Energy, Waste and the E
a Geochemical Pers

Special Publication reviewing procedures

The Society makes every effort to ensure that the scientific and production quality of its books matches that of its journals. Since 1997, all book proposals have been refereed by specialist reviewers as well as by the Society's Books Editorial Committee. If the referees identify weaknesses in the proposal, these must be addressed before the proposal is accepted.

Once the book is accepted, the Society has a team of Book Editors (listed above) who ensure that the volume editors follow strict guidelines on refereeing and quality control. We insist that individual papers can only be accepted after satisfactory review by two independent referees. The questions on the review forms are similar to those for *Journal of the Geological Society*. The referees' forms and comments must be available to the Society's Book Editors on request.

Although many of the books result from meetings, the editors are expected to commission papers that were not presented at the meeting to ensure that the book provides a balanced coverage of the subject. Being accepted for presentation at the meeting does not guarantee inclusion in the book.

Geological Society Special Publications are included in the ISI Index of Scientific Book Contents, but they do not have an impact factor, the latter being applicable only to journals.

More information about submitting a proposal and producing a Special Publication can be found on the Society's web site: www.geolsoc.org.uk.

It is recommended that reference to all or part of this book should be made in one of the following ways:

GIERÉ, R. & STILLE, P. (eds) 2004. *Energy, Waste, and the Environment: a Geochemical Perspective*. Geological Society, London, Special Publications, **236**.

EIKENBERG, J., BEER, H. & BAJO, S. 2004. Anthropogenic radionuclide emissions into the environment. *In*: GIERÉ, R. & STILLE, P. (eds) 2004. *Energy, Waste, and the Environment: a Geochemical Perspective*. Geological Society, London, Special Publications, **236**, 143–151.

GEOLOGICAL SOCIETY SPECIAL PUBLICATION NO. 236

Energy, Waste and the Environment: a Geochemical Perspective

EDITED BY

R. GIERÉ
Universität Freiburg, Germany

and

P. STILLE
ULP-École et Observatoire des Sciences de la Terre-CNRS,
Strasbourg, France

2004
Published by
The Geological Society
London

THE GEOLOGICAL SOCIETY

The Geological Society of London (GSL) was founded in 1807. It is the oldest national geological society in the world and the largest in Europe. It was incorporated under Royal Charter in 1825 and is Registered Charity 210161.

The Society is the UK national learned and professional society for geology with a worldwide Fellowship (FGS) of 9000. The Society has the power to confer Chartered status on suitably qualified Fellows, and about 2000 of the Fellowship carry the title (CGeol). Chartered Geologists may also obtain the equivalent European title, European Geologist (EurGeol). One fifth of the Society's fellowship resides outside the UK. To find out more about the Society, log on to www.geolsoc.org.uk.

The Geological Society Publishing House (Bath, UK) produces the Society's international journals and books, and acts as European distributor for selected publications of the American Association of Petroleum Geologists (AAPG), the American Geological Institute (AGI), the Indonesian Petroleum Association (IPA), the Geological Society of America (GSA), the Society for Sedimentary Geology (SEPM) and the Geologists' Association (GA). Joint marketing agreements ensure that GSL Fellows may purchase these societies' publications at a discount. The Society's online bookshop (accessible from www.geolsoc.org.uk) offers secure book purchasing with your credit or debit card.

To find out about joining the Society and benefiting from substantial discounts on publications of GSL and other societies worldwide, consult www.geolsoc.org.uk, or contact the Fellowship Department at: The Geological Society, Burlington House, Piccadilly, London W1J 0BG: Tel. +44 (0)20 7434 9944; Fax +44 (0)20 7439 8975; E-mail: enquiries@geolsoc.org.uk.

For information about the Society's meetings, consult *Events* on www.geolsoc.org.uk. To find out more about the Society's Corporate Affiliates Scheme, write to enquiries@geolsoc.org.uk.

Published by The Geological Society from:
The Geological Society Publishing House
Unit 7, Brassmill Enterprise Centre
Brassmill Lane
Bath BA1 3JN, UK

(*Orders*: Tel. +44 (0)1225 445046
      Fax +44 (0)1225 442836)
Online bookshop: http://bookshop.geolsoc.org.uk

The publishers make no representation, express or implied, with regard to the accuracy of the information contained in this book and cannot accept any legal responsibility for any errors or omissions that may be made.

British Library Cataloguing in Publication Data

A catalogue record for this book is available from the British Library.

ISBN 1-86239-167-X

Project managed by Techset Composition, Salisbury, UK
Printed by Cromwell Press, Trowbridge, UK.

Distributors

USA
AAPG Bookstore
PO Box 979
Tulsa
OK 74101-0979
USA
Orders: Tel. +1 918 584-2555
      Fax +1 918 560-2652
      E-mail bookstore@aapg.org

India
Affiliated East-West Press Private Ltd
Marketing Division
G-1/16 Ansari Road
Darya Ganj
New Delhi 110 002
India
Orders: Tels. +91 11 23279113, 23264180
      Fax +91 11 23260538
      E-mail affiliat@vsnl.com

Japan
Kanda Book Trading Company
Cityhouse Tama 204
Tsurumaki 1-3-10
Tama-shi, Tokyo 206-0034
Japan
Orders: Tel. +81 (0)423 57-7650
      Fax +81 (0)423 57-7651
      Email geokanda@ma.kcom.ne.jp

Contents

List of Abbreviations

AAS	Atomic absorption spectroscopy
ACT/An	Actinides
ACAA	American Coal Ash Association
AEC	Atomic Energy Commission
AFM	Atomic force microscopy
AFt/m	phase A : Al_2O_3 (substituted by F : Fe_2O_3) t: trisulphate, m: monosulphate
AGS	Acid generating salt
ANSTO	Australian Nuclear Science and Technology Organisation
APC	Air pollution control
AQM	Air quality modeling
ASTM	American Society for Testing and Materials
BA	Bottom ash
BE	Binding energy
BNFL	British Nuclear Fuels Ltd.
BOHC	Boron-oxygen hole center
BPP	Balti Power Plant
BWR	Boiling water reactor
CASH	Calcium aluminosilicate hydrate
CBM	Coal bed methane
CBO	Carbon burn-out
CBPC	Chemically-bonded phosphate ceramics
CC	Carbonate carbon
CCB	Coal combustion byproducts
CCM	Constant capacitance model
CCP	Coal combustion product
CDB	Citrate-dithionate-bicarbonate
CEC	Cation exchange capacity
CFBC	Circulating fluidized bed combustion
CFC	Chlorofluorocarbon
CFR	Cumulative fraction
CMM	Coal mine methane
COP	Coefficient of performance
CSH	Calcium silicate hydrate
CT	Collision theory
DDI	Distilled, deionized water
DISSOL	Thermodynamic simulation model
DLM	Diffuse layer model
DLVO	named after: Derjaguin, Landau, Vervey, Overbeek
DOC	Dissolved organic carbon
DOE	Department of Energy
DTA	Differential thermal analysis
DU	Depleted uranium

DWPF	Defense Waste Processing Facility
EBS	Engineered barrier system
EBR	Experimental breeder reactor
EDS	Energy dispersive spectroscopy
EDTA	Ethylene-diamine-tetraacetic acid
EDX	Energy dispersive X-ray
EDZ	Excavation-disturbed zone
EELS	Electron energy-loss spectroscopy
EGS	Enhanced geothermal system
EMPA	Electron microprobe analysis
ENEL	Ente Nazionale per l'Energia Elettrica
EPA	Environmental Protection Agency
EPMA	Electron probe microanalysis
EPP	Eesti power plant
EPRI	Electric Power Research Institute
ESEM	Environmental Scanning Electron Microscope
ESP	Electrostatic precipitator
ESRF	European Synchrotron Radiation Facility
EU	European Union
EXAFS	Extended X-ray absorption fine structure
FAA	Flame atomic absorption
FA	Fly ash
FBC	Fluidized bed combustion
FC	Filter cake
FEBEX	Full-scale engineered barriers experiment (in crystalline host rock)
FFFF	Flow-field flow fractionation
FGD	Flue gas desulphurization
FP	Fission products
FSU	Former Soviet Union
FT	Fourier transforms
FTIR	Fourier transformed infrared spectroscopy
FUETAP	Formed under elevated temperature and pressure
GCM	Glass-crystalline material
GEMS	Gibbs energy minimization selector (computer program)
GHG	Greenhouse gas
GLO	Guy-Lussac-Ostwald (Ostwald ripening)
GMT	Greenwich Mean Time
GSHP	Ground source heat pump
HA	Humic acid
HAP	Hazardous air pollutant
HC	Hydrocarbon (molecular weight: 13 g/mol)
HCFC	Hydrochlorofluorocarbon
HDPE	High-density polyethylene
HDR	Hot-dry-rock

HELP	Hydrological evaluation of landfill performance
HEU	Highly enriched uranium
HFO	Hydrous ferrous oxide or ferric hydroxide
HFR	Hot fractured-rock
HLW	High-level nuclear waste
HREE	Heavy rare earth elements (Gd-Lu)
HRL	Hard rock laboratory
HT	High temperature
HTGR	High-temperature gas-cooled reactor
HWR	Heavy water reactor

IAEA	International Atomic Energy Agency
IAP	Ion activity product
IBT	Ion beam thinned section
ICCM	Inductive cold-crucible melting
ICP	Inductively-coupled plasma
ICP-AES	Inductively-coupled plasma atomic emission spectrometry
ICP-MS	Inductively-coupled plasma mass spectrometry
IEX	Ion exchange
IGCC	Integrated coal gasification combined cycle
ILW	Intermediate-level nuclear waste
IMF	Inert matrix fuel
IPCC	Intergovernmental Panel on Climate Change
IQD	Interquartile distance
IR	Infrared

| JNREG | Joint Norwegian-Russian Expert Group |

LA-ICP-MS	Laser-ablation inductively-coupled plasma mass spectrometry
LDH	Layered double hydroxide
LEU	Low enriched uranium
LIBD	Laser-induced breakdown detection
LAW	Low-activity waste
LET	Linear energy transfer
LILW	Low- and intermediate-level nuclear waste
LIP	Lead-iron phosphate
LLNL	Lawrence Livermore National Laboratory
LLW	Low-level nuclear waste
LMA	Law of mass action
LMFBR	Liquid-metal-cooled fast-breeder reactor
LOI	Loss on ignition
LREE	Light rare earth elements (La-Sm)
L/S	Liquid-to-solid ratio (leachates)
LTA	Low-temperature ashing
LWR	Light water reactor

| MACT | Maximum achievable control technology |
| MAS-NMR | Magic-angle spinning nuclear magnetic resonance |

MCB	Microwave burn-out
MCC	Materials Characterization Center
MINEQL	Mineral equilibria (computer program)
MOX	Mixed-oxide (fuel)
MSWI	Municipal solid waste incinerator
μ-SXRF	Synchrotron-based X-ray microfluorescence
MW	Magnox waste glass (British Nuclear Fuel Public Ltd. Company)
NAA	Neutron activation analysis
NAAQS	National ambient air quality standards
NBO	Non-bridging oxygen
NCP	Non-carbonate portion
NEA	Nuclear energy agency
NEA-OECD	Nuclear Energy Agency-Organization for Economic Cooperation and Development
NEI	Nuclear Energy Institute
NETPATH	Interactive code for modelling net geochemical reactions along a flow path
NFC	Nuclear fuel cycle
NGCC	Natural gas combined cycle
NMR	Nuclear magnetic resonance
NO_x	$NO + NO_2$
NPP	Nuclear power plant
NRA	Nuclear reaction analysis
NRC	National Research Council
OPC	Ordinary Portland cement
PAH	Polycyclic aromatic hydrocarbon
PAN	Peroxyacetylnitrate
PBN	Peroxybenzoylnitrate
PCT	Product consistency test
PES	Plasma emission spectroscopy
PHREEQC	pH redox equilibrium calculations (computer program)
PIC	Product of incomplete combustion
PM	Particulate matter
PMATCHC	Program to manage thermochemical data, written in C^{++}
PM-10	Particulate matter with an aerodynamic diameter $<10\,\mu m$
PM-2.5	Particulate matter with an aerodynamic diameter $<2.5\,\mu m$
PRB	Powder River Basin
PUREX	Pu-U-recovery-extraction
PW	Purex waste
PWR	Pressurized water reactor
PZC	Point of zero charge
QA	Quality assurance
RBS	Rutherford backscattering spectroscopy
REE	Rare earth elements (lanthanides + Y)

ROBL	Rossendorf beamline
RSF	Radial structure function
SAEFL	Swiss Agency for Environment, Forests and Landscape
SAI	Strength activity index
SCR	Selective catalytic reduction
SCRW	Supercritical water reactor
SCM	Surface complexation model
SEM	Scanning electron microscopy
SEM-EDS	Scanning electron microscopy-energy dispersive spectroscopy
SFM	Soil forming material
SIMS	Secondary ion mass spectrometry
SIMFUEL	Simulated fuel
SNF	Spent nuclear fuel
SOHC	Silicon-oxygen hole center
SPC	Seasonal performance coefficient
SPFT	Single-pass flow-through
SSA	Specific surface area
SSS	Self-sustaining synthesis
STABCAL	Stability calculation software
STEM	Scanning transmission electron microscopy
TCLP	Toxicity characteristics leaching procedure
TDB	Thermodynamic database
TDF	Tyre-derived fuel
THC	Total hydrocarbon
TBP	Tri-n-butyl phosphate
TEM	Transmission electron microscopy
TLM	Triple layer model
TM-AFM	Tapping mode atomic force microscopy
TOC	Total organic carbon
TRLFS	Time-resolved laser fluorescence spectroscopy
TRU	Transuranic
TSP	Total suspended particles
TST	Transition state theory
TVS	Transportable vitrification system
UIC	Uranium Information Centre
UNSCEAR	United Nations Scientific Committee on the Effects of Atomic Radiation
USDOE	United States Department of Energy
USGS	United States Geological Survey
USPHS	United States Public Health Service
UV	Ultraviolet
VLNW	Very low-activity nuclear waste
VNCR	Valence coordination number ratio
VOC	Volatile organic compound
VSI	Vertical scanning interferometry

WATEQ   Water equilibria (computer program)
WHO   World Health Organisation
WIPP   Waste Isolation Pilot plant
WoCQI   World coal quality inventory

XANES   X-ray absorption near-edge spectroscopy
XAS   X-ray absorption spectroscopy
XAFS   X-ray absorption fine structure
XPS   X-ray photoelectron spectroscopy
XRD   X-ray diffraction
XRF   X-ray fluorescence

Referees

The editors are very grateful to the following people for their help in refereeing papers for this volume.

Energy, waste and the environment – a geochemical perspective: introduction

R. GIERÉ[1,2] & P. STILLE[3]

[1]*Institut für Mineralogie, Petrologie und Geochemie, Universität Freiburg, Freiburg, Germany*
(e-mail: giere@uni-freiburg.de)
[2]*Department of Earth and Atmospheric Sciences, Purdue University, West Lafayette, IN, USA*
[3]*ULP-EOST-CNRS, Centre de Géochimie de la Surface UMR 7517, Strasbourg, France*
(e-mail: pstille@illite.u-strasbg.fr)

Energy has played a key role in the development of civilizations around the globe. Since prehistoric times, Man has depended on energy sources for heating and cooking purposes. The main energy source at that time was wood, and wood burning continues to be of utmost importance in certain parts of the world. The need to secure energy sources has a direct interaction with the environment. Throughout history, this interaction has been detrimental to the environment, as documented, for example, by early deforestation in the Mediterranean area, where the growth of civilizations was linked to deforestation (Thirgood 1981). Today, the environmental impact resulting from energy production and consumption is more visible and more pronounced than ever before, as Man tries to satisfy an ever-growing energy demand. In the year 2001, the total world consumption of primary energy amounted to ~426 billion GJ (EIA 2003), an increase of more than 15% compared to 1992. By assuming a world population of 6.1 billion people in 2001, the per capita energy consumption is approximately 70 GJ. This figure, however, represents a global average only, and pronounced differences exist for various regions. Figure 1 demonstrates, for example, that in North America the per capita consumption of primary energy is four times greater than the global average, and nearly twice that of Western Europe. On the other hand, the per capita consumption in Africa is merely a third of the global average. This extreme geographical disparity in energy consumption is mirrored by the data for CO_2 emissions (Fig. 2), which reflect the fact that most of the energy consumed

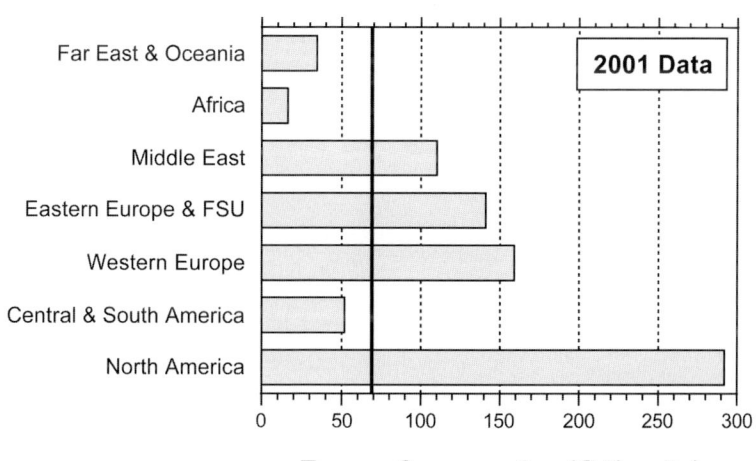

Fig. 1. World per capita consumption of primary energy for the year 2001. Thick vertical line represents global average; FSU = Former Soviet Union. Data from EIA (2003).

From: GIERÉ, R. & STILLE, P. (eds) 2004. *Energy, Waste, and the Environment: a Geochemical Perspective*. Geological Society, London, Special Publications, **236**, 1–5.
0305-8719/04/$15   © The Geological Society of London 2004.

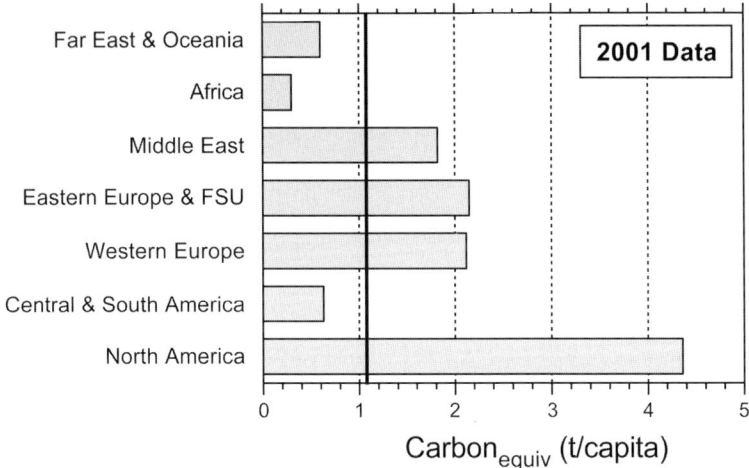

Fig. 2. World per capita carbon dioxide emissions (expressed as carbon equivalent) from the consumption and flaring of fossil fuels for the year 2001. Thick vertical line represents global average; FSU = Former Soviet Union. Data from EIA (2003).

globally is produced through combustion of fossil fuels. Fossil fuels are the source for ~86% of the total energy consumption (Fig. 3), of which petroleum is by far the most important source. These numbers dwarf the contributions of other types of energy, particularly those derived from renewable sources, such as biomass, sun, wind, and tides. The only exception to this pattern is hydroelectric power generation, which accounts for nearly 7% of the energy consumed globally, that is, almost identical to the global share of nuclear electric power (Fig. 3).

The historical data show that, while the total energy consumption increases steadily, the contributions from different sources do not increase at the same rate. For example, the largest growth during the past ten years was observed for energy derived from geothermal, solar, wind, wood, and waste sources, which together accounted for less than 1% of the energy consumption in 2001 (Fig. 3). The combined contributions from these types of energy have grown by more than 50%, whereas those from coal and natural gas only grew by 9% and 21%, respectively. The

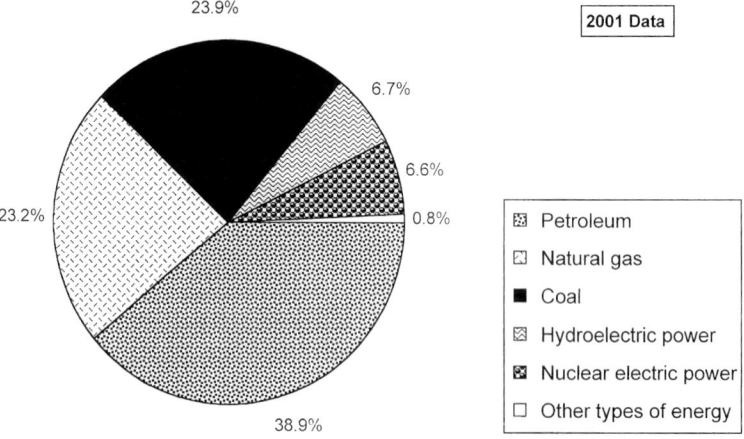

Fig. 3. World consumption of primary energy. The category 'Other types of energy' comprises data for geothermal, solar, wind, wood, and waste electric power. Data from EIA (2003).

increasing need for energy is associated with a constantly growing world population and an increasing demand for consumer goods. The relationship between world population and energy consumption is documented by the fact that the global per capita consumption has not changed significantly over the past few years (EIA 2003).

As a result of the increasing energy production in industrialized societies, the anthropogenic impact on the environment becomes more prominent and starts to affect natural geochemical and biochemical cycles at the surface of our planet. This impact reaches proportions that are comparable with those of evolutionary processes during the Earth's history. For example, the global CO_2 emissions from combustion and flaring of fossil fuels amount to an equivalent of ~6.5 Gt carbon per year, which corresponds to approximately 1 t carbon per person per year. The consequences of such enormous waste gas emissions, which result directly from energy production and consumption, become visible even on a global scale, with increasing atmospheric CO_2 concentrations, global warming, and climate change, which in turn is associated with increasing climate variability and storm events and increasing rates of erosion. The increasing flux of waste and its ominous interference with natural fluxes necessitates new strategies of global environmental management.

Much less known is the fact that energy production also generates large amounts of solid and liquid waste. These waste materials are extremely diverse: some need to be treated and isolated from the biosphere, because they are either toxic (e.g., fly ash and filter cake from high-temperature incineration of municipal solid waste) or radioactive (e.g., high-level waste from reprocessing of spent nuclear fuel). These materials must in general be solidified and/or stabilized (e.g., vitrified, embedded in cement) before they can be discarded in special waste deposits, which need to be constructed with multiple engineered barriers to avoid, or at least retard, migration of toxic and radioactive compounds into the biosphere. Other types of waste resulting from power generation accumulate in such large quantities (e.g., hundreds of Mt of coal combustion products per year) that simple disposal is in many cases no longer an option. Moreover, some waste materials contain valuable components (e.g., certain metals in fly ash and incinerator slags), which can be retrieved and recycled, thus saving precious primary resources.

This book focuses on geochemical approaches in immobilizing, isolating, or neutralizing waste derived from energy production and consumption.

It further addresses the importance of using some types of energy-derived waste as alternative raw materials for certain valuable components. Moreover, the book showcases research on how certain waste materials can be utilized for energy production, an increasingly important aspect of modern integrated waste management strategies. The main objectives are to:

(1) identify the most serious environmental issues related to various types of power generation and associated waste accumulation;
(2) demonstrate the importance of detailed geochemical characterization of waste;
(3) address the long-term stability and safe isolation of waste in the environment;
(4) document how toxic and radioactive components are encapsulated in nature for geologic time periods;
(5) present strategies, based on natural analogue materials, for the immobilization of toxic and radioactive waste components through mineralogical barriers;
(6) discuss modern procedures for reuse of waste or certain waste components;
(7) review the importance of geochemical modelling and kinetic concepts in describing and predicting the interaction between waste and the environment.

Owing to the breadth of the topic, only a few types of energy production are discussed. The choice of subjects is based primarily on availability of relevant scientific research, and on the desire to present examples from across the entire range of energy sources, that is, nuclear, fossil, renewable, and alternative fuel resources. Specifically, the aim is not to discuss all types of the less commonly used renewable energy resources (water, wind, sun, tides, waves), which are clearly preferable from an environmental standpoint. Rather, it is to focus attention on the most frequently discussed and criticized types of energy sources, that is, fossil and nuclear fuels. Each method of power generation, however, has significant impacts on the environment.

In introducing Part I, *The Nuclear Fuel Cycle*, Ewing discusses in a visionary way the future evolution of nuclear energy and points out that, even if the presently estimated reserves for uranium are completely utilized, the low-efficiency systems actually used and followed by direct disposal will lower the CO_2 increase by only 38 ppm. This suggests that, if nuclear power should have an increasing role in reducing the greenhouse gases, the development of advanced fuel cycles and waste management technologies, which do not at present exist on an industrial scale, have to be promoted. Among other problems arising from nuclear power generation is the management of radioactive waste. The long-term performance of nuclear waste repositories must be predicted over geological time scales, and its assessment requires not only detailed knowledge of the chemical and mineralogical characteristics of the waste matrix (e.g., glass, ceramics), as discussed in Part I, but also of its behavior during potential interactions with water that eventually intrudes an engineered barrier system. Laboratory experiments and detailed thermodynamic modelling provide the necessary basis for this understanding (see Part V). However, all laboratory experiments are performed for limited periods of time (a few years at the most) and, therefore, thermodynamic models have to be tested in natural, geological environments. This can be achieved by applying geochemical and mineralogical techniques to the study of so-called natural analogues and, as demonstrated in Part I, such investigations can serve to test the concept of repository design and to validate the geochemical models over geological time frames.

Unlike the nuclear fuel cycle, the fossil fuel cycle has a serious impact on the atmosphere and the climatic evolution, as outlined by Golomb & Fay in their introductory chapter to Part II of the book. These authors show that the rising atmospheric levels of greenhouse gases demand a strong reduction of CO_2 emissions from fossil fuel usage in order to fight against global warming and other climate changes. Nevertheless, the reduction of CO_2 emissions from fossil fuel usage is a very perplexing and arguably intractable environmental problem with which the modern urban–industrial society is faced. Possibly very soon, CO_2 collected directly from fossil fuel-combusting power plants can be routinely injected into deep geological formations in order to limit its emission into the atmosphere (see chapter by Saylor & Zerai). Another important aspect of the fossil fuel cycle is the large amount of waste produced through mining and combustion, particularly in the case of coal. To understand the environmental impacts of such waste products, it is essential to characterize them chemically and mineralogically, as shown in various papers in Part II.

The outlined perspectives of both nuclear and fossil fuel-derived energies are reason enough to promote alternative energy sources. As mentioned earlier, one of the chief renewable sources is water, and the generation of hydroelectric power is extensive on a global scale (Fig. 3), even though this type of energy is, like all others, associated with considerable environmental impacts (e.g., drowning of entire valleys, drying of river beds). Another important alternative energy source is contained in geothermal systems. In his introductory chapter to Part III, *The Geothermal Energy Cycle*, Arnórsson describes the overall environmental impact of developing geothermal systems and maintaining geothermal power plants. A major problem arising from the use of geothermal resources is that huge amounts of waste fluids are accumulating. Current trends to deal with this problem include reinjection of the spent fluids into the geothermal system. This topic, as well as the geophysical, geochemical, and mineralogical impacts of fluid injection, are discussed in this part of the book. An often neglected aspect of power generation in general is the large amount of heat that is wasted. Instead of discussing this subject for each of the various energy cycles, there is a special chapter on waste heat problems related to geothermal systems. The innovative solutions to waste heat problems described by Rybach & Kohl could, however, also be applied to other energy cycles.

In Part IV, *The Waste-to-Energy Cycle*, crucial geochemical aspects are discussed for this energy cycle, which gains in importance as landfill space becomes scarce and governments impose new regulations on the types of materials that can be landfilled. Moreover, many waste materials contain valuable chemical components or energy, or both, that could be extracted and, therefore, should not only be regarded as waste but also as a resource. This aspect of waste management is outlined for various waste materials, including municipal solid waste, scrap tyres, and mining waste. Those wastes that cannot be further utilized need to be discarded in special landfills, but typically require solidification and/or stabilization prior to disposal. Essential solidification/stabilization techniques as well as possible reuse options are discussed in Part IV of the book, and include vitrification, chemical stabilization in orthophosphate, and production of glass–ceramic or ceramic materials.

In the final section of the book, Part V, *Water–Waste Interaction*, various approaches to describe and predict the interaction between waste and water are presented. This part of the book does not only deal with landfills containing different types of municipal waste forms, but also with different nuclear waste forms. Of special interest are laboratory experiments on waste form corrosion and element speciation in aqueous media that represent realistic disposal environments. Such experiments are essential in reducing uncertainties about performance predictions. Moreover, thermodynamic modelling is, as outlined in this section of the book, a crucial component of an in-depth performance assessment. An approach combining all these individual techniques will dramatically improve our knowledge of the geochemical processes that take place at the waste–mineral–colloid–water interface and further our understanding of the migration behaviour of toxic heavy metals and radionuclides in the disposal environment.

This volume provides incentives for further development of sustainable fuel cycles through novel and interdisciplinary approaches to an Earth science-related topic. The issue of power generation and associated waste accumulation is of enormous political and social importance for people around the globe, and this book demonstrates that geological research is essential in providing solutions to some of the most pressing problems of today's society.

Many of the papers presented in this volume derive from presentations made at the 12th Annual V. M. Goldschmidt Conference held at Davos, Switzerland 18–25 August, 2002, and sponsored by The Geochemical Society, The European Association of Geochemistry, The Mineralogical Society of America, and ETH Zurich.

The Editors are grateful for the many discussions with, and helpful comments from, many of our colleagues, some of whom have also contributed to this volume. We are particularly grateful to all the reviewers, both the ones listed and those who wished to remain anonymous. Without their insight, hard work, and constructive suggestions, we would not have been able to compile a book on such diverse topics.

References

EIA 2003. Energy Information Administration, U.S. Department of Energy. World Wide Web Address: http://www.eia.doe.gov.

THIRGOOD, J. V. 1981. *Man and the Mediterranean Forest: A History of Resource Depletion.* Academic Press, London, New York, 194 p.

Environmental impact of the nuclear fuel cycle

RODNEY C. EWING

Departments of Geological Sciences, Materials Science & Engineering, and Nuclear Engineering & Radiological Sciences, University of Michigan, Ann Arbor, Michigan 48109, USA

(e-mail: rodewing@umich.edu)

Abstract: Nuclear power provides approximately 17% of the world's electricity, which is equivalent to a reduction in carbon emissions of ~0.5 Gt of C/year. This is a modest contribution to the reduction of global carbon emissions, ~6.5 Gt C/year. Most analyses suggest that in order to have a significant and timely impact on carbon emissions, carbon-free sources, such as nuclear power, would have to expand total energy production by a factor of three to ten by 2050. A three-fold increase in nuclear power capacity would result in a projected reduction in carbon emissions of 1 to 2 Gt C/year, depending on the type of carbon-based energy source that is displaced. This paper reviews the impact of an expansion of this scale on the generation of nuclear waste and fissile material that might be diverted to the production of nuclear weapons. There are three types of nuclear fuel cycles that might be utilized for the increased production of energy: open, closed, or a symbiotic combination of different reactor types (such as thermal and fast neutron reactors). Within each cycle, the volume and composition of the nuclear waste and fissile material depend on the type of nuclear fuel, the amount of burn-up, the extent of radionuclide separation during reprocessing, and the types of material used to immobilize different radionuclides. This chapter is a discussion of the relation between the different types of fuel cycles and their environmental impact.

As we enter the 21st century, the impact of human activity has reached a global scale. The exploitation of energy resources, 450 EJ/year (exajoules, 10^{18} joules), mainly carbon-based, has led to increases in atmospheric CO_2, CH_4, and N_2O, as well as other greenhouse gases, causing shifts in global climate patterns and an increase in global temperature during the 20th century (Houghton *et al.* 2001). Even as we debate the significance of future climate change and its effects, an array of 'advanced' technologies has been proposed either as substitutes for fossil fuels (Hoffert *et al.* 2002) or to reduce the energy demand by higher efficiency or low-carbon technologies (Rosenfeld *et al.* 2000). Although one may make optimistic assumptions about the potential for technological innovation and conservation to mitigate substantially the global impact of energy production, the immediate requirement is for solutions that can be put into place during the next fifty years, limiting the increase in atmospheric CO_2 concentrations to twice (550 parts per million) that of pre-industrial levels (275 ppm) (Fetter 2000). At present, the alternatives to fossil fuels for major energy production are limited to five possibilities:

nuclear fission, biomass, solar, wind, and decarbonized fossil fuels (i.e., the production of H from hydrocarbons) from which the carbon is sequestered (Fetter 2000). Unfortunately, no single energy source is a 'silver bullet' – each technology draws on a resource, each requires improvements in the implementing technologies, and all have an environmental impact. Of these five, only the nuclear option is presently deployed on a large scale, producing 17% of the world's electricity or 7.5% of the world trade primary energy (Grimston & Beck 2000). Additionally, during the 1990s, nuclear power was the fastest growing (energy output increasing 29.6%) of the major sources of global energy, as compared to natural gas (18.8%) and oil (12.1%) (Grimston & Beck 2000). Because nuclear power is so broadly used (used in more than 30 countries), there is a substantial basis for evaluating its cost and impact. However, such an analysis is difficult because there are different types of nuclear reactors, and there is not a single nuclear fuel cycle, but rather many variants of 'closed' and 'open' fuel cycles with different reprocessing and nuclear waste disposal technologies. The different fuel cycles reflect

From: GIERÉ, R. & STILLE, P. (eds) 2004. *Energy, Waste, and the Environment: a Geochemical Perspective.* Geological Society, London, Special Publications, **236**, 7–23.
0305-8719/04/$15 © The Geological Society of London 2004.

different strategies for the utilization of fissile nuclides, mainly ^{235}U and ^{239}Pu, and these different strategies have important implications for nuclear waste management and nuclear weapons proliferation. The 'once-through' open cycle treats the spent fuel as a waste without any attempt to reclaim the remaining ^{235}U or newly created ^{239}Pu, and the spent nuclear fuel (SNF) is directly disposed of in a geological repository. This is the US strategy. A closed fuel cycle with reprocessing retrieves approximately 99% of the fissile nuclides. However, the recovered fissile nuclides are only a supplement to the nuclear fuel that is mainly derived from newly mined ore. In this case, the high-level waste from reprocessing and the unreprocessed SNF are disposed of in a geological repository. The breeder reactor cycle creates more fissile material in the SNF than in the original fuel, and the breeder reactor cycle envisions multiple cycles of reprocessing. In 1977, President Carter decided to defer indefinitely reprocessing of SNF in the USA in order to have a more proliferation-resistant fuel cycle. In 1981, President Reagan lifted the ban on reprocessing, but he placed the financial responsibility for reprocessing on the private sector. By the mid-1980s, the commercial reprocessing of SNF had little attraction from a technical,

economic, regulatory, or policy perspective (Carter 1987). A recent, detailed analysis (Bunn *et al.* 2003) of the cost of reprocessing suggests that there is no financial incentive to pursue reprocessing, as compared with the simpler strategy of direct disposal of SNF.

This chapter is a discussion of the relations between the types of reactors and fuel cycles and their environmental impact. For the nuclear fuel cycle, the most obvious environmental impacts are related to the nuclear waste that is generated either from reprocessing of used nuclear fuel to reclaim fissile nuclides for another cycle of 'burning' in a reactor (the closed fuel cycle, Fig. 1) or the direct disposal of the SNF in a geological repository (the open fuel cycle, Fig. 2). The critical difference between the two fuel cycles is the use of reprocessing to reclaim fissile nuclides (^{235}U or ^{239}Pu). The spent fuel (UO_2) contains only about one atomic percent Pu and 3% fission products after typical burn-ups of 35 MegaWatt-days/kg (MWd/kg) of U (Hedin 1997). The most common reprocessing technology is the PUREX (Pu-U-Recovery-Extraction) process, which uses a solvent called tri-n-butyl phosphate (TBP) and liquid–liquid extraction combined with oxidation–reduction chemical reactions. The details of the reprocessing technique are critical

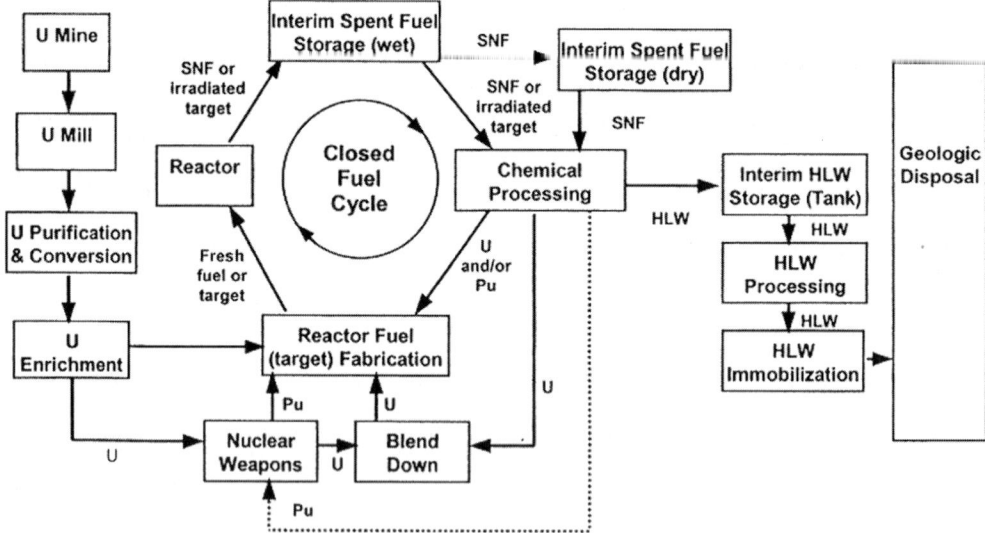

Fig. 1. Schematic illustration of the ideal closed nuclear fuel cycle (NRC 2003). In real practice, the reprocessing capacity does not match the generation rate of the spent nuclear fuel. Thus, the excess SNF must be placed in interim storage or disposed of in a geological repository. Under normal circumstances, the SNF will be in interim storage for just a few years. Also, note that excess material from nuclear weapons, e.g., highly enriched ^{235}U and ^{239}Pu, can be blended down to lower concentrations and used as a reactor fuel.

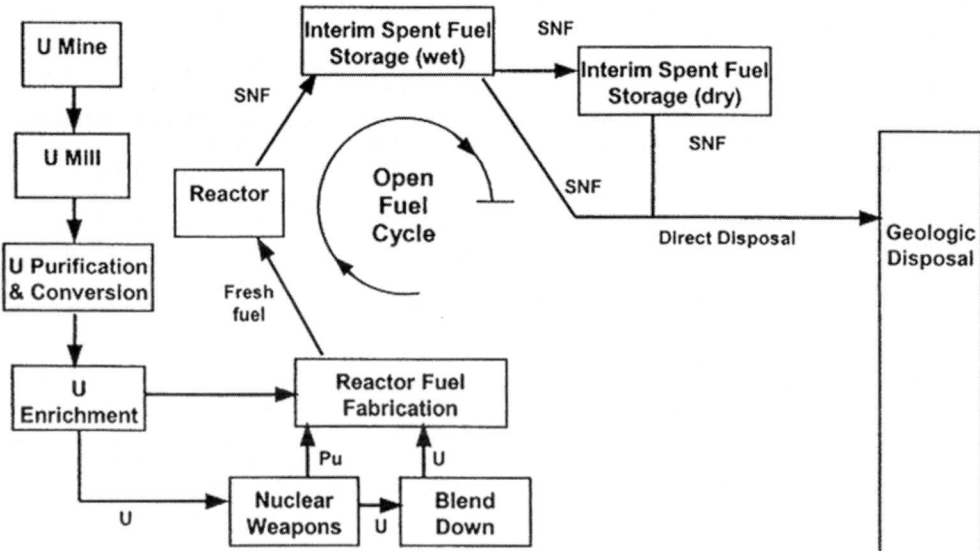

Fig. 2. Schematic illustration of the ideal open nuclear fuel cycle (NRC 2003). In this case, there is no reprocessing. Interim storage may last for tens of years so that the heat and radioactivity are much less prior to handling and final disposal. The spent fuel still contains fissile nuclides, such as ^{235}U and ^{239}Pu (generated by neutron capture reactions with ^{238}U).

to evaluating environmental impact, as this determines the waste stream compositions. Minimal reprocessing, as has been practised in the past, removes only U and Pu, leaving a complex mixture of fission products and the so-called 'minor' actinides, such as Np, Am, and Cm. These 'minor' actinides are important, as the annual production rates are not small: Np (3.4 metric tonnes (t)/y); Am (2.7 t/y); Cm (335 kg/y). A more ambitious reprocessing strategy could separate radionuclides according to their chemical properties, half-lives, and toxicity, so that each could be handled separately with different reprocessing, immobilization, and disposal schemes. This could be particularly advantageous for ^{237}Np, because it has a long half-life (2.1 million years) and is a major contributor to dose in analyses of the long-term performance of a geologic repository. A principal concern with reprocessing has always been the possibility of the diversion of fissile material, mainly ^{235}U and ^{239}Pu, for weapons production. However, other fissile nuclides, such as ^{237}Np and Am, may be separated during reprocessing. In 1997, the global inventory of ^{237}Np and ^{241,242,243}Am (the principle isotope of interest is ^{241}Am, which forms from the decay of ^{241}Pu) was estimated to be 80 t (enough for several thousand nuclear weapons), and this inventory grows at a rate of approximately 10 t/y (Albright

& Barbour 1999). The open fuel cycle avoids reprocessing, and the used fuel is sent for direct disposal in a geological repository. The very high radioactivity of the SNF is a deterrent to handling and reprocessing the fuel for fissile nuclides. However, the total activity of the fuel drops dramatically during the first few hundred years, as the main sources of activity are ^{137}Cs and ^{90}Sr, with half-lives of about 30 years. Thus, with increasing time, the inherent radioactive deterrent to reprocessing to reclaim the much longer lived fissile nuclides decreases. During the past decades, the relatively low price of uranium has made reprocessing uneconomical (Bunn et al. 2003), but it is still an important part of nuclear programs in countries such as France and Japan where nuclear power is a major source of electrical energy.

In addition to concerns for the proliferation of nuclear weapons, one may analyse the environmental impact of nuclear waste as the interface of the nuclear fuel cycle (NFC) with geochemical/hydrological cycles (Ewing 2002). The interface between these two cycles is the geological repository, although there is some release of radioactivity to the environment during mining of uranium ore, fabrication of nuclear fuels, and reprocessing (e.g., ^{85}Kr). There are several important points to be made with this schematic view (Fig. 3). (1) There is a great deal of flexi-

NUCLEAR FUEL CYCLES -- GEOCHEMICAL CYCLES

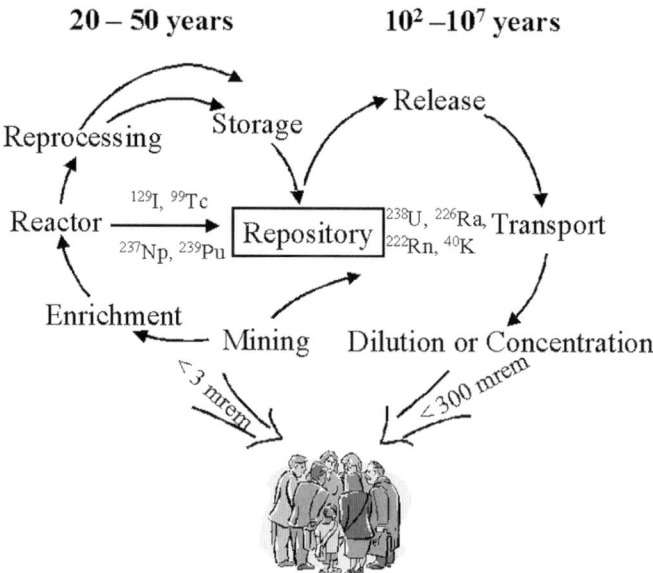

Fig. 3. Schematic illustration of the interface of the nuclear fuel cycle with geochemical/hydrological cycles. The geological repository is the interface for these two cycles. The principal sources of radioactivity (over the long term) are indicated by the radionuclides listed at the centre of each cycle. Total background exposures to radiation are less than 300 mrem/y. The total radiation exposure that can be attributed to the nuclear fuel cycle is less than 3 mrem/y.

bility in the nuclear fuel cycle for altering the volume and radionuclide composition of the waste (e.g., by using different types of nuclear fuels, different levels of burn-up, repeated cycles of reprocessing and transmutation). (2) The basic geochemical (e.g., redox conditions) and hydrological (e.g., saturated or unsaturated zone flow) properties of the repository determine the type and extent of release of the radionuclides to the environment. (3) The principal geological means of containment are slow release from the waste form and through the waste package materials, followed by slow transport (e.g., low flow rates or high sorption) and dilution. The most important decision in geological disposal is site selection – as the geology of the site determines the principal mechanisms for release, transport, and retardation. The important issues to address for geological disposal were clearly outlined over a quarter of a century ago (Brederhoeft *et al.* 1976). (4) The time scales for the NFC and the geochemical/hydrological cycles are vastly different, up to six orders of magnitude. Still, short-term changes in the NFC may have a profound effect on the types and extent of environmental impact. (5) Finally, the environmental impact of the NFC in terms of

annual exposure to radiation (<3 mrem $=$ 0.03 mSv) should be placed into the context of the exposure due to normal background (<300 mrem $=$ 3.00 mSv); however, such comparisons presume no failure either of nuclear power plants (NPP) or geological repositories.

For most energy producing systems, the main elements of the environmental impact can be captured in diagrams, more or less elaborate, similar to that in Fig. 3. However, for the NFC there are two other important environmental impacts, one negative and one positive: (1) the potential for the proliferation of nuclear weapons by the diversion of fissile material; and (2) reduced CO_2 emissions by the displacement of fossil fuel sources of energy. No other energy-producing system carries such profoundly important impacts on human beings and the environment. Thus, in addition to more traditional measures of environmental impact, the efficacy of any proposed NFC has to be evaluated in terms of the safeguards against proliferation of nuclear weapons and the reduction of CO_2 emissions. There is also the important issue of accidents at nuclear power plants. This chapter is too short to include a discussion of the environmental impact of the reactor accidents at the Three Mile Island (1978) and Chernobyl (1986) nuclear power plants or the probability of similar accidents occurring in the future. Here the range of opinion is wide, depending on whether one basis the analysis on past experience or predicted performance of advanced nuclear reactors (Sailor *et al.* 2000).

Thus, in the broadest sense, the issues related to nuclear power extend beyond the production of energy, but rather require an analysis of the potential of the NFC to impact the carbon cycle (e.g., reduce carbon emissions by substituting nuclear power for hydrocarbon-based sources of energy) and the effect of an active, growing nuclear power industry on the production of nuclear waste and potential for the proliferation of nuclear weapons. The ideal nuclear system produces a maximum amount of energy, substantially reduces carbon emissions, safely disposes of nuclear waste, and securely safeguards against proliferation of nuclear weapons. The connection between these cycles has become more evident, as observed by Carter & Pigford (1999):

'Some specialists around the world live by the argot and values of the non-proliferation community. There are others who live by the values and assumptions of nuclear waste disposal programs, thinking not in terms of rogue states, terrorists and hidden reprocessing labs, but rather of ground water travel times and radionuclide dispersion coefficients. It is now time for these two groups of specialists to cooperate closely in achieving their goals.'

Previous analyses

As the number of nuclear power plants (NPP) increased during the 1960s, there was a growing concern for the fate of the spent nuclear fuel (SNF) (Romer 1972). In 1976, there were two important studies. Pigford (1976), in connection with the Energy Policy Project of the Ford Foundation, completed an analysis of the 'Environmental Aspects of Nuclear Energy Production.' Pigford had already clearly outlined the challenge:

'The environmental acceptability of nuclear fission power plants rests upon the careful control of environmental effluents from each of the many diverse steps in the nuclear fuel cycle including uranium mining, fuel preparation, reactor operation, fuel reprocessing, and the storage and disposal of radioactive wastes. The nuclear fuel cycle involves a greater number of fuel process operations and waste treatment technologies than any present alternative power generation technology. The quantities of radioactivity pro-duced in nuclear fission are so great that they represent a very large theoretical potential for damage to the environment and to public health. However, the actual material quantities of these radioactive byproducts are relatively small and there are well-developed technologies to confine and control these radionuclides. As a result, the environmental impacts from nuclear fission power plants and their associated fuel cycles are the lowest of any of the currently available power-producing technologies dependent upon natural fuel resources.'

Budnitz & Holdren (1976) proposed a comprehensive, quantitative method of evaluating and comparing the environmental and social costs of energy producing systems (e.g., nuclear vs. hydrocarbon-based sources). However, having conceptualized the approach, they noted that there were important inadequacies in the analytical tools and the database available for the analysis and that there was a lack of clearly stated criteria to serve as a basis for comparison. Although optimistic about the possibility of meeting this challenge with '... vigorous (and often fervent) activity in environmental analysis,' they conclude:

'Finally, the situation that civilization has reached [is] the predicament where large-scale environmental disruptions are not only possible but perhaps likely, without having developed the knowledge to understand the possibilities in detail or to cope with them, gives reason to slow greatly the growth in energy consumption. Only such a slowdown can buy the time needed to obtain more knowledge of the threats and to develop and deploy more benign technologies.'

A quarter of a century after the publication of these two studies, their conclusions and approaches are now typical of the many similar analyses today. However, in 1976, these approaches were new and remarkably original and comprehensive. The conclusions of these authors are still echoed in today's discussion of the same issues – almost as if there has been no progress in resolving the basic issues or even in obtaining the fundamental data required for the analysis. There have, of course, been many recent and more elaborate analyses of these issues (Ansolabehere *et al.* 2003), and some of these studies will be the basis for discussion in this paper. However, it is instructive to review these first papers with the benefit of hindsight, as they add some perspective to the present review and its limitations.

Pigford's analysis (1976) is based on the projected growth in nuclear power capacity to the year 2000. This projection, of only a quarter of century, is one-half of the fifty year time-frame used to project the impact of energy producing systems on global climate change (Fetter 2000). Pigford (1976) used estimates from the US Atomic Energy Commission (AEC) that the 'most likely' additions to US generating capacity would be 700 GW of water reactors (pressurized water reactors (PWR) and boiling water reactors (BWR)); this amounts to seven hundred 1-GW reactors. In addition, the AEC expected that there would be an additional 100 GW of capacity from high-temperature gas-cooled reactors (HTGR) and that a liquid-metal-cooled fast-breeder reactor (LMFBR) would be introduced as a commercial power plant by 1986. The analysis presumed recovery of all Pu from water reactor fuel and diminished inventories of Pu toward the end of the century, as Pu would be consumed by fast-breeder reactors or as a mixed-oxide (MOX) fuel (U + Pu) in water reactors. The gas-cooled reactors would use fuels consisting of a mixture of ^{232}Th and ^{235}U (enriched to 93%), and due to the absence of ^{238}U, ^{239}Pu (created by (n,γ) reactions on ^{238}U), would not form. However, (n,γ) reactions on ^{232}Th would create fissile ^{233}U that would be reclaimed by reprocessing. Each of these fuel cycles and reprocessing schemes would lead to rather different nuclear waste streams; some, such as the radioactivity from ^{232}U (created by (n,2n)-reactions with ^{232}Th) would require special attention. Pigford's analysis assumed an average thermal energy generation of 33 MWd/kg U. The discharged fuel would be reprocessed after 150 days.

In the year 2000, the actual situation was very different: 103 nuclear power plants producing 72 GWy/y in the USA (GWy = the electrical energy corresponding to a power of 1 GWe maintained for a period of one year = 3.15 × 10^{16} J), and 434 reactors with a total capacity of 350 GW producing 2300 terawatt-hours (TWh) of electricity (approximately 260 GWy/y) world wide. There has been no reprocessing of commercially generated spent fuel in the USA (except for the Nuclear Fuel Services plant that operated from 1967 to 1972 at West Valley Plant in New York, which processed 600 t of irradiated fuel). Much of the commercially generated spent fuel has now been held in storage for over 20 years. Had the USA reached the 700 GWy/y production capacity, this would have resulted in an equivalent reduction of ∼0.7 GtC/y, compared to the 1.1 GtC/y required by the Kyoto Protocol. Also, the environmental impact in Pigford's analysis was judged mainly in the context of radiation exposure. Important figures, such as

the estimated risk of cancer due to exposure to low levels of ionizing radiation, have increased by a factor of more than five (0.8 × 10^{-2}/Sv vs. 5 × 10^{-2}/Sv) during the past 25 years. Atomic Energy Commission regulations (10 CFR 100) limited exposures to off-site individuals in the event of a loss-of-coolant accident to no more than 25 mrem. This is higher than the 15 mrem limit that the Environmental Protection Agency (EPA) has as its standard for Yucca Mountain, for which the maximum exposure is estimated for an individual at a distance of 20 km for a period up to 10 000 years. Thus, the standards that are the basis for the comparison have changed during the past 25 years, generally becoming more restrictive.

The approach by Budnitz & Holdren (1976) was more ambitious, using a method of materials accounting and critical pathway analysis to compare the social and environmental impacts and costs of very different energy-producing systems (i.e., coal, oil, gas, oil shale, tar sands, fission, fusion, solar, and geothermal). The impacts and costs are broadly defined as morbidity, mortality, economic loss, aesthetic losses, and undesirable social or political change. The actual methodology, mega-system analysis, required a considerable knowledge base and in principle could be expanded to include ecological and social systems. While one must admire the breadth of their approach, more than a quarter of a century later, the analysis of energy systems has yet to reach the goals outlined in their paper.

With the benefit of hindsight, and with high regard for the comprehensive and thoughtful quality of these previous analyses, we have to acknowledge that any analysis of overall environmental impact will be driven by the assumptions concerning future developments of the technologies, limited by a lack of fundamental data, complicated by a lack of or conflicting criteria of environmental impact, and fall far short of the comprehensive analysis that one would desire as part of the decision-making process. Thus, the discussion in this brief chapter must only be considered as illustrative and certainly the projected outcomes are not definitive.

Present and future situation

Nuclear power is a relatively new source of energy. The first electrical power production from a nuclear reactor occurred on 20 December, 1951, in Idaho at the Experimental Breeder Reactor Number 1 (EBR-1), and since that time nuclear science and technology have had a tremendous impact on the 20th century (Ewing

1999). In 2002, there were 433 nuclear power reactors in operation in 31 countries with a total capacity of approximately 350 GW (Grimston & Beck 2000). However, within the USA and the European Community there are no plants under construction or planned. At present, the main expansion of NPP construction is in Asia.

Inspired by the possibility of limiting carbon emissions by the increased use of nuclear power, there has been a resurgence of interest in the prospects for nuclear power and a growing number of analyses of its possible role in future energy production (Fetter 2000; Grimston & Beck 2000; Lake 2000; Magwood 2000; Adamov 2000; Ansolabehere *et al.* 2003). In the USA, there are major initiatives by the Department of Energy (DOE): an international program in Generation IV Nuclear Energy Systems and an Advanced Fuel Cycle Initiative. Present reactor systems, mainly advanced light water reactors (LWR), are considered the third generation (1995–2010) power plants. Generation III+ reactors will incorporate evolutionary designs and are considered to be deployable in the near-term. The Generation IV reactor systems will be the new systems that have the goals of being highly economical, passively safe, proliferation resistant, and generating minimal amounts of waste of lower toxicity (Lake 2000; Magwood 2000; DOE 2002). Among the systems selected for further consideration and development are gas-, lead-, or sodium-cooled fast-reactors, molten salt reactors, supercritical-water-cooled reactors or very high-temperature reactors. The Generation IV-A thermal reactors are envisioned to become available by 2020, and the Generation IV-B fast reactors would be available by 2040. In parallel with the development of new reactors, there is a programme to develop advanced closed fuel cycles. These new fuel cycles will involve the development of new types of fuel; actinide and fission product separation, such that radionuclides can be handled in different ways depending on disposal requirements or potential beneficial uses; advanced safeguards during reprocessing; and transmutation (by nuclear reactions) of long-lived actinides to shorter-lived or less toxic nuclides. The goals of this program are to reduce the volume of waste generated, the heat-load per unit of waste, and the radiotoxicity of the waste by transmutation to less toxic or shorter half-life radionuclides. Both the advanced reactors and advanced fuel cycles can be combined into a 'symbiotic' fuel cycle in which the fuel from thermal reactors (Generation IV-A) would be recycled once or

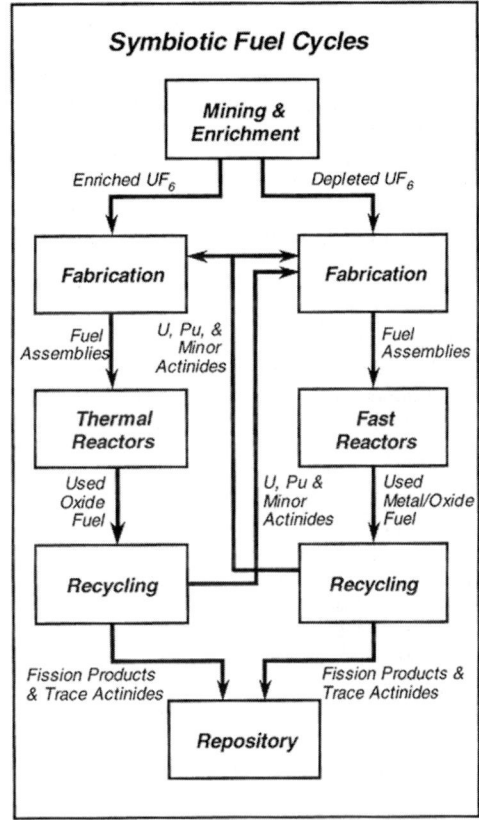

Fig. 4. Schematic illustration of a 'symbiotic' fuel cycle that utilizes both thermal and fast reactors. The fast reactors are used to 'burn' long-lived actinides (DOE 2002).

twice, and then fast reactors (Generation IV-B) would be used to transmute actinides during repeated cycles of reprocessing (Fig. 4). These are ambitious plans that could have a major effect on the environmental impact of the fuel cycle; however, there are strong economic and proliferation arguments against reprocessing (von Hippel 2001; Bunn *et al.* 2003). In Great Britain, there is speculation that British Nuclear Fuels Ltd. (BNFL) will soon shut down the Thorp (thermal oxide reprocessing) plant at Sellafield, even though this plant only opened in 1994 (*Science* 2003). At the same time, France maintains a large reprocessing facility at La Hague with a capacity of 1600 metric tonnes of heavy metal (tHM)/y, and Japan anticipates opening a major reprocessing plant in 2005 (at a cost of about US$20 billion). At present, the world-wide capacity to reprocess spent nuclear fuel is between one-third and one-half of the annual production rate of SNF ($\sim$10 000 tHM/y). At the same time, issues of radioactive waste disposal are of growing concern because after forty years of commercial nuclear power, no country has disposed of high-level nuclear waste (HLW). However, the present schedule in the USA calls for the receipt of HLW at Yucca Mountain in 2010, and other countries, such as Finland, anticipate opening their HLW repositories before 2020. Finally, concerns for the use of nuclear weapons, for example in a potential conflict between India and Pakistan, and the possible development and proliferation of nuclear weapons in North Korea, have made the expanded use of nuclear power a highly controversial subject. Thus, it is difficult to anticipate the future policies concerning nuclear power and reprocessing that will drive the decisions in individual nations. In some countries, such as India, the use of nuclear power plants to generate electricity, as well as fissile material for nuclear weapons, is inextricably intertwined (Perkovich 1999). Certainly, Asia is the region that will experience the most immediate expansion in nuclear power generation capacity. In the Asia-Pacific region, 23 new NPP were under construction in 1999 (Grimston & Beck 2000).

Nuclear waste

During the past fifty years, large quantities of nuclear waste have been created by the operation of NPP and nuclear weapons programs. The cumulative estimated cost in the USA of the clean-up from defence activities and the disposal of SNF and associated nuclear waste from commercial power generation is between 250 and 300 billion dollars in 1995 dollars (Crowley 1997). The wastes are usually classified as high-level (HLW), intermediate-level (ILW), and low-level waste (LLW). The definitions in terms of total activity and composition of each of the waste types varies from country to country, but the majority of the radioactivity is to be found in the HLW. The latter (97% of the radioactivity) consists mainly of SNF, the liquid wastes generated by the reprocessing of the SNF, and the solids used to immobilize the liquid waste. The ILW consists mainly of material produced during the fabrication of nuclear fuel assemblies, the production of nuclear weapons, and, to a lesser extent, reprocessing of fuel assemblies. In the USA, transuranic (TRU) waste is a special category containing more than 100 nCi/g of alpha-emitting transuranic isotopes (Z > 92), with half-lives greater than 20 years (except for HLW). The TRU waste is usually packaged in metal drums (55 gallons) lined with plastic that can be handled by direct contact, although some contain enough activity that they must be handled remotely. The Waste Isolation Pilot Plant (WIPP) in New Mexico is an operating repository for the TRU waste generated by defense programmes. In the USA, all radioactive waste that is not HLW or TRU is classified as LLW. Also, TRU with less than 100 nCi/g of TRU is considered to be LLW. In the USA, over two-thirds of the LLW was generated by activities associated with defence programmmes, consisting mainly of contaminated equipment and laboratory supplies (Ahearne 1997). The rest of the LLW is generated by nuclear power plants, medical procedures, and the fabrication of commercial products. The volumes of the LLW are large, but the activities are low (Table 1). The classification of the waste generally determines the disposal strategy. High-level and TRU waste require deep geological disposal with isolation for extended periods (>10 000 y). Intermediate- and low-level waste may be disposed of by shallow burial with protective barriers and monitoring for more limited periods (up to a thousand years).

During the past decade, a new waste type has emerged – the plutonium from dismantled nuclear weapons. Under the first and second Strategic Arms Reduction Treaties, as well as unilateral pledges made by the USA and Russia, thousands of nuclear weapons will be dismantled. Initially, this disarmament process will produce between 30 and 40 t, pure and impure, of weapons-grade plutonium in each country, as well as hundreds of tons of highly enriched ^{235}U (HEU). Still, this will leave well over 200 t of separated plutonium in nuclear weapons. An equally important problem is the fate of plutonium separated from commercially generated fuel originally destined for fabrication as a mixed-oxide (U + Pu) fuel. The largest inventories are in France (72 t, of which 33.6 t is foreign owned) and Great Britain (60 t, of which 6.1 t is foreign owned). The USA has only 5 t of civil plutonium. Japan has 5 t of Pu, but it has another 24.1 t held in other countries, mainly France (Albright et al. 1997). Global inventories of this 'civil' Pu are nearly 200 t, but it is now clear that not all of it will be fabricated into a MOX fuel (Ewing 2001). France probably has about 200 t of civil Pu already fabricated into MOX fuel for its 20 LWRs, but the inventory of civil plutonium continues to grow. This is an extremely important proliferation and environmental problem, as the bare critical mass of ^{239}Pu is less than 10 kg (Mark 1993), and such small volumes could be diverted to the production of a nuclear weapon. Addition-

Table 1. *Summary of estimated nuclear waste inventories in the USA in 2010**

Spent nuclear fuel (commercial)	61 800 tHM[‡]
	39 800 MCi
Spent nuclear fuel (weapons programs)	2500 tHM
High-level waste (reprocessing)	380 000 m^3
	2400 MCi
Buried waste (LLW)	6.2×10^6 m^3
	50 MCi
Excess nuclear materials:	
Highly enriched uranium	174 t[†]
Plutonium	
Weapons capable	38.2 t
Not weapons usable without processing	14.3 t
Depleted uranium as UF$_6$	700 000 t
^{137}Cs and ^{90}Sr separated from HLW in capsules as CsCl and SrF$_2$	90 GCi
Uranium mine & mill-tailings	438×10^6 m^3
	3000 MCi
Contaminated soil	30 to 80×10^6 m^3
Contaminated water	1800 to 4700×10^6 m^3

*Compiled from estimates in Crowley (1997); Ahearne (1997); Crowley & Ahearne (2002).
[†] t = metric tonnes; [‡] tHM = metric tonnes of heavy metal.

ally, Pu can cause acute health effects (Sutcliffe *et al.* 1995), and although these effects are often exaggerated in the press, this is inevitably a major public concern. One of the most active areas of research during the past decade has been to develop durable, solid matrices for the immobilization of Pu and other actinides (Ewing 2001; Ewing *et al.* 2004). Of equal importance are the stockpiles of highly enriched ^{235}U (HEU): 645 t (USA) and 1050 t (Russia) (Bunn & Holdren 1997). The HEU can be blended down to low enriched uranium (LEU) (4% ^{235}U) for use in LWRs. In 1993, the USA agreed to purchase 500 t of Russian weapons-origin uranium that is then blended down to LEU for use in commercial reactors. As of 2003, 193 t of bomb-grade HEU have been converted to 5700 t LEU for fuel in NPP. The HEU could have been used to produce over 7700 nuclear weapons.

In the USA, the total volume of all radioactive waste is 5.5 million m^3 (excluding contaminated soil and water) and the total radioactivity is approximately 30 GCi (Ahearne 1997). There has been a significant contamination of large volumes of soil and water by uranium mine- and as mill-tailings, LLW and ILW associated with nuclear weapons programmes (e.g., burial and injection into geological formations), and atmospheric testing of nuclear weapons. In the USA, over 70 million m^3 of soil and at least 1800 million m^3 of water have been contaminated by releases from DOE facilities that were used for weapons production (DOE 1997). Despite the large volumes of LLW and ILW and the associated contaminated soil and water,

LLW and ILW waste account for only a few percent of the total radioactivity generated by the NFC (Table 1). Thus, the discussion in this paper focuses on the fate of HLW and separated plutonium.

In 2002, there were approximately 150 000 t of SNF (slightly less than one-third of this amount is in the USA). Most of this fuel (90 000 t) is still at the 236 nuclear power stations (which together have 433 reactors) where it was originally generated in 36 different countries (Carter & Pigford 1999). In the USA, the SNF inventory in 2000 was approximately 42 300 tHM with a total activity of 32.6 GCi, increasing by slightly less than 2000 tHM per year. By the year 2020 the inventory will grow to approximately 77 100 tHM with an activity of 34.7 GCi (note, due to the decay of short-lived fission products, such as ^{137}Cs and ^{90}Sr, the original level of radioactivity, one month after removal from the reactor, drops rapidly to less than 1% of the original activity in 100 years) (Ahearne 1997). The legislated capacity for the proposed HLW repository at Yucca Mountain is 70 000 t of SNF equivalent. Almost all of the used fuel is from reactors used for producing commercial power, but approximately 2640 tHM is from reactors used to produce Pu for weapons, foreign and university research reactors, and naval reactors (mainly from submarines). Currently these HLW are stored at 131 sites in 39 states (Dyer 2003).

The other main type of HLW are the liquid and sludge (a mixture of liquid, suspended colloids, and solids) that resulted from reprocessing to reclaim fissile material, either for weapons production or MOX fuel. In the USA, approximately 380 000 m^3 (100 million gallons) of HLW have a total radioactivity of 960 MCi. The greatest volume, 340 000 m^3, is stored in tanks at Hanford, Washington, and Savannah River, South Carolina, and most of the balance is stored as a solid calcine in tanks at the Idaho National Engineering Laboratory. Over 99% of the present radioactivity is from non-actinide radionuclides, with half-lives of less than 50 years (reprocessing has removed much of the actinide content). After 500 years, the total activity will be substantially reduced, and the main radionuclides will be ^{238}Pu, ^{131}Sm, and ^{241}Am. After 50 000 years, most of the activity will be associated with long-lived actinides, such as ^{239}Pu and ^{240}Pu.

It is instructive to consider the production history and experience of a single site. During the 67 years that the Hanford site was in operation, some 67 t of Pu were produced by reprocessing 97 000 t of SNF (irradiated to lower burn-up to increase the yield of Pu) (Zorpette 1996). Using two chemical processing plants, some 2 million m^3 (500 million gallons) of highly radioactive and chemically toxic waste were produced (Crowley & Ahearne 2002). The composition of these waste sludges is exceedingly complex, consisting of nearly one-half of the elements in the periodic table, as well as a wide range of organic species and complex ions, such as oxalates, EDTA, organic complexes, and cyanides (Bunker et al. 1995). In addition, there are 76 000 m^3 of solid waste contaminated with actinides (mainly Pu), and an additional 1.2 million m^3 of LLW that have been buried on the site (Crowley & Ahearne 2002). At present, there are approximately 200 000 m^3 of HLW stored in 177 underground tanks (capacities range from 55 000 to 1.1 million gallons). The design lifetime for the first, single-shell tanks was 10 to 20 years, but these tanks have been in service between 17 and 59 years. Sixty-seven of the tanks are known or are suspected of having leaked, and estimates suggest that as much as 1.5 million gallons with one MCi of activity may have been released to the subsurface (Zorpette 1996).

In most countries, the HLW is destined for vitrification in a borosilicate glass. Many other materials, such as different glass compositions (e.g., phosphate glasses), polycrystalline ceramics (e.g., Synroc, glass ceramics, and concrete) and single phases (e.g., zircon, pyrochlore, and monazite), have been considered for the immobilization of HLW (Lutze & Ewing 1988a; Ewing et al. 1995; Donald et al. 1997). The most recent work has focused on the immobiliz-

ation of Pu (Ewing et al. in press). The type of waste form used for the immobilization of HLW and actinides has a profound effect on the extent of the environmental impact of the disposed waste, as there are three to five orders of magnitude difference in the release rates from the different waste forms (Lutze & Ewing 1988b).

Over 500 t of HLW have been vitrified in France and Germany. In the USA, the HLW at the Nuclear Fuel Services plant in West Valley Plant, New York, have been vitrified (300 two-ton canisters) and vitrification is ongoing at the Defense Waste Processing Facility (DWPF) at Savannah River, South Carolina ($\sim$1600 canisters by February 2004). A vitrification plant is under construction at Hanford, Washington. Vitrification of all of the HLW in the USA will generate approximately 20 000 canisters, which are destined for disposal at the geological repository at Yucca Mountain.

In the former Soviet Union and the present Russian Federation, nuclear waste continues to be generated by the weapons programme and commercial power generation. The most recent data are available in a study jointly sponsored by the USA and the Russian Academy of Sciences (NRC 2003). Within the Russian Federation there are approximately 14 000 tHM (5 GCi) that require disposition. Within its borders there are 30 operating nuclear power plants at ten locations that produce spent fuel at an annual rate of 850 tHM/year. The former Soviet Union also constructed 38 NPP in Eastern Europe, Ukraine, Finland, and Lithuania. The Russian Federation has confirmed its intention to honour the previous obligations of the former Soviet Union and take back the SNF from these reactors. Russian policy for the SNF is that most of it will be reprocessed as part of a closed fuel cycle. At the Mayak facility in the Urals, HLW from reprocessing has been immobilized into a Na-Al phosphate glass ($\sim$300 MCi). The Russians recognize that their aluminum-phosphate glass composition is not durable enough for long-term isolation, and a variety of new glass compositions and crystalline ceramics are under investigation as potential nuclear waste forms (Ewing & Wang 2002). The total activity of radioactive wastes at Minatom (Ministry of Atomic Energy of the Russian Federation) facilities is more than 2 GCi contained in more than 4×10^8 m^3 (NRC 2003). This does not include large volumes (at least 5×10^6 m^3 with more than 100 MCi of activity) injected into geological formations (NRC 2003). The total volume of liquid wastes from reprocessing is over 8×10^5 m^3 (including HLW and ILW). Environmental releases associated with reprocessing have been substantial. In the early days of the operation of the Mayak Plant 'B', the HLW were released directly into the Techa River at a rate of 1 kCi/day. The total amount of activity released to the Techa River between 1949 and 1956 was nearly 3 MCi, with detectable pollution stretching for hundreds of kilometres down river. After 1958, ILW (1.2 MCi) was placed in Lake Karachai. Liquid radioactive waste is still dumped into Lake Karachai. In 1957, a chemical explosion in a HLW tank released approximately 2 MCi over the surrounding area. The area surrounding the Pu-production reactors and reprocessing facilities at Mayak (near Chelyabinsk-65) must be considered as one of the most contaminated sites, due to nuclear activities, in the world (JNREG 1997).

As mentioned previously, the USA and Russia have excess plutonium from the dismantlement of nuclear weapons. In September 2000, the USA and Russia signed a Plutonium Management and Disposition Agreement that commits each country to dispose of 34 t of surplus Pu. Russia plans to fabricate a MOX fuel with the Pu and use it in existing or future reactors as part of its larger strategy to close the fuel cycle. The USA will also fabricate a MOX fuel to be used in existing commercial reactors, but there is presently no strategy for the fate of 5 to 7 t of so-called 'scrap' plutonium, which is too contaminated with impurities to be used in a MOX fuel.

One of the most pressing problems in Russia is the storage and disposal of the SNF from nuclear submarines and nuclear-powered surface ships (NRC 2003). With the end of the Cold War and the Soviet Union in 1991, 200 Soviet nuclear submarines were decommissioned. More than ten years later, over 100 submarines await dismantlement (Webster 2003). Storage facilities are in poor condition, and reactors (some with damaged fuel assemblies) remain in buoyed submarines in ports from the Kola Peninsula to Vladivostok. Even with the SNF removed, the reactors remain a major disposal problem because of their size and level of contamination. Eighty-nine of the removed reactors remain in their original compartments, held afloat in large circular buoys (Webster 2003). The largest SNF storage facility at Andreeva Bay on the northernmost coast of the Kola Peninsula (near the border with Norway) contains 21 000 spent fuel assemblies. Originally stored in cooling ponds that breached and released activity, the fuel assemblies are now stored in concrete silos. The immediate plan is to move the SNF assemblies to Mayak Chemical Combine in the Ural Mountains, more than 3200 km away, where they can be stored until reprocessing, but

this will still leave a substantial amount of contaminated soil, water, and concrete at the storage facility at Andreeva Bay.

As in the USA, most of the HLW (SNF or immobilized HLW fluids) in Russia is destined for geological disposal. Site investigations are under way in the area of the two reprocessing facilities (Mayak and Krasnoyarsk), as well as in the regions of nuclear navy bases in the Far East and Northwest.

Nuclear fuel cycles

In terms of assessing the environmental impact, the type of fuel cycle is probably the most important aspect of nuclear power generation. Broadly speaking, there are two types, open and closed. The most important difference is that with the closed fuel cycle there is reprocessing of the fuel to reclaim fissile material (Figs. 1 and 2). Reprocessing of the used nuclear fuel is always complicated by the issue of the possible diversion of fissile material and the proliferation of nuclear weapons. The type of fuel cycle is closely tied to the type of nuclear fuel (e.g., low-enriched uranium (LEU), high-enriched uranium (HEU), mixed-oxide (MOX), or inert matrix fuels (IMF)), the degree of burn-up (e.g., at lower burn-up the inventory of Pu is higher), the types of solids used for the immobilization of the waste (e.g., special waste forms for actinides and fission products), and the site selected for storage and geological disposal (e.g., the UO_2 in SNF is much less stable in an oxidizing geochemical environment than in a reducing environment). Although the choice between fuel cycles may be driven by economic considerations, with an open fuel cycle prevailing as long as the price of uranium is lower than the cost of reprocessing (von Hippel 2001; Bunn et al. 2003), other considerations may prevail. Thus, there are no simple criteria that can serve as a basis for the selection of a fuel cycle, as environmental, proliferation, and economic issues are closely tied to a nation's energy policy, and that depends on the energy resources that are available to each country. As an example, a once-through cycle using a PWR may create the lowest volumes of LLW and ILW, but will have very high volumes of mill-tailings and spent fuel (Ko et al. 2002). The PWR-MOX option with reprocessing has lower volumes of mill-tailings and spent fuel, but much higher volumes of LLW and ILW (Ko et al. 2002). From an economic point of view, the volume of waste is a consideration in estimating disposal costs, but may be relatively unimportant to estimating the environmental impact. As important as the volume of waste generated is the heat content of the waste generated by the decay of fission products. This heat not only affects the stability of the nuclear waste form (e.g., devitrification and accelerated corrosion of the borosilicate glass waste form at elevated temperatures), but it also affects repository design and performance. At the proposed Yucca Mountain repository in the USA, temperatures well over 200°C are envisioned for several hundred years. The elevated temperature has been incorporated into the design strategy in order to prevent water from coming into contact with the metallic waste packages. A reprocessing scheme in which actinides and fission products are separated into very specific compositions (e.g., separation of Tc from I) and immobilized in compositionally specific, durable waste forms (perhaps with high waste loadings) may substantially reduce the release of these radionuclides to the environment.

A recent MIT study has compared generic fuel cycles for a global growth scenario that is based on an expansion of worldwide nuclear power generating capacity by a factor of three (1000 GW) by the year 2050 (Ansolabehere et al. 2003). This would result in avoiding 1.8 GtC (if the displacement is from coal-burning plants), which is approximately 25% of the annual global emissions (Houghton 2000). The MIT study considered three fuel cycles: (1) a once-through cycle with direct geological disposal of the SNF (at different burn-ups); (2) a single cycle of reprocessing in which the plutonium is fabricated into a MOX fuel; and (3) a fully closed fuel cycle with a 'symbiotic' combination of thermal and fast reactors, the latter used to burn separated actinides (Fig. 4). An important aspect of the third type of fuel cycle is whether there is repeated reprocessing of the SNF from the thermal reactors, or whether repeated cycles of reprocessing are limited to the fast reactors. The general characteristics of each of these fuel cycles are summarized in Table 2. Although the data in Table 2 provide a very simplified view of the attributes of the different fuel cycles, one can see immediately that there are important differences in the volumes and types of radioactive waste that are generated. The once-through cycle followed by direct geological disposal generates the largest volumes of SNF, which contains substantial quantities of Pu (which can be viewed as either a source of energy or an environmental hazard). Going to higher burn-up reduces the volume of SNF generated per unit of energy and reduces, by approximately 100 t, the Pu in the spent fuel, because the fissioning of ^{239}Pu accounts for nearly one-third of the energy generated in a typical LWR (^{239}Pu is created by

Table 2. Selected characteristics of generic fuel cycles based on a projected deployment of 1500 GWe (gigawatts of electricity) per year in 2050

	Capacity (GWe)	Enrichment (% ^{235}U)	Burn-up (GWd/tHM)	SNF[†] (10³ tHM/y)	SNF reprocessed (10³ tHM/y)	HLW[¶] (vitrified) (t/y)	Pu (t/y)
Once-through (low burn-up)	1500	4.5	50	29.9	None	None	397
Once through (high burn-up)	1500	8.2	100	14.9	None	None	294
Closed cycle (MOX with one recyle)	1500						
UOX	1260	4.5	50	All reprocessed	25.1	2886[‡]	233[‡]
MOX	240	7.0 (Pu in MOX)	50	4.76	None		
Fully closed Hybrid cycle	1500						
LWR (thermal)	815	4.5	50	All reprocessed	16.24	1868[§]	0.7[‖]
Fast reactor	685	25.0	120	All reprocessed	4.69	539[§]	

*These figures are taken from Ansolabehere et al. (2003). The details, which are very important, of the calculations are given in the appendix of their study.
[†]The borosilicate glass contains 1292.6 t/y of fission products, 30.1 t/y of minor actinides, and 0.3 t/y of Pu.
[‡]Of the 233 t of Pu that are generated, 167 t of separated Pu are incorporated into MOX fuel, the reprocessing also generates 23 443 t/y of separated uranium.
[§]Mass of borosilicate glass estimated by author. This borosilicate glass contains 48.6 t/y of fission products, 1.1 t/y of plutonium and minor actinides.
[‖]Lost during reprocessing.
[¶]SNF = spent nuclear fuel; HLW = high-level waste.

(n,γ) reactions on ^{238}U). The proliferation risks are minimized because there is no separated Pu. The Pu in the SNF is protected from diversion by the strong radiation field generated by the fission products in the SNF. On the other hand, SNF, the waste form for all of the radionuclides, is not stable in an oxidizing environment, and so the properties of the geological repository become an essential, if not the most important, part of the strategy for nuclear waste containment. The closed fuel cycle (the MOX option with one recycle) has the advantage of more efficient use of the fissile radionuclides (^{235}U and ^{239}Pu), which are fabricated into MOX fuel. However, of the total amount of Pu (233 t) discharged in the spent uranium oxide (UOX) fuel, 167 t of the Pu will be separated. This amount of separated plutonium is equivalent to thousands of nuclear weapons (a nuclear weapon can be made with less than 10 kg of Pu). For both the open and closed fuel cycles, uranium, as well as Th-based, fuels may be utilized in different reactor types, such as heavy water reactors (HWR), supercritical water reactors (SCRW), and high-temperature and very high-temperature gas-cooled reactors (HTGR). However, the introduction of new fuels, new reactors, and advanced fuel cycles will require considerable resources and experience prior to deployment. The fully closed fuel cycle is one in which the actinides are separated from the spent fuel of the thermal reactors (once-through cycle) and incorporated into MOX fuel and burned in the fast neutron reactors. The fast neutron flux transmutes the actinides, and multiple cycles of reprocessing are used for the fast reactor fuel. This significantly reduces the plutonium and minor actinide inventories, transmuting the actinides to shorter-lived radionuclides. In this scheme, it is important to carefully balance the production of the actinides in thermal reactors with burning them in recycled MOX fuel. Because MOX fuel consists mainly of ^{238}U, additional ^{239}Pu is created by neutron capture and subsequent γ-decay reactions. The amount of Pu created can be reduced by burning the actinides in an inert-matrix fuel (i.e., one that does not contain ^{238}U, such as ZrO_2) that will provide more efficient burning of the actinides (Boczar et al. 1997; Oversby et al. 1997). In this case, reactors would burn a mixture of MOX and IMF (perhaps up to one-third of the core loading). This option substantially reduces the Pu and minor actinide content of irradiated IMF, and the zirconia (ZrO_2) is recognized as a durable, radiation-resistant waste form for direct disposal (Gong et al. 2000). The use of ZrO_2 as a fuel is under active investigation in Europe and Japan, but it may take more than a decade

to confirm its behaviour under actual reactor operating conditions. This fully closed fuel cycle should not be confused with breeder reactor fuel cycles in which breeder fast reactors are used to generate Pu, which is then burned as MOX fuel in thermal reactors. In this sequence, the inventory of Pu will increase.

Nuclear power and carbon emission

In 1997, the third Conference of the Parties (COP-3) produced the Kyoto Protocol. Although signed by then Vice President Al Gore, it has not been ratified by the Senate of the USA. To enter into force, the Protocol must be ratified by 55 parties representing at least 55% of the world's emissions of greenhouse gases (GHG) in 1990. As part of the Protocol, the developed countries must commit themselves to reducing their collective emissions of six GHGs to at least 5% below 1990 levels. The most prominent of these GHGs is CO_2, which accounts for nearly 65% of the warming effect (Houghton et al. 2001). The USA presently accounts for approximately 25% of the global emissions of CO_2, with 5% of the world's population. The Kyoto Protocol would require the USA to reduce present emissions by $\sim$22%, an annual reduction of 1.1×10^9 t_{CO_2}, equivalent to removing all the gasoline-powered vehicles from US roads (Loewen & León 2001). The USA produces nearly 20% of its electricity using nuclear power, and this is equivalent to avoiding the release of 6×10^8 t_{CO_2}, if this electricity had been produced from carbon-based fuels (Loewen & Léon 2001).

There is a pressing need for developing a timely strategy to reduce GHG emissions. Thus, a number of analyses are based on a goal of limiting the increase in CO_2 emissions to twice (550 ppm) the pre-industrial levels (275 ppm) by the year 2050 (Fetter 2000; Sailor et al. 2000). Present CO_2 levels are just over 360 ppm, increasing at an average rate of 1.5 ppm/y. This adds 3.3 GtC/y to the atmospheric reservoir, which is 750 GtC (Houghton et al. 2001). Models of CO_2 emissions suggest that strategies for reduction must be initiated in developed countries by 2010 in order to meet the goal of only doubling of the CO_2 concentration above the pre-industrial level (Wigley 1997). As previously discussed, of all the technologies presently capable of contributing to a major reduction in carbon emissions, nuclear power is one of the most promising, simply because the technology is already operating on a substantial scale, and in principle, it could be deployed more rapidly on a global scale. The Nuclear Energy Institute (NEI) maintains that the US nuclear power generating capacity can be increased to 10 GW by 2012 (equivalent to 0.022 GtC/y). The NEI supports a goal of adding 50 GW capacity (equal to approximately 50 new NPP) by 2020 (equivalent to a reduction of 0.1 GtC/y).

Analyses for the prospects of nuclear power have been presented by many, but two of the most detailed are by Fetter (2000) and Sailor et al. (2000). These analyses necessarily make many assumptions about future energy needs. Assuming a stabilization of CO_2 concentrations to approximately twice pre-industrial levels by 2050, and projecting a growth in world population to 9 billion (a 50% increase) and an increase in per capita energy consumption of 50%, the global energy demand in 2050 will be approximately 900 exajoules (10^{18} joules) per year (EJ/y) (Sailor et al. 2000). If nuclear power provides one-third of the projected energy requirement (300 EJ/y), and the balance is divided equally between conventional fossil fuels and 'decarbonized' fossil fuels, the 300 EJ from nuclear are roughly equivalent to 3300 GW-years (one GWy is the average annual energy output from a single large power plant) of capacity per year (present capacities are about 260 GWy/y). With this scenario, the projected 900 EJ/y of global energy use would still result in CO_2 emissions that would equal 5.5 GtC/y (present levels are approximately 6.6 GtC/y) (Sailor et al. 2000). Still, this would mean a more than a tenfold increase in nuclear power generating capacity requiring the construction of over 3000 NPP before 2050 (at present there are 439 operating nuclear generating units). The impact of this expansion in nuclear power generation capacity is difficult to anticipate because it depends critically on the types of reactors and fuel cycles that are used, as previously discussed. The figures previously cited from the MIT study (Ansolabehere et al. 2003) and tabulated in Table 2 were based on an increase by a factor of three of nuclear generating capacity by 2050 (1000 GWe). Still, one must expect that the most immediate deployment of new reactors will be of the Generation III+ type, not too different from the present water reactor technology, but with higher burn-up of the nuclear fuel. Thus, one may use the present technology as a basis for extrapolating the environmental impact and use the factors of 3 to 10 as the range of what has been considered for the increase in nuclear power production. On this basis, the annual increase in spent fuel production would be between 27 000 and 89 000 tHM. The higher number is more than the presently planned capacity (70 000 tHM

equivalent) for the proposed repository at Yucca Mountain. One approach to reducing the impact of the increased nuclear waste production would be to use reprocessing to minimize the volumes of waste produced and to utilize the fissile content of the SNF; however, this raises major issues related to the proliferation of nuclear weapons. A 1 GWy light water reactor produces 200 kg/y of Pu (enough for 20 nuclear weapons). If the nuclear energy capacity is increased to 3000 GW, then the annual production of Pu would be over 500 000 kg (Williams & Feiveson 1990). If one foresees a nuclear industry based on Pu-breeder reactors, the 3000 GW nuclear system would produce five million kilograms of plutonium per year (Williams & Feiveson 1990). Alvin Weinberg (2000) has related the reduction (avoided increase) in CO_2 content in the atmosphere to the amount of U consumed, that is, the percentage of U fissioned in nuclear power plants. A typical LWR without reprocessing has an efficiency (% of U fissioned) of only 0.5%, while a perfect breeder reactor cycle with reprocessing has an efficiency of 70%. Even if the presently estimated reserves for uranium (30×10^6 t) are completely utilized (Weinberg 2000), the low-efficiency system that we now use, LWR followed by direct disposal, will lower the CO_2 increase by only 38 ppm. Either there will have to be a shift to breeder reactors and reprocessing, or alternative sources of U must be found. All of these figures are speculative, but they do emphasize that an increase in the role of nuclear power in reducing carbon emissions must be substantial and go hand in hand with the development of advanced fuel cycles and waste management technologies that do not presently exist on an industrial scale.

Just as important as evaluating the performance of the nuclear fuel cycle, one must also consider the size of the fluxes and reservoirs of the carbon cycle. Present CO_2 emissions from fossil fuels and the production of cement are estimated to be 6.3 ± 0.4 GtC/y; emissions related to changes in land use (e.g., deforestation) are 1.6 ± 0.8 GtC/y (Schimel et al. 2001). At present, the reduction of CO_2 emissions that can be attributed to the use of nuclear power is 0.5 GtC/y. Thus, the uncertainties in the major fluxes in the carbon cycle are approximately the same as the present impact of nuclear power on CO_2 emissions (Sarmiento & Gruber 2002). To quote from Falkowski et al. (2000), 'Our knowledge is insufficient to describe the interactions between the components of the Earth system and the relationship between the carbon cycle and other biogeochemical and climatological processes.' We will need much more refined understanding of the carbon cycle and a more explicit description of the nuclear fuel cycle before one can quantify the impact of the fuel cycle on the carbon cycle.

Concluding remark

I realize that I have led the reader through a bewildering array of numbers and concepts. I am also acutely aware of the fact that I have failed to arrive at a set of final conclusions or recommendations. I think that this is to be expected, as the uncertainties in the analysis of the environmental impact of the nuclear fuel cycle remain large and even unconstrained in the absence of an explicit description of the type of fuel cycle to be used. This is made all the more evident by the failure of previous, although thorough, analyses (e.g., Pigford 1976) to anticipate the state of the nuclear power industry even 25 years into the future. It is also difficult to estimate the impact of the nuclear fuel cycle on the carbon cycle, because the uncertainties in our knowledge of the global carbon cycle remain large (Falkowski *et al.* 2000). Still, even in the face of the uncertainties for both of these cycles, nuclear and carbon, one must acknowledge the tremendous amounts of energy available from the non-carbon producing nuclear option. However, the scale of the increase in nuclear power production that will be required, up to a factor of 10, will lead to reprocessing of spent nuclear fuel, and this raises two issues: (1) the successful management and disposal of the radioactive waste, and (2) the potential proliferation of nuclear weapons. I remain optimistic about reaching a successful solution to the first problem and see no basis for optimism on the second.

This paper has benefited from the patience of two colleagues in different departments: Lynn Walter in Geological Sciences and Ron Fleming in Nuclear Engineering and Radiological Sciences, both at the University of Michigan. Lynn led me through the intricacies of the carbon cycle, while Ron gave me daily tutorials on reactor physics and the nuclear fuel cycle. It was an exhilarating experience to feel these two cycles come together in my mind. I also am grateful for the thorough reviews of this chapter by Mark Logsdon and Dan Golomb. A rather harsh, anonymous review was also much appreciated. Finally, I thank the editors, Peter Stille and Reto Gieré, for inviting me to write this chapter and the patience to wait for its completion. I am also grateful for the support of the John Simon Guggenheim Memorial Foundation for a fellowship that allowed me to stray far from my normal research path.

Figures 1 and 2 are reprinted with permission from End Points for Spent Nuclear Fuel and High-Level Radioactive Waste in Russia and the United States (2003) by the National Academy of Sciences and are reproduced courtesy of the National Academies Press, Washington, D.C.

References

ADAMOV, Y. O. 2000. Nuclear power may get its second wind in the 21st century. *Nuclear News*, **43**, 38–42.

AHEARNE, J. F. 1997. Radioactive waste: The size of the problem. *Physics Today*, **50**, 24–29.

ALBRIGHT, D. & BARBOUR, L. 1999. Separated neptunium and americium. *In*: ALBRIGHT, D. & O'NEIL K. (eds) *The Challenges of Fissile Material Control*. ISIS, Washington, DC, 85–96.

ALBRIGHT, D., BERKHOUT, F. & WALKER, W. 1997. *Plutonium and Highly Enriched Uranium 1996 World Inventories, Capabilities and Policies*. Oxford University Press, New York, 502 pp.

ANSOLABEHERE, S., DRISCOLL, M. *et al.* 2003. *The Future of Nuclear Power. An Interdisciplinary MIT Study*. MIT, 161 pp.

BOCZAR, P. G., GAGNON, M. J. N., CHAN, P. S. W., ELLIS, R. J., VERRALL, R. A. & DASTUR, A. R. 1997. Advanced CANDU systems for plutonium destruction. *Canadian Nuclear Society Bulletin*, **18**, 2–10.

BREDERHOEFT, J. D., ENGLAND, A. W., STEWART, D. B., TRASK N. J. & WINOGRAD, I. J. 1976. Geological disposal of high-level radioactive wastes – Earth-science perspectives. *U.S. Geological Survey Circular*, **779**, 28 pp.

BUDNITZ, R. J. & HOLDREN, J. P. 1976. Social and environmental costs of energy systems. *In*: HOLLANDER, J. M. & SIMMONS, M. K. (eds) *Annual Review of Energy*, vol. 1. Annual Reviews Inc., Palo Alto, CA, 553–580.

BUNKER, B., VIRDEN, J., KUHN, B. & QUINN, R. 1995. Nuclear materials, radioactive tank wastes. *In*: *Encyclopedia of Energy Technology and the Environment*, vol. 4. John Wiley & Sons, Inc., New York, 2023–2032.

BUNN, M. & HOLDREN, J. P. 1997. Managing military uranium and plutonium in the United States and the former Soviet Union: Current security challenges. *In*: SOCOLOW, R. H., ANDERSON, D. & HARTE, J. (eds) *Annual Review of Energy and the Environment*, **22**, 403–486.

BUNN, M., FETTER, S., HOLDREN, J. P. & VAN DER ZWAAN, B. 2003. *The Economics of Reprocessing vs. Direct Disposal of Spent Nuclear Fuel*. Project on Managing the Atom. Final Report, December, 2003, DE-FG26-99FT4028, 127 pp. World wide web Address: http://www.bcsia.ksg.harvard.edu/ BCSIA_content/documents/econ_reprocessing_m_ bunn.pdf

CARTER, L. J. 1987. *Nuclear Imperatives and Public Trust*. Resources for the Future, Inc., Washington, DC, 473 pp.

CARTER, L. J. & PIGFORD, T. H. 1999. The world's growing inventory of civil spent fuel. *Arms Control Today*, **January/February**, 8–14.

CROWLEY, K. D. 1997. Nuclear waste disposal: The technical challenges. *Physics Today*, **50**, 32–39.

CROWLEY, K. D. & AHEARNE, J. F. 2002. Managing the environmental legacy of U.S. Nuclear – Weapons Production. *American Scientist*, **90**, 514–523.

DONALD, I. W., METCALFE, B. L. & TAYLOR, R. N. J. 1997. The immobilization of high level radioactive wastes using ceramics and glasses. *Journal of Materials Science*, **32**, 5851–5887.

DYER, R. 2003. Nuclear has a lot going for it. *Nuclear News*, **46**, 36–37.

DOE (DEPARTMENT OF ENERGY), U.S., ENVIRONMENTAL MANAGEMENT PROGRAM. 1997. *Linking legacies – Connecting the Cold War Nuclear Weapons Production Processes to Their Environmental Consequences*. DOE/EM-0319, 230 pp.

DOE (DEPARTMENT OF ENERGY), U.S., NUCLEAR ENERGY RESEARCH ADVISORY COMMITTEE. 2002. *A Technology Roadmap for Generation IV Nuclear Energy Systems*. GIF-002-00, 91 pp.

EWING, R. C. 1999. Radioactivity in the 20th century. *In*: BURNS, P. C. & FINCH, R. (eds) *Uranium: Mineralogy, Geochemistry and the Environment, Reviews in Mineralogy*, vol. 38, Mineralogical Society of America, Washington, D.C., 1–21.

EWING, R. C. 2001. The design and evaluation of nuclear-waste forms: Clues from mineralogy. *Canadian Mineralogist*, **39**, 697–715.

EWING, R. C. 2002. Acceptance of Hawley Medal for 2002. *Canadian Mineralogist*, **41**, 241–242.

EWING, R. C., WEBER, W. J. & CLINARD, F. W., JR. 1995. Radiation effects in nuclear waste forms for high-level radioactive waste. *Progress in Nuclear Energy*, **29**, 63–127.

EWING, R. C. & WANG, L. M. 2002. Phosphates as nuclear waste forms. *In*: KOHN, M. J. U., RAKOVAN, J. & HUGHES, J. M. (eds) *Reviews in Mineralogy and Geochemistry*, **48**, Mineralogical Society of America, Washington, D.C., 673–699.

EWING, R. C., WEBER, W. J. & LIAN, J. (2004) Pyrochlore ($A_2B_2O_7$): A nuclear waste form for the immobilization of plutonium and 'minor' actinides. *Journal of Applied Physics*, **95**, 5949–5971.

FALKOWSKI, P., SCHOLES, R. J. *et al.* 2000. The global carbon cycle: A test of our knowledge of Earth as a system. *Science*, **290**, 291–296.

FETTER, S. 2000. Energy 2050. *The Bulletin of the Atomic Scientists*, **July/August**, 28–38.

GONG, W. L., LUTZE, W. & EWING, R. C. 2000. Zirconia ceramics for excess weapons plutonium waste. *Journal of Nuclear Materials*, **277**, 239–249.

GRIMSTON, M. C. & BECK, P. 2000. *Civil Nuclear Energy: Fuel of the Future or Relic of the Past?* The Royal Institute of International Affairs, Energy and Environment Programme, London, 126 pp.

HEDIN, A. 1997. Spent nuclear fuel – how dangerous is it? SKB Technical Report 97-13, Swedish Nuclear Fuel and Waste Management Co., Stockholm, Sweden, 60 pp.

HOFFERT, M. I., CALDEIRA, K. *et al.* 2002. Advanced technology paths to global climate stability: Energy for a greenhouse planet. *Science*, **298**, 981–987.

HOUGHTON, J. 2000. *Global Warming: The Complete Briefing*. Cambridge University Press, Cambridge, 251 pp.

HOUGHTON, J. T., DING, Y. *et al.* (eds) 2001. *Climate Change 2001: The Scientific Basis*. Cambridge University Press, 881 pp.

JNREG (JOINT NORWEGIAN–RUSSIAN EXPERT GROUP REPORT). 1997. *Sources contributing to radioactive contamination of the Techa River and areas surrounding the 'Mayak' production association, Urals, Russia*. Norwegian Radiation Protection Authority, 134 pp.

KO, W. I., KIM, H. D. & YANG, M. S. 2002. Radioactive waste arisings from various fuel cycle options. *Journal of Nuclear Science and Technology*, **39**, 200–210.

LAKE, J. A. 2000. Foundations for the fourth generation of nuclear power. *Nuclear News*, **43**, 32–35.

LOEWEN, E. P. & LEÓN, S. B. 2001. Implications of recent developments in global climate change policy — a radical suggestion. *Nuclear News*, **44**, 23–25

LUTZE, W. & EWING, R. C. (eds) 1988a. *Radioactive Waste Forms for the Future*. Elsevier Science Publishers, Amsterdam, 778 pp.

LUTZE, W. & EWING, R. C. 1988b. Summary and evaluation of nuclear waste forms. *In*: LUTZE, W. & EWING, R. C. (eds) *Radioactive Waste Forms for the Future*. Elsevier Science Publishers, Amsterdam, 699–740.

MAGWOOD, W. D., IV. 2000. Roadmap to the next generation of nuclear power systems: A vision for a powerful future. *Nuclear News*, **43**, 35–38.

MARK, J. C. 1993. Explosive properties of reactor-grade plutonium. *Science & Global Security*, **4**, 111–128.

NRC (NATIONAL RESEARCH COUNCIL) 2003. *End Points for Spent Nuclear fuel and High-Level Radioactive Waste in Russia and the United States*. The National Academies Press, Washington, DC, 137 pp.

OVERSBY, V. M., McPHEETERS, C. C., DEGUELDRE, C. & PARATTE, J. M. 1997. Control of civilian plutonium inventories using burning in a non-fertile fuel. *Journal of Nuclear Materials*, **245**, 17–26.

PERKOVICH, G. 1999. *India's Nuclear Bomb: The Impact on Global Proliferation*. University of California Press, Berkeley, 597 pp.

PIGFORD, T. H. 1976. Environmental aspects of nuclear energy production. *In*: HOLLANDER, J. M. & SIMMONS, M. K. (eds): *Annual Review of Energy*, vol. 1, Annual Reviews Inc., Palo Alto, CA, 515–559.

ROMER, R. H. 1972. Resource letter ERPEE-1 on energy: Resources, production, and environmental effects. *American Journal of Physics*, **40**, 805–829.

ROSENFELD, A. H., KAARSBERG, T. M. & ROMM, J. 2000. Technologies to reduce carbon dioxide emissions in the next decade. *Physics Today*, **November**, 29–34.

SAILOR, W. C., BODANSKY, D., BRAUN, C., FETTER, S. & VAN DER ZWANN, R. 2000. A nuclear solution to climate change. *Science*, **288**, 1177–1178.

SARMIENTO, J. L. & GRUBER, N. 2002. Sinks for anthropogenic carbon. *Physics Today*, **August**, 30–36.

SCIENCE NEWS SECTION. 2003. Britain set to pull plug on nuclear-fuel reprocessing. *Science*, **425**, 7–8.

SCHIMEL, D. S., HOUSE, J. I., *et al.* 2001. Recent patterns and mechanisms of carbon exchange by terrestrial ecosystems. *Nature*, **414**, 169–172.

SUTCLIFFE, W. G., CONDIT, R. H., MANSFIELD, W. G., MYERS, D. S., LAYTON, D. W. & MURPHY, P. W. 1995. A perspective on the dangers of plutonium. Center for Security and Technology Studies, Report CSTS-48-95, 13 pp.

VON HIPPEL, F. N. 2001. Plutonium and reprocessing of spent nuclear fuel. *Science*, **293**, 2397–2398.

WEBSTER, P. 2003. Haunted by Red October. *Science*, **301**, 1460–1463.

WEINBERG, A. M. 2000. Some necessary conditions for the rebirth of nuclear energy. *Nuclear News*, **43**, 47–48.

WILLIAMS, R. H. & FEIVESON, H. A. 1990. How to expand nuclear power without proliferation. *The Bulletin of the Atomic Scientists*, **46**, 1–8.

WIGLEY, T. M. L. 1997. Implications of recent CO_2 emission-limitation proposals for stabilization of atmospheric concentrations. *Science*, **390**, 267–270.

ZORPETTE, G. 1996. Hanford's nuclear wasteland. *Scientific American*, **May**, 88–97.

Reducing the long-term environmental impact of wastes arising from uranium mining

KAYE HART

Australian Nuclear Science and Technology Organisation (ANSTO),
PMB 1, Menai, NSW 2234, Australia (e-mail: kph@ansto.gov.au)

Abstract: Uranium mining and milling produces significant quantities of solid and liquid wastes. Solid wastes include waste rock arising from the mining activity and tailings produced by processing of the ore through the milling circuits. Of these, tailings have the most pronounced impact on the environment as they represent almost the entire mass of material processed in the milling circuits and also contain over 85% of the original radioactivity present in the ore. Modern rehabilitation schemes for tailings impoundments include a number of layers designed to prevent infiltration of water, oxidation of minerals, capillary rise of salts, and release of Rn, thereby minimizing the impact of the disposal of tailings on the environment. Although the effectiveness of these engineered barriers may degrade with time, it is possible, using evidence from natural systems and archaeological constructions, to provide evidence of the long-term isolation of the contaminants contained in the uranium tailings from the environment.

German scientist Martin Klaproth first identified uranium in 1789 as a constituent of ore derived from the Joachimsthal mines in Bohemia. Continued scientific study of this element by scientists including Röntgen, Becquerel, and the Curies (Goldschmidt 1989) led to the work of Rutherford and Solly, who elaborated the process of natural radioactive transmutation. Up until the 1900s, fluorescent uranium salts were in demand for colouring glass and for use in ceramic glazes. Uranium mining continued in the 1900s although, after the work of the Curies, U ores were principally mined to extract radium, which was in high demand for use as a medical isotope. Subsequently, during World War II, development of nuclear weapons renewed the importance of mining for U. After the war, US President Eisenhower, faced with an escalation in the development of nuclear weapons, proposed his 'Atoms for Peace' initiatives that supported the development of peaceful uses of nuclear energy including civilian nuclear power plants. These initiatives also included a call to create an international organization to promote the peaceful use of nuclear energy, which led to the founding of the United Nations' International Atomic Energy Agency (IAEA) in 1956 (IAEA 1996). As a result of these developments, the U mining and milling industry expanded to supply material to fuel the growing number of commercial nuclear power plants.

Current U production is primarily from Canada (32% of world supply) and Australia (19%), as shown in Table 1 with data of the Uranium Information Centre (UIC 2003). Other U supplies are obtained through initiatives designed to reduce stockpiles of materials suitable for use in weapons. This includes the down-blending of 'weapons-grade' uranium in the USA and Russia (UIC 2002). Uranium is

Table 1. *World U production from mines (UIC 2003)*

Country	U (t)	
	2001	2002
Canada	12 520	11 604
Australia	7756	6888
Niger	2920	3075
Russia (estimated)	2500	2900
Kazakhstan	2050	2800
Namibia	2239	2333
Uzbekistan	1962	1860
USA	1011	919
Others including; South Africa, Ukraine, China, Czech Republic, India, Brazil, Romania, Spain, France, and others	3408	3718
Total world mine production	36 366	36 097

From: GIERÉ, R. & STILLE, P. (eds) 2004. *Energy, Waste, and the Environment: a Geochemical Perspective*. Geological Society, London, Special Publications, **236**, 25–35.
0305-8719/04/$15   © The Geological Society of London 2004.

obtained from ores by either alkaline or acidic leaching processes. The major primary U ore mineral is uraninite (basically UO_2) or pitchblende ($U_2O_5 \cdot UO_3$, better known as U_3O_8), although a range of other uranium minerals is found in particular deposits. These include carnotite ($K_2(UO_2)_2V_2O_8 \cdot 3H_2O$), the davidite–brannerite–absite type U-titanates, and the euxenite–fergusonite-samarskite group (niobates of U and rare earths) (UIC 2001). Details of the processes commonly used to extract U are contained in Snodgrass (1990). Uranium mining and milling produces significant quantities of solid and liquid wastes. Solid wastes include waste rock arising from the mining activity and tailings produced by processing of the ore through the milling circuits. Tailings have the more pronounced impact on the environment, as they represent almost the entire mass of material processed in the milling circuits, and also contain over 85% of the original radioactivity present in the ore.

This chapter will present a description of the processes by which contaminants may be released from U mill tailings, and the results of studies that have been undertaken to reduce the impact of these releases on the environment. It is outside of the scope of this chapter to discuss the regulations that cover the handling of wastes from U mining and milling. These regulations are often country specific and very detailed, as they cover the design, monitoring, and rehabilitation of tailings impoundments. Consequently the reader is referred directly to regulatory publications (e.g., US Nuclear Regulatory Commission 2000, 2002a, b; IAEA 2002a, b).

Environmental impact of U tailings

Uranium ores contain ^{238}U (99.3 wt%) and ^{235}U (0.7 wt%) essentially in equilibrium with their progeny. As ^{235}U and its progeny present negligible hazards compared to ^{238}U, the focus of this chapter will be on ^{238}U and its progeny. The radioactivity contained within the tailings presents a hazard over geological time scales because of the long half-lives of the radionuclides present. (Note: The half-life of a radionuclide, $t_{1/2}$, is the time taken for its initial activity to decay by half. After 5 half-lives the activity of a radionuclide will decay to less than 0.7% of its original value, if it is not supported by a longer lived parent radionuclide.) When released from the mill, the tailings contain long-lived radionuclides arising from the decay of ^{238}U, viz. ^{230}Th ($t_{1/2} = 75\,400$ y) and ^{226}Ra ($t_{1/2} = 1620$ y). These progeny, originally at levels in equilibrium with the ^{238}U ($t_{1/2} = 4.5 \times 10^9$ y) contained in the ore, will

decay over very long periods of time until they reach equilibrium with the residual levels of ^{238}U in the tailings.

Tailings disposal schemes are designed particularly to control releases of ^{226}Ra and ^{222}Rn. Radium-226 is of concern because of its high mobility in the environment (Ritcey 1990). Moreover, if ^{226}Ra is ingested, its chemical similarity to Ca can result in exposures through its incorporation into the skeletal system. Short-lived ^{222}Rn is also of importance as it is continuously produced by the decay of ^{226}Ra. As a noble gas Rn can be dispersed readily from the emplaced tailings and have an impact on nearby communities through inhalation and its short-lived, alpha-emitting progeny.

Tailings differ from the U ore from which they are derived as they have undergone a number of physical and chemical processes that can include: crushing; leaching with chemicals, either strong acid (sulphuric or hydrochloric) or alkaline solutions (carbonate/bicarbonate solutions), depending on the mineralogy of the deposit; and contacting with other processing chemicals, including neutralization agents. All of these processes can lead to more accessible surfaces and to formation of host phases that can increase the release of radionuclides, heavy metals, and process chemicals when the tailings are contacted with infiltrating surface or ground water.

Consequently, tailings need to be isolated from the environment over hundreds of thousands of years to avoid release of radionuclides arising from: infiltrating groundwater; dispersion of tailings by wind and water erosion; and dispersal of radioactivity through the release of Rn (National Academies 1987). Consequently, disposal schemes for the tailings are designed to mitigate against these releases. Design conservatism is used to prolong the life and effectiveness of the rehabilitation scheme by increasing the thickness and number of barriers. Current strategies to isolate the tailings from the environment include multilayer covering systems and engineered underlying strata, augmented by passive controls, such as disposal below the surface and away from areas of groundwater recharge.

Disposal of U tailings

Until the 1970s, uranium tailings were stored at mine sites with little thought given to their impact on the environment and surrounding communities. Rehabilitation of U tailings emplaced before the 1970s has been the subject of major programs in the USA and other countries. Remediation of these tailings stockpiles has involved the movement of tailings into re-engineered structures, or the strengthening of structures in which the tailings had been placed, and treatment of ground and surface waters associated with the sites (Harries *et al.* 1997; National Academies 2002). Waugh *et al.* (2001) describe developments in the covering of tailings impoundments in the US since the 1970s, initially to achieve Rn attenuation, and then to provide low permeability designs to reduce infiltration and seepage of contaminated pore water. As this chapter focuses on the development of new facilities, it will not consider further the rehabilitation of existing tailings impoundments that were established before the 1970s.

Modern emplacements are characterized by engineered structures, commonly referred to as dams or impoundments, that include layers to control infiltration of groundwater, dispersal of the tailings by wind and water erosion, release of Rn, seepage from the tailings, and capillary rise of salts. These covering layers are typically installed at the end of the operating life of a U milling facility. Figure 1 shows details of the

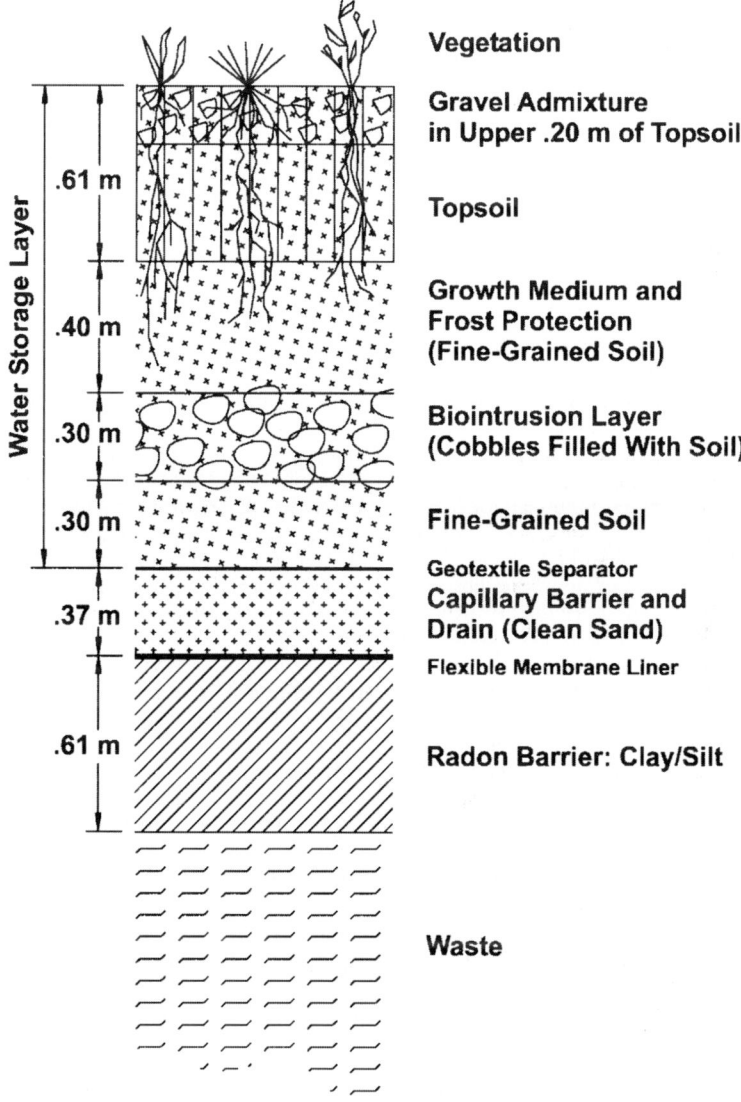

Fig. 1. Tailings covering system developed for the rehabilitation of the impoundment at Monticello, Utah. (Diagram from Waugh *et al.* 2001, used with permission)

engineered structures developed for the rehabilitation of the tailings impoundment at Monticello, Utah. Other designs for a range of climatic zones are given in Waugh *et al.* (2001). In areas where either wind or water erosion is a concern, the engineered structure is often 'armoured' by placing boulders and rocks over exposed surfaces. All of these engineering structures aim at reducing overall releases from the tailings.

As well as improved construction and rehabilitation techniques, other processes have been implemented to reduce the release of contaminants from U tailings impoundments. Neutralization of sulphuric acid leach solutions has been shown to reduce markedly the mobility of a wide range of contaminants. Results from experimental studies of Ring *et al.* (quoted in Snodgrass 1990) showed that neutralization of a sulphuric acid leach liquor/tailings slurry from a pH of $\sim$1.5 to 8.5 decreased the levels of ^{230}Th and ^{210}Pb by factors of 15 000 and 400, respectively. The behaviour of ^{226}Ra is more complicated, with levels increasing by a factor of about 10 during neutralization. The results from this laboratory study were very similar to those obtained for samples taken at the Narbalek mill in the Northern Territory of Australia. For ^{230}Th and ^{210}Pb, the decreases reflect the reduced solubility of these radionuclides at higher pH values. The increase in ^{226}Ra levels is explained by the increase in ^{226}Ra solubility that accompanies a decrease in sulphate concentration during neutralization, due to reduced levels of Ca in solution, possibly arising from the precipitation of Ca-sulphate (Snodgrass 1990). Longer term studies have shown that the levels of ^{226}Ra slowly decrease with time after neutralization as ^{226}Ra coprecipitates and/or sorbs onto fine material associated with the tailings (Levins *et al.* 1978). Other experimental studies (Levins *et al.* 1983; Dodson *et al.* 1985) showed that the concentrations of heavy metals were reduced by neutralization because of the formation of insoluble metal hydroxides at higher pH. These types of laboratory studies have supported the effectiveness of neutralization schemes to remove radionuclides and heavy metals from waste solutions and to limit the migration of these contaminants through the environment. The results have led to the use of neutralization techniques for waste solutions at most operating U mills (Dodson *et al.* 1985; Longmire & Thomson 1990). Other authors have further developed this philosophy to promote the use of neutralization as a reactive barrier to contaminants contained in seepage from the tailings (Spangler & Morrison 1991; Morrison *et al.* 1995).

Overall, there is agreement that tailings can be isolated from the environment in carefully engineered structures so that they will have minimum impact on the environment in the future. At each site, given the wide range of mineral contents, types of processing, and waste treatments, it is important to demonstrate that proposed rehabilitation strategies are applicable over very long timescales. The challenge is to develop techniques and results that enable the safety/effectiveness of the disposal method to be demonstrated over these timeframes, not only to the regulator but also to other stakeholders, including the public.

Long-term integrity of tailings disposal systems

The tailings disposal system needs to operate over both the short and long term to isolate the wastes from the environment. In the short term, while institutional controls are still in place, there is a primary reliance on the engineered barrier system. These barriers typically rely on compacted clay and soil layers to control water infiltration, Rn, and water seepage to prevent long-term release of contaminants. Over the long term, however, the viability of these barriers cannot be guaranteed. And, although the layers initially have the required integrity, their properties can degrade with time, through the impact of penetrations from vegetation and burrowing animals, drying/wetting cycles, and freeze/thaw cycles (Smith *et al.* 1997). With the degradation of the engineered barriers, the geochemical processes that help to stabilize components of the tailings and restrict releases to the wider environment increase in importance.

Barriers to limit seepage from a tailings impoundment have been investigated in a number of studies. For example, Gee *et al.* (1990) compared the types of liners available for tailings impoundments, viz. compacted clay, synthetic membranes, and natural sediments. None of these barriers is a panacea for isolating the wastes. Instead a combination of barriers optimized for short- and long-term performance is normally used. Gee *et al.* (1990) suggested that 1 m of compacted clay (typical permeability of 10^{-6} to 10^{-9} cm/s) provides an effective seal reducing the levels of contamination leaving the impoundment. However, these values for permeability are initial values and can vary with time depending on the physical and chemical interactions that can affect the engineered barrier's viability. These authors also noted that rock or soil strata can be effective natural geological barriers, and can, like clay liners, provide a longer term barrier to releases.

The long-term performance of synthetic liners is not known, and even in the short term can be affected by the presence of subsoil settlement, rock punctures, seam splitting, or entrapped air bubbles (Gee *et al.* 1990). The possible failure of synthetic liners creates an additional problem if they are the only barrier used in the tailings impoundment; instead, they should be seen as a short-lived barrier that is only one part of a multi-layer system.

Performance of the barrier system under operating conditions is not easy to predict. While some studies suggest that hydraulic properties of the clay layers change slowly, this depends on the interaction of the seepage and the barrier material. For example, Crim *et al.* (1979) studied the effect of a low pH environment on the effectiveness of liners for tailings dams. They found that as the pH of the leachate decreased to <1, there was an increase in permeability of the soil, by up to two orders of magnitude depending on the types of soil and the constituent clay minerals. This deterioration was attributed to the ion exchange of Na from sodium montmorillonite (a clay mineral) resulting in a decrease in bound water and consequently an increase in permeability. Other issues, including possible dissolution of minerals from the material, were not investigated. An additional concern was that the soils became more dispersive, that is, lost their ability to remain compacted. This increase in dispersivity is of major concern as it was identified as the cause for the loss of integrity at the Church Rock Tailings Dam (New Mexico, USA, 1979), which contributed to its catastrophic failure (Crim *et al.* 1979).

As well as the engineered barrier system, any releases from an impoundment will eventually come in contact with underlying sediments. These natural sediments often have a buffering capacity associated with Ca carbonate and other carbonate minerals in the soils, and this can be augmented by addition of neutralizing agents. These types of neutralization barriers can delay migration of contaminants both chemically, through reaction, and physically, by precipitates blocking pore space. Data of Gee *et al.* (1990) show that in a soil barrier augmented with lime (CaO), the permeability of the soil decreased from 3.5×10^{-6} to 4×10^{-8} cm/s after 24 pore volumes of acidic tailings solution had passed through the soil barrier. In the same untreated soil the permeability also reduced over the same number of eluted pore volumes but by a lesser amount, that is, from 4×10^{-6} to 1×10^{-6} cm/s. Although these results, which suggest that interactions between the soil layers and waste solutions act to reduce the mobility of contaminants, are encouraging, it should be noted that there are no data for the performance of these barriers over very long periods of time.

Soil layers placed on tailings impoundments may function well for years and decades, but problems exacerbated by variations in temperature (e.g., precipitation/drying and freeze/thaw cycles), soil erosion, subsidence, root intrusion, and animal intrusion may occur. These types of interactions have been recorded for various disposal sites, including: Grand Junction Colorado (U tailings); Hanford Reservation (radioactive waste); Idaho National Laboratory (radioactive waste); and Los Alamos (radioactive waste) (Smith *et al.* 1997). Data from these US Department of Energy (DOE) controlled sites indicate that as well as loss of cover integrity, plant roots have taken up radionuclides from wastes. Dreesen & Marples (1979) found elevated levels of ^{226}Ra, Mo, and Se in native vegetation growing on tailings from alkaline processing circuits. These levels of contaminants were of concern as they exceeded legislated levels for grazing animals. Consequently, modern disposal systems include layers to prevent intrusion of plants and animals ('bio-intrusion') as well as the capillary rise of salts. These layers are typically made up of thick (>0.5 m) beds of coarse gravels. Although an improvement to the original designs for covering of tailings impoundments, the effectiveness of these layers especially to withstand bio-intrusion has not been demonstrated over very long time periods.

Smith *et al.* (1997) suggested that additional information on the stability of earthen structures could be obtained by studying the fate of ancient tombs and ceremonial mounds. For example, the Mississippian culture in the eastern US, which flourished along the Mississippi River from about 1000 to 1500 CE, built flat-topped temple mounds sometimes co-located with round-topped burial mounds of compacted soil and loess. These mounds present an opportunity to study the effect of bio-intrusion on the stability of compacted soil layers. There are some limited data available from studies of these mounds that suggest that development of trees on the structure may promote slumping, whereas the presence of grasses alone seems to protect the structures (Smith *et al.* 1997). It is suggested that more detailed studies of these and similar structures (e.g., even older graves in Europe and Asia) could be used as analogues of engineered, compacted soil layers. These results could provide long-term data for modelling and predictive studies and for increasing stakeholder confidence in the use of such layers for isolating U tailings from the environment.

Ultimately, all of the approaches described above are relying on the long-term performance of covering and underlying layers in addition to the natural environment to immobilize the releases from the tailings. This type of approach has also been referred to as *natural attenuation* (National Academies 2000). In order to pursue this type of remediation approach it is important to develop a conceptual model of each site, which includes data for groundwater, contaminant location, concentrations, and types of processes that could affect the contaminants. The DOE has proposed using natural attenuation to clean up U and other groundwater contaminants at U mill tailings sites under the Uranium Mill Tailings Remediation Control Act (DOE 1996; National Academies 2000). However, the authors of the National Academies report felt that there was still a lack of scientific understanding of the processes associated with contaminant releases. In particular, they noted that there is a need to characterize the source term. This includes processes such as: advection; dispersion; phase transfer; chemical reactions, including acid–base, redox, precipitation, and dissolution reactions; as well as the impact of chemical sorption, hydrolysis, radioactive decay, and aqueous complexation (National Academies 2000).

Following on from the National Academies' recommendations, it is necessary to demonstrate, on a site-by-site basis, that our understanding of the impoundment systems is good enough to create confidence in stakeholders that the approach being used to remediate the tailings is suitable for now and the long term.

Importance of field and laboratory studies

Laboratory and field studies are needed to further our understanding of the behaviour of radionuclides and other contaminants in the tailings impoundment. This understanding also needs to be extended to the complex set of interactions between the tailings, infiltrating water, liners, and underlying sediments so that the effectiveness of the rehabilitation scheme can be evaluated.

A number of studies have shown that Ra mobility in processing circuits and tailings is controlled by sulphate levels. Snodgrass (1990) provided a detailed summary of studies that have been conducted on the distribution of ^{226}Ra in U processing circuits. The primary evidence for Ra mobility is that in the tailings, unlike the original crushed ore, Ra is concentrated in the fine size fraction, either associated with the fines or present as precipitates or co-precipitates with Ba and sulphate (Levins *et al.* 1978, 1983; Snodgrass 1990).

Laboratory studies on the leachability of Ra from tailings have also identified the importance of sulphate concentrations on the solubility of Ra. Leaching studies by Somot *et al.* (1997) have shown that in demineralized water 34–47 wt% of the gypsum is dissolved and 0.3–0.6% of the ^{226}Ra is solubilized (% of the total radioactivity associated with ^{226}Ra). Comparison with results obtained using stronger leachants suggests that the majority of Ra liberated when the tailings are contacted with demineralized water may be adsorbed locally from solution onto other minerals. Sorbed Ra appears to be strongly associated with amorphous Fe, Mn, and Al phases, as well as with pyrite. The silicate and resistate minerals in these tailings account for about 60–72 wt% of the tailings mass but contain only 3–7% of Ra and 7% of the U. Similarly, Bassot *et al.* (2000) found that ~70% of the ^{226}Ra present in the tailings had been sorbed onto oxyhydroxides present at trace levels in the tailings, with the remaining Ra associated with gypsum, barite, and silicate phases (including clays, feldspars, micas, quartz).

Laboratory studies have often focused on newly generated tailings. Additional studies have extended this work to samples from tailings dams in order to understand the effects of time on the mineral assemblages contained in the tailings. Drill hole samples of acid-leached tailings taken from the Écarpière tailings dam in France (Ruhlmann *et al.* 1996), were examined and found to have undergone significant diagenesis over the 30 years of storage in a covered tailings impoundment. New minerals were observed to have precipitated, including smectite, gypsum, barite, and minerals containing hexavalent U. These newly formed minerals were also found to incorporate heavy metals and radionuclides. It was further observed that some minerals, for example, sulphides, feldspars, biotite, and monazite had undergone dissolution. Leaching tests showed that releases of contaminants from the aged tailings were lower than those from fresh tailings. This result was attributed to the presence of smectite in the aged tailings, which provided surfaces onto which the radionuclides were adsorbed.

Ward *et al.* (1984) reported on studies of tailings from the Mary Kathleen Uranium mine in Queensland, Australia. Samples of tailings and associated pore water were taken from the tailings dam. Profiles obtained through the tailings dam showed that tailings at the base of the dam (18 m below the surface) were in secular equilibrium and that there does not appear to have been major losses of radionuclides from the dam. Accelerated leach tests were carried out to quantify chemical controls on releases from the tailings as a function of time and leachant composition. The results of these studies were used to predict that, after about 1000 years, only ~0.13 wt% of the ^{226}Ra would be released. Importantly, results from both the profile samples and the accelerated leach tests suggested that the high content of sulphate in the pore water was contributing to the relative immobility of ^{226}Ra, supporting the extrapolation of the results from the accelerated laboratory tests to predictions for the field.

Fernades *et al.* (1996) studied tailings in the impoundment of Brazil's U mining and milling facility at Poços de Caldas. These authors found that the distribution of heavy metals and radionuclides was non-uniform with depth through the tailings dam. This uneven distribution was attributed to segregation of fine-grained material, including precipitates of gypsum and metal hydroxides formed during neutralization. Other possible processes that may have occurred over time include pyrite oxidation and adsorption onto fine-grained material, clays, and Fe-oxides of major metals and trace radionuclides from neutralized pore water. Contaminants released from the surface layer were found to have undergone subsequent immobilization at depth in the tailings profile through reactions with unreacted lime added during the neutralization stage. Leaching studies of solids from the impoundment showed that near-surface samples have low releases, suggesting that readily leached elements have already been removed. Similar results have been obtained by Stanton (1973) for the redistribution of contaminants in the Mayday tailings dam in Colorado. His results showed that Pb, Zn, and Fe have been mobilized from the surface, as a result of oxidation of sulphides, and then redeposited in secondary phases at depth in the pile. Davé *et al.* (1982) also identified oxidation of pyrite as a major route for remobilization of contaminants from the tailings and, similarly to Stanton (1973) and Fernades *et al.* (1996), observed that there had been reprecipitation of the contaminants at depth in the tailings dam.

These laboratory and field studies have established that processes occurring in tailings dams are slow. Equilibrium between the tailings and pore waters has not necessarily been established over 20- to 30-year periods, emphasizing the slow kinetics of the processes occurring in impoundments. What is clear is that the range of interactions in the impoundments is complex, and additional approaches may be required to help extrapolate these results over geological time-scales.

Difficulties of modelling the complex nature and interactions of seepage from a U tailings impoundment have been summarized in Bradbury (2002), Zhu (2003), and Zhu & Burden (2001). These authors compared the K_d approach with more complicated models and highlight the need for site-specific data to support modelling predictions.* Zhu (2003) and Zhu & Burden (2001) noted that, although the chemical composition of the groundwater is well known, the mineralogy of underlying materials and the aquifer system and their evolution with time is not. This latter information is necessary to enable predictions to be reliably made about the effect of seepage on aquifer systems. Currently, modelling can provide a tool to evaluate different approaches for a particular site but predictions can only be as good as the data and knowledge of the site. To provide greater confidence in the results of modelling, a wider range of data, from a variety of sites and locales, including U deposits and other sources of naturally occurring radionuclides, are required.

Staff of the US Geological Survey are building geoenvironmental models from their mineral deposit models (Seal *et al.* 2002; Smith *et al.* 2002; Foley 2002; Hammarstrom & Smith 2002). Their work aims to provide a compilation of geological, geochemical, geophysical, hydrological, and engineering information to the behaviour of geologically similar mineral deposits. These models can also provide important information on the projected behaviour of mineral assemblages contained in the tailings. Currently, this approach is still being developed, and as such, is qualitative. However, in the longer term it is proposed to develop quantitative generic geoenvironmental models. The development of such highly integrated models should address some of the shortcomings of the currently available models for tailings systems, and enable long-term projections of releases from tailings dams to be made on a more viable basis.

Use of natural systems to provide long-term confidence in the disposal of U tailings

Natural systems have been studied to provide data to support the ability of geological repositories to isolate radioactive wastes (e.g.,

---

*The distribution coefficient K_d is defined by: $K_{d,i} = (C_i/S_i)$ where S_i and C_i represent the concentrations of component i bound to the solid matrix (mg/kg) and in groundwater (mg/L), respectively.

Smellie & Karlsson 1999). Results from studies of natural systems can also be applied to improve knowledge of processes within and transport of radionuclides from tailings impoundments, and importantly provide information over geological time-scales on essential geochemical processes. Geochemical processes that need to be studied further include: sorption, co-precipitation, precipitation, and movement of radionuclides from the fractures to the bulk-rock matrix occurring both near and at a distance from the tailings. In addition, these studies can provide input data to test models as well as information to gain public acceptance of the approaches being planned to reduce the environmental impact of the disposal of U tailings.

Examples of the use of natural systems to project long-term releases of radionuclides include research undertaken at the Koongarra U deposit (Australia), the Äspö underground facility (Sweden), and the Oklo U deposit (Gabon). There are many other examples of these types of studies (e.g., Bonotto 1998; Lozano et al. 2002; Marcos et al. 2000; Miller et al. 2003; Neymark et al. 2003; Read et al. 1998) that are too numerous to consider in detail in this chapter. All of these studies provide techniques that can be used to support predictions of long-term contaminant releases from natural systems located in similar geochemical environments to the tailings system being studied.

The Koongarra U deposit in the Northern Territory of Australia has been studied to evaluate the processes and mechanisms involved in the geochemical alteration of the primary ore zone, and to model the formation of the secondary U ore zone and dispersion fan (Duerden 1991; Duerden & Airey 1994). Studies of the distribution of the U in the dispersion fan (Murakami et al. 1991) have provided data on the fixation of U leached from the primary ore deposit. Their work has shown that, for this system, fractures are not only preferential pathways for groundwater movement but also contain secondary minerals with high sorption capacity for elements such as U. Even in the monsoonal climate, in which this deposit is located, a significant proportion of the uranium has not been released from the vicinity of the primary ore body.

Studies of the sorption of natural radionuclides from groundwater have established that minor minerals, such as clays coating the fractures in crystalline rock and Fe- and Mn-oxyhydroxides, are important sinks for radionuclides. For example, investigations carried out at the Äspö facility (Smellie & Karlsson 1999) showed qualitatively that rare earth elements (REEs), Sc, Th, U, Ra, and Ba have been scavenged by Fe-oxyhydroxide and calcite precipitates,

although it is not clear whether the process is sorption and/or co-precipitation. Their work also established that REEs, Sr, Th, and U had been incorporated into calcite.

Toulhoat et al. (1996) used data from the Oklo and Bangombé U deposits in Gabon to correlate groundwater evolution and controls over the dissolution of U. Their results showed that, in this case, redox is controlled by an equilibrium between Fe(II) in siderite and Fe(III) as $Fe(OH)_3$ and ferrihydrite. The results highlighted that buffering of the groundwater restricted the dissolution of U from the Bangombé deposit.

Long-term remobilization of radionuclides and contaminants from tailings cannot be studied unless use is made of a wide range of sources of information to help describe the radionuclide and contaminant host phases, the effect of infiltrating surface and groundwaters, and the ability of the natural system to immobilize any radionuclides released from the tailings. Studies of natural systems, in environments and locations similar to the site of the tailings impoundments, allow long-term data to be obtained. These data can be used to calibrate and verify models of the systems to provide greater credibility for the predictions of the long-term integrity of the disposal system. Further, the results can be used to identify the most important processes for each disposal site, enabling the design of the barrier system to be optimized to mitigate against releases.

Conclusions

Overall, there is international agreement that tailings can be isolated from the environment in carefully engineered structures so that they will have minimum impact on the environment in the future. However, to optimize the design of emplacement structures it is necessary to understand the geological environment in which the wastes are emplaced, including the geochemical processes that affect the mobility and distribution of contaminants. A wide range of natural processes can lead to immobilization of contaminants; however, these processes differ considerably from site to site and depend on a range of site-specific factors. Consequently, the information available for each site and each waste must be carefully collected and appraised. It is only where a full understanding of these processes is available that the data should be used in extrapolations to predict performance of waste systems over very long periods of time.

References

BASSOT, S., STAMMOSE, D., MALLET, C., FERREUX, J.-M. & LEFEBVRE, C. 2000. Study of the radium sorption/desorption on goethite, paper P-4b-258. *Proceedings of IRPA10*, 14–19 May 2000, Hiroshima, Japan. World Wide Web address: http://www.irpa.net/irpa10/.

BONOTTO, D. M. 1998. Generic performance assessment for a deep repository for low and intermediate level waste in the UK – a case study in assessing radiological impacts on the natural environment. *Journal of Environmental Radioactivity*, **66**, 89–119.

BRADBURY, J. W. 2002. *Insights from process-level modeling of contaminant transport from uranium mill tailings*. Annual Meeting: Geological Society of America, Denver, CO, United States, 2002.

CRIM, R. G., SHEPHERD, T. A. & NELSON, J. D. 1979. Stability of natural clay liners in a low pH environment. *Proc. Second Symposium on Uranium Mill Tailings Management*, Fort Collins, CO, USA, 19–20 November, 1979, 41–53.

DAVÉ, D. K., LIM, T. P. *et al.* 1982. Close-out concepts for the Elliot Lake Uranium Mining Operations, IAEA-SM-262/14. *Proceedings of an International Symposium on Management of Wastes from Uranium Mining and Milling*, co-organized by IAEA and OECD/NEA, Albuquerque NM, 10–14 May 1982, 215–229.

DODSON, M. E., OPITZ, B. E. & SHERWOOD, D. R. 1985. Two-reagent neutralization scheme for controlling the migration of contaminants from a uranium mill tailings disposal pond, PNL-SA-12629; CONF-850242, Management of uranium mill tailings, low-level waste and hazardous waste. *Proceedings of the Seventh Symposium*, Fort Collins, CO, 6–8 February 1985, 401–410.

DOE 1996. *Final Programmatic Environmental Impact Statement for the Uranium Mill Tailings Remedial Action Ground Water Project*, DOE/EIS-0198, DOE, Washington, USA.

DREESEN, D. R. & MARPLES, M. L. 1979. Uptake of trace elements and radionuclides from uranium mill tailings by four-wing Saltbush (Atripliex canescens) and Alkali Sacaton (Sporobolus airoides). *Proc. Second Symposium on Uranium Mill Tailings Management*, Fort Collins, CO, USA, 19–20 November 1979, 127–143.

DUERDEN, P. 1991. The Alligator Rivers Analogue Project – Characterisation of the development of the uranium dispersion fan. *Proc. Scientific Basis for Nuclear Waste Management XV*, 4–7 November, Strasbourg, France, published by the Materials Research Society, USA, 529–536.

DUERDEN, P. & AIREY P. 1994. *ARAP – The International Alligator Rivers Analogue Project: Background and Results*, OECD/NEA, Paris, France. World Wide Web Address: http://www.nea.fr/html/rwm/reports/1994/arap.pdf.

FERNADES, H. M., FRANKLIN, M. R., VEIGA, L. H. S., FREITAS, P. & GOMIERO, L. A. 1996. Management of uranium mill tailing: geochemical processes and radiological risk assessment. *Journal of Environmental Radioactivity*, **30**, 69–95.

FOLEY, N. K. 2002. Chapter E, Environmental geochemistry of platform carbonate hosted sulfide deposits. *In*: SEAL II, R. R. & FOLEY, N. K. (eds) *Progress on Geoenvironmental Models for Selected Mineral Deposit Types*. US Geological Survey Open-File Report 02-195, US Geological Survey, Denver CO, USA

GEE, G. W., PETERSON, S. R. & OPITZ, B. E. 1990. Optimization of liners for prevention of radium dispersion from tailings. *In*: *The Behaviour of Radium*. IAEA Technical Report Series 310, IAEA, Vienna, Austria, 163–194.

GOLDSCHMIDT, B. 1989. Uranium's scientific history 1789–1939. *Uranium Institute 14th International Symposium*, London, September 1989. (World Wide Web Address: http://ist-socrates.berkeley.edu/~rochlin/ushist.html.)

HAMMARSTROM, J. M. & SMITH, K. S. 2002. Chapter B, Geochemical and mineralogic characterization of solids and their effects on waters in metal-mining environments. *In*: SEAL II, R. R. & FOLEY, N. K. (eds) *Progress on Geoenvironmental Models for Selected Mineral Deposit Types*. US Geological Survey Open-File Report 02-195, US Geological Survey, Denver CO, USA.

HARRIES, J., LEVINS, D., RING, R. & ZUK, W. 1997. Management of waste from uranium mining and milling in Australia. *Nuclear Engineering and Design*, **176**, 15–21.

IAEA 1996. *History of the International Atomic Energy Agency: The First Forty Years*, IAEA, Vienna, Austria. World Wide Web Address: http://www-pub.iaea.org/MTCD/publications/PDF/Pub1032_web.pdf.

IAEA 2002*a. Monitoring and Surveillance of Residues from the Mining and Milling of Uranium and Thorium*. International Atomic Energy Agency, Safety Reports Series, no. 27, also STI/PUB/1146, IAEA, Vienna, Austria.

IAEA 2002*b. Safety Standards Series Ws-G-1.2, Management of Radioactive Waste from the Mining and Milling of Ores*, Safety Guide, International Atomic Energy Agency, Vienna, Austria.

LEVINS, D. M., RYAN, R. K. & STRONG, K. P. 1978. Leaching of radium from uranium tailings, CONF-780740; Management, stabilization and environmental impact of uranium mill tailings. *Proceedings of a Nuclear Energy Agency Seminar*, Albuquerque, NM, 24–28 July, 1978, 271–286.

LEVINS, D. M., RING, R. J. & HART, K. P. 1983. Mobilization of radionuclides and heavy metals in uranium mill and tailings dam circuits. *Proc. Scientific Workshop on the Environmental Protection in the Alligator Rivers Region*, sponsored by the Office of the Supervising Scientist, Jabiru, NT, Australia, 17–20 May, 10-1.

LONGMIRE, P. & THOMSON, B. 1990. Application of geochemical barriers for immobilizing hazardous constituents in uranium mill tailings, CONF-9004181 (Vol. 2); Remedial action under the environmental restoration and waste management five-year plan. *Proceedings of a Department of Energy Remedial Action Program Conference*, Albuquerque, NM, 16–19 April 1990, 68–108.

LOZANO, J. C., RODRÍGUEZ, P. B. & TOMÉ, F. V. 2002. Distribution of long-lived radionuclides of the ^{238}U series in the sediments of a small river in a uranium mineralized region of spain. *Journal of Environmental Radioactivity*, **63**, 153–171.

MARCOS, N., SUKSI, J., ERVANNE, H. & RASILAINEN, K. 2000. Fracture smectite as a long-term sink for natural radionuclides – indications from unusual U-series disequilibria. *Radiochimica Acta*, **88**, 763–766.

MILLER, W., HOOKER, P. & RICHARDSON, P. 2003. Natural analogue studies: Their application to a repository safety case. *Proceedings of the 10th International High-level Radioactive Waste Management Conference (IHLRWM)*, March 30 to April 3 2003, Las Vegas.

MORRISON, S. J., TRIPATHI, V. S. & SPANGLER, R. R. 1995. Coupled reaction/transport modeling of a chemical barrier for controlling uranium (VI) contamination in groundwater. *Journal of Contaminant Hydrology*, **17**, 347–363.

MURAKAMI, T., ISOBE, H., NAGANO, T. & NAKASHIMA, S. 1991. Uranium redistribution and fixation during chlorite weathering at Koongarra, Australia. *Proc. Scientific Basis for Nuclear Waste Management XV*, held 4–7 November, Strasbourg, France, published by the Materials Research Society, USA, 473–480.

NATIONAL ACADEMIES 1987. *Scientific Basis for Risk Assessment and Management of Uranium Mill Tailings*, Uranium Mill Tailings Study Panel, National Academy Press, Washington DC, USA, also DOE/DP/93032.

NATIONAL ACADEMIES 2000. *Natural Attenuation for Groundwater Remediation, Commission on Geosciences, Environment and Resources (CGER)*, National Academy Press, Washington DC, USA World Wide Web Address: http://www.nap.edu/books/0309069327/html/.

NATIONAL ACADEMIES 2002. *Remedial Action at the Moab Site – Now and for the Long Term*, National Academies, Washington DC, USA, June 2002, Letter Report.

NEYMARK, L. A., PACES, J. B. & AMELIN, Y. V. 2003 Reliability of U-Th-Pb dating of secondary silica at Yucca Mountain, Nevada. *10th International High-Level Radioactive Waste Management Conference*, March 30 to April 3 2003, Las Vegas, NV, 1–12.

READ, D., ROSS, D. & SIMS, R. J. 1998. The migration of uranium through clashach sandstone: The role of low molecular weight organics in enhancing radionuclide transport. *Journal of Contaminant Hydrology*, **35**, 235–248.

RITCEY, G. M. 1990. *Weathering Processes in Uranium Tailings and the Migration of Contaminants. In*: The Behaviour of Radium, IAEA Technical Report Series 310, IAEA, Vienna, 27–82.

RUHLMANN, F., PACQUET, A., REYX, J., THIRY, J., PAGEL, M. & SOMOT, S. 1996. Evidence of diagenetic processes in uranium mill tailings from the Ecarpiere Uranium Deposit (COGEMA – France). *Proceedings of the Waste Management Conference WM'96*, Tucson, AZ, 25–29 February.

SEAL I. R. R., FOLEY, N. K. & WANTY, R. B. 2002. Chapter A, Introduction to geoenvironmental models of mineral deposits. *In*: SEAL II, R. R. & FOLEY, N. K. (eds) *Progress on Geoenvironmental Models for Selected Mineral Deposit Types*. US Geological Survey, Denver CO, USA, US Geological Survey Open-File Report 02-195.

SMELLIE, J. A. T. & KARLSSON, F. 1999. The use of natural analogues to assess radionuclide transport. *Engineering Geology*, **52**, 193–220.

SMITH, E. D., LUXMOORE, R. J. & SUTER II, G. W. 1997. Natural physical and biological processes compromise the long-term performance of compacted soil caps. *In*: *Barrier Technologies for Environmental Management: Summary of a Workshop*. National Academy Press, Commission on Geosciences, Environment and Resources (CGER), 79–88. World Wide Web Address: http://books.nap.edu/books/0309056853/html/; http://www.infography.com/content/351566136978.html.

SMITH, K. S., CAMPBELL, D. L. *et al.* 2002. Chapter C, Toolkit for the rapid screening and characterization of waste piles on abandoned mine lands. *In*: SEAL II, R. R. & FOLEY, N. K. (eds) *Progress on Geoenvironmental Models for Selected Mineral Deposit Types*. US Geological Survey, Denver CO, USA, US Geological Survey Open-File Report 02-195.

SNODGRASS, W. J. 1990. The Chemistry Radium-226 in the Uranium Milling Process. *In*: *The Behaviour of Radium*, IAEA Technical Report Series 310. IAEA, Vienna, 5–25.

SOMOT, S., PAGEL, M. & THIRY, J. 1997. Spéciation du radium dans les résidus de traitement acide du minerai d'uranium de l'Écarpière (Vendée, France), Académie Sci. Paris. *Sciences de la terre et des Planètes*, **325**, 111–118.

SPANGLER, R. & MORRISON, S. J. 1991. Laboratory-scale tests of a chemical barrier for use at uranium mill tailings disposal sites, CONF-910981. *In*: WOOD, D. E. (ed) *Proceedings Environmental Remediation '91: Cleaning Up the Environment for the 21st Century*, Pasco, WA, 8–11 September, 739–744.

STANTON, M 1973 The role of weathering in trace metal distributions in subsurface samples from the Mayday Mine Dump near Silverton, Colorado. *In*: CHURCH, S. E. (ed) *The USGS Preliminary Release of Scientific Reports on the Acidic Drainage in the Animas River watershed San Juan County, Colorado*, US Geological Survey, Denver, CO, USA, 77–85.

TOULHOAT, P., GALLIEN, J. P. *et al.* 1996. Preliminary studies of groundwater flow and migration of uranium isotopes around the Oklo Natural Reactors (Gabon). *Journal of Contaminant Hydrology*, **21**, 3–17.

UIC (URANIUM INFORMATION CENTRE) 2001. *Geology of Uranium Deposits*. Nuclear Issues Briefing Paper 34. World Wide Web Address: http://www.uic.com.au/nip34.htm.

UIC (URANIUM INFORMATION CENTRE) 2002. *Military Warheads as a Source of Nuclear Fuel*. Nuclear Issues Briefing Paper 4. World Wide Web Address: http://www.uic.com.au/nip04.htm.

UIC (URANIUM INFORMATION CENTRE) 2003. *World Uranium Mining*, Nuclear Issues Briefing Paper 41. World Wide Web Address: http://www.uic.com.au/nip41.htm.

US NUCLEAR REGULATORY COMMISSION 2000. *Standard Review Plan for the Review of a Reclamation*

Plan for Mill Tailings Sites Under Title II of the Uranium Mill Tailings Radiation Control Act. US Nuclear Regulatory Commission, Washington DC, USA, NUREG-1620.

US NUCLEAR REGULATORY COMMISSION 2002*a*. US Nuclear Regulatory Commission, Washington DC, USA, U.S. NRC 10 CFR Part 40.

US NUCLEAR REGULATORY COMMISSION 2002*b*. *Design of Erosion Protection for Long-Term Stabilization*. Final Report U.S. Nuclear Regulatory Commission Office of Nuclear Material Safety and Safeguards Washington, DC 20555-0001, NUREG 1623, September 2002.

WARD, T. A., HART, K. P., MORTON, W. H. & LEVINS, D. M. 1984. Factors affecting groundwater quality at the rehabilitated Mary Kathleen Tailings Dam,

Australia. *Proceedings of the 6th Symposium on Uranium Mill Tailings Management*, Fort Collins, Colorado, 1–3 February 1984, 319–328.

WAUGH, W. J., SMITH, G. M., BERGMAN-TABBERT, D. & METZLE, D. R. 2001. Evolution of cover systems for the Uranium Mill Tailings Remedial Action Project, USA. *Mine Water and the Environment*, **20**, 190–197.

ZHU, C. 2003. A case against Kd-based transport models: Natural attenuation at a mill tailings site. *Computers & Geosciences*, **29**, 351–359.

ZHU, C. & BURDEN, D. S. 2001. Mineralogical compositions of aquifer matrix as necessary initial conditions in reactive contaminant transport models. *Journal of Contaminant Hydrology*, **51**, 145–161.

Nuclear waste forms

SERGEY V. STEFANOVSKY[1], SERGEY V. YUDINTSEV[2],
RETO GIERÉ[3] & GREGORY R. LUMPKIN[4]

[1]SIA Radon, 7th Rostovskii per. 2/14, Moscow 119121 Russia (e-mail: profstef@radon.ru)

[2]Institute of Geology of Ore Deposits, Staromonetny 35,
Moscow 119017 Russia (e-mail: syud@igem.ru)

[3]Institut für Mineralogie, Petrologie und Geochemie, Universität Freiburg,
Albertstrasse 23b, D-79104 Freiburg, Germany (e-mail: giere@uni-freiburg.de)

[4]ANSTO, Materials Division, Menai, NSW 2234, Australia; and Cambridge Centre for
Ceramic Immobilisation, Department of Earth Sciences, University of Cambridge,
Downing Street, Cambridge, CB2 3EQ, UK (e-mail: gregl@esc.cam.ac.uk)

Abstract: This review describes nuclear waste forms for high-level waste (HLW), that is, glasses, ceramics, and glass-ceramics, as well as for low- and intermediate-level waste (LILW), that is, cement, bitumen, glass, glassy slags, and ceramics. Ceramic waste forms have the highest chemical durability and radiation resistance, and are recommended for HLW and actinide (ACT) immobilization. Most radiation-resistant materials are based on phases with a fluorite-related structure (cubic zirconia-based solid solutions, pyrochlore, zirconolite, murataite). Glass is also a suitable matrix for HLW containing fission and corrosion products, and process contaminants such as Na salts. Within the framework of the HLW partitioning concept providing separation of short-lived (Cs, Sr) and long-lived (rare earth element-ACT) fractions, glass may be used for immobilization of the Cs–Sr-bearing fraction, whereas the rare earth–ACT fraction may be incorporated in ceramics. Glass-based materials or clay-based ceramics are the most promising LILW forms, but cement and bitumen may also be applied as matrices for low-level wastes (LLW).

High-level waste (HLW), intermediate-level waste (ILW), and low-level waste (LLW) are produced at all stages of the nuclear fuel cycle as well as in the non-nuclear industry, research institutions, and hospitals. The nuclear fuel cycle produces liquid, solid, and gaseous wastes. Moreover, spent nuclear fuel (SNF) is considered either as a source of U and Pu for re-use or as radioactive waste (Johnson & Shoesmith 1988), depending on whether the 'closed' ('reprocessing') or the 'open' ('once-through') nuclear fuel cycle is realized, respectively (Ewing, 2004).

Liquid HLW from reprocessing of SNF may consist of 50–60 elements, including about 90 radionuclides of 35 chemical elements of fission products (FP) and more than 120 radionuclides due to FP decay. The total activity of HLW may achieve 10^{16} Bq/m^3 (Nikiforov et al. 1985). The HLW elements can be divided into four groups: (1) fission products, such as ^{134,137}Cs, ^{90}Sr, ^{99}Tc, ^{129}I, ^{141,144}Ce, ^{147}Pm, ^{151}Sm, ^{152,154}Eu; (2) corrosion products, such as Fe, Al, Si, Mo, Zr, including the activated products ^{51}Cr, ^{54}Mn,

^{59}Fe, ^{58,60}Co, ^{122,124}Sb; (3) minor nuclear fuel components and transmutation products comprising α-emitters (^{235,238}U, ^{237}Np, ^{238,239}Pu, ^{241}Am, ^{242}Cm, ^{244}Cm); and (4) process contaminants, such as, Na, K, Ca, Mg, Fe, S, F, and Cl compounds.

In addition to HLW, reprocessing of SNF produces large amounts of low- and intermediate-level waste (LILW). The volume of LILW exceeds that of HLW by many times. The operation of nuclear power plants yields wastes such as activated coolant (water or steam), filter materials, pulps, regeneration solutions, and contour washing solutions. Non-nuclear fuel cycle institutional wastes represent solutions after regeneration of sorbents, various wastes of research laboratories, medical waste, solutions after decontamination of equipment and soils. Solid wastes from both nuclear power plants and non-nuclear fuel cycle facilities include contaminated equipment, ion-exchangers, lining and heat-insulating materials, laboratory dishes, various organic wastes, including ion-exchange resin,

From: GIERÉ, R. & STILLE, P. (eds) 2004. Energy, Waste, and the Environment: a Geochemical Perspective. Geological Society, London, Special Publications, **236**, 37–63.

cellulose, and biomaterials. Organic wastes are to be incinerated, forming incinerator ash or slag.

The inorganic wastes must be conditioned to durable waste forms. High-level wastes require long-term storage for $10^3 - 10^5$ years depending on composition and activity. The most dangerous wastes are those containing ACT and excess weapons Pu, derived from the dismantling of nuclear weapons. These types of waste must be stored for $10^5 - 10^6$ years. Glass and ceramic waste forms are suggested as suitable matrices for these wastes. A combined approach has also been developed on the basis of the application of glass–ceramics to the immobilization of HLW (Hayward 1988). Radionuclides partition between glass and crystals in such a way that Cs and other alkali elements enter the glass, whereas ACT and rare earth elements (REE) and possibly Sr are incorporated into crystalline phases. Recently, a HLW partitioning concept has been proposed (Actinide and Fission Products Partitioning and Transmutation 1999; Baestlé *et al.* 1999). In this scheme, HLW is to be partitioned into a short-lived Cs–Sr fraction and a long-lived REE–ACT fraction, which will be conditioned separately using borosilicate glass and crystalline ceramics, respectively.

Certain LILWs also need to be conditioned and stored for 300–500 years, and currently cement, bitumen, polymers, and composite materials are used as matrices (Sobolev & Khomtchik 1983; Dmitriev & Stefanovsky 2000). However, alternative waste forms, such as glass, glass–crystalline materials (glassy slags), and clay-based ceramics are also considered (Dmitriev & Stefanovsky 2000). Among the candidate HLW forms, glass and ceramics have been investigated in the greatest detail, and comprehensive reviews on their compositions, structure, physical properties, and geological behaviour are given, for example, by Hench *et al.* (1984), Ringwood (1985), Fielding & White (1987), Lutze & Ewing (1988), Laverov *et al.* (1994), Ewing *et al.* (1995), Merz & Walter (1996), Weber *et al.* (1997, 1998), and Ewing (1999, 2001).

Major requirements for nuclear waste forms

Nuclear waste forms have to be in compliance with the following requirements:

(1) High chemical durability;
(2) High radiation resistance (dependent on waste activity level);
(3) Long-term (thermodynamic) stability;
(4) Maximum waste volume reduction factor;
(5) Strong mechanical integrity;
(6) Appropriate thermal conductivity (for HLW forms);
(7) Appropriate viscosity and electric resistivity (for materials produced by electric melting);
(8) Homogeneous distribution of radionuclides (especially if fissile materials are present);
(9) Compatibility with geological environment;
(10) Simple, reliable, and safe production technology;
(11) Production at low temperature (T) to avoid losses of relatively volatile radionuclides (e.g., Cs, Ru);
(12) Resistance to biodegradation;
(13) Maximum difficulty to recover radioactive constituents from waste form (especially fissile materials, such as Pu).

No HLW form satisfies all the above requirements. For example, glass has high chemical durability and good radiation resistance, but the vitreous state is thermodynamically unstable and subject to devitrification. This process, taking place below the glass transformation temperature (T_g), may cause reduction of chemical durability and/or mechanical destruction. Devitrification may, however, be delayed if no aqueous phase is present, as documented by archaeological glasses. Ceramic material is an appropriate HLW form with respect to all parameters except that its production technology is not chosen yet and its compatibility with the geological environment is strongly dependent on the composition of both the ceramic and the geological medium. Nevertheless, both glass and ceramics satisfy the listed requirements to a large extent. LILW forms do not require such high chemical and physical standards. Major requirements include a maximum volume reduction factor, good chemical durability, and high mechanical integrity. Glass-based materials are the best solution in this case. Currently the most commonly used waste form for LLW is cement (Sobolev & Khomtchik 1983).

Chemical durability of waste forms

There are a number of methods for determining the chemical durability of waste forms. Different tests determine either the differential leach rate of individual constituents (in $g \cdot m^{-2} \cdot d^{-1}$) or the cumulative leaching (in $g \cdot m^{-2}$, or mol% or wt%). The differential leach rate is given by

$$L = A_i/(A_0 \times S \times t) \qquad (1)$$

in which A_i is the (thermodynamic) activity of the ith element in solution after the experiment; A_0 is the specific activity of the given element in the sample; S is the sample surface area; and t is the duration of the experiment. All tests may be divided into either static or dynamic according to the character of interaction between specimen and leachant. The most frequently applied methods for determining waste-form leachability are (Hespe 1971; US DOE 1981; IAEA 1985; ASTM 1994):

(a) *Materials Characterization Center (MCC) tests.* There are several MCC tests, performed under static (MCC-1), high-T (MCC-2), solubility-limited (MCC-3), low flow rate (MCC-4), and Soxhlet (MCC-5) conditions. The MCC-1 test is performed with monolithic samples at a specimen surface area (S) to leachant volume (V) ratio of 1:10, and under static conditions at 70 °C or 90 °C for 7 or 28 d. The other methods are normally applied for determining leach rates from glasses.

(b) *Single-Pass Flow-Through (SPFT) test.* The MCC and SPFT tests are used to determine leachability of glasses, glass–ceramics and ceramics.

(c) *Product Consistency Tests (PCT).* This test uses powdered material (149–74 μm) with $S/V = 2000$ m^{-1} at 90 °C for 7 days. The data are reported as elemental concentrations.

(d) *International Atomic Energy Agency (IAEA) test.* This test is performed under static conditions at 20 °C, with replacement of leachant every day for the first 7 d, and every week thereafter. It is used for LILW forms.

(e) *Toxicity Characteristics Leaching Procedure (TCLP).* This test uses 100 g of material with grain size <9.5 μm and requires a fluid to solid mass ratio of 20, mechanical shaking for 24 h, and filtering using a vacuum filtration apparatus. The results are reported as ppm elemental concentrations.

Owing to the use of these different methods, the results are often difficult to compare. Different procedures, however, are required to learn how waste-form materials behave under different conditions. Unless specified otherwise, all leach rates given in this chapter represent values determined with deionized water as the leachant.

Radiation resistance

The radiation stability of waste forms is determined by changes in their structure, chemical, and phase composition, and macroscopic properties such as, chemical durability, physical, and mechanical properties. These changes are due to atomic displacements, formation of point defects and ion radicals, accumulation of stored energy, formation and accumulation of He and O_2 bubbles, and radiolytic decomposition.

There are several techniques for the study of radiation resistance of actual waste forms, and they have different degrees of certainty (Ewing et al. 1995; Weber et al. 1997, 1998; Weber & Ewing 2002). One of these techniques is based on the investigation of radioactive U- and Th-containing minerals (natural analogue minerals; see Lumpkin et al. 2004). Maximum concentrations of $UO_2 + ThO_2$ in the minerals studied reached 58.2 wt% for brannerite, 31 wt% for pyrochlore, 23.5 wt% for zirconolite, ~10 wt% for zircon, ~6 wt% for perovskite, and 16 wt% for britholite (Alexandrova et al. 1966; Minerals 1967; Gieré et al. 1998, 2000; Lapina & Yudintsev 1999; Lumpkin et al. 1998a, 2000; Ewing et al. 1995; Ewing 1999). The cumulative α-decay dose in minerals is calculated as (Lumpkin & Ewing 1988):

$$D = 8N_1(e^{a_1 t} - 1) + 7N_2(e^{a_2 t} - 1)$$
$$+ 6N_3(e^{a_3 t} - 1) \qquad (2)$$

In this equation, D is the dose (α-decays/g); N_1, N_2, N_3 are the amounts (atoms/g) of ^{238}U, ^{235}U, ^{232}Th, respectively; a_1, a_2, a_3 are the decay constants of ^{238}U, ^{235}U, ^{232}Th (y^{-1}), respectively; and t is the age of the mineral (y). In the absence of isotopic data, the second term can either be ignored or a concentration ratio of ^{235}U = $1/139 \times {}^{238}$U can be assumed. Application of this method to natural glasses, however, is very limited because of their very low U and Th concentrations (typically <20 μg/g; Weber et al. 1997).

The second technique involves incorporation of short-lived ACT isotopes such as ^{238}Pu ($t_{1/2} = 87.7$ y) or ^{244}Cm ($t_{1/2} = 18.1$ y) into the structure of the phase to be studied. Normally, 0.2 to 3 wt% ACT are incorporated to achieve 10^{18} to 3×10^{19} α-decays/g in laboratory time periods of up to several years. This technique is most suitable for simulation of radiation damage in ACT waste forms, including those for Pu immobilization.

Another, more often applied, method is the irradiation of matrices with neutrons or charged particles (electrons, α-particles, heavy ions). Advantages of this method include: short time of irradiation (min or h); direct visibility of structural damage (observed *in situ* during irradiation using a transmission electron microscope); and the possibility of determination of radiation resistance of all phases at the temperature of the experiment. Owing to the short time of the experiments, the dose rate is higher by many orders of magnitude ($\sim 10^9$ times) compared with the dose rate of actual waste forms. Moreover, irradiation with α-particles and ions of light noble gasses (Ne, Ar, Kr) can be employed to simulate the effects of α-particles, whereas irradiation with heavy ions (Xe, Pb, Au) is an effective method to study α-recoil damage. The damage created by this technique, however, is restricted to a thin layer (typically between 0.1 and 1.0 μm) near the surface of the studied material, thus limiting the possibility of applying certain spectroscopic techniques. Similarly, electron irradiation can be used to study the effects of ionization and electron excitations from β-particles and γ-rays on glasses and ceramics; it may be applied to simulate radiation effects due to FPs. One disadvantage of the charged-particle irradiation is that all the phases in multiphase ceramics are irradiated, including those that do not contain ACTs. Therefore, this method is most suitable for glasses and single-phase ceramics with a homogeneous ACT distribution (Weber *et al.* 1997, 1998). Irradiation with fast neutrons produces significant numbers of atomic displacements, but it does not correctly simulate damage from α-particles and α-recoils and further, it does not simulate the accumulation of He from α-particles (Weber *et al.* 1997, 1998). Boron-containing materials, usually borosilicate glasses, may be irradiated with a thermal neutron flux to generate 1.78 MeV α-particles according to the $^{10}B(n, \alpha)^7$Li reaction (Weber *et al.* 1997). Materials doped with fissile isotopes (e.g., ^{235}U) may also be irradiated in a thermal neutron flux (Antonini *et al.* 1979; Vance & Pillay 1982) in order to provide a good simulation of spontaneous fission events in waste forms for HLW and Pu. However, spontaneous fission is a rare event, and its contribution to the total radiation damage in waste forms is insignificant.

To compare doses resulting from different types of radiation it is necessary to formulate a measure of radiation effects on the crystal structure: doses can be recalculated to the number of displacements per atom (dpa). A dose of 0.1 dpa, for example, means that one of ten atoms was displaced from its initial position. Equivalent values in dpa units may be calculated for different types of radiation from the effects of its interaction with the crystal lattice. To recalculate α-dose to dpa the following formula is used:

$$\text{Dose (dpa)} = N_d \times (D \times M)/(N_f \times N_A) \quad (3)$$

In this formula, N_d is the average number of atomic displacements per α-decay event; $D =$ accumulated α-dose; $M =$ molar mass of the compound; $N_f =$ number of atoms in the compound; and $N_A =$ Avogadro's number. The value of N_d is often assumed to be 1500, but computer codes can be used to estimate N_d and also to recalculate irradiation doses from heavy ions to dpa values (Ziegler *et al.* 1985).

Gamma-irradiation has been used to investigate effects on chemical durability of waste glasses (Lutze 1988) and to assess the effect on corrosion and defect formation in waste ceramics (Vlasov *et al.* 1987; Kulikov *et al.* 2001). Irradiation with γ-rays does not produce atomic displacements, but rather causes excitations and point defects and leads to formation of ion radicals. Therefore, this method may serve as a simulation for $\beta-\gamma$-radiation of FPs in the waste forms.

Glasses

Glass-based materials have been suggested as HLW forms due to the capability of their random network to accommodate ions with widely variable charges and radii. Other advantages of glass waste forms comprise: simple production technology (transferred from electric melting of industrial glasses); high chemical durability; and high radiation resistance due to the aperiodic structure, which minimizes the effect of ionizing radiation on the atomic arrangement (Lutze 1988; Weber *et al.* 1997).

Borosilicate glasses

HLW. Silicate glass was proposed for the first time in Canada in the 1950s for immobilization of acid HLW containing major FPs. Nepheline syenite was used as raw material, and the melting temperature of this glass was approximately 1350 °C, similar to those that are typical of silicate glasses in the system Na_2O-(K_2O)-CaO-MgO-Al_2O_3-SiO_2. At $T > 1200$ °C, however, Cs and Ru losses due to volatilization increase; therefore, this temperature may be regarded as an upper limit for vitrification of Cs-bearing waste.

To reduce the processing temperature it has been proposed to add boron oxide to the glass-forming system. The base system for borosilicate waste glasses is Na_2O-B_2O_3-SiO_2, and Li_2O may be added to further reduce the melting temperature, as well as the viscosity of the melts (Roderick *et al.* 2000). Normally, a base frit composition is chosen and then the frit is mixed with HLW calcine or slurry in either a mixer (Hamel *et al.* 1998) or directly in the melter (Masson *et al.* 1999). A number of boro-silicate frit and glass formulations were designed in various countries, mostly before 1987 (Lutze 1988), but only a few have been implemented so far. These baseline glasses are designed to achieve waste loadings of 10–30 wt%, and include (Table 1): the R7/T7 glass for French reprocessing waste; the SRL-165 and ATM-10 glasses for the Defense Waste Processing Facility at Savannah River (South Carolina) and the West Valley Demonstration Project (New York), respectively; special glasses for immobilization of various ACT wastes, that is, lanthanide-borosilicate glasses for excess weapons Pu, and Am-Cm-containing nitric acid solution (Fellinger *et al.* 1999); and U-enriched (Aloy *et al.* 2001) and Ca-Al-borosilicate ('boro-basalt', Russia) glasses for Pu-containing sludges.

Actinides are the most troublesome com-ponents of silicate glasses due to their low solu-bility in glass and high radiation hazard. Their solubility decreases with increasing atomic number and decreasing ionic radius at the same valence (Veal *et al.* 1987). Solubility limits for UO_3 and UO_2 in the binary alkali-silicate systems were determined to be ∼68 wt% and ∼34 wt%, respectively (Domine & Velde 1986). Solubili-ties of UO_2 and PuO_2 in complex borosilicate and aluminophosphate glasses were found to be ∼30 wt% and 5.6 wt%, respectively (Matyunin 2000). The PuO_2 concentration in lanthanide-borosilicate glass may reach ∼11 wt% (Bates *et al.* 1996; Bibler *et al.* 1996; Mesko *et al.* 1997; Chamberlain *et al.* 1997). The solubility limit for AmO_2 in Na-Li-borosilicate glass was found to be ∼2 wt% (Eller *et al.* 1985). More-over, significant attention must be paid to ensure homogeneous distribution of [239]Pu and other fissile isotopes to avoid concentration in local areas, where criticality could be reached (Matyunin *et al.* 2001).

One disadvantage of borosilicate glass is the low solubility of sulphates, molybdates, chro-mates, and halogenides, which may cause separa-tion of metastable phases (Fig. 1) at relatively low contents of these components (1–3 wt%, dependent on glass composition; Camara *et al.* 1980; Kawamoto *et al.* 1981; Stefanovsky 1989; Stefanovsky & Lifanov 1989). At higher concentrations, 'yellow phase' formation may occur (Morris & Chidley 1976; Stefanovsky & Lifanov 1988; Lutze 1988). The yellow phase, consisting of alkali and alkaline earth moly-bdates, sulphates, chromates, and halogenides, concentrates Cs and Sr radionuclides, and its presence increases leach rates of these radionuclides.

Most leach rate measurements of both matrix elements and radionuclides were performed at 90 °C using MCC-1 or PCT tests. According to these tests, leach rates range from 10^{-1} to $10\ \mathrm{g \cdot m^{-2} \cdot d^{-1}}$ (Lutze 1988). For example, the mass and elemental leach rates (in $\mathrm{g \cdot m^{-2} \cdot d^{-1}}$) for the PNL 76-68 glass containing 33 wt% waste oxides were determined at: mass – 0.42, Ca – 0.068, Cs – 1.03, Mo – 1.40, Na – 1.32, Sr – 0.075, B – 1.12, and Si – 0.73. These values are typical for borosilicate waste glass as measured by the MCC-1 procedure (90 °C, 28 d). Leach rates of Fe-group elements and ACTs under the same test conditions are con-siderably lower (10^{-3} and $10^{-4}\ \mathrm{g \cdot m^{-2} \cdot d^{-1}}$, respectively).

The effect of radiation on the structure of oxide glasses, including borosilicate nuclear waste glasses, depends on the type of radiation as distinguished on the basis of its linear energy transfer (LET) values. Radiation with low LET (γ-rays, β-particles, protons) causes excitation and ionization processes with for-mation of excitons (electron–hole pair), point defects, ion radicals, free electrons, and holes. Borosilicate glasses exhibit boron–oxygen hole centres (BOHC), silicon–oxygen hole centres (SOHC), and trapped electrons. Gamma- or elec-tron-irradiation may disrupt Si—O or B—O bonds with displacement of O atoms to inter-stitial positions. This may be followed by dimer-ization and molecular oxygen formation (DeNatale & Howitt 1985) or formation of O_2^- and O_3^- ions (Bogomolova *et al.* 1995, 1997). Radiation with high LET values (α-particles, α-recoil, fission fragments) produces atomic displa-cements and displacement cascades. Damage in the high-LET regime may cause volume changes of ±1.2% of the initial volume at a dose of $1-2 \times 10^{18}$ α-decays/g (corresponding to 1×10^9 Gy) and stored energy accumulation (up to ∼140 J/g at a dose of $0.1-0.3 \times 10^{18}$ α-decays/g, or 10^8 Gy). Other consequences include: He accu-mulation; formation of gas (probably oxygen) bubbles; increase in fracture toughness by 45%; and reduction in hardness and Young's modulus by 25 and 30%, respectively. Furthermore, there is an increase in Na diffusion by a factor of 2 to 5 below 150 °C as compared to initial values for unirradiated glasses. One of the most important radiation effects in waste glasses is the change in chemical durability. Data obtained from short-term testing of ACT-doped simulated nuclear waste glasses indicate that leach rates may increase by up to three times as a result of radiation effects from α-decay. Ion-, neutron-, and γ-irradiation to doses of up to 10^9 Gy increased the leach rate by a factor of up to four (Weber *et al.* 1997).

Table 1. *HLW glass compositions for industrial applications*

Type	R7/T7 (SON 68)	SRL-165 (Defence Waste Processing Facility)	ATM-10 (West Valley Demonstration Project)	Excess weapons plutonium	Borobasalt	Na–aluminophosphate glass
Country	France	USA	USA	USA	Russia	Russia
Waste	Raffinates from reprocessing of commercial LWR fuel at La Hague	Defence waste from the Savannah River Site	HLW from reprocessing of commercial SNF	Excess weapons plutonium	Excess weapons plutonium	HLW from reprocessing of SNF from water–water energetic reactors
Composition (in wt%)						
B_2O_3	16.9	8–9	11.2–14.8	10.4	22.2–28.0	—
SiO_2	55.0	48–52	39.5–48.4	25.8	29.0–33.7	—
Al_2O_3	5.9	4–5	5.6–7.1	19.0	8.1–9.6	20–24 (Al_2O_3 + Fe-group element oxides)
Fe_2O_3	—	11–13	10.7–13.5	—	5–6	—
CaO	4.9	—	0.2–0.6	—	14.4–16.6	—
Na_2O	11.9	12–13	7.1–8.6	—	1.9–5.7	24–27 (alkali oxides)
K_2O	—	—	4.1–5.3	—	0.5–0.9	—
Li_2O	2.4	3–4	3.3–4.2	—	—	—
MgO	—	~2	0.7–1.3	—	2.9–4.7	—
MnO	—	—	0.7–0.9	—	—	—
ZnO	3.0	—	—	—	—	—
SrO	—	—	—	2.2	—	—
TiO_2	—	—	0.7–0.9	—	0.5–0.9	—
P_2O_5	—	—	1.0–1.4	—	—	50–52
La_2O_3	—	—	—	11.0	—	—
Nd_2O_3	—	—	—	11.4	—	—
Gd_2O_3	—	—	—	7.6	—	—
ZrO_2	—	—	1.2–1.4	1.2	—	—
ThO_2	—	—	0.1–3.6	—	—	—
U_3O_8	~1	—	0.1–0.8	—	—	—
Waste loading	28 wt%, including 18.5 wt% of total fission products, ACTs, noble metals and Zr	Up to 33 wt%, including ^{137}Cs (~0.02 Ci/kg), ^{90}Sr (~0.2 Ci/kg), and ^{238}Pu (0.02–0.03 Ci/kg)	^{137}Cs and ^{90}Sr in concentrations of about 10 Ci/kg	11.4 wt% PuO_2	Up to 5 wt% PuO_2	Up to 10 wt% HLW oxides
References	Petitjean et al. 2002	Bibler et al. 1999; Marra et al. 1999	Barnes & Jain 1996; Jain & Barnes 1997; Palmer & Misercola 2003	Ramsey et al. 1995; Bibler et al. 1996; Mesko et al. 1997; Chamberlain et al. 1997; Meaker et al. 1997; Riley et al. 2000	Matyunin et al. 2001; Minaev et al. 2001	Vashman & Polyakov 1997

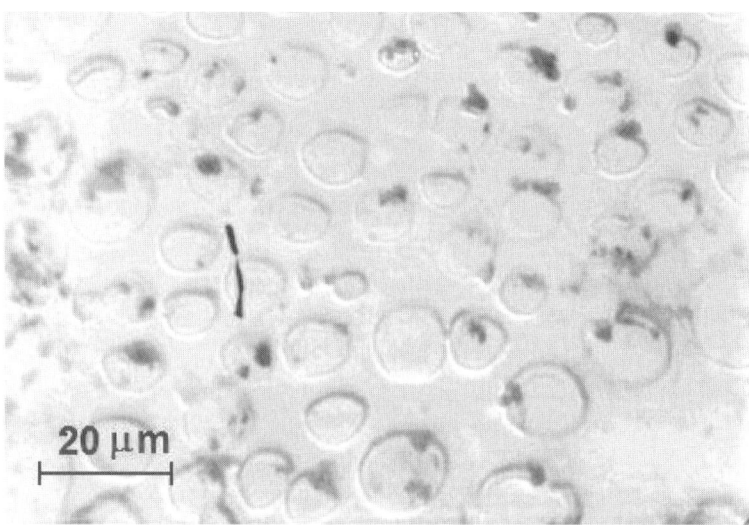

Fig. 1. Secondary electron micrograph of Na–borosilicate glass loaded with 30 wt% waste oxides and containing 0.9 wt% SO_3 (etching with 0.1 M HCl). Droplets with chemical durability lower than matrix are a sulphate-enriched phase (Stefanovsky 1989).

LILW. Vitrification of LILW was proposed in the late 1970s as an alternative to the bituminization process (Sobolev & Khomtchik 1983). Institutional and nuclear power plant wastes contain major amounts of Na-nitrate and minor amounts of Ca-, Mg-, Fe-nitrates and -carbonates. Some waste streams may additionally contain SO_4^{2-} and Cl^-, which can lead to yellow phase formation (Stefanovsky & Lifanov 1988). Operational wastes from nuclear power plants with a Russian water–water energetic reactor or pressurized water reactor (PWR) contain B, and thus, vitrification of these wastes may be performed with B-free fluxes (Lifanov *et al.* 1998, 2003). To vitrify Russian B-free institutional or channel fast-neutron reactor wastes, however, B-containing fluxes are required. Naturally occurring datolite, $CaBSiO_4(OH)$, is commercially available in Russia and has been proposed as a B source, whereas sandstone, loam clay, and bentonite have been chosen as glass-forming additives. The resulting materials are Na-Ca-borosilicate or B-free Na-aluminosilicate glasses (Sobolev *et al.* 1995a; Lifanov *et al.* 2003). The properties of these glasses are largely determined by the base oxides forming the matrix glass, because the radionuclide contents in LILW are at the level of 10^{-2}–10^{-5} wt% and they cannot affect the structure and properties of the glass. The major radioactivity in the LILW is due to β–γ-emitting [134,137]Cs and [90]Sr. Their leach rates (28-d IAEA test at 20 °C) from the glasses are 10^{-1}–10^{-2} g·m^{-2}·d^{-1}. Gamma-irradiation to a dose of 10^5 Gy (such a dose will be accumulated by solidified LILW during storage for ~300 y) caused only formation of point defects and ion radicals and did not affect the leach rate of radionuclides. Because these glasses are produced by electric melting, viscosity and electric resistivity are important parameters. They were found to be 4–7 Pa·s and 0.03–0.05 Ω·m, respectively (Dmitriev & Stefanovsky 2000).

Phosphate glasses

Development of phosphate-based glasses for HLW processing started in the early 1960s (Clark & Godbee 1963; Hatch *et al.* 1963). The main advantages of these glasses include low melting temperature (<1000 °C) and high solubilities of sulphates, chromates, molybdates, F, Cl, I, and oxides of multivalent elements, including ACT (Clark & Godbee 1963; Hatch *et al.* 1963; Brezhneva *et al.* 1976; Nikiforov *et al.* 1985; Stefanovsky *et al.* 1995; Vashman & Polyakov 1997; Badyal *et al.* 1999; Fujihara *et al.* 1999; Matyunin 1995, 2000). Phosphate melts, however, are extremely corrosive, requiring platinum pot lining to be produced, and the resulting phosphate-based glasses are more easily devitrified and have a lower chemical durability under hydrothermal conditions than borosilicate glasses (Brezhneva *et al.* 1976; Mukhamet-Galeyev *et al.* 1995).

The Russian *Na–aluminophosphate glass* contains mainly alkali oxides, Al_2O_3 and other oxides of the Fe group, P_2O_5, and up to 10 wt% other waste oxides including FPs and ACTs (Table 1). Leach rates determined by the IAEA procedure at room temperature for 7 d are $\sim 10^{-2}$ $g \cdot m^{-2} \cdot d^{-1}$ for Na, Cs, and Sr, and $\sim 10^{-3}$ $g \cdot m^{-2} \cdot d^{-1}$ for Pu. Glass annealing at 450 °C increases the Na leach rate by approximately two orders of magnitude due to crystallization of Na-phosphates and aluminophosphates with low chemical resistance. Gamma-irradiation of the glass to a dose of 10^8 Gy increases its surface damage depth from ~ 5 nm to ~ 100 nm after 10 min of water contact. The Na leach rate changes insignificantly with dose. ACT-doped glass demonstrated no crystallization or liquid–liquid phase separation at a dose of 2×10^{24} α-decays/m^3 ($\sim 8 \times 10^{17}$ α-decays/g). Physical properties of this glass are given in Table 2 (Vashman & Polyakov 1997).

Lead–iron phosphate (LIP) glass was proposed for immobilization of HLW with high Fe contents. The composition of the base glass corresponds to $Pb(PO_3)_2$ doped with 9 wt% Fe_2O_3. This glass may be prepared and poured at ~ 800 °C, and is very resistant to devitrification. The structure of this glass consists of polyphosphate chains cross-linked by octahedrally coordinated Fe^{3+} ions with a Fe–O distance of 0.19 nm. The coordination number of Pb^{2+} is 8 or 9, and the polyhedra are considerably disordered with an average Pb–O distance of 0.24 nm. Such a structure strengthens the polyphosphate network, thereby blocking the percolation of Pb and hydronium ions so that the chemical resistance of the overall structure in aqueous solutions is increased, leading to very high chemical durability. The net release of all elements from LIP glass loaded with commercial HLW is between 100 and 1000 times less than the corresponding release from the borosilicate glass PNL 76-68. For example, Cs, Sr, and U leach rates at 90 °C (MCC-1 test, 30 d) from the LIP glass loaded with defence HLW are <0.02, <0.02, and <0.001 $g \cdot m^{-2} \cdot d^{-1}$, respectively, whereas the corresponding values for the PNL 76-68 glass are ~ 1.0, 0.04, and 0.1 $g \cdot m^{-2} \cdot d^{-1}$. The LIP glass has a higher density (4.9 ± 0.2 g/cm^3) due to the high PbO content than both Na–aluminophosphate and borosilicate glasses, thus increasing the waste volume reduction factor. The LIP glass seems to be preferable for immobilization of defence HLW, because the HLW loading of 16 wt% is much higher than the 6.5 wt% in commercial HLW (Sales & Boatner 1988).

Production technology

Silicate glasses. Borosilicate glass has been used in France for conditioning of actual HLW since 1969. The currently implemented process at La Hague, the so-called two-step AVM process, uses a horizontal rotary calciner and an induction-heated metallic melter. The process has been in active operation since June 1978, and 8250 canisters of fully active glass have been produced so far (Masson *et al.* 1999; Petitjean *et al.* 2002). Currently, replacement of the induction-heated metallic melter by a cold-crucible melter is being considered (Petitjean *et al.* 2002). In the USA, the Department of Energy (DOE) recommended borosilicate glass as waste form for HLW in 1979 (Hench *et al.* 1984; Hamel *et al.* 1998). In March 1996, the Defence Waste Processing Facility began radioactive operations using a single-step vitrification process with slurry feeding into a Joule-heated ceramic melter (Marra *et al.* 1999). The West Valley Demonstration Project began production of borosilicate glass in June 1996 using the same process (Palmer & Misercola 2003). Borosilicate glasses are also considered as a primary HLW form in the United Kingdom, Germany, Belgium, Italy, Japan, and Korea (Lutze 1988). In Russia, immobilization of HLW in borosilicate glass has been suggested for implementation at the RT-2 plant of Krasnoyarsk Mining and Chemical Combine, Siberia (Aloy *et al.* 1998; Borisov *et al.* 2001), but the glass composition has not yet been chosen.

Vitrification of LILW is currently considered as a prospective process for waste dispositioning in various countries, including the USA (Whitehouse *et al.* 1995), whereby both borosilicate and B-free aluminosilicate glasses have been suggested. The Russian method of LILW vitrification is based on the inductive cold-crucible melting (ICCM) technology, developed at SIA Radon since the mid-1980s (Sobolev *et al.* 1995a,b). The full-scale vitrification plant with three cold crucibles and a total glass productivity of ~ 75 kg/h was put into active operation in 1998, producing borosilicate glass with a waste loading of 30–35 wt% LILW. Liquid waste is concentrated in a rotary film evaporator to a salt concentration of 1000–1100 g/L. This concentrate is mixed with datolite, bentonite, and sandstone to form a paste with a moisture content of 20–25 wt%, which is then fed into the cold crucibles. The molten glass is poured into 10 L containers, annealed in a tunnel furnace, and sent to a near-surface repository (Sobolev *et al.* 1995a, b). The stored glass is periodically sampled and

Table 2. *Comparison of properties of various HLW forms*

Properties	Borosilicate glass PNL 76–68	Russian Na-aluminophosphate glass	Synroc-C
Density, g/cm^3	2.6	2.5	4.35
Thermal conductivity, W/(m K)	1.1	0.7–1.5	2.1
Linear expansion coefficient, 1/K	8.1×10^{-6}	$(8-10) \times 10^{-6}$	10.5×10^{-6}
Specific heat capacity, J/(g K)	0.9	0.9–1.0	0.55
Glass transition temperature, K	725–775	700	–
Viscosity, Pa s (T, K)	5–20 (1375–1425)	3–7 (1125–1175)	0.5–1 (1550–1600)
Electric resistivity, Ω cm (T, K)	0.8–0.9 (1375–1425)	0.3–0.4 (1125–1175)	0.1–0.2 (1550–1600)
Flexural strength, MPa	68–74	40–46	110
Compressive strength, MPa	200–400	90–130	793
Young's modulus, GPa	82	54	134–203
Fracture toughness, MPa m$^{-0.5}$	0.77	–	1.83
Poisson's ratio	0.22	–	0.30
Microhardness	7.2	–	8.4
Leach rates, g/(m^2 day),*			
mass	0.42	–	0.005
Al	~0.01	–	<0.01
B	1.12	–	–
Ba	–	–	0.036
Ca	0.068	–	0.0065
Cs	1.03	0.35	0.033
Mo	1.40	–	0.12
Na	1.32	0.59	<0.2
Si	0.73	–	–
Sr	0.075	0.05	0.010
Ti	–	–	$<5 \times 10^{-5}$
Zr	–	–	$<6 \times 10^{-4}$
REE	0.01–0.05	–	10^{-4}–10^{-3}
ACT	0.01–0.05	$<10^{-3}$	2×10^{-5}–5×10^{-4}
References	Ringwood *et al.* 1988	Vashman & Polyakov 1997	Ringwood *et al.* 1988

*determined by MCC-1 test (90 °C, 7 d, deionized water)

examined with respect to occurrence of crystal-line phases and surface alteration. Long-term studies (up to 15 y) revealed a high chemical durability of the stored glasses (Sobolev *et al.* 1996).

The transportable vitrification system (TVS), operated in the USA, uses a Joule-heated ceramic melter (Whitehouse *et al.* 1995), which has been proposed for installation in other countries as well. However, a cold-crucible melter has also been considered for waste treat-ment in the USA, Italy, and Korea (Jouan *et al.* 1998).

Phosphate glasses. Currently, Na–alumino-phosphate glass is produced in a Joule-heated ceramic melter at the RT-1 plant of Production Association 'Mayak', Ural region, Russia. This plant has a waste capacity of up to 500 L/h, and by the end of 1995, the plant has treated 1974 t of HLW, or 2.4676×10^8 Ci (Glagolenko *et al.* 1996; Vashman & Polyakov 1997). The production of LIP glass is not yet implemented.

Ceramics

Single-phase ceramics

The hosts for ACT and REE immobilization are phases with a fluorite-derived structure (cubic zirconia-based solid solutions, pyrochlore, zirco-nolite, murataite), and zircon. The REEs and minor ACTs may be incorporated in perovskite, monazite, apatite–britholite, and titanite. Perov-skite and titanite are also hosts for Sr, whereas hollandite is a host phase for Cs and corrosion products. None of these ceramics is truly a single-phase material, and other phases such as silicates (pyroxene, nepheline, plagioclase), oxides (spinel, hibonite/loveringite, crichtonite), or phosphates may be present and incorporate some radionuclides and process contaminants. A brief description of the most important phases suitable for immobilization of ACTs and REEs is given below.

Pyrochlore, $^{VIII}A_2{}^{VI}B_2{}^{IV}X_6{}^{IV}Y$ *(Fd3m, Z = 8)* has a fluorite-derived structure, in which the *A* sites are occupied with large cations (Na, Ca, REE, ACT), and the *B* sites with lower-radius and higher-valence cations (Nb, Ta, Ti, Zr). Oxygen ions enter the *X* sites, and the *Y* sites are filled with additional ions or remain vacant.

The effect of α-decay on the pyrochlore structure $Gd_2Ti_2O_7$ doped with ^{244}Cm has been investigated in detail (Weber *et al.* 1998). The material was completely amorphized at a dose of 3.1×10^{18} α-decays/g. Amorphization is accompanied by volume expansion and increase of dissolution rate (by approximately a factor of

2.5), but leach rates of Gd and Cm may increase by factors of 20–50 due to incongruent dissol-ution. Samples of pyrochlore ceramic doped with 8.7 wt% ^{238}Pu and composed of mainly pyrochlore and minor brannerite (UTi_2O_6), rutile, and a fluorite-structure phase ((Pu,U)O_2) were kept for 6 months, and the volume expan-sion was found to be 1.2–1.6%. Most of the X-ray diffraction peaks have disappeared at a dose of 8.8×10^{17} α-decays/g. Another pyrochlore-based sample doped with ^{238}Pu has been amorphized at a similar dose of about 1×10^{18} α-decays/g, and the dissolution rate in water remains low ($\sim$0.04 g/m^2; Strachan *et al.* 2002).

Natural pyrochlores have been systematically studied as analogues of ACT host phases (Lumpkin *et al.* 1994, 1999, 2001). They were rendered amorphous at doses of 0.7–17 dpa, whereas synthetic samples became metamict at a significantly lower dose (0.3 dpa). Such a difference in irradiation doses causing complete structure disordering of natural and synthetic samples is due to radiation annealing of the lattice defects in minerals over geological time periods. This process is facilitated at elevated *T*, for example, due to metamorphism or natural geothermal gradients (Lumpkin *et al.* 1999).

The radiation resistance of the pyrochlore-structure phases has also been studied by ion irradiation (Smith *et al.* 1998; Weber & Ewing 2002; Chen *et al.* 2002; Lian 2002*b*). In all cases, the critical amorphization dose increases with *T* and with increasing Zr substitution for Ti in the series $Gd_2Ti_{2-x}Zr_xO_7$. Pyrochlores with $x \geq 1.8$ were not amorphized, even at 25 K. For example, $Gd_2Zr_2O_7$ is radiation resist-ant to a 1 MeV Kr$^+$ ion irradiation at 25 K at doses of up to 5 dpa. The radiation resistance of $A_2Zr_2O_7$ (A = La, Nd, Sm, Gd) increases with decreasing A-site cation radius, that is, from La^{3+} to Gd^{3+}. La$_2Zr_2O_7$ is the only zirco-nate pyrochlore that can be amorphized by ion-beam irradiation, and the critical amorphization temperature for La$_2Zr_2O_7$ irradiated by 1.5 MeV Xe$^+$ is $\sim$310 K (Chen *et al.* 2002). For a recent review on pyrochlore as a nuclear waste form for the immobilization of Pu, the reader is referred to Ewing *et al.* (2004).

Zirconolite, ideally $CaZrTi_2O_7$, incorporates ACTs and REEs by the isomorphic substitutions: $Ca^{2+} + Ti^{4+} = (ACT,REE)^{3+} + Me^{3+}$(Me = Al, Fe, Cr); $Ca^{2+} + Zr^{4+} = 2$ $(ACT,REE)^{3+}$; $Ca^{2+} + Ti^{4+} = ACT^{4+} + Me^{2+}$ (Me = Mg, Fe, Co, Mn); $Zr^{4+} + Ti^{4+} = (ACT,REE)^{3+} + Me^{5+}$ (Me = V, Nb, Ta); and $Zr^{4+} = ACT^{4+}$ (Gieré *et al.* 1998). There are several zirconolite polytypes with monoclinic (2M, 4M), trigonal (3T, 6T), and orthorhombic (3O) symmetry (Mazzi & Munno 1983; White 1984; Bayliss *et al.* 1989; Smith & Lumpkin 1993; Coelho *et al.* 1997; Begg *et al.* 2001). Both natural and synthetic zirconolites are leach-resistant, and even meta-mict samples with accumulated doses up to $\sim 10^{19}$ α-decays/g (~ 2 dpa) retain U and Th and their daughter products in the structure (Ringwood *et al.* 1988). Elemental release rates from zirconolite-based ceramics (7-d MCC-1 test, 90 °C) were found to be (in $g \cdot m^{-2} \cdot d^{-1}$): 10^{-1}–10^{-2} for Ca, 10^{-3}–10^{-4} for Nd, 10^{-4}–10^{-5} for U and Pu, 10^{-2} for Al, 10^{-6} for Ti, and 10^{-7} for Zr (Vance *et al.* 1995, 1996*a*; Hart *et al.* 2000). Zirconolite also exhibits exceptional corrosion resistance below 250 °C in various acidic and basic fluids over a wide range of pH. Between 250 and 500 °C, zirconolite is subject to partial corrosion in acidic and basic media, and significant corrosion occurs only above 500 °C (Gieré *et al.* 2001).

The radiation resistance of zirconolite has been tested with ^{238}Pu- and ^{244}Cm-doped and ion-irradiated samples (Weber *et al.* 1998; Weber & Ewing 2002). Amorphization of zirconolite occurred at doses corresponding to ~ 0.3–0.5 dpa at room temperature. Dose–age relationships have been determined by analytical transmission electron microscopy for the onset dose and critical amorphization dose of a suite of natural zirconolites (Lumpkin *et al.* 1994, 1998*b*).

Cubic fluorite-structure (Fm3m) zirconia-based solid solution, $(Zr,ACT,REE)O_{2-x}$, exhibits significant compositional flexibility to incorporate high concentrations of Pu, neutron absorbers, and impurities contained in Pu-bearing wastes (Gong *et al.* 1999). The phase has excellent radiation stability. No amorphization was observed under ion irradiation at room temperature to a dose corresponding to 200 dpa, and at 20 K to a dose of 25 dpa. Irradiation with I^{+} and Sr^{+} up to 300 dpa produced defect clusters in Y-stabilized zirconia, but did not cause amorphization. Amorphization

was achieved only after irradiation with Cs^{+} at room temperature to a dose of 1×10^{21} ions/m^{2}, or 330 dpa (Wang *et al.* 2001).

Murataite, a cubic phase with a fluorite-related structure ($F\bar{4}3m$ or $F432$ or $F3m$), exhibits wide compositional variation, and thus its exact formula is not established yet; the recently suggested formula is $^{VIII}A_3{}^{VI}B_6{}^{V}C_2O_{22-x}$ (Urusov *et al.* 2002). The natural variety has a structure characterized by a three-fold elementary fluorite unit cell with $a = 1.4863$ nm (Adams *et al.* 1974) or 1.4886 nm (Ercit & Hawthorne 1995), whereas the synthetic analogue has a unit-cell parameter of $a = 1.4576$ nm. Murataite has been found in Synroc-type ceramic containing simulated Russian HLW and produced by ICCM (Sobolev *et al.* 1997*c*); here, the murataite content was only ~ 5 vol%, but the phase accumulated about 40 wt% of the total U introduced. Several murataite polytypes with three- (3C), five- (5C), seven- (7C), and eightfold (8C) elementary fluorite unit cells (Fig. 2) were observed (Stefanovsky *et al.* 1999; Yudintsev *et al.* 2001; Kirjanova *et al.* 2002). Murataite exhibits very high chemical durability, with U and Pu leach rates of 10^{-6}–10^{-7} $g \cdot m^{-2} \cdot d^{-1}$ (Yudintsev *et al.* 2001). A special feature of the murataite-5C variety is a zoned distribution of waste elements, whereby the highest ACT concentration occurs in the core of the grains (Fig. 3, Table 3). In addition to the high leach resistance, a major advantage of murataite is its capability to simultaneously incorporate ACTs, REEs, corrosion products and some process contaminants (Na) in its structure. Therefore, it is a very promising host for immobilization of the ACT-REE fraction of partitioned HLW and ACT-bearing wastes with complex composition. The radiation resistance of murataite is comparable with that of pyrochlore: the critical amorphization dose under 1 MeV Kr^{+} ion irradiation was found to be 1.82–1.85×10^{18} ions/m^{2} (0.14–0.15 dpa), corresponding to $\sim 3 \times 10^{18}$ α-decays/g (Lian *et al.* 2002*a*).

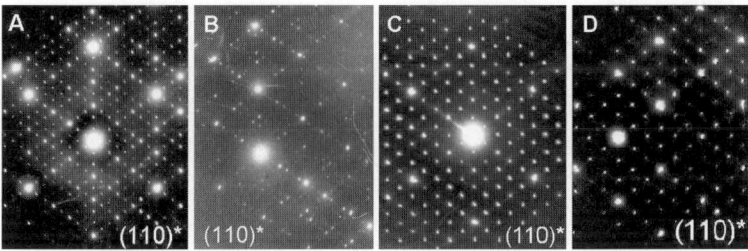

Fig. 2. Selected area electron diffraction patterns of the murataite varieties with (A) eight-, (B) seven-, (C) five-, and (D) threefold elementary fluorite unit cells.

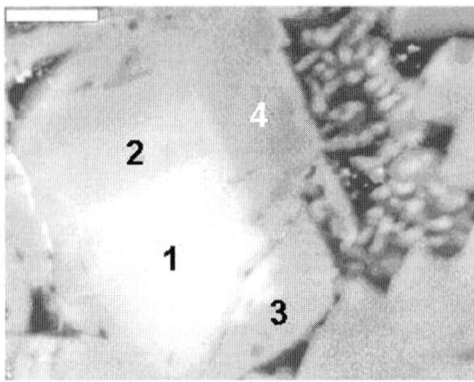

Fig. 3. Backscattered electron image of a zoned murataite-5C crystal in a Synroc-type ceramic containing 20 wt% HLW surrogate produced by ICCM at SIA Radon (Sobolev *et al.* 1997*c*). Scale bar: 5 μm. Numbers mark locations of analyses listed in Table 3. ACT concentrations in the core are approximately 10–20 times higher than at the edge.

Perovskite, $CaTiO_3$ (*Pnma*), may incorporate Sr instead of Ca, and trivalent ACT and REE and Fe-group elements via the substitution $Ca^{2+} + Ti^{4+} = REE^{3+} + Me^{3+}$ (Me = Al, Fe). Limited incorporation of tetravalent ACTs in the Ca site is also possible, whereby the solid solubility limit for tetravalent U and Pu is 0.01–0.1 atoms per formula unit (Vance *et al.* 2000). On the other hand, the trivalent REEs are capable of replacing Ca completely, thus forming a continuous solid solution, for example $CaTiO_3$–$GdAlO_3$ (Vance *et al.* 1996*c*). Trivalent Pu and Cm can also substitute for Ca. Perovskite is more leachable than zirconolite and pyrochlore (Ringwood *et al.* 1988).

Table 3. *Chemical composition of zoned murataite crystal (analysis spots shown in Fig. 3)*

wt%	1	2	3	4
Al_2O_3	0.8	1.6	3.7	2.8
SiO_2	0.2	0.5	0.2	0.3
K_2O	0.1	0.3	0.7	0.1
CaO	15.6	16.5	13.0	15.9
TiO_2	54.8	56.5	62.9	65.9
Cr_2O_3	0.3	–	0.2	–
MnO	3.6	4.1	5.7	5.6
FeO	0.2	–	0.1	–
NiO	0.2	–	–	0.2
ZrO_2	5.7	6.0	3.8	4.3
BaO	–	3.4	3.7	–
Ce_2O_3	6.0	2.9	2.7	3.7
UO_2	11.9	7.0	2.0	0.8
Total	99.4	98.8	98.7	99.6

Leach rates of ^{90}Sr, ^{147}Pm, ^{238}Pu, and ^{241}Am from perovskite ceramics are approximately $10^{-4}\,g\cdot m^{-2}\cdot d^{-1}$, whereas the leach rate of ^{238}U is higher by about two orders of magnitude.

Natural perovskites are normally crystalline; no traces of amorphization were found in samples with accumulated doses of up to 2.6×10^{18} α-decays/g. Significant damage was observed in perovskite containing ~6 wt% ThO_2 and with an accumulated dose of ~1×10^{19} α-decays/g (Lumpkin *et al.* 1998*a*). Perovskite was amorphized after irradiation with 1.5 MeV Kr^+ ions to a dose of ~1.8×10^{19} ions/m². The critical amorphization dose for perovskite is higher than for zirconolite by a factor of 2–4 (Smith *et al.* 1998).

Hollandite is one of the major Synroc phases incorporating Cs and corrosion products (Fe-group elements). Its generalized formula is $A_xB_yC_{8-y}O_{16}$, where $x \leq 2$. The A sites are occupied by large mono- or divalent cations (Na^+, K^+, Rb^+, Cs^+, Sr^{2+}, Ba^{2+}, Pb^{2+}), and various cations with a charge of +2 to +5 enter the B and C sites. Incorporation of Cs and Ba in hollandite is achieved under reducing conditions, thus suppressing formation of Cs and Ba molybdates and pertechnates, because Mo and Tc are reduced to their metallic state and form a separate phase, that is, a metal alloy (Ringwood *et al.* 1988).

Hollandite does not contain long-lived ACTs and, therefore, it is undergoing β–γ-irradiation from fission and corrosion products, but in multi-phase ceramics it can also be α-irradiated from neighbouring ACT-bearing phases. Irradiation by α-particles from external $^{238}PuO_2$ sources and heavy ions results in a volume expansion of 2–2.5% and transformation of tetragonal to monoclinic symmetry.

Zircon, $ZrSiO_4$ ($I4_1amd$, $Z = 4$), has been suggested as a promising host for ACTs (Ewing *et al.* 1995). Specific interest in zircon was enhanced by its occurrence in the Chernobyl lavas as an important ACT host phase (Ewing *et al.* 1995). However, later studies showed that the isomorphic capacity of zircon with respect to REEs and ACTs is very limited. The solubilities of $USiO_4$ and $ThSiO_4$ in zircon do not exceed 6 and 3 mol%, respectively (Ushakov 1998). Nevertheless, the radiation effects in zircon have been studied in detail using ^{238}Pu doping, heavy ion irradiation, and natural analogue studies (Ewing *et al.* 1995; Ewing 1999; Weber *et al.* 1998). The critical amorphization dose has been found to be $>8 \times 10^{18}$ α-decays/g.

Investigation of nuclide release from zircon with various U and Th contents as a function of α-dose showed that within a dose range of 10^{19}

to 10^{21} α-decays/g, the leach rate increased from 2.9×10^{-4} to 2.3×10^{-3} g·m^{-2}·d^{-1}. For fully metamict zircon, the leach rate was $<1.8 \times 10^{-2}$ g·m^{-2}·d^{-1}. Alpha-decay damage in zircon causes an increase in the unit-cell volume and in the total macroscopic swelling. The value of total macroscopic swelling at saturation was measured to be 16.6% and 18.4% for Pu-doped and natural zircons, respectively (Ewing et al. 1995).

Titanite, CaTiSiO$_5$ *(C2/c),* may incorporate Na$^+$, REE^{3+} and minor ACT in the Ca site, and Fe^{3+}, Al^{3+}, Nb^{5+} in the Ti site. The ThO$_2$ content of natural titanite may reach 0.28 wt% (Hayward 1988). Measured isomorphic capacities of titanite with respect to U^{4+}, Pu^{4+}, Pu^{3+}, Hf^{4+}, and Gd^{3+} are (in atoms per formula unit): 0.02–0.05 (Vance et al. 2000) or 0.07 (Stefanovsky et al. 2000b); 0.02; 0.05; 0.5; 0.3 (Vance et al. 2000) or 0.25 (Stefanovsky et al. 2000b), respectively. Due to limited solubility of ACTs and REEs in the structure, titanite is preferably considered as a host phase for these elements when their content is low, such as in titanite-based glass–ceramics developed for Canadian waste (Hayward 1988).

The radiation resistance of titanite was estimated both by the study of natural samples and by ion irradiation (Hayward 1988; Weber et al. 1998). Synthetic titanite irradiated with Ar$^+$ to a dose equivalent to 7×10^{18} α-decays/g was rendered completely amorphous, consistent with data for natural analogues (5×10^{18} alpha-decays/g).

Garnet-structure phases *(Ia3d, Z = 8)* with the general formula $^{VIII}A_3^{VI}B_2[XO_4]_3$, where $X = $ Si, Al, Fe, Ge, V, have also been proposed as host matrices. Natural silicate garnets contain small amounts of ACTs and REEs (typically <0.1 and <4 wt%, respectively; Minerals 1967). However, it is possible to produce garnet-structure aluminates and ferrates with high ACT and REE contents: the highest ThO$_2$, UO$_2$, and PuO$_2$ contents (16–20 wt%) were achieved in Gd–Zr–Fe garnet. An increase of the Gd content reduces the U content to 3–6 wt%, because both elements occupy the B sites in the garnet structure, although unlike Gd, U may also enter the A sites (Yudintsev 2001). Both silicate- and ferrite-aluminate garnets are susceptible to amorphization induced by ion irradiation below 1000 K: at room temperature, they have similar critical amorphization doses of about 0.2 dpa. The Si-free Ca–U–Zr–Fe–Al garnet (20.5 wt% UO$_2$) demonstrated the high-est radiation resistance, and its critical amorph-ization temperature was 890 K (Utsunomiya et al. 2002).

Phosphates and *silicophosphates,* such as apatite, Ca$_5$(PO$_4$,SO$_4$)$_3$(F,Cl,OH), and britholite, Ca$_{4-x}$(REE,ACT)$_{6+x}$(SiO$_4$)$_{6-y}$(PO$_4$)$_y$O$_2$, as well as monazite, REEPO$_4$, and kosnarite, NaZr$_2$(PO$_4$)$_3$, may be also considered as matrices for immobilization of various waste elements (Roy et al. 1982; Boatner & Sales 1988; Kryukova et al. 1991; Sheetz et al. 1994; Ewing et al. 1995; Wronkiewicz et al. 1996; Hawkins et al. 1997; Carpena et al. 1998; Aloy et al. 2002; Boatner 2002). Naturally occurring apatites contain limited amounts of ACTs (0.02 wt% U, 0.9 wt% Th; Heinrich 1958), but synthetic britholites may incorporate trivalent REEs and ACTs in significant concentrations. Synthetic britholite has been amorphized after 1.5 MeV Kr$^+$ irradiation at room temperature to a dose of 0.4 dpa. This phase may also be amorphized as a result of α-irradiation of incorporated ^{244}Cm (Weber et al. 1998).

Monazite is a mixed lanthanide orthophosphate proposed as a host phase for ACTs and REEs. Natural monazite contains <27 wt% UO$_2$ + ThO$_2$ and remains crystalline in spite of high accumulated α-doses (Boatner & Sales 1988; Weber et al. 1998). However, 1.5 MeV Kr$^+$ irradiation amorphizes monazite at a dose of 2.56×10^{18} ions/m^2 (Meldrum et al. 1996). Even fully amorphized monazite demonstrated low leachability, and the leach rate remained at the same level as that of unirradiated samples (Sales & Boatner 1988; Weber et al. 1998).

Multiphase ceramics

Synroc. Synroc is a titanate ceramic consisting of primarily zirconolite, hollandite, and perovskite, and of minor amounts of rutile, Magneli phases, hibonite/loveringite, metallic alloys, and phosphates. One variety of Synroc, Synroc-C, contains 20 wt% waste oxides from commercial power reactors, 57.0 wt% TiO$_2$, 5.4 wt% ZrO$_2$, 4.3 wt% Al$_2$O$_3$, 4.4 wt% BaO, and 8.9 wt% CaO. It consists of $\sim$30 wt% zirconolite, 30 wt% hollandite, 20 wt% perovskite, 15 wt% Ti oxides and Ca–Al titanates, and 5 wt% alloys and phosphates. Zirconolite is the major host for ACTs and a minor host for REEs, Sr and corrosion products (Fe-group elements), hollandite is the main host for Cs and Fe-group elements, and perovskite accommodates most of the Sr and REEs and some Fe-group elements and ACTs (Ringwood et al. 1988).

In addition to the conventional Synroc-C, other Synroc formulations were designed. The first Synroc variety, Synroc-A, contained the silicates kalsilite, leucite and celsian, and was produced via melting/crystallization. Synroc-D was specifically designed to immobilize the defence HLW at the Savanna River Site; because these wastes contain large amounts of process contaminants (Fe, Al, Mn, Ni, Si, Na), a special phase assemblage comprising zircono- lite ($\sim$16 wt%), perovskite ($\sim$11 wt%), spinel ($\sim$55 wt%), and nepheline ($\sim$18 wt%) was designed (Hench *et al.* 1984). Another variety, Synroc-E, consisted of Synroc-C incorporated in a rutile matrix. Synroc-F (for SNF) was dis- tinctly different from the other Synroc varieties and was composed of major pyrochlore, perov- skite, and uraninite, and minor hollandite (Ring- wood *et al.* 1988).

All Synroc varieties, except for Synroc-A, were produced via the hot-pressing route. Pro- duction of these waste forms via melting, especially cold-crucible melting, resulted in for- mation of additional alkali and alkali earth molyb- date phases, namely Cs and Ca molybdates (powellite) due to the oxidizing conditions during preparation (Knyazev *et al.* 1996). Other features of ICCM-produced Synroc ceramics are larger crystals and higher porosity and, as a result, lower density as compared to the hot- pressed samples (Fig. 4). In the late 1990s, Synroc–glass forms were suggested to accom- modate complex radioactive waste (e.g., alkali- bearing waste from Hanford, USA), and conven- tional glass melting technology was utilized for their production (Vance *et al.* 1996*b*, *d*). Within the framework of the HLW partitioning con- cept, special Synroc formulations were devel- oped in Australia (Hart *et al.* 1996) and Russia (Lashtchenova 1999); these varieties consist of major perovskite and hollandite, and minor Cs titanates, and were designed for immobilization of the Cs–Sr fraction of HLW.

Most of the detailed examinations were per- formed for the Synroc-C variety (Ringwood *et al.* 1988). Table 2 compares typical elemental leach rates (7-d MCC-1 test, 90 °C) from Synroc-C containing 10 wt% HLW calcine with those from borosilicate and aluminophosphate glasses, documenting that Synroc-C exhibits the lowest leach rates.

The high resistance of Synroc phases to α-decay damage was first recognized on the basis of esti- mates for natural zirconolites and perovskites, both containing substantial amounts of U and Th. The first quantitative evaluations of radiation sta- bility were obtained from irradiation experiments of Synroc and its constituents with fast neutrons in a nuclear reactor. The specimens, produced by cold pressing and sintering and hot pressing, were irradiated to a dose corresponding to 0.7 dpa, or 8×10^{18} α-decays/g (recalculated), resulting in larger volume expansion for the cold-pressed and sintered specimen than for the hot-pressed one. Accelerated tests were also per- formed using a ^{238}Pu-doped material (2 wt% and 5 wt%). X-ray diffraction revealed that the Pu- doped zirconolite and perovskite phases were ren- dered amorphous after accumulation of a dose of 2.8×10^{18} α-decays/g ($\sim$0.37 dpa). Volume expansion was found to be 6–8% (Ringwood *et al.* 1988).

Other multiphase ceramics. Numerous multi- phase ceramic formulations for conditioning of various wastes have been designed (Harker 1988). These so called 'tailored ceramics' were developed for immobilization of complex defence wastes at the Savannah River Plant and Rockwell Hanford Operation (Harker 1988). Tailored ceramics include ACT and REE hosts (fluorite-structure solid solutions, zirconolite,

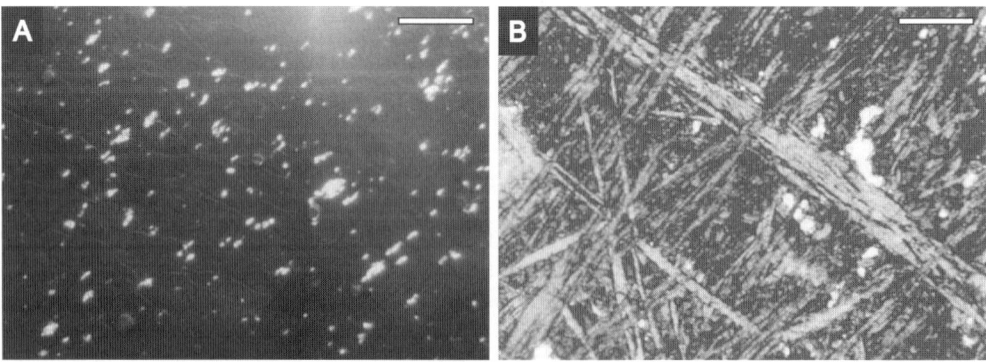

Fig. 4. Optical micrographs of (A) hot-pressed (at ANSTO) and (B) ICCM-produced (at SIA Radon) Synroc-C, loaded with 20 wt% HLW oxides. Scale bars: 500 μm (Sobolev *et al.* 1997*d*).

pyrochlore, perovskite, monazite, zircon), Sr and other alkaline earth hosts (magnetoplumbite, perovskite, hollandite), alkali hosts (nepheline, perovskite, interstitial glass), metal host phases (alloy), and non-FP host phases (spinel, corundum, rutile, pseudobrookite). Moreover, minor cubic murataite-type $(Zr(Ca,Mn)_2(Fe,Al)_4Ti_3O_{16})$ and loveringite-type $((Ca,Ce)(Ti,Fe,Cr,Mg)_{21}O_{38})$ phases were also found. The 'Supercalcine' composed of pollucite $(CsAlSi_2O_6)$, scheelite-structured Sr molybdate $(SrMoO_4)$, fluorite-structure solid solutions of REE $(REEO_2)$, and REE-rich apatite solid solution as well as zirconia (ZrO_2), corundum (Al_2O_3), spinel $(NiFe_2O_4)$, and various perovskites were designed to immobilize HLW from the PUREX (Pu–U-recovery–extraction) process.

In Russia, various polyphase ceramic compositions were developed in the late 1990s to immobilize the REE-ACT fraction of partitioned HLW. These ceramics are based on assemblages of perovskite, cubic oxide, pyrochlore, zirconolite, hibonite, loveringite, brannerite, and murataite (Stefanovsky et al. 1999, 2000a, 2001; Yudintsev et al. 1999, 2001). Moreover, it has been found that various garnet-containing assemblages (Burakov et al. 1999; Yudintsev 2001; Yudintsev et al. 2002) are promising matrices for ACTs, REEs, and Fe-group elements. The MCC-1 procedure revealed that leach rates were on the order of 10^{-5}–10^{-4} $g \cdot m^{-2} \cdot d^{-1}$ for ACTs and REEs, and 10^{-4}–10^{-2} $g \cdot m^{-2} \cdot d^{-1}$ for Fe-group elements. The lowest ACT leach rates were observed for murataite-containing ceramics (10^{-6}–10^{-5} $g \cdot m^{-2} \cdot d^{-1}$). The murataite (5C variety) grains exhibit ACT zoning (Fig. 3), with ACT concentrations in the core approximately 10–20 times higher than at the edge (Stefanovsky et al. 1999; Yudintsev et al. 2001). This zoning creates an additional barrier against leaching, but at the same time, it may cause a considerable problem by differential swelling due to α-decay damage (Chakoumakos et al. 1987). High radiation resistance of these ceramics has been confirmed by ion-irradiation tests up to a dose of 5×10^{18} $ions/m^2$ (Lian et al. 2002a; Utsonomiya et al. 2002).

Ceramic technologies

There are several possible technologies for production of ceramics, and all have been tested for inactive materials. The best-known technologies are based on high-T sintering of fine powders either under atmospheric pressure (cold pressing and sintering) or at high pressure (hot uniaxial or isostatic pressing). For example, the Synroc ceramics were produced by all three technologies; the hot uniaxial

pressing in bellows, originally developed at the Australian Nuclear Science and Technology Organisation (ANSTO), was implemented as production technology at the Synroc demonstration plant. Synroc powder is heated to 1150–1200 °C and uniaxially pressed at 14–21 MPa into 'pancakes', which can be stacked on top of one another in the disposal canister. To improve product quality, a special precursor is prepared using wet milling or sol-gel techniques (Ringwood et al. 1988). The cold pressing and sintering technology was developed at the US Lawrence Livermore National Laboratory for immobilization of excess weapons Pu in pyrochlore- and zirconolite-based ceramics (Brummond & Armantrout 1998). This process involves mixing of Pu oxide and a neutron absorber $(Gd_2O_3$ or $HfO_2)$ with ceramic additives $(TiO_2, ZrO_2, CaO, Al_2O_3)$ and cold pressing of the powder at 14–21 MPa in 300 g pucks ($\sim$9 cm in diameter, $\sim$1.6 cm thick, density $= 2.37$ g/cm^3, corresponding to 51% of the theoretical density), followed by heat treatment and a 4 h exposure to 1300 °C, and cooling for 14 h in a turned-off furnace. The final product (259 g pucks, diameter $= 6.86$ cm, thickness $= 1.57$ cm) has a density of 4.46 g/cm^3 ($\sim$95% of theoretical density).

Another technology for production of high-quality ceramics is based on the ICCM process. The cold crucible is a vessel fabricated from water-cooled Cu or stainless steel pipes, energized from a high-frequency generator operated within the range of $\sim$200 kHz to $\sim$13 MHz. The melt is heated by an electromagnetic field penetrating through the gaps between the crucible pipes. A special feature of the cold crucible is formation of an intermediate layer ('skull') between the melt and the cold-crucible pipes. This protects the cold-crucible walls from corrosion due to interaction with the melt and provides the following advantages relative to other melters: longer lifetime; higher achievable process temperature; no contact between melt, refractories and electrodes; smaller overall dimensions; and higher specific productivity because of the active hydrodynamic regime. The ICCM technology is currently being developed for the production of high-fusible glassy and ceramic waste forms in Russia (Sobolev et al. 1997a, b, c, 1998) and France (Jouan et al. 1996a, b). A number of polyphase ceramics based on zirconolite, pyrochlore, and murataite, as well as glass–ceramics have been produced by this method using laboratory- and bench-scale facilities and subsequently examined in detail.

An additional ceramic production method, developed mainly in Russia, is the self-sustaining synthesis (SSS). This process is based on redox reactions in oxide–metal mixtures, where the oxidizer is the oxide of a multivalent element in its highest oxidation state (MoO_3, WO_3, Fe_2O_3), peroxide (CaO_2) or nitrate, and the reducing agent is metallic (Ti, Zr, Al). Inert components used to produce the ceramic matrix are Ca, Ti, Zr, or calcine. An exothermic reaction is initiated in a local zone of batch mixture and then the burning process is self-sustained up to the formation of the ceramic or glass–ceramic. The SSS procedure was applied to production of zirconolite- and pyrochlore-based ceramics, including Pu-doped, and the leach resistance of these ceramics was comparable to ceramics produced by the cold pressing and sintering methods (Postnikov *et al.* 2001; Glagovsky *et al.* 2001).

Glass–ceramics

A number of glass–crystalline materials (GCM) have been proposed for immobilization of HLW. They can be classified in two groups: those produced by spontaneous crystallization, and those formed by induced (catalytic) crystallization.

Spontaneously crystallized GCM are composed of a vitreous phase and an uncontrolled amount of one or more crystalline phases. These GCM are called 'mineral-like materials' in the Russian literature, and 'glassy slags' in the English literature. They are normally produced by the ICCM method (Vlasov *et al.* 1987; Demine *et al.* 2001; Matyunin 2000; Matyunin *et al.* 2001), but may also be produced by the SSS technology (Glagovsky *et al.* 2001; Postnikov *et al.* 2001). These materials are very similar to glassy slags produced by plasma melting (Dmitriev *et al.* 1995; Feng *et al.* 1997), and contain 10–40 wt% crystalline phases distributed in a glassy matrix.

The GCM produced by *induced crystallization*, mostly via nucleating agents, are actual glass–ceramics. These GCM are composed of a target crystalline phase, which is embedded in interstitial glass and which amounts to <80 vol% of the material. The glass–ceramics may be produced using well-developed glass melting technologies in Joule-heated ceramic or cold-crucible melters (Hayward 1988; Dmitriev & Stefanovsky 2000) as well as in plasma furnaces (Dmitriev *et al.* 1995; Feng *et al.* 1997). The melts must then be either slowly cooled under controlled conditions, or quenched glasses must be annealed for a certain time to induce crystallization.

Best studied among these GCM materials is the titanite glass–ceramic (originally known as sphene-based glass–ceramic) developed in Canada (Hayward 1988). It is composed of titanite as the major crystalline phase (other minor phases may be present) and interstitial Na–aluminosilicate glass. Importantly, thermodynamic calculations indicate that titanite is stable in the Ca–Na–Cl brines typically encountered in the Canadian shield. Furthermore, analyses of natural specimens indicate that the titanite structure is able to accommodate ACTs, REEs, Na, Mn, Sr, Ba in the Ca site, and Fe and other transition metals in the Ti site. Both crystalline and metamict natural titanites are chemically durable, and the interstitial aluminosilicate glass also appears to be chemically durable. The titanite glass–ceramic is produced by melting at 1300–1400 °C, followed by cooling and thermal treatment (annealing) at 950–1050 °C for 1 h or more to induce crystallization. The resulting titanite concentrates REEs and corrosion products, whereas Cs^+ and U^{6+} remain in the vitreous phase. The Sr^{2+} ions are equally partitioned between titanite and the glass. Leach rates of major waste elements (Cs, Sr, U, La) are intermediate between those of borosilicate glass and Synroc. This glass–ceramic demonstrated good radiation stability and strong mechanical integrity.

A number of other glass–ceramics have been developed in various countries (Hayward 1988). At the Hahn–Meitner Institute, Germany, precursor borosilicate glasses melted at 1100–1400 °C were devitrified through a controlled heat treatment (typically at 530–720 °C), producing glass ceramics with the target phases celsian ($BaAl_2Si_2O_8$), eucryptite ($LiAlSiO_4$), spodumene ($LiAlSi_2O_6$), nepheline ($NaAlSiO_4$), perovskite ($CaTiO_3$), and diopside ($CaMgSi_2O_6$). To produce celsian glass–ceramics, precursor glasses were melted at 1175–1250 °C and then heat-treated in two stages, that is, at 600–650 °C (3 h) to create centres of crystallization, and at 800–850 °C (10–20 h) to crystallize the target phases (additional minor phases may be present). Leach rates of waste elements determined by the MCC-1 procedure were $\sim 1 \, g \cdot m^{-2} \cdot d^{-1}$. Irradiation with α-particles emitted from incorporated ^{239}Pu or ^{244}Cm increased the dissolution rate by 1–1.5 orders of magnitude due to radiation damage. Fresnoite ($Ba_2TiSi_2O_8$) glass–ceramics can be produced by melting at about 1200 °C and heat treatment at ~ 700 °C for nucleation, followed by heat treatment at 900–960 °C for crystallization of fresnoite and minor other phases.

Glass–ceramics based on natural or synthetic basalts, various zeolites, and apatite/britholite were also developed (Saidl & Ralkova 1966; Hayward 1988; Wronkiewicz *et al.* 1996; Sinkler *et al.* 2000; Nakazawa *et al.* 2001; Zhao *et al.* 2001; Tolstova *et al.* 2002).

Numerous studies have also been performed on glass–ceramics containing various Synroc phases, mainly zirconolite (Hayward 1988; Vance *et al.* 1996c, d; O'Holleran *et al.* 1997; Advocat *et al.* 1998; Lashtchenova & Stefanovsky 1998a, b; Loiseau *et al.* 2001; McGlinn *et al.* 2001; Martin *et al.* 2002). The precursor glasses were produced at 1300–1500 °C followed by slow cooling ($\sim$5 °C/h) or additional heat treatment at 900–1000 °C to produce the target phases. Actinides and REEs, and some Sr are incorporated into zirconolite, and Cs and the remaining Sr into the vitreous phase. As a rule, leach rates of ACTs and REEs from zirconolite glass–ceramics are between 10^{-4} and 10^{-2} g·m^{-2}·d^{-1}, and they are lower by 2–3 orders of magnitude than the Cs leach rate. The problem of Cs leaching is absent when glass-ceramics are used to immobilize separated ACTs.

A number of GCMs were developed on the basis of various natural rocks and industrial wastes for conditioning of LILW, mixed, and hazardous wastes. Such materials can be produced by smelting of various liquid and solid radioactive wastes (Kupfer *et al.* 1976; Komatsu *et al.* 1981; Palmer *et al.* 1981; Nomura *et al.* 1985; Lebeau & Girod 1987; Stefanovsky *et al.* 1992; Dmitriev *et al.* 1995; Lashtchenova *et al.* 1997) and by '*in situ*' vitrification (Ewing 1988; Timmons & Thompson 1996; McGlinn *et al.* 1998; Hartmann 2000). All these materials have chemical durabilities comparable to those of HLW glasses.

Cement

Portland cement has been extensively used since the early 1960s to immobilize LLW in many countries (Sobolev & Khomtchik 1983; McDaniel & Delzer 1988; Dmitriev & Stefanovsky 2000). Portland cement has a chemical composition of 64–67 wt% CaO, 21–25 wt% SiO$_2$, 4–8 wt% Al$_2$O$_3$, and 2–4 wt% Fe$_2$O$_3$, but minor alkali, Mg, Ti oxides, sulphates and phosphates may also be present. Major phases in the cement clinker are 3CaO·SiO$_2$ (alite, 42–60 wt%), 2CaO·SiO$_2$ (belite, 15–35 wt%), 3CaO·Al$_2$O$_3$ (tricalcium aluminate, 5–14 wt%), and 4CaO·Al$_2$O$_3$·Fe$_2$O$_3$ (brownmillerite, <5 wt%). The cement clinker is characterized by three parameters (weight concentration ratios), which must be kept within the following ranges: hydraulic modulus $m =$ CaO/(SiO$_2$ + Al$_2$O$_3$ + Fe$_2$O$_3$) = 1.9–2.4; silica modulus $n =$ SiO$_2$/(Al$_2$O$_3$ + Fe$_2$O$_3$) = 1.7–3.5; and alumina modulus $p =$ Al$_2$O$_3$/Fe$_2$O$_3$ = 1–3. The cement clinker is

mixed with water producing a grout. Mixing of dry clinker or grout and liquid and/or solid waste yields the cement waste form.

Currently, cement waste forms are used for immobilization of LLW and mixed wastes rather than HLW because of the low chemical and radiation resistance of the cement waste forms. The only exception is FUETAP (Formed Under Elevated Temperature And Pressure) concrete (McDaniel & Delzer 1988). Actual LLW normally contains NaNO$_3$ and other alkali salts, which compromise the properties of the final product as compared to those of waste-free cement. In particular, NaNO$_3$ retards the curing and reduces the mechanical integrity. An increase of the NaNO$_3$ concentration in LLW to 150 g/L reduces the compressive strength of the cement only insignificantly as compared to cement prepared with pure water (5–7 MPa). Higher NaNO$_3$ concentrations, however, result in strong reduction of the compressive strength (to <1 MPa). The appropriate waste:cement ratio is equal to 0.67 at an NaNO$_3$ concentration in LLW lower than 150 g/L. Leach rates of Cs and Sr from the cemented LLW as determined by the IAEA procedure (28 d, 20 °C) are relatively high at $\sim$10^2 g·m^{-2}·d^{-1} (Sobolev & Khomtchik 1983).

Major advantages of cement waste forms include very simple production and relatively low cost of both clinker and cementation technology. At the same time, however, there are numerous disadvantages of the conventional cement waste forms and cementation technology. The major disadvantage is that the waste volume increases by 10–30% relative to the initial material. Other drawbacks are low chemical, radiation, and freeze resistance, low mechanical integrity, and microbiological degradation. As a result, a number of attempts have been made to modify cement compositions in order to improve the waste-form properties. Cements have been modified by addition of clays, sand, rutile, plasticizers, milled incinerator bottom ash, fly ash, and other waste products (McDaniel & Delzer 1988; Jiang *et al.* 1993; Peri 1996; Dmitriev *et al.* 1999). Other cement waste forms have been designed on a non-Portland cement basis (Cougar *et al.* 1996; Siemer *et al.* 1997; Dmitriev *et al.* 1999). The FUETAP concrete, for example, consists of about 22 wt% cement, 11 wt% fly ash, 7.5 wt% clay, 28 wt% sand, 15 wt% simulated waste solids, 15 wt% water, and 0.75 wt% NaNO$_3$. This material has a density of 2 g/cm^3, a compressive strength of 60–100 MPa, a thermal conductivity of 1 W·m^{-1}·K^{-1}, and a porosity of 22–26%. Leach rates for Cs, Sr, and Pu were 10^{-2}, 1, and 10^{-4} g·m^{-2}·d^{-1}, respectively, and radiation stability was acceptable (McDaniel & Delzer 1988).

In Russia, extensive studies have been performed in regard to joint cementation of liquid and solid LLW, such as, incinerator ashes, contaminated soils, silts, and various other wastes accumulating through decommissioning of nuclear facilities (Sobolev & Khomtchik 1983; Dmitriev *et al.* 1999). The cementation process of radioactive silt (institutional or nuclear power plant waste) uses a fine-grained mixture of silt and/or ground CaO and Al$_2$O$_3$ additives, and dried radioactive waste salts. After heat treatment at 700–850 °C for 0.5–1 h, this material is transformed into a clinker with cement-like properties. After curing for 28 d, the cement exhibits a compressive strength of 20–35 MPa and a leach rate of 0.1–1 g·m^{-2}·d^{-1} (IAEA test, 20 °C), that is, lower by 2 to 3 orders of magnitude than for conventional cement waste forms (Dmitriev *et al.* 1999; Sobolev *et al.* 1999).

Cement is primarily considered as a matrix for LILW, and the main efforts are directed to improve the conventional cementation technology. One of the promising improvements is the application of high-energy mixers (e.g., inductive mixer with vortex layer) and vibration handling of the container during filling with high-penetrating cement grout (Sobolev *et al.* 1999).

Other waste forms

In addition to the waste forms described above, some other matrices for HLW and LILW were proposed. These include various recently developed composite materials, such as, glass composites (Sobolev *et al.* 1995*b*, 1997*b*); glass-encapsulated Ca-phosphate-based ceramics (Donald *et al.* 2002); glass-bonded sodalite (Esh *et al.* 1999; Pereira *et al.* 1999; Sinkler *et al.* 2000; Morss *et al.* 2000; Lambregts & Frank 2002; Jeong *et al.* 2002); apatite (Raicevic *et al.* 1999); silicotitanates (Nyman *et al.* 1999; Su *et al.* 1999); and complex Th-phosphates (Dacheux *et al.* 1999). Zeolites and zeolite-type compounds as well as phosphates were selected for immobilization of halogen-containing wastes due to their capability to accommodate halogens in their structure. Glass-bonded zeolites exhibit good radiation resistance to α-decay (Frank *et al.* 2002). Spent nuclear fuel can also be considered as a waste form (Johnson & Shoesmith 1988) as several countries show interest in direct disposal of SNF (for more details, see Buck *et al.*, 2004).

Recently, a new porous crystalline matrix ('Gubka') has been prepared on the basis of fly ash from power stations to incorporate complex ACT-containing wastes by means of repeated saturation–drying–calcining cycles. This matrix can accommodate up to ~50 wt% nitrate salts (after drying) and ~35 wt% calcine. The waste-loaded material can be consolidated by hot pressing with a 35% volume reduction (Aloy *et al.* 2000; Tranter *et al.* 2002).

Finally, a bitumen waste form should also be mentioned, as it has been widely applied for conditioning of LILW, particularly in Russia (Sobolev & Khomtchik 1983). The bituminization process is rather simple and consists of drying of the liquid waste, followed by mixing with molten bitumen. The bitumen waste form has many drawbacks, including: no waste volume reduction; fire danger; low chemical and radiation stability; low resistance to biodegradation; soil contamination with nitrates; however, bitumen as a waste form is still used in both Russia and Belgium. A recent study (Sobolev *et al.* 2000) has demonstrated that the bitumen waste form falls between the cement and glassy waste forms in terms of its chemical durability under near-surface repository conditions.

Outlook

To date, the most developed and commercially pursued HLW conditioning process is vitrification, which is implemented on a full-scale in the USA, France, and Russia. Borosilicate glasses are recognized to be more technologically and ecologically feasible as compared to phosphate-based glasses. Nevertheless, Russian HLW is vitrified at the Production Association 'Mayak' in the Ural region, yielding a Na–aluminophosphate glass, which is melted at lower temperature than borosilicate glass. Both borosilicate and aluminophosphate glasses are used for immobilization of non-partitioned HLW containing only traces of ACTs. To immobilize waste with a relatively high ACT content (e.g., ACT or REE-ACT fraction of HLW, whose separation is provided for by the HLW partitioning concept), glass is not a suitable matrix due to its thermodynamic instability and its insufficient chemical durability and radiation resistance. From this point of view, the ceramic materials are considered as an alternative to glass. The Synroc ceramic was designed for conditioning of non-partitioned HLW. Within the framework of the HLW partitioning concept, glass or special Synroc formulations may be used for immobilization of the short-lived Cs–Sr fraction, whereas crystalline waste forms based on zirconolite, pyrochlore, or perovskite are suggested as hosts for ACTs and REEs of the long-lived ACT-REE fraction. Extensive research was also performed on incorporation of excess weapons Pu, and phases with fluorite-derived structures (zirconolite, pyrochlore, zirconia-based solid solution) were recognized as most promising. Combined glass–ceramic forms also have a certain potential to be applied for conditioning of complex HLW and ILW.

Low- and intermediate-level waste is currently converted to cement and bitumen waste forms, but the existing trend is toward an increase in the radiation safety levels and reliability of immobilization. From this point of view, LILW should also be incorporated in vitreous and crystalline matrices. Current Russian and US experience on vitrification of such waste demonstrates the potential of the melting/vitrification process. Major advantages of vitrification over bituminization and cementation are a greater waste volume reduction, higher productivity, and higher durability of the final product.

The authors are very grateful to Dr Kenneth L. Nash (Washington State University), Dr Abdesselam Abdelouas (École des Mines de Nantes), and Dr Daniel Caurant (École nationale Supérieure de Chimie de Paris) for their constructive reviews and helpful suggestions.

References

ACTINIDE AND FISSION PRODUCTS PARTITIONING AND TRANSMUTATION 1999. *Proceedings of the Vth International Information Exchange Meeting*, Mol, Belgium, 25–27 November 1998, OECD, EUR 18898 EN.

ADAMS, J. W., BOTINELLY, T., SHARP, W. N. & ROBINSON, K. 1974. Murataite, a new complex oxide from El Paso County, Colorado. *American Mineralogist*, **59**, 172–176.

ADVOCAT, T., FILLET, C., MARILLET, J., LETURCQ, G., BOUBAIS, J. M. & BONNETIER, A. 1998. Nd-doped zirconolite ceramic and glass-ceramic synthesized by melting and controlled cooling. *Materials Research Society Symposium Proceedings*, **506**, 55–62.

ALEXANDROVA, I. T., GINSBURG, A. I., KUPRIYANOVA, I. I. & SIDORENKO, G. A. 1966. *Geology of Rare Elements Deposits. Rare Earth Silicates*. Nedra, Moscow (in Russian).

ALOY, A. S., ISKHAKOVA, O. A., KOLTSOVA, T. I. & TROFIMENKO, A. V. 1998. Glass waste form performance for disposal of the cesium and strontium concentrate resulting from the partitioning of HLW. *Materials Research Society Symposium Proceedings*, **506**, 901–906.

ALOY, A. S., ANSHITS, A. G., TRETYAKOV, A. A., KNECHT, D. A., TRANTER, T. J. & MACHERET, J.

2000. Development and testing of a new porous crystalline matrix (Gubka) for stabilizing actinide solutions. *Materials Research Society Symposium Proceedings*, **608**, 637–642.

ALOY, A. S., ISKHAKOVA, O. A., KOLTSOVA, T. I., TROFIMENKO, A. V. & JARDINE, L. J. 2001. Development and characterization of borosilicate glasses for immobilization of plutonium-containing sludges. *Materials Research Society Symposium Proceedings*, **663**, 189–198.

ALOY, A. S., KOVARSKAYA, E. N., KOLTSOVA, T. I., SAMOYLOV, S. E., ROVNY, S. I. & MEDVEDEV, G. M. 2002. Immobilization of Am-241 formed under plutonium metal conversion into monazite-type ceramic. *In*: LARDINE, L. J. & BORISOV, G. B. (eds) *Review of Excess Weapons Plutonium Disposition*. LLNL Contract Work in Russia, UCRL-ID-149341, 141–145.

ANTONINI, M., LANZA, F. & MANARA, A. 1979. Simulations of radiation damage in glasses. *In*: CHIKALLA, T. D. & MENDEL, J. E. (eds) *Proceedings of International Symposium on Ceramics in Nuclear Waste Management*, National Technical Information Service, CONF-790420, 289–294.

ASTM 1994. *Standard Test Methods for Determining Chemical Durability of Nuclear Waste Glasses: The Product Consistency Test (PCT)*. Standard C1285-94. ASTM, Philadelphia.

BADYAL, Y., KARABULUT, M. *et al.* 1999. The effect of uranium on the structure of iron phosphate glasses. *Materials Research Society Symposium Proceedings*, **556**, 297–304.

BAESTLÉ, L. H., WAKABAYASHI, T. & SAKURAI, S. 1999. Status and assessment report on actinide and fission product partitioning and transmutation, an OECD Nuclear Energy Agency review. *In*: Proceedings of the International Conference on Future Nuclear Systems Global '99 "Nuclear Technology – Bridging the Millenia", Jackson Hole, CD-ROM.

BARNES, S. M. & JAIN, V. 1996. Vitrification systems testing to ssupport radioactive glass production at the West Valley Demonstration Project. *In*: Waste Management '96 Conference, Laser Options, Inc., Tucson, CD-ROM.

BATES, J. K., ELLISON, A. J. G., EMERY, J. W. & HOH, J. C. 1996. Glass as a waste form for the immobilization of plutonium. *Materials Research Society Symposium Proceedings*, **412**, 57–64.

BAYLISS, P., MAZZI, F., MUNNO, R. & WHITE, T. J. 1989. Mineral nomenclature: zirconolite. *Mineralogical Magazine*, **53**, 565–569.

BEGG, B. D., DAY, R. A. & BROWNSCOMBE, A. 2001. Structural effects of Pu substitutions on the Zr-site in zirconolite. *Materials Research Society Symposium Proceedings*, **663**, 259–266.

BIBLER, N. E., RAMSEY, W. G., MEAKER, T. F. & PAREIZS, J. M. 1996. Durabilities and microstructures of radioactive glasses for immobilization of excess actinides at the Savannah River site. *Materials Research Society Symposium Proceedings*, **412**, 65–72.

BIBLER, N. E., FELLINGER, T. L., MARSHALL, K. M., CRAWFORD, C. L., COZZI, A. D. & EDWARDS, T. B. 1999. Demonstration of the defense waste processing facility vitrification process for Tank 42 radioactive sludge – Glass preparation and characterization. *Materials Research Society Symposium Proceedings*, **556**, 207–214.

BOATNER, L. A. 2002. Synthesis, structure, and properties of monazite, pretulite, and xenotime. *Reviews in Mineralogy and Geochemistry*, **48**, 87–121.

BOATNER, L. A. & SALES, B. C. 1988. Monazite. *In*: LUTZE, W. & EWING R. C. (eds) *Radioactive Waste Forms for the Future*. Elsevier Science Publishers, B.V., Amsterdam, 495–564.

BOGOMOLOVA, L. D., TEPLYAKOV, Y. G., STEFANOVSKY, S. V. & DMITRIEV, S. A. 1995. EPR of radiation centers in ion-implanted glasses simulating vitrified radioative wastes. *In*: SLATE, S., FEIZOLLAHI, F. & CREER, J. (eds) *Proceedings of the Fifth International Conference on Radioactive Waste Management and Environmental Remediation*. ASME, New York, **1**, 409–411.

BOGOMOLOVA, L. D., STEFANOVSKY, S. V., TEPLYAKOV, Y. G. & DMITRIEV, S. A. 1997. Formation of paramagnetic defects in oxide glasses during the bombardment of their surface with charged particles. *Materials Research Society Symposium Proceedings*, **465**, 657–664.

BORISOV, G. B., BALASHOV, A. V., MANSOUROV, O. A., NAZAROV, A. V. & VOLCHOK, Y. Y. 2001. Glass matrices for vitrification of Pu-containing sludge of MCC using microwave heating. 2001. *In*: JARDINE, L. J. & BORISOV, G. B. (eds) *Immobilization of Excess Weapons Plutonium in Russia: A Review of LLNL Contract Work*, UCRL-ID-143846, 57–60.

BRAMMOND, W. & ARMANTROUT, G. 1998. Ceramic process equipment for the immobilization of plutonium. *In*: *Waste Management '98 Conference*, Laser Options, Inc., Tucson, Rep. 65-05, CD-ROM.

BREZHNEVA, N. E., OZIRANER, S. G., MINAEV, A. A. & KUZTETSOV, D. G. 1976. Properties of phosphate and silicate glasses for solidification of radioactive wastes. *In*: *Management of Radioactive Wastes from the Nuclear Fuel Cycle*. IAEA, Vienna, **2**, 85–94.

BUCK, E. C., HANSON, B. D. & MCNAMARA, B. K. 2004. The geochemical behaviour of Tc, Np, and Pu in spent fuel in an oxidizing environment. *In*: GIERÉ, R. & STILLE, P. (eds). *Energy, Waste, and the Environment: a Geochemical Perspective*. Geological Society, London, Special Publications, **236**, 65–88.

BURAKOV, B. E., ANDERSON, E. B., KNECHT, D. A., ZAMORYANSKAYA, M. A., STRYKANOVA, E. E. & YAGOVKINA, M. A. 1999. Synthesis of garnet/perovskite-based ceramic for the immobilization of Pu-residue wastes. *Materials Research Society Symposium Proceedings*, **556**, 55–62.

CAMARA, B., LUTZE, W. & LUX, J. 1980. An investigation of the valency state of molybdenum in glasses with and without fission products. *In*: NORTHRUP JR, C. J. M. (ed) *Scientific Basis for Nuclear Waste Management*, Plenum Press, New York, **2**, 93–102.

CARPENA, J., AUDUBERT, F., BERNACHE, D., BOYER, L., DONAZZON, B., LACOUT, J. L. & SENAMAUD, N. 1998. Apatitic waste forms: Process overview. *Materials Research Society Symposium Proceedings*, **506**, 543–549.

CHAKOUMAKOS, B. C., MURAKAMI, T., LUMPKIN, G. R. & EWING, R. C. 1987. Alpha-decay-induced fracturing in zircon: the transition from the crystalline to the metamict state. *Science*, **236**, 1556–1559.

CHAMBERLAIN, D. B., HANCHAR, J. M. *et al.* 1997. Development and testing of a glass waste form for the immobilization of plutonium. *Materials Research Society Symposium Proceedings*, **465**, 1229–1236.

CHEN, J., LIAN, J., WANG, L. M., EWING, R. C., FARMER, J. M. & BOATNER, L. A. 2002. Structural alterations in titanate pyrochlores induced by ion irradiation: X-ray photoelectron spectrum interpretation. *Materials Research Society Symposium Proceedings*, **713**, 501–506.

CLARK, W. E. & GODBEE, H. W. 1963. Fixation of simulated highly radioactive wastes in glassy solids. *In*: *Treatment and Storage of High Level Radioactive Wastes*. IAEA, Vienna, 412–432.

COELHO, A. A., CHEARY, R. W. & SMITH, K. L. 1997. Analysis and structural determination of Nd-substituted zirconolite-4M. *Journal of Solid State Chemistry*, **129**, 346–359.

COUGAR, M. L. D., SIEMER, D. D. & Sheetz B. E. 1996. Vitrifiable concrete for disposal of spent nuclear fuel reprocessing waste at I.N.E.L. *Materials Research Society Symposium Proceedings*, **412**, 395–402.

DACHEUX, N., THOMAS, A. C., CHASSIGNEUX, B., PICHOT, E., BRANDEL, V. & GENET, M. 1999. Study of $Th_4(PO_4)_4P_2O_7$ and solid solutions with U(IV), Np(IV) and Pu(IV): Synthesis, characterization, sintering and leaching test. *Materials Research Society Symposium Proceedings*, **556**, 85–92.

DEMINE, A. V., KRYLOVA, N. V., POLUEKTOV, P. P., SHESTOPEROV, I. N., SMELOVA, T. V., GORN, V. F. & MEDVEDEV, G. M. 2001. High level liquid waste solidification using a "cold crucible" induction melter. *Materials Research Society Symposium Proceedings*, **663**, 27–34.

DENATALE, J. F. & HOWITT, D. G. 1985. The gamma-irradiation of nuclear waste glasses. *Radiation Effects*, **91**, 89–96.

DMITRIEV, S. A., STEFANOVSKY, S. V., KNYAZEV, I. A. & LIFANOV, F. A. 1995. Characterization of slag product from plasma furnace for unsorted solid radioactive waste treatment. *Materials Research Society Symposium Proceedings*, **353**, 1323–1332.

DMITRIEV, S. A., LIFANOV, F. A., VARLAKOV, A. P., KARLIN, S. V. & CHERNONOZSHKIN, V. N. 1999. Obtaining of an alkaline cementing material on the basis of radioactive silt and LRW of low and intermediate activity level. *In*: *Waste Management '99 Conference*. Laser Options, Inc., Tucson, CD-ROM.

DMITRIEV, S. A. & STEFANOVSKY, S. V. 2000. *Radioactive Waste Management*. Mendeleev University of Chemical Technology Press, Moscow (in Russian).

DOMINE, F. & VELDE, B. 1986. Preliminary investigation of the process governing the solubility of uranium in silicate melts. *Bulletin de Mineralogie*, **108**, 755–766.

DONALD, I. W., METCALFE, B. L. & GREEDHAREE, R. S. 2002. A glass-encapsulated ceramic wasteform for the immobilization of chloride-containing ILW: Formation of halite crystals by reaction between the glass encapsulant and ceramic host. *Materials Research Society Symposium Proceedings*, **713**, 287–293.

ELLER, P. G., JARVINEN, G. D., PURSON, J. D., PENNEMAN, R. A., RYAN, R. R., LYTLE, F. W. & GREEGOR, R. B. 1985. Actinide valences in borosilicate glass. *Radiochimica Acta*, **39**, 17–22.

ERCIT, T. S. & HAWTHORNE, F. C. 1995. Murataite, a UB12 derivative structure with condensed keggin molecules. *Canadian Mineralogist*, **33**, 1223–1229.

ESH, D. W., GOFF, K. M., HIRSCHE, K. T., BATTISTI, T. J., SIMPSON, M. F., JOHNSON, S. G. & BATEMAN, K. J. 1999. Development of a ceramic waste form for high level waste disposal. *Materials Research Society Symposium Proceedings*, **556**, 107–113.

EWING, R. C. 1988. Novel waste forms. *In*: LUTZE, W. & EWING, R. C. (eds) *Radioactive Waste Forms for the Future*. Elsevier Science Publishers, B.V., Amsterdam, 599–634.

EWING, R. C. 1999. Nuclear waste forms for actinides. *Proceedings of the National Academy of Sciences of the USA*, **96**, 3432–3439.

EWING, R. C. 2001. The design and evaluation of nuclear waste forms: Clues from mineralogy. *Canadian Mineralogist*, **39**, 697–715.

EWING, R. C. 2004. Environmental impact of the nuclear fuel cycle. *In*: GIERÉ, R. & STILLE, P. (eds) *Energy, Waste, and the Environment: a Geochemical Perspective*. Geological Society, London, Special Publications, **236**, 7–23.

EWING, R. C., WEBER, W. J. & CLINARD, JR., F. W. 1995. Radiation effects in nuclear waste forms for high level radioactive waste. *Progress in Nuclear Energy*, **29**, 63–127.

EWING, R. C., WEBER, W. J. & LIAN, J. 2004. Nuclear waste disposal—pyrochlore ($A_2B_2O_7$): Nuclear waste form for the immobilization of plutonium and "minor" actinides. *Journal of Applied Physics*, **95**, 5949–5971.

FELLINGER, A. P., BAICH, M. A. *et al.* 1999. Americium-curium vitrification process development (I). *Materials Research Society Symposium Proceedings*, **556**, 367–374.

FENG, X., EINZIGER, R. E. & ESCHENBACH, R. C. 1997. A direct single-step plasma arc – vitreous ceramic process for stabilizing spent nuclear fuels, sludges, and associated wastes. *Materials Research Society Symposium Proceedings*, **465**, 25–32.

FIELDING, P. E. & WHITE, T. J. 1987. Crystal chemical incorporation of high level waste species in aluminititanate-based ceramics: Valence, location, radi-

ation damage and hydrothermal durability. *Journal of Materials Research*, **2**, 387–414.

FRANK, S. M., BARBER, T. L. *et al.* 2002. Alpha-decay radiation damage study of a glass-bonded sodalite ceramic waste form. *Materials Research Society Symposium Proceedings*, **713**, 487–494.

FUJIHARA, H., MURASE, T., NISHI, T., NOSHITA, K., YOSHIDA, T. & MATSUDA, M. 1999. Low-temperature vitrification of radioiodine using $AgI-Ag_2O-P_2O_5$ glass system. *Materials Research Society Symposium Proceedings*, **556**, 375–382.

GIERÉ, R., WILLIAMS, C. T. & LUMPKIN, G. R. 1998. Chemical characteristics of natural zirconolite. *Schweizerische Mineralogische und Petrographische Mitteilungen*, **78**, 433–459.

GIERÉ, R., SWOPE, R. J., BUCK, E. C., GUGGENHEIM, R., MATHYS, D. & REUSSER, E. 2000. Growth and alteration of uranium-rich microlite. *Materials Research Society Symposium Proceedings*, **608**, 519–524.

GIERÉ, R., MALMSTRÖM, J. 2001. Durability of zirconolite in hydrothermal fluids: Implications for nuclear waste disposal. *Materials Research Society Symposium Proceedings*, **663**, 267–275.

GLAGOLENKO, YU, V., DZEKUN, E. G., DROZHKO, E. G., MEDVEDEV, G. M., ROVNY, S. I. & SUSLOV, A. P. 1996. Radioactive waste management strategy at production association 'Mayak'. *Issues of Radiation Safety*, 2, 3–10 (in Russian).

GLAGOVSKY, E. M., YUDINTSEV, S. V., KUPRIN, A. V., PELEVIN, L. P., KONOVALOV, E. E., VELICHKIN, V. I. & MYASOEDOV, B. F. 2001. A study of actinide crystalline matrices produced by self-sustaining high-temperature synthesis. *Radiochemistry*, **43**, 557–562 (in Russian).

GONG, W. L., LUTZE, E. & EWING, R. C. 1999. Zirconia – A ceramic for excess weapons plutonium wastes. *Materials Research Society Symposium Proceedings*, **556**, 63–70.

HAMEL, W. F., JR., SHERIDAN, M. J. & VALENTI, P. J. 1998. Vitrification at the West Valley Demonstration Project. *Radwaste Magazine*, 3, 27–42.

HARKER, A. B. 1988. Tailored ceramics. *In*: LUTZE, W. & EWING, R. C. (eds) *Radioactive Waste Form for the Future*. Elsevier Science Publishers B.V., Amsterdam, 335–392.

HART, K. P., VANCE, E. R., DAY, R. A., BEGG, B. D., ANGEL, P. J. & JOSTSONS, A. 1996. Immobilization of separated Tc and Cs/Sr in Synroc. *In*: MURPHY, W. M. & KNECHT, D. A. (eds) *Scientific basis for nuclear waste management XIX*. *Materials Research Society Symposium Proceedings*, **412**, 281–287.

HART, K. P., ZHANG, Y. *et al.* 2000. Aqueous durability of titanate ceramics designed to immobilize excess plutonium. *Materials Research Society Symposium Proceedings*, **608**, 353–358.

HARTMANN, T. 2000. Evaluation of phase- and element distribution after non-traditional *in situ* vitrification (NTISV) at Los Alamos National Laboratory on a simulated adsorption bed. *Materials Research Society Symposium Proceedings*, **608**, 619–624.

HATCH, L. P., WETH, G. C. & TUTHILL, E. J. 1963. Ultimate disposal of high level radioactive wastes – Fixation in phosphate glass with emphasis on the continuous mode of plant operation. *In: Treatment and Storage of High Level Radioactive Wastes.* IAEA, Vienna, 531–545.

HAWKINS, H. T., SHEETZ, B. E. & GUTRIE, JR., G. D. 1997. Preparation of monophasic (NZP) radiophases: Potential host matrices for the immobilization of reprocessed commercial high-level wastes. *Materials Research Society Symposium Proceedings,* **465**, 387–394.

HAYWARD, P. J. 1988. Glass-ceramics. *In:* LUTZE, W. & EWING, R. C. (eds) *Radioactive Waste Form for the Future.* Elsevier Science Publishers B.V., Amsterdam, 427–494.

HEINRICH, E. W. 1958. *Mineralogy and Geology of Radioactive Raw Materials.* McGraw-Hill Company, Inc., New York.

HENCH, L. L., CLARK, D. E. & CAMPBELL, J. 1984. High level waste immobilization forms. *Nuclear and Chemical Waste Management,* **5**, 149–173.

HESPE, E. D. 1971. Leach testing of immobilized radioactive waste solids. *Atomic Energy Review,* **9**, 195–207.

IAEA 1985. *Chemical Durability and Related Properties of Solidified High Level Waste Forms.* Technical Reports Series No. 257. IAEA, Vienna.

JAIN, V. & BARNES, S. M. 1997. Radioactive waste glass production at the WVDP. *In: Waste Management '97 Conference.* Laser Options, Inc., Tucson, CD-ROM.

JIANG, W., WU, X. & ROY, D. M. 1993. Alkali activated fly ash – slag cement based nuclear waste forms. *Materials Research Society Symposium Proceedings,* **294**, 255–260.

JEONG, S. Y., MORSS, L. R. & EBERT, W. L. 2002. Corrosion of glass-bonded sodalite and its components as a function of pH and temperature. *Materials Research Society Symposium Proceedings,* **713**, 413–420.

JOHNSON, L. H. & SHOESMITH, D. W. 1988. Spent fuel. *In:* LUTZE, W. & EWING, R. C. (eds) *Radioactive Waste Form for the Future.* Elsevier Science Publishers B.V., Amsterdam, 635–698.

JOUAN, A., MONCOUYOUX, J.-P., MERLIN, S. & ROUX, P. 1996a. Multiple applications of cold crucible melting. *Radwaste Magazine,* **3**, 77–81.

JOUAN, A., BOEN, R., MERLIN, S. & ROUX, P. 1996b. A warm heart in a cold body – melter technology for tomorrow. *In: Proceedings of the International Topical Meeting on Nuclear and Hazardous Waste Management SPECTRUM '96.* American Nuclear Society, Inc., La Grange Park, 2058–2062.

JOUAN, A., DO QUANG, R. & MERLIN, S. 1998. Industrial waste vitrification using the cold crucible melter. *In: Waste Management '98 Conference.* Laser Options, Inc., Tucson, CD-ROM.

KAWAMOTO, Y., CLEMENS, K. & TOMOZAWA, M. 1981. The effect on phase separation of the oxidation state of molybdenum in a $Na_2O-B_2O_3-SiO_2$ glass. *Physics and Chemistry of Glasses,* **22**, 110–114.

KIRJANOVA, O. I., STEFANOVSKY, S. V., YUDINTSEV, S. V. & NIKONOV, B. S. 2002. Effect of $CaO:Gd_2O_3$ and $CaO:UO_2$ ratios on phase composition in the $CaO-Gd_2O_3(UO_2)$ $MnO-TiO_2$ system. *Advanced Materials,* **5**, 38–45 (in Russian).

KNYAZEV, O. A., NIKONOV, B. S., OMELIANENKO, B. I., STEFANOVSKY, S. V., YUDINTSEV, S. V., DAY, R. A. & VANCE, E. R. 1996. Preparation and characterization of inductively-melted Synroc. *In: Proceedings of the International Topical Meeting on Nuclear and Hazardous Waste Management SPECTRUM '96.* ANS, Seattle, 2130–2137.

KOMATSU, F., SAWADA, Y., OHTSUKA, K. & OHUCHI, J. 1981. Development of a new solidification method for wastes contaminated by plutonium oxides. *In: Management of Alpha-Contaminated Wastes.* IAEA, Vienna, 325–337.

KRYUKOVA, A. I., ARTEMYEVA, G. YU., DEMARIN, V. T., ALFEROV, V. A. 1991. Cs-containing complex phosphates. Constitution. Cesium Leaching. *Radiochemistry,* **33**, 186–191 (in Russian).

KULIKOV, I. A., FILIN, V. M. & ANANINA, T. N. 2001. Radiation damage study of the U.S. specified Pu-containing ceramics. *In:* JARDINE, L. J. & BORISOV, G. B. (eds) *Immobilization of Excess Weapons Plutonium in Russia: A Review of LLNL Contract Work.* UCRL-ID-143846, 221–228.

KUPFER, M. J., SCHULZ, W. W., HOBBICK, C. W. & MENDEL, J. E. 1976. Glass forms for alpha waste management. *American Institute of Chemical Engineering Symposia Series,* **72**, 90–97.

LAMBREGTS, M. J. & FRANK, S. M. 2002. Preliminary studies of the disposition of cesium in glass-bonded sodalite waste form. *Materials Research Society Symposium Proceedings,* **713**, 373–380.

LAPINA, M. I. & YUDINTSEV, S. V. 1999. Study of natural zircon-xenotime assemblages for estimation of the actinide waste forms stability. *Materials Research Society Symposium Proceedings,* **556**, 785–792.

LASHTCHENOVA, T. N. 1999. Ph D thesis, SIA Radon, Moscow.

LASHTCHENOVA, T. N., LIFANOV, F. A. & STEFANOVSKY, S. V. 1997. Incorporation of radon incinerator ash in glass and glass crystalline materials. *Waste Management '97 Conference.* Laser Options, Inc., Tucson, CD-ROM.

LASHTCHENOVA, T. N. & STEFANOVSKY, S. V. 1998a. Immobilization of incinerator ash in Synroc–glass material. *In: Proceedings of the IT3 International Conference on Incineration & Thermal Treatment Technologies,* Salt Lake City, 603–607.

LASHTCHENOVA, T. N. & STEFANOVSKY, S. V. 1998b. Titanium–silicate based glass–crystalline wasteforms. *In:* CHOUDHARY, M. K., HUFF, N. Y. & DRUMMOND III, C. H. (eds) *Proceedings of the XVIII International Congress on Glass.* San-Francisco, CD-ROM.

LAVEROV, N. P., OMELIANENKO, B. I. & VELICHKIN, V. I. 1994. Geological aspects of radioactive waste disposal. *Geoecology,* **6**, 3–20 (in Russian).

LEBEAU, M.-J. & GIROD, M. 1987. Incorporation of simulated nuclear ashes in basalt: an experimental investigation. *American Ceramic Society Bulletin*, **66**, 1640–1646.

LIAN, J., YUDINTSEV, S. V., STEFANOVSKY, S. V., KIRJANOVA, O. I. & EWING, R. C. 2002a. Ion-induced amorphization of murataite. *Materials Research Society Symposium Proceedings*, **713**, 455–460.

LIAN, J., WANG, L. M., CHEN, J., EWING, R. C. & KUTTY, K. V. G. 2002b. Heavy ion irradiation of zirconate pyrochlores. *Materials Research Society Symposium Proceedings*, **713**, 507–512.

LIFANOV, F. A., KOBELEV, A. P. et al. 1998. Incorporation of intermediate-level liquid radioactive nuclear power plant wastes in glass and ceramics. *Proceedings of the IT3 International Conference On Incineration and Thermal Treatment Technologies*. Salt Lake City, 609–612.

LIFANOV, F. A., OJOVAN, M. I., STEFANOVSKY, S. V. & BURCL, R. 2003. Feasibility and expedience to vitrify NPP operational waste. *In: Waste Management '03 Conference*. Laser Options, Inc., Tucson, CD-ROM.

LOISEAU, P., CAURANT, D., BAFFIER, N., MAZEROLLES, L. & FILLET, C. 2001. Development of zirconolite-based glass-ceramics for the conditioning of actinides. *Materials Research Society Symposium Proceedings*, **663**, 179–187.

LUMPKIN, G. R. & EWING, R. C. 1988. Alpha-decay damage in minerals of the pyrochlore group. *Physics and Chemistry of Minerals*, **16**, 2–20.

LUMPKIN, G. R., HART, K. P., McGLINN, P. J., PAYNE, T. E., GIERÉ, R. & WILLIAMS, C. T. 1994. Retention of actinides in natural pyrochlores and zirconolites. *Radiochimica Acta*, **66/67**, 469–474.

LUMPKIN, G. R., COLELLA, M., SMITH, K. L., MITCHELL, R. H. & LARSEN, A. O. 1998a. Chemical composition, geochemical alteration, and radiation damage effects in natural perovskite. *Materials Research Society Symposium Proceedings*, **506**, 207–214.

LUMPKIN, G. R., SMITH, K. L., BLACKFORD, M. G., GIERÉ, R. & WILLIAMS, C. T. 1998b. The crystalline-amorphous transformation in natural zirconolite: evidence for long-term annealing. *Materials Research Society Symposium Proceedings*, **506**, 215–222.

LUMPKIN, G. R., DAY, R. A., McGLINN, P. J., PAYNE, T. E., GIERÉ, R. & WILLIAMS, C. T. 1999. Investigation of the long-term performance of betafite and zirconolite in hydrothermal veins from Adamello, Italy. *Materials Research Society Symposium Proceedings*, **556**, 793–800.

LUMPKIN, G. R., LEUNG, S. H. F. & COLELLA, M. 2000. Composition, geochemical alteration, and alpha-decay damage effects of natural brannerite. *Materials Research Society Symposium Proceedings*, **608**, 461–466.

LUMPKIN, G. R., EWING, R. C., WILLIAMS, C. T. & MARIANO, A. N. 2001. An overview of the crystal chemistry, durability, and radiation damage effects of natural pyrochlore. *Materials Research Society Symposium Proceedings*, **663**, 921–934.

LUMPKIN, G. R., SMITH, K. L., GIERÉ, R. & WILLIAMS, C. T. 2004. Geochemical behaviour of host phases for actinides and fission products in crystalline ceramic nuclear waste forms. *In*: GIERÉ, R. & STILLE, P. (eds). *Energy, Waste, and the Environment: a Geochemical Perspective*. Geological Society, London, Special Publications, **236**, 89–111.

LUTZE, W. 1988. Silicate glasses. *In*: LUTZE, W. & EWING, R. C. (eds) *Radioactive Waste Form for the Future*. Elsevier Science Publishers B.V., Amsterdam, 1–159.

LUTZE, W. & EWING, R. C. 1988. *Radioactive Waste Forms for the Future*. Elsevier Science Publishers B.V., Amsterdam.

MARRA, S. L., O'DRISCOLL, R. J., FELLINGER, T. L., RAY, J. M., PATEL, P. M. & OCCHIPINTI, J. E. 1999. DWPF vitrification – transition to the second batch of HLW radioactive sludge. *In: Waste Management '99 Conference*. Laser Options, Inc., Tucson, CD-ROM.

MARTIN, C., RIBET, I. & ADVOCAT, T. 2002. Alteration of a zirconolite glass–ceramic matrix under hydrothermal conditions. *Materials Research Society Symposium Proceedings*, **713**, 405–411.

MASSON, H., DESVAUX, J.-L., PLUCHE, E. & ROUX, P. 1999. The R7/T7 vitrification in La Hague: Ten years of Operation. *In: Waste Management '99 Conference*. Laser Options, Inc. Tucson, AZ, CD-ROM.

Matyunin, Yu. I. 1995. Investigation of plutonium and americium behavior at vitrification of simulated HLW with production of phosphate glass-like materials with various macrocompositions. *Radiochemistry*, **37**, 557–562 (in Russian).

MATYUNIN, Yu. I. 2000. PhD thesis, SIA Radon, Moscow.

MATYUNIN, Yu. I., ALEXEEV, O. A. & ANANINA, T. N. 2001. Immobilization of plutonium dioxide into borobasalt, pyroxene and andradite compositions. *In: GLOBAL 2001 International Conference on "Back End of the Fuel Cycle: From Research to Solutions"*, Paris, CD-ROM.

MAZZI, F. & MUNNO, R. 1983. Calciobetafite (new mineral of the Pyrochlore Group) and related minerals from Campi Flegrei, Italy; crystal structures of polymignite and zirkelite: comparison with pyrochlore and zirconolite. *American Mineralogist*, **68**, 262–276.

McDANIEL, E. W. & DELZER, D. B. 1988. FUETAP concrete. *In*: LUTZE, W. & EWING, R. C. (eds) *Radioactive Waste Form for the Future*. Elsevier Science Publishers B.V., Amsterdam, 565–588.

McGLINN, P. J., HART, K. P., DAY, R. A., HARRIES, J. R., WEIR, J. A. & THOMPSON, L. E. 1998. Scientific studies on the immobilization of Pu by ISV in field trials at maralinga, Australia. *Materials Research Society Symposium Proceedings*, **506**, 239–246.

McGLINN, P. J., ADVOCAT, T., LOI, E., LETURCQ, G. & MESTRE, J. P. 2001. Nd- and Ce-doped ceramic–glass composites: chemical durability under aqueous conditions and surface alteration in a moist clay medium at 90 °C. *Materials Research Society Symposium Proceedings*, **663**, 249–258.

MEAKER, T. F., PEELER, D. K., MARRA, J. C., PAREIZS, J. M. & RAMSEY, W. G. 1997. Actinide solubility in lanthanide borosilicate glass for possible immobilization and disposition. *Materials Research Society Symposium Proceedings*, **465**, 1281–1286.

MELDRUM, A., WANG, L. M. & EWING, R. C. 1996. Ion beam induced amorphization of monazite. *Nuclear Instruments and Methods in Physics Research B*, **116**, 220–224.

MERZ, E. R. & WALTER, C. E. 1996. *Disposal of Weapon Plutonium*. Kluwer Academic Publishers, Amsterdam.

MESKO, M. G., MEAKER, T. F., RAMSEY, W. G., MARRA, J. C. & PEELER, D. K. 1997. Optimization of lanthanide borosilicate frit compositions for the immobilization of actinides using a Plackett–Burman/simplex algorithm design. *Materials Research Society Symposium Proceedings*, **465**, 105–110.

MINERALS 1967. *The Directory*. Nedra, Moscow (in Russian).

MINAEV, A. A., KUZNETSOV, D. G., POPOV, I. B., KRAPUCHIN, V. B., ROVNIY, S. I., GUZHAVIN, V. I. & UFIMTSEV, V. P. 2001. Development of the method of radioactive wastes immobilization. *In*: *Waste Management '01 Conference*. Laser Options, Inc., Tucson, CD-ROM.

MORRIS, J. B. & CHIDLEY, B. E. 1976. Preliminary experience with the new Harwell Inactive Vitrification Pilot Plant. *In*: *Management of Radioactive Wastes from the Nuclear Fuel Cycle*. IAEA, Vienna, 241–256.

MORSS, L. R., STANLEY, M. I., TATKO, C. D. & EBERT, W. L. 2000. Corrosion of glass-bonded sodalite as a function of pH and temperature. *Materials Research Society Symposium Proceedings*, **608**, 733–738.

MUKHAMET-GALEYEV, A. P., MAGAZINA, L. O., LEVIN, K. A., SAMOTOIN, N. D., ZOTOV, A. V. & OMELIANENKO B. I. 1995. The interaction of Na-Al-P-glass (Cs, Sr-bearing) with water at elevated temperatures (70–250 °C). *Materials Research Society Symposium Proceedings*, **353**, 79–86.

NAKAZAWA, T., KATO, H., OKADA, K., UETA, S. & MIHARA, M. 2001. Iodine immobilization by Sodalite waste form. *Materials Research Society Symposium Proceedings*, **663**, 51–58.

NIKIFOROV, A. S., KULICHENKO, V. V. & ZHIKHAREV, M. I. 1985. *Conditioning of Liquid Radioactive Wastes*. Energoatomizdat, Moscow (in Russian).

NOMURA, I., NAGAYA, K. & HASHIMOTO, Y. 1985. Vitrification of low- and medium-level nuclear waste. *Transactions of the American Nuclear Society*, **49**, 74.

NYMAN, M., NENOFF, T. M., SU, Y., BALMER, M. L., NAVROTSKY, A. & XU, H. 1999. New crystalline silicotitanate (CST) waste forms: hydrothermal synthesis and characterization of Cs-Si-Ti-O

phases. *Materials Research Society Symposium Proceedings*, **556**, 71–76.

O'HOLLERAN, T. P., JOHNSON, S. G. *et al.* 1997. Glass–ceramic waste forms for immobilizing plutonium. *Materials Research Society Symposium Proceedings*, **465**, 1251–1258.

PALMER, C. R., MELLINGER, G. B. & RUSIN, J. M. 1981. Investigation of vitreous and crystalline ceramic materials for immobilization of alpha-contaminated residues. *In*: *Management of Alpha-Contaminated Wastes*. IAEA, Vienna, 339–354.

PALMER, R. A. & MISERCOLA, A. J. 2003. Waste form qualification experience at the West Valley Demonstration Project. *In*: *Waste Management '03 Conference*. Laser Options, Inc., Tucson, CD-ROM.

PASTUSHKOV, V. G., MOLCHANOV, A. V., SEREBRYAKOV, V. P., SMELOVA, T. V. & SHESTOPEROV, I. V. 2001. Technology and equipment based on induction melters with cold crucible for reprocessing active metal waste. *Materials Research Society Symposium Proceedings*, **663**, 59–64.

PEREIRA, C., HASH, M. C., LEWIS, M. A., RICHMANN, M. K. & BASCO, J. 1999. Incorporation of radionuclides from the electrometallurgical treatment of spent fuel into a ceramic waste form. *Materials Research Society Symposium Proceedings*, **556**, 115–120.

PERI, A. D. 1996. Influence of the titanium dioxide addition in matrix formulation on the Radwaste-Mortar matrix characteristics. *Materials Research Society Symposium Proceedings*, **412**, 429–433.

PETITJEAN, V., FILLET, C., BOEN, R., VEYER, C. & FLAMENT, T. 2002. Development of vitrification process and glass formulation for nuclear waste conditioning. *In*: *Waste Management '02 Conference*. Laser Options, Inc., Tucson, CD-ROM.

POSTNIKOV, A. YU., LEVAKOV, E. V., GAVRILOV, P. I., GLAGOVSKY, E. V. & KUPRIN, A. V. 2001. Self-sustaining high temperature synthesis of zirco-nolite-based materials for radioactive waste immobilization. *Physics and Chemistry of Materials Treatment*, **5**, 58–63 (in Russian).

RAICEVIC, S., PLECAS, I., LALOVIC, D. I. & VELJKOVIC, V. 1999. Optimization of immobilization of strontium and uranium by the solid matrix. *Materials Research Society Symposium Proceedings*, **556**, 135–142.

RAMSEY, W. G., BIBLER, N. E. & MEAKER, T. F. 1995. Compositions and durabilities of glasses for immobilization of plutonium and uranium. *In*: *Waste Management '95 Conference*. Laser Options, Inc., Tucson, CD-ROM.

RILEY, B. J., VIENNA, J. D. & SCHWEIGER, M. J. 2000. Liquidus temperature of rare earth-alumino-borosilicate glasses for treatment of americium and curium. *Materials Research Society Symposium Proceedings*, **608**, 677–682.

RINGWOOD, A. E. 1985. Disposal of high level nuclear wastes: A geological perspective. *Mineralogical Magazine*, **49**, 159–176.

RINGWOOD, A. E., KESSON, S. E., REEVE, K. D., LEVINS, D. M. & RAMM E. J. 1988. SYNROC. *In*: LUTZE, W. & EWING, R. C. (eds) *Radioactive*

Waste Forms for the Future. Elsevier Science Publishers B. V., Amsterdam, 233–334.

RODERICK, J. M., HOLLAND, D. & SCALES, C. R. 2000. Characterization and radiation resistance of a mixed-alkali borosilicate glass for high level waste vitrification. *Materials Research Society Symposium Proceedings*, **608**, 721–726.

ROY, R., VANCE, E. R. & ALAMO, J. 1982. [NZP] a new radiophase for ceramic nuclear waste forms. *Materials Research Bulletin*, **17**, 585–588.

SAIDL, J. & RALKOVA, J. 1966. Radioactive waste solidification by means of melting with basalt. *Atomic Energy*, **10**, 285–289 (in Russian).

SALES, B. C. & BOATNER, L. A. 1988. Lead-iron phosphate glasses. *In*: LUTZE, W. & EWING, R. C. (eds) *Radioactive Waste Forms for the Future*. Elsevier Science Publishers B. V., Amsterdam, 193–231.

SHEETZ, B. E., AGRAWAL, D. K., BREVAL, E. & ROY, R. 1994. Sodium zirconium phosphate (NZP) as a host structure for nuclear waste immobilization: A review. *Waste Management*, **14**, 489–505.

SIEMER, D. D., ROY, D. M., GRUTZECK, M. W., COUGAR, M. L. D. & SHEETZ, B. E. 1997. PCT leach tests of hot isostatically pressed (HIPed) zeolitic concretes. *Materials Research Society Symposium Proceedings*, **465**, 303–310.

SINKLER, W., O'HOLLERAN, T. P., FRANK, S. M., RICHMANN, M. K. & JOHNSON, S. G. 2000. Characterization of glass-bonded ceramic waste form loaded with U and Pu. *Materials Research Society Symposium Proceedings*, **608**, 423–430.

SMITH, K. L. & LUMPKIN, G. R. 1993. Structural features of zirconolite, hollandite and perovskite, the major waste-bearing phases in Synroc. *In*: BOLAND, J. N. & FITZ GERALD, J. D. (eds) *Defects and Processes in the Solid State: Geoscience Applications. The McLaren Volume*. Elsevier Science Publishers, B.V., 401–422.

SMITH, K. L., ZALUZEC, N. J. & LUMPKIN, G. R. 1998. The relative radiation resistance of zirconolite, pyrochlore, and perovskite to 1.5 MeV Kr^+ ions. *Materials Research Society Symposium Proceedings*, **506**, 931–932.

SOBOLEV, I. A. & KHOMTCHIK, L. M. 1983. *Conditioning of Radioactive Wastes at Centralized Sites*. Energoatomizdat, Moscow (in Russian).

SOBOLEV, I. A., DMITRIEV, S. A. *et al.* 1995a. Vitrification of intermediate level radioactive waste by induction heating. *In*: SLATE, S., FEIZOLLAHI, F. & CREER, J. (eds) *Proceedings of the Fifth International Conference on Radioactive Waste Management and Environmental Restoration ICEM'95*, Berlin, Germany. The American Society of Mechanical Engineers, New York, **1**, 1125–1127.

SOBOLEV, I. A., DMITRIEV, S. A. *et al.* 1995b. Waste vitrification: Using induction melting and glass composite materials. *In*: *Proceedings of the International Conference on Evaluation of Emerging Nuclear Fuel Cycle Systems – GLOBAL-1995*, Versailles, France, **1**, 734–740.

SOBOLEV, I. A., BARINOV, A. S., OJOVAN, M. I. & OJOVAN, N. V. 1996. Long term natural tests of NPP vitrified radioactive waste. *In*: *Waste Management '96 Conference*. Laser Options, Inc., Tucson, CD-ROM.

SOBOLEV, I. A., STEFANOVSKY, S. V., KNYAZEV, O. A., LASHTCHENOVA, T. N., VLASOV, V. I. & LOPUKH, D. B. 1997a. Application of the cold crucible melting for production of rock-type wasteforms. *In*: BAKER, R., SLATE, S. & BENDA, G. (eds) *Proceedings of the Sixth International Conference on Radioactive Waste Management and Environmental Remediation ICEM '97*. The American Society of Mechanical Engineers, New York, 265–269.

SOBOLEV, I. A., OJOVAN, M. I., KARLINA, O. K., STEFANOVSKY, S. V. & GALTSEV, V. E. 1997b. Processing and characterization of glass composites for ash immobilization. *Materials Research Society Symposium Proceedings*, **465**, 117–122.

SOBOLEV, I. A., STEFANOVSKY, S. V., IOUDINTSEV, S. V., NIKONOV, B. S., OMELIANENKO, B. I., & MOKHOV, A. V. 1997c. Study of melted Synroc doped with simulated high-level waste. *Materials Research Society Symposium Proceedings*, **465**, 363–370.

SOBOLEV, I. A., STEFANOVSKY, S. V., OMELIANENKO, B. I., IOUDINTSEV, S. V., VANCE, E. R. & JOSTSONS, A. 1997d. Comparative study of Synroc-C ceramics produced by hot-pressing and inductive melting. *Materials Research Society Symposium Proceedings*, **465**, 371–378.

SOBOLEV, I. A., DMITRIEV, S. A. *et al.* 1998. Application of the cold crucible technology in radioactive waste conditioning *In*: *Proceedings of the Eighth International Conference on High Level Radioactive Waste Management*, Las Vegas, 702–704.

SOBOLEV, I. A., LIFANOV, F. A., SAVKIN, A. E., VARLAKOV, A. P. & KOVALSKIY, E. A. 1999. Conditioning solid radioactive waste by using high-penetrating cement mortar. *In*: *Waste Management '99 Conference*. Laser Options, Inc., Tucson, CD-ROM.

SOBOLEV, I. A., BARINOV, A. S., OJOVAN, M. I., OJOVAN, N. V., STARTSEVA, I. V. & GOLUBEVA, Z. I. 2000. Long-term behavior of bitumen waste form. *Materials Research Society Symposium Proceedings*, **608**, 571–576.

STEFANOVSKY, S. V. 1989. The effect of sulfur oxide (VI) on phase separation in sodium borosilicate glasses. *Glass and Ceramic*, **3**, 10–11 (in Russian).

STEFANOVSKY, S. V. & LIFANOV, F. A. 1988. Phase separation at fixation of sulfate-containing radioactive waste in the $CaO-B_2O_3-SiO_2$ glass. *Radiochemistry*, **30**, 825–829 (in Russian).

STEFANOVSKY, S. V. & LIFANOV, F. A. 1989. Glasses for immobilization of sulfate-containing wastes. *Radiochemistry*, **31**, 129–134 (in Russian).

STEFANOVSKY, S., LIFANOV, F. & IVANOV, I. 1992. Glass forms for incinerator ash immobilization. *Proceedings of the XVI International Congress on Glass*, Madrid, **3**, 209–212.

STEFANOVSKY, S. V., IVANOV, I. A. & GULIN, A. N.
1995. Aluminophosphate glasses with high sulfate
content. *Materials Research Society Symposium
Proceedings*, **353**, 101–106.

STEFANOVSKY, S. V., YUDINTSEV, S. V.,
NIKONOV, B. S., OMELIANENKO, B. I. &
PTASHKIN, A. G. 1999. Murataite-based ceramics
for actinide waste immobilization. *Materials
Research Society Symposium Proceedings*, **556**,
121–128.

STEFANOVSKY, S. V., YUDINTSEV, S. V., NIKONOV, B. S.,
OCHKIN, A. V., CHIZHEVSKAYA, S. V.,
CHERNIAVSKAYA, N. E. 2000a. Phase compositions and elements partitioning in two-phase
hosts for immobilization of rare earth – actinide
high level waste fraction. *Materials Research
Society Symposium Proceedings*, **608**, 455–460.

STEFANOVSKY, S. V., YUDINTSEV, S. V.,
NIKONOV, B. S., OMELIANENKO, B. I. &
LAPINA, M. I. 2000b. Isomorphic capacity of syn-
thetic sphene with respect to Gd and U. *Materials
Research Society Symposium Proceedings*, **608**,
407–412.

STEFANOVSKY, S. V., YUDINTSEV, S. V.,
NIKONOV, B. S., LAPINA, M. I. & ALOY, A. S.
2001. Effect of synthesis conditions on phase
composition of Pyrochlore–Brannerite ceramics.
*Materials Research Society Symposium Proceed-
ings*, **663**, 315–324.

STRACHAN, D. M., SCHEELE, R. D. *et al.* 2002. Radiation damage in titanate ceramics for plutonium
immobilization. *Materials Research Society Symposium Proceedings*, **713**, 461–468.

SU, Y., BALMER, M. L., WANG, L., BUNKER, B. C.,
NYMAN, M., NENOFF, T. & NAVROTSKY, A. 1999.
Evaluation of thermally converted silicotitanate
waste forms. *Materials Research Society Symposium Proceedings*, **556**, 77–84.

TIMMONS, D. M. & THOMPSON, L. E. 1996. Geochemical and petrographic studies and the relationships
to durability and leach resistance of vitrified products from the in-situ vitrification process. *In: Proceedings of the International Topical Meeting on
Nuclear and Hazardous Waste Management
SPECTRUM'96*. American Nuclear Society, Inc.,
La Grange Park, 1026–1029.

TOLSTOVA, O. V., LASHTCHENOVA, T. N. &
STEFANOVSKY, S. V. 2002. Glassy materials from
basalt for intermediate-level wastes immobilization. *Glass and Ceramic*, **6**, 28–31 (in Russian).

TRANTER, T. J., ALOY, A. S., SAPOZHNIKOVA, N. V.,
KNECHT, D. A. & TODD, T. A. 2002. Porous crystalline silica (Gubka) as an inorganic support
matrix for novel sorbents. *Materials Research
Society Symposium Proceedings*, **713**, 907–913.

URUSOV, V. S., RUSAKOV, V. S. & YUDINTSEV, S. V.
2002. Valence state and structural positions of iron
atoms in synthetic murataite. *Doklady of the
Russian Academy of Sciences*, **384**, 527–534 (in
Russian).

US DOE 1981. *Nuclear Waste Materials Handbook
(Test Methods)*, Report DOE/TIC-11400. DOE
Technical Information Center, Washington, DC.

USHAKOV, S. V. 1998. PhD thesis, University of
St-Petersburg, Russia.

UTSONOMIYA, S., WANG, L. M., YUDINTSEV, S. V. &
EWING, R. C. 2002. Ion irradiation effects in synthetic garnets incorporating actinides. *Materials
Research Society Symposium Proceedings*, **713**,
495–500.

VANCE, E. R. & PILLAY, K. K. S. 1982. Fission frag-
ment damage in crystalline phases possibly
formed in solidified radioactive waste. *Radiation
Effects*, **62**, 25–38.

VANCE, E. R., BEGG, B. D., DAY, R. A. & BALL, C. J.
1995. Zirconolite-rich ceramics for actinide wastes.
*Materials Research Society Symposium Proceed-
ings*, **353**, 767–774.

VANCE, E. R., JOSTSONS, A., DAY, R. A., BALL, C. J.,
BEGG, B. D. & ANGEL, P. J. 1996a. Excess Pu disposition in zirconolite-rich Synroc. *Materials
Research Society Symposium Proceedings*, **412**,
41–47.

VANCE, E. R., CARTER, M. L., DAY, R. A., BEGG, B. D.,
HART, K. P. & JOSTSONS, A. 1996b. Synroc and
Synroc–glass composite waste forms for Hanford
HLW immobilization. *In: Proceedings of the Inter-
national Topical Meeting on Nuclear and Hazar-
dous Waste Management SPECTRUM '96*.
American Nuclear Society, Inc. La Grange Park,
2027–2031.

VANCE, E. R., DAY, R. A., ZHANG, Z., BEGG, B. D.,
BALL, C. J. & BLACKFORD, M. G. 1996c. Charge
compensation in Gd-doped CaTiO₃. *Journal of
Solid State Chemistry*, **124**, 77–82.

VANCE, E. R., DAY, R. A., CARTER, M. L. &
JOSTSONS, A. 1996d. A melting route to Synroc
for Hanford HLW immobilization. *Materials
Research Society Symposium Proceedings*, **412**,
289–295.

VANCE, E. R., CARTER, M. L., BEGG, B. D., DAY, R. A.
& LEUNG, H. F. 2000. Solid solubilities of Pu, U,
Hf and Gd in candidate ceramic phases for actinide
waste immobilization. *Materials Research Society
Symposium Proceedings*, **608**, 431–436.

VASHMAN, A. A. & POLYAKOV, A. S. 1997. *Phosphate
Glasses with Radioactive Wastes*. CNIIAtomin-
form, Moscow (in Russian).

VEAL, B. W., MUNDY, J. N. & LAM, D. J. 1987. Actinides in silicate glasses. *In: FREEMAN, A. J. &
LANDER, G. H. (eds) Handbook on the Physics
and Chemistry of the Actinides*. Elsevier Science
Publishers B.V., 271–309.

VLASOV, V. I., KEDROVSKY, O. L., POLYAKOV, A. S.
& SHISHTCHITZ, I. Y. 1987. Handling of liquid
radioactive waste from the closed nuclear fuel
cycle. *In: Back End of the Nuclear Fuel Cycle:
Strategies and Options*. IAEA, Vienna, 109–117.

WANG, L. M., ZHU, S., WANG, S. X. & EWING, R. C.
2001. Effects of cesium, iodine and strontium
ion implantation on the microstructure of cubic
zirconia. *Materials Research Society Symposium
Proceedings*, **663**, 293–300.

WEBER, W. J., EWING, R. C. *et al.* 1997. Radiation
effects in glasses used for immobilization of high
level waste and plutonium disposition. *Journal of
Materials Research*, **12**, 1946–1975.

WEBER, W. J., EWING, R. C. et al. 1998. Radiation effects in crystalline ceramics for the immobilization of high-level nuclear waste and plutonium. *Journal of Materials Research*, **13**, 1434–1484.

WEBER, W. J. & EWING, R. C. 2002. Radiation effects in crystalline oxide host phases for the immobilization of actinides. *Materials Research Society Symposium Proceedings*, **713**, 443–454.

WHITE, T. J. 1984. The microstructure and microchemistry of synthetic zirconolite, zirkelite and related phases. *American Mineralogist*, **69**, 1156–1172.

WHITEHOUSE, J. C., JANTZEN, C. M., VAN RYN, F. R. & DAVIS, D. H. 1995. Design and fabrication of a transportable vitrification system for mixed waste processing. *In*: *Proceedings of the Third Biennal Mixed Waste Symposium*. The American Society of Mechanical Engineers, Baltimore, MD, 8.3.1.

WRONKIEWICZ, D. L., WOLF, S. F. & DISANTO, T. S. 1996. Apatite- and monazite-bearing glass–crystal composites for the immobilization of low-level nuclear and hazardous wastes. *Materials Research Society Symposium Proceedings*, **412**, 345–352.

YUDINTSEV, S. V. 2001. Incorporation of U, Th, Zr, and Gd into the garnet-structured host. *In*: *Proceedings of the 8th International Conference on Radioactive Waste Management and Environmental Remediation*. The American Society of Mechanical Engineers, New York, CD-ROM.

YUDINTSEV, S. V., STEFANOVSKY, S. V. & EWING, R. C. 1999. Structural and compositional relationships in titanate-composed ceramics for actinide-bearing waste immobilization. *In*: *Proceedings of the 7th International Conference on Radioactive Waste Management and Environmental Remediation ICEM '99*. Nagoya, Japan, CD-ROM.

YUDINTSEV, S. V., STEFANOVSKY, S. V., NIKONOV, B. S. & OMELIANENKO, B. I. 2001. Phase and chemical stability of murataite containing uranium, plutonium and rare earths. *Materials Research Society Symposium Proceedings*, **663**, 357–366.

YUDINTSEV, S. V., LAPINA, M. I., PTASHKIN, A. G., IOUDINTSEVA, T. S., UTSONOMIYA, S., WANG, L. M. & EWING, R. C. 2002. Accommodation of uranium into the garnet structure. *Materials Research Society Symposium Proceedings*, **713**, 477–480.

ZIEGLER, J. F., BIERSACK, J. P. & LITTMARK, U. 1985. *The Stopping Range of Ions in Solids*. Pergamon Press, New York.

ZHAO, D., LI, L., DAVIS, L. L., WEBER, W. J. & EWING, R. C. 2001. Gadolinium borosilicate glass-bonded Gd-silicate apatite: A glass–ceramic nuclear waste form for actinides. *Materials Research Society Symposium Proceedings*, **663**, 199–206.

The geochemical behaviour of Tc, Np and Pu in spent nuclear fuel in an oxidizing environment

EDGAR C. BUCK, BRADY D. HANSON & BRUCE K. McNAMARA

Pacific Northwest National Laboratory, Richland, Washington, 99352, USA

(e-mail: edgar.buck@pnl.gov)

Abstract: Spent fuel from commercial nuclear reactors consists mainly of uranium oxide. However, the changes that occur during reactor operations have a profound effect on chemical and physical properties of this material. Heat build-up in the fuel pellet during reactor operations can cause redistribution of fission products. The fission products may aggregate in one of three types of precipitates; gaseous, metallic, or oxide, depending on the burn-up and in-core treatment. Radiation damage and variations in fission and neutron capture yields across the fuel pellets lead to Pu enrichment and increased porosity with increasing burn-up. A more porous surface may make the fuel more susceptible to oxidative dissolution. As the level of actinides and fission products increases, the fuel may become more resistant to oxidation. These changes may limit the usefulness of natural uraninite (UO_2) analogues for predicting the geological behaviour of spent fuel disposed in a high-level waste (HLW) repository. In this Chapter, an overview of spent fuel microstructure, radiolytic effects, and alteration processes is presented. Evidence for Np incorporation into U^{6+} phases, the nature of Pu surface precipitates on spent fuel, and evidence for the preferential removal of 4d-metals from ε-particles in corroded spent fuel is discussed. Understanding the potential mechanisms of radionuclide attenuation through sorption and/or incorporation requires techniques with both high spatial resolution and excellent elemental sensitivity.

Many countries, including those with active fuel reprocessing facilities, are planning to dispose spent nuclear fuel in a permanent geological repository. The resistance to radionuclide release to the biosphere is based primarily on the chemical stability of the radionuclide-bearing waste form. Reducing uncertainties in waste form corrosion models reduces uncertainties at all other points in repository performance assessment. Other than the proposed Yucca Mountain facility in Nevada, USA, all other proposed spent fuel repositories in the world are planned to be located in reducing environments. This is because UO_2 is thermodynamically unstable in oxidizing environments; however, the exceptionally low water infiltration rates (5 mm/y), unsaturated environment, remoteness, and sparse population, make Yucca Mountain a highly suitable site for a HLW repository. However, even under oxidizing environments, predicted corrosion rates for spent UO_2 fuel may be greatly exaggerated. This is partly because current models assume congruent release rates for all radionuclides and do not account for water flux and radiolysis effects. Furthermore, such conservative assumptions can lead to unnecessary and possibly expensive preventative measures for physically unrealistic processes at other

points in the repository performance assessment analysis.

Previous review articles on spent nuclear fuel and UO_2 corrosion have dealt with the details of uraninite alteration (Finch & Ewing 1992), spent fuel corrosion in oxidizing and reducing environments (Johnson & Shoesmith 1988; Wronkiewicz & Buck 1999; Shoesmith 2000), and the evolution of spent fuel microstructure (Poinssot *et al.* 2002). In this Chapter, the behaviour of three important radionuclides (^{237}Np, ^{99}Tc, and ^{239}Pu) is examined with examples of studies from spent fuel corrosion tests to illustrate the potential geochemical behaviour of these radionuclides.

Spent fuels vary in microstructure, and phase and elemental distribution depending on the in-core reactor operating conditions and reactor history. The chemical stability of spent U oxide fuel is described by local pH and Eh conditions, redox being the most important parameter. However, the redox system will also evolve with time as various radionuclides decay and the proportion of oxidants and reductants generated at the fuel/water interface changes with the altering α-, β-, γ-radiation field and with the generation of other corrosion products that can act as

From: GIERÉ, R. & STILLE, P. (eds) 2004. *Energy, Waste, and the Environment: a Geochemical Perspective*. Geological Society, London, Special Publications, **236**, 65–88.
0305-8719/04/$15 © The Geological Society of London 2004.

sinks for radiolytic species. Spent fuel may also undergo compositional and structural evolution before waste package failure and the potential for water contact that may further complicate weathering processes (Poinssot *et al.* 2002).

The majority of radionuclides (>90%) generated during in-reactor burning are retained in the U oxide matrix; hence, their release should be closely related to the corrosion rate of the matrix. The release of Cs and I are generally independent of the fuel corrosion rate, and the metallic elements (Mo, Ru, Tc, Rh, Pd), Xe–Kr gas bubbles, and perovskite phases ((Ba,Sr)ZrO₃) may only be partly controlled by the matrix corrosion rate.

Less is known about the fate of the fission and neutron capture products that could result in the precipitation of unique alteration phases depending on the availability of these species in the fuel matrix. Burns *et al.* (1997) theorized that many of the U(VI) alteration phases may be capable of incorporating the long-lived radiotoxic isotopes, including ^{237}Np, ^{99}Tc, and ^{239}Pu. In this chapter, we will discuss the evidence for Np incorporation into U(VI) phases and the behaviour of Pu in corroded spent nuclear fuel (SNF).

Evolution of aged spent fuel microstructure

Uranium oxide ceramics used in nuclear reactors have a small, uniform grain size (5–10 μm). A cross-section schematic diagram of a spent fuel pellet is shown in Figure 1. During reactor operation, the temperature at the centre-line of the fuel pellets may be as high as 1700 °C depending on the reactor design, operating conditions, and on the thermal conductivity of the fuel. The fuel is also exposed to an intense radiation field that can increase the mobility of the fission products. The elevated temperature can cause restructuring of the UO₂ grains and permit fission products that have a limited solid solubility in UO₂ to segregate to grain boundaries. The final disposition of all radionuclides depends on the in-reactor conditions; however, radionuclides can be separated into a number of distinct categories (Kleykamp 1985):

(1) Caesium and iodine that diffuse at different rates toward the fuel gap and grain boundary regions;
(2) Fission products that are incompatible with the uraninite (fluorite) structure and migrate

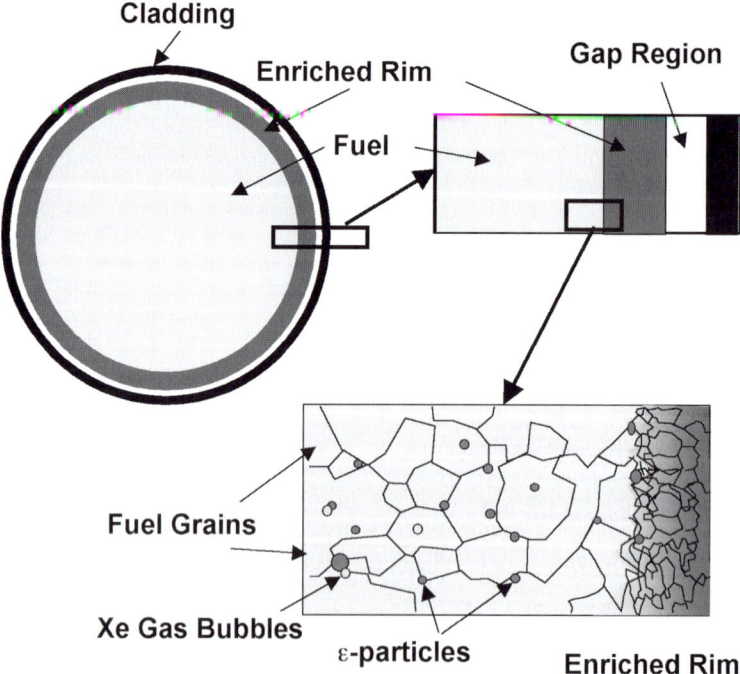

Fig. 1. Schematic cross-sectional views of SNF pellet showing an edge of an SNF with gap region, and a diagram of the grain structure showing ε-particles, gas bubbles, and the polygonized rim region.

Table 1. *Concentration of fission and neutron capture products (in ppm) in a low burn-up spent nuclear fuel*

Element	ppm	Element	ppm
Xe	5656.5	La	1269
I	258.7	Ce	2469.3
Cs	2605.3	Pr	1161
Sr	794.2	Nd	4189.8
Ba	1740.6	Sm	815.2
Se	57.7	Eu	154.5
Te	528.6	Gd	141.5
Zr	3639.4		
Mo	3497.5	Np-237	468
Tc	798.8	Pu (total)	9459
Ru	2404.9	Am (total)	484
Rh	483.5	Cm (total)	39
Pd	1684		
Ag	91.8		

to grain boundaries to occupy either fission gas bubbles (Xe and Kr) or metallic particles (Mo, Tc, Ru, Rh, Pd, Ag);

(3) Fission products, actinides, and lanthanides that are retained in the fuel matrix;

(4) Under some in-reactor conditions, other oxide phases, including perovskite phases (ideally ABO_3), may form. These are known to contain Ru, Zr, Ba, U, and Pu (Kleykamp 1985; Thomas *et al.* 1992)

The concentration of fission and neutron capture products in a light water reactor (LWR) fuel (30 MWd/kg U) are listed in Table 1 and presented in graphical form in Figure 2 (adapted from Oversby 1994).

Burn-up effects

Burn-up is not uniform across the fuel pellet. It increases dramatically at the pellet rim, resulting in the so-called 'rim effect' (Spino & Papaioannou 2000; Une *et al.* 2000). This effect is generally detrimental to the lifetime of the fuel in the reactor; however, the implications of rim formation have not been addressed with respect to long-term geological disposal. During irradiation of the fuel, neutrons are absorbed by ^{238}U leading to production of ^{239}Pu in the fuel. At the pellet edge, the resonance energies for absorption tend to be optimized for ^{239}Pu production. The ^{239}Pu may also undergo neutron capture or fission leading to a build-up of other transuranics and fission products. Metallic particles in the rim region tend to have higher concentrations of Pd because of enhanced ^{239}Pu fission. The rim region of high burn-up fuels is characterized by increased porosity and modification of the original UO_2 structure to form finer grains (polygonization). High burn-up means a greater initial concentration of fission gases; however, some have reported that the rim is characterized by a reduction in Xe concentration. This tends to suggest that bubble growth and migration may have resulted in interconnected porosity that may have allowed increased fission gas release. Increases in porosity within the rim will increase the surface area that can be contacted by moisture (Wasywich *et al.* 1992), and there may be an increase of the generation rate of α-radiolytic oxidants in the microporous regions (King *et al.* 1999). Both of these effects would tend to increase the rate of spent fuel corrosion and radionuclide release from the matrix.

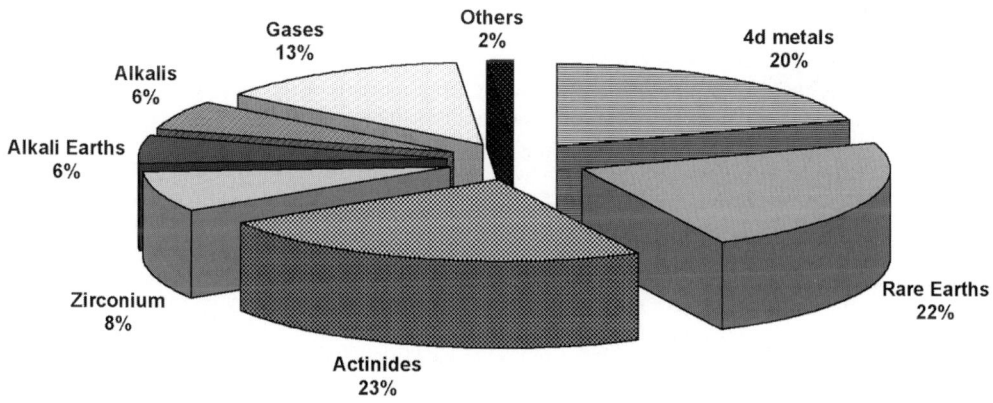

Fig. 2. Pie-diagram showing the major types of fission products and the actinides in their relative proportions (adapted from Oversby 1994).

Polygonization of the UO_2 grain structure in the rim region results in a significant size reduction of the original grains ($\sim 5-10$ μm) as they are subdivided into $\sim 0.15-0.3$ μm size grains (Matzke & Spino 1997; Thomas *et al.* 1992). The formation of submicrometre particles at the surface of a spent fuel pellet may increase the predisposition for primary colloidal particle release, if the fuel is contacted by large amounts of flowing water. Finn *et al.* (1994) reported polycrystalline UO_2 grains containing high levels of rare earth elements (REE) in colloidal particles from an SNF sample exposed to dripping water at the start of their tests. High levels of Am and Pu were also reported at this test sampling but decreased to minuscule levels at subsequent times (Finn *et al.* 1998).

Chemical substitution in the uraninite structure is known to lead to stabilization of the UO_2 matrix (Grandstaff 1976). In SNF, oxidation from $UO_{2.4}$ to U_3O_8 has been found to depend on burn-up (Hanson 1998). Hanson suggested that the oxidation-resistant fission products (e.g., Zr^{4+}, Gd^{3+}) and transuranics (e.g., Pu^{4+}, Am^{3+}) that occupy the U-site in the fluorite structure prevented oxidation beyond $UO_{2.4}$. However, such a relationship has not been clearly observable from dissolution tests on spent fuel. This may be a consequence of the β-,γ-radiolytic field and other competing processes. Grandstaff (1976) noted that uraninite dissolution rates decreased with increased substitution of REE and Th on the U-site in the fluorite structure. The reduction in dissolution rate was significant and, if a similar relationship occurs in spent fuels with increasing burn-up, it may provide a significant improvement to performance assessment models.

Dislocation structure and radiation effects

The extreme radiation field to which spent fuel is exposed during reactor operation results in the formation of numerous interstitial and vacancy defects. Recombination of these defects at reactor operation temperatures leads to the formation of dislocation loops and voids. The dislocation loops can redistribute the short-range accommodation strains into a long-range strain field, where the elastic energy per added interstitial decreases with increasing loop size. Therefore the interstitial loops in spent fuel may help relieve strain from fission-product incorporation into the fluorite structure. If such a mechanism occurs it suggests that there might be slight differences in the structure of spent fuels and simulated fuels (SIMFUELs). Radiation-enhanced diffusion could be a mechanism that could potentially lead to increased release of some fission products from UO_2 to grain boundaries, from which they may be leached by groundwater (Poinssot *et al.* 2002). In general, the changes that occur within spent fuel grains during reactor irradiation are likely to be dwarfed by the effects caused by moist-air oxidation on the surface microstructure.

The long-term consequences of α-decay will be He gas build-up, which may lead to grain boundary disintegration. Poinssot *et al.* (2002) have shown that the internal pressures generated over time in LWR fuels from He gas are insignificant; however, the pressures that could be produced by mixed oxide (MOX) fuels may need to be considered.

Fission gas release

Profiles through high burn-up fuels have indicated a significant reduction in Xe toward the edge of the fuel pellet (Walker *et al.* 1992; Yagnik *et al.* 1999). There may be several mechanisms for the fission gases, Xe and Kr, to migrate in a fuel matrix. These include grain boundary diffusion, diffusion along extended dislocations, or diffusion along the oxygen sublattice. The migration of fission gases in UO_2 leads to fission gas release and coarsening of Xe bubbles. Xenon may be trapped at lattice defects that increase in concentration as the irradiation proceeds. Childs (1963) argued that Xe and Kr diffusion is intimately connected with the diffusion of oxygen. The faster moving 'volatile' species are thought to be Te, I, and Xe in order of increasing mobility. The volatile radionuclides ^{133}Xe ($t_{1/2} = 5.245 \times 10^9$ y) and ^{85}Kr ($t_{1/2} = 10.72$ y) can be released instantly upon failure of the Zircaloy cladding.

Metallic fission products, Mo, Tc, Ru, Rh, and Pd, exist in irradiated LWR fuels as hexagonal Ru alloy metallic inclusions (termed ε-phases). These ε-phase particles range in diameter from a few micrometers to $5-20$ nanometers. An example of an ε-particle is shown in Figure 3*a*; it was found with transmission electron microscopy (TEM) in a high burn-up LWR fuel grain. The X-ray energy-dispersive (EDS) spectrum shows the presence of Mo, Tc, Ru, and Pd (Fig. 3*b*). A proportion of the Mo will exist in oxidized form within the fuel. The Mo concentration in the ε-phase decreases continuously during reactor irradiation due to an increase of the oxygen potential of the fuel with burn-up (Kleykamp 1985). This increase is because oxygen liberated following U fission cannot be completely balanced by the generated fission products. Fortner *et al.* (2003) have used X-ray absorption spectroscopy (XAS) to confirm the

(a)

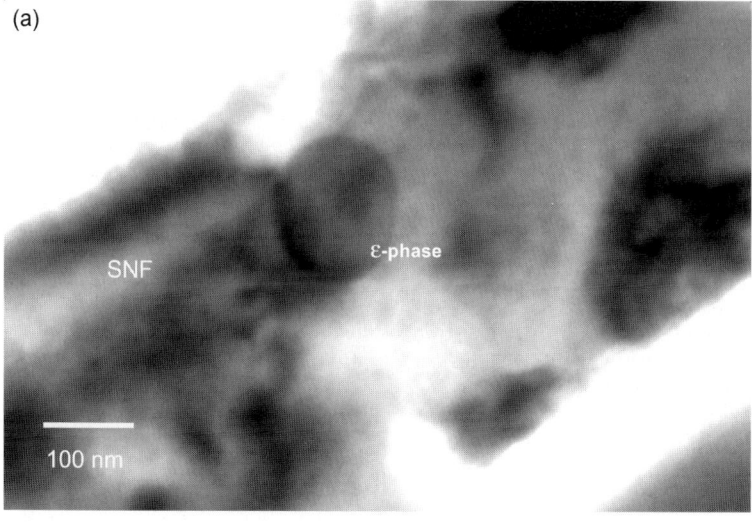

(b)

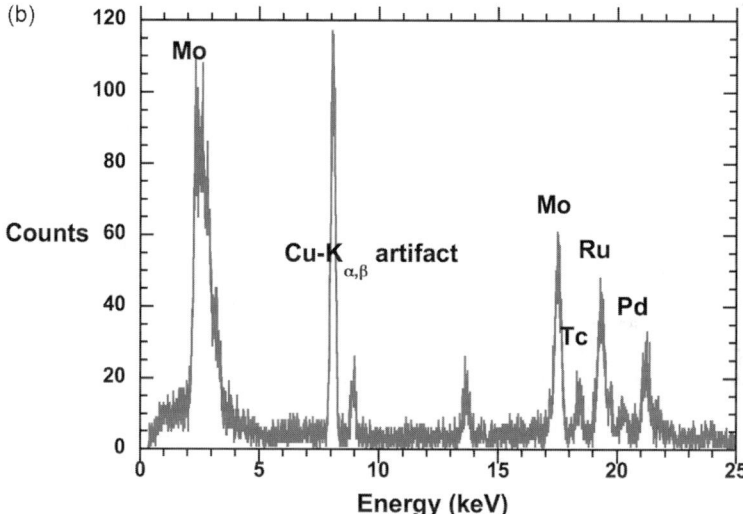

Fig. 3. (a) TEM Image of 4d-metal phase in high burn-up spent fuel; (b) EDS spectrum of the particle.

proportion of Mo–O bonds in spent fuel. These studies confirmed earlier work demonstrating the presence of both oxidized and metallic Mo. Molybdenum may also form a ternary oxide with Cs, Cs_2MoO_4, which has been observed on the fuel side of the fuel-cladding gap (Kleykamp 1985).

The exact size distribution of the metallic particles is unknown but these phases are important as they are the major residence site and source of ^{99}Tc, which, because of its long half-life and high solubility, once oxidized, dominates perform-

ance assessment calculations. Yet, the ε-particles are extremely chemically resistant and require a strong oxidation potential in order to dissolve; hence the release of ^{99}Tc from spent fuel may be considerable overestimated. The resistance of noble metal particles is well known in fuel reprocessing. Kessinger & Thompson (2002) reported during fuel dissolution tests that a finely divided material remained after nitric acid digestion of a spent fuel. It was supposed that this material was primarily the noble metal fission product. Similar observations of residue material from LWR fuel acid digestion have been reported by Adachi *et al.* (1990) and Kleykamp (1990), where the amount of undissolved residue solids increased disproportionately with burn-up. The only material identified with X-ray diffraction (XRD) was the hexagonal ε-phase containing the five metals Mo, Tc, Ru, Rh, and Pd.

Thomas *et al.* (1992) showed that fission gas bubbles in spent fuels nearly always occur in contact with the so-called 'five-metal' ε-phases, or noble metal particles. The 4d-group elements Mo through Pd form increasingly weaker bonds with oxygen, because of the smaller overlap between the oxygen s-orbital and the increasingly filled 4d-band. The composition of the metallic phases will vary in some instances but does correspond closely to anticipated fission yields. The rim region of the fuel pellet tends to contain increased levels of Pd, and this has been explained through the application of complex diffusion models. However, the high burn-up in the rim results in the production of ^{239}Pu, which has a slightly different fission product distribution of noble metals than ^{235}U. Fission of ^{239}Pu results in a higher yield for Pd. The increased Pd content in the metallic particles in high burn-up fuels makes them more resistant

to oxidation. In Figure 4, TEM-EDS analysis of a series of metallic phases reveals a significant range of compositions in the fuel.

Oxide phases

Segregation of fission product phases occurs readily in fuels that have operated at high temperatures. Axial and radial temperature gradients in the fuel pins result in thermally induced diffusion of materials. The segregation process is complicated by variable irradiation time and the changing oxygen-to-metal ratio and chemical potential of oxygen in the fuel. Kleykamp (1985) has identified Cs_2MoO_4, $Cs_2(U_{0.97}Pu_{0.03})_4O_{12}$, $(Ba,Sr)TeO_3$, BaO, Pu_2O_2Te, $Ba_{1-x}Sr_xO$, and $(Ba_{1-x-y}Sr_xCs_y)(U,Pu,Zr,Mo,REE)O_3$ (perovskite 'grey-phase' phase), in addition to metallic inclusions and fission gas bubbles in fuels that have operated at high temperatures. Perovskite phases were not reported in the LWR fuels examined by Guenther *et al.* (1988); however, Thomas *et al.* (1992) identified several secondary precipitates in LWR spent fuel (burn-up of ~45 MWd/kg U) during a TEM study. They characterized a face-centered cubic phase that was isostructural with UO_2 with the composition

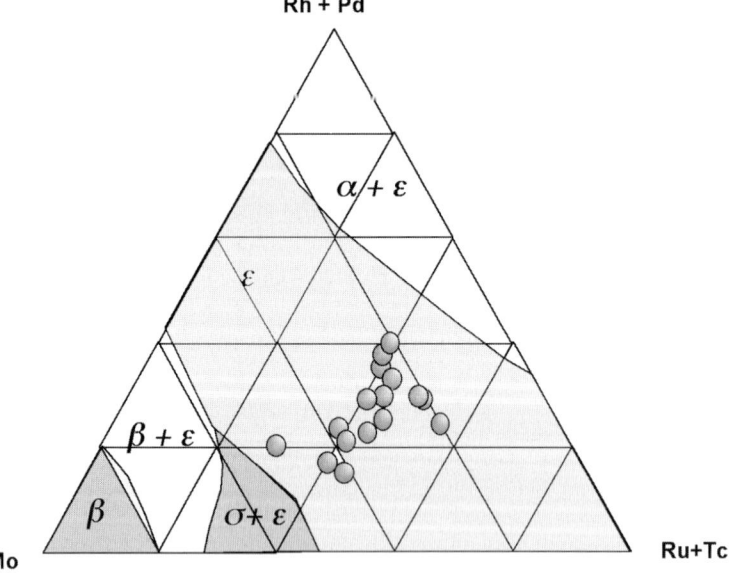

Fig. 4. EDS spectroscopy results of ε-particles from a high burn-up LWR fuel superimposed on the isothermal section of ternary phase diagram from Kleykamp (1985) at 1700 °C. These analyses show that there is distinct heterogeneity in the composition of metallic particles in the fuels. Hence, spot analysis of an individual ε-particle may not provide direct evidence of corrosion. The metallic system is dominated by the hexagonal close packing (ε) that occupies most of the phase space. The σ-space and the body centered cubic (β) space are limited to the Mo apex. The face centered cubic α-space occurs in the Pd-rich melt and is commonly observed in high burn-up fuels.

(Nd,Ce,Zr,U)O$_{2-x}$, and a Ba–zirconate phase that contained Sr, possibly similar to the perovskite phase identified in high-temperature spent fuels. The lattice parameter of the REE-enriched phase was slightly larger than that of UO$_2$. The TEM-EDS analysis did not indicate the presence of Pu at detectable levels, although, Thomas et al. (1992) reported enriched levels of Pu and REE in the rim region.

During aging of the fuel, all ^{90}Sr^{2+} will decay to ^{90}Zr^{4+}, and all ^{137}Cs$^+$ will have decayed to ^{137}Ba^{2+}. The effect of the change in charge may result in further compositional changes in the perovskite (grey-phases) and related oxides, making them possibly more susceptible to dissolution. In the perovskite phase, this additional Zr from Sr may not be accommodated on the A-site. These chemical changes make extrapolating results from present-day tests all the more difficult.

Chemically mobile radionuclides that can segregate from the fuel grains during reactor irradiation will be available for instantaneous release following waste package failure and contact with water. Thermodynamic calculations have suggested iodine should be present as CsI, and tellurium as Cs$_2$Te; however, neither of these phases has ever been observed experimentally on the surface of SNF (Kleykamp 1985). Axial γ-scans of spent fuels have confirmed that iodine migrates faster than Cs in the fuel during reactor operation. The different release rates reflect the fact that Cs is partly dissolved in the UO$_2$ matrix whereas iodine is not (Kleykamp 1985). Long-lived radioactive ^{129}I can therefore be released instantly following waste package failure. There is no indication that iodine is immobilized in any secondary phases associated with fuel corrosion. In contrast, caesium may be immobilized in alteration phases. Buck et al. (1997) identified a Cs-bearing U alteration phase during corrosion tests on spent fuel. Estimates of the amount of Cs available for instantaneous release (i.e., present in grain boundary and gap regions) are still a contentious issue. After a 20-year SNF corrosion test by Stroes-Gascoyne et al. (1997) only 5–7.7 wt% of the total calculated inventory of ^{137}Cs was leached. The researchers suggested that the total gap inventory for Cs may be overestimated.

Radiolysis

Owing to its high radiation field, fresh spent fuel in contact with water may undergo oxidative dissolution under both oxidizing and reducing environments through the formation of OH$^{\bullet}$ and H$_2$O$_2$ radiolysis products (Shoesmith 2000). Within 300 to 1000 years, this radiation field will decrease by 100 to 1000 times (Sunder et al. 1997), reducing the rate of reaction. Moreover, the production rate of OH$^{\bullet}$ radicals is lower in an α-radiation field compared to a β-,γ-field. The heterogeneous nature of spent fuel, particularly the occurrence of Pu-enriched rims (Matzke & Spino 1997), may continue to allow radiolytic species to form under moist-air (vapour) conditions even after the decay of the β-,γ-field (King et al. 1999). Performance assessment codes for HLW repositories require a rate for spent nuclear fuel corrosion; yet, in over 25 years of fuel research, such a relationship has been difficult to obtain. Oversby (1999) summarized much of the scientific community's data on UO$_2$ fuel corrosion and demonstrated the difficulty in obtaining a mechanistic dependency for the spent fuel corrosion rate. New models that include dependence on the effective generation rates of H$_2$O$_2$ from α-radiolysis in an aged fuel/canister environment are providing improvements in our models for spent fuel behaviour in a geological repository (King et al. 1999).

The HLW repository environment will be a dynamic redox system owing to the time-dependent generation of radiolytic oxidants and reductants and the corrosion of Fe-bearing canister materials (Spahiu et al. 2002; Pérez del Villar et al. 2002). After 300 to 1000 years, the β-,γ-radiolytic field will be negligible; however, the α-field may still create an oxidizing environment at the fuel/water interface for tens of thousands of years (King et al. 1999). As oxidants are produced, they must diffuse to the fuel surface to react. At this time, they can be destroyed by Fe-oxide corrosion products that will lower the effective G-value (molecules generated per 100 eV energy deposited) for H$_2$O$_2$ production. This results in the formation of a radiolytic zone (Sunder et al. 1997), outside of which radiolytic products cannot increase the oxidation potential nor increase the corrosion rate. The occurrence of studtite, [UO$_2$(O$_2$)(H$_2$O)$_2$](H$_2$O)$_2$, in nature indicates that natural uraninite can generate enough radiolytic oxidants to cause a change in the UO$_2$ paragenesis (Finch & Ewing 1992; Burns & Hughes 2003). The results from natural systems imply that even aged fuel may be susceptible to radiolytic processes as long as the temperature is low and iron corrosion products are not present. Sunder et al. (1997) explored the effect of radiolysis on fuel corrosion and determined that most of the radiolytic effects would be due to the intense β-,γ-field that decays over the first 1000 years of emplacement. New mixed potential models are now looking at α-emitters concentrated at the spent fuel

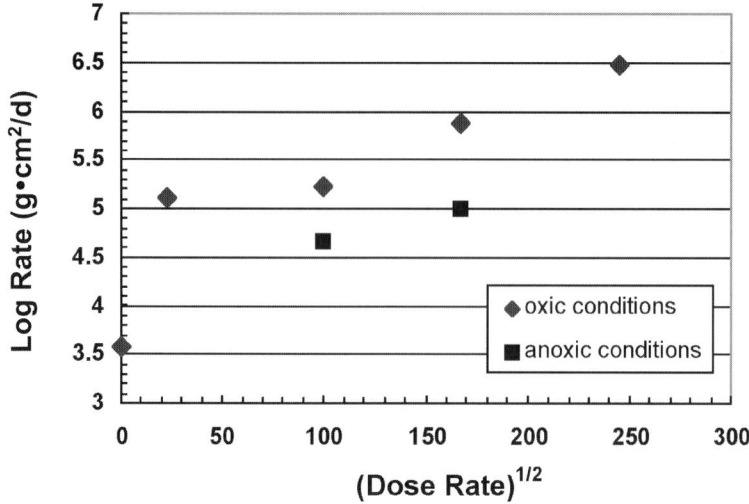

Fig. 5. Plot showing the effect of radiation field intensity on UO_2 dissolution rate (data extracted from Christensen & Sunder 1996). Dissolution rates were obtained by electrochemical measurements. A significant enhancement in the reaction rate is observed with dose and in the presence of oxygen.

surface. Under stagnant water conditions or thin water films, α-radiolysis may have a significant impact on the local oxidation potential. At sufficiently high dose rates, the corrosion rate is directly proportional to the square root of the dose rate (Shoesmith, 2000) (Fig. 5).

The absence of an observable oxygen dependence in flow-through tests by Gray *et al.* (1992) on spent fuel at 25 °C and 75 °C may be due to the competitive effects of temperature, oxygen, and radiolysis products. At the high temperature, the main radiolytic oxidant, peroxide (H_2O_2), is rapidly decomposed (Stefanic & LaVerne 2002). The thermal decomposition reaction rate constant for H_2O_2 in aqueous solution has been estimated to be 6.5×10^5 $\exp(-71 \text{ kJ} \cdot \text{mol}^{-1}/RT)\text{s}^{-1}$ (where R is the gas constant, and T is the temperature in degrees Kelvin). This may have led to the oxygen partial pressure parameter becoming rate-determining. At low temperature, H_2O_2 is not decomposed as readily, resulting in the radiolytic field becoming the rate-determining effect. As the lifetime of radiolytic species is very short and the effective diffusion lengths of the radiolysis products are on the order of $0.1-1 \mu$m, the flow rate in a flow-through test is too low to disrupt radiolytic-assisted corrosion of the fuel surface. The role of temperature in studtite formation can be demonstrated by observations of extensive studtite formation in the K-Basins at Hanford, Washington (Abrefah *et al.* 1998), where the water temperature is held at 10 °C; whereas, at 90 °C on Hanford 'N' (ninth) reactor corroding U-metal fuel particles do not generate studtite. If the fuel in a geological repository is held at temperatures above 60 °C, it is unlikely that there will be significant H_2O_2 build-up, and thus there will be little or no studtite formation.

In SNF corrosion tests, there has been a tendency to use the release of more soluble species Tc, Cs, and Mo as markers for fuel corrosion (Finn *et al.* 2002). As none of these elements are present in the UO_2 matrix, this approach may not reveal the actual fuel matrix corrosion rate. Furthermore, short-term leaching tests may not expose possible diffusion-limited ($t^{1/2}$) release of gap and grain boundary species and assume excessive rates of reaction based on initial fast release rates. The microstructure, radiation field, and composition will change over time, so that tests carried out on fuel today may not be relevant to fuel behaviour 300 to 1000 years from now, once the high β-,γ-field has decayed.

Secondary U mineral paragenesis

Weathering of natural uraninite (Finch & Ewing 1992; Finch & Murakami 1999) and synthetic UO_2 (Wronkiewicz *et al.* 1992, 1996) under oxidizing conditions can result in the formation of a series of U^{6+} secondary phases, such as U^{6+} oxide hydrates (e.g., dehydrated-schoepite, $UO_3 \cdot 0.8H_2O$), alkali U^{6+} oxide hydrates (e.g., compreignacite, $K_2[(UO_2)_6O_4(OH)_6] \cdot 8H_2O$), U^{6+} silicates (e.g., uranophane, $Ca(UO_2)_2(SiO_3)_2$

Table 2. *Potential U minerals of interest to spent fuel behavior in a geological repository*

Mineral	Formula
Uranium oxide	UO_2
Natural Uraninite	$[U_{1-x-y-z-u}^{4+}U_x^{6+}(Th^{4+})_uREE_y^{3+}M_z^{2+}]O_{2+x-(0.5y)-z}$
Uranyl oxide-hydrates	
Dehydrated schoepite	$UO_3 \cdot 0.8H_2O$
Ianthinite	$U^{4+}(UO_2)_5(OH)_{14} \cdot 3H_2O$
Meta-schoepite	$(UO_2)_4O(OH)_6 \cdot 5H_2O$
Schoepite	$[(UO_2)_8O_2(OH)_{12}](H_2O)_{12}$
U^{6+} peroxides	
Meta-studtite	$UO_4 \cdot 2H_2O$
Studtite	$[(UO_2)(O_2)(H_2O)_2](H_2O)_2$
Becquerelite group	
Becquerelite	$Ca[(UO_2)_6O_4(OH)_6] \cdot 8H_2O$
Billietite	$Ba[(UO_2)_6O_4(OH)_6] \cdot 8H_2O$
Calciouranoite	$(Ca,Na,Ba)U_2O_7 \cdot 2H_2O$
Clarkeite	$(NaCa_{0.5}Pb_{0.5})[(UO_2)O(OH)](H_2O)_{0-1}$
Compreignacite	$K_2[(UO_2)_6O_4(OH)_6] \cdot 8H_2O$
Curite	$Pb_3[(UO_2)_8O_8(OH)_6] \cdot 3H_2O$
Fourmarierite	$Pb_{1-x}[(UO_2)_4O_{3-2x}(OH)_{4+2x}] \cdot 4(H_2O)$
Protasite	$Ba[(UO_2)_3O_3(OH)_2] \cdot 3H_2O$
Uranyl silicates	
Soddyite	$(UO_2)_2SiO_4 \cdot 2H_2O$
Uranosilite	$(Mg,Ca)_4(UO_2)_4(Si_2O_5)_{5.5}(OH)_5 \cdot 13H_2O$
Haiweeite group (U : Si = 1 : 3)	
Haiweeite	$Ca(UO_2)_2(Si_2O_5)_3 \cdot 5H_2O$
Weeksite	$K_2(UO_2)_2(Si_2O_5)_3 \cdot 4H_2O$
Uranophane group (U : Si = 1 : 1)	
Boltwoodite	$K(H_3O)[(UO_2)(SiO_4)] \cdot 0.5H_2O$
Kasolite	$Pb(UO_2)(SiO_4) \cdot H_2O$
Sklodowskite	$Mg(H_3O)_2[(UO_2)(SiO_4)]_2 \cdot 4H_2O$
Swamboite	$(UO_2)_{0.33}[(UO_2)(SiO_4)]_2 \cdot 6H_2O$
β-uranophane	$Ca(UO_2)_2(SiO_3)_2(OH)_2 \cdot 5H_2O$
Uranyl carbonates	
Andersonite	$Na_2Ca(UO_2)(CO_3)_3 \cdot 18H_2O$
Rutherfordine	UO_2CO_3
Wyartite	$CaU^{5+}(UO_2)_2(CO_3)O_4(OH)(H_2O)_7$
Uranyl phosphates and arsenates	
Autunite	$Ca[(UO_2)(PO_4)]_2 \cdot 11H_2O$
Bassetite	$Fe^{2+}[(UO_2)(PO_4)] \cdot 8H_2O$
Chernikovite	$(H_3O)_2(UO_2)(PO_4)_2 \cdot 6H_2O$
Meta-autunite	$Ca(UO_2)_2(PO_4)_2 \cdot 6\text{-}8H_2O$
Meta-torbenite	$Cu[(UO_2)(PO_4)]_2 \cdot 8H_2O$
Renardite	$Pb[(UO_2)(PO_4)]_2 \cdot 7H_2O$
Saléeite	$Mg[(UO_2)(PO_4)]_2 \cdot 10H_2O$
Sklodowskite	$(H_2O)_2Mg[(UO_2)(PO_4)]_2 \cdot 4H_2O$
Torbenite	$Cu[(UO_2)(PO_4)]_2 \cdot 10H_2O$
Uranyl molybdates	
Calcurmolite	$Ca(UO_2)_3(MoO_4)_3(OH)_2(H_2O)_{11}$
Iriginite	$U(MoO_4)_2(OH)_2(H_2O)_3$
Umohoite	$(UO_2)(MoO_2)(OH)_2(H_2O)_2$

(Continued)

Table 2. Continued

Mineral	Formula
Uranyl vanadates	
Carnotite	$K_2(UO_2)_2(VO_4)_2 \cdot 3H_2O$
Uranyl sulphates	
Deliensite	$Fe(UO_2)_2(SO_4)_2(OH)_2 \cdot 3H_2O$
Na–Zippeite	$Na_4(UO_2)_6(SO_4)_3(OH)_{10} \cdot 4H_2O$
Schroeckingerite	$NaCa_3(UO_2)(CO_3)_3(SO_4)F \cdot 10H_2O$
Uranopilite	$(UO_2)_6(SO_4)O_2(OH)_6(H_2O)_6 \cdot 8H_2O$
Uranyl selenites and tellurites	
Cliffordite	$UO_2(Te_3O_7)$
Derriksite	$Cu_4[(UO_2)(SeO_3)_2](OH)_6$
Demesmaekerite	$Pb_2Cu_5[(UO_2)(SeO_3)_3]_2(OH)_6(H_2O)_2$
Guilleminite	$Ba[(UO_2)_3(SeO_3)_2O2](H_2O)_3$
Marthozite	$Cu[(UO_2)_3(SeO_3)_2O_2](H_2O)$
Moctezumite	$PbUO_2(TeO_3)_2$
Actinide orthosilicates	
Coffinite	$U(SiO_4)_{4-x}(OH)_{4x}$

$(OH)_2 \cdot 5H_2O$) in silicate bearing solutions (see Table 2 for a list of U minerals). If phosphate or sulphate is present in a repository environment, possibly due to the introduction of sulphur-bearing steels or phosphate-engineered barriers, U sulphates such as Na–zippeite [$Na_4(UO_2)_6$ $(SO_4)_3(OH)_{10} \cdot 4H_2O$], and U phosphates such as meta-autunite [$Ca(UO_2)_2(PO_4)_2 \cdot 6\text{-}8H_2O$] might form.

Laboratory tests with UO_2 under oxidizing conditions

Under oxidizing conditions in the presence of water or moisture, UO_2 is not thermodynamically stable. The rate of UO_2 oxidation in air is slow owing to the slow diffusion of oxygen; however, in the presence of water, the oxidation process is far more rapid (Wasywich et al. 1992). Many spent fuel dissolution models assume that spent fuel will be completely converted to U^{6+} phases within 100 to 1000 years after waste package failure and exposure to aerated water under oxidizing conditions (Budnitz et al. 1999). Even if UO_2 undergoes complete alteration to U secondary phases, some radionuclides may be held up either by these secondary U phases formed through alteration, sorption on canister corrosion products, or through the formation of other precipitates. Initial phases to precipitate from the oxidative dissolution of UO_2 are U^{6+} oxyhydroxides (schoepite). However, there is mounting evidence for the importance of studtite (McNamara et al. 2003; Burns & Hughes 2003) under conditions where thin films of water are present at ambient temperatures.

Scanning electron microscopy (SEM) images of euhedral meta-schoepite crystals precipitated after reaction with water for one year are shown in Figure 6a. The banding on the crystal may be due to drying in the vacuum chamber. After two years of reaction, the original schoepite started to re-dissolve (Fig. 6b). This may have occurred owing to slight changes in the solution chemistry. The corrosion of the schoepite occurred along specific crystallographic directions.

The U-bearing phases studtite and meta-studtite, $UO_4 \cdot 2H_2O$, are the only known peroxide minerals (Burns & Hughes 2003). Studtite forms during the build-up of α-generated H_2O_2 on the surface of natural uraninite (Walenta 1974; Deliens & Piret 1983; Finch & Ewing 1992; Burns & Hughes 2003). With increasing oxidant concentrations or radiation field, studtite and meta-studtite have been found to be more prevalent than schoepite-related structures in UO_2 tests (Amme 2002; Satonnay et al. 2001) and also in spent UO_2 fuel tests (McNamara et al. 2003). The radiolytic field with spent fuel may create an additional complexity for predicting UO_2 paragenesis in a geological repository.

UO_2 corrosion under reducing conditions

Most repository sites under consideration for commercial spent fuel disposal are in reducing environments, such as in Boom clay formations of the Mol site, Belgium, where UO_2 is thermodynamically stable. In oxygen-free conditions, Spahiu et al. (2002) have shown that fuel in an

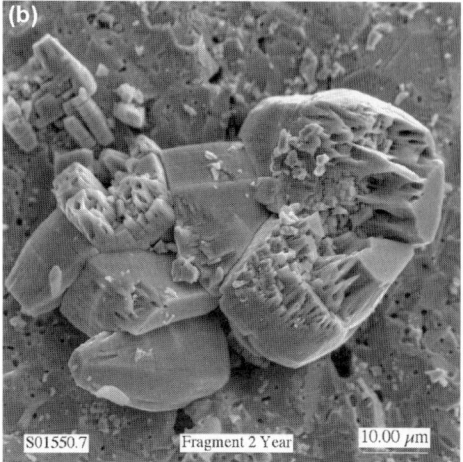

Fig. 6. SEM images of meta-schoepite: (**a**) Precipitation of meta-schoepite after 1 y of reaction; (**b**) After 2 y, preferential corrosion of the U^{6+} phase occurred along specific crystallographic directions.

environment containing Fe generated hydrogen. This pressurized H_2 substantially hindered the further reaction of spent fuel in spite of radiolysis effects. Under reducing conditions, uraninite can undergo coffinitization, resulting in the formation of coffinite, $U(SiO_4)_{4-x}(OH)_{4x}$ (Speer 1982).

Laboratory tests with SNF under oxidizing conditions

Studtite and meta-studtite have been identified by McNamara *et al.* (2003) on the surface of corroding commercial spent fuel replacing the meta-schoepite precipitated earlier (Fig. 7). As the H_2O_2 concentration increased with time, studtite became the dominant alteration phase. Abrefah *et al.* (1998) also noted the preponderance of studtite on Hanford's N-reactor fuel extracted from storage pools. It is likely that the high temperature used in tests with spent fuel (Finn *et al.* 1994; Finch *et al.* 1999) may have prevented the formation of studtite by thermal destruction of radiolytic H_2O_2. Burns (personal communication 2003) has noted that studtite can easily be mistaken for uranophane and can only be distinguished through XRD analysis. It is possible that studtite and/or meta-studtite may have been missed during analysis of some natural analogue sites, such as, Peña Blanca, Mexico. Indeed, phase characterization at Peña Blanca was mainly by visual inspection (Pearcy *et al.* 1994).

In a 20-year corrosion test by Stroes-Gascoyne *et al.* (1997), oxidative dissolution of the fuel

Fig. 7. SEM image of studtite from corroded SNF. Studtite replaces meta-schoepite, which is not visible in the image.

resulted in the precipitation of a U oxide phase, identified as schoepite. However, SEM images of the alteration phases reveal phases that were rod-shaped, similar to those shown in Figure 7. Long-term tests by Wronkiewicz *et al.* (1992, 1996) on UO_2 exposed to dripping aerated water were able to establish the paragenesis of UO_2 alteration. A similar paragenesis for spent U oxide fuel has been presented by Finch *et al.* (1999). However, it is evident from other fuel corrosion studies that the paragenesis of spent U oxide fuel may be more complex and dependent on temperature as well as solution composition. Unusual phases, including Cs–Mo U^{6+} oxides and studtite (Stroes-Gascoyne *et al.* 1997; Buck *et al.* 1997; McNamara *et al.* 2003) suggest that other alteration pathways may be possible. The complete paragenesis for fuel alteration and disposition of individual radionuclides in both oxidizing and reducing environments remains unresolved.

Anthropogenic U sources

There are many examples of U contamination in the environment, from U-mining sites to U production facilities (Bertsch *et al.* 1994; Abdelouas *et al.* 1999; Buck *et al.* 1996) and disposed U ordnance (Salbu *et al.* 2003). The major U-bearing phases found in soils at the former U-processing site at Fernald in Ohio, USA, were Ca–meta-autunite, U–Ca-oxide, uraninite, and uranium (IV) meta-phosphate (U_3PO_4) (Buck *et al.* 1996). Bertsch *et al.* (1994) and Allen *et al.* (1994) used XAS to determine the oxidation state of U in bulk soil samples. The position of the U–L_{III} edge indicated that 80% of the U was hexavalent.

Salbu *et al.* (2003) used micro-XAS to examine oxidation of depleted uranium (DU) munitions. Interestingly, these studies revealed the presence of UO_2 and U_3O_8 but no U^{6+} oxide hydrate phases. Brock *et al.* (2003) examined the corrosion of DU penetrators in an arid environment. Using SEM, they observed aggregates of tabular, hexagonal schoepite and meta-schoepite crystals with clay/silt particles that were coated with amorphous silica. Brock *et al.* (2003) suggested that as the schoepite/meta-schoepite phases were coated with amorphous silica/clays, further dissolution was inhibited.

In most of these cases of anthropogenic sources of U, the only radionuclide present is U. However, at the former gaseous diffusion site of Portsmouth in Ohio, USA, the recycling of U from nuclear reactors resulted in small amounts of ^{237}Np, ^{239}Pu, and ^{99}Tc being introduced into the production process (BJC/PORTS 2000). The Portsmouth plant received approximately 90 000 t of recycled U containing an estimated 3.6 ppb of ^{239}Pu, 0.2 ppm of ^{237}Np, and 7.3 ppm of ^{99}Tc. At the concentrations in U from fuel reprocessing facilities, the radiological impact of these impurities is negligible. However, the chemical processes in the plant tended to concentrate these impurities either in the U product or in reaction byproducts. Over a period of several years, kilogramme quantities of ^{99}Tc and ^{237}Np were introduced into the plant. Examination of U contamination at this site might provide useful information on the behaviour of ^{237}Np and ^{99}Tc.

Weathering and the redistribution of radionuclides

In laboratory tests with SNF, Buck *et al.* (1997) demonstrated that a U^{6+} phase structurally related to billietite, $Ba[(UO_2)_6O_4(OH)_6] \cdot 8H_2O$, incorporated Cs, Ba, and Mo. Kim *et al.* (2002) synthesized Nd- and Ce-bearing U(VI) phases, becquerelite and ianthinite, although the results suggested surface sorption rather than incorporation. Obtaining direct evidence for incorporation of minor and trace elements into U^{6+} phases has proven difficult as most instrumental methods are unable to distinguish between sorption and incorporation. Further work is required to establish the ability for U^{6+} phases to incorporate and/or sorb radionuclides in various environments. Se-79 is another long-lived (1.1×10^6 y) fission product (although present at extremely trace levels in SNF) that in oxygenated solutions may become incorporated into U phases (Chen *et al.* 1999). Many U^{6+} selenite minerals have been identified in nature (see Table 2) and it seems highly probably that SeO_3 may substitute for (SiO_3OH) in a variety of U^{6+} silicate minerals under the appropriate conditions. Chen *et al.* (2000) have determined that incorporation of Tc into U^{6+} phases is unlikely owing to the underbonding at the U^{6+} site that will destabilize the structure.

Neptunium and Pu are the primary actinides of concern for the geological disposal of spent fuel owing to their long half-lives (2.41×10^4 y for ^{239}Pu; 2.14×10^6 y for ^{237}Np). Both these actinides occur in two or more oxidation states. Neptunium and Pu will be present as Np(IV) and Pu(IV), respectively, in spent fuel substituting for U in the UO_2 fluorite structure. However, the studies of Kleykamp (1985) and Thomas *et al.* (1992) suggest that transuranic elements may also be present in various other oxide and/or zirconate phases. Owing to its high solubility and low sorption affinity, Np(V) is considered to be the most environmentally mobile actinide species (Runde *et al.* 2002). Burns *et al.* (1997) have suggested that the similarities in the NpO_2^+ and the UO_2^{2+} ions may allow significant solid solubility of Np(V) into U^{6+} secondary minerals.

Plutonium and REE behaviour

Detailed analysis of corroded spent fuel reacted under moist-air conditions revealed a precipitate with an anomalously high concentration of Pu (Buck *et al.* 2004). These regions were patchy and appear to be about 100 nm thick (Fig. 8). The Pu-enriched regions did not possess the microstructure typical of a U(VI) alteration phase but were crypto-crystalline. Analysis with TEM-EDS and electron energy-loss spectroscopy (EELS) indicated that the U:Pu ratio was about 1 : 0.3–0.2 compared to an expected ratio of 1 : 0.02–0.04 in the unaltered spent fuel (Fig. 9). Such EELS techniques are almost completely analogous to XAS and, like XAS, the near-edge structure of absorption edges can be used to determine the chemical nature of an element. The EELS techniques generally use much lower energy absorption edges than XAS and, when combined with TEM, have much higher spatial resolution.

The Pu-enriched region also contained Am, REE, Zr, and Ru; however, americium and the REE were present at lower than expected levels. The depletion of REE and Am in the precipitated Pu-layer is interesting as these elements are usually considered to be good analogues for Pu chemistry. The chemistry of Pu has been recently complicated by the discovery of PuO_{2+x} (Haschke *et al.* 2000; Conradson *et al.*

2004). Direct evidence for the occurrence of Pu(V) in PuO_{2+x} has now been obtained with XAS. The precipitation of a superoxide PuO_{2+x} can release Pu(V) to solution. Later disproportionation of Pu(V) eventually results in the precipitation of Pu(OH)$_4$ solids. Such a mechanism would explain the partitioning of Pu from the REE, and the EELS results that suggested that the oxidation state of the Pu in the precipitate phase may be Pu(V) (Buck *et al.* 2004); however, further work is necessary to confirm this finding. The EELS results also indicated that the U was oxidized relative to U(IV). These results were consistent with the electron diffraction analysis of the residue region that matched with U_3O_8. These discrete altered regions were only visible with TEM and could not be detected with XRD or SEM. Wasywich *et al.* (1992) also detected oxidized U using X-ray photo-electron spectroscopy at the surface of UO$_2$ exposed to moist-air conditions.

Similar Pu-rich regions were identified by SEM of solids remaining from an acidic (pH = 4) flow-through test (Fig. 10). Observation of similar Pu-rich precipitates in the acid flow through tests supports the contention that the Pu enrichment is not an artifact of the starting material.

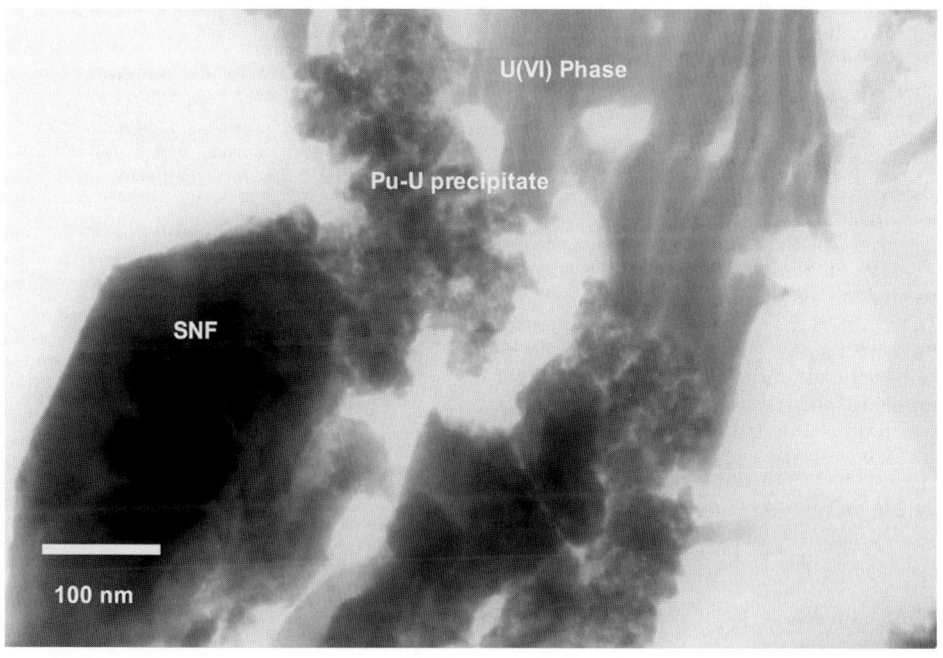

Fig. 8. Transmission electron micrograph of corroded SNF exposed to moist-air conditions for 3 years. A thin layer of Pu-rich precipitate was observed on the weathered fuel surface along with uranyl oxide hydrate and Cs–Mo uranyl oxide hydrate alteration phases (adapted from Buck *et al.* 2004).

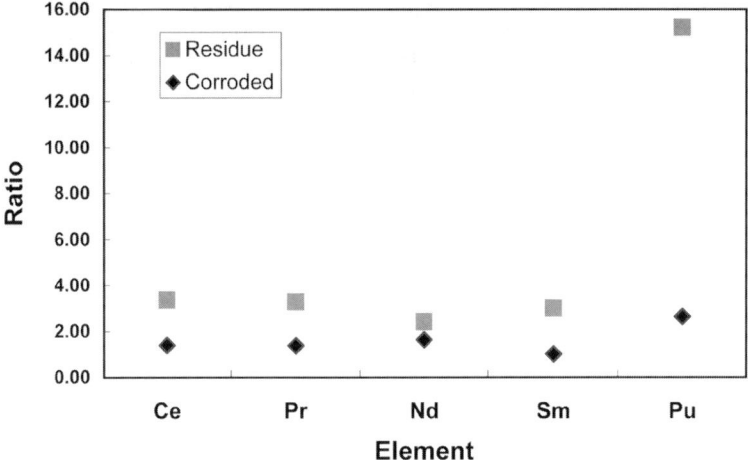

Fig. 9. Plot using EELS signal ratios to demonstrate the large enrichment of plutonium compared to rare earths in the interfacial region of a corroded SNF (45 MWd/kg U) exposed to moist-air for 3 y (adapted from Buck *et al.* 2004). The residue and corroded fuel values have been divided by the expected values based on SNF inventories from Guenther *et al.* (1988).

Examination of the surface of the reacted particle demonstrated that there had been a dramatic increase in the surface area of the particle. There were no large differences in contrast with back-scattered electron imaging. A few areas were observed that contained relatively high concentrations of Zr, Mo, Ru, and Pu. Although, 4d-metals (Mo, Tc, and Ru) can be observed during SEM examination of SNF, regions enriched in Pu and/or Zr have not been observed in fuel. However, these regions did not possess a distinctive morphology or contrast. The majority of the sample remained U-rich with little evidence of other elements.

Haschke *et al.* (2000) have demonstrated the occurrence of a superoxide $PuO_{2+x}(s)$ that reportedly formed between 25 and 350 °C and contained Pu(VI). Runde *et al.* (2002) reported XAS evidence for the existence of Pu(V) instead of Pu(VI). Fine structure analysis for the Pu solid precipitated at 90 °C in a Yucca Mountain groundwater was compared with data for a $PuO_{2(s)}$ standard. The measured Pu–O distance of ~0.185 nm is characteristic for Pu(V) and agrees well with the measured Pu–O bond distance of the $Pu(V)O_2^+(aq)$ ion at 0.181 nm. The Pu–O bonds in hexavalent Pu compounds are much shorter. Other details in the X-ray absorption fine structure spectra supported the assignment of Pu as Pu(V). At ambient temperatures, oxidation of Pu(IV) was extremely slow and formation of Pu(V) was not observed.

Neptunium behaviour

Several research groups are examining the potential for Np incorporation or association with U^{6+} phases that would lead to improvements in performance assessment models (Chen 2003). A fundamental problem addressing the issue of Np behaviour in weathered waste forms is the low level of this actinide in the starting material. Typical spent fuels contain about 400–600 ppm [237]Np (see Table 2), although this amount will increase with the decay of [241]Am. To date, the only method available with the necessary sensitivity combined with high spatial resolution to probe the submicrometre sized phases, has been EELS on a TEM. However, operation of this technique is extremely difficult because of the specific energy region that must be used. To date, only a few examples of possible Np incorporation into an alteration phase have been found (Buck *et al.* 2003). There has been no experimental evidence from fuel tests on Np incorporation into the U^{6+} silicate secondary phases β-uranophane and boltwoodite, $K(H_3O)[(UO_2)(SiO_4)] \cdot 0.5H_2O$; however, Np-bearing uranophane has been synthesized (Douglas *et al.* 2003; Buck *et al.* 2003; Burns *et al.* 2004). Preliminary data from Buck *et al.* (1998) suggested that Np was being sequestered in a U^{6+} oxide-hydrate phase during unsaturated corrosion tests on spent fuel, confirming the theoretical prediction of Burns *et al.* (1997). However, significant experimental problems with the detection of Np in a U

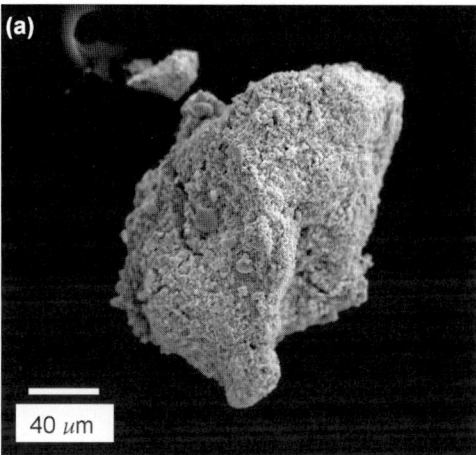

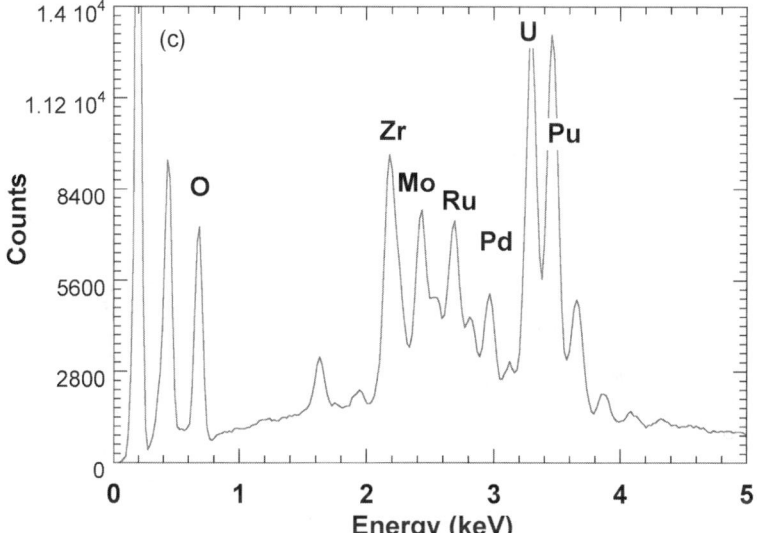

Fig. 10. SEM image of a corroded SNF particle from an acidic regime flow-through test exhibiting patches of Pu enrichment: (a) Low-magnification SEM image of particle from the low pH flow-through test; (b) High-magnification image of the Pu-rich region; (c) EDS spectrum of Pu-enriched region.

matrix challenges the veracity of these results and additional proof of this process will be necessary before solubility models can include the possibility of Np incorporation and/or uptake into U^{6+} phases.

An indication that Np can enter U^{6+} phases has been observed in a number of different U^{6+} phases. Np-doped studtite has been synthesized and examined with EELS and infrared spectroscopy (Buck *et al.* 2003). The Np was held in the pentavalent oxidation state with $NaNO_2$. A uranyl nitrate (0.5 g) solution was reacted

with 1 M H_2O_2. To produce a Np-bearing U solid, a solution of Np(V) was added to the reacting uranyl nitrate solution. The resulting solid that precipitated was red-brown (pure U end-member studtite is white). A green colouration remained in the solution, indicating the presence of some residual Np(V). Characterization of the Np-doped U phase was then performed with optical microscopy, XRD, infrared spectroscopy, and electron microscopy and compared to a pure studtite phase. The EELS analysis provided the best evidence for incorporation of Np into

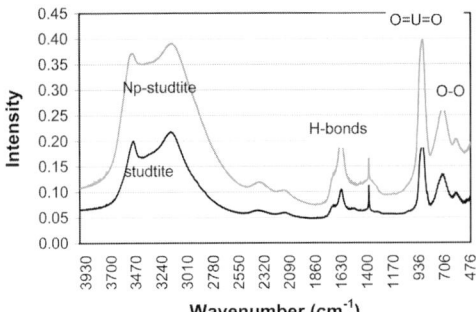

Fig. 11. Infrared spectrum of synthetic studtite and Np-doped studtite.

this phase. The analysis was performed on the actinide N-edges, which, although very weak, are at a lower energy and easier to obtain in radiation sensitive materials. X-ray diffraction of the U end-member and the Np-doped material were both consistent with studtite. No other minor phases were detected. Further evidence for the purity of the U-only and Np-doped solids was obtained from infrared spectroscopy (Fig. 11). The high level of Np associated with the studtite phase suggests that Np must be incorporated into the studtite structure, possibly as Np(VI).

Radiochemical analysis of meta-studtite from an SNF test indicated that ^{237}Np was associated with the U^{6+} peroxide (Table 3). Plutonium, Cm, and Am were found to be in lower concentration in the secondary U phases; however, 5–6% of the available Pu and Am, based on reactor code estimates, was co-precipitated with the U phase. The radiochemical data for the collected alteration products from the SNF samples (Table 3) are presented as µg-analyte per g U for comparison to the reactor inventory code calculation for 30 MW/d burn-up fuel at 20 y (extracted from Guenther *et al.* 1988).

Douglas *et al.* (2003) prepared Np-doped U^{6+} silicate phases under different pH conditions. In Figure 12, EELS spectra from a synthetic uranophane crystallite clearly illustrate the presence of

Np (Buck *et al.* 2003). Uranium and Np were identified by determining the energy gap ($\Delta E_{M4,5}$) between the M$_4$ and M$_5$ absorption edges (U: $\Delta E_{M4,5} = 176$ eV; Np: $\Delta E_{M4,5} = 184$ eV) recorded by EELS. The Np–M$_5$ (3665 eV) edge can be subject to interference from the plural scattering from U–M$_5$ (3552 eV) and the U–O$_{4,5}$ edge (95–111 eV). However, the Np–M$_4$ edge (3850 eV) is not subject to similar interference and is separated from any plural events on the U–M$_4$ edge by 7–8 eV. One of the samples had a nominal bulk Np concentration measured at 6300 ppm, whereas the second sample had 1300 ppm of Np. While the U peaks remain at similar intensities for the two samples, the Np signal has decreased for the sample with lower bulk Np concentration. Direct evidence for Np incorporation in these studies with the U^{6+} silicates is debatable; however, the relatively high levels of Np and the consistency of the Np/U intensity ratios measured by EELS supports incorporation rather than sorption.

Technetium behaviour

Technetium (^{99}Tc) is a major dose contributor to repository performance assessment calculations owing to its long half-life (2.13×10^5 y), and its high solubility when oxidized. Technetium has been used as a marker for the rate of fuel matrix dissolution (Finn *et al.* 2002) even though it exists primarily in ε-particles in the fuel, which have very different dissolution characteristics to UO$_2$. Corroded particles of fuel from unsaturated drip tests were examined with TEM to look for compositional evidence for the preferential removal of Tc. Evidence of weathering was not obvious from TEM examination of the particles (Fig. 13). However, the composition of the particles did not match that reported during electron microscopy characterization of the uncorroded fuel (Guenther *et al.* 1988; Thomas *et al.* 1992). In Figure 14, TEM-EDS analyses of several particles have been

Table 3. *Radiochemical analysis of corrosion products from spent fuel after 1.5 y of immersion in water*

Sample	Total U (µg)	^{237}Np (µg/g U)	^{239}Pu (µg/g U)	^{241}Am (µg/g U)	$^{243+244}$Cm (µg/g U)
Collected solids	623	330	943	32.3	0.62
Reactor code calculation		345	12 200	516	15.8
% Uptake		96%	8%	NV	4%

NV = Not valid, fuel age not taken into account in reactor inventory code results.

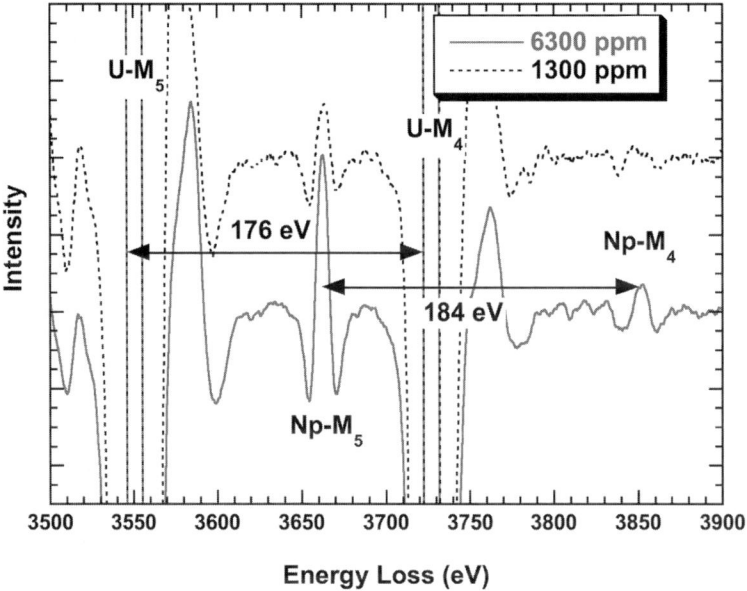

Fig. 12. Electron energy-loss spectrum of uranophane solids containing 1300 and 6300 ppm Np. The double arrows on the figure are 176 eV and 184 eV wide, demonstrating the presence of Np in the solid (extracted from Buck *et al.* 2003).

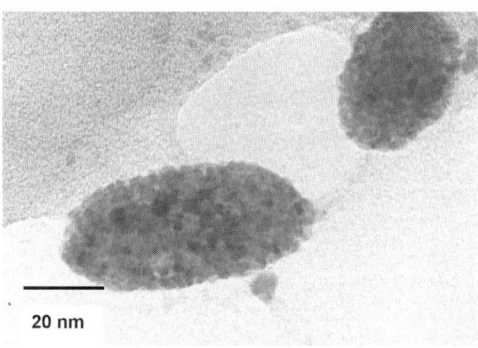

Fig. 13. TEM image of corroded ε-phase particle exhibiting preferential surfacial corrosion.

plotted on a ternary diagram similar to that shown in Figure 4. Most of the particles examined were extremely small, around 20–50 nm in diameter. In most cases, the Mo concentration appeared suppressed relative to Ru, suggesting preferential removal of this metal. Shoesmith (2000) has argued that for Tc or Ru to be removed from ε-particles, the oxidation potential exceeds that which can be accounted for by H_2O_2 alone, suggesting that only the intense β-,γ-radiolytic field can lead to the dissolution of these types of phases. However, it is not clear

that metallic particles exhibiting relatively high concentrations of Pd and/or Rh are necessarily an indication of corrosion. Thomas *et al.* (1992) reported TEM-EDS results of 4d-metals in the ε-phases, where the distribution of metals was very close to that expected on the basis of reactor inventory codes for the five elements. However, they also reported the occurrence of α-Pd particles in the SNF pellet centre. Earlier it was shown that the distribution of 4d-metals in metallic particles from a higher burn-up fuel varied considerably, possibly due to reactor operation conditions. This makes the case for preferential removal of Tc more difficult to confirm. A better method for looking for evidence of corrosion of the metallic particles is to profile across an individual corroded particle. In Figure 15, spot analyses indicated the removal of Mo but not Tc. The levels of Ru, Rh, and Pd appeared to be unchanged. As these particles are often only around 5–50 nm in diameter, this type of analysis is difficult and cannot be accomplished with techniques that do not have nanometer spatial resolution.

It is important to have accurate data on the composition of uncorroded fuel grains to be confident that observations on corroded particles reflect real corrosion. Based on the work by Thomas *et al.* (1992) and Guenther *et al.* (1988), we should not be seeing such high levels of Pd in

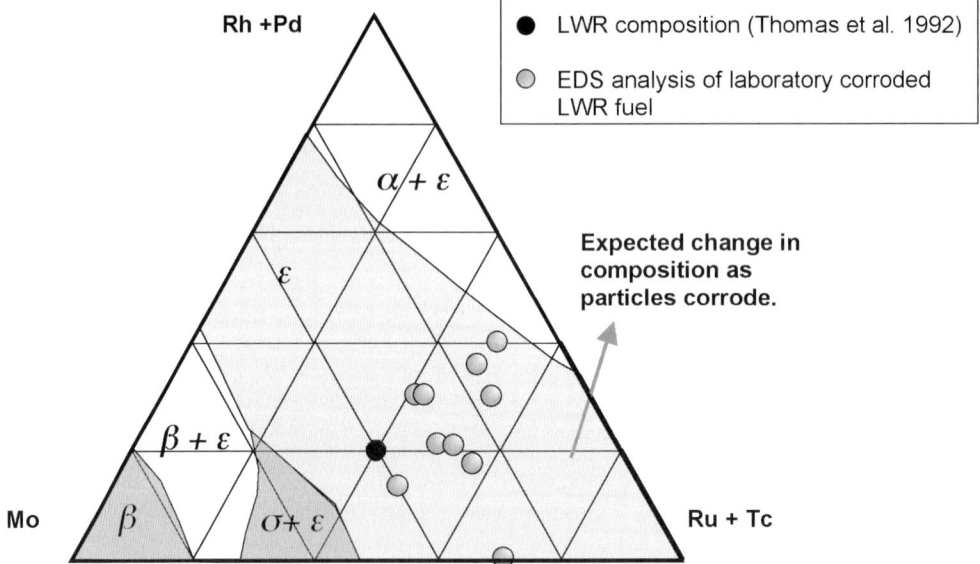

Fig. 14. Ternary diagram similar to that in Figure 4 showing the distribution of compositions of corroded metallic particles from an LWR fuel sample. Corrosive loss of Mo makes the average composition more Pd-rich.

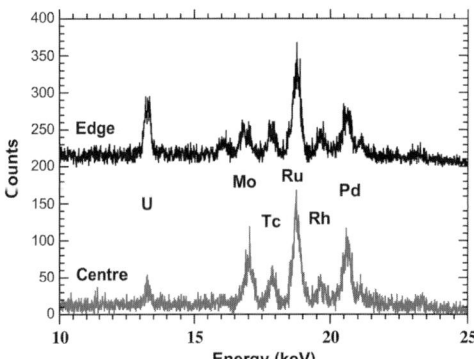

Fig. 15. EDS analysis across a 50 nm diameter ε-particle indicating preferential removal of Mo at the edge but little evidence for removal of Tc. The edge spectrum shows U from the neighbouring oxide. This behaviour is consistent with the oxidation potential of the 4d-group metals.

the ε-metal phase in the LWR fuel. The reported ratio Mo/(Ru + Pd) in uncorroded LWR spent fuel (30 MWd/kg U) is about one for the 4d-metal phases. In the particles found in the corroded fuel samples, this ratio was found to be much lower for all of the analysed particles. These results suggest that the ε-metal phases examined exhibit preferential removal of Mo during corrosion. However, if Mo is released preferentially to Tc, it would be inaccurate to use

these elements to determine the dissolution rate of spent fuel.

Natural analogues

Natural analogues for spent fuel have been reviewed extensively elsewhere (Finch & Ewing 1992, Murakami *et al.* 1997; Janeczek 1999; Pérez del Villar *et al.* 2002) for the geochemical behaviour of U. It is important to use appropriate natural analogues to validate geochemical models developed to predict the migration of radionuclides in a future radioactive waste geological repository system. In this section, a brief overview will be given of the natural fission reactors at Oklélobondo and Bangombé (Oklo) in Gabon (Brookins 1990; Hidaka & Holliger 1998; Stille *et al.* 2003); the analogue of the proposed Yucca Mountain repository at Peña Blanca, Mexico (Pearcy *et al.* 1994); the Cigar Lake deposit in Saskatchewan, Canada (Fayek *et al.* 1997); the Mina Fe in Salamanca, Spain (Pérez del Villar *et al.* 2002); and studies at the Colorado Plateau in the U.S. (Zhao & Ewing 2000).

The Oklo natural reactors in Gabon are the best natural analogues for assessing the geological behaviour of fission products and actinides (see also Gauthier-Lafaye *et al.* 2004). Elements that were compatible with the U ore structure were retained, whereas elements that were incompatible in the pitchblende (UO_{2+x}) structure were lost by diffusion into the surrounding rocks. Alkalis and alkaline earths were immobilized in clay minerals, secondary carbonates, and sulphates. This site continues to be a useful source for geochemical data on ^{99}Tc, ^{237}Np, and ^{239}Pu. The Oklo region reactors have been studied extensively (Hidaka & Holliger 1998; Janeczek 1999; Jensen et al. 2002).

The Bamgombé site in the Oklo region is of particular interest as it is located at shallow depths. Microprobe analysis of uraninite in contact with U^{6+} phases indicates that the alteration resulted in increased concentrations of Si, P, S, Zr, Ce, and Nd. The dissolution of accessory apatite, monazite and sulphides resulted in the retardation of U through the formation of secondary U^{6+} sulphates and phosphates, including phosphatian coffinites, $U(HPO_4)_2 \cdot 2H_2O$, and Fe–uranyl phosphate hydrates (i.e., bassetite, $Fe^{2+}[(UO_2)(PO_4)] \cdot 8H_2O$). Jensen et al. (2002) investigated the mineralogy of the U deposits at Bamgombé and compared their observations with thermodynamic predictions based on groundwater conditions. The primary U minerals at Bamgombé are uraninite and minor coffinite. The U^{6+} minerals include: fourmarierite, Pb_{1-x} $[(UO_2)_4O_{3-2x}(OH)_{4+2x}] \cdot 4(H_2O)$; bassetite, possibly associated with phosphatian coffinite and/or chernikovite, $(H_3O)_2(UO_2)(PO_4)_2 \cdot 6H_2O$; torbernite, $Cu[(UO_2)(PO_4)]_2 \cdot 10H_2O$; Ce-rich françoisite, $(Nd,Y,Sm,Ce)(UO_2)_3(PO_4)_2O(OH)$ $(H_2O)_6$; and uranopilite, $(UO_2)_6(SO_4)O_2(OH)_6$ $(H_2O)_6 \cdot 8H_2O$. Autunite, $Ca[(UO_2)(PO_4)]_2 \cdot 11H_2O$, has also been reported. Eh-pH diagrams predict that phosphatian coffinite and $UOF_2 \cdot H_2O$ are the only stable U^{4+} phases, and that uranopilite, torbernite, and bassetite will become stable during oxidative alteration. Soddyite, $(UO_2)_2$ $SiO_4 \cdot 2H_2O$, has been predicted to form at the expense of coffinite, but soddyite has not been identified at Bangombé. Jensen et al. (2002) have demonstrated that if SO_4^{2-} and PO_4^{3-} are excluded from geochemical models at Bamgombé, thermodynamic models fail to predict the observed U phase assemblages.

At Oklo, reactor zones 10, 13, and 16 have been of special interest because of their depth-mitigated surface weathering (Hidaka & Holliger 1998). Savary & Pagel (1997) showed that elements that migrated outside of the reactor zones 10 and 16, were usually elements not affected by changes in redox conditions (i.e., Rb, Sr, Ba, Cs, Kr, Xe, I, and Br). The geochemical behaviour of ^{240}Pu and ^{237}Np was assessed from determining the disposition of ^{232}Th and ^{209}Bi, respectively. Hidaka & Holliger (1998) suggest that there was little chemical fraction-ation of ^{90}Sr and ^{90}Zr. Given the relatively short half-life for ^{90}Sr, this result is not unexpected. ^{90}Zr is produced both during ^{235}U fission and from the decay of ^{90}Sr.

The fissogenic Ru, Rh, Pd, and Te in the reactor zones at Oklo were found to be incorporated into metallic aggregates and, interestingly, were difficult to dissolve for analysis, requiring first attack with peroxide, followed by dissolution with 6 M HCl. Similar metallic phases are found in irradiated U oxide from a commercial nuclear reactor and are also acid-resistant Pt-group metal particles (Mo, Tc, Ru, Rh, Pd). Fissogenic ^{99}Ru is produced from the decay of its long-lived precursor ^{99}Tc ($t_{1/2} = 2.1 \times 10^5$ y). It is likely that ^{99}Tc would be more chemically active than Ru. Chemical fractionation between ^{99}Ru and ^{101}Ru has been used by Hidaka & Holliger (1998) to establish the degree of ^{99}Tc mobility. Analyses indicated that the ^{99}Ru/^{101}Ru ratio in the metallic inclusions was greater than the ratio in the surrounding rock. Of the 730 kg of ^{99}Tc produced at Oklo, 60–85 wt% was retained in the reactor zone pitchblende. Of that which migrated, a majority was found within 10 m of the reactor zones (Brookins 1990).

In contrast, fissiogenic Mo, which is sensitive to redox conditions, was shown to migrate. In dissolution experiments with spent fuels, Mo appears to be the most easily leached element in the metallic particles (see Fig. 14). All fissogenic REE were well retained in the U matrices of the Oklo reactor zones. The geochemical behaviour of Np has been deduced from the disposition of ^{209}Bi. Significant differences in the ^{209}Bi/^{238}U ratio have been observed in some Oklo reactor zones, indicating migration of ^{237}Np; however, it is possible that some (or all) of this migration may reflect the transport of the decay product ^{209}Bi. Evidence for incorporation of Np into U^{6+} phases has not been established at Oklo.

Analysis of the ^{235}U/^{238}U level can provide an indication of the fate of Pu at the site. After leaching apatite crystals in clay deposits at the reactor zone 16 site, the remaining ^{235}U/^{238}U ratio was even higher than the initial value, suggesting the original ^{239}Pu may have been associated with the more leach-resistant phosphate phase (Hidaka & Holliger 1998). The REE were also associated with phosphates and sulphates: florencite, (La,Ce,Nd,Sm,Ca,Sr) $(Al,Fe)_3[(P(O,OH)_4]_2(OH)_6$; rhabdophane; crandallite $(Ca,REE,Pb,Sr)(Al,Fe)(Si,P,As)O_4(OH)_{5-6}$; and Ce–françoisite (Stille et al. 2003). Most of the REE minerals such as rhabdophane and florencite are enriched in the lighter REEs as the stability of REE–PO_4 complexes increases with decreasing atomic number (Pérez del Villar *et al.* 2002). Although REE phases are abundant in the reactor zones, the REE concentrations of the water in contact with these sediments are very low (<17 ppt). Similar Pu and REE behaviour has been observed in corrosion tests on HLW borosilicate glasses that contain trace amounts of phosphorus (Buck & Bates 1999). Confidence in waste form corrosion tests can be established by linking natural analogue and waste form behaviour. In nature, Th and the REE often behave in the same manner (Wood *et al.* 1997), whereas U chemistry is appreciably different. This behaviour has been duplicated in laboratory tests with nuclear waste materials where Pu is often found associated with Th and REE. However, these trends may not apply to spent fuel. The observation of enriched levels of Pu and depressed levels of REE suggests that the behaviour of Pu may be more complex. It is possible that the anomalous Pu chemistry may be a result of the stronger radiation field and the ensuing higher fuel surface oxidation potential.

Cigar Lake is a 1.3 Ga old U deposit that has remained relatively intact for at least 10 000 years (see also Gauthier-Lafaye *et al.* 2004). There is no surface evidence for the deposit, and its geometric arrangement is similar to that envisaged for deep HLW repositories. The U matrix is surrounded by a clay zone and buffered by several hundred metres of host rock. The U mineralization is comprised mainly of uraninite surrounded by impermeable illite and Mg–chlorite. The presence of iron(II) sulphide contributed to the stability of the orebody (Pérez del Villar *et al.* 2002). XRD analysis has revealed the presence of coffinite, which has replaced uraninite and pitchblende (UO_{2+x}) along fractures (Fayek *et al.* 1997). The coffinite was characterized by a high SiO_2 content and occasionally high P, Ti, and Zr contents.

Studies at the Nopal I U deposit in the Peña Blanca District, Chihuahua, Mexico (Pearcy *et al.* 1994) and the U deposit at Shinkolowbwe, Zaire (Finch & Ewing 1992) show that the primary uraninite altered to a suite of secondary U minerals similar, but not identical, to those observed in corrosion tests with UO_2 (Wronkiewicz *et al.* 1992) and SNF (Finch *et al.* 1999). Although these natural U deposits can tell us something about the possible fate of U, they cannot provide information on the important long-lived radionuclides ^{129}I, ^{99}Tc, ^{237}Np, and ^{239}Pu.

The Mina Fe deposit at Salamanca, Spain, presents an interesting potential analogue for waste package interactions with SNF. The oxidized zone has a large abundance of Fe(III) and Mn– Fe oxyhydroxides (Pérez del Villar *et al.* 2002).

Secondary U and REE minerals include autunite, Ce–phosphate, and Ld–Nd phosphates. The geochemical behaviour can be explained through pyrite oxidation that increases acidity and releases sulphate and Fe(III), that would allow oxidative dissolution of the U ore, possibly precipitating uranopilite. When the pH increased at sites more distant from pyrite dissolution, U(VI) was hydrolysed and eventually co-precipitated with Fe^{3+}-oxyhydroxides.

Zhao & Ewing (2000) examined altered uraninite from the Colorado Plateau with quantitative electron microprobe analysis in order to determine the fate of trace elements, including Pb, Ca, Si, Th, Zr, and REE, during corrosion under oxidizing conditions. The alteration phases identified included schoepite, calciouranoite, uranophane, fourmarierite, a Fe-rich U phase, and coffinite. The primary uraninites and alteration phases generally had low trace element contents. The electron microprobe analyses indicated that the trace elements preferentially entered the secondary U phases. Alteration caused the loss of U, Pb, and Zr, and incorporation of Si, Ti, Ca, and P into U phases.

The reactors at Oklo and Bangombé provide great confidence in the feasibility of radioactive waste isolation. The reactors operated for 800 000 years and after two billion years, most of the radionuclides (except for iodine that has migrated away completely) and/or their decay products have migrated only a few metres.

Conclusions

Most current performance assessment models assume conservatively that as the UO_2 matrix corrodes, the important radionuclides (^{129}I, ^{99}Tc, ^{237}Np, ^{239}Pu) will be released congruently even though some of these elements are present in different phases and are enriched to different levels throughout the spent fuel. All current SNF experiments are compromised by their high β-, γ-radiation field that results in the generation of powerful oxidants in the presence of water (e.g., OH$^•$ and H_2O_2) at the fuel. As a consequence, models based on spent fuel tests are biased and use unrealistically high dissolution rates that force increased reliance on engineered barriers.

Future models need to account for several complex and potentially competing factors, including:

(1) The time-dependent generation of radiolytic oxidants and reductants at the spent fuel/water/canister interfaces;

(2) The evolution of spent fuel microstructure (e.g., increased porosity) and microtextures

(e.g., the size distribution and the changing oxidation resistance of 4d-transition metal particles) before waste package failure;

(3) The role of secondary mineral precipitation and possible immobilization of released radionuclides in these alteration products under different conditions.

A high radiation field can change the nature of precipitating U^{6+} alteration phases (e.g., precipitation of Np-bearing meta-studtite) as well as the disposition of the radionuclides (Amme 2000; Satonnay et al. 2001; McNamara et al. 2003). Given the occurrence of studtite as alteration phase on natural uraninite due to α-radiolysis (Finch & Ewing 1992; Burns & Hughes 2003), it is apparent that this phase may still dominate the SNF paragenesis. Some laboratory tests on SNF may have inadvertently prevented studtite from precipitating by decomposing radiolytic H_2O_2 with hot water (75–90 °C). If the surface of the SNF in a geological repository remains above 60 °C, studtite is unlikely to form.

As the fuel ages, this radiation field will decrease dramatically; however, weathering may lead to the precipitation of low-solubility α-emitters at the corroding surface, so retaining the high radiolytic field for tens of thousands of years (King et al. 1999). The presence of Fe corrosion products in a repository environment will also impact the effective G-values for H_2O_2 production at time periods where α-radiolysis dominates (King et al. 1999). Unless more realistic assessment for radiolytic production rates under repository conditions are developed, models will continue to use exaggerated corrosion rates for SNF.

Indications from both microscopic analyses of metallic particles from corrosion tests and evidence from the Oklo natural reactors indicate that performance assessment calculations should not assume ^{99}Tc is easily mobilized. It is entirely inappropriate to use ^{99}Tc release as a marker for fuel corrosion because Tc is not located in the fuel matrix. The TEM examinations of corroded ε-particles have shown that Mo is preferentially leached from these phases, a behaviour that is similar to the one observed at Oklo. It is interesting to note that laboratory dissolution of billion-year old 4d-metallic particles for a chemical analysis required a corrosive mix of peroxide and acid (Hidaka & Holliger 1998) similar to the experience at SNF reprocessing plants. It is doubtful that the oxidation potential at the surface of an aged fuel will be sufficient to move Tc(0) from the ε-metal particles.

The evidence for radionuclide (e.g., Np, Se, Pu) incorporation into U^{6+} phases still needs to be established under a variety of conditions, and improvements are needed in analysing the radionuclides associated with these phases. Improved understanding of the chemical processes that occur at the waste form/water interface will result in less conservative models that would remove the need for expensive and unnecessary engineered barriers, increasing scientific credibility and public acceptance in a HLW repository.

Experimental work and analysis was performed at Pacific Northwest National Laboratory (PNNL) and Argonne National Laboratory. Radiochemical analysis was performed by Mr Charles Z. Soderquist (PNNL). Scanning electron microscopy was performed by Mr Bruce W. Arey (PNNL) and X-ray diffraction by Dr Evan D. Jensen (PNNL). Neptunium-bearing U(VI) phases were provided by Dr Judah I. Friese (PNNL) and Mr Matthew Douglas (Washington State University, Pullman, WA, USA). Work supported by the US Department of Energy under Contract DE-AC06-76RLO 1830. This text has benefited greatly from reviews by Dr Reto Gieré, Dr David W. Shoesmith, and Dr Lawrence Johnson.

References

ABDELOUAS, A., LUTZE, W. & NUTTALL, H. E. 1999. Uranium contamination in the subsurface: Characterization and remediation. In: BURNS, P. C. & FINCH, R. (eds) Uranium: Mineralogy, Geochemistry and the Environment. Mineralogical Society of America Reviews in Mineralogy, 38, 433–473.

ABREFAH, J., MARSCHMAN, S. C. & JENSEN, E. D. 1998. Examination of the Surface Coatings Removed from K-East Fuel Elements. Pacific Northwest National Laboratory Report, PNNL-11806.

ADACHI, T., OHNUKI, M. et al. 1990. Dissolution study of spent PWR fuel: Dissolution behavior and chemical properties of insoluble residues. Journal of Nuclear Materials, 174, 60–71.

ALLEN, P. G., BERG, J. M. et al. 1994. Determining Uranium Speciation in Fernald Soil by Molecular Spectroscopic Methods. Los Alamos National Laboratory Report, LA-12799-PR.

AMME, M. 2002. Contrary effects of the water radiolysis product H_2O_2 upon the dissolution of nuclear fuel in natural ground water and deionized water. Radiochimica Acta, 90, 399–406.

BERTSCH, P. M., HUNTER, D. B., SUTTON, S. R., BAJT, S. & RIVERS, M. L. 1994. In situ chemical speciation of uranium in soils and sediments by micro x-ray absorption spectroscopy. Environmental Science and Technology, 28, 980–984.

BJC/PORTS. 2000. Recycle Uranium Mass Balance Project, Portsmouth, OH, Site Report. Bechtel Jacobs Company LLC, BJC/PORTS-139/R1.

BROCK, A. L., BUCK, B., JOHNSON, W. & ULERY, A. L. 2003. Corrosion of depleted uranium in an arid environment: Characterizaion with XRD, SEM/EDS, and microprobe analyses. Geological

Society of America Annual Meeting, Seattle, 2–5 November 2003.

BROOKINS, D. G. 1990. Radionuclide behavior at the Oklo nuclear reator, Gabon. *Waste Management*, **10**, 285–296.

BUCK, E. C., BROWN, N. R. & DIETZ, N. L. 1996. The nature of contaminant uranium phases at the Fernald site in Ohio. *Environmental Science and Technology*, **30**, 81–88.

BUCK, E. C., WRONKIEWICZ, D. J., FINN, P. A. & BATES, J. K. 1997. A new uranyl oxide hydrate phase derived from spent fuel alteration. *Journal of Nuclear Materials*, **249**, 70–76.

BUCK, E. C., FINCH, R. J., FINN, P. A. & BATES, J. K. 1998. Retention of neptunium in uranyl alteration phases formed during spent fuel corrosion. *Materials Research Society Symposium Proceedings*, **506**, 87–94.

BUCK, E. C. & BATES, J. K. 1999. Microanalysis of colloids and suspended particles from nuclear waste glass alteration. *Applied Geochemistry*, **14**, 635–653.

BUCK, E. C., MCNAMARA, B. K., DOUGLAS, M. & HANSON, B. D. 2003. *Possible Incorporation of Neptunium in Uranyl (VI) Alteration Phases*. Pacific Northwest National Laboratory Report, PNNL-14277.

BUCK, E. C., FINN, P. A. & BATES, J. K. 2004. Electron energy-loss spectroscopy of anomalous plutonium behavior in nuclear waste materials. *Micron*, **35**, 235–243.

BUDNITZ, B., EWING, R. C., MOELLER, D. W., PAYER, J., WHIPPLE, C. & WITHERSPOON, P. A. 1999. *Peer Review of the Total System Performance Assessment-Viability Assessment Final Report*. Total System Performance Assessment Peer Review Panel, Las Vegas, Nevada.

BURNS, P. C., MILLER, M. L. & EWING, R. C. 1997. Incorporation mechanisms of actinide elements into the structures of U^{6+} phases formed during the oxidation of SNF. *Journal of Nuclear Materials*, **245**, 1–14.

BURNS, P. C. & HUGHES, K. A. 2003. Studtite, $[(UO_2)(O_2)(H_2O)_2](H_2O)_2$: The first structure of a peroxide mineral. *American Mineralogist*, **88**, 1165–1168.

BURNS, P. C., DEELY, K. M., SKANTHAKUMAR, S. 2004. Neptunium incorporation into uranyl compounds that form as alteration products of spent nuclear fuel: Implications for geologic repository performance. *Radiochim. Acta*, **92**, 151–159.

CHEN, F., BURNS, P. C. & EWING, R. C. 1999. ^{79}Se: geochemical and crystallo-chemical retardation mechanisms. *Journal of Nuclear Materials*, **275**, 81–94.

CHEN, F., BURNS, P. C. & EWING, R. C. 2000. Near-field behavior of ^{99}Tc during oxidative alteration of spent nuclear fuel. *Journal of Nuclear Materials*, **278**, 225–232.

CHEN, Y. 2003. Using reactive transport modeling to evaluate the source term at Yucca Mountain. *Computers & Geosciences*, **29**, 385–397.

CHILDS, B. G. 1963. Fission product effects in uranium dioxide. *Journal of Nuclear Materials*, **9**, 217–244.

CHRISTENSEN, H. & SUNDER, S. 1996. An evaluation of water layer thickness effective in the oxidation of UO_2 fuel due to radiolysis of water. *Journal of Nuclear Materials*, **238**, 70–77.

CHRISTENSEN, H. & SUNDER, S. 2000. Current state of knowledge of water radiolysis effects on spent nuclear fuel corrosion. *Nuclear Technology*, **131**, 102–123.

CONRADSON, S. D., ABNEY, K. D. *et al.* 2004. Higher order speciation effects on plutonium L_3 X-ray absorption near edge spectra. *Inorganic Chemistry*, **43**, 116–131.

DELIENS, M. & PIRET, P. 1983. Meta-studtite, $UO_2 \cdot 2H_2O$, a new mineral from Shinkolobwe, Zaire. *American Mineralogist*, **68**, 456–458.

DOUGLAS, M., CLARK, S. B., BUCK, E. C. & HANSON, B. D. 2003. Neptunium association with uranium (VI) silicate solid phases. *Migration'03*, Gyeongju (Kyongju), Korea, 21–26 September 2003. Abstract volume.

FAYEK, M., JANECZEK, J. & EWING, R. C. 1997. Mineral chemistry and oxygen isotopic analyses of uraninite, pitchblende and uranium alteration minerals from the Cigar Lake deposit, Saskatchewan, Canada. *Applied Geochemistry*, **12**, 549–565.

FINCH, R. J. & EWING, R. C. 1992. The corrosion of uraninite under oxidizing conditions. *Journal of Nuclear Materials*, **190**, 133–156.

FINCH, R. J., BUCK, E. C., FINN, P. A. & BATES, J. K. 1999. Oxidative corrosion of spent UO_2 fuel in vapor and dripping groundwater at 90 °C. *Materials Research Society Symposium Proceedings*, **556**, 431–438.

Finch. R. J. & MURAKAMI, T. 1999. Systematics and paragenesis of uranium minerals. *In*: BURNS, P. C. & FINCH, R. (eds) Uranium: Mineralogy, Geochemistry and the Environment. *Mineralogical Society of America, Reviews in Mineralogy*, **38**, 91–180.

FINN, P. A., BUCK, E. C., GONG, M., HOH, J., EMERY, J. W., HAFENRICHTER, L. D. & BATES, J. K. 1994. Colloidal products and actinides species in leachate from SNF. *Radiochimica Acta*, **66/67**, 189–195.

FINN, P. A. FINCH, R., BUCK, E. & BATES, J. 1998. Corrosion mechanism of spent fuel under oxidizing conditions. *Materials Research Society Symposium Proceedings*, **506**, 123–131.

FINN, P. A., TSAI, Y. & CUNNANE, J. C. 2002. Corrosion tests of LWR fuels – Nuclide release. *Materials Research Society Symposium Proceedings*, **713**, 607–614.

FORTNER, J. A., KROPF, J. A., FINCH, R. J., MERTZ, C. J., JERDEN, J. L., GOLDBERG, M. M. & CUNNANE, J. C. 2003. X-ray absorption spectroscopy of trace elements in spent nuclear fuel. *ANS/ENS International Winter Meeting*, 16–20 November, New Orleans, Louisana. Abstract volume.

GAUTHIER-LAFAYE, F., STILLE, P. & BROS, R. 2004. The Gabon and Cigar Lake uranium ore deposits. *In*: GIERÉ, R. & STILLE, P. (eds). *Energy, Waste, and the Environment: a Geochemical Perspective*. Geological Society, London, Special Publications, **236**, 123–134.

GRANDSTAFF, D. E. 1976. A kinetic study of the dissolution of uraninite. *Economic Geology*, **71**, 1493–1506.

GRAY, W. J., LEIDER, H. R. & STEWARD, S. A. 1992. Parametric study of LWR spent fuel dissolution kinetics. *Journal of Nuclear Materials*, **190**, 46–52.

GUENTHER, R. J., BLAHNIK, D. E., CAMPBELL, T. K., JENQUIN, U. P., MENDEL, J. E., THOMAS, L. E. & THORNHILL, C. K. 1988. *Characterization of Spent Fuel Approved Testing Material ATM-106*. Pacific Northwest Laboratory Report, Richland, WA 99352, PNL-5109–106.

HANSON, B. D. 1998. *The Burnup Dependence of Light Water Reactor Spent Fuel Oxidation*. Pacific Northwest National Laboratory Report, PNNL-11929.

HASCHKE, J. M., Allen T. H. & MORALES, L. A. 2000. Reaction of plutonium dioxide with water: Formation and properties of PuO_{2+x}. *Science*, **237**, 285–287.

HIDAKA, H. & HOLLIGER, P. 1998. Geochemical and neutronic characteristics of the natural fossil fission reactors at Oklo and Bamgombe, Gabon. *Geochimica et Cosmochimica Acta*, **62**, 89–108.

JANECZEK, J. 1999. Systematics and paragenesis of uranium minerals. *In*: BURNS, P. C. & FINCH, R. (eds) Uranium: Mineralogy, Geochemistry and the Environment. *Mineralogical Society of America, Reviews in Mineralogy*, **38**, 321–392.

JENSEN, K. A., PALENIK, C. S. & EWING, R. C. 2002. U^{6+} phases in the weathering zone of the Bangombé U-deposit: observed and predicted mineralogy. *Radiochimica Acta*, **90**, 761–770.

JOHNSON, L. H. & SHOESMITH, D. W. 1988. Spent fuel, Chapter 11. *In*: LUTZE, W. & EWING, R. C. (eds) *Radioactive Waste Forms for the Future*. Elsevier, New York, NY, 635–698.

KESSINGER, G. F. & THOMPSON, M. C. 2002. *Dissolution of Dresden reactor fuel*. Westinghouse Savannah River Company Report, WSRC-TR-2002-00448.

KIM, C. W., WRONKIEWICZ, D. J., FINCH, R. J. & BUCK, E. C. 2002. Incorporation of cerium and neodymium in a uranyl hydroxide solid. *Materials Research Society Symposium Proceedings*, **713**, 663–670.

KING, F., KOLAR, M. & SHOESMITH, D. W. 1999. Modeling the oxidative dissolution of UO_2. *Materials Research Society Symposium Proceedings*, **556**, 463–470.

KLEYKAMP, H. 1985. The chemical state of fission products in oxide fuels. *Journal of Nuclear Materials*, **131**, 221–246.

KLEYKAMP, H. 1990. Post irradiation examinations and composition of the residues from nitric acid dissolution experiments of high burn-up LWR fuel. *Journal of Nuclear Materials*, **171**, 181–188.

MATZKE, H. & SPINO, J. 1997. Formation of the rim structure in high burnup fuel. *Journal of Nuclear Materials*, **248**, 170–179.

MCNAMARA, B. K., BUCK, E. C. & HANSON, B. D. 2003. Observation of studtite and meta-studtite on spent fuel. *Materials Research Society Symposium Proceedings*, **757**, 401–408.

MURAKAMI, T., OHNUKI, T., ISOBE, H. & SATO, T. 1997. Mobility of uranium during weathering. *American Mineralogist*, **82**, 888–899.

OVERSBY, V. M. 1994. Nuclear waste materials. *In*: FROST B. R. T. (ed.) *Material science and technology, a comprehensive treatment, Nuclear Materials*, VCH, Weinheim, Germany, **10B**, 392–442.

OVERSBY, V. M. 1999. *Uranium Dioxide, SIMFUEL, and Spent Fuel Dissolution Rates – A Review of Published Data*. Swedish Nuclear Fuel and Waste Management Co., Stockholm, Sweden, TR-99-22.

PEARCY, E. C., PROKRYL, J. D., MURPHY, W. M. & LESLIE, B. W. 1994. Alteration of uraninite from the Nopal I deposit, Peña Blanca District, Chihuahua, Mexico, compared to degradation of SNF in the proposed U.S. high level nuclear waste repository at Yucca Mountain, Nevada. *Applied Geochemistry*, **9**, 713–732.

PÉREZ DEL VILLAR, L., BRUNO, J. *et al.* 2002. The uranium ore from Mina Fe (Salamanca, Spain) as a natural analogue of processes in a spent fuel repository. *Chemical Geology*, **190**, 395–415.

POINSSOT, C., LOVERA, P. & FAURE, M.-H. 2002. Assessment of the evolution with time of the instant release fraction of spent nuclear fuel in geological disposal conditions. *Materials Research Society Symposium Proceedings*, **713**, 615–623.

RUNDE, W., CONRADSON, S. D. EFURD, D. W., LU, N., VANPELT, C. E. & TAIT, C. D. 2002. Solubility and sorption of redox sensitive radionuclides (Np,Pu) in J-13 water from the Yucca Mountain site: comparison between experiment and theory. *Applied Geochemistry*, **17**, 837–853.

SALBU, B., JANSSENS, K., LIND, O. C., PROOST, K. & DANESI, P. R. 2003. Oxidation states of uranium in DU particles from Kosovo. *Journal of Environmental Radioactivity*, **64**, 167–173.

SATONNAY, G., ARDOIS, C. CORBEL, C. LUCCHINI, C. F., BARTHE, M. F., GARRIDO, F. & GOSSET, D. 2001. Alpha-radiolysis effects on UO_2 alteration in water. *Journal of Nuclear Materials*, **288**, 11–19.

SAVARY, V. & PAGEL, M. 1997. The effects of water radiolysis on local redox conditions in the Oklo, Gabon, natural fission reactors 10 and 16. *Geochimica et Cosmochimica Acta*, **61**, 4479–4494.

SHOESMITH, D. W. 2000. Fuel corrosion processes under waste disposal conditions. *Journal of Nuclear Materials*, **282**, 1–31.

SPAHIU, K., EKLUND, U.-B., CUI, D. & LUNDSTRÖM, M. 2002. The influence of near field redox conditions on spent fuel leaching. *Materials Research Society Symposium Proceedings*, **713**, 633–638.

SPEER, J. A. 1982. Actinide orthosilicates. *In*: RIBBE, P. H. (ed) Orthosilicates. *Reviews in Mineralogy*, **5**, 113–135.

SPINO, J., & PAPAIOANNOU, D. 2000. Lattice parameter changes associated with the rim-structure formation in high burnup UO_2 fuels by micro X-ray diffraction. *Journal of Nuclear Materials*, **281**, 146–162.

STEFANIC, I. & LAVERNE, J. A. 2002. Temperature dependence of the hydrogen peroxide production in the γ-radiolysis of water. *Physical Chemistry*, **A106**, 447–452.

STILLE, P., GAUTHIER-LAFAYE, F. *et al.* 2003. REE mobility in groundwater proximate to the natural fission reactor at Bangombé (Gabon). *Chemical Geology*, **198**, 289–304.

STROES-GASCOYNE, S., JOHNSON, L. H., TAIT, J. C., MCCONNELL, J. L. & PORTH, R. J. 1997. Leaching of used Candu fuel: Results from a 19-year leach test under oxidizing conditions. *Materials Research Society Symposium Proceedings*, **465**, 511–518.

SUNDER, S., SHOESMITH, D. W. & MILLER, N. H. 1997. Oxidation and dissolution of nuclear fuel (UO$_2$) by the products of the alpha radiolysis of water. *Journal of Nuclear Materials*, **244**, 66–74.

THOMAS, L. E., BEYER, C. E. & CHARLOT, L. A. 1992. Microstructural analysis of LWR spent fuels at high burnup. *Journal of Nuclear Materials*, **188**, 80–89.

UNE, K., HIRAI, M., NOGITA, K., HOSOKAWA, T., SUZAWA, Y., SHIMIZU, S. & ETOH, Y. 2000. Rim structure formation and high burnup fuel behavior of large-grained UO$_2$ fuels. *Journal of Nuclear Materials*, **278**, 54–63.

WALENTA, K. 1974. On studtite and its composition. *American Mineralogist* **59**, 166–171.

WALKER, C. T., KAMEYAMA, T., KITAJIMA, S. & KINOSHITA, M. 1992. Concerning the microstructure changes that occur at the surface of UO$_2$ pellets on irradiation to high burnup. *Journal of Nuclear Materials*, **188**, 73–79.

WASYWICH, K. M., HOCKING, W. H., SHOESMITH, D. W. & TAYLOR, P. 1992. Differences in oxidation behavior of used CANDU fuel during prolonged storage in moisture-saturated air and dry air at 150 °C. *Nuclear Technology*, **104**, 309–329.

WOOD, S. A., VAN MIDDLESWORTH, P., GIBSON, P. & RICKETTS, A. 1997. The mobility of the REE, U, and Th in geological environments in Idaho and their relevance to radioactive waste disposal. *Journal of Alloys Compounds*, **249**, 136–141.

WRONKIEWICZ, D. J., BATES, J. K., GERDING, T. J., VELECKIS, E. & TANI, B. S. 1992. Uranium release and secondary phase formation during unsaturated testing of UO$_2$ at 90 °C. *Journal of Nuclear Materials*, **190**, 107–127.

WRONKIEWICZ, D. J. BATES, J. K., WOLF, S. F. & BUCK, E. C. 1996. Ten-year results from unsaturated drip tests with UO$_2$ at 90 °C: implications for the corrosion of spent nuclear fuel. *Journal of Nuclear Materials*, **238**, 78–95.

WRONKIEWICZ, D. J. & BUCK, E. C. 1999. Uranium mineralogy and the disposal of spent nuclear fuel. *In*: BURNS, P. C. & FINCH, R. (eds) Uranium: Mineralogy, Geochemistry and the Environment. *Mineralogical Society of America, Reviews in Mineralogy*, **38**, 475–497.

YAGNIK, S. K., MACHIELS, A. J. & YANG, R. L. 1999. Characterization of UO$_2$ irradiated in the BR-3 reactor. *Journal of Nuclear Materials*, **270**, 65–73.

ZHAO, D. & EWING, R. C. 2000. Alteration products of uraninite from the Colorado Plateau. *Radiochimica Acta*, **88**, 739–749.

Geochemical behaviour of host phases for actinides and fission products in crystalline ceramic nuclear waste forms

GREGORY R. LUMPKIN[1,2], KATHERINE L. SMITH[2],
RETO GIERÉ[3] & C. TERRY WILLIAMS[4]

[1]Cambridge Centre for Ceramic Immobilisation, Department of Earth Sciences,
University of Cambridge, Downing Street, Cambridge CB2 3EQ, UK
(e-mail: gregl@esc.cam.ac.uk)
[2]Australian Nuclear Science and Technology Organisation, PMB 1,
Menai, NSW 2234, Australia
[3]Department of Earth and Atmospheric Sciences, Purdue University,
West Lafayette, IN 47907-1397, USA and Institut für Mineralogie, Petrologie und Geochemie,
Universität Freiburg, Albertstrasse 23B, D-79104 Freiburg, Germany
[4]Department of Mineralogy, The Natural History Museum, Cromwell Road,
London SW7 5BD, UK

Abstract: A number of polyphase or single-phase ceramic waste forms have been considered as options for the disposal of nuclear waste in geological repositories. Of critical concern in the scientific evaluation of these materials is their performance in natural systems over long periods of time (e.g., 10^3 to 10^6 years). This paper gives an overview of the aqueous durability of the major titanate host phases for actinides (e.g., Th, U, Np, Pu, Cm) and important fission products (e.g., Sr and Cs) in alternative crystalline ceramic waste forms. These host phases are compared with reference to some basic acceptance criteria, including the long-term behaviour determined from studies of natural samples. The available data indicate that zirconolite and pyrochlore are excellent candidate host phases for actinides. These structures exhibit excellent aqueous durability, crystal chemical flexibility, high waste loadings, and well-known processing conditions. Although both pyrochlore and zirconolite become amorphous due to alpha-decay processes, the total volume swelling is only 5–6% and there is no significant effect of radiation damage on aqueous durability. Hollandite also appears to be an excellent candidate host phase for radioactive Cs isotopes. Brannerite and perovskite, on the other hand, are more prone to alteration in aqueous fluids and have a lower degree of chemical flexibility. With the exception of hollandite, many of the properties of these potential host phases have been confirmed through studies of natural samples.

The immobilization and long-term disposal of nuclear wastes is a world-wide issue and one of the greatest challenges facing modern society today. In the USA, for example, large amounts of high-level wastes (HLW) have been generated from the operation of commercial nuclear power stations and the nuclear weapons programme. The main sources of HLW are spent fuel from commercial nuclear power stations, liquid wastes from the reprocessing program of the 1960s, wastes generated from the production of nuclear weapons, and weapons-grade plutonium resulting from nuclear disarmament treaties between the USA and Russia. In 1994, approximately 400 000 cubic meters of HLW was stored in more than 200 underground tanks at the major United States Department of Energy (USDOE) sites (Ewing et al. 1995; Weber et al. 1998).

The current policy of the USA allows for the direct disposal of the spent fuel from commercial power generation, whereas most of the HLW from the defence programme is intended for immobilization using borosilicate glass. Borosilicate glass is also the currently accepted material for many countries that continue to reprocess their commercial spent fuel (e.g., England, France, Japan, and Russia). Nevertheless, a significant fraction of the existing 'legacy' wastes are characteristically very complex in physical

From: GIERÉ, R. & STILLE, P. (eds) 2004. Energy, Waste, and the Environment: a Geochemical Perspective. Geological Society, London, Special Publications, **236**, 89–111.
0305-8719/04/$15 © The Geological Society of London 2004.

form and chemical composition. These complex waste materials, together with scrap plutonium, and the fission products and actinides generated from the various partitioning strategies, may be better suited for existing or new types of high-performance crystalline waste forms or glass–ceramics (Donald *et al.* 1997; Trocellier 2000, 2001; Fillet *et al.* 2002, 2004).

In the existing strategy for HLW disposal using spent fuel and borosilicate glass, these HLW materials are part of a multibarrier concept relying heavily on the container, backfilling materials, and the repository itself for prevention of radionuclide migration into the environment. There are a number of alternative crystalline waste forms that may be capable of providing a much higher level of chemical durability as the best defence against radionuclide migration away from the repository. Some of these waste forms are illustrated in Table 1. Materials such as tailored ceramics (Harker 1988), the Synroc titanate waste forms (Fielding & White 1987; Ringwood *et al.* 1988), and their special purpose derivatives (e.g., pyrochlore, zirconolite, and hollandite) are reasonably well developed and have been the subject of extensive dissolution testing and radiation damage studies. Pyrochlore is the major component of Synroc-F, a polyphase ceramic designed for partially reprocessed nuclear fuel (Ball *et al.* 1989) and later

appeared as the principle host phase for excess weapons Pu and U in a crystalline titanate ceramic form. Zirconolite has also been proposed as an ideal host phase for actinides due to a combination of crystal chemical flexibility and very high durability in aqueous fluids (Vance 1994). In view of the current 'partitioning' strategies for actinides and fission products, it has become apparent that hollandite may provide an excellent host material for separated radioactive Cs for similar reasons of crystal chemistry and durability (e.g., Kesson 1983; Cheary 1988).

As noted by Stewart *et al.* (2003), any existing or new high-performance waste form must meet several requirements in order to reach final consideration for use in a repository. These requirements are (1) a high level of durability in aqueous fluids, (2) crystal chemical flexibility allowing the material to incorporate impurity elements on crystallographic sites of the chosen host phases, (3) acceptable waste loadings, and (4) a reliable and cost-effective means of processing. In view of recent trends, one might be tempted to add radiation resistance as a fifth criterion. However, we prefer a more general statement: (5) the material must possess acceptable physical and mechanical properties including thermal behaviour, hardness, radiation damage induced swelling, cracking, and so on (Donald *et al.* 1997; Trocellier 2000; Stefanovsky *et al.*, 2004,

Table 1. *A selection of crystalline ceramic waste forms, applications, and mineralogy*

Waste form	Application	Mineralogy
Tailored ceramics	High Al SRP waste, 86 wt% loading	Magnetoplumbite, spinel, uraninite $\pm$ corundum, perovskite, pseudobrookite
	High Al RHO Purex SRP waste, 60 wt% loading	Magnetoplumbite, zirconolite, spinel, pseudobrookite, perovskite, rutile, loveringite
	SRP Composite HLW, 60 wt% loading	Magnetoplumbite, spinel, nepheline, uraninite, corundum
	SRP Composite HLW, 60 wt% loading	Nepheline, spinel, zirconolite, perovskite, murataite, magnetoplumbite
	Barnwell HLW, up to 57 wt% loading	Pyrochlore, perovskite, monazite, fluorite, intermetallic phases
Synroc-C	Purex type HLW from reprocessed spent fuel, 10–20 wt%	Ba–Cs–hollandite, perovskite, rutile, zirconolite, Al oxides, intermetallic phases
Synroc-C	Japanese type HLW from reprocessed spent fuel, 10–20 wt%	Ba–Cs–hollandite, perovskite, rutile, zirconolite, freudenbergite, loveringite, Al oxides, intermetallic phases
Synroc-D	US defence waste, 60–70 wt%	Zirconolite, perovskite, spinel, nepheline
Synroc-E	HLW from reprocessed spent fuel, 5–7 wt%	Rutile, Cs–hollandite, perovskite, zirconolite
Synroc-F	Conversion of spent fuel, 50 wt%	Pyrochlore, perovskite, uraninite $\pm$ hollandite
Pyrochlore	Pu, U, other ACTs, up to 35 wt%	Pyrochlore $\pm$ brannerite, baddeleyite, rutile
Zirconolite	Pu, U, other ACTs, up to 25 wt%	Zirconolite $\pm$ baddeleyite, rutile
Hollandite	Separated radioactive Cs, 5–10 wt% easily achieved	Ba–Cs–hollandite $\pm$ rutile

for details). Ultimately, it is the overall balance sheet that is important and there may be cases where a material that becomes amorphous due to radiation damage may excel in other areas and vice versa. Although studies of natural systems provide the only means of confirming the long-term behaviour of a given waste form phase, this concept is seldom considered at a level on par with the criteria listed above (see Crovisier *et al.*, 2004). In fact, such studies have closely paralleled the experimental research and development programmes over the past 30 years and extensive data sets now exist for many of the candidate waste form phases (Lumpkin 2001). The purpose of this paper is to review the available data for the major crystalline titanate ceramics and their mineral analogues.

Geochemical alteration

Pyrochlore

The structure of pyrochlore is considered to be an anion-deficient derivative of the fluorite structure type with a doubled *a* cell parameter and change in space group from *Fm3m* to *Fd3m* (Subramanian *et al.* 1983; Chakoumakos 1984; Smith & Lumpkin 1993). Minerals of the pyrochlore group conform to the general formula $A_{2-m}B_2X_{6-w}Y_{1-n} \cdot pH_2O$, where A represents cations in eightfold coordination, B represents cations in sixfold coordination, and X and Y are anion sites. The basic structural element of pyrochlore is the framework of corner-sharing octahedra. Within this framework, continuous tunnels are arranged parallel to the <110> directions. Both the A-site cations and Y-site anions are located in these tunnels. In synthetic systems, some A-site cation exchange capacity has been demonstrated in defect pyrochlores, in which the values of *m* in the general formula can be quite large.

Natural pyrochlore typically crystallizes under magmatic conditions in granitic pegmatites, nepheline syenite pegmatites, and carbonatites (in both calcite- and dolomite-rich varieties). The composition of common pyrochlore usually approaches the stoichiometric form (Na, Ca, REE, U)$_2$(Nb, Ta, Ti)$_2$O$_6$(F, OH, O), but the structure type is extremely flexible in terms of the sheer number of elements that can be incorporated, and is particularly amenable to the incorporation of actinides. Natural samples are known to contain up to 30 wt% UO$_2$, 9 wt% ThO$_2$, and 16 wt% REE$_2$O$_3$, an important consideration for the issue of nuclear criticality. However, as shown by the general formula, the crystal chemistry of pyrochlore is complicated

by the potential for vacancies at the A-, X-, and Y-sites (*m* = 0.0–1.7, w = 0.0–0.7, and n = 0.0–1.0) and the incorporation of water molecules (*p* = 0–2) in the vacant tunnel sites. The total water content of the natural defect pyrochlores may be as high as 10–15 wt% H$_2$O (with speciation as both water molecules and OH groups).

Numerous investigations have demonstrated that the pyrochlore structure type is susceptible to alteration via reaction with aqueous fluids over a range of P–T–X conditions. These studies have been recently reviewed in detail by Lumpkin (2001) and will only be summarized briefly here. A number of authors have studied the behaviour of Nb- and Ta-rich pyrochlore group minerals in natural systems (Lumpkin *et al.* 1986; Wise & Černý 1990; Ohnenstetter & Piantone 1992; Lumpkin & Ewing 1992, 1995, 1996; Lumpkin & Mariano 1996; Wall *et al.* 1996; Williams *et al.* 1997; Chakhmouradian & Mitchell 1998; Lumpkin 1998; Nasraoui *et al.* 1999). These authors found that many samples of pyrochlore exhibit geochemical alteration effects under a range of P–T–X conditions from weathering environments to late-stage magmatic and hydrothermal conditions associated with igneous intrusions. The alteration patterns of pyrochlore group minerals generally follow the coupled substitution schemes listed below (where □ represents a vacancy):

$$^A\square^Y\square \longrightarrow {}^ACa^YO$$
$$^ANa^YF \longrightarrow {}^ACa^YO$$
$$^ANa^YOH \longrightarrow {}^ACa^YO$$
$$^ANa^YF \longrightarrow {}^A\square^Y\square$$
$$^ACa^YO \longrightarrow {}^A\square^Y\square$$
$$^ACa^XO \longrightarrow {}^A\square^X\square$$

From top to bottom, these exchange mechanisms generally reflect the changes that occur from high- to low-temperature environments, although the details vary with rock type and specific P–T–X conditions. At higher temperatures (~300–650 °C, <400 MPa) in highly evolved late-stage magmatic fluids, Ca enrichment is commonly observed (via the filling of vacancies and replacement of Na–F by Ca–O pairs). The main effect of alteration at moderate temperatures under hydrothermal conditions (~200–350 °C, <200 MPa) is the loss of Na and F, typically combined with some cation exchange for Sr, Ba, REE, and Fe to produce a hydrated pyrochlore near AB$_2$O$_6 \cdot$H$_2$O in stoichiometry. Additional removal of Na, F, Ca, and O may occur in low-temperature hydrothermal or weathering environments. These defect pyrochlores tend to exhibit maximum numbers of A-site, Y-site, and X-site vacancies, maximum hydration levels, and may take up large cations such as K, Sr, Cs, Ba, Ce, and Pb in certain environments. An example of this type of alteration is shown in Figures 1 and 2. Note that the alteration effects illustrated here are largely confined to the Na–Ca–F pyrochlores, (Na,Ca,REE,U)$_2$(Nb,Ta,Ti)$_2$O$_6$(F,OH,O), in which the REE and U contents usually remain unchanged (e.g., Lumpkin & Ewing 1992, 1996). Under extreme lateritic weathering, fragmentation and subsequent dissolution of pyrochlore can occur,

with precipitation of Nb and other B-site cations in submicron amorphous material consisting predominantly of Fe, Al, and P, with variable Si, Ca, and Ti. The available results suggest that Th is retained within micron-size secondary phases, whereas U is relatively mobile and may be released from U-rich zones in altered pyrochlore crystals (Wall *et al.* 1999).

A variety of alteration effects have been observed in completely metamict Ti- and U-rich pyrochlores. In the samples from Adamello, Italy, the pyrochlore occurs as overgrowths on zoned zirconolite grains and contains 29–34 wt% UO$_2$. Electron microscopy and microanalytical work has revealed that these pyrochlore samples have

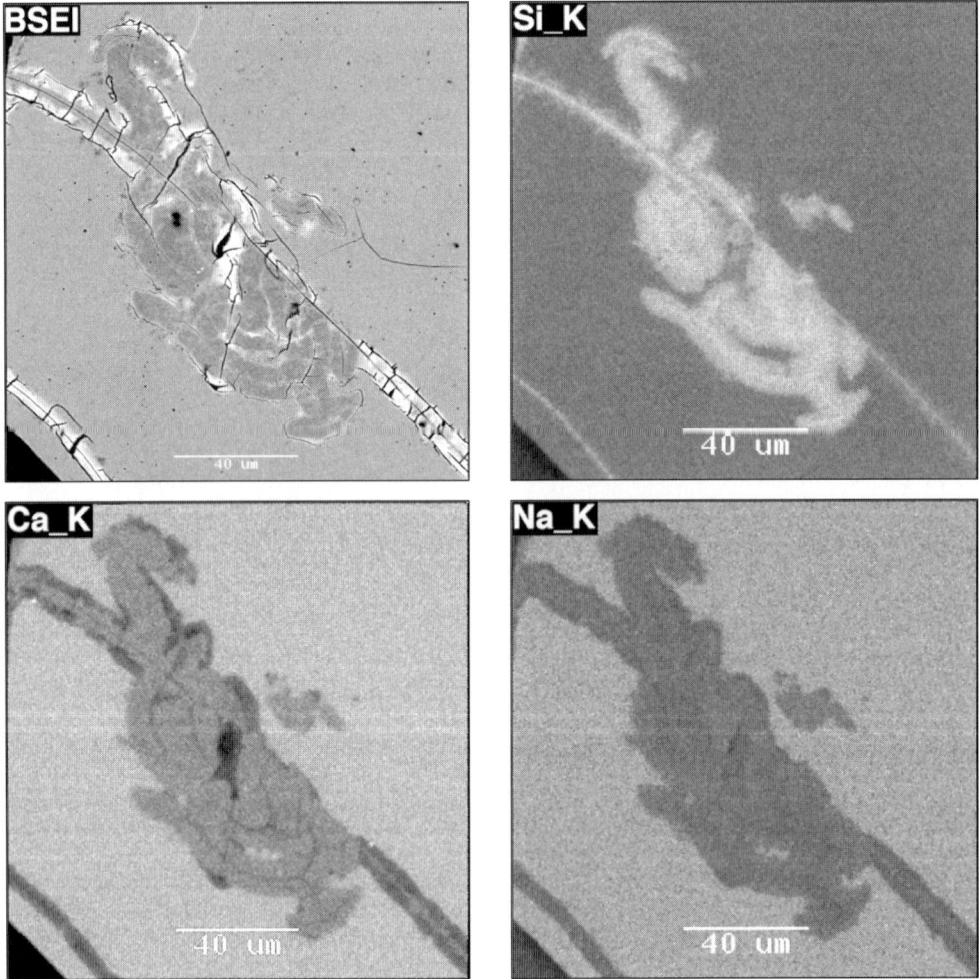

Fig. 1. SEM backscattered electron image, Si X-ray map, Ca X-ray map, and Na X-ray map of alteration in pyrochlore from Vishnevogorskii, Russia. Note the loss of Na and Ca and incorporation of Si along cracks and the dark area near the middle of the backscattered image.

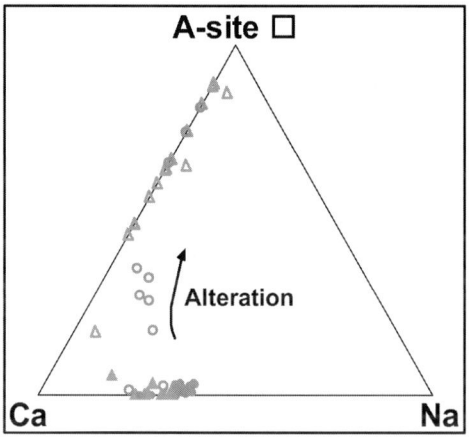

Fig. 2. Triangular plot of Ca, Na, and A-site vacancies in pyrochlore from Vishnevogorskii calculated from structural formulae normalized to two B-site cations. Compositions of primary pyrochlore plot near the Ca–Na join (solid symbols). Compositions of the altered areas move toward the Ca-vacancy join, then along this join toward the top of the triangle.

only suffered a minor late-stage hydration event as evidenced by lower backscattered electron image contrast around the rims of the grains. Results of this study demonstrate retention of U and Th for time periods of 40 Ma, even though the cumulative total alpha-decay dose is on the order of 3 to $4 \times 10^{16} \alpha \, \text{mg}^{-1}$ (Lumpkin *et al.* 1999). In two samples from Bancroft, Ontario, Canada, Lumpkin & Ewing (1996) had previously concluded that the major result of alteration was hydration, with only minor changes in elemental composition, apart from the precipitation of galena due to mobility of radiogenic Pb. In contrast to these examples, the Ti- and U-rich pyrochlores from Madagascar exhibit a range of alteration effects beginning with a relatively low temperature style of alteration that initially proceeds by removal of Na, Ca, and F as shown in the list above. The alteration is usually accompanied by hydration together with minor increases in Al, K, Mn, Fe, Sr, and Ba. At this stage, U and Th remain relatively unaffected by the alteration process (stage 1). If the Ca content drops below about 0.2–0.3 atoms per formula unit, these Ti-rich pyrochlores exhibit various levels of recrystallization to liandratite + uranpyrochlore + rutile (or anatase) and partial dehydration (stage 2). This may be accompanied by loss of up to 20–30% of the original amount of U and local redistribution of the radiogenic Pb.

The effect of radiation damage on the alteration of 440 Ma old pyrochlore from Mozambique

has been studied by Gieré *et al.* (2000, 2001). These pyrochlore crystals exhibit distinctive chemical zoning, characterized by a U-free core and a U-rich rim (up to 17 wt% UO_2). Following uplift and cooling, groundwater penetrated these fractured crystals and led to the deposition of clay minerals along both fractures and cleavage planes. This low-temperature process also led to chemical alteration of the pyrochlore, but only within the zone of the U-rich rim. During this alteration, which is the result from exposure to tropical conditions, Na, Ca, and F were removed from the pyrochlore, leading to increased A-site vacancies (up to about 1.8 A-site vacancies per formula unit). The alteration also led to localized redistribution of radiogenic Pb and to hydration, but U remained immobile. The low-temperature alteration effects are only observed in the U-rich rim, which is largely amorphous and characterized by abundant microfractures, suggesting that the susceptibility of pyrochlore to low-temperature alteration is enhanced by the microstructural damage caused by the α-decay of U (also see Ohnenstetter & Piantone 1992).

Zirconolite

The structure of zirconolite (Smith & Lumpkin 1993) is also considered to be an anion-deficient derivative of the fluorite structure type. The structure can be viewed as a volumetrically condensed, layered version of pyrochlore with reduced symmetry and several polytypic forms (monoclinic 2M or 4M, both with space group $C2/c$; orthorhombic 3O with space group *Acam*, hexagonal 3T with space group $P3_121$). The chemical composition of zirconolite 2M corresponds to $CaZrTi_2O_7$, but in nature it typically deviates from this end-member composition due to extensive substitution of REE, Th, and U for Ca and Nb, Fe, and Mg for Ti (Gieré *et al.* 1998; Bellatreccia *et al.* 1999). Zirconolite may incorporate up to 24 wt% UO_2, 22 wt% ThO_2, and 32 wt% REE_2O_3 in natural systems. In natural samples Zr is subject to only limited substitution by other elements (e.g., minor amounts of Y, REE, U, and Ti). Experimental work has shown that extensive substitution of REE, Th, and U generally results in a polytypic phase transformation from monoclinic 2M to trigonal 3T or from monoclinic 2M to monoclinic 4M. It is often difficult to determine the polytype of natural zirconolite due to extensive radiation damage; however, the polytypes 2M, 3O, and 3T are known from a few samples with low alpha-decay dose (Sinclair & Eggleton 1982; Mazzi & Munno 1983; Bellatreccia *et al.* 2002).

Isotopic age dating work by Oversby & Ring-wood (1981) and electron microscopy studies by Ewing *et al.* (1982) demonstrate that zirconolite remained a closed system with respect to U, Th, and Pb for up to 650 Ma with little, if any, evidence for geochemical alteration. Lumpkin *et al.* (1994), Gieré *et al.* (1994), and Hart *et al.* (1996) have described the alteration of metamict zirconolite from the 2060 Ma carbonatite complex of Phalaborwa, South Africa, in some-what greater detail. Electron microprobe ana-lyses, element mapping, and backscattered electron images demonstrate that the alteration is localized along cracks and resulted in the incorporation of Si and loss of Ti, Ca, and Fe. However, in these samples the REE, Th, and U contents remained relatively constant across the alteration zones. Radiogenic Pb appears to have been mobile and precipitated mainly within the altered areas as galena (PbS), similar to obser-vations reported by Lumpkin & Ewing (1996) for metamict U- and Ti-rich pyrochlores.

At higher temperature and pressure in mag-matic, hydrothermal, or metamorphic systems, zirconolite may be altered by dissolution–repre-cipitation or replacement mechanisms. Gieré & Williams (1992) have described zoned zircono-lites from Adamello, Italy, which exhibit cor-rosion and replacement by a new generation of zirconolite together with loss of Th and U to a hydrothermal fluid. Importantly, a thermody-namic analysis of the mineral assemblages was performed in this work and indicated that the zir-conolite crystallization and corrosion occurred at 500–600 °C in a reducing hydrothermal fluid rich in H_2S, HCl, HF, and P and relatively poor in CO_2. Pan (1997) has also described the break-down of zirconolite to a new mineral assemblage consisting of zircon, sphene, and rutile in meta-morphosed (∼600 °C) ferromagnesian silicate rocks at Manitouwadge, Canada. This reaction can be expressed as follows (modified slightly from Pan 1997):

$$CaZrTi_2O_7 + 2SiO_2 = ZrSiO_4 + CaTiSiO_5 + TiO_2$$

Bulakh *et al.* (1998) and Williams *et al.* (2001) have reported on the alteration of zirconolite from carbonatites. In these examples, the zirco-nolite has been replaced along cracks and within micron-sized domains by an unidentified Ba–Ti–Zr–Nb–ACT silicate phase, suggesting that zirconolite may not be stable in the presence of hydrothermal fluids rich in Si. Figure 3 pre-sents an example of this type of alteration, where one may see additional evidence for enhanced alteration along zones enriched in Th and U. In

contrast to pyrochlore, however, where complete breakdown of this mineral can occur in extreme tropical lateritic weathering environments (see above), zirconolite appears relatively resistant to dissolution, with only minor dissolution fea-tures occurring at the margins (Williams *et al.* 2001).

Brannerite

Brannerite, ideally UTi_2O_6, has a crystal struc-ture based on a distorted array of hexagonal close packed oxygens. The structure is monocli-nic, space group $C2/m$, and consists of layers of Ti octahedra connected by columns of U octahe-dra (Szymański & Scott 1982). Natural and syn-thetic brannerites can incorporate substantial amounts of Ca, REE, Th, and other elements. In both cases, the incorporation of lower valence elements on the A-site may be charge balanced by oxidation of some U^{4+} to U^{5+} and/or U^{6+} ions (Vance *et al.* 2000). Lumpkin *et al.* (2000) have examined brannerite samples from different localities by SEM-EDX, showing that the unaltered areas of the samples contain up to 7 wt% CaO, 8 wt% REE_2O_3, 7 wt% PbO, and 15 wt% ThO_2. Low but variable amounts of Al_2O_3, SiO_2, MnO, FeO, NiO, and Nb_2O_5 were also found in these samples.

Electron microscopy studies show that most natural brannerites with ages greater than about 10 Ma are fully amorphous due to alpha-decay damage and are commonly altered by natural aque-ous fluids (Lumpkin 2001). Alteration usually occurs around the rim of the brannerite or along cracks (Fig. 4a), and may involve the formation secondary such as anatase, galena, and thorite. As reported for pyrochlore and zirconolite, the galena may precipitate due to the combined effects of radiogenic Pb migration and the pre-sence of S species in the aqueous fluid. Altered brannerite typically loses U and the concentration may decrease to approximately 1 wt% UO_2 in the most heavily altered areas (Fig. 5). The loss of U may be compensated in part by the incorporation of up to 18 wt% SiO_2 and 16 wt% FeO, together with significant amounts of P_2O_5, As_2O_5, and Al_2O_3 in certain examples. In some cases, the associated rock or mineral matrix surrounding the brannerite may be highly fractured, providing pathways for the migration of U, evidenced by the precipitation of secondary U minerals (Lumpkin *et al.* 2000). An example of this process is illus-trated in Fig. 4b.

Hollandite

The crystal structure of hollandite, $A_{1.1-1.7}B_8O_{16}$, is similar to that of rutile and consists of edge-

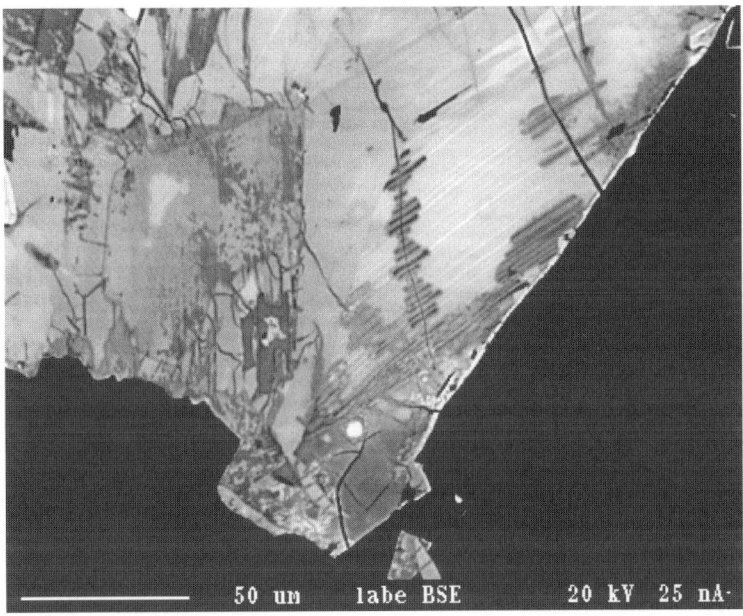

Fig. 3. SEM backscattered electron image of alteration in zirconolite from the Afrikanda alkaline complex, Kola Peninsula, Russia. This crystal exhibits complex magmatic zoning, late-stage replacement by an unknown Ba–Zr–Ti–silicate phase, and preferential alteration along cracks and Th–U-rich zones.

sharing chains of octahedra connected via corner sharing to form a three-dimensional framework. Hollandite, however, has two octahedral chains connected by edge-sharing instead of the single chain found in rutile, resulting in a rather large 2×2 tunnel capable of accommodating large A-site cations like K, Rb, Cs, and Ba (Kesson & White 1986a, b; Cheary 1987). These cations exhibit various ordering sequences over the available tunnel sites, commonly resulting in superlattice peaks in X-ray or electron diffraction patterns. In general, the superlattice is incommensurate with the sublattice as defined by the octahedral cation repeat distance, but may be commensurate for A-site occupancy values of 1.200, 1.333, and 1.500 atoms per formula unit. The space group is typically $I4/m$ or $C2/m$ depending upon the A/B cation radius ratio. Numerous synthetic samples have been produced with Ti, Mn, Mo, noble metals, or Sn as the most common major elements and with Mg, Al, V, Fe, Co, Ni, Zn, and Sb, among others, as minor elements on the B-site. The composition of hollandite in titanate-based waste forms is generally given as $(Ba_x Cs_y)[(Ti,Al)^{3+}_{2x+y} Ti^{4+}_{8-2x-y}]O_{16}$ in which charge compensation for Ba and Cs is usually provided by Al or Ti^{3+} (Myhra et $al.$ 1988a; Carter et $al.$ 2002; Bart et $al.$ 2003).

In natural samples (cryptomelane group), typical B-site cations are Ti, V, Cr, Fe, Mn^{2+}, and especially tetravalent Mn^{4+}. The closest natural analogue for synthetic titanium hollandite is the mineral priderite, which occurs in the leucite lamproites of Western Australia and elsewhere and has a composition of approximately $(Ba,K)_{1.2-1.6}[Ti,Fe,Mg]_8O_{16}$ (Sinclair & McLaughlin 1982; Post et $al.$ 1982). Unfortunately, priderite is a relatively rare mineral and there have been no detailed studies of the alteration of this phase in natural systems.

Perovskite

Perovskite is an ABX$_3$ structure type based around a framework of corner-sharing, octahedral B-site cations. The large A-site cations occupy the center of a large cavity formed by eight B-site octahedra and are coordinated to 12 oxygens in the ideal cubic structure. Most perovskites are distorted via octahedral tilting and generally have lower symmetry. In nature, only near end-member SrTiO$_3$ is cubic; most other compositions are orthorhombic (Mitchell 1996). The major A-site cations in natural perovskites are Na, Ca, Sr, and REE, with minor amounts of K, Ba, and U. The concentration of Th is usually low, but may reach levels as high as 18 wt% ThO$_2$ in certain alkaline rocks (Mitchell & Chakhmouradian 1998a). The B-site of natural perovskite is occupied predominantly by Ti, Fe, and Nb, together

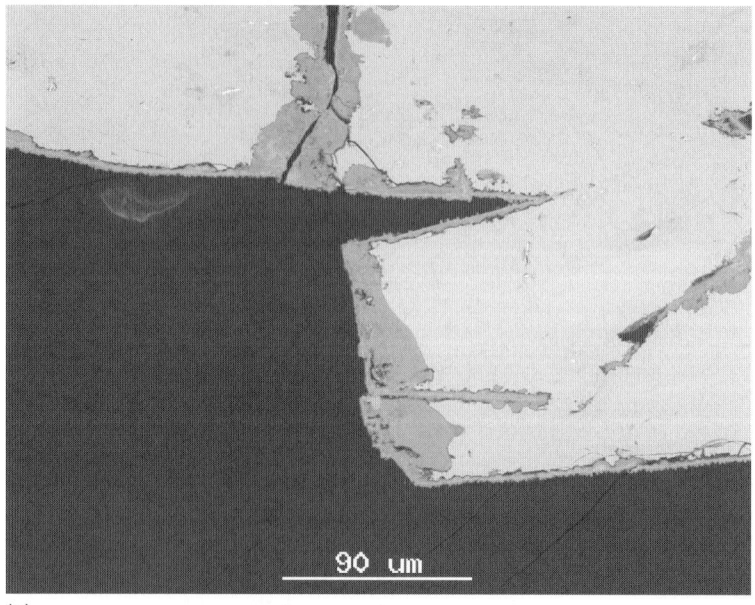

(a)

(b)

Fig. 4. SEM backscattered electron images of alteration in natural brannerite: (**a**) Part of a large brannerite specimen showing minor alteration around the rim of the crystal and along fractures extending into the interior; (**b**) Brannerite crystals showing extensive alteration along their rims, together with the presence of U-rich phases along cracks in the host rock.

with minor amounts of Mg, Al, Zr, and Ta, depending on bulk rock chemistry and mineralogy.

In low-temperature environments, perovskite commonly breaks down to one or more polymorphs of TiO_2, in the process releasing Ca and possibly other elements into the aqueous fluid. This is well illustrated by the alteration of perovskite to anatase, cerianite, monazite, and

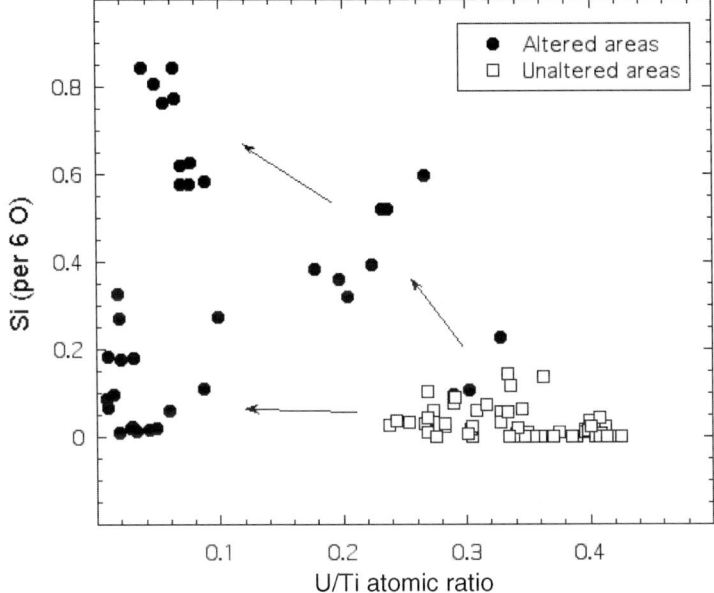

Fig. 5. Scatter plot showing the number of Si atoms per formula unit and the U/Ti atomic ratio in a suite of natural brannerite samples, ($\square$, unaltered areas; $\bullet$, altered areas). In unaltered brannerite, the U/Ti ratio lies below the ideal value of 0.5 due to extensive substitution of Th, Ca, and minor REEs for U on the A-site.

crandallite group minerals during severe weathering of carbonatites in Brazil (Mariano 1989). Using electron microscopy, Banfield & Veblen (1992) have proposed that the perovskite–anatase reaction mechanism involves 'topotactic inheritance' of layers of the perovskite Ti–O framework. In hydrothermal systems, Mitchell & Chakhmouradian (1998b) have described the alteration of perovskite to kassite, anatase, titanite, calcite, and ilmenite in the presence of a CO_2- and SiO_2-rich fluid phase at temperatures of 250–600 °C in alkaline rocks. In general, the results reported above are consistent with the thermodynamic properties of perovskite and related minerals. Nesbitt *et al.* (1981) have shown that perovskite is unstable with respect to sphene, rutile, calcite, and quartz in many hydrothermal fluids and ground waters at 25–300 °C.

Na-bearing perovskite, or loparite (ideally $Na_{0.5}REE_{0.5}TiO_3$), is also subject to hydrothermal alteration in alkaline igneous rocks (Lumpkin *et al.* 1998; Chakhmouradian *et al.* 1999). The primary result of this alteration is removal of Na from the original perovskite, producing a phase which is now amorphous ('metaloparite') due to radiation damage and has a composition of approximately $REETi_2O_{6-x}$ $(OH,F)_x \cdot H_2O$. It is entirely possible that this amorphous phase was originally lucasite, a

mineral that is isostructural with kassite, $CaTi_2O_4$ $(OH)_2 \cdot H_2O$. The simplified reaction relationships for perovskite, kassite, and lucasite are:

$$2H^+ + H_2O + 2CaTiO_3$$
$$= CaTi_2O_4(OH)_2 \cdot H_2O + Ca^{2+}$$

$$H^+ + H_2O + 2Na_{0.5}REE_{0.5}TiO_3$$
$$= REETi_2O_5(OH) \cdot H_2O + Na^+$$

In both cases, the replacement product is usually reported as having a distinctly fibrous or prismatic morphology. The higher level of radiation damage in metaloparite (lucasite?) could be due to a difference in the critical amorphization dose of loparite and the alteration product, provided that the alteration event occurred soon after crystallization of the loparite. For example, Smith *et al.* (1998) and Lumpkin *et al.* (1998) have shown that the critical amorphization dose of perovskite structure types may be as much as a factor of five greater than the critical dose of other Nb–Ta–Ti minerals (e.g., pyrochlore and zirconolite).

Perovskites also exhibit reaction relationships with phosphate minerals and exotic alkali–titanium–silicate minerals under late-stage magmatic conditions. In a systematic study of loparite mineral chemistry in the Lovozero alkaline complex, Russia, Kogarko *et al.* (2002) have identified several reactions between loparite and a late-stage alkaline melt, where loparite was corroded and replaced by apatite, monazite, and the rare minerals lamprophyllite, lomonosovite, mosandrite, nenadkevichite, steenstrupine, and vitusite. During this stage of the evolution of the Lovozero peralkaline magma, concentrations of volatile components (F, H_2O) and alkalis reached very high levels, and reacted with loparite.

Experimental data

Polyphase ceramics

The basic dissolution tests carried out on polyphase Synroc formulations have been carried out in the temperature range 25–200 °C and in deionized water, following the standard MCC-1 and MCC-2 procedures, with periodic replacement of the aqueous fluid. Some additional tests have been performed using carbonate, saline, or silicate solutions instead of deionized water. Much of the early work on the polyphase ceramics, including the effects of composition and processing parameters on elemental release rates, has been reviewed by Ringwood *et al.* (1988) and will not be repeated in detail here. The release rates of soluble elements like Al, Ca, Sr, Mo, Cs, and Ba exhibit a non-linear decrease with time, typically from 10^{-1} g/m²/d after one day to 10^{-3} g/m²/d or lower after about 90 days at temperatures of 70–100 °C. Release rates of the rare earth elements are about an order of magnitude lower than the soluble elements and Ti and Zr are lower still, often reaching detection limits within a few days. Experiments performed at 90 °C reveal that the dissolution of Synroc-C has a weak dependence on pH, possibly with a shallow minimum near pH = 7. Over a temperature range of 45–300 °C, experiments indicate an increase in release rates by a factor of 25 and activation energies in the range 15–30 kJ/mol.

In subsequent work, Smith *et al.* (1991) showed that release rates of Cs, Mo, and Sr decrease rapidly with time in deionized water at 150 °C, reaching values below 10^{-3} g/m²/d after 336 days. The release rates are lower by a factor of about 2–5 for experiments performed at 70 °C in deionized water. Interestingly, experiments were also run at 70 °C in bicarbonate and silicate solutions, and the release rates for Cs and Mo were lower still, a clear indication of the possible effects of real aqueous fluids. Further analysis of the data by Smith *et al.* (1992) showed that Al and Mo possessed the highest release rates in water at 150 °C, with about

0.1% of the original amount of these elements lost after 532 days. By way of comparison, about 0.06–0.07% Ca and Sr, 0.02–0.05% Ce and Nd, and 0.01–0.03% of the Cs and Ba were released after 532 days of testing. The chemical data are consistent with the maximum alteration depths observed by electron microscopy: 0.5 μm for intermetallic alloys, 0.3 μm for Al-rich phases, and 0.2 μm for perovskite. Using data for total mass loss, Smith *et al.* (1992) determined that the average alteration rate of Synroc is approximately 10^{-1} nm/d. It is possible to estimate the alteration depths of some of the phases in Synroc-C from elemental release rate data. This is illustrated in Fig. 6, where the cumulative alteration depth is plotted against time for two different temperatures. These results indicate that there is a significant increase in the alteration depth after increasing the temperature from 70 °C to 150 °C in pure water. The estimated alteration depths after 1 year at 150 °C are 350 nm for Al-rich oxides, 160 nm for perovskite, and 13 nm for hollandite (Fig. 6*b*) and are in good agreement with surface SEM and cross-sectional TEM observations (Smith *et al.* 1992).

At temperatures below 100 °C the major alteration product on the surface of Synroc consists of an amorphous or poorly crystalline Ti–O–H film derived mainly from the dissolution of perovskite (Murakami 1985; Kastrissios *et al.* 1987; Myhra *et al.* 1988*b*, 1988*c*; Solomah & Matzke 1989; Lumpkin *et al.* 1991, 1995; Smith *et al.* 1997*a*). Some dissolution of Mo-rich intermetallic particles is also observed at this temperature. Electron microscope observations made following experiments at 150 °C clearly reveal the alteration of perovskite to anatase + brookite, oxidative dissolution of intermetallic phases, dissolution of minor Al-rich phases, and the presence of scattered TiO_2 crystals on hollandite. Interestingly, monazite has also been observed as a secondary alteration product at 150 °C, incorporating REEs from perovskite and P from dissolving intermetallic grains. Other secondary phases at this temperature include at least two Al–O–H phases and a poorly crystalline Fe–O–H material (Lumpkin *et al.* 1995). Using cross-sectional TEM methods to examine the interface between perovskite and secondary TiO_2 phases, Smith *et al.* (1997*a*) have recently suggested that there may be a thin zone in the upper surface of the perovskite, approximately 40 nm thick, which is depleted in Ca and Sr.

Additional work has been carried out on the behaviour of actinides in Synroc-type ceramics (Ringwood *et al.* 1988; Reeve *et al.* 1989; Hart *et al.* 1992; Smith *et al.* 1996; 1997*b*).

(a)

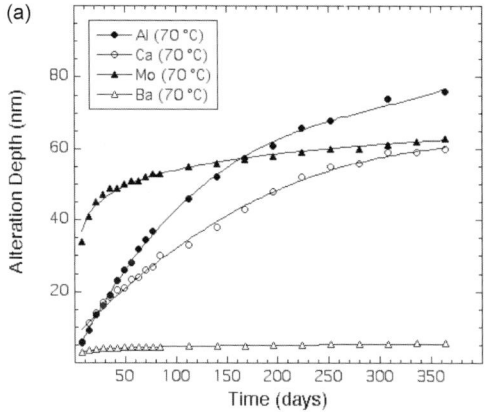

(b)

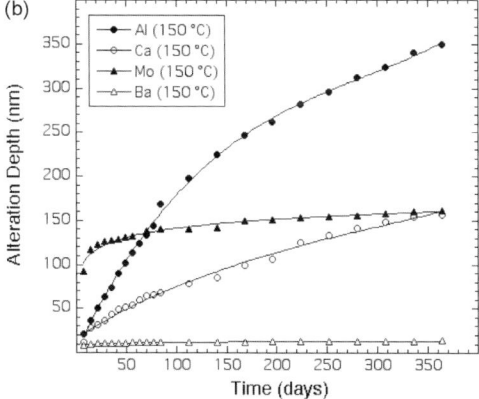

Fig. 6. Plots of the cumulative alteration depth versus time for Synroc-C in deionized water at 70 °C (**a**) and 150 °C (**b**). These plots use Al, Ca, Mo, and Ba as indicator elements for the alteration depth of Al-rich oxide phases, perovskite, intermetallic phases, and hollandite, respectively.

Importantly, the early studies summarized by Ringwood *et al.* (1988) suggested that the release rates of the actinides are approximately two orders of magnitude lower than the more soluble elements such as Ca, Sr, Mo, Cs, and Ba. This has been generally confirmed by Reeve *et al.* (1989), who report release rates of less than 10^{-4} g/m^2/d for ^{237}Np and less than 10^{-5} g/m^2/d for ^{239}Pu, ^{241}Am, and ^{244}Cm in short-term tests (less than 60 days). For experiments performed in buffered solutions of pH $= 2-10$ at 70 °C using Synroc containing Pu, Np, and Am, Hart *et al.* (1992) have used the EQ3/6 code to demonstrate that the solubility of Pu controls the behaviour in all experiments except for pH $= 2$. The solubility limits of Am and Np were not exceeded under any of the experimental conditions. The Pu and Np release

rates decrease from a maximum of 10^{-3} g/m^2/d down to approximately 10^{-6} g/m^2/d after more than 2000 days in pure water at 70 °C (Smith *et al.* 1996, 1997*b*). For short-term tests, the results of Smith *et al.* (1997*b*) suggest that the Pu release rate increases by about one order of magnitude in silicate and carbonate solutions. Although surface characterization of the samples tested for long periods of time revealed the presence of abundant secondary phases (anatase, brookite, Fe-oxides, etc.), actinide elements were not detected (<0.2 wt%) within these alteration products.

The alteration of several titanate ceramics in pure water at 90 °C has been investigated by Leturcq *et al.* (2001). These experiments were performed under conditions of high surface area to volume ratio and lasted for over one year without replacement of the solution. Starting materials included melted Synroc-like materials and hot pressed Synroc-C. This study reported the normalized elemental mass losses, defined by the equation:

$$NL(i) = (C_i \times 1000)/(W_i \times S/V)$$

where C_i is the steady-state concentration in ppb of element i in solution, W_i is the weight percentage of element i in the starting material, S is the surface area of the specimen, and V is the volume of the liquid. Results demonstrate that steady-state concentrations were obtained after several days in solution. Normalized mass losses were typically between 10^{-1} and 10^{-4} g/m^2 for relatively soluble elements (Al, Ca, Mo, Sr, Cs, Ba) and between 10^{-4} and 10^{-7} g/m^2 for the less soluble elements (Ti, Zr, REEs). Thermodynamic calculations suggested that the 'cessation of alteration' is not explained by the solubility limits of the primary phases in the ceramics. The results appear to be consistent with the development of a layer of secondary material (Ti–Zr hydroxides) on the surface of the ceramic that retards the alteration process after a short period of time. The estimated thickness of this layer is about 3–5 nm, but unfortunately this thickness was stated as being less than the sensitivity of the characterization methods available to the authors (Leturcq *et al.* 2001).

In an important study of accelerated radiation damage effects in polyphase Synroc samples doped with ^{244}Cm, Mitamura *et al.* (1992, 1994) studied the behaviour of Mo, Sr, Ca, Cs, Ba, and Cm for up to 56 days in distilled water at 90 °C. These ceramics were doped with either Na-free simulated Purex waste (PW-4b) or simulated waste containing 1.65 wt% Na$_2$O (e.g., JW-A type). We should note here that the two different waste streams lead to a real difference in the physical and chemical response to radiation damage. Up to the maximum dose achieved during this study (1.3×10^{15} α/mg), the PW-4b samples showed a consistent decrease in density with increasing dose. A single specimen loaded with JW-A type waste, however, showed a significant change in the dose–density behaviour at a dose of 8.5×10^{14} α/mg related to the onset of cracking (see fig. 5 in Mitamura *et al.* 1994). For the longest testing period, the release rates of Ca and Sr in the PW-4b samples were within error of one another and increased by less than a factor of 10 with increasing dose, from approximately 3×10^{-4} g/m²/d to 2×10^{-3} g/m²/d. The fact that Ca and Sr are released at approximately the same rate points to perovskite as the major source of these elements in solution, even though the damage rate of zirconolite appears to be much higher than that of perovskite according to the X-ray diffraction results.

Pyrochlore ceramics

A variety of pyrochlore ceramics have been studied using flow-through experiments in pH buffered solutions and a range of temperatures below 100 °C (Shoup *et al.* 1997; Icenhower *et al.* 2000; Roberts *et al.* 2000; Zhang *et al.* 2001*a, b*). Shoup *et al.* (1997) examined the behaviour of a pyrochlore ceramic, $Er_{1.78}Ce_{0.22}Ti_2O_7$, in three different aqueous fluids: WIPP type brine, 0.1 M NaCl, and 0.1 M HCl at 50 °C. All elements were below detection limits in the NaCl solution and only Er was detected at a level of 10 ppm in the WIPP brine. For acidic conditions, the authors reported 1375 ppm Er, 150 ppm Ce, and only 3 ppm Ti in solution. The Er/Ce ratio in solution is very close to that of the starting material; however, the measured amount of Ti in solution is well below the level of Er and Ce, possibly due to precipitation of secondary phases on the surface.

Icenhower *et al.* (2000) investigated $Gd_2Ti_2O_7$ and $Lu_2Ti_2O_7$ monoliths and $(Ca,Gd,Ce,Hf)_2(Ti,Mo)_2O_7$ powders at 90 °C and pH = 2–12. Release rates for Gd, Ce, and Ti at pH = 2 were approximately $1–4 \times 10^{-3}$ g/m²/d and $4–7 \times 10^{-4}$ g/m²/d for the $Gd_2Ti_2O_7$ and $Lu_2Ti_2O_7$ monoliths, respectively. The releases were essentially the same within experimental error. Experiments with the more complex sample in powdered form yielded release rates of $6–9 \times 10^{-5}$ g/m²/d for Gd and Ce using a flow rate of 2 mL/d. Using a higher flow rate of 10 mL/d caused higher release rates of $2–4 \times 10^{-5}$ g/m²/d. Release rates for all of the samples approached steady-state conditions after about 90 days. The release of Mo was used to monitor the dissolution of

$(Ca,Gd,Ce,Hf)_2(Ti,Mo)_2O_7$ in neutral to basic solutions. Near steady-state Mo concentrations (0–4 ppb range) yield apparent dissolution rates of $2–6 \times 10^{-5}$ g/m²/d for both samples at pH = 6–8 and flow rate = 2 mL/d. In total, the results for $(Ca,Gd,Ce,Hf)_2(Ti,Mo)_2O_7$ indicate a weak pH dependence with a minimum at pH = 7. Similar pyrochlore-rich ceramics were studied by Hart *et al.* (2000), who showed that the release rate of Pu dropped from approximately 10^{-3} g/m²/d to 10^{-5} g/m²/d or less after nearly one year in pure water at 90 °C (Fig. 7). The release rates of U and Gd in these experiments were higher than Pu by factors of about 10 and 100, respectively.

The kinetics of U release from a pyrochlore, $(Ca,Gd,Ce,Hf,U)_2Ti_2O_7$, containing 24 wt% UO_2 has been investigated by Roberts *et al.* (2000) and Zhang *et al.* (2001*b*). Although the results of these two studies differ in detail, they both report that the pH dependence follows a shallow V-shaped pattern with a minimum near neutral pH (Fig. 8). The activation energy (E_a) for the release of U ranges from about 3 to 20 kcal/mol below the minimum pH and from 3 to 12 kcal/mol above the minimum pH value. The release rates for U, converted from the limiting rate constants given by Zhang *et al.* (2001*b*), range from 6×10^{-7} g/m²/d to 7×10^{-5} g/m²/d for all experimental conditions (e.g., $T = 25–75$ °C and pH = 2–12).

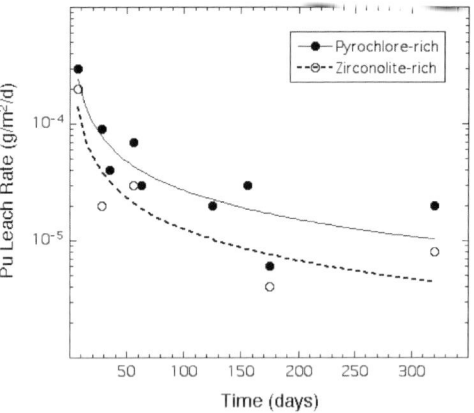

Fig. 7. Plot of the Pu leach rate as a function of time for pyrochlore- and zirconolite-rich LLNL-type waste forms in experiments performed at 90 °C in pure water (after Hart *et al.* 2000). Power law curve fits illustrate the normal trends for these materials: initial rapid decrease in leach rates followed by slower release rates, which decrease to 10^{-5} g/m²/d or less after time periods of several months to one year.

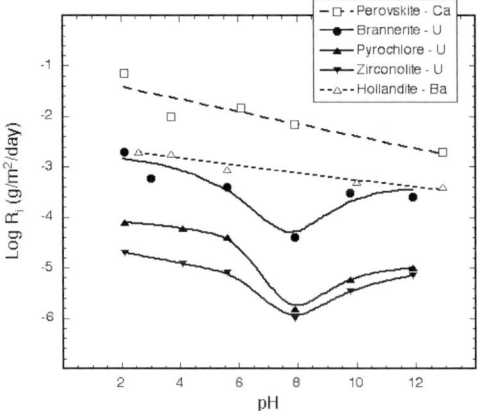

Fig. 8. Plots showing the dependence of the elemental release rate on pH for U in brannerite, pyrochlore, and zirconolite at 75 °C (after Zhang *et al.* 2001*b*). Also shown are data for Ca in perovskite and Ba in hollandite (after McGlinn *et al.* 1995; Carter *et al.* 2002). Linear fits are shown for perovskite and hollandite, but the other curves are weighted fits used to illustrate the trends.

Weber and co-workers were among the first investigators to conduct dissolution tests on single-phase pyrochlore samples ($Gd_2Ti_2O_7$) doped with ^{244}Cm (Weber *et al.* 1986). In this work, the dissolution tests were limited to annealed, fully crystalline samples and fully amorphous samples and were exercised at 90 °C in pure water for 14 days. The experiments revealed weight losses of 0.02% and 0.05% for the crystalline and amorphous pyrochlore samples, respectively. The results of this study also indicated that the release rates of Cm and the radioactive daughter product ^{240}Pu increased by factors of 17 and 49, respectively, as a consequence of amorphization. It is instructive to compare the results of actinide doping experiments with the analogous situation wherein samples are irradiated using heavy ions and then subjected to dissolution tests. The irradiation of bulk pyrochlore using 2 MeV Au ions has been shown to produce an amorphous layer to depths of approximately 300–400 nm (Begg *et al.* 2001). Subsequent dissolution tests on crystalline and amorphous samples performed at 90 °C in a solution of nitric acid at pH = 2 gave different results, depending on the A-site cation in a series of $A_2Ti_2O_7$ pyrochlores with A = Y, Lu, and Gd. The process of amorphization in these samples caused the release rates of Y, Lu, and Gd to increase by factors of 0, 3–5, and 10–15, respectively. Stewart *et al.* (2003) recently reported similar results for a Ce-doped LLNL-type pyrochlore waste form irradiated with

heavy ions. Dissolution of this material at 90 °C in a solution with pH = 1.75 suggested that the irradiation caused an increase in the release rates of Ca, Gd, and Ti by a factor of about 2–12 over those of the unirradiated sample. The results of both studies indicate that the elemental release rates of Ca and REEs are generally 1–2 orders of magnitude higher than Ti. As noted above, this may be due to precipitation of Ti-rich secondary phases on the surface of the sample.

Significant new results are now available (Icenhower *et al.* 2003) for synthetic pyrochlore samples with compositions corresponding to the LLNL Baseline Formulation with Ce or Pu–U, single-phase pyrochlore with Ce or Pu–U, and single-phase $Gd_2Ti_2O_7$ for comparison. The baseline and single phase Pu–U pyrochlores were fabricated with ^{238}Pu or ^{239}Pu in order to compare samples with high and low alpha-decay dose under identical conditions. The experiments were designed for single-pass flow through conditions allowing for a range of flow rate to surface area ratios, while at the same time minimizing experimental problems (e.g., effect of F from teflon, effect of atmospheric CO_2 on pH, and effect of radiolysis at low flow rates). Results for solutions buffered at pH = 2 and temperatures of 85–90 °C reveal that a steady-state, forward reaction rate can be determined for high values of the flow rate to surface area ratio. Under these conditions, Icenhower *et al.* (2003) report release rates of 1.3×10^{-3} g/m^2/d for the LLNL Baseline pyrochlore formulations and 2×10^{-4} g/m^2/d for the single-phase pyrochlores. However, the most important result of this work is that the dissolution rates of the crystalline and X-ray amorphous (^{238}Pu-doped) samples are the same, within the experimental errors.

Zirconolite ceramics and natural samples

Prior to 1988, dissolution studies had only been reported on natural samples and the results were complicated by the presence of inclusions (Ringwood *et al.* 1988). Nevertheless, the release rate of Ca from fully amorphous zirconolite from Sri Lanka, averaged over temperatures of 95 and 200 °C, appeared to decrease from about 10^{-1} g/m^2/d to 3×10^{-3} g/m^2/d after about 2 weeks. McGlinn *et al.* (1995) studied the pH dependence of single-phase zirconolite in pure flowing water at 90 °C. Results showed that after 43 days the release rates for Ca decrease with increasing pH for all samples, although there is a lot of scatter, especially where the data are close to the detection limit. Over the duration of the experiments, the Ca rate of release dropped from about 10^{-2} to 10^{-3} g/m^2/d. Preferential release of Ca over Ti (by a factor of 10–450) occurred in all samples over the entire pH range. However, SEM work failed to show conclusive evidence for alteration products on the surface of the zirconolite. It is worth noting here that the Ca release rate of the amorphous natural zirconolite, extrapolated to 43 days, is approximately 2×10^{-3} g/m^2/d, suggesting that the alpha-decay induced amorphization has little influence on the long-term dissolution rate.

We have included here, for comparison, the results of a study of zirconolite-rich Synroc nominally composed of 80 wt% Ce- or Pu-doped zirconolite plus 10 wt% hollandite and 10 wt% rutile (Hart *et al.* 1998). Inclusion of this study in this section is significant because the two additional phases are both highly durable in their own right and the experiments were conducted at two different temperatures (90 and 200 °C) and in three different aqueous solutions (pure water, silicate, and brine). The authors found no major differences in the release rates of Ca, Ce, Hf, Ti, Zr, Pu, and Gd apart from those for Ce and Ti, which appeared to be somewhat higher in the brine. On average, for all elements, the increase in temperature caused the release rates to increase by a factor of approximately seven. Release rates were generally below 10^{-2} g/m^2/d for Ca, 10^{-3} g/m^2/d for Ce and Gd, and 10^{-4} g/m^2/d for Ti, Zr, Hf, and Pu (except for the brine at 200 °C, which gave a Ti release rate of 2×10^{-3} g/m^2/d). Hart *et al.* (2000) also determined the release rate of Pu in an LLNL-type zirconolite ceramic. After nearly one year in pure water at 90 °C the release rate of Pu decreased from 2×10^{-3} g/m^2/d to less than 10^{-5} g/m^2/d (Fig. 7).

The dissolution kinetics of zirconolite have now been determined as a function of pH using pure water in single-pass flow through tests at temperatures of 75 °C and lower (Roberts *et al.* 2000; Zhang *et al.* 2001*b*). These authors have independently studied a Ce-, Gd-, and Hf-doped zirconolite containing about 16 wt% UO$_2$ and the results of the two studies are similar. Release rates determined by Zhang *et al.* (2001*b*) indicate that Ti and U are released from zirconolite at about the same rate after about 20 days following an initial period where U is released at a somewhat faster rate than Ti. The limiting rate constants, K, are equivalent to U release rates of 6.4×10^{-7} to 1.3×10^{-5} g/m^2/d for zirconolite over the entire pH range of 2–12 and temperature range of 25–75 °C. The dissolution rate of zirconolite is characterized by a shallow V-shaped pattern with a minimum near pH = 8, similar to the results obtained for pyrochlore (Fig. 8).

Malmström and coworkers (Malmström *et al.* 1999, 2000; Malmström 2000) have investigated the performance of several zirconolite compositions under hydrothermal conditions (150–700 °C, 50–200 MPa) in fluids containing different concentrations of HCl, NaOH, H$_3$PO$_4$, silicate, or carbonate, in addition to pure water. Starting materials consisted of Ca$_{0.99}$Zr$_{1.07}$Ti$_{1.94}$O$_7$ and Ca$_{0.8}$U$_{0.2}$ZrTi$_{1.6}$Al$_{0.4}$O$_7$ with 1–3 wt% excess CaZrTi$_2$O$_7$, and single-phase samples of Ca$_{0.8}$Nd$_{0.2}$ZrTi$_{1.8}$Al$_{0.2}$O$_7$, Ca$_{0.85}$Ce$_{0.1}$Gd$_{0.1}$Zr$_{0.85}$Hf$_{0.1}$Ti$_{1.9}$Al$_{0.1}$O$_7$, and Ca$_{0.85}$U$_{0.1}$Gd$_{0.1}$Zr$_{0.85}$Hf$_{0.1}$Ti$_{1.8}$Al$_{0.2}$O$_7$. The results of these experiments demonstrate that zirconolite is most highly reactive in the NaOH-bearing fluids, but temperatures in excess of 500 °C are required to produce a continuous alteration layer consisting of perovskite + calzirtite at 50 MPa or perovskite + baddeleyite at 200 MPa. In the case of HCl, similar temperatures are required to produce an alteration layer consisting of rutile and anatase. Somewhat surprisingly, SEM observations revealed that the silicate and carbonate fluids had no visible effect on the zirconolite surface after experimental runs at 550 °C and 50 MPa. Only limited reaction was observed in pure water or H$_3$PO$_4$ fluids at the same temperature and pressure, with rutile and monazite appearing as products on the surface.

Brannerite ceramics

Vance *et al.* (2000) presented some preliminary results on the release of U from brannerite in flowing solutions buffered to three different pH values at 70 °C. After 60 days of testing, the U release rates were approximately 8×10^{-3}, 2×10^{-3}, and 10^{-4} g/m^2/d for solutions with pH = 2.1, 9.8, and 7.9, respectively. In addition to basic dissolution tests, many of these laboratory experiments were specifically designed to shed light on the kinetics and mechanisms of brannerite alteration. The aforementioned results have been extended to additional pH values and temperatures by Zhang *et al.* (2001*b*). For temperatures of 20–50 °C and pH values in the range of 2–12, these authors report limiting rate constants for U equivalent to elemental release rates of about 2×10^{-3} to 2×10^{-6} g/m^2/d. The lowest rate was obtained at 20 °C and pH = 5.6. However, the pH dependence was examined in detail at 70 °C and the results show a shallow V-shaped pattern similar to that of pyrochlore and zirconolite, although the release rates of U from these two phases are

1–2 orders of magnitude less than that of brannerite (Fig. 8).

In particular, Zhang *et al.* (2003) investigated the dissolution of brannerite as a function of pH, phthalate, and bicarbonate concentration in aqueous fluids at low temperature. Their results show that the initial uranium release rate is weakly dependent on phthalate concentration with the reaction order of 0.19 and then approaches a constant value after about one hour. There is no obvious rate dependence of titanium on phthalate concentration and the uranium release rate is 2–3 orders of magnitude greater than titanium. Furthermore, experimental results show that the phthalate ion increases titanium solubility. In bicarbonate solutions, however, the uranium release rate is strongly dependent on bicarbonate concentration with the initial reaction order of ~0.5, increasing to ~0.7 after 7 days. Therefore, the concentration of bicarbonate enhances the dissolution of brannerite under alkaline conditions (cf. Szymański and Scott, 1982). In the same experiments, the release of titanium is not dependent upon bicarbonate concentration, either in release rate or equilibrium concentration (~3 ppb), indicating that unlike phthalate, bicarbonate does not interact strongly with titanium, either on the solid surface or in solution.

Further experiments by Zhang *et al.* (2003) using ion beam thinned sections (for subsequent TEM examination) show that for a pH 2 solution, an apparent preferential release of uranium occurs (Fig. 9). Results from TEM show that the IBT specimen tested in pH 2 solution at 90 °C for four weeks has large areas of unaltered brannerite as well as a relatively small amount of secondary phase. The primary rutile grains appear to have been partially etched. Energy dispersive X-ray (EDX) spectroscopy indicates that the secondary phase is TiO_2 with differing amounts of uranium and trace amounts of other elements. Selected area electron diffraction confirms that the secondary phase is polycrystalline, probably anatase and/or brookite. In the case of pH 11 solution, the less linear and lower release of uranium than titanium suggests that some dissolved uranium may absorb back onto the secondary phase or even the sample itself. Overall, the release rates of U and Ti from the IBT specimen in pH 11 solution are nearly the same. The IBT specimen exposed to pH 11 solution at 90 °C for four weeks shows large areas of a fibrous secondary phase associated with the original brannerite. Selected area electron diffraction indicates that the secondary phase is amorphous. Energy dispersive X-ray spectroscopy shows that the Ti-rich secondary phase contains varying amounts of uranium and trace amounts of other elements. Also, XPS analyses indicate the existence of oxidized U^{5+} and U^{6+} species on specimens both before and after the dissolution tests, and U^{6+} was the dominant component on the specimen tested in the pH 11 solution.

Hollandite ceramics

Experimental studies of the polyphase ceramics noted above demonstrate that hollandite is one of the most durable titanate phases in aqueous solutions. Pham *et al.* (1994) carried out experimental work on synthetic Ba–hollandite doped with Cs and containing Al on the B-site for charge balance. These authors suggested that, following the initial release of Cs and Ba from reactive surface sites the first few monolayers of the structure rapidly dissolved due to the release of Al and consequent precipitation of Al-OH species, driving solution pH to lower values. However, the alteration process was mediated via the formation of a continuous Al- and Ti-rich surface layer. Further evidence for selective removal of Ba and enrichment of Al and Ti on the surface of hollandite tested at 250–300 °C was presented by Myhra *et al.* (1988*b*). These conclusions were largely based on the different release rates of Ba (10^{-1} g/m^2/d), Al (7×10^{-3} g/m^2/d), and Ti ($<8 \times 10^{-4}$ g/m^2/d) after 14 days of dissolution testing, combined with XPS analyses of the altered surfaces.

Recent work by Carter *et al.* (2002) on the alteration of hollandite samples in pure water at 90 °C and 150 °C demonstrates that the release rates of Ba and Cs are nearly identical. In these same experiments, Al and Ti are below detection limits at 90 °C and only Al is detected at 150 °C, but only by a factor of 2–3 above the detection limit. These authors also examined the pH dependence of the release rates at 90 °C, finding that the release of Ba decreases linearly (Fig. 8) from about 2×10^{-3} g/m^2/d at pH = 2.5 to 4×10^{-4} g/m^2/d at pH = 12.9. SEM work revealed the presence of nodular secondary phases on the surface of the hollandite at both temperatures. This was confirmed by XTEM work on cross-sections of the material, which identified both Ti-rich and Al-rich nodules in a ratio of about ten to one, respectively. Furthermore, XPS analysis of the hollandite surface after testing at 150 °C for 7 days showed an increase in the Al/Ti ratio from 0.26 to 0.47, consistent with the presence of an Al-rich layer approximately 0.5 nm thick. XPS results also show a decrease in the Ba/Ti ratio from 0.095 to 0.072 and this is equivalent to removal of about 25% of the Ba within 5 nm of the surface.

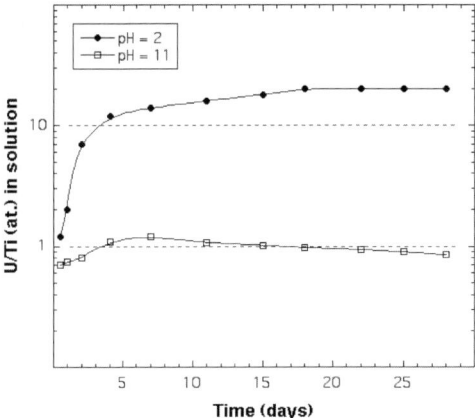

Fig. 9. Graph showing the change in U/Ti atomic ratio in the fluid (normalized to the stoichiometry of the starting material) after dissolution of synthetic brannerite acidic and basic fluids (after Zhang *et al.* 2003). The experiment at pH = 2 reveals a strong preferential release of U, reaching a steady-state normalized U/Ti atomic ratio of about 20 after several weeks.

The effects described above by Carter *et al.* (2002) are further illustrated in Fig. 10. Here we plot variations in the concentrations of Al_2O_3, Fe_2O_3, Cs_2O, and BaO determined by AEM from replicas of hollandite removed from the surface of Synroc-C tested at 150 °C in pure water for up to

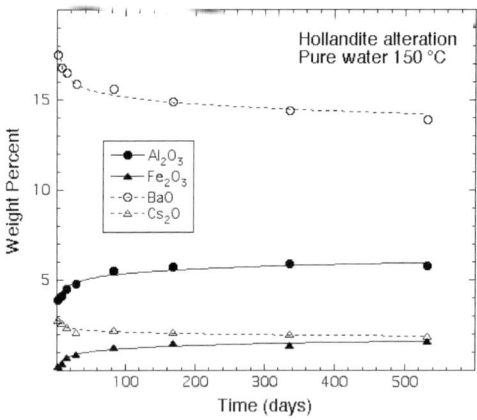

Fig. 10. Change in composition of thin hollandite grains, lifted from the surface of Synroc-C, as determined by AEM after leaching in pure water at 150 °C for up to 532 days. The measured concentrations of BaO and Cs_2O decrease whereas the concentrations of Al_2O_3 and Fe_2O_3 increase with time. There is also a slight increase in the TiO_2 concentration out to 532 days (not shown). See text and Fig. 11 for further discussion and an illustration of hollandite alteration.

532 days. These analyses indicate that there is an accumulation of Al–Fe–O–H material on the surface of hollandite up to about 150–200 days of testing. If the hollandite grains have thicknesses of 25–75 nm, typical values for AEM analysis, then the increases in Al_2O_3 and Fe_2O_3 content correspond to a surface layer thickness of approximately 1–3 nm. The AEM analyses also reveal significant decreases in the Cs_2O and BaO content, the magnitude of which cannot be fully explained by the Al–Fe–O–H layer. Therefore, it can be postulated that the tunnel cations have been released to a depth of 8–16 nm depending upon the actual hollandite crystal thickness used for the AEM analyses. The overall alteration mechanism is illustrated in Figure 11 and is generally consistent with the observations of Carter *et al.* (2002) and the hollandite alteration depth estimated from elemental release data.

Perovskite ceramics and natural samples

Nesbitt *et al.* (1981) performed a detailed analysis of the thermodynamic and kinetic stability of perovskite. Thermodynamic calculations and data for natural groundwaters and hydrothermal

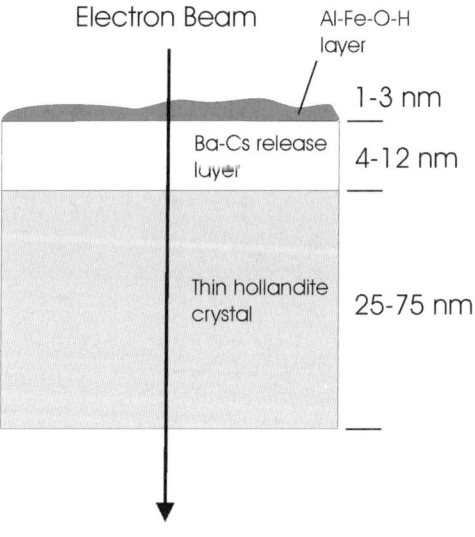

Fig. 11. Schematic drawing of the alteration of hollandite, as deduced from material recovered on TEM replicas (see Fig. 10). The model assumes a thickness of 50–100 nm for the hollandite grains and a uniform thickness of Al–Fe–O–H material built up on the surface during the alteration process. The decrease in the measured concentration of Ba, for example, cannot exceed that required by the surface layer alone and requires additional Ba release to a depth of approximately 8–16 nm depending on the actual grain thickness.

fluids (up to 300 °C) revealed that perovskite is unstable with respect to titanite, titanite + quartz, quartz, rutile, or rutile + calcite. The groundwaters used in this study represented a range of compositions emanating from dunites, peridotites, serpentinites, rhyolites, granites, and limestones. This work included measurements of the dissolution rates of two natural perovskites and synthetic $SrTiO_3$ and $BaTiO_3$ samples in pure water at 25–300 °C. Results of the experimental work gave elemental release rates in the range of 5×10^{-2} to 8×10^{-1} $g/m^2/d$ for Ca, 4×10^{-2} to 3×10^{-1} $g/m^2/d$ for Sr, and 3×10^{-2} to 4×10^{-1} $g/m^2/d$ for Ba. The Ca release rates were obtained from a single crystal from Magnet Cove, Arkansas. Using a powdered sample of perovskite from Australia, lower release rates of 10^{-3} to 10^{-2} $g/m^2/d$ for Ca have been reported. Kamizono et al. (1991) have also examined the behaviour of $CaZrO_3$ in acidic (HCl, pH = 1) and near neutral (deionized water, pH = 5.6) solutions at 90 °C. The release rates of Ca and Zr were determined to be 10^{-3} $g/m^2/d$ and 2×10^{-4} $g/m^2/d$, respectively, in the experiment with pH = 1. Release rates were about two orders of magnitude lower in the experiment using deionized water (see figs 1 and 2 in Kamizono et al. 1991). As noted above, the perovskite contained small amounts of Ce, Nd, and Sr as simulated waste elements. These elements had unusually high release rates of about 10^{-1} $g/m^2/d$.

Surface analytical studies of synthetic perovskite after experiments conducted at 150–250 °C in silica saturated aqueous fluids were reported by Myhra et al. (1987). These authors, using a combination of AES, SEM-EDX, and XPS techniques, identified the presence of a surface reaction layer ranging in thickness from a few monolayers to several hundred nanometers, depending on the experimental conditions. In spite of the presence of SiO_2 and CO_2 in solution (the system was open to the atmosphere), the alteration layer was composed mainly of crystalline TiO_2, a thin siliceous layer, and possible calcium carbonate or hydroxide species. Following this work, Myhra et al. (1988b) reported additional results for $CaTiO_3$ and $BaTiO_3$ tested at 300 °C in pure water. They determined release rates after 14 days of 10^{-2} $g/m^2/d$ for Ca and 3×10^{-2} $g/m^2/d$ for Ba from the two samples, respectively. Surface analytical work showed that Ca and Ba were depleted to a depth of about 200 nm and that oxygen was enriched near the surface, consistent with the release of Ca and Ba to solution and the formation of a crystalline TiO_2 layer.

Pham and co-workers have performed similar experiments at temperatures of 20–100 °C.

These authors showed that near surface decreases in the Ca/Ti ratio determined by XPS are accompanied by the formation of an amorphous Ti-rich layer up to 10 nm thick, as observed by TEM (Pham et al. 1989; Turner et al. 1989). The authors proposed a base catalysed hydrolysis and ion exchange model to account for their observations, whereby surface Ca^{2+} is released to solution via exchange with H^+ and Ti–O–Ti surface species are converted to Ti–OH species via reaction with OH^- and H_2O. The overall reaction can be written as follows (Pham et al. 1989):

$$CaTiO_3 + (6 - x)H^+$$
$$= Ca^{2+} + Ti(OH)_x^{(4-x)+} + (3 - x)H_2O$$

This reaction indicates that the aqueous Ca^{2+}/H^+ ratio and the presence of Ti–OH surface species control the dissolution of perovskite. The presence of the amorphous Ti–OH surface film at low temperatures may be due to kinetic factors, as crystalline TiO_2 phases are thermodynamically favoured at low temperatures (Pham et al. 1989). McGlinn et al. (1995) investigated the pH dependence of the release of Ca from two perovskite samples: end-member $CaTiO_3$ and $Ca_{0.78}Sr_{0.04}Nd_{0.18}Ti_{0.82}Al_{0.18}O_3$. Results of this study, performed at 90 °C with pH ranging from 2.1 to 12.9, demonstrated that the Ca release rates generally decrease with increasing pH. After 43 days of testing, the Ca release rate for the end-member perovskite decreased from just below 10^{-1} $g/m^2/d$ at pH = 2.1 to 2×10^{-3} $g/m^2/d$ at pH = 12.9. Similar results were obtained for the doped perovskite, in which the Ca release rate decreased from 4×10^{-2} $g/m^2/d$ at pH = 2.1 to less than 2×10^{-3} $g/m^2/d$ at pH = 12.9. These results are summarized in Figure 8. In SEM imaging the development of 'agglomerated, submicron, titanaceous particles' are clearly shown on the perovskite surface after testing under acidic conditions.

Discussion and conclusions

Using the criteria listed in the introduction, we now provide a brief summary of the performance of these materials. With reference to criterion number 5, physical properties, we restrict out discussion here to the effects of radiation damage on the crystalline structure. For materials destined for geological disposal, the long-term performance in natural systems over long periods of time (e.g., 10^3 to 10^6 years) is of critical concern. Therefore, we believe that it is necessary to consider one more criterion: (6) natural samples exist and allow for an assessment of the long-term behaviour. Using these criteria as a guide, we illustrate below some of the advantages and disadvantages of these waste form phases.

Starting from the requirement of aqueous durability, the data summarized in this paper clearly demonstrate that perovskite is the least durable phase present in the polyphase waste forms such as Synroc-C. As described above, the low-temperature alteration of perovskite to anatase and other phases in natural systems is well known. These results are consistent with the available thermodynamic data and are generally corroborated by laboratory experiments on single-phase and polyphase samples. Most of these experiments have been executed at temperatures of 70–150 °C in pure water, with fewer studies concerning the performance in sodium chloride, bicarbonate, and silicate solutions, or at temperatures above and below the stated range. Characterization studies have shown that perovskite readily reacts with aqueous fluids to form an amorphous Ti–O–H film at temperatures below about 100 °C or crystalline TiO_2 polymorphs at higher temperatures. Therefore, from a natural systems and experimental viewpoint, perovskite is not the best host phase for either ACTs or Sr in waste forms destined for a geological repository. Note, however, that these comments largely apply to ACT-poor perovskite compositions in the system $CaTiO_3-SrTiO_3-REETiO_3-REEAlO_3$ with $CaTiO_3$ as the dominant component.

The kinetics and mechanisms of brannerite alteration in aqueous fluids are now reasonably well established through experiments covering a range of temperature, fluid composition, and pH. Although generally highly resistant to total dissolution due to the low solubility of the TiO_2 component, brannerite is susceptible to chemical alteration by ion exchange and recrystallization processes. However, the overall influence of this phase on waste form performance, at least in terms of U release, should be minimal as long as the volume proportion is kept low (e.g., below approximately 10 vol%). Furthermore, recent experimental work has shown that end-member brannerite, UTi_2O_6, is energetically stable with respect to the oxide assemblage $UO_2 + 2TiO_2$, with $\Delta H_{f-ox}^0 = -7.7$ kJ/mol at 25 °C (Helean *et al.* 2003). End-member $PuTi_2O_6$ is metastable with respect to the oxides $PuO_2 + TiO_2$; however, Helean *et al.* (2003) estimate that inclusion of up to 30% brannerite of composition $U_{0.8}Pu_{0.2}Ti_2O_6$ will not significantly alter the thermodynamic stability of the waste form.

Pyrochlore and zirconolite have excellent release rates with minimal variation as a function of pH, they are highly flexible in terms of their ability to incorporate impurities, they are well established with regard to processing, and they can accommodate substantial waste loadings. Both the aqueous performance and radiation effects determined experimentally are confirmed through studies of natural samples. The general chemistry and alteration mechanisms of natural pyrochlores in the system $NaCaNb_2O_6F-NaCaTa_2O_6F-Ca_2Nb_2O_7-Ca_2Ta_2O_7-REE_2Ti_2O_7-CaUTi_2O_7$ are well documented. Alteration processes are dominated by ion exchange effects, but these are largely restricted to the Nb–Ta endmembers due to the nature of the structure and bonding (Lumpkin 2001, p. 160). Some evidence for dissolution and reprecipitation of pyrochlore has been presented, but only in the most extreme geological environments. In fact, the world's largest Nb ore deposits are formed through the dissolution of the host rocks and concentration of pyrochlore in overlying laterite horizons. Studies of natural samples suggest that zirconolite is even more durable that pyrochlore, consistent with experimental evidence and the nature of the crystal structure. Although titanate pyrochlore and zirconolite are rendered amorphous by alpha-decay processes, the *total volume expansion is low* (5–6%) and the *effect of amorphization on aqueous durability is negligible* (e.g., Icenhower *et al.* 2003). Of the two phases, the performance of zirconolite appears to be somewhat better than that of pyrochlore in both the experimental and natural systems. Additionally, zirconolite is one of the few host phases subjected to experimentation at elevated temperature and pressure and in a range of fluid compositions, and the results indicate that this mineral is highly resistant to aqueous attack up to about 250–400 °C in a range of fluid compositions. The titanate pyrochlore and zirconolite-rich waste forms both appear to be thermodynamically stable relative to the oxides or to assemblages containing perovskite + oxides (e.g., Putnam *et al.* 1999; Helean *et al.* 2002).

Finally, hollandite is the only non-actinide host phase considered here, and is the only Cs host phase that has been extensively studied. Results summarized herein indicate that hollandite has excellent aqueous durability, crystal chemical flexibility, well-known processing parameters, and more than adequate waste loadings. Although natural analogues exist for hollandite, they have not been examined in the context of nuclear waste disposal, thus we have no details on the long-term behaviour of this important Cs host phase.

We thank J. P. Icenhower and T. Advocat for constructive reviews of this manuscript. Further credit must be given to

M. G. Blackford, B. C. Chakoumakos, M. Colella, R. A. Day, R. C. Ewing, K. P. Hart, A. Jostsons, A. N. Mariano, E. R. Vance, and F. Wall, among others, for their significant contributions to this work. Natural samples provided by the American Museum of Natural History, Australian Museum, Australian National University, ETH Zürich, Harvard University, The Natural History Museum have formed much of the basis of this work. We acknowledge funding support over many years from the Australian Nuclear Science and Technology Organisation the United States Department of Energy. The Cambridge Centre for Ceramic Immobilisation is supported by British Nuclear Fuels Limited and the Cambridge-MIT Institute.

References

BALL, C. J., BUYKX, W. J. et al. 1989. Titanate ceramics for the stabilization of partially reprocessed nuclear fuel elements. Journal of the American Ceramic Society, 72, 404–414.

BANFIELD, J. F. & VEBLEN, D. R. 1992. Conversion of perovskite to anatase and TiO_2 (B): A TEM study and the use of fundamental building blocks for understanding relationships among the TiO_2 minerals. American Mineralogist, 77, 545–557.

BART, F., LETURCQ, G. & RABILLER, H. 2003. Iron-substituted barium hollandite ceramics for cesium immobilization. 105th American Ceramic Society Symposium Proceedings, Nashville, TN, USA, 27 April–2 May.

BEGG, B. D., HESS, N. J., WEBER, W. J., DEVANATHAN, R., ICENHOWER, J. P., THEVUTHASAN, S. & McGRAIL, B. P. 2001. Heavy-ion irradiation effects on structures and acid dissolution of pyrochlores. Journal of Nuclear Materials, 288, 208–216.

BELLATRECCIA, F., DELLA VENTURA, G., CAPRILLI, E., WILLIAMS, C. T. & PARODI, G. C. 1999. Crystal-chemistry of zirconolite and calzirtite from Jacupiranga, São Paulo (Brazil). Mineralogical Magazine, 63, 649–660.

BELLATRECCIA, F., DELLA VENTURA, G., WILLIAMS, C. T., LUMPKIN, G. R., SMITH, K. L. & COLELLA, M. 2002. Non-metamict zirconolite polytypes from the feldspathoid-bearing alkali-syenitic ejecta of the Vico volcanic Complex (Latium, Italy). European Journal of Mineralogy, 14, 809–820.

BULAKH, A. G., NESTEROV, A. R., WILLIAMS, C. T. & ANISIMOV, I. S. 1998. Zirkelite from the Sebl'yavr carbonatite complex, Kola peninsula, Russia: and x-ray and electron microprobe study of a partially metamict mineral. Mineralogical Magazine, 62, 837–846.

CARTER, M. L., VANCE, E. R., MITCHELL, D. R. G., HANNA, J. V., ZHANG, Z. & LOI, E. 2002. Fabrication, characterisation, and leach testing of hollandite, $(Ba,Cs)(Al,Ti)_2Ti_6O_{16}$. Journal of Materials Research, 17, 2578–2589.

CHAKHMOURADIAN, A. R. & MITCHELL, R. H. 1998. Lueshite, pyrochlore and monazite-(Ce) from apatite–dolomite carbonatite, Lesnaya Varaka complex, Kola Peninsula, Russia. Mineralogical Magazine, 62, 769–782.

CHAKHMOURADIAN, A. R., MITCHELL, R. H., PANKOV, A. V. & CHUKANOV, N. V. 1999. Loparite and 'metaloparite' from the Burpala alkaline comples, Baikal Alkaline Province (Russia). Mineralogical Magazine, 63, 519–534.

CHAKOUMAKOS, B. C. 1984. Systematics of the pyrochlore structure type, ideal $A_2B_2X_6Y$. Journal of Solid State Chemistry, 53, 120–129.

CHEARY, R. W. 1987. A structural analysis of potassium, rubidium and caesium substitution in barium hollandite. Acta Crystallographica, B43, 28–34.

CHEARY, R. W. 1988. The immobilisation of cesium in synroc. Materials Science Forum, 27/28, 397–406.

CROVISIER, J. L., ADVOCAT, T. & DUSSOSSOY, J. L. 2004. Relevance of analogues for long-term prediction. In: GIERÉ, R. & STILLE, P. (eds). Energy, Waste, and the Environment: a Geochemical Perspective. Geological Society, London, Special Publications, 236, 113–121.

DONALD, I. W., METCALFE, B. L. & TAYLOR, R. N. J. 1997. The immobilisation of high level radioactive wastes using ceramics and glasses. Journal of Materials Science, 32, 5851–5887.

EWING, R. C., WEBER, W. J. & CLINARD, F. W., JR. 1995. Radiation effects in nuclear waste forms for high-level radioactive waste. Progress in Nuclear Energy, 29, 63–127.

EWING, R. C., HAAKER, R. F., HEADLEY, T. J. & HLAVA, P. F. 1982. Zirconolites from Sri Lanka, South Africa and Brazil. In: TOPP, S. V. (ed) Scientific Basis for Nuclear Waste Management V. Elsevier, New York, 249–256.

FIELDING, P. E. & WHITE, T. J. 1987. Crystal chemical incorporation of high level waste species in aluminotitanate-based ceramics: valence, location, radiation damage, and hydrothermal durability. Journal of Materials Research, 2, 387–414.

FILLET, C., ADVOCAT, T. et al. 2002. Des matrices sur mesure pour les radionucléides à vie longue. Les researches pour la gestion des déchets nucléairs ISSN 0298-6248. CLEFS CEA, 46, 64–67.

FILLET, C., ADVOCAT, T., BART, F., LETURCQ, G. & RABILLER, H. 2004. Titanate ceramics for separated long-lived radionuclides. Compte rendu de l'Académie des Sciences, in press.

GIERÉ, R. & WILLIAMS, C. T. 1992. REE-bearing minerals in a Ti-rich vein from the Adamello contact aureole (Italy). Contributions to Mineralogy and Petrology, 112, 83–100.

GIERÉ, R., WILLIAMS, C. T. & LUMPKIN, G. R. 1998. Chemical characteristics of natural zirconolite. Schweiz. Mineral. Petrogr. Mitt., 78, 433–459.

GIERÉ, R., GUGGENHEIM, R. et al. 1994. Retention of actinides during alteration of aperiodic zirconolite. ICEM 13 Paris, 1269–1270.

GIERÉ, R., SWOPE, R. J., BUCK, E. C., GUGGENHEIM, R., MATHYS, D. & REUSSER, E. 2000. Growth and alteration of uranium-rich microlite. In: SMITH, R. W. & SHOESMITH, D. W. (eds) Scientific Basis for Nuclear Waste Management XXIII. Materials Research Society Symposium Proceedings, 608, 519–524.

GIERÉ, R., BUCK, E. C., GUGGENHEIM, R., MATHYS, D., REUSSER, E. & MARQUES, J. 2001. Alteration of

Uranium-rich Microlite. *In*: HART, K. P. &
LUMPKIN, G. R. (eds) Scientific Basis for Nuclear
Waste Management XXIV. *Materials Research
Society Symposium Proceedings*, **663**, 935–944.

HARKER, A. B. 1988. Tailored ceramics. *In*: LUTZE, W.
& EWING, R. C. (eds) *Radioactive Waste Forms
for the Future*. North-Holland, Amsterdam,
335–392.

HART, K. P., GLASSLEY, W. E. & McGLINN, P. J.
1992. Solubility control of actinide elements
leached from Synroc in pH-buffered solutions.
Radiochimica Acta, **58/59**, 33–35.

HART, K. P., LUMPKIN, G. R., GIERÉ, R.,
WILLIAMS, C. T., McGLINN, P. J. & PAYNE, T. E.
1996. Naturally occurring zirconolites – analogues
for the long-term encapsulation of actinides in
Synroc. *Radiochimica Acta*, **74**, 309–312.

HART, K. P., VANCE, E. R. *et al.* 1998. Leaching behaviour of zirconolite-rich synroc used to immmobi-
lise 'high-fired' plutonium oxide. *In*: McKINLEY,
I. G. & McCOMBIE, C. (eds) Scientific Basis for
Nuclear Waste Management XXI. *Materials
Research Society Symposium Proceedings*, **506**,
161–168.

HART, K. P., ZHANG, Y. *et al.* 2000. Aqueous durability of titanate ceramics designed to immobilise
excess Pu. *In*: SMITH, R. W. & SHOESMITH, D. W.
(eds) Scientific Basis for Nuclear Waste Management XXIII. *Materials Research Society Symposium Proceedings*, **608**, 353–358.

HELEAN, K. B., NAVROTSKY, A. *et al.* 2002. Enthalpies of formation of Ce–pyrochlore, $Ca_{0.93}Ce_{1.00}$
$Ti_{2.035}O_{7.00}$, U–pyrochlore, $Ca_{1.46}U_{0.23}^{4+} U_{0.46}^{6+}Ti_{1.85}$
$O_{7.00}$ and Gd–pyrochlore, $Gd_2Ti_2O_7$: three
materials relevant to the proposed waste form for
excess weapons plutonium. *Journal of Nuclear
Materials*, **303**, 226–239.

HELEAN, K. B., NAVROTSKY, A. *et al.* 2003. Enthalpies of formation of U–, Th–, Ce–brannerite:
implications for plutonium immobilisation.
Journal of Nuclear Materials, **320**, 231–244.

ICENHOWER, J. P., McGRAIL, B. P., SCHAEF, H. T. &
RODRIQUEZ, E. A. 2000. Dissolution kinetics of
titanium pyrochlore ceramics at 90 °C by single-
pass flow-through experiments. *In*: SMITH, R. W.
& SHOESMITH, D. W. (eds) Scientific Basis for
Nuclear Waste Management XXIII. *Materials
Research Society Symposium Proceedings*, **608**,
373–378.

ICENHOWER, J. P., STRACHAN, D. M., LINDBERG,
M. M., RODRIQUEZ, E. A. & STEELE, J. L. 2003.
*Dissolution Kinetics of Titanate-Based Ceramic
Waste Forms: Results from Single-Pass Flow
Tests on Radiation Damaged Specimens*. Pacific
Northwest National Laboratory, Report No.
PNNL-14252.

KAMIZONO, H., HAYAKAWA, I. & MURAOKA, S. 1991.
Durability of zirconium-containing ceramic waste
forms in water. *Journal of the American Ceramic
Society*, **74**, 863–864.

KASTRISSIOS, T., STEPHENSON, M., TURNER, P. S. &
WHITE, T. J. 1987. Hydrothermal dissolution of
perovskite: implications for synroc formulation.
Journal of the American Ceramic Society, **70**, C-
144–C-146.

KESSON, S. E. 1983. The immobilisation of cesium in
synroc hollandite. *Radioactive Waste Management
and the Nuclear Fuel Cycle*, **4**, 53–72.

KESSON, S. E. & WHITE, T. J. 1986a. $[Ba_xCs_y]$
$[(Ti,Al)_{2x+y}^{3+}Ti_{8-2x-y}^{4+}]O_{16}$ Synroc-type hollandites I.
Phase chemistry. *Proceedings of the Royal
Society of London*, A **405**, 73–101.

KESSON, S. E. & WHITE, T. J. 1986b. $[Ba_xCs_y]$
$[(Ti,Al)_{2x+y}^{3+}Ti_{8-2x-y}^{4+}]O_{16}$ Synroc-type hollandites
II. Structural chemistry. *Proceedings of the Royal
Society of London*, A **408**, 295–319.

KOGARKO, L. N., WILLIAMS, C. T. & WOOLLEY, A. R.
2002. Chemical evolution of loparite through the
layered, peralkaline Lovozero complex, Kola
Peninsula, Russia. *Mineralogy and Petrology*, **74**,
1–24.

LETURCQ, G., ADVOCAT, T., HART, K., BERGER, G.,
LACOMBE, G. & BONNETIER, A. 2001. Solubility
study of Ti,Zr-based ceramics designed to immobilize long-lived radionuclides. *American Mineralogist*, **86**, 871–880.

LUMPKIN, G. R. 1998. Rare-element mineralogy and
internal evolution of the Rutherford #2 pegmatite,
Amelia County, Virginia: a classic locality
revisited. *Canadian Mineralogist*, **36**, 339–353.

LUMPKIN, G. R. 2001. Alpha-decay damage and aqueous
durability of actinide host phases in natural systems.
Journal of Nuclear Materials, **289**, 136–166.

LUMPKIN, G. R. & EWING, R. C. 1992. Geochemical
alteration of pyrochlore group minerals: Microlite
subgroup. *American Mineralogist*, **77**, 179–188.

LUMPKIN, G. R. & EWING, R. C. 1995. Geochemical
alteration of pyrochlore group minerals: Pyrochlore
subgroup. *American Mineralogist*, **80**, 732–743.

LUMPKIN, G. R. & EWING, R. C. 1996. Geochemical
alteration of pyrochlore group minerals: Betafite
subgroup. *American Mineralogist*, **81**, 1237–1248.

LUMPKIN, G. R. & MARIANO, A. N. 1996. Natural
occurrence and stability of pyrochlore in carbonatites, related hydrothermal systems, and weathering
environments. *In*: MURPHY, W. M. & KNECHT,
D. A. (eds) Scientific Basis for Nuclear Waste
Management XIX. *Materials Research Society
Symposium Proceedings*, **412**, 831–838.

LUMPKIN, G. R., CHAKOUMAKOS, B. C. & EWING, R. C.
1986. Mineralogy and radiation effects of microlite
from the Harding pegmatite, Taos County, New
Mexico. *American Mineralogist*, **71**, 569–588.

LUMPKIN, G. R., SMITH, K. L. & BLACKFORD, M. G.
1991. Electron microscope study of Synroc
before and after exposure to aqueous solutions.
Journal of Materials Research, **6**, 2218–2233.

LUMPKIN, G. R., SMITH, K. L. & BLACKFORD, M. G.
1995. Development of secondary phases on
synroc leached at 150 °C. *In*: MURAKAMI, T. &
EWING, R. C. (eds) Scientific Basis for Nuclear
Waste Management XVIII. *Materials Research
Society Symposium Proceedings*, **353**, 855–862.

LUMPKIN, G. R., LEUNG, S. H. F. & COLELLA, M. 2000.
Composition, geochemical alteration, and alpha-
decay damage effects of natural brannerite. *In*:
SMITH, R. W. & SHOESMITH, D. W. (eds) Scientific

Basis for Nuclear Waste Management XXIII. *Materials Research Society Symposium Proceedings*, **608**, 359–365.

LUMPKIN, G. R., HART, K. P., MCGLINN, P. J., PAYNE, T. E., GIERÉ, R. & WILLIAMS, C. T. 1994. Retention of actinides in natural pyrochlores and zirconolites. *Radiochimica Acta*, **66/67**, 469–474.

LUMPKIN, G. R., COLELLA, M., SMITH, K. L., MITCHELL, R. H. & LARSEN, A. O. 1998. Chemical composition, geochemical alteration, and radiation effects in natural perovskite. *In*: MCKINLEY, I. G. & MCCOMBIE, C. (eds) Scientific Basis for Nuclear Waste Management XXI. *Materials Research Society Symposium Proceedings*, **506**, 207–214.

LUMPKIN, G. R., DAY, R. A., MCGLINN, P. J., PAYNE, T. E., GIERÉ, R. & WILLIAMS, C. T. 1999. Investigation of the long-term performance of betafite and zirconolite hydrothermal veins from Adamello, Italy. *In*: WRONKIEWICZ, D. J. & LEE, J. H. (eds) Scientific Basis for Nuclear Waste Management XXII. *Materials Research Society Symposium Proceedings*, **556**, 793–800.

MALMSTRÖM, J. C. 2000. Zirconolite: experiments on the stability in hydrothermal fluids. *Beiträge zur Geologie der Schweiz, Geotechnische Serie*, **93**, 130 pp.

MALMSTRÖM, J. C., REUSSER, E., GIERÉ, R., LUMPKIN, G. R., DÜGGELIN, M., MATHYS, D. & GUGGENHEIM, R. 1999. Zirconolite corrosion in dilute acidic and basic fluids at 180–700 °C and 50 MPa. *In*: WRONKIEWICZ, D. J. & LEE, J. H. (eds) Scientific Basis for Nuclear Waste Management XXII. *Materials Research Society Symposium Proceedings*, **556**, 165–172.

MALMSTRÖM, J. C., REUSSER, E. *et al.* 2000. Formation of perovskite and calzirtite during zirconolite alteration. *In*: SMITH, R. W. & SHOESMITH, D. W. (eds) Scientific Basis for Nuclear Waste Management XXIII. *Materials Research Society Symposium Proceedings*, **608**, 475–480.

MARIANO, A. N. 1989. Economic geology of rare earth minerals. *In*: LIPIN, B. R. & MCKAY, G. A. (eds) Geochemistry and mineralogy of rare earth elements. *Reviews in Mineralogy*, **21**, 309–337.

MAZZI, F. & MUNNO, R. 1983. Calciobetafite (new mineral from the pyrochlore group) and related minerals from Campi Flegrei, Italy: crystal structure of polymignite and zirkelite: comparison with pyrochlore and zirconolite. *American Mineralogist*, **68**, 262–276.

MCGLINN, P. J., HART, K. P., LOI, E. H. & VANCE, E. R. 1995. pH Dependence of the aqueous dissolution rates of perovskite and zirconolite at 90 °C. *In*: MURAKAMI, T. & EWING, R. C. (eds) Scientific Basis for Nuclear Waste Management XVIII. *Materials Research Society Symposium Proceedings*, **353**, 847–854.

MITAMURA, H., MATSUMOTO, S. *et al.* 1992. Aging effects on curium-doped titanate ceramic containing sodium-bearing high-level nuclear waste. *Journal of the American Ceramic Society*, **75**, 392–400.

MITAMURA, H., MATSUMOTO, S. *et al.* 1994. α-Decay damage effects in curium-doped titanate ceramic containing sodium-free high-level nuclear waste. *Journal of the American Ceramic Society*, **77**, 2255–2264.

MITCHELL, R. H. 1996. Perovskites: a revised classification scheme for an important rare earth element host in alkaline rocks. *In*: JONES, A. P., WALL, F. & WILLIAMS, C. T. (eds) *Rare Earth Minerals, Chemistry, Origin and Ore Deposits*. Chapman and Hall, London, 41–76.

MITCHELL, R. H. & CHAKHMOURADIAN, A. R. 1998a. Th-rich loparite from the Khibina alkaline complex, Kola Peninsula: isomorphism and paragenesis. *Mineralogical Magazine*, **62**, 341–353.

MITCHELL, R. H. & CHAKHMOURADIAN, A. R. 1998b. Instability of perovskite in a CO$_2$-rich environment: examples from carbonatite and kimberlite. *Canadian Mineralogist*, **36**, 939–952.

MURAKAMI, T. 1985. Crystalline product on surface of synroc after long leaching. *Journal of Nuclear Materials*, **135**, 288–291.

MYHRA, S., WHITE, T. J., KESSON, S. E. & RIVIÉRE, J. C. 1988a. X-ray photoelectron spectroscopy for the direct identification of Ti valence in [Ba$_x$Cs$_y$][(Ti,Al)$^{3+}_{2x+y}$Ti$^{4+}_{8-2x-y}$]O$_{16}$ hollandites. *American Mineralogist*, **73**, 161–167.

MYHRA, S., SMART, R. ST. C. & TURNER, P. S. 1988b. The surfaces of titanate minerals, ceramics and silicate glasses: surface analytical and electron microscope studies. *Scanning Microscopy*, **2**, 715–734.

MYHRA, S., DELOGU, P., GIORGI, R. & RIVIÉRE, J. C. 1988c. Scanning and high-resolution Auger analysis of zirconolite/perovskite surfaces following hydrothermal treatment. *Journal of Materials Science*, **23**, 1514–1520.

MYHRA, S., BISHOP, H. E., RIVIÉRE, J. C. & STEPHENSON, M. 1987. Hydrothermal dissolution of perovskite (CaTiO$_3$). *Journal of Materials Science*, **22**, 3217–3226.

NASRAOUI, M., BILAL, E. & GIBERT, R. 1999. Fresh and weathered pyrochlore studies by Fourier transform infrared spectroscopy coupled with thermal analysis. *Mineralogical Magazine*, **63**, 567–578.

NESBITT, H. W., BANCROFT, G. M., FYFE, W. S., KARKHANIS, S. N. & NISHIJIMA, A. 1981. Thermodynamic stability and kinetics of perovskite dissolution. *Nature*, **289**, 358–362.

OHNENSTETTER, D. & PIANTONE, P. 1992. Pyrochlore-group minerals in the Beauvoir peraluminous leucogranite, Massif Central, France. *Canadian Mineralogist*, **30**, 771–784.

OVERSBY, V. M. & RINGWOOD, A. E. 1981. Lead isotopic studies of zirconolite and perovskite and their implications for long range synroc stability. *Radioactive Waste Management*, **1**, 289–307.

PAN, Y. 1997. Zircon- and monazite-forming metamorphic reactions at Manitouwadge, Ontario. *Canadian Mineralogist*, **35**, 105–118.

PHAM, D. K., MYHRA, S. & TURNER, P. S. 1994. The surface reactivity of hollandite in aqueous solution. *Journal of Materials Research*, **9**, 3174–3181.

PHAM, D. K., NEALL, F. B., MYHRA, S., SMART, R. ST. C. & TURNER, P. S. 1989. Dissolution mechanisms of

$CaTiO_3$ – solution analysis, surface analysis and electron microscope studies – implications for Synroc. *In*: LUTZE, W. & EWING, R. C. (eds) Scientific Basis for Nuclear Waste Management XII. *Materials Research Society Symposium Proceedings*, **127**, 231–240.

POST, J. E., VON DREELE, R. B. & BUSECK, P. R. 1982. Symmetry and cation displacements in hollandites: structure refinements of hollandite, cryptomelane and priderite. *Acta Crystallographica*, **B38**, 1056–1065.

PUTNAM, R. L., NAVROTSKY, A., WOODFIELD, B. F., BOERIO-GOATES, J. & SHAPIRO, J. L. 1999. Thermodynamics of formation for zirconolite ($CaZrTi_2O_7$) from T = 298.15 K to T = 1500 K. *Journal of Chemical Thermodynamics*, **31**, 229–243.

REEVE, K. D., LEVINS, D. M., SEATONBERRY, B. W., RYAN, R. K., HART, K. P. & STEVENS, G. T. 1989. Fabrication and leach testing of synroc containing actinides and fission products. *In*: LUTZE, W. & EWING, R. C. (eds) Scientific Basis for Nuclear Waste Management XII. *Materials Research Society Symposium Proceedings*, **127**, 223–230.

RINGWOOD, A. E., KESSON, S. E., REEVE, K. D., LEVINS, D. M. & RAMM, E. J. 1988. Synroc. *In*: LUTZE, W. & EWING, R. C. (eds) *Radioactive Waste Forms for the Future*. North-Holland, Amsterdam, 233–334.

ROBERTS, S. K., BOURCIER, W. L. & SHAW, H. F. 2000. Aqueous dissolution kinetics of pyrochlore, zirconolite and brannerite at 25, 50, and 75 °C. *Radiochimica Acta*, **88**, 539–543.

SHOUP, S. S., BAMBERGER, C. E., HAVERLOCK, T. J. & PETERSON, J. R. 1997. Aqueous leachability of lanthanide and plutonium titanates. *Journal of Nuclear Materials*, **240**, 112–117.

SINCLAIR, W. & EGGLETON, R. A. 1982. Structure refinement of zirkelite from Kaiserstuhl, West Germany. *American Mineralogist*, **67**, 615–620.

SINCLAIR, W. & MCLAUGHLIN, G. M. 1982. Structure refinement of priderite. *Acta Crystallographica*, **B38**, 245–246.

SMITH, K. L. & LUMPKIN, G. R. 1993. Structural features of zirconolite, hollandite and perovskite, the major waste-bearing phases in synroc. *In*: BOLAND, J. N. & FITZ GERALD, J. D. (eds) *Defects and Processes in the Solid State: Geoscience Applications, The McLaren Volume*. Elsevier, Amsterdam, 401–422.

SMITH, K. L., ZALUZEC, N. J. & LUMPKIN, G. R. 1998. In situ studies of ion irradiated zirconolite, pyrochlore and perovskite. *Journal of Nuclear Materials*, **250**, 36–52.

SMITH, K. L., HART, K. P., LUMPKIN, G. R., MCGLINN, P., LAM, P. & BLACKFORD, M. G. 1991. A description of the kinetics and mechanisms which control the release of HLW elements from synroc. *In*: ABRAJANO, T., JR. & JOHNSON, L. H. (eds) Scientific Basis for Nuclear Waste Management XIV. *Materials Research Society Symposium Proceedings*, **212**, 167–174.

SMITH, K. L., LUMPKIN, G. R., BLACKFORD, M. G., DAY, R. A. & HART, K. P. 1992. The durability of Synroc. *Journal of Materials Research*, **190**, 287–294.

SMITH, K. L., BLACKFORD, M. G., LUMPKIN, G. R., HART, K. P. & ROBINSON, B. J. 1996. Neptunium-doped Synroc: partitioning, leach data and secondary phase development. *In*: MURPHY, W. M. & KNECHT, D. A. (eds) Scientific Basis for Nuclear Waste Management XIX. *Materials Research Society Symposium Proceedings*, **412**, 313–319.

SMITH, K. L., COLELLA, M. *et al.* 1997a. Dissolution of Synroc in deionised water at 150 °C. *In*: GRAY, W. J. & TRIAY, R. (eds) Scientific Basis for Nuclear Waste Management XX. *Materials Research Society Symposium Proceedings*, **465**, 349–354.

SMITH, K. L., LUMPKIN, G. R., BLACKFORD, M. G., HAMBLEY, M., DAY, R. A., HART, K. P. & JOSTSONS, A. 1997b. Characterisation and leaching behavior of plutonium-bearing Synroc-C. *In*: GRAY, W. J. & TRIAY, R. (eds) Scientific Basis for Nuclear Waste Management XX. *Materials Research Society Symposium Proceedings*, **465**, 1267–1272.

SOLOMAH, A. G. & MATZKE, HJ. 1989. Leaching studies of synroc crystalline ceramic waste forms. *In*: LUTZE, W. & EWING, R. C. (eds) Scientific Basis for Nuclear Waste Management XII. *Materials Research Society Symposium Proceedings*, **127**, 241–248.

STEFANOVSKY, S. V., YUDINTSEV, S. V., GIERÉ, R. & LUMPKIN, G. R. 2004. Nuclear waste forms. *In*: GIERÉ, R. & STILLE, P. (eds) *Energy, Waste, and the Environment: a Geochemical Perspective*. Geological Society, London, Special Publications, **236**, 37–63.

STEWART, M. W. A., BEGG, B. D. *et al.* 2003. Ion irradiation damage in zirconate and titanate ceramics for plutonium disposition. *Proceedings of ICEM '03: The 9th International Conference on Radioactive Waste Management and Environmental Remediation*, in press.

SUBRAMANIAN, M. A., ARAVAMUDAN, G. & SUBBA RAO, G. V. 1983. Oxide pyrochlores – a review. *Progress in Solid State Chemistry*, **15**, 55–143.

SZYMAŃSKI, J. T. & SCOTT, J. D. 1982. A crystal structure refinement of synthetic brannerite, UTi_2O_6, and its bearing on rate of alkaline-carbonate leching of brannerite in ore. *Canadian Mineralogist*, **20**, 271–279.

TROCELLIER, P. 2000. Immobilisation of radionuclides in single-phase crystalline waste forms: a review on their intrinsic properties and long term behaviour. *Ann. Chim. Sci. Mat.*, **25**, 321–337.

TROCELLIER, P. 2001. Chemical durability of high level nuclear waste forms. *Ann. Chim. Sci. Mat.*, **26**, 113–130.

TURNER, P. S., JONES, C. F., MYHRA, S., NEALL, F. B., PHAM, D. K. & SMART, R. ST. C. 1989. Dissolution mechanisms of oxides and titanate cermics – electron microscope and surface analytical studies. *In*: DUFOUR, L.-C., MONTY, C. & PETOT-ERVAS, G. (eds) *Surfaces and Interfaces of Ceramic*

Materials. Kluwer Academic Publishers, Dordrecht, 663–690.

VANCE, E. R. 1994. Synroc: a suitable waste form for actinides. *Materials Research Society Bulletin*, **XIX**, 28–32.

VANCE, E. R., WATSON, J. N. *et al.* 2000. Crystal chemistry, radiation effects and aqueous leaching of brannerite, UTi_2O_6. *Ceramic Transactions*, **107**, 561–568.

WALL, F., WILLIAMS, C. T. & WOOLLEY, A. R. 1999. Pyrochlore in niobium ore deposits. *In*: STANLEY, C. J. (ed) *Mineral Deposits: Processes to Processing*. Balkema Publishers, Rotterdam, Vol. 1, 687–690.

WALL, F., WILLIAMS, C. T., WOOLLEY, A. R. & NASRAOUI, M. 1996. Pyrochlore from weathered carbonatite at Lueshe, Zaire. *Mineralogical Magazine*, **60**, 731–750.

WEBER, W. J., WALD, J. W. & MATZKE, HJ. 1986. Effects of self-radiation damage in Cm-doped $Gd_2Ti_2O_7$ and $CaZrTi_2O_7$. *Journal of Nuclear Materials*, **138**, 196–209.

WEBER, W. J., EWING, R. C. *et al.* 1998. Radiation effects in crystalline ceramics for the immmobilization of high-level nuclear waste and plutonium. *Journal of Materials Research*, **13**, 1434–1484.

WILLIAMS, C. T., WALL, F., WOOLLEY, A. R. & PHILLIPO, S. 1997. Compositional variation in pyrochlore from the Bingo carbonatite, Zaire. *Journal of African Earth Sciences*, **25**, 137–145.

WILLIAMS, C. T., BULAKH, A. G., GIERÉ, R., LUMPKIN, G. R. & MARIANO, A. N. 2001. Alteration features in natural zirconolites from carbonatites. *In*: HART, K. P. & LUMPKIN, G. R. (eds) Scientific Basis for Nuclear Waste Management XXIV. *Materials Research Society Symposium Proceedings*, **663**, 945–952.

WISE, M. A. & ČERNÝ, P. 1990. Primary compositional range and alteration trends of microlite from the Yellowknife pegmatite field, Northwest Territories, Canada. *Mineralogy and Petrology*, **43**, 83–98.

ZHANG, Y., HART, K. P. *et al.* 2001a. Durabilities of pyrochlore ceramics designed for the immobilisation of surplus plutonium. *In*: HART, K. P. & LUMPKIN, G. R. (eds) Scientific Basis for Nuclear Waste Management XXIV. *Materials Research Society Symposium Proceedings*, **663**, 325–332.

ZHANG, Y., HART, K. P. *et al.* 2001b. Kinetics of uranium release from Synroc phases. *Journal of Nuclear Materials*, **289**, 254–262.

ZHANG, Y., THOMAS, B. S., LUMPKIN, G. R., BLACKFORD, M., ZHANG, Z., COLELLA, M. & ALY, Z. 2003. Dissolution of synthetic brannerite in acidic and alkaline fluids. *Journal of Nuclear Materials*, **321**, 1–7.

Relevance of analogues for long-term prediction

JEAN LOUIS CROVISIER[1], THIERRY ADVOCAT[2] & JEAN LUC DUSSOSSOY[2]

[1]*ULP-EOST-CNRS, Centre de Géochimie de la Surface UMR 7517, Strasbourg, France*
(e-mail: jlc@illite.u-strasbg.fr)
[2]*CEA-VRH/DEN-DIEC/SCDV, Marcoule BP17171, Bagnols/Cèze, France*

Abstract: The long-term consequences on the environment of materials containing potentially toxic elements must be assessed on the basis of experimental data obtained over short
time-scales and from models of the phenomena involved over several tens of thousands of
years. Predicting the future is only possible through the study of collection of past events to
infer a possible long-term behaviour. However, not everything we would like to predict has
already occurred in the past, nor is it necessarily observable under perfectly similar circumstances. Considering natural or artificial analogues permits us to study materials that, even if
not homologous, are similar or equivalent to some of the properties of the unknown
materials. Examples presented in this paper illustrates the fact that some reactions observed
in short-term experiments can be validated over the long term only by examining such
natural analogues.

Society today produces increasing quantities of
materials containing elements potentially toxic
for the environment. Most of these materials
come from conventional industry, incineration
of household waste, and nuclear activities, and
consist mainly of clinker, cement, or glass.
Although their corrosion resistance is identified
and quantified by a wide range of experimental
tests performed at laboratory scale, their long-
term consequences on the environment must be
assessed on the basis of experimental data
obtained over short time-scales and from models
of the phenomena involved (migration of heavy
metals, effect on the pH of the surrounding
media, and so on).

The scientific issue is thus one of predicting
the long-term behaviour of these materials and
their impact on the environment over several
tens of thousands of years in the case of some
nuclear waste and – a point that is often overlooked – for all eternity in the case of some
other wasteforms such as lead or cadmium
arising from industrial activities that are generally considered more mundane (e.g., chemical
industries). The long-term behaviour cannot be
inferred exclusively from laws using parameters
measured experimentally in the laboratory. The
geological environment in which these materials
may be placed is complex, and many of the parameters are subject to variation: for example the
nature of the fluids, climatic changes, and so on.
Most of the parameters are interdependent:
radionuclide behaviour, for example, cannot be

understood independently of phenomena such
as co-precipitation or adsorption of other major
or minor system constituents.

This short chapter does not address all the
issues related to predicting the long-term behaviour of materials. Our goal is simply to evoke
a few key concepts relevant to the use of
natural analogues.

Predicting the future

Predicting the future is an age-old dream of
humanity. Little progress has been made in this
area since Aristotle (384–322 BC). Fortunately,
he had already laid down the mechanisms of
the syllogism, of induction and deduction that
today form the basis for computer logic. The
question: 'Will the sun rise tomorrow?' was
already answered by Aristotle: 'Probably, since
it rose yesterday, and the day before, and has
risen in the same way for a very long time.' Is
this a proof? No, it is an empirical generalization.
By induction, based on a collection of past events
(or events observed in the present), we infer a
probable future or a possible behaviour. The
degree of probability depends, of course, on
the number of cases observed and on their
effective equivalence. For example, by observing
the temperature-dependent volume and pressure
changes of a gas, we can inductively infer the
ideal gas law with a high degree of probability.
Conversely, the observation that 45% of a sample of French voters support a particular

From: GIERÉ, R. & STILLE, P. (eds) 2004. *Energy, Waste, and the Environment: a Geochemical
Perspective*. Geological Society, London, Special Publications, **236**, 113–121.

presidential candidate does not allow us to infer with the same degree of certainty that this candidate will receive about 45% of the actual votes.

Moreover, this method cannot be generalized since not everything we would like to predict has already occurred in the past, nor is it necessarily observable under perfectly similar circumstances. This is the case in particular for new waste containment matrices. The problem of the specific nature of unknown materials has been widely discussed in the literature (Brookins 1976; Ewing 1979; Ewing & Jercinovic 1987; Petit 1992). The idea proposed by these authors is to study materials that, even if not homologous, are similar or equivalent to some of the properties of the unknown materials. Hence the concept of 'natural analogues'. These analogues, if not identical, must at least have some common properties (e.g., composition, structure, or complexity level) with the new material (Petit 1992). For example, Roman cements 2000 years old could be considered suitable analogues for today's cement matrices. Some authors have studied stained glass windows in medieval cathedrals or ancient basaltic glasses as analogues for radioactive waste containment glass. In 1976, Brookins suggested that the natural reactor at Oklo (Gabon) could be considered as a geological laboratory to investigate nuclear waste disposal.

Such studies are sometimes criticized because they appear to be based on the concept of a global analogy between the specific example studied and a waste repository site (e.g., Miller *et al.* 1994). Clearly no geological site fully resembles a radioactive waste repository site. Confusion also seems to arise from the fact that analogy has sometimes been considered as a particular form of reasoning distinct from conventional logic. We maintain that this is not the case: there is and has been only one logic from Plato and Aristotle to Russel (1913).

The following discussion limited to basaltic glasses illustrates the fact that natural analogues are irreplaceable for the development of predictive models.

Basaltic glasses as analogues for glass waste

Activation energies

Since the 1980s, several authors have investigated the initial alteration of basaltic glass at temperatures ranging from 3 °C to 300 °C (Crovisier *et al.* 1985, 1989a; Gislason & Eugster 1987; Guy & Schott 1989; Daux *et al.* 1997; Techer 1999; Techer *et al.* 2000). The reported values are plotted logarithmically versus the reciprocal of the temperature in Fig. 1, forming a linear relation throughout the considered temperature range:

$$\ln(r_0) = \ln(A) + \frac{-E_a}{RT}$$

where r_0 is the initial alteration rate (mole$\cdot$m^{-2} s^{-1}), A the pre-exponential factor (mole$\cdot$m^{-2} s^{-1}), R the ideal gas constant, and T the temperature (K). The activation energy E_a (J$\cdot$mol^{-1}) is the product of $s \times R$, where s is the slope of the regression line. The literature data are corrected to the same pH value of 8.4, using Guy & Schott data on the evolution of r_0 with pH for temperatures ranging from 50 to 200 °C (Guy & Schott 1989). The activation energy of

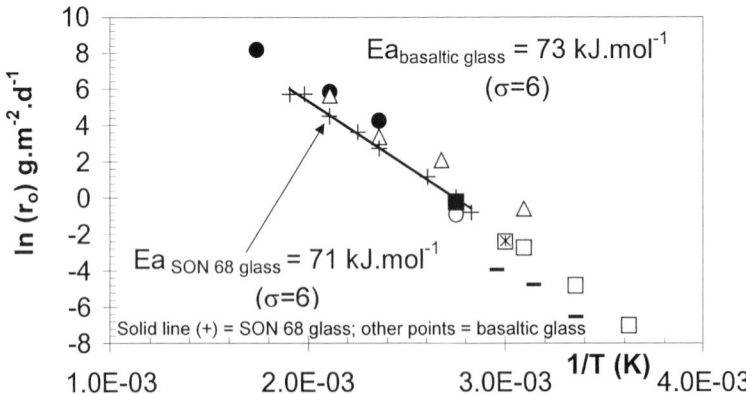

Fig. 1. Calculated activation energies of radioactive waste containment glass and basaltic glasses (Techer, 1999; Techer *et al.* 2000).

the initial basaltic glass alteration reaction is constant between 3 and 300 °C at a value of 72.4 kJ·mol^{-1} ($2s = 8$). This is comparable to the value of 71 kJ·mol^{-1} obtained earlier for the SON68 nuclear glass dissolution reaction (Delage & Dussossoy 1991) (Fig. 1). The initial alteration of basaltic glass could thus be characterized by mechanisms similar to those involved during SON68 glass alteration. The similarity of the activation energies is important for comparing the alteration kinetics of the two materials: the initial dissolution of both silicates under equivalent conditions is governed by surface reactions and not by diffusion.

Alteration layers

Another property common to both basaltic glasses and waste containment glasses is the very rapid formation of surface alteration layers consisting of the products of the glass/solution reaction. This property has already been highlighted by several authors (Ewing 1979; Malow & Ewing 1981; Malow *et al.* 1984; Byers *et al.* 1985, 1987*a*, *b*; Grambow *et al.* 1985; Lutze *et al.* 1985; Ewing & Jercinovic 1987; Jercinovic & Ewing 1987; Crovisier *et al.* 1989*a*, *b*, 1992*a*; Petit 1992; Petit & Côme 1994; Steinmann *et al.* 1999; Techer 1999; Techer *et al.* 2000, 2001*a*, *b*). Figure 2 shows an example of alteration layers viewed with an electron microscope. A porous zone is visible at the interface between the glass and the overlying phyllosilicates. These layers generally trap a large fraction

of the potentially toxic elements either in the structure of the secondary minerals or by adsorption (Chapman *et al.* 1984; Murakami *et al.* 1988; Advocat 1991; Crovisier *et al.* 1992*b*; Abdelouas *et al.* 1995, 1997; Gong *et al.* 1998; Techer 1999; Gauthier *et al.* 1999; Advocat *et al.* 2001; De Putter *et al.* 2002). It is interesting to note that these alteration layers are also observed on the surface of ancient basaltic glasses altered in marine or continental environments (von Waltershausen 1845; Nayudu 1964; Bonatti 1965; Moore 1965; Honnorez 1972, 1981; Jakobsson 1972; Eggleton & Keller 1982; Crovisier *et al.* 1989*a*, 1992*b*; Jercinovic *et al.* 1990; Le Gal 1999; Le Gal *et al.* 1999). The importance of studying these natural samples is immediately obvious, especially with regard to the rare earth elements whose behaviour resembles that of some radioactive elements. Recent work in this area has shown that the rare earth elements are trapped in such basaltic alteration layers (Steinmann & Stille 1998; Steinmann *et al.* 1999, 2001).

From the laboratory to the natural environment by calculation

The following example concerns a basaltic glass altered experimentally in seawater at temperatures between 25 and 90 °C for durations ranging from a few minutes to 30 days (Crovisier *et al.* 1982; Thomassin & Touray 1982). Regardless of the temperature, the first secondary mineral observed was a hydroxycarbonate from

Basaltic glass

Nuclear glass

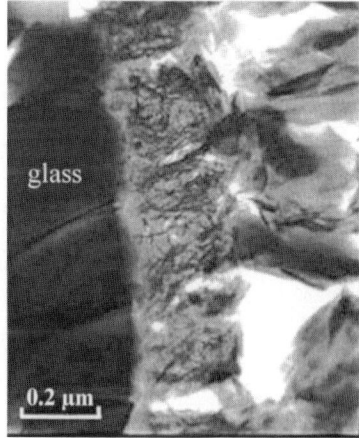 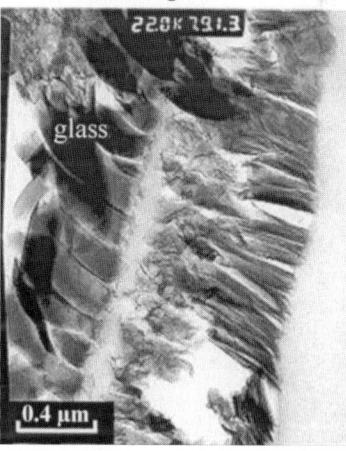

Fig. 2. Alteration layers observed on the surface of basaltic and nuclear glasses, comprising a porous zone and phyllosilicates: (**a**) tholeiitic basalt glass altered 120 days at 50 °C (Crovisier *et al.* 1987); (**b**) MW nuclear glass altered 5.7 years at 90 °C (Curti *et al.* in preparation).

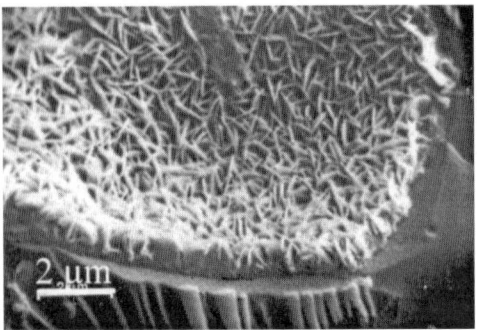

Fig. 3. Hydrotalcite $Mg_6Al_2CO_3(OH)_{16} \cdot 4H_2O$ formed experimentally by alteration of basaltic glass in seawater at 50 °C (30 days).

the hydrotalcite family $Mg_6(Al, Fe)_2CO_3 (OH)_{16} \cdot 4H_2O$), forming an alteration layer on the glass surface (Fig. 3).

These experimental findings raised a problem: hydrotalcite had never been described in the natural environment as a basaltic glass alteration product. The mineral association generally observed is a saponite phyllosilicate and zeolites (Honnorez 1967, 1972, 1978, 1981; Melson & Thompson 1973; Hoffert *et al.* 1978; Kastner 1979; Karpoff *et al.* 1980; Mevel 1980). Two hypotheses were proposed to account for this situation: (1) the experimental medium was too simple compared with the natural environment, or (2) hydrotalcite is unstable over the long term. We tested the second hypothesis using the DISSOL thermodynamic simulation model (Fritz 1975, 1981). This model is derived from work by Garrels & Thompson (1962), Helgeson (1968), and Helgeson *et al.* (1971). They are based on a constant equilibrium approach. The models describe reaction paths when dissolving a reactant, here the glass, iteratively incrementing the progress variable ζ and checking the aqueous phase for saturation with regard to minerals from a data bank, and then precipitating or dissolving the required mass of appropriate mineral phase to maintain equilibrium.

The results are shown in Fig. 4. Figure 4a indicates the saturation values ($\log Q/K$) of the

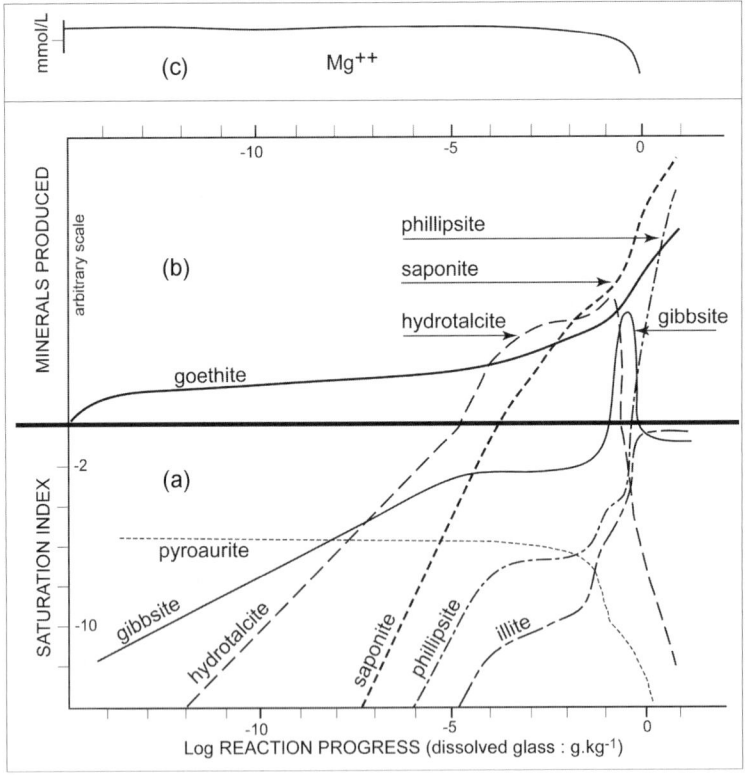

Fig. 4. Calculated estimate of the long-term stability of hydrotalcite in seawater at 60 °C: (**a**) log ions activity product/equilibrium constant (log IAP/K); (**b**) quantity of minerals formed; (**c**) magnesium concentration in solution.

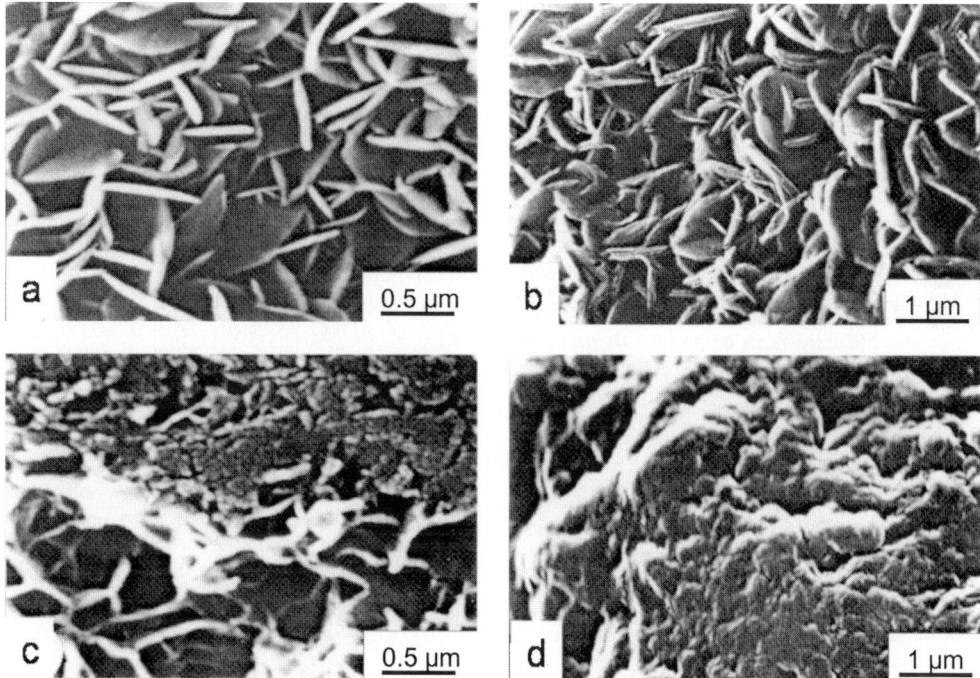

Fig. 5. Hydrotalcite crystals formed by alteration of basaltic glass in seawater at 50 °C: (**a**) 10 days, (**b**) 30 days, (**c**) 120 days, (**d**) 600 days.

principal minerals tested. Note that an iron hydroxide is the first mineral assumed to form, followed by hydrotalcite and finally saponite and phillipsite (Fig. 4b). The hydrotalcite is then assumed to dissolve when the magnesium concentration diminishes in solution (Fig. 4c); it remains largely undersaturated as the reaction progresses. The model thus allowed us to conclude that hydrotalcite is probably not present in the marine environment because of its long-term instability when saponite forms, and not because the experimental medium was oversimplified. The model also predicts the formation of phillipsite at the most advanced stages of reaction progress. We were thus able to model the paragenesis observed in the marine environment (Honnorez 1981).

Laboratory experiments of longer duration confirmed the model prediction. Figure 5 is made of scanning electron micrographs showing the surface condition after 10, 30, 120, and 600 days at 50 °C. The crystals gradually become covered by a deposit and are entirely masked after 600 days. Transmission electron microscope examination of an ultramicrotome cross-section of the layer formed after 600 days shows (1) exfoliation of the outer hydrotalcite

crystals and (2) their replacement by a silicate similar to saponite (Fig. 6).

This provides experimental evidence confirming the model predictions: hydrotalcite is not stable over the long term and our experimental

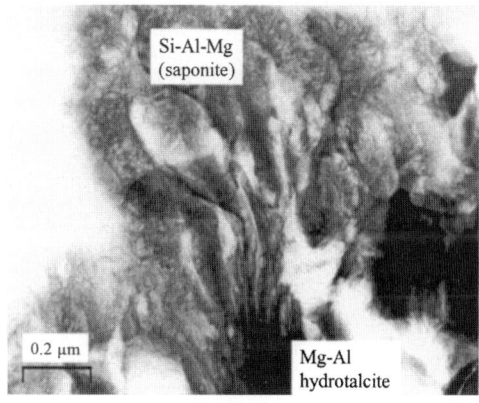

Fig. 6. Outer region of the alteration layer formed after 600 days. Note that the outer hydrotalcite crystals are replaced by a silicate.

protocol is not oversimplified compared with the natural environment.

Nevertheless, hydrotalcite is not a rare mineral. It has been observed earlier as a nuclear glass alteration product in MgCl brines (Abdelouas et al. 1993, 1994). It is also described as an alteration mineral of slags and cements (Mascolo 1973, 1986; Mascolo & Marino 1980; Faucon et al. 1996).

Conclusions

Predicting the long-term behaviour of materials (performance assessment) is a vast issue that cannot be covered in a brief discussion. The work presented here simply illustrates the fact that some reactions observed in short-term experiments can be validated over the long term only by examining natural analogues. It also demonstrates that models are powerful tools for testing such hypotheses. In this chapter we have seen why hydrotalcite is observed in laboratory alteration of basaltic glass but not in the oceanic environment. This illustrates the fact that the construction of safety models is a three-pronged endeavour, involving experimentation, natural analogues, and numerical modeling.

This study has been supported by grants from the European Commission (GLASTAB project, contract FIKW-CT-2000-00007). This is EOST/CGS contribution no. 2003. 402-UMR7517. Our thanks to Greg Lumpkin, an anonymous reviewer, and Peter Stille for the improvement of the manuscript.

References

ABDELOUAS, A., CROVISIER, J. L., LUTZE, W. & BERNOTAT, W. 1993. Formation de l'hydrotalcite au cours de l'altération du verre nucléaire R7T7 dans une saumure à 190°C. Comptes Rendus Académie des Sciences, Paris, II., 317, 1067–1072.

ABDELOUAS, A., CROVISIER, J. L., LUTZE, W., FRITZ, B., MOSSER, A. & MÜLLER R. 1994. Formation of hydrotalcite-like compounds during R7T7 nuclear waste glass and basaltic glass alteration. Clays and Clay Minerals, 42, 526–533.

ABDELOUAS, A., CROVISIER, J. L., LUTZE, W., MÜLLER, R. & BERNOTAT, W. 1995. Structure and chemical properties of surface layers developed on R7T7 simulated nuclear waste glass altered in brine at 190°C. European Journal of Mineralogy, 7, 1101–1113.

ABDELOUAS, A., GRAMBOW, B., MULLER, R., CROVISIER, J.-L., LUTZE, W. & DRAN, J.-C. 1997. Surface layers on a borosilicate nuclear waste glass corroded in MgCl₂ solution. Journal of Nuclear Materials, 240, 100–111.

ADVOCAT, T. 1991. Les mécanismes de corrosion en phase aqueuse du verre nucléaire R7T7. Approche expérimentale. Essai de modélisation thermodynamique et cinétique. PhD thesis, Université Louis Pasteur, Strasbourg, 213 p.

ADVOCAT, T., JOLLIVET, P., CROVISIER, J. L. & DEL NERO, M. 2001. Long-term alteration mechanisms in water for SON68 radioactive borosilicate glass. Journal of Nuclear Materials, 298, 55–62.

BONATTI, E. 1965. Palagonite, hyaloclastites and alteration of volcanic glass in the ocean. Bulletin of Volcanology, 28, 257–269.

BROOKINS, D. G. 1976. Shale as a repository for radioactive waste: the evidence from Oklo. Environmental Geology, I, 225–269.

BYERS, C. D., EWING, R. C. & JERCINOVIC, M. J. 1987a. Experimental alteration of basalt glass applied to the alteration of nuclear waste glass. Advances in Ceramics, 20, 733–744.

BYERS, C. D., JERCINOVIC, M. J. & EWING, R. C. 1987b. A study of natural glass analogues as applied to alteration of nuclear waste glass. Report no. NUREG/CR-4842 ANL-86-46, Argonne National Laboratory, 150 p.

BYERS, C. D., JERCINOVIC, M. J., EWING, R. C. & KEIL, K. 1985. Basalt glass: an analogue for the evaluation of the long-term stability of nuclear waste form borosilicate glasses. In: JANTZEN, C. M., STONE, J. A. & EWING R. C. (eds) Scientific Basis for Nuclear Waste Management VIII. Material Research Society Symposium Proceedings, 44, 583–590.

CHAPMAN, N. A., MCKINLEY, I. G. & SMELLIE, J. A. T. 1984. The potential of natural analogues in assessing systems for deep disposal of high-level radioactive waste. Technical Report SKB 84-16, Swedish Nuclear Fuel and Waste Management Co., Stockholm, 103 p.

CROVISIER, J. L., EBERHART, J. P., THOMASSIN, J. H., JUTEAU, T., TOURAY, J. C. & EHRET, G. 1982. Interaction eau de mer-verre basaltique à 50°C. Formation d'un hydroxycarbonate et de produits silicatés amorphes (Al, Mg) et mal cristallisés (Al, Fe, Mg). Etude en microscopie électronique et par spectrométrie des photoélectrons (ESCA). Comptes Rendus Académies des Sciences, Paris, 294, 989–994.

CROVISIER, J. L., FRITZ, B., GRAMBOW, B., & EBERHART, J. P. 1985. Dissolution of basaltic glass in seawater: experiments and thermodynamic modelling. In: WERME, L. O. (ed) Scientific Basis for Nuclear Waste Management IX. Material Research Society Symposium Proceedings, 50, 273–280.

CROVISIER, J. L., HONNOREZ, J. & EBERHART, J. P. 1987. Dissolution of basaltic glass in seawater: Mechanism and rate. Geochimica et Cosmochimica Acta, 51, 2977–2990.

CROVISIER, J. L., ADVOCAT, T., PETIT, J. C. & FRITZ, B. 1989a. Alteration of basaltic glass in Iceland as a natural analogue for nuclear waste glasses: geochemical modelling with DISSOL. In: LUTZE, W. & EWING, R. C. (eds) Scientific Basis for Nuclear Waste Management XII. Material Research Society Symposium Proceedings, 127, 41–48.

CROVISIER, J. L., ATASSI, H., DAUX, V., HONNOREZ, J., PETIT, J. C. & EBERHART, J. P. 1989b. A new insight into the nature of the leached layers formed on basaltic glasses in relation to the choice of constraints for long term modelling. *In*: LUTZE, W. & EWING, R. C. (eds) Scientific Basis for Nuclear Waste Management XII. *Material Research Society Symposium Proceedings*, **127**, 41–48.

CROVISIER, J. L., HONNOREZ, J., FRITZ, B. & PETIT, J. C. 1992a. Dissolution of subglacial volcanic glasses from Iceland: laboratory study and modelling. *Applied Geochemistry*, **SI 1**, 55–81.

CROVISIER, J. L., VERNAZ, E., DUSSOSSOY, J. L. & CAUREL, J. 1992b. Early phyllosilicates formed by alteration of R7T7 glass in water at 250°C. *Applied Clay Science*, **7**, 47–57.

CURTI, E., CROVISIER, J.-L., KARPOFF, A.-M. & MUNIER, I. Long-term glass alteration of two simulated nuclear waste glasses (MW and SON68): kinetics and geochemical results. (in preparation).

DAUX, V., GUY, C., ADVOCAT, T., CROVISIER, J.-L., & STILLE, P. 1997. Kinetic aspects of basaltic glass dissolution at 90°C: role of aqueous silicon and aluminium. *Chemical Geology*, **142**, 109–126.

DELAGE, F. & DUSSOSSOY, J. L. 1991. R7T7 glass initial dissolution rate measurements using a high-temperature soxhlet device. In: ABRAJANO, T., JR. & JOHNSON, L. H. (eds) Scientific Basis for Nuclear Waste Management XIV. *Material Research Society Symposium Proceedings*, **212**, 41–47.

DE PUTTER, T., ANDRE, L., BERNARD, A., DUPUIS, C., JEDWAB, J., NICAISE, D. & PERRUCHOT, A. 2002. Trace element (Th, U, Pb, REE) behaviour in a cryptokarstic halloysite and kaolinite deposit from Southern Belgium: importance of 'accessory' mineral formation for radioactive pollutant trapping. *Applied Geochemistry*, **17**, 1313–1328.

EGGLETON, R. A. & KELLER, J. 1982. The palagonitization of limburgite glass – A TEM study. *Neues Jahrbuch für Mineralogie: Monatshefte*, **7**, 321–336.

EWING, R. C. 1979. Natural glasses: analogues for radioactive waste forms. *In*: MCCARTHY, G. J. (ed) *Scientific Basis for Nuclear Waste Management I*. Plenum Publishing Corp., New York, 57–68.

EWING, R. C. & JERCINOVIC, M. J. 1987. Natural analogues: Their application to the prediction of the long-term behavior of nuclear waste forms. *In*: BATES, J. K. & SEEFELDT, W. B. (eds) Scientific Basis for Nuclear Waste Management X. *Material Research Society Symposium Proceedings*, **84**, 67–86.

FAUCON, P., LE BESCOP, P., ADENOT, F., BONVILLE, P., JACQUINOT, J. F., PINEAU, F. & FELIX, B. 1996. Leaching of cement: study of the surface layer. *Cement and Concrete Research*, **26**, 1707–1715.

FRITZ, B. 1975. Etude thermodynamique et modélisation des réactions entre minéraux et solutions.

Application à la géochimie des altérations et des eaux continentales. *Sciences Géologiques Mémoires*, **41**, 152.

FRITZ, B. 1981. Etude thermodynamique et modélisation des réactions hydrothermales et diagénétiques. *Sciences Géologiques Mémoires*, **65**, 197.

GARRELS, R. M. & THOMPSON, M. E. 1962. A chemical model for sea water at 25°C and one atmosphère total pressure. *American Journal of Science*, **260**, 57–66.

GAUTHIER, A., THOMASSIN, J. H. & LE COUSTUMER, P. 1999. Rôle de matériaux zéolitiques lors de l'altération expérimentale du verre nucléaire R7T7. *Comptes Rendus Académie des Sciences, Paris, II*, **329**, 331–336.

GISLASON, S. R. & EUGSTER, H. P. 1987. Meteoric water–basalt interactions. I: A laboratory study. *Geochimica et Cosmochimica Acta*, **51**, 2827–2840.

GONG, W. L., WANG, L. M., EWING, R. C., VERNAZ, E., BATES, J. K. & EBERT, W. L. 1998. Analytical electron microscopy study of surface layers formed on the French SON68 nuclear waste glass during vapor hydration at 200°C. *Journal of Nuclear Materials*, **254**, 249–265.

GRAMBOW, B., JERCINOVIC, M. J., EWING, R. C. & BYERS, C. D. 1985. Weathered basalt glass: a natural analogue for the effects of reaction progress on nuclear waste glass alteration. *In*: WERME, L. O. (ed) Scientific Basis for Nuclear Waste Management IX. *Material Research Society Symposium Proceedings*, **50**, 263–272.

GUY, C. & SCHOTT, J. 1989. Multisite surface reaction versus transport control during the hydrolysis of a complex oxide. *Chemical Geology*, **78**, 181–204.

HELGESON, H. C. 1968. Evaluation of irreversibles reactions in geochemical processes involving minerals and aqueous solutions. I. Thermodynamic relations. *Geochimica et Cosmochimica Acta*, **32**, 853–877.

HELGESON, H. C., JONES, J. A., MUNDT, T., BROWN, T. H., NIGRINI, A., LEEPER, R. H. & KIRKAHM, D. H. 1971. *Path I and data bank PATDAT*. Report no. 1000 Computer Library, U.C Berkeley.

HOFFERT, M., KARPOFF, A. M., CLAUER, N., SCHAFF, A., COURTOIS, C. & PAUTOT, G. 1978. Néoformations et alteriès dans trois faciès volcano sédimentaires du Pacifique Sud. *Oceanologica Acta*, **1,2**, 187–202.

HONNOREZ, J. 1967. *La palagonitisation: l'altération sous marine du verre volcanique basique de Palagonia (Sicile)*. Thèse, Université Libre de Bruxelles, 227 p.

HONNOREZ, J. 1972. *La palagonitisation: l'altération sous marine du verre volcanique basique de Palagonia (Sicile)*. Vulkaninstitut Immanuel Friedlaender. Birkhäuser Verlag, Basel, vol. **9**, 131 p.

HONNOREZ, J. 1978. Generation of phillipsites by palagonitization of basaltic glass in seawater and the origin of K-rich deep-sea sediments. *In*: SAND, S. B. & MUMPTON, F. A. (eds) *Natural Zeolites,*

Occurrence, Properties, use, Pergamon Press, Oxford, 245–258.

HONNOREZ, J. 1981. The aging of the oceanic crust at low temperature. *In*: EMILIANI, C. (ed) *The sea*. Wiley Interscience, New York, 525–587.

JAKOBSSON, S. P. 1972. *On the Consolidation and Palagonitization of the Tephra of the Surtsey Volcanic Island, Iceland*, Surtsey Program Reports VI, 1–8.

JERCINOVIC, M. J. & EWING, R. C. 1987. *Basaltic Glasses from Iceland and the Deep Sea: Natural Analogues to Borosilicate Nuclear Waste-Form Glass*. Swedish Nuclear Fuel and Waste Management Co, Stockholm, Technical Report JSS 88–01, 221 p.

JERCINOVIC, M. J., KEIL, K., SMITH, M. R. & SCHMITT, R. A. 1990. Alteration of basaltic glasses from north-central British Columbia, Canada. *Geochimica et Cosmochimica Acta*, **54**, 2679–2696.

KARPOFF, A. M., PETERSCHMITT, I. & HOFFERT, M. 1980. *Mineralogy and Geochemistry of Sedimentary Deposits on Emperor Seamounts, Sites 430, 431 and 432: Authigenesis of Silicates, Phosphates and Ferromanganese Oxides*. Initial Reports Deep Sea Drilling Project, LV, US Gov. Print. Office, 463–489.

KASTNER, M. 1979. Zeolites. *In*: RIBBE, P. H. (ed) *Review in Mineralogy: Marine Minerals*. Mineralogical Society of America, Blacksborg, 111–122.

LE GAL, X. 1999. *Etude de l'Altération de Verres Volcaniques du Vatnajökull (Islande). Mécanismes et bilans à basse température*. PhD thesis, Université Louis Pasteur, Strasbourg, 153 p.

LE GAL, X., CROVISIER, J.-L., GAUTHIER-LAFAYE, F., HONNOREZ, J. & GRAMBOW, B. 1999. Meteoric alteration of Icelandic volcanic glass. long-term changes in the mechanism. Comptes Rendus de l'Académie de Sciences, Paris – Série IIa. *Sciences de la Terre et des Planètes*, **329**, 175–181.

LUTZE, W., MALOW, G., EWING, R. C., JERCINOVIC, M. J. & KEIL, K. 1985. Alteration of basalt glasses: implications for modelling the long-term stability of nuclear waste glasses. *Nature*, **314**, 252–255.

MALOW, G. & EWING, R. C. 1981. Nuclear waste glasses and volcanic glasses: a comparison of their stabilities. *In*: MOORE, J. G. (ed) *Scientific Basis for Nuclear Waste Management III*. Plenum Publishing Corp., New York, 315–322.

MALOW, G., LUTZE, W. & EWING, R. C. 1984. Alteration effects and leach rates of basaltic glasses: implications for the long-term stability of nuclear waste form borosilicate glasses. *Journal of Non-Crystalline Solids*, **67**, 305–321.

MASCOLO, G. 1973. Hydration products of synthetic glasses similar to blast furnace slag. *Cement and Concrete Research*, **3**, 207–213.

MASCOLO, G. 1986. Hydrotalcite observed in mortars exposed to sulfate solutions. A discussion. *Cement and Concrete Research*, **16**, 610–612.

MASCOLO, G. & MARINO, O. 1980. MgO-bearing phases in the hydration products of slag cement.

VII International Congress on the Chemistry of Cement, Paris, Vol. II, Sect. III, 59–62.

MELSON, W. G. & THOMPSON, G. 1973. Glassy abyssal basalts, Atlantic sea floor near St Paul's rocks: petrography and composition of secondary clay minerals. *Geological Society of America Bulletin*, **84**, 703–716.

MEVEL, C. 1980. *Mineralogy and Chemistry of Secondary Phases in Low Temperature Altered Basalts from Deep Sea Drilling Project Legs, 51, 52 & 53*. Initial Report Deep Sea Drilling Project LI, LII, LIII., US Gov. Print. Office., 1299–1317.

MILLER, B., ALEXANDER, R., CHAPMAN, N. A., MCKINLEY, I. G. & SMELLIE J. A. T. 1994. Natural analogues revisited. *Fourth International Conference on the Chemistry and Migration Behavior of Actinides and Fission Products in the Geosphere*, Charleston, 545–549.

MOORE, J. G. 1965. Petrology of deep-sea basalts near Hawaii. *American Journal of Science*, **263**, 40–53.

MURAKAMI, T., EWING, R. C. & BUNKER, B. C. 1988. Analytical electron microscopy of leached layers on synthetic basalt glass. *In*: APTED, M. J. & WESTERMAN, R. E. (eds) Scientific Basis for Nuclear Waste Management XI. *Material Research Society Symposium Proceedings*, **112**, 737–748.

NAYUDU, R. Y. 1964. Palagonite tuffs (hyaloclastites) and the products of post eruptive processes. *Bulletin Volcanologique*, **27**, 391–420.

PETIT, J. C. 1992. Reasoning by analogy. Rational foundation of natural analogue studies. *Applied Geochemistry, S.I.*, **1**, 9–14.

PETIT, J. C. & CÔME, B. 1994. L'analogie au service du stockage des déchets radioactifs. *Clefs CEA, Paris*, **28**, 24–35.

RUSSELL, B. 1913. Theory of knowledge. *In*: VRIN, D. (ed) *Théorie de la connaissance*, Le manuscrit de 1913. Paris, 2002, 254 p.

STEINMANN, M. & STILLE, P. 1998. Strongly fractionated REE patterns in salts and their implications for REE migration in chloride-rich brines at elevated temperatures and pressures. *Comptes Rendus Académie des Sciences, Paris, II*, **327**, 173–180.

STEINMANN, M., STILLE, P., BERNOTAT, W. & KNIPPING, B. 1999. The corrosion of basaltic dykes in evaporites: Ar–Sr–Nd isotope and rare earth elements evidence. *Chemical Geology*, **153**, 259–279.

STEINMANN, M., STILLE, P., MENGEL, K. & KIEFEL, B. 2001. Trace element and isotopic evidence for REE migration and fractionation in salts next to a basalt dyke. *Applied Geochemistry*, **16**, 351–361.

TECHER, I. 1999. *Apports des Analogues Naturels Vitreux à la Validation des Codes de Prédiction du Comportement à Long Terme des Verres Nucléaires*. Report CEA-R-5880, Commissariat à l'Energie Atomique, Marcoule, France, 206 p.

TECHER, I., ADVOCAT, T., LANCELOT, J. & LIOTARD, J.-M. 2000. Basaltic glass: alteration mechanisms and analogy with nuclear waste glasses. *Journal of Nuclear Materials*, **282**, 40–46.

TECHER, I., ADVOCAT, T., LANCELOT, J. & LIOTARD, J.-M. 2001*a*. Dissolution kinetics of basaltic glasses: control by solution chemistry and protective effect of the alteration film. *Chemical Geology*, **176**, 235–263.

TECHER, I., LANCELOT, J., CLAUER, N., LIOTARD, J.-M. & ADVOCAT, T. 2001*b*. Alteration of a basaltic glass in an argillaceous medium: The Salagou dike of the Lodève Permian Basin (France). *Geochimica et Cosmochimica Acta*, **65**, 1071–1086.

THOMASSIN, J. H. & TOURAY, J. C. 1982. L'hydrotalcite, un hydroxy-carbonate transitoire précocement formé lors de l'interaction verre basaltique/eau de mer. *Bulletin de Minéralogie* **105**, 312–319.

VON WALTERSHAUSEN, S. 1845. Ueber die Submarinen vulkanischen Ausbrücke in der tertiär Formation des Val di Nito im Vergleich mit verwandten Erscheinungen am Aetna. *Göttingen Studien*, **I**, 371–431.

Special cases of natural analogues:
The Gabon and Cigar Lake U ore deposits

F. GAUTHIER-LAFAYE[1], P. STILLE[1] & R. BROS[2]

[1]ULP-EOST-CNRS, Centre de Géochimie de la Surface UMR 7517, Strasbourg, France
(e-mail: gauthier@illite.u-strasbg.fr)

[2]JNC, Waste Isolation Research Division, Tokai-mura, Ibaraki-ken, Japan

Abstract: The Gabon and Cigar Lake uranium deposits may be used as natural analogues for nuclear waste because they both provide information on actinide immobilization in a geological system over a very long period of time. These deposits contain high-grade uranium ores (40–60% UO_2) embedded in a clay matrix. Futhermore, the Oklo and Bangombé uranium deposits in Gabon were natural nuclear fission reactors that operated 2000 Ma ago. This offers the unique opportunity to study the behaviour of fission products and actinides in a geological environment. The geological stability of the sites is one of the most important parameters that can explain the preservation of the natural reactors of Gabon and of the Cigar Lake deposit. The mineralogy of the hosted rocks and the chemical composition of the fluids are likewise important parameters. It can be shown that clays, Fe-sulphides and organic matters provide effective redox buffering and consequently play an important role in the long-term preservation of the ores. Adsorption of actinides and fission products on clays and Ti-oxides is also an important process for retention of these elements. This process is even more efficient when it is followed by the formation of coffinite, $USiO_4 \cdot nH_2O$. Coffinitization is common at the Cigar Lake deposit and in the natural reactors of Gabon. Coffinitization may be related to hydrothermal alteration (Cigar Lake) and to weathering (Bangombé).

The repository analogy

The uranium deposits of Oklo and Bangombé in Gabon and Cigar Lake in Canada (Saskatchewan province) are similar in the sense that they both present high-grade uranium ores (40 to 60% UO_2) embedded in a clay matrix. Both belong to the Proterozoic and are located in sandstones. These two deposits may be used as natural analogues for nuclear waste because they are good examples for actinide immobilization, in a geological system over a very long period of time. We can learn a lot about the main geological, mineralogical, and geochemical parameters that allowed the preservation of these sites. Oklo and Bangombé uranium deposits are, however, unique because they contain natural nuclear fission reactors that worked 2000 Ma ago, leaving in the geological environment natural spent fuels that can be studied in detail. This offers the unique opportunity to study the behaviour of fission products and actinides in a geological environment. These fission products and actinides were originally located in uraninites embedded in clay layers. Analogies between these natural reactors and high-level nuclear wastes exist with respect to the role of clays, secondary mineral phases (phosphates, oxides, and so on) and organic matter in the containment of fission products. These natural nuclear fission reactors (hereafter reactors) are located at depths ranging between 250 and 12 m and are therefore submitted to different weathering conditions, the deepest being the best preserved to any surface alteration, whereas the upper one in the Bangombé deposit is subject to intense weathering and is involved in a laterite profile that develops just above it (Fig. 1).

The natural nuclear reactors of Oklo and Bangombé

The uranium deposits of the Franceville basin in Gabon are the oldest known high-grade uranium deposits (Gauthier-Lafaye & Weber 1989). They are unique because they are the only known places in the world containing natural nuclear fission reactors. Fission reactions took place spontaneously 2000 Ma ago (Gancarz 1978; Ruffenach 1978) and have been sustained for 100 000 to 500 000 years. The reactors now contain high-grade uranium ores with a high content of fission products and their decay end members. Therefore, these deposits are natural analogues for places where spent nuclear fuels

From: GIERÉ, R. & STILLE, P. (eds) 2004. *Energy, Waste, and the Environment: a Geochemical Perspective*. Geological Society, London, Special Publications, **236**, 123–134.
0305-8719/04/$15 © The Geological Society of London 2004.

have been stored in a geological environment over a very long period of time (i.e., 2.0 Ga).

Uranium deposits are located at the top of an up to 1000 m thick sedimentary sequence of mainly sandstones and conglomerates. This formation is overlain by black shales with organic carbon content ranging between 1 and 15% (Gauthier-Lafaye 1986). Most of the uranium is in the form of uraninite and is closely associated with migrated hydrocarbons that fill the secondary porosity of the sandstones (Gauthier-Lafaye & Weber 1981). Uranium mineralization occurred during diagenesis, when the mineralized FA formation was 3000 m deep and when oxidized uranium-bearing fluids met reduced fluids associated with the hydrocarbons in tectonic traps. The mineralization occurred 2050 ± 40 Ma ago (Gancarz 1978). Fifteen natural nuclear fission reactors were discovered in two deposits of the Franceville basin. Fourteen were located in the Oklo deposit at 100 to 260 meters depth but are now mined out. One however, has been preserved in the small deposit of Bangombé (30 km from Oklo). This reactor is located in the groundwater table at only 12 m depth and is therefore exposed to weathering (Fig. 1b). Extensive geological, mineralogical, and geochemical descriptions of the reactors exist (Gauthier-Lafaye *et al.* 1989, 1996).

The size of the reactors is quite variable. In length, the biggest reactor has dimensions of 12×18 m and has a thickness of 20 to 50 cm (Fig. 1a). The core of the reactors consists of a 5 to 20 cm thick layer of uraninite embedded in clays (illite and chlorite). Clays around the reactors result from the hydrothermal alteration of the host sandstone during the fission reactions. This alteration occurred at a temperature close to 400 °C in the core. Temperature decreased drastically toward the vicinity with a thermal gradient of 100 °C/m (Pourcelot & Gauthier-Lafaye 1999). The uranium content of the core ranges between 40 and 60%. Accessory minerals are mainly sulphides (pyrite and galena), hematite and phosphates (mainly hydroxyapatite).

The Cigar Lake uranium deposit

The Cigar Lake uranium deposit is located in northern Saskatchewan in the eastern part of the Athabasca basin (Fouques *et al.* 1986; Bruneton 1987, 1993; Cramer & Smellie 1994). Similar to the Gabon ore deposits, the primary mineralization is hosted by a hydrothermally altered sandstone (Athabasca formation), above the contact with an Archean metamorphic basement at a depth of about 430 m (Fig. 2). The ore formation occurred 1360 Ma to 1550 Ma

ago at a depth of 3000 m (Cumming & Krstic 1992; Alexandre & Kyser 2003).

The mineralization consists primarily of uraninite $UO_{2,x}$, with subordinate coffinite $USiO_4$ (Sunder *et al.* 1988). Similar to Oklo, the U ore is embedded in a clay matrix, which is shaped like an irregular lens measuring approximately 2000 m in length, 20–100 m in width and 1–20 m in thickness. The average grade of the mineralization is 8 wt% U, with local enrichments of up to 55 wt% U. The ore contains high amounts of sulphides, arsenides and sulph-arsenides. Uranium mineralization occurred when hydrothermal fluids passing through the crystalline basement became reducing in contact with graphite-bearing metapelitic rocks. The uranium minerals were deposited where reducing fluids interacted with more oxidizing diagenetic U-bearing solutions of the red-bed type sandstones. The alteration caused the reduction of Fe(III) from hematite to Fe(II) in a marcasite/pyrite rich zone, which forms a plume of 'bleached' sandstones above the axis of the mineralization. High-temperature alteration also caused the dissolution of quartz from the sandstones, leading to an enrichment in clays toward the high-temperature side and reprecipitation of mobilized silica toward the cooler side, which forms a quartz-cemented zone above the mineralization. The clay zone is 5–30 m thick (predominantly illite) and encloses the high-grade uranium mineralization. Thus, Oklo reactors and Cigar Lake U deposit show similar phenomena of hydrothermal alteration of the sandstones as well as clay formation around the U ore. The clay-rich halo has played an important role in the long-term preservation of the main ore (Cramer & Smellie 1994). It has low hydraulic conductivities (4 mm/y; Winberg & Stevenson 1994), compared with altered and unaltered sandstones of the overlying main aquifer (4–7 m/y) and effectively sealed the ore zone from bulk groundwater flow through this zone. The clay halo and the altered sandstones enriched in Fe-sulphides (pyrite, marcasite) provided effective redox buffering.

Retention of fission products and actinides in the Gabon reactors

It has been recognized that some fission and decay products had almost completely been retained, whereas others had been almost totally lost from the vicinity of the reactors (Gauthier-Lafaye *et al.* 1996) (Fig. 3). Elements that had been retained essentially within the core of the reactors are Pu, Th, REE, Zr, Ru, Rh, and Pd. These

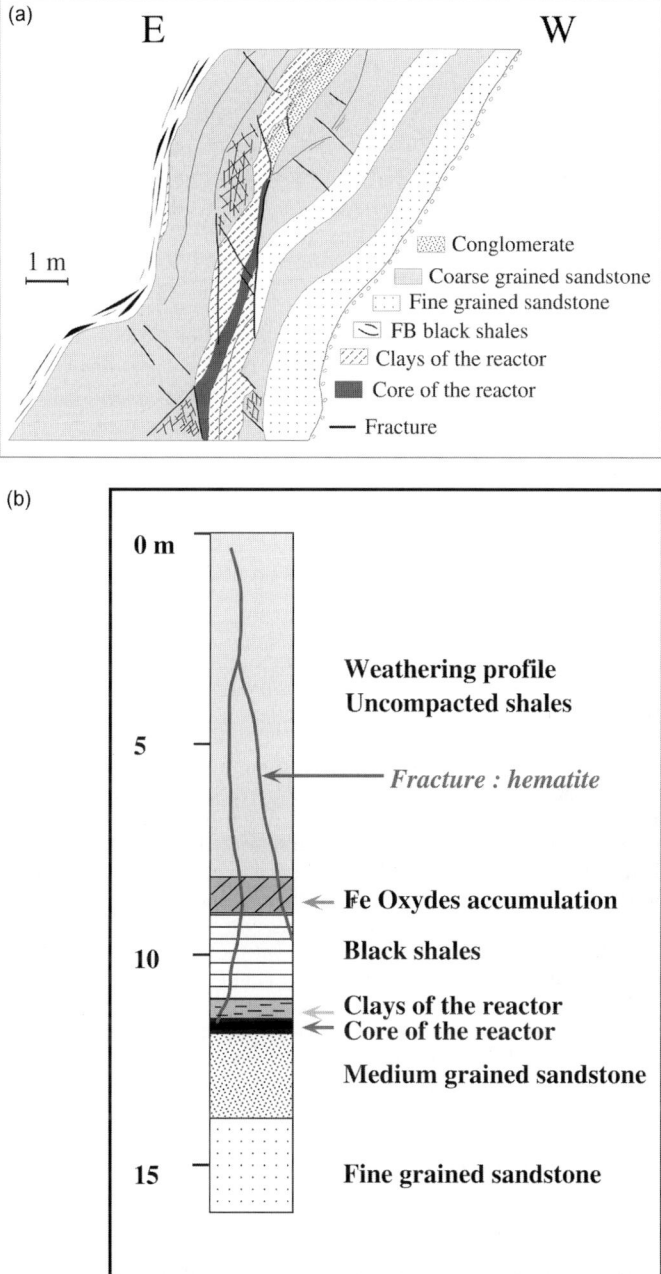

Fig. 1. (a) Cross-section of reactor 9 in the Oklo uranium deposit; (b) bore-hole BAX3 crossing the reactor of Bangombé.

elements have low solubilities in groundwater, especially under the reducing conditions that pertained in the uranium deposits of Oklo. Other fission products, such as rare gases, I, Cd, alkali metals, and alkaline earth (e.g., Sr), appeared to have migrated out of the reactors. These elements either are gaseous or have high solubility in groundwater, even under reducing conditions.

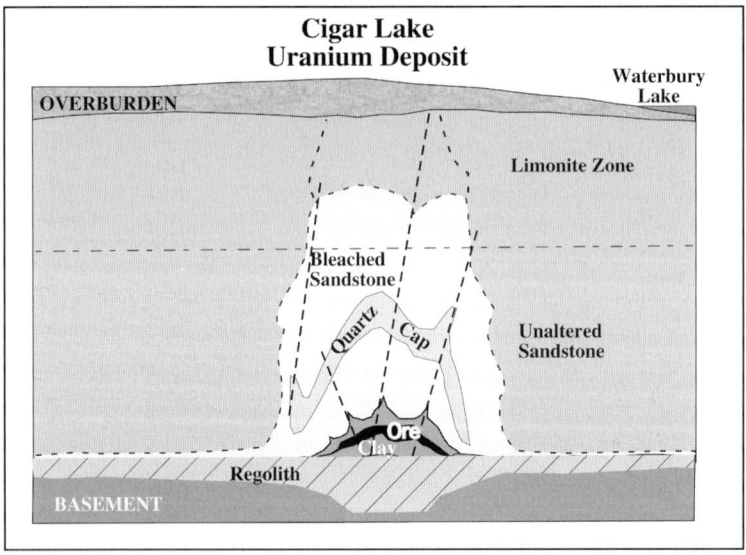

Fig. 2. Schematic cross-section of the Cigar Lake deposit (Cramer & Smellie 1994).

Detailed analyses of specific isotopes that were formed during the fission reactions provided information on the behaviour of their precursors. Analyses of isotopes such as ^{235}U (Bros *et al.* 1993), ^{99}Ru (Hagemann *et al.* 1974; Gancarz *et al.* 1980; Curtis *et al.* 1989; Hidaka *et al.* 1993, 1999), ^{90}Zr (Hidaka *et al.* 1994), and ^{209}Bi (Gauthier-Lafaye *et al.* 1996; Hidaka & Holliger 1998), whose precursors are ^{239}Pu, ^{99}Tc, ^{90}Sr, and ^{237}Np, respectively, indicated that these fission elements and actinides migrated during the fission reaction or soon thereafter.

The most efficient matrix for retention of actinides and fission products is the uraninite mineral. However, it has been shown that other matricies such as apatite, clay minerals, zirconium silicates, and oxides (Fe, Mn) may also be important in the retention of fission products and actinides. For example, Pu was stored in apatite (Bros *et al.* 1996) and chlorite (Bros *et al.* 1993) in the core of the reactor 10. In the core of the reactors, between uraninite grains, 20–200 μm-sized metallic aggregates containing fissiogenic Ru, Rh, and Te associated with As, Pb, and S were found. These aggregates also exist in spent fuels of water-pressured type reactor plants, suggesting their analogy with spent fuels.

Uraninites at Oklo and Bangombé have a very constant chemical composition (Jensen *et al.* 1997) (Table 1). Their 'impurity' contents are: SiO_2: 0.7–1%; FeO: 0.3–0.7%; CaO: 1.5–1.7%. Lead content is very low for strongly altered uraninites and reaches up to 25% in well-preserved uraninites from the deep-seated reactors. Most uraninites suffered coffinitization, that is, the transformation of $U(IV)O_{2-x}$ to U(IV)-silicate. Coffinitization is possibly related to the process of oxidative alteration of uraninites by water radiolysis and the later precipitation of U(IV)-silicate again under reducing conditions.

Recent migrations of actinides and fission products were mainly studied in the Bangombé reactor located at 12 m depth within the groundwater table. Therefore, this reactor is subjected to weathering and chemical exchange with the groundwater (Fig. 1*b*). Dissolution of uraninites has been proven in the reactor core where secondary U(VI) minerals occur (idiomorphic U, Ca, Fe, Al, LREE phosphates; U (Al, Pb) and U (Fe, Ti) silicates) (Salah *et al.* 2000; Perez del Villar *et al.* 2000) and in the groundwater with $^{235}U/^{238}U$ ratios lower than 0.0072 (Louvat *et al.* 2000). The original shape and size of the reactor have probably been strongly modified during this weathering process. However, the fact that uraninite is still present after a long period of time suggests that some geochemical processes should maintain the dissolution phenomenon of uraninite at a very low rate. The stability of uraninite results from the buffering effect of Fe(II)/Fe(III) equilibrium controlled by various mineral associations such as siderite/ferrihydrite, daphnite/ferrihydrite, magnetite/nontronite or chlorite/ferrihydrite

Element	Behavior of fission products and actinides at Oklo				
	Retention				
	Core of the reactor			Argillaceous rocks surrounding the reactors	
	Uraninites	Inclusions	Migration	Clays	Others
Cs			+		
Rb			+		
Sr			+		
Ba			+		
Mo		+	+		
Tc		+		+	
Ru		+		+	?
Rh		+		+	?
Pd		+		+	?
Y	+			+	
Nb	?				
Zr	+			+	**Heavy.miner/Oxide**
Te					
REE light	+			+	**apatite**
Ce					
Pb					
Pb	+				**galena**
Bi					
Th	+			+	**Heavy miner.**
U	+			+	**apatite**
Np	?				
Pu	+			+	**apatite**

+	+	

Close relation to no relation between the minerals and the chemical elements

Fig. 3. Retention of the various fission products and actinides by different mineral phases.

and by the hydrolysis of graphite, which is encountered within the 'argiles de pile' in the close vicinity of the reactor (Madé et al. 2000). The reducing environment that is generally maintained inside the reactor can explain the conservation of the uranium in spite of the leaching by oxidizing groundwater. These apparent redox-buffering conditions may change with time (i.e., due to changes in the level of the water table), explaining the occurrence of minerals representative of reducing (uraninite, sulphides) and oxidizing (uranophosphates, Fe-oxides, P-rich coffinites) conditions in the same samples.

Mechanisms of retention of uranium during weathering include precipitation of new U-bearing minerals and adsorption processes (Salah et al. 2000; Perez del Villar et al. 2000). The vertical and lateral flow of groundwater is responsible for the oxidation and dissolution of primary sulphides, leading to acidic solutions that facilitated the oxidation and dissolution of uraninite. The resulting uranyl cations migrated and precipitated as uranyl minerals, mainly phosphates, silicates, silico-phosphates. In certain local conditions, reduction of these uranyl cations allowed precipitation of coffinite with a high content of P and LREE. Adsorption of uranium, together with P, mainly occurs on Fe-oxyhydroxides, but this kind of uranium retention seems less efficient than the precipitation, at least in the close vicinity to the reactor. On the other hand, in the weathered FB formation, U adsorption on clays and Fe–Mn–oxyhydroxides minerals may be the dominant process of uranium retention (Salah 2000).

Rare earth element distribution patterns of the sandstones enclosing the U ore and Sm–Nd isotope data of groundwater suggest that fissiogenic REE were mobilized due to dissolution of uraninite and migrated over at least 3 m but less than 25 m (Bracke *et al.* 2001; Stille *et al.* 2003). Detailed studies on the mineralogy of the reactor and on the elemental and isotopic compositions of U, Sm, and Nd in groundwater samples show that newly precipitated P–LREE-coffinites in the core of the reactor allow the retention of most of the dissolved uranium and fissiogenic LREE that were liberated by the dissolution of primary uraninites (Stille *et al.* 2003). This observation is in agreement with experiments that show that the UO_2 stability in the solid phase and U(VI) predominance in the aqueous phase are thermodynamically compatible, particularly under slightly acidic to neutral pH conditions and in transitional redox waters (Casas *et al.* 1998).

Mechanisms of U retention at Cigar Lake

At the Cigar Lake deposit uraninite is also the most important U-bearing mineral and contains similar impurity elements, which can be related to alteration (Table 1). Among others, radiogenic Pb has been identified. PbO reaches high concentrations of up to 13.5 wt%. Lead is not compatible with the uraninite structure due to its large ionic radius and may exsolve from uraninite (Janeczek & Ewing 1992). Release of Pb includes uraninite dissolution and diffusion loss at uraninite grain boundaries. Coffinitization of uraninite is also responsible for Pb removal as indicated by the low Pb contents of Si-bearing uraninite. Owing to its relative immobility in groundwater (Mann & Deutscher 1980) and the high S^{2-} activity under reducing conditions prevailing at the Cigar Lake deposit, a large portion of Pb released from uraninite precipitated in the form of galena, which occurs as small single grains or large euhedral crystals.

The production of Pb by radioactive decay of U may also affect the oxidation state of uraninite by a process called 'auto-oxidation' (Frondel 1958; Finch & Murakami 1999). Radiogenic Pb^{4+} is a strong oxidizer and due to its instability in the presence of U^{4+}, it is reduced to Pb^{2+} while simultaneously U^{4+} oxidizes to U^{6+} leading to elevated U^{6+}/U^{4+} ratios in uraninite. The U^{6+}/U^{4+} ratios determined from XPS spectra of uraninite from Cigar Lake can be high (0.3) despite the reducing conditions at the depth of the deposit (Sunder *et al.* 1996) (Table 1). Similarily the U^{6+}/U^{4+} ratios in uraninite from the deep Oklo reactors 10 and 13 vary between 0.10 and 0.19 (Sunder *et al.* 1992, 1994) and point to the presence of U^{6+} in uraninite.

Electron microprobe traverses were performed on cubic and massive uraninite in order to find elemental transfers in response to alteration of uraninite. The results show no significant gradient of concentration for U, Ca, Si, and Pb. Sometimes, Ca is enriched at the border of the crystal and inversely Pb is depleted. A similar feature has been observed on uraninites of the reactor 10 at Oklo, showing a zoning with Pb-rich and Ca-poor zones (Gauthier-Lafaye *et al.* 1996). The dissolution of uraninite seems to be congruent and controlled by the hydrolysis of the bulk matrix rather than by selective dissolution controlled by exchange between elements of uraninite and hydrogen of the solution. While the reaction progresses, secondary coffinite forms at the surface of uraninite (including the edges of microfractures). It is apparent that the enrichment in Si, P, Ca and Y and Pb depletion around massive uraninite is the result of the alteration by Si–P–Y–Ca-rich solutions and subsequent coffinitization.

After uraninite, coffinite ($USiO_4 \cdot nH_2O$) is the most important U-bearing mineral at Cigar Lake. It often occurs as a secondary alteration product of uraninite along microfractures, cracks, and boundaries of uraninite grains. Along with uraninite dissolution, coffinitization of uraninite is another process responsible for the release of U.

Chemical analyses of coffinite show several types of coffinite defined by various substitutions with other phases such as zircon ($ZrSiO_4$), hafnon ($HfSiO_4$), thorite ($ThSiO_4$), thorogummite ($ThSiO_4 \cdot nH_2O$), xenotime YPO_4, ningyoite (U, Ca, REE)$(PO_4)_2 1-2H_2O$ as well as synthetic actinide phases such as $PuSiO_4$ and $NpSiO_4$ (Speer 1982; Janeczek & Ewing 1996; Ewing 1999; Finch & Murakami 1999; Bros *et al.* 2003). Similar to the coffinites at Oklo, the Cigar Lake coffinites may contain up to 8 wt% P_2O_5, 3 wt% CaO, and more than 0.5 wt% ThO_2, ZrO_2, Ce_2O_5.

The occurrence of Ca, REE, and P can be explained by a limited solid solution between coffinite and ningyoite, (U, Ca, Ce)$_2(PO_4)_2 \cdot 1-2H_2O$ according to the substitution:

$$Si^{4+} + U^{4+} = P^{5+} + Ca^{2+} + REE^{3+}$$

Ningyoite is an orthorhombic hexagonal phosphate (Hikichi *et al.* 1989). Because of structural differences, a complete solid solution between coffinite and ningyoite is impossible, whereas substantial solid solution between coffinite and xenotime is likely to occur (Finch & Murakami 1999). Such P–REE-rich coffinite has already been described for the Gabon natural nuclear reactors (Janeczek & Ewing 1996; Bros *et al.* 2001, 2003).

In clay samples Zr–Th-rich coffinite was found around remnants of zircon. It is likely that it is the result of solid solution with zircon, $ZrSiO_4$ and thorite, $ThSiO_4$, which are isostructural with coffinite (Finch & Murakami 1999; Jensen & Ewing 2001). The presence of phosphorus and sulphur in coffinite suggests that both elements substituted for Si in the coffinite structure. A previous study at the Bangombé site in Gabon has clearly shown that coffinites are most important secondary minerals for the retention of fissiogenic lanthanides and actinides (Stille *et al.* 2003).

The adsorption of U onto phyllosilicates (chlorite and illite) is also an important retardation mechanism. Illite is the dominant clay in bulk rocks and <2 μm fractions. It occurs at all levels in the massive ore and massive clays. The illite crystallinity points to high-grade diagenetic to anchimetamorphic conditions. Low potassium content (1.2–4.1 wt% K_2O) and low cation oxide totals (70–88%) determined by electron microprobe suggest the presence of molecular water in interlayer sites. Chlorite is the second major clay mineral in the clay halo. Chlorite has a Fe–Mg composition with FeO and MgO concentrations of 12.3–34.4% and 8.9–24.2%, respectively, or displays an aluminous character (21.8–25.5 wt% Al_2O_3) with low Mg (0.6–3 wt% MgO) and Fe (5.8–7.2 wt% FeO) contents. Such peraluminous chlorites (Percival & Kodama 1989; Percival *et al.* 1993) are associated with hydrothermal alteration in high-grade U deposits or polymetallic deposits (Alysheva *et al.* 1977; Bros *et al.* 1993; Gauthier-Lafaye *et al.* 1996). The textural association of chlorite and coffinite strongly suggests an early stage of U adsorption onto the surface of existing chlorite followed by the formation of coffinite (Fig. 4). Studies of the clay halo at the Oklo and Bangombé natural reactors have shown that chlorites are efficient actinide scavengers (Bros *et al.* 1993; Eberly *et al.* 1995, 1996; Gauthier-Lafaye *et al.* 1996; Bros *et al.* 2003) and revealed the sorption of significant amounts of U (100–1000 ppm) onto Fe–Mg chlorite (Bros *et al.* 2003). Along with U, chlorite has also the potential to sorb Pu. This was demonstrated by an experiment in which thin sections from the Bangombé deposit remained in contact with a solution containing 10^{-6}M ^{239}Pu (Eberly *et al.* 1996). After exposure, the distribution of Pu by autoradiography indicated that adsorption of Pu was much more effective onto chlorite than illite and kaolinite. Another mechanism of U fixation is the sorption onto clays during growth, followed by the precipitation of U minerals. At the initial stage, both Fe-illite and Al-chlorite are associated to small micron-sized inclusions we assume to be the uraninite that forms between the clay sheets. At an intermediate stage, the growth of uraninite within clay sheets leads to an elongated 'finger'-shaped mineral (Fig. 4*b*). At the last stage, the matrix of clays is almost completely invaded by the U mineralization as shown by the clear contrast in back-scattered electron images, but the layered structure of clays is preserved (Fig. 4*c*, *d*). Highly mineralized Fe-illite and Al-chlorite are therefore characterized by elevated U contents ranging from 2.7 to 10.9 wt%.

Rutile, TiO_2, is abundant in the clay halo and forms large agglomerates (100–500 μm) embedded in an illite matrix. Single 10–20 μm-long TiO_2 tubular rods are often rimmed by coffinite that fills the void between crystals. As for chlorite, this paragenetic association suggests an initial stage of U adsorption onto rutile followed by the development of coffinite enhanced by the presence of Si in groundwater. The agglomerate itself is also surrounded by coffinite, indicating that coffinitization might have continued after the formation of rutile. Similar mineralogical associations were observed for the Bangombé and Oklo natural nuclear reactors, where euhedral Ti-oxides grains (likely anatase, another polytype of TiO_2) are rimmed by secondary uraninite and coffinite (Janeczek 1999; Bros *et al.* 2003) in hydrothermal clays. The affinity of U for Ti-oxide is being used industrially to selectively remove U from seawater (Yui, personal communication).

Al-phosphate minerals were identified in massive clays above and beneath the high-grade ore. The average composition of these aluminous hydroxy phosphates is $Cra_{24-28} Flo_{24-30} Goy_{44-48}$ in which coefficients correspond to the percentages of Ca-crandallite, REE-florencite and Sr-goyazite poles, respectively. Strontium predominates, with concentrations ranging from 9 to 11%. Only light REE were detected whereas Sm and heavier REE were not detected. La and Ce are dominant, with concentrations ranging from 3.4 to 3.8% (La) and 3.8 to 4.1% (Ce). Nd displays lower concentrations, 0.7–1.1 wt% Nd_2O_3. The presence of Si (1.3–1.9 wt% SiO_2) and S (2.1–3.0 wt% SO_3) can be explained by the substitution of P by SiO_4 and SO_4. In the natural reactors of Gabon, crandallite group minerals were identified in

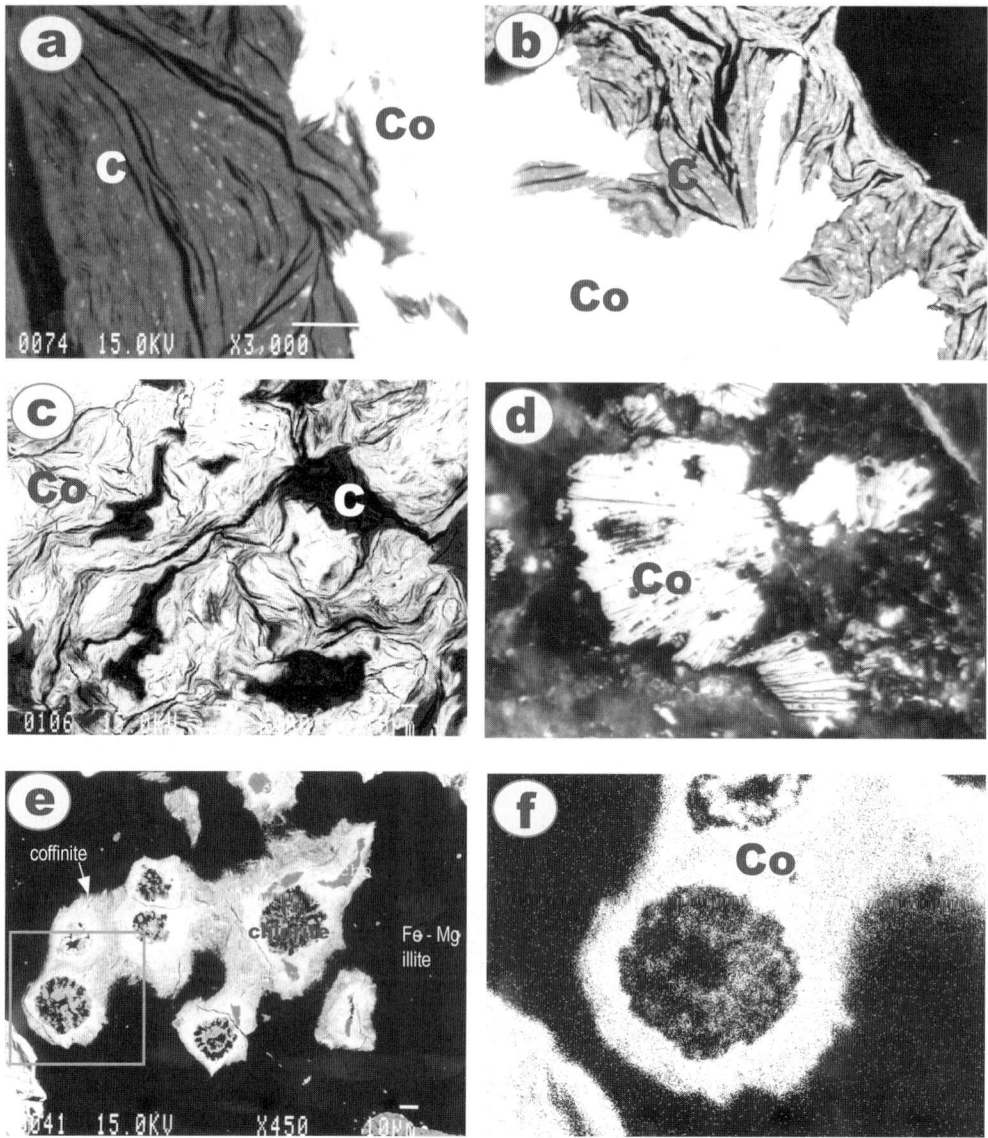

Fig. 4. Relation between clays and coffinite at Cigar Lake (**a, b, c, e** and **f**) and at Oklo (**d**): (**a**) to (**d**) uranium progressively invades the clay phase until complete replacement in (**d**); (**e**) coffinite replacing a chlorite mineral (precipitation of coffinite around a chlorite flake); (**f**) map of the distribution of uranium corresponding to the rectangle in (**e**). (C = clay mineral; Co = coffinite.)

illite within and beneath the reactor of Bangombé (Janeczek & Ewing 1996) and in the core of reactor 13 (Gauthier-Lafaye *et al.* 1996; Dymkov *et al.* 1997).

Conclusions

The study of the uranium deposit of Cigar Lake and of the natural nuclear fission reactors of Oklo and Bangombé shows that the geological location, the mineralogy of the hosted rocks, and the chemistry of the fluids are important parameters for the preservation of the uranium deposits and of the natural nuclear reactors and that this information could be used to design and select materials for high-level waste disposal.

The geological stability of the sites is one of the most important parameters that can explain the preservation of the natural reactors of Gabon and of the Cigar Lake deposit for such a long period of time. Both uranium deposits in Gabon and Canada are located in very stable cratons. However, other analogies with nuclear wastes can be found:

(1) Both the Gabon reactors and the Cigar Lake U deposit show similar phenomena of hydrothermal alteration of the ore-containing sandstones as well as of clay formation around the U ore. This clay halo and the altered sandstones enriched in Fe-sulphides provide effective redox buffering and have, together with the organic matter, an important role in the long-term preservation of the main ores. It also emphasizes the role of an illite-based clay layer that impeded groundwater penetration, sealing the deposit from aqueous migration of fission products, despite the relatively low swelling property of illite compared to bentonite, which will be used in HLW repositories.

(2) High amounts of Ca (and potentially Ra), Sr, and LREE were co-precipitated by Al-phosphates from the crandallite group, which crystallized in the clay halo. Clay minerals (mainly chlorite and to a lower extent illite) and Ti-oxide were found to have sorbed significant amounts of uranium. Sorption onto mineral surfaces was followed by the formation of coffinite, $USiO_4 \cdot nH_2O$, rimming clay, and rutile particles.

(3) Uraninite crystals retain most of the actinides produced by the fission reactions and most of the fission products that have ionic radii close to that of uranium. When uraninite becomes hydrothermally altered or transformed during supergene weathering, that is, in the weathered zone of the Bangombé reactor, the reduced conditions in the close vicinity of the U ore allows its precipitation in newly formed Si–P–REE–uranium minerals (coffinite).

(4) The formation of coffinite is an important factor for the long-term preservation of radioactive material because it retards the mobility of actinides and fissiogenic lanthanides. Coffinitization of uraninite and co-precipitation of U-bearing phosphate minerals like Fe-uranyl phosphate hydroxide hydrates seems to be mainly related to the weathering of U ores and to the interaction with meteoric fluids, as observed for the Bangombé reactor in Gabon. The Si, necessary for the transformation of uraninite into coffinite, might originate from simultaneous alteration of the surrounding clay minerals. The radio-nuclides adsorbed on the surface of the altering clay minerals become incorporated into the structure of the newly formed coffinite. Thus, in the performance of a clay barrier, coffinite is among the secondary silicate minerals a very important sink for radionuclides.

The formation of coffinite under repository conditions is consistent with a scenario involving the failure of the canister, congruent glass dissolution and release of Si, U, and other actinides and their reincorporation in precipitating coffinite. Although the temperatures were somehow higher at Cigar Lake than those expected in repository conditions, it is well known that coffinite can form at various temperatures in natural conditions, in low-temperature ($<100\,°C$) to high-temperature ($>200\,°C$) systems. Coffinite can form at low Eh and relatively high Si activity ($>10^{-3}$ mol/L). The concentration of mobilized radionuclides was probably high at Cigar Lake and they display solubility-limited transport behaviour. This contrasts with conditions expected in repositories in which radionuclides would be released in such diluted concentrations that they would remain below solubility limits. However, supersaturation conditions leading to coffinitization may occur in stagnant groundwater zones or in dead-end pores, for example. Despite the potentially important role of coffinitization in understanding radionuclide transport and retardation, little is known about the thermodynamic constant and the kinetics of formation of this mineral.

Table 1. *Chemical compositions (wt%) of low-pressured water reactor (LWR) spent fuel and uraninite from Oklo reactors 10, 16 and Cigar Lake*

Oxide	LWR*	Cigar Lake	Oklo 10[†]	Oklo 16[†]
UO_2	95.4	79.7	77.4	90.5
PuO_2	0.9	Traces	Traces	Traces
AcO_2	0.1	—	—	—
FP[§]	3.6	—	0.3–0.5	nd
PbO	—	10.5	18.7	5.6
ThO_2	—	—	0.1	0.1
CaO	—	0.4	0.3	1.2
FeO	—	0.3	0.1	0.5
SiO_2	—	0.4	0.3	0.3
U^{6+}/U^{4+}	0.001	0.02–0.75[‡]	0.10–0.19	nd

*Burn-up of 35 MWd/kg, 10 years after discharge from the reactor.
[†]Janeczek 1999.
[‡]Sunder *et al.* 1994.
[§]FP, fission products.
nd, not determined

The authors would like to thank M. Cuney, R. Gieré, and an anonymous reviewer for the very helpful comments. This study was partly financially supported by JNC, Waste Isolation Research Division, Tokai-mura, Japan. This is EOST/CGS contribution no. 2004.601-UMR7517.

References

ALEXANDRE, P. & KYSER, K. 2003. Geochronology of the Paleoproterozoic basement-hosted unconformity-type uranium deposits in northern Saskatchemwan, Canada. *In*: CUNEY, M. (ed) *Proceedings of 'Uranium Geochemistry' 2003 International Conference*, Nancy-France, 13–16 April 2003, p. 37–40.

ALYSHEVA, E. I., RUSINOVA, O. V. & CHEKVAIDZE, V. B. 1977. *Sudoite from Polymetal Deposit of Rudnyy*. Academy of Sciences, USSR, Doklady Earth Science Section, 236, 1–6, 167–169.

BRACKE, G., SALAH, S. & GAUTHIER-LAFAYE, F. 2001. Weathering process at the natural fission reactor of Bangombé. *Environmental Geology*, 40, 403–408.

BROS, R., TURPIN, L., GAUTHIER-LAFAYE, F., HOLLIGER, P. & STILLE, P. 1993. Occurrence of naturally enriched [235]U: implications for plutonium behaviour in natural environments. *Geochimica Cosmochimica Acta*, 57, 1351–1356.

BROS, R., CARPENA, J., SERE, V. & BELTRITTI, A. 1996. Occurence of Pu and fissiogenic rare earth elements in hydrothermal apatites from the fossil natural nuclear reactor 16 of Oklo (Gabon). *Radiochimica Acta*, 74, 277–282.

BROS, R., HIDAKA, H., KAMEI, G. & OHNUKI, T. 2001. Retardation of fissiogenic REE in clays surrounding the Oklo reactor 2 (Gabon). *In*: *Proceedings of Atomic Energy Research Society of Japan Meeting*, Sapporo, September 2001, 932.

BROS, R., KAMEI, G. & OHNUKI, T. 2003. Mechanisms of mobilization and retardation in the Oklo reactor zone 2 (Gabon) – Inferences from U, REE, Zr, Mo and Se isotopes. *Applied Geochemistry*, 18, 1807–1824.

BROS, R., GAUTHIER-LAFAYE, F., STILLE, P., UENO, K., YOSHIKAWA, H. & YUI, M. 2003. Mechanisms of transport and retardation in clays enclosing the Cigar Lake high-grade uranium deposit (Saskatchewan, Canada). *In*: *Proceedings of Uranium Geochemistry 2003 International Conference*, Nancy, 13–16 April 2003, 87–90.

BRUNETON, P. 1987. Geology of the Cigar Lake uranium deposit (Saskatchewan, Canada). *In*: GILBOY, C. F. & VIGRASS, L. W. (eds) *Economic Minerals of Saskatchewan*. Saskatchewan Geological Society, Special Publication, Regina, SK, Canada, 8, 99–119.

BRUNETON, P. 1993. Geological environment of the Cigar Lake uranium deposit. *Canadian Journal of Earth Science*, 30, 653–673.

CASAS, I., DE PABLO, J. *et al.* 1998. The role of Pe, pH, and carbonate on the solubility of UO_2 under nominally reducing conditions. *Geochimica et Cosmochimica Acta*, 62, 2223–2231.

CRAMER, J. J. & SMELLIE, J. 1994. *Final Report of the AECL/SKB Cigar Lake Analog Study*. Technical Report AECL-10851, COG-93-147, SKB TR 94-04, 393 p.

CUMMING, G. L. & KRSTIC, D. 1992. The age of unconformity-related uranium mineralization in the Athabasca basin, northern Saskatchewan. *Canadian Journal of Earth Science*, 29, 1623–1639.

CURTIS, D., BENJAMIN, T. *et al.* 1989. The Oklo nuclear reactors: cumulative fission yields and nuclear characteristic of reactor 9. *Earth and Planetary Science Letters*, 89, 193–206.

DYMKOV, Y. M, HOLLIGER, P., PAGEL, M., GORSHKOV, A. & ARTYUKHINA, A. 1997. Characterization of a La–Ce–Sr–C aluminous hydroxy phosphate in nuclear zone 13 in the Oklo uranium deposit (Gabon). *Mineralium Deposita*, 32, 617–620.

EBERLY, P., JANECZEK, J. & EWING, R. C. 1995. Precipitation of uraninite in chlorite-bearing veins of the hydrothermal alteration zone (argiles de pile) of the natural nuclear reactor at Bangombé, Republic of Gabon. *Proceedings of Material Research Society Symposium*, 353, 1195–1202.

EBERLY, P., EWING, R. C., JANECZEK, J. & FURLANO, A. 1996. Clays at the natural nuclear reactor at Bangombé, Gabon: Migration of actinides. *Radiochimica Acta*, 74, 271–275.

EWING, R. C. 1999. Nuclear waste forms for actinides. *Proceedings of National Academy of Sciences colloquium 'Geology, Mineralogy, and Human Welfare'*, 8–9 November 1998, Irvine, CA, USA, 96, 3432–3439.

FINCH, R. & MURAKAMI, T. 1999. Systematics and paragenesis of uranium minerals. *In*: BURNS, P. C. & FINCH, R (eds) *Uranium: Mineralogy, Geochemistry and the Environment. Reviews in Mineralogy, Mineralogical Society of America*, 38, Washington DC, USA, 91–179.

FOUQUES, J. P., FOWLER, M., KNIPPING, H. D. & SCHIMANN, K. 1986. The Cigar Lake uranium deposit – Discovery and general characteristics. *In*: EVANS, E. L. (ed) *Uranium Deposits of Canada*, Special Volume. Canadian Institute of Mining and Metallurgy, Regina, SK, Canada, 33, 218–229.

FRONDEL, C. 1958. Systematic mineralogy of uranium and thorium. *US Geological Survey Bulletin*, 1064, 400.

GANCARZ, A. J. 1978. U–Pb age (2.05×10^9 years) of the Oklo uranium deposit. *In*: *Proceedings of the Technical Committee Meeting*, Paris, 19–21 December 1975. Natural Fission Reactors, IAEA, Vienna, 513–520.

GANCARZ, A., COWAN, G., CURTIS, D. & MAEK, W. 1980. 99Tc, Pb, and Ru migrations around the Oklo natural fission reactors. *Scientific Basis For Nuclear Waste Management*, 2, 601–608.

GAUTHIER-LAFAYE, F. 1986. Les gisements d'uranium du Gabon et les réacteurs d'Oklo. Modèle métallogénique de gites à fortes teneurs du Protérozoique inférieur. *Mémoire Sciences Géologiques*, 78, 206.

GAUTHIER-LAFAYE, F. & WEBER, F. 1981. Les concentration uranifères du Francevillien du Gabon:

leur association avec des gîtes à hydrocarbures fossiles du Protérozoïque inférieur. *Comptes Rendus de l'Académie des Sciences, Paris*, **292**, 69–74.

GAUTHIER-LAFAYE, F. & WEBER, F. 1989. The Francevillian (Lower Proterozoic) uranium ore deposits of Gabon. *Economic Geology*, **84**, 2267–2285.

GAUTHIER-LAFAYE, F., WEBER, F. & OHMOTO, H. 1989. Natural fission reactors of Oklo. *Economic Geology*, **84**, 2286–2295.

GAUTHIER-LAFAYE, F., HOLLIGER, P. & BLANC, P. L. 1996. Natural fission reactors in the Franceville basin, Gabon: A review of the conditions and results of 'critical event' in a geological system. *Geochimica Cosmochimica Acta*, **60**, 4831–4852.

HAGEMANN, R., LUCAS, M., NIEF, G. & ROTH, E. 1974. Mesures isotopiques du rubidium et du strontium et essais de mesure de l'âge de la minéralisation de l'uranium du réacteur naturel d'Oklo. *Earth and Planetary Science Letters*, **23**, 170–176.

HIDAKA, H., SHINOTSUKA, K. & HOLLIGER, P. 1993. Geochemical behaviour of ^{99}Tc in the Oklo natural fission reactors. *Radiochimica Acta*, **63**, 19–22.

HIDAKA, H., SUGIYAMA, T., EBIHARA, M. & HOLLIGER, P. 1994. Isotopic evidence for the retention of 90Sr inferred from excess 90Zr in the Oklo natural fission reactors: implication for geochemical behaviour of fissiogenic Rb, Sr, Cs and Ba. *Earth and Planetary Science Letters*, **122**, 173–182.

HIDAKA, H. & HOLLIGER, P. 1998. Geochemical and neutronic characteristics of the natural fossil fission reactors at Oklo and Bangombé, Gabon. *Geochimica and Cosmochimica Acta*, **62**, 89–108.

HIDAKA, H., HOLLIGER, P. & GAUTHIER-LAFAYE, F. 1999. Tc/Ru fractionation in the Oklo and Bangombé natural fission reactors (Gabon). *Chemical Geology*, **155**, 323–333.

HIKICHI, Y., MURAYAMA, K., OHSATO, H. & NOMURA, T. 1989. Thermal changes of rare earth phosphate minerals. *Journal of the Mineralogical Society of Japan*, **19**, 117–126.

JANECZEK, J. & EWING, R. C. 1992. Dissolution and alteration of uraninite under reducing conditions. *Journal of Nuclear Materials*, **490**, 133–156.

JANECZEK, J. & EWING, R. 1996. Florencite-(La) with fissiogenic REE from a natural fission reactor at Bagombé, Gabon. *American Mineralogist*, **81**, 1263–1269.

JANECZEK, J. 1999. Mineralogy and geochemistry of natural fission reactors in Gabon. *In*: BURNS, P. C. & FINCH, R (eds) *Uranium: Mineralogy, Geochemistry and the Environment. Reviews in Mineralogy.* Mineralogical Society of America, Washington DC, USA, **38**, 321–392.

JENSEN, K. A., EWING, R. C. & GAUTHIER-LAFAYE, F. 1997. Uraninite: a 2 Ga spent nuclear fuel from the natural fission reactor at Bangombé in Gabon, West Aftrica. *Proceedings of Material Research Society Symposium*, **465**, 1209–1218.

JENSEN, K. A. & EWING, R. C. 2001. The Oklelobondo natural fission reactor, southeast Gabon: Geology, mineralogy and retardation of nuclear reaction products. *Geological Society of America Bulletin*, **113/1**, 32–62.

LOUVAT, D., LOT, K., MICHELOT, J.-L., SMELLIE, J. & TUNIZ, C. 2000. Environmental isotope data of water and dissolved species used in model development, calibration and testing at Bagombé and Okélobondo. *In*: *Proceeding of the Second Joint EC-CEA Workshop on the Oklo-Natural Analogue Phase II Project*, Helsinki, 16–18 June 1998. European Commission, Nuclear Science and Technology, EUR19116 EN, 391–398.

MADÉ, B., LEDOUX, E., SALIGNAC, A.-L., LE BOURSICAUD, B. & GURBAN, I. 2000. Modélisation du transport réactif de l'uranium autour du réacteur naturel de Bangombé (Oklo, Gabon). *C R de l'Académie des Sciences, Paris*, **331**, 587–594.

MANN, A. W. & DEUTSCHER, R. L. 1980. Solution chemistry of lead and zinc in water containing carbonate, sulfate and chloride ions. *Chemical Geology*, **29**, 293–311.

PERCIVAL, J. B. & KODAMA, H. 1989. Sudoite from Cigar Lake, Saskatchewan. *The Canadian Mineralogist*, **27**, 633–641.

PERCIVAL, J. B., TORRANCE, J. K. & BELL, K. 1993. Clay mineralogy and isotope geochemistry of the alteration halo at the Cigar Lake uranium deposit. *Canadian Journal of Earth Sciences*, **30**, 689–704.

PÉREZ DEL VILLAR, L., COZAR, J. S., PARDILLO, J. & LABAJOS, M. A. 2000. Mineralogical characterization of fracture filling in the far field of the Bagombé natural reactor (Gabon): implications on the migration/retention processes involving natural radionuclides and the analogue elements. *In*: LOUVAT, D. (ed) *OKLO Working Group Proceedings of the Final Meeting OKLO-Natural Analogue Phase II Project Held in Cadarache, France*, 20 to 21 May 1999. European Commission, Nuclear Science and Technology, EUR 19137EN, 45–74.

POURCELOT, L. & GAUTHIER-LAFAYE, F. 1999. Hydrothermal and supergen clays of the Oklo natural reactors: conditions of radionuclide release, migration and retention. *Chemical Geology*, **157**, 155–174.

RUFFENACH, J. C. 1978. Etude des migrations de l'uranium et des terres rares sur une carotte de sondage et application à la détermination de la date des réactions nicléaires. *In*: *Proceedings of the Technical Committee Meeting*, Paris, 19–21 December 1975. Natural Fission Reactors, IAEA, Vienna, 441–471.

SALAH, S. 2000. *Weathering Processes at the Natural Nuclear Reactor of Bangombé (Gabon). Identification and Geochemical Modeling of the Retention and Migration Mechanisms of Uranium and Rare Earth Elements*. PhD thesis, Université Louis Pasteur, Strasbourg, France.

SALAH, S., GAUTHIER-LAFAYE, F., DEL NERO, M., BRACKE, G. 2000. FB formation: Mineralogy of the pelites and description of the weathering profile. *In*: LOUVAT, D. (ed) *OKLO Working Group Proceedings of the Final Meeting OKLO-Natural Analogue Phase II Project Held in*

Cadarache, France, 20–21 May 1999. European Commission, Nuclear Science and Technology, EUR 19137EN, 75–90.

SPEER, J. A. 1982. Actinide orthosilicates. *In*: RIBBE, P. H. (ed) *Reviews in Mineralogy, Orthosilicates*. Mineralogical Society of America, Washington DC, USA, **5**, 67–135.

STILLE, P., GAUTHIER-LAFAYE, F. *et al.* 2003. REE mobility in groudwater proximate to the natural fission reactor at Bangombé (Gabon). *Chemical Geology*, **198**, 289–304.

SUNDER, S., TAYLOR, P. & CRAMER, J. J. 1988. XPS and XRD studies of uranium rich minerals from Cigar Lake, Saskatchewan. *Materials Research Society Symposia Proceedings*, **112**, 465–472.

SUNDER, S., SHOESMITH, D. W., CHRISTENSEN, H. & MILLER, N. H. 1992. Oxidation of UO_2 fuel by the products of gamma radiolysis of water. *Journal of Nuclear Materials*, **190**, 78–86.

SUNDER, S., MILLER, N. H. & DUCLOS, A. M. 1994. XPS and XRD studies of samples from the natural fission reactors in the Oklo uranium deposits. *Materials Research Society Symposia Proceedings*, **333**, 631–638.

SUNDER, S., CRAMER, J. J. & MILLER, N. H. 1996. Geochemistry of the Cigar Lake deposit: XPS studies. *Radiochimica Acta*, **74**, 303–307.

WINBERG, A. & STEVENSON, D. 1994. Compilation of hydrogeological data and 2D and 3D conceptualisation and 2D modelling of the groundwater flow conditions in and around the Cigar Lake uranium deposit. *In*: CRAMER, J. J. & SMELLIE, J. A. T. (eds) *Final Report of the AECL/SKB Cigar Lake Analog Study*. Atomic Energy of Canada Limited Report, AECL-10851 Pinawa (Manitoba), Canada, 104–142.

Basaltic dykes in evaporites: a natural analogue

MARC STEINMANN[1] & PETER STILLE[2]

[1]*Département de Géosciences, Université de Franche-Comté, Besançon, France*
(e-mail: marc.steinmann@univ-fcomte.fr)

[2]*ULP-EOST-CNRS, Centre de Géochimie de la Surface UMR 7517, Strasbourg, France*
(e-mail: pstille@illite.u-strasbg.fr)

Abstract: We present rare earth element (REE) data of basalt and salt samples from central Germany where basaltic dykes of Tertiary age crosscut Upper Permian rock and potash salt. The glassy rims of the dykes can be considered as a natural analogue for the corrosion of nuclear waste glass in a salt repository, whereas the REE data from the salt can serve as an analogue for radionuclide migration in salt next to a leaking nuclear waste repository because the light rare earths (LREE) have a geochemical behavior similar to that of some actinides.
Our basalt data demonstrate mobility and fractionation of the REE during postintrusive circulation of salt brines. The processes controlling this behavior of the REE were dissolution and reprecipitation of phosphate minerals. The salt data show that a small portion of the REE has left the basalt during postintrusive fluid circulation and migrated into the salt where a strong depletion of the LREE can be observed with increasing distance from the basalt contact. This fractionation is most probably due to precipitation of LREE-enriched accessory minerals such as apatite. In analogy to this, a similar behavior might be expected from actinides such as Am and Cm, which would in the case of a leaking salt nuclear waste repository probably be immobilized when phosphate minerals are present in the backfill material.

The repository analogy

Salt deposits are potential host rocks for the disposal of high-level radioactive waste. In this chapter we will present data from the Werra–Fulda district (northern Germany) where Upper Permian (Zechstein) salt is crosscut by numerous basalt dykes of Miocene age.

The chilled margins of the dykes are glassy and can therefore, as proposed by Ewing (1979), be used as a natural analogue for the long-term behaviour of vitrified high-level radioactive waste (HLW) glass in a salt environment. Several studies have shown that similar alteration products form on both nuclear waste and natural glasses in spite of differences in the chemical composition between natural basalt and synthetic HLW borosilicate glass (Malow et al. 1984; Byers et al. 1985; Crovisier et al. 1992; Lutze & Grambow 1992; Petit 1992; Abdelouas et al. 1994; Daux et al. 1997).

The unique setting of the Werra–Fulda district allows not only the study of basalt alteration, but also the ability to trace basalt-derived elements within the salt. The study site is thus, on one hand, a natural analogue for the long-term corrosion expected for nuclear waste glass in contact with salt brines, and on the other hand, a natural analogue for radionuclide migration in salt next to a leaking salt repository. We will focus mainly on the behaviour of the rare earth elements (REE) during basalt alteration and fluid migration in the salt. The REE are of particular interest for natural analogue studies because they are considered to be chemical analogues for the actinides Am and Cm (Choppin 1983; Krauskopf 1986; Seaborg 1993; Silva & Nitsche 1995). The basalt–salt reaction zones have been studied in detail previously (Steinmann & Stille 1998; Steinmann et al. 1999, 2001) and the reader is referred to these earlier publications for analytical details and raw data. The aim of the present article is to give an overview on REE behaviour at basalt–salt contacts and to contribute to radionuclide behaviour during HLW glass corrosion and radionuclide migration in salt formations.

The study site

The evaporites of the Werra–Fulda potash district were deposited during the Upper Permian (Zechstein). The series has a total thickness of

From: GIERÉ, R. & STILLE, P. (eds) 2004. *Energy, Waste, and the Environment: a Geochemical Perspective*. Geological Society, London, Special Publications, **236**, 135–141.
0305-8719/04/$15 © The Geological Society of London 2004.

up to 400 m. The most important salt is halite (NaCl), followed by thinner horizons of potash salt and anhydrite (CaSO$_4$). The series contains two important potash salt horizons, mainly consisting of carnallite (KMgCl$_3$·6H$_2$O), which are mined at depths of about 800 m. The thickness of these horizons varies between 2 and 10 m.

In the Tertiary, basaltic dykes intruded the evaporite. Small flakes of newly formed clay minerals cover the basalt surface at the contact with the evaporite. In the underground mines, the basalt is exposed as subvertical dykes, which can be followed horizontally over several kilometres. All basalt–salt contacts discussed here are located in underground mines in the two potash salt horizons mentioned above.

Near the basalt contacts, in a zone of less than 3 m width, the potash salt has been transformed mainly into halite and sylvite (KCl) by fluids saturated in NaCl. These fluids accompanied the basaltic melt and became saturated in NaCl during upward migration by the partial dissolution of an underlying, up to 100 m thick, halite formation (Knipping & Herrmann 1985; Gutsche 1988; Knipping 1989). During the metamorphism of the potash salt, the NaCl-fluids were mixed with fluids originating from the potash salt and the resulting mixture subsequently altered the basalt and transported basalt-derived trace elements such as Sr and REE into the salt (Steinmann *et al.* 1999, 2001). The study of the alteration of the basalt presented here is based on a detailed profile across a dyke of about 50 cm width (Fig. 1), whereas trace element migration in the salt has been studied in another outcrop in a 4 m long horizontal salt profile (Fig. 2).

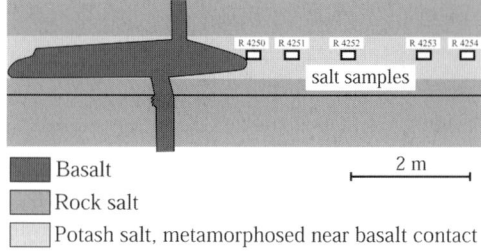

Basalt

Rock salt

Potash salt, metamorphosed near basalt contact

Fig. 2. The salt profile studied here is situated next to a basalt dyke, which is locally enlarged to an apophysis. The sampling points in the salt are indicated by the small rectangles numbered R 4250 to R 4254.

Mobility of the REE during basalt corrosion

In order to obtain information about the exchange processes between the basalt and the salt, leaching experiments were performed with 1.5 M HCl on basalt powder samples that had previously been washed in bi-distilled water in order to remove salt minerals. The H$_2$O-washed whole rock samples, as well as the resulting leachate and residue fractions have been analysed for major elements, REE, and for Sr isotopes.

A key problem in the present natural analogue study is the distinction between chemical variations related to trace element migration during basalt alteration and variations due to magmatic fractionation and other syn-intrusive processes. The detailed evaluation of the available data has shown that the chemical and isotopic composition of the HCl residues is largely controlled by fractional crystallization and syn-intrusive assimilation of salt. In contrast, the chemical composition of the leachates is strongly modified by post-intrusive alteration (Steinmann *et al.* 1999).

For both HCl leachates and HCl residues, a positive correlation between ΣREE and P$_2$O$_5$ has been found suggesting that phosphate is an important REE carrier in both cases (Fig. 3). The positive correlation between La/Yb and P$_2$O$_5$ furthermore indicates that the REE carrier phase is enriched in the light REE (LREE, La–Nd). Such an LREE enrichment is characteristic for phosphate minerals such as apatite, which is in agreement with the suggestion that phosphate is the principal REE carrier in the leachable phase and in the residues.

A plot of the average REE distribution patterns of the leachates and residues normalized to the H$_2$O-washed whole rock data shows that they have complementary patterns (Fig. 4). Furthermore the data show that leaching removed

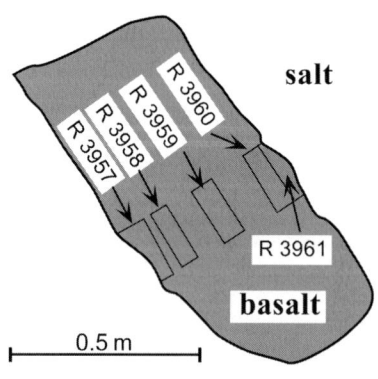

Fig. 1. Detailed cross-section of the basalt dyke with sampling points of samples R 3957 to R 3960.
The dyke is conserved as an isolated tectonic boudin embedded in potash salt.

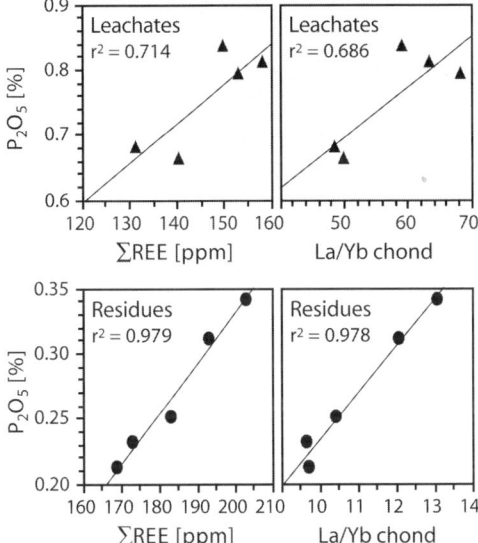

Fig. 3. Correlations of $\sum$REE and chondrite normalized La/Yb ratios with P_2O_5 for the HCl leachates and residues of the basalt samples.

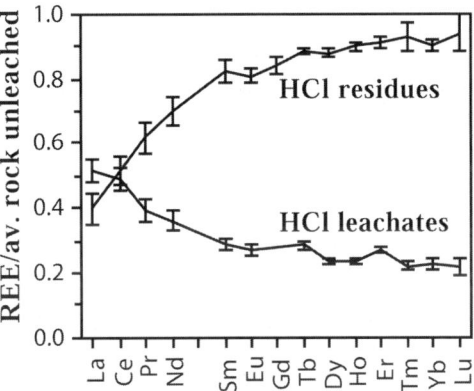

Fig. 4. Average REE concentrations with standard deviations of the 5 HCl leachates and residues normalized to the average concentrations of the H_2O-washed whole rock.

almost 50% of the total REE content, in the case of La even more. The leachable phosphate is more enriched in the LREE than the residual one. Such an enrichment of the LREE is typical for a secondary Al-phosphate such as florencite or rhabdophane and has been described in several weathering profiles (Duddy 1980; Tazaki et al. 1987; Banfield & Eggleton 1989; Braun et al. 1990, 1993; Prudêncio et al. 1993; Macfarlane et al. 1994) and in hydrothermal deposits (Nasraoui et al. 2000). A comparative study of weathered and unweathered phosphate by Tazaki et al. (1987) has shown that weathered phosphate is characterized by a less ordered lattice structure. This could in our case explain why leaching with HCl removed in a very selective way essentially secondary and not primary phosphate.

The only primary phosphate mineral in the basaltic groundmass is an idiomorphic acicular apatite, which we therefore regard as the most probable primary REE carrier in the residues. However, the REE concentrations of this apatite were below the detection limit of the EDS device of our SEM.

The secondary phase that contained the leachable REE was identified in alteration rims around small salt inclusions in the peripheral parts of the dyke where numerous phosphate grains with high REE concentrations occur in association with secondary sheet silicates (Fig. 5). This secondary phosphate could be identified as REE–chlorapatite [$Ca_2La_{0.5}CeNd_{0.5}(PO_4)_3Cl$] and its high REE content is demonstrated by EDS element maps (Fig. 6).

In summary, the study of the basalt dykes shows that the REE of the primary basalt have been mobilized during basalt alteration, most probably by dissolution of primary apatite. This mobilization was strongest in the intermediate samples R 3958 and R 3960 (Fig. 1) and the mobilized REE were subsequently immobilized by precipitation of secondary apatite.

Migration of the REE in the salts

In order to trace the migration of basalt-derived REE in the salt, REE distribution patterns (Fig. 7) and Nd isotopic compositions (Fig. 8) have been determined in a salt horizon adjacent to a basalt dyke (Fig. 2). The flat REE distribution patterns and the almost basaltic Nd isotopic composition of the salt samples collected at the basalt–salt contact point to a basaltic origin of the REE for this sample. With increasing distance from the contact, the patterns are more and more depleted in Ce, Pr, Nd, Sm, and Eu and the Nd isotopic compositions are slightly shifted towards lower ε_{Nd} values, which, however, still remain above values typical for continental crust or Permian seawater (Stille et al. 1996, and citations therein). This evolution of the REE distribution patterns and the Nd isotopic compositions could basically be due to mixing between a basalt and a salt end member or, alternatively, it could have been fractionation of the REE during migration in the salt that modified the REE patterns.

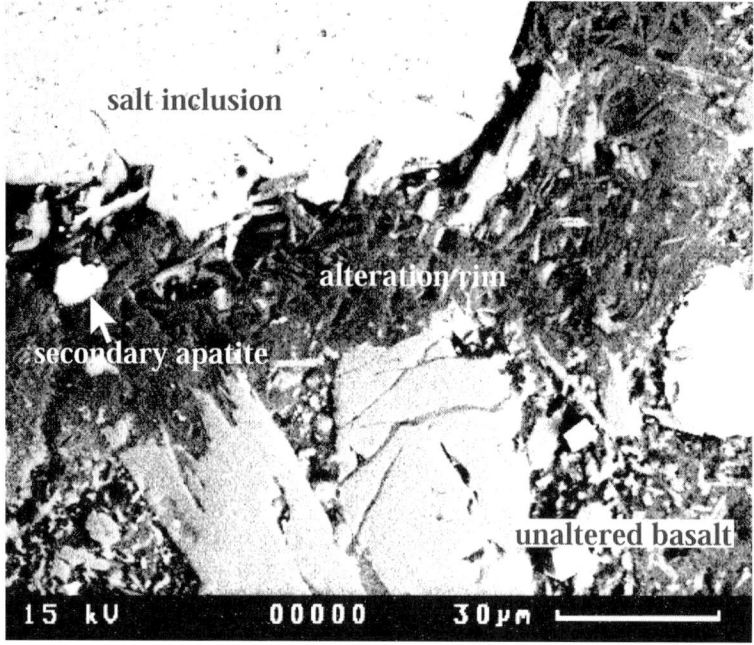

Fig. 5. SEM photograph showing a secondary apatite crystal within a rim of altered basalt around a salt inclusion. The other minerals contained in the rim are sheet silicates and hematite.

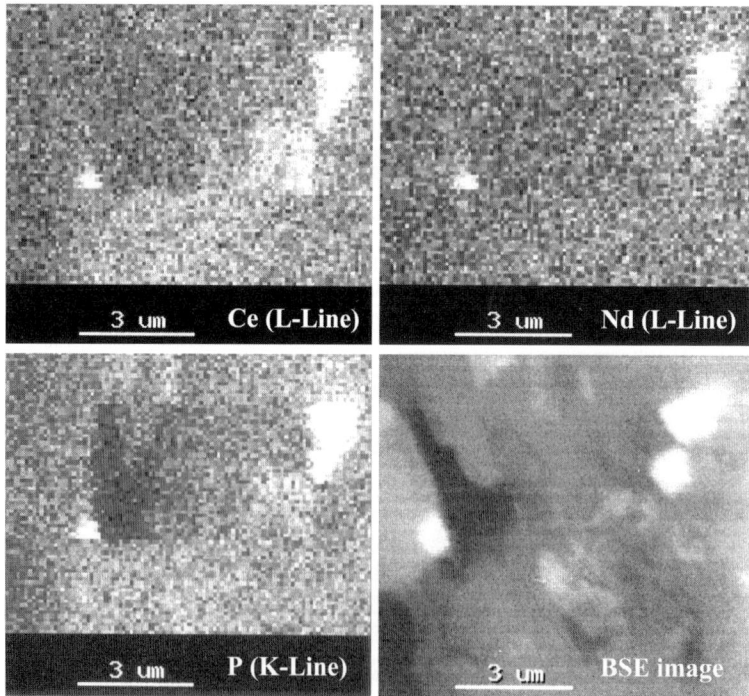

Fig. 6. Element maps of Ce, Nd, and P demonstrating the high REE content of the secondary apatite. The elevated REE concentrations lead to the high contrast of the apatite in the backscattered electron (BSE) image.

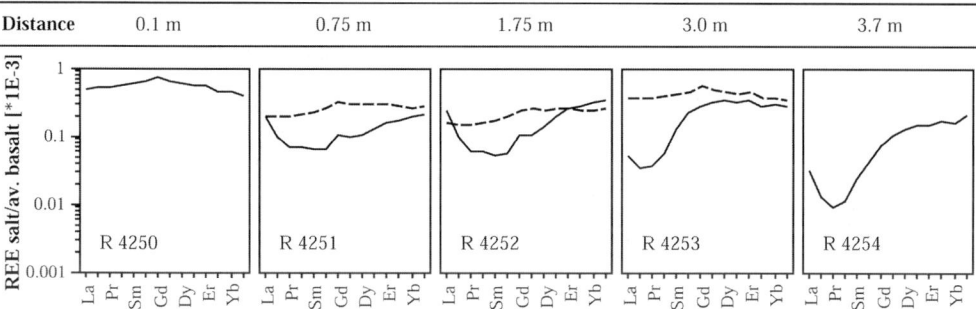

Fig. 7. Basalt normalized REE distributions of the salt samples R 4250 to R 4254 from 0.1 to 4.7 m from the basalt contact. The solid lines show the measured REE distributions. Note the depletion of Ce, Pr, Sm, and Eu appearing with distance. The dashed lines are calculated mixing patterns discussed in the text. See Fig. 2 for sample locations.

The Nd isotopic compositions of the salt give valuable information to test these two hypotheses. The lack of an alignment in the Nd-mixing diagram (not shown) is a first indication that the observed modification of the REE patterns cannot be due to mixing (Steinmann *et al.* 2001). In addition, hypothetical REE patterns have been calculated for the three intermediate samples R 4251, R 4252, and R 4253 by using a binary mixing model and the sample R 4250 from the basalt contact and the most distant sample R 4254 as end members. First, the percentage of REE derived from the contact sample R 4250 has been calculated for each of the three intermediate samples on the basis of their Nd isotopic compositions and the Nd isotopic compositions of the two end members from equation (1):

$$\% \ REE_{R\,4250}$$

$$= 100 \times \left(\frac{{}^{143}Nd/{}^{144}Nd_{sample} - {}^{143}Nd/{}^{144}Nd_{R\,4254}}{{}^{143}Nd/{}^{144}Nd_{R\,4250} - {}^{143}Nd/{}^{144}Nd_{R\,4254}} \right)$$

$$(1)$$

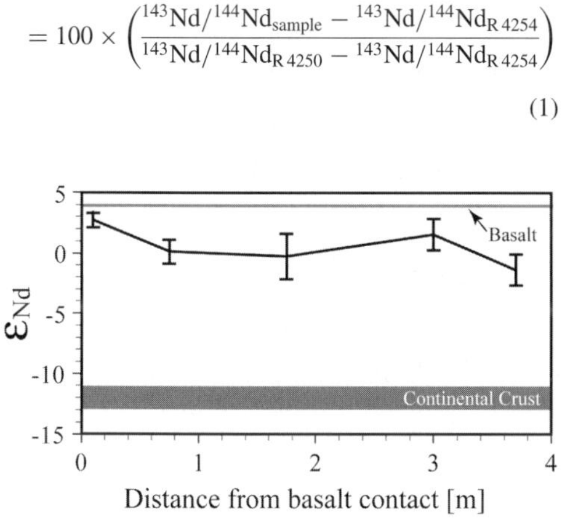

Fig. 8. The evolution of the Nd isotopic compositions with distance in the salt profile together with the compositions of average basalt and average continental crust. The error bars represent ± 2 sigma mean values of the individual measurements.

Afterwards, the percentage calculated in equation (1) has been used together with the REE patterns of the two end members to calculate a hypothetical REE mixing pattern for each of the three intermediate samples according to equation (2):

$$REE_{sample}$$
$$= \{\% \ REE_{equation\,(1)} \times REE_{R\,4250}\}$$
$$+ \{(100 - \% \ REE_{equation\,(1)})$$
$$\times REE_{R\,4254}\} \qquad (2)$$

The hypothetical REE mixing patterns calculated for the three intermediate samples are shown as dashed lines in Fig. 7. They are almost flat and show no similarities with the observed patterns. We therefore conclude that the observed evolution of the REE distributions cannot be attributed to mixing, but must be due to a fractionation process.

This with distance increasing fractionation of the salt patterns is probably due to preferential immobilization of the REE by coprecipitation with a secondary mineral phase rather than to better solubility in the fluid by solution complexation (Steinmann *et al.* 2001). This secondary mineral phase could, in analogy to the finding in the previous section on REE immobilization during basalt corrosion, be apatite. This hypothesis is ascertained by a comparison with experimental data on REE fractionation by phosphate precipitation (Fig. 9, Byrne *et al.* 1996). The experimental data in fact show that apatite precipitation preferentially immobilizes Ce, Pr, Nd, Sm, and Eu, which is in agreement with our traverse in the salt. This finding is also in agreement with *in situ* observation and modelling of Byrne and Kim (1993) and Johannesson *et al.* (1995), who found that phosphate precipitation is an important process leading to fractionation of the rare earths in surface water and groundwater.

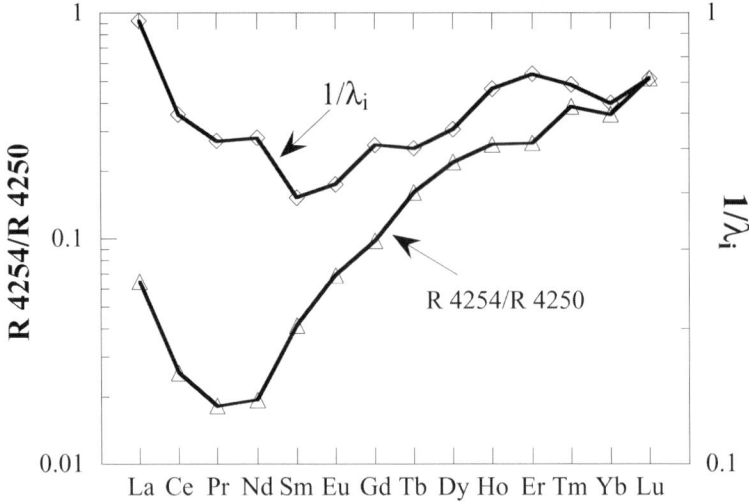

Fig. 9. Comparison of REE fractionation in the salt with fractionation factors λ_i determined by Byrne *et al.* (1996) from experimental data for REE coprecipitation with phosphate. REE fractionation in the salt is illustrated by the ratio of samples R 4254/R 4250, which are the two extreme samples of the salt profile. The λ_i values are plotted in reciprocal form in order to show REE fractionation in the remaining solution from which phosphate has precipitated.

A key question for the present natural analogue is the temperature at which REE migration and fractionation occurred in the salt. The temperature of the basaltic melt was around 1150 °C (Knipping 1989). However, model calculations based on conductive heat transport suggest that the temperatures in the salt never exceeded 800 °C, even directly at the basalt contact, and that they never attained 200 °C at 3 m distance from a basalt dyke of 1–1.8 m thickness (Knipping 1989). This is also supported by the presence of relicts of primary carnallite, for which the thermal stability limit is 167.5 °C (Braitsch 1971), in salt less than 1 m from basalt contacts (Knipping & Herrmann 1985; Knipping 1989). All these data suggest that REE migration in the salt occurred only in close proximity to the basalt at temperatures above 200 °C, whereas temperatures were already lower at 1–3 m from the contact. These temperatures can be regarded as maximum temperatures because the study of the basalt has shown that mobilization of the REE occurred after intrusion, at a moment when cooling of the basalt and the adjacent salt had already started. This decrease of temperature with distance might, together with increasing pH, have been one of the key parameters that triggered apatite precipitation within the salt, which in turn led to immobilization and fractionation of the REE.

Concluding remarks

The study of the basaltic dykes in evaporites demonstrates that dissolution and precipitation of phosphate minerals is a key process for the control of REE mobility and REE fractionation. In the present case, all REE found in secondary apatite in the basalt and in the salt are derived from the dissolution of primary magmatic apatite during basalt corrosion. This loss of REE from the basalt to the salt was not sufficient to lower significantly the REE concentrations of the basalt and it could only be detected by the analysis of the salt. The absolute quantity of REE transferred from the basalt into the salt, however, cannot be quantified because we have no three-dimensional control on the REE concentrations around the basalt apophysis.

Our results show that coprecipitation of the REE with phosphate removed Ce, Pr, Nd, Sm, and Eu more easily from the brine than other REE. This finding might be of importance for the mobility of trivalent Am and Cm in a radioactive waste salt repository, because for these elements, owing to their almost identical ionic radii, an almost analogous geochemical behaviour is expected as for Sm and Nd (Choppin 1983; Krauskopf 1986). These radionuclides would, in the case of a leaking HLW salt repository, probably be retained when phosphate minerals are present in the backfill material.

M. Cuney and S. Church are thanked for their constructive reviews. This is EOST/CGS contribution no. 2004.602-UMR7517.

References

ABDELOUAS, A., CROVISIER, J. L., LUTZE, W., FRITZ, B., MOSSER, A. & MÜLLER, R. 1994. Formation of hydrotalcite-like compounds during R7T7 nuclear waste glass and basaltic glass alteration. *Clays and Clay Minerals*, **42**, 526–533.

BANFIELD, J. F. & EGGLETON, R. A. 1989. Apatite replacement and rare earth mobilization, fractionation, and fixation during weathering. *Clays and Clay Minerals*, **37**, 113–127.

BRAITSCH, O. 1971. *Salt Deposits. Their Origin and Composition.* Springer-Verlag, Berlin-Stuttgart-Heidelberg-New York, 297 pp.

BRAUN, J. J., PAGEL, M., HERBILLON, A. & ROSIN, C. 1993. Mobilization and redistribution of REEs and thorium in a syenitic lateritic profile: A mass balance study. *Geochimica et Cosmochimica Acta*, **57**, 4419–4434.

BRAUN, J. J., PAGEL, M., MULLER, J. P., BILONG, P., MICHARD, A. & GUILLET, B. 1990. Ce anomalies in lateritic profiles. *Geochimica et Cosmochimica Acta*, **54**, 781–795.

BYERS, C. D., EWING, R. C. & KEIL, K. 1985. Basalt glass: an analogue for the evaluation of the long-term stability of nuclear waste form borosilicate glasses. *In*: JANTZEN, C. M., STONE, J. A. & EWING, R. C. (eds) *Scientific Basis for Nuclear Waste Management VIII. Mat. Res. Soc. Symp. Proc.*, Pittsburgh, Pennsylvania, 44.

BYRNE, R. H. & KIM, K.-H. 1993. Rare earth precipitation and coprecipitation behavior: The limiting role of PO_4^{3-} on dissolved rare earth concentrations in seawater. *Geochimica et Cosmochimica Acta*, **57**, 519–526.

BYRNE, R. H., LIU, X. & SCHIJF, J. 1996. The influence of phosphate coprecipitation on rare earth distributions in natural waters. *Geochimica et Cosmochimica Acta*, **60**, 3341–3346.

CHOPPIN, G. R. 1983. Comparison of the solution chemistry of the actinides and the lanthanides. *Journal of the Less-Common Metals*, **93**, 232–330.

CROVISIER, J.-L., HONNOREZ, J., FRITZ, B. & PETIT, J.-C. 1992. Dissolution of subglacial volcanic glasses from Iceland: laboratory study and modelling. *Applied Geochemistry*, **Suppl. Issue No. 1**, 55–81.

DAUX, V., GUY, C., ADVOCAT, T., CROVISIER, J.-L. & STILLE, P. 1997. Kinetic aspects of basaltic glass dissolution at 90 °C: Role of aqueous silicon and aluminium. *Chemical Geology*, **142**, 109–126.

DUDDY, I. R. 1980. Redistribution and fractionation of rare earth and other elements in a weathering profile. *Chemical Geology*, **30**, 363–381.

EWING, R. C. 1979. Natural glasses: Analogues for radioactive waste forms. *In*: MCCARTHY, G. (ed) *Scientific Basis for Nuclear Waste Management.* Plenum Press, New York, 57–68.

GUTSCHE, A. 1988. *Mineralreaktionen und Stofftransporte an einem Kontakt Basalt-Hartsalz in der Werra-Folge des Werkes Hattorf.* Unpubl. Diploma Thesis, Georg-August-Universität Göttingen.

JOHANNESSON, K. H., LYONS, W. B., STETZENBACH, K. J. & BYRNE, R. H. 1995. The solubility control of rare earth elements in natural terrestrial waters and the significance of PO43- and CO32- in limiting dissolved concentrations: A review of recent information. *Aquatic Geochemistry*, **1**, 157–173.

KNIPPING, B. 1989. *Basalt Intrusions in Evaporites. Lecture Notes in Earth Sciences 24.* Springer-Verlag, Berlin, 132 pp.

KNIPPING, B. & HERRMANN, A. G. 1985. Mineralreaktionen und Stofftransporte an einem Kontakt Basalt-Carnallitit im Kalisalzhorizont Thüringen der Werra-Serie des Zechsteins. *Kali und Steinsalz*, **9**, 111–124.

KRAUSKOPF, K. B. 1986. Thorium and rare-earth elements as analogues for actinide elements. *Chemical Geology*, **55**, 323–335.

LUTZE, W. & GRAMBOW, B. 1992. The effect of glass corrosion on near field chemistry. *Radiochimica Acta*, **58/59**, 3–7.

MACFARLANE, A. W., DANIELSON, A., HOLLAND, H. D. & JACOBSEN, S. B. 1994. REE chemistry and Sm–Nd systematics of late Archean weathering profiles in the Fortescue Group, Western Australia. *Geochimica et Cosmochimica Acta*, **58**, 1777–1794.

MALOW, G., LUTZE, W. & EWING, R. C. 1984. Alteration effects and leach rates of basaltic glasses: implications for the long-term stability of nuclear waste form borosilicate glasses. *Journal of Non-Crystalline Solids*, **67**, 305–321.

NASRAOUI, M., TOULKERIDIS, T., CLAUER, N. & BILAL, E. 2000. Differentiated hydrothermal and meteoric alterations in the Lueshe carbonatite complex (Democratic Republic of Congo) identified by a REE study combined with a sequential acid-leaching experiment. *Chemical Geology*, **165**, 109–132.

PETIT, J. C. 1992. Reasoning by analogy: rational foundation of natural analogue studies. *Applied Geochemistry*, **Suppl. Issue 1**, 9–11.

Prudêncio, M. I., BRAGA, M. A. S. & GOUVEIA, M. A. 1993. REE mobilization, fractionation and precipitation during weathering of basalts. *Chemical Geology*, **107**, 251–254.

SEABORG, G. T. 1993. Overview of the actinide and lanthanide (the f) elements. *Radiochimica Acta*, **61**, 115–122.

SILVA, R. J. & NITSCHE, H. 1995. Actinide environmental chemistry. *Radiochimica Acta*, **70/71**, 377–396.

STEINMANN, M. & STILLE, P. 1998. Strongly fractionated REE patterns in salts and their implications for REE migration in chloride-rich brines at elevated temperatures and pressures. *Comptes Rendus de l'Académie des Sciences Paris, Série II a*, **327**, 173–180.

STEINMANN, M., STILLE, P., BERNOTAT, W. & KNIPPING, B. 1999. The corrosion of basaltic dykes in evaporites: Ar–Sr–Nd isotope and REE evidence. *Chemical Geology*, **153**, 259–279.

STEINMANN, M., STILLE, P., MENGEL, K. & KIEFEL, B. 2001. Trace element and isotopic evidence for REE migration and fractionation in salts next to a basalt dyke. *Applied Geochemistry*, **16**, 351–361.

STILLE, P., STEINMANN, M. & RIGGS, S. R. 1996. Nd isotope evidence for the evolution of the paleocur-rents in the Atlantic and Tethys Oceans during the past 180 Ma. *Earth and Planetary Science Letters*, **144**, 9–20.

TAZAKI, K., FYFE, W. S. & DISSANAYAKE, C. B. 1987. Weathering of apatite under extreme conditions of leaching. *Chemical Geology*, **60**, 151–162.

Anthropogenic radionuclide emissions into the environment

J. EIKENBERG, H. BEER & S. BAJO

Division for Radiation Safety and Security, Paul Scherrer Institute,
CH-5232 Villigen, Switzerland (e-mail: jost.eikenberg@psi.ch)

Abstract: Two sources contribute essentially to the presence of anthropogenic radioisotopes in the environment: (i) release from nuclear materials with a major fraction derived from the nuclear bomb testing period during the period 1950–1963 and (ii) emissions from the nuclear industry, such as waste waters from U-mine tailing or nuclear fuel reprocessing plants. This chapter focuses mainly on the major source responsible for global surface contamination, that is, radioisotope aerosol deposition after release into the atmosphere. The atmospheric emissions were caused mainly via surface atomic bomb tests and reactor accidents, with the Chernobyl reactor catastrophe as the most important contribution. In contrast with most fission products, almost all actinides (which are produced via neutron capture reactions) are rather long lived and can be measured in environmental samples with high precision. Some of the actinides (i.e., U, Pu, Cm) consist of various isotopes (e.g., ^{238,239,240,241}Pu) and hence can be used to distinguish between soil contamination from different emission sources. Approaches are therefore presented for calculating the different fractions of contaminants by use of characteristic isotopic signatures.

In the first years after the detection of nuclear fission by Hahn, Strassmann and Meitner in 1938, the major focus on using atomic energy was the production and testing of nuclear weapons in the terrestrial atmosphere (Eisenbud & Gesell 1997). Between 1950 and 1963, when an agreement to ban atomic bomb testing in the atmosphere was signed by the majority of nuclear weapon producing nations, more than 6 PBq (1 PBq = 10^{15} Bq) of ^{239}Pu (that did not undergo fission during the bomb explosion), 900 PBq of fissiogenic ^{137}Cs, and 600 PBq of ^{90}Sr were released into the atmosphere (Table 1; UNSCEAR 2000). This source still contributes predominantly to the presence of anthropogenic radioisotopes that can be found in the environment (mainly in the northern hemisphere; Volchok & Kleinmann 1971). In addition, with the development of the nuclear power industry from 1970 onwards, further emissions occurred to a minor extent from properly operating reactors (waste water), but in most cases from accidents that destroyed reactors, particularly the Chernobyl accident in 1986 (Buzulukov & Dobrynin 1993; NEA-OECD 1996). These are the major sources via aerosol deposition from the atmospheric pathway, but there are further significant releases from nuclear reprocessing plants (mainly isotopes of U, Pu, Am, and fission products) and the uranium mining industry (U, Th, Ra, Po). These radionuclide contaminants propagate mainly via aquatic pathways,

and since most of the long-lived radioisotopes undergo strong interaction with soil minerals or river/sea sediments, the concentrations of several radio-pollutants in the sediments show maximum values around the vicinity of the emission site (e.g., Irish Sea sediment pollution from the Sellafield fuel processing plant or Russian emissions into the Arctic Sea (Black 1984; Johnes *et al.* 1991; Khodakovsky 1994)) but not on a global scale as presented below.

Release from nuclear weapon tests

In the late 1950s to the early 1960s more than 500 atmospheric atomic bombs tests were conducted at different test sites in the northern hemisphere of our planet (e.g., Pacific Ocean atolls such as the Marshall Islands and other sites such as Novaja-Semlja, Severjana-Semlja, Kamchatka, Semipalatinsk, the Gobi Desert and so on). Table 2 summarizes the countries involved, the number of tests performed, and the total energy released into the atmosphere during these tests. The radioactive debris from these nuclear detonations can be divided roughly into three fractions: (i) larger 'hot particles', which fall out within a few hours and near to the detonation site; (ii) a smaller fraction, which is dispersed into the troposphere and behaves like larger aerosoles, that is, deposition is mainly due to washout by rainfall within weeks after the detonation; and, finally, (iii) fine-grained

From: GIERÉ, R. & STILLE, P. (eds) 2004. *Energy, Waste, and the Environment: a Geochemical Perspective*. Geological Society, London, Special Publications, **236**, 143–151.
0305-8719/04/$15   © The Geological Society of London 2004.

Table 1. *Total atmospheric release of long-lived dose-relevant fission radioisotopes and actinides from the atomic bomb tests and the Chernobyl reactor explosion*[¶]

	Totally released atmospheric input (Bq)		Soil surface activity concentration (Bq/m^2)		
Isotope	Atomic bomb tests*	Chernobyl accident[†]	Chernobyl 30 km zone[‡,§]	Western Europe	
				Chernobyl component[‡]	Atomic bomb component[‡,‖]
^{90}Sr	6.2×10^{17}	2.2×10^{17}	$(0.5-5) \times 10^5$	$<1 \times 10^2$	$(0.1-2) \times 10^3$
^{134}Cs	$<1 \times 10^{14}$	1.7×10^{17}	$(0.1-2) \times 10^5$	$(0.1-4) \times 10^3$	$<1 \times 10^0$
^{137}Cs	9.5×10^{17}	2.8×10^{17}	$(0.2-2) \times 10^6$	$(0.2-4) \times 10^4$	$(0.4-8) \times 10^3$
^{238}Pu	3.3×10^{14}	1.0×10^{15}	$(0.5-5) \times 10^3$	$<1 \times 10^1$	$(0.3-6) \times 10^0$
$^{239+240}$Pu	1.1×10^{16}	2.1×10^{15}	$(0.1-1) \times 10^4$	$<1 \times 10^1$	$(0.1-2) \times 10^2$
^{241}Pu	1.4×10^{17}	1.7×10^{17}	$(0.5-5) \times 10^4$	$<1 \times 10^1$	$(0.3-6) \times 10^2$
^{241}Am	$<1 \times 10^{14}$	1.5×10^{14}	$(0.1-1) \times 10^4$	$<1 \times 10^1$	$(0.4-8) \times 10^1$

*From UNSCEAR (1982, 2000), reference year 1963; [†]From Buzulukov and Dobrynin, reference year 1986; [‡]Reference period: 1990–2000; [§]From Swiss Commission on Atomic Protection (1995); [‖]From Hoelgye & Filgas (1995), Bunzel *et al.* (1995), Swiss Commision on Atomic Protection (1995), Geering *et al.* (2000), Eikenberg *et al.* (2002, 2003). [¶]The last three columns summarize typical values of soil surface concentrations in the immediate vicinity of the Chernobyl reactor and less affected areas such as western Europe.

aerosol-sized particles, which enter the stratosphere (Eisenbud & Gsell 1997). This smaller fraction, which does not undergo rain-out initially, has stratospheric residence times of several months (Krey & Krajewski 1970) and is therefore globally distributed over the entire Earth. The fraction of debris entering the stratosphere arises mainly from the large fission–fusion–fission bombs that release tremendous energies exceeding 500 kilotons TNT (1 ton TNT equals to 4.2×10^{12} J).

Nuclear reactor catastrophes

The Chernobyl accident, April 1986

Besides the input of nuclear fuel material via atomic bomb detonation, reactor accidents also contributed to atmospheric release of anthropogenic radioisotopes, with the Chernobyl reactor catastrophe in the year 1986 as the major source.

The explosion of the graphite reflector-type reactor happened as a consequence of various mistakes on a criticality experiment that was performed although the reactor operated on a non-stable basis at that time (NEA-OECD 1996). The major cause for the disaster was a construction/operation error during movement of the neutron-absorbing elements into the reactor core during a short-term shut-down. Instead of achieving subcritical conditions, the reactor became highly supercritical within a few seconds, causing rapid development of extreme heat in the fuel material (>2000 °C) and ignition of the graphite reflector along with release of hydrogen and carbon monoxide. These gases reacted with the oxygen in the surrounding air and two explosions caused such a massive increase of the gas pressure that the whole reactor exploded. The amounts of the most dose-relevant radioisotopes that were emitted into the atmosphere are listed in Table 1.

Table 2. *Estimated yields of atmospheric nuclear weapon tests (From UNSCEAR 1982)*

Country	Testing period	No. of tests	Estimated yield (MT TNT)*	
			Fission	Total
United States	1945–1962	193	72	139
USSR	1949–1962	142	111	358
United Kingdom	1952–1953	21	11	17
France	1960–1974	45	11	12
China	1964–1980	22	13	21
Total		423	218	547

*One ton TNT equals 4.2×10^{12} joules (J).

Compared to the sum of all previously performed atmospheric bomb tests, the values for ^{90}Sr, ^{137}Cs, and $^{239+240}$Pu from the Chernobyl accident are in the order of 10% (Table 1) and much higher for almost non-fissible actinides such as ^{238}Pu and ^{241}Am.

Fallout of hot particles in the vicinity of the reactor caused such a considerable contamination of the soil surface (e.g., ^{137}Cs up to 10^6 Bq/m^2) that the population had to be evacuated within a zone of 30 km distance from the reactor. The residence time of the relatively long-lived fission isotopes ^{137}Cs and ^{90}Sr ($t_{1/2}$ of both isotopes ~30 years) as well as of the actinides in the topsoil layers are in the order of tens of years (Hardy 1974; Alberts et al. 1980; Hodge et al. 1996) such that up to now some 100 000 people still cannot return into their original domiciles, although intensive decontamination attempts were performed to remediate the most affected regions. Typical soil surface contamination in the 30 km zone around the reactor and, for comparison, in considerably less affected areas such as western Europe are additionally listed in Table 1. The table shows clearly that the contribution of ^{137}Cs from the Chernobyl plume is significant even 2000 km away from the emission location; however, the total surface contamination is at least two orders of magnitude below the level within the 30 km exclusion zone.

The Windscale reactor accident, October 1957

The Windscale graphite reactor was built on the British west coast 30 km north of the city of Lancaster at the Sellafield nuclear fuel reprocessing plant. The accident was caused by the release of initially undetected residual heat during the routine annealing procedure of the graphite reflector, during which the energy was generated using nuclear heat. Insufficient core instrumentation was responsible for the failure to detect overproduction of heat early enough to prevent reaction of hot metallic uranium and graphite with air. The following combustion of the reactor caused release of 5×10^{13} Bq of ^{137}Cs (Clarke 1974), that is, a quantity that is about four orders of magnitude less than the amount released during the explosion of the Chernobyl reactor. While the committed dose (E$_{50}$) from ^{137}Cs on the population living in the vicinity of the site remained rather low (E$_{50}$ < 1–5 mSv per person), more significant doses were calculated from uptake of short-lived ^{131}I (8 days half-life) into the human thyroid via uptake from the food chain. This pathway was as follows: significant release of the highly volatile radioiodine (>10^{15} Bq)

during the accident, dry deposition and fallout contamination of grassland, and uptake by grazing cows and incorporation by consumption of milk. Maximum committed doses, particularly for children at Sellafield, were estimated to be in the order of 100 mSv (Dunster et al. 1958). Such a dose exceeds about five times the ICRP-recommended maximum dose for employees exposed to ionizing radiation at the workplace (ICRP 1994).

Ignition of the satellite SNAP 9A, April 1964

A navigational US satellite (Transit 5BN powered by SNAP-9A) equipped with a highly radioactive ^{238}Pu power generator (total activity 6.3×10^{14} Bq) was unsuccessfully launched into the atmosphere on 17 April 1964 (Krey 1967). Instead of gaining orbital velocity at a stationary distance from Earth, the satellite dropped back to Earth and ignited in the atmosphere above the Indian Ocean. The stratospheric fallout residence time of the radioactive debris from this accident was approximately 14 years (Krey & Krajewski 1970). Similar to the fallout from large >500 kilotons atomic bombs, this debris was also distributed over the entire Earth (about two-thirds in the southern hemisphere, one-third in the northern hemisphere).

The Palomares airplane accident, May 1966

A local contamination of soil with Pu isotopes and ^{241}Am from an (unexploded) atomic bomb impact was caused by collision of two US army aeroplanes during a mid-air refuelling operation in the Palomares area, southern Spain (details in Garcia-Olivares & Iranzo 1997; Montero & Sanchez 2001). The collision destroyed both planes and four thermonuclear bombs fell in the area, three onto the soil surface and one into the Mediterranean sea. Owing to the heavy impact, radioactive material was released into the environment, part of it as aerosols that were transported further downwind. In total, a region of 226 ha was contaminated from this accident.

The accident at Three Mile Island, March 1979

The accident at unit II of the two pressurized-water nuclear power stations at Three Mile Island (located outside the town of Middletown, Pennsylvania, USA) occurred only three months after starting the commercial production of electricity via steam generators. The accident was initiated by malfunctioning of the feedwater pumps that supplied the reactor's steam generators, leading to an automatic shutdown of the whole reactor (Rogovin & Frampton 1980). Consecutively, heat from the residual radioactivity in the core caused temperature and pressure of the reactor coolant to rise and a relief valve on top of the pressurizer to open. This valve should have closed in seconds, but was stuck in the open position, causing a significant loss of reactor coolant water. Although a safeguard system could have injected fresh water to the hot reactor core, the operators disregarded this option and the fuel started to melt. Radioactivity entered the reactor building basement, but, the pressure in the whole containment did not exceed a threshold value at which the reactor would have exploded. The explosion was avoided at the last moment because a few hours after the start of the accident the crew was able to open a backup valve and pump fresh coolant water into the reactor chamber. Since the reactor was already shut down the temperature in the reactor chamber dropped with an increasing degree of flooding and fortunately insignificant radioactivity was released into the environment. Although this accident did not cause notable radiation effects on individuals living in the vicinity of the reactor, this event initiated an intensive discussion in the USA on the radiological safety of nuclear power reactors.

Isotopic methods to distinguish between radionuclide contamination from different sources

This section summarizes isotope approaches to distinguish between different aerosol-derived soil contamination sources. Radioisotope data on soil samples from Switzerland and surrounding countries in western Europe are taken as examples for such an investigation. Typical long-lived products from nuclear fallout debris are isotopes of Pu (^{238}Pu, ^{239}Pu, ^{240}Pu, ^{241}Pu), ^{241}Am, ^{244}Cm and fissiogenic ^{137}Cs and ^{90}Sr. In more contaminated samples trace amounts of short-lived ^{134}Cs from the Chernobyl event may also still be detected using extreme low-level counting techniques (L'Annunziata 1998). If the isotopic signature from the different atmospheric emission sources (bomb testing, reactor explosions, ignited satellites, etc.) shows a different pattern, the isotope composition can be useful for distinguishing between such sources and a few approaches are presented in the following.

Apart from locally contaminated sites, typical present-day environmental levels of $^{239+240}$Pu in topsoil layers of the northern hemisphere vary between 0.1 and a few Bq/kg depending on the soil type, vegetation, and altitude (Hardy *et al.* 1973; Hardy 1974; Hoelgye & Filgas 1995; Hodge *et al.* 1996; Eisenbud & Gesell 1997; Michel *et al.* 2002). For instance, Pu isotope measurements performed by Geering *et al.* (2000) in Switzerland yielded significantly higher contributions of atmospheric fallout Pu in the Alps while the northern lower altitudes were less affected. Soils from forest ecosystems also show higher levels of fallout debris than cultivated areas because of the higher filtration effect of forests by aerosol deposition on leaves compared to farmland, which has a less specific surface area of plants (Smith 2001). Typical activity concentrations of ^{238}Pu, $^{239+240}$Pu, and ^{241}Am from various sites in Switzerland are shown in Fig. 1. The data for $^{239+240}$Pu scatter are typically between 0.1 and 3 Bq/kg with an average of 0.5 Bq/kg, while the activity concentration of ^{238}Pu is considerably lower. This isotope was mainly released into the atmosphere by ignition and disintegration of the satellite SNAP-9A in 1964 (6.3×10^{14} Bq) and to a minor extent from nuclear tests (3.3×10^{14} Bq; UNSCEAR 1982). Although the contribution of ^{238}Pu on total Pu is insignificant in the soil samples presented in Figure 1, a good correlation between the isotopes of Pu is indicated and the ^{238}Pu/$^{239+240}$Pu ratio of 0.03 ± 0.01 calculated via regression analysis (Eikenberg *et al.* 2002, 2003) agrees well with that obtained from other sites in the northern hemisphere (Hoelgye & Filgas, 1995; Hodge *et al.* 1996; Michel *et al.* 2002).

Figure 1 also displays the ^{241}Am vs. $^{239+240}$Pu data of the same samples, which also strongly correlate, and the regression line through the data yields a result for ^{241}Am/$^{239+240}$Pu of 0.38 ± 0.04 (reference date: 1 January 2003). This result is within uncertainty identical to the year 2003 ^{241}Am/$^{239+240}$Pu ratio of 0.36 calculated via the ^{241}Pu/^{241}Am progenitor/progeny relationship assuming (i) meaningless initial ^{241}Am after the atomic weapon tests (Fig. 2), and additionally considering (ii) that the initial ^{241}Pu(0)/$^{239+240}$Pu(0) ratio resulting from nuclear weapon testing in the 1960s was about 13 (UNSCEAR 2000). The ^{241}Pu decay and ^{241}Pu/^{241}Am progenitor/progeny relationships (discussed in detail in Eikenberg *et al.* 2002, 2003) are given by:

$$A_{241Pu}(t) = A_{241Pu}(0) \times e^{-\lambda_{241Pu} t} \quad (1)$$

where $A(t)$ is the remaining activity with time following radioactive decay (λ_{241Pu}) and $A(0)$ is the

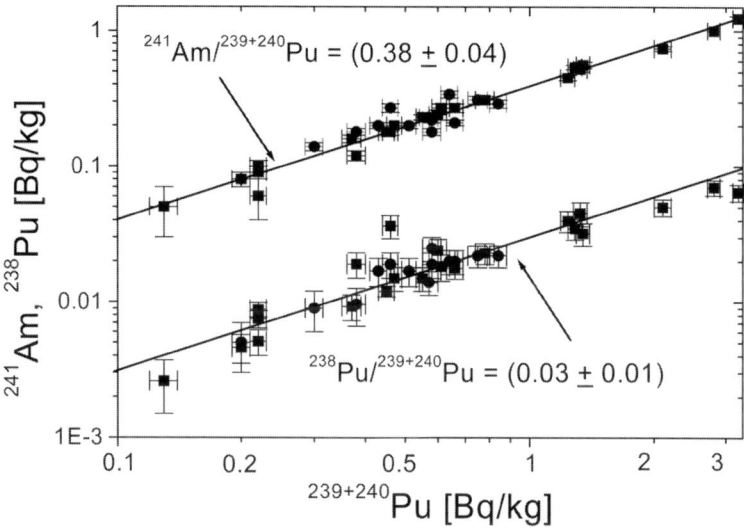

Fig. 1. Correlation diagram with ^{238}Pu vs. $^{239+240}$Pu and ^{241}Am vs. $^{239+240}$Pu obtained on Swiss soil samples from various locations (circles, data from Eikenberg *et al.* (2002); squares, data from Geering *et al.* (2000); reference date, 1 January 2003).

initial activity for ^{241}Pu (i.e., in the year 1963). For the ingrowing progeny (^{241}Am) in the coupled decay/ingrowth relationship the following equation holds:

$$A_{241Am}(t) = A_{241Pu}(0) \times \frac{\lambda_{241Am}}{\lambda_{241Am} - \lambda_{241Pu}}$$

$$\times \left(e^{-\lambda_{241Pu}\,t} - e^{-\lambda_{241Am}\,t}\right) \qquad (2)$$

This relationship is graphically illustrated in Fig. 2 with the relative activity development as a function of time. The calculated and measured ^{241}Am/$^{239+240}$Pu ratio is additionally in good agreement with data from Bunzl *et al.* (1995) and shows clearly that ^{241}Am measured in soil samples in central and western Europe can be well explained via support from decaying ^{241}Pu initially released during the surface A-bomb testing period. It can be concluded consequently

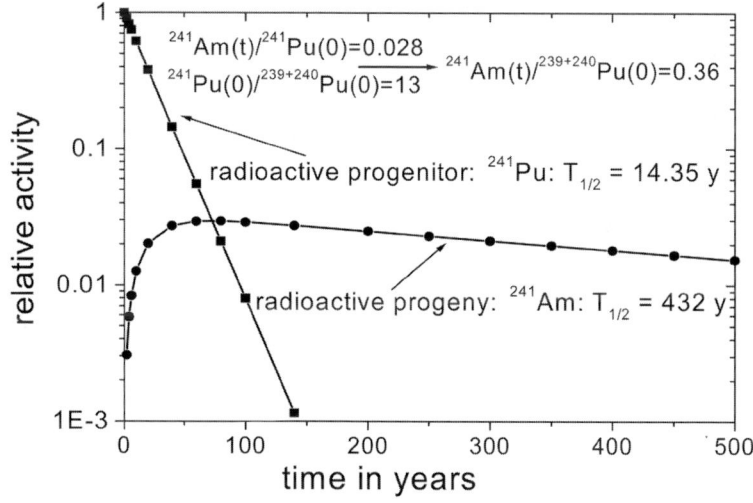

Fig. 2. Progeny/progenitor relationship for the ^{241}Pu/^{241}Am couple with time.

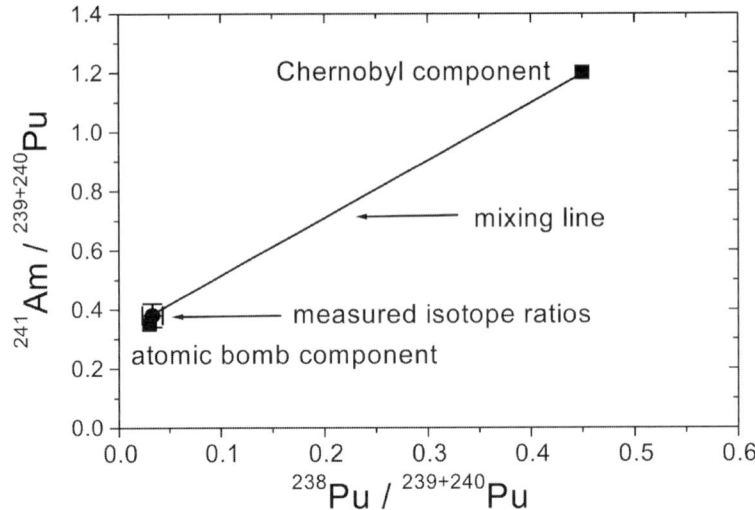

Fig. 3. Graphical illustration of a mixing relationship between two end-member components in a three-isotope normalization diagram (^{241}Am/$^{239+240}$Pu vs. ^{238}Pu/$^{239+240}$Pu). It is indicated that the measured anthropogenic actinide isotope ratios are, within uncertainty, identical to the atomic weapon component, therefore clearly indicating that actinide fallout from the Chernobyl accident was insignificant in western Europe.

that the contribution of heavy particles from the Chernobyl accident in 1986 is not significant in western Europe.

For distinguishing between two components three-isotope normalization plots such as on Fig. 3 are useful. Figure 3 and Table 3 indicate that reactor components (such as Chernobyl reactor fuel material) are characterized by higher levels of ^{238}Pu and ^{241}Am compared to $^{239+240}$Pu because of the longer neutron-flux exposure time of fuel material in a reactor in comparison to nuclear bomb explosions. The measured ^{238}Pu/$^{239+240}$Pu and ^{241}Am/$^{239+240}$Pu isotope ratios in western Europe (Hoelgye & Filgas 1995; Bunzl *et al.* 1995; Hodge *et al.* 1996; Geering *et al.* 2000; Michel *et al.* 2002; Eikenberg *et al.* 2003) are within uncertainty identical to the average actinide isotope ratios resulting from the surface

atomic bomb tests. Table 3 lists also the ^{241}Pu/$^{239+240}$Pu ratios of 1.8 measured in central Europe (Eikenberg *et al.* 2002, 2003) and these values additionally agree with the data from UNSCEAR (2000) for global bomb fallout (^{241}Pu/$^{239+240}$Pu = 1.9, decay corrected at reference date 1 January 2003). The use of the ^{241}Pu/$^{239+240}$Pu isotope ratio is, however, not as sensitive as the above listed actinide isotope ratios because ^{241}Pu is a weak β-emitter, that is, difficult to analyse (L'Annunziata 1998).

Disregarding the significant contribution of actinides from the Chernobyl accident to the inventory of Pu in western Europe, and inserting the average $^{239+240}$Pu/^{137}Cs ratio of 0.03 reported for North American soils (Hodge *et al.* 1996; i.e., pure atomic bomb component) as the initial pre-Chernobyl Pu/Cs ratio, then the frac-

Table 3. *Anthropogenic Pu and Am in soil samples of the northern hemisphere: measured mean, and calculated values for the following isotope ratios (reference date 1 January 2003)*

	^{241}Pu/$^{239+240}$Pu	^{241}Am/$^{239+240}$Pu	^{238}Pu/$^{239+240}$Pu
Measured ratios	1.8 ± 0.5*	0.38 ± 0.04*	0.03 ± 0.01*
Atomic weapon component	1.9[†]	0.36[†]	0.03[‡]
Chernobyl component	40[§]	1.2[§]	0.45[§]

*Measured values from Eikenberg *et al.* (2002, 2003), counting uncertainties 2σ; [†]Calculated with the initial ^{241}Pu/$^{239+240}$Pu ratio of 13 (UNSCEAR 2000) at reference date, 1 January 2003; [‡]Literature data for regions that were not contaminated by Chernobyl hot particles (i.e., actinide fallout) (Hoelgye *et al.* 1995; Hodge *et al.* 1996; Geering *et al.* 2000; Michel *et al.* 2002); [§]Data for Chernobyl from Buzulukov & Dobrynin (1993); Decay corrected at reference date.

tions of ^{137}Cs from both events can be roughly estimated using the following relationships:

$$R_A = \left(\frac{^{239+240}Pu}{^{137}Cs}\right)_A = 3 \times 10^{-2} \quad (3)$$

where R_A equals the $^{239+240}Pu/^{137}Cs$ ratio of the atomic weapon component. Also,

$$R_m = \frac{^{239+240}Pu_A}{^{137}Cs_A + ^{137}Cs_C} \quad (4)$$

where R_m equals the measured $^{239+240}Pu/^{137}Cs$ ratio (Eikenberg et al. 2002, 2003), that is, measured $^{137}Cs_m$ is composed of $^{137}Cs_A$ (fraction from the bomb fallout) and $^{137}Cs_C$ (fraction from the Chernobyl fallout). Combining equations (3) and (4) yields the ratio between the components for the case of Cs:

$$R_{Cs} = \frac{^{137}Cs_A}{^{137}Cs_C} = \frac{R_m}{R_A - R_m} \quad (5)$$

or for the fraction of the bomb component from the calculated two-component ratio R_{Cs}:

$$f_A^{^{137}Cs} = \frac{R_{Cs}}{1 + R_{Cs}} \quad (6)$$

A more precise result for both components is, however, obtained by the measured $^{134}Cs/^{137}Cs$ ratio. This is an advantage, because both isotopes are from the same element, that is, they behave chemically identically, while Pu and Cs do not have the same transport and transfer properties. This approach was applied several times to assess the excess radio-caesium contamination in European countries after the Chernobyl event (Kliment & Bucina 1990; Livens et al. 1992). The approach is as follows: Since ^{134}Cs is a short-lived radionuclide ($t_{1/2} = 2.06$ years) all potential contribution of this isotope from older sources has decreased completely, meaning that measured ^{134}Cs is derived solely from the Chernobyl accident (production of ^{134}Cs during bomb testing was meaningless anyway). Inserting, in addition, the initial $^{134}Cs/^{137}Cs$ ratio for aerosols in the Chernobyl plume after the explosion, it is possible to calculate the contribution of ^{137}Cs from both nuclear sources at any time as long as ^{134}Cs can be monitored. As shown in Table 4, more than 80% of surface-bound ^{137}Cs in western Europe can be attributed to fallout of very fine aerosols from the Chernobyl accident in 1986. In this case the relationships are as follows:

$$R_C = \left(\frac{^{134}Cs}{^{137}Cs}\right)_C (t) = \left(\frac{^{134}Cs}{^{137}Cs}\right)_C (0)$$
$$\times e^{(\lambda_{137} - \lambda_{134})t} \quad (7)$$

$$R_m = \left(\frac{^{134}Cs}{^{137}Cs}\right)_m = \frac{^{134}Cs_C}{^{137}Cs_C + ^{137}Cs_A} \quad (8)$$

and therefore for the ratio between both components (R_{Cs}) in the case of Cs,

$$R_{Cs} = \frac{^{137}Cs_A}{^{137}Cs_C} = \frac{R_C - R_m}{R_m} \quad (9)$$

A similar consideration can be made with respect to ^{90}Sr by use of the measured (as well as reported initial) $^{90}Sr/^{137}Cs$ ratio and the fractional distribution of ^{137}Cs in both nuclear components (as derived from $^{134}Cs/^{137}Cs$). However, in contrast to ^{137}Cs, the calculation yields that the amount of ^{90}Sr from the bomb tests exceeds by far that resulting from the Chernobyl source, indicating that released ^{90}Sr was fixed on more coarse-grained particles with significantly lower residence time in the atmosphere (all results and equations are summarized in Table 4).

Applications in radioecology

Most of the anthropogenic radioisotopes that were widely distributed on the Earth's surface were deposited as fine-grained aerosols and undergo all geochemical relevant processes in the biosphere such as adsorption/desorption, hydrolysis, oxidation/reduction, and are also available for plant uptake and consequently contribute to the dose to man. Such dose calculations are performed in the frame of radioecology, which is the investigation of the behaviour of radionuclides in the environment with a special focus on soil/plant/human trans-fer to assess doses of ionizing radiation in man. The three major fields of radioecology research are:

(1) Investigation or transport of radionuclides through natural and agricultural ecosystems to various receptors such as plants, animals and humans;

(2) Study of the effects of environmental radioactivity on plants and animals, particularly at the population and community levels of biological organization;

(3) Uses of radioisotopes as tracers to understand dynamic ecological processes.

Particularly after the Chernobyl accident, a large effort was made to assess doses to man from all types of radiation and pathways. Various

Table 4. *Calculated fractions of ^{137}Cs and ^{90}Sr in Swiss soil samples contaminated by fallout from previous (40 years ago) atmospheric atomic bomb testing (component A) and from the 1986 Chernobyl accident (component C) (reference date: 1 January 2003)*

	End component activity ratio (R_B or R_C)	Measured activity ratio§ (R_m)	Bomb to Chernobyl fallout ratio (R_{Cs} and R_{Sr})
1	$R_A = \left(\dfrac{^{239+240}\text{Pu}}{^{137}\text{Cs}}\right)_A = 3 \times 10^{-2*}$	$R_m = \dfrac{^{239+240}\text{Pu}_A}{^{137}\text{Cs}_A + {}^{137}\text{Cs}_C} = 4.8 \times 10^{-3}$	$R_{Cs} = \dfrac{^{137}\text{Cs}_A}{^{137}\text{Cs}_C} = \dfrac{R_m}{R_A - R_m} = 0.18$
2	$R_C = \left(\dfrac{^{134}\text{Cs}}{^{137}\text{Cs}}\right)_C = 4 \times 10^{-3\dagger}$	$R_m = \dfrac{^{134}\text{Cs}_C}{^{137}\text{Cs}_A + {}^{137}\text{Cs}_C} = 1.4 \times 10^{-2}$	$R_{Cs} = \dfrac{^{137}\text{Cs}_A}{^{137}\text{Cs}_C} = \dfrac{R_C - R_m}{R_m} = 0.15$
3	$R_B = \left(\dfrac{^{90}\text{Sr}}{^{137}\text{Cs}}\right)_A = 6 \times 10^{-1\ddagger}$	$R_m = \dfrac{^{90}\text{Sr}_A + {}^{90}\text{Sr}_C}{^{137}\text{Cs}_A + {}^{137}\text{Cs}_C} = 7.4 \times 10^{-2}$	$R_{Sr} = \dfrac{^{90}\text{Sr}_A}{^{90}\text{Sr}_C} = \dfrac{R_A}{7.7 \times R_m - R_A} > 20$

*From Hodge *et al.* (1996); †From NEA-OECD (1996); ‡From UNSCEAR (2000); §From Eikenberg *et al.* (2002, 2003); All values decay corrected at reference date, 1 January 2003.

numerical models were developed considering the complex matrix of radioisotope movement and transfer to soil/plant/animal/man. Among various models two computer codes gained international acceptance: (i) ECOSYS, which is a general program for the standard biosphere; and (ii) FORESTPATH, which uses generic differential equations to describe the transfer of radionuclides in forest compartments (van der Stricht & Kirchman 2001). Therefore, on the basis of the knowledge of even complex diffusion processes in air, water, and soil, radioecology allows the determination of the radioisotope transfer processes in natural surroundings, calculation of intake of radionuclides by humans, and finally, predicts committed doses from radioisotope incorporation. Another issue of radioecology is focusing on the remediation of contaminated sites via complete topsoil layer removal.

Conclusions

Soil contamination with long-living radioisotopes, particularly actinides, has to be seriously considered for the assessment of committed doses to man from ionizing radiation following environmental releases. The major contribution to soil contamination so far was caused by atmospheric releases from the nuclear weapon testing period about 50 years ago and from the Chernobyl nuclear reactor accident in April 1986. Contribution from both components can still be measured in many European countries. Therefore the use of the isotopic pattern of elements such as Pu, Am, and Cm provides a powerful tool to calculate the respective fraction from different emission sources. This is also of great advantage for monitoring trace levels of radioisotope releases from nuclear facilities in the frame of national radioisotope emmission surveillance programmes.

References

ALBERTS, J. J., BOBULA, C. M. & FARRAS, D. T. 1980. A comparison of the distribution of industrially released ^{238}Pu and ^{239,240}Pu in temperate, northern United States soils. *Journal of Environmental Quality*, **9**, 592–596.

BLACK, F. 1984. *Investigation of the Possible Increased Incidence of Cancer in West Cumbria. Report of the Independent Advisory Group.* Her Majesty's Stationery Office, London.

BUNZL, K., FLESSA, H., KRACKE, W. & SCHIMMACK, W. 1995. Association of fallout $^{239+240}$Pu and ^{241}Am with various soil components in successive layers of grassland soil. *Environmental Science Technology*, **29**, 2513–2518.

BUZULUKOV, YU. P. & DOBRYNIN, YU. L. 1993. Release of radionuclides during the Chernobyl accident. *In*: MERVIN, S. & BALONOV, M. (eds) *The Chernobyl Papers*. Research Enterprises Inc., Richland, Washington 99353.

CLARKE, R. H. 1974. An analysis of the Windscale accident using the WERRIE code. *American Nuclear Science Engineering*, **1**, 73–74.

DUNSTER, H. J., HOWELLS, H. & TEMPLETON, W. L. 1958. District surveys following the Windscale incident, October 1957. *In: Proceedings of the Second UN Conference on Peaceful Uses of Atomic Energy*. United Nations, New York.

EIKENBERG, J., BAJO, S. *et al.* 2002. *Spezialnuklid-Analysen fuer die Beweissicherung ZWILAG 1997–2001: Final Report of the Immission Measurements PSI Technical Report No 02–15*. Paul Scherrer Institut, CH-5232 Villigen, 40 pp.

EIKENBERG, J., BAJO, S., GANN, C., RUETHI, M. & BUTTERWECK, G. 2003. A rapid procedure for determining ^{239,240}Pu and ^{241}Pu in environmental samples using α/β-LSC. *In*: MÖBIUS, S., NOAKES, J., & SCHÖNHOFER, F. (eds) *Proceedings LSC-2001, Advances in Liquid Scintillation Spectrometry Radiocarbon*, 351–361.

EISENBUD, M. & GESELL, T. 1997. *Environmental Radioactivity*. Academic Press, San Diego, New York, Toronto.

GARCIA-OLIVARES, A. & IRANZO, C. E. 1997. Resuspension and transport of plutonium in the Palomares area. *Journal of Environmental Radioactivity*, **37**, 101–114.

GEERING, J. J., FROIDEVAUX, P., SCHMITLER, T., BUCHILLIER, T. & VALLEY, J. F. 2000. Mesures de plutonium et d'americium dans l'environnement. *In*: VOELKLE, H. & GOBET, M. (eds) *Environmental Radioactivity and Radiation Exposure in Switzerland*. Swiss Inspectorate for Public Health, Annual Report 1999, Section B7.2.1–8.

HARDY, E. P., KREY, P. W. & VOLCHOCK, H. L. 1973. Global inventory and distribution of fallout plutonium, *Nature*, **241**, 444–446.

HARDY, E. P. 1974. *Depth Distribution of Global Fallout of ^{90}Sr, ^{137}Cs and ^{239,240}Pu in Sandy Loam Soil*. Fallout Program Quarterly Summary Report HASL-286. US Atomic Energy Commission, New York.

HODGE, V., SMITH, C. & WHITING, J. 1996. Radio-cesium and plutonium: still together in background soils after more than thirty years. *Chemosphere*, **32**, 2067–2075.

HOELGYE, Z. & FILGAS, R. 1995. Inventory of ^{238}Pu and ^{239,240}Pu in the soil of Czechoslovakia in 1990. *Journal of Environmental Radioactivity*, **27**, 181–189.

ICRP 1994. International Commission of Radiological Protection, ICRP Publications 61, Recommendations of the Commission. ICRP, Vol. 21, No. 4

JOHNES, S. R., FULKNER, M. J., MCKEEVER, J. & STEWERT, T. H. 1991. Aspects on population exposure consequent on discharges of radionuclides to the environment from the nuclear reprocessing plant at Sellafield at Cumbria. *Radiation Protection Dosimetry*, **36**, 199–204.

KHODAKOVSKY, I. K. 1994. Radionuclide sources of arctic contamination. *Arctic Research United States*, **8**, 262–265.

KLIMENT, V. & BUCINA, I. 1990. Contamination of food in Czechoslovakia by caesium radioisotopes from the Chernobyl accident. *Journal of Environmental Radioactivity*, **12**, 167–178.

KREY, P. W. 1967. Atmospheric burn-up of a plutonium-238 generator. *Science*, **158**, 769–771.

KREY, P. W. & KRAJEWSKI, B. 1970. Comparison of atmospheric transport model calculations with observation of radioactive debris. *Journal of Geophysical Research*, **75**, 2901–2908.

LIVENS, F. R., FOWLER, D. & HORILL, A. D. 1992. Wet and dry deposition of ^{131}I, ^{134}Cs and ^{137}Cs at an upland site in northern England. *Journal of Environmental Radioactivity*, **16**, 243–254.

L'ANNUNZIATA, M. F. 1998. *Handbook of radioactivity analysis*. Academic Press, New York, 771 pp.

MICHEL, H., BARCI-FUNEL, G., DALMASSO, J., ARDISSON, G., APPLEBY, P. G., HAWORTH, E. & EL-DAOUSHY, F. 2002. Plutonium and americium inventories in atmospheric fallout and sedimentory coves from Blelham Tarn, Cumbria, UK. *Journal of Environmental Radioactivity*, **59**, 127–137.

MONTERO, P. R. & SANCHEZ, A. M. 2001. Plutonium contamination from accidental release or simply fallout: study of soils at Palomares (Spain). *Journal of Environmental Radioactivity*, **55**, 157–165.

NEA-OECD 1996. Chernobyl Ten Years On: Radiological and Health Impact. OECD Nuclear Energy Agency, Paris Cx, 112 pp.

ROGOVIN, M. & FRAMPTON, G. J. 1980. *A Sequence of Physical Events. Three Mile Island*, Vol. II, Part 2, pp 309–340, Report to the Commissioners and the Public, Special Inquiry GROUP, US Nuclear Regulatory Commission, Washington, DC.

SMITH, F. B. 2001. Dispersion and transfer in the terrestrial environment. *In*: E. VAN DER STRICHT, E. & KIRCHMANN, R. (eds) *Radioecology*. Fortemps, Liege, Belgium.

SWISS COMMISSION ON ATOMIC PROTECTION 1995. *10 Years After the Chernobyl Accident – a Swiss Perspective*. EDMZ-Report 805.567, 3000 Bern.

UNSCEAR 1982. *Sources and Effects of Ionising Radiation*. UNSCEAR 1982 Report to the General Assembly, United Nations Scientific Committee on the Effects of Atomic Radiation, New York.

UNSCEAR 2000. *Sources and Effects of Ionising Radiation*. UNSCEAR 2000 Report to the General Assembly, United Nations Scientific Committee on the Effects of Atomic Radiation, New York.

VAN DER STRICHT, R. & KIRCHMANN, R. 2001. *Radioecology: Radioactivity and Ecosystems*. Fortemps, Liège, Belgium.

VOLCHOK, H. L. & KLEINMANN, L. 1971. *Worldwide Deposition of Sr-90 Through 1970*. Report HASL-243. US Atomic Energy Commission, New York.

Atmospheric impact of the fossil fuel cycle

D. S. GOLOMB[1] & J. A. FAY[2]

[1]University of Massachusetts Lowell, Lowell, MA 01854, USA

(e-mail: dan_golomb@uml.edu)

[2]Massachusetts Institute of Technology, Cambridge, MA 02139, USA

Abstract: Fossil fuels supply about 86% of the global primary energy consumption for transportation, industrial, commercial, and residential uses. While a minor part of fossil fuels is used as a raw material for the chemical industry, the vast majority is combusted providing heat, mechanical, and electric energy for the urban–industrial society. Owing to the combustion of fossil fuels, copious quantities of pollutants are emitted into the air and impact on the local, regional, and global air quality. It is technologically possible to significantly reduce the emissions of most pollutants, albeit at considerable energy and economic penalty. However, the reduction of emissions of carbon dioxide, one of the principal products of fossil fuel combustion, poses technological, economic, societal, and political problems of enormous magnitude. In recent decades it has become evident that rising levels of atmospheric carbon dioxide (and other greenhouse gases) will cause global warming and other climate changes.

A minor part of mined fossil fuels is used as a raw material for the chemical industry (e.g., plastics, synthetic fabrics, carbon black, ammonia, and fertilizers). The major part supplies the energy needs for modern society. Fossil fuels supply about 86% of global primary energy consumption (39% oil, 24% coal, and 23% natural gas), providing energy for transportation, electricity generation, and industrial, commercial, and residential uses (EIA 2001). Coal, and to a lesser extent oil, combustion leaves a significant amount of solid waste. The treatment of solid waste from fossil fuel combustion is treated in different chapters of this book. In this chapter we focus on air emissions of fossil fuel combustion, and their impact on human health and the environment.

The air emissions of fossil fuel combustion are dispersed and diluted within the atmosphere, eventually falling or migrating to the surface of the Earth or ocean at various rates. Until recently, most attention was focused on the so-called primary pollutants of fossil fuel combustion that are harmful to human health: oxides of sulphur and nitrogen, carbon monoxide, suspended particles (including soot), heavy metals, and products of incomplete combustion. These pollutants are most concentrated in urban or industrialized areas close to large or multiple sources. However, the primary pollutants may interact with each other, and with atmospheric constituents and sunlight, forming secondary pollutants that disperse far beyond the urban–industrial areas. The secondary pollutants can affect regions of continental size. Typical examples are tropospheric ozone and other photo-oxidants, acid rain, and regional haze.

It is technologically possible to reduce significantly the emission of primary pollutants (and the ensuing secondary ones) by orders of magnitude, albeit by incurring substantial energy and economic costs. In a complex advanced industrial economy this requires an extensive regulatory regime that imposes emission standards for primary pollutants and for precursors of secondary pollutants.

In recent decades it has become evident that rising levels of atmospheric carbon dioxide (and other greenhouse gases) have already warmed the Earth's surface slightly, and it is predicted that with continuous and increasing use of fossil fuels, the warming trend will increase. Most of the increase in carbon dioxide (CO_2) levels is a direct consequence of fossil fuel combustion. More importantly, this emission accumulates in the atmosphere for about a century before being removed to ocean and land storage, unlike ordinary primary and secondary pollutants that last at most a few weeks before being removed. Although atmospheric CO_2 is non-toxic to human health, and in fact necessary to support biological life, its growth threatens to change global climate and thereby perturb the biosphere.

The energy potential of fossil fuels can only be fully realized by the oxidation of its carbon and

From: GIERÉ, R. & STILLE, P. (eds) 2004. Energy, Waste, and the Environment: a Geochemical Perspective. Geological Society, London, Special Publications, **236**, 153–167.
0305-8719/04/$15   © The Geological Society of London 2004.

hydrogen content to gaseous CO_2 and water vapour. Technological processes can capture the carbon dioxide and sequester it underground or in the ocean, albeit at substantial economic and energy penalty. The necessity to reduce both ordinary combustion pollutants and CO_2 will substantially alter the nature of future fossil fuel use around the globe.

The goal of this article is to describe the characteristics of fossil fuel generated air pollutants, their transport and fate in the atmosphere, and their effects on human health and the environment. The technology for reducing air emissions is generally beyond the scope of this article, although a brief overview will be given on the ways and means for the reduction of CO_2 emissions in order to ameliorate global warming.

Health and environmental effects of air pollutants

Air pollutants, when they exceed certain concentrations, can cause acute or chronic diseases in humans, animals, and plants. They can impair visibility, cause climatic changes, and damage materials and structures. Table 1 lists some of the effects of air pollutants on human health, fauna and flora, and structures and materials. The order of listing does not follow any particular ranking; some individuals or plants are more sensitive to one kind of pollutant than to another. The listed pollutants are designated by the US Environmental Protection Agency (EPA) as criteria pollutants. These are the pollutants for which dose–response relationships are fairly well known from clinical and epidemiological studies. Regulatory agencies in most developed countries prescribe *ambient* concentration standards for criteria air pollutants. The standards are set at a level below which it is estimated that no harmful effect will ensue on human health. On the other hand, if the standards are exceeded, increased human mortality and morbidity is expected. For example, in the USA the EPA set National Ambient Air Quality Standards (NAAQS) for five pollutants: carbon monoxide (CO), nitrogen oxides (NO_x, the sum of NO and NO_2), ozone (O_3), sulphur dioxide (SO_2), and particulate matter (PM). The standards are listed in Table 2. In the past, particulate matter was regulated regardless of size (the so-called total suspended particles, TSP). Since 1978 only particles with an aerodynamic diameter less than 10 μm are regulated. This is called PM-10. More recently, the US EPA proposed to promulgate a standard for even smaller particles, PM-2.5, but this proposal has not yet been approved by the US judiciary. The reason for regulating only small particles is that these particles can be lodged deep in the alveoli of the lungs, and hence are most detrimental to our health, whereas the larger particles are filtered out in the upper respiratory tract.

In order to meet ambient standards, *emission* standards have to be set for principal emitters, called *sources*. Stationary sources include power plants, incinerators, steel, cement, paper and pulp factories, chemical manufacturers, and refineries. Mobile sources include automobiles, trucks, locomotives, ships, and aircraft. As an example, Table 3 lists the US emission standards for fossil-fueled steam generators, which are fossil-fueled power plants and industrial boilers. Table 4 lists the US emission standards for passenger cars and light trucks. The emission standards are set in a manner that air quality modeling shows that after dispersion in the atmosphere the ambient standards in the vicinity of the sources will not be exceeded. The emission standards are maintained by employing emission reduction control technologies for specific sources. For example, for coal- and oil-fired power plants, industrial boilers, and municipal solid waste incinerators, the particle (fly ash) emission reduction technologies include electrostatic precipitators or fabric filters. For SO_2 control the commonly applied technologies include limestone ($CaCO_3$) based 'scrubbers' either in the wet or dry form. For partial NO_x control the low-NO_x burner can be used. However, for more complete removal of NO_x the more effective (and more expensive) flue gas denitrification technologies must be used, based on the selective catalytic reduction of NO_x by ammonia or urea. For automobile emission control the three-way catalytic converter is used in most developed countries. Unfortunately, the catalytic converter does not work on diesel-powered automobiles and trucks, so these vehicles remain the principal sources of urban air pollution.

The description of the workings of the control technologies is beyond the scope of this article. However, it is worth noting that many of these technologies create substantial amounts of solid or liquid waste that needs to be disposed of properly. For example, in the USA the total amount of fly ash produced from coal combustion alone is about 57 Mt/y (Kalyoncu 2000). About one-third is utilized as secondary raw material (e.g., for aggregate and asphalt), but the rest is usually disposed of in landfills. The wet and dry scrubbers for SO_2 control produce a sludge or dry waste that finds little secondary use, and a large amount is disposed in landfills.

Table 1. *Effects of criteria air pollutants on human health, fauna and flora, and structures and materials*

Pollutant	Health effect	Effects on fauna and flora	Effects on structure and material
O$_3$ and photo-oxidants	Pulmonary oedema, emphysema, asthma, eye, nose, and throat irritation, reduced lung capacity.	Vegetation damage, necrosis of leaves and pines, stunting of growth, photosynthesis inhibitor, probable cause of forest die-back, suspected cause of crop loss.	Attack and destruction of natural rubber and polymers, textiles and materials.
CO	Neurological symptoms, impairment of reflexes and visual acuity, headache, dizziness, nausea, confusion. Fatal in high concentrations because of irreversible binding to haemoglobin.	NA	NA
NO$_x$	Pulmonary congestion and oedema, emphysema, nasal and eye irritation.	Chlorosis and necrosis of leaves. Precursor to acid rain and photo-oxidants.	Weathering and corrosion.
SO$_2$	Bronchoconstriction, cough.	Cellular injury, chlorosis, withering of leaves and abscission. Precursor to acid rain: acidification of surface waters with community shifts and mortality of some aquatic organisms. Possible effect on uptake of Al and other toxic metals by plant roots.	Weathering and corrosion. Defacing of monuments.
Particulate matter	Non-specific composition: bronchitis, asthma, emphysema. Composition dependent: brain and neurological effects (e.g., Pb, Hg), toxigens (e.g., As, Se, Cd), throat and lung cancer (e.g., coal dust, coke oven emissions, polycyclic aromatic hydrocarbons, Cr, Ni, As).	NA	Soiling of materials and cloth. Visibility impairment due to light scattering of small particles.

Adapted from Wark *et al.* (1998).

Table 2. *US 2000 National Ambient Air Quality Standards (NAAQS)* [‡]

Pollutant	Standard	
	ppm	μg/m³
Carbon monoxide (CO)		
8 h average	9	10 mg/m³
1 h average	35	40 mg/m³
Nitrogen dioxide (NO₂)		
Annual arithmetic mean	0.053	100
Ozone (O₃)*		
3 y average of annual fourth highest daily maximum 1 h concentration	0.12	236
Particulate matter, diameter <10 μm (PM-10) [†]		
Annual arithmetic mean		50
Arithmetic mean of 24 h 99th percentile, averaged over 3 y		150
Sulphur dioxide (SO₂)		
Annual arithmetic mean	0.03	80
24 h average	0.14	365

*Proposed new standard: 3 y average of annual 4th highest daily max; 8 h concentration, 0.08 ppm = 157 μg/m³.
[†]Proposed additional new standard for particulate matter, diameter <2.5 μm (PM-2.5): 3 y annual arithmetic mean, 15 μg/m³; arithmetic mean of 24 h 98th percentile averaged over 3 y, 65 μg/m³.
[‡]Data from EPA (2003*a*).

In addition to the above criteria pollutants, one finds in the air a host of other gaseous and particulate pollutants, generally designated as hazardous air pollutants (HAPs), or simply *air toxics*. The US EPA identified 189 HAPs. Of course, not all HAPs are related to fossil fuel usage. Examples of fossil fuel HAPs are products of incomplete combustion (PIC), volatile organic compounds (VOC), polycyclic aromatic hydrocarbons (PAH), toxic metals (e.g., Hg, Pb, Cd,

Table 3. *US emission standards, called new source performance standards, for fossil fuel steam generators with heat input >73 MW* [†]

Pollutant	Fuel	g/GJ Heat input
SO₂	Coal	516
	Oil	86
	Gas	86
NOₓ	Coal (bituminous)	260
NOₓ	Coal (subbituminous)	210
NOₓ	Oil	130
NOₓ	Gas	86
PM*	All	13

*PM, particulate matter. For PM emissions an opacity standard also applies, which allows a maximum obscuration of the background sky by 20% for a 6 minute period.
[†]Data from EPA (2003*b*).

Se, As, V). Many of the fossil fuel related HAPs are found as condensed matter on particles emitted by stationary sources (e.g., fly ash from power plants) or mobile sources (e.g., exhaust smoke from trucks). While the HAPs may be more deleterious to our health than the criteria pollutants (some of them are carcinogens), it is difficult to establish a dose–response relationship. Therefore, instead of setting emission standards for HAPs, the US EPA mandates that specific control technologies be installed on major emitting sources. These are called maximum achievable control technologies (MACT). For example, MACTs for volatile organic compounds include secondary incineration, chemical absorption, or physical adsorption. For the capture of Hg vapour, the designated MACT is activated carbon injection into the flue gas. The activated carbon particles physically adsorb the mercury vapour. Of course, most of the MACT technologies transfer the air pollutants to a solid or liquid phase, thus presenting another hazardous waste disposal problem.

Local and regional air pollution

Advection and dispersion of air pollutants

When air pollutants exit the smoke stack or exhaust pipe (called the *sources*), they are advected by winds and dispersed by turbulent diffusion. Winds blow from high pressure toward low pressure cells at speeds that depend on the pressure gradient. Because of the Coriolis force, wind trajectories are curvilinear in reference to fixed Earth coordinates, although within a relatively short (few to tens of km) distance, wind trajectories can be approximated as linear. Winds have a horizontal and vertical component. Over flat terrain the horizontal component predominates; in mountainous and urban areas with tall buildings, the vertical component can be significant, as well at the land/sea interface.

In the bottom layer of the atmosphere, called the troposphere, usually the temperature declines with altitude, on the average about 10 °C per km. Because of the negative temperature gradient, upper parcels of the air are denser than lower ones, so they tend to descend, while lower parcels buoy upward, giving rise to eddy or turbulent mixing. The more negative the temperature gradient, the stronger the turbulence, and the faster the dispersion of a pollutant introduced into the troposphere. Wind shears also cause turbulence. A wind shear exists when in adjacent layers of the atmosphere winds blow in different directions and/or speeds.

Table 4. *US federal vehicle emission standards* [‡]

Model year	Light-duty vehicles (auto)				Light-duty trucks (gasoline)			
	HC* (g/km)	CO (g/km)	NO_x (g/km)	PM (g/km)	HC (g/km)	CO (g/km)	NO_x (g/km)	PM (g/km)
1968	2.0	20.5						
1971[†]	2.9	29.2	2.49					
1974	2.11	24.2	1.86					
1977	0.93	9.32	1.24					
1978	0.93	9.32	1.24		1.24	12.4	1.93	
1979	0.93	9.32	1.24		1.06	11.2	1.43	
1980	0.25	4.35	1.24		1.06	11.2	1.43	
1981	0.25	2.11	0.62		1.06	11.2	1.43	
1982	0.25	2.11	0.62	0.37	1.06	11.2	1.43	
1985	0.25	2.11	0.62	0.37	0.50	6.21	1.43	0.99
1987	0.25	2.11	0.62	0.12	0.50	6.21	1.43	1.62
1988	0.25	2.11	0.62	0.37	0.50	6.21	0.75	1.62
1994	0.25	2.11	0.25	0.05	0.50	3.42	0.60	0.06

*HC, Hydrocarbon (molecular weight 13 g/mol).
[†]Test method changed in 1971.
[‡]Data from EPA (2003c).

Occasionally, the temperature increases with altitude, for example during radiative cooling of the Earth surface, such as may occur during cloudless nights. This is called an inversion. During an inversion, there is little turbulence, and a pollutant will disperse very slowly. Inversions may also occur aloft, that is, a negative temperature gradient exists at the bottom, followed by a positive gradient above. The layer up to the altitude at which the inversion occurs is called the mixing layer, and the altitude at the inversion is called the mixing height. The shallower the mixing layer, the greater chances for air pollution episodes to develop, because pollutants emitted near the ground are confined to the shallow mixing layer. This often occurs in cities such as Los Angeles, Houston, Atlanta, Salt Lake City, Denver, Mexico City, Sao Paulo, Athens, Madrid, Rome, Istanbul, and others. Some of these cities are surrounded by mountain chains. In the basin of the mountain chain, or in the valley, the mixing layer is shallow, and the winds in the layer are usually weak, leading to poor ventilation. During the morning rush hour, pollutants are emitted into the shallow mixing layer, where they are concentrated because of the 'lid' imposed by the inversion aloft. Later in the day, when solar radiation breaks up the inversion, the pollutants disperse to higher altitude, and the pollutant concentration is diluted a little. The photo in Fig. 1 shows a pollution episode in Los Angeles. The mixing height extends only to the middle of the tall building. The lower part is obscured by the smog, the upper part, which is above the mixing layer, is in relatively clear air.

The spatial and temporal extent of air pollution episodes is determined by the lifetime of the pollutant in question. Some pollutants have a short lifetime, and therefore impact only the immediate area where they are emitted. For example, CO is oxidized into CO_2 within a few kilometres from the emission sources. Larger particles (larger than about 10 μm in diameter) settle on the ground within minutes to hours, so they disperse at relatively short distances from the emission sources. Small particles (less than a few μm in diameter) settle on the ground within days to weeks, so they travel long distances. They are of special concern because of their impact upon human health. Other pollutants have a long lifetime – for example, CH_4 (few years), CO_2 (tens of years), chlorofluorocarbons (CFCs) (hundreds of years) – so they tend to stay in the atmosphere for a very long time, enveloping the whole Earth (Seinfeld & Pandis 1998).

In addition to chemical transformation (e.g., $CO \rightarrow CO_2$; $SO_2 \rightarrow H_2SO_4$; $NO_x \rightarrow HNO_3$) and settling, the lifetime of a pollutant is also dependent on its removal by precipitation. Falling hydrometeors (rain drops or snow flakes) scavenge ('washout') the pollutants from the air. Hydrophilic pollutants (e.g., inorganic vapours, sulphate and nitrate particles) are efficiently scavenged by hydrometeors; others (e.g., organic vapours, CFCs, soot, organic particles) are not scavenged efficiently, so they stay in the atmosphere longer. Precipitation is the ultimate cleanser of the atmosphere. We have all experienced the cleansing property of a heavy thunderstorm in summer. Before the storm, the air is hazy and polluted; after it, the air is transparent and clean.

Fig. 1. Los Angeles smog. Note the low inversion beneath which the smog accumulates. Above the inversion the air is relatively clear. (Photo by the South Coast Air Quality Management District.)

Air quality modeling

The estimation of pollutant concentration in space and time is called air quality modeling (AQM). There are models pertaining to an individual source, for example, a smoke stack of a single power plant, or to aggregate sources, such as an oil refinery, a steel manufacturing complex, or a multitude of sources within an urban area, including commercial, residential, industrial, and vehicular sources. Some models are of the *trajectory* type, where the pollutant plume is followed as it is advected by the wind over the terrain and dispersed laterally and vertically by turbulent diffusion. Others are *grid-type* models, where the terrain is divided into grids. To each grid is assigned an emission strength of a single or multiple pollutants. The emission strengths for single or gridded sources are obtained from local or national environmental regulatory agencies. To each grid are assigned terrain conditions, such as land or sea, topography, forest, prairie, farmland, or desert.

For both trajectory and grid models, meteorological conditions are of utmost importance, as they determine the transport, dispersion, transformation of primary to secondary pollutants, and washout by precipitation of the pollutants. Meteorological data are available from numerous weather stations operating around the world, especially in the more developed countries. The weather stations measure and record surface and upper air winds, atmospheric pressure, humidity, precipitation, solar irradiation, temperature on the ground, and the temperature gradient in the atmosphere. The measurements are usually rendered twice daily at 0000 and 1200 Greenwich Mean Time (GMT), so that measurements are synchronized all over the world. Past and present weather data are available from national repositories, for example, in the USA, from the National Weather Service in Asheville, North Carolina.

Other important inputs to AQM are the transformation and deposition processes of air pollutants. Examples of air quality modeling where transformation and deposition processes need to be considered are acid deposition, regional haze, and photo-oxidants.

Acid deposition

While commonly called acid rain, acid deposition is a better term because deposition can occur both in the wet and dry form. The ingredients of acid deposition are sulphuric and nitric acids. The primary pollutants are SO_2 and NO_x, which is the sum of NO and NO_2 molecules. Both SO_2 and NO_x result from fossil fuel combustion. Sulphur is a ubiquitous ingredient of coal and petroleum. When these fuels are burned in air, SO_2 is emitted from the smokestack or the exhaust pipe of a vehicle. Coal and petroleum also contain nitrogen in their molecular make-up, resulting in NO_x emissions when these fuels are combusted in air. In addition, some NO_x is formed from the recombination of air oxygen and nitrogen at the high flame temperatures. So, even the combustion of natural gas, which has no nitrogen in its molecular make-up, produces some NO_x.

Primary emitted SO_2 is transformed to sulphuric acid by a series of steps. First, it is oxidized to HSO_3 by the ubiquitously present OH radical in the atmosphere. Next, further oxidation to H_2SO_4 can occur in the gaseous or aqueous phase (i.e., within cloud droplets) by atmospheric oxidants, such as OH, HO_2, or O_3. Similarly, primary emitted NO_x is transformed in the gaseous or aqueous phase by the same atmospheric oxidants (Seinfeld & Pandis 1998). The resulting acids can be either deposited in the dry phase on land or water, a process called dry deposition, or scavenged by a falling hydrometeor, resulting in wet deposition. Acid deposition modeling is fairly successful in predicting the amount of acid deposition given the emission strength of the precursors SO_2 and NO_x (see, for example, Fay et al. 1985). From these models it was concluded that acid deposition within a geographic domain is approximately linearly dependent on SO_2 and NO_x emission strength in that domain, so a certain percentage of either precursor emission reduction results in a proportional deposition reduction of sulphuric or nitric acid. Indeed, in countries and continents where serious curtailments of precursor emissions have been made, a proportional reduction in acid deposition occurred. In the USA, as a result of reducing emissions of SO_2 by approximately one-half since the enactment of the Clean Air Act of 1990, sulphuric acid deposition has also declined by approximately one-half (Butler et al. 2001). Nitric acid deposition has not fallen appreciably, because the control of NO_x is much harder to accomplish than the control of SO_2, especially from dispersed sources such as commercial/residential boilers and furnaces, automobiles, and diesel trucks.

Regional haze

Small particles (also called fine particles or aerosols) less than $1-2$ μm in diameter settle very slowly on the ground. Small particles can be either in the solid or liquid phase. For example, fly ash or smoke particles are solid, mist is liquid. Small particles can be of natural origin (e.g., volcanic dust, windblown soil dust, forest and brush fires) or of anthropogenic origin (e.g., fly ash, diesel truck smoke). They can be emitted as primary particles or formed by transformation and condensation from primary emitted gases. They can travel hundreds to thousands of kilometres from their emitting sources. The particles can envelope vast areas, such as the northeastern or southwestern USA, southeastern Canada, western and/or central Europe, and southeastern Asia. Satellite photographs often show continental areas covered with a blanket of haze. The haze may extend far out over the ocean. This phenomenon is called regional haze. Small particles are efficient scatterers of sunlight. Light scattering prevents distant objects from being seen. This is called visibility impairment. Figure 2 shows an encroaching haze episode on a mountain chain in Vermont. As the haze thickens, distant mountain peaks are no longer visible, and eventually neighbouring peaks disappear. Increasing concentration of particles in urbanized parts of the continents causes the loss of visibility of the starlit nocturnal sky. On these days, small stars, less than a fifth order of magnitude, rarely can be seen from populated areas of the world.

The composition of fine particles varies from region to region, depending on the precursor emissions. In the northeastern USA, central Europe, and southeastern Asia, more than half of the composition is made up of sulphate particles, due to the combustion of high-sulphur coal and oil. The rest is made up of nitrate particles, carbonaceous material (elemental and organic carbon), and crustal matter (fugitive particles from soil, clay, and rock erosion).

Photo-oxidants

The family of photo-oxidants includes tropospheric ozone, O_3 (the 'bad' ozone), ketones, aldehydes and nitrated oxidants, such as peroxyacetylnitrate (PAN) and peroxybenzoylnitrate (PBN). The modeling of photo-oxidants is more complicated than that of acid deposition (NRC 1991). Here, the primary precursor is NO_x, which as mentioned before, is emitted as a result of fossil fuel combustion. A part of NO_x is the NO_2 molecule, which splits (photodissociates)

Fig. 2. Regional haze in the Vermont Green Mountains. Note as the pollution episode progresses, adjacent peaks are no longer visible. (Photo by the Vermont Agency of Environmental Conservation.)

by solar ultraviolet (UV) and blue photons into NO and atomic oxygen. The photodissociation rate is dependent on solar irradiation, which in turn is dependent on latitude, season, time of day, and cloudiness. Atomic oxygen combines with molecular oxygen to form O_3. The NO that is formed in the photodissociation is quickly re-oxidized into NO_2 by peroxy-radicals, RO_2, present in the polluted atmosphere. The peroxy-radicals are formed from VOCs that are emitted as a consequence of incomplete combustion of fossil fuels, or from evaporation and leakage of liquid fossil fuels and solvents. The VOCs are oxidized in a complicated sequence of photochemical reactions to the peroxy-radicals. The oxidation rates of VOCs are also dependent on solar irradiation and on the specific VOC molecule. Long- and branch-chained hydrocarbons (e.g., n-octane and iso-octane) are more reactive than short- and straight-chained ones (e.g., methane and ethane); unsaturated hydrocarbons (e.g., ethene) are more reactive than saturated ones (e.g., ethane). Aromatic hydrocarbons (e.g., benzene) are more reactive than aliphatic ones (e.g., hexane), and so on.

Thus, photo-oxidants have really two kinds of precursors, NO_x and VOCs, which makes abatement of these secondary pollutants, as well as their modeling, so complicated. First of all, as mentioned above, complete NO_x emission control is difficult to accomplish because in addition to large stationary sources, NO_x is emitted from a myriad of dispersed sources, such as home and commercial furnaces and boilers, automobiles, trucks, off-road vehicles, aircraft, locomotives, and ships. In principle, anthropogenic VOCs could be substantially controlled, for example, by ensuring complete combustion of the fossil fuel, or with the catalytic converter on automobiles. However, not all VOCs are of anthropogenic origin. Trees and vegetation emit copious quantities of VOCs, such as terpenes and pinenes, which are pleasant smelling, but they do participate in photochemical reactions that produce photo-oxidants. Even though great effort and expenses are being made in many developed countries to control precursors, photo-oxidant concentrations over urban–industrial continents have improved only slightly, if at all. In less developed countries that do not have the means of controlling photo-oxidant precursors, their concentrations are on a steady increase.

Global air pollution

Some air pollutants, because of their prolonged lifetime in the atmosphere, advect and disperse from the emission sources and envelope the whole planet. Examples are the stratospheric O_3-destroying CFCs and the global-warming-causing greenhouse gases.

Stratospheric ozone depletion

Stratospheric O_3 depletion, commonly known as the ozone hole, is caused by the release into the atmosphere of certain manmade substances that destroy the ozone (the 'good' ozone) at high altitude. Because of the thinning of the ozone layer,

more solar UV rays reach the surface of the Earth, which may cause a higher incidence of sunburn and skin cancer. The ozone hole is not caused directly by fossil fuel usage, but indirectly by using appliances such as refrigerators, air conditioners, chillers, and heat pumps. The compressors of these appliances are filled with CFCs that leak into the atmosphere during their lifetime, or are vented during their destruction before being used as scrap iron. The insulation foam of these devices also contains CFCs, which were used in the foam-blowing process. The CFCs are leaking slowly out from the foam cells of still-in-use appliances or discarded ones.

The ozone-destroying CFCs are *fully halogenated* hydrocarbons. For example, CFC-11 is CCl_3F, CFC-12 is CCl_2F_2, CFC-113 is CCl_2FCClF_2. These compounds are non-toxic and non-flammable, and their thermodynamic properties are ideally suited for the compression/expansion cycle in cooling and heat pump appliances. However, CFCs are chemically very inert, so when they are vented into the atmosphere, they do not react with atmospheric constituents. They diffuse unscathed first into the troposphere, then penetrate slowly into the stratosphere. There, the solar UV radiation photodissociates these compounds, liberating free chlorine atoms (the C–Cl bond is weaker than the C–F bond). The chlorine atoms react with atmospheric O_3 to form chlorine oxide, which in turn reacts with atmospheric atomic oxygen regenerating chlorine atoms:

$$Cl + O_3 \longrightarrow ClO + O_2 \qquad (1)$$
$$ClO + O \longrightarrow Cl + O_2 \qquad (2)$$

The cycle repeats itself, so that one Cl atom can destroy thousands of O_3 molecules (Seinfeld & Pandis 1998). It is estimated that so far about 10% of the stratospheric ozone has been depleted. Because of the Montreal Convention of 1987 and its Amendment of 1992, fully halogenated CFCs are no longer manufactured legally in the world. Unfortunately, these CFCs are very long lived (in the order of hundreds of years), so the ozone hole will only be slowly filled in by natural production of O_3 in the stratosphere (IPCC 2001).

The phasing-out of fully halogenated CFCs is a success story of international cooperation in ameliorating an environmental hazard created by anthropogenic activities. In part, this success was achieved because chemical manufacturers found ready substitutes for the fully halogenated CFCs, the *partially halogenated* hydrochlorofluorocarbons (HCFCs), for example HCFC-22,

$CHClF_2$. These HCFCs do react with atmospheric hydroxyl radicals, shortening their lifetime so that they do not reach the stratosphere. The problem with the HCFCs is that they cannot be used in older appliances that were designed for CFCs. When CFCs will no longer be found in the market, the older appliances will need to be replaced by new ones designed for HCFCs.

Global warming

Of all environmental effects of fossil fuel usage, global warming, including its concomitant climate change, is the most perplexing, potentially most threatening, and arguably most intractable problem. It is caused by the ever-increasing accumulation in the atmosphere of CO_2 and other gases, such as CH_4, nitrous oxide (N_2O), CFCs, as well as anthropogenic aerosols, collectively called greenhouse gases (GHG).

The term greenhouse effect is derived by analogy to a garden greenhouse. There, a glass-covered structure lets in the Sun's radiation, warming the soil and plants that grow in it, while the glass cover restricts the escape of heat into the ambient surroundings by convection and radiation. Similarly, the Earth atmosphere lets through most of the Sun's radiation, which warms the Earth's surface, but the GHGs trap outgoing terrestrial infrared (IR) radiation, keeping the Earth surface warmer than if the GHGs were absent from the atmosphere.

A schematic of the greenhouse effect is represented in Fig. 3. At the left of the schematic is the atmospheric temperature structure, which defines the various 'spheres'. In the troposphere the temperature decreases on the average by 10 °C per km. This sphere is thoroughly mixed by thermal and mechanical turbulence, and contains most of the air pollutants mentioned in a previous section. In the stratosphere the temperature increases steadily. Because of the positive temperature gradient this sphere is very stable with very little turbulence and mixing. At about 50–60 km height the temperature declines again with altitude, giving rise to the mesosphere. Finally, above 90–100 km the temperature gradient reverses itself, and becomes positive. The highest sphere is called the thermosphere. Its temperature is dependent on solar irradiation, and varies from several hundred degrees Kelvin at solar minimum to up to 2000 K at solar maximum. On the average, though, the Earth atmosphere has a negative temperature gradient of about -6.5 °C/km.

The average solar radiation that impinges on the top of the atmosphere is about 343 W/m^2. This is the annual, diurnal, and spatial average irradiation. Of this irradiation, currently about 30% (103 W/m^2) is immediately reflected into space by land, ocean, icecaps, and clouds, and scattered into space by atmospheric molecules. (That is the reason the sky is blue. Atmospheric molecules preferentially scatter the blue portion of the solar spectrum.) The reflected and scattered sunlight is called *albedo*. The albedo may not remain constant over time. With increased melting of the icecaps, and increased cloud cover, in part due to anthropogenic influences, the albedo may change over time. The remaining 70% of solar irradiation heats the Earth surface, its land and oceans. Currently, the global average surface temperature is 288 K (about 15 °C). A body (the so-called *black body*) that is heated to 288 K radiates 390 W/m^2. The Earth surface radiation occurs in the far IR and is called *earthshine*. A part of the earthshine is reflected back to the Earth surface by clouds and aerosols; another part is first absorbed by certain gaseous molecules, then re-radiated back to the surface. The absorption/re-radiation occurs by poly-atomic molecules, including water vapour and the GHGs CO_2, CH_4, N_2O, O_3, CFC, and others. The reflection and re-radiation to the Earth surface of the outgoing terrestrial IR radiation is causing the Earth surface to become warmer than it would be merely by solar irradiation. This is essentially the greenhouse effect. With increasing concentrations of anthropogenic GHGs and aerosols, the Earth surface temperature will increase. It should be emphasized that the greenhouse effect is not due to trapping of the incoming solar radiation, but the outgoing terrestrial IR radiation. In fact, due to the trapping of earthshine by GHGs the upper layers of the atmosphere (primarily the stratosphere) will become *colder* (Mitchell 1989).

The extent of global warming can be predicted by radiative transfer models. These models include the radiative properties of the GHGs and their distribution in the Earth atmosphere, as well as the temperature and pressure gradients in the atmosphere. There is general agreement among the models as to the extent of surface warming due to GHG absorption/re-radiation, called radiative forcing. Based on the models, the International Panel on Climate Change (IPCC, 2001) projects that the average Earth surface temperature will increase as shown in Fig. 4. The middle, 'best', estimate predicts a rise of the Earth surface temperature by the end of the 21st century of about 2 °C; the 'optimistic' estimate predicts about 1 °C, and the 'pessimistic' estimate predicts about a 3 °C rise. The optimistic estimate relies on slowing of CO_2 and other GHG emissions; the pessimistic estimate on

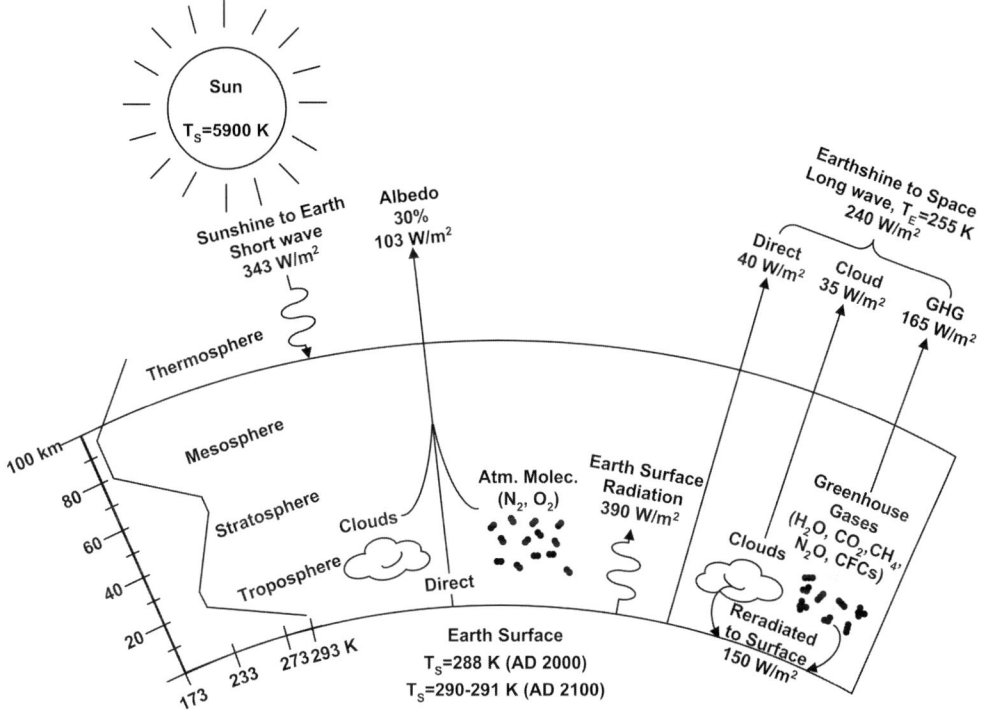

Sun-Earth-Space Radiative Equilibrium

Fig. 3. Schematic of the greenhouse effect.

'business-as-usual', that is, on continuing rate of growth of CO_2 and other GHG emissions, and the best estimate is somewhere in between. If the GHG concentrations increase in the atmosphere at their current rate, then by the year 2100, CO_2 will contribute about 65% to global warming, CH_4 15%, N_2O 10%, and CFCs about 5–10%. (By international conventions CFCs are being phased out entirely, but due to their very long lifetime in the atmosphere, they will still contribute to global warming by the year 2100.)

In addition to radiative forcing, global warming may be enhanced by the so-called feedback effects. For example, water vapour is a natural GHG. When the temperature of the ocean surface increases, the evaporation rate will increase. As a consequence, the average water vapour content of the atmosphere will increase. This causes more absorption of the outgoing infrared radiation and more global warming. Furthermore, increased evaporation may cause more cloud formation. Clouds can also trap outgoing terrestrial radiation,

again increasing global warming. Melting icecaps and glaciers decrease the reflection of incoming solar radiation (reduced albedo), which also increases global warming. The prediction of the feedback effects is more uncertain than the prediction of radiative forcing, but generally it is assumed that the feedback effects may double the surface temperature increase due to radiative forcing alone.

Has the surface temperature already increased due to anthropogenic activities? Figure 5 plots the global average surface temperature over the last century and a half. Even though there are large annual fluctuations, the smoothed curve through the data points indicates an upward trend of the temperature. From 1850, the start of the industrial revolution, to date, the global average surface temperature has increased by about 0.5–1 °C. This is in accordance with radiative models that predict such a trend considering the increase of GHG concentrations over that period.

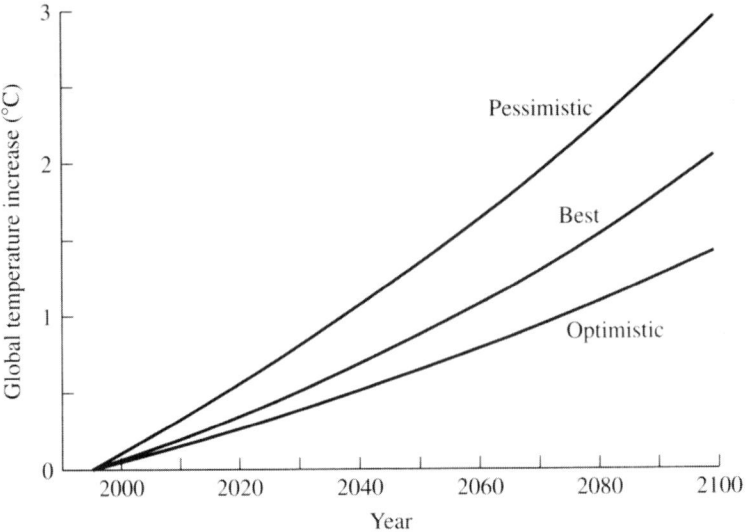

Fig. 4. Projected trend of the Earth surface temperature increase. Upper curve, pessimistic scenario with no emission curtailment; lower curve, optimistic scenario with significant emission curtailment; middle curve, in-between scenario. (Adapted from IPCC 2001.)

Other effects of global warming

As a consequence of increased GHG concentrations in the atmosphere, the Earth surface temperature may rise as discussed in the previous section. The surface temperature rise may cause several ancillary effects on global climate and hydrogeology, which in turn will affect human habitat, welfare and the ecology.

Sea level rise. With increasing surface temperatures the average sea level will rise because of three factors: melting of polar ice caps, receding of glaciers, and thermal expansion of the ocean surface waters. Combining all three factors,

it is estimated that by the end of the next century the average sea level may be 30–50 cm higher than it is today. This can seriously affect low-lying coastal areas, such as the Netherlands in Europe, Bangladesh in Asia, and low-lying islands in the Pacific and other oceans (IPCC, 2001).

Climate changes. Predicting global and regional climatic changes as a consequence of average surface temperature rise is extremely difficult and fraught with uncertainties. It is expected that regional temperatures, prevailing winds, storm and precipitation patterns will change, but where

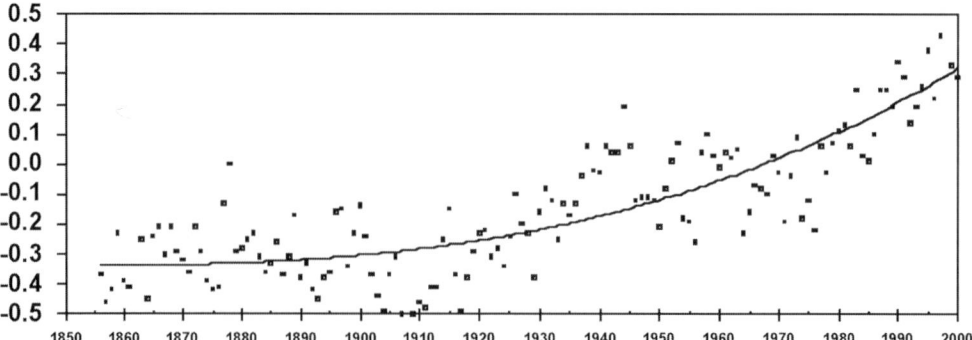

Fig. 5. Global average surface temperature trend 1850–2000. Scale on y-axis (°C) is indexed to 1970 global average surface temperature. (Adapted from Carbon Dioxide Information Analysis Center 2000.)

and when, and to what extent changes will occur, is a subject of intensive investigation and modeling on the largest available computers, the so-called supercomputers. Climate is not only influenced by surface temperature changes, but also by biological and hydrological processes, and by the response of ocean circulation, which are all coupled to temperature changes. It is expected that temperate climates will extend to higher latitudes, probably enabling the cultivation of grain crops further toward the north than at present. However, crops need water. On average, the global evaporation and precipitation balance will not change much, although at any instant more water vapour (humidity) may be locked up in the atmosphere. However, precipitation patterns may alter, and the amount of rainfall in any episode may be larger than it is now. Consequently, the runoff (and soil erosion) may be enhanced, and areas of flooded watersheds may increase. Hurricanes and typhoons spawn in waters that are warmer than 27 °C, in a band from 5 to 20° north and south latitude. As the surface waters become warmer, and the latitude band expands, it is very likely that the frequency and intensity of tropical storms will increase.

The sea level and climatic changes may cause a redistribution of agricultural and forestry resources, a considerable shift in population centres, and incalculable investments in habitat and property protection.

Greenhouse gas concentrations trends

Currently, about 6.8 Gt/y of carbon (25 Gt/y CO_2) is emitted into the atmosphere by fossil fuel combustion. Another 1.5 ± 1 Gt/y is emitted due to deforestation and land use changes, mainly artificial burning of rainforests in the tropics, and logging of mature trees, which disrupts photosynthesis. Figure 6 plots the trend of atmospheric concentrations of CO_2, measured consistently at Mauna Loa, Hawaii, since 1958. Currently, the average CO_2 concentration is about 370 parts per million by volume (ppmv). The plot shows seasonal variations due mainly to assimilation/respiration of CO_2 by plants, but there is a steady increase of the average concentration at a rate of approximately 0.4%/y. If that rate were to continue into the future, a doubling of the current CO_2 concentration would occur in about 175 years. However, if no measures are taken to reduce CO_2 emissions, then due to the population increase, and the concomitant enhancement of fossil fuel use, the rate of growth of CO_2 concentration will increase more than 0.4%/y, and the doubling time will be achieved sooner.

Methane emissions are in part due to fossil energy usage, because it leaks from gas pipes, storage tanks, tankers, and coal mine shafts. Anthropogenic emissions of CH_4 from fossil fuel usage amount to about 100 Mt/y. However, CH_4 is also emitted from municipal waste landfills, sewage treatment, biomass burning, cultivated rice paddies, enteric fermentation of cattle, and other anthropogenic activities, so that the total amount of CH_4 emissions is about 400 Mt/y. Currently, the average atmospheric concentration of CH_4 is about 1.7 ppmv, growing at about 0.6%/y. Nitrous oxide is a minor product of combustion of fossil fuels. Currently, the concentration of N_2O is about 0.3 ppmv, growing by about 0.25%/y. Although CFCs are not directly associated with fossil fuel usage, they are, however, emitted inadvertently from energy-using devices, such as refrigerators, air conditioners, chillers, and heat pumps. Current concentrations of the various CFCs are about 0.5 parts per billion by volume (ppbv), and their concentrations are slowly declining due to the phase-out of production of CFCs.

What can be done about global warming

Global warming could be ameliorated by reducing significantly the emissions of GHGs into the atmosphere. Most GHG emissions are a consequence of fossil fuel use. While it is important to reduce the emissions of all GHGs, the greatest preventative measure would come from reducing CO_2 emissions. Emission reductions of CO_2 can be accomplished by a combination of several of the following approaches.

Demand side conservation and efficiency improvements. This includes less space heating and better insulation, less air conditioning, use of more fluorescent instead of incandescent lighting, more energy efficient appliances, process modification in industry, and very importantly, more fuel-efficient automobiles. Some measures may even incur a negative cost, that is, consumer savings by using less energy, or at least a rapid payback period for the investment in energy-saving devices.

Supply side efficiency measures. Here we mean primarily increasing the efficiency of electricity production. Natural gas combined cycle power plants (NGCC) emit less CO_2 than single-cycle coal-fired power plants; first, because natural gas emits about one-half the amount of CO_2 per fuel heating value than coal, and secondly, because the thermal efficiency of combined-cycle power plants is in the 45–50% range compared with the 35–38% range of single-cycle plants. In the

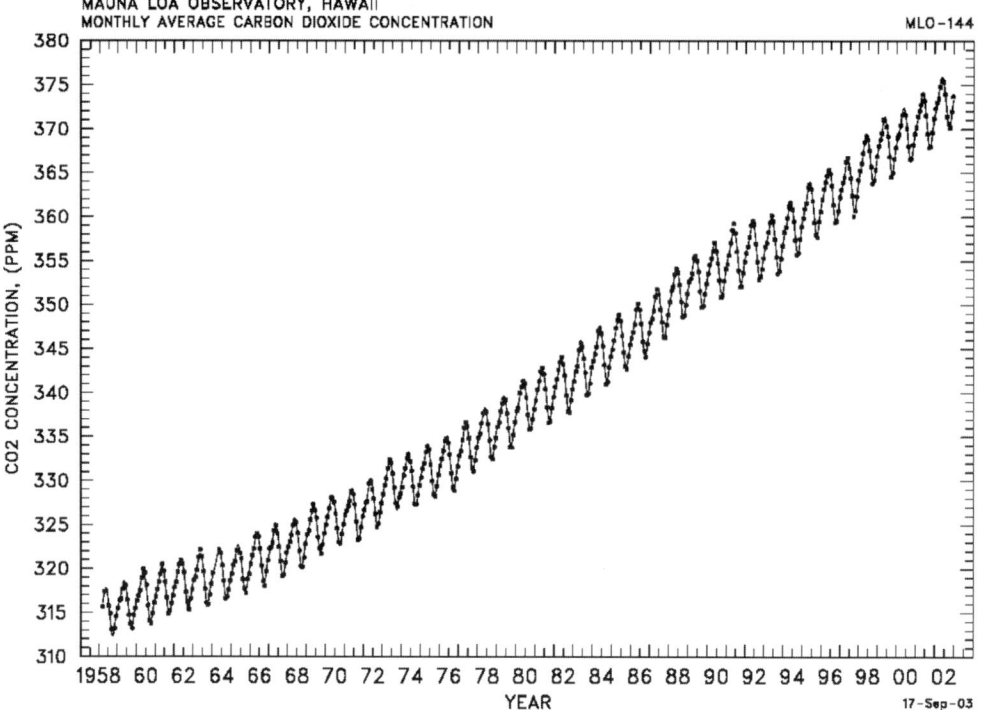

Fig. 6. Carbon dioxide concentration trend 1958–2002. (Adapted from Carbon Dioxide Information Analysis Center 2000.)

future, integrated coal gasification combined-cycle power plants (IGCC) may come on line with a thermal efficiency in the 40–45% range reckoned on the basis of coal heating value. Furthermore, IGCCs may enable the capture of CO_2 at the gasification stage, with subsequent sequestration of the captured CO_2 in geologic and deep ocean repositories (see Saylor & Zerai, 2004).

Shift to non-fossil energy sources. The choices here are agonizing, because the largest impact could be made by shifting to nuclear and hydro-electricity, both presently very unpopular and fraught with environmental and health hazards. The shift to solar, wind, geothermal, and ocean energy is popular, but because of their limited availability and intermittency, and because of their larger cost compared to fossil energy, a substantial shift to these energy sources cannot be expected in the near future.

None of these options can prevent global warming alone. They have to be taken in combination and on an incremental basis, starting with the least expensive ones and progressing to the more expensive ones. Even if the predictions of

global climate change were to turn out to be exaggerated, the fact that fossil fuel usage entails many other environmental and health effects, and the certainty that fossil fuel resources are finite, makes it imperative that we curtail fossil energy usage as much as possible.

Summary and conclusions

The use of fossil fuels to supply energy for the use of the world's population has resulted in the release to the atmosphere of troublesome chemical byproducts that present harm to humans and other natural species. These effects can be localized (near the emission source), can extend to large regional areas (involving subcontinents), and can even cover the globe, from pole to pole. A large portion of the human population is exposed to one or more of these environmental effects.

The scientific understanding of how fossil fuel use causes these effects is well advanced, providing quantitative means for explaining what is currently observed and predicting what changes will occur in the future from projected future fuel consumption. These projections provide a basis for modifying the amount and character of future energy supply so as to ameliorate harmful environmental consequences.

The technological systems that employ fossil fuel energy have been developed to lessen the amounts of harmful emissions, albeit at significant energy and economic cost. Further improvements can be expected, but at increasing marginal cost. The most severe emission control problem, in terms of economic and energy cost, is that of CO_2, a major contributor to global warming. The implementation of policies by national governments and international bodies to curtail the use of fossil energy and the concomitant emissions of CO_2 will become a growing task for mankind in this century.

The authors wish to thank Drs R. Gieré and R. Hinrichs for their careful reviews of this paper and the valuable comments they made, which hopefully we addressed in the revised manuscript.

References

BUTLER, T. J., LIKENS, G. E. & STUNDER, B. J. B. 2001. Regional-scale impacts of Phase I of the Clean Air Act Amendments in the USA: the relation between emissions and concentrations, both wet and dry. *Atmospheric Environment*, **35**, 1015–1028.

CARBON DIOXIDE INFORMATION ANALYSIS CENTER 2000. *Trends Online*. Oak Ridge National Laboratory, Oak Ridge, TN. http://cdiac.esd.ornl.gov

EIA 2001. *International Energy Annual* 2001. US Department of Energy, Energy Information Administration, Washington, DC 20585. Available at http://www.eia.doe.gov.

EPA 2003a. *National Ambient Air Quality Standards*. US Environmental Protection Agency, Washington, DC 20460. Available at http://www.epa.gov/ttn/naaqs/

EPA 2003b. *Code of Federal Regulations Title 40: Protection of Environment, Part 60: Standards of Performance for New Stationary Sources*. US Environmental Protection Agency, Washington, DC 20460. Available at http://www.epa.gov/docs/epacfr40/chapt-I.info/60tc.htm

EPA 2003c. *Code of Federal Regulations Title 40: Protection of Environment, Part 85: Control of Air Pollution From Mobile Sources*. US Environmental Protection Agency, Washington, DC 20460. Available at http://www.access.gpo.gov/nara/cfr/cfrhtml_00/Title_40/40cfr85_00.html

FAY, J. A., GOLOMB, D. & KUMAR, S. 1985. Source apportionment of wet sulfate deposition in Eastern North America. *Atmospheric Environment*, **19**, 1773–1782.

IPCC 2001. *Intergovernmental Panel on Climate Change, Climate Change 2001: The Scientific Basis*. Cambridge University Press, Cambridge CB2 IRP, UK.

KALYONCU, R. S. 2000. *Coal Combustion Products, Metals and Minerals, Minerals Yearbook*. US Geological Survey, Reston, VA.

MITCHELL, J. F. B. 1989. The greenhouse effect and climate change, *Review of Geophysics*, **27**, 115–139.

NRC 1991. *National Research Council, Rethinking the Ozone Problem in Urban and Regional Air Pollution*. National Academy Press, Washington, DC.

SAYLOR, B. & ZERAI, B. 2004. Injection and trapping of carbon dioxide in deep saline aquifers. *In*: GIERÉ, R. & STILLE, P. (eds). *Energy, Waste, and the Environment: a Geochemical Perspective*. Geological Society, London, Special Publications, **236**, 285–296.

SEINFELD, J. H. & PANDIS, S. N. 1998. *Atmospheric Chemistry and Physics: From Air Pollution to Climate Change*. John Wiley and Sons, New York, NY, USA.

WARK, K., WARNER, C. F. & DAVIS, W. T. 1998. *Air Pollution: Its Origin and Control*, 3rd edn. Addison-Wesley, Menlo Park, CA, USA.

Environmental impacts of coal mining and associated wastes: a geochemical perspective

PAUL L. YOUNGER

*Hydrogeochemical Engineering Research and Outreach (HERO),
School of Civil Engineering and Geosciences, University of Newcastle,
Newcastle upon Tyne, UK*

Abstract: Ever since the commencement of industrial-scale coal mining (in northeast England around 1600), substantial environmental impacts have been recorded as arising from both the mined voids and from the wastes left behind at the surface. In the early days of coal mining, complaints about such impacts were strident, as the newly established industry adversely affected long-established agricultural interests. When the coal trade had come to dominate regional economies in mining districts, its negative impacts came to be accepted as a necessary byproduct of the generation of coal-based wealth. It has only been since large-scale mine closures began to take place in the major coal-mining economies of the developed world during the last few decades that the negative impacts of coal mining have once more been deemed unacceptable. The environmental impacts arising from coal mining activities are fundamentally attributable to the exposure of reduced earth materials (especially coal, pyrite, siderite, and ankerite) to the oxidizing power of the Earth's atmosphere. The consequences range from the spontaneous combustion of coal to the release of acidic waters from pyrite oxidation. A typology of the known impacts arising from mine voids and wastes in coal mining districts has been developed, which recognizes many subcategories of impacts under five major headings: air pollution, fire hazards, ground deformation, water pollution, and water resource depletion. A robust understanding of geochemical processes is key to understanding how these impacts arise, and to developing sustainable mitigation strategies. The application of the newly developed typology is illustrated using the case of the Shilbottle Coalfield (Northumberland, UK). Although few demonstrable impacts have arisen in the categories of air pollution, fire hazards, or ground deformation, major problems of water pollution have required both preventative and remedial interventions. For the flooded underground voids, these took the form of a pump-and-treat system, whereas emissions of leachates from surface spoil heaps have necessitated the installation of an innovative 'hybrid' passive treatment system, comprising a permeable reactive barrier, oxidation ponds, and a wetland. Inverse geochemical modelling has clarified the linkages between the various types of water encountered in the coalfield, providing a baseline geochemical understanding upon which future investigations of remedial system sustainability can be based.

The mining and processing of coal has supplied most of the fuel for the world's industrialized societies for more than two centuries, and looks set to retain a leading share of the global energy generation market for many decades yet to come. The undeniable benefits that coal-powered electricity generation has brought to society have not been without their costs. In this chapter, the environmental debit sheet of coal mining is examined from a geochemical perspective. The four principal purposes of this chapter are to:

(1) Outline the history of how perceptions of the environmental impacts of coal mining have evolved over the centuries;

(2) Proffer a typology of the environmental impacts attributable to coal mining, distinguishing between those arising from bodies of mine waste and those arising from the mined voids themselves;

(3) Highlight the geochemical background to these impacts; and

(4) Highlight observations arising from (2) and (3) by means of an illustrative case study.

It should be noted that this chapter does *not* address the following topics:

(1) Environmental conditions *within* working coal mines, which are relevant to the

From: GIERÉ, R. & STILLE, P. (eds) 2004. *Energy, Waste, and the Environment: a Geochemical Perspective*. Geological Society, London, Special Publications, **236**, 169–209.
0305-8719/04/$15   © The Geological Society of London 2004.

health of workers. This is a specialist topic on which a vast socio-medical literature already exists (e.g., Seixas *et al.* 1992; Buchanan & Brenkley 1994; Love *et al.* 1997; Grayson 1999; Bloor 2000, 2002; Finkelman *et al.* 2002), and it will therefore only be discussed in this paper in the context of explaining the evolution of public consciousness on issues of pollution and public health associated with coal mining.

(2)   The contribution to the global greenhouse gas budget made by direct methane releases from coal mines (Williams & Mitchell 1994), which are known to account for a significant proportion of total atmospheric emissions of this gas (cf. Breas *et al.* 2002), nor the lesser methane releases now thought to occur through fractured strata overlying coal mines (Thielemann *et al.* 2001).

(3)   Atmospheric impacts of coal burning for power generation and other industrial purposes; these are summarized by Golomb & Fay (2004).

(4)   Potential environmental hazards associated with solid residues arising from coal-burning power stations (Donahoe 2004; Glasser 2004; Groppo *et al.* 2004; Spears & Lee 2004; Tishmack & Burns 2004).

(5)   Beneficial re-uses of abandoned mine sites (Harrison 2000), mine wastes (Skarzynska 1995*a*, *b*; Banks *et al.* 2004) and mine waters (Banks *et al.* 1996).

Rather, this paper is intended to be a synthesis from a geochemical perspective of the various impacts of coal mine voids and mine wastes on the surrounding environment, that is on the habitats and urban environments which adjoin active and abandoned mine sites. Since many of the potential negative impacts of coal mining are effectively mitigated in the case of most working mines (which in most jurisdictions are subject to environmental management regulations of varying degrees of stringency), this paper inevitably includes more information on the environmental problems associated with abandoned (including reclaimed) mine sites, rather than on those associated with active mines.

Before proceeding, it is necessary to define some of the key terminology used in this paper to denote various features of mine sites. Table 1 provides the necessary glossary. For further explanation of the mining and mineral processing techniques that give rise to these features, the interested reader is referred to text books concerning mining engineering (Hartman &

Mutmansky 2002) and mine water management (Younger *et al.* 2002).

The evolution of environmental consciousness in relation to coal mining

Although the burning of coal is much maligned in these days of concern over climate changes, there is no doubting its importance to the history of mankind. Although coal was used for metallurgical purposes in ancient times in both China and the Roman Empire, such usage was on relatively modest scales (Lynch 2002). The first time that coal mining began to take place on a modern industrial scale was actually in the early 17th century, in the Tyneside district of northeast England (Levine & Wrightson 1991). This area was the first to experience 'carboniferous capitalism' simply because it was the only place in Britain where major, naturally navigable rivers (i.e., the Rivers Tyne and Wear) pierced the heart of a major coalfield. From these rivers, coal was shipped via the North Sea to what was then world's greatest demand centre, the City of London. Thus England's most northerly major coalfield led the way in large-scale coal mining two centuries before the advent of rail transport allowed economic competition from the various inland coalfields sited much closer to London. It is impossible to overstate the importance of emergence of the Tyneside coal industry for the evolution of the material culture with which all who live in industrialized countries are now familiar: river-dredging technology, railways, steam locomotives, pumping engines, steam-powered boats, turbines, and even the electric light bulb were all originally developed in the milieu of innovation stimulated by the Tyneside coal trade.

However, if the coal industry was the dynamo of invention throughout the first industrial revolution, it soon revealed itself also as a nuisance to its neighbours. Thus we find that as early as April 1620, residents of the Tyneside village of Whickham were making legal depositions in protest at the drying up of wells and springs due to underdrainage by recently driven coal mine adits. They also complained that '. . . two hundred acres and above [*of good meadow land had been*] quyte [*sic*] spoiled and cankered with the water that issueth out of the colewaists [*sic*] . . .' Similar complaints were also raised in respect of the 'unwholesome, cankered and infectious' water that flowed from the drainage channels (i.e., adits) of the mines, which were deemed to have polluted the receiving land surface so badly that grass would no longer

Table 1. *Glossary of selected coal mining terms used in this paper, with selected cross-referenced entries for other synonymous terms used in literature cited by this paper*

Term	Definition
Abandoned	Denotes a *mine* or part of a *mine* that is no longer being worked
Active	Denotes a *mine* or part of a *mine* that is currently being worked to produce coal
Adit	Subhorizontal tunnel connecting the *workings* of an *underground mine* to the ground surface, which is normally used to provide a gravity drain to the *workings*, but often also as access for personnel and materials (synonym: *drift*, meaning (a))
Backfill	Fractured waste rock deposited into a former *opencast mine void* (sedimentologically similar to *spoil* and to *goaf*)
Bing	See *spoil heap*
Blackdamp	O_2-deficient air, typically rich in CO_2 (synonym: *stythe*)
Blower	Significant point source of gas entering *mine workings*
Bord	*Mine* roadway (or dead-end room) from which *run-of-mine* coal has been extracted
Bord-and-pillar	A 'supported' method of underground coal mining in which coal is extracted from a rectilinear network of *bord*s, with *pillar*s of intact coal left in place to support the roof. (synonyms: *pillar-and-stall*, *room-and-pillar*, *stoop-and-room*)
Colliery	Coal mine, especially an *underground mine* for coal
Culm bank	See *spoil heap*
Deep mine	See *underground mine*
Dinting	Excavation of floor strata below a thin coal *seam* to improve the overall working height of roadways, thus improving access (see also *ripping*)
Drift	(a) An *adit*, which may be further qualified as a 'haulage drift', 'access drift', or 'drainage drift', depending on its principal function
	(b) Any tunnel in an *underground mine* that connects one set of *workings* with another
Drift mine	*Deep mine* in which the workings are principally accessed by an *adit*
Dust explosion	Accidental explosion in an *underground mine* due to the explosive ignition of fine particles of coal suspended in the *mine atmosphere*
Face	Excavated wall of coal
Feeder	Significant point-source of water entering mine *workings*
Finings	Fine-grained *waste rock* material removed from coal in a *washery*, and deposited from suspension in a pond (synonyms: *slimes*, *tailings*)
Finings pond	Pond in which *finings* are settled out of suspension (synonym: *slimes dam*)
Fire damp	Explosive mixture of methane and air encountered in mine *workings*
Goaf	Brecciated mass of roof strata that has collapsed into an abandoned *void* in an *underground mine* (sedimentologically similar to *backfill* and to *spoil*)
Gob	Synonym of *goaf* used in Yorkshire and the eastern USA (where it is also used to denominate *spoil*)
Gob fire	Fire in *goaf*
Gob pile	See *spoil heap*
Highwall	Highest wall of intact strata in an *opencast* mine
Longwall	'Caving' method of mining involving the removal of all coal from a number of large *panel*s, into which the ceiling is subsequently left to fall, forming *goaf*
Loose wall	Slope of a pile of *backfill* in a working *opencast* mine *void*; the loose wall typically faces the current *highwall*
Low wall	Lowermost edge of intact strata around the perimeter of an *opencast* void
Maingate	One of two parallel roadways along the long edges of a *longwall panel*, which provide access to the *face* during working (see also *tailgate*)
Mine	Any discrete set of *workings* for coal, whether a *surface mine* or an *underground mine*
Overburden	See *waste rock* (meaning (b))
Opencast mine	*Surface mine* in which *overburden* is cast immediately behind the working bench, onto the latest worked-out bench behind the advancing excavators (synonym: *strip mine*)
Panel	Discrete volume of coal within a single *seam*, which is extracted in a *longwall* underground mining operation; panels often comprise the entire thickness of the seam, and are typically rectangular in plan, with total areas of several tens of hectares (typically 25–40 ha)
Pillar	Intact mass of coal (usually rectangular in plan) left in place in an *underground mine* to support the roof
Pillar-and-stall	See *bord-and-pillar*

(continued)

Table 1. *Continued*

Term	Definition
Restored	Adjective describing a former mine site that has been relandscaped (and usually also *revegetated*) following the end of mining
Revegetated	Body of *opencast backfill* or *colliery spoil* upon which vegetation has been established
Ripping	Extraction of roof strata above a thin coal *seam* to improve headroom for access purposes (see also *dinting*)
Run-of-mine	Adjective describing the raw coal product (as excavated from the working *bord*s or *panel*s) prior to processing by screening, crushing and/or washing; 'run-of-mine coal' therefore consists of a mixture of coal (the predominant component) together with fragments of *waste rock*
Seam	Discrete bed of coal
Seat-earth	Shale bed immediately underlying a coal *seam*, often containing fossilized roots from the coal-forming plants
Shaft	Vertical or subvertical tunnel providing access from the surface to the *workings* of an *underground mine*
Shaft pillar	*Pillar* of unworked coal left in place for a predetermined radius around a *shaft* in order to ensure the structural integrity of the *shaft*
Slimes	Predominantly South African term for *finings*
Slimes dam	See *finings pond*
Soil-forming material (SFM)	Earth material (typically *spoil*, *overburden* containing large quantities of mudstone), which can be conditioned (by mixing with compost) such that it evolves over a period of a few years to form a semi-natural soil, which is a key step in developing *revegetated* mine sites
Spoil	*Waste rock* removed from a *colliery* and deposited at the Earth's surface (sedimentologically similar to *backfill* and to *goaf*)
Spoil heap	Deposit of spoil, typically in the form of a large pile of rock fragments (synonyms: *bing* (Scotland), *culm bank* (India), *gob pile* (Eastern USA), *waste rock pile* (Australia, Canada)
Strip mine	See *opencast mine*
Stythe	See *blackdamp*
Surface mine	Any mine in which excavation takes place in the open air (in the case of coal mining considered in this paper, this term is largely synonymous with *opencast mine*)
Tailgate	One of two parallel roadways along the long edges of a *longwall panel*, which provide access to the *face* during working (see also *maingate*)
Tailings	Analogous term to *finings* principally used in metalliferous mining, but occasionally applied without further qualification also to coal *washery finings*
Underground mine	Any mine in which extraction occurs in underground galleries (synonym: *deep mine*)
Void	Any excavated opening in a mine (surface or underground)
Washery	A coal-processing plant in which *finings* are removed from coal by the action of water (usually aided by flotation agents)
Waste rock	(a) Rock removed accidentally with coal and separated from the coal by later processing (b) Barren rock (*overburden*) removed in the course of gaining access to the coal (approximately synonymous with *spoil* and *backfill* once it has been disposed of) (Sedimentologically similar to *goaf*)
Water make	The flow rate of water entering a given *mine* or discrete body of *workings*
Workings	A collective term for the *voids* belonging to a particular mine

grow there, resulting in streams that were so polluted that even the beasts of the field would refuse to drink their waters or to eat grass grown by such streams (Levine & Wrightson 1991, pp. 110–116).

The same district also yields what are probably the earliest reliable accounts of uncontrolled burning of coal seams, both dating from the 1640s (Galloway 1898, pp. 133, 163). During the Anglo-Scottish Covenant Wars, a retreating detachment of English soldiers set fire to their camp in the vicinity of Whickham in order to deny its availability to approaching Scottish forces. This surface fire then ignited an underlying coal seam, which proceeded to burn for several years, giving rise to sporadic displays of flame at the ground surface along its line of outcrop through the adjoining village. In about 1643, careless use of a candle in a working mine in the village of Benwell set fire to a coal seam, which went on to burn out of control for more than 50 years, leading to the collapse of overlying houses and the loss of substantial coal reserves. This fire gave rise to gaseous emissions rich in NH_3 and S, some of which formed white precipitates around aureoles where the fumes reached the surface. Both of these instances of burning gave rise to sustained complaints, relating both to their hazards for the public and to the threat they posed to adjoining coal mines.

As the coal industry continued to expand, nascent environmental complaints such as these gradually came to pale into insignificance in comparison with concerns relating to the mortal hazards encountered by the miners themselves in the course of their daily work. By the middle of the 19th century, a long and sad catalogue of mining disasters had impressed the perils of mining on the popular consciousness (Duckham & Duckham 1973; Younger 2004). The overall breakdown of workplace mortalities in the UK coal-mining industry up to the middle of the 20th century is summarized in Table 2. The breakdown of such figures for other European countries and the eastern USA are unlikely to differ substantially. Table 2 gives the lie to the widespread misconception that explosions were the principal cause of workplace death in the coal-mining industry. In reality, roof falls and accidents with machinery and explosives were far more important overall, although as they tended to strike day by day, claiming one or two lives at a time, they never received the same publicity as major explosions. As it deals strictly with workplace mortalities, Table 2 also cannot reveal the massive loss of life associated with lung diseases such as silicosis, pneumoconiosis, emphysema, and asthma, which resulted

Table 2. *Summary of causes of workplace death in UK coal mines since 1850**

Cause	% Total fatalities attributed to given cause
Roof and wall collapses ('falls of ground')	50
Accidents with underground haulage equipment	25
Accidents with explosives	10
Explosions of methane and/or coal dust	3
Inundation by inrushes of water/liquefied ground	1.5
Other (e.g., falling down shafts, asphyxiation by unexpected bad air, unidentified causes)	10.5

*See Richards 1951; Younger 2003.

in death *outside* the work place (e.g., Bloor 2000; 2002).

The growing awareness of such problems during the 19th and 20th centuries prompted trade union agitation for stricter controls on the health and safety of the mining workforce (Bloor 2002). Despite the continued persistence of non-negligible health risks for coal miners (e.g., Love *et al.* 1997), especially those engaged in underground mining (e.g. Seixas *et al.* 1992), the development of generally effective health and safety regulations can justly be regarded as one of the principal social achievements of the industrial age (e.g. Richards 1951; Seixas *et al.* 1992; Buchanan & Brenkley 1994; Grayson 1999; Borm 2002; Seal & Bise 2002; Li *et al.* 2002).

Although the wider damage to the environment caused by coal mining was already evident locally by the early 17th century, by the time coal mining had become the main economic activity in its host regions this damage came to be accepted as something of a necessary evil. In response to the most severe problems, such as those associated with mining subsidence affecting houses and other property, compensation systems were eventually put in place to ensure that reparations were relatively swift and (usually) equitable (e.g., Bell & Genske 2001). Nevertheless, to outsiders the prevailing environmental conditions in colliery villages usually seemed intolerably bad, and they found it difficult to comprehend the stoical attitudes of locals towards the dirt and disruption (Doyle 2001). One classic example of an account by a horrified outsider is that of Priestley (1934) who experienced the pervasive air pollution in the colliery village of Shotton (County Durham, UK) due to the perpetual burning of the local colliery 'tip' or spoil heap: '... Imagine a village consisting of a few shops, a public house and a clutter of dirty little houses, all at the base of what looked at first like an active volcano. This volcano was the notorious Shotton "tip", literally a man-made smoking hill [*which*] towered to the sky ... its black dark bulk steaming and smoking at various levels .... The atmosphere was thickened with ashes and sulphuric fumes; like that of Pompeii, as we are told, on the eve of its destruction. I do not mean that by standing in one particular place you could find traces of ash in the air and could detect a whiff of sulphur. I mean that the whole village and everybody in it was buried in this thick reek, was smothered in ashes and sulphuric fumes'.

It was only as coal mining lost its economic stranglehold in its traditional heartlands that outspoken environmental advocacy commenced. For ex-mining communities the conversion from tacit approvers of environmental damage to campaigners on behalf of Mother Nature was often abrupt and radical (e.g., Kemp & Griffiths 1999). Unsurprisingly, environmental radicalization was most marked where proposals were made for surface mining of coal in the vicinity of villages, which had only recently lost their livelihoods due to the closure of deep mines (Beynon *et al.* 2000). Ill-will was further fostered by the fact that surface mining was undertaken by civil engineering contractors, who generally brought in their own staff from further afield and rarely made any effort to recruit local residents (Beynon *et al.* 2000). The great boom in surface coal mining following World War II, which was stimulated by war-time technological developments in heavy-duty track-laying plant (Grimshaw 1992), involved the conspicuous and wholesale disturbance of large volumes of coal-bearing ground on a scale rarely experienced previously in any sector of the mining industry.

It was precisely in the context of public policy debates over the proliferation of surface coal mining that the long-latent public health concerns in relation to coal mining were finally given voice. In contrast to the extended discussions in the 19th century of employee health issues in underground coal mines, the many debates over the merits of opencast developments in the second half of the 20th century barely mentioned the occupational health of the surface miners. Rather, the focus was squarely on the impact of the opencast mine on the local environment, and in particular the health and quality of life of nearby residents (Beynon *et al.* 2000). Typical concerns raised by objectors to opencasting related to noise, vibrations, airborne dust, disruption of natural drainage and the loss of cherished landscape features such as hedgerows and mature woodlands.

Similar issues arise in relation to the spoil heaps arising from underground mining operations (e.g., Bell 1996), which can also yield dust to the wind, and often require extensive landscaping and stimulation of soil-forming processes if they are not to remain barren wastelands of infertile rock debris (e.g., Backes *et al.* 1986, 1987; Rimmer & Younger 1997). Furthermore, spoil heaps and bodies of opencast backfill often give rise to polluted leachates, which are typically contaminated with Fe, and are sometimes also acidic. Other metals such as Al and Zn are frequently present in solution at concentrations known to be acutely toxic to aquatic invertebrates and fish. Surprisingly, but perhaps reflecting a lack of public understanding of groundwater hydrology, the generation of such leachates has rarely been raised as a prime concern by campaigners against opencast. In the post-closure period of opencast mines, however, and in the case of many deep-mine spoil heaps, leachate pollution can be the most long-lived environmental impact of all (e.g., Younger *et al.* 2002).

Water pollution problems are not restricted to opencast backfill and spoil heaps: far more abundant volumetrically in most major coalfields are similar polluted waters emanating from abandoned underground voids (e.g., Younger 2001). It is only in the 1990s, with the closure of entire coalfields in the UK, that pollution from abandoned mine voids has risen in public consciousness to be identified as a major, long-term environmental impact of coal mining (e.g., CCC 1995). A similar growth in consciousness is now occurring in France, Germany, Spain, and other countries, as entire coalfields are being abandoned (Younger 2002a). All of this is occurring when the European Union's Water Framework Directive is coming into force.

The mineralogy and geochemistry of coal and associated rocks: pollutant sources and buffers

It is impossible to rationally assess the environmental risks associated with mining a given sequence of coal-bearing strata unless the assessor has a good grasp of the mineralogy and geochemical behaviour of those strata. Although such assessments are beset with complications, in practice a range of techniques are commonly applied to this end. These techniques range from simple analytical assays known as 'acid–base accounting', through various types of simulated weathering experiments (including 'humidity cell tests') to detailed mineralogical characterizations coupled with predictive geochemical modelling. An excellent summary and critical review of these approaches has been recently presented by Banwart *et al.* (2002), and hence no further discussion of the relative merits of the various techniques is given here. Rather, in the following paragraphs, a few pointers are provided to those aspects of the mineralogy and geochemistry of coal-bearing strata, which must be understood before any risk assessment can be confidently attempted. The relative importance of these aspects is explored practically, by means of geochemical modelling, in the case study that concludes this chapter.

In essence, all of the environmental problems associated with coal mining and the disposal of the lithic wastes to which it gives rise can be attributed to a single cause: the incompatibility between coal-bearing strata (which are essentially geochemically reduced) and the oxidizing nature of the surface/near-surface atmosphere. Coal mining results in the introduction of considerable imbalances in the dynamic redox equilibria, which have typically developed over geological time at the interface between the atmosphere and the geosphere in natural (unmined) systems (e.g., Strömberg & Banwart 1994). The bulk of the biosphere thrives precisely at the geosphere–atmosphere interface, with many microbes drawing energy from their participation in the redox transformations of C, O, N, S, and other nutrients at this interface. By means of coal mining, components of the geosphere, which were hitherto stable under the markedly reducing conditions at depth, are suddenly exposed to O_2, giving rise to degradation of redox-sensitive minerals.

Coal itself is a naturally reduced geochemical entity that undergoes substantial changes in the presence of atmospheric O_2 (Thomas 2002). The various coal macerals vary in their susceptibility to oxidation, but vitrinite and inertinite are both known to oxidize significantly. Oxidation of the coal leads to break down into fine-grained particles, which are then prone to hydration. In some cases, oxidation leads inexorably to the spontaneous combustion of the coal, which occurs because the sorption of O_2 gas onto fresh coal surfaces is a strongly exothermic process. Where the supply of O_2 is great enough for extensive heating of the coal surface to occur, but still sufficiently limited that the heat is not dissipated safely into the atmosphere, the coal can reach ignition point (Thomas 2002). This is the root cause of uncontrolled fires in coal seams and coal mine wastes (see below).

The *in situ* burning of coal within spoil results in substantial changes in its mineralogy, geochemistry, and mechanical properties. Visually, the most obvious change is that burnt colliery spoil is red in colour (as opposed to the greys and blacks of unburnt spoil), due to the high-temperature ($\leq 1000\,°C$) oxidation of all Fe-bearing minerals in the strata to form hematite (Fe_2O_3). Large fused masses of oxidized rock up to 10 m in diameter are frequently found in burnt spoil (Rimmer & Younger 1997), and these clearly have far greater mechanical strength than the pre-existing loose waste rock. Gaseous emissions from burning zones in coal comprise mixtures of noxious oxides of C, S, and N, as well as smoke particles containing polycyclic aromatic hydrocarbons (PAHs) and other potentially toxic organic compounds (Lee *et al.* 1981).

The risk of spontaneous combustion of coal is known to be exacerbated by the presence of pyrite (FeS_2), the oxidation of which is strongly exothermic. Pyrite swells on oxidation (Bell 1996), promoting the disintegration of the coal and shale clasts within which it occurs, and rendering them ever-more permeable to O_2. The first step in pyrite oxidation is the seven-electron oxidation reaction, which raises S from its average redox state within pyrite of -1 up to the $+6$ state, which it maintains in the SO_4^{2-} molecule. With such a profound shift in oxidation state, the maintenance of electroneutrality demands hydrolysis, which results in the release of protons (H^+) into solution:

$$FeS_{2(s)} + 3.5\,O_{2(g)} + H_2O$$
$$\longrightarrow\ Fe^{2+}_{(aq)} + 2\,SO^{2-}_{4(aq)} + 2H^+_{(aq)} \quad (1)$$

This reaction (along with others in the overall chain of pyrite oxidation) results in this process being one of the most acid-generating in Nature (e.g., Appelo & Postma 1993; Backes *et al.* 1993).

Since pyrite oxidation has been the subject of numerous in-depth reviews in recent years (e.g., Backes *et al.* 1986; Evangelou 1995; Younger *et al.* 2002; Rimstidt & Vaughan 2003) no detailed exposé is warranted here. However, a few words on the occurrence of pyrite in coal-bearing sequences are appropriate in this context. It is well known that pyrite is present at greatest concentrations in those coals (and adjacent shales) that were deposited and/or underwent early diagenesis in the presence of marine or brackish, tidal waters (Williams & Keith 1963; Caruccio & Ferm 1974; Casagrande 1987; Morrison *et al.* 1990; Younger 1994, 2000). High total S contents therefore correlate quite closely with stratigraphic proximity between a given coal seam and the nearest bed in the sequence containing marine fossils (most notably bivalves of *Lingula sp.*). Total S contents are of economic importance in the marketing of coal, and hence data on this parameter are widely available, typically being measured as part of a suite of determinations termed 'ultimate analysis' (Thomas 2002). Total S contents in coals vary within rather restricted limits: few coals will have less than 0.1 percent by weight (wt%) total S, and contents in the range 0.5–1.5 wt% S are typical of most marketable coals. 'High-S' coals typically have total S contents between 3 and 5 wt%, with some extreme examples reaching 10 wt% S (for instance, in marine-influenced coals of Jurassic and Cretaceous age in Europe). Given that a considerable component of the total S in any coal is actually 'organic S' bound within the coal macerals (which does not contribute to acidification of mine waters) it is important to determine what proportion of the total S is actually attributable to the presence of pyrite (i.e., what is the wt% of 'pyritic S'). Although direct determinations of pyritic S are occasionally made, pyritic S content data are far less common than total S data. It is therefore desirable to be able to estimate the likely pyritic S content of a sample of coal-bearing strata for which only the total S content is known. The following 'rules' (derived from Casagrande 1987) are useful in this regard:

(1) Organic S is a major component of total S in *all* coals, and where total S is less than about 0.5 wt%, organic S probably accounts for nearly all of the S present;
(2) As the total S rises significantly above 0.5 wt%, pyrite comes to gradually account for more and more of the total S content, although even in the highest-S coals, pyritic S seldom accounts for more than 50% of total S.

Armed with the above information, it is possible to make a preliminary assessment of the pollution potential of specific coal seams and associated strata simply on the basis of their total S content (see Younger 2000 for further discussion).

Equations such as equation (1) above imply that the oxidative dissolution of pyrite is congruent, directly liberating Fe^{2+}, SO_4^{2-}, and H^+ to solution. However, in the common circumstance that water is insufficiently abundant to immediately transport the oxidation products away from the mineral surfaces, pyrite oxidation more commonly results initially in the accumulation of various hydroxysulphate evaporite minerals. These minerals form efflorescent crusts, typically white and yellow in colour, on the surfaces of pyrite-rich coals and mudstones (Fig. 1), and they effectively 'store' the oxidation products in a readily soluble form until some hydrological event delivers sufficient water to dissolve and transport them away. Because pyrite often occurs in mudstones, where Al-bearing clay minerals are in contact with acidic pyrite oxidation waters, Al is frequently released from the clays and is also stored in these hydroxysulphate phases. When these minerals finally dissolve, they result in abrupt and extreme increases in dissolved acidity. For this reason, they have been termed 'acid generating salts' (AGS) (Bayless

Fig. 1. Efflorescent 'acid generating salts' present within underground workings of the former Blenkinsopp Colliery (Northumberland, UK). The base of the hammer handle is resting on the contact between the seat-earth and the overlying coal seam. The coal seam is visible in the top third of the photograph, but is entirely obscured down to its contact with the seat-earth by the efflorescent salts, which form a thorough lightly coloured covering on the exposed seam beneath an undulating upper limit which passes a few cm above the hammer head.

& Olyphant 1993). A wide range of AGS have been identified temporarily storing acidity on surface exposures of coal mine spoil (e.g., Bayless & Olyphant 1993), whence they release dissolved acidity whenever a significant rainfall event follows a dry spell during which the AGS have been accumulating. They are also commonly found in underground coal workings (e.g., Younger 2000), where they typically store acidity until the dewatering pumps are finally withdrawn and the workings containing the AGS flood. Table 3 lists some of the AGS commonly encountered in coal mine settings.

Pyrite is not the only Fe^{2+}-bearing mineral that undergoes oxidative weathering when coal-bearing strata are exposed to the atmosphere through mining. The Fe-rich carbonate minerals

siderite, $FeCO_3$, and ankerite, $Ca(Mg,Fe)(CO_3)_2$, are also important reservoirs of Fe^{2+}. Siderite is a relatively common mineral in the shales and sandstones of Carboniferous coal-bearing sequences (e.g., Spears 1989), in which it usually occurs in the form of ellipsoidal concretions (Fig. 2a) orientated with their long axes parallel to bedding. In some cases the density of siderite concretions in specific mudstone horizons is so high that they have historically been worth mining as Fe ores (Hemingway 1974). The fact that these siderite masses could often be won from the same mines, that produced the coal needed to smelt them, meant that the capital costs of developing them were relatively modest, which in part explains why (despite their relatively low Fe grade) these ores were instrumental in the development of the European iron and steel industries

Table 3. *Fe and Al hydroxysulphate minerals, so-called 'acid generating salts' (AGS), which effectively store the products of pyrite oxidation in solid form until later submergence and dissolution*[*]

Mineral phase	Formula
Fe-rich minerals	
Aluminocopiapite	$Fe(II)Fe(III)_2Al_2(SO_4)_6(OH)_2 \cdot 20H_2O$
Copiapite	$Fe(II)Fe(III)_4(SO_4)_6(OH)_2 \cdot 20H_2O$
Coquimbite	$Fe(III)_2(SO_4)_3 \cdot 9H_2O$
'Ferrosic hydroxide'[†]	$Fe(II)_4Fe(III)_2(OH)_{12}(SO_4 \cdot 3H_2O)$
('Interlayered Green Rust'[†])	
Jarosite	
Hydronium Jarosite	$(H_3O^+)Fe(III)_3(SO_4)_2(OH)_6$
Natrojarosite	$NaFe(III)_3(SO_4)_2(OH)_6$
Jarosite	$KFe(III)_3(SO_4)_2(OH)_6$
Kornelite	$Fe(III)_2(SO_4)_3 \cdot 7H_2O$
Melanterite	$Fe(II)SO_4 \cdot 7H_2O$
Rhomboclase	$(H_3O^+)Fe(III)(SO_4)_2 \cdot 3H_2O$
Römerite	$Fe(II)Fe(III)_2(SO_4)_4 \cdot 14H_2O$
Rozenite	$Fe(II)SO_4 \cdot 4H_2O$
Schwertmannite	$Fe_8O_8(OH)_6SO_4$
Szomolnokite	$FeSO_4 \cdot H_2O$
Al-rich minerals	
Alunite	
Amorphous	$KAl_3(SO_4)_2(OH)_6$ (am)
Crystalline	$KAl_3(SO_4)_2(OH)_6$ (c)
Alunogen	$Al_2(SO_4)_3 \cdot 17H_2O$
Basaluminite	
Amorphous	$Al_4SO_4(OH)_{10} \cdot 5H_2O$ (am)
Crystalline	$Al_4SO_4(OH)_{10} \cdot 5H_2O$ (c)
Halotrichite	$Fe(II)Al_2(SO_4)_4 \cdot 22H_2O$
Pickeringite	$MgAl_2(SO_4)_4 \cdot 22H_2O$

[*]It should be noted that other sulphate minerals, especially gypsum ($CaSO_4 \cdot 2H_2O$), epsomite ($MgSO_4 \cdot 7H_2O$) and kieserite ($MgSO_4(H_2O)$, also frequently occur within the same white–yellow–pale green efflorescent crusts which contain these AGS, storing excess SO_4 beyond that associated with Fe and Al. Adapted from Younger *et al.* (2002), with additional observations from the author, and mineralogical information from Bevins (1994) and Nordstrom (1999).
[†]Note: these are not formal mineral names, but are used in the literature to describe this poorly defined phase.

(a)

(b)

Fig. 2. Carbonate minerals, which play important roles in the hydrogeochemistry of coal mines and coal mine wastes: (**a**) A typical siderite concretion (from the mudstone overlying the Halifax Hard Seam at Bullhouse, West Yorkshire, UK); long axis of concretion is 11 cm; (**b**) Patchy films of creamy-white ankerite on a cleat surface of coal (from the Main Seam, Ravensworth Grange Opencast Coal Mine, Durham, UK); maximum width of hand specimen is 10.5 cm.

(and thus the Industrial Revolution as a whole) in the first half of the 19th century (Lynch 2002).

In terms of the environmental impacts of coal mine drainage, siderite plays an ambiguous role (see, for instance, Morrison *et al.* 1990, and works cited therein). On the one hand, the acidity released by pyrite oxidation can be transported into the presence of siderite, which will dissolve to neutralize it:

$$FeCO_{3(s)} + H^+_{(aq)} \longrightarrow Fe^{2+}_{(aq)} + HCO^-_{3(aq)} \quad (2)$$

Further neutralization (i.e., consumption of H^+) occurs when the Fe^{2+} released by reaction (2) is oxidized to Fe^{3+} (e.g., Singer & Stumm 1970):

$$Fe^{2+}_{(aq)} + 0.25\,O_{2(g)} + H^+_{(aq)} \longrightarrow Fe^{3+}_{(aq)} + 0.5\,H_2O$$

$$(3)$$

However, when the Fe^{3+} produced in reaction (3) undergoes hydrolysis, three moles of H^+ are released for every mole of Fe^{3+} hydrolysed:

$$Fe^{3+}_{(aq)} + 3\,H_2O \longrightarrow Fe^{3+}(OH)_{3(s)} + 3\,H^+_{(aq)} \quad (4)$$

If the overall budget of protons consumed and released by reactions (2) through (4) is calculated, it is at once evident that siderite dissolution has no net neutralizing capacity for acidic waters (see also Morrison *et al.* 1990; Younger 1998). Thus siderite dissolution can only provide local neutralization, and only where O_2 is absent from the system. Moreover, reaction (2) suggests that siderite can be a net source of dissolved Fe in coal mine drainage waters, a point that is corroborated by evidence from geochemical modelling of certain neutral pH coal mine and ironstone mine waters in the UK (e.g., Younger 1995, 2002*b*).

In contrast to siderite, ankerite does not commonly occur in any great quantities in the mudstones and sandstones of coal-bearing sequences (Hawkes & Smythe 1937; Smythe & Dunham 1947). Rather, ankerite tends to occur in the form of patchy films lining the surfaces of the many cleats (i.e., micro-joints) that cut the coal itself (Fig. 2b). Cleat is considered to develop relatively late in the post-burial history of coal-forming peats (Thomas 2002), and thus ankerite is typically a product of late-stage diagenesis. As Thomas (2002) has recently pointed out, most sedimentological studies of coal-bearing sequences focus on the depositional fabrics of the clastic sediments rather than the coals themselves. Furthermore, insofar as diagenesis is considered at all in such studies, the focus is on its early stages, and there is thus a tendency to ignore post-cleat mineralization phases. For both of these reasons, ankerite occurrence tends to be underreported in the scientific literature relating to coal-bearing sequences. However, the preferential location of ankerite on coal cleat surfaces means that it is ideally located to participate in rock–water interactions, and it therefore often makes a marked contribution to the hydrochemical evolution of groundwaters flowing through coal-bearing strata, most notably in the elevation of dissolved Mg^{2+} concentrations.

Given that siderite dissolution has no overall neutralization effect on acidic waters, it is important to assess whether the same applies to ankerite. The dissolution of one mole of stoichiometric ankerite consumes two moles of protons:

$$Ca(Mg,Fe)(CO_3)_{2(s)} + 2\,H^+_{(aq)}$$
$$\longrightarrow Ca^{2+} + 0.5\,Mg^{2+} + 0.5\,Fe^{2+}_{(aq)}$$
$$+ 2\,HCO^-_{3(aq)} \quad (5)$$

When the 0.5 moles of Fe^{2+} liberated in (5) are oxidized to form $Fe^{3+}_{(aq)}$, a further 0.5 moles of protons are consumed (according to reaction (3)). The ultimate hydrolysis of this Fe^{3+} to form solid $Fe(OH)_3$ (according to reaction (4)) will yield 1.5 moles of H^+. Thus the overall proton balance, $\sum(H^+)$, can be summarized as: $\sum(H^+) = -2 - 0.5 + 1.5 = -1$. Hence, for every mole of stoichiometric ankerite dissolved, there is a net consumption of one mole of proton acidity. Given that most ankerites contain rather less Fe^{2+} than Mg^{2+} (Smythe & Dunham 1947), in the majority of cases, the net consumption of protons will be greater than one. Thus, in contrast to the case of siderite, ankerite *does* possess net neutralization potential for acidic waters.

Besides these Fe^{2+}-rich carbonate minerals, those carbonate minerals that are most common in the geological record as a whole, that is, calcite and dolomite, tend to be rather underrepresented in most coal-bearing sequences. Calcite cements and discrete limestone beds are extremely rare in the Westphalian (=Pennsylvanian) Coal Measures of northwestern Europe. However, limestone is more common in the stratigraphically equivalent coalfields of northern Spain (Central Asturian Coal Basin), Nova Scotia, and parts of Appalachia, reflecting more widespread marine incursions into depositional basins in those areas during the late Carboniferous. Having noted above the correlation between marine diagenesis and high pyritic S content in coals and shales, if marine influences were strong enough to allow accumulation of limestone this can offer significant buffering of the acidity, which can be released during the extraction and burning of marine-influenced coals.

Quartz is one of the most common minerals in the sedimentary rocks typically associated with coal. A relatively inert mineral, quartz does not significantly affect the acid–alkali balance of pore waters, and to that extent it is 'environmentally neutral'. On the other hand, quartz is the principal agent of silicosis, which is one of the most severe respiratory health problems associated with coal mining.

The mudstones and sandstones associated with coals contain many aluminosilicate minerals of geochemical significance, most notably clays (predominantly kaolinite and illite; Thomas 2002) and muscovite. The various clay minerals are generally rather benign in their environmental geochemistry, acting as powerful sorbents, which serve to limit the mobility of ecotoxic metals at circum-neutral pH (Stumm & Morgan 1996). Along with the other aluminosilicate minerals, many clays will dissolve in the presence of acidic waters, providing valuable buffering capacity in situations where no carbonates are available (Banwart et al. 2002; Younger et al. 2002). It should be noted, however, that dissolution of these minerals in very acidic mine waters is the principal source of dissolved Al, which is strongly ecotoxic. Since dissolved Al is trivalent, it hydrolyses in a manner similar to Fe^{3+} (reaction (4) above):

$$Al^{3+}_{(aq)} + 3H_2O \longrightarrow Al(OH)_{3(s)} + 3H^+_{(aq)} \quad (6)$$

The solid phase resulting from this reaction typically forms a white precipitate, which on account of its low density often forms impressive masses of white foam where the water, from which it is precipitating, is subjected to turbulent eddies. Generally, this white precipitate is X-ray amorphous, although if it is allowed to settle for periods in excess of six months, crystalline forms of $Al(OH)_3$, most notably gibbsite, are

often identified. (Other Al-rich secondary minerals are sulphates, formed by evaporation of mine waters which have undergone reaction (6); see Table 3.)

A typology of the environmental impacts attributable to coal mining

Figure 3 summarizes the principal environmental compartments that can in theory be affected by coal mining, as well as the principal potential environmental stresses that can arise from active and abandoned mine voids on the one hand, and from mine waste management facilities (i.e., spoil heaps and washery finings ponds) on the other. Further details on the mechanisms and impacts of these environmental stresses are given in Table 4 (for coal mine voids) and Table 5 (for coal mine wastes).

To the impacts listed in Tables 4 and 5, one might also have added issues of land use and landscape degradation, which are inherent in certain aspects of mining. For instance, surface mining for coal inevitably results in the temporary, but wholesale, disruption of the pre-existing land use. Where the pre-existing land use was undesirable (e.g., a derelict industrial site) opencast mining and subsequent restoration can result in a substantial improvement in land use for the long term (e.g., Harrison 2000). In some cases,

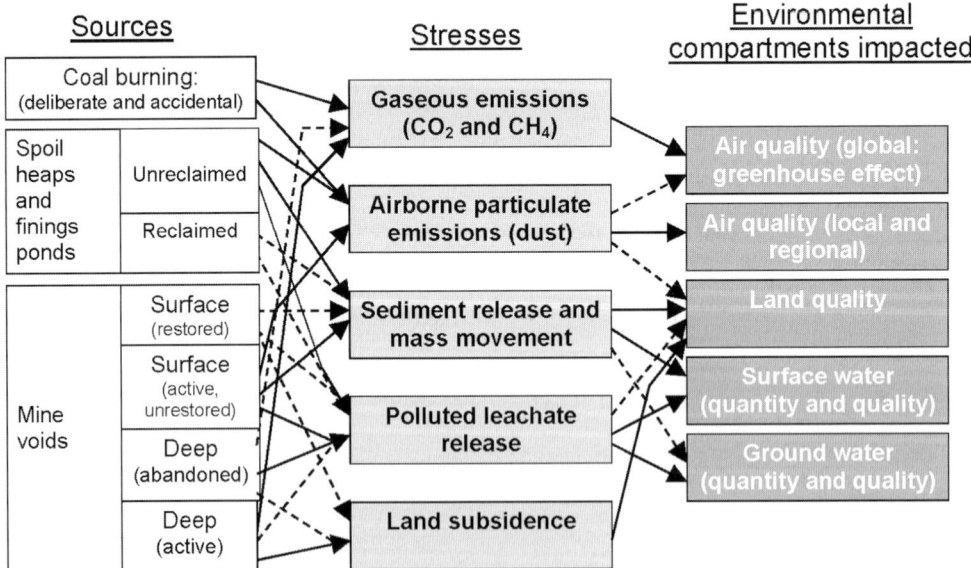

Fig. 3. Flow diagram illustrating the principal environmental compartments affected by, and the principal environmental stresses that can arise from, active and abandoned coal mine voids and coal mine waste deposits.

Table 4. A generalized typology of the external environmental impacts arising from coal mine voids

Hazard category	Hazard source	Pathway/mode of occurrence	Impacts on principal receptors (with references for examples/further information)
Air pollution	Flammable mine gas*	Emitted from adits/shafts and through fractured overlying strata	Explosion risk in confined, inhabited spaces (Robinson 2000)
	Asphyxiating mine gas*	Emissions from adits/shafts	Asphyxiation risk in confined, inhabited spaces (Robinson 2000)
	Combustion fumes*	Emitted from old mine openings and fire-induced ground fissures	Widespread smog, leading to diseases (respiratory, skin and eye), asphyxiation risks in confined spaces above point sources, and to reductions in visibility (affecting transport safety) (Stracher et al. 2002)
	Dust from excavation zones	Wind-borne transport of dust out of the working area	Neighbouring communities (dirt accumulation requiring frequent cleansing); possible links to respiratory diseases (Pless-Mulloli et al. 2000a; Ghose & Majee 2001)
Fire hazards	Open freely-burning fires in underground voids*	Flames and exhaust fumes pass to surface via ground fissures and mine openings	Localized risks to human health from burns (e.g., victims fall into burning zone) and/or asphyxiation by smoke (especially where fumes enter confined spaces); more diffuse problems of smog, leading to diseases (respiratory, skin and eye) and to reductions in visibility (affecting transport safety) (Dunrud 1984; Stracher et al. 2002; Thomas 2002)
	Concealed burning zones within coal seams*	Exhaust fumes pass to surface via ground fissures along and above seam outcrop	
Ground deformation	Subsidence caused by longwall extraction	Usually manifest in relatively broad settlement of ground over areas in excess of 25–50 ha, typically in the form of a shallow closed basin	Damage to property, including fracturing of buildings and disruption of surface drainage, causing flooding of previously dry land; effects develop slowly and thus pose little or no threat to human/animal life (Bell & Genske 2001; Younger et al. 2002)
	Collapse of previously meta-stable old bord-and-pillar workings	Typically results in steep-sided vertical collapse structures (crown holes), approximately circular in plan and ranging from 0.5 m to more than 50 m diameter	Acute surface collapse, known to threaten human/animal life directly on occasions (Bell & Genske 2001); can also disrupt surface infrastructure (utilities, highways) and can cause collapse of buildings. Interception of surface waters also possible (Thomas 2002; Younger et al. 2002)
	Collapse of old shafts (previously capped)		

(continued)

Table 4. *Continued*

Hazard category	Hazard source	Pathway/mode of occurrence	Impacts on principal receptors (with references for examples/ further information)
	Ground fracturing arising from underground seam fires	Results in linear or gently curved open fissures, with apertures typically between several mm and several m	Direct damage to properties, and a hazard to health if wide enough for humans to accidentally fall into burning zone (Dunrud 1984)
Water pollution	Siltation of watercourses	Silt-laden runoff from working areas entering nearby natural water bodies	Benthic smothering (by silt or ochre precipitates) prevents photosynthesis and thus destroys base of local food chain; direct toxicity to fish where Al and Zn high and/or pH low; unsightly appearance of affected streams detracts from amenity value (e.g., Jarvis & Younger 1997; Diamond & Serveiss 2001)
	Acidic and/or ferruginous drainage entering surface watercourses	Water rich in Fe and other metals, sometimes also with low pH (≥ 2.5)	
	Saline water discharge to surface watercourses	Highly mineralized water discharged to streams from actively pumped workings	Salinization of surface waters, which can render water unsuitable for potable supply/irrigation uses (e.g., Helios Rybicka 1996)
	Aquifer pollution	Highly mineralized waters flow from abandoned workings into overlying aquifers	Extensive plumes of contaminated water in overlying aquifers can render water unsuitable for potable supply/irrigation uses (e.g., Younger 2002a)
Water resource depletion	Drawdown of water table due to pumping of active workings	Drying up of natural streams, springs, and/or lowering of the water table in adjoining aquifers	Depletion of water resources in one catchment, often transferring them to another depending where dewatering effluent is disposed (Younger et al. 2002)
	Disruption of natural flow paths (surface and subsurface) due to mine subsidence		Impacts on surface waters often permanent, whereas subsurface impacts often temporary (Younger et al. 2002: Booth 2002)

*All of these hazard sources contribute to global greenhouse gas accumulations (see Golomb & Fay 2004) in addition to more localized impacts listed.

Table 5. *A generalized typology of the external environmental impacts arising from coal mine wastes*

Hazard category	Hazard source	Pathway	Impacts on principal receptors (with references for examples/further information)
Air pollution	Combustion fumes from burning spoil*	Emitted from burning zones within spoil heap	Locally intense smog, leading to diseases (respiratory, skin, and eye), asphyxiation risks in confined spaces above point sources, and to reductions in visibility (affecting transport safety) (Stracher et al. 2002)
	Dust ablation from bare spoil	Wind-borne transport of dust from bare spoil/dry finings pond surfaces	Effects on neighbouring communities: dirt accumulation requiring frequent cleansing (Beynon et al. 2000); possible links to respiratory diseases (Pless-Mulloli et al. 2000b; Ghose & Majee 2001)
Fire hazards	Combustion of spoil heaps/backfill*	Multiple emissions of smoke and dense noxious gases to surrounding air	Localized risks to human health from burns (e.g., victims fall into burning zone) and/or asphyxiation by smoke (especially where fumes enter confined spaces) (Stracher et al. 2002) (See also entry above under 'Air pollution; combustion fumes from burning spoil')
	Ground fracturing arising from underground fires in spoil/backfill	Results in linear or gently curved open fissures, with apertures typically between several mm and several metres	Direct damage to properties, and a hazard to health if wide enough for humans to accidentally fall into burning zone (Stracher et al. 2002; Thomas 2002)
Ground deformation	Differential subsidence of opencast backfill	Typically results in shallow, Steep-sided vertical collapse hollows (≤5 m diameter) above actively eroding pipe structures	Can cause damage to structures (Charles & Skinner 2001) and to crops, farm animals and equipment (where reclaimed spoils used for arable agriculture); also pose a danger of injury to humans and animals that may fall into hollows accidentally; can intercept surface runoff, thus depleting surface water resources (Groenewold & Rehm 1982)
	Landslides of unstable spoil, often triggered by excessive pore water pressures	Mudflows/debris-flows that prograde downhill at high velocities, destroying obstacles in their path and depositing thick mantles of spoil on the surface	These can be extremely devastating where they enter inhabited areas, killing large numbers of people (McLean & Johnes 2002); even where their paths are utterly rural they can damage infrastructure and devastate terrestrial and aquatic flora and fauna (Siddle et al. 1996)
Water pollution	Silt-laden runoff from spoil/dry finings surfaces	Rill-erosion of spoil surface by surface runoff, especially where spoil is low permeability/rainfall is intense	Very rapid runoff typical of many spoil surfaces (Nicolau 2002), both restored and unrestored, and is often accompanied by excessive silt mobilization, which can blanket the benthos of receiving watercourses, contributing to faunal impoverishment (e.g., Garcia-Criado et al. 2002)

(continued)

Table 5. *Continued*

Hazard category	Hazard source	Pathway	Impacts on principal receptors (with references for examples/further information)
	Acidic and/or ferruginous surface runoff from spoil/dry finings	Sporadic events in which highly contaminated runoff is released to adjoining watercourses	Benthic smothering by ochre precipitates prevents photosynthesis and thus destroys base of local food chain; direct toxicity to fish where Al and Zn high and/or pH low; unsightly appearance of affected streams detracts from amenity value (e.g. Jarvis & Younger 1997; Diamond & Serveiss 2001; Younger *et al.* 2002)
	Acidic and/or ferruginous leachates from perched water table systems within spoil heaps/finings ponds	Perennial flows of contaminated water to adjoining watercourses	
	Saline pore water leaching from freshly deposited spoil	Leaching of original pore water, which is often highly saline (even in low-S spoils)	Localized salinization of surface waters, with detrimental effects on the local aquatic fauna (Zielinski *et al.* 2001; Garcia-Criado *et al.* 2002)

*Both of these hazard sources contribute to global greenhouse gas accumulations (see Golomb & Fay 2004) in addition to more localized impacts listed.

however, the restoration of viable soils and re-creation of authentic ecosystems on former opencast sites can prove very difficult (e.g., Backes *et al.* 1986; Rimmer & Younger 1997). More permanent land use change is associated with the deposition of spoil from deep mines onto a pre-existing land surface. In such cases, a new landform is created, which is likely to remain in place for centuries or even millennia. Although some of the environmental impacts associated with unreclaimed spoil are listed in Table 5, broader aesthetic considerations are in most cases a major driver for restorative action.

A cursory reading of Fig. 3 and Tables 4 and 5 reveals that the full array of *possible* environmental impacts attributable to the working of coal mines and the disposal of the wastes associated with them is dauntingly wide. In practice, however, very few of these possible impacts will actually come to pass during the full life-cycle of any one coal mining operation. Even in cases where a very wide range of impacts are detectable, it is only very rarely that more than one or two of these impacts will be sufficiently severe as to merit preventative/remedial measures. In the following section, further detail is provided on the *geochemical* aspects of those impacts listed in Tables 4 and 5, which either arise most frequently in practice, or are especially severe in their impacts on the rare occasions when they do arise.

Geochemical aspects of environmental stresses and impacts

Air pollution hazards

Flammable mine gases. The explosion risks associated with flammable mine gases are infamous as the cause of the most devastating coal mine disasters. The geochemical origins of methane (CH_4) in coal-bearing strata have been extensively studied, both as an aid to developing effective measures to prevent underground explosions and in the context of exploration for coalbed CH_4 resources. In virgin coal beds, CH_4 occurs as what is termed an 'occluded gas', which is held in the unmined coal in the following three ways (in order of decreasing total abundance; Thomas 2002):

(1) Sorbed onto coal surfaces;
(2) Present as free pockets of gas within pores (especially fracture pores associated with cleat);
(3) Dissolved in water present within the pore system.

Mining relieves the *in situ* stress on the coal, radically changing the pore pressure regime. This allows rapid escape of any CH_4 present in free pockets, followed by a more gradual release of CH_4 by desorption from the coal and exsolution from pore water, which occurs as the pore gases begin to equilibrate with the mine atmosphere. As CH_4 is lighter than air, where it moves en masse into the mine voids it tends to rise to the ceiling of any workings, where it becomes trapped in any domed areas. Where CH_4 is present in very high concentrations ($\gg 15\%$ of the total air volume) it will not spontaneously ignite. However, where CH_4 comprises between 5 and 15% of the total volume of a mixture with air, a substantial explosion risk exists, with maximum explosivity occurring at a CH_4 content of 9.5 vol% (Coulshed 1951).

Asphyxiating mine gases. As was noted above, many of the minerals found in coal mine environments oxidize briskly when in contact with atmospheric O_2:

(1) Coal sorbs molecular O_2 from the mine atmosphere, partially oxidizing and releasing CO_2 gas.
(2) The weathering of siderite (reaction (2)) and ankerite (reaction (5)) in the presence of O_2 can sometimes lead to dissociation of dissolved HCO_3^-, resulting in the release of CO_2 to the atmosphere; a significant proportion of the atmospheric O_2 can also be trapped in the solid form as secondary ferric (hydr)oxide precipitates (reactions (3) and (4)).
(3) Pyrite oxidation transfers atmospheric O_2 to the solid phase (as hydroxysulphate minerals; Table 3) and or the dissolved phase (as SO_4^{2-}; reaction (1)), thereby reducing the atmospheric concentration of the gas.

Furthermore, many mine waters contain CO_2 well in excess of concentrations explicable by equilibration with the atmosphere. Although some of this CO_2 may be derived from oxidation of C by sulphate-reducing bacteria (e.g., Younger 1995), most of it is likely due to closed-system dissolution of ankerite, siderite, and other carbonate minerals. When these CO_2-charged waters encounter the atmosphere (in the mine workings or at the surface), they rapidly de-gas, releasing large quantities of CO_2 into the air. If the supply of O_2 in the mine atmosphere is not replenished by ventilation at a rate equal to or greater than that at which it is consumed by reaction with such minerals, O_2-deficient air quickly develops. It is a

common experience in the coal mining industry that O_2-deficient air can be expected to be present wherever underground mine voids are left unventilated. This includes confined spaces adjoining well-ventilated roadways: for instance, the author has twice encountered dangerously low levels of O_2, and toxically elevated levels of CO_2, while attempting to sample mine water feeders in short headings only a few metres off the main ventilation circuit. Hence it is imperative that underground coal mine voids should only be entered when they are known to be securely ventilated. Gas monitoring boreholes in colliery spoil heaps have also revealed that O_2-deficient air is commonly encountered at depths of only a few metres below ground surface, especially where the spoil contains much fine-grained sediment.

Carbon dioxide is heavier than air, which results in it blanketing the floor within mine voids. This tendency needs to be taken into account in planning mine safety, as breathable air above a mantle of O_2 deficient air can give a false sense of security where a sloping mine roadway is being travelled. Blankets of O_2-deficient air can also hug the ground surface around abandoned mine shafts and adit portals, and this should be borne in mind when planning fieldwork in the vicinity of such features.

The transfer of O_2 to dissolved SO_4^{2-} by means of the pyrite oxidation reaction is frequently only the first step of a series of further reactions, which culminate in the bacterially mediated reduction of SO_4^{2-} to sulphide, which typically occurs in anoxic, ponded water within mine systems (e.g., Younger 1995). If such water is disturbed by mining, it will tend to de-gas rapidly, releasing potentially lethal concentrations of hydrogen sulphide (H_2S) gas into the mine atmosphere. The first recorded fatality from an event of this type occurred in 1835, when a miner using a hand-drill to probe flooded old workings about 110 m below ground level in Hartley Colliery (Northumberland, UK) was killed when H_2S fumes rushed from the drill-hole into the confined space where he was working (Hair 1844, p. 40). More recently, the objectionable odour of H_2S has been noted repeatedly in the vicinity of certain mine water discharges (most notably the Six Bells Colliery site in South Wales, UK), giving rise to complaints from local residents, which necessitated the installation of ventilation and chemical oxidation facilities, so that H_2S is quickly converted to odourless SO_2 gas, which is then simply dispersed in the air.

Ignition (accidental or deliberate) and/or spontaneous combustion of coals clearly leads to extreme demands for O_2 from the mine atmosphere. Although CO_2 is commonly released from open, freely burning fires that enjoy continual access to O_2, where the supply of O_2 becomes limited a far more dangerous gaseous product is frequently formed: carbon monoxide (CO). This gas is extremely toxic (it can be lethal to humans following exposures of only a few minutes at atmospheric concentrations as low as 0.5 vol%), and is the principal cause for concern in relation to human safety in the exhaust streams emanating from underground coal fires. Migration of CO from hidden zones of burning in abandoned workings into nearby residential properties has claimed the lives of a number of unsuspecting victims (Robinson 2000).

Airborne dust from excavation zones and bare spoil/backfill surfaces. Lung diseases including silicosis (arising from adverse reactions to inhaled quartz particles), pneumoconiosis (lesions arising from inhaled coal particles), pulmonary fibrosis, and related conditions have long been well-known diseases of miners, and have given rise to extensive regulations and practices intended to reduce their incidence, particularly amongst the workforce of underground mines (e.g., Seixas et al. 1992; Buchanan & Brenkley 1994; Finkelman et al. 2002). Workers in opencast coal mines are far less likely to develop these illnesses than their underground counterparts (Love et al. 1997), although the risks are still non-negligible for workers employed in the most dusty opencast jobs (e.g., drilling, excavation, machine loading) (Love et al. 1997; Ghose & Majee 2002).

As surface mining for coal expanded in Europe and North America during the second half of the 20th century, there was a corresponding growth in anxiety over possible detrimental effects of airborne dusts from such operations on the health of residents in nearby towns and villages (e.g., Temple & Sykes 1992; Munro & Crompton 1999). Such concerns continue to be expressed as opencast coal mining expands in developing countries (e.g., Ghose & Majee 2001). The key issue is the risk of exposure to respirable dusts, especially particulate matter less than 10 μm in diameter (PM_{10}) (see Pless-Mulloli et al. 2000a; King et al. 2000; Borm 2002). Detailed sampling of PM_{10} around a number of opencast coal sites around the world (Pless-Mulloli et al. 2000a), coupled with careful statistical analyses of public health records for communities located near to and far away from such sites (e.g., Pless-Mulloli et al. 2000a, b, 2001; Howel et al. 2001) have recently shed considerable light on the magnitude of such risks. Although it was found that residential areas fringing opencast coal sites do receive greater mean concentrations of PM_{10} than communities located far away from such sites, mineralogical studies have shown that most of the excess PM_{10} is attributable to clay minerals (e.g., Pless-Mulloli et al. 2000b; Jones et al. 2002), which are considerably less likely to cause lung disease than quartz. Furthermore, despite having far lower specific gravities (G_s) than quartz or shale, coal particles ($G_s = 1.3$, compared with G_s values around 2.65 for quartz and shale) were not found to form a significant fraction of the captured PM_{10} in the areas surrounding opencast sites (Jones et al. 2002). This finding supports the general supposition that the processes of overburden stripping and mining generate a far greater proportion of airborne dust than does the winnowing of coal stockpiles by the wind.

In terms of the number of consultations made with local physicians, the respiratory health of children living in communities close to opencast sites was very similar to that of children living in communities distant from such sites (Pless-Mulloli et al. 2001; Howel et al. 2001). Nevertheless, in four-fifths of the cases studied, there was a small but statistically significant increase in consultations for respiratory, skin and eye conditions in those living close to opencast sites, albeit the absolute numbers of consultations were still small. Overall, the evidence gathered to date supports the contention that such increments in exposure to PM_{10} as may arise in local communities on account of opencast coal extraction are highly unlikely to have any detectable effects on health (Pless-Mulloli et al. 2000a, b, 2001; Howel et al. 2001).

It is noteworthy that regulatory pressures are beginning to mount to consider controlling particles smaller than 2.5 μm. This is because many particles around 10 μm in diameter can be coughed out by humans, whereas those smaller than 2.5 μm tend to stay in the lungs, and therefore have the potential to cause most harm. No specific studies of $PM_{2.5}$ distributions in the vicinity of opencast mines have yet been published, but it is likely that such studies will need to be undertaken in the near future in Europe, where tightening of regulatory controls is already advancing towards this more radical benchmark.

Fire hazards

Uncontrolled fires in coal seams are a significant hazard in coalfields world wide, and are particularly prevalent at the present time in China, India, Indonesia, South Africa, Australia, and parts of the USA (eastern Pennsylvania, Colorado, Wyoming) (Prakash *et al.* 2001). In China, uncontrolled burning of coal seams is estimated to consume around 200 Mt of coal per annum (Anon 2003), a figure which is of the same order of magnitude as the total annual coal consumption of many industrialized countries. Quite apart from the obvious mortal dangers posed by the fire itself, these uncontrolled blazes also cause land subsidence and give rise to major smoke hazards. They also represent a major waste of important natural energy resources and contribute to the global greenhouse gas budget without rendering any useful service to humanity in the process. Many coal fires burn for decades before they finally exhaust their fuel supplies, and they can develop over such vast areas that they can only really be quantified using remote sensing techniques (e.g., Prakash *et al.* 2001).

The ignition of uncontrolled fires in coal mining areas can occur in a number of ways (Stracher *et al.* 2002):

(1) by spontaneous combustion of coal, due to the exothermic process, whereby O_2 is sorbed to the surface of freshly exposed coal (discussed above);

(2) deliberate or accidental exposure to naked flames, including those associated with explosions of CH_4 and/or coal dust, and those associated with forest fires;

(3) the generation of sparks where drilling or cutting equipment strikes a nodule or a clast of hard, siliceous rock;

(4) by means of lightning strikes, either directly on seam outcrops or by the transmission of electrical charges through overburden to pockets of explosive CH_4–air mixtures 200 m or more below the ground surface (e.g. Novak & Fisher 2001).

Propagation of fires, whether within *in situ* coal sequences, underground workings, or spoil heaps, depends on a continued supply of O_2 to the burning zone: where the O_2 supply is abundant, the principal gaseous product will be CO_2, whereas the far more hazardous CO is generated where the fire burns in an atmosphere containing limited O_2 (e.g., Robinson 2000). Other common gaseous products include the oxides and dioxides of N and S (typically represented as NO_x and SO_x), unoxidized N_2 gas, and (where the fire is burning under the same low-O_2 conditions that give rise to CO gas) free gaseous S_2 (e.g., Stracher 1995; Stracher *et al.* 2002). In addition to gaseous products, smoke arising from the free burning of coal also includes a range of fine-grained particles of C and various trace metals (e.g., Linak & Wendt 1994).

Exhaust streams emerging from underground burning zones in coal and coal wastes have frequently been observed to precipitate mineral condensates at their points of surface emergence. The process is essentially one of rapid cooling of gaseous mixtures leading to redox transformations (Stracher 1995). The earliest historic account of this phenomenon, relating to a seam fire that commenced accidentally in about 1643, was noted earlier in this text. During the 1670s, with the fire still burning, the earliest scientific investigations of these condensates were made. They were all white in colour, and were identified, using what were then state-of-the-art chemical analysis techniques, to be 'flour of sulphur and sal ammoniac' (Galloway 1898, p. 163). The most recent studies of the same phenomenon concur with these early findings (Stracher *et al.* 2002), noting the accumulation of both orthorhombic S and salammoniac (NH_4Cl) around seam fire fumaroles. A wide range of 'trace' condensates have also been detected using modern analytical techniques, including elemental selenium, various compounds of Se and As, galena (PbS), gypsum ($CaSO_4 \cdot 2H_2O$), and mullite ($Al_6Si_2O_{13}$) (Stracher *et al.* 2002). Where previously burned zones are later flooded, the dissolution of these condensates can give rise to elevated dissolved concentrations of ammonia and As (Younger, unpublished data).

Ground deformation

Land subsidence is one of the most widespread negative impacts of underground coal mining, so much so that its management and mitigation has prompted the enactment of laws and binding codes of practice in many countries, for instance in the UK, Germany, the USA, and South Africa (NCB 1975; Dunrud 1984; Bell & Genske 2001). The causative processes are generally well understood, and are primarily physical in nature, and thus outside the scope of this paper. However, geochemical processes do make a contribution to *some* aspects of subsidence. For instance, it is well known that ground deformation processes occur far more readily in the presence of water than under dry conditions (e.g., Siddle *et al.* 1996; Younger *et al.* 2002). In large measure, the explanation for this phenomenon is physical, relating to grain displacements wrought by the action of pressurized pore waters and/or to the lubrication of failure planes by excess water (Siddle *et al.*

1996; McLean & Johnes 2000). Nevertheless, certain geochemical processes exacerbate these conditions, most notably mineral ion exchange and oxidative weathering. Freshly excavated mudstones and sandstones in many coal-bearing sequences contain waters in which the major cation is Na^+ (reflecting their deep subsurface origin, in the presence of deep basinal brines or highly evolved meteoric waters of the Na–HCO_3 facies). Reflecting this, surface sorption sites in many coal measures rocks are occupied by Na^+. Where these rocks are subjected to leaching by shallow-sourced groundwaters rich in Ca^{2+} and/or Mg^{2+} (generally derived from weathering of calcite and/or ankerite), ion exchange occurs, with these larger divalent cations occupying the surface sorption sites and displacing the monovalent Na ions into solution. The valency differences between Na^+ and Ca^{2+}/Mg^{2+} result in substantial reductions in the strength of particle–particle adhesions in mudstones and clay-rich sandstones (Taylor & Spears 1970; Taylor 1973, 1975; Bell 1996). Where the rock in question contains expansive, mixed-layer clays, this exchange process can even extend to interlayer bonds within the clay crystal structure, leading to intraparticle expansion. These processes serve to weaken the host rocks and promote their disintegration to fine-grained sediments. This tendency is further exacerbated where pyrite is present within waste rock or drained coal washery finings: the liberation of protons in the pyrite oxidation reaction (1) can lead to the destruction of natural grain–grain cements, resulting in the disintegration of sedimentary clasts. Extreme levels of acidity can lead to direct dissolution of clay minerals, a process that liberates dissolved Al in large quantities, which later hydrolyses according to reaction (4), giving rise to diagnostic white foams in spoil leachates. Additional processes such as 'air breakage' contribute still further to the breakdown of originally competent rock clasts (Badger et al. 1956; Bell 1996). By means of all of these processes, a spoil heap that was originally composed of coarse fragments of relatively competent rock (with correspondingly large pores allowing free drainage), can over a period of months to years be gradually weathered to a water-retaining mass of disaggregated mud and sand (Bell 1996). Clearly, the latter is far weaker than the former in geotechnical terms. Thus slopes that were stable at the time of spoil deposition may become unstable once weathering has become sufficiently pervasive (Groenewold & Rehm 1982). Rapid slope failures in coal spoil heaps have been recorded on numerous occasions (e.g. Siddle et al.

1996). In the worst instance, these have caused substantial loss of human life, most notably in the Aberfan disaster in South Wales in 1966, when a substantial part of a large, steep-sided colliery spoil heap failed catastrophically, and slid down a mountainside into the adjoining village, engulfing several homes and part of a primary school (McLean & Johnes 2000). A total of 144 people died in the Aberfan disaster, of whom 116 were schoolchildren, buried alive in their classrooms along with their teachers. Recent psychiatric research has revealed the persistence of post-traumatic stress syndrome in survivors of this incident, nearly forty years later (Morgan et al. 2003).

In the majority of cases, of course, ground deformation on spoil heaps does not occur on such a dramatic scale. Although it is clearly impossible to avoid some settlement of spoil following initial deposition (see Younger et al. 2002, p. 49), most spoil settles relatively evenly, with total settlement rarely exceeding a few percent of the total spoil thickness. However, as water in spoil heaps tends to follow highly preferential pathways associated with rubbly zones, significant erosion can occur even in very old spoil heaps, leading to the development of soil pipes, which collapse to form sinkholes (e.g., Groenewold & Rehm 1982; Younger et al. 2002). Clearly the disintegration of spoil by chemically assisted weathering contributes to the availability of loose sediment, which is vulnerable to piping.

Similarly within underground workings, seat-earths that formed relatively sound floors to roadways when dry may become dangerously weak when wetted. Cyclic wetting and drying of clay-rich seat-earths exploits the listric surfaces naturally present in these rocks, with complete disintegration occurring after only a few cycles (Bell 1996). Disintegration will occur even more rapidly in cases where the seat-earth is pyritic (Younger 2002c), such that acid-generating salts have been accumulating in and on the seat-earth during the period prior to flooding (as shown in Fig. 1). With the seat-earth thus weakened, pillars of coal that were left in place to support the roof can punch into the floor strata, leading to void migration and, ultimately, surface subsidence (e.g., Younger 2002c).

The increase in volume of expansive clays during wetting does not only lead to collapse and subsidence. A major geodesic study of the abandoned Limburg Coalfield in the southern Netherlands revealed that wetting of the strata as the deep coal mines flooded actually led to a general raising of the ground surface by several centimetres (Bekendam & Pottgens 1995). This has been ascribed to the wetting of expansive clays in the sequence.

Water pollution

Drainage is the most important pathway for environmental contamination associated with mining. Accordingly, a very extensive literature exists on this topic, and this will not be rehearsed in great detail here. Pyrite oxidation processes have been reviewed in detail by Evangelou (1995), with a significant recent update being provided by Rimstidt & Vaughan (2003). Broader discussions of the evolution of the acidity–alkalinity balance in coal mine settings are provided by Younger (1998) and Banwart *et al.* (2002). Similar processes also occur during the operation of non-conventional coal exploitation techniques, most notably coalbed methane production, giving rise to similarly polluted waters (Clarke 1996). All geochemical processes are mediated by transport phenomena, which are dependant on the hydrogeology of mine water pollution source-zones (see Younger & Robins 2002) and by the limnology and hydrodynamics of receiving watercourses. Some of the more impressive examples of closely juxtaposed source and receiving zones are the acidic pit lakes, which dot the landscape in areas such as Saxony (Germany) (e.g., Geller *et al.* 1998), where high-S coals have been surface-mined in the absence of sufficient overburden to allow for restoration of the mine void to a terrestrial landform. The complex interactions between hydrogeology and geochemical reactions account for the bewildering variety of water types found in mining districts (e.g., Banks *et al.* 1997). Besides pyrite oxidation processes, the gradual flushing of saline pore waters from recently disturbed rock, which was previously of low permeability, is also important in many deep mines and spoils deposited from these mines (e.g., Zielinski *et al.* 2001). Although much of the mine water literature tends to focus on dissolved load contaminants, it is important to bear in mind that erosion of bare spoil can give rise to high sediment loadings in streams draining active and restored coal mine sites (Tiwary 2001). Rejuvenated rill erosion can even arise on steep slopes within otherwise revegetated mine sites (e.g., Nicolau 2002). Finally, activities inherent in the working of mines can give rise to non-mining specific forms of pollution, such as spillages of oils (e.g., Tiwary 2001). Without adequate mitigation, the cocktail of dissolved and suspended pollutants can take a heavy toll on aquatic biota. Recent advances in elucidating the geochemical processes that

govern the ecotoxicity of polluted mine waters are described by Jarvis & Younger (1997), Kelly (1999), Diamond & Serveiss (2001), and Garcia-Criado *et al.* (2002).

Distilling the insights from the literature cited above, together with further specific results of research undertaken in Europe over the last decade, it is now possible to outline a state-of-the-art conceptual model of water pollution issues associated with coal mining. At least in the major coalfields of Europe, acidic coal mine drainage accounts for a relatively modest proportion of the total inventory of polluted mine waters. Acidic drainage is predominantly associated with spoil heaps containing pyritic shales, and with shallow, freely draining drift mines in high-S strata. While some abandoned, flooded deep mines *will* emit acidic waters in the initial period following the completion of flooding, over a period of time the acidity will abate until some long-term 'asymptotic' level of acidity release is reached (Glover 1983). This period of time, referred to as 'the first flush', typically takes about four times as long as the time it took for the mine to flood in the first place (Younger 2000). Long-abandoned, flooded deep mines still tend to release large volumes of water, which, although generally of circum-neutral pH, are often still significantly contaminated with Fe^{2+} and Mn^{2+} (e.g., Wood *et al.* 1999). Because the oxidation and hydrolysis of Fe^{2+} leads to precipitation of ochre deposits that thoroughly cloak stream beds and prevent light penetration to the benthos, just as much ecological damage is caused by mine waters that are of neutral pH, and still display elevated Fe concentrations, as is caused by the more acidic discharges (Jarvis & Younger 1997).

Water resource depletion

Being a wholly physical process, this impact will not be discussed in detail in this geochemically oriented review. The interested reader is referred to Younger *et al.* (2002).

Geochemical mitigation measures

Air pollution hazards

Flammable and asphyxiating mine gases. For both flammable and asphyxiating mine gases, the first step in mitigating hazards is the recognition that gases behave in a reasonably reliable manner. Fortunately, all of the hazardous mine gases will mix with and disperse irreversibly into ordinary air if they are subjected to sufficient mixing. The release of hazardous gases into the mine atmosphere, and thence to the surface atmosphere, occurs most briskly when atmospheric pressure is low; consequently mine gas hazards tend to be less marked during sunny weather. In the absence of vigorous mixing with ambient air, flammable mine gases tend to accumulate against the ceiling in underground workings, whereas asphyxiating gases tend to blanket the floor.

Bearing these facts in mind, it is clear that the key to minimizing mine gas hazards is the operation of adequate ventilation systems, which both introduce enough air to the workings to dilute the hazardous gases and pass through the workings with sufficient turbulence to entrain and mix with the mine gases. Although it is entirely possible for natural ventilation to suffice, particularly where mines are located in mountainous country and are well connected to the open atmosphere at a number of points of different elevation, in many cases adequate ventilation will demand the use of electrical fans. Such applications are commonplace in deep mining; they are also used strategically in areas with abandoned mine workings, to prevent the dangerous build-up of hazardous gases in confined spaces (Robinson 2000). In cases where CH_4 is the principal mine gas of concern, it may be cost-effective to safely tap it using boreholes, and lead it to the surface where it may be safely used to fire electricity-generating plant. Such combustion of CH_4, which yields CO_2, is preferable to merely venting it to the atmosphere. This is because CH_4 is a far more potent 'greenhouse gas' than CO_2. CH_4 has been extensively intercepted during the working of mines in France, Germany, and the UK. Although CH_4 release rarely continues at a high rate in the absence of ongoing fracturing of fresh strata by continued mining, the post-closure CH_4 yield of some collieries remains sufficiently high to warrant continued capture and use of the gas. This approach to CH_4 use has recently been labelled 'coal mine methane' (CMM), to distinguish it from coal bed methane (CBM) production from unmined seams. The production of CMM is now proceeding significantly in the Ruhr Basin of Germany. However, expansion of the CMM market in the UK has been stalled by governmental policies that classify CMM as an ordinary fossil fuel, and thus subject it to unfavourable tarrifs aimed at discouraging CO_2 emissions. The result of this misguided policy is that CH_4 continues to vent to the atmosphere in any case, wasting the energy it contains and causing greater environmental damage than it would in the form of CO_2.

In emergency situations, where a mine ventilation system is compromised and underground fires are releasing dangerous quantities of CO into the mine atmosphere, chemical oxidation of the CO to less-harmful CO_2 is the usual practice. This is achieved using a 'self-rescuer' apparatus, which is essentially a mask, into which the mine atmosphere is drawn by the respiration of the wearer. As mine atmosphere enters the 'self rescuer' it passes over a $CuMn_2O_4$ catalyst (known as 'Hopcalite'; see Vepřek et al. 1986), which expedites the oxidation of CO to CO_2. Normally, self-rescuers contain sufficient Hopcalite to allow for 60 to 90 minutes of use, which will generally be sufficient time to allow the wearer to escape from the mine, or at least to reach a secure 'fresh air base' where the ventilation system continues to operate.

Airborne dust from excavation zones and bare spoil/backfill surfaces. Perhaps the most effective mitigation measure for airborne dust is the enactment of well-designed planning controls, such that the location and timing of major excavation activities are designed with nuisance avoidance in mind. In some cases this may mean limiting the 'take' of a surface mining operation so that it does not impinge within several hundred metres of sensitive residential areas. A variety of physical barriers may also be used to help minimize 'fugitive' (i.e., off-site) airborne dust emissions (Ghose & Majee 2001). The most common barrier is a bund (known as a 'battle bank'), formed by stacking stored soil and overburden around the working area; this deflects air flow and thus helps to cut down on fugitive emissions. Baffle banks can also help minimize the visual intrusiveness of the working areas, and additionally help minimize noise problems associated with mining operations. Trees and bushes can be planted on both the baffle bank and around all site approach roads, where they serve to intercept dust from the moving air and store it on leaf surfaces until the next period of rain washes the dust back into the site where it belongs.

These precautions may not be entirely sufficient during dry periods. The spreading of water on haul roads and working floors is the principal operational practice aimed at minimizing airborne dust release. Mathematical models of dust release from opencast mine haul roads have recently been developed, which allow forward planning of water-spreading frequencies for summer and winter conditions in opencast coal mines in specific climatic settings (Thompson & Visser 2002). Under the driest of conditions, it may be necessary for water spreading vehicles to spray a given section of haul road every twenty minutes.

Multifactor models have been developed recently to support the estimation of emission rates from opencast coal mining sites (Chaulya *et al.* 2002). The modelling approach is based on empirical relationships derived from airborne particulate sampling in the vicinity of individual activities typical of opencast coal mines (e.g., drilling, overburden loading, tipping) on seven study sites. Verification of the model was undertaken by predicting and measuring airborne emissions at another opencast mine, which had not been studied during model derivation. The agreement between predicted and observed values fell in the range 85.6–99.9%, which is considered to be a robust vindication of this activity-level predictive approach.

For abandoned opencast sites, and for large spoil heaps associated with underground coal mines, landscaping and revegetation are clearly the key to long-term prevention of dust entrainment by the wind (see below).

Fire hazards

As uncontrolled fires in coal seams and coal wastes cannot occur in the absence of O_2, one of the most obvious approaches to extinguishing such fires is to exclude O_2 from the burning zone. However, there are very few circumstances in which the precise loci of O_2 ingress are known, and even fewer where 100% effective seals can be constructed. Instances are known where preferential compaction of 'hot spots' (i.e., localized pockets of burning) within spoil heaps has resulted in the extinction of fires, presumably due to constriction of O_2 ingress pathways (Bell 1996). In other cases it has been found necessary to augment preferential compaction with the injection of an enclosing grout curtain to cut off deep-seated lateral pathways for O_2 ingress. Another approach that has been trialled (although without success to the best knowledge of the author) is to cap the spoil with dolomite or limestone fragments, on the assumption that the CO_2 released by these minerals when they are heated might replace O_2 in the gaseous stream entering the heap (M. Foster, formerly of British Coal, personal communication, 1993). The possibility that this CO_2 might be thus converted to the more dangerous CO appears not to have been considered by advocates of this approach. In some cases, the only sure way of removing a risk of spoil combustion is to systematically excavate all combustible sections of the heap one after another, either smothering them

with clay in a controlled manner in a new burial pit (e.g., Poremba 2003), or else allowing them to complete their burning more rapidly than they would *in situ*, before finally re-burying the characteristically red burnt spoil after combustion has finally ceased.

Clearly, these spoil heap management options are not readily applicable to underground fires in old mines and *in situ* coal seams, and alternative approaches are therefore warranted. Perhaps the simplest cases occur where the burning zone lies above a water table that has been depressed by mine dewatering: a temporary cessation of mine water pumping can allow water levels to rise and extinguish the fire. Where this option is not available, alternative 'drowning' approaches (such as infusion of liquid N) may be feasible, but are unlikely to be economically viable in the cases of all but a handful of small fires. It should be noted that simply spraying water over the outcrop zone of a burning seam is unlikely to be helpful, and may even make matters worse, given that partial wetting of freshly exposed coal surfaces can actually catalyse the exothermic O_2 sorption process responsible for many cases of spontaneous combustion (Thomas 2002). Containment of coal seam fires in active mines has been tackled for many years by the construction of air-tight seals in major roadways (e.g., Coulshed 1951). Of course, O_2 can find other pathways to the fire via fissures in the rockmass, so such measures are generally conceived as 'containment measures' rather than as fire extinction. In abandoned mine workings, injection of special grout mixtures can be used to create air-tight, low-combustibility zones around in-seam fires (Anon 2003). Where the geology allows, one of the most sure ways to combat an uncontrolled fire is to deny it fuel, which means rapid excavation of a 'fire break' within the coal seam at some position 'down-burn' over the currently burning zone. With no coal available at this point, the fire will not be able to progress further once it finally arrives at that position. Anti-inflammatory substances can also be injected into the fire break where appropriate. However, in many situations seams will be too deep for the application of surface mining approaches to fire break installation, and too hazardous to re-enter using subsurface mining methods. In such cases, there are few feasible strategies other than patience. In some parts of the world, little more can be done than to monitor and wait until fires become extinguished naturally.

Prevention is, as ever, better than cure, and thus management approaches that seek to prevent spontaneous combustion in the first place are far to be preferred to reactive fire-fighting once a blaze has already started (Coulshed 1951). Working from the known fact that different coals vary greatly in their susceptibility to spontaneous combustion, Singh et al. (2002) devised and tested a comprehensive spontaneous combustion management plan for application to collieries in South Queensland, Australia. This involves classifying all local coals according to their propensity for spontaneous combustion, and then planning the mine layout and ventilation system in such a manner as to allow for both ready detection of the early signs of a 'heating' (i.e., incipient burning), which is indicated by measurable increases in the CO content of the mine atmosphere, and for the selective sealing-off of portions of the mine in which a heating has developed into an uncontrolled fire.

Ground deformation

Prevention of ground deformation in spoil can largely be achieved simply by enforcing a robust water management strategy for the heap. Capping of spoil with compacted clay (or even artificial liners) can be an effective means of reducing infiltration to the body of the heap, although it will result in a localized zone with a high potential for generating surface runoff (and hence flooding). Installation of drainage pipe networks within spoil heaps can limit the accumulation of water within the spoil, and thus help to minimize both weathering rates and adversely high pore water pressures. However, as such drainage is rarely built into heaps during deposition, it can be a costly remedial measure. Reinforcement of critical failure-prone slopes, by the insertion of geotextiles and/or the construction of retaining walls, may be a less costly option where it is desirable to avoid wholesale reworking of the spoil.

In underground coal mine voids, the best strategies for avoidance of ground deformation associated with geochemical changes is to design for rapid flooding of voids, avoiding extended periods in which water can be flowing in open channel mode through zones with sensitive seat-earths, and also minimizing the occurrence of repeated wetting–drying cycles. Clearly, at the outer limit of a saturated zone in old coal workings, at the so-called 'tail of the water' within a system of mine workings, there will always be a 'tidal zone' in which the water level fluctuates at least seasonally. In the best interests of maintaining long-term ground stability, there are good reasons for controlling water levels such that the tail of the water lies in a zone where the seat-earths or other floor strata are least susceptible to loss of strength during weathering.

Water pollution

Siltation of watercourses due to sediment transport from areas of bare spoil can be addressed actively by the use of silt traps, which are essentially sedimentation basins sized for hydraulic retention times on the order of 3–4 h (Younger et al. 2002. pp. 286–287). Passive prevention of sediment release is usually approached by the establishment of a dense vegetation cover, in order to 'knit' the soil together and make it less susceptible to erosion by overland flow (Fig. 4). Although such approaches are often successful in areas with temperate climates, the high-intensity/short-duration convective storms typical of tropical and Mediterranean climates often overwhelm the stabilizing capacity of vegetation (Fig. 4a), so that sediment export continues to occur long after re-vegetation has been completed (e.g., Nicolau 2002). Avoidance of sediment release in such settings can only be assured by re-grading slopes to shallow final gradients, and incorporating ponds and lakes into the surface runoff system.

Mitigation measures to minimize the discharge of saline mine waters to freshwater surface watercourses range from the application of desalination technologies (such as reverse osmosis) to the implementation of 'flow balancing' strategies, in which water is temporarily stored (either within disused mine voids or in purpose-built reservoirs) until stormy conditions give rise to high dilution capacities in the surface watercourses into which the saline waters are then released (e.g., Gandy & Younger 2002).

The amelioration of acid waters in pit lakes developed in areas of former lignite mining is a significant technical challenge, for the in situ treatment of the entire water body present in one of these lakes would not in itself make any difference to the long-term prognosis, given that acidified groundwaters will continue to make their way into the lake, mixing with and displacing the previously treated waters. The search for more sustainable approaches is currently focusing on the stimulation of what will hopefully develop into self-sustaining biological systems within these lakes, harnessing processes such as anaerobic microbial respiration, which have the capacity to consume acidity. This is a complex and challenging area of research; for further details of some of the key issues, the interested reader is referred to Geller et al. (1998).

Fig. 4. Two contrasting spoil heaps in the Gardanne Coal Basin, Provence, France: (**a**) Unrestored heap near Domainé de la Salle, showing largely bare spoil with characteristic development of scree and rills (cf. Nicolau 2002), with the only vegetation being a few self-seeding Mediterranean pines; (**b**) Restored, thoroughly afforested spoil heap at Meyreuil, barely recognizable as a former mine waste disposal site.

The mitigation of surface water pollution by the discharge of acidic and/or ferruginous mine waters typically requires the use of treatment technologies. These have been extensively documented recently, and extensive guidelines for their design and implementation have been established (Younger *et al.* 2002; PIRAMID Consortium 2003); hence only a brief summary is given here. There are two principal approaches to mine water treatment, 'active' and 'passive', which are best viewed as complementary rather than competing.

'Active treatment' refers to the application of conventional wastewater treatment processes, which typically require ongoing inputs of electrical power and/or chemical reagents in a closely controlled process (which usually demands frequent operator attention). The classic approach to active treatment of acidic and/or ferruginous mine drainage involves three steps:

(1) *Oxidation* (usually by means of a simple cascade open to the atmosphere), which helps to convert soluble Fe^{2+} to far less soluble Fe^{3+}, as well as allowing pH to rise by venting excess CO_2 (where present), until equilibrium with the atmospheric CO_2 content is attained.

(2) *Dosing with alkali* (usually 'hydrated lime' ($Ca(OH)_2$), and less frequently caustic soda (NaOH)), which serve both to raise the pH (thus lowering the solubility of most problematic metals) and to supply OH^- ions for the rapid precipitation of metal hydroxide solids.

(3) *Accelerated sedimentation*, usually by use of a clarifier or lamellar plate thickener,

often aided by the addition of flocculants and/or coagulants. Current practice in the industry favours the re-circulation of an aliquot of $Fe(OH)_3$ sludge into the influent of the sedimentation unit, which has been shown to promote the attainment of higher densities in the final sludge. This practice is called the 'high density sludge' process and it typically yields sludges with 25–30 vol% solids, as opposed to the 5 vol% solids contents typically obtained without re-circulation.

The term 'passive treatment system' in the context of mine waters is defined by the European Commission's PIRAMID project as follows: 'a water treatment system that utilises naturally available energy sources such as topographical gradient, microbial metabolic energy, photosynthesis and chemical energy, and requires regular but infrequent maintenance to operate successfully over its design life' (PIRAMID Consortium 2003). The working definition of 'infrequent' in this context is currently around six-monthly. Types of passive system currently in use include (in order of decreasing frequency): (1) aerobic, surface flow wetlands (reed-beds); (2) compost wetlands with significant surface flow; (3) mixed compost/limestone systems, with predominantly subsurface flow (so-called reducing and alkalinity producing systems); (4) subsurface reactive barriers treating acidic, metalliferous ground waters; (5) closed-system limestone dissolution systems for Zn removal from alkaline waters; and (6) roughing filters for the aerobic treatment of ferruginous mine waters where there is no room for a surface wetland. Each of the above technologies is appropriate for a different kind of mine water, or for specific hydraulic circumstances.

Relatively few cases have been reported to date in which the sensitive receptor for water draining from flooded mine workings is a major aquifer (Younger & Adams 1999). However, what little evidence exists to date reinforces the general supposition that such pollution is so persistent that it is almost impossible for it to be adequately remediated over time-scales of relevance to human institutions. *In situ* remedial technologies generally applicable to groundwaters, such as the use of permeable reactive barriers (e.g., Starr & Cherry 1994), are only likely to be applicable in cases where aquifers are extremely shallow. Accordingly, the only published application of this type of technology to coal mine site problems has been deployed at depths not exceeding 5 m, to intercept and treat acidic leachate emanating from the toe of a disused spoil heap (Amos & Younger 2003; Younger

et al. 2003). In general, permeable reactive barriers are not a feasible technology for *in situ* remediation of polluted groundwaters in consolidated rock aquifers at greater depths (tens to hundreds of metres below the ground surface). The aquifers overlying the major Carboniferous coalfields of Europe and Asia fall precisely into this category, and therefore alternatives to permeable reactive barrier technology are needed. Given that clean-up of aquifers contaminated with mine waters is so problematic, risk-based preventative strategies are *de rigeur* in this context. In some cases, a risk assessment for a coalfield beneath a major aquifer might conclude that the extent of any future contamination is unlikely to be so great as to affect the utility of the aquifer. In such a case, the most cost-effective approach may simply amount to precautionary monitoring in case the predictions begin to prove inaccurate. In other cases, the risks may be so great that nothing less than a perpetual pump-and-treat option, holding water levels in the mine workings below the lowest feasible decant route to the aquifer, may be the optimal solution.

Water resource depletion

Being a wholly physical process, mitigation measures for this impact are also primarily physical, although it should be noted that the benefits of restoring flows in watercourses that were previously depleted can have important water quality implications, such as provision of useful dilution for sewage and other forms of pollution (e.g., Banks *et al.* 1996).

Reclamation/restoration: soil, vegetation, and ecosystem establishment

Numerous mentions have been made in the foregoing text of the use of land reclamation/restoration techniques to abate some of the negative geochemical impacts of coal mine wastes. The objective is generally to move from a stark, sparsely vegetated waste-rock pile (Fig. 4a) to a thoroughly vegetated landform (Fig. 4b). A brief explanation of how this can be achieved is therefore warranted here. Much of the following summary is drawn from the thorough review of this topic presented by Rimmer & Younger (1997), to which the interested reader is referred for further details.

Although the challenges and methodologies are essentially the same in both cases, the term 'reclamation' is typically applied to applications concerning spoil heaps formed by deep mining operations, whereas 'restoration' is the preferred term in the context of returning an opencast mine to a non-mining use. Until relatively recently, with the focus of much reclamation/restoration being on the establishment of agricultural or forestry as the main after-use, the principal activities have been limited to three steps:

(1) *Preparation of the final landform*: this often includes spreading the spoil over larger areas than it previously occupied, in order to arrive at surface profiles that are less obtrusive in the enclosing landscape (e.g., Roy 2003). Drainage infrastructure is typically incorporated into the spoil at this point, taking care to allow for significantly higher proportions of storm runoff than would be anticipated on natural soils in the vicinity;

(2) *Soil establishment:* either with or without the use of imported topsoil. Some of the geochemical implications of soil importation and/or *in situ* soil formation from spoil precursors are problematic (PIRAMID Consortium 2003); and

(3) *Long-term maintenance of the revegetated area*, which will normally be carried out by farmers or foresters who have little background in the intricacies of spoil as an Earth material. Relatively little attention is devoted to this topic in the literature, beyond documentation of geotechnical stability problems, which can arise during later agricultural use of restored spoils (Groenewold & Rehm 1982).

To these three activities, a further one is now often added in cases where the objective is not agricultural after-use but the establishment of areas of high nature conservation value, namely *ecosystem establishment*. Still in its relative infancy, to date this has most often involved attempts to create species-rich grasslands. Although such efforts can obviously draw upon transferable skills from the agricultural and forestry sectors, the aim of achieving significantly greater species diversity ($20-25$ species/m^2) than one would expect in an agricultural sward ($2-3$ species/m^2) is challenging in practice. The fact that most soils formed from mine spoil tend to contain less N than natural soils might be considered a benefit in this context, as it exerts ecological pressures that potentially open the soil system to colonization by a range of plants that are better adapted to such conditions. However, the N content can be so low that many of these plants struggle to become established, leading to dominance by clover and other plants capable of fixing atmospheric N. 'Ecosys-

tem establishment' should in most cases extend beyond ensuring species diversity among plants to include also: soil microbes (Frouz 2002), the activities of which are crucial to the long-term turnover of soil nutrients; worms, including nematodes (Hanel 2002), and earthworms (Pizl 2001); isopods, microarthropods, spiders, millipedes, and insects (Tajovsky 2001; Wanner & Dunger 2002); and higher animals, particularly small mammals and ground-nesting birds.

In terms of geochemistry the key challenges lie in avoiding the various pitfalls that beset the establishment of viable soils on re-shaped spoil. Where good quality topsoil can simply be imported, most of these problems can be side-stepped, although soil structure is inevitably lost during translocation. In many cases, insufficient topsoil will be available at reasonable cost, and alternatives must be explored. Most modern opencast operations, foreseeing the shortage of importable topsoil, go to some lengths to selectively strip the pre-existing soil before mining commences and store it separately from rock overburden. On the face of it, this seems like a good idea, but it is beset with geochemical hitches (Rimmer & Younger 1997; Ghose 2001). For instance, anaerobic conditions often become established deep within soil stores, leading to the bacterial reduction of soil nitrates to ammonia. The longer a body of topsoil is stored, the more these adverse geochemical reactions tend to completeness, such that topsoils that have been stored for some decades become incapable of supporting macrophyte growth without major rehabilitation (Ghose 2001). When the topsoil is dug up and spread on the land, accumulated ammonia may be rapidly leached, leading to surface water pollution problems (e.g., Rimmer & Younger 1997).

A popular alternative to importation or re-spreading of stored topsoil is active soil formation. This is achieved by selecting suitable 'soil forming materials' (SFMs) from within the body of spoil, such as low-S, friable mudstones, and subjecting them to physical comminution and sorting until they possess acceptable physical properties. In many cases, so-called 'soil amendments' (i.e., imported organic matter, such as sewage sludge, paper mill wastes, piggery slurries, and composted manures) are added to SFMs in order to mimic the form and function of natural humus. The intention is that, as the added organic matter gradually decays, it will be replaced by true humus developing naturally from the decay of root and leaf litter from the established vegetation. As with re-spread topsoils, however, the use of soil amendments must be undertaken with care if it is not to lead to problems of nutrient washout, with undesirable consequences of eutrophism in receiving water-courses (e.g., Salazar *et al.* 2002).

Case study: the Shilbottle–Whittle Coalfield, Northumberland, UK

Purpose and focus of case study

This case study is intended to illustrate the application of the typology of environmental impacts of coal mines and coal wastes developed in this paper to a real case. To keep the presentation concise, the example selected is a relatively small ($\sim$40 km^2) coalfield located in northeast England. Of course, most of the world's major coalfields are much larger than this, ranging up to several thousand km^2 in area, and it is therefore important not to overinterpret the significance of the findings for this one illustrative example. The case study concludes with the results of a geochemical modelling exercise, designed to elucidate the principal controls on the nature and severity of the most serious environmental impact associated with the working of this coalfield.

Geological setting

The Shilbottle Coalfield lies within the Northumberland Basin, one of the largest of a number of sedimentary basins of Carboniferous age in northwest Europe (Johnson 1992). The Northumberland Basin contains more than 4.5 km of sedimentary fill, displaying an upward evolution in character from predominantly fluviatile sediments (predominantly sandstones and mudstones) in the basal sequences (Viséan stage: Arundian–Holkerian substages), through mixed fluviatile/marine facies in the middle sequences (Viséan: Asbian–Brigantian substages), to deltaic facies, with only occasional evidence of short-lived marine incursions, in the uppermost sequences (Serpukhovian stage) (Johnson 1992). The Shilbottle Coalfield exploited coal seams in the Middle Limestone Group (Viséan: Brigantian) sequence. The local stratigraphy is summarized in Fig. 5, which shows strata from the Shilbottle Seam seat-earth up to the strata overlying the Hazon Seam. This is the main sequence of strata affected by hydrogeological and mechanical disruptions induced by mining. Beneath the seat-earth of the Shilbottle Seam the sequence continues principally as low-permeability shales of marine origin, which formed an effective base to the mined hydrological circulation system.

Coal quality information is only available for the most widely worked of the three local coal seams, that is, the Shilbottle Seam. The Shilbot-

tle Seam is generally between 0.7 and 0.9 m thick in this area. The seam produced what was widely regarded as the finest household coal in the north of England (Carruthers *et al.* 1930); a high volatile content (39–45 wt% d.a.f. [dry ash-free]; Carruthers *et al.* 1930; Temple *et al.* 1945) meant it was easy to ignite and burnt freely. It also reportedly left little ash (ultimate analyses recorded around 4.8 wt% total ash). From a modern geo-environmental perspective, however, the Shilbottle coal has the disadvantage of a medium to high total S content: Carruthers *et al.* (1930) quote a total S content of 2.8 wt%, although unpublished Coal Authority records from samples analysed during the height of production in the mid-1960s reveal wt% total S values ranging from 2.5 to 4.3, with an average around 3. Intriguingly, Temple *et al.* (1945, p. 25) commented that the Shilbottle Seam 'usually contains heavy deposits of nodular pyrites which are now [*i.e., during World War II*] separated and marketed separately'. The immediate roof and floor strata of the Shilbottle Seam are recorded on several geological logs for mine shafts and boreholes as being mudstones of various types. Unpublished mine records held by the Coal Authority describe the floor strata as 'dark grey mudstone' and the roof strata as 'pyritic shale' and as 'dark grey mudstone with a 1/2″ layer of pyrites adjacent to the top coal'. This is clearly a high-S environment, reflecting the strong influence of marine waters during diagenesis of the coal.

Structurally, the Shilbottle Coalfield is relatively simple. The strata strike approximately 45°N and dip about 10° to the SE. Antithetic extensional faults disturb seam continuity locally (Fig. 6), and a single major fault system striking approximately 85°N (cropping out in the vicinity of Whittle; Fig. 6) down-throws the coal to the north by some 200 m.

Mining history

Although occasional efforts were made to work coal profitably from the two higher seams in the local succession (Fig. 5), the disappointing thicknesses of the Townhead and Hazon seams (both around 0.45 m; Carruthers *et al.* 1930; Fowler 1936) frustrated all such ventures. Thus almost all of the mining in the Shilbottle Coalfield occurred in the Shilbottle Seam alone. Figure 6 shows the known distribution of workings in the Shilbottle Seam, which were originally developed by a number of separate mining operations accessed by numerous drifts and shallow shafts. Although it is likely that mining commenced much earlier, public records

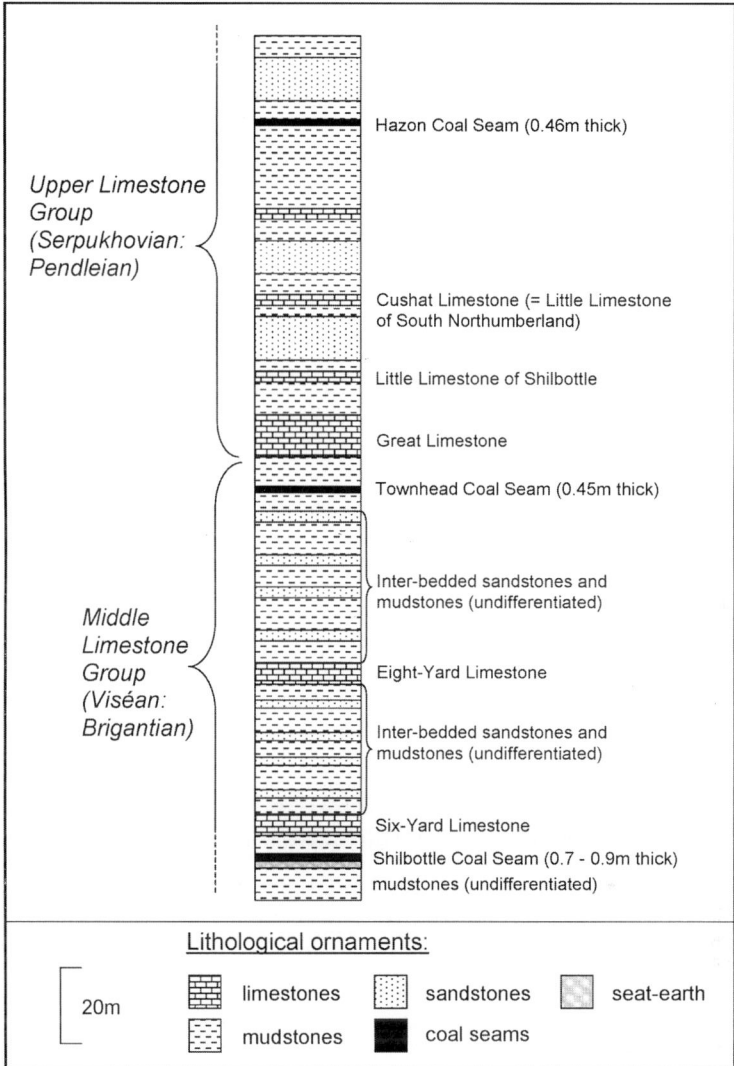

Fig. 5. Summary stratigraphic log for the 200 m interval of coal-bearing strata of Lower Carboniferous age in the Shilbottle Coalfield, Northumberland. Collated from two borehole records in the vicinity of Hazon, in the centre of the Coalfield: the Hazon Ventilation Borehole (record held by the Coal Authority) and the Hazon Lee Water Bore (Fowler 1936, p. 51). The chronostratigraphic classification has been updated from the scheme initially established by Johnson (1992).

of mining extend reliably back to 1869 for the Shilbottle, Newton and Framlington Collieries. Gradual exhaustion of reserves in the Framlington and Newton collieries, coupled with shaft collapses in the Healeycote Colliery, led to the centralization of all production in Shilbottle and Whittle collieries by the mid-20th century. Further details of the history of mining in the Shilbottle–Whittle complex are given by Tuck (1993). In terms of mine wastes, pyritic mudstone roof strata removed from the mine as part of routine 'ripping' operations were deposited in substantial spoil heaps at Shilbottle Grange Pit and (to a lesser extent) at Whittle. The highly visible pyrite content of the Shilbottle spoil earned the heap at that site the local sobriquet of 'the Brass Heap' ('brass' being a dialect name for pyrite).

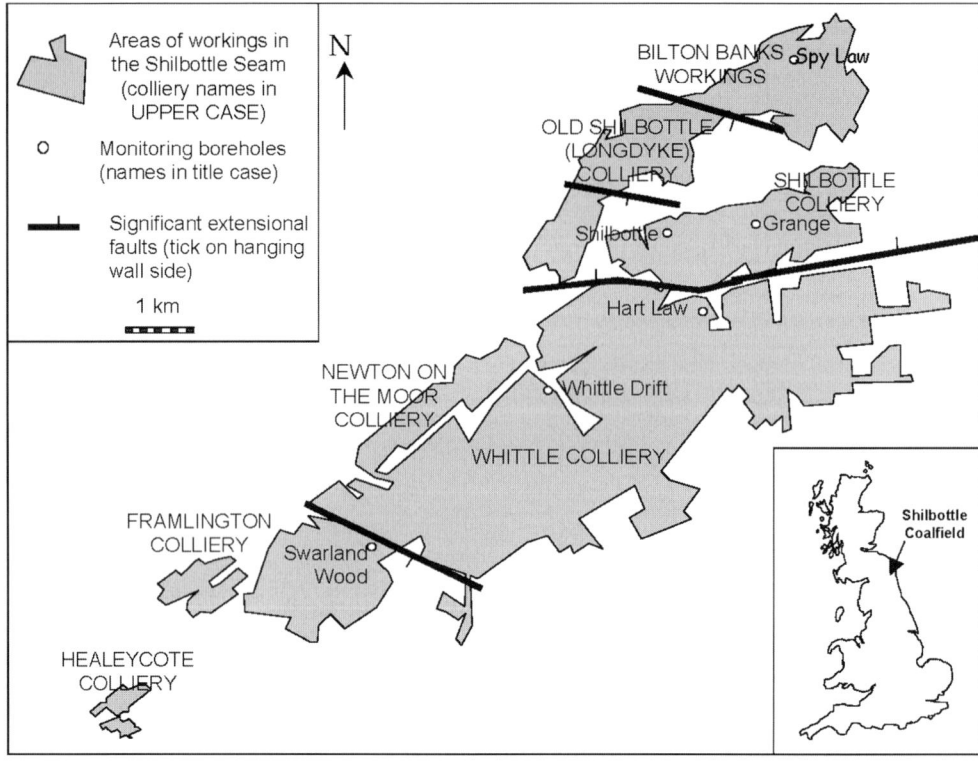

Fig. 6. Map showing the distribution of underground workings in the Shilbottle Seam, Shilbottle Coalfield, Northumberland (UK), traces of extensional faults that had sufficient throw to disrupt mining, and positions of monitoring boreholes referred to in Table 7 and associated text. Developed from abandonment plans and geological data held by the Coal Authority.

Typological examination of environmental impacts for the Shilbottle Coalfield

Table 6 summarizes the observed and/or historically recorded impacts of both mine voids and mine spoils in the Shilbottle Coalfield over its entire life-cycle to date. It is clear from Table 6 that in this particular coalfield the predominant environmental impact is, and has long been, water pollution, originating both from the substantial spoil heap at Shilbottle Grange and (more recently) from the now-flooded deep mine workings. In line with the generic conceptual model presented above, the spoil leachates are strongly acidic, whereas the deep mine waters are of circum-neutral pH, albeit still heavily contaminated with Fe. Mitigation measures, in line with the earlier recommendations of this paper, have now been successfully implemented. For the remediation of polluted leachates emanating from the Shilbottle Grange spoil heap an innovative 'hybrid' passive treatment system has been installed, comprising a shallow reactive barrier

that debouches into a series of oxidation ponds, which flow onwards to a pre-existing aerobic wetland (Younger *et al.* 2003). This system has reduced the impact of the spoil leachates on the adjoining watercourse by more than 95%. For the flooded underground mine voids, the mitigation measures took the form of a pump-and-treat system, with a purpose-drilled borehole at Whittle Drift delivering the polluted water to a treatment system comprising an aeration cascade, settlement ponds, and a series of three aerobic wetlands (Nuttall 2003). This system has successfully achieved the Coal Authority's previously stated intention of preventing the negative impacts of an uncontrolled mine water discharge in the first place (Parker 2000).

That subsidence has not been reported to be a significant problem in this coalfield perhaps reflects the sparsely populated nature of the overlying countryside. However, given that the UK's main N–S rail and trunk routes cross the coalfield, there has clearly been scope for observable impacts to arise. The lack of reported subsidence

Table 6. *Summary of environmental impacts of the Shilbottle Coalfield, Northumberland (UK) over its full life-cycle**

Hazard category	Hazard source	Pathway	Impacts on principal receptors (with references for examples/further information)
Air pollution	Flammable and/or asphyxiating mine gases	Emitted from adits/shafts and through fractured overlying strata	No recorded explosions; insufficient CH_4 released to warrant exploitation; little or no impact
	Combustion fumes from burning spoil	Emitted from burning zones within spoil heap	No record of any such impact in this coalfield
	Dust ablation from bare spoil	Wind-borne transport of dust from bare spoil/dry finings pond surfaces	Predominantly westerly winds meant that such dust as was entrained did not fall on residential areas; thus no negative impacts noted
Fire hazards	Combustion of spoil heaps/backfill or freely burning fires in underground voids	Multiple emissions of smoke and dense noxious gases to surrounding air	Spoil did burn in the past at both Whittle and Shilbottle Grange sites, producing red burnt shale (used in shaft filling); no negative impacts on communities reported from periods of spoil combustion
Ground deformation	Ground fracturing arising from underground fires in spoil or workings	Open fissures, with apertures typically between several mm and several metres	No incidents recorded in this coalfield
	Differential subsidence of opencast backfill or collapse of old shafts or bord-and-pillar workings	Typically results in shallow, steep-sided vertical collapse hollows (≤ 5 m diameter) above actively eroding pipe structures	No incidents recorded in this coalfield
	Subsidence caused by longwall extraction	Usually manifest in relatively broad settlement of ground over areas in excess of 25–50 ha, typically in the form of a shallow closed basin	Although this form of subsidence certainly occurred, it did not give rise to environmental/public safety problems in this coalfield
	Landslides of unstable spoil, often triggered by excessive pore water pressures	Mudflows/debris-flows that prograde downhill at high velocities, destroying obstacles in their path	No such incidents occurred in this coalfield
Water pollution	Siltation of watercourses	Silt-laden runoff from working areas/spoil entering nearby natural water bodies	Sufficient problem at Whittle that major sedimentation ponds were constructed in 1986; some silt-laden runoff persists from Whittle spoils

(*continued*)

Table 6. *Continued*

Hazard category	Hazard source	Pathway	Impacts on principal receptors (with references for examples/further information)
	Acidic and/or ferruginous drainage entering surface watercourses	Water rich in Fe and other metals, sometimes also with low pH (≥ 2.5)	Major problem at the Shilbottle Grange site prior to installation of remediation system in 2002 (Younger *et al.* 2003). Minor uncontrolled discharges (circum-neutral pH, Fe < 10 mg/L) from Bilton Banks and Hazon Coal workings. Anticipated grave mine water discharge at Whittle (Parker 2000) was averted by the Coal Authority's timely construction of a major pump-and-treat facility at Whittle Drift, which came into permanent use in May 2002 (Nuttall 2003)
	Saline water discharge to surface watercourses	Highly mineralized water discharged to streams from actively pumped workings/spoil leachates	Apart from salinity of acidic leachates at Shilbottle Grange, no problems with non-acidic salinity in this coalfield
	Aquifer pollution	Highly mineralized waters flow from abandoned workings into overlying aquifers	It is highly likely that a SO_4-rich plume of ground water now exists in some of the limestone aquifers to the east of Whittle/Shilbottle, but as these are not exploited by boreholes at present there is no evidence for such a plume
Water resource depletion	Drawdown of water table due to pumping of active workings	Drying up of natural streams, springs, and/or lowering of the water table in adjoining aquifers	Pumping of Whittle Colliery intercepted karst drainage in the 8-Yard Limestone, which effectively underdrained the Swarland Burn throughout the working life of the mine, often leaving the burn completely dry downstream of Swarland village. No doubt pumping also affects other overlying limestone aquifers (Fig. 5) but no piezometric data exist to substantiate this
	Disruption of natural flow paths (surface and subsurface) due to mine subsidence		Not considered to be a contributory factor in this coalfield

*During coal production and in the post-closure period to date (January 2004).

is likely due to a combination of careful management of mine layout in more recent decades, assisted by limitations imposed on settlement by the presence of some of the thicker limestones and sandstones overlying the Shilbottle Seam (Fig. 5), which will have been able to limit void migration to a considerable degree (cf. NCB 1975).

Modelling investigations of the principal environmental impact in the coalfield

Given the geochemical focus of this paper, it is appropriate to focus in a little more detail on the geochemical processes responsible for the principal environmental impact arising from this coalfield, that is, water pollution (Table 6). In this section, a brief account is given of how geochemical modelling using the well-known package PHREEQC (Parkhurst & Appelo 1999) has been used to discriminate between a number of alternative conceptual models of how observed water qualities have evolved. Table 7 lists the typical analyses of waters in this coalfield upon which the analysis is based.

Before embarking upon geochemical modelling, it is important that the key elements of the hydrology of the study system be adequately understood. In this case, the requisite understanding can be based firmly on prior hydrological monitoring and modelling, various aspects of which have already been documented (Younger & Adams 1999; Adams & Younger 2001). In fact, in reviewing earlier predictions made in 1998 (Younger & Adams 1998), four years before the pump-and-treat facilities were brought into permanent service (Nuttall 2003),

we have a very rare opportunity to undertake retrospective analysis of predictive hydrogeological modelling. In the 1998 work, predictive modelling of the flooding of the Shilbottle Seam workings was undertaken using an innovative model that coupled a pipe-network code (representing major mine roadways) to a three-dimensional variably saturated subsurface flow model (Adams & Younger 2001). Comparison of the 1998 predictions with subsequent observations reveal a pleasing correspondence. For instance, the median post-rebound discharge rate from the workings was predicted to be 1.7 ML/d, with a range varying from 0.8 ML/d following a lengthy drought period, to 2.6 ML/d for the peak flows following a very wet year. This median figure matches precisely the rate of pumping that has since been found necessary to maintain water levels in the mined system at around 50 m aOD [above Ordnance Datum] (Nuttall 2003). The 1998 modelling also predicted that water levels would reach the 50 m aOD level (at which pumping would become necessary to prevent uncontrolled discharge) by around May 2002. In the event, the newly constructed pump-and-treat system was tested in January–February 2002, and brought into permanent operation in May 2002 (Nuttall 2003). The geochemical modelling presented below is therefore based on an unusually well-substantiated hydrogeological model.

The analysis that follows has been developed using the inverse modelling capabilities of PHREEQC, in which we 'account for the chemical changes that occur as a water evolves along a flow path [as] one aqueous solution is assumed to mix with [another], and [/or] to react with minerals and gases to produce the observed

Table 7. *Summary of representative analyses of selected mine waters in the Shilbottle Coalfield**

	'Mature' Whittle drift borehole	'Whittle recharge' water	Spy Law borehole	Shilbottle Grange spoil heap leachate
Temperature (°C)	13	9	11.6	16
pH	7.7	6.5	5.27	3.95
Alkalinity	510	195	78	0
Na	727	15	458	<1
Mg	198	62	406	2160
Al	0.01	0.01	1.4	424
SO_4	2730	341	4460	13 075
Cl	62	52	46	41
K	29.9	4.6	45	7.1
Ca	248	143	450	28
Mn	1.5	1.2	11.3	268
Fe	29	0.34	456	888

*All values in mg/L except for pH, temperature and alkalinity (mg/L as $CaCO_3$ equivalent). (Data sources: Coal Authority and University of Newcastle.)

composition of a second aqueous solution' (Parkhurst & Appelo 1999). In doing so, PHREEQC calculates mixing fractions for the aqueous solutions and mole transfers of the gases and minerals into/out of solution, which are consistent with the change in chemistry between the 'initial' and 'final' waters along the specified flow-path. To configure PHREEQC for a specific simulation it is necessary to define a number of 'phases' (i.e., minerals or gases) that it is considered might reasonably take part in the geochemical reactions affecting the waters under investigation. For mineral phases, it is possible to simulate both precipitation/dissolution reactions and sorption/desorption processes. Having defined the phases, the final requirement of the code is to specify 'balances', that is, chemical elements for which a mole-transfer balance is sought (which must be present both in the water analyses used and in the specified phases).

In this case study, the selected phases are pyrite, amorphous FeS, calcite (present in limestones in the roof strata; Fig. 5), dolomite (possibly also present in the limestones), siderite (which occurs as nodules in roof-strata mudstones), ankerite (present on coal cleats in the Shilbottle Seam), melanterite and potassium–jarosite (representing the hydroxysulphate minerals; see Table 3), amorphous ferric hydroxide (i.e., the 'ochre' commonly observed in these workings, forming by precipitation from ferruginous mine waters), and gypsum (a mineral known to precipitate subaqueously from mine waters with SO_4 contents in excess of about 2500 mg/L at ambient groundwater temperatures in this region, and with which most of the mine waters in the district are known to be in equilibrium). In addition, sorption reactions were included in some of the simulations, to contribute to the mole transfer balances for Ca, Na, and Fe.

The 'initial' and 'final' waters used in these simulations were selected from among the analyses listed in Table 7. Although a number of simulations were made for various different flowpaths deemed likely on the basis of the hydrogeological model, the presentation here is restricted solely to investigations of the evolution of the 'mature' pumped waters at Whittle Drift Borehole (Fig. 6) from a range of possible precursor waters. The 'mature' Whittle Drift pumped water is of Ca–Na–SO_4 facies, with a pH around 7.5 and ≤45 mg/L dissolved Fe. The possible precursor waters were identified as follows:

(1) *Whittle 'recharge' water*: water found at the top of the water column in the Whittle

Drift when the newly constructed treatment system was tested in January and February of 2002 (Nuttall 2003). This water, of Ca–Mg–HCO_3 facies, is similar in quality to individual roof drippers sampled during the operation of the mine (unpublished Coal Authority data), and clearly represents incoming water, which has thus far been largely in contact with the overlying limestones (Fig. 5).

(2) *Spy Law Borehole Water*: samples of water pumped from the Spy Law borehole (Fig. 6), which represent the poorest quality water (in terms of elevated Fe and SO_4) sampled anywhere in flooded portions of the Shilbottle seam workings. Clearly no direct flow-path is likely to exist from the Spy Law Borehole itself to the Whittle Drift Borehole, due to the intervention of several barriers of unworked coal between the two, reflecting the difficulties of continuing mining through the faults marked on Fig. 6. However, there is no *a priori* reason to suppose that waters of similarly poor quality do not exist in portions of the workings that are interconnected to the Whittle Drift Borehole, but which have not yet been accessed by monitoring wells.

(3) *The extremely acidic Shilbottle Grange spoil heap leachates*: although these waters have clearly evolved in the near-surface environment rather than within the mine workings, they are generated by the same rock materials, which are certainly present in the old underground workings. Given that it is far easier to sample spoil leachates than deep mine waters, it is often tempting to use the former as analogues for inaccessible waters deep underground. By testing the credibility of these leachates as precursors for the 'mature' Whittle Drift pumped waters, we have a powerful test of the wisdom of succumbing to this temptation.

The results of various attempts to explain the provenance of the 'mature' Whittle waters in terms of the above precursor waters and mineral phases are outlined below. Assuming that 'Whittle recharge' water is the sole precursor, several attempts were made to model the evolution of the 'mature' Whittle water by reaction of the 'Whittle recharge' waters with various minerals. Where the mineral suite included calcite, dolomite, gypsum, pyrite and K-jarosite, but not ankerite, the resulting models were of poor credibility: they invoke coupled precipitation of calcite (although dissolution is far more likely) and substantial dissolution of gypsum and dolomite (neither of which is very likely, given the lack of primary gypsum in the strata and the low dissolution rate of dolomite relative to calcite). Iron concentrations are explained reasonably well by dissolution of both pyrite and the Fe hydroxysulphate secondary minerals. If dolomite is replaced by ankerite in the mineral assemblage, then the latter begins to play an important role as a source for dissolved Fe, as well as accounting for the increase in dissolved Mg between the 'recharge' and 'mature' waters. Evaporative water losses are predicted to vary between zero and 33%, which is an intuitively reasonable range. However, a number of predictions of poor credibility remain, including the dissolution of gypsum and precipitation of melanterite and calcite.

If it is assumed that Spy Law-type water is the sole precursor, as with the foregoing case, the resulting models are not especially appealing. Although credible rates of dissolution are predicted for pyrite (2.4×10^{-3} moles per kilogramme of water (mol/kgw)) and calcite (1.0×10^{-2} mol/kgw), less credible aspects of these models include the dissolution of ferric hydroxide (unlikely under the observed pH conditions) and *precipitation* of ankerite and the hydroxysulphates (both of which would be expected to dissolve rather than precipitate).

In the light of these experiences, simulations using the extremely acidic Shilbottle Grange spoil leachate waters as a sole precursor were not even attempted: it is clear that such highly contaminated waters could only feasibly play a role in the evolution of 'mature' Whittle waters through mixing with less aggressive waters.

A series of simulations involving mixing of the different precursors were undertaken, in the hope of finding more satisfactory explanations for the evolution of 'mature' Whittle waters. The most logical results were obtained from simulations in which Spy Law water was modelled as mixing with Whittle recharge-type water to produce 'mature Whittle' water. These models predicted dissolution of calcite (4.8×10^{-3} mol/kgw), pyrite (1.5×10^{-4} mol/kgw) and the hydroxysulphate mineral jarosite (5.4×10^{-4} mol/kgw), with an upper limit on dissolved SO_4^{2-} concentrations being imposed by the precipitation of gypsum (2.4×10^{-3} mol/kgw). In more than 70% of models obtained, excess Fe is predicted to be removed from solution by sorption to clay minerals.

Simulation of the mixing of Shilbottle Grange spoil leachates with the Whittle recharge water resulted in models which invoke tiny fractions of spoil leachate in the final mixture (<1 ppm). This serves to demonstrate that very little of the water in the underground mine voids can be as poor in quality as the Shilbottle spoil leachates. Accordingly, this study supports the contention that spoil leachate quality is *not* a very useful guide to the quality of waters in the mine voids from which the spoil was originally excavated.

Conclusions

Environmental impacts associated with the mining of coal, arising from the mined voids themselves and from the wastes left behind at the surface, have been recorded ever since the onset of industrial-scale extraction around 1600. In the early days of coal mining, complaints about such impacts were strident, as the newcomer industry adversely affected long-established agricultural interests. As the coal trade came to dominate regional economies in mining districts, so its negative impacts came to be accepted stoically, finally passing unremarked). Only with the onset of large-scale mine closures have the negative impacts of coal mining come to be deemed unacceptable once more. The environmental impacts arising from coal mining activities are fundamentally attributable to the exposure of profoundly reduced Earth materials (especially coal, pyrite, siderite, and ankerite) to the oxidizing power of the atmosphere. A typology of the known impacts arising from mine voids and wastes in coal mining districts has been developed, which recognises many subcategories of impacts under five major headings: air pollution, fire hazards, ground deformation, water pollution, and water resource depletion. A robust understanding of geochemical processes is key to understanding how these impacts arise, and to developing sustainable mitigation strategies. The application of the newly developed typology is illustrated using the case of the Shilbottle Coalfield (Northumberland, UK). The geography and stratigraphy of this coalfield have ensured that very few demonstrable impacts have arisen in the categories of air pollution, fire hazards, or ground deformation. However, major problems of water pollution relating to both the flooded underground voids and emissions of leachates from surface spoil heaps have necessitated the installation of substantial remediation works (pump-and-treat allied with passive treatment). Inverse geochemical modelling has clarified the linkages between the various types of water encountered in the coalfield, providing a baseline geochemical understanding upon which future investigations of remedial system sustainability can be constructed.

I am grateful to Dr C. Nuttall for the collation of much of the data discussed in the Case Study, and for the initial drafting of the plan from which Fig. 6 has been derived. A. Elliott of Northumberland County Council, and K. Parker of the Coal Authority, helpfully provided data held by their respective organizations, which were used in the Case Study. Dr T. Pless-Mulloli kindly helped me find my way into the medical literature. This work was in part financially supported by the European Commission 5th Framework Research Project 'Environmental Regulation of Mine Waters in the EU' (ERMITE), contract number EVK1-CT-2000-0078 (see www.minewater.net/ermite).

References

ADAMS, R. & YOUNGER, P. L. 2001. A strategy for modeling ground water rebound in abandoned deep mine systems. *Ground Water*, **39**, 249–261.

AMOS, P. W. & YOUNGER, P. L. 2003. Substrate characterisation for a subsurface reactive barrier to treat colliery spoil leachate. *Water Research*, **37**, 108–120.

ANON 2003. *A Global 'Catastrophe' – Coal Fires Threaten Environment, Human Health*. American Association for the Advancement of Science, Annual Meeting February 2003, Internet article. http://www.aaas.org/news/releases/2003/0213fire.shtml.

APPELO, C. A. J. & POSTMA, D. 1993. *Geochemistry, Groundwater and Pollution*. A. A. Balkema Publishers, Rotterdam. 536 pp.

BACKES, C. A., PULFORD, I. D. & DUNCAN, H. J. 1986. Studies on the oxidation of pyrite in colliery spoil. I. The oxidation pathway and inhibition of the ferrous–ferric oxidation. *Reclamation and Revegetation Research*, **4**, 279–291.

BACKES, C. A., PULFORD, I. D. & DUNCAN, H. J. 1987. Studies on the oxidation of pyrite in colliery spoil. II. Inhibition of the oxidation by amendment treatments. *Reclamation and Revegetation Research*, **6**, 1–11.

BACKES, C. A., PULFORD, I. D. & DUNCAN, H. J. 1993. Seasonal variation of pyrite oxidation rates in colliery spoil. *Soil Use and Management*, **9**, 30–34.

BADGER, C. W., CUMMINGS, A. D. & WHITMORE, R. L. 1956. The disintegration of shale. *Journal of the Institute of Fuel*, **29**, 417–423.

BANKS, D., SKARPHAGEN, H., WILTSHIRE, R. & JESSOP, C. 2004. *Heat Pumps as a Tool for Energy Recovery from Mining Wastes*. *In*: GIERÉ, R. & STILLE, P. (eds) *Energy, Waste, and the Environment: a Geochemical Perspective*. Geological Society, London, Special Publications, **236**, 499–513.

BANKS, D., YOUNGER, P. L. & DUMPLETON, S. 1996. The historical use of mine-drainage and pyrite-oxidation waters in central and eastern England, United Kingdom. *Hydrogeology Journal*, **4**, 55–68.

BANKS, D., YOUNGER, P. L., ARNESEN, R.-T., IVERSEN, E. R. & BANKS, S. D. 1997. Mine-water chemistry: the good, the bad and the ugly. *Environmental Geology*, **32**, 157–174.

BANWART, S. A., EVANS, K. A. & CROXFORD, S. 2002. Predicting mineral weathering rates at field scale for mine water risk assessment. *In*: YOUNGER, P. L. & ROBINS, N. S., (eds) *Mine Water Hydrogeology and Geochemistry*. Geological Society, London, Special Publications, **198**, 137–157.

BAYLESS, E. R. & OLYPHANT, G. A. 1993. Acid-generating salts and their relationship to the chemistry of groundwater and storm runoff at an abandoned mine site in southwestern Indiana, USA. *Journal of Contaminant Hydrology*, **12**, 313–328.

BEKENDAM, R. F. & POTTGENS, J. J. 1995. Ground movements over the coal mines of southern Limburg, the Netherlands, and their relation to rising mine waters. *International Association of Hydrological Sciences*, **234**, 3–12.

BELL, F. G. 1996. Dereliction: colliery spoil heaps and their rehabilitation. *Environmental and Engineering Geoscience*, **2**, 85–96.

BELL, F. G. & GENSKE, D. D. 2001. The influence of subsidence attributable to coal mining on the environment, development and restoration: some examples from western Europe and South Africa. *Environmental and Engineering Geoscience*, **7**, 81–99.

BEVINS, R. E. 1994. *A Mineralogy of Wales*. National Museum of Wales, Cardiff, 146 pp.

BEYNON, H., COX, A. & HUDSON, R. 2000. *Digging up Trouble. The Environment, Protest and Opencast Coal Mining*. Rivers Oram Press, London, 306 pp.

BLOOR, M. 2000. The South Wales Miners Federation, miners' lung and the instrumental use of expertise, 1900–1950. *Social Studies of Science*, **30**, 125–140.

BLOOR, M. 2002. No longer dying for a living: collective responses to injury risks in South Wales mining communities, 1900–47. *Sociology*, **36**, 89–105.

BOOTH, C. J. 2002. The effects of longwall mining on overlying aquifers. *In*: YOUNGER, P. L. & ROBINS, N. S. (eds) *Mine Water Hydrogeology and Geochemistry*. Geological Society, London, Special Publications, **198**, 17–45.

BORM, P. J. A. 2002. Particle toxicology: from coal mining to nanotechnology. *Inhalation Toxicology*, **14**, 311–324.

BREAS, O., GUILLOU, C., RENIERO, F. & WADA, E. 2002. The global methane cycle: Isotopes and mixing ratios, sources and sinks. *Isotopes In Environmental and Health Studies*, **37**, 257–379.

BUCHANAN, D. J. & BRENKLEY, D. 1994. Green coal mining. *In*: HESTER, R. E. & HARRISON, R. M. (eds) *Mining and its Environmental Impact*. Royal Society of Chemistry, London, 71–95.

CARUCCIO, F. T. & FERM, J. C. 1974. Paleoenvironment – predictor of acid mine drainage problems. *Proceedings of the 5th Coal Mine Drainage Research Symposium*, National Coal Association (USA), Kentucky, 5–9.

CARRUTHERS, R. G., BURNETT, G. A. & ANDERSON, W. 1930. *The Geology of the Alnwick District. Memoirs of the Geological Survey of England*

and Wales, Explanation of Sheet 6. HMSO, London. 144 pp.

CASAGRANDE, D. J. 1987. Sulphur in peat and coal. In: SCOTT, A. C. (ed) Coal and Coal-Bearing Strata: Recent Advances. Geological Society, London, Special Publications, 32, 87–105.

CCC 1995. Poisoned Legacy: Water Pollution from Abandoned Mines. A Strategy for Action from the Coalfield Communities Campaign. Coalfield Communities Campaign, Barnsley, UK, 10 pp.

CHARLES, J. A. & SKINNER, H. D. 2001. Compressibility of foundation fills. Proceedings of the Institution of Civil Engineers – Geotechnical Engineering, 149, 145–157.

CHAULYA, S. K., CHKRABORTY, M. K., AHMAD, M., SINGH, R. S., BONDYOPADHAY, C., MONDAL, G. C. & PAL, D. 2002. Development of empirical formulae to determine emission rate from various opencast coal mining operations. Water Air and Soil Pollution, 140, 21–55.

CLARKE, L. B. 1996. Environmental aspects of coalbed methane production, with emphasis on water treatment and disposal. Transactions of the Institution of Mining and Metallurgy, (Section A), 105, A105–A113.

COULSHED, A. J. G. 1951. Fires. In: MASON, E. (ed) Practical Coal Mining for Miners (2nd edition). Virtue and Company, London, 280–288, vol. I.

DIAMOND, J. M. & SERVEISS, V. B. 2001. Identifying sources of stress to native aquatic fauna using a watershed ecological risk assessment framework. Environmental Science and Technology, 35, 4711–4718.

DONAHOE, R. J. (2004). Secondary Mineral Formation in Coal Combustion Byproduct Disposal Facilities: Implications for Trace Element Sequestration. In: GIERÉ, R. & STILLE, P. (eds) Energy, Waste, and the Environment: a Geochemical Perspective. Geological Society, London, Special Publications, 236, 641–658.

DOYLE, A. 2001. The Colliery Aesthetic: Cultural Responses at the End of Industry. Unpublished PhD thesis, Department of Sociology, University of Durham, UK, 278 pp.

DUCKHAM, H. & DUCKHAM, B. F. 1973. Great Pit Disasters: Great Britain 1700 to the Present Day. David and Charles, Newton Abbot, 227 pp.

DUNRUD, C. R. 1984. Coal mine subsidence – western United States. In: HOLZER, T. L. (ed) Man-Induced Land Subsidence. Geological Society of America, Boulder, Reviews in Engineering Geology, VI, 151–194.

EVANGELOU, V. P. 1995. Pyrite Oxidation and its Control: Solution Chemistry, Surface Chemistry, Acid Mine Drainage. CRC Press, Florida, 293 pp.

FINKELMAN, R. B., OREM, W. et al. 2002. Health impacts of coal and coal use: possible solutions. International Journal of Coal Geology, 50, 425–443.

FOWLER, A. 1936. The Geology of the Country Around Rothbury, Amble and Ashington. Memoirs of the Geological Survey of England and Wales, Explanation of Sheets 9 and 10. HMSO, London, 159 pp.

FROUZ, J. 2002. The effect of soil microfauna on litter decompostion and soil organic matter accumulation during soil formation in spoil heaps after brown coal mining. Ekologia-Bratislava, 21, 363–369.

GALLOWAY, R. L. 1898. Annals of Coal Mining and the Coal Trade, Volume 1 (First Series, up to 1835). Colliery Guardian Company Ltd, London, 534 pp. (Reprinted 1971 by David & Charles, Newton Abbot.)

GANDY, C. J. & YOUNGER, P. L. 2002. A Review of Saline Mine Waters in the Silesian Region and Methods for Managing Them. 1st IMAGE-TRAIN Case Study. European Commission 5th Framework RTD Programme, project EVK1-2001-80002, 48 pp. www.image-train.net.

GARCIA-CRIADO, F., FERNÁNDEZ-ALAEZ, M. & FERNÁNDEZ-ALAEZ, C. 2002. Relationship between benthic assemblage structure and coal mining in the Boeza River Basin (Spain). Archiv für Hydrobiologie, 154, 665–689.

GELLER, W., KLAPPER, H. & SALOMONS, W. (eds) 1998. Acidic Mining Lakes. Acid Mine Drainage, Limnology and Reclamation. Springer, Heidelberg, 435 pp.

GHOSE, M. K. 2001. Management of topsoil for geoenvironmental reclamation of coal mining areas. Environmental Geology, 40, 1405–1410.

GHOSE, M. K. & MAJEE, S. R. 2001. Air pollution due to opencast coal mining and its control in Indian context. Journal of Scientific and Industrial Research (India), 60, 786–797.

GHOSE, M. K. & MAJEE, S. R. 2002. Assessment of the status of work zone air environment due to opencast coal mining. Environmental Monitoring and Assessment, 77, 51–60.

GLASSER, F. P. 2004. Coal Combustion Wastes: Characterisation, Reuse and Disposal. In: GIERÉ, R. & STILLE, P. (eds) Energy, Waste, and the Environment: a Geochemical Perspective. Geological Society, London, Special Publications, 236, 211–222.

GLOVER, H. G. 1983. Mine water pollution – an overview of problems and control strategies in the United Kingdom. Water Science and Technology, 15, 59–70.

GOLOMB, D. S. & FAY, J. A. 2004. Atmospheric Impact of the Fossil Fuel Cycle. In: GIERÉ, R. & STILLE, P. (eds) Energy, Waste, and the Environment: a Geochemical Perspective. Geological Society, London, Special Publications, 236, 153–167.

GRAYSON, R. L. 1999. Mine health and safety: industry's march towards continuous improvement – the United States experience. In: AZCUE, J. M. (ed) Environmental Impacts of Mining Activities. Emphasis on Mitigation and Remedial Measures. Springer, Heidelberg, 83–100.

GRIMSHAW, P. N. 1992. Sunshine Miners. Opencast Coalmining in Britain, 1942–1992. British Coal Opencast, Mansfield, Notts, 113 pp.

GROENEWOLD, G. H. & REHM, B. W. 1982. Instability of contoured surface-mined landscapes in the northern Great Plains: causes and implications.

trip mine spoils – western North Dakota. *Reclamation and Revegetation Research*, **1**, 161–176.

GROPPO, J., ROBL, T. & HOWER, J. C. 2004. *The Beneficiation of Coal Combustion Ash. In*: GIERÉ, R. & STILLE, P. (eds). *Energy, Waste, and the Environment: a Geochemical Perspective*. Geological Society, London, Special Publications, **236**, 247–262.

HAIR, T. H. 1844. *A Series of Views of the Collieries in the Counties of Northumberland and Durham, with Descriptive Sketches and a Preliminary Essay on Coal and the Coal Trade by M ross*. Davis Books, Newcastle upon Tyne, 52 pp (Republished 1987).

HANEL, L. 2002. Development of soil nematode communities on coal-mining dumps in two different landscapes and reclamation practices. *European Journal of Soil Biology*, **38**, 167–171.

HARRISON, S. R. 2000. Opportunities, not threats: how today's coal-mining industry can resolve problems created by past mining. *Transactions of the Institution of Mining and Metallurgy (Section A: Mining Technology)*, **109**, A157–A164.

HARTMAN, H. L. & MUTMANSKY, J. M. 2002. *Introductory Mining Engineering*. John Wiley & Sons Inc, New York, 584 pp.

HAWKES, L. & SMYTHE, J. A. 1937. Ankerites of the Northumberland coal-field. *Mineralogical Magazine*, **24**, 65–75.

HELIOS, RYBICKA, E. 1996. Environmental impact of mining and smelting industries in Poland. *In*: APPLETON, J. D., FUGE, R. & McCALL, G. J. H. (eds) *Environmental Geochemistry and Health*. Geological Society, London, Special Publications, **113**, 183–193.

HEMINGWAY, J. E. 1974. Ironstones. *In*: RAYNER, D. H. & HEMINGWAY, J. E. (eds) *The Geology and Mineral Resources of Yorkshire*. Yorkshire Geological Society, Leeds, 329–335.

HOWEL, D., PLESS-MULLOLI, T., KING, A., STONE, I. & MEREFIELD, J. 2001. Consultations of children living near open-cast coal mines. *Environmental Health Perspectives*, **109**, 567–571.

JARVIS, A. P. & YOUNGER, P. L. 1997. Dominating chemical factors in mine water induced impoverishment of the invertebrate fauna of two streams in the Durham Coalfield, UK. *Chemistry and Ecology*, **13**, 249–270.

JOHNSON, G. A. L. (ed) 1992. *Robson's Geology of North East England*. (The Geology of North East England, Second Edition. Published as volume 56 of the Transactions of the Natural History Society of Northumbria.) Natural History Society of Northumbria, Newcastle Upon Tyne, 391 pp.

JONES, T., BLACKMORE, P., LEACH, M., BERUBE, K., SEXTON, K. & RICHARDS, R. 2002. Characterisation of airborne particles collected within and proximal to an opencast coalmine: South Wales, U.K. *Environmental Monitoring and Assessment*, **75**, 293–312.

KELLY, M. G. 1999. Effects of heavy metals on the aquatic biota. *In*: PLUMLEE, G. S. & LOGSDON, M. J. (eds) *The Environmental Geochemistry of Mineral Deposits. Part A: Processes, Techniques*

and Health Issues. Reviews in Economic Geology, Volume 6A. Society of Economic Geologists, Littleton, CO, 363–371.

KEMP, P. & GRIFFITHS, J. 1999. *Quaking Houses. Art, Science and The Community: A Collaborative Approach to Water Pollution*. Jon Carpenter Publishing, Charlbury, Oxon, 142 pp.

KING, A. M., PLESS-MULLOLI, T., MEREFIELD, J. & STONE, I. 2000. TEOMs and the volatility of UK non-urban PM_{10}: a regulatory dilemma? *Atmospheric Environment*, **34**, 3211–3212.

LEE, M. L., NOVOTNY, M. V. & BARTLE, G. D. 1981. *Analytical Chemistry of Polycyclic Aromatic Hyrocarbons*. Academic Press, New York.

LEVINE, D. & WRIGHTSON, K. 1991. *The Making of an Industrial Society. Whickham 1560–1765*. Clarendon Press, Oxford, 456 pp.

LI, H. F., WANG, M. L., SEIXAS, N., DUCATMAN, A. & PETSONK, E. L. 2002. Respiratory protection: associated factors and effectiveness of respirator use among underground coal miners. *American Journal of Industrial Medicine*, **42**, 55–62.

LINAK, W. P. & WENDT, J. O. L. 1994. Trace-metal transformation mechanisms during coal combustion. *Fuel Processing Technology*, **39**, 173–198.

LOVE, R. G., MILLER, B. G. *et al.* 1997. Respiratory health effects of opencast coalmining: A cross sectional study of current workers. *Occupational and Environmental Medicine*, **54**, 416–423.

LYNCH, M. 2002. *Mining in World History*. Reaktion Books Ltd, London, 350 pp.

McLEAN, I. & JOHNES, M. 2000. *Aberfan. Government and Disasters*. Welsh Academic Press, Cardiff, 250 pp.

MORGAN, L., SCOURFIELD, J., WILLIAMS, D., JASPER, A. & LEWIS, G. 2003. The Aberfan disaster: 33-year follow-up of survivors. *British Journal of Psychiatry*, **182**, 532–536.

MORRISON, J. L., SCHEETZ, B. E., STRICKLER, D. W., WILLIAMS, E. G., ROSE, A. W., DAVIS, A. & PARIZEK, R. R. 1990. Predicting the occurrence of acid mine drainage in the Alleghenian coal-bearing strata of western Pennsylvania; an assessment by simulated weathering (leaching) experiments and overburden characterization. *In*: CHYI, L. L. & CHOU, C. L. (eds) *Recent Advances in Coal Geochemistry*. Geological Society of America, Special Paper, **248**, 87–99.

MUNRO, J. F. & CROMPTON, G. K. 1999. Health effects of respirable dust from opencast coal mining. *Proceedings of the Royal College of Physicians, Edinburgh*, **29**, 11–15.

NCB 1975. *The Subsidence Engineers' Handbook*. National Coal Board, Mining Department, London, 111 pp.

NICOLAU, J. M. 2002. Runoff generation and routing on artificial slopes in a Mediterranean–continental environment: the Teruel coalfield, Spain. *Hydrological Processes*, **16**, 631–647.

NORDSTROM, D. K. 1999. Efflorescent salts and their effects on water quality and mine plugging. *In*: FERNÁNDEZ-RUBIO, R. (ed) *Mine, Water and Environment*. (Proceedings of the International Congress of the International Mine Water

Assocation, Sevilla, Spain, 13–17 Sept 1999). International Mine Water Association, Madrid, 543–550, vol. II.

NOVAK, T. & FISHER, T. J. 2001. Lightning propagation through the earth and its potential for methane ignitions in abandoned areas of underground coal mines. *IEEE Transactions on Industry Applications*, **37**, 1555–1562.

NUTTALL, C. A. 2003. Testing and performance of a newly constructed full-scale passive treatment system at Whittle Colliery, Northumberland. *Land Contamination and Reclamation*, **11**, 105–112.

PARKER, K. 2000. Mine water – the role of the Coal Authority. *Transactions of the Institution of Mining and Metallurgy (Section A: Mining Technology)*, **109**, A219–A223.

PARKHURST, D. L. & APPELO, C. A. J. 1999. *User's Guide to PHREEQC (Version 2) – a Computer Program for Speciation, Batch-Reaction, One-Dimensional Transport, and Inverse Geochemical Calculations*. US Geological Survey, Denver (CO), Water-Resources Investigations Report 99-4259, 312 pp.

PIRAMID CONSORTIUM (2003) *Engineering Guidelines for the Passive Remediation of Acidic and/ or Metalliferous Mine Drainage and Similar Wastewaters*. European Commission 5th Framework RTD Project no. EVK1-CT-1999-000021 'Passive *in-situ* remediation of acidic mine/industrial drainage' (PIRAMID). (Freely downloadable from: www.piramid.org). University of Newcastle Upon Tyne, Newcastle Upon Tyne, UK, 164 pp.

PIZL, V. 2001. Earthworm succession in afforested colliery spoil heaps in the Sokolov region, Czech Republic. *Restoration Ecology*, **9**, 359–364.

PLESS-MULLOLI, T., KING, A., HOWEL, D., STONE, I. & MEREFIELD, J. 2000a. PM$_{10}$ levels in communities close to and away from opencast coal mining sites in Northeast England. *Atmospheric Environment*, **34**, 3091–3101.

PLESS-MULLOLI, T., KING, A., HOWEL, D., STONE, I., MEREFIELD, J., BESSELL, J. & DARNELL, R. 2000b. Living near opencast coal mining sites and children's respiratory health. *Occupational and Environmental Medicine*, **57**, 145–151.

PLESS-MULLOLI, T., HOWEL, D. & PRINCE, H. 2001. Prevalence of asthma and other respiratory symptoms in children living near and away from opencast coal mining sites. *International Journal of Epidemiology*, **30**, 556–563.

POREMBA, M. 2003. Reclamation in Gateshead: creating landscapes for people. *In*: MOORE, H. M., FOX, H. R. & ELLIOTT, S. (eds) *Land Reclamation – Extending the Boundaries*. Balkema, Lisse (NL), 101–107.

PRAKASH, A., FIELDING, E. J., GENS, R., VAN GENDEREN, J. L. & EVANS, D. L. 2001. Data fusion for investigating land subsidence and coal fire hazards in a coal mining area. *International Journal of Remote Sensing*, **22**, 921–932.

PRIESTLEY, J. B. 1934. *English Journey: Being a Rambling but Truthful Account of What One Man Saw During the Autumn of the Year 1933*. Heinemann/Victor Gollancz, London, 422 pp.

RICHARDS, H. P. 1951. Accidents. *In*: MASON, E. (ed) *Practical Coal Mining for Miners* (second edition). Virtue and Company, London, 337–341, vol. I.

RIMMER, D. L. & YOUNGER, A. 1997. Land reclamation after coal-mining operations. *In*: HESTER, R. E. & HARRISON, R. M. (eds) *Contaminated Land and its Reclamation*. Thomas Telford, London, 73–90.

RIMSTIDT, J. D. & VAUGHAN, D. J. 2003. Pyrite oxidation: a state-of-the-art assessment of the reaction mechanism. *Geochimica et Cosmochimica Acta*, **67**, 873–880.

ROBINSON, R. 2000. Mine gas hazards in the surface environment. *Transactions of the Institution of Mining and Metallurgy (Section A: Mining Technology)*, 109, A228–A236.

ROY, S. 2003. *Pyrite Oxidation in Coal-Bearing Strata: Controls on in situ Oxidation as a Precursor of Acid Mine Drainage Formation*. Unpublished PhD thesis, Department of Geological Science, University of Durham, UK, 439 pp.

SALAZAR, M., POCH, R. M. & BOSCH, A. D. 2002. Reclamation of steeply-sloping coal spoil banks under Mediterranean semi-arid climate. *Australian Journal of Soil Research*, **40**, 827–845.

SEAL, A. B. & BISE, C. J. 2002. Case study using task-based, noise-exposure assessment methods to evaluate miner noise hazards. *Mining Engineering*, **54**, 44–48.

SEIXAS, N. S., ROBINS, T. G., ATTFIELD, M. D. & MOULTON, L. H. 1992. Exposure–response relationships for coal-mine dust and obstructive lung-disease following enactment of the Federal Coal-Mine Health and Safety Act of 1969. *American Journal of Industrial Medicine*, **21**, 715–734.

SIDDLE, H. J., WRIGHT, M. D. & HUTCHINSON, J. N. 1996. Rapid failures of colliery spoil heaps in the South Wales Coalfield. *Quarterly Journal of Engineering Geology*, **29**, 103–132.

SINGER, P. C. & STUMM, W. 1970. Acid mine drainage: the rate limiting step. *Science*, **167**, 1121–1123.

SINGH, R. N., SHONHARDT, J. A. & TEREZOPOULOS, N. 2002. A new dimension to studies of spontaneous combustion of coal. *Mineral Resources Engineering*, **11**, 147–163.

SKARZYNSKA, K. M. 1995a. Reuse of coal mining wastes in civil engineering – part 1: properties of minestone. *Waste Management*, **15**, 3–42.

SKARZYNSKA, K. M. 1995b. Reuse of coal mining wastes in civil engineering – part 2: application of minestone. *Waste Management*, **15**, 83–126.

SMYTHE, J. A. & DUNHAM, K. C. 1947. Ankerites and chalybites from the northern Pennine orefield and the north-east coalfield. *Mineralogical Magazine*, **28**, 53–74.

SPEARS, D. A. 1989. Aspects of iron incorporation into sediments with special reference to the Yorkshire ironstones. *In*: YOUNG, T. P. & TAYLOR, W. E. G. (eds) *Phanerozoic Ironstones*, Geological Society Special Publication No 46. Geological Society, London, 19–30.

SPEARS, D. A. & LEE, S. 2004. Geochemistry of Leachates from Coal Ash. In: GIERÉ, R. & STILLE, P. (eds) Energy, Waste, and the Environment: a Geochemical Perspective. Geological Society, London, Special Publications, 236, 619–639.

STARR, R. C. & CHERRY, J. A. 1994. In situ remediation of contaminated groundwater: the funnel-and-gate system. Ground Water, 32, 465–476.

STRÖMBERG, B. & BANWART, S. A. 1994. Kinetic modelling of geochemical processes at the Aitik mining waste rock site in northern Sweden. Applied Geochemistry, 9, 583–595.

STRACHER, G. B. 1995. The anthracite smokers of eastern Pennsylvania: $PS_{2(g)}$–T stability disgram by TL analysis. Mathematical Geology, 27, 499–511.

STRACHER, G. B., TAYLOR, T. P. & PRAKASH, A. 2002. Coal fires: a synopsis of their origin, remote sensing detection and thermodynamics of sublimation. In: SHANNON, S. (ed) Environmental Technology for Mining: Case Histories of Mine Reclamation and Regulation. Robertson Geo-Consultants Inc., Reno, Nevada, 1–8. (Available on-line at: www.infomine.com/technology/enviromine/case_hist/coal%.../Stracher_et_al.htm)

STUMM, W. & MORGAN, J. J. 1996. Aquatic Chemistry (3rd edition). J. Wiley and Sons, New York.

TAJOVSKY, K. 2001. Colonisation of colliery spoil heaps by millipedes (Diplopoda) and terrestrial isopods (Oniscidea) in the Sokolov region, Czech Republic. Restoration Ecology, 9, 365–369.

TAYLOR, R. K. 1973. Compositional and geotechnical characteristics of a hundred year old colliery spoil heap. Transactions of the Institution of Mining and Metallurgy (Section A), 82, A1–A14.

TAYLOR, R. K. 1975. English and Welsh colliery spoil heaps – mineralogical and mechanical relationships. Engineering Geology, 9, 39–52.

TAYLOR, R. K. & SPEARS, D. A. 1970. The breakdown of British Coal Measures rocks. International Journal of Rock Mechanics and Mining Science, 7, 481–501.

TEMPLE, J. M. F. & SYKES, A. M. 1992. Asthma and open cast mining. British Medical Journal, 305, 396–397.

TEMPLE, F. C., ALLAN, W. et al. 1945. Northumberland and Cumberland Coalfields Regional Survey Report (Northern 'A' Region. Report to the Ministry of Fuel and Power. HMSO, London, 59 pp.

THIELEMANN, T., KROOSS, B. M., LITTKE, R. & WELTE, D. H. 2001. Does coal mining induce methane emissions through the lithosphere/atmosphere boundary in the Ruhr Basin, Germany? Journal of Geochemical Exploration, 74, 219–231.

THOMAS, L. 2002. Coal Geology. John Wiley & Sons Ltd, Chichester, 384 pp.

THOMPSON, R. J. & VISSER, A. T. 2002. Benchmarking and management of fugitive dust emissions from surface-mine haul roads. Transactions of the Institution of Mining and Metallurgy, Section A: Mining Technology, 111, A28–A34.

TISHMACK, J. K. & BURNS, P. E. 2004. The chemistry and mineralogy of coal and coal combustion products. In: GIERÉ, R. & STILLE, P. (eds) Energy, Waste, and the Environment: a Geochemical Perspective. Geological Society, London, Special Publications, 236, 223–246.

TIWARY, R. K. 2001. Environmental impact of coal mining on water regime and its management. Water Air and Soil Pollution, 132, 185–199.

TUCK, J. T. 1993. The Collieries of Northumberland, Vol. 1. Trade Union Printing Services, Newcastle upon Tyne, 109 pp.

VEPŘEK, S., COCKE, D. L., KEHL, S. & OSWALD, H. R. 1986. Mechanism of the deactivation of Hopcalite catalysts studied by XPS, ISS and other techniques. Journal of Catalysis, 100, 250–263.

WANNER, M. & DUNGER, W. 2002. Primary immigration and succession of soil organisms on reclaimed opencast coal mining areas in eastern Germany. European Journal of Soil Biology, 38, 137–143.

WILLIAMS, A. & MITCHELL, C. 1994. Methane emissions from coal mining. In: HESTER, R. E. & HARRISON, R. M. (eds) Mining and Its Environmental Impact. Royal Society of Chemistry, London, 97–109.

WILLIAMS, E. G. & KEITH, M. L. 1963. Relationship between sulfur in coals and the occurrence of marine roof beds. Economic Geology, 58, 720–729.

WOOD, S. C., YOUNGER, P. L. & ROBINS, N. S. 1999. Long-term changes in the quality of polluted minewater discharges from abandoned underground coal workings in Scotland. Quarterly Journal of Engineering Geology, 32, 69–79.

YOUNGER, P. L. 1994. Minewater pollution: The revenge of Old King Coal. Geoscientist, 4, 6–8.

YOUNGER, P. L. 1995. Hydrogeochemistry of minewaters flowing from abandoned coal workings in the Durham Coalfield. Quarterly Journal of Engineering Geology, 28, S101–S113.

YOUNGER, P. L. 1998. Coalfield abandonment: geochemical processes and hydrochemical products. In: NICHOLSON, K (ed), Energy and the Environment. Geochemistry of Fossil, Nuclear and Renewable Resources. Society for Environmental Geochemistry and Health, McGregor Science, Aberdeen, 1–29.

YOUNGER, P. L. 2000. Predicting temporal changes in total iron concentrations in groundwaters flowing from abandoned deep mines: a first approximation. Journal of Contaminant Hydrology, 44, 47–69.

YOUNGER, P. L. 2001. Mine water pollution in Scotland: nature, extent and preventative strategies. Science of the Total Environment, 265, 309–326.

YOUNGER, P. L. 2002a. Coalfield closure and the water environment in Europe. Transactions of the Institution of Mining and Metallurgy (Section A: Mining Technology), 111, A201–A209.

YOUNGER, P. L. 2002b. The importance of pyritic roof strata in aquatic pollutant release from abandoned mines in a major, oolitic, berthierine–chamosite–siderite iron ore field, Cleveland, UK. In: YOUNGER, P. L. & ROBINS, N. S. (eds) Mine

Water Hydrogeology and Geochemistry. Geological Society, London, Special Publications, **198**, 251–266.

YOUNGER, P. L. 2002c. Deep mine hydrogeology after closure: insights from the UK. *In*: MERKEL, B. J., PLANER-FRIEDRICH, B. & WOLKERSDORFER, C. (eds) *Uranium in the Aquatic Environment. Proceedings of the International Conference Uranium Mining and Hydrogeology III and the International Mine Water Association Symposium*, held in Freiberg, Germany, 15–21 September 2002. Springer–Verlag, Berlin, 25–40.

YOUNGER, P. L. 2003. Passive *in situ* remediation of acidic mine waste leachates: progress and prospects. *In*: MOORE, H. M., FOX, H. R. & ELLIOTT, S. (eds) *Land Reclamation – Extending the Boundaries.* Balkema, Lisse (NL), 253–264.

YOUNGER, P. L. 2004. 'Making water': the hydrogeological adventures of Britain's early mining engineers. *In*: MATHER, J. D. & TORRENS, H. S. (eds) *Two Hundred Years of British Hydrogeology.* Geological Society, London, Special Publication (in press).

YOUNGER, P. L. & ADAMS, R. 1998. *Improved Modelling of Abandoned Coalfields: 5th Six-monthly Progress Report.* Environment Agency (England & Wales) R&D Contract no. B04(95)2. University of Newcastle, UK. 55pp.

YOUNGER, P. L. & ADAMS, R. 1999. *Predicting Mine Water Rebound.* Environment Agency R&D Technical Report W179, Bristol, UK, 108 pp.

YOUNGER, P. L. & ROBINS, N. S. (eds) 2002. *Mine Water Hydrogeology and Geochemistry.* Geological Society, London, Special Publications, **198**, 396 pp.

YOUNGER, P. L., BANWART, S. A. & HEDIN, R. S. 2002. *Mine Water: Hydrology, Pollution, Remediation.* Kluwer Academic Publishers, Dordrecht, ISBN 1-4020-0137-1, 464 pp.

YOUNGER, P. L., JAYAWEERA, A. *et al.* 2003. Passive treatment of acidic mine waters in subsurface-flow systems: exploring RAPS and permeable reactive barriers. *Land Contamination and Reclamation*, **11**, 127–135.

ZIELINSKI, R. A., OTTON, J. K. & JOHNSON, C. A. 2001. Sources of salinity near a coal mine spoil pile, north-central Colorado. *Journal of Environmental Quality*, **30**, 1237–1248.

Coal combustion wastes: characterization, reuse and disposal

F. P. GLASSER

School of Engineering and Physical Sciences, University of Aberdeen, Meston Building, Meston Walk, Aberdeen, UK (e-mail: f.p.glasser@abdn.ac.uk)

Abstract: Coal extraction and combustion produce a number of waste streams, some of which can be utilized. Fly ash from the combustion of pulverized coal is characterized; it finds application as a partial replacement for Portland cement in concrete. Relatively few uses exist for other wastes, much of which is stockpiled, stored, or landfilled. These leave a legacy for future generations that must be managed to minimize geochemical impacts on air, soil, and groundwater quality.

Coal combustion provides a substantial fraction of world energy production, both thermal and electrical. In addition, coal and its semi-processed derivative, coke, are cheap chemical reductants essential to the extraction of a variety of metals from their ores (e.g., Fe, Mn, Cu). The market for coal is likely to increase over the next decade only slowly in developed countries but more rapidly in developing countries with suitable domestic resources (e.g., China, India).

The impact of coal extraction, distribution, and combustion is considerable. Long-established coal-producing regions are often burdened with legacy deposits arising from coal extraction (see Younger, 2004). Moreover, combustion emits both gases (CO_2, SO_2, NO_x) and fine particulates ($\sim 1\%$ of combustion ash) into the atmosphere (see Golomb & Fay, 2004). However, the large amounts of solid wastes arising as a consequence of combustion and extraction form the main subject of this chapter.

Modern coal combustion employs two principal techniques: combustion in a fluidized bed or pulverization, followed by combustion of fine particles suspended in moving air. Figure 1 shows a schematic of pulverized coal combustion, a process much used in steam-raising plants. Each process produces a characteristic residue; fluidized bed combustion gives rise mainly to a clinker-like or granular product, whereas pulverization, followed by combustion, produces mainly a much finer, micrometre-sized ash residue. Pulverization also yields a coarser fraction, the so called 'bottom ash', which is periodically removed without difficulty. However, the finer 'fly ash' has to be recovered by filtration and electrostatic precipitation. Commercially, fly ash has

the greatest economic potential for reuse. Recent estimates by Kalyoncu (2000) are that 57 Mt of fly ash are produced annually in the USA alone.

However, it should be recalled that in some applications, coal combustion residues are not recovered as separate fractions; instead, they are incorporated directly in another product or byproduct. For example, Portland cement is often produced using coal as the fuel. The mineral fraction of the coal reacts with the feed, mainly limestone and shale, and is incorporated directly into the final product, termed cement clinker. The chemical composition of the feedstock has to be adjusted to take account of the amount and chemistry of the ash to enable the optimum overall batch composition to be achieved (Lea 1988). The ash undergoes complete mineralogical and microstructural reconstitution in the course of pyroprocessing to form Portland cement and is no longer recognizable as ash. In the metallurgical industries, coal is heated in retorts to expel volatiles. The resulting hard, but spongy residue, termed coke, is used as the reductant for Fe-oxides; mineral impurities in the coke combine with those in the ore and flux, and the resulting molten oxide melt, termed slag, has a composition that reflects the three principal sources of impurities: coking coal, ore, and flux (normally limestone and dolomite). Slag is technically a byproduct of iron making, but, as will be shown, is recovered for utilization.

However, overall utilization of coal combustion products is low, with the result that coal extraction and combustion yields a copious supply of waste materials. These present serious ecological and environmental problems of management and stewardship.

From: GIERÉ, R. & STILLE, P. (eds) 2004. *Energy, Waste, and the Environment: a Geochemical Perspective*. Geological Society, London, Special Publications, **236**, 211–222.
0305-8719/04/$15   © The Geological Society of London 2004.

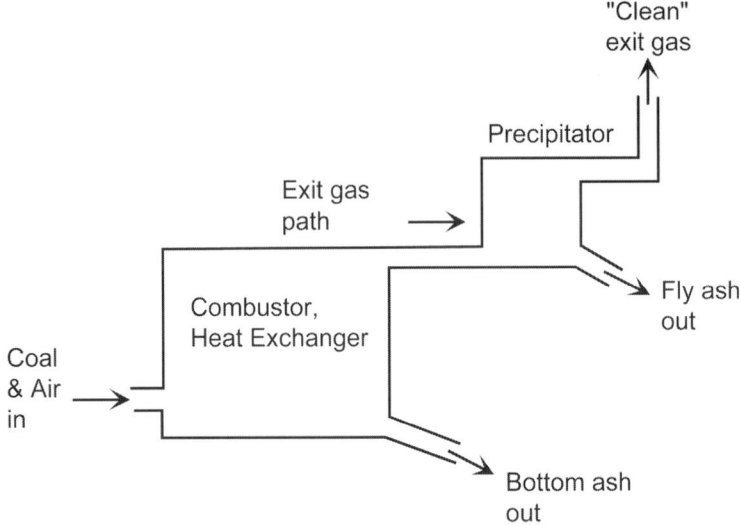

Fig. 1. Combustion pathways utilizing pulverized coal. Exit gas is released directly to the atmosphere.

Coal and its inorganic constituents

Coal is comprised mainly of carbon, hydrogen, and oxygen. It is generally regarded as arising from the accumulation and alteration of vegetable matter in the absence, or near absence, of free oxygen. Because of its commercial and geological importance, numerous coal classification schemes exist, and legal standards govern its thermal value, grading, and analysis (see Tishmack & Burns, 2004). Our present concern is with the mineral matter, which may range from a few wt% upwards.

The mineral matter in coal arises from various sources. Detrital minerals are derived from fossil soil as well as from wind- and water-derived particulates contained within coal, in partings and in material bounding the coal seam. Detrital constituents include quartz, felspar (often more or less altered), and clay minerals, of which kaolinite is perhaps ubiquitous. Other resistate minerals are often present in small amounts: mica, zircon, rutile, or other TiO_2 polymorphs, and hydrated Fe-oxides. Secondary minerals are formed by percolation. In the course of time, water percolates coal, and mineral constituents present in the water react with coal and its primary minerals. These reactions are often mediated by microbiological action, for example, resulting in formation of CO_2, H_2S, and sulphur in oxidized speciations. The nature of mineral formation depends on geological setting and local hydrology of the coal in its original depositional environment: calcite is common

in some coals. Sulphur present in coal tends to be reduced to S^{2-}, on account of the chemically reducing local environment, and is insolubilized by reaction, typically by formation of minerals, such as the FeS_2 polymorphs marcasite and pyrite. The grain size of the mineral matter is generally fine, typically of μm to sub-μm size with few exceptions. These exceptions include calcite and sulphides, both of which may form nodules and concretions up to several cm across. However, no categoric statements can be made; many coals contain sulphide, which is very fine-grained and, on that account, difficult to remove in the course of coal beneficiation.

Modern coal mining and processing has as its objective removal or reduction of sulphides and other non-combustible mineral constituents, which lower the recoverable heat content. The success achieved by these beneficiation methods is variable; for example, relatively large lumps of sulphides may be removed, but the finer-grained, well-distributed sulphides are often not accessible without the necessity of very fine comminution, with the result that valuable methane trapped within the coal escapes, thereby reducing the residual heat content and creating a potential explosion hazard. Finely ground coal product also forms an explosive dust when in air suspension and is prone to spontaneous combustion in storage. Hence much coal has to be burnt with relatively little beneficiation and, if pulverized, is immediately combusted.

Coal combustion

During coal combustion, temperatures reach 1200–1400 °C in fluidized-bed conditions, and somewhat higher, perhaps 1400–1800 °C, in the course of gas-suspension combustion of pulverized coal. Under these latter conditions, extensive fusion of ash occurs. Owing to normally oxidizing conditions, sulphur oxidizes to SO_2, which partially escapes unless a separate scrubber is installed. In general, the entire sequence of combustion from ignition to combustion and subsequent partial cooling occurs rapidly, within hundreds of milliseconds. As a consequence, the mineral matter only partially equilibrates to the high-temperature environment, and the high-temperature imprint is effectively quenched during cooling.

Figure 2 shows a projection of mineral ash compositions on a ternary grid having as its end-members $(CaO + MgO)$, $(Al_2O_3 + Fe_2O_3)$ and SiO_2. This projection generally includes >90% of ash components expressed in mol%. Although ash compositions spread over the area shown, depending on coal provenance, two distinct population clusters occur. These clusters give rise to terms 'Class F' and 'Class C' ashes (Tishmack & Burns, this volume; Helmuth 1987). Broadly, Class-F ashes comprise a population with low-lime (CaO) contents, whereas

Class-C ashes are lime-rich. It was at one time believed that coal combustion ash fell into one of these two categories, but today it is known that ash compositions can lie anywhere in the zone delineated in Fig. 2, even though examples of intermediate ash types (between classes F and C) are perhaps less common.

The impacts of thermal cycling on ash mineralogy are partially explained by relating the bulk composition of the ash to those of appropriate reference minerals (see Fig. 2). The phase content of fly ash reflects both the high-temperature excursion as well as the persistence of detrital minerals in coal (Table 1). During combustion, virtually all detrital minerals become unstable at the high peak temperatures. The extent to which reaction occurs reflects opposing tendencies, that is, the demands of equilibrium, the limitations on reaction imposed by granulometry and kinetics of reaction and of dissolution in the melt, as well as the large intergranular differences in composition. The persistence and dissolution of minerals in the melt phase at peak temperatures is controlled in part by particle size, by the degree of undersaturation of the melt with respect to each mineral phase, as well as by the kinetics of reaction at mineral–melt interfaces. Thus, while strong thermodynamic driving forces exist to attain a

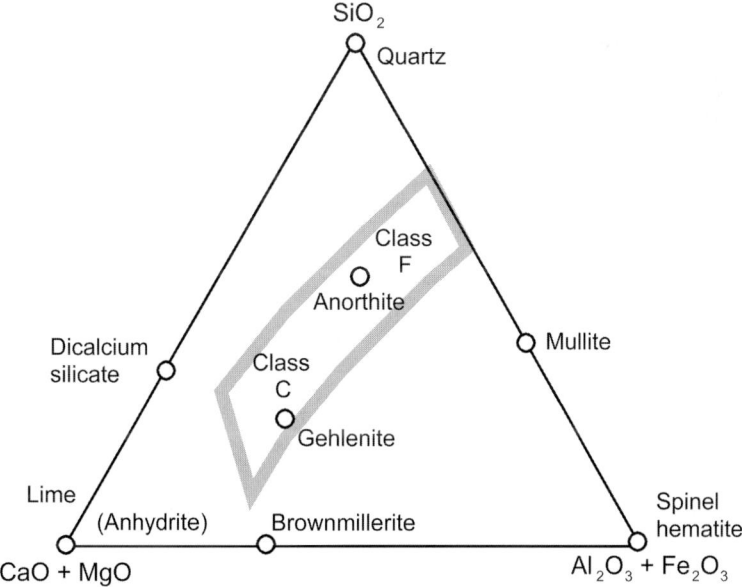

Fig. 2. Pseudoternary diagram showing projected compositions in mol% of coal combustion fly ashes and of phases present in ash. Polymorphs of dicalcium silicate include larnite. Bulk composition of fly ash lies within the field bounded by the shaded thick line.

Table 1. *Origin of mineral constituents in fly ash*

Ash type	Detrital	High temperature	Quench features
Aluminosilicate ash (Class F)	Quartz and unburnt carbon	Melt, mullite, Fe_2O_3, Fe_3O_4	Glass-in-glass phase separation: secondary mullite
Ca-rich ash (Class C)	Quartz (relatively uncommon)	Glass (minor), CaO, MgO, Fe_2O_3, Fe_3O_4, Ca_2SiO_4, $Ca_3Al_2O_6$, $Ca_{12}Al_{14}O_{33}$, $Ca_2(Fe,Al)O_5$, $Ca_2Al_2SiO_7$, $CaAlSi_2O_8$	Crystallization of melt to yield Ca-aluminates, silicates and ferrite; minor glass persists

high-temperature equilibrium, the duration of heating of a Class-F (aluminosilicate) ash is typically too brief to attain complete reaction, with the result that quartz tends to persist. As melt develops, it wets residual quartz strongly. However, the melt often approaches silica saturation at least locally; a thin envelope of siliceous glass develops at the quartz–melt interface through which diffusion must occur if reaction is to continue. This is a comparatively slow process: for example, experience of relevant commercial glass compositions (e.g., window and container glasses) supports the contention that dissolution of quartz is kinetically sluggish. On the other hand, quartz–albite–kaolinite mixes begin eutectic melting above ~1100 °C, at atmospheric pressure and much melt forms (Levin *et al.* 1964). Melting reactions also compete with internal reorganization within the separate minerals: for example, heating kaolinite yields initially a dehydroxylated phase, termed metakaolin, while eventually, above ~1000 °C, mullite, nominally $Al_6Si_2O_{13}$, and melt will form. Fe-oxides, depending on temperature and oxygen fugacity, do not melt but instead form either or both spinel and hematite with loss of hydroxide, evolved as water, and oxygen. Because Fe-oxides have generally poor solubility in siliceous melts, they too, like quartz, tend to persist into the cooled product. The ash product is thus chemically and physically inhomogeneous across a broad range of length scales (Kilgour & Diamond 1988).

During cooling, the high-temperature phases persist. The melt phase undercools, and becomes rigid, that is, it forms a glass (Hemmings & Berry 1988). The composition of the glass may vary locally, because melting is too rapid to allow the glass to homogenize. On account of its high content of solid inclusions, it is not easy to determine the average composition of the glass in commercial fly ashes. Table 2 shows the bulk composition of a US National Bureau of Standards reference fly ash, described by Kanare (1986), together with high-resolution nanoscale analysis of the glass

Table 2. *Analysis of reference fly ash SRM 2689 (U.S. National Bureau of Standards*)[†]*

wt%	Bulk fly ash	Glassy phase
Na_2O	0.34	—
MgO	1.01	0.1
Al_2O_3	24.45	22.6
$SiO2$	51.74	50.1
SO_3	—	0.2
K_2O	2.65	3.9
CaO	3.05	3.8
TiO_2	1.25	2.1
Fe_2O_3	13.32	16.8
L.O.I.[‡]	1.76	—
Total	99.30	99.6

*Now NIST, National Institute of Science and Technology.
[†]The bulk analysis is by standard wet chemical methods. The glassy phase was analysed by analytical transmission electron microscopy. Adapted from Qian & Glasser (1988) and Qian *et al.* (1988).
[‡]L.O.I. = Loss on ignition.

phase averaged over a number of inclusion-free regions. Relative to the bulk composition of the fly ash, the glass is somewhat enriched in K_2O and, in all probability, Na_2O, although the detector system used was not sufficiently sensitive for Na to permit accurate analysis.

The glassy phase undergoes spontaneous phase separation during cooling (Roth *et al.* 1987). It is perhaps surprising in view of the high temperatures required for melting that this should occur, but the phenomenon is well known from laboratory studies as well as from its occurrence in fused or partially fused refractories. The liquid phase develops local structures spontaneously in the course of cooling and, with decreasing temperature, these structures increasingly tend to diverge in composition, with the result that a broad range of aluminosilicate melts develop a two-phase structure: one melt is SiO_2-rich, the other Al_2O_3-rich. Phase separation occurs on a nanoscale so that even high-resolution analyses, for example, by analytical transmission electron microscopy, are insufficient to resolve the domains for analysis;

Table 2 thus averages the compositions of the domains. Figure 3 shows a temperature–composition equilibrium diagram of the Al_2O_3–SiO_2 system. Melting relationships are well established: fusion begins above $\sim 1630\,°C$ in the pure $Al_2O_3 \cdot SiO_2$ binary system (cf. eutectic point in Fig. 3), but will be somewhat lowered in fly ash because of the presence of fluxing oxides, such as alkalis, and Mg- and Fe-oxides.

A wide range of melt compositions undergo phase separation upon cooling. Two mechanisms are observed: classical phase separation (dashed curve) as well as by spinodal (dot–dashed curve) mechanisms. Both the position of these regions and the onset temperature of phase separation reported in the literature vary (e.g., Roth et al. 1987), but there is no doubt that the phenomenon occurs amongst the glassy phase of aluminosilicate fly ash despite the presence of other oxides that tend to promote miscibility.

In the course of cooling commercial aluminosilicate fly ash, phase separation is often followed by partial crystallization of the alumina-rich amorphous phase. Primary crystallization at high temperature yields coarser mullite, often revealed in micrographs of acid-etched particles, whereas the subsequent crystallization of mullite through devitrification during cooling yields finer crystals (Kilgour & Diamond 1988). Thus two generations of mullite often occur, distinguished by origin and size, as well as textural relationships to the boundaries between immiscible regions. The consequences of phase separation and mullite crystallization to reactivity of glass are unknown, but presumably strain energy arises at interfaces owing to mismatch of thermal coefficients of expansion; given the nanoscale at which immiscibility occurs, the resulting interface area potentially available for stress accumulation must be very large. The writer speculates that the additional strain energy thus created enhances the reactivity of the glass phase.

Lime-rich fly ashes exhibit large variations in both chemistry and mineralogy. First, as with the mineral and chemical composition of all fly ashes, large variations may occur from grain to grain (Stevenson & Huber 1987; McCarthy et al. 1987). This variation extends to the distribution

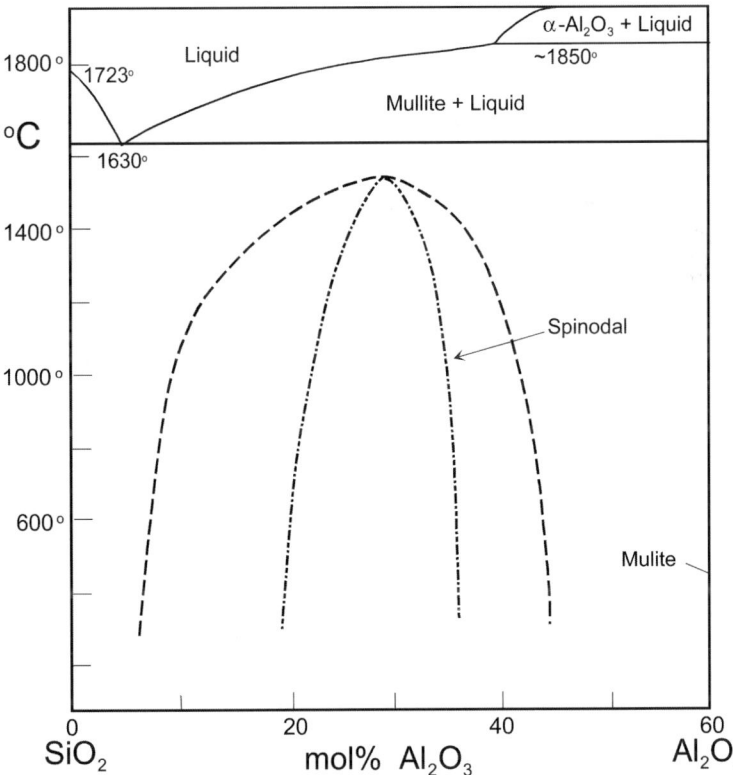

Fig. 3. Phase diagram of a portion of the SiO_2–Al_2O_3 system, showing regions of phase separation of undercooled melts. The projected composition of Class-F fly ash glasses lies in the range of 10–40% mol% Al_2O_3. Adapted from Roth et al. (1987).

of minor and trace elements, as shown for a Class-F fly ash by Gieré et al. (2003). Moreover grains may experience different thermal histories and may only partially equilibrate in the course of combustion. Secondly, although much melting may occur at high temperatures, the low-silica melts are not good glass formers and normally crystallize in whole or in part during cooling of the ash. As a result, some minerals pass through the combustion stage essentially unaffected while others form as a direct consequence of the high-temperature excursion followed by cooling. Examples of local equilibria induced by combustion include free lime, formed by thermal decomposition of $CaCO_3$, and the Fe-oxides hematite and magnetite, formed by oxidation of Fe-sulphides and dehydration of ferrous and ferric oxyhydrates. However, reaction also occurs between mineral phases within individual particles, resulting in formation of phases that may be rare in nature and are not primary constituents of coal. Examples of such phases include: dicalcium silicate, β-Ca_2SiO_4 (larnite); Ca-aluminates, Ca_3AlSiO_6 and $Ca_{12}Al_{14}O_{33}$ (mayenite); melilite solid solutions close to gehlenite, Ca_2AlSiO_7; anorthite, $CaAlSi_2O_8$; and brownmillerite, $Ca_2(Al,Fe)O_5$. Metastable phase coexistence is commonly observed (e.g., free lime and anorthite).

A noteworthy feature of the mineralogy of high-Ca fly ash is that many of the phases formed are reactive towards water with development of a cementitious bond, for example: CaO, β Ca_2SiO_4, $Ca_3Al_2O_6$, and $Ca_{12}Al_{14}O_{33}$. Indeed, several of these phases occur in Portland cement, a manufactured product comprised of, in decreasing order of abundance, Ca_3SiO_5, Ca_3SiO_4, and roughly equal amounts of $Ca_3Al_2O_6$ and $Ca_2(Fe,Al)_2O_5$. Thus the Ca-rich ashes may be cementitious in their own right, although the strength attained and rate of strength gain are normally too slow for commercial use. The ash reactivity is improved by adding activators, such as portlandite, $Ca(OH)_2$, gypsum, $CaSO_4 \cdot 2H_2O$, and Portland cement. Zenieris & Lagutos, (1988) described the use of high fly ash content mixes in conjunction with activators to form low-strength subfoundations.

In general, lack of equilibrium during the pyro-processing stage and heterogeneity of composition make it impractical to calculate the phase content of fly ash from its chemical analysis. Two other factors, common to all types of fly ash, further complicate evaluation of the potential to use fly ash. One is the presence of unburnt carbon, which should preferably not exceed a few wt%; the other is the presence of alkali-sulphates (see Groppo et al., 2004). When used in cementitious formulations, free carbon increases the water demand of the mix. Alkali sulphates have high vapour pressures at coal combustion temperatures and partially evaporate (Barry & Glasser 2000). During cooling, the vapour stream may become supersaturated with respect to Na- and/or K-sulphate; under these conditions, and depending on temperature, a condensate may form on ash particles consisting of molten or solid sulphates. The equilibria and kinetics of the sulphur cycle are probably similar to conditions in a cement kiln, the thermodynamics of which were explored by Barry & Glasser (2000). As the vapour stream cools, the surfaces of fly ash particles act as nucleation centres, as a result of which the outer surfaces of fly ash concentrate alkali-sulphates. As the sulphates are potentially water-soluble when fly ash is used as a cement replacement material, they rapidly dissolve in mix water and, if present in sufficient quantity, the alkali and sulphate can affect the normal hydration properties of the cement. Soluble sulphates also affect ash behaviour in disposal (see below).

Fly ash morphology

The shape and size of fly ash particles are important in application areas. Various classifications exist for fly ash; Table 3 shows a morphological classification (Helmuth 1987). Ca-rich fly ashes, typically having high crystalline contents, tend to have rather irregular morphologies with occasional agglomerates, types 4 through 8 (Table 3) are thus most frequently encountered. Aluminosilicate ashes also exhibit a range of morphologies, typically types 1 through 3. The hollow structures of aluminosilicate ashes arise when gases (SO_2, CO_2, H_2O) are actively evolved while fusion occurs; the rather viscous aluminosilicate melt lends itself to bloating with formation of bubbles. Figure 4 shows an example of a multiple, thin-walled structure formed in an aluminosilicate fly ash. These complex multiwalled structures, termed plerospheres, may comprise an appreciable volume fraction of certain ashes (see also Tishmack and Burns, 2004). Simpler, single-shelled hollow spheres, termed cenospheres, are also common. Coal fly ashes with a high quantity of hollow particles are often beneficiated by immersion in water-filled ponds, where the cenospheres float. Thus, they can be removed by a rotating skimmer arm similar to those used in sewage works to skim off froths. The low-density product is washed, graded and dried, and has found application in insulating products, in lightweight concrete materials, and as fillers for organic polymeric components. In recent decades, however, the advent of bloated volcanic

Table 3. *Morphological classification of reference fly ash particles**

Designation	Description of morphology	Characteristic size range (μm)
Type 1	Spherical or rounded, often with high glass content. Particles may be solid or hollow, colourless.	0–20
Type 2	As above, but light brown to black. Mainly solid with a few cenospheres.	5–30
Type 3	Rounded, colourless. Spongy texture and high glass content. May include cenospheres.	10–200
Type 4	Irregular, tan to brown. Partially crystalline.	10–100
Type 5	Sintered agglomerates of smaller particles of all types.	50–500
Type 6	Black, carbon-rich.	20–200
Type 7	Angular, crystalline, pale colour perhaps indicating quartz.	10–100
Type 8	As above, but red-brown; probably hematite/magnetite.	5–50

*After Helmuth (1987).

glass has supplanted the use of fly ash concentrates.

Use of fly ash

Although many Ca-rich fly ashes are self-cementing, strength gain is usually too slow for most practical applications. Moreover, aluminosilicate rich ashes are essentially unreactive with water. The most important application of fly ash is as a partial replacement for Portland cement. In this application, Portland cement furnishes much of the early strength, up to ~1 month, while the high alkalinity of cement chemically activates the ash such that its slow reaction with cement components and water contribute increasingly to long-term strength gain. Replacement levels of fly ash up to 30–40 wt% of cement content are often specified in the belief that replacement enhances the performance of concrete in the environment (Helmuth 1987; Malhotra 1986, 1987). However, many national codes permit lower levels of supplementary cementing materials, up to about 10 wt%, to be marketed as 'Portland cement' without special labelling. Fly ash competes with ground limestone and granulated iron blast-furnace slag as a supplementary cementing material in commercial Portland cement formulations.

The reason why fly ash is an attractive supplementary cementing material is because the blends formulated with aluminosilicate-rich

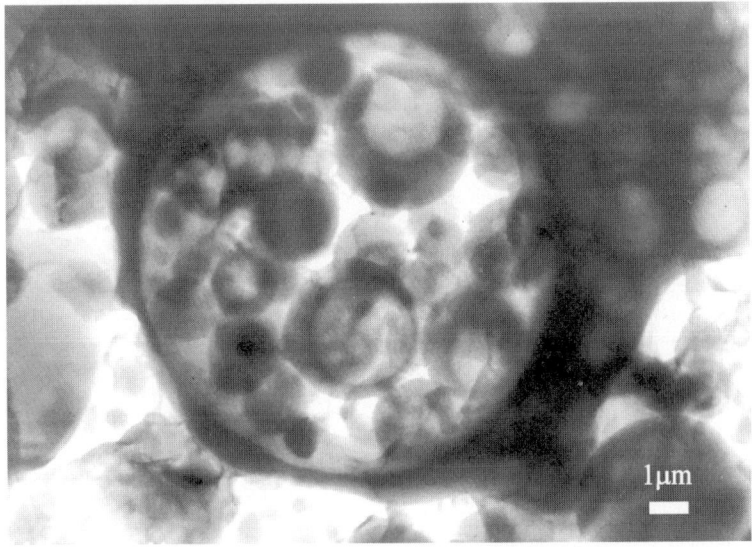

Fig. 4. Plerospheres in coal combustion fly ash. The large plerosphere in the centre of the field of view has broken open to reveal smaller nests of plerospheres and cenospheres inside. Transmission electron micrograph courtesy of Dr. E. E. Lachowski.

ashes are easily handled and achieve satisfactory mix rheology of fresh concretes, especially at low water contents. When hardened, the resulting concretes made with low water contents achieve reduced permeability and, as noted, give better protection to embedded steel and have improved resistance to penetration by aggressive chemicals in the service environment. These qualities are, however, only ensured if the resulting concrete mix is well mixed and well cured; water is required for the chemical reactions associated with hydration, and it is essential that added mix water should not be lost prematurely by drying. Since fly ash reacts rather slowly, relative to modern Portland cement, fly ash cement blends may require longer cure times relative to equivalent concretes formulated without fly ash. Other advantages are claimed for ash-rich blends: the aluminosilicate hydrates formed spontaneously in the course of reaction promote absorption of alkalis from the cement pore fluids; the alkali sorption mechanism has been quantified by Glasser & Hong (2002). Alkali-sensitive mineral aggregates may be accommodated in fly ash concrete with less risk of subsequent expansion and distress than if the concrete were made with plain Portland cement (Macphee & Glasser 1996). Ash-rich blends do not normally exhibit greater strengths than concretes of equivalent grade but made without ash. Indeed early strengths, 1–14 d, may be reduced at high levels of cement replacement. However, besides the improved long-term durability of fly ash-rich blends, they also limit the temperature rise resulting from rapid release of the heat of hydration typical of plain Portland

concretes and thereby minimize thermally induced cracking: hydration of cement is strongly exothermic and on that account, large concrete constructions have to be carefully planned to avoid thermal excursions with crack development during cooling. While fly ash blends do not free designers from the constraint of avoiding excess temperature rise, the magnitude of the problem can be reduced by replacing cement in part by fly ash.

Cement is a highly specified and relatively homogeneous material on a scale >100 μm, approximately. Fly ash, on the other hand, is notably inhomogeneous. Partly on that account it has proven difficult to develop adequate tests of the performance of cement–fly ash blends. Speed of reaction, with resultant development of strength, termed 'pozzolanic reactivity', is a key engineering parameter. However, no agreed definition of reactivity or its measurement exists and, as a result, the suitability of rapid tests of pozzolanic activity are the subject of considerable controversy. While test formulations can be cured and their properties, for example, compressive strength, measured as a function of age, this is a slow process, requiring perhaps months of curing and elaborate statistical control, given the inherently poor repeatability of individual determinations. Therefore rapid tests have long been sought. Such tests need in the first instance to be benchmarked against specimens normally made with plain cement or standard mortars and concretes in which cement is diluted by inert material aggregate. Table 4 illustrates some of the proposed rapid test methods and their limitations. The Feret–Florentin method allows for

Table 4. *Tests of pozzolanic activity of fly ash**

Test designation	Basis of test	Comment[†]
Vicat	Rate of absorption of calcium measured indirectly from a saturated $Ca(OH)_2$ solution in contact with excess $Ca(OH)_2$ (solid) by determining residual $Ca(OH)_2$.	Residual free portlandite is measured by DTA, TG, or solvent extraction. Correlation with strength gain is poor.
Blaime's method, modified by Sestini and Santarelli	Sequential extraction with HCl and KOH. Soluble Al and Si are analysed.	Poor reproducibility and poor correlation with strength gain.
Lea's conductimetric test	Electrical conductivity of a solution of $Ca(OH)_2$ in a cell with fly ash is measured periodically.	Amount of reaction small: reproducibility is poor. Correlation with mortar strengths are also poor.
Feret–Florentin method	Reactivity measured of 0.5 g fly ash and 0.5 g $Ca(OH)_2$ in 50 mL water. Amounts of Al, Si, and Fe thus solubilized into acid solution are determined.	Results are corrected for median particle diameters. Resulting correlations with strength are good for ash particles in the range 7–46 μm.

*Adapted from Helmuth (1987).
[†]DTA, differential thermal analysis; TGA, thermo-gravimetric analysis.

the impact of granulometry and, on that account, is arguably the most realistic test. Typically, a good reactive fly ash will be found to have a median particle diameter less than that of the Portland cement with which it is blended. The fly ash grains tend to pack in the interstices of the cement particles, thus ensuring optimum contact with the activator (cement) and best possible densification of the still-wet mix.

Use of slags

Brief mention may also be made of other products whose compositions are affected by the mineral content of coal, for example iron blast-furnace slag. Iron ore contains various impurities, depending on source and extent of pre-processing and beneficiation. Silica is a common contaminant present in ore, which also may contain phosphorus. The function of the iron blast furnace is to reduce Fe-oxide to metallic iron. Coke is the preferred reductant, as its physical structure allows combustion air to permeate the charge (Richardson 1974). However, the coke also contributes a considerable content of impurity minerals. These do not appear as an ash, but instead react with the mineral components of the ore and added fluxes, typically limestone and dolomite, to form a fluid melt, termed slag. At temperatures between $\sim 1250\,°C$ and the peak operating temperature of $\sim 1550\,°C$, carbon-saturated Fe metal is also molten: it accumulates at the bottom of the furnace from whence it is periodically removed. Slag floats on the molten Fe as a second, immiscible liquid, which is also completely molten, or almost so. The slag plays an important role in the process: to dissolve as much as possible of the impurities, principally sulphur and phosphorus from the metal. Its physical chemistry is well known, with the result that the slag is made as rich in CaO as is consistent with the remaining liquid at operating temperatures (Richardson 1974). To achieve optimum slag performance consistent with having minimum slag volume, it is therefore essential to factor in the mineral content of the coke.

Slag utilization is favoured by quenching the molten slag to glass, using air or water as the quenching medium. The resulting glass is much fractured by quenching and is readily ground in a cement mill, quenched slag glass being somewhat softer than clinker. The resulting blend of slag and Portland cement, containing up to 60–70 wt% ground glassy slag, finds ready acceptance in concrete making. Thus utilization of suitable slags approaches 100% in the European market. Once dumped, however, slag readily looses its reactive properties, so old dumps cannot in general be reworked: the product weathers readily and has a relatively short storage life if left unprotected from reaction with water and atmospheric carbon dioxide (Glasser, unpublished).

Flue gas desulphurization product

The exit gases from coal combustion frequently carry much of the S, originally present in coal, as SO_2. Depending on the nature of the combustion process and combustion chemistry, some of the S may react with alkali in the vapour stream to condense as alkali sulphates, for example, K_2SO_4, Na_2SO_4, and their solid solutions, but most of the SO_2 remains in the vapour; at lower temperatures it may partially oxidize to SO_3. Removal of SO_2 and SO_3 is the subject of numerous patents. Broadly, the main commercial process utilizes injection of powdered lime into the vapour stream. In the presence of steam, reaction occurs forming mainly $CaSO_4$ as a fine particulate solid (see Tishmack & Burns, this volume). This is removed by scrubbing the cooling vapour stream with water, which, when filtered, gives rise to mixtures of gypsum, $CaSO_4 \cdot 2H_2O$, $CaSO_4 \cdot 0.5H_2O$ (hemihydrate), and anhydrite, $CaSO_4$. Hemihydrate is the main raw material for production of gypsum plaster and, provided anhydrite can be reduced or eliminated from the product, the fine-grained calcium sulphate is often suitable for introduction into slurries used to make gypsum products such as drywall plasterboard. Other uses include intergrinding with Portland cement clinker, typically at 1–4 wt%, to retard its initial set. However, accumulation of desulphurization products typically much exceeds demand and the product is often disposed to landfill.

Disposal

The disposal of coal combustion residues must take into account the nature and amount of the products to be disposed, as well as the nature of the disposal environment. It is a characteristic of coal utilization that many historic wastes exist that lack, or almost totally lack, characterization data. Only in recent decades, with concern over the environmental impacts of present and future accumulation of wastes, have characterization studies been made. The principal types of waste are described in Table 5.

Studies of the long-term evolution of disposed wastes are often site specific and most studies have been of relatively brief duration, months, or at most a few years. Fine particulate wastes are subject to rapid erosion by rain and, during dry spells, transport by wind. For this reason, wastes are often ponded as slurries. Earth or clay bunds are constructed and the resulting barriers used to contain the slurried combustion product. Over the course of years or decades, the fine particulates gradually compact and consolidate so that the contents can eventually be allowed to drain and dry, thus achieving self-consolidation. Alternatively, the ash may be mounded. The core of ash mounds is usually formed from the fine particulate materials with a cover of coarser material, for example, fly ash core with a bottom ash cover. Upon abandonment, it may be appropriate to add a final layer of soil and plant the mound.

The concern is that the combustion process mobilizes the major/minor and trace element concentrations of combustion product and that fresh water (streams, lakes) as well as groundwater may become contaminated by leachate. In general, sites for ash disposal are physically close to the combustion site. Thus, the location is chosen primarily on the basis of economics, and the disposal sites exhibit a wide range of local hydrological regimes and are, of course, subject to prevailing climatic conditions.

Although coal combustion residues are not particularly hazardous or toxic, the scale of the problem should not be underestimated. A large coal-fired electricity-generating plant may, over a 25–40 year lifetime, consume upwards of 10^8 t of coal; depending on ash content, this might yield $>10^7$ t of material for disposal. Large masses of disposed combustion material may thus have substantial geochemical impacts that cumulate with time, and we are at present unable to give reliable estimates of the future behaviour and impacts of disposal.

Present leaching behaviour can be characterized, although the applicability of many leach tests with fly ash is doubtful. Special tests are often more appropriate. For example, Fig. 5 shows a simple packed bed, through which leachant percolates and which acts like a lysimeter: leachate can be sampled periodically through the

valve without disturbing the test. Scope also exists for renewing the leachant, if desired.

Analyses of leachates are generally benchmarked with reference to drinking water standards. Coal combustion products will, of course, frequently have a geochemical signature characteristic of the coal provenance and combustion technology. Nevertheless, some general conclusions can be drawn from leachate analyses. Species with high leach rates, exceeding drinking water standards, include among major elements Na, K, and Al. Among minor and trace elements, Cr, Se, As, and Mn often exceed permitted limits (Andrada et al. 1990). However, concentrations of leached elements also change with time, and among the elements whose concentration increases/tends to increase with time are Na, K, As, B, Cr, Li, Mo, and Se.

In general, the speciation of the leachable elements in the ash is unknown. Moreover, a series of weathering reactions occur during and after disposal. The nature and severity of reactions is modified by the water regime as well as biological action. Microbiological regimes are sensitive to the environment, and bioactivity in the waste is also modified by access to oxygen and groundwater levels. For example, chromium is most readily mobilized in the oxidizing zone of wastes, where oxidation to Cr(VI) yields species that are, in general, more soluble than Cr(III). Oxidation is prevalent in air-permeable materials, whereas flooded or ponded ash remains anaerobic. Thus in flooded environments, the regime is anaerobic and Fe remains chemically reduced and methane may be produced, whereas oxygenated environments typically result in CO_2 production and oxidation of Fe^{2+} to Fe^{3+}. Many deposits fluctuate between aerobic and anaerobic conditions seasonally.

It is also important to control pH or effluent discharges. Low pHs are often associated with production of sulphuric and sulphurous acids.

Table 5. *Waste arising from coal production and use*

Nature of wastes	Key features relevant to behaviour
Coal mining and processing wastes	Characteristically close to coal source. Contain much rock. Free carbon and S^{2-} present. Spontaneous combustion hazard. Oxidation of S^{2-} to S^{6+} results in acid (H_2SO_4) drainage with leaching
Coal, coking and allied coal tar industries	Heavily contaminated land and dumps. High content of toxic or hazardous organics.
Coal gasification processes	Heated lignitic residues liable to leaching.
Bottom ash and fluidized bed ashes	Coarse, cindery agglomerates liable to leaching. Neutral to alkaline.
Fly ash	Characteristically much finer than bottom ash. Mobility of fine particulates as dusts may be a problem.
Flue gas desulphurization product	Gypsum, lime and other impurities. Nominally alkaline, but if sulphide is present, it may oxidize causing transition to an acidic regime.

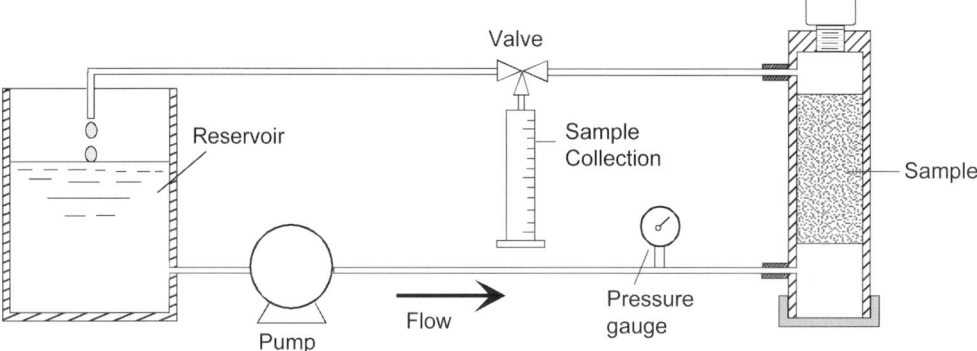

Fig. 5. Schematic of the assembly used in flow-leaching experiments.

Not only do acidic discharges leach metals from ash, but they may also liberate metals normally bound onto or into mineral structures of geomaterials, including minerals in flow paths and aquifers. Consequently, an important part of disposal strategy lies in estimating the future potential for acid production. Insofar as acid production is associated with oxidation, for example, of sulphide sulphur to SO_3^{2-} and SO_4^{2-} species, acid production can be limited by preventing or reducing free access to oxygen, either gaseous or dissolved in percolating water. A variety of strategies have evolved to control oxygen access, for example, impermeable clay caps and barriers or static flooding are widely used.

The thermodynamic incompatibility of many of the solid phases present with each other as well as their local environment, results in formation of secondary minerals. Although the secondary materials may comprise only a small volume fraction of the waste, they (1) tend to increase in amount with time, as weathering processes proceed, (2) typically form at grain surfaces and are thus physically liable to react with percolating gas or liquids, and (3) may exhibit sites suitable for sorption or crystallochemical incorporation of trace elements (see Donahoe, 2004). Frequently observed secondary minerals include jarosite and ettringite: the former is known to sorb ions such as Mn and As, whereas ettringite can form solid solutions, in which SO_4 is replaced by CrO_4 (Kumarathasan et al. 1990).

While the geochemical impacts of weathering in the disposal environment are far from being understood, it is apparent that considerable scope exists for managing waste disposal so as to achieve low leach ratios and enhance incorporation of toxic or hazardous species in secondary minerals. The management aspects include siting and construction as well as controlling the amount of percolating water. Thus the impacts of disposal cannot be eliminated, but can be managed so as to mitigate environmental impacts.

Concluding remarks

Coal will continue in the foreseeable future as a major source of energy, and thus coal combustion waste will continue to be generated. However, problems associated with near-surface and shallow disposal of combustion products have already arisen and, as the geochemical 'footprint' of disposal sites continues to spread, affecting surface and groundwater quality, disposal practices are likely to increasingly attract critical attention.

'Clean coal' production will undoubtedly help reduce combustion waste, especially if reject material is returned to the coal extraction site. Some ash, especially fly ash, finds useful application, but scope exists for innovative research to improve utilization of coal combustion products. For example, tailoring ash properties to specific applications offers scope for increased utilization, provided tailoring can be done economically.

The combustion process activates mineral ash with the result that leachates extract relatively high proportions of elements whose concentrations in potable water are limited. We are as yet some way from understanding the speciations of these elements in combustion waste as well as the geochemical evolution of waste in its disposal environment. Preliminary studies show that the design, construction, and operation of disposal sites have a major influence on releases. The underlying geochemical processes are at present only known in outline and provide a fascinating field for interdisciplinary studies.

Achieving a balanced presentation in limited space has presented the writer with a challenge. The constructive suggestions of referees has helped to achieve this goal.

References

ANDRADA, A., COEANEGRACHT, Y. M. A., HOLLMAN, G. G., JANSEN-JURKOVICOVA, M., PETERSEN, M. S., VRIEND, S. P. & SCHUILING, R. D. 1990. Leaching characteristics of fly ash after four years of natural weathering. *Materials Research Society Symposium Proceedings*, **178**, 105–114.

BARRY, T. I. & GLASSER, F. P. 2000. Calculation of Portland Cement clinkering reactions. *Advances in Cement Research*, **12**, 19–28.

DONAHOE, R. J. 2004. *Secondary Mineral Formation in Coal Combustion Byproduct Disposal Facilities: Implications for Trace Element Sequestration. In*: GIERÉ, R. & STILLE, P. (eds). *Energy, Waste, and the Environment: a Geochemical Perspective.* Geological Society, London, Special Publications, **236**, 641–658.

GIERÉ, R., CARLETON, L. E. & LUMPKIN, G. R. 2003. Micro- and nanochemistry of fly ash from a coal-fired power plant. *American Mineralogist*, **88**, 1853–1865.

GLASSER, F. P. & HONG, S.-Y. 2002. Alkali sorption by C–S–H and C–A–S–H gels. Part II: role of alumina. *Cement and Concrete Research*, **32**, 1101–1111.

GOLOMB, D. S. & FAY, J. A. 2004. Atmospheric Impact of the Fossil Fuel Cycle. In: GIERÉ, R. & STILLE, P. (eds) *Energy, Waste, and the Environment: a Geochemical Perspective.* Geological Society, London, Special Publications, **236**, 153–167.

GROPPO, J., ROBL, T., JAMES, C. & HOWER, J. C. 2004. The beneficiation of coal combustion ash. In: GIERÉ, R. & STILLE, P. (eds) *Energy, Waste, and the Environment: a Geochemical Perspective.* Geological Society, London, Special Publications, **236**, 247–262.

HELMUTH, R. 1987. *Fly Ash in Cement and Concrete.* Portland Cement Association, Skokie IL, ISBN 0-89312-085-5.

HEMMINGS, R. T. & BERRY, E. E. 1988. On the glass in coal fly ashes: recent advances. *Materials Research Society Symposium Proceedings*, **113**, 3–38.

KALYONCU, R. S. 2000. *Coal Combustion Products, Metals and Minerals Yearbook.* US Geological Survey, Reston, VA. Vol. 1, 20.1–20.5.

KANARE, H. M. 1986. Preparation and characterisation of three new NBS fly ash standard reference materials. *Materials Research Society Symposium Proceedings*, **65**, 159–160.

KILGOUR, C. L. & DIAMOND, S. 1988. The internal structure of a low-calcium fly ash. *Materials Research Society Symposium Proceedings*, **113**, 65–73.

KUMARATHASAN, P., MCCARTHY, G. J., HASSETT, D. J., & PFLUGHOEFF-HASSETT, D. F. 1990. Oxyanion-substituted ettringites: synthesis and characterisation and their potential role in immobilisation of As, B, Cr, Se and V. *Materials Research Society Symposium Proceedings*, **178**, 83–104.

LEA, F. M. 1988. Portland cement: classification and manufacture; 'The burning of Portland cement'. In: HEWLITT, P. C. (ed) *Lea's Chemistry of Cement and Concrete.* Arnold, London, ISBN 0-340-56589-9, 4th edition, 25–94, 195–240.

LEVIN, E. M., ROBBINS, C. R. & MCMURDIE, H. F. 1964. *Phase Diagrams for Ceramists.* The American Ceramic Society, Columbus, OH. Figs 828–851, 279–284.

MCCARTHY, G. J., MANZ, O. E., JOHANSEN, D. M., STEINWAND, S. J. & STEVENSON, R. J. 1987. Correlations of chemistry and mineralogy of western US fly ash. *Materials Research Society Symposium Proceedings*, **86**, 109–112.

MACPHEE, D. E. & GLASSER, F. P. 1996. A modelling approach to the prediction of pore fluid alkalinity in concrete. In: SHAYAN, A. (ed) *Alkali Reaction in Concrete.* Blackie and Sons, Glasgow, 792– In: GIERÉ, R. & STILLE, P. (eds) *Energy, Waste, and the Environment: a Geochemical Perspective.* Geological Society, London, Special Publications, **236**, 796.

MALHOTRA, V. M. 1986. *Fly Ash, Silica Fume, Slag and Natural Pozzolans in Concrete.* American Concrete Institute Special Publication, SP-91. Detroit MI (Zvoli), Library of Congress 85-73563.

MALHOTRA, V. M. 1987. *Supplementary Cementing Materials for Concrete.* Ministry of Supply and Services Canada, Ottawa, ISBN 0-660-12550-1.

QIAN, J. C. & GLASSER, F. P. 1988. Bulk composition of the glassy phase in some commercial FPA's. *Materials Research Society Symposium Proceedings*, **113**, 39–44.

QIAN, J. C., LACHOWSKI, E. E. & GLASSER, F. P. 1988. Microstructure and chemical variation in Class F fly ash glass. *Materials Research Society Symposium Proceedings*, **113**, 45–53.

RICHARDSON, F. D. 1974. *Physical Chemistry of Melts in Metallurgy.* Academic Press, London/ New York, 2 volumes, ISBN 0-12-58-7901-06.

ROTH, R. S., DENNIS, J. R. & MCMURDIE, H. F. 1987. *Phase Diagrams for Ceramists.* The American Ceramic Society, Vol. VI, Figs 6442–6448, 146–152.

STEVENSON, R. J. & HUBER, T. P. 1987. SEM study of chemical variations in Western US fly ash. *Materials Research Society Symposium Proceedings*, **86**, 99–108.

TISHMACK, J. K. & BURNS, P. E. 2004. *The Chemistry and Mineralogy of Coal and Coal Combustion Products. In*: GIERÉ, R. & STILLE, P. (eds) *Energy, Waste, and the Environment: a Geochemical Perspective.* Geological Society, London, Special Publications, **236**, 223–246.

YOUNGER, P. L. 2004. *Environmental Impacts of Coal Mining and Associated Wastes: A Geochemical Perspective. In*: GIERÉ, R. & STILLE, P. (eds) *Energy, Waste, and the Environment: a Geochemical Perspective.* Geological Society, London, Special Publications, **236**, 169–209.

ZENIERIS, P. & LAGUTOS, J. G. 1988. Fly ash as a binder in aggregate base courses. *Materials Research Society Symposium Proceedings*, **113**, 231–240.

The chemistry and mineralogy of coal and coal combustion products

JODY K. TISHMACK[1] & PERRE E. BURNS[2]

[1]*Agricultural and Biological Engineering, Purdue University, West Lafayette, IN, USA*
(e-mail: tishmack@purdue.edu)
[2]*Department of Agronomy, Purdue University, West Lafayette, IN, USA*

Abstract: The generation of electricity by thermal conversion of coal results in significant volumes of solid waste. Most of these materials are disposed of in surface impoundments near coal-fired power plants or in active coal mines. Disposal rates vary from country to country. In the USA, 60–70 wt% of these materials are disposed. Although these materials rarely meet the definition of hazardous, leachate tests on some coal ash have shown them capable of producing elevated concentrations of some regulated metals. As a result of the well-documented environmental concerns posed by coal combustion, and the disposal of coal combustion products (CCPs), a large body of research has focused on characterizing the mechanisms of mobilization and attenuation of trace elements in coal and its ash. However, groups opposed to coal combustion or unregulated disposal of ash overlook the value of these materials as well-proven replacements for aggregate, cement, or soil in numerous engineering and agricultural applications. An extensive body of knowledge has been gathered describing the variability and versatility of these materials. It has been shown that proper utilization or management of ash requires a good understanding of its chemical, physical, and mineralogical properties. This chapter is intended to provide a broad overview of the chemistry and mineralogy of coal, and the combustion products that are formed when coal is burned.

Each year the US electric utility industry generates over 1500 billion kWh of electricity from the combustion of over 800 Mt of coal (Wang & Sweigard 1996). The combustion of this coal leads to the generation of approximately 110 Mt of coal combustion products (CCPs) annually (ACAA 2000). The total mass of CCPs produced in the USA now ranks among non-fuel mineral resources such as crushed stone, sand, and gravel.

The vast quantities of CCPs produced annually have spurred considerable interest in the usage of CCP as construction and building materials, backfill for mining operations, and more recently as liming agents and soil amendments in agricultural settings. Despite a large body of research in the literature, only about 30 wt% of the CCPs produced annually are utilized in the USA (ACAA 2000). The remaining 70 wt% are disposed in saturated ash impoundments or ash landfills. Possible environmental concerns related to coal combustion include the disposal of enormous volumes of ash, pollution of water resources neighboring unlined disposal facilities by metal constituents derived from ash residues, and the emission of gaseous oxides (primarily SO_2, NO_x, and CO_2), and volatile trace metals to the atmosphere during the combustion process.

A thorough understanding of the chemical and mineralogical composition of CCPs is necessary for proper management of these materials. This chapter will cover: (1) the composition of coal; (2) the formation of CCPs; (3) the physical, chemical, and mineralogical characteristics of CCPs; (4) characterization of North American fly ashes; (5) hydrated minerals in fly ash/water pastes; (6) sulphur scrubbing products; and (7) environmental impact of CCPs.

Coal composition

The American Society for Testing and Materials (ASTM) classification of coals includes four rankings, which generally correspond to the age of coal. These rankings, in order of increasing energy content, include lignite, subbituminous, bituminous, and anthracitic coals (ASTM 2003*a*). Lignite is the youngest class of coal and is characterized as having a relatively low C content (resulting in energy contents as low as 1.37×10^4 MJ/t). Subbituminous and bituminous coals are of intermediate age, and have

From: GIERÉ, R. & STILLE, P. (eds) 2004. *Energy, Waste, and the Environment: a Geochemical Perspective*. Geological Society, London, Special Publications, **236**, 223–246.
0305-8719/04/$15   © The Geological Society of London 2004.

energy contents approximating 1.9×10^4 and 2.53×10^4 MJ/t, respectively. Subbituminous and bituminous coals, as well as lignite in some states, make up the vast majority of coal used for the production of electricity in the USA. Anthracitic coals are the hardest in texture, contain typically greater than 90 wt% C, and are highly valued due to their purity. The energy content in anthracitic coal is relatively high, approximately 2.43×10^4 MJ/t; however, because of its increasingly limited availability, the use of anthracitic coal for the production of electricity has been virtually eliminated.

The primary component of coal is carbonaceous material resulting from the accumulation and decay of plant matter in marine or freshwater environments and marshes (Hessley *et al.* 1986). As plant matter accumulates it becomes humified and may eventually be consolidated into coal through a process called coalification. In the organic matrix, C is the major element by weight, with smaller amounts of H, O, N, and S, and many trace elements. The abundance of these trace elements is highly variable, but based on the reported trends in the affinity of elements for the organic fraction of coal (Table 1), elements such as B, Ge, Be, Ti, and V are expected to exist primarily within the organics in coal.

During coalification distinct changes in the chemical composition of the organic fraction of coal have been observed. A large consensus has been reached that the organic fraction of coal is primarily composed of aromatic compounds. As coalification progresses, the degree of aromaticity increases concentrating organic carbon into increasingly denser forms. As this consolidation

of C occurs, the weight percent of O and H in coal decrease (Valkovic 1983). Oxygen diminishes from about 25 wt% in lignite to nearly 0 wt% in anthracite, and H decreases from approximately 5 wt% in lignite to 2–3 wt% in anthracite. In contrast to O and H, the percentage of N and S show little to no correlation with the age (Valkovic 1983).

Coal contains detrital minerals that were deposited along with the plant material, and authigenic minerals that were formed during coalification. The abundance of mineral matter in coal varies considerably with its source, and is reported to range between 9.05 and 32.26 wt% (Valkovic 1983). Minerals found in coal include (Table 2): aluminosilicates, mainly clay minerals; carbonates, such as, calcite, ankerite, siderite, and dolomite; sulphides, mainly pyrite (FeS_2); chlorides; and silicates, principally quartz. Trace elements in coal are commonly associated with one or more of these minerals (see Table 2).

The distribution of major, minor, and trace elements and their forms vary with coal type. Miller & Given (1986) studied the association of inorganic elements in lignite coal, in particular the organic/inorganic interactions. They separated a coal sample by gravimetric methods, and then further subdivided these fractions using ammonium acetate and HCl extractions. Their analyses revealed that Ca, Mg, Na, K, Sr, Ba, and Mn were present largely or partly in ion-exchangeable form. They found appreciable amounts of K (illite), Ba (sulphate, carbonates), and Mn present in mineral forms. Some of the Al appeared in organic associations. Ti, which often occurs in sediments by substitution in clays, was found in substantial amounts in the organic fraction as acid-soluble and acid-insoluble organic chelates. Several trace elements were enriched in the coal fractions with the lowest specific gravity, indicating that Be, Sc, Cr, Y, Yb, V, Ni, Cu, and Zn were primarily associated with the organic matter, a situation more commonly found in lignite than bituminous coal (Miller & Given 1986).

Conditions that are conducive to the accumulation and decay of plant residues, and therefore conducive to the formation of coal, are typically associated with water-saturated and reducing environments. Consequently, a large portion of the trace elements associated with the mineral fraction of coal are expected to occur in reduced forms, primarily as sulphides or carbonates. However, because of its abundance compared to sulphur, it is unlikely that the complete reduction of iron oxides to iron sulphide would ever occur. Therefore, the presence of Fe oxides, and trace

Table 1. *Elemental affinity for the organic fraction of coal**

Element	Valkovic (1983)	Zubovic *et al.* (1960)
Germanium	100	87
Vanadium	100	76
Beryllium	75–100	82
Boron	75–100	77
Titanium	75–100	78
Nickel	0–75	59
Chromium	0–100	55
Cobalt	25–50	53
Molybdenum	50–75	40
Copper	25–50	34
Tin	0	27
Zinc	50	0

Sources: Valkovic (1983); Zubovic *et al.* (1960).
*% of total elemental content that exists in the organic fraction.

Table 2. *Mineral constituents in coal*

Mineral	Chemical formula	Trace element associations
Silicates		
Kaolinite	$Al_2Si_2O_5(OH)_4$	Lithophile elements:
Illite	$KAl_2(AlSi_3O_{10})(OH)_2$	Li, F, B, Cr, Sr, Ba, I, Ti,
Mixed Illite–Smectite	$Al_2Si_4O_{10}(OH)_2 \cdot H_2O$	Br, Mn
Chlorite	$Mg_5Al(AlSi_3O_{10})(OH)_8$	
Quartz	SiO_2	
Feldspar	$(K,Na)AlSi_3O_8$	
Oxides		
Hematite	Fe_2O_3	Siderophile elements:
Magnetite	Fe_3O_4	Ni, Co, Mo, Mn, P, Ge,
Rutile	TiO_2	Pt, Ti, Sn
Anatase	TiO_2	
Sulphides		
Pyrite	FeS_2	Chalcophile elements:
Marcasite	FeS_2	As, Cu, Co, Pb, Hg, Zn,
Pyrrhotite	$Fe_{1-x}S$	Mo, Cd, Se
Sulphates		
Gypsum	$CaSO_4 \cdot 2H_2O$	Na, Sr, Ba, Pb
Jarosite	$KFe_3(SO_4)_2(OH)_6$	
Thenardite	Na_2SO_4	
Carbonates		
Calcite	$CaCO_3$	Mn, Zn, Sr
Dolomite	$CaMg(CO_3)_2$	
Siderite	$FeCO_3$	
Ankerite	$Ca(Mg,Fe,Mn)(CO_3)_2$	
Chlorides		
Sylvine	KCl	
Halite	$NaCl$	

Sources: Valkovic (1983); Harvey & Ruch (1986).

elements associated with Fe oxides, will persist in coal despite the strongly reducing conditions.

Formation of coal combustion products

Pulverized-coal combustion is the predominant technology for the generation of electricity in the world. It involves pulverizing the raw coal into fine particles and injecting them with air into a furnace ('boiler') where they are burned at temperatures often exceeding 1500 °C. The heat released during combustion boils water in tubes located along the furnace walls. Turbine generators use the steam from the boiling water to produce electricity. Although coal combustion processes vary significantly, the waste streams generated in most modern coal combustion facilities are similar.

Figure 1 illustrates a generalized process diagram for the combustion of coal and the formation of CCP. Coal is first fed through a conveyor/hopper system to a coal crusher. Within the crusher the coal is pulverized and then fed to a furnace. As coal particles are heated within the furnace they swell and become porous, with the degree of swelling dependent on both the type of coal and the combustion conditions (Wu & Chen 1987). Most of the organic material in the coal is combusted and liberated as CO_2, leaving behind inorganic minerals and uncombusted organic compounds. Volatile hydrocarbons and the inorganic elements associated with them are vaporized, along with highly volatile mineral matter. Approximately one-fifth of the ash falls to the bottom of the furnace and is collected as coarser textured 'bottom ash' or as slag (Helmuth 1987). The rest of the solids, called 'fly ash', are transported to a low temperature zone (200 °C) where they consolidate into fine-textured, sphere-shaped fly ash particles that contain both crystalline and non-crystalline phases (Helmuth 1987).

The amount of mineral matter contained in coal is highly variable and, in many cases, coal

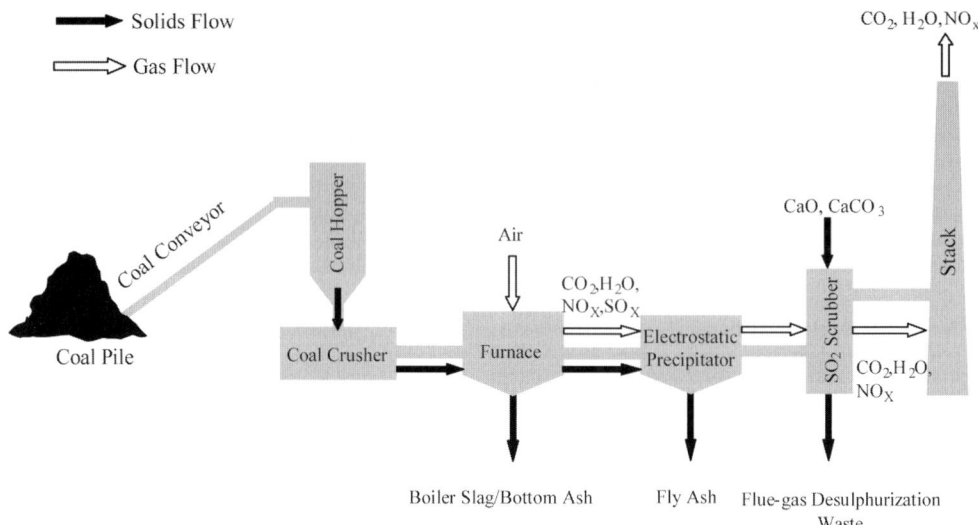

Fig. 1. Generalized coal combustion process.

companies clean or remove as much of it as is economically feasible before the coal is sold to electricity-generating stations. Most of the resulting utility coal contains between 5 and 10 wt% mineral impurities, which is reported as wt% ash in a proximate coal analysis. Higher contents of ash in coal can cause slag formation on boiler tubes, a significant problem for power plants because of the reduction in heat transfer.

Minerals that decompose at or below combustion temperatures (primarily carbonates and sulphides), or have elements that sublime at combustion temperatures are likely to be vaporized into the furnace gases. A comparison of boiling temperatures of selected elements to that of the operating temperatures of modern fluidized-bed combustion units (800–900 °C) suggests that elements including C, O, H, S, N, P, Na, K, As, Cd, Pb, Hg, and Se are likely to be present in vapour form in furnace gases.

Approximately 70–90 wt% of mineral matter in coal undergoes some form of thermal transformation during the combustion process. Mineral matter and residual organics that are not vaporized are burned by heterogeneous oxidation (Wu & Chen 1987). The rate of oxidation of small particles ($<100 \, \mu m$) is controlled by chemical kinetics, while the diffusion of oxygen to the particle surface limits the rate of combustion of larger particles (Flagan & Friedlander 1976). Although coal combustion is largely an oxidation process, if air is not vigorously mixed with burning coal particles within the combustion furnace, localized reducing zones can exist

(Katrinak & Zygarlicke 1995). Therefore, even though mineral constituents in ash are predominantly converted to oxidized forms, the presence of reduced forms of minerals cannot be ruled out, particularly within large particles.

Silicate and aluminosilicate minerals are the principal non-combustible phases present in pulverized coal and, with the exception of quartz, are wholly transformed upon ignition, when they are raised to the melting temperature range (1100–1600 °C; Hubbard *et al.* 1984). Bound water and P released during thermal decomposition of the aluminosilicate mineral illite, fluxes partial melting of these particles (Hubbard *et al.* 1984). The melt thus formed is frothed by simultaneous production of CO_2 from Fe-catalysed oxidation of carbonaceous matter. This produces hollow, gas-filled glass-walled fly ash particles called 'cenospheres'. The mineral kaolinite has a very high ash-fusion temperature due to the absence of fluxing ions (Hubbard *et al.* 1984). Its lattice is destroyed without melting, forming cenosphere-free fly ash aggregates.

As exhaust gases and fly ash particles are vented from the furnace, they quickly begin to cool, leading to the condensation and adsorption of volatilized elements onto the surfaces of fly ash particles entrained in the gas stream (Kaakinen *et al.* 1975). Under high-temperature combustion conditions certain elements, including S, are enriched on the surface of particles (Davison *et al.* 1974; Smith 1980). A vaporization–condensation process is the primary mechanism responsible for the enrichment of volatile trace metals onto the surfaces of fly ash particles, especially on the small-sized fraction (Kaakinen et al. 1975; Smith et al. 1979; Smith 1980; Summers et al. 1983; Van der Sloot & Nieuwendijk 1985; Gieré et al. 2003).

Upon leaving the furnace, exhaust gases are diverted through a series of pollution control devices. Baghouses or electrostatic precipitators remove fly ash particles entrained in the exhaust gases. In the case of electrostatic precipitators, ash particles in the gases are imparted with a negative charge by high-voltage electrodes. The negatively charged ash particles are then attracted to positively charged collecting surfaces, effectively removing more than 99% of the volume of ash particles from the gas stream.

Exhaust gases are treated to minimize SO_2 emissions into the atmosphere. Removal of SO_2 is achieved by the utilization of a scrubber system and will be discussed below in greater detail. Although numerous variations of both wet and dry scrubbers are commercially available, most rely on the adsorption of SO_2 onto Al-, Na-, or Ca-based adsorbents. Ca-based adsorbents, such as lime (CaO) or calcite ($CaCO_3$), are usually preferred due to their low cost and high availability.

Based on the above description of the coal combustion process several conclusions become apparent. First, the type and amount of ash accumulated during coal combustion greatly depends on the mineralogy of the coal being used, the combustion process, and the presence of emission control devices. Secondly, the chemical forms in which elements are found in ash are affected by coal combustion process variables such as combustion temperature and the mode of combustion (e.g., pulverized-coal fired, fluidized bed, cyclone, stoker). Lastly, the amount of CCPs accumulated by power plants is predominantly a consequence of the presence of emission control devices. The latter is supported by the fact that the total amount of CCPs produced in the US has increased significantly since the use of electrostatic precipitators became prevalent in the early 1970s (Simsiman et al. 1987).

Physical characteristics of fly ash and bottom ash

Beyond similarities in bulk density, the physical properties of fly ash and bottom ash are very distinct (Table 3). The diameter of fly ash particles is generally several orders of magnitude smaller than that of bottom ash particles, leading to

Table 3. *Physical characteristics of fly ash and bottom ash from coal combustion*

Property	Fly ash	Bottom ash
Particle size range (mm)	10^{-4}–10^{-1}	10^{-1}–10^{1}
Mean particle diameter (μm)	20–80	500–700
Saturated hydraulic conductivity (cm/s)	10^{-6}–10^{-4}	10^{-3}–10^{-1}
Specific gravity	1.59–3.1	2.17–2.78
Dry bulk density (g/cm^3)	1.0–1.6	0.74–1.6
Surface area (m^2/g)	0.2–3.06	0.4

Sources: Summers et al. (1983); Valkovic (1983); Hostetler et al. (1989).

saturated hydraulic conductivities of fly ash that are one to three orders of magnitude lower than bottom ash. The small diameter of fly ash particles also results in their larger surface areas.

The morphologies of fly ash and bottom ash particles are also quite different. Bottom ash primarily consists of random conglomerations of fused subangular particles, with smaller amounts of angular fragments of unmodified quartz, feldspar, and aluminosilicates. Fly ash consists of particles resulting from the cooling of molten droplets of fused material contained within the exhaust gases leaving the combustion furnace. Fusion of silicate minerals results in the formation of glassy spherical particles (Fig. 2). The majority of these spherical particles are classified into two groupings. Hollow spheres packed with smaller spheres are called plerospheres; hollow and empty spheres are called cenospheres. The large range in specific gravity of fly ash is a result of these unique hollow sphere morphologies.

Chemical composition of fly ash and bottom ash

The chemical composition of CCPs varies with coal origin and rank; however, the major elemental constituents of all coal ash residues are O, Si, Al, Fe, and Ca, along with lesser amounts of Mg, S, and C. The relative abundance of constituents that typically make up more than 1% of the total mass of fly ash and bottom ash are summarized in Table 4. These elements are found in the ash because of their lower volatility and the short time the particles actually remain in the furnace during combustion (Helmuth 1987). Both crystalline and non-crystalline compounds form on the surface of fly ash particles when elements react with oxygen in the flue gases, and through

Fig. 2. Scanning electron micrograph of typical coal fly ash particles.

Table 4. *Chemical composition of fly ash and bottom ash from coal combution*

Chemical component	Fly ash (wt%)	Bottom ash (wt%)
SiO_2	45.5–57	20–60
Al_2O_3	18–33.7	10–35
Fe_2O_3	6.85–16	5–35
CaO	2.8–10	1–20
MgO	1–5.5	0.3–4
SO_3	0.75–3.3	0.1–12

Sources: Braunstein *et al.* (1977); Valkovic (1983); Wang & Sweigard (1996).

condensation/crystallization within melt droplets (Smith 1980).

Ca- and Mg-oxides (lime, periclase [MgO]) may occur as small crystals embedded within the glass or they may be located on its surface depending on temperature and furnace conditions where the minerals formed. Low-temperature minerals such as anhydrite ($CaSO_4$) may form on the surface of the fly ash grains after they have left the high-temperature zones in the furnace (Linton *et al.* 1977; Soroczak *et al.* 1987; Ainsworth *et al.* 1993; Fishman *et al.* 1999).

Glass is amorphous vitreous silicate or aluminosilicate that is formed when liquid material is cooled rapidly, locking the elements into a disordered, non-crystalline structure. Because of its high content of silicate glass, fly ash has cementitious or pozzolanic properties (depending on available lime) and is widely used in concrete as a replacement for either fine aggregate or for Portland cement (Helmuth 1987). Both bottom ash and fly ash are suitable for use in structural fills, and a standard guide for design and construction has been developed (ASTM 2003b).

Bituminous coal ash is often higher in pyritic Fe than subbituminous or lignite ashes, which are often higher in Ca and/or other basic elements. As a result of the increased content of pyritic Fe, bituminous coal ashes may form acidic solutions when mixed with water. Conversely, subbituminous and lignite coals often form basic solutions when mixed with water due to their lower content of pyrite and/or their high content of basic cations. Therefore the pH of ash leachate can range from acidic to basic values depending on coal source (see Donahoe, this volume).

Few comprehensive classification schemes for CCP exist. The American Society for Testing and Materials (ASTM 1994) classifies two catgories of fly ash (Class F and Class C) based upon chemical and physical properties of the fly ash (the total amount of $Si + Al + Fe$, sulphate, loss on ignition). This classification system was developed for the use of fly ash as an admixture in concrete. More recently, new classification schemes have been developed that place emphasis on textural descriptions, the form of carbon (or 'char'), and the surface properties of fly ash (Hower & Mastalerz 2001). These new classification schemes for fly ash may be the result of growing concern over mercury emissions from coal-fired boilers. Studies have shown that mercury adsorption onto the surface of fly ash particles is a function of both the total carbon content and the gas temperature at the point of fly ash collection (Hower et al. 2000).

Mineralogy of fly ash

Fly ash is a heterogeneous material at the micrometer scale, and even at the nanometer scale (Hemmings & Berry 1986; Qian et al. 1988, 1989; Enders 1995; Gieré et al. 2003). The chemical composition and assemblage of minerals in a given fly ash sample is characteristic of the coal source from which it was derived, and the boiler conditions. Fly ash is composed of 75–90 wt% Si–Al-bearing glass, which is highly variable in its composition. Glass

composition depends on the rate of quenching and on the amount of substitution or modification by other ions in the glass such as Fe, Ca, Na, K, and Mg. Subbituminous and lignite coals generally contain higher concentrations of Ca and Mg either as Ca- and Mg-bearing minerals (carbonate and clay minerals) or in the organic fraction of the coal. The reactivity of fly ash glass in high-pH solutions, such as those found in concrete pore water, depends on the long-range order of the glass matrix. Ions such as Ca, Mg, K, and Na act as network modifiers in the glass, reducing long-range order and increasing its reactivity, or the rate at which it can be dissolved (Diamond 1984; Hemmings & Berry 1988). Tikalsky (1989) provides an excellent review and discussion of fly ash glass formation.

Diamond (1984) conducted a study of fly ash glass using X-ray diffraction (XRD) and found that the position of the diffuse scattering maximum, or 'glass hump', in XRD patterns provided information about the composition of the glass. The centre of the glass maximum reflects the position of the Bragg diffraction peak for the compound that would have crystallized from the molten material had it been allowed to cool slowly. The position of the diffuse scattering maximum, therefore, can reflect differences in glass composition. The glass in low-Ca (<5 wt% CaO) fly ashes has a maximum at $23°$ 2-theta (Cu Kα radiation) indicative of a siliceous glass, while more moderate-Ca fly ashes (between 10 and 20 wt% CaO) have a scattering maximum nearer $28°$ 2-theta. As the total Ca content increases above 20 wt% an asymmetrical glass hump forms at about $32°$ 2-theta, presumably due to the formation of Ca aluminate-rich glass.

Hemmings and Berry (1988), studying fly ashes with a CaO content of approximately 10 wt%, identified two types of glass, which they referred to as Types I and II. Type I glass (similar to the low-Ca glass in Diamond's findings) is generally higher in Si and Al. Type II glass contains more Ca and other alkali earth elements, such as Mg and K.

Although fly ash is largely non-crystalline glass, the major and minor elements in fly ash (O, Si, Al, Fe, Ca, Mg, and S) form several different phases that have been characterized by XRD in many research projects (Diamond 1984; McCarthy 1988; Bergeson et al. 1988; Thedchanamoorthy & McCarthy 1989; Ainsworth et al. 1993). The phases of major elemental constituents that have been commonly identified in fly ash include: quartz; mullite, $Al_6Si_2O_{13}$; tricalcium aluminate, $Ca_3Al_2O_6$

(commonly referred to as C_3A in cement short-hand notation); alite, Ca_3SiO_5; belite, Ca_2SiO_3; magnetite, Fe_3O_4; hematite, Fe_2O_3; lime; anhydrite; $Ca_4Al_6O_{12}SO_4$; periclase; melilite, $Ca_2(Mg,Al)(Al,Si)_2O_7$; merwinite, $Ca_3Mg(SiO_4)_2$; and thenardite, Na_2SO_4. A list of these minerals is provided in Table 5 along with their principal XRD peak positions.

Much of the quartz in the fly ash originates from the coal as silt- and sand-sized particles, and it remains in the ash because it survives thermal transformation during the combustion process (Helmuth 1987). Small amounts of vola-tilized Si may also oxidize to form very fine crys-tals of quartz within the fly ash glass (Diamond 1984; Hubbard et al. 1984). Although bitumi-nous coal ash may contain more than 50 wt% analytical SiO_2, only 5–10 wt% of it is present in the form of quartz (McCarthy et al. 1990). Some Si is present in the mineral mullite, but the majority of it is in the amorphous glass phase.

Mullite is the principal Al-bearing mineral in low-Ca bituminous fly ash. It originates by direct crystallization from clay minerals or by devitrification of glass upon cooling (Hubbard et al. 1984). Despite a careful search, Hubbard et al. (1984) did not find distinct grains of mullite in any of the fly ashes they studied. They concluded that as the clay minerals in coal are progressively heated they lose structural water between 200 and 900 °C, followed by crystallization of mullite and cristobalite before melting at 1100–1600 °C. However, other researchers observed mullite needles near the surface of etched fly ash particles (Hulett & Weinberger 1980; Joshi et al. 1985) as well as within the bulk matrix (Warren & Dudas 1986).

Mullite is almost twice as abundant in low-Ca fly ash when compared to high-Ca fly ash, mainly due to differences in the Al content of the clay minerals associated with the coal (McCarthy et al. 1990). Using atomic absorption spectroscopy (AAS) and scanning electron microscopy/electron microprobe analyses (SEM/EMPA) Stevenson & Huber (1987) found a corre-lation between the elemental composition of ash particles and the clay mineral species in the raw coal. They concluded that the geologic origin of the coal had a significant impact on the microche-mical composition of the ash.

The molar ratio of Al : Si in clay minerals is variable, kaolinite (0.85), mica-illite (0.61), and smectitic clays (0.35) (McCarthy et al. 1987). Bituminous coal tends to have higher concentrations of kaolinite, while lignite and subbituminous coals contain mica-illite or smec-titic clays. McCarthy et al. (1987) predicted that the higher the Al : Si ratio of the clay minerals the

more mullite will form in the fly ash. However, variations in clay minerals are not the only reason for differences in mullite content of the ash. Fly ash with higher Ca content also contains less mullite because some of the Al combines with Ca to form C_3A.

The principal Fe-oxide minerals in coal ash are magnetite and hematite. Magnetite results from the volatilization and oxidation of Fe-bearing minerals in the coal, mainly pyrite (FeS_2). Similar to quartz, hematite crystals in coal survive the brief exposure to high temperatures, while pyrite is oxidized to form magnetite with the release of SO_2. Approximately one-third to one-half of the Fe present in fly ash is in the form of magnetite and hematite (McCarthy et al. 1990), which are largely inert. The rest of the Fe in the fly ash is contained in the glass phase and becomes available when the glass dis-solves (Helmuth 1987).

In addition to quartz, mullite, and hematite, moderate- and high-Ca fly ashes (>15 wt% CaO) contain several Ca-bearing minerals that are reactive in the presence of water. The most common Ca-aluminate mineral in fly ash is C_3A, which occurs along with Ca-aluminate-rich glass. Smaller amounts of the Ca-silicate min-erals (C_2S and C_3S) are probably present in most high-Ca fly ash (>20 wt% CaO), but, due to several overlapping peaks, these minerals are difficult to identify or quantify using XRD (Tishmack et al. 1999).

The principal Mg-phase, periclase, contains about half of the Mg present in subbituminous and lignite fly ash. Magnesium is also found in melilite and merwinite, neither of which are known to react significantly with water. McCarthy et al. (1987) found that the periclase in fly ash/water pastes did not hydrate to form the mineral brucite ($Mg(OH)_2$) even after one year. Brucite is associated with deleterious expan-sion in concrete, and therefore its formation in fly ash is of concern when ash is used as an admixture in concrete. However, experience has shown that high-Mg content fly ash/water pastes (>5.0 wt% MgO) do not expand in the ASTM autoclave expansion soundness test (McCarthy et al. 1990). Miller & Given (1986) suggested that periclase in fly ash originates from the Mg in the organic portion of the coal, while its origin in Portland cement is from dolomite.

Most of the S in fly ash is in the form of anhy-drite, which forms as a solid phase on the surface of the subbituminous and lignite fly ash particles (McCarthy et al. 1989). Sulphur accumulates on the surface of fly ash particles from reactions between Ca, Na, O, and SO_2 as the fly ash

Table 5. *Selected phases identified in coal fly ash*

Name	Abbreviation used in XRD patterns	Chemical formula	2-theta (Cu $K\alpha$) position and intensity of the strongest Bragg diffraction peaks*
Klein's phase	Kl	$Ca_4Al_6O_{12}SO_4$	**23.7**$_x$, 33.8$_3$, 41.7$_2$
Anhydrite	Ah	$CaSO_4$	**25.5**$_x$, 31.4$_4$, 38.6$_2$, 40.8$_2$, 48.7$_2$
Mullite	Mu	$Al_6Si_2O_{13}$	**26.3**$_x$, 16.5$_5$, 26.0$_9$, 33.3$_4$,35.3$_5$, 40.9$_6$
α-Quartz	Qz	SiO_2	**26.7**$_x$, 20.9$_2$
Calcite		$CaCO_3$	**29.4**$_x$, 39.3$_2$, 43.1$_2$, 47.6$_2$
Melilite	Ml	$Ca_2MgSi_2O_7$	**31.1–31.5**$_x$, 28.9–29.2$_2$, 51.9–52.1$_2$
Thenardite		Na_2SO_4	**32.2**$_x$, 19.1$_8$, 28.1$_5$, 29.0$_6$, 33.9$_5$, 38.6$_3$, 48.8$_4$
Alite	C$_3$S	Ca_3SiO_4	**32.2**$_x$, 34.3$_8$, 32.7$_6$, 29.4$_5$, 41.4$_4$, 52.1$_4$
Belite	C$_2$S	Ca_2SiO_4	**32.7**$_x$, 32.1$_9$, 40.8$_x$, 40.2$_3$
Tricalcium aluminate	C$_3$A	$Ca_3Al_2O_6$	**33.2**$_x$, 47.5$_4$, 59.4$_3$
Merwinite	Mw	$Ca_3Mg(SiO_4)_2$	**33.3**$_x$, 32.5$_3$, 33.6$_7$, 33.8$_5$, 40.8$_2$, 47.6$_4$, 48.5$_2$
Hematite	Hm	Fe_2O_3	**33.3**$_x$, 24.3$_3$, 35.8$_5$, 41.0$_3$, 49.6$_4$, 54.3$_6$
Magnetite	Mg	Fe_3O_4	**35.5**$_x$, 30.1$_3$, 43.1$_1$, 53.4$_1$
Lime	Lm	CaO	**37.4**$_x$, 32.2$_3$, 53.9$_5$
Periclase	Pc	MgO	**42.9**$_x$, 37.0$_1$
Hydration products			
Strätlingite		$Ca_2Al_2SiO_7 \cdot 8H_2O$	**21.3**$_x$, 7.1$_5$,14.1$_4$, 20.2$_3$, 31.1$_6$, 34.3$_4$, 37.8$_4$, 38.0$_5$, 48.2$_3$
Ettringite		$Ca_6Al_2(SO_4)_3(OH)_{12} \cdot 26H_2O$	**9.1**$_x$, 15.8$_8$, 21.2$_x$, 18.9$_6$, 32.3$_7$, 35.0$_9$, 40.9$_7$
Monosulphoaluminate		$Ca_4Al_2(SO_4)(OH)_{12} \cdot 6H_2O$	**10.0**$_x$, 20.0$_9$, 22.3$_2$, 37.1$_2$, 40.4$_3$, 41.2$_2$
Gypsum		$CaSO_4 \cdot 2H_2O$	**11.6**$_x$, 20.7$_x$, 29.1$_8$, 31.1$_5$, 33.4$_4$

*XRD peak position; 100% peak shown in bold.

particles leave the hot part of the furnace and enter low-temperature zones (Smith 1980; Fishman et al. 1999). Small amounts of anhydrite have been observed in fly ash with CaO contents as low as 12 wt% (McCarthy et al. 1989). In moderate-Ca fly ash, anhydrite accounts for most of the S, but Na in coal will also help to capture SO_2 in the boiler, resulting in the formation of thenardite (McCarthy et al. 1989). In high-Ca fly ash (>25 wt% CaO), Klein's phase, $Ca_4Al_6O_{12}SO_4$, is found and contains soluble S (McCarthy et al. 1989). A small amount of gypsum, $CaSO_4 \cdot 2H_2O$, is sometimes observed in moderate- and high-Ca fly ashes that contain anhydrite and have been exposed to atmospheric moisture. The soluble anhydrite in fly ash tends to be reactive and readily converts into gypsum.

Advancements in analytical technology have helped to obtain new information about the minerals present in these materials. Tapping mode atomic force microscopy (TM-AFM) was used to image the surface of high-Ca fly ash (Bosbach & Enders 1998). Combined with single-particle X-ray analysis, these authors identified nanometer-sized anorthite, gehlenite, anhydrite, and magnetite crystals on the surface of the fly ash spheres. The TM-AFM revealed that the glass matrix in fly ash is heterogeneous even on the subnanometer scale.

Characterization of North American fly ash

A database of chemical, mineralogical, and physical characteristics of North American fly ashes was assembled from the analysis of more than 178 samples of North American fly ash (McCarthy et al. 1989). These fly ashes were derived from the combustion of five principal

coal sources in the USA: the Appalachian Region, the Illinois Basin, the Gulf Coast Region, the Fort Union Region, and the Powder River Basin (EIA 1995). The bulk chemical composition of a typical fly ash from each coal source (Table 6) reflects the source of the coal. Fly ash from bituminous coal is classified as Class F, from subbituminous coal as Class C, and fly ash from lignite can be either Class F or Class C.

The Appalachian Basin and Illinois Basin bituminous coals produce Class-F fly ashes that mainly contain Si, Al, and Fe ($SiO_2 + Al_2O_3 + Fe_2O_3$ <85 wt%) and only small amounts of CaO (<5 wt%; see Table 6). The principal phases in Class-F fly ash are quartz, mullite, magnetite, and minor hematite (Fig. 3). The silicate-rich glass in the Appalachian Basin fly ash forms a broad 'hump' in the diffraction pattern centered near 22° 2-theta, the location for the major Bragg diffraction peak of crystobalite, a silicate mineral (Diamond 1984). In fly ash richer in Al, Al substitutes for Si in the glass structure, and the position of the glass hump in, for example, the Illinois Basin fly ash, shifts to smaller d-spacing (25° to 26° 2-theta) nearer the main diffraction peak of mullite (Diamond 1984; Diamond & Olek 1987; Hemmings & Berry 1988).

Illinois Basin coal has relatively high concentrations of sulphide minerals (pyrite, marcasite) and contains moderate to high amounts of S (1.5–6.0 wt% S). The resulting fly ashes tend to contain higher concentrations of magnetite. Coal companies can remove some S from the coal by various physical and chemical means, but utilities burning high-S coal must remove SO_2 from flue gases.

There are two sources of lignite mined in the USA, the Gulf Coast Region in Texas and the Fort Union basin in North Dakota. Combustion of Gulf Coast lignite forms a low- to

Table 6. *Select elemental composition of common fly ashes resulting from combustion of coal from the major US coal basins*

Chemical component in wt%	Bituminous coal		Lignite coal		Subbituminous coal
	Appalachian Region Class F	Illinois Basin Class F	Gulf Coast Class F	Fort Union Class C	Powder River Basin Class C
SiO_2	50	57.4	57	45.8	33.8
Al_2O_3	28.7	28.2	21	15.2	18.7
Fe_2O_3	9.4	6.12	7.8	7.2	4.8
Total	**88.1**	**91.7**	**85.8**	**68.2**	**57.3**
CaO	1.7	0.8	9.3	18.3	26.5
SO_3	0.3	0.15	0.2	1.4	1.8
MgO	1	1.3	2.4	5.5	5.4
Na_2O	0.33	0.82	0.37	1.08	2.04

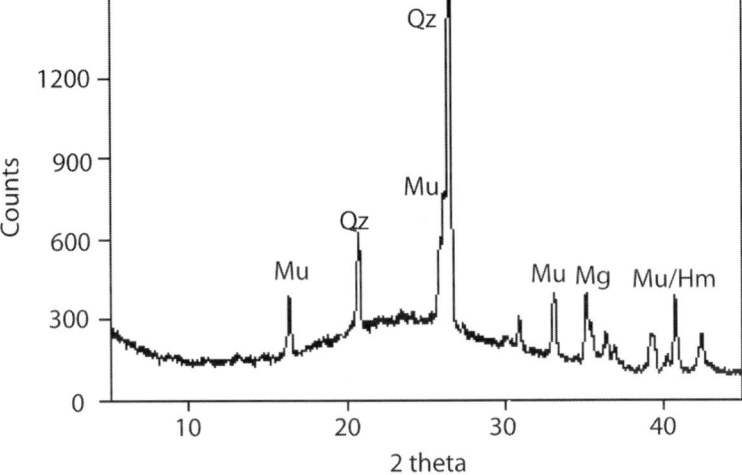

Fig. 3. X-ray diffractogram of Class-F bituminous coal fly ash. Analytical conditions: diffraction data were collected using a Philips X-ray powder diffractometer (45 kV/30–40 mA; CuKα; theta-compensating variable divergence slit; diffracted-beam graphite monochromator; scintillation detector) automated with an MDI/Radix Databox. The scan parameters were typically 0.02° step size for 1 s count times over a range of 5–60° 2-theta. All data were analysed and displayed using a data reduction and display code (JADE) from Materials Data Inc., Livermore, CA.

moderate-Ca Class-F fly ash (7–12 wt% CaO), whereas Fort Union lignite forms a moderate-to high-Ca Class-C fly ash (15–24 wt% CaO). Gulf Coast lignite produces Class-F fly ash similar in composition to bituminous Class-F fly ash, but with higher Ca-glass. Diamond (1984) observed a shift in the glass maximum, which generally became broader and centered at 28° 2-theta, the position of a major Bragg diffraction peak for anorthite, $CaAl_2Si_2O_8$. The glass in the Gulf Coast lignite fly ash has a higher pozzolanic reactivity compared to eastern bituminous fly ash, and makes a very good mineral admixture in concrete (Tikalsky 1989).

Gulf Coast lignite Class-F fly ash is generally low in SO_3 (<0.5 wt%) and in the presence of water typically will not react to form hydrated minerals (although there have been some exceptions). One such sample containing 9.3% CaO was observed to form small amounts (<2 wt%) of ettringite, $Ca_6Al_2(SO_4)_3OH_{12} \cdot 26H_2O$ (Solem & McCarthy 1992).

Fort Union lignite is low in S (<0.5 wt% SO_3) and it forms Class-C fly ash that contains Ca-and/or Mg-bearing phases such as lime, anhydrite, C_3A, periclase, melilite; and merwinite. As Ca and Mg concentrations in the coal increase, so too does the amount of Ca- and Mg-bearing minerals in the fly ash. At the lower range of CaO concentrations (15 wt%), only a small percentage of the total CaO is

present as the mineral lime and C_3A. As CaO concentrations in Fort Union fly ash reach the upper range (about 22 wt%), there is a larger amount of Ca-bearing phases present due to the reaction between Ca and Al, S, and/or Si in the boiler forming melilite, merwinite, and/or anhydrite. The XRD pattern (Fig. 4) shows the Ca-rich glass present in this type of fly ash, with its diffuse scattering maximum centered between 28° and 31° 2-theta. The fly ash sample contained quartz, magnetite (minor hematite), periclase, C_3A, and anhydrite (only the major peak for each mineral is identified in the XRD pattern).

Lime provides the high pH necessary to activate pozzolanic reactions, making high-Ca fly ash 'self cementing' (Solem and McCarthy 1992). Fort Union fly ash is widely used as a pozzolan in North Dakota, but replacement levels are generally no more than 10–15 wt% due to their slower strength gain (McCarthy *et al.* 1990). Fly ashes with thenardite present tend to form significant amounts of ettringite when hydrated (10–12 wt%; Solem & McCarthy 1992).

The largest volume of coal mined in the USA comes from the Powder River Basin (PRB) located in Wyoming and Montana (EIA 1995). Combustion of the PRB coal produces a high-Ca Class-C fly ash (22–32 wt% CaO) that is widely used as a replacement for Portland cement in concrete. An XRD pattern of this

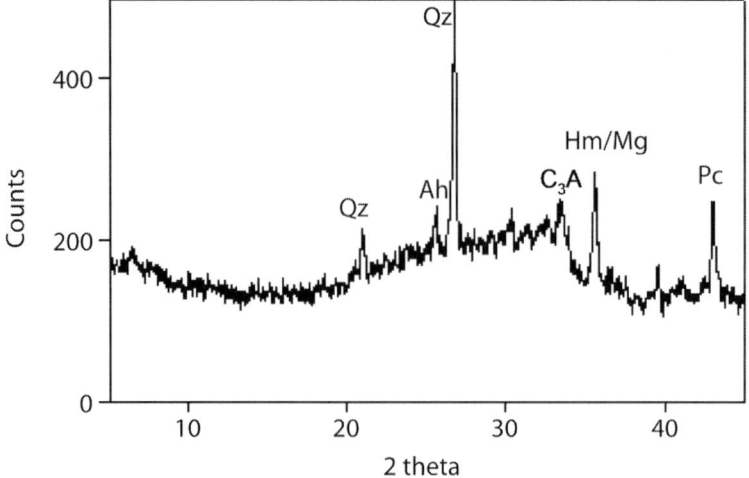

Fig. 4. X-ray diffractogram of Class-C lignite fly ash (from Fort Union coal). For analytical conditions, see Fig. 3.

type of fly ash (Fig. 5) shows the Ca-rich glass that forms a broad hump centered around 32° 2-theta, the location of the major Bragg reflection for C_3A.

The fly ash resulting from combustion of PRB coal contains more Ca than that from the Fort Union lignite (Table 6). In addition to quartz, mullite, and hematite, the PRB fly ash has greater concentrations of Ca-aluminate-rich glass and Ca-bearing phases including lime, C_3A, melilite, alite (C_3S in cement shorthand notation), belite (C_2S), and/or Klein's phase (Solem-Tishmack *et al.* 1990; Solem & McCarthy 1992; Solem-Tishmack 1993). (Alite and belite peaks are difficult to identify with XRD because of several overlapping peaks in the 32° 2-theta range.) As Ca increases so too

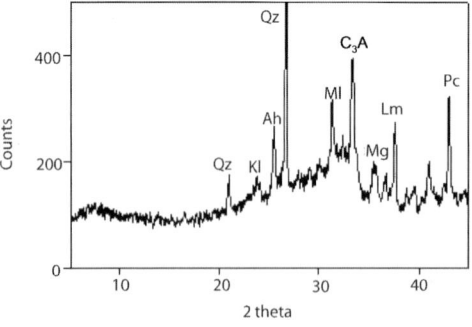

Fig. 5. X-ray diffractogram of Class-C subbituminous fly ash (from Powder River Basin coal). For analytical conditions, see Fig. 3.

does Mg, generally resulting in more periclase in PRB fly ashes compared with Fort Union fly ash. A comparison of their diffraction patterns (Figs 4 and 5) shows that the PRB fly ash has a larger glass hump, as well as sharper and stronger lime, anhydrite, C_3A, and periclase peaks. Also, SEM micrographs of both types of fly ash show that the Fort Union fly ash (Fig. 6) contains fewer small-sized particles than the PRB fly ash (Fig. 7). Bergeson *et al.* (1988) performed particle size separation and mineralogical characterization and found that the minerals lime, anhydrite, and Klein's phase were enriched in the fine-sized fraction of the fly ash. Tishmack *et al.* (1999) provided a detailed characterization of PRB fly ashes and their hydration behaviour in Portland cement.

Based on the above description of North American fly ashes several conclusions become apparent. The hydration behaviour, or reactivity, of fly ash largely depends on the amount of Ca present. Low-Ca bituminous Class-F fly ash (<10 wt% CaO) will not react with water unless an external source of lime is supplied to initiate dissolution of the glass, that is, pozzolanic reactions. Lignite fly ash with moderate- to high-Ca contents may form either Class-F or Class-C fly ash. Lignite Class-F fly ash, containing more Ca in the glass, is more reactive as pozzolans compared with bituminous Class-F fly ash. Pozzolanic reactivity and cementitious properties increase with CaO concentrations. As CaO increases above 15 wt%, so too does the amount of Ca-bearing minerals in the fly ash. High-Ca fly ash (>20 wt% CaO), such as the PRB fly ash,

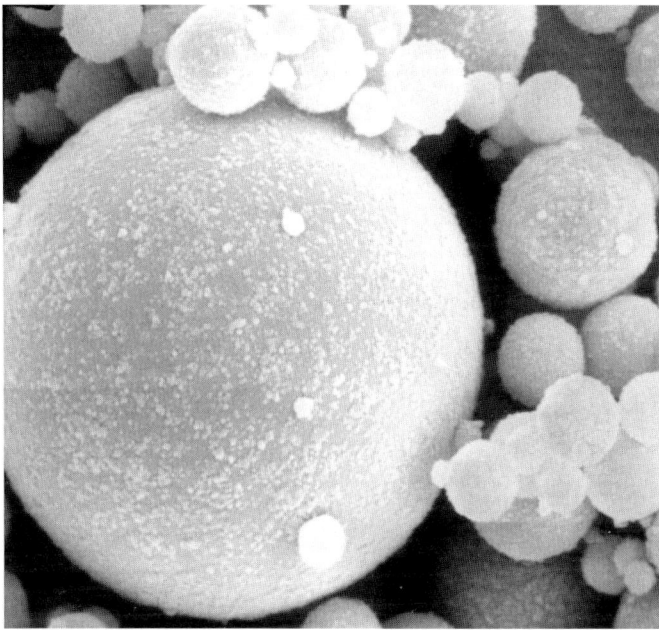

Fig. 6. Scanning electron micrograph of moderate-Ca Class-C lignite fly ash (from Fort Union coal).

contains several different types of water soluble Ca-, Al-, and S-bearing minerals that make this type of fly ash highly cementitious when it comes in contact with water (Tishmack *et al.* 1999).

Hydrated minerals in fly ash/water pastes

The principal hydration product identified by XRD in moderate-Ca Fort Union fly ash/water

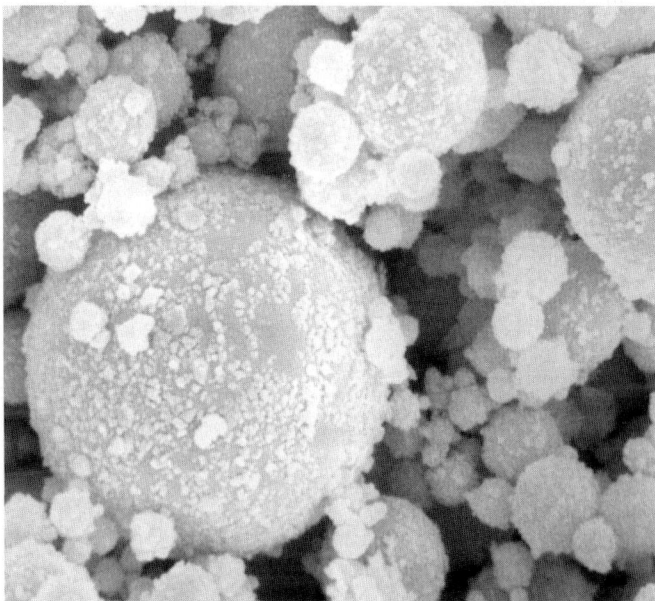

Fig. 7. Scanning electron micrograph of high-Ca Class-C subbituminous fly ash (from Powder River Basin coal).

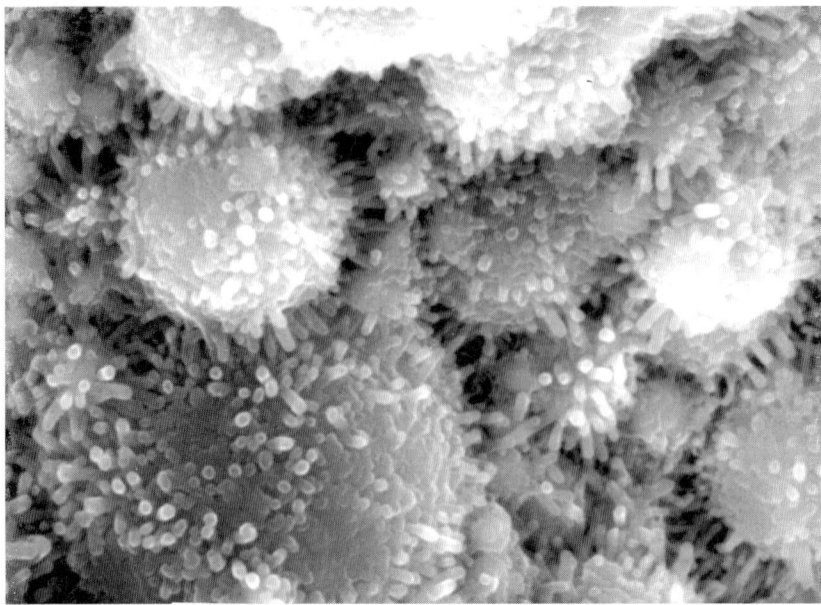

Fig. 8. Scanning electron micrograph of hydrated Class-C lignite fly ash (from Fort Union coal). Ettringite crystals are the small bumps seen on the surface of the fly ash particles.

pastes is ettringite (Fig. 8). Paste made from PRB fly ash and water (shown in Fig. 9) contains several minerals including ettringite, monosulphate, strätlingite ($Ca_2Al_2SiO_7 \cdot 8H_2O$), and probably C–S–H gel. The formation of these minerals in high-Ca fly ash has been studied in detail, particularly the formation of the mineral ettringite and its impact on cementitious

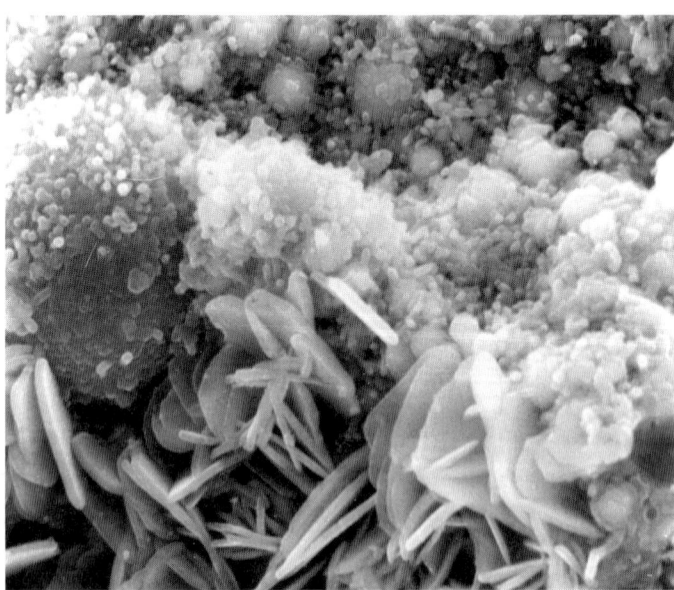

Fig. 9. Scanning electron micrograph of hydrated Powder River Basin high-Ca fly ash. The phases identified by XRD included ettringite (bumps on surface), monosulphate, and strätlingite (the latter are both platy).

systems (McCarthy & Solem-Tishmack 1994; Tishmack *et al.* 1999).

Ettringite is one of the principal hydration products that form in hydrated CCP containing more than 15 wt% CaO . It is the first Ca-sulphoaluminate hydrate to form in fly ash–water pastes as well as in Portland cement concrete (Lea 1971). The name ettringite causes confusion because it is both the name of a mineral having a distinct composition, as well as the name of a group of six minerals having the 'ettringite' structure (Pöllmann *et al.* 1989). Identification of ettringite is made more complicated because these minerals can have several different hydration states as well as various elemental substitutions in their structure, giving rise to many different forms (Pöllmann *et al.* 1989). The water content of ettringite is variable (32–24 moles of water per formula unit), and its crystals can lose up to 8 moles of water per formula unit without showing structural change (Moore & Taylor 1970); however, it strongly influences the XRD peak height (Tishmack 1999). It has a characteristic needle-like shape and commonly forms as secondary pore fillings in concrete and mortar, also known as 'secondary ettringite' (Mehta 1973; Monteiro & Mehta 1985; Day 1992).

Ettringite has been widely studied in cement and concrete due to the deleterious expansion, which has been attributed to its formation after the concrete has hardened (Lea 1971; Mehta 1973; Dunstan 1980; Cohen 1983; Monteiro & Mehta 1985; Tikalsky 1989; Day 1992; Tishmack *et al.* 1999). Ettringite also causes expansion in some types of high-S CCPs (see below).

Monosulphoaluminate, $Ca_4Al_2(SO_4)(OH)_{12} \cdot 6H_2O$, is similar to ettringite but with one-third the S (Hampson & Bailey 1982). In cement notation it is referred to as an AFm phase (aluminate ferrite mono-substituted), more widely known as 'monosulphate' (Lea 1971). Monosulphate and ettringite are the two common S-bearing compounds in PRB fly ashes as well as in ordinary Portland cement concrete (Lea 1971). Monosulphate is one of several AFm phases that form in hydrated Portland cement and have a layered structure and platy crystal habit (Lea 1971; Hampson & Bailey 1982). Its presence in PRB fly ash/water pastes as well as in Portland cement concrete indicates a low $S:C_3A$ ratio (Tishmack *et al.* 1999). Monosulphate will readily convert into ettringite if additional sulphate becomes available, such as from the ingress of S-bearing water, and its conversion to ettringite is associated with deleterious expansion in concrete ('delayed ettringite formation') (Lea 1971; Mehta 1973). As with ettringite,

monosulphate can have several different hydration states, and there are various substituted phases in which carbonate replaces sulphate in the structure (Highway Research Board 1972; Hampson & Bailey 1982).

Strätlingite (or Strätling's compound) is also sometimes referred to as gehlenite hydrate (Lea 1971; Highway Research Board 1972; MacDowell 1991). It is a calcium aluminate silicate hydrate, probably an AFm phase with S occurring as the interlayer ion (Lea 1971). It forms platy crystals that have been reportedly found in lime–pozzolan and cement–pozzolan mixtures (Highway Research Board 1972). Its occurrence in fly ash–water pastes has only been observed in pastes made from PRB fly ashes that contained high concentrations of Ca-aluminosilicate-rich glass and lower S (Solem & McCarthy 1992; Tishmack 1999).

The Ca-aluminate-rich glass present in PRB high-Ca fly ashes undergoes pozzolanic reactions and forms various hydration products including a Ca–Al-hydrate mineral sometimes referred to as hydrogarnet (MacDowell 1991). According to the Highway Research Board (1972) 'garnet' is a general name for cubic Ca–, Mg–, Fe–, or Mn–aluminum silicates, and 'hydrogarnet' $(Ca_3Al_2(SiO_4)_{3-x}(OH)_{4x})$ is the name used for end-members of the solid–solution series that contain appreciable crystalline water. Members of the garnet–hydrogarnet solid–solution series occur sporadically and to a limited extent in Portland cement concrete but more frequently in lime–pozzolan systems (Lea 1971; Highway Research Board 1972). They occur most frequently as a result of low-SO_3 availability or high temperatures.

One such Ca-aluminate hydrate mineral was observed in fly ash–water pastes (Tishmack 1999). The fly ash analysed in the study was washed with deionized water and Na–EDTA to selectively remove the soluble S-bearing minerals prior to being blended with additional CaO and water. The XRD analysis of the hydrated pastes showed peaks that closely matched the pattern of hydrogarnet with one notable exception, a strong peak at 32.1° 2-theta. The phase may be a tetracalcium aluminate hydrate prototype AFm mineral noted to form in systems that have a low $S:C_3A$ composition (Highway Research Board 1972). It may also be possible that the phase had a composition close to that of hydrogarnet with some substitution of Si for Al.

The hydration behaviour of high-Ca fly ashes, such as PRB fly ash, makes these materials a valuable replacement for Portland cement in concrete. It is commonly added at levels as high as

25 wt%, but research has shown they can even be used at higher replacement levels (as much as 45 wt%) without adversely affecting the 28-day compressive strength of the concrete (Malhotra 1994). PRB fly ash contains C_3S, C_2S, and C_3A, the principal phases in Portland cement. Although C–S–H gel is not typically identifiable with XRD due to its low crystallinity, it probably accounts for a major portion of the observed strength gain in fly ash pastes made from high-Ca fly ash and water (Solem-Tishmack et al. 1990). It is the presence of these reactive minerals along with free lime, that makes high-Ca fly ashes a valuable replacement for Portland cement in concrete and soil stabilization (Helmuth 1987; Solem-Tishmack 1993; Malhotra 1994). However, PRB fly ashes that contain >25 wt% CaO may cause durability problems in concrete because of limited SO_3 availability and the subsequent formation of secondary ettringite (Mehta 1973; Dunstan 1980; Tikalsky 1989; Day 1992; Malhotra 1994; Tishmack et al. 1999).

Sulphur-scrubbing products

The US Clean Air Act (CAA) introduced in 1970 was developed in order to maintain healthy air quality by controlling air pollution. The 1990 amendments to the CAA renewed and intensified national efforts to reduce air pollution in the USA. Besides CO_2, the two principal gaseous pollutants from coal combustion are SO_2 and NO_x. S- and N-compounds in the coal volatilize and react with O in the furnace, forming gaseous emissions that contain SO_2 and NO_x. A number of clean coal combustion technologies have been developed that remove SO_2 from the stack gases (International Energy Agency 1985). This can be accomplished by controlling combustion parameters, such as air (oxygen)/fuel ratios, temperature, and fuel feed rate, or by adding sorbent (chemicals that react with SO_2) directly into the combustion zone. Combustion cleaning methods include low-NO_x burners, gas reburning, flue gas desulphurization, and fluidized-bed combustion technology (US Department of Energy 1992). A detailed description of these technologies is beyond the scope of this chapter.

Flue gas desulphurization (FGD) is post-combustion technology that cleans gases after they leave the boiler or combustion zone. The basic approach to cleaning flue gas is to react alkaline minerals with the gaseous SO_2. Calcium is the principal alkaline sorbent used, although Mg or K are also used. Slurry made from limestone and water, or lime and water,

can be sprayed directly into hot flue gases forming a dry particulate material (spray-dryer technology). Flue gases can also be directed into a vessel, in which alkaline slurry is used to capture the SO_2 gases. These so-called 'wet scrubbers' produce a sludge that must be dewatered and stabilized before disposal (Baker Environmental Consulting Engineers 1995). Both of these types of FGD technology (spray dryer or wet scrubber) are used in conjunction with conventional pulverized-coal fired furnaces.

Scrubbers significantly change the chemistry and mineralogy of the resulting CCPs. Residues from scrubber systems contain high concentrations of Ca and S (between 30 and 60 wt%). They may contain lime, calcite, dolomite, anhydrite, gypsum, hannebachite ($CaSO_3 \cdot 1/2H_2O$), and/or portlandite ($Ca(OH)_2$). Residues of FGD vary in their chemical and mineralogical composition depending on a number of variables including: boiler type; coal feedstock; sorbent type; feed rate; residence time; airflow rate; and collection and handling operations.

Sulphite FGD contains hannebachite, which is formed at lower oxygen concentrations in the wet scrubbing system according to the following:

$$CaO + SO_2 + \frac{1}{2}H_2O \longrightarrow CaSO_3 \cdot \frac{1}{2}H_2O$$

Sulphate FGD contains gypsum and is formed when air is added into the system, increasing oxygen concentrations according to the following:

$$CaO + SO_2 + \frac{1}{2}O_2 + 2H_2O \longrightarrow CaSO_4 \cdot 2H_2O$$

Most sulphite FGD residues are combined with lime and fly ash to make a dry granular material that has better physical handling properties than unstabilized residues. These materials are generally called fixated scrubber sludge and have been used in road base and structural fill (Baker Environmental Consulting Engineers 1995). The added lime and fly ash creates pozzolanic reactions and results in the formation of C–S–H gel, improving the long-term strength gain in these materials (Perri & Laskey 1988).

The sulphate FGD residues often contain unreacted sorbent, that is, CaO or $Ca(OH)_2$, which acts as in internal source of lime and produces a high pH (12.5) when these materials come in contact with water. The amount of unreacted sorbent varies considerably depending on whether or not the power plant recycled the residues back through the scrubbers (McCarthy et al. 1989). Research has shown that these high-Ca, high-S materials can be used effectively as soil amendments to improve drainage, reduce erosion, and increase plant nutrients (Rechcigl 1995). They may also be used as soil amendments after being combined with organic byproducts (Stratton & Rechcigl 1998). Also, FGD gypsum is used widely as a replacement for gypsum in the manufacture of wallboard.

Fluidized-bed combustion (FBC) is a different technology from conventional pulverized-coal-fired boilers. A mixture of powdered rather than pulverized coal and a sorbent (usually limestone) are injected directly into the furnace and burned while suspended on a bed of high-velocity air. The turbulent mixing of coal and sorbent causes the solids to behave as a pseudo-fluid, enhancing heat generation and transfer at much lower temperatures (800–900 °C) compared to conventional pulverized-coal-fired furnaces (1500 °C) (US Department of Energy 1992).

These temperatures are below the fusion temperature for most silicate minerals, thus no glass phases are formed in these byproducts, making them significantly different from conventional pulverized-coal fly ash and bottom ash (Solem-Tishmack et al. 1990; Hemmings & Berry 1995). They are also well below the threshold where thermally induced NO_x is formed. Thus fluidized-bed combustion reduces both SO_2 and NO_x formation. However, the addition of large volumes of sorbent (as much as half a ton limestone per ton of coal) may be required, thus increasing the volume of CCPs by as much as four times the volume produced by conventional coal-fired boilers (International Energy Agency 1985).

An excellent study of FBC ash and its hydration behaviour was presented by Berry et al. (1991). The FBC ash consists of coarse-grained spent bed material and fine-grained 'fly ash' or baghouse ash that should not be confused with pulverized-coal fly ash because it is neither spherical nor glassy. In some cases, inert bed materials such as sand or shale are used to fluidize the coal and limestone particles. Clay minerals heated to 800 °C are dehydroxylated but not melted, resulting in the formation of X-ray amorphous phases, which produce a diffuse scattering maximum (glass hump) in the XRD pattern that resembles the glass in pulverized-coal fly ash. The spent bed contains sand-sized grains of reacted and unreacted sorbent and small amounts of unburned coal. The minerals generally identified in FBC residues are anhydrite, lime, quartz, and/or hematite (Solem-Tishmack et al. 1990). Under reducing conditions they may also contain Ca-sulphide phases and, upon conditioning with water, the senior author

has detected the presence of hydrogen sulphide gas.

Use of FBC ash in structural fill is limited compared to fly ash, bottom ash, and FGD products due to the expansion problems that occur when FBC ash first comes in contact with water (Hemmings & Berry 1995; Solem-Tishmack et al. 1990). Berry et al. (1994) investigated the possibility of blending FBC ash with low-Ca fly ash in order to improve particle bonding, increase strength, and reduce expansion. The engineering performance and environmental impact of an embankment constructed from these FBC products was monitored for several years in a study at Purdue University (Deschamps 1998). Hydrated FBC ash formed large amounts of ettringite and had a tendency to heave and swell for several years after placement in the field. The swelling behaviour was reduced when ash was allowed to weather (i.e., allowed to undergo sufficient hydration) prior to placement in the fill. The mixture of compacted FBC ash and fly ash had much higher stiffness compared to compacted soil and has demonstrated a very high resistance to penetration, as measured during coring and split-spoon sampling.

Long-term stability of FGD and FBC materials placed in landfills was reported by McCarthy et al. (1997). Five types of these materials from four different disposal sites in the USA were characterized by XRD. In settings, where moisture was available during or after placement of materials in the landfill, the materials changed chemically, physically, and mineralogically over time in a process analogous to diagenesis of buried sediments. At three of the sites, the transformations resulted in a matrix dominated by ettringite and thaumasite (an ettringite mineral with substitution of Si for Al and carbonate for sulphate). At a site that contained a high-Na CCP, an assemblage of Na-rich phases formed, including a zeolite and a zeolite-related nosean–hauyne phase (McCarthy et al. 1997). Mineral transformations in most of the materials resulted in decreased strength and increased permeability after only a few years of placement in the landfill. Results suggested, however, that blending of scrubber materials with fly ash to increase the proportion of C–S–H formation, combined with controlling subsequent moisture additions (i.e., capping) could minimize deleterious diagenesis.

Environmental impact of CCP

Most of the mass of CCP (99 wt%) is made up of Si, Al, Fe, Ca, K, Mg, Na, O, P, and Ti; the same elements that make up the composition of natural soil. It is the 1 wt% of trace elements that pose the greatest environmental concern, and research on trace elements in coal ash has historically focused on addressing their capacity to adversely impact the environment (Eary *et al.* 1990; Mattigod *et al.* 1990). There is concern over the emission of toxic trace elements, either in vapour form or enriched onto escaping fly ash particles, into the atmosphere. This is because the concentrations of several toxic trace elements (e.g., As, Cd, Pb, Se) are reported to be concentrated into the smallest respirable particles (Natusch *et al.* 1974; Smith 1980; Gieré *et al.* 2003).

As a result of the well-documented environmental concerns posed by coal combustion, and the disposal of CCPs, a large body of research has focused on characterizing the mechanisms of mobilization and attenuation of trace elements in coal and its ash. Based on their reported distribution in the solid phases of both source coals and coal ash, knowledge of the thermal transformations that occur to major mineral constituents during coal combustion, and a limited number of studies that have identified discrete solid phases of trace elements, a conceptual model of the chemical and mineralogical characteristics of trace elements in coal ash has been developed.

The concentrations of most trace elements in coal ash are typically higher than those found in the Earth's crust, coal, and soil (Table 7), indicating the coal combustion process tends to enrich the wastes in trace elements (Eary *et al.* 1990). Combustion at 800–900 °C causes a series of reactions involving nearly every solid contained within coal. The most important of these reactions, with respect to the geochemical behaviour of trace elements contained within coal, include: (1) the vaporization of volatile elements associated with the organic fraction and thermally unstable mineral matter within coal; (2) the condensation of elements volatilized during combustion onto the surfaces of fly ash particles; (3) the oxidation of sulphide minerals to metal oxides and inorganic salts, with the release of $SO_{2(g)}$ and $SO_{3(g)}$; and (4) the decomposition of carbonate minerals into metal oxides and inorganic salts, with the release of CO_2.

As was previously mentioned, trace elements that sublime at temperatures below those attained during coal combustion (e.g., As, Se, Hg, Zn), and are associated with thermally unstable solid phases (in particular organic matter and sulphide minerals), are subject to vaporization into furnace gases. Once these gases, and fly ash particles entrained in the gases, are vented from the combustion furnace they quickly cool, leading to the condensation of volatilized elements onto the

Table 7. *Concentrations (in mg/kg) of minor elements in solid coal combustion residues and natural sediments*

Element	Boiling temp (°C)	Fly ash*	Bottom ash*	Coal[†]	Soil[‡]	Surface enrichment[§]
Hg	357	0.01–12	0.01–4	0.01–1.6	n.d.	100
As	614	2–1385	0.02–168	0.5–106	1–50	98
Se	685	0.2–130	0.1–10	0.4–8	0.1–2	93
Zn	907	14–3500	4–1800	0–5600	10–300	100
Sr	1382	30–7600	170–6440	100	50–1000	32
Sb	1587	0.8–131	<10	0.2–9	n.d.	n.d.
Pb	1740	3–2120	0.4–1100	4–218	2–200	76
Ba	1870	1–13800	110–9360	150	100–3000	66
Mn	2061	25–3000	56–1940	6–181	20–3000	38
Sn	2602	8–56	<9–90	0–51	2–200	n.d.
Cr	2671	4–900	0.2–5820	0–610	1–1000	58
Ni	2913	1.8–4300	10–2900	0.4–104	5–500	9
Cu	2927	19–2200	4–930	1.8–185	2–100	64
Ti	3287	1310–10100	1540–1300	20–3200	1000–10000	n.d.
V	3407	12–1180	12–540	0–1281	20–500	80
B	3927	10–5000	2–513	1.2–356	2–100	n.d.
U	3927	0.8–30.4	5–26	<10–1000	n.d.	n.d.
Mo	4639	1–236	1–443	0–73	0.2–5	85

*Eary *et al.* (1990); Ainsworth & Rai (1987); Valkovic (1983); Adriano *et al.* (1980).
[†]Valkovic (1983); Torrey (1978).
[‡]Lindsay (1979).
[§]Hansen & Fisher (1980).
n.d., not detectable

surfaces of fly ash particles at varying rates (Natusch *et al.* 1974; Smith 1980). Through this vaporization–condensation mechanism, enrichment of volatile elements onto fly ash particles increases with decreasing particle size, due to the increased surface area/volume ratio of the smaller particles. The most volatile elements, which are the last to condense, are more strongly enriched onto the smallest particles (Natusch *et al.* 1974; Smith 1980; Eary *et al.* 1990; Gieré *et al.* 2003). Condensates of these volatile elements are predicted, and in some cases observed, to principally form relatively soluble metal oxides and simple ionic salts, primarily metal sulphates (Eary *et al.* 1990).

In addition to volatile trace elements, non-volatile chalcophile elements and trace elements that form associations with carbonate minerals (e.g., Ba, Sr) are also subject to thermal transformations that may enhance their accessibility to the environment, although to a far lesser degree than volatile elements. Both sulphides and carbonates are unstable at combustion temperatures, and readily decompose with the release of SO_2, SO_3, and CO_2. Although a large portion of the trace elements associated with these mineral domains are incorporated within relatively insoluble spinels (primarily magnetite) formed from the decomposition of pyrite and ankerite (Mattigod 1982; Lauf *et al.* 1982; Wang & Sweigard 1996), portions of these trace elements are converted to relatively soluble oxide and sulphate compounds.

Iron oxides present in coal are generally stable for the relatively short period of time that they are exposed to combustion temperatures. Therefore, siderophile elements (e.g., Ni, Co, Mo, Pt, Pd, Au) that are incorporated within iron oxides are also expected to remain stable, and escape any significant thermal transformation reactions (Burns 2003). Similarly, lithophile elements (e.g., Ba, B, Cr, Mn, Sr, V) that are initially found in association with silicates and aluminosilicates in coal are expected to be incorporated within the glassy fraction of coal ash upon thermal transformation of their parent minerals (Burns 2003).

Leachability of trace elements from CCP is a function of their type and hydration behaviour. The leachability of metals from lignite and sub-bituminous CCP is lower than from bituminous CCP due to the formation of hydrated minerals. Leachability of Se and B from Fort Union and PRB CCP is probably controlled by metal substitution within the ettringite crystal structure (Hassett & Hassett 1988; Hassett *et al.* 1989; Kumatharasan *et al.* 1990; Solem-Tishmack

et al. 1990; McCarthy *et al.* 1992; Solem-Tishmack 1993). The alkaline, high-pH solution generated by high-Ca CCP and the formation of C–S–H gel reduces solubility of Pb.

Trace elements on the surfaces of fly ash particles that are accessible to humans through air, soil, water, can affect health in several ways. The pathways by which metals from CCP may cause harm include: (1) soil deposition and resulting plant uptake of metals and subsequent movement into the food chain; (2) direct ingestion of soil by animals or humans; (3) leaching of metals from CCP to water systems and uptake by plants, animals, or humans; and (4) inhalation of dust (from soil) or respirable ash particles (Ryan & Bryndzia 1997).

The US EPA characterizes As, Be, Sb, Cd, Cr, Cu, Pb, Hg, Ni, Se, Ag, Tl, and Zn as 'priority metals' because of their potential hazardousness to human health. However, the environmental fate and effect of only a few metals (As, B, Cd, Cr, Cu, Mo, Ni, Pb, and Zn) have been studied extensively (Rechcigl 1995). For a given metal the potential to cause harm depends on the identifiable risk pathway, which is different for different metals. One pathway usually provides the highest probability of adverse affects to some receptor and is, therefore, the limiting pathway (Ryan & Bryndzia 1997). The most toxic elements to humans are Hg, Pb, Cd, Ni, and Co. Some of the principal limiting pathways for various metals are: the direct ingestion of Pb-contaminated soil by children; plant phytotoxicity from Cu, Zn, Ni; food-chain concentration and transfer of Cd and Hg to humans; and food-chain transfer of Se and Mo to livestock (Ryan & Bryndzia 1997).

The mobility of metals in soil solutions is controlled by several processes: (1) desorption or dissolution (rate depends on the solubility of metal–mineral form); (2) diffusion (depends on speciation of metal, soil oxidation/reduction potential, and pH); (3) sorption or precipitation (depends on soil solution concentration and rhizosphere effects); and (4) translocation in the plants (depends on plant species, soil solution concentration, and competing ions) (McBride 1994). Overall, a soil with a large clay component, oxides, or humus, tends to strongly retain most of the trace metals present (McBride 1995).

Modern analytical methods can detect some concentration of most elements in soils. The concentration varies as a function of several factors including: the age of the soil, its parent material, climate and weathering regimes, vegetation, and human activity. Whether an element is present naturally in soil or water, or has been introduced by pollution, its potential harmfulness is related to its bioavailablity or lability in the environment (McBride 1994). Bioavailablity of metals is difficult to measure. Neither the total quantity of an element in a soil, nor the extractability of that element from the soil by aggressive laboratory reagents, enable us to determine potential toxicity, because these measurements have not been shown to correlate with a plant-available pool of the element (McBride 1995).

Phytotoxicity is often a limiting mechanism in and of itself because it results in limited plant growth, seed and vegetative production and therefore, limited food-chain transfer to higher life forms (Ryan & Bryndzia 1998). Trace metals in CCP deposited on soil are not concentrated soluble metal salts, and they constitute only a small percentage of the fly ash mass (Rechcigl 1995; Ryan & Bryndzia 1998). In addition, surface trace metals on fly ash are only leachable at elevated concentrations for a short period of time due to their relatively low concentrations in these materials (Hassett & Hassett 1988). Short-term toxicity to plants or microorganisms is most closely related to the free metal cation concentration in soil solutions. Metal uptake over the long term depends to a large extent on the total metal concentration in soil solutions and on the ability of the soil to maintain this concentration (McBride 1995).

Rarely are CCP capable of sustaining elevated concentrations of trace metal species when exposed to in any type of solution (HCl, alkaline hydroxide, or even neutral water), However, when excessively large volumes of ash are continuously placed in saturated impoundments or landfills this situation can lead to movement of metals into local water resources. Several elements have been documented to leach from ash residues in concentrations exceeding water quality standards. For example, the US Environmental Protection Agency (EPA) found that 86% of groundwater samples taken near ash landfills contained arsenic levels more than 10 times the USEPA's new health standard of 0.01 mg/L (EPA 1988). Additionally, Hopkins *et al.* (2000) reported increases in the occurrence of malformations of bullfrog tadpoles from wetlands near CCP impoundments in South Carolina. Although the specific causes of these malformations were unclear, the researchers noted elevated concentration of teratogens (an agent that can cause malformations of an embryo or fetus) such as Se, Cd, and Cu in the tadpoles. Only recently have studies attempted to bridge the gap between the large amount of empirical data available and the development of a conceptual model that would allow for the prediction of the interactions between CCPs and water, and relate this to a given management practice. This is an area where further research is warranted.

Summary

The world's use of coal for electric power generation produces large volumes of CCP that could be safely utilized in many applications. The USA is the third largest coal consumer in the world and produces more than 100 Mt of CCP annually, only 30 wt% of which are used beneficially. Around the world much of the coal ash generated is disposed of in on-site landfills near electricity-generating stations. Growing concerns about groundwater protection are leading regulators to consider more stringent landfill protection at power plants. Power companies are looking for ways to recycle coal ash because it can be more economical than building new landfills.

CCP vary widely in their chemical and mineralogical composition, therefore proper utilization or disposal of these materials depends on detailed characterization in order to provide quality control, standardization, and regulatory guidance. New methods of characterizing these materials show that they are inhomogeneous even on the subnanometer scale. The relative content of mineral constituents in coal ash residues is known to vary, depending on the mineralogy of the source coal and the process variables during coal combustion. However, the dominant mineral constituents in coal and the thermal transformation reactions occurring during coal combustion are very predictable.

The chemical composition of coal ash is typically made up of O, Si, Ca, Al, and Fe, along with lesser amounts of Mg, S, C, and various trace elements. Crystals may occur embedded within the glass, or attached to the surface of fly ash particles, where they form through condensation reactions leading to surface enrichment of some elements. The common minerals found in CCB that contain less than 15 wt% CaO are made up of Si, Al, and Fe. They are unreactive and will not form hydration products. Coal ash with CaO contents greater than 15 wt% contain varying amounts of Ca-bearing minerals that are soluble in water and together with their more reactive Ca-rich glass form various hydrated minerals. Ca-content, therefore, is a controlling factor in the reactivity of these materials when they contact water, and therefore, must be considered in the management of these materials. The formation of delayed ettringite in concrete may limit the use of fly ash having

>25 wt% CaO as a replacement for Portland cement.

The compounds that result from FGD technologies have very different bulk chemical and mineralogical characteristics than those from conventional combustion. They still contain Si, Al, and Fe, but they contain as much as 50–60 wt% Ca and S, much of which is soluble in water. Although many types of FGD products have been used successfully as engineered fill, they may also be utilized as soil amendments, particularly when combined with organic byproducts. Use of FBC ash in structural fill is limited compared to fly ash, bottom ash, and FGD products due to the expansion problems that occur when FBC ash first comes into contact with water.

Regardless of the source coal or specific process variables, normal high-temperature ashing of coal leads to a series of thermal transformation reactions that will dictate the composition of coal ash residues. The combined influence of these thermal transformations results in the production of coal ash with predictable chemical and mineralogical characteristics. The formation of ash residues with similar and predictable mineralogy suggests that the composition of water contacting these residues will also be similar and predictable. However, only recently have studies attempted to bridge the gap between the large amount of empirical data available and the development of a conceptual model that would allow for the prediction of the interactions between CCPs and water, particularly with respect to the environmental impact of ash disposal.

We would like to thank the reviewers of this chapter, Dr J. Hower (Center for Applied Energy Research, University of Kentucky), Dr M. Brownfield (USGS Denver), and one anonymous reviewer. Their questions, comments, and suggestions were greatly appreciated and helped us to expand the focus of the chapter and improve the quality of this review.

References

ACAA 2000. *American Coal Ash Association – 1999 Coal Combustion Product Production and Use.* ACAA, Alexandria, VA.

ADRIANO, D. C., PAGE, A. L., ELSEEWI, A. A., CHANG, A. C. & STRAUGHAN, I. 1980. Utilization and disposal of fly ash and other coal residues in terrestrial ecosystems: A review. *Journal of Environmental Quality*, **9**, 333–344.

AINSWORTH, C. C. & RAI, D. 1987. *Chemical Characterization of Fossil Fuel Wastes.* EPRI EA-5321. Electric Power Research Institute, Palo Alto, CA.

AINSWORTH, C. C., MATTIGOD, S. V., RAI, D. & AMONETTE, J. E. 1993. *Detailed Physical, Chemical, and Mineralogical Analyses of Selected Coal and Oil Combustion Ashes.* Electric Power Research Institute, Palo Alto, CA.

ASTM 1994. *ASTM C618: Standard Specification for Fly Ash and Raw or Calcined Natural Pozzolan for Use as a Mineral Admixture in Portland Cement Concrete.* ASTM International. World Wide Web Address: http//:www.astm.org.

ASTM 2003a. *ASTM D388-99e1: Standard Classification of Coals by Rank.* ASTM International. World Wide Web Address: http//:www.astm.org.

ASTM 2003b. *ASTM E2277-03: Standard Guide for Design and Construction of Coal Ash Structural Fills.* ASTM International. World Wide Web Address: http//:www.astm.org.

BAKER ENVIRONMENTAL CONSULTING ENGINEERS 1995. *FGD By-product Disposal Manual, 4th Edition,* EPRI TR-104731. Electric Power Research Institute, Palo Alto, CA.

BERGESON, K. L., SCHLORHOLTZ, S. & DEMIREL, T. 1988. *Development of a Rational Characterization Method for Iowa Fly Ash.* Final report, Iowa Department of Transportation project HR-286, ERI project 1847, ISU-ERI 86-450.

BERRY, E. E., HEMMINGS, R. T. & CORNELIUS, B. J. 1991. *Commercialization Potential of AFBC Concrete: Part 2. Mechanistic Basis for Cementing Action.* Report GS-7122, Electric Power Research Institute, Palo Alto, CA.

BERRY, E. E., HEMMINGS, R. T., ZHANG, M., CORNELIUS, B. J. & GOLDEN, D. M. 1994. Hydration in high-volume fly ash concrete binders. *ACI Materials Journal*, **91**, 382–389.

BOSBACH, D. & ENDERS, M. 1998. Microtopography of high-calcium fly ash particle surfaces. *Advances in Cement Research*, **10**, 17–23.

BRAUNSTEIN, H. M., COPENHAVER, E. D. & PFUDERER, H. A. 1977. *Environmental, Health, and Control Aspects of Coal Conversion: An Information Overview,* Vol. 1. Oak Ridge National Laboratory, Contract No. W-7405-eng-26.

BURNS, P. E. 2003. *Identifying Mechanisms for the Mobilization and Attenuation of Elemental Constituents from Weathered Alkaline Coal Ash Using Geochemical Modeling.* PhD dissertation. Purdue University, West Lafayette, IN.

COHEN, M. D. 1983. Theories of expansion in sulfoaluminate-type expansive cements: schools of thought. *Cement and Concrete Research*, **13**, 809–818.

DAVISON, R. L., NATUSCH, D. F. S., WALLACE, J. R. & EVANS, C. A. 1974. Trace elements in fly ash: Dependance of concentration on particle size. *Environmental Science and Technology*, **8**, 1107–1112.

DAY, R. L. 1992. *The Effect of Secondary Ettringite Formation on the Durability of Concrete: A Literature Analysis.* Research and Development Bulletin RD108T, Portland Cement Association, Skokie, IL.

DESCHAMPS, R. J. 1998. Using fluidized bed combustion and stoker ash as structural fill: A case study.

Journal of Geotechnical and Geoenvironmental Engineering, **124**, 1120–1127.

DIAMOND, S. 1984. On the glass present in low-Ca and high-Ca fly ash. *Cement and Concrete Research*, **13**, 459–464.

DIAMOND, S. & OLEK, J. 1987. Some properties of contrasting end-member high calcium fly ashes. *Materials Research Society Symposium Proceedings*, **86**, 315–324.

DONAHOE, R. J. (2004). *Secondary Mineral Formation in Coal Combustion Byproduct Disposal Facilities: Implications for Trace Element Sequestration. In*: GIERÉ, R. & STILLE, P. (eds) *Energy, Waste, and the Environment: a Geochemical Perspective.* Geological Society, London, Special Publications, **236**, 641–658.

DUNSTAN, E. R. 1980. A possible method for identifying fly ashes that will improve the sulfate resistance of concrete. *Cement, Concrete, & Aggregates*, **2**, 22–30.

EARY, L. E, RAI, D., MATTIGOD, S. V. & AINSWORTH, C. C. 1990. Geochemical factors controlling the mobilization of inorganic constituents from fossil fuel combustion residues: II. Review of minor elements. *Journal of Environmental Quality*, **19**, 202–214.

ENDERS, M. 1995. Microanalytical characterization (AEM) of glassy spheres and anhydrite from a high-calcium lignite fly ash from Germany. *Cement and Concrete Research*, **25**, 1369–1377.

EIA 1995. *Energy Information Administration Coal Data: A Reference.* DOE/EIA-0064(93). US Department of Energy, Washington, DC.

EPA 1988. *US Environmental Protection Agency Report to Congress: Wastes from the Combustion of Coal by Electric Utility Power Plants.* Office of Solid Waste and Emergency Response, Washington, DC. Report No. EPA/530-SW-88-002.

FISHMAN, N. S., RICE, C. A., BREIT, G. N. & JOHNSON, R. D. 1999. Sulfur-bearing coatings on fly ash from a coal-fired power plant: composition, origin, and influence on ash alteration. *Fuel*, **78**, 187–196.

FLAGAN, R. C. & FRIEDLANDER, S. K. 1976. *Particle Formation in Pulverized Coal Combustion: A Review.* Paper presented at the Eighty-Second National Meeting of the American Institute of Chemical Engineers, Atlantic City, NJ.

GIERÉ, R., CARLETON L. E. & LUMPKIN, G. R. 2003. Micro- and nanochemistry of fly ash from a coal-fired power plant. *American Mineralogist*, **88**, 1853–1865.

HAMPSON, C. J. & BAILEY, J. E. 1982. On the structure of some precipitated calcium alumino-sulfate hydrates. *Journal of Materials Science*, **17**, 3341–3346.

HANSEN, L. D. & FISHER, G. L. 1980. Elemental distribution in coal fly ash particles. *Environmental Science and Technology*, **14**, 1111–1117.

HARVEY, R. D. & RUCH, R. R. 1986. Mineral matter in Illinois and other U.S. coal. *In*: VORRES, K. S. (ed) *Mineral Matter and Ash in Coal. American Chemical Society Symposium Series*, 301, Washington, DC.

HASSETT, D. J. & HASSETT, D. F. 1988. Fixation of leachable elements in composite waste forms from North Dakota lignite coal conversion ash. *Materials Research Society Symposium Proceedings*, **113**, 333–340.

HASSETT, D. J., PFLUGHOEFT-HASSETT, D. F., KUMARATHASAN, P. & MCCARTHY, G. J. 1989. Ettringite as an agent for the fixation of hazardous oxyanions. *Proceedings of the 12th Annual Madison Waste Conference on Municipal and Industrial Waste*, 471–482.

HELMUTH, R. 1987. *Fly Ash in Cement and Concrete.* Portland Cement Association, Skokie, IL.

HEMMINGS, R. T. & BERRY, E. E. 1986. Speciation in size and density fractionated fly ash. *Materials Research Society Symposium Proceedings*, **65**, 91–104.

HEMMINGS, R. T. & BERRY, E. E. 1988. On the glass in coal fly ashes: recent advances. *Materials Research Society Symposium Proceedings*, **113**, 3–38.

HEMMINGS, R. T. & BERRY, E. E. 1995. Cementitious properties of atmospheric fluidized bed combustion residues and their use in engineering applications. *Proceedings of the Industrial Waste Conference*, **49**, 591–596.

HESSLEY, R. K., REASONER, J. W. & RILEY, J. T. 1986. *Coal Science: An Introduction to Chemistry Technology and Utilization.* John Wiley & Sons, New York, NY.

HIGHWAY RESEARCH BOARD 1972. *Guide to Compounds of Interest in Cement and Concrete Research.* Special Report 127. Highway Research Board, Division of Engineering, National Research Council, Washington, DC.

HOPKINS, W. A., CONGDON, J. D. & RAY, J. K. 2000. Incidence and impact of axial malformations in bullfrog larvae developing in sites impacted by coal combustion byproducts *Environmental Toxicology and Chemistry*, **19**, 862–868.

HOSTETLER, C. J., ERIKSON, R. L. FRUCHTER, J. S. & KINCAID, C. T. 1989. *FASTCHEM^{TM} Package, Volume 1: Overview and Application to a Chemical Transport Problem.* Electric Power Research Institute, EA-5870, Vol. 1, Research Project 2485-2.

HOWER, J. C. & MASTALERZ, M. 2001. An approach toward a combined scheme for the petrographic classification of fly ash. *Energy and Fuels*, **15**, 1319–1321.

HOWER, J. C., FINKELMAN, R. B., RATHBONE, R. F. & GOODMAN, J. 2000. Intra- and inter-unit variation in fly ash petrography and mercury adsorption: examples from a western Kentucky power station. *Energy and Fuels*, **14**, 212–216.

HUBBARD, F. H., McGILL, R. J., DHIR, R. K. & ELLIS, M. S. 1984. Clay and pyrite transformations during ignition of pulverized coal. *Mineralogical Magazine*, **48**, 251–256.

HULETT, L. D. & WEINBERGER, A. J. 1980. Some etching studies of the microstructure and composition of large aluminosilicate particles in fly ash from coal-burning power plants. *Environmental Science and Technology*, **14**, 965–970.

INTERNATIONAL ENERGY AGENCY 1985. *The Clean Use of Coal: A technology Review.* Organization for Economic Co-operation and Development (OCED), Paris, France.

JOSHI, R. C., NATT, G. S., DAY, R. L. & TILLEMAN, D. D. 1985. Scanning electron microscopy and x-ray diffraction analysis of various size fractions of fly ash. *Materials Research Society Symposium Proceedings*, **1**, 31–39.

KAAKINEN, J. W., JORDEN, R. M. LAWASANI, M. H. & WEST, R. E. 1975. Trace element behavior in coal-fired power plant. *Environmental Science and Technology*, **9**, 862–869.

KATRINAK, K. A. & ZYGARLICKE, C. J. 1995. Size-related variations in coal fly ash composition as determined using automated scanning electron microscopy. *Fuel Processing Technology*, **44**, 71–79.

KUMARATHASAN, P., McCARTHY, G. J., HASSETT, D. J. & PFLUGHOEFt-HASSETT, D. F. 1990. Oxyanion substituted ettringites: synthesis and characterization; and their role in immobilization of As, B, Cr, Se and V. *Materials Research Society Symposium Proceedings*, **178**, 83–104.

LAUF, R. J., HARRIS, L. A. & RAWISTON, S. S. 1982. Pyrite framboids as the source of magnetite spheres in fly ash. Environmental Science and Technology, **16**, 218–220.

LEA, F. M. 1971. *The Chemistry of Cement and Concrete*, 3rd ed. Chemical Publishing Co., New York.

LINDSAY, W. L. 1979. *Chemical Equilibria in Soils*. John Wiley & Sons, Inc., New York.

LINTON, R. W., WILLIAMS, P., EVANS, C. A. & NATUSCH, D. R. S. 1977. Determination of the surface predominance of toxic trace elements in airborne particles by ion microprobe mass spectrometry and Auger electron spectroscopy. *Analytical Chemistry*, **49**, 1514–1521.

MACDOWELL, J. F. 1991. Strätlingite and hydrogarnet from calcium aluminosilicate glass cements. *In*: *Special Cement Advanced Properties, Mater. Res. Soc. Symp. Proc.*, **179**, 159–179.

MALHOTRA, V. M. 1994. *Fly Ash in Concrete*. Canada Center for Mineral and Energy Technology (CANMET), Ottawa, Canada.

MATTIGOD, S. V. 1982. Characterization of fly ash particles. *Scanning Electron Microscopy*, **2**, 611–617.

MATTIGOD, S. V., RAI, D., EARY, L. E. & AINSWORTH, C. C. 1990. Geochemical factors controlling the mobilization of inorganic constitutents from fossil fuel combustion residues: I. Review of major elements. *Journal of Environmental Quality*, **19**, 188–201.

McBRIDE, M. B. 1994. *Environmental Chemistry of Soils*. Oxford University Press, New York.

McBRIDE, M. B. 1995. Toxic metal accumulation from agricultural use of sludge. Are U.S. EPA regulations protective? *Journal of Environmental Quality*, **24**, 5–18.

McCARTHY, G. J. 1988. X-ray powder diffraction for studying the mineralogy of fly ash. *In*: *Fly Ash and Coal Conversion By-Products: Characterization, Utilization and Disposal. Materials Research Society Symposium Proceedings*, **113**, 75–86.

McCARTHY, G. J. & SOLEM-TISHMACK, J. K. 1994. Hydration mineralogy of cementitious coal combustion by-products. *In*: *Advances in Cement and Concrete, Proceedings of an Engineering Foundation Conference*. Materials Engineering Division, ASCE, Durham, NH.

McCARTHY, G. J., MANZ, O. E., JOHANSEN, D. M., STEINWAND, S. J. & STEVENSON, R. J. 1987. Correlations of chemistry and mineralogy of Western U.S. fly ash. *Materials Research Society Symposium Proceedings*, **86**, 109–112.

McCARTHY, G. J., SOLEM, J. K. *et al.* 1989. Database of chemical, mineralogical and physical properties of North American low-rank coal fly ash. *In*: NESS, H. M. (ed) *Proceedings of the 15th Biennial Low-Rank Fuels Symposium*, DOE/METC-90/6109, 555–563.

McCARTHY, G. J., SOLEM, J. K., MANZ, O. E. & HASSETT, D. J. 1990. Use of a database of chemical, mineralogical and physical properties of North American fly ash to study the nature of fly ash and its utilization as a mineral admixture in concrete. *Materials Research Society Symposium Proceedings*, **178**, 3–34.

McCARTHY, G. J., HASSETT, D. J. & BENDER, J. A. 1992. Synthesis, crystal chemistry and stability of ettringite, a material with potential applications in hazardous waste immobilization. *Materials Research Society Symposium Proceedings*, **245**, 129–140.

McCARTHY, G. J., BUTLER, R. D., GRIER, D. G., ADAMEK, S. D., PARKS, J. A. & FOSTER, H. J. 1997. Long-term stability of landfilled coal combustion by-products. *Fuel*, **76**, 607–703.

MEHTA, P. K. 1973. Mechanism of expansion associated with ettringite formation. *Cement and Concrete Research*, **3**, 1–6.

MILLER, R. N. & GIVEN, P. H. 1986. The association of major, minor, and trace inorganic elements with lignites. I. Experimental approach and study of a North Dakota lignite. *Geochimica et Cosmochimica Acta*, **50**, 2033–2043.

MONTEIRO, P. J. M. & MEHTA, P. K. 1985. Ettringite formation on the aggregate–cement paste interface. *Cement and Concrete Research*, **15**, 378–380.

MOORE, A. E. & TAYLOR, H. F. W. 1970. Crystal structure of ettringite. *Acta Crystallographica*, **B26**, 386–392.

NATUSCH, D. F. S., WALLASCE, J. R. & EVANS, C. A. 1974. Toxic trace elements: Preferential concentration in respirable particles. *Science*, **183**, 202–204.

PERRI, J. S. & LASKEY, J. S. 1988. *Advanced SO₂ Control By-product Utilization Laboratory Evaluation. Electric Power Research Institute Report*, CS-6044. Palo Alto, CA.

PÖLLMANN, H., KUZEL, H. J. & WENDA, R. 1989. Compounds with ettringite structure. *Neues Jahrbuch für Mineralogie, Abhandlungen*, **160**, 133–158.

QIAN, J. C., LACHOWSKI, E. E. & GLASSER, F. P. 1988. Microstructure and chemical variation in Class F fly ash glass. *Materials Research Society Symposium Proceedings*, **113**, 45–53.

QIAN, J.C., LACHOWSKI, E.E. & GLASSER, F.P. 1989. The microstructure of National Bureau of Standards reference fly ashes. Materials Research Society Symposium Proceedings, 136, 77-85.

RECHCIGL, J. E. (ed) 1995. *Soil Amendments and Environmental Quality*. CRC Press, Boca Raton, FL.

RYAN, J. A. & BRYNDZIA, L. T. 1997. Fate and potential effects of trace elements: issues in co-utilization of by-products. *In*: BROWN, S., ANGLE, J. S. & JACOBS, L. (eds) *Beneficial Co-utilization of Agricultural, Municipal, and Industrial By-products*. USDA Agricultural Research Service, Kluwer Academic Publishers, Netherlands, 219–233.

SIMSIMAN, G. V., CHESTERS, G. A. & ANDERS, W. 1987. Effect of ash disposal ponds on groundwater quality at coal-fired power plant. *Water Research*, **21**, 417–426.

SMITH, R. D. 1980. The trace element chemistry of coal during combustion and the emissions from coal-fired plants. *Progress in Energy and Combustion Science*, **6**, 53–119.

SMITH, R. D., CAMPBELL, J. A. & NIELSON, K. K. 1979. Concentration dependence upon particle size of volatilized elements in fly ash. *Environmental Science and Technology*, **13**, 553–558.

SOLEM, J. K. &. MCCARTHY, G. J. 1992. Hydration reactions and ettringite formation in selected cementitious coal conversion by-products. *Material Research Society Symposium Proceedings*, **245**, 71–79.

SOLEM-TISHMACK, J. K. 1993. *Use of Coal Conversion Solid Residuals for Stabiliztion/Solidification*. Masters thesis, Agronomy Department, North Dakota State University.

SOLEM-TISHMACK, J. K., MCCARTHY, G. J., DOCKTOR, B., EYLANDS, K. E., THOMPSON, J. S. & HASSETT, D. J. 1990. High-calcium coal conversion by-products: Engineering properties, ettringite formation, and potential application in solidification and stabilization of selenium and boron. *Cement and Concrete Research*, **25**, 658–670.

SOROCZAK, M. M., EATON, H. C. & TITTLEBAUM, M. E. 1987. An ESCA and SEM study of changes in the surface composition and morphology of low-calcium coal fly ash as a function of aqueous leaching. *Materials Research Society Symposium Proceedings*, **86**, 37–47.

STEVENSON, R. J. & HUBER, T. P. 1987. SEM study of chemical variations in western U.S. fly ash. *Materials Research Society Symposium Proceedings*, **86**, 99–108.

STRATTON, M. L. & RECHCIGL, J. E. 1998. Agronomic benefits of agricultural, municipal, and industrial by-products and their co-utilization: an overview. *In*: BROWN, S., ANGLE, J. S. & JACOBS, L. (eds)

Beneficial Co-utilization of Agricultural, Municipal, and Industrial By-products. USDA Agricultural Research Service, Kluwer Academic Publishers, Netherlands, 9–34.

SUMMERS, D. V., RUPP, G. L. & GHERINI, S. A. 1983. *Physical-chemical Characteristics of Utility Solid Wastes*. Electric Power Research Institute, EA-3236. Palo Alto, CA.

THEDCHANAMOORTHY. A. & MCCARTHY, G. J. 1989. Semi-quantitative XRD analysis of fly ash using rutile as an internal standard. *Advances in X-Ray Analysis*, **32**, 569–576.

TIKALSKY, P. J. 1989. *The Effects of Fly Ash on the Sulfate Resistance of Concrete*. PhD dissertation, University of Texas at Austin, August.

TISHMACK, J. K. 1999. *Characterization of High-Calcium Fly Ash and Its Influence on Ettringite Formation in Portland Cement Pastes*. PhD dissertation, School of Civil Engineering, Purdue University, West Lafayette, IN.

TISHMACK, J. K., OLEK, J. & DIAMOND, S. 1999. Characterization of high-calcium fly ashes and their potential influence on ettringite formation in cementitious systems. *Cement, Concrete, and Aggregates*, **21**, 82–92.

TISHMACK, J. K., OLEK, J. & DIAMOND, S. 2001. Characterization of pore solutions expressed from high-calcium fly ash–water pastes. *Fuel*, **80**, 815–819.

TORREY, W. 1978. Coal ash utilization: fly ash, bottom ash and slag. *Pollution Technology Review*, **48**, 136–145.

US DEPARTMENT OF ENERGY 1992. *Clean Coal Technology Demonstration Program: Program Update 1991*. Assistant Secretary for Fossil Energy, Washington, DC, Report No. DOE/FE-247P.

VAN DER SLOOT, H. A. & NIEUWENDIJK, B. J. T. 1985. Release of trace elements from surface-enriched fly ash in seawater. *Wastes Ocean*, **4**, 449–465.

VALKOVIC, V. 1983. *Trace Elements in Coal: Volumes I and II*. CRC Press, Inc. Boca Raton, FL.

WANG, D. & SWEIGARD, R. J. 1996. Characterization of Fly Ash and Bottom Ash from a Coal-Fired Power Plant. International Journal of Surface Mining, Reclamation and Environment, 10, 181–186.

WARREN, C. J. & DUDAS, M. J. 1986. Mobilization and Attenuation of Trace Elements in Artificially Weathered Fly Ash. Electric Power Research Institute Report, EA-4747, Palo Alto, CA.

WU, E. J. & CHEN, K. Y. 1987. *Chemical Form and Leachability of Inorganic Trace Elements in Coal Ash*. Electric Power Research Institute Report, EA-5115, Palo Alto, CA.

ZUBOVIC, P., STADNICHENKO, T. & SHEFFEY, N.B. 1960. *The Association of Minor Elements with Organic and Inorganic Phases of Coal*. US Geological Survey Professional Paper, 400-B, Washington, DC.

The beneficiation of coal combustion ash

JOHN GROPPO, THOMAS ROBL & JAMES C. HOWER

University of Kentucky, Center for Applied Energy Research, Lexington, KY, USA

(e-mail: groppo@caer.uky.edu)

Abstract: Pulverized coal combustion ash is an important source of strategic materials in the USA. Bottom ash is used as a source of aggregate for use in concrete and masonry units (blocks). It is processed primarily to improve its grading. A top size is removed and the finest sizes are removed via wet or dry screens. Pyrite and rock may also be present with the ash. These materials can be removed by spiral concentrators and jigs. In some cases high-quality/high-value lightweight aggregates are produced from stored bottom ash. Fly ash is used as a pozzolanic additive to Portland cement concrete. In addition to partially replacing the cement, it contributes substantially to the durability of the concrete. The advent of low-NO_x burners and selective catalytic reduction (SCR) supported ammonia injection has altered the character of the fly ash, particularly the Class F, low-Ca type, by generally increasing the amount of unburned carbon. This contaminant adsorbs air-entrainment reagents and can decrease the resistance of the concrete to freeze–thaw damage. Over the past decade several technologies and approaches have been developed to remove the carbon from the fly ash, including: air classification; electrostatic separation; and fluidized-bed combustion. Other approaches such as microwave heating also show promise. Froth flotation has been successfully applied to wet ash. The amount of ash that is beneficiated has increased to a current level of about 1 million tons per year in the USA, which is expected to grow in time due to the need for predictable materials with constant characteristics. The primary environmental advantage of ash beneficiation is that it enables the use of combustion ash that would otherwise be disposed as waste. High-quality, consistent products can be generated, thus increasing the usefulness and acceptance of these processed products in both traditional and emerging markets. By doing so, the amount of ash that is utilized will be increased, thus reducing the amount of ash that is disposed, while conserving other resources such as aggregate and sand for other uses not applicable to combustion ash.

Pulverized coal combustion produces hundreds of millions of tons of ash each year. Significant amounts of this ash are used beneficially. Globally, the amount used varies considerably, ranging from only 7% in India to ~30% in the USA, to complete utilization in the Netherlands (Meij & van den Berg, 2001). The most important use of this material is for pozzolanic additives in concrete to replace and augment cement, as aggregate in concrete masonry units, as a major component in autoclaved cellular concrete, and in flowable fills. Changes in the combustion characteristics of the ash, as dictated by the need to burn coal more cleanly, have resulted in changes in its composition. Low-NO_x burners generally produce more carbon, and the advent of selective catalytic reduction (SCR) of NO_x also leaves a small residue of ammonia salts, primarily ammonium bisulphate and sulphate. These changes represent an overall decrease in the quality of the ash.

Beneficiation processes to remove contaminants and improve the overall quality of ash have been developed. The amount of ash that is currently beneficiated is hard to determine precisely; however, we estimate that it is over 1 million tons in the USA annually, and growing.

Classification and composition of fly and bottom ash

For the purpose of this discussion, coal combustion ash consists of two distinct products: bottom ash and fly ash. The distinction between bottom ash and fly ash is how they form and exit from the boiler. Bottom ash is removed from the bottom of the boiler, whereas fly ash exits in the flue gas, where it is subsequently collected by a variety of devices. This distinction is important because the beneficiation options available for either of these products, as well as potential end uses, are different.

Objectives of beneficiation

Beneficiation is the practice of improving raw material properties in order to increase their

From: GIERÉ, R. & STILLE, P. (eds) 2004. *Energy, Waste, and the Environment: a Geochemical Perspective*. Geological Society, London, Special Publications, **236**, 247–262.
0305-8719/04/$15   © The Geological Society of London 2004.

usefulness. The term is perhaps most commonly applied to mineral mixtures, where the practice consists of concentrating desirable components and separating them from less desirable components. Mineral beneficiation or mineral processing, as it is sometimes referred to, has been practised for hundreds of years in order to concentrate the useful components of an ore. The extent to which mineral beneficiation is carried out is dependent on the complexity of the ore and the level of concentration that is required. One of the simplest beneficiation schemes is to remove undesirable fines from a sand deposit by size classification or to screen crushed gravel to the appropriate sizes. An example of the other extreme would be the selective concentration of Cu, Pb, Zn, Ag, and Au from a complex sulphide ore. Regardless of the specific application, effective beneficiation provides usable product(s) and added value from raw materials that are less valuable.

The term ash beneficiation, for the purposes of this review, is applied to adding value and usefulness to coal combustion ash. The primary reason that ash beneficiation is necessary is that recent changes in combustion practice have rendered combustion ash less usable than in previous generations. For example, bottom ash was once commonly used to produce concrete masonry units. So common was the practice of using the coarse ash residue from stoker boilers, frequently called cinders, that the masonry units, or blocks, were called cinder blocks. Unfortunately, changes in the quality of the ash over the years made the practice less common. However, a resurgence is currently taking place, made possible by beneficiating the ash to remove the less desirable components and provide a high-quality, consistent product.

Environmental impacts of ash beneficiation

An unavoidable consequence of coal combustion is the generation of byproducts, specifically bottom ash and fly ash. These constituents are comprised primarily of the inorganic components of coal remaining after combustion. Common practice is to store or impound these byproducts in landfills or ponds frequently located on property adjacent to the utility. Since these storage facilities have limited capacity, at some point in time they are filled to capacity. It is then necessary to create new storage capacity by excavating accumulated ash and transporting it either to a permanent disposal site or identifying some beneficial reuse such as a structural fill.

If neither of these options are available, it is necessary to construct additional ash storage capacity.

Excavating existing disposal sites adds significantly to the cost of ash disposal. While this approach creates additional ash storage capacity, the costs associated with excavating, transporting, and placing reclaimed ash may preclude consideration of this option. Construction of new disposal sites bears considerable costs as well, such as land purchase, permitting, and site preparation. The impact of ash beneficiation is that processed ash products can be generated from combustion ash and marketed, thus reducing the amount of ash in storage. If sufficient processed ash can be marketed as useful products, the need to create additional storage capacity is reduced, or potentially even eliminated with an ideal goal of utilizing all of the ash generated at a utility station. This scenario is technically and economically feasible, provided that consistent, high-quality ash products can be produced. The role of ash beneficiation is to provide processed products that meet these criteria.

The beneficial impact of ash utilization is not merely limited to economic considerations; a proactive ash utilization programme constitutes good environmental stewardship. For example, the need for raw materials, such as aggregate, in the construction industry is ever increasing. Aggregate can be derived from quarry operations, but may also be generated from specific size fractions of bottom ash, provided that the processed bottom ash meets the necessary quality specifications. By utilizing bottom ash, the amount of ash disposed is reduced while raw aggregate from the quarry can be conserved for other uses not applicable for bottom ash. Another example is pozzolan derived from fly ash. High-quality fly ash can be substituted for 20–30 wt% of Portland cement in concrete, greatly improving some concrete quality criteria, such as permeability and workability without compromising strength. Since the use of cement is thereby reduced, an environmental benefit is realized with respect to the energy and raw materials consumed, along with the emissions produced by cement manufacture.

As previously mentioned, the ultimate goal of ash utilization is to market all of the ash generated at a utility station. Given the broad variations in ash characteristics, such as size, physical and chemical properties, it is unlikely that total utilization can be realized with a single ash market. Effective beneficiation can separate highly variable ash into a range of consistent, high-quality products, such as lightweight aggregate, high-carbon fuel and pozzolan.

With a range of products available, the marketing potential is significantly broadened and the utilization potential is increased.

What makes a good quality bottom ash?

Currently, the most common market for bottom ash is as lightweight aggregate, a component in lightweight concrete masonry units. The first commercial-scale bottom ash beneficiation system to recover lightweight aggregate from stored bottom ash was installed in 1997 at Santee Cooper's Winyah Generating Station in Georgetown, South Carolina (Edens 1999). The facility has the capacity to produce as much as 200 000 t/y lightweight aggregate. Since then, several other bottom ash beneficiation plants have been installed. The specific purpose of these plants is to produce lightweight aggregate, compliant with the specifications outlined by the American Society for Testing and Materials (ASTM) in procedure C-331. The specifications are primarily targeted at gradation with a top size of 9.5 mm, which can usually be readily attained by adequate screening. Another important criterion is bulk density. While ASTM specifications limit the bulk density to 1120 kg/m^3, most block manufacturers prefer a bulk density of less than 800 kg/m^3. When the bulk density of bottom ash is as low as 640 kg/m^3 or less, friability is generally too high to be used in this application. If the bulk density exceeds 1120 kg/m^3, bottom ash can be marketed as construction fill, as is currently being done at LG&E's Mill Creek facility in Louisville, Kentucky.

Perhaps the most limiting quality issue is the presence of pyrite in the bottom ash. Utility coal grinding or pulverizing equipment is frequently designed to reject hard particles such as rock, hard coal, and pyrite. If the pulverizer rejects are co-disposed with the bottom ash, potential use as lightweight aggregate is severely limited. Even if co-mingled pyrite constitutes a small portion of the bottom ash, the potential of the pyrite to cause iron staining in the final product can be substantial. An example of this type of staining is shown in Fig. 1, where microscopic analysis of the stained portion revealed binary coal/pyrite fragments.

What makes a good quality fly ash?

One of the most significant uses of fly ash is as a mineral admixture in Portland cement concrete. This use alone accounted for over 56 wt% of the fly ash utilized in the USA in 2001 (American Coal Ash Association 2002). Fly ash is particularly well suited to this application because of its pozzolanic properties. As defined by the American Concrete Institute, a pozzolan is a glassy silicate or aluminosilicate that does not possess cementitious properties when mixed with water, but will react with aqueous calcium hydroxide at ordinary temperatures to form a cement. The term 'pozzolan' comes from the neighbourhood of Pozzoli, located in Naples, Italy, where natural pozzolan was first mined for use in Roman Cement.

The chemistry of Portland cement is beyond the scope of this review; suffice it to say that it is made up of Ca-rich glass and four poorly crystalline phases, all of which, upon mixing with water, undergo hydration reactions and form cementitious calcium aluminosilicate hydrate (CASH) gels, along with a large amount of $Ca(OH)_2$, or portlandite. For example, the hydration of 1 mole of tricalcium silicate, that is, alite, the most important phase in Portland cement, forms about 1.5 moles of $Ca(OH)_2$ for every mole reacted as illustrated in the idealized equation (1). The calcium hydroxide can react with a pozzolanic glass (equation 2), such as fly ash, to form additional cementitious CASH gels:

$$2\,(3\,CaO \cdot SiO_2) + 6\,H_2O$$
$$\rightarrow 3\,CaO \cdot 2\,SiO_2 \cdot 3\,H_2O + 3\,Ca(OH)_2 \quad (1)$$
$$Ca(OH)_2 + SiO_2 \cdot Al_2O_3 + OH$$
$$\rightarrow CaO \cdot SiO_2 \cdot Al_2O_3 \cdot xH_2O \quad (2)$$

By consuming the $Ca(OH)_2$ the pozzolan makes it unavailable for sulphation and carbonation reactions and increases the density of the concrete, which lowers its permeability and reduces the risk of chloride attack.

The standard specifications for pozzolans from fly ash are outlined in ASTM C-618, which designates their chemical and physical characteristics. As a pozzolanic additive, fly ash is classified by its chemistry. The USA produces several distinct types. Class F ash is generally produced from the combustion of bituminous coal as mined in the Appalachian, Illinois and Eastern Interior Coal Basins. As defined under ASTM C-618, the content of $SiO_2 + Al_2O_3 + Fe_2O_3$ in Class F ash must exceed 70 wt%. Class C ash must have a minimum $SiO_2 + Al_2O_3 + Fe_2O_3$ of 50 wt% and is rich in CaO. This Class C ash (the 'C' is for calcium) is primarily derived from subbituminous coal and lignite mined in the Western USA. For both Class F and Class C ashes, the maximum allowable amounts of SO_3, moisture and loss on ignition (LOI) are 5.0 wt%, 3.0 wt%, and 6.0 wt%, respectively.

Class F and Class C ash may be better differentiated by their calcium oxide concentrations.

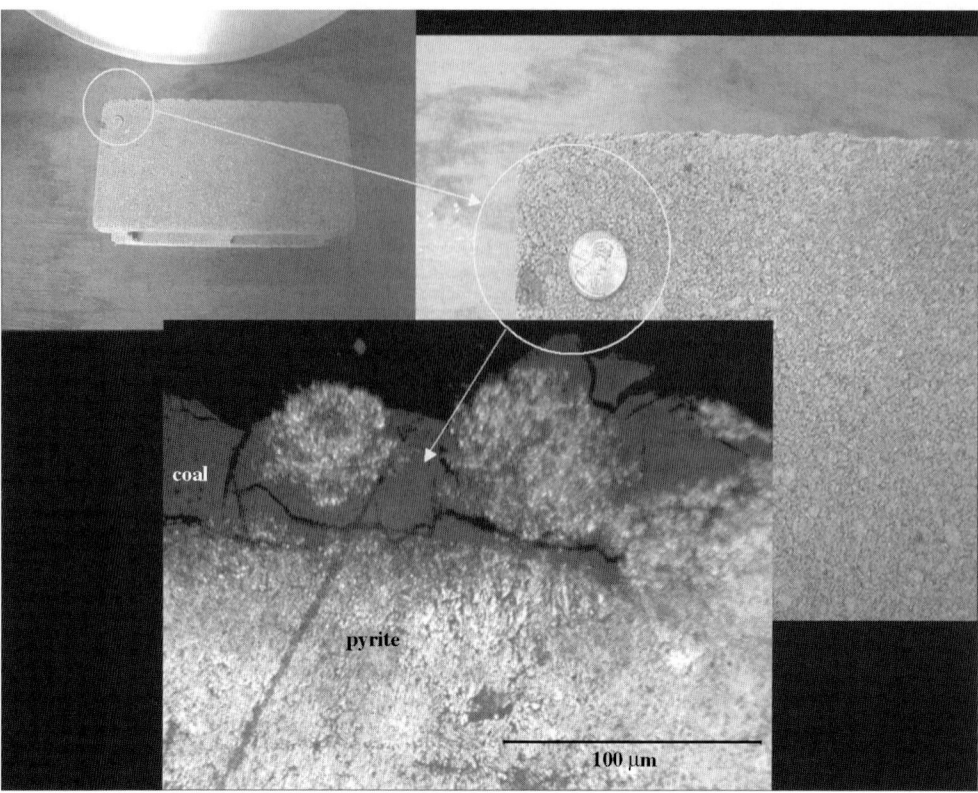

Fig. 1. Photograph of iron staining on concrete block produced by pyrite.

Thomas *et al.* (1999) proposed that ash be classified according to CaO content as follows: Class F, <8 wt% CaO; Class CI, 8–20 wt% CaO; and Class CH, >20 wt% CaO. These CaO ranges have a bearing on their behaviour in concrete, and these subdivisions are used in this review, where appropriate, along with the general term Class C ash defined by ASTM. Class F ash has pozzolanic properties; Class CI ash is similar to Class F ash but is also cementitious. Class CH is more aggressively cementitious and readily forms $Ca(OH)_2$. It is more likely to contribute to the heat of hydration of the cement.

There is little that can be done to affect the chemical composition of the ash as far as SiO_2, Al_2O_3, and Fe_2O_3 are concerned. The chemical composition of the post-combustion ash is primarily a function of the chemical composition of the non-combustible matter in the coal feed to the boiler. The moisture specification is primarily for storage and handling considerations since fly ash used as a pozzolan has historically been stored, transported, and metered as a dry product. There is no technical reason that fly ash must be dry for use as a concrete admixture.

In Class F ashes, the LOI is essentially unburned carbon (Fig. 2), present as a consequence of combustion inefficiencies. The unburned carbon is particularly problematic because of its propensity to adsorb air-entraining agents. The latter are essentially surface-active agents added to entrain air into the concrete mix. Fluctuating levels of unburned carbon make proper dosing of air-entraining agents difficult and hence contribute to poor control of the amount of air in the final concrete. This variable amount of air can adversely affect durability to such a degree that most state highway departments, major consumers of concrete, specify maximum LOI concentrations of 3–4 wt%, rather than the 6 wt% allowed by ASTM C-618. For this reason, LOI reduction has been one of the focus areas of fly ash beneficiation for the past decade. However, reducing LOI alone does not ensure the production of high-quality pozzolan.

The major physical requirements specified in ASTM C-618 pertain to fineness and strength activity index (SAI). The fineness requirement is that no more than 34 wt% be retained on a

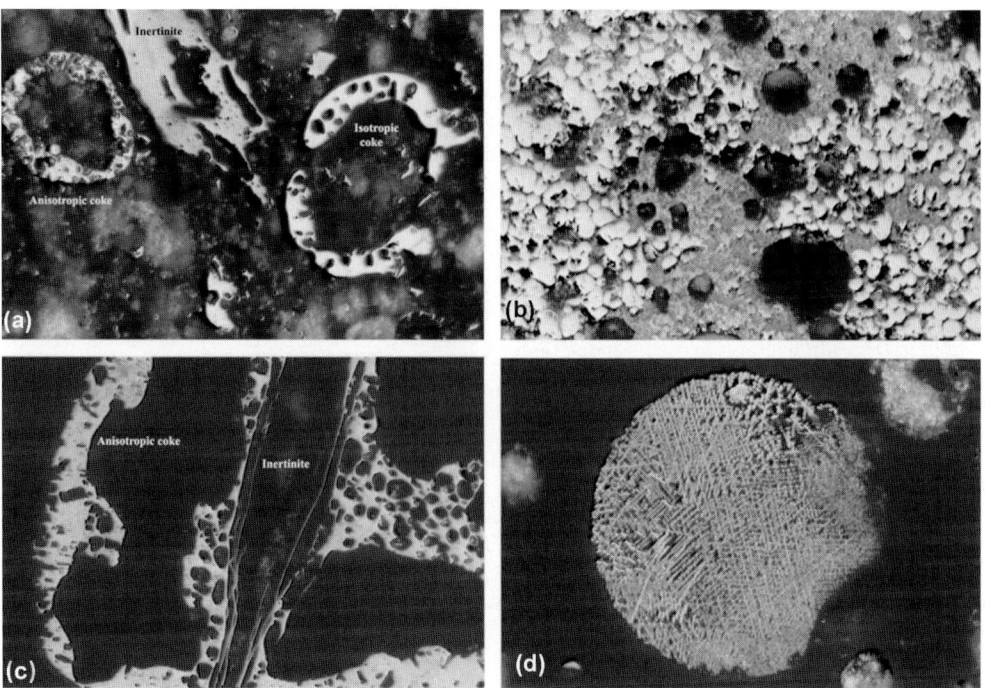

Fig. 2. (**a**) Discrete isotropic coke, anisotropic coke, and inertinite particles in fly ash from an eastern US bituminous coal (reflected light, oil immersion optics, 250 μm on long axis of photographs); (**b**) Anisotropic coke attached to inertinite in fly ash from an eastern US bituminous coal – the coke develops from vitrinite in the feed coal and the inertinite is an original, essentially unaltered, constituent of the coal; (**c**) Massive and fine spinel in bottom ash from an eastern US bituminous coal; (**d**) Spinel (light-coloured rods) in spherical glass in fly ash from an eastern US bituminous coal.

325 mesh (45 μm) screen. The SAI is a ratio of the strength achieved at 20 wt% substitution in mortar cubes to the strength achieved with a control mix comprised of Portland cement and no pozzolan. The substitution mix must achieve at least 75% of the strength of the control mix after 7 or 28 days of curing.

A relationship exists between fineness and SAI. While 34 wt% fineness meets the ASTM C-618 specifications, finer ash samples have been shown to achieve higher compressive strengths. While sizing on a 325 mesh screen (>45 μm) can be indicative of the amount of fine materials present, it does not directly measure the reactive component of the ash, which is in the very finest fractions (e.g., <5 μm). The presence of this ultra-fine material has a profound impact on strength development of the concrete, as the surface area available for pozzolanic cementitious reactions is contained in the finest fractions. For example, the surface area of a typical Class F fly ash, with a mean particle size (D_{50}) of 24 μm will have over 90% of the total surface area accounted for in the <5 μm fraction. This is shown in Table 1 (Robl & Groppo 2002), which compares the strengths of concrete achieved for three different ashes at a substitution of 20 wt%. Ashes H and M have mean particle sizes of 3.8 and 2.3 μm, respectively, compared to 25 μm for ash C. The control sample was comprised of Portland cement with no pozzolan. The strength development of the ultra-fine ash samples was significantly higher than that of the conventional ash, matching the

Table 1. *Effect of concrete curing time on SAI (% of control strength) for ashes with differing <10 μm proportions*

	D_{50} (μm)	Curing time				
		1 Day	3 Days	7 Days	28 Days	56 Days
Ash H	3.8	84.9	103.6	103.6	116.7	137.4
Ash M	2.3	90.7	100.0	100.9	118.8	147.1
Ash C	25.0	69.8	76.9	70.7	76.8	90.9
Control	–	100.0	100.0	100.0	100.0	100.0

control strength after as little as three days and achieving significantly higher strengths after 28 and 56 days. The results for the conventional Class F ash (Ash C) demonstrated the delay in strength development that is typical for conventional Class F ash. These results clearly show the relationship between size distribution and concrete strength while illustrating the limitation of using fineness as a quality specification.

Bottom ash processing

Bottom ash is removed from the bottom of a boiler in essentially two different ways depending on the boiler design: wet and dry. With either of these methods, the bottom ash may be quenched after removal in a water bath to facilitate transport to the ash disposal site. In most cases, a clinker grinder is used to reduce the top size of the ash after it is removed from the boiler to minimize transport difficulties. A schematic classification of the beneficiation options that are currently in practice is shown in Fig. 3. The products resulting from these processes include graded aggregate, lightweight aggregate, and asphalt components.

Dry bottom ash processing

For dry bottom ash, the primary option is to screen to the appropriate size for the end use. The degree of screening is ultimately determined by the customer, but is primarily intended to control the top size and minimize the amount of fines. Incremental screen sizes can be included if necessary on a multi-deck screen to meet

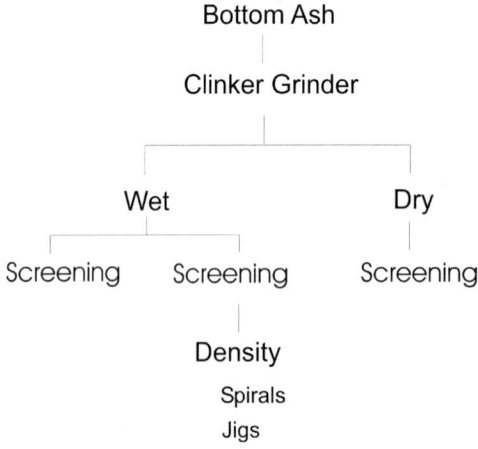

Fig. 3. Schematic flow chart of beneficiation options for bottom ash.

gradation requirements, but this will typically require additional handling to blend the desired proportions. Oversize material can also be crushed and fed back into the system. While wet screening is generally accepted to be more efficient than dry screening, on the coarser screens used for bottom ash processing, the inefficiencies inherent to dry screening can be remedied by oversizing the screens used.

Wet bottom ash processing

Similarly, screening can be practised on wet ash with what is generally accepted as higher capacities as well as higher screening efficiencies for a given screen area. An example of a vibrating wet screen is shown in Fig. 4. The primary advantage of wet screening is that spray bars can be used to increase screening efficiencies, which is particularly important for minimizing fine particles in a coarse product. Using wet sprays effectively washes fines from coarser particles while also breaking up loosely held agglomerates. Since bottom ash has a rough texture, removal of fine particles from the surface pockets and pores may require the addition of spray water.

If wet screening is to be incorporated into the processing scheme, bottom ash can be pumped directly into the processing plant provided sufficient sump capacity is available there to accommodate the excess water since sluicing ash from the boiler is normally accomplished at low pulp density (i.e., <5 wt% solids). The same is true for dredge recovery from ash ponds. Dredges used for excavating ash can operate successfully at pulp densities as high as 30 wt% solids, although lower pulp densities are more common. A more desirable option may be to use higher solids loading from stockpiled or decanted bottom ash. In this arrangement, long-arm track hoes or backhoes excavate ash and store it in a stockpile adjacent to the processing plant. Water drains from the stockpiled ash to facilitate solids handling and transport. For these installations, material is fed to a slurry tank by an inclined conveyor where water is added and the slurry is pumped onto the screen deck. The slurrying and pumping actions tends to de-agglomerate the ash and liberate fines, facilitating effective screening. Although this feed arrangement requires the additional handling step of stockpiling ash prior to processing, it does provide the advantage of added control of the feed slurry pulp density, which, in turn, enables higher solids throughput.

Density separation is also applicable to wet bottom ash processing and can be required in

(a)

(b)

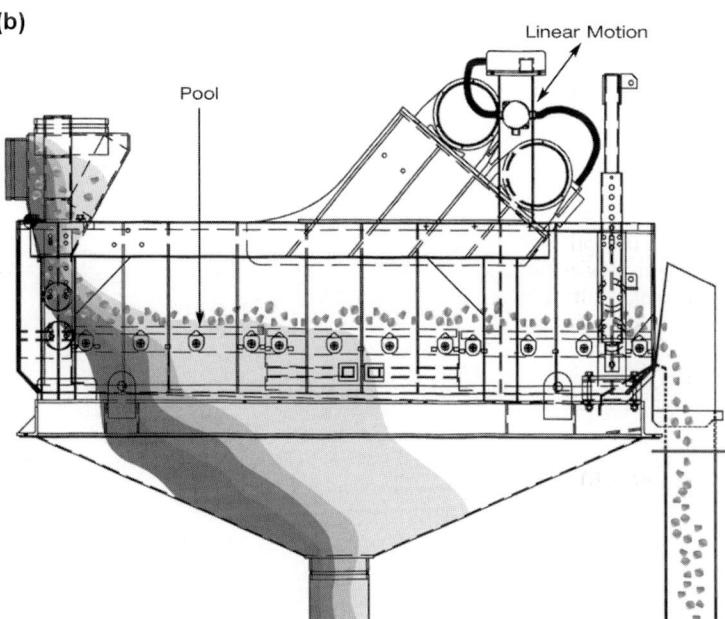

Fig. 4. Photograph of vibrating linear motion screen (top) and diagram of linear motion screen showing flow of materials (bottom). Source: Derrick Corporation (2003).

order to produce certain products such as light-weight aggregate. Removal of dense contaminants (i.e., rock, dense ash, pyrite) will improve product quality by reducing the bulk density of the lightweight aggregate and ensuring uniformity. Removal of low-density contaminants (i.e., coal, char, vegetation) may be just as critical to meet product specifications. The most commonly

applied density separation approach is to use concentrating spirals. These devices provide the capability of separating particles of similar size based on differences in density while they will also separate particles of similar density based on differences in size. For this reason, it is important that the material to be processed be closely sized by effective screening prior to spiral concentration in order to achieve effective density separations.

Photographs of spiral concentrators are shown in Fig. 5. The spiral consists of a banked trough, or race, wrapped around a support post. The feed slurry is introduced at the top of the spiral and the slurry washes down the trough by gravity and a density gradient is achieved across the spiral profile. Finer, lighter particles report to the outside edge of the trough, whereas coarser, denser particles remain on the inside edge; intermediate density, or middling particles remain in the middle of the trough. Adjustable splitters at the base of the spirals divert the product streams to the appropriate subsequent processing step. Most spirals produce three products in this manner and middling particles are frequently re-circulated back to the spiral feed. In some spirals, such as the one shown in Fig. 5, an additional splitter is included on the inside of the trough approximately three turns from the top. This additional splitter diverts and rejects undesirable heavy particles that have accumulated on the inside of the race and prevent them from being washed back into the other product streams.

Another option for density separation is a jig. The jigging mechanism involves pulsing air, water, or both through a bed of particles, which are thus rearranged. After several cycles of pulsing, layers of differing density result, and the layers are separated as the particles exit the device. An example of a fine particle jig is shown in Fig. 6. In this particular example, the feed slurry is introduced at the top of the device along with water, if necessary. Air is then pulsed through a screen deck to develop a jigging bed of particles stratified by density. Low-density particles overflow via a launder, whereas high-density particles accumulate at the bottom of the jig tank and are removed through a control valve to maintain the slurry level in the jig. Once commonly used in coal preparation, these devices are being applied for bottom ash processing to provide separations not achievable with spirals. An example is processing high-density ash co-mingled with pyrite. While the high-density ash is not suitable for use as lightweight aggregate, it is frequently marketable as construction sand, provided pyrite is removed. The specific gravity difference between pyrite (s.g. $= 4.5–5$) and high-density ash (s.g. $= 3–3.5$) is not amenable for separation using spirals, but jigs have demonstrated that these two components can be separated effectively.

Fly ash processing

Fly ash is removed from the boiler in the flue gas. Before the flue gas is vented to the atmosphere

(a)   (b)

Fig. 5. Photographs of spiral concentrator showing overall view and detail of trough. Source: Outokumpu (2003).

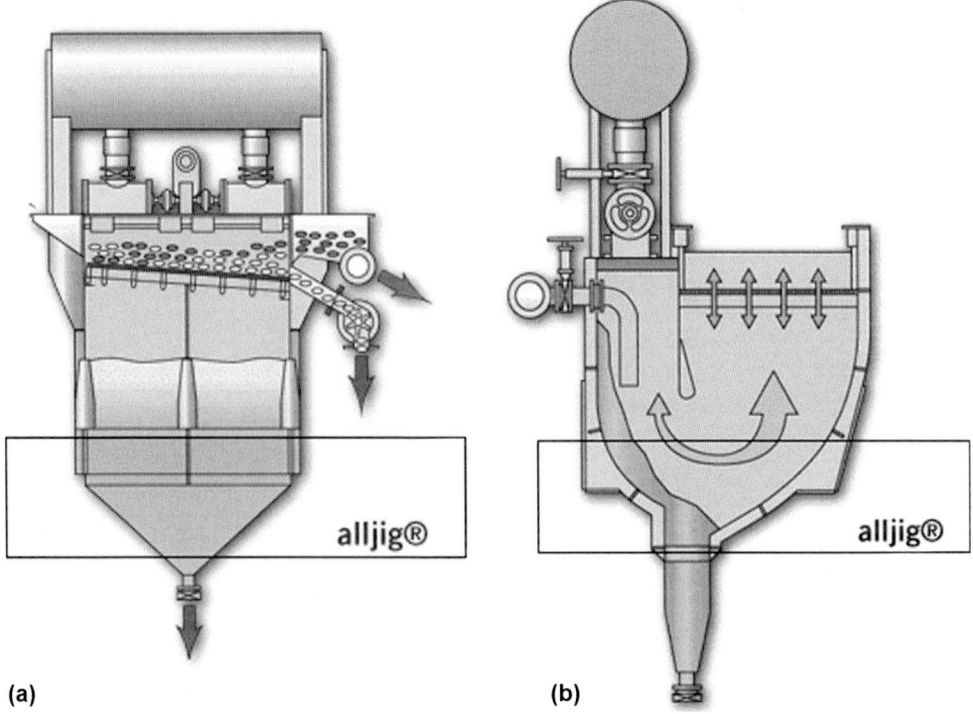

Fig. 6. Front and side views of density separation achieved with a jig. Source: Allmineral (2003).

through the smoke stacks, it is necessary to collect sufficient fly ash to meet emission opacity requirements. This is accomplished with a variety of devices including cyclones, electrostatic precipitators, and bag houses. With each case, fly ash is collected dry and stored in hoppers for removal. To reduce the amount of dust, fly ash is removed from the hoppers either as a slurry through a slurry reduction system where it is pumped to an impoundment or pond, or it is moisturized prior to transport to a landfill.

A schematic flow chart of the beneficiation options that are currently in practice is shown in Fig. 7. These processing schemes are primarily employed for the production of pozzolan, but other products may result, such as carbon fuel and mineral grade filler. The fly ash beneficiation option applicable to a specific site is dependent on many factors, but the primary consideration is whether the fly ash is wet or dry. Once ash has been wetted, flotation is the only practical beneficiation option. It may be technically feasible to use thermal processes on damp ash, but the amount of heat required will be significantly increased, thus decreasing the economic value

and environmental benefits of reusing the ash. Similarly, electrostatic separation processes have been shown to be effective on ashes that have been stored wet, but in order to utilize this separation process, the feed must be dry. The additional step of drying the ash prior to processing diminishes the economic value and environmental benefit.

Air classification

Air classification is perhaps the simplest fly ash processing option and is normally employed to improve the fineness of the ash (i.e., remove coarse particles). A typical cyclone classifier uses centrifugal force to separate fine particles from an air stream. The particles enter tangentially into a cylindrical chamber dispersed in an air stream and centrifugal force forces the coarser particles to the wall of the cylinder while the air stream and finer particles spiral to an inner vortex. The air exits from the inner core via an outlet port while the particles slide down the chamber walls and exit the bottom.

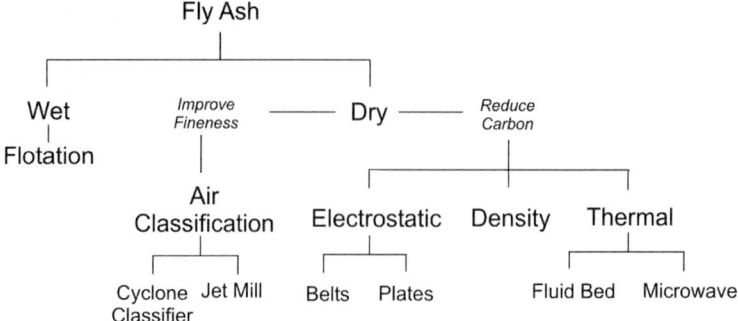

Fig. 7. Schematic flow chart of beneficiation options for fly ash.

When air classification is employed for ash beneficiation purposes, the mechanism is similar, but a high-performance cyclone is used. An example is shown in Fig. 8. Devices vary considerably in the feed arrangement used, but most employ some mechanism to effectively distribute the ash particles across the profile of the cyclone, such as a rotating cage or distribution plate. The velocity of the feed is controlled to produce the desired size separation whereby fine particles and most of the air migrate to the inner vortex while coarse ash exits the bottom. Air classifiers have been used for many years to control pozzolan fineness simply by removing coarse ash particles, particularly when LOI is not a problem. Air classification is usually not effective for reducing LOI, particularly if the carbon is fine. Some coarse carbon particles may be rejected with the coarse ash, but since the density of the carbon particles is lower than

that of the ash particles, carbon may actually be concentrated with the fine ash in the vortex (Groppo *et al.* 1995, 1996).

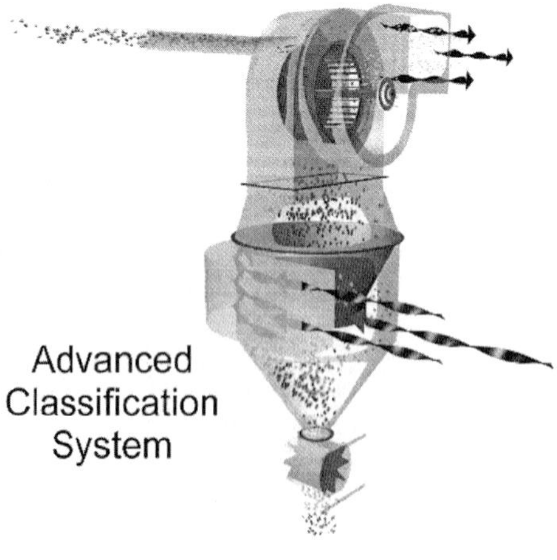

Another classification technique addresses the difficulties presented by coarse carbon in air classification by combining grinding with air classification in a device called a jet mill (Fig. 9). The principle is to use high-velocity air to accelerate the particles and induce autogenous grinding. Since the carbon particles have a higher grindability (i.e., are softer) than the ash particles, the carbon is ground very finely and is rejected in the vortex while the ash exits the bottom of the mill. The manufacturer reports that this technique is also applicable to damp ash. While the jet mill has been demonstrated to reject carbon as ultra-fine particles, some fine ash may inevitably also be lost. Since the pozzolan is collected in the coarse product stream, this technique will do little to improve fineness. In order to reject coarse ash particles, a secondary classification step may be necessary.

Density separation

Density separation processes exploit the differences in densities (ρ) of the particles to be separated. Fly ash is comprised primarily of two components: aluminosilicate ash particles ($\rho \approx 2.1-2.4 \, \text{g/cm}^3$) and unburned carbon particles ($\rho \approx 1.8-2.0 \, \text{g/cm}^3$) with minor amounts of iron-rich spinel, commonly referred to as magnetite ($\rho \approx 3.0-4.5 \, \text{g/cm}^3$). Under ideal conditions, particles can be efficiently separated based on differences in particle densities alone. However, in continuous density separation processes, particle size differences induce separation inefficiencies; that is, smaller, high-density particles behave similarly to larger, low-density particles. For this reason, narrowing the size

Fig. 8. Air classifier. Source: RSG, Inc. (2003).

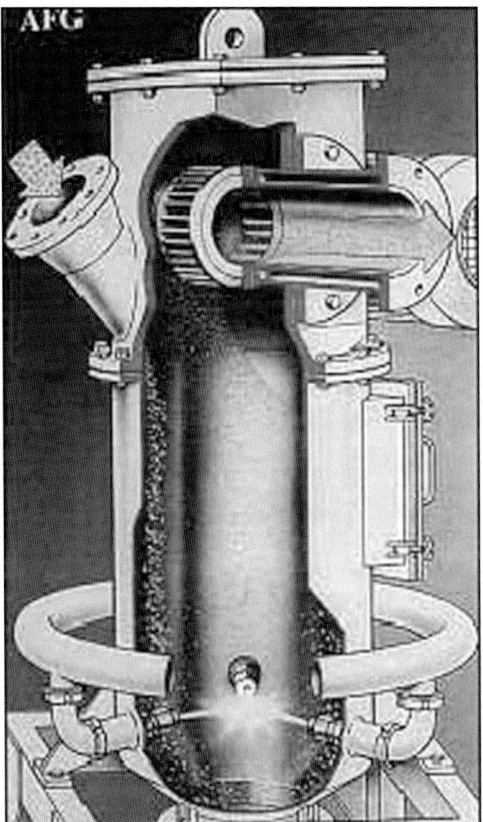

**Courtesy of Hosokawa Micron
Powder Systems, Summit, New Jersey**

Fig. 9. Jet mill. Source: Pittsburgh Mineral and
Environmental Technology, Inc. (2003).

distribution of the particles to be separated will
increase the efficiency of the density separation.
Many fly ashes indeed have a narrow size distribution and thus may be amenable to effective
density separation.

One density-based method is the Dry Tribo-
Separation process where separation relies not
only on density, but also on the difference in fric-
tion coefficients of various particle sizes and
shapes (Eiderman *et al.* 2000). Dry material is
fed onto the surface of a rotating conical bowl
and a combination of centrifugal, frictional, and
gravitational forces separates the ash into two products; one that is enriched in carbon and one that
is carbon-lean. This process is currently under
development but has demonstrated the potential
to separate particles by size as well as by shape.

Another density-based separation method
is described by Levy *et al.* (1999). In this
technique, ash is added at one end of an inclined
table and fluidized by air passing through a distributor. Particles segregate as they move up the
inclined table because the fluidizing air transports low-density carbon-rich particles to the
top of the bed while higher density carbon-lean
products migrate to the bottom of the bed. The
separation is primarily based on density, although
the particle sizes also play an important role.
Results have shown that this technique is particularly effective for recovering coarse (>45 µm)
carbon. The authors note that process efficiency
is improved when a high-intensity acoustic field
was applied above the bed. As with the Dry Tribo-
Separation process, this approach is also in the
development stages.

Electrostatic separation

Electrostatic separation, for the purposes of this
discussion, is accomplished by exploiting the
differences in electrical properties between ash
and carbon particles. This technique has been
employed in a variety of mineral separations for
many years, with the major commercial application being the separation of heavy minerals
from sands. Particles are electrically charged
and, depending on their conductivity, will gain
or lose electrons and thus become differentially
charged. Separation occurs by diverting the particles to electrodes of opposite charge. A major
limitation to widespread use of this technique
in mineral processing is that the feed must be perfectly dry. This, however, is not at all a limitation
for fly ash beneficiation since most fly ash is collected dry. Thus, electrostatic separation is similar
to the operation of electrostatic precipitators used
to remove fly ash particulates from flue gas.

In electrostatic precipitators, ash particles are
charged by a conventional power source and
removed from the flue gas by attraction to oppositely charged plates where the particles are collected. The precipitators are typically designed
to allow the flue gas to pass through a series of
electrodes and vertical plates. The power source
creates an electrostatic charge on the electrodes
and an opposite electrostatic charge on the plates
so that particles carried by the flue gas are
charged by electrodes and attracted to the plates.
Each particle within the effluent stream is
imparted with a negative charge and is retained
on a positively charged plate. Typically, collection of the particulates occurs with successive
stages because negatively charged particles
move across the plates, some particles may be
more difficult to charge than others, and some
particles may lose their charge and require
recharging. Gas velocity may also affect the

collection efficiency since lower velocities may allow more time for the charged particles to adhere to the plates and reduce re-entrainment in the flue gas. A patent by Gillen & Mills (1998) describes control of the electrostatic separators to recover a high-purity carbon product in the initial stage, or field, of electrostatic precipitators. Remaining ash particles are collected in subsequent fields. Thus, the electrostatic precipitators can, in some cases, be used to achieve selective recovery of carbon.

In electrostatic separation, particle charging is accomplished by inter-particle contact, where charges are transferred between particles by differences in electron affinity. Carbon particles, having a lower electron affinity, lose electrons and become positively charged, while the ash particles gain electrons and become negatively charged. This differential charge between the constituent species is the basis for the separation mechanism.

The primary difference between the various electrostatic separation techniques is the mechanical device that is used to achieve the physical separation of the oppositely charged particles. The first commercial application of this technique was reported by Separation Technologies, Inc. (STI) (Bittner *et al.* 1997). The STI separator (Fig. 10) works by feeding ash from the raw ash silo through a vibrating screen to distribute the ash uniformly into a thin gap (0.635–1.9 cm) between two parallel plane electrodes. The particles are then swept up by a moving open mesh belt and conveyed in opposite directions, depending on their charge. The belt moves particles adjacent to each electrode toward opposite ends of the separator, which is approximately 6.1 m long. A different application of this

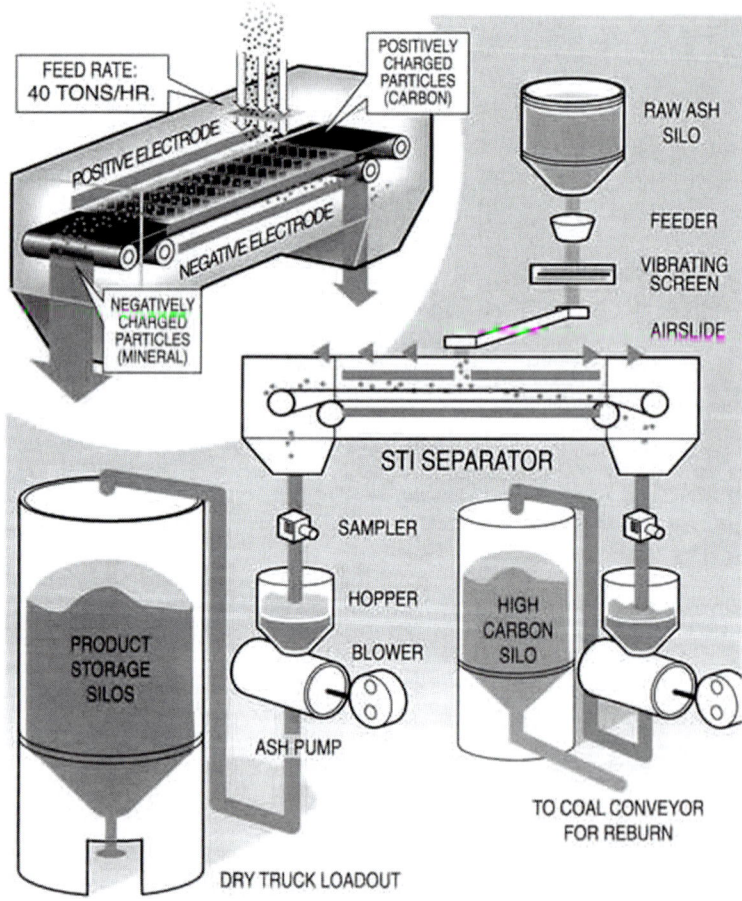

Fig. 10. STI separator. Source: Separation Technologies, Inc. (2003).

technique is described by Rao *et al.* (1999) where electrostatic separation was used to remove Fe-enriched particles for ceramic and refractory applications rather than pozzolan.

Similar approaches have been proposed by others using different separation devices (e.g., Ban *et al.* 1996, 1997; Kim *et al.* 2000, 2001; Soong *et al.* 2001, 2002). One example is a device developed at the University of Kentucky Center for Applied Energy Research and is marketed by Solvera Controls, which utilizes plates and pneumatic ash transport. Ash particles are differentially charged as described previously, and deflected toward oppositely charged plates, thus achieving a carbon/ash separation.

Thermal processing

Perhaps the most effective means of removing unburned carbon from fly ash is to simply burn the carbon using thermal processes (Boyd & Cochran 1994). Given the low proportion of carbon in fly ash (typically 4–10 wt%), conventional combustion processes would not be adequate without the use of supplemental fuel. However, using a fluidized bed presents the opportunity for effective thermal decomposition of unburned carbon. Such is the basis for Progress Material's Carbon Burn-Out (CBO) process, where start-up fuel is used to heat the ash to 460 °C, the auto-ignition temperature of the residual carbon in fly ash. Once the ash is heated and the desired carbon reduction is achieved, low-carbon ash is removed from the fluidized bed, cooled, and stored for use as a pozzolan. Heated ash that is not removed from the process is proportioned back to the bed to maintain bed temperature (Keppeler 2001). Once the bed temperature is heated to the auto-ignition temperature of the residual carbon (460 °C), no additional fuel is necessary to maintain bed temperature provided there is a sufficient amount of unburned carbon in fresh ash introduced into the process. For this reason, CBO is particularly well suited for treating ashes with higher LOI.

Another approach to thermal processing is to use microwave energy. Carbon has very high microwave receptivity, even when present in very low concentrations. The Microwave Burn-Out (MCB) process uses microwave energy to oxidize carbon to CO_2, thus reducing its concentration to the desired value in the ash. The MCB process operates at the auto-ignition temperature of the unburned carbon, using microwave energy and air to supplement the heat required to maintain the temperature necessary to oxidize the carbon. The reactor is a fluidized bed and the finished product is removed as it overflows the reactor (Tranquilla & MacLean 2001).

Froth flotation

Since froth flotation is such a versatile separation method for fine particles, it is not surprising that results of numerous investigations have been reported (Hemmings & Berry 1986; Grunewald & Otterstetter 1989; Gray *et al.* 2001) and several patents issued (Hurst & Styron 1978; Hwang 1991) on the application of this technique for fly ash beneficiation, particularly with respect to the use of the technique to reduce unburned carbon to acceptable levels. The incorporation of hydraulic classification into the process flowsheet to reject coarse ash (i.e., >75 μm) provides a means of improving fineness (Groppo *et al.* 1998). Thus, the combination of classification and flotation provides the capability to produce a pozzolan that is tailor-made to user requirements.

The application of flotation exploits the hydrophobicity of unburned carbon. Although carbon particles possess little or no natural hydrophobicity, various additives (i.e., oils) can be used to selectively adsorb onto carbon and render carbon particles hydrophobic. Air bubbles produced in an agitated tank effectively collect the carbon particles and transport them to the top of the slurry vessel where a stable froth is formed and removed. The slurry remaining in the vessel is essentially a low-carbon pozzolan slurry, which then must be thickened and filtered. Drying is typically required by most consumers, but this is primarily for transportation and handling reasons. There is no technical reason that the pozzolan cannot be used as a damp cake since water is added in the concrete mix design. It is important to note that with respect to ash beneficiation, froth flotation is primarily applicable to low-Ca or Class F ashes. While flotation itself is capable of working effectively at the higher pH values that would be encountered with Class C ashes (pH = 9–11), the presence of CaO may adversely affect the cementitious properties of the ash, particularly if the processed or unprocessed ash was stored with any water.

Perhaps the most significant advantage of using froth flotation to beneficiate fly ash is the ability to process ash that has been impounded in disposal ponds and landfills. This is advantageous for several reasons. First, it effectively de-couples ash beneficiation from ash production. There is no need for the utility to alter disposal practices or install additional ash handling equipment in order to accommodate ash processing since the feed to the processing plant

can be excavated from the disposal pond. Secondly, the storage requirements are significantly
reduced. Dry ash processing options require
that the ash remain dry. If there is a seasonal
decrease in pozzolan demand, it is necessary to
provide dry storage for processed ash or
dispose of it into an ash pond or landfill. This
is particularly important in colder climates
where ash production peaks when construction
provides little or no demand for the pozzolan.
Similarly, if the dry processing option is not
available for maintenance or other reasons, ash
must be disposed. Thirdly, since ash is sluiced
to the disposal pond, classification occurs in the
pond itself, providing extensive areas of finer
ash to be processed (Tyra *et al.* 2003). Last, but
certainly not least, this process provides the
potential of reclaiming old ash storage areas.
This is particularly advantageous since it can
delay or avoid the construction of new disposal
areas, in addition to providing the opportunity
to recover high-quality ash that has been previously disposed rather than marketed.

Ammonia removal processes

Utilities using post-combustion SCR-supported
ammonia injection for NO_x control as well as
those using ammonia conditioning to improve
electrostatic precipitator performance will
produce fly ash that contains ammonia compounds. The ammonia is primarily physically
adsorbed onto the fly ash particles as sulphate
and bisulphate species. In many cases, the
residual ammonia levels are quite low
(<50 ppm); however, elevated concentrations
can occur as the catalyst ages or due to mechanical problems with the ammonia injection
system. While elevated ammonia concentrations
in fly ash do not negatively impact pozzolanic
properties, it can reduce ash marketability due
to odour concerns. For this reason, several
processes have been developed to remove or
reduce the amount of ammonia in fly ash.

Thermal decomposition of the primary
ammonia species present on fly ash occurs at
temperatures higher than 434 °C, thus the
thermal beneficiation processes described previously (i.e., CBO and MCB) effectively remove
ammonia. Several other thermal processes are
also described in the literature (Harald & Ruegg
1994; Fisher *et al.* 1997), but unlike CBO and
MCB, they address the removal of ammonia
only and not the carbon reduction.

Chemical treatment can also be used to release
ammonia as a gas by elevating the pH. The
various treatments differ primarily in the
amounts of lime and water that are used

(Epperly & Sprague 1991; Martin *et al.* 1993;
Gasiorowski *et al.* 2000). A different chemical
approach is to oxidize ammonia by the addition
of hypochlorite to the fly ash prior to use as a
cement additive (Minkara 2003). Wet beneficiation of fly ash with flotation, although not a
specific chemical treatment, will also remove
ammonia due to the high solubility of the
ammonia salt compounds present.

Summary

Ash from pulverized coal combustion is a strategic material that has many critical applications
from a source of aggregate to the most important
source of pozzolan for addition to Portland
cement concrete. Environmental control measures on the emissions of coal combustion have
resulted in a loss of quality for these materials.
In response we have seen the advent of beneficiation processes applying both proven and new
technologies to produce high-quality consistent
products from these materials. Currently we
estimate that about one-fifth of all ash products
marketed are processed through some form of
beneficiation method. We expect that the demand
for quality and consistency will continue and the
relative amount of process ash products will
increase in the future.

The authors wish to acknowledge the contributions and
efforts of reviewers H. Ban (University of Alabama),
Y. Soong (US Department of Energy), and
S. Gasiorowski (Separation Technologies, Inc.).

References

ALLMINERAL 2003. *Introducing the Allmineral Alljig.*
    Allmineral LLC, Alpharetta, GA. World Wide
    Web Address: www.allmineral.com.
AMERICAN COAL ASH ASSOCIATION 2002. *2001 Coal
    Combustion Product (CCP) Production and Use
    (Short Tons).* World Wide Web Address: http://
    www.acaa-usa.org / PDF / ACAA2001CCPSurvey.
    pdf. (accessed June 2003).
BAN, H., LI, T. X., HOWER, J. C., SCHAEFER, J. L. &
    STENCEL, J. M. 1997. Dry triboelectrostatic benefi-
    ciation of fly ash. *Fuel,* **76**, 801–805.
BAN, H., LI, T. X., SCHAEFER, J. L. & STENCEL, J. M.
    1996. Characterizing dry triboelectrostatic benefi-
    ciation of coal and fly ash using recovery analysis.
    In: CHIANG, S. H. (ed) *Coal-Energy and the
    Environment. Thirteenth Annual International
    Pittsburgh Coal Conference,* September 1996,
    Pittsburgh, PA, **2**, 873–878.
BITTNER, J., GASIOROWSKI, S., TONDU, E. &
    VASILIAUSKAS, A. 1997. STI fly ash separation
    system 10% in, 1% out; 160,000 tons of STI ash
    in the new England ready mix concrete market.
    Proceedings: 1997 International Ash Utilization

Symposium, 20–22 October 1997, Lexington, KY, 630–636.

BOYD, T. J. & COCHRAN, J. 1994. Fly ash beneficiation by carbon burnout. *Proceedings of the American Power Conference*, **56**, 937–942.

DERRICK CORPORATION 2003. World Wide Web Address: http://www.derrickcorp.com/products/wetsizing/floline.htm (accessed December 2003).

EDENS, T. F. 1999. Recovery and utilization of pond ash. *Proceedings: 1999 International Ash Utilization Symposium*, 18–20 October 1999, Lexington, KY, 304–310.

EIDERMAN, B., VOSKOBOINIK, M., LEVY, H. & SOSKINE, M. 2000. Operational and financial advantages of a new technology for fly ash separation. *Proceedings: 2000 Conference on Unburned Carbon on Utility Fly Ash*, US Department of Energy, Office of Fossil Energy, National Energy Technology Laboratory, 51–55.

EPPERLY, W. R. & SPRAGUE, B. N. 1991. *Method and Composition for the Reduction of Ammonia Emissions from Non-Acidic Residue*. US Patent No. 5,069,720.

FISHER, B. C., BLACKSTOCK, T. & HAUKE, D. 1997. Fly ash beneficiation using an ammonia stripping process. *Proceedings: 12th International Symposium on Coal Combustion By-Product (CCB) Management and Use*, Vol. 2, Paper 65, pp. 1–8.

GASIOROWSKI, S. A., HRACH, JR. & FRANK, J. 2000. *Method for Removing Ammonia from Ammonia Contaminated Fly Ash*. US Patent No. 6,077,494.

GRAY, M. L., CHAMPAGNE, K. J., SOONG, Y. & FINSETH, D. H. 2001. Parametric study of the column oil agglomeration of fly ash. *Fuel*, **80**, 867–871.

GILLEN, J. E. & MILLS, R. E. 1998. *Method for Obtaining Devolatilized Bituminous Coal from the Effluent Streams of Coal Fired Boilers*. US Patent No. 5,779,764.

GROPPO, J. G., BROOKS, S. M. & KREISER, C. 1995. Fly Ash Beneficiation by Air Classification. *124th Annual American Institute of Mining and Metallurgical Engineers Meeting*, Denver, CO, 6–9 March 1995, 95–196.

GROPPO, J. G., ROBL, T. L., GRAHAM, U. M. & MCCORMICK, C. J. 1996. Selective beneficiation for high loss-on-ignition fly ash. *Mining Engineering*, **48**, 51–53.

GROPPO, J. G., ROBL, T. L. & MCCORMICK, C. J. 1998. *Method for Improving the Pozzolanic Character of Fly Ash*. US Patent No. 5,817,230.

GRUNEWALD, K. & OTTERSTETTER, H. 1989. Investigations into the beneficiation of fly ash from power plants using flotation. *Bundesministerium für Bildung, Wissenschaft und Kultur*, **42**, 61–63.

HARALD, W. & RUEGG, H. 1994. *Process for the Thermal Treatment of Solids Which Arise in the Purification of Flue Gases*. US Patent No. 5,501,161.

HEMMINGS, R. T. & BERRY, E. E. 1986. *Evaluations of Plastic Filler Applications for Leached Fly Ash*. Electric Power Research Institute Report No. CS4765.

HURST, V. J. & STYRON, R. W. 1978. *Fly Ash Beneficiation Process*. US Patent No. 4,121,945.

HWANG, J. Y. 1991. *Wet Process for Fly Ash Beneficiation*. US Patent No. 5,047,145.

KEPPELER, J. 2001. Carbon burnout, an update on commercial applications. *Proceedings: 2001 International Ash Utilization Symposium*, 20–24 October 2001, Lexington, KY, cd-rom.

KIM, J.-K., CHO, H.-C. & KIM, S.-C. 2001. Removal of unburned carbon from coal fly ash using a pneumatic triboelectrostatic separator. *Journal of Environmental Science and Health, Part A Toxic/Hazardous Substances and Environmental Engineering*, **36**, 1709–1724.

KIM, J.-K., CHO, H.-C., KIM, S.-C. & CHUN, H.-S. 2000. Electrostatic beneficiation of fly ash using an ejector-tribocharger. *Journal of Environmental Science and Health, Part A: Toxic/Hazardous Substances and Environmental Engineering*, **35**, 357–377.

LEVY, E., HERRERA, C., COATES, M. & AFONSO, M. 1999. Beneficiation of fly ash using a fluidized bed separator. *Proceedings: 13th International Symposium on Use and Management of Coal Combustion Products*, American Coal Ash Association, Paper No. 17, January 2003, St. Petersburg, FL.

MARTIN, W. J., Martin, J. E., HORLER, S. & NIKOLAUS, T. 1993. *Process for Extracting of Disposing of Ammonia or Ammonia Compounds from Dust Mixtures*. US Patent No. 5,211,926.

MEIJ, R. & VAN DEN BERG, J. 2001. Coal Fly Ash Management in Europe: Trends, Regulations and Health & Safety Aspects. Keynote Address, *2001 International Ash Utilization Symposium*, 20–24 October 2001, Lexington, KY.

MINKARA, R. 2003. Technologies to Mitigate Ammonia Slip. *Proceedings: 13th International Symposium on Use and Management of Coal Combustion Products*, American Coal Ash Association, Paper No. 73, January 2003, St. Petersburg, FL.

OUTOKUMPU 2003. Minerals Processing Technology. World Wide Web Address: http://www.outokumpu.fi/mineralprocessing/phy-7-3-4.htm. (accessed June 2003).

PITTSBURGH MINERAL & ENVIRONMENTAL TECHNOLOGY, INC. 2003. World Wide Web Address: http://www.pmet-inc.com/(accessed June 2003).

RAO, R. B., CHATTOPADHYAY, P. & BANERJEE, G. N. 1999. Removal of iron from fly ash for ceramic and refractory applications. *Magnetic and Electrical Separation*, **10**, 21–27.

ROBL, T. L. & GROPPO, J. G. 2002. The production of superpozzolan from coal fired utility ash ponds. *In*: DHIR, R. K., DYER, T. D. & HALLIDAY, J. E. (eds) *Sustainable Concrete Construction*. Thomas Telford Publishing, London, England, 2002.

RSG INC. 2003. World Wide Web Address: http://www.airclassify.com/ (accessed June 2003).

SEPARATION TECHNOLOGIES, INC. 2003. World Wide Web Address: http://www.stiash.com/L1/power-sti_technology.html. (accessed June 2003).

SOONG, Y., SCHOFFSTALL, M. R., GRAY, M. L., KNOER, J. P., CHAMPAGNE, K. J., JONES, R. J. & FAUTH, D. J. 2002. Dry beneficiation of high loss-on-ignition fly ash. *Separation and Purification Technology*, **26**, 177–184.

SOONG, Y., SCHOFFSTALL, M. R. & LINK, T. A. 2001. Triboelectrostatic beneficiation of fly ash. *Fuel*, **80**, 879–884.

THOMAS, M. D. A, SHEHATA, M. H. & SHASHIPRA-KASH, S. G. 1999. The use of fly ash in concrete: Classification by composition. *Cement Concretes and Aggregates*, **21**, 105–110.

TRANQUILLA, J. & MACLEAN, J. 2001. Microwave Carbon Burnout (MCB), Gas Products and Developments of Specific Metallic Elements. *Proceedings: 2001 International Ash Utilization Symposium*, 20–24 October 2001, Lexington, KY.

TYRA, M. A., GROPPO, J. G., ROBL, T. L. & MINSTER, T. 2003. Using digital mapping techniques to evaluate beneficiation potential in a coal ash pond. *International Journal of Coal Geology*, **54**, 261–268.

The chemistry and mineralogy of waste from retorting and combustion of oil shale

O. M. SAETHER[1], D. BANKS[2], U. KIRSO[3], L. BITYUKOVA[4] & J. E. SORLIE[5]

[1]Geological Survey of Norway, Trondheim, Norway

(email: ola.sather@ngu.no)

[2]Holymoor Consultancy, Chesterfield, Derbyshire, UK

[3]National Institute of Chemical Physics and Biophysics, Tallinn, Estonia

[4]Institute of Geology, Tallinn University of Technology, Tallinn, Estonia

[5]Norwegian Geotechnical Institute, Oslo, Norway

Abstract: Oil shales are comprised of clastic, organic, carbonate and minor sulphide fractions. The relative proportions of these fractions influence the composition and potential environmental impact of wastes produced by mining, combustion, and retorting of these shales. Mining produces spoils and gangues, which may or may not produce acidic leachates, depending on relative proportions of sulphide and carbonate minerals. Combustion of oil shale for power production produces slag and ash, which, when deposited as huge heaps and plateaux, can generate highly alkaline leachates. Power plants also emit acid-producing and greenhouse gases (SO_2, NO_x, and CO_2) and particulate matter. Emitted particulate matter is dominated by basic oxides (e.g., CaO), the fallout of which tends to neutralize any tendency to environmental acidification. Estonian oil shales are highly unfavourable in terms of greenhouse gas emissions: CO_2 is generated both by decomposition of carbonates and oxidation of organic carbon, and 0.029 t/GJ carbon (C) is emitted for an energy yield of only 9 GJ/t shale. Finally, retorting of oil shales produces organically contaminated condensate water and semi-coke solid residue, leachates from which can contain several hundred mg/L phenols.

The role of oil shale in energy production is relatively unknown to most people, in comparison with, for example, petroleum and coal. Its contribution to the global energy budget of our industrial age is very small, as its use as an energy source is geographically limited. Production of energy or (shale) oil from oil shales is now restricted to a few countries, such as Estonia, China, and Brazil, although more widespread usage historically has taken place in, for example, Scotland, Sweden, France, Canada, and the USA (Russell 1990). World reserves of shale oil or energy from oil shales are huge, however, conservatively estimated at 350 Gt shale oil (equivalent to 2600×10^9 barrels of oil = 410×10^9 Sm^3 o.e. (Standard cubic metre oil equivalents)), of which 227 Gt shale oil (or 2100×10^9 barrels = 340×10^9 Sm^3 o.e.) are located in the USA, and only 0.52 Gt shale oil (or 3.9×10^9 barrels = 0.62×10^9 Sm^3 o.e.) in Estonia (Dyni 2000) and 0.54 Gt shale oil (or 4×10^9 barrels = 0.64×10^9 Sm^3 o.e.) in the Fushun and Maoming deposits of China (Xianglin & Qian 1986; Qian, personal communication 2003). Smaller deposits occur in, for example,

Russia, Turkey, and South Africa. For comparison, the Norwegian oil and gas resources that have been produced between 1971 and 2002 amount to 3.3×10^9 Sm^3 o.e. (Norwegian Ministry of Oil and Energy 2003), the world's total oil and gas resources are estimated at 478×10^9 Sm^3 o.e. (3008×10^9 barrels), and the coal and lignite resources of the world at 3708×10^9 Sm^3 o.e. (Press & Siever 1997). It might be argued that oil shales represent a significant strategic energy source for periods when the international distribution of energy is unstable or difficult to predict.

In addition to their energetic value, oil shales have also been used as sources for other materials, such as alumina, ammonium sulphate, phosphate, sodium carbonate, S, U, V, and Zn (Murray 1974). In the early 17th century, potassium aluminium sulphate was extracted from the Alum Shales in Sweden for use in the tanning and textile industry. It was only by the 19th century that hydrocarbons were being extracted from these Alum Shales while, in the 1960s, their content of U and V was being exploited (Dyni 2000).

From: GIERÉ, R. & STILLE, P. (eds) 2004. *Energy, Waste, and the Environment: a Geochemical Perspective*. Geological Society, London, Special Publications, **236**, 263–284.
0305-8719/04/$15   © The Geological Society of London 2004.

Ash waste from combustion of oil shale is used in the cement industry, but other applications analogous to those found for ash from coal-fired power plants should be investigated (Hanni 1996; Manz 1999; Hall & Livingstone 2002; Swanepoel & Strydom 2002; Woolard *et al.* 2002).

What are oil shales?

Oil shale is 'a kerogen-bearing, finely laminated brown or black shale that will yield liquid or gaseous hydrocarbons on distillation' (American Geological Institute 1972). Usually only oil shales expected to yield more than 40 L/t shale upon Fisher assay are considered reserves (Qian, personal communication 2003). The term *oil shale* is a misnomer as it is neither a true shale nor does it generally have oil in it. Oil shales, or subsets thereof, may also be referred to (although not always strictly synonymously) as alum shale, bituminite, cannel coal, gas coal, kerosene shale, or kukersite (American Geological Institute 1972; Urov & Sumberg 1999). The organic component, called kerogen, is typically insoluble in most common organic solvents (e.g., carbon disulphide), and temperatures above 350 °C are required to decompose it in the absence of oxygen to condensable shale oil, gas, and a solid semi-coke residue. This process is known as *retorting*. Unocal Inc. of California, USA, produced 0.6×10^6 t (4.5×10^6 barrels = 0.72×10^6 Sm^3 o.e.) of oil from oil shale in a large-scale experimental mining and retorting facility in northwestern

Colorado during the period 1980–1991, with an average yield of ~110 L shale oil per tonne of exploitable oil shale (Dyni 2000). The facility has a capacity of ~1 600 000 L shale oil per day (10 000 barrels/d; Press & Siever 1997). However, most of the oil shale mined in the world today is utilized as feedstock for production of energy, both thermal and electrical. In such power plants, the temperatures reach up to 1500 °C. As an example, Estonian energy production accounts for about 70% of the world's oil shale consumption (Ots & Uus 2002).

Genesis and composition of oil shales

Oil shales are found on most continents (Fig. 1) in rocks of Cambrian to Neogene age (Russell 1990; Urov & Sumberg 1999). Depending on their original depositional environment, Hutton (1987) divided oil shales into three basic groups: (1) terrestrial, (2) lacustrine, and (3) subtidal or marine.

The Tertiary Fushun Group in northeastern China is underlain by bituminous coal, which grades upwards into 40–190 m thick oil shale with interbedded coal seams of 0.5–0.8 m thickness, and is overlain by green to greenish-grey calcareous shale, indicating a lacustrine origin (Xianglin & Qian 1986; Russell 1990). The exploitable Mahogany zone of the Tertiary Green River Formation in the USA is another example of a lacustrine oil shale (Smith 1974; Surdam & Wolfbauer 1975; Desborough 1978). The kerogen in the Estonian Ordovician kukersites is classified as being of marine

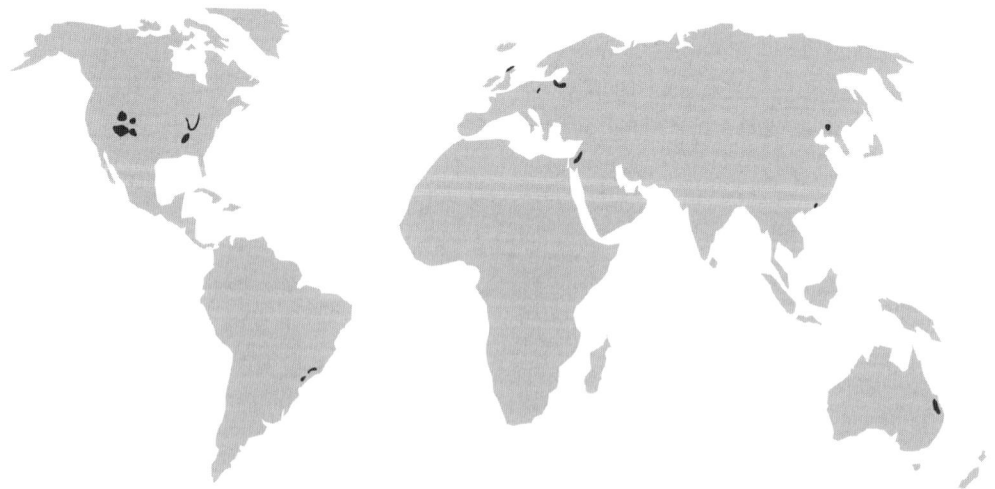

Fig. 1. Main oil shale deposits in the world (in black). (Simplified from Duncan & Swanson 1965 and Dyni 2000).

origin by Hutton (1995) and Lille *et al.* (2002, 2003), but Kattai *et al.* (2000) suggested it is more characteristic of lacustrine oil shales and sapropelic coal. This variety of depositional environments of oil shales is reflected in their mineralogical and geochemical composition and, especially, in the biotic derivation of their organic matter. Bacterial processes were probably important during the deposition and early diagenesis of most oil shales. Oil shales usually contain authigenic minerals such as calcite, dolomite, and pyrite, formed during biogenic and diagenetic processes (Dyni 2000). They also contain three major types of macerals (Hutton 1987): *telalginite*, structured organic matter derived from thick-walled or colonial unicellular algae; *lamalginite*, distinct laminae, displaying few or no recognizable biologic structures, but may include thin-walled colonial or unicellular algae; and *bituminite*, largely amorphous material, which lacks significant fluorescence under the microscope. Some oil shales may also contain macerals derived from land plants, such as inertinite and vitrinite (Dyni 2000).

Waste products from processing of oil shale

Associated with their potential as an energy resource, the use of oil shales in power production and retorting poses environmental challenges. The waste products generated by the oil shale industry can be divided into the following categories:

(1) *Mining wastes and overburden.* Estonian kukersite oil shales occur in limestone strata (Schmidt 1858, 1881), yielding a carbonate-rich spoil with significant potential for acid neutralisation. Other oil shales, for example, the siliceous Estonian Dictyonema shales, which contain only traces of carbonate compared with the kukersites, are associated with sulphides, such as pyrite, and may thus generate acidic leachate due to pyrite oxidation (Puura & Pihlak 1998; Puura *et al.* 1999).

(2) *Semi-coke, or coke ash residue.* This waste material accumulates when oil shale is retorted. The residues and their leachates contain organic contaminants, most importantly phenolic species, for example, phenol, cresols, xylenols, and resorcinols (Raidma 1994; Kahru *et al.* 1999).

(3) *Liquid wastes from processing of shale oil.* The liquids include condensate water from retorting and waste water from 'washing'

of shale oil. These 'technical' waters contain organic contaminants, primarily phenols (Kahru *et al.* 1999, 2000; Tuvikene *et al.* 1999).

(4) *Bottom ash and fly ash from combustion of oil shale in power plants.* The expulsion of acidic gases during combustion leaves the remaining ashes rich in basic components (e.g., alkaline earth oxides), which can generate highly alkaline leachates when disposed in heaps or landfills. These leachates might be rich in metals or metalloids and other trace elements (e.g., As, B, F, Mo, Se), which are mobile under alkaline conditions and might be harmful to the ecosystem (Lindsay 1979; Saether 1980; Stollenwerk 1980; Stollenwerk & Runnells 1981). When performing environmental risk assessments, it is important to know the host phases of these elements in the ashes to understand their release mechanisms during leaching with water (Saether *et al.* 1981, 1984; Pets 1999).

(5) *Particulate emissions into the atmosphere from combustion.* Depending on installed filtration and scrubbing systems, fine particles (often <45 μm) may not be collected efficiently by electrostatic precipitators (ESP; Paat & Traksmaa 2002) and are deposited on foliage or the ground (Ots *et al.* 2000), or might be inhaled by vertebrates, including humans.

(6) *Emission of combustion gases, such as SO_2, NO_x, and CO_2.* These gases are potential sources of acid precipitation and might contribute to climate change (Häsänen *et al.* 1997).

Northeastern European oil shales

History

The largest oil shale resources in the world used industrially today are located in northeastern Estonia and northwestern Russia (Fig. 2). Oil shales have been known here since at least the late 18th century, but large-scale exploration for oil shale deposits and subsequent exploitation by continuous mining started during World War I to supply Petrograd (subsequently Leningrad, now St. Petersburg) with fuel. Following World War II, the Soviet Union processed the oil shales to provide fuel for the Baltic Sea Navy and to generate gas for Leningrad (Reinsalu 1998; Kattai *et al.* 2000). During the 1950s, the shales were introduced as a source for electrical power generation. New mines were developed to provide fuel for the Tallinn, Kohtla-Järve

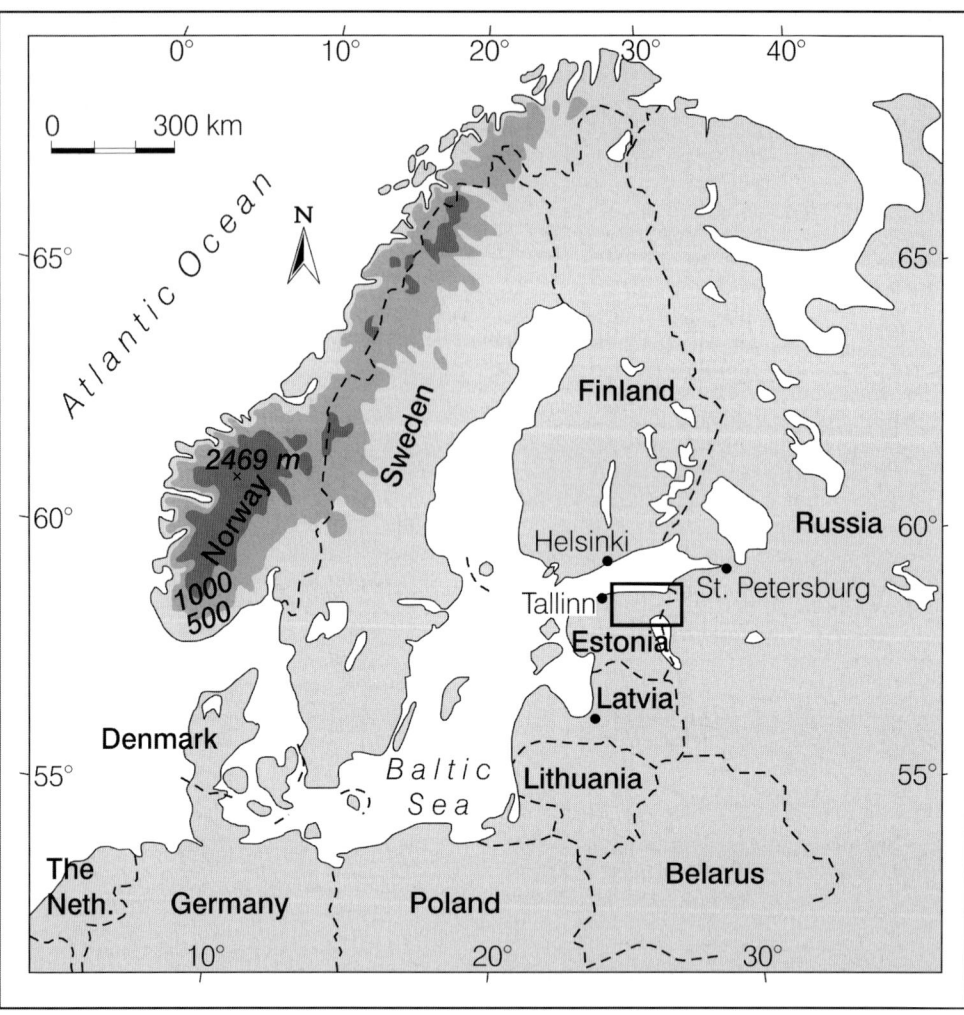

Fig. 2. Location of Estonia and of the area containing the oil shales in Estonia and Russia, as detailed in Fig. 3. (Modified from J. Donner, The Quarternary History of Scandinavia, 1995, Cambridge University Press, Fig. 1.1).

and Ahtme power stations, feedstock for the Kohtla-Järve and Kivióli chemical plants, and fuel for the 'Kunda-Nordic Tsement' cement plant (Fig. 3). From the late 1950s to the 1970s, the massive Balti and Eesti power plants (subsequently referred to as BPP and EPP) near the city of Narva (Fig. 3) were developed; they are the largest operations based on oil shale in the world today. Up to 90% of the shale mined annually in Estonia is consumed by power plants, mainly BPP and EPP, but also by smaller ones at Kohtla-Järve and Sillamäe. Chemical refineries, producing fuel oil, aromatic hydrocarbons, phenol products, epoxy adhesive, antiseptics, antifreeze, and mastics, consume most of the remainder (>8%). Around 2% is

consumed by the 'Kunda-Nordic Tsement' cement factory (Eesti Pólevkivi Ltd. 2003). By the year 2000 an estimated cumulative amount of 920 Mt of oil shale had been mined in Estonia (Table 1).

Occurrence

The northeastern European oil shale deposits crop out on the northern coast of Estonia (Figs 2 and 3) with a strike N100–120 °E and a dip of c. 3 m/km towards the SSW. They occur as two stratigraphic units (Bauert & Kattai 1997; Kattai & Lokk 1998): (1) Middle Ordovician oil shale, also known as 'kukersite', from the German 'Kuckers' (Kattai *et al.* 2000); and

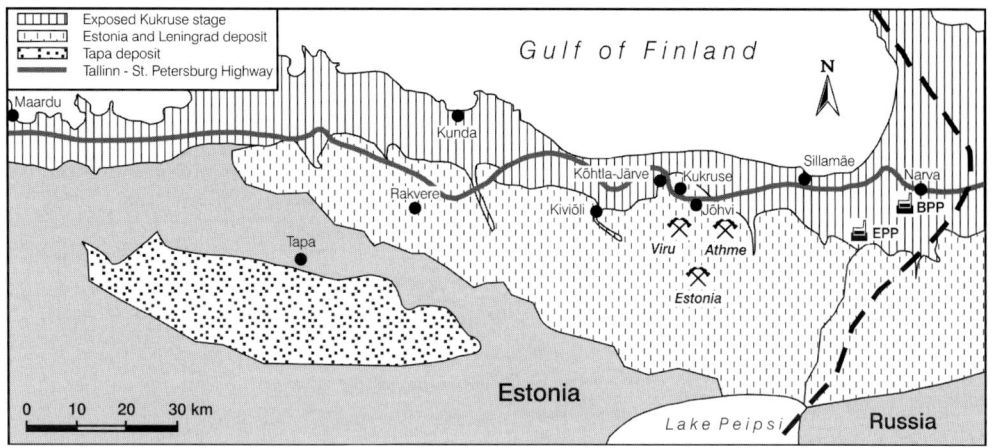

Fig. 3. Geological sketch indicating areal distribution of the Estonia–Leningrad and Tapa oil shale deposits and the location of the main production and consumption facilities. (Courtesy Bauert & Kattai 1997, Fig. 203, p. 314).

Table 1. *Cumulative statistics on mined quantities and waste generated in the Estonian oil-shale industry**

Type of material	Volume (Mm^3)	Weight (Mt)
Surface rock removed from quarries	1500	860
Oil shale mined from mines	350	560
Oil shale mined from quarries	225	360
Gangue (limestone with oil shale residue)	86	150
Semi-coke	80	100
Oil shale ash from power plants	140	230

*After Toomik & Liblik 2000.

(2) Lower Ordovician Dictyonema shale, also referred to as Dictyonema argillite, graptolitic argillite, alum shale.

The Dictyonema shale, with a seam thickness of 1–8 m, contains by far the largest reserves, and underlies an area of c. 11 000 km^2 in the northern and northeastern parts of Estonia at a depth of 10–90 m. These shales have been explored and exploited for mineral resources, for example, Mo and U (Pukkonen & Rammo 1992). However, their low energy content (5–8 MJ/kg), high sulphur content (2–6 wt%) and depth render them unsuitable for energy production or processing (Petersell 1997).

The younger kukersite shale, which economically is the most important oil shale resource of Estonia, occurs in two deposits. The Estonia deposit, located in the northeast of Estonia, underlies an area of 3000 km^2 and comprises up to 50 laterally continuous kukersite seams with a thickness between several centimetres and 0.9 m. The deposit is regarded as extending from surface outcrops down to 100–120 m depth (Kattai & Lokk 1998). The commercial stratigraphic interval consists of seven seams with a thickness of 5–60 cm (Kattai 2003). The Tapa deposit, located to the SW of the Estonia deposit, underlies an area of 1,150 km^2 and occurs 50 to 160 m below the surface. It has a seam thickness of less than 2.3 m and is currently unexploited (Kattai & Lokk 1998).

The kukersites mined from the Estonia deposit have a light chocolate-brown colour in dry condition (10YR 6/4, i.e., light yellowish brown: Munsell Color Company 1994), with a conchoidal fracture oblique to the lamination. The density is less than 2.1 g/cm^3, but varies depending on the content of clastic material and carbonate minerals versus organic matter.

Chemical and mineralogical composition of Estonian kukersite

The chemical composition of natural oil shale is highly dependent on the proportions of organic, carbonate, and clastic components present (Table 2). According to Pokonova & Fainberg (1985), the inorganic component of kukersite consists of (in vol%): 65.9% carbonates, 8.5% quartz, 8.5% feldspar, 5.0% clay minerals, 3.1% sulphides, and 0.8% sulphates, that is, similar to the values cited in Table 3. Torpan (1954) identified calcite (CaCO$_3$), lime (CaO), magnesite (MgCO$_3$), quartz (SiO$_2$), orthoclase (KAlSi$_3$O$_8$), gypsum (CaSO$_4$·2H$_2$O), and biotite

Table 2. *Components of Estonian kukersite oil shale**

Component	Volume %
Organic material	10–60
Carbonate minerals	20–70
Clastic particles	15–60

*After Kattai & Puura 1988.

Table 3. *Mineralogical composition of Estonian kukersite oil shale**

Mineral	Volume %
Calcite	58.2
Dolomite	12.6
Quartz	11.8
Feldspar	4.0
Illite	10.0
Pyrite	3.4

*After Bondar & Koel 1998.

$(K(Mg,Fe^{2+})_3(Al,Fe^{3+})Si_3O_{10}(OH)_2)$ in kukersite. Dilaktorsky & Galibina (1955) additionally reported limonite $(FeO(OH) \cdot nH_2O)$, muscovite $(KAl_2(AlSi_3)O_{10}(OH)_2)$, zircon $(ZrSiO_4)$, rutile and anatase (both TiO_2), glauconite, and tourmaline. Using X-ray diffractometry (XRD), Utsal (1984) and Vingisaar *et al.* (1984) identified chlorite, siderite $(FeCO_3)$, and pyrite (FeS_2).

Within the raw oil shale, carbonates are the predominant minerals (Table 3), whereas (alumino) silicates are the most important components of the non-carbonate residue (Table 4).

Table 4. *Chemical composition (in wt%) of the sandy clay (<250 μm), non-carbonate portion (NCP) of typical Estonian kukersite after dissolution in acid* and of ashed kukersite oil shale [†]*

	Kukersite (NCP)	Ash (range)	Ash (average)
CaO	0.7	30–60	41.5
SiO_2	59.8	20–50	30
Al_2O_3	16.1	5–15	9.5
Fe_2O_3	2.8	2–9	5.5
TiO_2	0.7	0.2–1.2	0.7
MgO	0.4	1–6	3
Na_2O	0.8	0.1–0.5	0.2
K_2O	6.3	2–6	3.5
FeS_2	9.3	–	–
SO_3	0.5	3.0–6.5	4.5
H_2O	2.6	–	–

*After Arro *et al.* 1998.
[†]Bauert & Kattai 1997.

In ashed kukersite oil shale, however, CaO is a dominant component, most likely due to pyrolysis of calcium carbonate:

$$CaCO_3 \text{ (solid)} \rightarrow CaO \text{ (solid)} + CO_2 \text{ (gaseous)}$$

Similarly, the MgO content of the ashes is largely derived from thermal decomposition of magnesian carbonates. This process also expels water from hydrated minerals present in the oil shale. The Fe-sulphide content of the original shales is oxidized to ferric oxides and sulphates in the ash (expressed as SO_3 in Table 4).

The bulk Ca and S contents of the raw Estonian kukersite oil shale are rather high (Table 5), considerably larger than those of the Green River oil shale in Colorado (Mahogany Zone, cf. Table 8). However, the Green River oil shale is richer in Mg, Na, K, and Al, as well as in most of the trace metals and metalloids.

Based on the characteristics of the organic component, the kukersite shales are most likely of marine origin, and its telalginite is derived from the colonial micro-organism *Gloeocapsomorpha prisca* (Derenne *et al.* 1990; Lille *et al.* 2002, 2003). The kerogen component is dominated by phenols, mainly alkyl-1,3-benzenediols (Lille *et al.* 2003). The elemental composition of kerogen from Estonian kukersite (Table 6) shows low levels of nitrogen and medium to high levels of sulphur, and is classified as a Type II/I kerogen in a 'van Krevelen'-diagram (Tissot & Welte 1978; Derenne *et al.* 1990; Zhmur 1990; Koel 1999; Lille *et al.* 2003).

Exploitation

Oil shale deposits in Estonia are exploited either in open pits (typically at a depth <30 m) or underground mines (if depth >30 m) in the ratio 47 wt% to 53 wt% (Eesti Põlevkivi Ltd. 2003). The largest underground mine, the 'Estonia' mine, reaches depths of 40–70 m; the smaller 'Viru' mine is still in operation, whereas the 'Ahtme' mine was closed by 2002 (for locations, see Fig. 3). There are two open pit mines in operation, 'Narva' and 'Aidu', with a combined production of 5.5 Mt in 2002 (Eesti Põlevkivi Ltd. 2003) and yielding shales with calorific values of 8.3–9.0 MJ/kg (Koel 1999).

The total reserves of oil shale in Estonia are estimated at about 5000 Mt, of which 1500 Mt are regarded as being readily accessible, that is, mineable (Veidermaa 2002). Annual production peaked at 31 Mt in 1980 (Kattai & Lokk 1998), and is now 12–13 Mt/y (Adamson & Jostov 2002). These production levels point to sufficient feedstock for another hundred years at the current

Table 5. *Elemental concentrations determined in raw kukersite oil shale and in ash after combustion from Estonia deposit, Estonia,* compared with elemental concentrations in the Earth's crust[†] and Estonian soil[‡]*

Element	Unit	BPP (raw oil shale)	BPP (ash cyclone)	EPP (raw oil shale)	EPP (ash cyclone)	Earth's crust	Estonian natural soil (n = 16)
Ca	wt%	15.9	35.2	16.9	41.9	2.95	0.75
Mg	wt%	1.69	2.84	1.51	3.45	1.35	0.27
Na	wt%	0.029	0.054	0.022	0.041	2.57	0.78
K	wt%	0.85	1.55	0.90	1.18	2.87	2.08
Al	wt%	1.74	3.55	1.35	2.88	7.74	4.35
Fe	wt%	1.85	3.15	1.63	2.55	3.09	1.45
Mn	mg/kg	214	493	306	654	527	449
As	mg/kg	10	10	16	50	2	3.5
B	mg/kg	71	118	55	73	—	—
Ba	mg/kg	45	147	59	148	668	393
Cd	mg/kg	0.2	0.2	0.3	0.3	0.1	—
Co	mg/kg	2.6	5	2.7	5	12	5.2
Cr	mg/kg	15	26	16	21	35	23
Cu	mg/kg	6	16	7	9	14	7
Mo	mg/kg	3	4	5	6	4.4	—
Ni	mg/kg	13	26	13	25	19	8.1
Pb	mg/kg	19	32	27	45	17	15
S	mg/kg	18600	15600	16700	13100	953	—
Sb	mg/kg	0.3	0.4	0.4	0.6	0.3	—
Se	mg/kg	1	2	—	—	—	—
V	mg/kg	19	43	22	38	53	29
Zn	mg/kg	27	51	76	93	52	41

*Häsänen *et al.* (1997), determined by inductively coupled plasma mass spectrometry (ICP-MS), except Al, Ba, Ca, Fe, Mg, Na, P, and S, which were determined by inductively coupled plasma atomic emission spectrometry (ICP-AES).
[†]Wedepohl (1995), determined by X-ray fluorescence (XRF).
[‡]Reimann *et al.* (2003), determined by XRF.

Table 6. *Elemental composition of kerogen in Estonian kukersite oil shale**

Element	wt%
C	76.0–77.5
H	9.4–9.9
S	1.2–2.0
N	0.2–0.5
O	9.0–11.0
Cl	0.5–0.9
H/C (molar ratio)	1.48

*After Arro *et al.* 1998.

rate. All oil shale currently consumed in Estonia is enriched by flotation before processing.

Usage in power generation

Power plants based on oil shale contribute 60% of Estonia's stationary energy, consuming up to 90% of the Estonian oil shale production. During the period 2001–2002, 80.7 wt% of the oil shale production was consumed by AS Narva Elektrijaamad (also referred to as Narva power plants), which, since 1999, consists of the Balti and Eesti power plants located near the city of Narva (Figs 3 and 4). Running at full capacity, the Narva power plants consumed 22 Mt of oil shale (Ots 1992), producing about 16–17 TWh of electricity per year. The Kohtla-Järve and Sillamäe power stations (Eesti Põlevkivi Ltd. 2003) consumed 2.9 and 0.9 wt%, respectively, for heat production. Much of the remaining 15 wt% was consumed by the AS Viru Keemia Group chemical fuel refineries at Kohtla-Järve.

Usage in oil production and refining

Chemical retorting and refining of oil shale commenced in Estonia in 1922–1924 at factories in Kiviõli and Kohtla-Järve. Currently AS Narva Elektrijaamad also has major retorting operations next to the BPP and EPP. Over the past two decades, two main methods have been used in producing shale oil (Veidermaa 2002): the 'Galoter' process (or 'UTT-3000' process),

Fig. 4. Balti power plant run by AS Narva Elektrijamaad with ash plateaux and pond with ultra-alkaline leachate in foreground. The two tallest smoke-stacks are 180 m (Photo: Courtesy AS Narva Elektrijamaad, from www.powerplant.ee/eng/photos.php).

which has a relatively higher energy efficiency than the Kiviter process (see below), but it requires bulky and less reliable equipment; and the '*Kiviter*' *process*, which retorts with internal heating and achieves oil yields constituting 15–17 wt% of the organic matter, but the energy efficiency is low, and potentially hazardous semi-coke is formed as a byproduct (Veidermaa 2002).

The 'Galoter' process was first implemented in an oil shale retorting plant of the solid-heat-carrier type, constructed adjacent to the EPP in the late 1970s. There are currently eleven of these units in operation as part of AS Narva Elektrijaamad, each with a capacity of 3000 t oil shale per day. Since the early 1980s, more than half of the equipment has been reconstructed and the chemical efficiency has reached up to 80%. The process is based on introducing dried oil shale into an aerofountain drier where it is mixed with hot (590–650 °C) shale ash produced by combustion of oil shale semi-coke (at 760–810 °C with O_2 deficiency; Golubev 2003). These plants produce fuel-grade shale oil and raw materials for bitumen and antiseptics, and can also use technical rubber waste and petroleum washes as fuel (Pokonova & Fainberg 1989).

The 'Kiviter' process used by the AS Viru Keemia Group is a direct heat retorting method based on vertical gas generation. Most of the oil shale refining in Estonia is accomplished by the 'Kiviter' process: in 1993, this process was used to refine 1.993 Mt, whereas only 0.5 Mt were refined with the 'Galoter' process (Kattai 2003).

During the last few years, the AS Viru Keemia Group has initiated the implementation of new solid-heat-carrier technologies, for example, the 'Alberta Taciuk' process (Purga 2002; Ritcey 2003). In 2002, the chemical refineries of AS Viru Keemia Group at Kohtla-Järve (Viru Õlitööstus; now the largest such operation in Europe) and the T. R. Tamme Auto OÜ consumed about 10 wt% and 3.6 wt%, respectively, of Estonia's oil shale production (Eesti Põlevkivi Ltd. 2003).

Solid wastes of the Estonian oil shale industry

The majority of the industrial waste in Estonia originates from the oil shale sector, and includes waste materials resulting from mining operations, power generation, and production of chemicals. In 1997, waste from other branches of the economy accounted for only 2.6 wt% of the total hazardous waste generated. The oil shale wastes are considered to be hazardous due to their often high alkalinity and high content of potentially toxic organic compounds and trace elements. Most of the disposal of solid industrial wastes via landfills is geographically restricted to northeastern Estonia, reflecting production based on oil shale in this region (Toomik 1998; Toomik & Liblik 2000).

Mining wastes. A historical byproduct of oil-shale mining in Estonia is limestone gangue, which was dumped to form approximately 30 mounds (previously referred to as 'terricones') of varying form and volume with heights ranging from 10 to 60 m (Table 1). One of these, 'Mt. Kukruse', has become a popular tourism site (Eesti Põlevkivi Ltd. 2003). In addition to limestone, these mounds contain between 8 vol% (recent mounds) and 15 vol% (older mounds) of oil-shale residue, which has undergone spontaneous *in situ* ignition.

Residues from refining and retorting. Retorting of oil shales produces a charcoal-coloured, non-decarbonized, spent shale waste referred to as semi-coke, which contains significant amounts of organic compounds (e.g., hydrocarbons, phenols) and sulphides. A total of 0.63 Mt semi-coke accumulated in Estonia during 1999, and all the disposal mounds together contain 70–80 Mt of semi-coke covering 180–200 ha of land (Fig. 5). The height of these mounds is up to 120 m and the wastes constitute a serious source of pollution through leaching, especially as they are not equipped with any form of basal liner or waterproof foundation. Various avenues of recycling semi-coke are presently being examined (Arro *et al.* 2002).

The composition of the waste depends on the conditions of retorting; higher temperatures, for example, lead to greater decarbonation. A Fisher assay retorting of typical Estonian kukersite oil shale yielded 64.8 wt% ash, 28.1 wt% CO_2, 7.6 wt% C, and 1.5 wt% total S (Arro *et al.* 1998). The chemical composition of semi-coke generated from typical Estonian kukersite is shown in Table 7. A thorough analysis of major, minor, and trace elements in raw and retorted (at 480–650 °C) oil shale from the Green River Formation (Mahogany Zone), in Colorado, was performed and published by Fruchter *et al.* (1980), and Saether (1980) determined the elemental concentrations of natural oil shale from the same location after low-temperature ashing (Table 8). Similar comprehensive data on retorted oil shale from Estonia has, to our knowledge, not yet been published. The concentrations of trace elements reported for retorted Green River oil shale (Table 8) are three to ten times higher than those determined for Estonian semi-coke (Table 7). However, the different concentration levels are partly a result of the different analytical methods used, that is, total elemental content (Table 8) versus 4-acid extractable or leached elemental content (Table 7). Direct

Fig. 5. View of two mounds of semi-coke in the vicinity of Kivioli. The mound in the foreground is about 100 m tall (Photo: Courtesy L. Michelson, from http://virumaa.kolhoos.ee/discuss/msg).

Table 7. *Chemical compositon* of semi-coke and ash waste from Kiviõli, Kohtla-Järve, and Narva areas in Estonia*

Elements	Unit	Kiviõli semi-coke			Kohtla-Järve		Narva	
		Fresh	<20 y old	>40 y old	Semi-coke fresh	Ash from powerplant	Ash from cyclone	Ash from filter
SiO_2	wt%	32.3	34.3	39.4	30.5	37.8	22.4	17.2
TiO_2	wt%	0.24	0.22	0.27	0.31	0.36	0.42	0.47
Al_2O_3	wt%	9.15	8.32	10.5	11.9	12.7	12.7	17.3
Fe_2O_3	wt%	8.26	8.26	7.66	7.95	7.86	8.89	8.83
MnO	wt%	0.04	0.04	0.03	0.05	0.05	0.07	0.05
MgO	wt%	2.30	1.59	0.75	3.68	4.61	3.60	3.47
CaO	wt%	30.8	30.5	25.2	23.6	25.8	42.4	32.8
Na_2O	wt%	0.21	0.14	0.15	0.19	0.35	0.28	0.31
K_2O	wt%	4.26	3.16	4.67	5.37	6.38	4.50	9.78
P_2O_5	wt%	0.18	0.18	0.22	0.23	0.25	0.32	0.38
SO_3	wt%	4.89	5.39	0.27	4.57	2.55	4.37	9.44
C_{org}	wt%	7.34	7.84	10.9	11.7	1.20	0.02	0.02
Ag	µg/kg	90	58	68	123	90	82	123
B	mg/kg	10	0.2	0.2	0.6	0.1	1.3	1.0
As	mg/kg	4	8	10	11	14	12	40
Cd	mg/kg	0.1	0.1	0.1	0.1	0.1	0.2	0.3
Cr	mg/kg	36	34	56	43	44	38	70
Cu	mg/kg	6	6	7	11	9	11	10
Hg	µg/kg	<0.1	<0.1	<0.1	<0.1	<0.1	5	6
Mo	mg/kg	5	4	4	5	5	5	10
Ni	mg/kg	22	21	22	23	21	23	31
Pb	mg/kg	38	42	44	42	37	43	97
Se	µg/kg	20	3	0.8	24	3	18	82
V	mg/kg	26	27	34	34	36	34	61
Zn	mg/kg	25	27	20	37	32	58	89

*Elemental concentrations determined by inductively coupled plasma mass spectroscopy (ICP-MS) in 4-acid extract (HNO_3–$HClO_4$–HF–HCl), except for B, Hg, and Se, which were determined in aqueous leachates (ratio solid/water $= 1/10$) by ICP-MS. Percent oxides are calculated from elemental concentrations. Samples collected by the authors.

comparison of the trace element content of semi-coke from the two different types of oil shales must thus be done with caution.

The environmental risk leachates from the retort residue represent on the ecosystem is not only related to the absolute content of various potentially toxic trace elements in the semi-coke, which in most cases may not be significantly higher than the content of the natural oil shales. It also depends on the leachability of these elements when the residue comes in contact with water, and on the elemental speciation. Important factors to consider in this context are:

(1) The degree of decarbonation of the waste. Leachate alkalinity is generally proportional to the degree of decarbonation.
(2) Mineralogical microfacies. The leachability of trace elements in aqueous solutions depends on their mineralogical residence (Saether *et al.* 1984).
(3) Particle surface morphology and area. During retorting at temperatures 480–650 °C, the pulverized shale becomes friable and the surface area may increase by up to one order of magnitude (Saether 1980).
(4) Solid/water ratio, and water throughput. Different elements within any given residue will tend to exhibit different concentration profiles as a function of leaching time or of the number of pore-volumes that have passed through the waste (Stollenwerk 1980).

Saether (1980), Saether & Runnells (1980), and Stollenwerk & Runnells (1981) studied the leaching of various retorting residues from US Green River oil shales. Alkaline pH values were established even before 25% of the first pore-volume had passed through the waste, and were maintained at high levels for the first 15 pore-volumes. The main potentially toxic elements mobilized under such alkaline conditions were found to be As, B, F, Mo, and Se. In contrast, base metals such as Cu, Ni, Pb, and Zn will be immobilized under such alkaline conditions (Baes & Mesmer 1976; Bell 1976).

Table 8. *Elemental concentrations determined in raw (i.e., feedstock),* retorted,* and after low-temperature ash (LTA)[†] of oil shale from Green River Formation, Mahogany Zone, (GRFMZ), USA*

Element	Unit	GRFMZ (Feed-stock)	GRFMZ (Retorted)	GRFMZ (LTA) ($n = 10$)
Ca	wt%	10.7	13.9	10.6
Mg	wt%	3.59	4.32	—
Na	wt%	1.68	2.15	—
K	wt%	1.66	1.94	2.46
Al	wt%	3.89	4.83	—
Fe	wt%	2.02	2.56	2.50
Mn	mg/kg	319	420	309
As	mg/kg	41.6	59.8	50.9
B	mg/kg	94	107	—
Ba	mg/kg	483	593	513
Cd	mg/kg	0.64	0.9	1.9
Co	mg/kg	9	11.1	—
Cr	mg/kg	39.7	49.6	—
Cu	mg/kg	40.3	56.3	156
Hg	mg/kg	0.089	0.035	—
Mo	mg/kg	24	32.7	44.9
Ni	mg/kg	24.2	32.1	37.1
Pb	mg/kg	26.5	36.2	42.1
S	mg/kg	5730	6780	—
Sb	mg/kg	2.09	2.63	—
Se	mg/kg	2.7	3.4	4.0
V	mg/kg	139	139	—
Zn	mg/kg	62.6	86.2	117

*Fruchter *et al.* (1980), determined by X-ray fluorescence (XRF), except Al and Na by neutron activation analysis (NAA), Mg by flame atomic absorption (lithium borate fusion) (FAA), and B by plasma emission spectroscopy (sodium carbonate fusion) (PES); [†]Saether (1980), determined by XRF after low-temperature ashing (LTA) of raw oil shale samples ($n = 10$).

As regards organic contaminants, leachates from semi-coke contain compounds such as phenols, for example, cresols, resorcinols, and xylenols, which occur at mg/L concentrations. Indeed, Kahru *et al.* (2002) found total phenols at concentrations up to 380 mg/L in semi-coke dump leachates. Phenols also volatilize from such leachates, depending on temperature and pH (Kundel & Liblik 2000). Atmospheric phenol concentrations of 4–50 $\mu g/m^3$ have been observed in the proximity of leachate ponds (Koel 1999). Generally, aliphatic hydrocarbons, carboxylic acids, and organo-nitro and organo-sulpho compounds do not occur at elevated concentrations in leachates from Estonian semi-coke (Koel 1999).

Ash and slag from combustion. Owing to their high mineral content, combustion of oil shales results in vast quantities of ash (Table 1). Oil shale ash dumps form massive plateaux around the Estonian electric power stations near Narva (Fig. 4). The bulk of the ash waste is carried as slurry in pipes from the power stations to sedimentation basins up onto disposal mounds with heights up to 30–40 m, covering an area of over 20 km^2. New ash slurry is continuously disposed into sedimentation ponds located on the top of existing mounds. Thus the height and area of the mounds tend to increase. The cumulative length of the perimeters of these sedimentation pools is almost 10 km, and there is a hazard of alkaline water breaking through the walls of the ponds and reaching adjacent surface and groundwaters. The ash deposits tend to be highly alkaline, and often generate alkaline leachates when brought in contact with water, because the expulsion of acidic gases (SO_2, CO_2) during the combustion process leaves the ashes of kukersite oil shales enriched in the oxides of base cations, such as CaO (lime CaO).

The ash generated per tonne raw feedstock shale during combustion is different in terms of both amount and composition depending on where in the flue-line it accumulates. For example, Häsänen *et al.* (1997) distinguished between bottom ash – 39 wt%, cyclone ash – 32 wt%, electrostatic precipitator ash – 19 wt%, fly ash (>6 μm) – 1.5 wt%, and fly ash (<6 μm) – 0.5 wt%, and listed the elemental concentrations for 30 major, minor, and trace elements determined in each type at BPP and EPP. Table 5 shows elemental concentrations of the raw feedstock shale and of the combustion ash from the major cyclones of these two plants. It should be noted that the concentrations of many of the potentially toxic trace elements are even higher in the electrostatic precipitator ash and the fly ash than in ash from the main cyclone (Häsänen *et al.* 1997). The different contents of trace elements in different types of ashes is also evident from Table 7, in which it is shown that ash from the filter at the Narva power plants contains higher concentrations of most of the potentially toxic trace elements compared with ash from the major cyclone. Furthermore, the trace element content of ash from the power plant at Kohtla-Järve is comparable to that of semi-coke, whereas the content of organic carbon and sulphur is significantly lower.

Table 9 again demonstrates the high CaO content in the fine fraction (30–90 μm) of the combustion ash, due to the high carbonate content associated with the oil shale feedstock. In the very finest fraction (<30 μm), however, Si and Al contents are relatively elevated, at the expense of CaO, presumably due to the presence of clays in this fraction.

Paat & Traksmaa (2002) studied the mineralogical composition of oil shale ash by X-ray diffraction (XRD) and mapped changes in phase content along the slag and ash handling system in one of the boilers (K-3A) of EPP and also in the flue stack from the BPP. The ash was

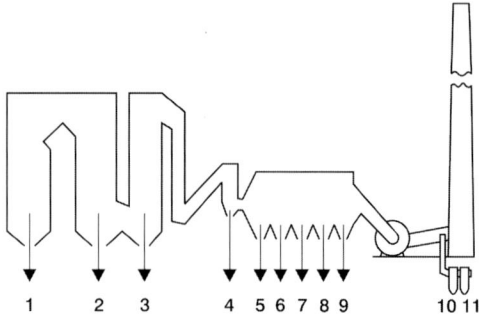

Fig. 6. Sketch of boiler (K-3A) system at the Eesti power plant, showing locations of ash sampling. Samples from locations 10 and 11 were taken from the Balti power plant (modified from Paat & Traksmaa 2002, fig. 1, p. 374). Furnace bottom (1); gas duct: superheater (2), economizer (3), cyclone (4), electrostatic precipitators (ESP): prechamber (5), field I (6), field II (7), field III (8), field IV (9), flue: cyclone (10) and cloth filter (11). (©*Oil Shale.*)

sampled under normal operating conditions, that is, with steam production at 235 t/h, $P_{H_2O} = 13.2$ MPa, and $T = 535$ °C, at locations 1–9 in the EPP (see Figs 6, 7, and 8; Table 10). Locations 10 and 11 in Figs 6, 7 and 8 show sampling points at BPP for fly ash at the cyclone (10) and the flue cloth filter (11), respectively.

Coarse residue with grain sizes 45 μm–1 mm, or slag, from the EPP constitutes the major fraction in the furnace ash and the ash from the superheater, economizer, and cyclone (locations 1–4; Fig. 6, Table 10). The grain size distribution of the gas duct ashes varies widely, being coarser in the economizer adjacent to the major cyclone (Table 10). Some of the coarse ash from the major cyclone carries over into the ash from the electrostatic precipitator, but is reduced from Field I through IV.

The six main phases identified by XRD in the non-magnetic portion of the ashes were lime (CaO), quartz, calcite, anhydrite (CaSO₄), larnite (Ca₂SiO₄), and periclase (MgO). The distribution of these phases in ashes from different sampling points was assessed semi-quantitatively, based on counts per second (cps) of the most prominent XRD peak in the diffractograms. If the relative content of minerals in the ash is considered as a whole throughout the EPP, it can be seen that lime is the predominant mineral, especially in the furnace ash (location 1; Figs 7 and 8). In the superheater ash (location 2), the amount of lime was considerably smaller, increased towards the prechamber of the electrostatic precipitator (location 3; Figs 7 and 8), and

Table 9. *Physical properties and chemical composition of ash residues from combustion of Estonian kukersites in power plants**

Property	Fine fraction	Finest fraction
Specific surface area (m²/kg)	50–120	320–500
Particle size (μm)	30–90	10–30
CaO (wt%)	46–58	28–35
SiO₂ (wt%)	20–28	30–35
Al₂O₃ (wt%)	6–8	10–12
SO₂ (wt%)	7–12	12–15
CO₂ (wt%)	2–3	1–2
Fe₂O₃ (wt%)	4–6	4–5
MgO (wt%)	3–4	2–3
Na₂O (wt%)	0.1	1–2
TiO₂ (wt%)	0.4–0.5	0.5–0.6
K₂O (wt%)	1–2	4–6

*After Hanni 1996.

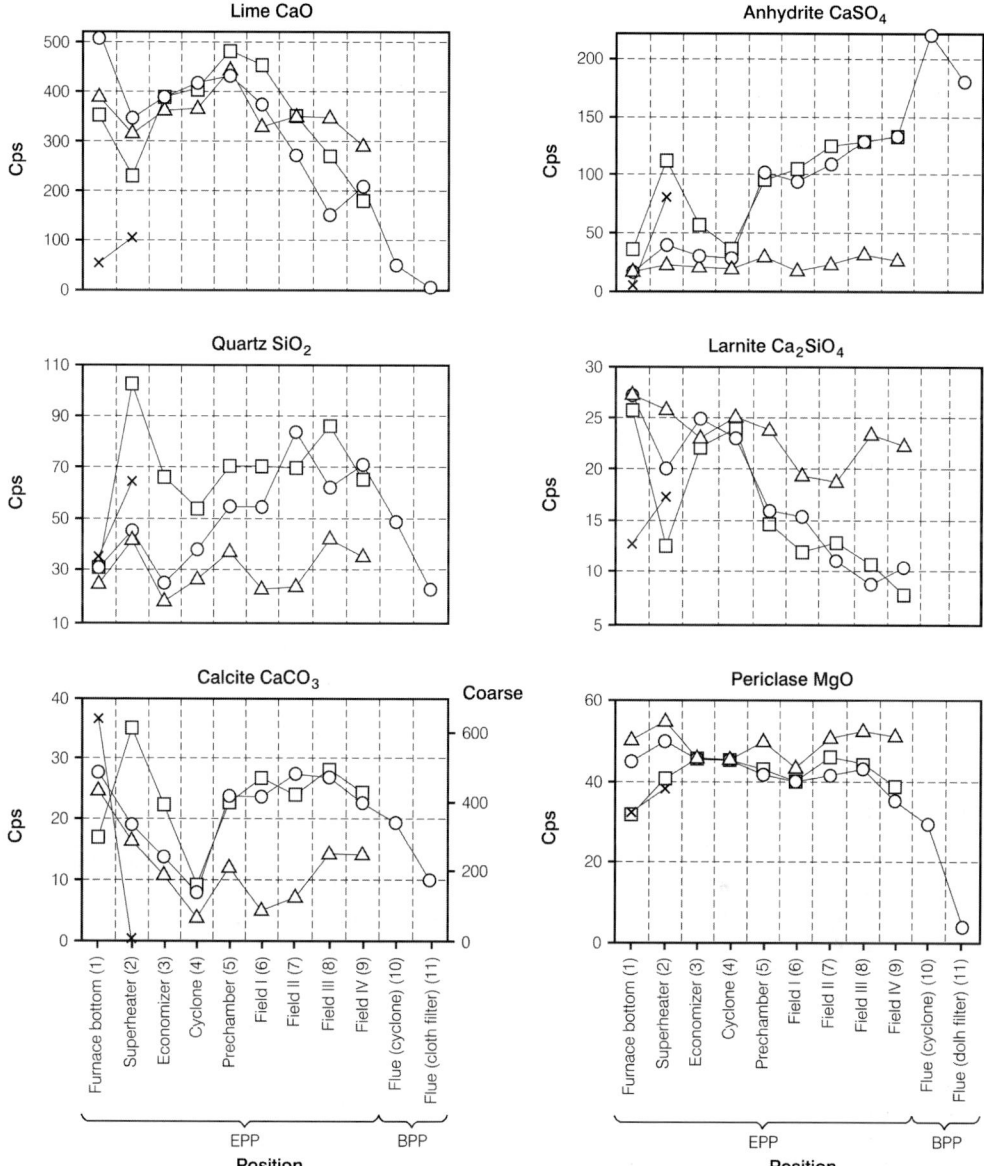

Fig. 7. The content of six main phases in different ashes of the Eesti and Balti power plants: ○, whole ash; □, fine fraction (<45 μm); △, coarse fraction (45 μm–1 mm); ×, coarsest fraction (>1 mm). Modified from Paat & Traksmaa (2002, fig. 3, p.378). (©*Oil Shale.*)

then decreased rapidly in ashes from one field to the next in the electrostatic precipitator (locations 6–9). Anhydrite exhibits the opposite trend, reaching a minimum content in the ash collected in the cyclone (location 4; Figs 7 and 8) and maximum in the flue cyclone (location 10). The content of quartz in the whole ash fluctuates significantly along the combustion path,

but increases markedly from the economizer (location 3) onwards in the EPP. In samples from the final cyclone and flue cloth filter of the BPP, the quartz content is successively lower. The calcite content of the whole ash decreases from the furnace bottom (location 1) to the cyclone (location 4), but shows a small increase in the whole ash from the electrostatic

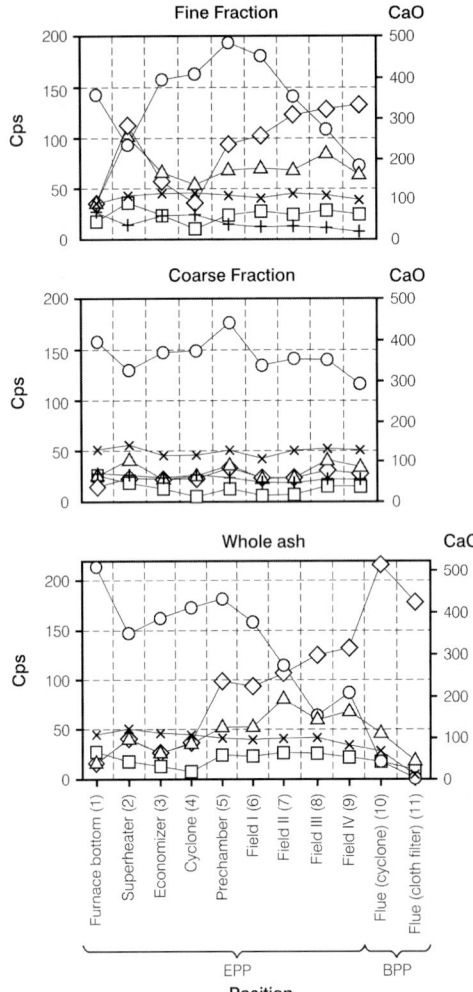

Fig. 8. The composition of: whole ash; fine fraction (<45 μm) of ash; and coarse fraction (45 μm–1 mm) of ash; ○, lime; ◇, anhydrite; △, quartz; ×, periclase; □, calcite; +, larnite. Modified from Paat & Traksmaa (2002, fig. 4, p. 379). Note that cps scale for CaO (at right) is different from scale at left. (©*Oil Shale.*)

Table 10. *Weight percent of ash particles in different grain-size fractions based on dry sieving at different locations in the boiler at Eesti power plant (locations 1–9 indicated in Fig. 6)**

Ash type	Fine fraction <45 μm	Coarse fraction 45 μm–1 mm	Coarsest fraction >1 mm
Furnace bottom (1)	6.4	93.6	<0.1
Gas duct:			
Superheater (2)	19.2	80.0	0.8
Economizer (3)	4.8	95.2	—
Cyclone (4)	33.8	66.2	—
Electrostatic precipitators (ESP's):			
Prechamber (5)	94.4	5.6	—
Field I (6)	93.6	6.4	—
Field II (7)	97.4	2.6	—
Field III (8)	98.4	1.6	—
Field IV (9)	99.6	0.4	—

*After Paat & Traksmaa 2002.

(location 1) in the EPP, but is significantly lower in the flue cyclone and flue cloth filter of the BPP (locations 10 and 11).

Although oil shale fly ash has been used for soil amelioration, mainly because of its relatively high content of easily soluble Ca and K (Kikas 1988; Turbas 1992), it may also contain concentrations of potentially harmful trace elements and heavy metals, and thus may adversely affect plants and soil quality (Pets *et al.* 1985, 1995; Pets & Haldna, 1995; Pets 1999). These authors found elevated concentrations of Co, Cu, Hg, Ni, and Sb, as well as Pb and Zn, in ashes from electrostatic precipitators, cyclones, and steam heaters, relative to natural uncombusted oil shale. These results are confirmed by Häsänen *et al.* (1997). Additionally, the concentrations of As, Ba, Br, Ce, Hg, Th, Zn, U, and Yb in the ashes exceed the average concentrations estimated for the Earth's crust (Clarke 1924; Pets & Haldna 1995), whereas As, Cd, Cu, Ni, Pb, V, and Zn exceed the average concentrations of these elements in the continental crust as estimated by Wedepohl (1995; Table 5). Results from a geochemical survey of agricultural soils from circum-Baltic countries (Reimann *et al.* 2003) also confirm that the background soil concentrations of As, Ni, Pb, V, and Zn are significantly lower than those in Estonian oil shale cyclone and filter ash (Table 5).

Our own data for the chemical composition of Estonian combustion ashes (Table 7) based on 4-acid extracts and aqueous leachates of these (ratio of solid/water = 1/10) show that Hg

precipitators, and successively smaller in the flue cyclone and flue cloth filter of the BPP (locations 10 and 11, Figs 7 and 8). Hematite (Fe_2O_3) and magnetite (Fe_3O_4) are present in small amounts and could only be detected in concentrates made by applying strong magnets (Paat & Traksmaa 2002). The largest amounts of larnite are found in the furnace bottom with another maximum in the economizer whereafter it decreases, especially in the fine fraction (location 3; Fig. 7). Periclase shows a slight decrease in ash from the furnace bottom occurs enriched in the ashes from the EPP (Narva) at concentrations of over 5 $\mu g/kg$, with Cr occurring at 40–70 mg/kg, Pb at 40–100 mg/kg, and Zn at 60–90 mg/kg. These are about twice the concentrations found for the same elements in low-temperature and 550 °C ash from the Ermelo coal deposit in South Africa (Wang et al. 2003), and are similar (Pb, Zn) or higher (Cr) than those reported for coal ash and fly ash by Itkonen & Jantunen (1989). The concentration of hazardous organic compounds is much lower in combustion ash than in the semi-coke wastes.

Atmospheric emissions from oil shale combustion

Pulverized oil shale is the main fuel for Estonian power stations, and atmospheric emissions therefrom have been studied in detail by Aunela et al. (1995), Häsänen et al. (1997), and Jalkanen (2000). The main emissions are acidic or greenhouse gases (SO_2, NO_x, CO_2) and large amounts of airborne particulate matter that escape the trapping devices in the smoke stack (i.e., the part of fly ash <50 μm in diameter).

Gaseous emissions. Flue gases of BPP and EPP have been continuously monitored for SO_2 and NO_x (Aunela et al. 1995). Average emissions of NO_x are in the range of 90–100 mg/MJ, and the total NO_x emissions during the year 2000 were about 9000 t (Eesti Energia 2002), down from 14 000 t during 1992 (Häsänen et al. 1997). Average emissions of SO_2 were also high, ranging between 820 and 1360 mg/MJ for different boilers. This reflects the S content in oil shale feedstock at the BPP and EPP, averaging 1.7–1.8 wt%, of which about two-thirds occurs as sulphide, one-third as organic S and about one-twentieth as sulphate (Häsänen et al. 1997). This distribution of S species is similar to that found in Colorado oil shale (Stollenwerk & Runnells 1981). Emissions of SO_2 during the year 2000 were ~70 000 t (Eesti Energia 2002), compared to 140 000 t SO_2 during 1992 reflecting the reduced consumption of oil shale (Häsänen et al. 1997). During 2000, CO_2 emissions from oil shale combustion were ~10 Mt and constituted 67% of the total CO_2 emissions in Estonia (Eesti Energia 2002). As CO_2 is produced by decomposition of carbonate minerals as well as by oxidation of organic C, the C emission from oil shale combustion is relatively high at 29.1 t/TJ. Without decomposition of the carbonate, this emission rate would be about 22 t/TJ (Austrian Energy Agency 2003).

Particulate emissions. Particulate atmospheric emissions from oil shale combustion include particles of varying composition and size. In terms of composition, oil shale fly ash is comprised mainly of oxides, especially CaO (Pets et al. 1985; Õispuu & Rootamm 1994; Õispuu et al. 1994). At elevated temperatures, the reaction between CaO and SO_2 is rapid and produces $CaSO_4$. In Estonia, it is estimated that BPP and EPP, at full capacity of 3000 MW, have produced more than 300 Mt of ash since their opening in the 1970s, including about 5 Mt fly ash emitted directly to the atmosphere (Õispuu & Rootamm 1994). Current emissions (2000) are estimated at ~47 000 t/y (Eesti Energia 2002). Average particle concentrations in the flue gases were estimated at 2250 mg/m^3, but the actual range was 1730–3320 mg/m^3 for emissions from the BPP (Aunela et al. 1995, 1998).

Häsänen et al. (1997) cite results of chemical analysis of major, minor and trace elements in different size fractions of emitted fractions of oil shale fly ash from Estonia. In their study, the major constituents in emissions during 1992 from two oil-shale-fuelled power plants were Ca (36 000 t/y), K (9700 t/y), Al (8400 t/y), Mg (3000 t/y), Fe (4500 t/y), and Na (250 t/y). The elements Se, Tl, V, and Cd were enriched in the coarse fraction (i.e., larger than 6 μm diameter) of fly ash of both power plants, whereas Sb, U, Cr, Zn were enriched in the finer fraction. The concentrations of Co, Ni, and Sr were similar in both fractions. Thus, the chemical composition of various size fractions of fly ash varies, as does the proportion of each element emitted in the gaseous or particulate phase. For example, Häsänen et al. (1997) found that for the Narva power plants 40–50% of S and Se and 20–35% of As, Cd, and Tl are released in the gas phase, whereas over 90% of the halogens Cl, Br, and I were retained in the ash fraction. In a similar study of Canadian coal ash, Goodarzi et al. (2002) found that the elements Ca, Cu, Ga, Mg, Na, Pb, and Zn, which partially vaporize during combustion and then condense along the route of the flue gases, are being deposited on the particulate matter.

Organic contaminants. The concentration of polynuclear aromatic hydrocarbons (PAH) in the particulate phase of flue gases of oil-shale-combusting thermal power plants has been estimated to range from 0.04 to 3.16 mg/m^3 (Aunela et al. 1995). The solvent-extractable fraction (<1.5 wt%) from fly ash particles collected from Narva power plant smog chambers included several PAHs (phenanthrene, fluoranthene, pyrene, benz(a)-anthracene, chrysene, benzo(b)-fluoranthene, benzo(k)-fluoranthene, benzo(a)-pyrene), some of which are considered mutagenic and/or carcinogenic (Teinemaa *et al.* 2002). Analysis of soils in the Kohtla-Järve and Ida-Virumaa counties of northeastern Estonia demonstrated that contamination by PAHs was most likely introduced as fly ash particles (Liblik & Kundel 1996; Trapido 1999). Indeed, PAH concentrations in soil samples from Kohtla-Järve were significantly higher than in some soils from other study areas. The accumulation of PAHs in plants has also been demonstrated by Junninen (2001).

Environmental impact. Contamination by fallout from downwind transport of emitted particulate matter from the oil-shale power stations has been demonstrated throughout northeastern Estonia, as far as c. 140 km northwards to Finland and as far as 100 km to the southwest within Estonia (Jalkanen *et al.* 2000). Indeed, Teinemaa *et al.* (2002) and Jalkanen *et al.* (2000) examined the morphology of particles in power station fly ash (see Fig. 9) and found similarities in structure, size and chemical composition of particles collected from moss surfaces in northeastern Estonia and southeastern Finland. Although the emissions of SO_2 and NO_x from the BPP and EPP are generally similar to, or lower than, those from conventional

oil- and coal-fired power plants, the particulate emissions, which were estimated to be 186 000 t/y in 1992 (Häsänen *et al.* 1997), amounted to three times the total particulate emissions from Finnish sources during 1993 (Jalkanen *et al.* 2000). The CaO emissions (although highly variable on a daily basis; Jalkanen *et al.* 2000) are so large that they have a marked neutralization effect on both the SO_2 in the flue gases and the environmental acidification over large areas of northeastern Estonia and even southeastern Finland (Kaasik & Sõukand 2000). Indeed, airborne particulate Ca emissions from the Narva power plants represent approximately 5% of the European total (Jalkanen *et al.* 2000).

Some degree of fractionation as function of distance from the power station smoke stack is to be expected: coarse particles will fall out in the immediate vicinity of the power station, whereas fine fly ash will be transported further, and gaseous emissions might be expected to be transported the furthest. Thus, from the point of view of environmental health, not only the chemical composition of emitted particles and aerosols, but also their size, is relevant (Teinemaa *et al.* 2002). As particulate matter is dominated by basic oxides (e.g., CaO) and gaseous emissions by acidic gases (e.g., CO_2, SO_2), this fractionation will influence the pH of

800nm 25900X

Fig. 9. Scanning electron microscope image of the typical fly ash particles. The samples were taken in a simulated experiment, 130 minutes after generation of the fly ash aerosol (Photo: Courtesy Teinemaa *et al.* 2002) (©*Elsevier*).

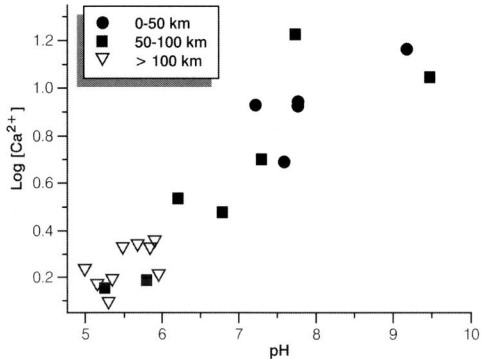

Fig. 10. Concentration of Ca^{2+} (mg/L) vs. pH in snow melt water taken at different distances from the BPP and EPP (Adapted from Teinemaa *et al.* 2003) ($^{©}$*Oil Shale*).

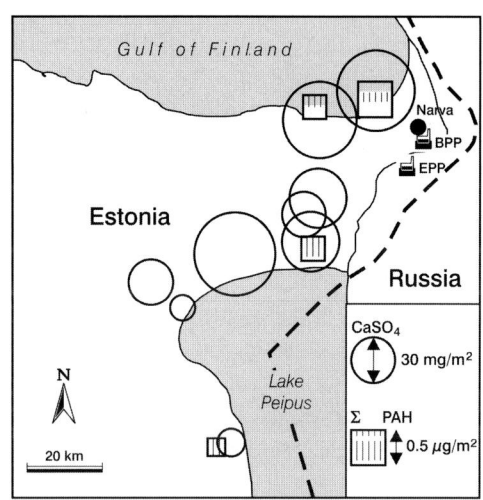

Fig. 11. Map showing the integrated flux (mg/m^2/d) of typical mineral components (CaSO$_4$) and organic trace compounds (total PAH) at sampling points of particulate emissions as an indication of pollution load (Modified from Teinemaa *et al.* 2003) ($^{©}$*Oil Shale*).

surface waters. Figure 10 demonstrates that the pH of snow meltwater decreases with increasing distance from oil-shale-fired power plants. One could thus argue that the acidic emissions (both locally derived and, possibly, transported from elsewhere in Central Europe) appear to be neutralized to a certain extent by the deposition of basic oxides and base cations from power plant emissions. The future installation of a new generation of electrostatic precipitators at power plants will influence the particle size distribution of emissions. It may be speculated that, as this will decrease emissions of potentially neutralizing basic oxide particles, it could have an impact that may not be wholly beneficial. A similar conclusion was reached in the case of the metallurgical industry on the Kola Peninsula (Reimann *et al.* 2000; Kashulina *et al.* 2003).

Paalme *et al.* (1990) studied the bulk deposition chemistry in the environment around the Estonian oil shale combustion area (Fig. 11: see also Kirso *et al.* 2002). Cations (i.e., calcium, potassium, sodium), anions (i.e., chloride, sulphate, nitrate), and *n*-hexane-extractable PAHs were analysed in snow meltwater samples collected from 21 sampling stations in northeastern Estonia. It was found that the characteristic products of oil shale combustion, that is, Ca^{2+} and SO_4^{2-}, accounted for over 92% of the major cations and 90% of the major anions in the snow. Correlation coefficients of $r = 0.86$ and 0.92 were noted for Ca^{2+} vs. SO_4^{2-} and Ca^{2+} vs. ($SO_4^{2-} + 2Cl^-$), respectively. A high degree of correlation ($r = 0.83$) was also noted between Ca^{2+} vs. total PAHs in snow samples taken 150 km to the south of the thermal power plants. The deposition fluxes of Ca^{2+} and PAHs decreased with distance from the power plants. The average Ca^{2+} deposition flux 90 km

south of BPP and EPP, was 2.7 mg/m^2/d, compared with 10.7 mg/m^2/d in the immediate vicinity of the plants. The deposition of total PAHs in the area immediately surrounding the power plants was c. 0.47 μg/m^2/d, as compared with only 0.16 μg/m^2/d 90 km to the southwest (Teinemaa *et al.* 2003).

Conclusions

Oil shales comprise clastic, carbonate, organic, and minor sulphide fractions. Each of these fractions generates several types of environmentally harmful wastes, as evidenced by information from the world's largest industrial consumer of such shales: Estonia. In general, mining wastes and spoil may yield acidic or alkaline leachates, depending on the relative proportions of sulphide and carbonate minerals present.

Combustion of oil shales releases the greenhouse gas CO_2, derived from oxidation of organic matter and decomposition of carbonates. If carbonates are present in high proportions, this renders the oil shales inefficient in terms of energy per unit of CO_2 emitted. Furthermore, oil shale combustion emits acidic gases (NO_x and SO_2) derived both from inorganic sulphides and organically bound N and S. Although the emissions of CO_2, SO_2, and NO_x from combustion of oil shales are at the same level or lower than those from oil- or coal-based power plants with comparable capacity, the combustion of oil shales also yields particulate emissions (potentially enriched in a variety of metals, metalloids, and organics) at a rate 20 to 50 times higher (Häsänen *et al.* 1986; Pets *et al.* 1995; Sloss *et al.* 1996). These emissions have the potential to alter environmental pH in the vicinity of power stations, although the direction of change in pH will depend on the relative proportions of acidic gases or aerosols and basic phases (e.g., particulate oxides) deposited (Reimann *et al.* 2000; Kaasik & Sõukand 2000; Kashulina *et al.* 2003). In Estonia, the prevalence of basic oxides in particulate fallout results in a tendency to elevate pH in the vicinity of power plants. Combustion also results in ashes and slags, which are typically rich in oxides such as CaO and generate high-pH leachates.

Retorting of oil shales to produce shale oil results in wastes (condensate water and solid semi-coke residue) that are heavily contaminated with organic compounds, especially phenolic compounds. Semi-coke leachate is typically alkaline (Kundel & Liblik 2000) and can contain several hundred mg/L phenol in Estonia, in addition to potentially toxic heavy metals and trace elements, for example, As, B, F, Mo, and Se, which might be mobilized during leaching by water. Volatilization of phenols from leachate lagoons can also impact atmospheric quality.

With consumption of fossil fuels allegedly outstripping discovery of new resources (e.g., Deffeyes 2001), it could be argued that oil shales may represent a viable energy alternative for oil-poor countries, provided they are prepared for potential conflicts with international environmental agreements intended to regulate national emissions of greenhouse gases. Nevertheless, due to their modest energy yield, with typical heat contents of 8–9 GJ/t shale (Estonian oil shale, Öpik 1998) compared to 17–24 GJ/t in coal, the economic competitiveness of oil shale must be doubted. Interest in the oil shales of the western USA as a strategic reserve increased after the oil embargo of 1973, when the price of oil doubled, but was found to be commercially unviable in the 1980s. If oil shale should be considered as raw material for shale oil it must contain enough organic matter to yield more energy than it requires to process the rock. A pyrolysis temperature of 500 °C requires approximately 1 GJ/t rock. Thus, assuming an energy content of kerogen of 40 GJ/t, a kerogen content of less than 2.5 wt% would result in a net energy inefficiency (Burger 1973). In fact, the organic content needs to be 8–10 wt% (i.e., yielding 40–50 L/t oil), before it becomes economical as raw material for synthetic fuel (Tissot & Welte 1978).

Geographical, economical, and political aspects will heavily influence future consumption of oil shale. From an environmental viewpoint, the most favourable remediation strategies will favour (1) minimizing gaseous emissions and reducing particulate emission (while being mindful of the latter's potential to neutralize the former), (2) reducing the amount of leaching from ash and semi-coke waste dumps, and (3) investigating further recycling of solid wastes, which is necessary to reduce the environmental hazards caused by the solid waste from the oil shale industry.

The authors wish to acknowledge the constructive comments of D. Crawford-Brown (University of North Carolina, Chapel Hill), J. Dyni (USGS, Denver), R. Gieré (Purdue University, Lafayette), and an anonymous reviewer. Communications with Jialin Qian (Petroleum University, Beijing) and K. Stollenwerk (USGS, Denver) are gratefully appreciated. I. Lundquist and T. E. Finne (Geological Survey of Norway) were of invaluable assistance in getting the figures and tables copy-ready.

Technical note

1 Sm^3 = 6.29 barrels crude oil = 0.84 t of, for example, shale oil
1 barrel crude oil = 159 L = 0.134 t
1 t oil is equivalent to 42 GJ energy
1 GJ = 277.8 kWh = 0.2778 MWh
1 t oil shale may yield up to about 0.1 t shale oil

References

ADAMSON, A. & JOSTOV, M. 2002. Estonian oil shale mining. *Abstracts of the Symposium on Oil Shale*, 18–21 November 2002, Tallinn, 11.

AMERICAN GEOLOGICAL INSTITUTE 1972. *Glossary of Geology.* American Geological Institute, Washington, DC, 805 pp. + one appendix.

ARRO, H., PRIKK, A., & PIHU, T. 1998. Calculation of composition of Estonian oil shale and its combustion products on the basis of heating value. *Oil Shale*, **15**, 329–340.

ARRO, H., PRIKK, A., PIHU, T. & ÖPIK, I. 2002. Utilization of semi-coke of Estonian shale oil industry. *Oil Shale*, **19**, 117–125.

AUNELA, L., Häsänen, E. *et al.* 1995. Emissions from Estonian oil shale power plants. *Oil Shale*, **12**, 165–177.

AUNELA, L., FRANDSEN, F. & HÄSÄNEN, E. 1998. Trace metal emission from the Estonian oil shale fired power plant. *Fuel Processing Technology*, **57**, 1–24.

AUSTRIAN ENERGY AGENCY 2003. *Energy and the Environment (Estonia Country Profile).* World Wide Web Address: http://www.eva.ac.at/(en)/ enercee/est/environment.htm.

BAES, C. F. & MESMER, R. E. 1976. *The Hydrolysis of Cations.* Wiley-Interscience, New York.

BAUERT, H. & KATTAI, V. 1997. Kukersite oil shale, *In*: RAUKAS, A. & TEEDUMÄE, A. (eds) *Geology and Mineral Resources of Estonia*, Tallinn, Estonian Academy, 313–327.

BELL, A. V. 1976. Waste controls at base metal mines. *Environmental Science and Technology*, **10**, 130.

BONDAR, E. & KOEL, M. 1998. Application of supercritical fluid extraction to organic geochemical studies of oil shale. *Fuel*, **77**, 211–213.

BURGER, J. 1973. L'exploitation des pyroschistes ou schistes bitumineux. *Revue de l'Institut Français du Pétrole*, **28**, 315–372.

CLARKE, F. W. 1924. *The Data of Geochemistry. 5th edition*. United States Geological Survey Bulletin (Monograph) B 0770, United States Geological Survey, Reston, VA, 841 pp.

DEFFEYES, K. S. 2001. *Hubbert's Peak: The Impending World Oil Shortage*. Princeton University Press, Princeton, NJ, 208 pp.

DERENNE, S., LARGEAU, C., CASADEVALL, E., SINNINGHE DAMSTE, J. S., TEGELAAR, E. W. & DELEEUW, J. W. 1990. Characterization of Estonian Kukersite by spectroscopy and pyrolysis: Evidence for abundant alkyl phenolic moieties in an Ordovician, marine, Type II/I kerogen. *In*: DURAND, B. & BEHAR, F. (eds) *Advances in Organic Geochemistry, Part 2. Molecular Geochemistry. Proceedings of the 14th International Meeting on Organic Geochemistry, 18–22 September, 1989, Paris, France. Organic Geochemistry*, **16**, 873–888.

DESBOROUGH, G. A. 1978. A biogenic–chemical stratified lake model for the origin of oil shale of the Green River Formation: An alternative to the playa-lake model. *Geological Society of America Bulletin*, **89**, 961–971.

DILAKTORSKY, N. L. & GALIBINA, E. A. 1955. The processes of mineral formation during the heating of oil shale ash and hardening of hydrate products of burning. *In*: *Oil Shale Ash Materials in Building*. Estonian Academy, Tallinn, 31–46 (in Russian).

DUNCAN, D. C. & SWANSON, V. E. 1965. Organic rich shale of the United States and world land areas. *United States Geological Survey Circular*, **523**, 12 pp.

DYNI, J. R. 2000. *Oil Shale. Report of the Energy Minerals Division of the American Association of Petroleum Geologists*. World Wide Web Address: http://emd.aapg.org/technical_areas/oil_shale.cfm. [Accessed 22 July 2004].

EESTI ENERGIA 2002. *Eesti Energia AS Annual Report 2001/2002*. World Wide Web Address: http://www.energia.ee/documents/ed20248dbc548d5.pdf. [Accessed 22 July 2004].

EESTI PÕLEVKIVI LTD. 2003. World Wide Web Address: http://www.ep.ee/eng/Environment/Index.html (in English).

FRUCHTER, J. S., WILKERSON, C. L., EVANS, J. C., SANDERS, R. W. 1980. Elemental partitioning in an aboveground oil shale retort pilot plant. *Environmental Science and Technology*, **14**, 1374–1381.

GOLUBEV, N. 2003. Solid oil shale heat carrier technology for oil shale retorting. *Oil Shale*, **20**, 324–332.

GOODARZI, F., PEEL, W. P., HUGGINS, F. E., BROWN, J. R., CHARLAND, J.-P. & PERCIVAL, J. 2002. Chemical and mineralogical characteristics of milled coal, ashes, and stack-emitted material from unit no. 5, Battle River coal-fired power station, Alberta, Canada. *Geological Survey of Canada Bulletin*, **570**, 148 p.

HALL, M. L. & LIVINGSTONE, W. R. 2002. Fly ash quality, past, present and future, and the effect of ash on the development of novel products. *Journal of Chemical Technology and Biotechnology*, **77**, 234–239.

HANNI, R. 1996. Energy and valuable material byproduct from firing Estonian oil shale. *Waste Management*, **16**, 97–99.

HÄSÄNEN, E., POHJLOLA, V., HAKKALA, M., ZILLIACUS, R. & WICKSTRØM, K. 1986. Emissions from power plants fuelled by peat, coal, natural gas and oil. *The Science of the Total Environment*, **54**, 29–51.

HÄSÄNEN, E., AUNELA-TAPOLA, L. *et al.* 1997. Emission factors and annual emissions of bulk and trace elements from oil shale fuelled power plants. *The Science of the Total Environment*, **198**, 1–12.

HUTTON, A. C. 1987. Petrographic classification of oil shales. *International Journal of Coal Geology*, **8**, 203–231.

HUTTON, A. C. 1995. Organic petrography of oil shale. *In*: SNAPE, C. (ed) *Composition, Geochemistry and Conversions of Oil Shales*. Kluwer Academic Publishers, Doordrecht, Boston, London, 17–33.

ITKONEN, A. O. & JANTUNEN, M. J. 1989. The properties of fly ash and fly ash mutagenicity. *In*: CHEREMISINOFF, P. N. (ed) *Encyclopedia of Environmental Control Technology. Volume 1: Thermal Treatment of Hazardous Wastes*. Butterworth-Heinemann, Burlington, MA.

JALKANEN, L., MÄKINEN, A., HÄSÄNEN, E. & JUHANOJA, J. 2000. The effect of large anthropogenic particulate emissions on atmospheric aerosols, deposition and bioindicators in the eastern Gulf of Finland region. *The Science of the Total Environment*, **262**, 123–136.

JUNNINEN, H. 2001. *Monitoring and Source Apportionment of Polycyclic Aromatic Hydrocarbons in Estonian Soil*. Master's thesis, Kuopio University, Finland, 63 pp.

KAASIK, M. & SÕUKAND, Ü. 2000. Balance of alkaline and acidic pollution loads in the area affected by oil shale combustion. *Oil Shale*, **17**, 113–128.

KAHRU, A., PÕLLUMAA, L., REIMAN, R. & RÄTSEP, A. 1999. Predicting the toxicity of oil-shale industry waste-water by its phenolic composition. *ATLA* (Alternatives to laboratory animals/published by Fund for the replacement of animals in medical experiments), **27**, 359–366.

KAHRU, A., PÕLLUMAA, L., REIMAN, R. & RÄTSEP, A. 2000. Microbiotests for the evaluation of the pollution from the oil shale industry. *In*: PERSOONE, G., JANSSEN, C. & DE COEN, W. (eds) *New Microbiotests for Routine Toxicity Screening and*

Bio-monitoring, Kluwer Academic/Plenum, New York, 357–365.

KAHRU, A., MALOVERJAN, A., SILLAK, H. & PÕLLUMAA, L. 2002. The toxicity and fate of phenolic pollutants in the contaminated soils associated with the oil-shale industry. *Environmental Science and Pollution Research International (Special Issue)*, **1**, 27–33.

KASHULINA, G., REIMANN, C. & BANKS, D. 2003. Sulphur in the arctic environment (3): environmental impact. *Environmental Pollution*, **124**, 151–171.

KATTAI, V. & LOKK, U. 1998. Historical review of the kukersite oil shale exploration in Estonia. *Oil Shale*, **15**, 102–110.

KATTAI, V. & PUURA, V. 1988. Commercial zonation of the Estonia oil shale deposit. *In: Proceedings of the International Conference on Oil Shale and Shale Oil*, 16–19 May 1988, Beijing, China, 51–58.

KATTAI, V., SAADRE, T. & SAVITSKI, L. 2000. *Eesti Põlevkivi: Geoloogia, Ressurss, Kaevandamistingimused*. Tallinn, Eesti Geoloogiakeskus, 226 pp. + 22 plates, with summary in English p. 179–194 and Russian p. 195–224.

KATTAI, V. 2003. *Põlekivi-Õlikivi*. Eesti Geoloogiakeskus, Tallinn (in Estonian) 162 pp.

KIKAS, V. H. 1988. Mineral part of oil shale kukersite and their use. *Oil Shale*, **5**, 15–27.

KIRSO, U., TEINEMAA, E. & IRHA, N. 2002. Environmental problems of use of solid wastes in agriculture. *Abstracts of the Symposium on Oil Shale*, 18–21 November 2002, Tallinn, Estonia, 72–73.

KOEL, M. 1999. *Estonian Oil Shale*. Review compiled for Oil Shale Journal (Tallinn). World Wide Web Address: http://www.kirj.ee/oilshale/Est-OS.htm.

KUNDEL, H. & LIBLIK, V. 2000. Emission of volatile phenols from stabilization ponds of oil shale ash dump leachate. *Oil Shale*, **17**, 81–94.

LIBLIK, V. & KUNDEL, H. 1996. Pollution sources and formation of air contamination multi-componental concentration fields of organic substances in North-Eastern Estonia. *Oil Shale*, **13**, 43–63.

LILLE, Ü., HEINMAA, I., MÜÜRISEPP, M. & PEHK, T. 2002. Investigation of kukersite structure using NMR and oxidative cleavage: On the nature of phenolic precursors in the kerogen of Estonian kukersite. *Oil Shale*, **19**, 101–116.

LILLE, Ü., HEINMAA, I. & PEHK, T. 2003. Molecular model of Estonian kukersite kerogen evaluated by ^{13}C Mas NMR spectra. *Fuel*, **82**, 799–804.

LINDSAY, W. L. 1979. *Chemical Equilibria in Soils*. Wiley-Interscience, New York.

MANZ, O. E. 1999. Coal fly ash: a retrospective and future look. *Fuel*, **78**, 133–136.

MUNSELL COLOR COMPANY 1994. *Munsell Soil Color Charts*. Munsell Color Company, Baltimore, Maryland.

MURRAY, D. K. (ed) 1974. Energy Resources of the Piceance Creek Basin, Colorado. *25th Field Conference*, Rocky Mountain Association of Geologists, Denver, Colorado, 302 pp.

NORWEGIAN MINISTRY OF OIL AND ENERGY 2003. *Fakta 2003 – Norsk Petroleumsvirksomhet* (in

Norwegian). Oslo, Olje-og Energidepartementet, 203 pp.

ÕISPUU, L. & ROOTAMM, R. 1994. Environmental pollution by burning pulverized oil-shale. *Transactions of Tallinn Technical University*, **739**, 70–85 (in Estonian).

ÕISPUU, L., RANDMANN, R. & ROOTAMM, R. 1994. Components of oil-shale ash deposited on ashfields, consumed or ejected with exhaust gases. *Transactions of Tallinn Technical University*, **739**, 94–113 (in Estonian).

ÖPIK, I. 1998. Future of Estonian oil shale energy sector. *Oil Shale*, **16**, 295–301.

OTS, A. 1992. Formation of air-polluting compounds while burning oil shale. *Oil Shale*, **9**, 63–75.

OTS, K., RAUK, J. & MANDRE, M. 2000. The state of the forest ecosystem in the area of oil shale mining and processing. 2. Morphological characteristics of Norway Spruce. *Oil Shale*, **17**, 168–183.

OTS, A. & UUS, M. 2002. Estonia oil shales as power fuel. *Abstracts of the Symposium on Oil Shale*, November 18–21, 2002, Tallinn, Estonia, 12–13.

PAALME, L., VOLL, M., URBAS, E., PALVADRE, R., JOHANNES, I. & KIRSO, U. 1990. Influence of oil shale region on the atmospheric pollution of Lake Peipsi. *Proceedings of the Estonian Academy of Science, Chemistry*, **39**, 18–27 (in Estonian).

PAAT, A. & TRAKSMAA, R. 2002. Investigation of the mineral composition of Estonian oil-shale ash using X-ray diffractometry. *Oil Shale*, **19**, 373–386.

PETERSELL, V. 1997. Dictyonema argillite. *In*: RAUKAS, A. & TEEDUMÄE, A. (eds) *Geology and Mineral Resources of Estonia*. Estonian Academy, Tallinn, 327–331.

PETS, L. 1999. Probable modes of occurrence of trace elements in oil shale ashes of power plant. *Oil Shale*, **16**, 464–472.

PETS, L. & HALDNA, Ü. 1995. Microelements in Estonian and Green River (USA) oil shales: a quantitative comparison. *Oil Shale*, **12**, 239–245.

PETS, L. I., VAGANOV, P. A., KNOT, I., HALDNA, J. L., SHWENKE, G., SHNIR, K. & JUGA, R. J. 1985. Microelements in the ash of kukersite oil shale from Baltic Electric Power Station. *Oil Shale*, **2**, 379–390.

PETS, L., VAGANOV, P. & RONGSHENG, Z. 1995. A comparative study of remobilization of trace elements during combustion of oil shale and coal at power plants. *Oil Shale*, **12**, 129–138.

POKONOVA, Y. & FAINBERG, V. 1985. *Slantsehimiya. Itogi nauki I techniki. Seriya: Technologiya Organicheskih Veshestv*. (Chemistry of Oil Shale. Results of Science and Technology. Series: Technology of Organic Matter). Moscow, Vol. 10.

POKONOVA, Y. & FAINBERG, V. 1989. The production of carbon materials from shale oil. *Fuel Science and Technology International*, **7**, 951–967.

PRESS, F. & SIEVER, R. 1997. *Understanding Earth*, 2nd edition. W. H. Freeman & Company, New York, USA, 682 pp.

PUKKONEN, E. & RAMMO, M. 1992. Distribution of molybdenum and uranium in the Tremadoc graptolitic argillite (Dictyonema shale) North-Western

Estonia. *Bulletin of the Geological Survey of Estonia*, **2**, 3–15.

PURGA, J. 2002. The Alberta Taciuk Process-VKG developments for Estonian oil shale. *Abstracts of the Symposium on Oil Shale, 18–21 November 2002, Tallinn, Estonia*, 29.

PUURA, E. & PIHLAK, A. 1998. Oxidation of Dictyonema shale in Maardu mining waste dumps. *Oil Shale*, **15**, 239–267.

PUURA, E., NERETNIEKS, I. & KIRSIMÄE, K. 1999. Atmospheric oxidation of the pyritic waste rock in Maardu, Estonia. 1. Field study and modelling. *Environmental Geology*, **39**, 1–19.

RAIDMA, E. 1994. Diglycidyl ethers from oil shale alkyl resorcinols. Part 1. Initial raw phenols and the structure of reaction products. *Oil Shale*, **11**, 241–249.

REIMANN, C., BANKS, D. & DE CARITAT, C. 2000. Impacts of airborne contamination on regional soil and water quality: the Kola Peninsula, Russia. *Environmental Sciences and Technology*, **34**, 2727–2732.

REIMANN, C., SIEWERS, U. *et al.* 2003. *Agricultural Soils in Northern Europe: A Geochemical Atlas*. Geologisches Jahrbuch, Sonderhefte, Reihe D, Heft SD 5. Schweizerbart'sche Verlagsbuchhandlung, Stuttgart, 279 pp., ISBN 3-510-95906-X.

REINSALU, E. 1998. Is Estonian oil shale beneficial in the future? *Oil Shale*, **15**, 97–101.

RUSSELL, P. L. 1990. *Oil Shales of the World – Their Origin, Occurrence and Exploitation*. Pergamon Press, New York, 753 pp.

SAETHER, O. M. 1980. *The Geochemistry of Fluorine in Green River Oil Shale and Oil-Shale Leachates*. PhD dissertation, University of Colorado, USA, 232 pp.

SAETHER, O. M., RUNNELLS, D. D. 1980. The geochemistry of fluorine in oil shale leachates. *In*: *Proceedings of the 13th Annual Oil Shale Symposium, Colorado School of Mines, 16–18 April 1980*. (Extended abstract, 7 p.)

SAETHER, O. M., RUNNELLS, D. D., RISTINEN, R. A. & SMYTHE, W. 1981. Fluorine; its mineralogical residence in oil shales of the Mahogany Zone of the Green River Formation, Piceance Creek Basin, Colorado, USA. *Chemical Geology*, **31**, 169–184.

SAETHER, O. M., RUNNELLS, D. D. & MEGLEN, R. R. 1984. Trace and minor elements in Green River Oil Shale (Colorado, USA), concentrated by differential density centrifugation. *Chemical Geology*, **47**, 1–14.

SCHMIDT, F. 1858. *Untersuchungen über die silurische Formation von Ehstland, Nord-Livland und Oesel*. Archiv für Naturkunde Liv-, Ehst- und Kurlands, Serie I (Mineralogische Wissenschaften, nebst Chemie, Physic und Erdbeschreibung). Band II. Dorpat, 1–250, 1 Karte.

SCHMIDT, F. 1881. *Revision der ostbaltischen Silurischen Trilobiten nebst geognostischer übersicht des ostbaltischen Silurgebiets. Abteilung I: Phacopiden, Cheiruriden und Encrinuriden*. Memoires de l'Academie Imperiale des Sciences de St. Petersbourg, VII'e Série: XXX,

No. 1, Imperial Academy of Sciences, St. Petersburg, 237 pp.

SLOSS, L. L., SMITH, I. M. & ADAMS, D. M. B. 1996. Pulverized coal ash – requirements for utilization, IEACR/88. IEA Coal Research, London, UK, 88 p.

SMITH, J. W. 1974. Geochemistry of oil-shale genesis in Colorado's Piceance Creek Basin, in Rocky Mountain. *In*: MURRAY, K. D. (ed) *Association of Geologists Guidebook for 25th Annual Field Conference*, Rocky Mountain Association of Geologists, Denver, CO, 71–79.

STOLLENWERK, K. S. 1980. *Geochemistry of Leachate from Retorted and Unretorted Colorado Oil Shale*. PhD dissertation, University of Colorado, USA, 280 pp.

STOLLENWERK, K. S. & RUNNELLS, D. D. 1981. Composition of leachate from surface-retorted and unretorted Colorado oil shale. *Environmental Science and Technology*, **15**, 1340–1346.

SURDAM, R. C. & WOLFBAUER, C. A. 1975. Green River Formation, Wyoming: A playa-lake complex. *Geological Society of America Bulletin*, **86**, 336–345.

SWANEPOEL, J. C. & STRYDOM, C. A. 2002. Utilisation of fly ash in a geopolymeric material. *Applied Geochemistry*, **17**, 1143–1148.

TEINEMAA, E., KIRSO, U., STROMMEN, M. R. & KAMENS, R. M. 2002. Atmospheric behaviour of oil-shale combustion fly ash in a chamber study. *Atmospheric Environment*, **36**, 813–824.

TEINEMAA, E., KIRSO, U., STROMMEN, M. R. & KAMENS, R. M. 2003. Deposition flux and atmospheric behavior of oil shale combustion aerosols. *Oil Shale*, **20**, 429–440.

TISSOT, B. P. & WELTE, D. H. 1978. *Petroleum Formation and Occurrence*. Springer Verlag, New York, 538 pp.

TOOMIK, A. 1998. Environmental heritage of oil shale mining. *Oil Shale*, **15**, 170–183.

TOOMIK, A. & LIBLIK, V. 2000. Põlevkivi kujundab maastikku [Oil shale designed the landscape – in Estonian]. *Eesti Loodus* [Estonian Nature], **4**, Apr. 2000.

TORPAN, B. K. 1954. *Chemical and Mineralogical Composition of Seams and Interbeds Seams of Kukersite*. Tallinn Technical University, Tallinn, Series A, N **57**, 22–31 (in Russian).

TRAPIDO, M. 1999. Polycyclic aromatic hydrocarbons in Estonian soil: contamination and profiles. *Environmental Pollution*, **105**, 67–74.

TURBAS, E. 1992. Use of oil shale ashes as a lime fertilizer in Estonia. *Oil Shale*, **9**, 301–309 (in Russian).

TUVIKENE, A., HUUSKONEN, S., KOPONEN, K., RITOLA, O., MAUER, Ü. & LINDSTRÖM-SEPPÄ, P. 1999. Oil shale processing as a source of aquatic pollution: monitoring of the biologic effects in caged and feral freshwater fish. *Environmental Health Perspectives*, **107**, 745–752.

UROV, K. & SUMBERG, A. 1999. Characteristics of oil shales and shale-like rocks of known deposits and outcrops. *Oil Shale*, **16**, 64 pp.

UTSAL, K. 1984. The use of the X-ray diffraction method for the complex study of the matter composition of oil shale. *Oil Shale*, **1**, 69–79.

VEIDERMAA, M. 2002. Estonian oil shale: reserves and usage. *Abstracts of the Symposium on Oil Shale*, 18–21 November 2002, Tallinn, Estonia, 8.

VINGISAAR, P., KATTAI, V. & UTSAL, K. 1984. The composition of kukersite in the Baltic oil shale basin. *Proceedings Estonian Academy of Geological Sciences*, **33**, 55–62.

WANG, J., SHARMA, A. & TOMITA, A. 2003. Determination of the modes of occurrence of trace elements in coal by leaching coal and coal ashes. *Energy and Fuels*, **17**, 29–37.

WEDEPOHL, K. H. 1995. The composition of the continental crust. *Geochimica et Cosmochimica Acta*, **59**, 1217–1232.

WOOLARD, C. D., STRONG, J. & ERASMUS, C. R. 2002. Evaluation of the use of modified coal ash as a potential sorbent for organic waste streams. *Applied Geochemistry*, **17**, 1159–1164.

XIANGLIN, H. & QIAN, J. (eds) 1986. *Shale Oil Industry in China*. The Hydrocarbon Processing Press, Beijing, China.

ZHMUR, S. I. 1990. Origin of Ordovician oil shales in the Baltic syneclise, paper 2: kukersites. *Lithology and Mineral Resources*, **24**, 244–251.

Injection and trapping of carbon dioxide in deep saline aquifers

BEVERLY Z. SAYLOR & B. ZERAI

Department of Geological Sciences, Case Western Reserve University,
Department of Geological Sciences, Cleveland, OH, USA
(e-mail: bzs@case.edu)

Abstract: Carbon dioxide (CO_2) collected from the waste streams of point sources can be injected into deep geologic formations in order to limit the emission of greenhouse gases to the atmosphere. Deep saline aquifers provide the largest potential subsurface storage capacity for injected CO_2. Once injected, free CO_2 can be retained in deep aquifers for long time periods by slow-moving, downward-directed formation waters. Over time, the injected CO_2 will dissolve in the formation waters and, through reactions with formation minerals, may be converted to carbonate minerals, resulting in permanent sequestration. Factors that influence the mass of CO_2 that can be injected and stored in free or aqueous form, and as mineral phases, are reviewed and applied to estimate storage capacity of the Rose Run Sandstone, a saline aquifer beneath eastern Ohio, USA. It is estimated that 30 years of CO_2 emissions from five of Ohio's largest coal-fired power plants can be injected into the Rose Run Sandstone and, over time, converted to aqueous and, ultimately, mineral phases.

Growing concern about human-induced climate change is driving technology development aimed at slowing the buildup of greenhouse gases in the atmosphere. Improved efficiency and increased use of alternative energy to reduce the use of fossil fuels almost certainly will be part of the solution. A third prospect is to increase the storage of waste CO_2 in reservoirs other than the atmosphere. Potential reservoirs include the terrestrial biosphere (Seneviratne 2003), the oceans (Drange et al. 2001), mineralized forms at the Earth's surface (Lackner 2002), and the deep underground (Holloway 2001). Injection of CO_2 deep underground is particularly promising because geological formations have the potential to store large volumes of CO_2, safely, and to retain the CO_2 in the subsurface for thousands to millions of years. Technology is in development to capture CO_2 from the waste streams of point sources, such as power plants, and inject it into depleted oil and gas reservoirs, deep unmineable coal seams, and deep saline aquifers.

Carbon dioxide is less dense than formation fluids. Once injected, some CO_2 dissolves into the pore water, and the remainder rises buoyantly as a separate phase. Escape of the buoyant CO_2 to the surface is a significant concern because it could be hazardous, for example by contaminating shallow potable aquifers, and it would

compromise storage (Saripalli & McGrail 2002; Klusman 2003). Safe, long-term storage requires that the CO_2 be immobilized or otherwise prevented from migrating upward. Closed structures immobilize buoyancy-driven flow in oil and gas reservoirs and demonstrably have retained oil, natural gas, and naturally occurring CO_2 in the subsurface for millions of years (Pearce et al. 1996). The CO_2 injected into deep coal seams is immobilized by adsorption onto the coal itself. Given a low permeability cap rock, deep saline aquifers theoretically can trap CO_2 hydrodynamically by slow-moving, downward-directed formation waters (Bachu et al. 1994), by solubility in the pore waters (Weir et al. 1995), and by reactions with minerals and pore waters that convert CO_2 to carbonate minerals (Gunter et al. 1993).

Technology for injecting CO_2 has long been used in the oil and gas industry for enhanced oil recovery. In addition, a pilot study of CO_2 injection for enhanced methane recovery from deep coal seams has been underway since 1996 (Gale & Freund 2001). These technologies potentially can provide economic return while storing CO_2 in the subsurface for long time periods. Injection of CO_2 into depleted oil and gas reservoirs, in particular, is a reasonable first approach to CO_2 storage because the infrastructure is largely in place. However, the storage capacity may not be sufficient to meet

From: GIERÉ, R. & STILLE, P. (eds) 2004. *Energy, Waste, and the Environment: a Geochemical Perspective*. Geological Society, London, Special Publications, **236**, 285–296.
0305-8719/04/$15 © The Geological Society of London 2004.

long-term needs. The technology also exists for injecting CO_2 in deep saline aquifers and a pilot project is under way as part of efforts to limit CO_2 release from oil production in the Sleipner West field of the North Sea (Gale *et al.* 2001). Deep saline aquifers provide no economic return for CO_2 injection, but have the largest potential storage capacity.

Deep aquifers are widespread, are geographically associated with fossil fuel sources, and, because it is not necessary to identify and inject directly into closed structural traps, are likely to have large storage volumes and suitable injection sites in close proximity to power-plant sources of CO_2 (Hitchon *et al.* 1999). Deep aquifers potentially have CO_2 storage capacities sufficient to hold many decades worth of CO_2 emissions, but estimates of global capacity are poorly constrained, varying from 300 to 10 000 Gt CO_2 (Holloway 2001). The variation reflects different assumptions about the effectiveness of trapping mechanisms. The low estimate counts only CO_2 that could be stored as an immiscible phase in closed structures within aquifers. The high number assumes closed structures are not necessary and that CO_2 can be stored through a combination of hydrodynamic, solubility, and mineral trapping.

This review article summarizes the factors that influence the storage of CO_2 in deep aquifers. A case study of expected mineral–brine–CO_2 reactions in the Rose Run Sandstone, a deep aquifer and oil- and gas-containing formation in the Appalachian Basin area of eastern Ohio, USA, is presented. Geochemical reactions between CO_2, brine, and formation minerals are emphasized in the example because these reactions determine the ultimate fate of CO_2.

Mechanisms for aquifer storage of CO_2

Hydrodynamic trapping

Hydrodynamic trapping refers to storage of free CO_2 in the pore spaces of sedimentary layers and the transport of that CO_2 away from the surface by regional groundwater flow (Bachu *et al.* 1994). Free CO_2 is the main form of storage during injection, which can last 30–50 years. The injected CO_2 is subject to injection-related hydrodynamic gradients and to buoyancy forces that cause it to form a plume that rises and spreads laterally. The CO_2 will rise until it meets a confining layer that impedes vertical ascent causing the CO_2 to accumulate as a cap. The buoyant force of the CO_2 cap will depend on the difference in density between the CO_2 and the brine and also on the dip of the confining layer. Provided a near horizontal confining layer and relatively small density difference,

the CO_2 theoretically will travel laterally with the downward-directed regional groundwater flow (Bachu *et al.* 1994). However, faults or other high-permeability zones in the stratigraphic seal could provide past path escape routes to the surface (Saripalli & McGrail 2002). The over-pressuring required for reasonable rates of CO_2 injection and the buoyancy forces exerted by the CO_2 cap can widen small fractures, exacerbating the risk for CO_2 escape (Saripalli & McGrail 2002; Klusman 2003).

Carbon dioxide can exist in three different states under the pressure and temperature conditions of deep saline aquifers: liquid, gas, and supercritical (Fig. 1).

Supercritical CO_2 behaves like a gas, filling all the volume available, but has a density that varies with pressure and temperature from less than 200 kg/m^3 to more than 900 kg/m^3 (Angus *et al.* 1976). To reduce costs associated with injection and to limit the buoyancy forces and maximize the mass of free CO_2 that can fill a given pore volume, CO_2 should be injected in a supercritical state (Bachu 2002). Early studies assumed an average surface temperature of 10 °C, a geothermal gradient of 25 °C/km, and a hydrostatic pressure gradient of 10 MPa/km to determine the depth at which the critical point for CO_2 (31.1 °C and 7.38 MPa) is reached (Holloway & Savage 1993). These studies set 800 m as a minimum depth for the confining layer. A depth corresponding to a pressure of 10 MPa, on the order of 1000 m, may be better because it avoids large CO_2 density gradients in the vicinity of the critical point (Holloway 2001).

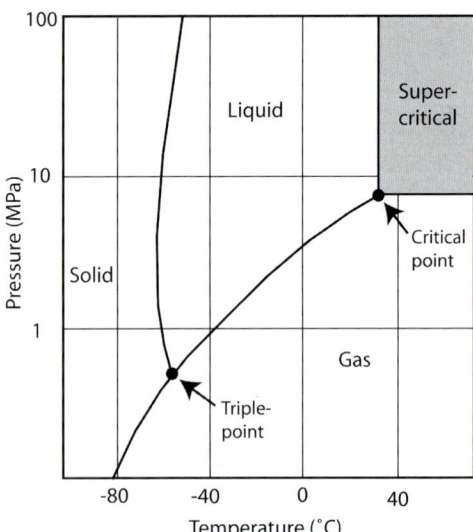

Fig. 1. Phase diagram for CO_2.

In reality, the depths in the Earth at which the pressure and temperature conditions for super-critical CO$_2$ are reached vary depending on climate conditions and the geology of the basin. In warm, overpressured basins, the critical point can lie as shallow as 400 m and the pressure–temperature distribution can be such that CO$_2$ passes with depth directly from a gas to a super-critical state. In cold basins, the depth at which the pressure and temperature conditions of the critical point are reached is likely to lie below the gas-to-liquid phase change. All things being equal, old, cold, stable basins are better for hydrodynamic trapping of CO$_2$ because higher CO$_2$ densities are reached at shallower depths and gradients in the density of CO$_2$ are more easily avoided (Bachu 2000). Old foreland and continental basins are best suited for hydrodynamic trapping because they tend to be cold, stable, and close to hydrostatic pressure, and have erosion- or topography-driven, down-dip-directed regional flow regimes (Bachu 2000).

At 10 MPa and 35 °C, CO$_2$ has a density of approximately 700 kg/m^3. Under these conditions, a cubic meter of sandstone with 10% porosity contains approximately 70 kg of CO$_2$ if the pore space is completely filled by CO$_2$. However, saturation of CO$_2$ is not complete, and some brine remains in the invaded pore spaces (Saripalli & McGrail 2002; Pruess et al. 2003). In addition, non-uniform flow of CO$_2$ bypasses parts of the aquifer entirely. Darcy-flow based analytical and numerical solutions are used to evaluate some of these effects by simulating the advance of the CO$_2$ front over time-scales of decades to hundreds of years and over lateral distances of tens to hundreds of kilometers. To account for the extreme changes in density and viscosity of CO$_2$ with pressure and temperature, these models must incorporate experimentally constrained equations of state (Adams & Bachu 2002).

Initially the free CO$_2$ is distributed in radially decreasing concentrations in zones around the injection site (Fig. 2a; van der Meer 1996). Nearest the injection site lies a zone of near completely saturated pores, containing isolated beads of trapped brine, some of which evaporate into the CO$_2$ (Pruess et al. 2003). The middle zone contains mixed brine and CO$_2$ (Saripalli & McGrail 2002; Pruess et al. 2003). In the outer zone CO$_2$ is present only as aqueous species. Following injection, CO$_2$ saturations around the injection site are predicted to decrease over tens of years as the free CO$_2$ rises buoyantly, spreads laterally, and dissolves into the brine (Weir et al. 1995). Over time-scales of hundreds of years, dispersion, diffusion, and dissolution can reduce the concentration of both free and aqueous CO$_2$ to near zero (McPherson & Cole 2000).

Pruess et al. (2003) used the Darcy-flow based Buckley–Leverette two-phase displacement theory to analytically solve for average saturation of injected CO$_2$ under a range of conditions. Under Buckley–Leverette conditions, the radius of the region swept by the CO$_2$ front increases with duration of injection and the CO$_2$ saturation decreases along that radius, but the average CO$_2$ saturation in the region is time independent. Assuming a homogeneous aquifer and uniformly swept region, Pruess et al. (2003) found that average saturation is most sensitive to permeability, and for a range of rock types calculated a saturation range of 20–40%, with higher average saturations corresponding to rocks with higher permeabilities. Under these conditions, a rock with 10% porosity has a storage capacity of 14–28 kg of CO$_2$ per cubic meter. These maximum values ignore flow patterns that may cause the CO$_2$ to bypass large parts of the aquifer. For example, taking into account buoyancy effects, van der Meer (1995) estimated only 6% average saturation by CO$_2$. In addition, over certain ranges of viscosity ratio and injection velocity, the displacement front between CO$_2$ and brine can develop fractal fingers (Sahimi 1995) rather than advancing as straight front as modeled by Darcy-flow. Fractal fingers may reduce the volume of the aquifer accessed by CO$_2$ to as low as 1%. Formation heterogeneity, including low-permeability layers and lateral discontinuities in permeable rock types can compartmentalize the aquifer vertically and laterally, reducing the amount of the aquifer accessible to the flow of CO$_2$. However, aquifer compartmentalization can also work in the other direction, limiting buoyancy and fingering effects and increasing the access of CO$_2$ (Johnson et al. 2001). Any aquifer bypassing that does occur, due either to buoyancy effects, fingering, or formation heterogeneity, may be partially compensated by higher saturations in the layers swept by CO$_2$ (Pruess et al. 2003).

Solubility trapping

Solubility trapping refers to the CO$_2$ that dissolves into the brine. The CO$_2$–brine solution has a density greater than brine alone, preventing buoyant flow of the CO$_2$ toward the surface, even along high-permeability vertical pathways such as faults.

Most models of solubility trapping assume instantaneous equilibrium between the brine and free CO$_2$. The solubility of CO$_2$ varies as a function of pressure, temperature, and salinity in a manner described by a modified Henry's law (Pruess & Garcia 2002):

$$\Phi X_{CO_2(free)}P = K_H X_{CO_2(l)} \tag{1}$$

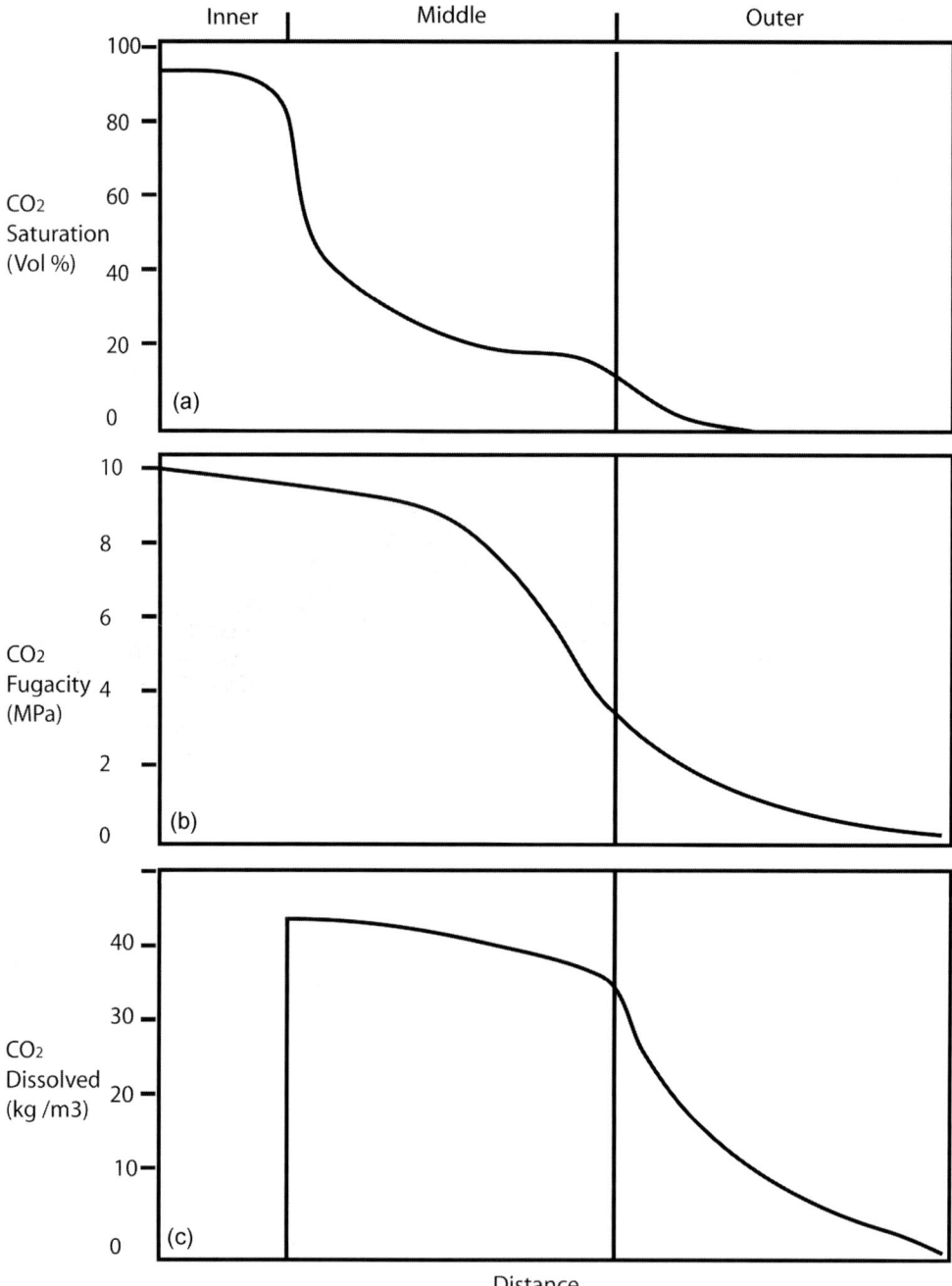

Fig. 2. Schematic variation of (a) CO_2 saturation, (b) CO_2 fugacity, and (c) dissolved CO_2 with radial distance from injection site.

where X_{CO_2}(free) is the mole fraction of free CO_2, P is total pressure and the product is equal to the partial pressure of CO_2. X_{CO_2}(l) is the mole fraction solubility of CO_2 in the brine. The fugacity coefficient, Φ, accounts for the non-linear increase in the solubility of CO_2 with increasing P and T. The modified Henry's coefficient, K_H, corrects for non-ideal solution behaviour with increasing temperature and mole fraction dissolved salt.

Under typical cold deep-aquifer conditions of 10 MPa, 35 °C, and 10% mass fraction salt, the solubility of CO_2 in brine is approximately 40 kg/m^3 of brine. A rock layer with 10% porosity has a storage capacity of as much as 4 kg CO_2/m^3. Under the conditions of interest for aquifer storage, CO_2 solubility increases approximately linearly by about 10 kg/m^3 for each 10 MPa increase in pressure over the range of 10–30 MPa (Pruess et al. 2003). It decreases by an average of approximately 10 kg/m^3 for each 25 °C rise of temperature over the range of 35–100 °C (Pruess et al. 2003). Thus under hydrostatic pressure gradients 10 MPa/km and average geothermal gradients of 25 °C/km, the effects of increasing pressure and temperature with depth on the solubility of CO_2 essentially cancel each other out. However, around the injection site CO_2 pressure increases to as much as twice hydrostatic pressure and CO_2 solubility increases with it (Fig. 2b, c). The solubility of CO_2 decreases by about 30% as salinity increases from zero to saturated NaCl (Møller et al. 1998, Pruess et al. 2003; Bachu & Adams 2003). The solubility of CO_2 is sensitive not only to the mass of salt dissolved, but also to the particular kind of salt. However, there are few experimental constraints on this effect.

The extent to which CO_2 dissolves into the brine is influenced by the migration of the CO_2 front and by the rate of dispersion and diffusion of CO_2. Fractal fingering and buoyancy flow, which tend to limit the storage of free CO_2, may increase solubility trapping by increasing the surface area of the brine–CO_2 contact, allowing more rapid dissolution. In addition, diffusion of CO_2 into the brine can set up reverse density gradients that lead to convective mixing and increased rate of dissolution of free CO_2 (Lindeberg & Wessel-Berg 1997). Through the processes of dispersion, diffusion, and dissolution, aqueous CO_2 becomes the dominant form of CO_2 storage in aquifers over time periods of tens to hundreds of years following injection (Weir et al. 1995; Law & Bachu 1996; Lindeberg 1997; McPherson & Cole 2000). Given a large enough storage aquifer and sufficient time for dilution, the mass fraction of aqueous CO_2 also approaches zero (McPherson & Cole 2000).

Mineral trapping

Mineral trapping is the fixing of CO_2 in carbonate minerals as a result of geochemical reactions among aquifer brines, formation minerals, and aqueous species of CO_2. The density of CO_2 in calcite is 1250 kg CO_2/m^3 calcite. In a rock with 10% porosity and the pores completely filled with calcite the storage capacity would be 125 kg CO_2/m^3. However, the mass of CO_2 sequestered as carbonate minerals is sensitive to formation mineral and aquifer brine composition, P, T, and brine–rock ratio. Time is also important because mineral trapping reactions take hundreds to thousands of years and more to complete (Gunter et al. 1997).

Four kinds of models are available to simulate mineral–brine–CO_2 reactions: equilibrium, path of reaction, kinetic, and reactive transport models. Mineral equilibrium models and path of reaction models are used to calculate equilibrium solid phases and solution compositions for a given set of reactants based on a data set of equilibrium constants and activity coefficients. Equilibrium models calculate only the final state. Path of reactions models also calculate transitional phases along the way. These models do not provide information on the amount of time it takes to reach equilibrium or transition states. Kinetic models consider the rates of reactions. Widely available geochemical modelling codes such as PATHARC (Hitchon 1996), SOLMINEQ (Kharaka et al. 1988), and Geochemists Workbench (Bethke 1996) have been used for equilibrium, path of reaction, and kinetic simulations of CO_2 storage in aquifers. Because these models have no transport component, these studies simulate closed-system batch conditions and do not take into account migration of CO_2 through the aquifer (Gunter et al. 1993, 1996, 1997). Studies using full-scale reactive transport codes to simulate the flow, dissolution, and reaction of CO_2 are just becoming available (e.g., Johnson et al. 2001; Xu et al. 2003). In addition, experimental studies are investigating the kinetics of mineral–brine–CO_2 reactions in mineral separates and rocks to refine and test model reliability (Kaszuba et al. 2003; Liu et al. 2003).

Early studies by Gunter et al. (1993) recognized three general cases for mineral–brine–CO_2 reactions: (1) reactions with mafic minerals, including anorthite feldspar (CaAlSi$_2$O$_8$); (2) reactions with alkali feldspar (solid solution between albite [NaAlSi$_3$O$_8$] and orthoclase [KAlSi$_3$O$_8$]); and (3) reactions with carbonate minerals. They concluded that the most promising reactions for mineral trapping involve mafic minerals, which provide divalent cations (Fe^{2+}, Mg^{2+}, Ca^{2+}) for precipitation of carbonate. Mafic minerals such as olivine, pyroxene, and amphibole have the highest CO_2-fixing potential (Pruess et al. 2003), but are rare in sedimentary basins. The most common sedimentary mineral sources of divalent cations are anorthite, mica-group minerals, especially glauconite, and clays, such that a general equation for mineral trapping in aquifers is

anorthite + micas + clays + CO_2 + H_2O

$$= quartz + kaolinite + calcite$$

$$+ dolomite + siderite \qquad (2)$$

The mineral trapping takes place in three steps as demonstrated by the example of anorthite dissolution:

$$CO_2(aq) + H_2O = H_2CO_3 \qquad (3)$$

$$H_2CO_3 = H^+ + HCO_3^- $$

$$CaAl_2Si_2O_8 + 2H^+ + H_2O$$

$$= Ca^{2+} + Al_2Si_2O_5(OH)_4 \qquad (4)$$

$$Ca^{2+} + HCO_3^- = CaCO_3 + H^+ \qquad (5)$$

which leads to the net reaction:

$$\underset{anorthite}{CaAl_2Si_2O_8} + CO_2(aq) + 2H_2O$$

$$= \underset{calcite}{CaCO_3} + \underset{kaolinite}{Al_2Si_2O_5(OH)_4} \qquad (6)$$

First the aqueous CO_2 dissociates in water, producing carbonic acid, then the acid attacks the anorthite, leaching Ca^{2+} and neutralizing the acid, and finally, calcium carbonate precipitates.

Reactions with alkali feldspars do not provide divalent cations for the precipitation of carbonate minerals and initially were thought to be of little significance for mineral trapping (Gunter et al. 1997). However, more recent work indicates that dissolution of alkali feldspars contributes to the fixing of CO_2 as the sodium alumino-carbonate mineral dawsonite, $NaAlCO_3(OH)_2$ (Johnson et al. 2001). In this case, the Na necessary for dawsonite precipitation is available in abundance in the brine, but dissolution of alkali feldspar provides a source of aluminum and neutralizes the acidic CO_2 according to (Johnson et al. 2001):

$$KAlSi_3O_8 + Na^+ + CO_2(aq) + 2H_2O$$

$$= NaAlCO_3(OH)_2 + 3SiO_2 + K^+ \qquad (7)$$

The feasibility of mineral trapping of CO_2 in dawsonite is demonstrated by the Bowen–Gunnedah–Sydney Basin in Australia, which has abundant diagenetic dawsonite that formed in response to magmatic CO_2 (Baker et al. 1995). In addition, abundant dawsonite in the

Green River Formation of Colorado and in Pleistocene ash beds at Olduvai Gorge Tanzania was formed by reactions of aqueous carbonate species with nepheline in the sediment (Smith and Milton 1966).

Dissolution of carbonate minerals does not lead to mineral trapping of CO_2 (Gunter et al. 1993). However, carbonate dissolution, and other mineral precipitation–dissolution reactions can impact sequestration capacity by altering the permeability of the aquifer near the injection site.

Case study: CO_2 storage in the Rose Run Sandstone, Eastern Ohio

Geologic and geographic suitability

Ohio is a populous and heavily industrialized state with high CO_2 emissions from its coal-fired electric plants. The largest power plants are concentrated in the eastern part of the state (Fig. 3), where coal is mined from the Appalachian Basin. Cambrian sandstones of the Mount Simon Sandstone, the Kerbal Formation, and the Rose Run Sandstone are deep saline aquifers beneath Ohio that have potential for storing injected CO_2 (Fig. 4). These formations and other potential injection targets are the subject of a recently initiated research project by the US Department of Energy and Battelle aimed at characterizing the rock formations using seismic surveys and collection of a 3000 m core (Gupta & Dooley 2003). The site for the study is American Electric Power's Mountaineer plant located in West Virginia along the Ohio River (Fig. 3).

The Rose Run Sandstone is the best characterized of the Cambrian sandstones because it is also an oil and gas reservoir (Fig. 3). It is also the only one of the Cambrian sandstones that is known to retain its sandstone composition in the eastern part of the state rather than passing laterally into carbonate. The Rose Run Sandstone is a sandy layer in the middle of the Knox Dolomite (Fig. 4), which across much of eastern Ohio lies at depths suitable for injection of supercritical CO_2 (Fig. 3). The Rose Run Sandstone was deposited in a passive margin phase of the Appalachian Basin and consists of interbedded layers of carbonate, primarily dolostone, and sandstone (Fig. 5). The sandstone is compositionally mature, consisting largely of quartz. Subordinate reactive minerals are the alkali feldspars and locally abundant glauconite (Fig. 5). Dolomite and quartz are the dominant cements (Janssens 1973; Riley et al. 1993).

Erosional truncation by the Knox unconformity restricts the Rose Run Sandstone to the

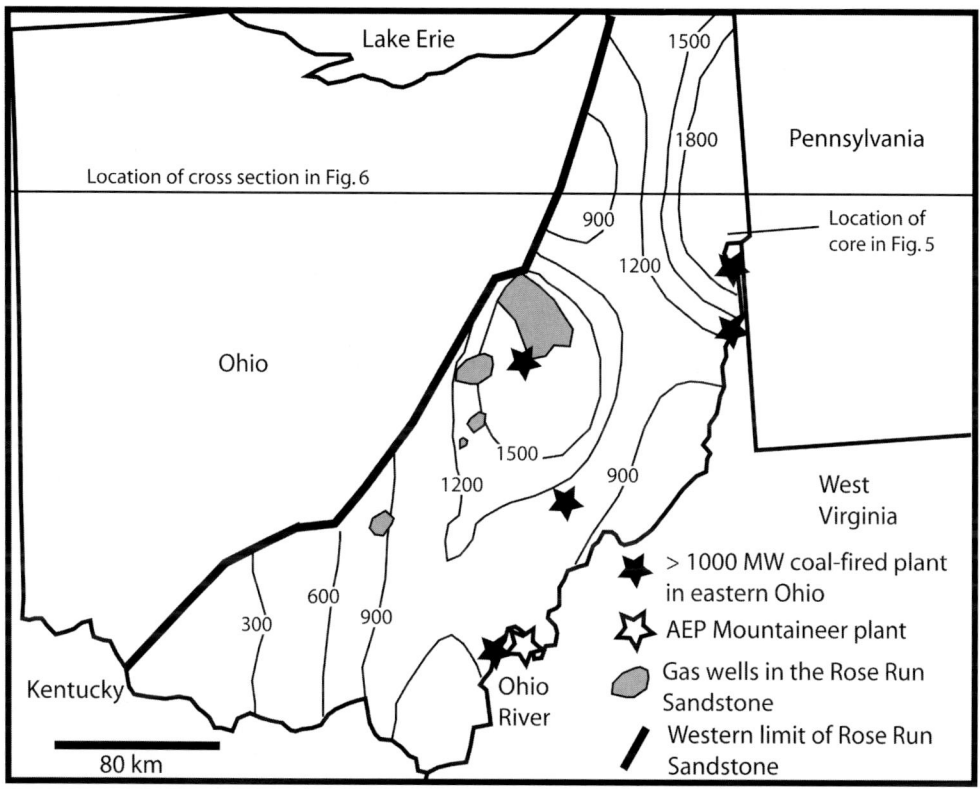

Fig. 3. Map showing depth to Knox Unconformity where it overlies the Rose Run Sandstone in eastern Ohio. (Contour interval in metres; AEP, American Electric Power.)

eastern third of Ohio (Fig. 6). It forms a wedge that thickens eastward from 60 m in easternmost Ohio to more than 200 m in central Pennsylvania, where laterally equivalent formations are brought to the surface by thrust sheets of the Appalachian Mountains. The subcrop belt, where the Rose Run Sandstone intersects the Knox unconformity, is the principal locale for oil and gas production in the unit and great care must be taken not to interfere with this production while injecting CO$_2$ for aquifer storage. However, most of the large power plants are located east of this belt, along the Ohio River (Fig. 3), so that interference with oil and gas production can be avoided.

The Knox Unconformity is immediately overlain by the thin Wells Creek Formation (Fig. 4), followed by Mohawkian limestone and Cincinnatian carbonate and shale, which together form a low-permeablility confining unit that totals more than 1000 m (Gupta & Bair 1997) in thickness (Fig. 6). Regional stratigraphic dip

is gentle, approximately 5°, down to the east and southeast. Regional flow is down dip (Gupta & Bair 1997).

Geochemical modelling of mineral trapping

Batch equilibrium modelling was conducted using The Geochemist's Workbench (Bethke 1996) to simulate multicomponent mineral–brine–CO$_2$ reactions resulting from injection of CO$_2$ into the Rose Run Sandstone. Only reactions with sandstone layers were considered. CO$_2$ fugacity ($fCO_2 = \Phi X_{CO_2(\text{free})} P = K_H X_{CO_2(l)}$) was varied by 2 MPa increments from 10 to 2 MPa as a simple way to investigate possible variation in reaction products as a consequence of decreasing CO$_2$ pressure and solubility with increasing distance from the injection site. A CO$_2$ fugacity of 10 MPa was selected to represent a location near the injection site. Setting partial pressure equal to total pressure and assuming a CO$_2$ fugacity coefficient of 0.45

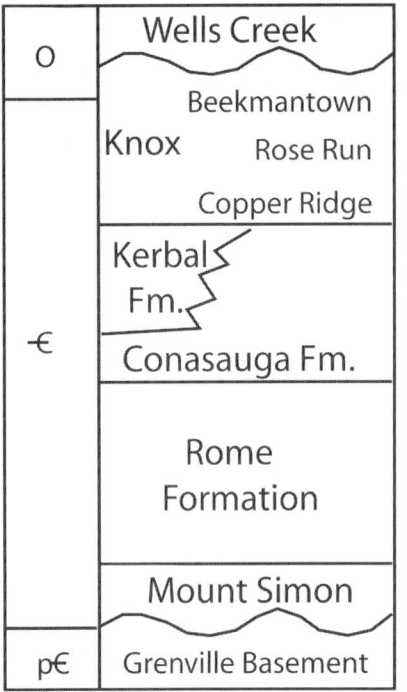

Fig. 4. Cambrian stratigraphy beneath eastern Ohio (Courtesy of Janssens 1973).

from Pruess & Garcia (2002), fCO_2 of 10 MPa corresponds to a CO_2 pressure of 22 MPa, or approximately 85% of lithostatic pressure (26 MPa) at 1000 m depth. Injection pressure must be kept below lithostatic pressure to avoid pressure-induced rock fracturing, which could induce earthquakes and open pathways for the escape of CO_2. Low fCO_2 values represent locations far from the injection site or long after injection has ceased where the CO_2 is present primarily as aqueous phases. This approach does not take into account the decrease in the saturation of the pores by free CO_2 with distance from the injection site. Nor does it consider the impact of flow of CO_2 from the injection site or the rates of geochemical reactions.

Temperature was set at 35 °C. Brine–rock mass ratio was set to 0.4 : 10, which corresponds to a porosity of approximately 10%. The mineral content and brine compositions were set to measured values (Table 1). Debye–Hückel equations were used to correct activity coefficients for saline solutions. The brine was allowed to come to equilibrium with the CO_2, then the 10 kg of sandstone was added and equilibrium assemblages were computed a second time.

The simulations indicate precipitation of the carbonate minerals calcite ($CaCO_3$), dolomite

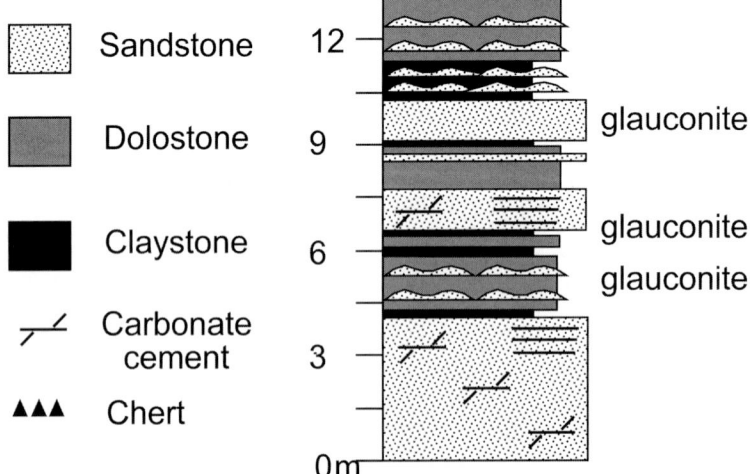

Fig. 5. Measured section of core through part of the Rose Run Sandstone. For location of core, see Fig. 3.

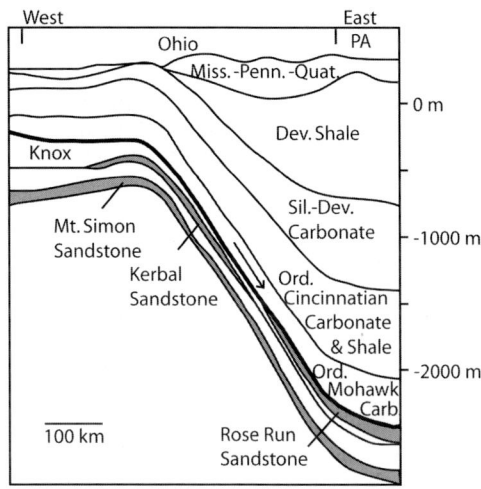

Fig. 6. Cross-section for central Ohio. Cambrian sandstones that are candidates for CO₂ storage are shaded (after Gupta & Bair 1997) (PA, Pennsylvania; for location of cross-section, see Fig. 3).

$(CaMg(CO_3)_2)$, siderite $(FeCO_3)$, strontianite $(SrCO_3)$, and dawsonite $(NaAlCO_3(OH)_2)$. The quantity of strontianite and dolomite formed at equilibrium varies little with initial fCO_2 and is less than 0.05 moles per 10 kg of rock reacted under all conditions studied. The quantities of the other carbonate minerals that formed vary with fCO_2 (Fig. 7). The quantity of calcite precipitated increases from 0.01 moles at 10 MPa fCO_2 to more than 0.2 moles per 10 kg of rock reacted for all other conditions of fCO_2 studied. The quantities of dawsonite and siderite formed at equilibrium decrease with decreasing fCO_2. Dawsonite only forms at fCO_2 above 6 MPa, whereby a maximum of 0.5 moles per 10 kg of

rock reacted forms at fCO_2 equal to 10 MPa (Fig. 7). A maximum of 0.67 moles of siderite form per 10 kg of rock reacted at fCO_2 equal to 10 MPa. There is a net loss of siderite at fCO_2 equal to 2 MPa. The total amount of CO₂ trapped as carbonate minerals per 10 kg of rock reacted varies from a maximum of 1.22 moles (54 g) at 10 MPa to 0.2 moles (9 g) at 2 MPa. The corresponding range of effective storage capacity if reactions reach equilibrium is 14 to 3 kg CO₂ per cubic metre of sandstone reacted. In a rock with 10% porosity, the carbonate minerals would fill at most 12% of the pore space.

Estimation of storage capacity

Table 2 compares the density in pure form and the effective storage density (kg CO_2/m^3 rock) for the different mechanisms of trapping CO₂ estimated for the Rose Run Sandstone. Although the density of CO₂ in free, aqueous, and mineral forms differs by two orders of magnitude, there is substantial overlap of estimated effective storage density for the formation. Stated another way, according to these estimates, all the free CO₂ that can be injected into a formation, if converted to aqueous or mineral form would cover close to the same volume of the aquifer as that covered by the injection plume. For the 60 m thick Rose Run Sandstone the volume necessary to store 30 years of emissions from a 1000 MW power plant (330 Mt) has a radius of 15–34 kms. Since the five largest coal-fired power plants in Eastern Ohio are spaced on the order of 50 miles apart (80 km) the Rose Run Sandstone has the potential to receive 30 years of their CO₂ emissions and to store them over millennia as a negatively buoyant aqueous solution and, ultimately, as immobile carbonate mineral.

Summary and future work

Estimates of storage capacity based on simple flow and equilibrium geochemical models indicate that the Rose Run Sandstone, by itself, potentially can store 30 years of emissions from the five largest coal-burning power plants in eastern Ohio. Ultimately the injected CO₂ can dissolve into the brine and be converted to the stable, immobile, carbonate mineral phases, primarily siderite, dawsonite, and calcite.

The mineral content of the Rose Run Sandstone is similar to other deep formations that are being considered for sequestration of CO₂, including other Cambrian sandstones deep beneath Ohio (Janssens 1973) and the glauconitic sandstone in the Alberta Basin of Canada (Gunter *et al.* 1993). Like the Rose Run, these

Table 1. *Rose Run mineral content and brine composition data*

Mineral	wt%	Species	ppm (mass)
Quartz	83	Cl^-	191203
Orthoclase	10	Na^+	60122
Kaolinite	3	Ca^{2+}	37600
Albite	2	Mg^{2+}	5880
Siderite	1	K^+	3354
Annite (for glauconite)	1	Sr^{2+}	455
		SO_4^{2-}	326
		HCO_3^-	122
		Rb^+	5
		$SiO_2(aq)$	3
		Al^{3+}	2

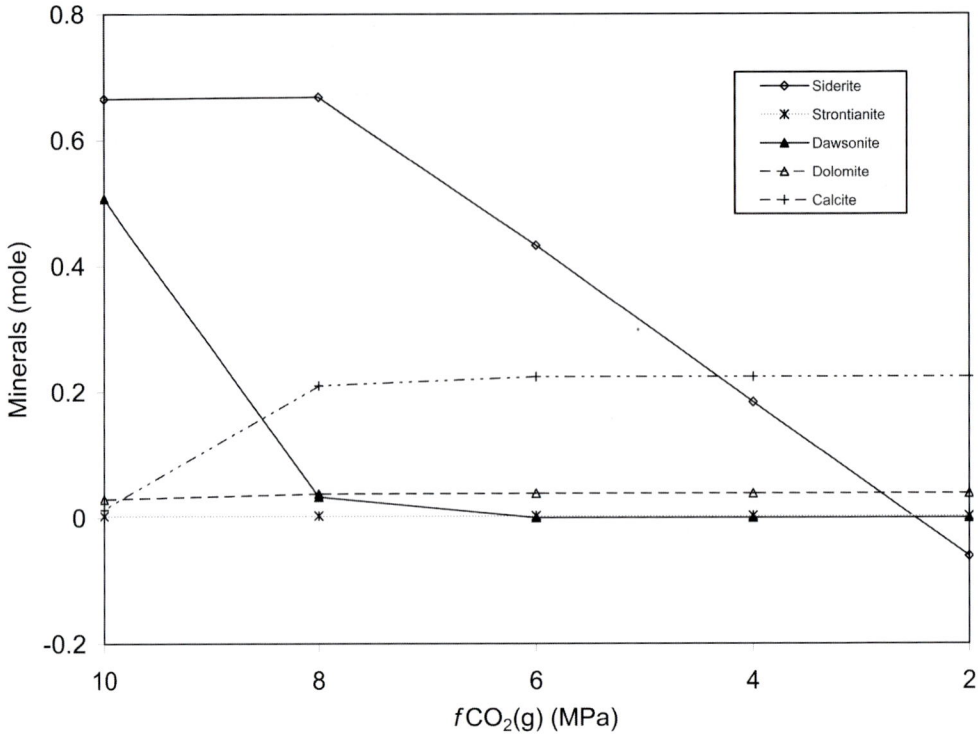

Fig. 7. Carbonate minerals formed per 10 kg of reacted rock at equilibrium as function of fCO$_2$.

sandstone layers were deposited in old, cold, stable basins, and consequently are well suited for hydrodynamic and solubility trapping, but have a paucity of the mafic minerals that are best suited for mineral trapping reactions. The potential that alkali feldspar and glauconite dissolution can provide the cations necessary for dawsonite and siderite precipitation makes these widespread aquifers better candidates for CO$_2$ storage.

A number of factors can alter the estimates of storage potential. These are areas of future work that apply to the Rose Run Sandstone and other potential sites for CO$_2$ injection. They include the following.

(1) Formation heterogeneity, both in terms of porosity and permeability structure and mineral content, was not considered in any of the estimates. This heterogeneity influences how much of the aquifer is accessed by free CO$_2$, how quickly that CO$_2$ dissolves into the brine, and what proportion of the reactive minerals reacts with

Table 2. *Capacity of different mechanisms to store CO$_2$*

CO$_2$ trapping mechanism	CO$_2$ density at 10 MPa, 35 °C	Average storage density in aquifer at 10 MPa, 35 °C, 10% porosity	Radius and height of storage volume for 330 Mt CO$_2$
Hydrodynamic	700 kg/m^3	20% saturation: 14 kg/m^3 6% saturation: 5 kg/m^3	$H = 60$ m $R = 11–18$ km
Solubility	40 kg/m^3	4 kg/m^3	$H = 60$ m $R = 21$ km
Mineral	CO$_2$ in calcite 1250 kg/m^3	10 MPa fCO$_2$: 14 kg/m^3 2 MPa fCO$_2$: 3 kg/m^3	$H = 30$ m (sandstone layers only) $R = 15–34$ km

the dissolved CO$_2$. Some questions that need to be addressed include: what is the impact of low permeability, but highly reactive carbonate layers on the flow and dissolution of CO$_2$ through the Rose Run Sandstone? What is the impact of the uneven distribution of glauconite, which is concentrated in layers, on the effectiveness of mineral trapping reactions?

(2) Faults, fractures, and plugged well holes, which all have the potential to form vertical escape routes through the cap rock to the surface, were not considered. Large faults and plugged well holes can be identified through mapping and high-resolution seismic, and possibly can be avoided. Small fractures induced by injection and buoyancy pressures are more difficult to avoid. Experimental and modelling research is needed to evaluate the risks of leakage through large and small vertical pathways.

(3) Results from equilibrium modelling indicate that the extent of mineral trapping depends strongly on the fugacity of CO$_2$. Consequently, the extent of mineral trapping is sensitive to the rate of mineral–brine–CO$_2$ reactions relative to the rate of flow and dispersion of CO$_2$ away from the site of injection. Reactions must be fast enough to reach carbonate phase saturation before the CO$_2$ is overly diluted by outward radial flow, dispersion, and diffusion. The rates of reaction and the factors that influence the rates of reaction must be better constrained.

Research on the carbon storage potential of the Rose Run Sandstone is supported by Ohio Coal Research Consortium Grant OCRC3-00-4.C4-1. The authors are grateful to J. Friedmann (University of Maryland) and S. Bachu (Alberta Energy and Utilities Board, Canada) and R. Gieré for their helpful reviews.

References

ANGUS, S., ARMSTRONG, B. & REACH, K. M. 1976. *International Thermo-Dynamics Tables of Fluid-State Carbon Dioxide.* Pergamon Press, Oxford.

ADAMS, J. J. & BACHU, S. 2002. Equations of state for basin geofluids: algorithm review and intercomparison for brines. *Geofluids*, **2**, 257–271.

BACHU, S. 2000. Sequestration of CO$_2$ in geological media: criteria and approach for site selection in response to climate change. *Energy Conversion and Management*, **41**, 953–970.

BACHU, S. 2002. Sequestration of CO$_2$ in geological media in response to climate change: road map for site selection using the transform of the geological space into CO$_2$ phase space. *Energy Conversion and Management*, **43**, 87–102.

BACHU, S. & ADAMS, J. J. 2003. Sequestration of CO$_2$ in geological media in response to climate change: capacity of deep saline aquifers to sequester CO$_2$ in solution. *Energy Conversion and Management*, **44**, 3151–3175.

BACHU, S., GUNTER, W. D. & PERKINS, E. H. 1994. Aquifer disposal of CO$_2$: hydrodynamic and mineral trapping. *Energy Conversion and Management*, **35**, 269–279.

BAKER, J. C., BAI, G. P., GOLDING, S. D., HAMILTON, P. J. & KEENE, J. 1995, Continental-scale magmatic carbon-dioxide seepage recorded by dawsonite in the Bowen–Gunnedah–Sydney Basin system, eastern Australia. *Journal of Sedimentary Research*, **65**, 522–530.

BETHKE, C. M. 1996. *Geochemical Reaction Modeling: Concepts and Applications.* Oxford University Press, New York, NY, USA, 397 p.

DRANGE H., ALENDAL, G. & JOHANNESSEN, O. M. 2001. Ocean release of fossil fuel CO$_2$: a case study. *Geophysical Research Letters*, **28**, 2637–2640.

GALE, J. & FREUND, P. 2001. Coal-bed methane enhancement with CO$_2$ sequestration worldwide potential, *Environmental Geosciences*, **8**, 210–217.

GALE, J., NIELS, P. C., Cutler, A. & TORP, T. A. 2001. Demonstrating the potential for geological storage of CO$_2$: the Sleipner and GESTCO projects. *Environmental Geosciences*, **8**, 160–165.

GUNTER, W. D., PERKINS, E. H. & MCCANN, T. J. 1993. Aquifer disposal of CO$_2$-rich gases: reaction design for added capacity. *Energy Conversion and Management*, **34**, 941–948.

GUNTER, W. D., BACHU, S. *et al.* 1996. Technical and economic feasibility of CO$_2$ disposal in aquifers within the Alberta sedimentary basin, Canada. *Energy Conversion and Management*, **37**, 1135–1142.

GUNTER, W. E., PERKINS, E. H. & WIWCHAR, B. 1997. Aquifer disposal of CO$_2$-rich greenhouse gases: extension of the time scale of experiment for CO$_2$-sequestering reactions by geochemical modelling. *Mineralogy and Petrology*, **59**, 121–140.

GUPTA, N. & BAIR, E. S. 1997. Variable-density flow in the midcontinent Basin and Arches Region of the United States. *Water Resources Research*, **33**, 1785–1802.

GUPTA, N. & DOOLEY, J. 2003. *New Project on Geologic Storage of CO$_2$. Greenhouse Issues*, 64. World Wide Web Address: http://www.ieagreen.org.uk/jan64.htm

HITCHON, B. (ed) 1996. *Aquifer Disposal of Carbon Dioxide: Hydrodynamics and Mineral Trapping –* Proof of Concept. Geoscience Publishing Ltd., Alberta, Canada.

HITCHON, B., GUNTER, W. D., GENTZIS, T. & Bailey R. T. 1999. Sedimentary basins and greenhouse gases: a serendipitous association. *Energy Conversion and Management*, **40**, 825–843.

HOLLOWAY, S. 2001. Storage of fossil fuel-derived carbon dioxide beneath the surface of the Earth. *Annual Reviews in Energy and the Environment*, **26**, 145–166.

HOLLOWAY, S. & SAVAGE, D. 1993. The potential for aquifer disposal of carbon dioxide in the UK. *Energy Conversion and Management*, **34**, 925–932.

JANSSENS, A. 1973. *Stratigraphy of the Cambrian and Lower Ordovician Rocks of Ohio*. Ohio Department of Natural Resources, Division of the Geological Survey, 64, Columbus, OH, USA, 197.

JOHNSON, J. W., NITAO, J. J., STEEFAL, C. I. & KNAUSS, K. G. 2001. Reactive transport modeling of geologic CO_2 sequestration in saline aquifers: the influence of intra-aquifer shales and the relative effectiveness of structural, solubility, and mineral trapping during prograde and retrograde sequestration. *First National Conference on Carbon Sequestration*. World Wide Web Address: http://www.netl.doe.gov/publications/proceedings/01/carbon_seq/P28.pdf

KASZUBA, J. P., JANECKY, D. R. & SNOW, M. G. 2003. Carbon dioxide reaction processes in a model brine aquifer at 200 degrees C and 200 bars: implications for geologic sequestration of carbon. *Applied Geochemistry*, **18**, 1065–1080.

KHARAKA, Y. K., GUNTER, W. D., AGGARWAL, P. K., PERKINS, E. H. & DEBRAAL, J. D. 1988. *SOLMNEQ.88. A Computer Program for Geochemical Modelling of Water–Rock Reactions*. United States Geological Survey, Water-Resources Investigations Report 88-4227.

KLUSMAN, R. W. 2003. Evaluation of leakage potential from a carbon dioxide EOR/sequestration project. *Energy Conversion and Management*, **44**, 1921–1940.

LACKNER, K. S. 2002. Carbonate chemistry for sequestering fossil carbon. *Annual Review of Energy and the Environment*, **27**, 193–232.

LAW, D. H. S. & BACHU, S. 1996. Hydrogeological & numerical analysis of CO_2 disposal in deep aquifers in the Alberta Sedimentary Basin. *Energy Conversion and Management*, **37**, 1167–1174.

LINDEBERG, E. 1997. Escape of CO_2 from aquifers. *Energy Conversion and Management*, **38**, 235–240.

LINDEBERG, E. & Wessel-BERG, D. 1997. Vertical convection in an aquifer under a gas cap of CO_2. *Energy Conversion and Management*, **38**, 5229–5234.

LIU, L., SUTO, Y., BIGNALL, G., YAMASAKI, N. & HASHIDA, T. 2003. CO_2 injection to granite and sandstone in experimental rock/hot water systems. *Energy Conversion and Management*, **44**, 1399–1410.

MCPHERSON, B. J. O. L. & COLE, B. S. 2000. Multiphase CO_2 flow, transport and sequestration in the Powder River Basin, Wyoming. *Journal of Geochemical Exploration*, **69–70**, 65–69.

MØLLER, N., GREENBERG, J. P. & WEARE, J. H. 1998. Computer modeling for geothermal systems: predicting carbonate and silica scale formation, CO_2 breakout, and H_2S exchange. *Transport in Porous Media*, **33**, 173–204.

PEARCE, J. M., HOLLOWAY, S., WACKER, H., NELIS, M. K., ROCHELLE, C. & BATEMAN, K. 1996. Natural occurrences as analogues for the geological disposal of carbon dioxide. *Energy Conversion and Management*, **37**, 1123–1128.

PRUESS, K. & GARCIA, J. 2002. Multiphase flow dynamics during CO_2 disposal in aquifers. *Environmental Geology*, **42**, 282–295.

PRUESS, K., XU, T., APPS, J. & GARCIA, J. 2003. Numerical modeling of aquifer disposal of CO_2. *SPE Journal*, **8**, 49–60.

RILEY, R. A., HARPER, J. A., BARANOSKI, M. T., LAUGHREY, C. D. & CARLTON, R. W. 1993. *Measuring and Predicting Reservoir Heterogeneity in Complex Deposystems: The Late Cambrian Rose Run Sandstone of Eastern Ohio and Western Pennsylvania*. Appalachian Oil and Natural Gas Research Consortium, Morgantown, West Virginia.

SAHIMI, M. 1995. *Flow and Transport in Porous Media and Fractured Rock. From Classical Models to Modern Approaches*, VCH Verlagsgesellschaft, Weinheim, Germany.

SARIPALLI, P. & MCGRAIL, P. 2002. Semi-analytical approaches to modeling deep well injection of CO_2 for geological sequestration. *Energy Conversion and Management*, **43**, 185–198.

SENEVIRATNE, G. 2003. Global warming and terrestrial carbon sequestration. *Journal of Biosciences*, **28**, 653–655.

SMITH, J. W. & MILTON, C. 1966. Dawsonite in the Green River Formation of Colorado. *Economic Geology and the Bulletin of the Society of Economic Geologists*, **61**, 1029–1042.

VAN DER MEER, L. G. H. 1995. The CO_2 storage efficiency of aquifers. *Energy Conversion and Management*, **36**, 513–518.

VAN DER MEER, L. G. H. 1996. Computer modeling of underground CO_2 storage. *Energy Conversion and Management*, **37**, 1155–1160.

WEIR, G. J., WHITE, S. P. & KISSLING, W. M. 1995. Reservoir storage and containment of greenhouse gases. *Energy Conversion and Management*, **36**, 531–534.

XU, T., APPS, J. A. & PRUESS, K. 2003. Reactive geochemical transport simulation to study mineral trapping for CO_2 disposal in deep arenaceous formations. *Journal of Geophysical Research*, **108 B2**, 2071.

Environmental impact of geothermal energy utilization

STEFÁN ARNÓRSSON

Science Institute, University of Iceland, Reykjavík, Iceland (e-mail stefanar@raunvis.hi.is)

Abstract: Presently geothermal energy accounts for 0.25% of the annual world-wide energy consumption. Extraction of heat from geothermal reservoirs involves drillings. Exploitation of geothermal resources is generally far less a cause of pollution than fossil-fuel combustion, making it preferable as a source for power and heat. Geothermal energy exploitation has locally some adverse environmental effects. Possible effects include scenery spoliation, drying out of hot springs, soil erosion, noise pollution, and chemical pollution of the atmosphere and of surface- and groundwaters. Exploitation may enhance seismic activity and cause land subsidence. The development of geothermal fields must sometimes be evaluated against the value of tourism. Various measures have been successfully employed to reduce the adverse environmental effects of geothermal energy utilization. The most important ones are directional drilling and injection of spent geothermal fluid. Future use of geothermal resources should specifically consider further use of geothermal heat pumps, extraction of chemicals from spent geothermal fluids to yield useful byproducts, integrated multiple use of high-temperature systems, and furthering of technology to develop hot-dry-rock systems including those occurring in the roots of active volcanic geothermal systems.

Geothermal systems consist of a body of hot aqueous fluid and hot rock (sometimes hot rock only) within the top few kilometres of the Earth's crust and are characterized by a particular rock–hydrological situation. The ultimate source of heat to these systems is the Earth's gravity field and decay of radioactive elements within the Earth. Typically, geothermal systems develop by deep convection of groundwater. The convection may be driven by hydraulic head or by the density difference of hot water within the geothermal system and cooler groundwater outside it (density-driven convection). The convecting water transports heat from deeper to shallower levels in the Earth's crust. Thus, geothermal systems are recognized as surface or shallow-level thermal anomalies. The geothermal fluid usually consists of ancient to modern precipitation but sometimes the source water is seawater, or a mixture of meteoric water and seawater. A small magmatic component may be present in some geothermal fluids. During its convection, the fluid composition is altered through interaction with the enclosing rock. The fluid may also receive chemical components from a degassing magma heat source.

Convective geothermal systems have been classified as volcanic (also termed high-temperature or high-enthalpy) and non-volcanic (low-temperature or low-enthalpy). By definition, high-T systems have temperatures of $>180\,°C$ above 1000 m depth, but in low-T systems temperature is $<180\,°C$ above 1000 m depth (Fridleifsson 1979). The volcanic geothermal systems are associated with tectonically active areas of Quaternary to Recent volcanism and are characteristically located by converging or diverging plate boundaries, but also within hot spots. They have a magmatic heat source. Convection in non-volcanic systems is characteristically associated with geologically young tectonic fractures within which the groundwater convects. Such systems have been classified as tectonic systems (Goff & Janik 2000). The heat source is hot rock at depth. Two other types of non-volcanic geothermal systems have been identified: geopressurized and hot-dry-rock (HDR) systems (Goff & Janik 2000). Geopressurized systems occur in sedimentary basins, where subsidence and accumulation and compaction of water-bearing sediments have generated an overpressurized pore fluid. HDR systems are represented by heat stored in low-porosity, impermeable rocks at varying depths and of varying temperatures.

Utilization of geothermal energy is of two types, electric power generation and direct use of the heat. In geothermal electric power plants, steam is used to drive turbines in much the same way as in conventional fossil fuel plants. The main difference is that in fossil fuel plants the steam is generated by boiling water by

From: Gieré, R. & Stille, P. (eds) 2004. *Energy, Waste, and the Environment: a Geochemical Perspective*. Geological Society, London, Special Publications, **236**, 297–336.
0305-8719/04/$15

burning of coal, oil, or gas, but in the case of geothermal power plants, the steam is extracted from the ground. Electric power generation by geothermal energy is largely based on utilization of high-enthalpy systems, although binary plants can use reservoirs with temperatures as low as 100 °C. Binary geothermal power plants are based on vaporizing a low-boiling-point fluid by the geothermal fluid using a heat exchanger. The vapour of the low-boiling-point fluid is conveyed to the turbine to drive it. When leaving the turbine, the binary fluid is condensed, and returned to the heat exchanger to be re-heated.

Direct uses of geothermal heat are very varied. They are largely based on exploitation of low-enthalpy convection systems. Uses include balneotherapy, space heating, and many agricultural and industrial uses. The potential use of geothermal resources is dictated by fluid temperature as described by the Líndal diagram (Fig. 1).

The utilization of geothermal reservoirs involves drilling and extraction of fluid from the reservoir at a rate, which usually far exceeds the natural discharge rate. Often, hot water in

sub-boiling reservoirs is extracted by pumping. In the case of high-T (high-enthalpy) reservoirs, on the other hand, discharge from wells is always self-sustained.

Exploitation of geothermal energy is generally far less a cause of pollution than fossil-fuel combustion. Therefore, replacement of fossil-fuel energy by geothermal energy is environmentally beneficial. Geothermal energy is, however, not a clean source of energy and its utilization may have some adverse effects on the local environment. The environmental impact of direct uses of low-T (low-enthalpy) geothermal resources is generally trivial. However, unfavourable chemical composition of the hot water in some low-enthalpy reservoirs is of concern and reservoir pressure drawdown, which occurs in response to the exploitation, is known to cause hot springs to dry out. Additionally, road building and construction work may be regarded as environmentally negative.

The environmental impact of high-enthalpy reservoir exploitation is greater and more varied than that of low-enthalpy reservoirs. Increased

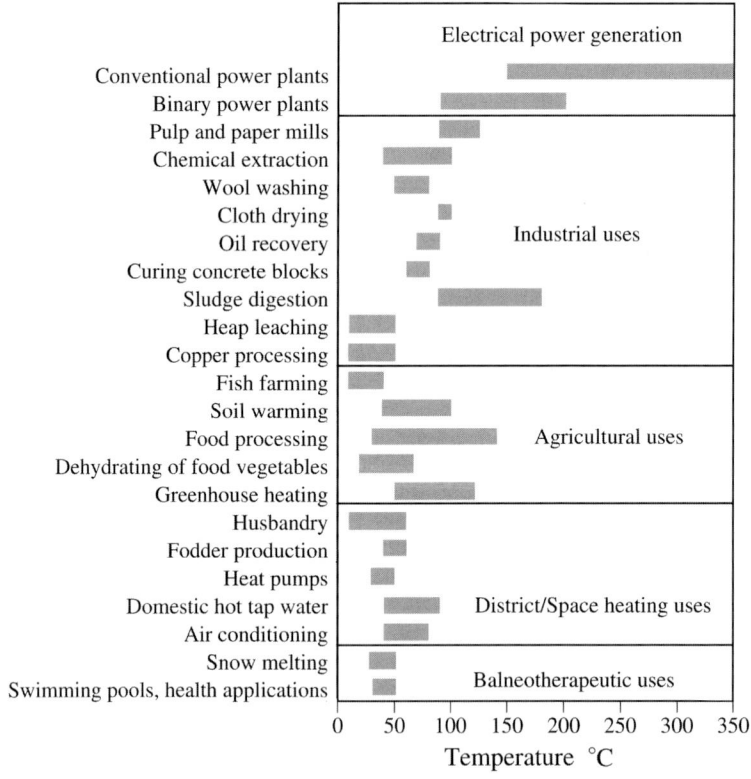

Fig. 1. The Líndal diagram, which shows the potential uses of geothermal energy in relation to the temperature of the fluid.

fluid withdrawal, which leads to pressure decline in the high-enthalpy reservoirs, tends to enhance boiling and may in this way enhance fumarole activity but cause hot springs to dry out. Seismic activity may be enhanced (e.g., Sherburn *et al.* 1990; Smith *et al.* 2000; Brandsdóttir *et al.* 2002). Yet, it may be difficult to establish to what extent seismic activity may be enhanced by exploitation of high-enthalpy geothermal reservoirs because many such reservoirs are located in areas that are tectonically active. Land subsidence is known to occur as a consequence of fluid withdrawal from some high-enthalpy reservoirs (Allis *et al.* 1997, 1998; Allis 2000; Eysteinsson 2000; Glowacka *et al.* 2000; Lee & Bacon 2000). The fluid brought to the surface from high-enthalpy geothermal reservoirs often contains constituents that may have adverse effects on surface- and groundwater and the atmosphere, if not disposed of properly. Drillings, road building, and other construction work are known to have caused extensive soil erosion and scenery spoliation in fields that are located in rugged terrain. Operation of electric power plants is accompanied with significant visual and noise pollution.

Development of a geothermal field may have an adverse effect on tourism, at least if natural manifestations have an attraction. Decision on the benefit of development of a geothermal resource must at times be balanced against the value of tourism.

Many steps have been taken during the last few decades to reduce the environmental impact of geothermal energy utilization. They include: (1) directional drillings, which aim at decreasing scenery spoliation, visual pollution, and soil erosion; and (2) injection of condensate and waste water into the bedrock, which reduces the chemical pollution of local surface- and groundwaters and impedes land subsidence. Furthermore, H_2S abatement from steam has been implemented in some fields to reduce atmospheric pollution. Technologies have also been developed to remove As, B, and Hg from steam, thus reducing atmospheric pollution with respect to these elements.

The efficiency of conventional use of geothermal steam for electric power generation is poor. Only about 10–12% of the heat in the steam discharged from drillholes is converted into electric energy and, in the case of high-enthalpy liquid-dominated reservoirs, the water brought to the surface with the steam is not utilized. In contrast to conventional electric power generation, direct use of geothermal heat is efficient. In Iceland, two high-T fields are presently being exploited for combined electric power generation and production of hot water for space heating. Such a combined scheme permits an efficient exploitation of the energy from liquid-dominated high-T geothermal fluids. The possibility of such use depends on the availability of surface- and/or groundwater for heating and the market for such heated water. High-T water cannot be used directly for space heating, or other direct uses, due to its chemical composition.

Use of geothermal resources

On a global scale, geothermal resources constitute a small, yet rapidly growing, energy resource. It is very important for some countries (Tables 1 and 2) and an important renewable energy source. In the year 2000, geothermal energy constituted about 0.25% of the annual world-wide energy consumption (Fridleifsson 2001). At that time the installed electric power generating capacity was 7974 MW (Huttrer 2001), and direct energy use was 190 699 TJ (Lund & Freeston 2001; IGA 2003). Direct uses of geothermal heat include balneology, swimming, space heating, geothermal heat pumps, greenhouse heating, aquaculture pond heating, various agricultural and industrial applications, snow melting, and air conditioning (Fig. 1).

In 2000 a total of 21 countries were generating electric power from geothermal resources (Huttrer 2001). Five countries accounted for 78.6% of the total globally installed power capacity, and 10 countries for 96.9% of the total (Fig. 2, Table 1). In all, 58 countries were using geothermal heat directly (Lund & Freeston 2001) (Fig. 3, Table 2). Various countries have presented plans for increasing the use of geothermal resources for electric power generation. The total planned increase from 2000 to 2005 is 3440 MW (Huttrer 2001), but the reliability of the plans of the various countries has not been evaluated here. The number above, however, reflects the existence of much interest and effort in increasing electric power generation from geothermal resources.

In 1960, at the time of the United Nations Rome Conference on New Sources of Energy, the total globally installed geothermal electric power was 420 MW (Armstead 1975*a*). In 1970 and 1975, the figure had risen to 675 and 1310 MW, respectively (Armstead 1975*a*), and in 1995 to 6833 MW (Huttrer 2001) (Fig. 4). In the period 1970–2000, the increase in the installed geothermal electric power has been steady, on average 253 MW annually.

Direct use of geothermal energy has grown very much during the last few years. In 1995 and 2000 the estimated energy usage was 112 441

Table 1. *Electricity production by geothermal steam in the world by country in the year 2000**[*]

Country	Installed capacity (MW)	Additional planned capacity by 2005 (MW)	Percentage of country's power generation
Australia	0.2		
China	29.2		
Costa Rica	142.5	20	10.21
El Salvador	161		20.00
Ethiopia	8.5	3	
France (Guadeloupe)	4.2	15.8	2.00
Guatemala	33.4		
Iceland	170	170[†]	14.73
Indonesia	589.5	180	5.12
Italy	785	161	1.68
Japan	546.9		0.40
Kenya	45	120[‡]	8.41
Mexico	855	250	3.16
New Zealand	437	15	6.08
Nicaragua	70		
Philippines	1909	626	21.52
Portugal[§]	25	21	16.00
Russia (Kamchatka)	23	200	
Thailand	0.3		
Turkey	20.4	500	
USA	2228	148	0.40
Total	7974	2429.8	

[*]Based on Huttrer (2001).
[†]A new 30 MW unit at Nesjavellir is already in operation (2001).
[‡]82 MW are already in operation and 38 planned. T. Mwangi (pers. comm.).
[§]Azores only.

and 190 699 TJ, respectively, which corresponds to a growth of 70% in a five-year period (Lund & Freeston 2001). The estimated growth rate for bathing and swimming (from 15 742 TJ in 1995 to 79 546 TJ in 2000) is excessive, because the 1995 evaluation did not include much of the capacity and energy for this usage due to uncertainty in reported data (Lund & Freeston 2001). It is, however, clear that the use of geothermal heat pumps for space heating has experienced a very large growth rate in this period. Capacity has increased by 185% and energy utilization by 59% (Lund & Freeston, 2001). Published information is not available to assess globally the increase in direct use of geothermal energy prior to 1995.

In the year 2000, bathing accounted for almost 42% of the direct use of geothermal heat. Use for space heating was second with approximately 23%, heat pumps third (12%), and greenhouses fourth (9%; see Fig. 5). The countries with the largest utilization were China, Japan, USA, Iceland and Turkey with 63% of the total.

Direct use of geothermal heat has developed rapidly in China during the last decade. In the year 2000, integrated direct use of geothermal

heat was 37 908 TJ. The emphasis is on the replacement of coal-fired projects with geothermal heat. According to Zhang *et al.* (2000), direct use of hot water at the end of 1998 consisted of 7 900 000 m^2 of space heating, 700 000 m^2 of greenhouses, 3 000 000 m^2 of aquaculture ponds, and over 1600 sites of spa. No heat pumps are currently operated in China.

Traditionally, use of geothermal water in Japan for bathing and swimming is important. Lund & Freeston (2001) estimate the geothermal energy utilization for this sector as 22 030 TJ in 2000, whereas Sekioka & Yoshii (2000) give a much lower value. The estimate of Lund & Freestone (2001) is based on the same methodology as they used for other countries and is therefore consistent with data given in Table 2. In the year 2000, the energy consumption for space heating in Japan from geothermal was 2953 TJ (Sekioka & Yoshii 2000). Other important uses include greenhouses (654 TJ/y; 1 TJ/y = 0.278 GWh), fish farming (577 TJ/y), and snow melting (495 TJ/y). Other uses amounted to 225 TJ/y.

Direct use of geothermal energy involving heat pumps is the most important in the USA, accounting for 59.1% of the direct use of geothermal heat

Table 2. *Direct use of geothermal energy in the world by country at the end of 1999*

	Power (MWt)	Energy (TJ/y)
Algeria	100	1586
Argentina	25.7	449
Armenia	1	15
Australia	34.4	351
Austria	255.3	1609
Belgium	3.9	107
Bulgaria	107.2	1637
Canada	377.6	1023
Caribbean Islands	0.1	1
Chile	0.4	7
China	2282	37908
Colombia	13.3	266
Croatia	113.9	555
Czech Republic	12.5	128
Denmark	7.4	75
Egypt	1	15
Finland	80.5	484
France	326	4895
Georgia	250	6307
Germany	397	1568
Greece	57.1	385
Guatemala	4.2	117
Honduras	0.7	17
Hungary	472.7	4086
Iceland	1469	20170
India	80	2517
Indonesia	2.3	43
Israel	63.3	1713
Italy	325.8	3774
Japan	1167	26933
Jordan	153.3	1540
Kenya	1.3	10
Korea	35.8	753
Lithuania	21	599
F.Y.R. of Macedonia	81.2	510
Mexico	164.2	3919
Nepal	1.1	22
Netherlands	10.8	57
New Zealand	307.9	7081
Norway	6	32
Peru	2.4	49
Philippines	1	25
Poland	68.5	275
Portugal	5.5	35
Romania	152.4	2871
Russia	308.2	6144
Slovak Republic	132.3	2118
Slovenia	42	705
Sweden	377	4128
Switzerland	547.3	2386
Thailand	0.7	15
Tunisia	23.1	201
Turkey	820	15756
United Kingdom	2.9	21
USA	3766	20302
Venezuela	0.7	14
Yemen	1	15
Yugoslavia	80	2375
Total	15145	190699

in 1999, or 12 000 TJ (Lund & Boyd 2000). In the period 1990–2000, the use of heat pumps has about doubled. Other important direct uses of geothermal heat in the USA include aquaculture (2800 TJ/y), bathing and swimming (2500 TJ/y), space heating (1480 TJ/y), and greenhouses (1130 TJ/y).

In Iceland space heating accounts for the larger part of the direct use of geothermal heat (77.4%, or 15 600 TJ/y in the year 2000). Approximately 86% of the inhabitants of the country use geothermal heat for space heating (Ragnarsson 2000). Other uses in 2000 were as follows: various industrial uses 1600 TJ/y; bathing and swimming pools 1100 TJ/y; greenhouses and soil heating 790 TJ/y; fish and animal farming 650 TJ/y; and snow melting 410 TJ/y (Ragnarsson 2000).

Most of the recent development in Turkey has occurred in space heating. However, balneological uses are the most important, accounting for 65.5% of the total in the year 2000, or 10 314 TJ. Other uses are: space heating 4327 TJ/y and greenhouses 1115 TJ/y (Batik *et al.* 2000).

Geothermal electric power plants

Typical modern geothermal electric power plants are essentially composed of two major parts: (1) production wells with a steam gathering system, and injection wells with a waste fluid disposal piping system; and (2) a power house with turbines and generators, and cooling towers (Fig. 6). In liquid-dominated fields, steam separators are also required as part of the steam gathering system, but in vapour-dominated fields they are not required. (Note: the function of the steam separator is to separate the water and steam discharged from a wet-steam well. The steam is conveyed to the turbine through the steam gathering system. The water is generally disposed of but sometimes used as a direct source of heat, such as for space heating.) In conventional power plants about two-thirds of the steam is condensed in the cooling towers and about one-third escapes into the atmosphere. Other escape routes for steam and gas are:

(1) Gas ejectors: (an ejector is a vent for removing non-condensable gases from the condensing unit located downstream from the turbine. Condensate from the cooling towers is pumped to the condenser unit to condense the steam leaving the turbine. In this way, low pressure is created downstream from the turbine, which leads to improved efficiency to generate electric

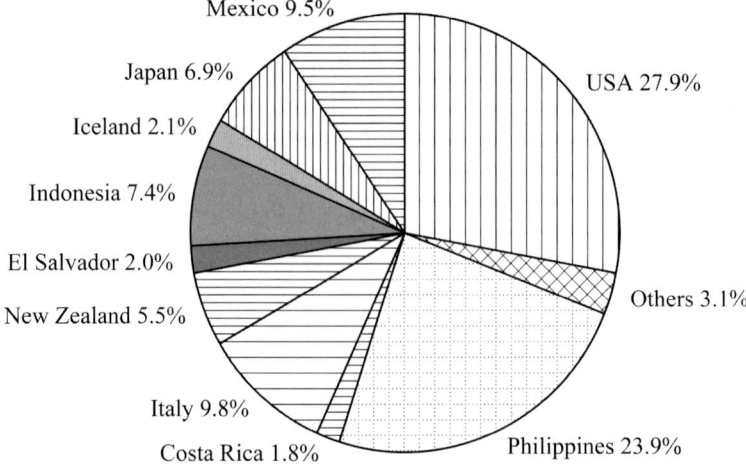

Fig. 2. Percentage installed geothermal electric power by country in the year 2000. The installed power was 7974 MW (see also Table 1). The USA had the highest installed capacity, or 27.9% of 7974 MW, followed by Philippines with 23.9%.

power by the work performed by the steam. To maintain low pressure in the condenser, the non-condensable gases must be removed from it and this happens by their discharge (together with some steam) through the ejectors.)
(2) Drains and pots: drains and pots are typically located at low points on the steam gathering system. Through valves on their lower side, any water that forms in the steam pipes by condensation of steam can be removed.

(3) Excess steam from wells not absorbed by the plant.
(4) Silencers.

Further, steam and gas may escape into the atmosphere from wells undergoing tests, and from blow-out drillholes. Some gas may also escape in solution in the surplus condensate flowing from the cooling towers.

Many older power plants still do not have injection wells and a waste fluid piping system. Various methods have been employed to

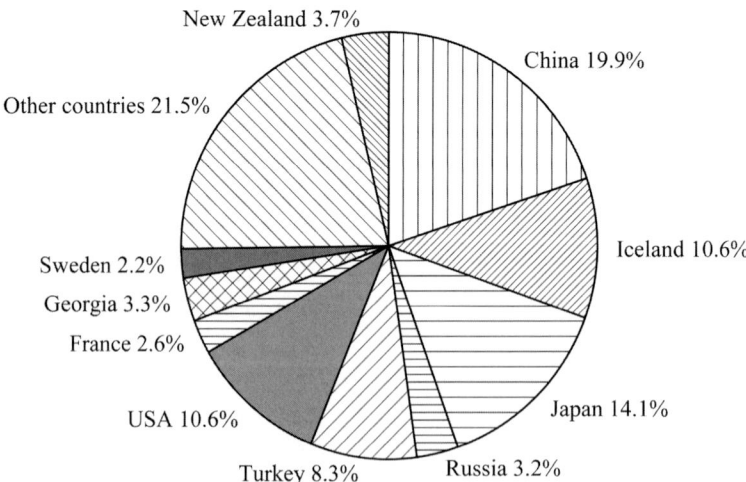

Fig. 3. Percentage of direct use of geothermal energy by country in the year 2000. The total energy usage was 190 699 TJ (see also Table 2). China is the largest user of geothermal heat with 19.9% of the total, followed by Japan, USA, and Iceland.

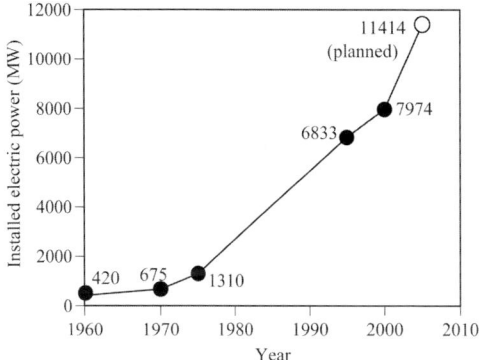

Fig. 4. Installed geothermal electric power by year.

dispose of the waste fluid from such plants. They include infiltration (Krafla, Iceland; Olkaria, Kenya), evaporation (Cerro Prieto, Mexico), and discharge into nearby rivers (Wairakei, New Zealand). The disposal method favoured today is injection into deep wells. It is discussed in some detail in a later section of this chapter (see also Panichi 2004). At Krafla the condensate from the cooling towers and the water from steam separators has been disposed of into a lava field to the south of the power station. The permeability of the lava is sufficiently high for all the waste water to infiltrate. Presently, a part of the waste water at Krafla is injected. At Olkaria in Kenya, all the waste water from the power plant was disposed of by infiltration

during the first 15 years of operation. Presently, however, a part of the waste water is injected into a deep well (Kubo 2003). This injection has helped in maintaining reservoir pressures and has in fact improved the performance of some of the producing wells. The infiltration method at Olkaria involves conveying the separated water from the silencers by each wellhead into a conditioning pond located next to each silencer. The residence time of the water in the conditioning ponds is about two hours. During this time, dissolved silica in excess of amorphous silica solubility polymerizes. From the conditioning pond, the water is conveyed in open ditches to an infiltration pond. By allowing the excess silica to polymerize in the ponds, precipitation of amorphous silica in the ditches from these conditioning ponds and in the infiltration pond is practically inhibited, but if such precipitation had occurred in the infiltration pond, it was feared that the bottom of the pond would gradually become impermeable by amorphous silica deposit. The excess condensate from the cooling towers is conveyed in open ditches directly to the infiltration pond. This disposal method for the waste water from the Olkaria power station has been quite successful. It is inexpensive but not as favourable environmentally as injection. At Cerro Prieto in Mexico, the climate is sufficiently hot and dry for evaporation of the excess condensate and the waste water from the power plant (Mercado 1975). However, the evaporation ponds expand during the cooler season but shrink

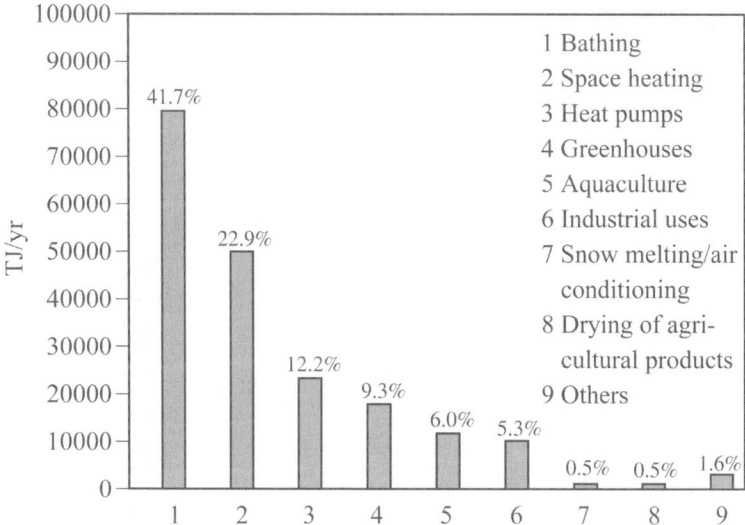

Fig. 5. World direct use of geothermal heat by type in the year 2000. The numbers above each column show % of the energy usage. The total use for this year was 190 699 TJ.

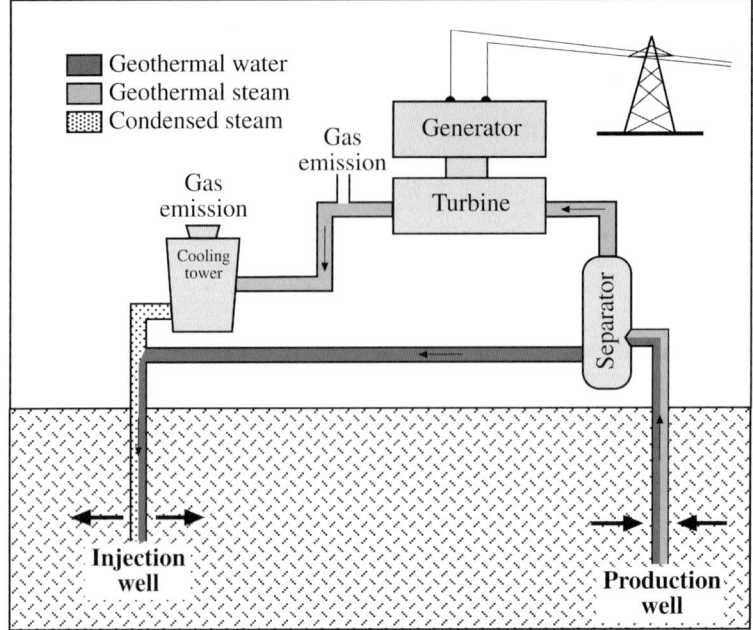

Fig. 6. Simplified schematic layout of a classic geothermal power plant. The main escape routes for steam are from the cooling tower and gas ejectors, which are located just downstream from the turbine.

during the hotter season. When they shrink, fine silica and salt dust that forms on the bottom of the evaporation ponds blows away. This dust has been a nuisance. At Wairakei in New Zealand, the waste water from the wellhead separators has been conveyed in open concrete ditches for a distance of several kilometres to the Waikato river (Ellis & Mahon 1977). An environmental impact study demonstrated that this method of disposal caused both some thermal and chemical pollution of the river water (Axtmann 1975a). Injection tests have been carried out since 1978, and plans exists for injecting 60% of the waste water, as it is considered to be important both for environmental reasons and for maintaining reservoir pressures (Carey 2000).

The Larderello vapour-dominated field was the first geothermal field to be exploited for electric power generation. Production started in 1904 (Zancani 1975). Initially, extraction of B from the steam was coupled with the power production. Development of the field was initially slow but many new turbines were installed in 1939–1940, at which time the installed capacity was 250 MW. By 1964 it had increased 490 MW (Villa 1975). In 1995, the installed capacity had decreased to 460 MW (Allegrini *et al.* 1995),

but risen again to 547 MW by the end of 1999. In that year the energy production from Larderello was 3020 GWh (Cappetti *et al.* 2000), which is very close to the average since 1951 (Minissale 1991). Deep drillings have verified that the Larderello field and the neighbouring Travale–Radicondoli field belong to the same geothermal area. Its lateral extent is 400 km^2 for the 300 °C isotherm at 3000 m below sea level (Barrell *et al.* 1995). The geothermal reservoir is hosted by meta-sedimentary rocks of limestone, marls, quartzite, and phyllite. A deep-seated granitic batholith, which may constitute the heat source to the geothermal system, has been identified under the area (Gianelli & Puxeddu 1992) and there exists seismic evidence for a magmatic body at 7 km depth (Fiordelisi *et al.* 1995). The T of the exploited geothermal reservoir is generally 240–250 °C; however, a maximum of 427 °C has been recorded at 3300 m depth (Ruggieri & Gianelli 1995). D'Amore & Truesdell (1979) interpreted chemical and ^{2}H and ^{18}O isotopic compositional variations at Larderello as resulting from pre-exploitation lateral steam flow from central boiling zones towards the margin of the system. Hydrologic balance studies carried out by Petracco & Squarci (1975) indicate that natural infiltration into the Larderello reservoir is only about one-third of the mass of fluid extracted from the reservoir at that time. Banwell (1975) states that limited additional steam can be extracted from the reservoir by further drillings. In order to dispose of condensate and to compensate for decreasing yield of wells and decreasing reservoir pressure, injection has been exercised at Larderello (e.g., Baldi *et al.* 1980; D'Amore *et al.* 1987; Minissale 1991). Injection has improved the performance of production wells located in the vicinity of injection wells (Panichi, this volume). Evidence from δ^2H and $\delta^{18}O$ indicates mixing between the injected fluid and a 'deep' steam component (D'Amore *et al.* 1987). During early injection tests at Larderello, it was evaluated that 85% of the injected condensate was recovered as superheated steam (Bertrami *et al.* 1985).

The first liquid-dominated geothermal field exploited for electric power generation is Wairakei in New Zealand. The first turbine (6.5 MW) was commissioned in 1958 (Carey 2000). By 1963, the installed capacity had increased to 199.6 MW. Three pressure systems have been operated at Wairakei supplying steam to turbines with different inlet pressures. In 1982, the high-pressure steam supply system was decommissioned due to falling reservoir pressures, decreasing the installed capacity to 157.2 MW (Carey 2000). Electricity generation has remained about constant in the period 1965–1996, 7000–8000 GWh/y. The mass withdrawn from the reservoir annually in the period 1960–2000 is roughly 45 Mt. This amounts to liquid water stored in 2.4 km^3 of rock with 10% porosity. During the early years of production at Wairakei, the so-called Eastern and Western borefields were exploited. With time, production has shifted westwards and production from the Eastern field has been replaced by production from the Te Mihi field, which is in the west of the Wairakei field (Clotworthy 2000). An anomaly of low bedrock resistivity indicates that the lateral extent of the Wairakei field is about 25 km^2 (Clotworthy 2000). Reservoir temperatures are about 240–260 °C. Deep liquid pressures have declined by about 25 bars over the production period (Clotworthy 2000). Initially the decline was sharp, but after 1972 pressure has declined more slowly and remained stable since 1985. The pressure decline has been about uniform within the bedrock resistivity boundary but measurements in deep drillholes outside it show little pressure decline, indicating that the resistivity boundary is a hydrological boundary. The stabilization of reservoir pressure, despite about constant mass production from the

reservoir, is due to recharge of cooler water into the reservoir. The pressure drawdown in the reservoir has enhanced boiling, leading to the formation of a steam cap above the liquid reservoir over much of the field. Shallow wells have been drilled in the Te Mihi area to produce from the steam cap (Clotworthy 2000). Cooler water recharge into the production field is evidenced by decline in the Cl concentrations of well discharges (Glover & Bacon 2000). The weighted average decline from 1580 ppm in 1960 to 1375 ppm in 1997 indicates that 27% of well discharge was from cool inflow water in 1997 (Glover & Bacon 2000). Carbon dioxide concentrations in well discharges have decreased more with time than those of Cl, presumably due to boiling and separation in the reservoir of the steam and water phases (Kissling *et al.* 1996). Numerical simulation studies indicate that the geothermal reservoir can sustain its present production until 2050, assuming injection of 60 000 t/d, which is 60% of the water flow from the wellhead separators (O'Sullivan *et al.* 1998).

At Nesjavellir in Iceland a high-T geothermal reservoir is being exploited for both power generation and space heating. Production of hot water for space heating started in 1990 and electric power generation in 1998 (Gunnarsson *et al.* 1992; Ballzus *et al.* 2000a; Kjartansson 2000). By 2001 installed electric capacity was 90 MW and thermal capacity 200 MW (Ballzus *et al.* 2000b). High-P steam is used to drive the turbines. The separated water, together with the steam coming from the turbines, is passed through heat exchangers to heat up fresh groundwater that is subsequently piped for 27 km to the city of Reykjavík and the neighbouring communities for space heating (Ballzus *et al.* 2000b; Fig. 7). The geothermal water and steam are not suitable for direct use due to high contents of dissolved solids and gases, in particular SiO_2 ($\sim$750 ppm) and H_2S ($\sim$100 ppm). The heat exchangers have a twofold role. They have the function of cooling towers used in conventional geothermal power plants, that is, to condense steam. At the same time, they are used to heat up fresh water for space heating. As a result of the design at Nesjavellir, emission of steam into the atmosphere is limited and, together with injection of the spent fluid, which will be exercised in the near future, surface installations are largely a closed loop through which the geothermal fluid passes. Only the gases in the steam escape into the atmosphere, as the condensed steam, after passing through the turbines and heat exchangers, is exposed to the atmosphere before it is mixed with the geothermal water

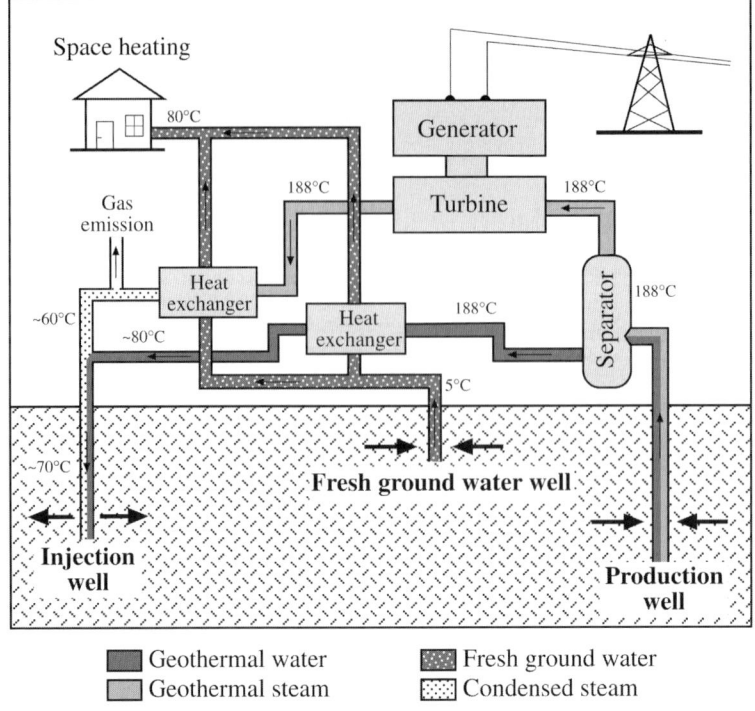

Fig. 7. Simplified schematic layout of the Nesjavellir plant, Iceland. Presently, this plant is capable of generating 90 MW electric and 300 MW thermal power.

from the steam separators. If the gases were disposed of properly, or extracted to be used for some industries, the arrangement at Nesjavellir would offer a solution that is environmentally much more feasible than that adopted in conventional geothermal power plants. The exploitation involves effective extraction of heat from the geothermal water and steam and practically no escape of fluids on the ground and into the atmosphere. The economy of this scheme depends on the availability of surface- or groundwater for the heat exchangers and a market for the heated water.

In the Nesjavellir power plant, groundwater, initially at 5 °C, is heated in the plant to produce 80 °C hot water for space heating. The utilized hot water is disposed of at an average temperature of about 40 °C. The idea has been put forward that improved utilization of the geothermal resource at Nesjavellir could involve recycling the 40 °C waste water to the geothermal plant for re-heating. Improved utilization of the geothermal resource would, accordingly, involve construction of a 'return' piping system from Reykjavík to Nesjavellir, a distance of about 27 km. The heated fresh water presently

being pumped from the Nesjavellir power plant to Reykjavík cools by only 1–2 °C in the 27 km long pipeline.

Space heating and related installations

Direct uses of geothermal heat, both from high- and low-T systems, are very varied, depending on the application and the reservoir temperature. However, constructions needed for direct use of low-enthalpy geothermal resources are generally much simpler than those necessary for the exploitation of high-enthalpy resources for electric power generation. It is beyond the scope of this chapter to describe all types of installations linked to direct uses of geothermal heat. A brief description only will be given of installations for the most important direct uses: space heating, bathing and swimming, and greenhouses.

In some instances, particularly in the case of bathing and swimming, direct use of geothermal heat is based on utilization of natural hot spring discharges. Such usage has minimal environmental impact. However, the far larger part of direct use of low-enthalpy geothermal resources relies on extracting the water from drill holes.

Typically, such extraction induces a pressure drawdown in the reservoir, which is known to cause hot springs to dry out. The piping system used to convey water from production wells to pumping stations or storage tanks is entirely, or for the most part, underground. The piping system that distributes the hot water to individual buildings is with few exceptions underground as well. Wellheads may also be underground.

Spent low-enthalpy water that is exploited for bathing and swimming, space heating, and greenhouses is generally disposed of into the sewer system of the respective community, or it may be recycled. Reykjavík Energy, the company supplying hot water for space heating to the city of Reykjavík, Iceland, and the neighbouring communities (170 000 inhabitants) utilizes four low-enthalpy geothermal reservoirs with temperatures ranging from 75 to 130 °C, in addition to the Nesjavellir high-enthalpy field discussed in the previous section. Some of the spent water from the low-enthalpy fields, which is at about 40 °C, is mixed with 130 °C water pumped from drill holes in one of the utilized low-enthalpy fields to produce water at 80 °C which is of the appropriate T for space heating. The water exploited by Reykjavík Energy is of low salt content, (150–300 mg/L dissolved solids), alkaline and not corrosive. Scaling problems have not been encountered. The water is, therefore, suitable for direct use.

In the Dogger geothermal aquifer of the Paris Basin, hot (47–85 °C), saline, H_2S-bearing brine has been exploited for space heating since 1970 (Rojas et al. 1987). The installed capacity of the district heating plants in the Paris Basin was 295 MW thermal in the year 2000 (Lund & Freeston 2001). Problems caused by corrosion and sulphide scaling have been encountered (Fouillac et al. 1990). Owing to the unfavourable composition of the geothermal fluids, water from one production well is passed through heat exchangers at the surface and then disposed of into an injection well. The closed loop of a production well, a heat exchanger, and an injection well has been referred to as the doublet concept (LaPlaige et al. 2000).

Geo-hydrological features of geothermal systems

High-T systems

High-T geothermal systems are of two types, vapour-dominated systems and liquid-dominated systems (White et al. 1971). Liquid-dominated systems are much more abundant than the vapour-dominated ones. In vapour-dominated systems, cracks, fractures, and faults in the reservoir are filled with vapour, but micropores are filled with liquid. Temperature and pressure (P) vary insignificantly with depth in these systems. The temperature in all drilled vapour-dominated systems is around 240 °C, which corresponds to the maximum enthalpy of steam. In liquid-dominated systems, liquid water forms a continuous phase in channelways, so P increases with depth as dictated by the density of liquid water. Such systems may be boiling, that is, containing two phases (water and steam), but others are sub-boiling. Measured temperatures in deep wells (>2000 m) in liquid-dominated systems are frequently over 300 °C and sometimes as high as 350 °C.

Productive wells in vapour-dominated systems discharge steam only. Characteristically, wells in liquid-dominated systems discharge a mixture of water and steam. The steam forms essentially by depressurization boiling of the reservoir water. It is, however, not uncommon that some wells drilled into liquid-dominated systems discharge dry steam or possess a higher steam to water ratio than can be accounted for by depressurization boiling. The reason for this is partial or complete immobilization of the water in the aquifer due to the effects of capillary pressure and relative permeability.

The mass flow rate of water through a body of rock is proportional to the absolute permeability of this rock, the pressure gradient, and the water density, but inversely proportional to its viscosity. (Note, absolute permeability is the cross-sectional area of connected pores in a rock; its SI unit is m^2, but the unit darcy is most often used, where 1 darcy $= 10^{-12}$ m^2.) When two phases are flowing through rocks, such as steam and water in many geothermal systems, the relative permeability of each phase represents the volume fraction of the absolute permeability occupied by it. The mass flow rate of each phase is determined by this relative permeability and the pressure gradient, which generally is different for the two phases, as well as their intrinsic properties, density and viscosity. The pressure of liquid water is generally lower than that of the steam. The difference is due to capillarity, sometimes called capillary pressure. The cause of this pressure is adhesive forces between mineral surfaces and the pore fluid and they are stronger for water than for steam. In other words, mineral surfaces are preferentially wetted by water. The effect of capillary pressure becomes stronger in formations with small pores and fractures, that is, when permeability is low. Typical relative permeability functions are depicted in Fig. 8. They show the relationship between liquid saturation (fraction of water by

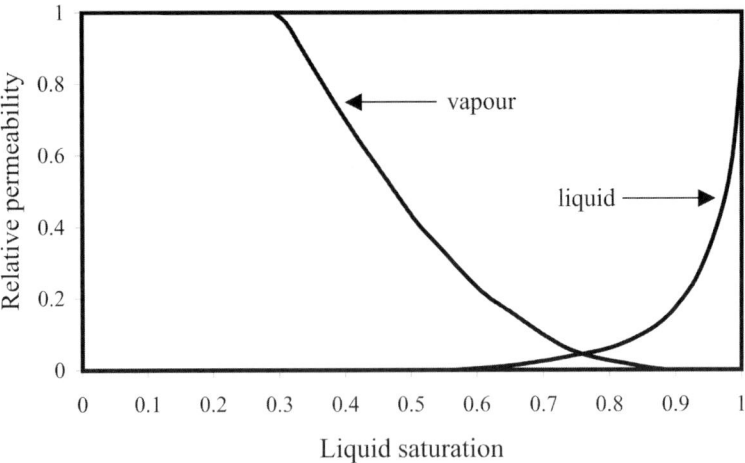

Fig. 8. Typical relative permeability functions (from Pruess 2002).

volume) and relative permeability. From this figure one can read that water is immobile when water saturation is below about 0.6.

Because a considerable quantity of water is always brought to the surface through production wells in liquid-dominated systems, which needs to be disposed of, their exploitation is not as friendly environmentally as exploitation of vapour-dominated systems.

An important property of high-T systems is the reservoir formation porosity and the mass fractions of water and steam occupying the pores. These mass fractions and the formation porosity determine how large a fraction of the heat of the system is stored in the fluid and how large a fraction in the rock. Usually the quantity of heat stored in the reservoir rock is considerably larger than that stored in the fluid. This is particularly the case for vapour-dominated systems. For a liquid-dominated system at 250 °C with 10% porosity and no steam, the quantity of heat stored in the fluid in 1 m^3 of reservoir rock is about 87 MJ, but that stored in the rock is some 500 MJ. The corresponding number for the heat stored in steam in a vapour-dominated system is only 5.6 MJ.

To exploit the heat stored in the rock, it is important to have cooler water recharging the reservoir. The recharging water gains heat by flowing through the hot reservoir rock towards producing wells and it extracts heat from this rock in the process. The recharge of the water may be natural. Alternatively, it may be injected waste fluid.

From the above discussion, it is seen that the long-term performance of a geothermal reservoir depends on reservoir rock porosity, recharge, and

distribution of permeable channelways within the reservoir rock that affect how easily the heat can be extracted from this rock by conductive transfer into the cooler recharging water. The low quantity of heat stored in unit volume of steam means that vapour-dominated reservoirs will not have a long life-time without recharge, if the volume of micropores in the reservoir rock, which is occupied by liquid water, is small. This appears to have been the case for, at least, parts of The Geysers field in California, which was manifested by rapid pressure decline in the reservoir and subsequent improvement after injection was started (Dellinger 1997). In the vapour-dominated field of Larderallo in Italy, reservoir performance was very significantly improved by injection (Panichi 2004).

During steam production from vapour-dominated systems, heat stored in the reservoir rock is always exploited to some extent, even if there is no recharge into the reservoir. This occurs through evaporation of water held in micropores by capillary forces. In the same way, heat will also be extracted from the reservoir rock of liquid-dominated systems, if the relative permeability effect causes some phase segregation in producing aquifers. In both cases the steam formed by boiling of micropore water will flow into producing wells. When this happens, the latent heat of vaporization is provided by the reservoir rock. For water at 250 °C the latent heat of vaporization is higher than that of liquid water by a factor of about 1.65.

Natural recharge of cooler ground water to high-T geothermal systems may cause changes in the composition of the fluid discharged from producing wells (Panichi 2004). Typically, the Cl concentration in well discharges decreases (Glover & Bacon 2000; Gudmundsson & Arnórsson 2002) (Fig. 9) because cold groundwater is characteristically lower in Cl than geothermal waters. The gas content may also decrease. These changes tend to reduce the chemical environmental impact of the

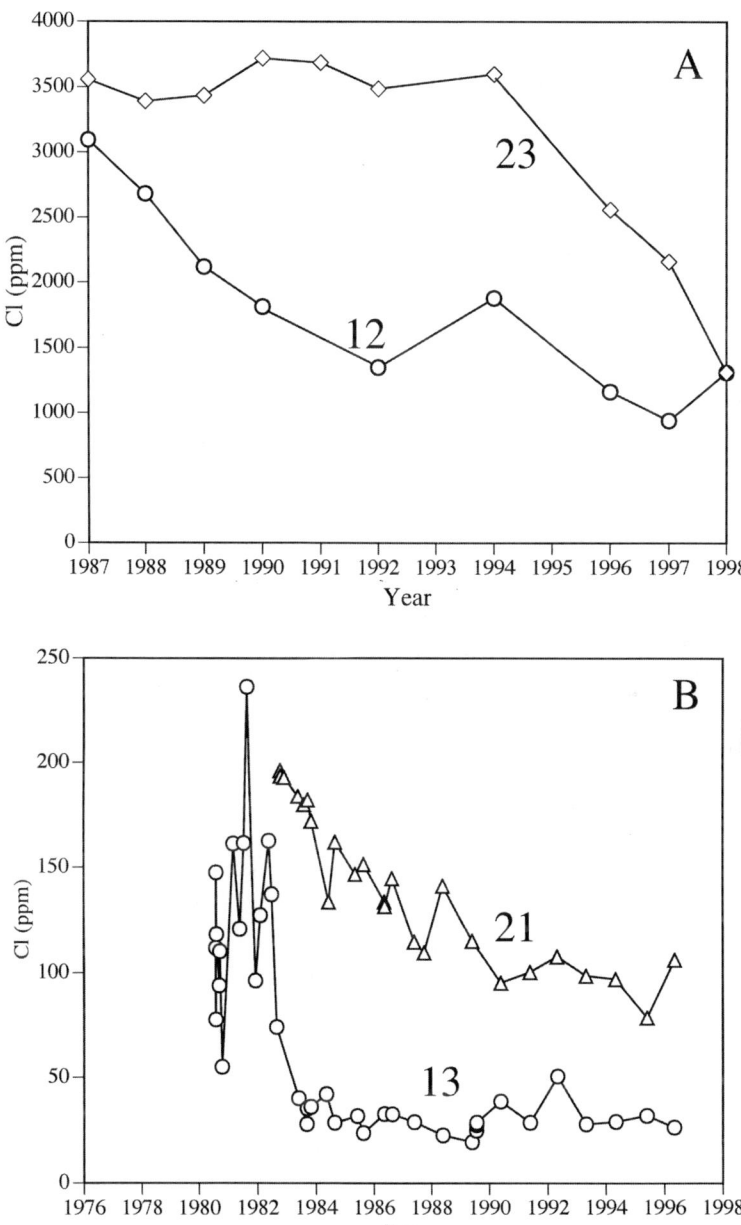

Fig. 9. Decline in Cl concentrations with time in water at 10 bars absolute separation pressure discharged from wells at (**a**) Momtombo, Nicaragua, and (**b**) Krafla, Iceland. The cause of the decline is cooler water recharge into producing aquifers. Numbers indicate well numbers. From Arnórsson (1996) and Gudmundsson & Arnórsson (2002).

exploitation of high-T geothermal resources. Injection of waste water from liquid-dominated reservoirs will cause the salt content of the aquifer water to increase through successive evaporation by repeated cycling through the reservoir of the spent water. The increase in the salt content of the water may in the long run lead to increased scale formation and in this way cause operational problems.

Low-T systems

Low-T (non-volcanic), tectonic systems are typically found in areas of active seismicity where temperature gradients are high. Most of the low-T systems in Iceland, which are exploited for space heating, are of this type. Formation of faults creates secondary permeability, in which groundwater convects to form a geothermal system. Hydrostatic pressure at depth in a fracture, where the geothermal water rises, may be lower than in the formation outside the fracture due to the difference in the density of the hotter and cooler water columns in the fracture and outside it, respectively. This pressure difference will lead to flow of water into the fracture from the enclosing rock. Thus, the geothermal water consists essentially of two components; a relatively old groundwater that has been residing in the bedrock and relatively young water circulating in the tectonic fracture. The drawdown caused by production from tectonic systems enhances convection in the permeable fractures. Experience in some fields such as Ellidaár in Reykjavík, Iceland, has revealed that cooling of the water in such systems is excessive, when permeable formations overlie the fractured and globally less permeable reservoir rock (Arnórsson *et al.* 1992). The drawdown in the fractures causes rapid recharge from the overlying groundwater in the overlying permeable formations. In the case that the impermeable rock reaches the surface, injection is beneficial to sustain pressure in the fractures (Axelsson *et al.* 2000).

Geopressurized systems, which occur in sedimentary basins, require two coincident factors for their formation: (1) sediments of high porosity and permeability, and (2) elevated regional heat flow (Muffler 1975). The best-known examples of geopressurized systems are around the Gulf of Mexico (Poli 1975) and in Hungary (Boldizsár 1970). In the Gulf of Mexico area, temperatures as high as 200 °C and pressures of 1000 bars have been encountered in drill holes. The water is very saline and often associated with methane. Poli (1975) considered that the resource could be developed for power generation and methane extracted as a byproduct. In Hungary, Boldizsár (1970) estimated the lateral extent of the geopressurized reservoir to be 4000 km^2, containing water at 60–200 °C. The geothermal reservoir water in Hungary is not so saline but relatively rich in CO_2. It is extensively exploited.

Hot-dry-rock systems, which are represented by heat stored in low-porosity, impermeable rocks at varying depths and of varying temperatures, have not yet been economically exploited. However, experimental systems have been developed in the USA (e.g., Robertson-Tait *et al.* 2000), France (e.g., Baria *et al.* 2000; Gender *et al.* 2000; Baldeyrou-Bailly *et al.* 2004), England (e.g., Richards *et al.* 1992, 1994; Bruel 1995), and Japan (e.g., Tenma *et al.* 2000; Hayashi & Minamiura 2000). To extract heat from these systems, it is necessary to create artificial permeability by hydraulic fracturing, and to use some wells to inject cold water and others to produce hot water. The hot water is generated by flow along fractures through the hot rock from injection to production wells. One of the problems encountered in exploiting HDR systems is that the temperature of the produced hot water falls relatively rapidly, such as at the Rosemanowes site in Cornwall, England (Nichol & Robinson 1990). A detailed account of HDR systems is also given by Baldeyrou-Bailly *et al.* (2004).

Chemical composition of geothermal fluids

The chemical composition of geothermal fluids is widely variable (Table 3). The dissolved solids content of geothermal water ranges from a few hundred mg/L (ppm) to as much as 30% by weight. The total gas content of geothermal steam can be as high as 20% by volume, although values between 0.2 and 1 vol% are more typical (Table 4). The concentrations of dissolved solids and gases in geothermal fluids are determined by their sources of supply to the fluid and by the formation of hydrothermal minerals, which remove dissolved components from the fluid. The sources of dissolved solids and gases include (1) the rock through which the fluid percolates, (2) magma, and (3) dissolved matter in the recharging water.

The dissolved solids and gases in geothermal fluids have been divided into two types, reactive and conservative components (Giggenbach 1991). Temperature-dependent equilibria between solution and hydrothermal minerals fix the aqueous concentrations of the *reactive* components, at least if temperatures exceed some 100 °C (e.g., Giggenbach 1980, 1981; Arnórsson *et al.* 1983).

Table 3. *Major component composition of separated well waters from wet-steam wells and one sample from a hot water well**

Location[†]	Rock type	Aquifer temperature (°C)	Sampling pressure (bars)	Discharge enthalpy (kJ/kg)	pH/°C[§]	SiO₂	B	Na	K	Ca	Mg	CO₂	SO₄	H₂S	Cl	F	Total dissolved solids	Source[¶]
Cerro Prieto 5, Mexico	Sediments	289	0.0	1790	8.99/19	790	14.5	6950	1670	395	0.45	31.0	5.0	5.0	12900	1.90	22748	1
Krafla 15, Iceland	Basalt	265	11.7		7.46/20	797	2.0	206	38.3	1.60	0.0049	43.6	198.9	82.8	24.5	1.58	1300	2
Krafla 20, Iceland	Basalt	280	9.8		7.70/20	815	1.7	214	43.7	0.94	0.006	207.8	38.8	53.4	114.3	1.29	1371	3
Ngawha 9, New Zealand	Sediments	228	0.0	980		559	1250	1193	107	3.50	0.12	167.0	25.0	1.0	1700.0	1.20	4953	4
Námafjall 12, Iceland	Basalt	250	28.4	1712	7.63/26	447	4.9	111.8	16.8	0.40	0.0024	13.2	14.4	110.0	41.5	0.47	646	2
Nisyros 2, Greece	?	279	18.0	1607	4.47/19	718	55	23754	3224	8594	52.5	637	24.6	1.2	56007	1.76	92865	5
Ohaaki 28, New Zealand	Andesite	274	0.0	1315	8.12/20	784	52	910	155	1.3	0.07	191	30.0	—	1414	6.1	3483	6
Olkaria 25, Kenya	Basalt-trachyte	260	0.0	2516	9.15/25	641	5.5	522	94	1.1	0.11	150	28.0	2.12	671	70.0	2135	7
Reykjanes 8, Iceland	Basalt	248	19.0	1150	6.38/20	631	8.7	11150	1720	1705	1.44	63.1	29.3	2.2	22835	0.21	38124	8
Salton Sea 1, California	Sediments[||]	300	0.0		5.2/20	400	390	50400	17500	28000	233	7100	5.4	16	155000	15.0	256784	9
Tongonan 202, Philippines	Andesite	312	0.0	1395	7.08/25	1034	235	6750	1710	211	0.08	67.1	19.0	10.0	12390	—	22395	10
Wairakei 24, New Zealand	Rhyolite	248	0.0	995	7.70/15	557	26.2	1256	200	26.7	0.02	64.1	34.2	—	2183	6.90	4334	4
Zunil D-1, Guatemala	Granodiorite	296	9.0	1380	8.13/25	896	36.2	866	230	3.73	0.22	29	25.0	2.9	1506	4.63	35880	11
Geysir, Iceland	Basalt + rhyolite	87			9.41/25	534	1.05	238	22.6	0.77	0.0038	143.3	96.3	2.98	124.3	8.85	1123	12
Landmannalaugar, Iceland	Rhyolite	96			9.08/26	190	7.69	382	13.6	14.4	0.0087	9.5	16.3	5.59	602.6	10.6	1243	12
Laugaból, S-Iceland	Basalt	98			9.56/23	279	0.58	120	5.74	1.40	0.0014	38.4	74.0	4.56	40.4	1.98	549	12
Varmahlíð, 1 Iceland[‡]	Basalt	89			9.47/24	123	0.47	75.1	1.90	1.72	0.0020	31.7	50.7	2.8	27.4	2.20	304	13
Yellowstone, Wyoming, USA	Rhyolite	93			8.61/120	243	3.98	331	9.45	1.00	0.001		20.7	1.09	324	31.60	965	14

*Concentrations are in ppm.

[†]Number indicates well number.

[‡]Hot water well.

[§]Temperature at which pH is measured.

[||]Includes evaporites.

[¶]1: Truesdell *et al.* (1981); 2: Nehring & D'Amore (1984); 3: Gudmundsson & Arnórsson (2002); 4: Arnórsson *et al.* (1999); 5: Bjarnason & Arnórsson (1986); 6: Hedenquist (1990); 7: Karingithi (2002); 8: Arnórsson *et al.* (1983); 9: Muffler & White (1968); 10: Arnórsson & Gunnlaugsson (1985); 11: Arnórsson (1995); 12: Unpublished data of the Science Institute, University of Iceland; 13: Arnórsson *et al.* (2002); 14: Ball *et al.* (1998).

Table 4. *Gas concentrations in steam from selected geothermal wells*[†]

Location	Sampling pressure (bars)	CO_2	H_2S	H_2	CH_4	N_2	O_2	Ar	Vol% of gas in steam	Source[*]
Cerro Prieto, Mexico	0.0	195.2	18.9	11.5	12.8	1.49	0	0.035	0.432	1
Krafla 15, Iceland	11.7	250.0	37.5	21.2	0.044	1.88	0.034	0.034	0.560	2
Nagwha 9, New Zealand	0.0	160.3	1.87	0.5	4.84	0.59	0.003	0.004	0.303	4
Námafjall 12, Iceland	28.4	142.0	51.4	89.5	0.782	2.47	0.05	0.050	0.516	2
Nisyros 2, Greece	18.0	306.6	5.27	0.22	1.48	0.28	—	0.030	0.565	5
Ohaaki 28, New Zealand	0.0	744.4	11.56	1.36	10.92	6.45	0		1.396	6
Olkaria 25, Kenya	0.0	78.3	5.61	4.33	0.17	2.44	0.006		0.164	7
Reykjanes 8, Iceland	10.0	103.7	3.13	0.22	0.11	0.65	0		0.194	8
Tongonan 202, Philippines	0.0	96.9	3.61						0.181	10
Wairakei 24, New Zealand	0.0	31.1	0.47	0.07	0.03	0.22	0		0.057	4
Zuni D-1, Guatemala	0.0	157.9	5.78	0.25	0.12	1.51	0		0.298	11

[*]See footnotes in Table 3.
[†]Concentrations in mmol/kg steam.

By contrast, the fluid concentrations of *conservative* components are determined by their supply to the fluid. Once in solution they remain there. Waters that have percolated through sediments containing evaporites contain the highest concentrations of dissolved solids, as, for example, the Salton Sea brines in southern California (White *et al.* 1963), whereas waters in basalt tend to be lowest (see Table 3).

Relatively limited primary mineral dissolution is sometimes needed to saturate geothermal waters with many common hydrothermal minerals, which remove reactive components from solution as they precipitate. The concentrations of these components may not be higher in geothermal waters than in surface- and groundwaters, at least if supply of Cl^- to the geothermal water is limited. By contrast, conservative component concentrations are typically much higher in geothermal fluids than in surface- and groundwaters, because the former have reacted more extensively with the enclosing rock due to their higher T and longer residence time underground.

Many chemical components in geothermal fluids have high mobility; that is, a relatively large fraction of the component that enters the fluid phase remains there. The fluid concentrations of these components are largely controlled by their supply to the fluid. Components of this type include As, B, CO_2, Hg, and sulphide-S. It is thus not surprising to have high concentrations of these components in geothermal fluids, which can extract them from either the enclosing rock or acquire them from a degassing magma heat source.

Chloride (Cl^-) is the major anion of most geothermal waters and the principal conservative constituent. This element only forms soluble salts with cations, which are abundant in natural waters. Increase in aqueous Cl^- concentrations is accompanied by an equal increase (in terms of charge) in the cation concentrations. However, at equilibrium with a specific hydrothermal mineral assemblage at a given T and P, the relative activities of the cations are constant. This is exemplified for Na^+/H^+ and $Ca^{+2}/(H^+)^2$ activity ratios in Fig. 10. The consequence of the equilibrium between cations and hydrothermal minerals is thus that any cation activity ratio is unaffected by the aqueous Cl^- concentration although individual cation activities are. In line with this observation, it is not unexpected that many cation-forming metals, some of which are of environmental concern, occur in high concentrations in geothermal brines, but in low concentrations in geothermal waters of low salt content (Table 5).

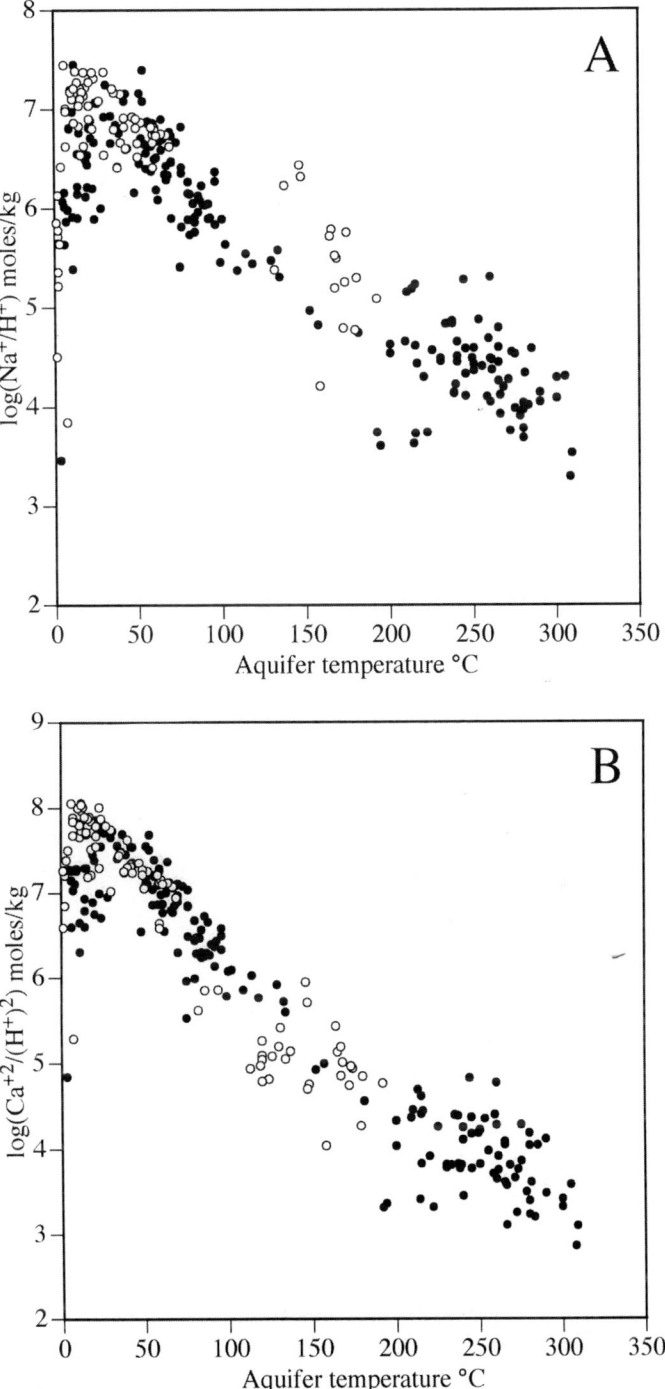

Fig. 10. Na$^+$/H$^+$ (**a**) and Ca^{+2}/(H$^+$)^2 (**b**) activity ratios in selected springs (circles) and geothermal aquifer waters (dots). Individual ion activity ratios are in moles/kg. The data for springs are from Yellowstone, Wyoming, USA, and Iceland (many different fields). The drillhole data are from Guatemala (Zunil), Iceland (many different fields), Japan (Onuma, Fusime), New Zealand (Ngawha, Wairakei), and Nicaragua (Momotombo).

Table 5. *Trace element content (in ppb) of selected geothermal waters (1—6)**

Atomic number	Element	1	2	3	4	5	6
3	Li		215000	467	55.7	50	3960
4	Be			0.248	0.664		<0.001
5	B	1730	390000	1001	8274	576	3980
13	Al	1169	41000	717	125	315	0.28
15	P	1.20		2.78	811	0.58	—
21	Sc			0.0049	0.0062		—
22	Ti	0.13		0.92	0.151	0.03	—
23	V			7.1	0.0396	0.883	<0.02
24	Cr	0.474		0.0433	0.01	0.103	<0.09
25	Mn	0.94	1560000	1.21	7.3	0.063	<0.12
26	Fe	27.0	2090000	7.85	7.32	6.4	0.0232
27	Co	<0.005		0.0013	0.0026	<0.005	<0.04
28	Ni	0.35		0.387	0.044	<0.02	<0.04
29	Cu	<0.05	9000	0.204	0.17	<0.020	<0.14
30	Zn	0.27	850000	0.72	0.584	0.06	<0.01
31	Ga			9.62	6.84	4.96	—
32	Ge			24.4	25.2	14.1	—
33	As	48	13000	86.2	237	20.2	0.86
34	Se			1.79	12.4	0.015	—
35	Br		120000			136	
37	Rb		147000	180	83.9	25.4	—
38	Sr	4.42	413000	1000	630	3	8.0
39	Y			0.0049	0.0396		—
40	Zr			0.0427	0.0656		—
41	Nb			0.0058	0.0038		—
42	Mo	1.04		41	0.147	25	—
44	Ru			<0.001	<0.001		—
45	Rh			<0.002	<0.015		—
46	Pd			<0.007	<0.003		—
47	Ag	—	1000	0.0307	0.0016	<0.002	—
48	Cd	<0.003		<0.005	<0.001	0.0128	<0.04
50	Sn			0.0221	0.0128	<0.020	—
51	Sb			4.3	41	0.4	—
52	Te			0.0221	0.0034		—
53	I		19000	8.69	49.3	42.9	—
55	Cs		17000	12.2	16.4	0.37	—
56	Ba	0.73	235000	0.106	4.73	0.12	<40
57	La			0.0109	0.0335		—
58	Ce			0.0092	0.0691		—
59	Pr			0.0027	0.01		—
60	Nd			0.0055	0.0382		—
62	Sm			0.0043	0.0083		—
63	Eu			0.0002	0.0024		—
64	Gd			0.0004	0.0081		—
65	Tb			0.0001	0.0014		—
66	Dy			0.0009	0.0073		—
67	Ho			0.0004	0.0014		—
68	Er			0.0004	0.0036		—
69	Tm			<0.00005	0.00043		—
70	Yb			0.0004	0.0394		—
71	Lu			<0.0001	0.00044		—
72	Hf			0.0011	0.0016		—
73	Ta			0.0003	0.0003		—
74	W			20.4	58.4	7.95	—
75	Re			0.0053	0.0002		—
77	Ir			<0.00005	<0.00005		—
78	Pt			<0.0005	<0.0005		—
79	Au	—		0.0185	0.0029	<0.0005	—

(continued)

Table 5. *Continued*

Atomic number	Element	1	2	3	4	5	6
80	Hg	<0.001				0.004	
81	Tl			0.109	0.0041	0.022	—
82	Pb	0.061	91000	0.050	0.012	<0.010	<0.15
83	Bi			<0.0005	<0.0005		—
90	Th			0.0013	0.0017	<0.002	—
92	U			0.001	0.0008	<0.0005	—

*1: Krafla, well 20, Iceland; Arnórsson *et al.* (1992). 2: Salton Sea well 1; Muffler & White (1968). 3: Geysir, Iceland; unpublished data of the Science Institute, University of Iceland. 4: Landmannalaugar, Iceland; unpublished data of the Science Institute, University of Iceland. 5: Laugaból, South-Iceland; unpublished data of the Science Institute, University of Iceland. 6: Yellowstone, Wyoming, USA; Bell *et al.* (1998).

Chemical pollution

The most adverse environmental effect of geothermal energy utilization is chemical pollution, from gaseous components in steam that are discharged into the atmosphere and from aqueous components in spent water that may mix with surface- and groundwaters. This problem has been reduced by injecting into drill-holes both the separated water and the steam condensate. Injection is specifically discussed in a separate section below.

Airborne pollutants

Various poisonous chemical components in geothermal steam escape into the atmosphere from electric power plants via ejector exhausts, cooling towers, silencers, and drains and traps. These compounds include H_2S, B, Hg, As, and Rn. Other compounds of environmental concern in geothermal steam, although not poisonous, include CO_2 and CH_4. Apparently, not much attention has been paid to airborne poisons in geothermal steam other than H_2S. This noxious gas has an unpleasant smell when present in low and harmless concentrations. When more strongly concentrated, H_2S can paralyse the olfactory nerves and thus becomes odourless, and eventually lethal.

Carbon dioxide. Available data indicate that the concentrations of CO_2 in >100 °C geothermal reservoir waters are generally controlled by equilibrium with various mineral buffers (Arnórsson & Gunnlaugsson 1985), at least when not affected by long-term exploitation. Aqueous CO_2 concentrations increase with increasing T (Fig. 11a): at 200 °C, they are around 3.5 mmoles CO_2/kg of steam, whereas the corresponding number at 300 °C is around 35 mmoles/kg. In the range of about 230–300 °C, the mineral buffer in systems hosted by basaltic to silicic volcanics is clinozoisite + prehnite + quartz + calcite (Arnórsson

& Gunnlaugsson 1985). The concentration of CO_2 in steam discharged from wet-steam wells is higher than that of the parent fluid, frequently between 50 and 300 mmoles/kg but values as high as 1000 mmoles/kg are not uncommon (Table 4). The concentrations of CO_2 in steam from wet-steam wells depend on (1) the CO_2 concentration in the parent geothermal water, (2) the steam fraction, which has formed by depressurization boiling, (3) the reservoir steam fraction, if present, and (4) the boiling processes, which lead to the steam formation (see, e.g., Arnórsson *et al.* 1990). Observations by the author indicate that the CO_2 concentrations in the steam initially produced from the vapour-dominated fields of the Geysers (California) and Larderello (Italy) correspond to those in equilibrium with liquid water, which in turn is in equilibrium with the same mineral buffer, which controls CO_2 concentrations in reservoir waters of liquid-dominated geothermal systems (see above).

The concentration of CO_2 in steam of well discharges is known to change with exploitation (Axtmann 1975a; Glover & Bacon 2000). It may decline as a consequence of recharge of cooler water into producing aquifers coupled with insufficient supply of CO_2 from other sources (magma, the reservoir rock) for maintaining equilibrium with a specific mineral buffer. Boiling in producing aquifers, enhanced by exploitation-induced pressure drawdown, and separation of water and steam during lateral flow in the aquifer, with only the water flowing into producing wells, may also lead to a decrease in the CO_2 of well discharges (Kissling *et al.* 1996). Such a decrease is accompanied by an increase in the quantity of CO_2 emitted from fumaroles and/or the formation of a steam cap over the liquid-dominated reservoir, as is, for example, the case at Svartsengi in Iceland and Wairakei in New Zealand (Ármannsson 2003; Clotworthy 2000). Emission of CO_2 from a geothermal reservoir can be very much enhanced by exploitation

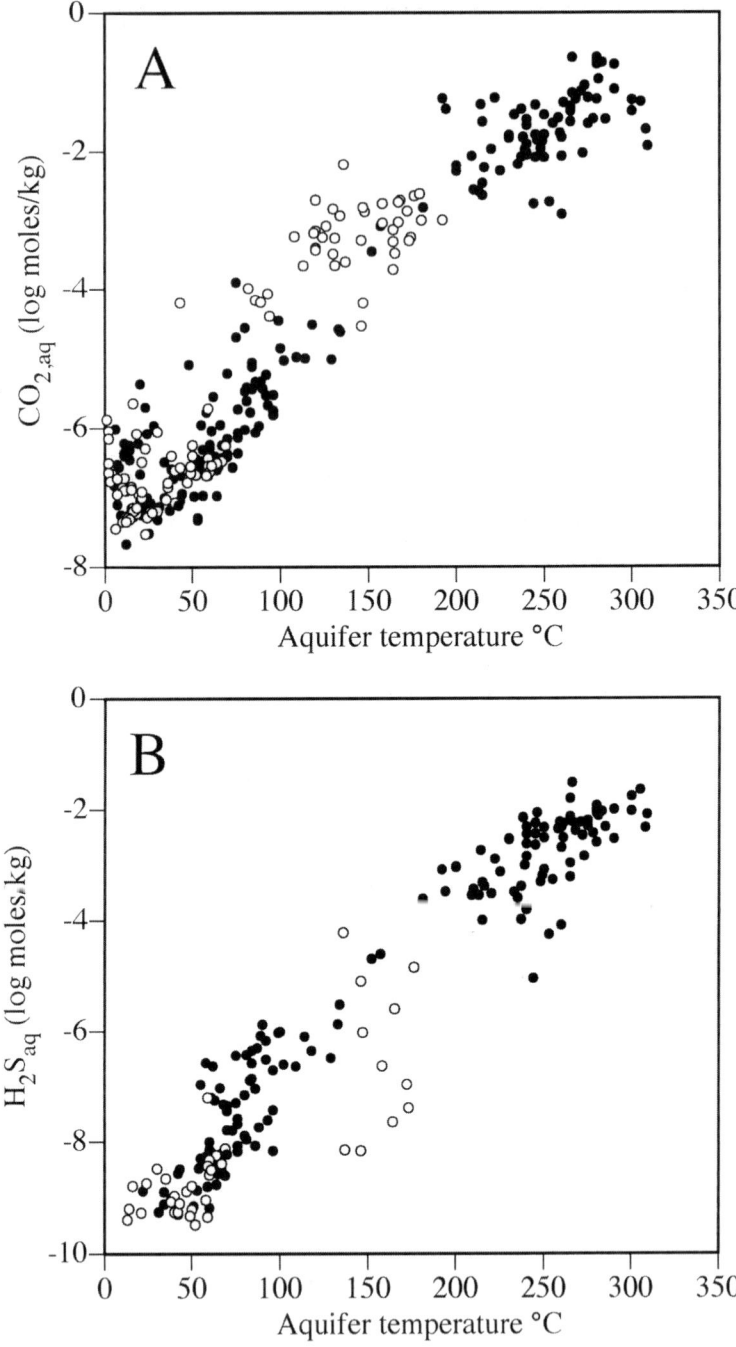

Fig. 11. Carbon dioxide (**a**) and hydrogen sulphide (**b**) concentrations in selected springs (circles) and geothermal aquifer waters (dots). The data on springs are from Yellowstone, Wyoming, USA, and Iceland (many fields). The drill hole data are from Guatemala (Zunil), Iceland (many fields), Japan (Onuma, Fusime), New Zealand (Ngawha, Wairakei), and Nicaragua (Momotombo).

relative to the initial natural emission, at least during the early years of exploitation, when steam caps develop in response to production. The enhanced CO_2 emission may occur through fumaroles, but also through wells, which produce from the steam cap.

Figure 12 shows the quantity of CO_2 emitted into the atmosphere from two geothermal power plants in Iceland and how it compares with CO_2 emission from fossil fuel plants. At Krafla, the quantity of CO_2 emitted per MW-year (MWy) is 840 t, which is only about 10% of that from a typical coal-fired power plant (8760 t/MWy, Ármannsson & Kristmannsdóttir 1992). The

comparison is even more favourable in the case of the Nesjavellir power plant, which emits only about 62 t/MWy (Gíslason 2000).

Methane. The concentrations of CH_4 in geo-thermal fluids are quite variable (see Table 4). In some geothermal systems, the CH_4 aquifer fluid concentration is apparently controlled by attainment of equilibrium for the reaction:

$$CH_4 + 2\,H_2O = CO_2 + 4\,H_2 \qquad (1)$$

In other systems, equilibrium by this reaction does not prevail, as this redox reaction is

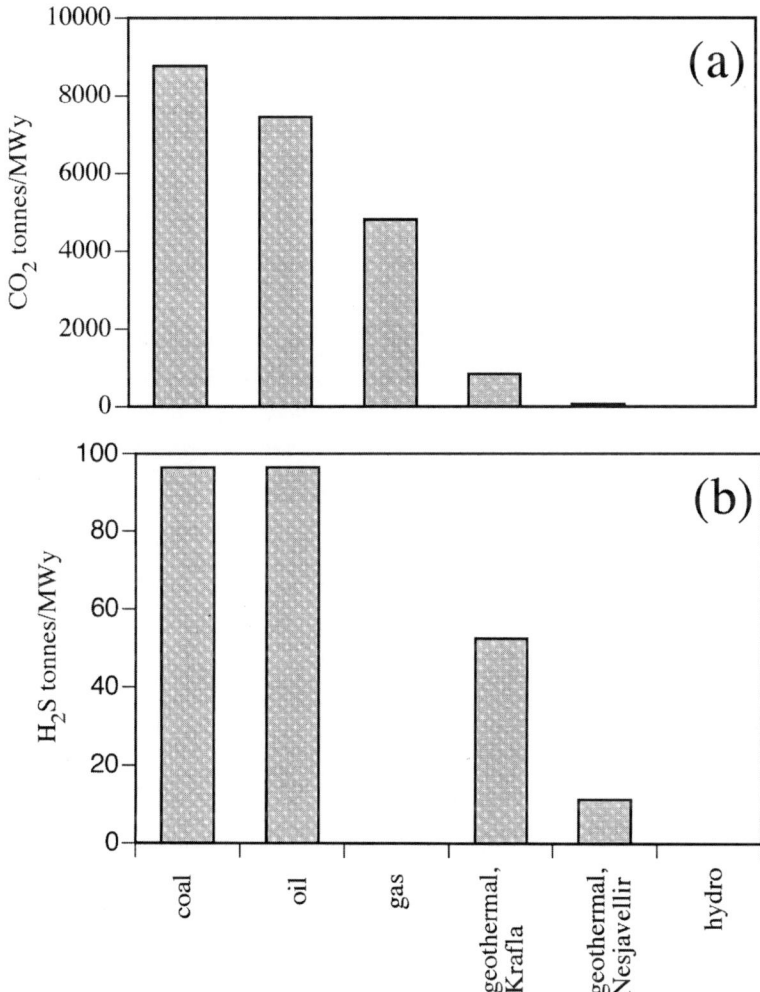

Fig. 12. Carbon dioxide (**a**) and hydrogen sulphide (**b**) emissions from some power plant types. Based on Ármannsson & Kristmannsdóttir (1992) and Gíslason (2000).

extremely slow, particularly at low temperatures (e.g., Giggenbach 1982). When equilibrium is not attained, the aquifer fluid concentrations of CH_4 are probably governed by the supply of CH_4 to the fluid. Methane concentrations in geothermal fluids tend to be highest in systems hosted by marine sediments, such as at Ngwaha (New Zealand) and Cerro Prieto (Mexico) (Table 4). The concentrations of CH_4 in geothermal steam are affected by the same parameters as those listed above for CO_2. Methane is, on a molal basis, 21 times more effective in absorbing infrared radiation than CO_2 (Andrews et al. 1996). Its atmospheric emission from geothermal power plants may therefore be of as much concern as emission of CO_2 with respect to the greenhouse effect.

Hydrogen sulphide. Like CO_2 concentrations, the concentrations of H_2S in aquifer waters of undisturbed geothermal systems are considered to be controlled by specific mineral buffers (Arnórsson & Gunnlaugsson 1985). In high-T waters (230–300 °C) of low salt content, which have a relatively high pH and are highly reducing, the mineral buffer is considered to be pyrite + pyrrhotite + epidote + prehnite (Arnórsson & Gunnlaugsson 1985). In waters of higher salinity, which are less reducing than waters of low salt content (Stefánsson & Arnórsson 2002), the respective mineral buffer may consist of pyrite + magnetite + epidote + prehnite (Arnórsson & Gunnlaugsson 1985; Ragnarsdóttir et al. 1984). Other buffers may also be involved. Figure 11b shows how H_2S concentrations in geothermal aquifer waters vary with T. The scatter of the data points at each T, at least, when above about 200 °C, suggests that different mineral buffers may be involved in controlling aqueous H_2S concentrations. The scatter may also be caused by variation in the composition of one or more of the minerals constituting these buffers, in particular the epidote.

The concentrations of H_2S in steam discharged from wet-steam wells of high-T geothermal systems typically lie in the range of 2–20 mmoles/kg, but they may be as high as 50 mmoles/kg (Table 4). In a particular area, H_2S in steam discharged from fumaroles is generally lower than in steam from wells. This has been attributed to reactions between steam and rock in upflow zones (Arnórsson & Gunnlaugsson 1985). In steam from wells, the H_2S concentrations depend on the same factors as those listed above for CO_2.

Owing to its poisonous nature, H_2S in the atmosphere has negative effects on vegetation. In the Italian geothermal field of Travale–Radicondoli in southern Tuscany, high quantities of S were found in leaves in an area in the vicinity of the geothermal power plant and the concentrations correlate negatively with distance from the plant (Bussotti et al. 1997). The S content was positively associated with reduction in the leaf size. Assuming a threshold limit of 0.3 ppm H_2S in the air, Gallegos-Ortega et al. (2000) concluded that the present levels of atmospheric H_2S in the vicinity of the Cerro Prieto geothermal power plant, Mexico, do not pose a risk for public health in the neighbouring towns. Studies at Olkaria in Kenya (Marani et al. 2000) revealed that variations in atmospheric H_2S concentrations could be correlated with changes in weather parameters and distance from the geothermal power plant. Concentrations in excess of 7.5 ppm could be expected occasionally close to the power plant, but at a distance of about 500 m, they were always below 0.3 ppm.

Abatement of H_2S has been considered and implemented at several geothermal power plants. There is a plethora of methods available for such abatement. Only a few of them will be mentioned here. The subject has been summarized recently by Sanopoulos & Karabelas (1997). The H_2S in the steam may be removed either upstream from the turbine or downstream. Upstream escape routes for H_2S are drains and pots and steam stacking. Downstream, steam escapes mainly from the gas ejectors and the cooling towers. Upstream methods include (1) continuous condensation and reboiling of the steam, and (2) scrubbing with alkali (NaOH). The latter method is relatively expensive. It can be used downstream as well. Other downstream methods include absorption of part of the H_2S from steam ejectors into citric acid and oxidation of the rest of the H_2S by air into SO_2. The SO_2 is subsequently reacted with the absorbed H_2S to produce elemental sulphur (Claus reaction). Another downstream method involves catalytic oxidation of the H_2S with air and H_2O_2. According to Sanopoulos & Karabelas (1997) the most attractive abatement methods for H_2S include (1) scrubbing with alkali, (2) use of steam re-boilers, and (3) compression and use of the reactant BIOX (chlorine- and bromine-stabilized biological oxidizing agents; see Hoyer et al. 1991) as a catalyst to oxidize H_2S to H_2SO_4.

Arsenic boron and mercury. As discussed in the following section, As, B, and Hg are typically enriched in geothermal fluids, as compared to surface- and groundwaters (Sakamoto et al. 1988; Ballantyne & Moore 1988). Being quite fugitive, Hg partitions significantly into the steam phase at low T. Higher temperatures are required for such partitioning of As and B. Mercury concentrations in geothermal steam may be as high as 0.5 ppm. Steam in vapour-dominated fields, such as at Larderello in Italy, may be high in B. The B exists as boric acid (H_3BO_3) in the steam. In fact the geothermal steam at Larderello was exploited for its B content before it was utilized for power production.

Owing to the high concentrations of As, B, and Hg in geothermal fluids, hot spring and fumarole discharges are known to contaminate stream and river waters (Robinson et al. 1995; Nimick et al. 1998; Webster 1999), accumulate in vegetation (Loppi & Bonini 2000; Loppi 2001) and in fish (Axtmann 1975b; Kim 1995). Elevated concentrations of As, B, and Hg have been reported around geothermal power plants, both in the atmosphere (Bargagli et al. 1997) and in vegetation (Loppi et al. 1999). At the Mt. Amiata geothermal power plant in Italy, Hg emission rates are $3-4$ g/h per MW (Bacci et al. 2000). In Japan, the estimated yearly Hg output from geothermal fumaroles is estimated as 1.4 t, which is only a small part (2%) of the total amount of Hg released annually to the environment in Japan; the remaining 98% come mostly from various industrial activities (Nakagawa 1999).

Various chemical methods have been developed with the purpose of reducing atmospheric and terrestrial pollution from geothermal power plants with respect to As, B, and Hg. It has been demonstrated that As can be removed from the water by co-precipitation with calcium silicate (Rothbaum & Anderton 1975) and Fe(III)-hydroxide (Yanagese et al. 1983; Decarlo & Thomas 1985). Mercury can be removed by S-impregnated and virgin activated carbon via the formation of HgS (Vitolo & Seggiani 2001). Neither of these chemical methods have, however, been realized in exploited geothermal systems. To reduce the environmental impact of B on vegetation the most promising method is injection (Einarsson et al. 1975; Satman et al. 2000). However, a method using ion-exchange has been developed and considered to solve a B-pollution problem at Kizildere in Turkey (Badruk & Kabay 2003).

Waterborne pollutants

Elements that sometimes are present in poisonous concentrations in geothermal waters are As, B, and H_2S. Other elements that may be present in harmful concentrations include Al, F, ammonia (NH_3), and various heavy metals. Additionally, geothermal fluids may exhibit a very high salt content. High concentrations of heavy metals are associated with high-T brines, such as those at Salton Sea in California (White et al. 1963) and on the island of Milos, Greece (Thórhallsson et al. 1984). High B and As concentrations are found in many geothermal systems associated with andesitic volcanism. Examples include Mt. Apo in the Philippines and Achuapan in El Salvador. Boron-rich geothermal waters are known to form upon reaction with marine sediments, such as at Ngwaha in New Zealand. The source of the B is the illite of the marine sediments. Waters of geothermal systems hosted by basaltic rock are low in B and As, and heavy metals, but relatively high in both H_2S and Al. This is exemplified by Krafla, Iceland (see Table 3).

Boron and arsenic. Geothermal waters of high-T systems in andesitic and silicic volcanics at converging plate boundaries may contain As and B in concentrations as high as 50 and 1000 ppm, respectively, although concentrations between 1 and 10 ppm for As and $10-50$ ppm for B are more typical (Ellis & Mahon 1977; Ballantyne & Moore 1988). By contrast, high-T fluids hosted in basaltic rocks on the diverging plate boundary in Iceland are much lower in As and B (Arnórsson 2003). The concentrations of As and B in geothermal fluids are positively related to their concentrations in the enclosing rock. Indeed, the source of these elements to the water is the primary minerals of the enclosing rock. Being quite mobile, only a small part of the As and B dissolved from the rock is removed from solution again by incorporation into secondary minerals that precipitate from the fluid (see, e.g., Ellis & Mahon 1964, 1967; Arnórsson 2003).

High concentrations of B in soil water may be deleterious for certain crops. The most sensitive crops can tolerate no more than $0.5-1.0$ ppm B (McKee & Wolf 1971). Boron concentrations in geothermal waters are not expected to exceed acceptable limits for stock watering and aquatic life, as indicated by data provided by McKee & Wolf (1971). During the early development stages of the Achuapan geothermal field, El Salvador in $1970-1971$, injection tests were performed with the purpose of solving an environmental problem by surface disposal of well water rich in B (Einarsson et al. 1975). At the Kizildere geothermal field in Turkey, the waste water from the steam separators has been disposed of into the nearby Buyuk Menderes River. The river water is used for irrigation. The adverse effects of the high B content of the geothermal water has called for injection of the waste water back into the geothermal reservoir (Satman *et al.* 2000) but removal of the B from the water using ion-exchange has also been considered (Badruk & Kabay 2003).

Epidemiological evidence has verified that arsenic is one of the most carcinogenic and toxic substances in surface- and groundwaters. By European Union (EU) standards and United States Public Health Service (USPHS) Drinking Water Standards the maximum allowable As concentration in drinking water is 0.05 ppm, and 0.01 ppm for aquaculture, but as already mentioned As concentrations in geothermal waters may be as high as 50 ppm. From a toxicity assessment perspective, it is not sufficient to quantify total dissolved As, because its chemical form must be determined (Buyuktuncel *et al.* 1997). Trivalent As is harmful, whereas As^{5+} is not. In geothermal waters, As is largely trivalent, although upon contact with the air, it is relatively rapidly oxidized to As^{5+} (Yokoyama *et al.* 1993; Gihring *et al.* 2001).

Hydrogen sulphide. In boiled, relatively saline high-T waters (>1000 ppm Cl), total sulphide-S concentrations are characteristically below 5 ppm (as H_2S), but they may be as high as 100 ppm in high-T waters of low salt content hosted by basaltic rocks. In low-T waters dissolved sulphide-S is most often below 2 ppm. Its concentration decreases with decreasing T (Fig. 11b). It is the H_2S species that is poisonous, not total sulphide-S. Thus, the S-species distribution in the water is an important parameter. The fraction of sulphide-S present as H_2S is determined by the water pH. For waters with a pH of 7, as measured at 25 °C, H_2S accounts for about 50% of the total sulphide-S. In dilute high-T waters, which build up a high pH (>9) by boiling, practically all the sulphide-S is present as HS^-. If high-pH, sulphide-rich geothermal water mixes with surface- or groundwater so that the pH of the mixed water is significantly lowered, the H_2S concentration of the mixed water may exceed 1 ppm depending on the dilution effect. Such concentrations, or even lower, have been reported to be toxic to many types of fish within a period of 24 hours (McKee & Wolf 1971). For this reason, H_2S-bearing water from geothermal power plants, whether separated water or condensate, may have harmful effects on aquatic life and water quality if disposed of on the surface, where it can seep into the soil and mix with surface- and shallow groundwater.

Aluminium and fluoride. In geothermal waters, which have equilibrated with low-albite (NaAl Si_3O_8) and quartz (SiO_2), or its microcrystalline variety, chalcedony, Al concentrations decrease at any specific temperature with increasing Na concentrations, that is, increasing water salinity. This can be seen from the following reactions:

$$SiO_2 + 2\,H_2O = H_4SiO_4^0 \qquad (2)$$
$$NaAlSi_3O_8 + 2\,H_2O = Na^+ + Al(OH)_4^-$$
$$+ 3\,H_4SiO_4^0 \qquad (3)$$

and the observation that Na^+ and $Al(OH)_4^-$ constitute the most important Na- and Al-bearing aqueous species (Arnórsson & Andrésdóttir 1999). Stefánsson & Arnórsson (2000) observed that geothermal reservoir waters generally closely approach equilibrium with low-albite when T is above about 100 °C. Equilibrium with chalcedony is attained for waters with temperatures below 180 °C, but with quartz at higher temperatures (Arnórsson 1975). With increasing T, at a particular salinity (Na concentration), Al aqueous concentrations increase because low-albite solubility increases much more with rising temperature than quartz solubility (see, e.g., Arnórsson & Stefánsson 1999).

High-T waters of low salinity, which also contain low Na (<200 ppm) may contain as much as $2-3$ ppm Al. By EU standards, recommended Al concentration in domestic waters is 0.05 ppm, and the permissible limit is 0.2 ppm. For aquaculture it is less. Thus, the Al content of high-T geothermal waters may far exceed the permissible limit. Low-T waters of low salt content may also exceed the permissible limit, if their pH exceeds about 9.

Fluoride concentrations in geothermal waters hosted by silicic volcanics are controlled by fluorite solubility (Nordström & Jenne 1977). In geothermal waters in basalt, ion-exchange equilibria involving OH^- and F^- and an OH-bearing silicate mineral probably control the aqueous activity of F (Arnórsson *et al.* 1983). Fluoride concentrations of $10-20$ ppm are common in waters in silicic volcanics (Nordström & Jenne 1977; Arnórsson 1985), although values as high as $50-80$ ppm have been reported in the Olkaria field in Kenya (Karingithi 2002). In basaltic environments, F concentrations in geothermal waters typically lie in the range of 0.5 to 2 ppm depending on T and salt content of the water. Aqueous F concentrations in silicic rock environments far exceed the permissible limits for water for domestic use of 1.5 ppm by EU standards. As described by McKee & Wolf (1971), there are numerous articles describing the effects of F-bearing water on dental enamel. It may also cause skeletal damage, for example, on sheep, as is well known from the effects of some volcanic eruptions in Iceland (Óskarsson 1982).

Scaling

Many types of scales are known to form from geothermal fluids (e.g., Ellis & Mahon 1977; Arnórsson 1981). They include amorphous silica (e.g., Yanagese et al. 1970; Kuboda & Aosaki 1975; Klein 1995; Takeuchi et al. 2000), calcium carbonate (e.g., Arnórsson 1978a, 1989; Ramos-Candelaria et al. 2000; Lee & Bacon 2000), iron silicate and various metallic sulphides (e.g., Arnórsson 1981; Gallup 1989; Kristmannsdóttir 1989; Gallup et al. 1995; Hardardóttir et al. 2001), Al- and Mg-silicates (e.g., Thórhallsson et al. 1975; Kristmannsdóttir, 1989; Hauksson et al. 1995; Gallup 1997; Simmons & Brown 2000), and rare anhydrite (e.g., Kristmannsdóttir 1989). Scale formation is largely confined to liquid-dominated geothermal systems with temperatures in excess of some 200 °C. However, scales of Mg-silicates, which are largely amorphous or poorly crystalline, are confined to low-T waters and geothermally heated groundwaters used for space heating (Hauksson et al. 1995). Scale formation may occur in production wells, surface piping, and injection wells. It has caused operational problems; it affects the effectiveness in the utilization of the geothermal resource or calls for the use of chemical inhibitors to reduce the rate of scale formation. In extreme cases, intense scale formation observed during testing of exploration wells has led to cessation of further development.

The most common type of troublesome scale is that of amorphous silica and calcium carbonate. Scales of various metallic sulphides is the rule rather than the exception. By far the most abundant sulphide scale consists of iron sulphides. They include pyrite, marcasite, and pyrrhotite (Kristmannsdóttir 1989), but sulphide scale of other metals have also been observed, such as Cu, Pb, and Zn (White et al. 1963; Gallup 1989; Gallup et al. 1995; Hardardóttir et al. 2001; Reyes et al. 2002). Sulphide scales are often poorly crystalline and they may be amorphous to X-rays. Moreover, the sulphide-bearing scales are known to be enriched in various elements such as Ag, As, Au, Cd, and Mn. Reyes et al. (2002) observed that scales at Rotokawa, New Zealand, also contained elevated concentrations of Hg, Sb, and Se, which were incorporated in pyrite. The quantity of sulphide scale formation is generally very limited and may in fact be beneficial rather than troublesome as the scale forms a stable protective film on the inside surface of casings and pipe work that inhibits corrosion. Intense and serious sulphide scale formation is only known to occur from saline and/or acid fluids, such as at Salton Sea, California (White et al. 1963), Krafla (Kristmannsdóttir 1989) and Reykjanes (Hardardóttir et al. 2001), Iceland, and on the Greek island of Milos (Thórhallsson et al. 1984). Scale of anhydrite is rare and apparently confined to acid–sulphate–chloride waters.

Scale formation occurs largely in response to cooling and degassing of the geothermal water, as it boils by depressurization. Some of the scale-forming minerals have prograde solubility (the solubility increases with increasing temperature), such as amorphous silica and metallic sulphides. Others, such as calcite, aragonite, and metallic sulphides have pH-dependent solubility. Cooling can cause the geothermal water to become oversaturated with minerals with prograde solubility. Degassing, which occurs as a consequence of boiling, tends to produce oversaturated water with respect to minerals with pH-dependent solubility. The degassing causes the pH of the water to increase strongly by transfer of the 'acid gases', CO_2 and H_2S, to the steam phase. The kinetics (rate of formation) of the scale-forming mineral reactions determines how troublesome the scale formation may be, but not so much the degree of oversaturation that is produced by cooling and degassing. Also, the availability (aqueous concentration) of the scale-forming components is very important. For example, the content of Fe and various heavy metals is very low in geothermal waters of low salt content. This sets a limit to the quantity of metallic sulphides that can precipitate from the water. Boiled and degassed geothermal waters are oversaturated with many minerals that are known to be present as hydrothermal minerals in geothermal reservoirs, such as quartz, albite, and K-feldspar. These minerals are, however, not known to form scales in geothermal wells, the reason being slow kinetics. On the other hand, calcite precipitates readily from slightly oversaturated solution. If degassing causes excessive calcite oversaturation, aragonite, rather than calcite, tends to form. Metallic sulphides apparently precipitate readily from solution.

Geothermal aquifer waters are close to saturation with some scale-forming minerals (calcite, pyrite) but undersaturated with others (amorphous silica, amorphous metallic sulphides). Only the slightest degassing suffices to produce calcite oversaturated water. By contrast, extensive cooling may be required to produce amorphous-silica oversaturation. As solubility constants are not known for the amorphous or poorly crystalline sulphide phases, which precipitate from geothermal waters, it is not known how much cooling and degassing is required to render the water oversaturated with these phases.

Geothermal fluids are generally strongly reducing and not corrosive, but there are many exceptions. Many warm geothermal waters ($<60\ °C$) contain dissolved O_2, which makes them corrosive. It is common that geothermal waters contain free CO_2 in significant concentrations, which also makes them corrosive. Water and steam may also contain high boron (e.g., Zancani 1975), ammonia and hydrochloric acid (Truesdell et al. 1989; Izquierdo et al. 2000), all of which are corrosive. Corrosion problems may be solved by material selection, chemical treatment, or scrubbing of undesirable components from the water or steam. Acid, high-T geothermal waters that have been encountered by deep drillings in several fields deserve specific attention. They include Mt. Labo and Palinpinon in Philippines (Hermoso et al. 1998; Maturgo et al. 2000; Matsuda et al. 2000), several fields in Kyushu in Japan (Kiyota et al. 1996; Maturgo et al. 2000) and Krafla, Iceland (Kristmannsdóttir 1989). In the aquifer, the pH of these waters is buffered by bisulphate (HSO_4^-/SO_4^{2-}) and has a value of 5–6. The dissociation constant for HSO_4^-($HSO_4^- = H^+ + SO_4^{2-}$) increases strongly with decreasing T. Thus cooling, which occurs as a consequence of depressurization boiling, causes most of the HSO_4^- in the aquifer water to dissociate into SO_4^{2-} and H^+, decreasing the pH of the water in the process to values as low as 2 at the wellhead. Owing to the highly corrosive nature of their fluid, such acid geothermal wells have not been exploited. The author knows, however, that injection in the Palinpinon field in Philippines has raised the pH of the aquifer water of 'acid wells' making it possible to produce from them.

Below, amorphous silica and calcite scale formation is discussed in some detail because scale formation of these minerals is the most common. Space does not permit more detailed handling of scaling and corrosion problems associated with the utilization of geothermal fluids.

Silica

Dissolved silica is one of the most troublesome products of liquid-dominated high-T geothermal fields. In the reservoir, aqueous silica concentrations are controlled by quartz solubility according to reaction (2) above (Mahon 1966; Fournier & Rowe 1966; Arnórsson 1975). Boiling of high-T water, which lowers its T and increases its silica concentration due to steam formation, may lead to amorphous silica oversaturation (Fig. 13). This silica phase precipitates relatively easily from oversaturated solution and may form scales in well bores, surface equipment and injection wells, thus creating operational problems. Quartz does not form because of the slow kinetics of its precipitation reaction, as supported by the quartz solubility experiments of Morey et al. (1962) and Rimstidt (1997), although the boiled geothermal water is much more oversaturated with quartz than with amorphous silica.

In high-T geothermal reservoir waters, the dominant silica species is invariably $H_4SiO_4^0$ (Arnórsson et al. 1983). Upon boiling and CO_2 and H_2S degassing of dilute high-T waters, the pH of the boiled water may be sufficiently raised to cause a substantial fraction of the dissolved silica to ionize to form $H_3SiO_4^-$ ($H_4SiO_4^0 = H_3SiO_4^- + H^+$). The formation of this species counteracts the effects of decreasing temperature and steam formation on amorphous silica saturation, thus lowering the temperature at which amorphous silica saturation will be reached.

Many methods have been applied to avoid problems linked with amorphous silica scaling. The one that has probably been most widely used involves adjusting pressure (therefore also T) in steam separators, surface equipment, and pipings to injection wells above the T of amorphous silica saturation. This temperature is related to the aquifer temperature of production wells, as may be read from Fig. 13, and may exceed $200\ °C$ for reservoirs with temperatures in excess of $300\ °C$. The above described method to avoid amorphous silica scaling means poor exploitation of the heat brought to the surface through production wells. Other methods that have been applied to avoid troublesome amorphous silica deposition include evaporation of the waste water in ponds, use of inhibitors (Shimizu et al. 2000), acidification, silica polymerization, and removal of silica from solution ('brine clarification'). Still other methods have been attempted, such as precipitation of the silica with lime to form a useful calcium silicate (Rothbaum & Anderton 1975). Silica has also been successfully precipitated from the high-T geothermal seawater at Reykjanes, Iceland, by bubbling CO_2 through the flashed brine. The mechanism of the reaction is not known. Inhibition of silica scale formation by acidification can be achieved by biologically oxidizing H_2S in geothermal steam to sulphuric acid and mixing the steam condensate with the separated water (Takeuchi et al. 2000). Adjustment of steam separator pressure, acidification, and the

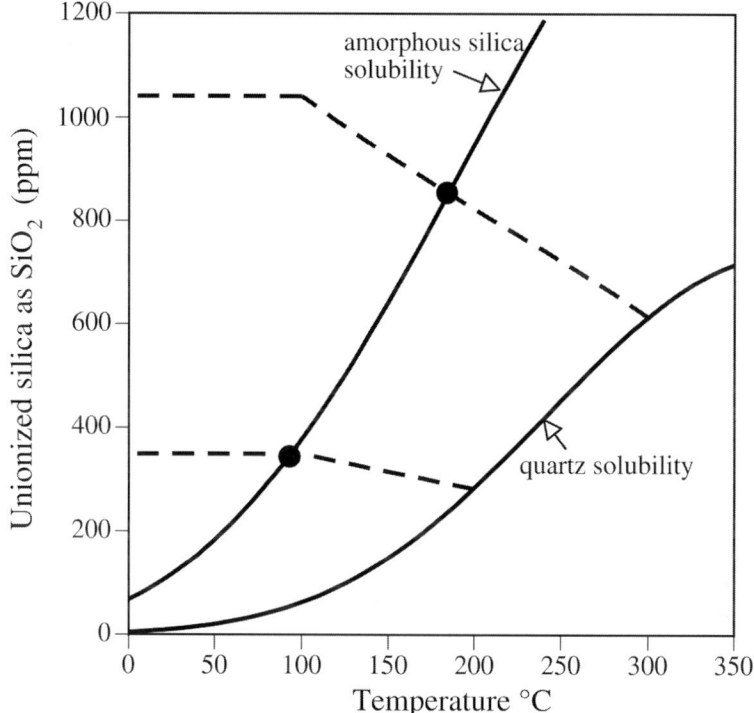

Fig. 13. Quartz and amorphous silica solubility vs. temperature along the vapour saturation curve. The dashed lines show the silica concentration in water initially in equilibrium with quartz during adiabatic boiling to 100 °C and subsequent cooling. The increase in aqueous silica concentrations during boiling is the consequence of steam formation. Amorphous silica saturation (shown by the dots) is attained at 188 °C in the case of the 300 °C aquifer water, but at 94 °C in the case of the 200 °C aquifer water. It was assumed that the pH of the water is not raised sufficiently during boiling to cause significant ionization of the aqueous silica. If some ionization had occurred, amorphous silica saturation would be reached at lower temperatures than those indicated in Fig. 13.

use of inhibitors can be used to avoid silica scaling in production wells, surface equipment, and injection wells. The other methods listed above can only be applied to solve silica scaling problems in surface equipment and injection wells.

In the Palinpinon geothermal field, Philippines, amorphous silica scaling has been avoided by adjusting steam separator pressures and injection water temperatures above amorphous silica saturation temperatures. Silica polymerization has been successively applied at Olkaria, Kenya, to avoid amorphous silica deposition. Silica molecules (monomeric silica) in solution ($H_4SiO_4^0$) in excess of amorphous silica solubility tend to react with each other to form colloidal silica (polymerize). The rate of silica polymerization varies with many factors, including pH, temperature, water salinity, and the degree of amorphous silica oversaturation (e.g., Gunnarsson & Arnórsson 2003). Completion of the polymerization reaction,

that is, when the polymerization reaction has reduced monomeric silica concentration to that corresponding to equilibrium with amorphous silica, may range from a few minutes for saline geothermal waters to a day or more for geothermal water of low salt content. In saline geothermal water, silica colloids grow sufficiently fast and to such a size that they settle by gravity. In waters of low salt content, on the other hand, the colloidal silica may remain suspended in solution for years (Arnórsson, unpublished work). Experience has shown that colloidal silica precipitates much more slowly from solution than monomeric silica. Therefore, the rate of silica scale formation can be reduced much, even inhibited for all practical purposes, by allowing the dissolved silica in excess of amorphous silica solubility to polymerize, either in tanks or in conditioning ponds.

At Olkaria, temperatures in wellhead separators are above saturation with respect to amorphous silica. Flashing (frequently used in the geothermal industry to indicate explosive boiling of the geothermal water upon depressurization) in silencers by each wellhead, however, leads to oversaturation of the water entering the weirboxes. (The tank into which water flows from an atmospheric silencer is called a weirbox. The overflow from the weirbox is frequently through a V-notch so the water flow rate can be easily estimated by measuring the water level behind the V-notch.) At each weirbox a conditioning pond has been constructed where the water resides for 2–3 hours before being conveyed in open ditches to an infiltration pond. In the conditioning ponds, practically all the aqueous silica in excess of amorphous silica solubility polymerizes. About 20 years of experience has demonstrated that very insignificant amounts of silica precipitates in the open ditches and infiltration ponds. At one time, however, the author observed extensive amorphous silica deposition from hot waste water, which flowed directly from the weirbox of a well into a nearby stream bed after torrent rain had destroyed the conditioning pond for that well. This observation proves that amorphous silica precipitation can be reduced to practically nil by allowing aqueous silica in excess of amorphous silica solubility to polymerize in special conditioning ponds.

In the Salton Sea area, California, silica in the hyper-saline brine (Table 3) has been removed from solution in special 'brine clarification' tanks. A sludge containing precipitates of silica and various sulphides is injected into hot brine in the tanks. The precipitates in the sludge act as nuclei for precipitation of additional silica and the rate of precipitation is sufficiently high for removal of aqueous silica to <100 ppm. This method most likely only applies to very saline waters.

At Nesjavellir, Iceland, hot water from the steam separators at 188 °C is conveyed through 'capillary heat exchangers' and cooled in 10–15 s to 90 °C. Experience over the past three years indicates that this residence time in the heat exchangers is not sufficiently long to allow onset of amorphous silica precipitation. At 180 °C the separated water, which contains around 770 ppm SiO_2, is close to saturation with respect to amorphous silica but when leaving the heat exchangers it is highly oversaturated (see Fig. 13). At Nesjavellir, treatment involving polymerization of the silica in the water leaving the capillary heat exchangers is being considered, before injecting the cooled geothermal water. The experience at Nesjavellir suggests that capillary heat exchangers may successfully allow extraction of heat from the separated hot water before disposal without

amorphous silica precipitation. This is a simple method, which at the same time is environmentally friendly, as it does not involve the use of any chemicals. The economy of the technique at Nesjavellir depends on the availability of cold water to be heated and the market for such water. The technology of cooling is, however, of much interest, as it appears to open up the possibility of effective heat extraction from silica-rich high-T geothermal waters without having operational problems resulting from silica scale formation. It remains to be demonstrated, however, if this technology applies to saline geothermal waters as well. At Nesjavellir, the separated hot water is very low in dissolved solids, similar to that at Námafjall (see Table 3).

Silica deposition from flashed high-T geothermal waters may not be negative altogether. At Svartsengi in Iceland, the 240 °C reservoir brine, which has two-thirds the salinity of seawater (Arnórsson 1978b), is double flashed. The steam from the first flash step is passed through turbines and then condensed in heat exchangers. The separated brine is flashed for a second time in low-P separators (0.3 bars) to a temperature of 70 °C (Arnórsson et al. 1975). The brine from this second flash is disposed of into a large infiltration pond. The aqueous silica polymerizes rapidly in the brine to form polymers that grow to sufficient size to settle by gravity. The gelly precipitate of silica colloids is a useful industrial product. Bathing in the brine has also proved to have healing ability for psoriasis patients (Ólafsson 1996).

Calcium carbonate

Geothermal reservoir waters are close to calcite saturation. Degassing of such water, which occurs in response to boiling, leads to a sharp rise in the pH of the water. This in turn causes the activity of the carbonate ion (CO_3^{2-}) to increase much, leading to oversaturation (Arnórsson 1989):

$$CaCO_{3,s} = Ca^{2+} + CO_3^{2-} \qquad (4)$$

The solubility of calcite increases with decreasing T. Thus cooling, which occurs during depressurization boiling, counteracts the effect of CO_2 degassing with respect to the state of calcite saturation. During the early stages of boiling, the effect of degassing dominates, but after the water has been largely degassed with respect to CO_2, the effect of further cooling by boiling takes over, decreasing steadily the degree of oversaturation. For many waters sufficient cooling may lead to calcite undersaturation (Fig. 14).

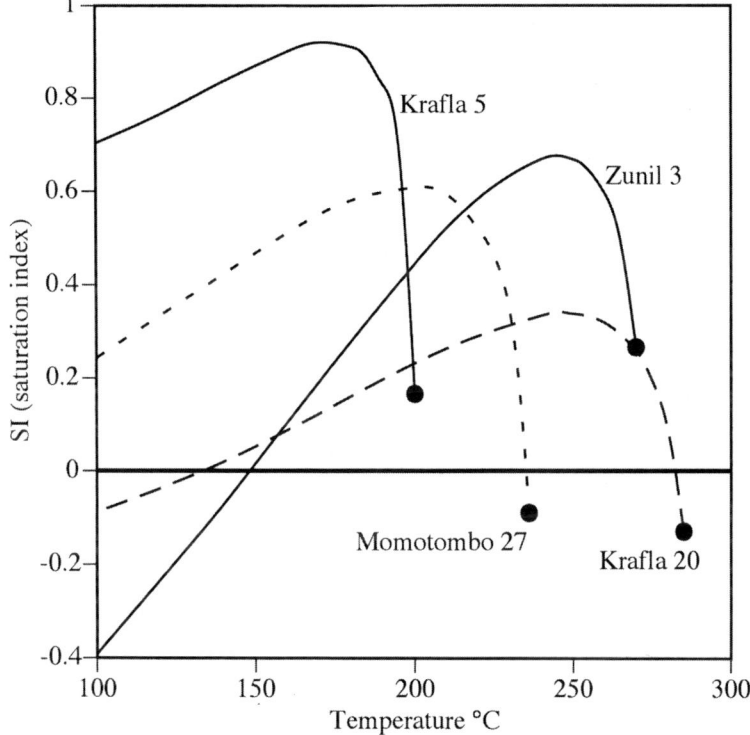

Fig. 14. Change in calcite saturation state during single-step adiabatic boiling (vapour saturation pressure) of aquifer water from four wet-steam wells in three areas: Krafla, Iceland; Momotombo, Nicaragua; and Zunil, Guatemala. A positive SI-value corresponds to oversaturation and a negative value to undersaturation. An SI of zero corresponds with equilibrium. SI is on a log scale so an SI-value of +1 indicates tenfold oversaturation. The numbers indicate well numbers. The calculated SI-values for the aquifer waters (dots) depart a little from equilibrium. In view of all errors involved in the calculation of the SI-values, the departure from equilibrium is, however, not significant. Note that variations in SI-values are more accurately calculated than absolute values.

Many observations have shown that calcium carbonate deposition in wells is only troublesome when the first level of boiling is in the well and that the deposition is strongest just above this level. The deposit is most often calcite. However, aragonite, which is slightly more soluble than calcite, has also been identified. It seems that aragonite forms at the expense of calcite, when CO_2 degassing causes strong oversaturation (Arnórsson 1989).

The observed formation of calcite scales in geothermal wells is consistent with calcite saturation calculations (see Fig. 14). The intensity of calcite deposition is largely determined by two factors, the temperature of the water and its salt content (Arnórsson 1978a). In relation to temperature, calcite scale formation is most troublesome around 200 °C and it decreases at both higher and lower T. The cause is the temperature variation in the solubility of CO_2. It is at a minimum around 200 °C. Thus boiling, when it starts around 200 °C, leads to more rapid degassing than boiling, which starts at higher and lower temperatures. The intensity of calcite scale formation increases with increasing salt content of the water. The relative concentrations of Ca and carbonate carbon (CC) in geothermal waters depend on the salt content of the water (Arnórsson et al. 1983). At any particular temperature, waters of low salt content have low Ca concentrations (1–10 ppm) and relatively low Ca/CC ratios ($\ll 1$). This ratio increases with increasing salt content. The availability of Ca is a limiting factor for the amount of calcite precipitation, explaining the observed relationship between the salt content of geothermal waters and the intensity of calcite scale formation.

Experience in many parts of the world has shown that calcium carbonate scale formation is only a problem in producing wells, when boiling starts in the well. It is not a problem when the flashing point is in the aquifer. The reason is that the cross-sectional area of pores in the formation, even at a small distance from the well, must far exceed the cross-sectional area of the wellbore. Hence, the aquifer rock can cope with calcium carbonate scale formation for a much longer time than wellbores, probably longer than the life-time of a geothermal power plant (Arnórsson 1989).

Three methods have been applied to cope with calcite scales in wet-steam wells. They include the use of inhibitors (e.g., Ramos-Candelaria et al. 2000; Buning et al. 2000), periodic mechanical cleaning (Argueta 1995), and dissolving the scale with acid (e.g., Líndal & Kristmansdóttir 1989; Evanoff et al. 1995). All of these methods have proved successful. However, mechanical cleaning is only acceptable if the rate of scale formation is not too rapid so that cleaning once or twice a year is sufficient to maintain full well output.

It is not uncommon for calcite to precipitate from CO_2-rich low-enthalpy geothermal waters. The cause is CO_2 degassing, which occurs as a consequence of depressurization. The scaling problem may be solved by the use of inhibitors. Two additional methods have been used to solve calcium carbonate scaling problems associated with the use of low-enthalpy CO_2-rich waters. One involves enhancement of calcium carbonate precipitation under controlled conditions by air bubbling and seeding 50–70 °C geothermal water with aragonite (El Fil et al. 2003). The other method, which has been operating for four years at Haedarendi in Iceland, involves placing a heat exchanger into the well to cool 164 °C water to about 110 °C. In this way, fresh, hot water is produced for space heating and the geothermal water is sufficiently cooled to make it calcite undersaturated after degassing (Bjarnason, personal communication).

Byproducts

Extraction of various gases and solid waste products resulting from exploitation of geothermal resources has been considered and implemented on a small scale. Thus, B was extracted as a byproduct from steam at Larderello during the early years of electric power production (Villa 1975). Carbon dioxide has been recovered from steam, both in Iceland and Turkey (Ragnarsson 2000; Mertoglu et al. 2003). In Iceland, a project is ongoing to recover H_2 from steam for use as fuel. This project also considers the possibility to extract H_2S from the steam for production of H_2 by electrolysis (Árnason,

personal communication). Probably, silica has received the most attention. As already mentioned, Rothbaum & Anderton (1975) described how silica could be removed from geothermal water by precipitating it with lime to form a calcium silicate that could be used as a building material. In Svartsengi, Iceland, silica precipitate has been used on a small scale for the production of powder make-up. Recently, production of glass from geothermal waste water products has been considered. At Cerro Prieto, Mexico, >99% pure SiO_2 has been produced from the waste water residue. This silica can be used as a raw material in the ceramic industry, for example, in the production of soda-lime (Diaz et al. 1999, 2000). In the high-T geothermal sea water system at Reykjanes, Iceland salt was extracted from the brine by evaporating it with geothermal steam (Líndal 1975). The salt was used for the domestic fishing industry as well as low-Na health salt for export. The salt plant was in operation for almost 20 years but it was closed down in 1994 (Ragnarsson 2000). A part of the plant was restarted in 1999, and full operation is planned.

Scenery spoliation and soil erosion

Surface manifestations in geothermal areas are often valued as beauty spots and represent a tourist attraction. Development of such areas needs to be balanced against the value of preservation for national heritage and tourism. Withdrawal of reservoir fluid through drillholes leads to a pressure decline in the reservoir and may cause hot springs and geysers to dry out. Reservoir depressurization may also enhance boiling at shallow depth and in this way increase fumarolic activity. It is difficult to predict these changes upon exploitation although attempts have been made (Shimada et al. 2000; Tokita et al. 2000a, b).

Many geothermal power plants, including the steam supply system, have not been constructed to minimize visual pollution. Yet, some systems have been well camouflaged into the surroundings. It is important to employ architects to design geothermal power plants, as is the case, for example, at Svartsengi and Nesjavellir in Iceland. Making use of such expertise for power plant design is clearly very beneficial for reduction of scenery spoliation.

Many geothermal fields are located in rugged terrain. The artificial levelling of ground on steep slopes for roads may aggravate soil erosion, particularly by heavy rain, by creating steep gradients in road cuts and by removing vegetation (e.g., Read & Campbell 1975).

Directional drilling

Directional drilling, rather than vertical drilling, into geothermal reservoirs is commonly practised today. Its purpose is to reduce scenery spoliation. Moreover, directional drilling reduces the demand for building of roads and drillpads, which is particularly advantageous in areas of rugged topography. In such settings, roads unavoidably incise steep slopes, which may lead to severe soil erosion by torrential rain. Several wells may be drilled from the same pad. In comparison with vertical drillings, directional drillings allow more flexibility in siting of wells.

Directional wells are initially drilled vertically and deviated subsequent to cementation of the production casing. The depth of this casing may be 1000 m, even more, depending on targeted well depth, lithology, and the depth level of permeable horizons. Drilling of many wells from the same pad may not be successful if most of the producing aquifers occur not far below the bottom of the production casing, in which case the wells interfere with each other.

Directional drilling is some 20% more expensive than vertical drilling of a well of the same length. However, the extra cost of drilling directionally may be saved on a shorter steam gathering system, shorter roads, and fewer drillpads. The steam gathering system becomes shorter if many wells are drilled directionally from the same drillpad, as compared to more scattered wellheads in a production field with vertical wells only. Directional drilling is also advantageous for side-tracking dry (unproductive) wells, or poor producers to improve their performance. The drilling of the first $\sim$1000 m and the cementing of surface-, anchor-, and production-casing accounts for as much as 70% of the total cost of a typical geothermal well drilled to 2000–2500 m total depth (length).

Injection

The first effort to inject spent fluid from a high-enthalpy well back into the geothermal reservoir was made in the Achuapan field in El Salvador around 1970 during the early stages of its development (Einarsson et al. 1975). The reason for this effort was environmental. Surface disposal was not acceptable due to the high B content ($\sim$50 ppm) of the separated water from the well. Early results of these injection tests were discussed at the United Nations Geothermal Symposium in Pisa in 1970, and it was considered a subject worthy of study (Armstead 1975b). The geothermal community was reluctant, however, to accept this method of disposal of spent geothermal

water and condensate as a solution to environmental problems. It was feared that the injected waste fluid could enter aquifers of production wells and deteriorate their performance. Today injection is the rule for disposal of spent fluid from geothermal power plants rather than the exception. Generally speaking, it is favourable, both environmentally and economically. The advantages of injection are essentially threefold. Firstly, injection helps maintaining reservoir pressures and thus reducing the decline in flow rate from producing wells, which in turn increases their operational lifetime and the lifetime of the reservoir. Secondly, injection is favourable in comparison with surface disposal of waste water and condensate, because such water frequently contains constituents that have adverse effect on the environment. Thirdly, injection favours utilization of the heat stored in the rock of geothermal systems, which usually is much larger than the heat stored in the fluid of the system, when exploitation begins. The cooled waste water gains heat as it percolates through the reservoir rock into the aquifers of producing wells. In this way heat in the reservoir rock is utilized. Of course, natural recharge of cold water will have the same effect.

The development of an injection strategy requires consideration of many variables (Benoit et al. 2000). They include: (1) an injection scheme, (2) where to inject and how deep, (3) whether or not the steam condensate should be mixed with the waste water before injection, (4) reservoir fluid cooling, (5) waste water chemistry, and (6) possible treatment of this water to prevent scaling in injection wells. Injection calls for intensified monitoring studies of the response of the reservoir to the production load. Such studies include tracer tests, and monitoring of changes in reservoir temperature and pressure, and in well discharge compositions (see, e.g., Panichi 2004). Injection of fresh water has been implemented in a few fields, particularly at the Geyser in California, to counteract reservoir pressure decline (Dellinger 1997).

Injection into shallow or deep drillholes outside the geothermal reservoir has been considered and implemented in some fields (e.g., Seastres et al. 2000; Clotworthy 2000), as well as injection into the reservoir itself (e.g., Tokita et al. 2000a, b; Dellinger 1997). Moreover, exploration wells and old damaged production wells have also been used for injection (Puente & Rodriguez 2000). The first approach reflects reluctance to inject back into the reservoir because of the risk of thermal breakthrough, that is, the relatively cool injected fluid may flow rapidly into the aquifer of producing wells and deteriorate their performance. Indeed such cooling has been observed (e.g., Alincastre *et al.* 2000).

Geothermal reservoir rocks are typically fractured and therefore exhibit variable and anisotropic permeability. For that reason it is neither possible to predict with confidence how an injection well may perform with respect to its injectivity nor with respect to which way the injected fluid will flow once it is in the reservoir. Because of this complication, the success of injection varies between fields and it is anticipated that a special injection scheme must be developed for each field depending on its characteristics, mainly the three-dimensional distribution of permeability and the waste fluid composition. Injection may require drilling of special wells. Alternatively, wells drilled for the purpose of production may not have adequate yield but can be used successfully as injection wells. When this is the case, no special wells need to be drilled for injection purposes, which reduces road building and therefore scenery spoliation.

In some fields, decreased injectivity of the injection wells has created problems, such as at Hatchubaru (Kiyota *et al.* 2000) and Takigami (Goto 2000) in Japan. This problem has been attributed to silica scaling. At Hatchubaru, the rate of silica deposition has been reduced by acidifying the brine to a pH of 5 (Kiyota *et al.* 2000). Decreased injectivity was also observed in an injection well at Sumikava, Japan, when injecting separated water. The condition was improved by lowering the silica concentration in the spent water by mixing it with condensate before injection (Kumagai & Kitao 2000). At Ohaaki, New Zealand, injectivity and pressure transient tests showed clearly that hot water injection into colder rock caused a thermally induced reduction of permeability near the well bore. The reduction of permeability was reversed when cold water was injected (Clotworthy 2000). During drilling and completion tests (tests carried out on a well after it has been drilled to final depth but before the drill-rig is dismounted are called completion tests and include T and P logging, injectivity tests, and sometimes other tests, such as caliper logging), fluid circulation losses are commonly observed at deep levels in geothermal wells. Yet, the chemistry of the well discharge, when flowed, indicates that these deep (and hot) loss zones do not produce into the well (Gudmundsson & Arnórsson 2002). The reason is probably that permeability was induced by thermal contraction of the rock, when the cold water was injected. Upon thermal recovery of the rock, this permeability was destroyed. This observation, and that described by Clotworthy

(2000) above, indicates that, generally speaking, injection is favourable into rock, which is hotter than the injected fluid. Scaling of amorphous silica in injection wells is known to deteriorate injectivity. To avoid such scaling, various methods have been applied. They include the use of inhibitors, dilution of the waste water with condensate, acidification, and silica polymerization.

Changes in well fluid chemistry have occurred as a consequence of injection. Chloride concentrations have increased, such as at Cerro Prieto, Mexico (Puente & Rodriguez 2000) and Sumikawa, Japan (Kumagai & Kitao 2000). In the long run, repeated circulation of injected brine and its successive evaporation will cause the salinity of the water discharged from wells to increase, which may ultimately lead to scaling and corrosion problems. In some areas, the concentration of gases like H_2S has decreased but that of the atmospheric gases, N_2 and Ar, has increased (Arriaga *et al.* 2000)

Condensed steam from the cooling towers or heat exchangers is acid because of the presence of dissolved CO_2 and H_2S. Oxidation of H_2S into sulphuric acid may also contribute to the acidity. The acid condensate is corrosive, making common casing material in injection wells inadequate. This problem may be overcome by mixing the condensate with separated water. Alternatively, addition to the condensate of caustic soda (NaOH) to raise its pH may solve the corrosion problem. From an environmental point of view use of caustic soda is undesirable. Its production and transportation to the geothermal site, and subsequent use, may lead to some spilling, for example, if car accidents occur, causing vegetation to be killed, harming animal life and contaminating groundwater.

Hot-dry-rock systems

Use of HDR systems has received considerable attention during the last two decades or so (Baldeyrou-Bailly *et al.* 2004). The amount of heat stored within the uppermost few kilometres of the Earth's crust is enormous. Projects aiming at developing technology to exploit heat from these impermeable systems have focused on areas with a relatively high T gradient (for continental crust), but away from regions of active volcanism and where thermal manifestations are lacking. The technology considered involves the drilling of two kinds of wells, injection and production wells. Permeability is created by hydro-fracturing. Injected water flows into the fractures, picks up heat from the hot rock as it flows through it, and is pumped to the surface via the production wells.

Hot-dry-rock systems in the roots of high-T volcanic geothermal systems should be given more attention. At present, production wells drilled into such systems are typically 2000–2500 m deep. Below this depth, and above the magma heat source, one can expect the presence of a large volume of very hot rock, which probably has generally low permeability. Injection and production tests of wells drilled into one of the Cornish granites at Rosemanowes, south-west-England, as a part of an HDR development project, have demonstrated that permeability can be thermally induced. Observations during injection tests in New Zealand indicate the same (Clotworthy 2000), as well as information on deep circulation losses during drilling and completion tests of geothermal wells in many parts of the world. One may speculate that drilling of very deep wells (3–5 km) into the roots of volcanic geothermal systems, and injection of cold water into these wells, will lead to the production of very hot fluid at deep levels that could ascend by buoyancy into shallower production wells, thus contributing to their output and maintaining reservoir pressures. This process would reduce recharge of cooler water into the reservoir from shallow zones, but at the same time it could improve the performance of production wells and increase their lifetime. If this happened, deep injection would increase the thickness of the exploitable reservoir, allowing more power (heat) to be produced per unit surface area of a geothermal field. Exploitation of heat from the roots of high-T geothermal systems, where temperatures may be 350–500 °C, could be expected to be associated with scaling problems. Chemical reaction rates, including mineral deposition, are expected to be much higher at these high temperatures than at temperatures commonly found today in exploited high-T geothermal reservoirs (200–300 °C). Therefore, potential scaling problems need to be assessed, as well as the process of thermally induced permeability, before embarking on a programme to develop the roots of high-T systems for electric power generation or other uses.

Conclusions

Geothermal energy is a rapidly growing energy resource. In the year 2000, it accounted for 0.25% of the annual world-wide energy consumption. Exploitation of geothermal resources is generally far less a cause of pollution than fossil-fuel combustion. It has, however, some adverse local effects on the environment, including: scenery spoliation; drying out of hot springs; soil erosion; noise pollution; and chemical pollution of the atmosphere and of surface- and groundwaters. Exploitation of geothermal resources may also enhance seismic activity and cause land subsidence. Various measures have been successfully employed to reduce the adverse environmental effects of geothermal energy utilization. The most important ones are directional drilling and injection of spent geothermal fluid.

I wish to extend my sincere thanks to Dr. B. Fritz, Prof. C. Panichi, Dr. H. Kristmannsdóttir and Dr. S. Loppi, who reviewed the manuscript and made many useful suggestions. I am particularly grateful to one of the editors of this book, Dr. R. Gieré, for his thorough review and many suggestions, which have improved the manuscript very much, yet made it substantially longer.

References

ALINCASTRE, R. S., SAMBRANO, B. M. G. & NOGARA, J. B. 2000. Geochemical evaluation of the reservoir response to exploitation of the Mindanao-1 geothermal production field, Philippines. *In: Proceedings World Geothermal Congress 2000*, Kyushu–Tohoku, Japan, 28 May–10 June 2000, 2483–2488.

ALLEGRINI, G., CAPPETTI, G. & SABATELLI, F. 1995. Geothermal development in Italy: Country update report. *In: Proceedings World Geothermal Congress*, Florence, 18–31 May, 201–208.

ALLIS, R. G. 2000. Review of subsidence at Wairakei field, New Zealand. *Geothermics*, **29**, 355–478.

ALLIS, R. G., CAREY, B., DARBY, D., READ, S. A. L., ROSENBERG, M. & WOOD, C. P. 1997. Subsidence at Haaki field, New Zealand. *In: Proceedings 19th New Zealand Geothermal Workshop*, 9–15.

ALLIS, R. G., ZHAN, X. & CLOTWORTHY, A. 1998. Predicting future subsidence at Wairakei field, New Zealand. *Geothermal Resources Council Transactions*, **22**, 43–47.

ÁRMANNSSON, H. & KRISTMANNSDÓTTIR, H. 1992. Geothermal environmental impact. *Geothermics*, **21**, 869–880.

ÁRMANNSSON, H. 2003. CO_2 emission from geothermal plants. *In: Proceedings of International Conference on Multiple Integrated Uses of Geothermal Resources*, Reykjavík, 14–17 September, S12, 56–62.

ANDREWS, J. E., BRIMBLECOMBE, P., JICKELLS, T. D. & LISS, P. S. 1996. *An Introduction to Environmental Chemistry*. London, Blackwell Science, 209 pp.

ARGUETA, G. G. M. 1995. *Rehabilitation of Geothermal Wells With Scaling Problems*. United Nations University Geothermal Training Programme Report, 1995, 207–240.

ARMSTEAD, H. C. H. 1975a. Summary of Section VIII. Electricity production. *In: Proceedings Second United Nations Symposium on the Development and Use of Geothermal Resources*, San Francisco, 20–29 May, 1975, ciii–cx.

ARMSTEAD, H. C. H. 1975b. Summary of Section V. Environmental factors and waste disposal. *In: Proceedings Second United Nations Symposium on the Development and Use of Geothermal Resources*, San Francisco, 20–29 May, 1975, lxxxvii–xciv.

ARNÓRSSON, S. 1975. Application of the silica geothermometer in low-temperature hydrothermal areas in Iceland. *American Journal of Science*, **275**, 763–784.

ARNÓRSSON, S. 1978a. Precipitation of calcite from flashed geothermal waters in Iceland. *Contributions to Mineralogy and Petrology*, **66**, 21–28.

ARNÓRSSON, S. 1978b. Major element chemistry of the geothermal sea-water at Reykjanes and Svartsengi, Iceland. *Mineralogical Magazine*, **42**, 209–220.

ARNÓRSSON, S. 1981. Mineral deposition from Icelandic geothermal waters, environmental and utilization problems. *Journal of Petroleum Technology*, **33**, 181–187.

ARNÓRSSON, S. 1985. The use of mixing models and chemical geothermometers for estimating underground temperatures in geothermal systems. *Journal of Volcanological and Geothermal Research*, **23**, 299–335.

ARNÓRSSON, S. 1989. Deposition of calcium carbonate minerals from geothermal waters – theoretical considerations. *Geothermics*, **18**, 33–39.

ARNÓRSSON, S. 1995. *Geothermal Investigations of Geothermal Resources in* Guatemala. International Atomic Energy Agency, Project Report GUA/8/009-06, 20 pp.

ARNÓRSSON, S. 1996. *Interpretation of Geochemical and Isotopic Data from Well Discharges in the Momotombo Geothermal Field, Nicaragua With Recommendations on Monitoring Studies*. International Atomic Energy Agency Report (Project NIC/8/008-03), May 1996, 24 pp.

ARNÓRSSON, S. 2003. Arsenic in surface- and up to 90 °C ground waters in a basalt area, N-Iceland: processes controlling its mobility. *Applied Geochemistry*, **18**, 1297–1312.

ARNÓRSSON, S. & ANDRÉSDÓTTIR, A. 1999. The dissociation constants of Al-hydroxy complexes at 0–350 °C and P_{sat}. *In: Proceedings of the Sixth International Symposium on Geochemistry of the Earth's Surface*, Reykjavík, 425–428.

Arnórsson, S. & GUNNLAUGSSON, E. 1985. New gas geothermometers for geothermal exploration – calibration and application. *Geochimica et Cosmochimica Acta*, **49**, 1307–1325.

ARNÓRSSON, S. & STEFÁNSSON, A. 1999. Assessment of feldspar solubility constants in water in the range 0° to 350 °C at vapor saturation pressures. *American Journal of Science*, **299**, 173–209.

ARNÓRSSON, S., BJÖRNSSON, S., JÓHANNESSON, H. & GUNNLAUGSSON, E. 1992. *The Production Characteristics of the Reykjavík District Heating Service low-Temperature Geothermal Systems*. *Árbók Verkfrædingafélags Íslands 1991/1992*. Reykjavík, 344–366. (In Icelandic.)

ARNÓRSSON, S., BJÖRNSSON, S., MUNA, Z. W. & OJIAMBO, S. B. 1990. The use of gas chemistry to evaluate boiling processes and initial steam fractions in geothermal reservoirs with an example from the Olkaria field, Kenya. *Geothermics*, **19**, 497–514.

ARNÓRSSON, S., ELÍASSON, J. & GUDMUNDSSON, B. TH. 1999. *40 MW Geothermal Power Plant in Bjarnarflag: Evaluation of the Impact on Ground Water and Natural Geothermal Manifestations*. Science Institute Report RH-26–99, 36 pp. (In Icelandic.)

ARNÓRSSON, S., GUNNARSSON, I., STEFÁNSSON, A., ANDRÉSDÓTTIR, A. & SVEINBJÖRNSDÓTTIR, Á. E. 2002. Major element chemistry of surface- and ground waters in basaltic terrain, N-Iceland. *Geochimica et Cosmochimica Acta*, **66**, 4015–4046.

ARNÓRSSON, S., GUNNLAUGSSON, E. & SVAVARSSON, H. 1983. The chemistry of geothermal waters in Iceland. II. Mineral equilibria and independent variables controlling water compositions. *Geochimica et Cosmochimica Acta*, **47**, 547–566.

ARNÓRSSON, S., RAGNARS, K. *et al.* 1975. Exploitation of saline high-temperature water for space heating. *In: Proceedings Second United Nations Symposium on the Development and Use of Geothermal Resources*, San Francisco, 20–29 May, 1975, 2077–2082.

ARRIAGA, M. C. S., TELLO, M. R. L. & SAMANIEGO, F. V. 2000. Geochemical evolution of the Los Azufres, Mexico, geothermal reservoir. Part II. Non-condensible gases. *In: Proceedings World Geothermal Congress 2000*, Kyushu–Tohoku, Japan, 28 May–10 June 2000, 2227–2233.

AXELSSON, G., FLOVENZ, O. G. *et al.* 2000. Thermal energy extraction from the Laugaland geothermal systems in N-Iceland. *In: Proceedings World Geothermal Congress 2000*, Kyushu–Tohoku, Japan, 28 May–10 June 2000, 3027–3032.

AXTMANN, R. C. 1975a. Chemical aspects of the environmental impact of geothermal power. *In: Proceedings Second United Nations Symposium on the Development and Use of Geothermal Resources*, San Francisco, 20–29 May, 1975, 1323–1327.

AXTMANN, R. C. 1975b. Environmental impact of a geothermal power plant. *Science*, **187**, 795–803.

BACCI, E., GAGGI, C., LANZILLOTTI, E., FERROZZI, S. & VALLI, L. 2000. Geothermal power plants at Mt. Amiata (Tuscany–Italy): mercury and hydrogen sulphide deposition revealed by vegetation. *Chemosphere*, **40**, 907–911.

BADRUK, M. & KABAY, N. 2003. Removal of boron from Kizildere–Denizli geothermal brines using ion-exchange method. *In: Proceedings International Conference on Multiple Integrated Uses of Geothermal Resources*, Reykjavík, 14–17 September, S14, 8–13.

BALDI, P., BERTINI, G., CALORE, C., CAPETTI, G., CATALDI, C., CELATI, R. & SQUARCI, P. 1980. Selection of dry wells in Tuscani for stimulation tests. *In: Proceedings 2nd DOE-ENEL Workshop for Cooperative Research in Geothermal Energy*, Berkeley, 98–115.

BALDEYROU-BAILLY, A., SURMA, F. & FRITZ, B. 2004. Geophysical and mineralogical impacts of fluid injection in a geothermal system: the Hot Fractured Rock site at Soultz-sous-Forêts, France. *In*: GIERÉ, R. & STILLE, P. (eds) *Energy, Waste and the Environment: a Geochemical Perspective*. Geological Society, London, Special Publication 236, 355–367

BALL, J. W., NORDSTROM, D. K., CUNNINGHAM, K. M., SCHOONEN, M. A. A., XU, Y. & DEMONGE, M. 1998. *Water-Chemistry and On-Situ Sulfur Speciation Data for Selected Springs in Yellowstone National Park, Wyoming, 1994–1995*. USGS Open-file report 98–574, 35 pp.

BALLANTYNE, J. M. & MOORE, J. N. 1988. Arsenic geochemistry in geothermal systems. *Geochimica et Cosmochimica Acta*, **52**, 475–483.

BALLZUS, C., FRÍMANNSSON, H., GUBBARSSON, G. I. & HROLFSSON, I. 2000*a*. The geothermal plant at Nesjavellir, Iceland. *In*: *Proceedings World Geothermal Congress 2000*, Kyushu–Tohoku, Japan, 28 May–10 June 2000, 3109–3114.

BALLZUS, C., FRÍMANNSSON, H. & MAACK, R. 2000*b*. Modular development of the Nesjavellir power plant for flexibility. *In*: *Proceedings World Geothermal Congress 2000*, Kyushu–Tohoku, Japan, 28 May–10 June 2000, 3115–3120.

BANWELL, C. J. 1975. Geothermal energy and its uses: Technical, economic, environmental, and legal aspects. *In*: *Proceedings Second United Nations Symposium on the Development and Use of Geothermal Resources*, San Francisco, 20–29 May 1975, 2257–2267.

BARGAGLI, R., CATENI, D., NELLI, L., OLMASTRONI, S. & ZAGARESE, B. 1997. Environmental impact of trace element emissions from geothermal power plants. *Archives of Environmental Contamination and Toxicology*, **33**, 172–181.

BARIA, R., BAUMGÄRTNER, J., GÉRARD, A. & GARNISCH, J. 2000. The European HDR programme: Main targets and results of the deepening of the well GPK2 to 5000 m. *In*: *Proceedings World Geothermal Congress 2000*, Kyushu–Tohoku, Japan, 28 May–10 June 2000, 3643–3652.

BARRELL, A., CAPPETTI, G. & STEFANI, G. 1995. Results of deep drilling in the Larderello–Travale/Radicondoli geothermal area. *In*: *Proceedings World Geothermal Congress*, Florence, 18–31 May, 1275–1278.

BATIK, H., KOÇAK, A., AKKUS, I., SIMSEK, S., MERTOGLU, O., DOKUZ, I. & BAKIR, N. 2000. Geothermal energy utilization development in Turkey – present geothermal situation and projections. *In*: *Proceedings World Geothermal Congress 2000*, Kyushu–Tohoku, Japan, 28 May–10 June 2000, 85–91.

BENOIT, D., JOHNSON, S. & KUMATAKA, M. 2000. Development of an injection augmentation program at the Dixie Valley, Nevada geothermal field. *In*: *Proceedings World Geothermal Congress 2000*, Kyushu–Tohoku, Japan, 28 May–10 June 2000, 819–824.

BERTRAMI, R., CALORE, C., CAPETTI, G., CELATI, R. & D'AMORE, F. 1985. A three-year recharge test by reinjection in the central area of Larderello field: Analysis of production rate. *Geothermal Resources Council Transactions*, **9**, 293–298.

BETTAGLI, N. & BIDINI, G. 1996. Larderello–Farinello–Valle Secolo geothermal area: Energy analysis of the transportation network and of the electric power plants. *Geothermics*, **25**, 3–16.

BJARNASON, J. O. & ARNÓRSSON, S. 1986. Nisyros Geothermal Development. *Nisyros Well* NIS-2. *Production Characteristics and Fluid Composition*. Unpublished National Energy Authority Report OS-86038, Reykjavík, Iceland, 97 pp. (In Icelandic.)

BOLDIZSÁR, T. 1970. Geothermal energy production from porous sediments in Hungary. *Geothermics*, **2**, 99–109.

BRANDSDÓTTIR, B., FRANZSON, H., EINARSSON, P., ÁRNASON, K. & KRISTMANNSDÓTTIR. H. 2002. Seismic monitoring during an injection experiment in the Svartsengi geothermal field, Iceland. *Jökull*, **51**, 43–52.

BRUEL, D. 1995. Heat extraction modeling from forced fluid flow through stimulated fractures rock masses: Applications to the Rosemanowes hot-dry rock reservoir. *Geothermics*, **24**, 361–374.

BUNING, B. C., NORIEGA, M. T., SARMIENTO, Z. F. & MALATE, R. C. M. 2000. Experimental injection set-up for downhole chemical dosing. *In*: *Proceedings World Geothermal Congress 2000*, Kyushu–Tohoku, Japan, 28 May–10 June 2000, 3033–3038.

BUSSOTTI, F., CENNI, E., COZZI, A. & FERRETTI, M. 1997. The impact of geothermal power plants on forest vegetation. A case study at Travale (Tuscany, Central Italy). *Environmental Monitoring and Assessment*, **45**, 181–194.

BUYUKTUNCEL, E., BEKTAS, S., SALIH, B., EVIRGEN, M. M. & GENC, O. 1997. Arsenic speciation in geothermal waters by HPLC/GFAAS and HPLC/HGAAS methods. *Fresenius Environmental Bulletin*, **6**, 494–501.

CAPPETTI, G., PASSALEVA, G. & SABATELLI, F. 2000. Italy country update report 1995–1999. *In*: *Proceedings World Geothermal Congress 2000*, Kyushu–Tohoku, Japan, 28 May–10 June 2000, 109–116.

CAREY, B. 2000. Wairakei 40 plus years of generation. *In*: *Proceedings World Geothermal Congress 2000*, Kyushu–Tohoku, Japan, 28 May–10 June 2000, 3145–3149.

CLOTWORTHY, A. 2000. Response of Wairakei geothermal reservoir to 40 years of production. *In*: *Proceedings World Geothermal Congress 2000*, Kyushu–Tohoku, Japan, 28 May–10 June 2000, 2057–2062.

D'AMORE, F. & TRUESDELL, A. H. 1979. Models for steam chemistry at Larderello and the Geysers. *In*: *Proceedings 5th Stanford Geothermal Engineering Workshop*, Stanford, 283–297.

D'AMORE, F., FANCELLI, R. & PANICHI, C. 1987. Stable isotope study of reinjection processes in the Larderello geothermal field. *Geochimica et Cosmochimica Acta*, **51**, 857–867.

DECARLO, E. H. & THOMAS, D. M. 1985. Removal of arsenic from geothermal fluids by adsorptive bubble flotation with colloidal ferric hydroxide. *Environmental Science & Technology*, **19**, 538–544.

DELLINGER, M. 1997. The Lake County Geysers effluent pipeline & injection project. *Geothermal Resources Council Bulletin*, **26**, 218–223.

DIAZ, C., GRACIA, H., ZAYAS, M. E., ESPINOZA, F. J. & VALLE-FUENTES, F. J. 2000. Producing optical glass with geothermal waste. *American Ceramic Society Bulletin*, **79**, 57–59.

DIAZ, C., VALLE-FUENTES, F. J., ZAYAS, M. E. & AVALOS-BORJA, M. 1999. Cordierite glass–ceramic from geothermal waste. *American Ceramic Society Bulletin*, **78**, 62–64.

EINARSSON, S. S., VIDES, R. A. & CUÉLLAR, G. 1975. Disposal of geothermal waste water by reinjection *In: Proceedings Second United Nations Symposium on the Development and Use of Geothermal Resources*, San Francisco, 20–29 May 1975, 1349–1363.

EL FIL, H., MANZOLA, A. S. & BENMOR, M. 2003. Decarbonation of geothermal waters by seeding with aragonite crystals coupled with air bubbling. *Applied Geochemistry*, **18**, 1137–1148.

ELLIS, A. J. & MAHON, W. A. J. 1964. Natural hydrothermal systems and experimental hot water/rock interactions. *Geochimica et Cosmochimica Acta*, **28**, 1323–1357.

ELLIS, A. J. & MAHON, W. A. J. 1967. Natural hydrothermal systems and experimental hot water/rock interactions. Part II. *Geochimica et Cosmochimica Acta*, **31**, 519–538.

ELLIS, A. J. & MAHON, W. A. J. 1977. *Chemistry and Geothermal Systems*. Academic Press, New York. 392 pp.

EVANOFF, J., YEAGER, V. & SPIELMAN, P. 1995. Stimulation and damage removal of calcium carbonate scaling in geothermal wells: A case study. *In: Proceedings World Geothermal Congress*, Florence, 18–31 May, 2481–2486.

EYSTEINSSON, H. 2000. Elevation and gravity changes at geothermal fields on the Reykjanes Peninsula, SW-Iceland. *In: Proceedings World Geothermal Congress, 2000*, Kyushu–Tohoku, Japan, 28 May–10 June 2000, 559–564.

FIORDELISI, A., MACKIE, R. L., MADDEN, T., MANZELLA, A. & RIEVEN, S. A. 1995. Application of the magnetotelluric method using a remote–remote reference system for characterizing deep geothermal system. *In: Proceedings World Geothermal Congress*, Florence, 18–31 May, 893–897.

FOUILLAC, C., FOUILLAC, A. M. & CRIAUD, A. 1990. Sulphur and oxygen isotopes of dissolved sulphur species in formation waters from the Dogger geothermal aquifer, Paris Basin, France. *Applied Geochemistry*, **5**, 415–427.

FOURNIER, R. O. & ROWE, J. J. 1966. Estimation of underground temperatures from the silica content of water from hot springs and wet-steam wells. *American Journal of Science*, **264**, 685–697.

FRIDLEIFSSON, I. B. 2001. Geothermal energy for the benefit of the people. *Renewable & Sustainable Energy Reviews*, **5**, 299–312.

FRIDLEIFSSON, I. B. 1979. Geothermal activity in Iceland. JÖKULL, **29**, 47–56.

GALLEGOS-ORTEGA, R., QUINTERO-NUNEZ, M. & GARCIA-CUETO, O. R. 2000. H_2S dispersion model at Cerro Prieto geothermoelectric power plant. *In: Proceedings World Geothermal Congress 2000*, Kyushu–Tohoku, Japan, 28 May–10 June 2000, 579–584.

GALLUP, D. L. 1989. Iron-silicate scale formation and inhibition at the Salton Sea geothermal field. *Geothermics*, **18**, 97–103.

GALLUP, D. L. 1997. Aluminum silicate scale formation and inhibition: Characterization and laboratory experiments. *Geothermics*, **26**, 483–499.

GALLUP, D. L., FEATHERSTONE, J. L., REVERENTE, J. P. & MESSER, P. H. 1995. Line Mine: A process for mitigating injection well damage at the Salton Sea, California (USA) geothermal field. *In: Proceedings World Geothermal Congress*, Florence, 18–31 May, 2403–2408.

GENDER, A., TRAINEAU, H., LEDÉSERT, B., BOURGINE, B. & GENTIER, S. 2000. Over 10 years of geological investigations within the HDR Soultz Project, France. *In: Proceedings World Geothermal Congress 2000*, Kyushu–Tohoku, Japan, 28 May–10 June 2000, 3707–3712.

GIANELLI, G. & PUXEDDU, M. 1992. Geologic comparison between Larderello and The Geysers geothermal fields. *Abstracts of the 29th IGC*, Kyoto, 3/3, 853.

GIGGENBACH, W. F. 1980. Geothermal gas equilibria: *Geochimica et Cosmochimica Acta*, **44**, 2021–2032.

GIGGENBACH, W. F. 1981. Geothermal mineral equilibria. *Geochimica et Cosmochimica Acta*, **45**, 393–410.

GIGGENBACH, W. F. 1982. C-13 exchange between CO_2 and CH_4 under geothermal conditions. *Geochimica et Cosmochimica Acta*, **46**, 159–165.

GIGGENBACH, W. F. 1991. Chemical techniques in geothermal exploration. *In:* D'AMORE F. (ed) *Application of Geochemistry in Geothermal Reservoir Development*. UNITAR/UNDP Centre on Small Energy Resources, Rome, 119–144.

GIHRING, T. M., DRUSCHEL, G. K., MCCLESKEY, R. B., HAMERS, R. J. & BANFIELD, J. F. 2001. Rapid arsenite oxidation by Thermus aquaticus and Thermus thermophilus: Field and laboratory investigations. *Environmental Science and Technology*, **35**, 3857–3862.

GÍSLASON, G. 2000. Nesjavellir co-generation plant, Iceland. Flow of geothermal steam and non-condensable gases. *In: Proceedings World Geothermal Congress 2000*, Kyushu–Tohoku, Japan, 28 May–10 June 2000, 585–590.

GLOVER, R. & BACON, L. 2000. Chemical changes in natural features and well discharges at Wairakei, New Zealand. *In: Proceedings World Geothermal Congress 2000*, Kyushu–Tohoku, Japan, 28 May–10 June 2000, 2081–2086.

GLOWACKA, E., GONZALES, J. & NAVA, F. A. 2000. Subsidence in Cerro Prieto geothermal field, Baja California, Mexico. *In: Proceedings World Geothermal Congress 2000*, Kyushu–Tohoku, Japan, 28 May–10 June 2000, 591–596.

GOFF, F. & JANIK, C. J. 2000. Geothermal systems. *In:* SIGURDSSON, H., HOUGHTON, B., MCNUTT, S. R., RYMER, H. & STIX, J. (eds) *Encyclopedia of Volcanoes*. Academic Press, New York, 817–834.

GOTO, H. 2000. The decrease of capacity in re-injection wells in the Takigami field, Japan. *In: Proceedings World Geothermal Congress 2000*, Kyushu–Tohoku, Japan, 28 May–10 June 2000, 3059–3063.

GUDMUNDSSON, B. TH. & ARNÓRSSON, S. 2002. Geochemical monitoring of the Krafla and Námafjall geothermal areas, N-Iceland. *Geothermics*, **31**, 195–243.

GUNNARSSON, I. & ARNÓRSSON, S. 2003. Silica scaling: The main obstacle in efficient use of high-temperature geothermal fluids. *In: Proceedings International Geothermal Conference on Multiple Integrated Uses of Geothermal Resources*, Reykjavík, 14–17 September, S13, 30–36.

GUNNARSSON, A., STEINGRÍMSSON, B. S., GUNNLAUGSSON, E., MAGNÚSSON, J. & MAACK, R. 1992. Nesjavellir co-generation power plant. *Geothermics*, **21**, 559–583.

HARDARDÓTTIR, V., KRISTMANNSDÓTTIR, H. & ÁRMANNSSON, H. 2001. *Scale Formation in Wells RN-9 and RN-8 in the Reykjanes Geothermal Field, Iceland. Water–Rock Interaction (WRI-10).* Balkema, Rotterdam, 851–854.

HAUKSSON, T., THÓRHALLSSON, S., GUNNLAUGSSON, E. & ALBERTSSON, A. 1995. Control of magnesium silicate scaling in district heating systems. *In: Proceedings World Geothermal Congress*, Florence, 18–31 May, 2487–2490.

HAYASHI, K. & MINAMIURA, K. 2000. Growth process of the reservoir in the Yunomori HDR/HWR test site – an approach based on a model of reservoir growth in the shear mode. *In: Proceedings World Geothermal Congress 2000*, Kyushu–Tohoku, Japan, 28 May–10 June 2000, 2263–2268.

HEDENQUIST, J. W. 1990. The thermal and geochemical structure of the Broadlands–Ohaaki geothermal system, New Zealand. *Geothermics*, **19**, 151–185.

HERMOSO, D. Z., MEJORADA, A. V. & RAE, A. J. 1998. Determination of the nature of acidic fluids in the Palinpinon geothermal field, Philippines through the use of $\delta^{34}S$ in sulphates and sulphides. *In: Proceedings 19th Annual PNOC-EDC Geothermal Conference*, 65–75.

HOYER, D., KITZ, K. & GALLUP, D. 1991. Salton Sea unit 2: Innovations and successes. *Geothermal Resources Council Transactions*, **15**, 355–361.

HUTTRER, G. W. 2001. The status of world geothermal power generation 1995–2000. *Geothermics*, **30**, 1–27.

IGA 2003. International Geothermal Association World Wide Web Address: iga.igg.cnr.it/index.php

IZQUIERDO, G., ARELLANO, V. M., ARAGÓN, A., PORTUGAL, E. & MARTINEZ, I. 2000. Fluid acidity and hydrothermal alteration at the los Humeros geothermal reservoir Puebla, Mexico. *In: Proceedings World Geothermal Congress 2000*, Kyushu–Tohoku, Japan, 28 May–10 June, 1301–1306.

KARINGITHI, C. W. 2002. *Hydrothermal Mineral Buffers Controlling Reactive Gases Concentrations in the Greater Olkaria Geothermal System, Kenya.* MSc thesis, University of Iceland, 94 pp.

KIM, J. P. 1995. Methylmercury in rainbow-trout (oncorhynchus-mykiss) from lakes Okareka, Okaro, Rotomahana, Rotorua and Tarawera, North-Island, New Zealand. *Science of the Total Environment*, **164**, 209–219.

KISSLING, W. M., BROWN, K. L., O'SULLIVAN, M. J., WHITE, S. P. & BULLIVANT, D. P. 1996. Modelling chloride and CO_2 chemistry in the Wairakei geothermal reservoir. *Geothermics*, **25**, 285–305.

KIYOTA, Y., MATSUDA, K. & SHIMADA, K. 1996. Characteristics of acid water on the Otake–Hatchubaru geothermal field. *In: Proceedings 17th Annual PNOC-EDC Geothermal Conference*, 131–135.

KIYOTA, Y., HIROWATARI, K., TOKITA, H., HARUGUCHI, K. & UOGATA, K. 2000. Evaluation on geothermal injection treatment by pH modification. *In: Proceedings World Geothermal Congress 2000*, Kyushu–Tohoku, Japan, 28 May–10 June 2000, 3077–3082.

KJARTANSSON, G. 2000. Power plant process at Nesjavellir based on experimental tests. *In: Proceedings World Geothermal Congress 2000*, Kyushu–Tohoku, Japan, 28 May–10 June 2000, 1355–1359.

KLEIN, C. W. 1995. Management of fluid injection in geothermal wells to avoid silica scaling at low levels of silica oversaturation. *In: Proceedings World Geothermal Congress*, Florence, Italy, 18–31 May, 2451–2456.

KRISTMANNSDÓTTIR, H. 1989. Types of scaling occurring by geothermal utilization in Iceland. *Geothermics*, **18**, 183–190.

KUBO, B. M. 2003. Environmental management at Olkaria geothermal project, Kenya. *In: Proceedings International Geothermal Conference on Multiple Integrated Uses of Geothermal Resources*, Reykjavík, 14–17 September, S12, 72–80.

KUBODA, K. & AOSAKI, K. 1975. Reinjection of geothermal hot water at the Otake geothermal field. *In: Proceedings Second United Nations Symposium on the Development and Use of Geothermal Resources*, San Fransisco, California, 20–29 May, 1379–1383.

KUMAGAI, N. & KITAO, K. 2000. Reinjection problems encountered in Sumikawa geothermal power plant, Japan. *In: Proceedings World Geothermal Congress 2000*, Kyushu–Tohoku, Japan, 28 May–10 June 2000, 2683–2688.

LAPLAIGE, P., JAUDIN, F., DESPLAN, A. & DEMANGE, J. 2000. The French geothermal experience. Review and perspectives. *In: Proceedings*

World Geothermal Congress 2000, Kyushu–Tohoku, Japan, 28 May–10 June 2000, 283–295.

LEE, S. & BACON, L. 2000. Operational history of the Ohaaki geothermal field, New Zealand. *In: Proceedings World Geothermal Congress 2000*, Kyushu–Tohoku, Japan, 28 May–10 June 2000, 3211–3216.

LÍNDAL, B. 1975. Development of industry based on geothermal energy, geothermal brine, and sea water in the Reykjanes Peninsula, Iceland. *In: Proceedings Second United Nations Symposium on the Development and Use of Geothermal Resources*, San Francisco, 20–29 May 1975, 2223–2228.

LÍNDAL, B. & KRISTMANNSDÓTTIR, H. 1989. The scaling properties of the effluent water from Kizildere power station, Turkey, and recommendation for a pilot plant in view of district heating applications. *Geothermics*, **18**, 217–223.

LOPPI, S. 2001. Environmental distribution of mercury and other trace elements in the geothermal area of Bagnore (Mt. Amiata, Italy). *Chemosphere*, **45**, 991–995.

LOPPI, S. & BONINI, I. 2000. Lichens and mosses as biomonitors of trace elements in areas with thermal springs and fumarolic activity (Mt. Amiata, Central Italy). *Chemosphere*, **41**, 1333–1336.

LOPPI, S., GIOMARELLI, B. & BARGAGLI, R. 1999. Lichens and mosses as biomonitors of trace elements in a geothermal area (Mt. Amiata, central Italy). *Cryptogamie Mycologie*, **20**, 119–126.

LUND, J. W. & BOYD, T. L. 2000. Geothermal direct-use in the United States update: 1995–1999. *In: Proceedings World Geothermal Congress 2000*, Kyushu–Tohoku, Japan, 28 May–10 June 2000, 297–305.

LUND, J. W. & FREESTON, D. H. 2001. World-wide direct uses of geothermal energy 2000. *Geothermics*, **30**, 29–68.

MAHON, W. A. J. 1966. Silica in hot water discharged from drillholes at Wairakei, New Zealand. *New Zealand Journal of Science*, **9**, 135–144.

MARANI, M., TOLE, M. & OGALO, L. 2000. Concentrations of H_2S in air around the Olkaria geothermal field, Kenya. *In: Proceedings World Geothermal Congress 2000*, Kyushu–Tohoku, Japan, 28 May–10 June 2000, 649–661.

MATSUDA, K., SHIMADA, K. & KIYOTA, Y. 2000. Development of study methods for clarifying formation mechanism and distribution of acid geothermal-fluid. Case studies of geothermal areas in Kyushu, Japan. *In: Proceedings World Geothermal Congress 2000*, Kyushu–Tohoku, Japan, 28 May–10 June 2000, 1425–1430.

MATURGO, O., ZAIDE-DELFIN, M., LAYUGAN, D. & CATANE, J. P. 2000. Characteristics of the volcanic-hydrothermal system in Mt. Labo, Philippines: Implications to development. *In: Proceedings World Geothermal Congress 2000*, Kyushu–Tohoku, Japan, 28 May–10 June 2000, 1431–1436.

MCKEE, J. E. & WOLF, H. W. 1971. *Water Quality Criteria*. State Water Resources Control Board, California, Publication 3-A, 548 pp.

MERCADO, S. 1975. Cerro Prieto geothermoelectric project: Pollution and basic protection. *In: Proceedings Second United Nations Symposium on the Development and Use of Geothermal Resources*, San Francisco, 20–29 May 1975, 1394–1398.

MERTOGLU, O., BAKIR, N. & KAYA, T. 2003. Geothermal applications in Turkey. *Geothermics*, **32**, 419–428.

MINISSALE, A. 1991. The Larderello geothermal field – A review. *Earth Science Reviews*, **31**, 133–151.

MOREY, G. W., FOURNIER, R. O. & ROWE, J. J. 1962. The solubility of quartz in water in the temperature interval from 25 to 300 °C. *Geochimica et Cosmochimica Acta*, **26**, 1029–1043.

MUFFLER, L. J. P. 1975. Tectonic and hydrologic control of the nature and distribution of geothermal resources. *In: Proceedings Second United Nations Symposium on the Development and Use of Geothermal Resources*, San Francisco, 20–29 May 1975, 499–507.

MUFFLER, L. J. P. & WHITE, D. E. 1968. Origin of CO_2 in the Salton Sea geothermal system, southeastern CALIFORNIA, U.S.A. *XXIII International Geological Congress*, Prague, **17**, 185–194.

NAKAGAWA, R. 1999. Estimation of mercury emissions from geothermal activity in Japan. *Chemosphere*, **38**, 1867–1891.

NEHRING, N. L. & D'AMORE, F. 1984. Gas chemistry and thermometry of the Cerro Prieto, Mexico, geothermal field. *Geothermics*, **13**, 75–84.

NICHOL, D. A. C. & ROBINSON, B. A. 1990. Modelling of the heat extraction from the Rosemanowes HDR reservoir. *Geothermics*, **19**, 247–257.

NIMICK, D. A., MOORE, J. N., DALBY, C. E. & SAVKA, M. W. 1998. The fate of geothermal arsenic in the Madison and Missouri rivers, Montana and Wyoming. *Water Resources Research*, **34**, 3051–3067.

NORDSTRÖM, D. K. & JENNE, E. A. 1977. Fluorite solubility equilibria in selected geothermal waters. *Geochimica et Cosmochimica Acta*, **41**, 175–188.

ÓLAFSSON, J. H. 1996. The blue lagoon in Iceland and psoriasis. *Clinics in Dermatology*, **14**, 647–651.

ÓSKARSSON, N. 1982. On chemical reactions in volcanic gases. *In: Eldur i Nordri*. Societas Scientiarum Islandica, Reykjavík, 275–281. (In Icelandic.)

O'SULLIVAN, M. J., BULLIVANT, D. P., MANNINGTON, W. I. & FELLOWS, S. E. 1998. Modelling of the Wairakei–Tauhara geothermal system. *In: Proceedings 20th New Zealand Geothermal Workshop*, University of Auckland, 59–66.

PANICHI, C. 2004. Geochemical impact of re-injecting geothermal waste waters: Example Larderello, Italy. *In: GIERÉ, R. & STILLE, P. (eds) Energy, Waste and the Environment: a Geochemical Perspective. Geological Society*, London, Special Publication **236**, 337–354.

PETRACCO, C. & SQUARCI, P. 1975. Hydrological balance of Larderello geothermal region. *In: Proceedings Second United Nations Symposium on the Development and Use of Geothermal*

Resources, San Francisco, 20–29 May 1975, 521–530.

POLI, M. 1975. Relevance of geothermal energy in today's energy situation. *In*: *Proceedings Second United Nations Symposium on the Development and Use of Geothermal Resources*, San Francisco, 20–29 May 1975, 2339–2341.

PRUESS, K. 2002. *Mathematical Modeling of Fluid Flow and Heat Transfer in Geothermal Systems.* An introduction in five lectures held at the United Nations University Geothermal Programme, Reykjavík, Iceland. Report to the Earth Science Division, Lawrence Berkeley National Laboratory, University of California, 83 pp.

PUENTE, H.G & RODRIGUEZ, M. H. 2000. 28 years of production at Cerro Prieto geothermal field. *In*: *Proceedings World Geothermal Congress 2000*, Kyushu–Tohoku, Japan, 28 May–10 June 2000, 855–859.

RAGNARSSON, A. 2000. Geothermal developments in Iceland 1995–1999. *In*: *Proceedings World Geothermal Congress 2000*, Kyushu–Tohoku, Japan, 28 May–10 June 2000, 363–375.

RAGNARSDÓTTIR, K. V., WALTHER, J. V. & ARNÓRSSON, S. 1984. Description and interpretation of the composition of fluid and alteration mineralogy in the geothermal system at Svartsengi, Iceland. *Geochimica et Cosmochimica Acta*, **48**, 1535–1553.

RAMOS-CANDELARIA, M., CABEL, A. C. JR., BUNING, B. C. & NORIEGA, M. T. JR. 2000. Calcite inhibition field trials at the Mindanao geothermal production field (MGPF), Philippines. *In*: *Proceedings World Geothermal Congress 2000*, Kyushu–Tohoku, Japan, 28 May–10 June 2000, 2171–2176.

READ, M. J. & CAMPBELL, G. E. 1975. Environmental impact of development in the Geysers geothermal field, USA. *In*: *Proceedings Second United Nations Symposium on the Development and Use of Geothermal Resources*, San Francisco, 20–29 May 1975, 1399–1410.

REYES, A. G., TROMPETTER, W. J., BRITTEN, K. & SEARLE, J. 2002. Mineral deposits in the Rotokawa geothermal pipelines, New Zealand. *Journal of Volcanology and Geothermal Research*, **119**, 215–239.

RICHARDS, H. G., PARKER, R. H. *et al.* 1994. The performance of the experimental hot dry rock geothermal reservoir at Rosemanowes, Cornwall (1985–1988). *Geothermics*, **23**, 73–109.

RICHARDS, H. G., SAVAGE, D. & ANDREWS, J. N. 1992. Granite–water reactions in an experimental Hot Dry Rock geothermal reservoir: Rosemanowes test site, CORNWALL, U. K. *Applied Geochemistry*, **7**, 193–222.

RIMSTIDT, J. D. 1997. Quartz solubility at low temperatures. *Geochimica et Cosmochimica Acta*, **61**, 2553–2558.

ROBERTSON-TAIT, A., KLEIN, C. & McLARTY, L. 2000. Utility of the data gathered from the Fenton Hill Project for development of enhanced geothermal systems. *In*: *Proceedings World Geo-*

thermal Congress 2000, Kyushu–Tohoku, Japan, 28 May–10 June 2000, 3847–3852.

ROBINSON, B., OUTRED, H., BROOKS, R. & KIRKMAN, J. 1995. The distribution and fate of arsenic in the Waikato river system, North Island, New Zealand. *Chemical Speciation and Bioavailability*, **7**, 89–96.

ROJAS, J., MENJOS, A., MARTIN, J. C., CRIAUD, A. & FOUILLAC, C. 1987. Development and exploitation of low enthalpy geothermal systems: example of the Dogger in the Paris Basin, France. *In*: *Proceedings of the 12th Workshop on Geothermal Reservoir Engineering*, Stanford University, 203–212.

ROTHBAUM, H. P. & ANDERTON, B. H. 1975. Removal of silica and arsenic from geothermal discharge waters by precipitation with useful calcium silicates. *In*: *Proceedings Second United Nations Symposium on the Development and Use of Geothermal Resources*, San Francisco, 20–29 May 1975, 1417–1425.

RUGGIERI, G. & GIANELLI, G. 1995. Fluid inclusion data from the Carboli 11 well, Larderello geothermal field, Italy. *In*: *Proceedings World Geothermal Congress*, Florence, 18–31 May, 1087–1091.

SAKAMOTO, H., KAMADA, M. & YONEHARA, N. 1988. The contents and distributions of arsenic, antimony, and mercury in geothermal waters. *Bulletin of the Chemcal Society of Japan*, **61**, 3471–3477.

SANOPOULOS, D. & KARABELAS, A. 1997. H_2S abatement in geothermal plants: Evaluation of process alternatives. *Energy Resources*, **19**, 63–77.

SATMAN, A., SERPEN, U. & MIHCAKAN, M. 2000. Assessment of reinjection trials in Kizildere geothermal field. *In*: *Proceedings World Geothermal Congress 2000*, Kyushu–Tohoku, Japan, 28 May–10 June 2000, 1695–1700.

SEASTRES, J. S. JR., SALONGA, N. D., SAW, V. S. & MAXONI, D. A. 2000. Reservoir management strategies to sustain the full exploitation of greater Tongonan geothermal field, Philippines. *In*: *Proceedings World Geothermal Congress 2000*, Kyushu–Tohoku, Japan, 28 May–10 June 2000, 2863–2868.

SEKIOKA, M. & YOSHII, M. 2000. Country update report of geothermal direct uses in Japan. *In*: *Proceedings World Geothermal Congress 2000*, Kyushu–Tohoku, Japan, 28 May–10 June 2000, 433–437.

SHERBURN, S., ALLIS, R. & CLOTWORTHY, A. 1990. Microseismic activity at Wairakei and Ohaaki geothermal fields. *In*: *Proceedings 12th New Zealand Geothermal Workshop*, University of Auckland, 51–55.

SHIMADA, K., INUYAMA, F. & TOKITA, H. 2000. Hot spring interference study for predicting hot spring changes in geothermal fields. *In*: *Proceedings World Geothermal Congress 2000*, Kyushu–Tohoku, Japan, 28 May–10 June 2000, 757–762.

SHIMIZU, K., SUZUKI, E. *et al.* 2000. Prevention of scale adhesion using surface treatment of coating with organic compound at geothermal power station. *In*: *Proceedings World Geothermal Congress 2000*, Kyushu–Tohoku, Japan, 28 May–10 June 2000, 3299–3301.

SIMMONS, S. F., & BROWN, P. R. L. 2000. Hydrothermal minerals and precious metals in the Broadlands-Ohaaki geothermal system: Implications for understanding low-sulfidation epithermal environments. *Economic Geology and the Bulletin of the Society of Economic Geologists*, **95**, 971–999.

SMITH, B., BEALL, J. & STARK, M. 2000. Induced seismicity in the Geysers field, California, USA. *In*: *Proceedings World Geothermal Congress 2000*, Kyushu–Tohoku, Japan, 28 May–10 June 2000, 2887–2892.

STEFÁNSSON, A. & ARNÓRSSON, S. 2000. Feldspar saturation state in natural waters. *Geochimica et Cosmochimica Acta*, **64**, 2257–2584.

STEFÁNSSON, A. & ARNÓRSSON, S. 2002. Gas pressures and redox reactions in geothermal fluids in Iceland. *Chemical Geology*, **190**, 251–271.

TAKEUCHI, K., FUJIOKA, Y., HIROWATARI, K., KUSABA, S. & SUZUKI, H. 2000. Scale prevention method by pH modification using advanced bioreactor. *In*: *Proceedings World Geothermal Congress 2000*, Kyushu–Tohoku, Japan, 28 May–10 June 2000, 3623–3626.

TENMA, N., YAMAGUCHI, T., TEZUKA, K. & KARASAWA, H. 2000. A study of the pressure–flow response of the Hijiori reservoir at the Hijiori HDR test site. *In*: *Proceedings World Geothermal Congress 2000*, Kyushu–Tohoku, Japan, 28 May–10 June 2000, 3917–3920.

THÓRHALLSSON, S., ÁRMANNSSON, H. & HAUKSSON, T. 1984. *Milos Geothermal Development. Milos M-2 Production Test*. National Energy Authority Report, December 1984, 93 pp.

THÓRHALLSSON, S., RAGNARS, K., ARNÓRSSON, S. & KRISTMANNSDÓTTIR, H. 1975. Rapid scaling of silica in two district heating systems. *In*: *Proceedings Second United Nations Symposium on the Development and Use of Geothermal Resources*, San Francisco, 20–29 May, 1445–1449.

TOKITA, H., HARUGUCHI, K. & KAMENOSONO, H. 2000a. Maintaining the rated power output of the Hatchubaru geothermal field through an integrated reservoir management. *In*: *Proceedings World Geothermal Congress 2000*, Kyushu–Tohoku, Japan, 28 May–10 June 2000, 2263–2268.

TOKITA, H., TAKAGI, H. *et al.* 2000b. Development and verification of a method to forecast hot springs interference due to geothermal power exploitation. *In*: *Proceedings World Geothermal Congress 2000*, Kyushu–Tohoku, Japan, 28 May–10 June 2000, 725–730.

TRUESDELL, A. H., HAIZLIP, J. R., ÁRMANNSSON, H. & D'AMORE, F. 1989. Origin and transport of chloride in superheated geothermal steam. *Geothermics*, **18**, 295–304.

TRUESDELL, A. H., THOMPSON, J. M., COPLEN, T. B., NEHRING, N. L. & JANIK, C. J. 1981. The origin of the Cerro Prieto brine. *Geothermics*, **10**, 225–238.

VILLA, F. P. 1975. Geothermal plants in Italy: Their evolution and problems. *In*: *Proceedings Second United Nations Symposium on the Development and Use of Geothermal Resources*, San Francisco, 20–29 May 1975, 2061–2064.

VITOLO, S. & SEGGIANI, M. 2001. Mercury removal from geothermal exhaust gas by sulfur-impregnated and virgin activiated carbons. *Geothermics*, **31**, 431–442.

WEBSTER, J. G. 1999. The source of arsenic (and other elements) in the Marbel–Matingao river catchment, Mindanao, Philippines. *Geothermics*, **28**, 95–111.

WHITE, D. E., ANDERSON, E. T. & GRUBBS, D. K. 1963. Geothermal brine well: Mile deep drill hole may tap ore-bearing magmatic water and rocks undergoing metamorphism. *Science*, **139**, 919–922.

WHITE, D. E., MUFFLER, L. J. P. & TRUESDELL, A. H. 1971. Vapor-dominated hydrothermal systems compared with hot-water systems. *Economic Geology*, **66**, 75–97.

YANAGESE, T., SUGINORA, Y. & YANAGESE, K. 1970. The properties of scales and methods to prevent them. *Geothermics*, **Special Issue 2**, 1619–1624.

YANAGESE, K., YOSHINAGA, T. & KAWANO, K. 1983. Arsenic removal from geothermal water by coprecipitation with iron(III) hydroxide. *Bunseki Kagaku*, **32**, T111–T116.

YOKOYAMA, T., TAKAHASHI, Y. & TARUTANI, T. 1993. Simultaneous determination of arsenic and arsenious acids in geothermal water. *Chemical Geology*, **103**, 103–111.

ZANCANI, C. F. A. 1975. Summary of thirty years' experience in selecting thermal cycles for geothermal power pants. *In*: *Proceedings Second United Nations Symposium on the Development and Use of Geothermal Resources*, San Francisco, 20–29 May 1975, 2069–2074.

ZHANG, Z., WANG, J., REN, X., LIU, S. & ZHU, H. 2000. The state-of-the-art and future development of geothermal energy in China. Country update report for the period 1996–2000. *In*: *Proceedings World Geothermal Congress 2000*, Kyushu–Tohoku, Japan, 28 May–10 June 2000, 505–507.

Geochemical impact of re-injecting geothermal waste waters: example, Larderello, Italy

COSTANZO PANICHI

Istituto di Geoscienze e Georisorse, CNR, Pisa, Italy (e-mail: panichi@igg.cnr.it)

Abstract: The impacts of re-injection of geothermal waste waters on the physico-chemical characteristics of the fluids of the Larderello vapour-dominated geothermal field in Tuscany (Italy) were monitored using deuterium and oxygen-18 as 'natural' tracers. The observed variations were due mainly to mixing between the re-injected and deep components. Large isotopic fractionations occur at depth during the evaporation of the re-injected water in the reservoir and may affect the evaluation of the recovered fluid using a simple mixing model. Stable isotope and gas/steam ratios are closely correlated in the fluid collected from the monitored wells. Maps showing the isotopic distribution of the steam and the distribution of the gas content have been generated for the whole Larderello geothermal field. Evaluation of data collected before and after re-injection reveals a good distinction between natural inflow induced by exploitation, and artificial recharge with waste fluid. Gases other than N_2 in the wells affected by injection are substantially diluted since condensates are formed by evaporation. Isotopic variations of H_2, CH_4, and CO_2 have been used to describe the disequilibrium conditions among gas components. This disequilibrium is induced by the settling of a liquid plume at the producing level in the reservoir.

Geothermal power generation is often considered as a clean alternative to fossil fuel or nuclear power stations. Although chemical contamination of the environment may occur through gas, steam, and bore or cooling water discharge, impacts can be minimized or even eliminated by careful management. The chemical composition of the fluid discharged is determined by the composition of the reservoir fluid, and the operating conditions used for power generation. Reservoir fluid chemistry is different for different fields. For example, geothermal fluids of the Salton Sea (USA), which is hosted by sediments including evaporite deposits, are acid and highly saline (pH < 5, [Cl] = 155 000 ppm; Henley *et al.* 1984). At the other extreme, those of the Nesjavellir, Krafla, and Namafjall fields, hosted by volcanic rocks in Iceland, are alkaline and of very low salinity (mostly <100 ppm Cl; Henley *et al.* 1984).

Most bore waters include high concentrations of at least one of the following chemical contaminants (Table 1): lithium (Li), boron (B as H_3BO_3), arsenic (As), hydrogen sulphide (H_2S), mercury (Hg), and sometimes ammonia (NH_3). If released into a river or lake, these contaminants can potentially damage aquatic life, terrestrial plants, and/or human health. The disposal of highly saline bore waters can also have an adverse effect on water quality.

In natural geothermal environments, the impact of such contaminants may be controlled by precipitation from the atmosphere (rain), leading to contamination of soil and/or vegetation. For example, Hg and As are precipitated in silica sinter, and NH_3 is readily taken up by soils. Boron can have a serious impact on vegetation (see Arnórsson, 2004). Steam condensate will typically have higher concentrations of H_2S, Hg, NH_3, and, to a lesser extent, B. Consequently these contaminants can become concentrated in the cooling water discharge (see Table 1).

At vapour-dominated fields, such as Larderello (Italy) and The Geysers (USA), gas in steam will be the most important effluent from an environmental perspective. The bulk gas emissions will be escaping from the exhausts of the power station, often through a cooling tower. Although consisting of mainly carbon dioxide (CO_2), the geothermal gases may contain very high concentrations of H_2S. The impact of H_2S discharge comprises an unpleasant odour, equipment corrosion, eye irritation, and respiratory damage to humans. When dissolved in water aerosols (e.g., fog), H_2S reacts with atmospheric oxygen to form more oxidized sulphur-bearing compounds. Although some of these oxidized sulphur compounds have been identified as components of 'acid rain', a direct link between acid rain and geothermal activity has not been proven (Reed & Renner 1995). Geothermal steam may also contain hydrocarbons, such as, methane (CH_4) or ethane (C_2H_6), as well radon (Rn) in

From: GIERÉ, R. & STILLE, P. (eds) 2004. *Energy, Waste, and the Environment: a Geochemical Perspective*. Geological Society, London, Special Publications, **236**, 337–354.
0305-8719/04/$15   © The Geological Society of London 2004.

Table 1. *Contaminant concentrations (ppm) in selected geothermal fluids and gases, and in a world average freshwater*

	Li	B	As	Hg	H_2S	NH_3
Freshwater	0.003	0.01	0.002	0.00004	—	0.04
Deep well water						
Salton Sea (USA)	215	390	12	0.006	16	386
Cerro Prieto (Mexico)	—	19	2.3	0.00005	0.16	127
Wairakei (New Zealand)	14	30	4.7	0.0002	1.7	0.2
Steam (s) or non-condensable gases (ncg)						
The Geysers, s	—	16	0.019	0.005	540	700
The Geysers, ncg	—	—	—	—	222	52
Cerro Prieto, s	—	—	—	0.04	—	—
Cerro Prieto, ncg	—	—	—	—	350	190
Wairakei, s	—	0.23	—	0.002	52	4

addition to H_2, N_2, and the noble gases (Henley *et al.* 1984).

Waste water disposal

The optimum utilization of a geothermal resource requires a good knowledge of the hydrological characteristics of the reservoir. One of the methods used to improve exploitation of geothermal reservoirs is re-injection of the waste geothermal fluids into the reservoir. This procedure, indeed, compensates for pressure drawdown caused by fluid production from wells. This operational strategy was initiated as a disposal method for purely environmental reasons, but theoretical and practical studies have shown that re-injection is also a powerful technique for increasing the longevity of geothermal resources, to maximize the amount of energy that can be extracted from a given reservoir, and to avoid ground subsidence due to depletion of thermal fluids in the geothermal reservoir (see Arnórsson, 2004).

During the development of the Ahuachapan geothermal field in El Salvador in the late 1960s, the injection of the waste brine from producing wells was, for the first time, considered a viable method of disposal, both for economic and environmental reasons.

For the purpose of fluid re-injection, either special wells must be drilled or non-productive wells must be used. If an injection well is located within the field of producing wells, there is a risk that relatively cold water from the injection well will flow too rapidly into the aquifer of a producing well, thus deteriorating its performance or even destroying it. In the long run, injection of waste brine into a producing reservoir may have some negative effects.

In order to avoid irreversible negative effects of re-injection, some tests must be carried out prior to launching a full-scale re-injection. Natural and artificial tracers are, in this respect, of considerable importance as they can provide information on processes taking place in the reservoir and on their evolution with time. Such tracers make it possible to calculate the proportion of the injected water in the fluid recovered at the production well, and can reveal existence of preferential flow paths of the injected water in the reservoir.

Chemical compounds, which have been used as tracers, include sodium fluorosceinate, rhodamin-WT, and halogen salts of potassium and magnesium. In some geothermal fields tritium-enriched water is used, for example, at The Geysers (Gulati *et al.* 1978) and at Ahuachapan (Einarsson *et al.* 1975). One of the drawbacks of using this tracer, however, is that it destroys the natural tritium balance in the reservoir. At Larderello, the natural abundance of tritium has been described, and is still used to study the recharge of the field by meteoric water (Celati *et al.* 1973; Panichi *et al.* 1974; Panichi & Gonfiantini 1978). Iodine-125 and iodine-131 have also been used as artificial tracers (Adams 1995). The stable isotopes of oxygen and hydrogen were suitable to trace quantitatively the underground pathways of the re-injected water within the Larderello geothermal field (Nuti *et al.* 1981).

Gambill (1990), Beall *et al.* (1989), and Beall (1993) have used geochemical data such as deuterium isotope and/or NH_3 concentration in non-condensable gases as natural tracers to estimate the amount of injected water recovered in various part of The Geysers geothermal field. In one injection experiment in the Svartsengi field (Iceland), an elevated level of N_2 was observed in some of the production wells. In this case, an aerated groundwater was used for re-injection, and the nitrogen in the groundwater served as a natural tracer in the experiment (Gudmundsson 1983).

Injection benefits were observed in well fields with low reservoir pressure (low boiling temperature), high steam enthalpy (dry reservoir with large heat contents), and high fracture density (large surface area for heat transfer). Several steam field operators have found that water injection into the vapour-dominated reservoir can be beneficial if performed properly (Cappetti *et al.* 1982; Bertrami *et al.* 1985; Gambill 1990; Enedy *et al.* 1991; Goyal & Box 1992). The positive contributions of water injection include: providing reservoir pressure support; maintaining steam production rate; reducing makeup well requirements; and increasing reserves and lifetime of the field by recovering a portion of approximately 90% of the heat that is stored in the rocks of the vapour-dominated systems.

In the oil industry, re-injection has been recognized and widely used for decades to increase the sweeping effect in the reservoir and, hence, to improve the recovery of hydrocarbons. That view is gradually gaining recognition in the geothermal industry, as a large number of studies conducted on re-injection have concluded that re-injection should be beneficial for the management of geothermal reservoirs.

In the past, the main purpose of re-injection to geothermal fields was to dispose of the waste water. At present, it is considered desirable to recover most of the re-injected water in the fluids delivered from the production wells. This change in attitude is based on the experience that, in most cases, it is possible to recover only a small fraction of the geothermal energy stored in the reservoir rock unless re-injection is practised in the exploitation of the geothermal resource. Re-injection experiments have been reported for up to 40 geothermal fields, and re-injection is an integrated part of field operations in about 20 fields worldwide (Stefansson 1997). This author revealed that re-injection into geothermal fields is not a widespread method for reservoir management. In many fields, re-injection is required for geothermal waste water disposal, and fields under development usually have plans for re-injection.

Injection underground does not necessarily require injection into the geothermal reservoir from which production is derived, because this process carries with it the risks of potential cooling of the production wells and possible adverse impacts on the geochemical characteristics of geothermal fluids. The waste fluid may also be injected into an aquifer other than the geothermal reservoir simply to avoid environmental impact due to surface disposal. Moreover, injection in an aquifer at a shallower level than that of the producing reservoir saves drilling costs. In such a case, obviously, reservoir pressure support, heat scavenging, or mitigation of ground subsidence cannot be expected; instead, the wellhead injection pressure may become impractically high over time because the waters injected into the aquifer are not subject to depletion. In fact, the injection pressure may become so high as to create seismic activity, ground heaving, leakage of the injection fluid to the surface, or even groundwater contamination. Such problems have occurred, or are suspected to have occurred, in some commercial projects in the USA, even if no more than one of these problems has affected a single project (Sanyal *et al.* 1995).

Compositional changes in reservoir fluids and recovery factor

The injected fluids usually have a chemical composition that is different from that of the *in situ* reservoir fluids, particularly when steam has been separated from the liquid phase and only the separated water is re-injected. The steam phase will be enriched in non-condensable gases, whereas the separated water has a higher concentration of dissolved solids, but a lower gas content. Changes in the chemistry of the produced fluids are to be expected in all fields where re-injection of separated water has been applied. When the same water is recycled, successive steam loss will increase its salinity, which may in turn lead to scaling problems (see Arnórsson, 2004).

In the Palimpinon geothermal field in the Philippines, where waste fluid has been injected for some years, re-injected fluid has already returned into the production wells, as indicated by an increase in the salinity of the water produced. Harper & Jordan (1985) report an increase in Cl concentration in a large number of production wells of the Palimpinon field. This response to re-injection was observed only three years after re-injection operations started. The increase in Cl is different for each of the production wells and reflects the amount of the re-injected fluid returning to each of the wells.

As the chemical composition of both the injected and produced fluids is monitored with time, the changes in chemical composition can be used to determine the fraction of injected fluid in the flow from each production well. Re-injection experience at Dixie Valley in Nevada (Benoit 1992) revealed increments of

Cl concentrations in production wells similar to those in Palimpinon, and the collected data document injection returns in all production wells of the field. The recovery of injectate in production wells varies between 18 and 100%.

The 'recovery factor' is defined as the ratio of additional steam provided by injection to the amount of water injected over the same period of time. Additional steam is the steam produced at the new decline rate (or improvement rate) due to injection minus the steam production calculated at a decline rate without re-injection. The recovery factor defined on the basis of production data may be different from that defined on the basis of geochemical data if considered on a well-by-well basis. However, the combined recovery from all production wells affected by one or more injection wells should agree when applying both methods given sufficient time, since (1) the total amount of boil water should appear as steam in production wells and be reflected in the production data, and (2) the steam originally to be produced from a given well but replaced by injection-derived steam should eventually be produced in other wells.

Recovery factor calculations may not be possible under certain situations, including: (1) a decrease in steam flow rate due to water breakthrough; (2) scale deposits in the well-bore and/or fracture conduits; (3) a fluctuating flow rate; and (4) the completion of additional production wells in injection-affected areas, which will have an impact on the decline rates of nearby production wells.

Case study: The Larderello geothermal field, Italy

The Larderello geothermal field, located 80 km southeast of Pisa in Tuscany, Italy (Fig. 1), produces superheated steam from a vapour-dominated reservoir. The thermal fluids are derived from different productive horizons located inside lithologic units that differ from each other according to depth. The upper productive horizon ('shallow reservoir') is constituted mainly of carbonate rocks and evaporites of Triassic age. The permeable levels of the crystalline basement, mainly comprising phyllites, quartzites, and gneiss of Late Palaeozoic age, form the 'deeper reservoir'. The metamorphic units of the basement have been reached by drill holes during the 1970s, and are, at present, the subject of intensive exploration, both in the western and eastern portions of the exploited field, where depths of 2500–4500 m below ground level have been reached (Cappetti et al. 1995).

Between 1926 and 1940, 111 wells (82% productive) reaching the top of the carbonate reservoir were drilled in an area of less than 4 km^2 (Fig. 1). Steam output gradually increased up to 1500 t/h. Between 1940 and 1950, the exploited area almost doubled to 7 km^2 and the total steam production arose to 2200 t/h in spite of the decrease occurring in the older exploited zones. During the next 30 years the exploited area was extended to 180 km^2, whereby drilling was performed outside of the high structures of carbonates. The new, deeper and more widely spaced (1 well/km^2) wells, notwithstanding the less favourable permeability conditions, enabled the reaching of a steam output of 3000 t/h in the early 1960s. After this period, exploitation of the suitable areas of the crystalline basement became the main goal for the exploitation of the Larderello field and was successfully conducted, even though the percentage of the drilling success dropped to 38%. The presence of steam inside these deeper horizons suggests the existence of a lateral continuity between the early productive areas and some peripheral basins (Barelli et al. 2000). This means that the total area of the Larderello geothermal district is over 400 km^2, practically doubled with respect to the conventional surface extension generally attributed to the Larderello field ('the classical areas').

The geothermal power plants in the Larderello vapour-dominated field produce large quantities of condensate, on the order of 900 kg/s (Ente Nazionale per l'Energia Elettrica [ENEL], personal communication). These waters have high contents of NH_3 and boric acid (H_3BO_3) and are therefore potential pollutants of the groundwater and the surface water used for local water supply. The chemical composition of the fluids from wells of the Larderello geothermal field is shown in Table 2. The data refer to wells drilled between 1930 and 1940 in several subunits of the field (for locations, see Fig. 1). The fluid composition in these wells remained almost unchanged during the entire production history, and thus represents the initial conditions of the field, the so-called *reference cores*.

Temporal variations in gas compositions, mainly due to depressurization of the field caused by extensive exploitation of the reservoir, have been observed in each of the subunits. These variations have been described for the entire exploitation history of the Larderello field before initiation of the re-injections, that is, from 1926 to 1979 (Scandiffio et al. 1995).

A mechanism of simple mixing between two endmembers has been proposed to explain the observed chemical and isotopic compositions of the steam produced in the field. The two

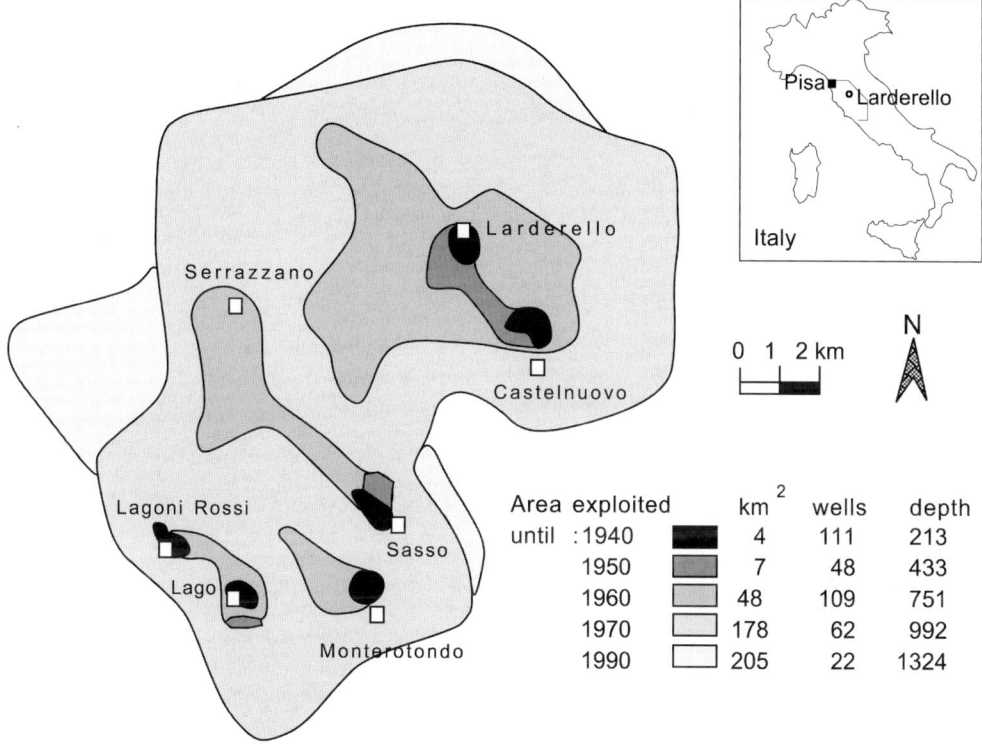

Fig. 1. Development of the Larderello geothermal field from 1926 to 1990.

endmembers are:

(1) A primary steam produced inside the reference cores. Typical isotopic values are $\delta D = -40\%_o$ and $\delta^{18}O = -1\%_o$ (all δ values given in this paper are $\%_o$ relative to V-SMOW), owing to extensive water–rock interaction undergone by the liquid phase in the reservoir. This interaction caused the original meteoric water to become enriched in ^{18}O, and to acquire the gas content as reported in Table 2.

(2) A secondary steam derived from the boiling of freshwater introduced into the geothermal reservoir after depressurization. Because of its short residence time at depth (Celati *et al.* 1973), this water interacts only lightly with the rock. Consequently, the steam contains small amounts of gases and its isotopic composition keeps typical meteoric values of $\delta D = -40\%_o$ and $\delta^{18}O = -7\%_o$. The occurrence of secondary steam can be traced back to the early 1960s and represents a natural recent recharge of the field.

Table 2. *Fluid composition of some Larderello subunits (in mmol/mol of total fluid)*

Gas	Larderello	Castelnuovo	Monterotondo	Lago
H_2O	979.95	974.69	990.25	982.95
CO_2	18.992	23.817	8.519	15.282
H_2	0.412	0.288	0.329	0.584
H_2S	0.428	0.430	0.234	0.463
CH_4	0.312	0.260	0.371	0.370
N_2	0.181	0.161	0.095	0.110
NH_3	0.212	0.227	0.107	0.107
H_3BO_3	0.115	0.127	0.092	0.136

Wells in the southern zones of the Larderello, Castelnuovo, Lago, and Monterotondo subunits (Fig. 1) produce steam with different degrees of mixing between the two endmembers. In the triangular diagrams of Fig. 2, where the relative concentrations of H_2O, CO_2, and H_2 are reported, this mixing process results in a shift of the data points of the reference cores towards the H_2O corner. The dilution of original steam may be explained by the fact that carbonate rocks crop out only south of the field, and new recent meteoric water may infiltrate here and reach the reservoir. On the other hand, the wells located in the northern zones of the field produce steam samples that are enriched in CO_2, moving towards the CO_2 corner starting from the reference core values. Two different mechanisms may be responsible for the high concentration of CO_2 in the gas extracted in northern zones of the Larderello geothermal field. The first requires that the CO_2 origin is internal to the system due to metasomatic interactions between water and rocks of the reservoir, assuming that steam is formed by liquid boiling between 200 and 300 °C. As the liquid evaporates, releasing CO_2 (and other gases) into the vapour phase, a new amount of this gas is formed in the residual liquid by feldspar alteration in the metamorphic basement (Calore *et al.* 1990).

On the other hand, a linear combination of 10% of an equilibrium boiling (single-step steam separation) and a Rayleigh distillation

(continuous steam separation) at 90% of the liquid phase in the reservoirs may also explain the compositional differences among the several zones of the Larderello geothermal field. The results of this process at 250 °C are reported in Fig. 3, where the theoretical path of the steam produced during such boiling, starting from a liquid system that has the composition of the Larderello reference core, is shown. It can be observed that within 6% of evaporation (corresponding to the early stages of exploitation) a strong enrichment of CO_2 in the vapour phase occurs (at 8% of evaporation, the composition of the resulting vapour phase is the same as initially assumed for the dissolved gases in the original liquid). As the evaporated fraction increases (later stages of production) and the residual liquid becomes almost devoid of gases, the representative points of the produced steam will move along line B in Fig. 3. The same figure shows, as comparison, the linear path of the fluid obtained by simple mixing of primary and secondary steam (line C). According to the boiling process the reference core composition (A) could represent either an early stage of boiling or, more probably, the result of complete evaporation of the original liquid water under local conditions of full liquid–vapour equilibrium.

Already before re-injection commenced, significant temporal variation in fluid composition had occurred and in certain zones, substantial differences existed in fluids of wells a few

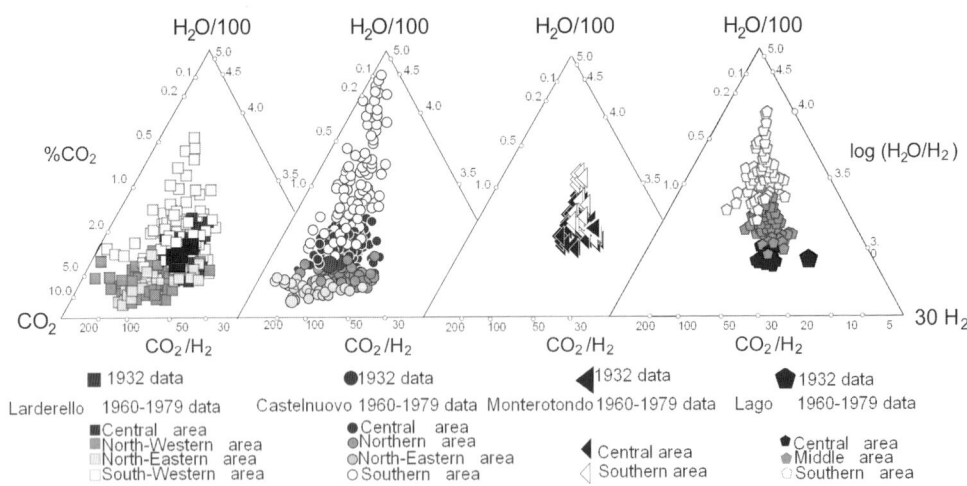

Fig. 2. Triangular diagrams describing the relative concentration of H_2O–CO_2–H_2 in the steam produced before re-injection in the four main subunits of the Larderello field. Different symbols represent different locations of producing wells in the same subunit. Largest symbols in each diagrams represent the chemical composition of the *reference core* within a single area (see text for discussion). Data for Castelnuovo result from the largest amount of meteoric water inflow observed in the entire basin.

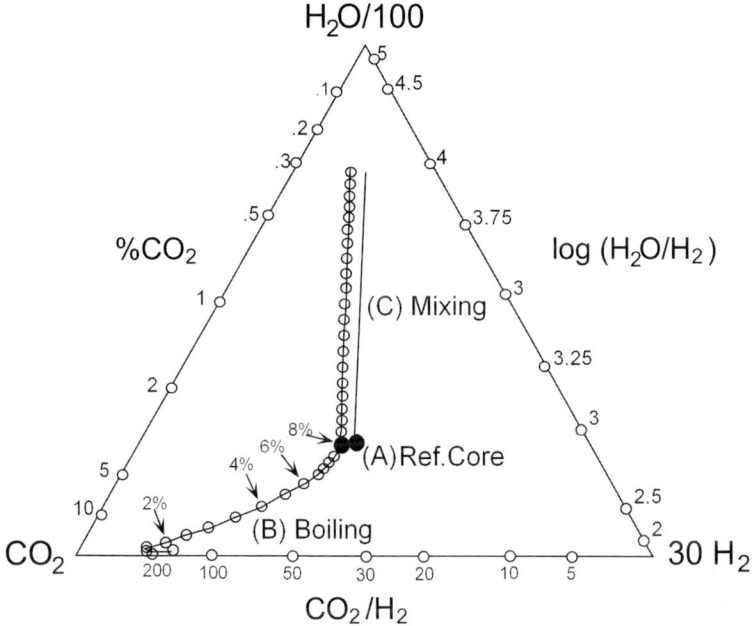

Fig. 3. Computed trend of steam produced according to a boiling process, which is a linear combination of 10% of a single-step steam separation and 90% of a continuous steam separation (Rayleigh distillation). The temperature was taken as 250 °C. (A) represents the reference core of the Larderello area. Two different processes that may involve the original fluid are shown: boiling and mixing. The effects of progressive boiling and mixing are represented by curves (B) and (C), respectively. Dots represent a complex evaporation constituted by a linear combination of an equilibrium boiling ('single-step separation' type) at 10% and a Raleigh distillation ('continuous steam loss' type) at 90%. Each dot corresponds to an evaporation step of 0.5% by weight. The linear path of fluid obtained by mixing between primary and secondary steam is represented by (C).

hundred metres apart. A knowledge of such variations prior to re-injection was required to better understand the chemical and isotopic variations induced by injection of waste water.

At the Larderello geothermal field, re-injection began in the 1970s to dispose of the steam condensate discharged from the power plants, in compliance with new environmental legislation, which prohibits the dispersion of waste waters on the surface, as was common practice at that time. The fear that long-term injection of cold water might damage the field production characteristics led to the choice of wells located relatively far from the exploitation areas, that is, in marginal zones, which were already affected by the inflow of meteoric water towards the geothermal field. Subsequently, studies and field tests were conducted in order to assess the potential of using re-injection to recover the thermal energy stored in the rocks, paying special attention to those areas characterized by high permeability and by a degree of superheating of the steam inside the reservoir (Giovannoni et al. 1981). Of several examined areas, Valle

Secolo was the most promising, because intensive exploitation of that area has gradually depleted the fluid inside the reservoir and caused the reservoir pressure to fall from 3 to 0.5 MPa, although the reservoir temperature remained at 240–250 °C. Under these conditions, a large fraction of thermal energy remains stored in the reservoir rocks. Water injection tests in this superheated area started in 1979 and demonstrated that an enhancement of heat recovery through the boiling of injected water was possible. In 1984, re-injection of water discharged by power plants became an important part of the exploitation strategy of this area. Analysis of the well production parameters confirmed that a large part of the re-injected water has been recovered as superheated steam, with a significant increase in steam flow rate and reservoir pressure. In light of the positive results, a supplementary injection program was planned and put into operation in 1994 in order to increase the secondary heat recovery in the Valle Secolo area and other selected zones of the Larderello geothermal field (Barelli et al. 1995).

This geothermal system is now regarded has a heat mine, from which the heat is recovered by means of an artificial water flow. The exploitation strategy is therefore progressively moving from a strategy typical of hydrothermal systems to one typical of hot-dry-rock (HDR) systems. An HDR system is any system where injection is necessary to extract the heat at a commercial rate for a prolonged period (Baldeyrou-Bailly *et al.*, 2004).

The use of natural tracers

Prior to re-injection, the steam produced from wells of the entire Larderello geothermal field, had $\delta^{18}O$ values ranging between -7.0 and $-1.0‰$, and δD between -37 and $-42‰$ (Panichi *et al.* 1974). After leaving the turbine, the condensate is piped to the cooling towers. A relatively large fraction of this water (approximately 30 wt%) is discharged into the atmosphere in the form of steam. As a result of this evaporation, the residual water, which is subsequently re-injected, is strongly enriched in heavy isotopes due to large fractionation factors for evaporation at about 70 °C. The isotopic composition of the injected water is, in fact, characterized by $\delta^{18}O$ values of $+4$ to $+5‰$ and by δD values between $+3$ and $+7‰$ (Panichi *et al.* 1995). As the isotopic composition of the injected water differs from that of the local groundwater, the system can be systematically monitored throughout the re-injection test.

In 1979, re-injection of spent steam condensate into the geothermal reservoir at Larderello has been monitored with measurements of both the isotopic composition of steam and of gas/steam ratios in the production wells surrounding the re-injection well (Giovannoni *et al.* 1981; Nuti *et al.* 1981; Bertrami *et al.* 1985). Immediately after re-injection began, it was noted that the isotopic composition and the gas/steam ratio of the steam produced in the monitored wells changed and varied over the period covered by the tests (Fig. 4). Injection is reflected by an increase in both the $\delta^{18}O$ and δD values of the discharged steam and by a decrease in its gas content, showing that the injected water is rapidly, at least partly, vaporized. The close correlation found between stable isotopic composition of the steam and gas/steam ratios suggests that mixing between re-injected condensate and a deep steam component takes place at depth.

Assuming a complete evaporation of the re-injected water, the recovered discharge from each productive well (Q_i) is given by:

$$Q_i = Q_{wh} \times (X_{wh} - X_r)/(X_i - X_r) \qquad (1)$$

In this equation, i is re-injected contribution; r is steam from reservoir, deep contribution; wh is total amount of steam at well head; and X is gas/steam ratio (G/S) or $\delta^{18}O$ or δD according to the used tracer. For the 1979–1982 period, Bertrami *et al.* (1985) calculated a recovery of about 100% from G/S, whereas using the isotopic composition yielded a value of 85%. Between October 1991 and March 1992, the computed recovery was 100% according to G/S, and up to 90% using stable isotope ratios. In the following period (April–September 1992), these values were lower, about 90% using G/S, and 50–70% using the isotope ratios (Panichi *et al.* 1995). The decreased recovery at the end of 1992 is due to a change in the re-injection sites: in this period, the wells used for re-injection were deeper and located farther away from those that were monitored (Panichi *et al.* 1995).

Theoretical returns obtained using the gas content always exceed the ones based on the isotopic data, but they are less reliable because the reference values for the deep steam gas content (X_r) refer to pre-1979. However, the isotopic values are also affected by uncertainties; the isotopic composition of re-injected water (X_i) is variable with averages of $\delta^{18}O = 5 \pm 2‰$ and $\delta D = 0 \pm 10‰$. Moreover, some fractionation occurs as a consequence of the existing residual plume (D'Amore *et al.* 1987).

Inspection of Fig. 5 suggests that the processes taking place in the reservoir during re-injection are more complex than assumed in simple mixing. A simple mixing between the deep steam of composition W and a complete evaporation of the re-injected water of composition R will produce a fluid with isotopic compositions intermediate between these two values (dashed line). The deviations observed suggest that all points plot as straight mixing lines between the composition of the original deep reservoir steam and the composition of steam produced by partial evaporation of the re-injected water. This means that low-temperature injected water causes the formation of a liquid plume under the re-injection well. This water is fractionated during boiling and the newly formed steam then mixes with the deep reservoir fluid. Data for samples collected (open squares) both during and after injection plot along a line that suggests mixing of deep steam with about 60% of the steam phase of the injected water ($f_w = 0.40$).

In contact with the reservoir rocks, indeed, the fluid obtained is the result of a simple mixing in different proportions of the primary and secondary steam. This will usually happen in the early

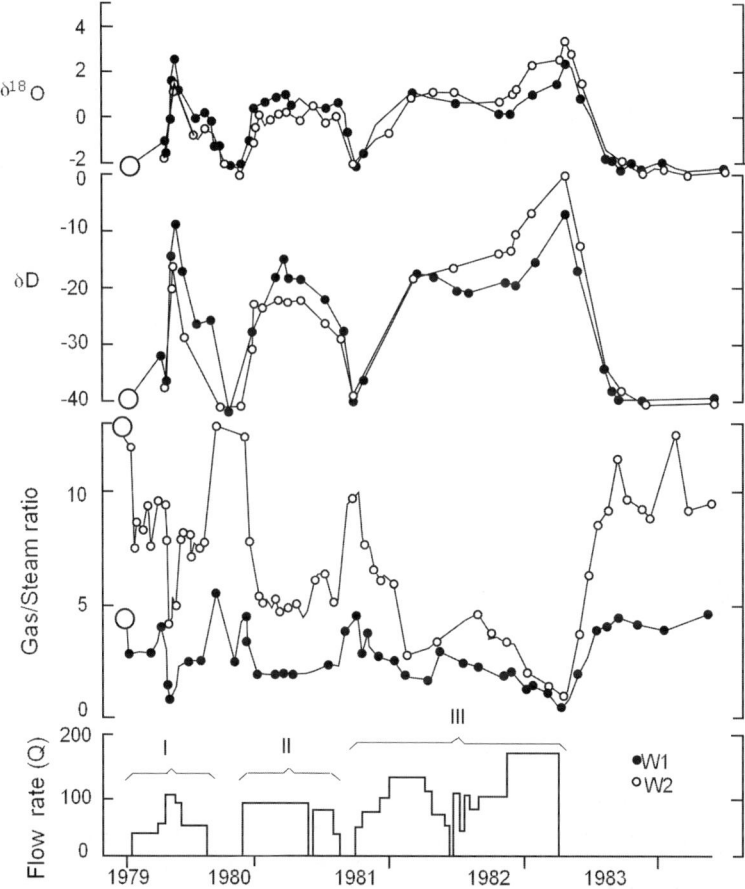

Fig. 4. Temporal variations in isotopic composition and gas/steam ratio (G/S) monitored in two wells (W1, W2) of the central area in the Larderello geothermal field, located near the re-injection well. Undisturbed pre-injection values (first point of each series, shown as large open circle) and flow rate of injected water (Q) are also shown. I, II and III refer to the three re-injection periods considered.

stages of the re-injection. As the amount of re-injected water increases, partial boiling of the injected water occurs, which produces a fractionation between the newly formed steam and the residual liquid. Consequently, during injection a liquid plume forms beneath the injection well, despite injection into shallow superheated zones.

The drawback of tracer methodologies that use stable isotope compositions of water is that water–rock interaction will eventually modify the original oxygen isotopic composition of the re-injected water, if it remains in the reservoir for a long time. Moreover, isotopic re-equilibration may take place between the water and the gas species when the injected fluid comes into contact with large quantities of reservoir gas. Finally, the isotopic composition

of the steam produced from injected water may differ from that of waste water because of isotopic fractionations occurring in the reservoir caused by incomplete boiling of injected liquid.

Gas chemistry

Figure 6 shows some $X_{H_2O}-X_{CO_2}-X_i$ diagrams, where $i = H_2$, CH_4, and N_2 in the fluids discharged from three productive wells located in the 'classic zone' of the Larderello field. Comparison of steam collected before and after re-injection proves that wells are strongly affected by re-injection. The higher the vapour from the re-injected water in the produced fluid, the more the representative points move towards increasing H_2O contents. Since the waste condensate is gas-depleted, except for a small

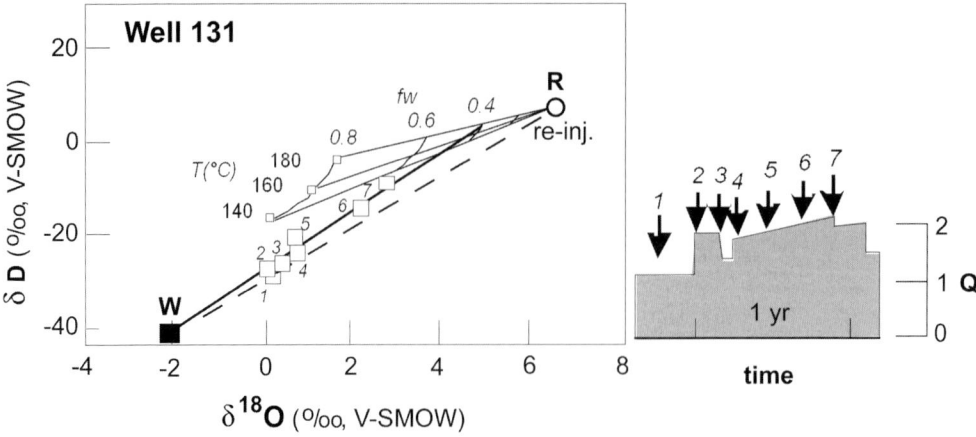

Fig. 5. Variation of the $\delta^{18}O$ and δD values of the fluid delivered from well 131, located 500 m away from the re-injection site, during an injection test conducted in the peripheral area of Serrazzano in the Larderello field (open squares). The figure also shows graphically, and in arbitrary units, the flow rate Q of water re-injected into the well as a function of time, and the position of each sample collected. Theoretical isotopic pattern of the steam produced by re-injected water, assuming continuous steam separation at depth, is also reported. Since the actual evaporation temperature and the fraction of residual water are unknown, calculations were made for three different temperatures (140, 160, and 180 °C) and fractions (f_w) of residual liquid water after boiling. Dashed line represents the hypothetical mixing between deep geothermal steam (W) and completely evaporated re-injected water (R).

amount of N_2, the overall effect can be explained as dilution of the original deep steam.

Unlike the other gases, the distribution of N_2 suggests that the re-injected water is the main source of N_2 in the monitored well fluids, as shown by the top right diagram in Fig. 6. Here, the H_2O/N_2 ratio at zero contribution of deep steam, corresponds to that of air-saturated groundwater at 20 °C (labelled as GDWR).

The atmospheric gas contribution to the fluids discharged after re-injection is shown in Fig. 7, where the relative contents of N_2, He, and Ar for samples collected before and after re-injection are plotted. Injected and/or infiltrated waters previously equilibrated with the atmosphere are expected to introduce 2.4×10^{-4} mmol/mol of Ar into the deep reservoir (Mazor 1978). This value is very similar to the average content of the fluid produced before re-injection (1.6×10^{-4}, Scandiffio et al. 1995). However, the two endmembers differ in N_2 contents; in the injectate, the theoretical N_2/Ar ratios are between 84 and 38 (values for air and air-saturated groundwater, respectively), whereas in the original fluid, this ratio is about 800 (Scandiffio et al. 1995).

Radiogenic He is present in the deep fluid, but absent in the secondary steam produced by re-injected waters because the very short residence time of the latter prevents any substantial interaction with the reservoir rocks. The trend observed in Fig. 7 seems to be the result of

simple mixing between deep and meteoric gas components introduced by the re-injected waste water. The full dots shown in Fig. 7 refer to values obtained for gas emerging anywhere in the Larderello geothermal system before re-injection (i.e., prior to 1979), whereas the open dots represent later values. The Ar/He ratios increase more than a hundredfold compared to the pre-injection period. In general, the data reveal that Ar is a gas component that is very sensitive to re-injection. The usefulness of this tracer should be at least of the same order as the isotopic composition of the steam or the G/S value.

Other components, such as B and NH_3 have also been used to trace recharges into the geothermal reservoirs. The injected concentrations of H_3BO_3 range from 0.1 to 0.25 mmol/mol of total fluid, those of NH_3 from 0.2 to 0.4 mmol/mol (Fig. 8). The liquid/vapour distribution coefficients play a fundamental role for these condensates because, during boiling, NH_3 is concentrated in the steam relative to H_3BO_3 due to the different values of the distribution coefficients. In Fig. 8, the composition of steam delivered from a single well shows chemical variations with respect to the composition of water before re-injection (black circles). Mixing of deep steam and a new vapour causes H_3BO_3 depletion, but strong NH_3 enrichment along different patterns describing single-step steam

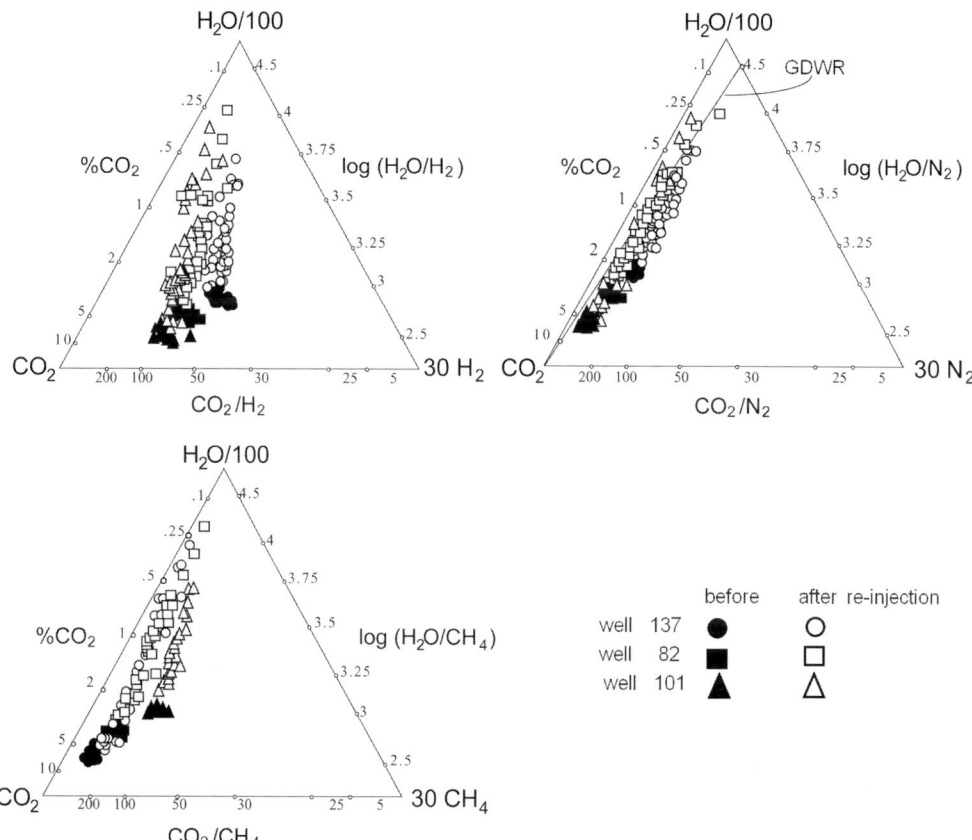

Fig. 6. Relative concentrations of $CO_2-H_2O-X_i$ ($X_i = H_2$, N_2, CH_4) components in three wells located at different distances from the injection sites, before and after re-injection (full and open symbols, respectively). All wells are strongly affected by the injected fluids. GDWR indicates the N_2/Ar ratios in groundwater equilibrated with the atmosphere at 25 °C.

separation at temperatures ranging from 240 to 280 °C. Two different set of boiling curves are reported in Fig. 8, because the chemical composition of the re-injected waters change over time (empty and grey squares), and consequently the chemical composition of steam samples change in good agreement with the variations expected by a single-step steam separation process.

Chemical and isotopic variations with time

It is advisable to draw maps for well fields showing concentrations of a particular chemical element or isotope, or a ratio, at a given time in order to document changes with time. Comparison of maps from different times facilitates evaluation of the spatial distribution of changes and helps in locating where recharge into an exploited reservoir may be occurring. In August 1992, an isotopic survey was carried out in the field to distinguish between the contributions of natural and re-injected inflow. Figure 9 shows the large difference in isotopic composition of the steam before and after the re-injection of spent waters in the active field. Before re-injection, the isotopic composition of the vapour produced was characterized by $\delta D = -40 \pm 2‰$ and by $\delta^{18}O$ ranging between about -5 and $0‰$ (Fig. **9a**). The natural recharge ($\delta D = -40‰$; $\delta^{18}O = -7‰$) tends to modify only the ^{18}O content of the steam. On the other hand, the re-injected waters are very enriched in both heavy isotopes (δD and $\delta^{18}O$ values both about $+5‰$; grey dots in Fig. **9b**), and the mixing with the original deep steam determines the variations observed for both isotopes (open circles). The steam delivered from the entire field thus becomes strongly modified in its isotopic composition with the introduction into the geothermal reservoir of the waste waters.

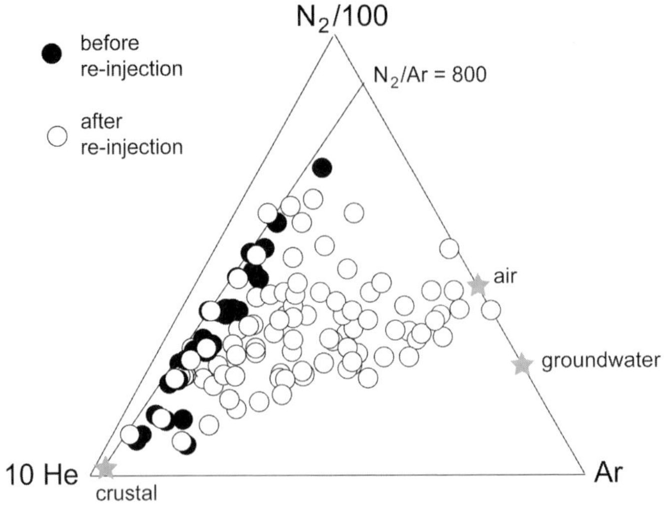

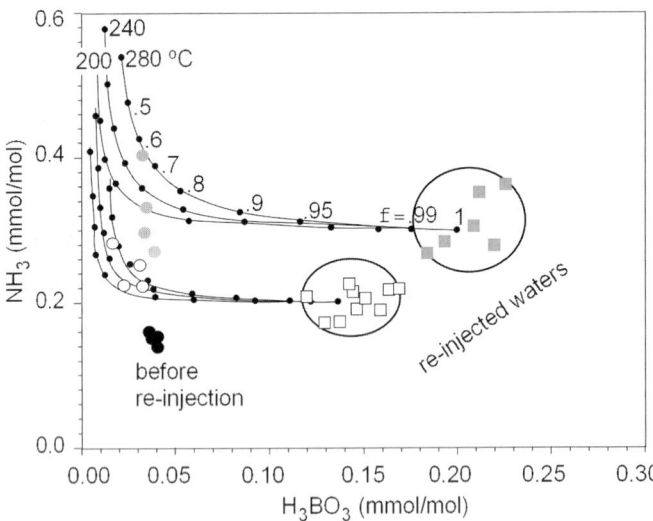

Fig. 7. Relative concentrations (mmol/mol on a water-free basis) of N_2, He, and Ar components of the gas mixtures of the steam produced in different periods of exploitation. Composition of crustal fluids, together with N_2/Ar ratios for atmosphere and air-saturated groundwater are reported for comparison (Giggenbach & Goguel 1989). The N_2/Ar ratio of 800 represents the ratio found in geothermal fluids before re-injection.

Figure 10 shows a picture of the status of the characteristics of the fluids emerging from the entire Larderello field based on all systematic analyses of major gas components collected during ten years of re-injection (Panichi *et al.* 1995). Figure 10a shows areas of the geothermal field where the chemical composition of the gas mixtures did not show changes between the

Fig. 8. Ammonia and boric acid concentrations of the fluids delivered from the same well at different times (between August 1983 and April 1985). The composition of the re-injected waters (squares) and steam produced before re-injection (black dots) are reported with the variations expected by a single-stage separation process. Two groups of isotherms at 200, 240, and 280 °C are drawn starting from two average values of the chemical composition of waste waters injected at different times.

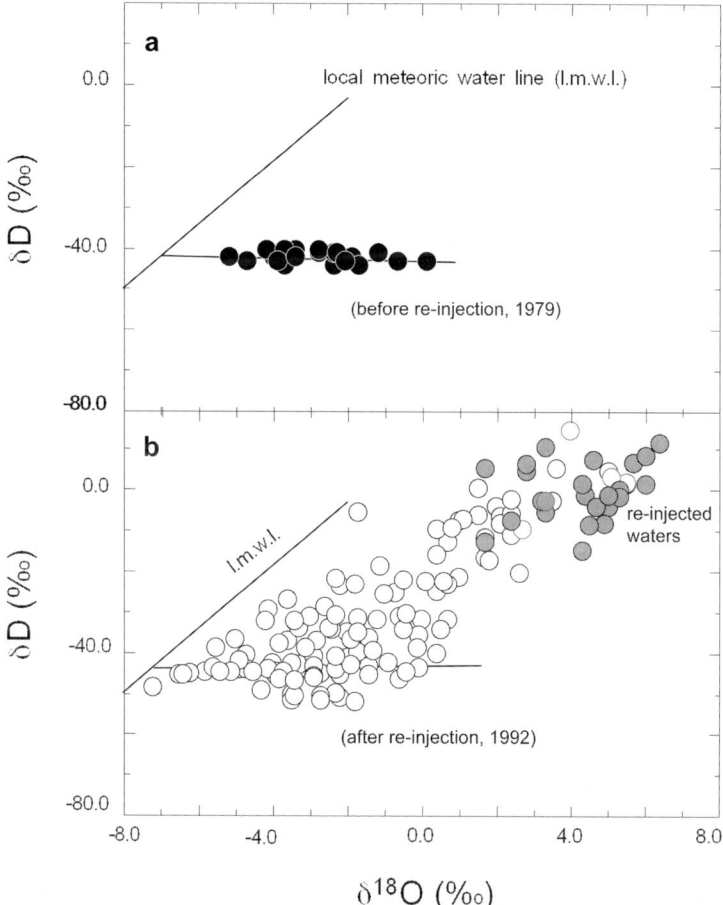

Fig. 9. Variation of $\delta^{18}O$ and δD values of steam delivered at the Larderello field (**a**) before 1979, and (**b**) in 1992, that is, before and after re-injection. l.m.w.l. represents local meteoric water line.

beginning of exploitation and the present (reference cores), and zones of the field whose produced fluids have gas concentrations diluted by appreciable amounts of surface water (gas-depleted) that reach the reservoir after decreasing of reservoir pressure as a consequence of the exploitation (Scandiffio *et al.* 1995) or by re-injection of waste fluids. Different sources of freshwater inflow are recognizable in the maps, but cannot be discriminated on a gas chemistry basis because the variation is due to a substantial dilution of the original deep gaseous mixtures.

Cold water recharge may neither cause immediate cooling of the producing aquifer nor of the discharged fluid, because the recharging waters gain heat by contact with the hot reservoir rock on passage to discharging wells. Thus, cold water may be particularly difficult to detect in boiling reservoirs. However, examination of the distribution of the values for $\delta^{18}O$ and δD in

the field made it possible to draw Fig. 10b, where the natural inflow with respect to the re-injected one is labelled by both hydrogen and oxygen isotopes.

The agreement between the two maps based on chemical and isotopic data is very good. The slight differences can be explained by changes of re-injection sites and by the different rate of response of the two methods, on which these maps are based.

Disequilibrium conditions in the reservoir

Isotopic disequilibrium between gases induced by re-injection has been reported by Bolognesi *et al.* (1990). Oxygen, H, and C isotopes of the fluid discharged by wells of a S–N oriented transect crossing the whole geothermal field reveal that after re-injection, samples show different isotopic temperatures from those obtained

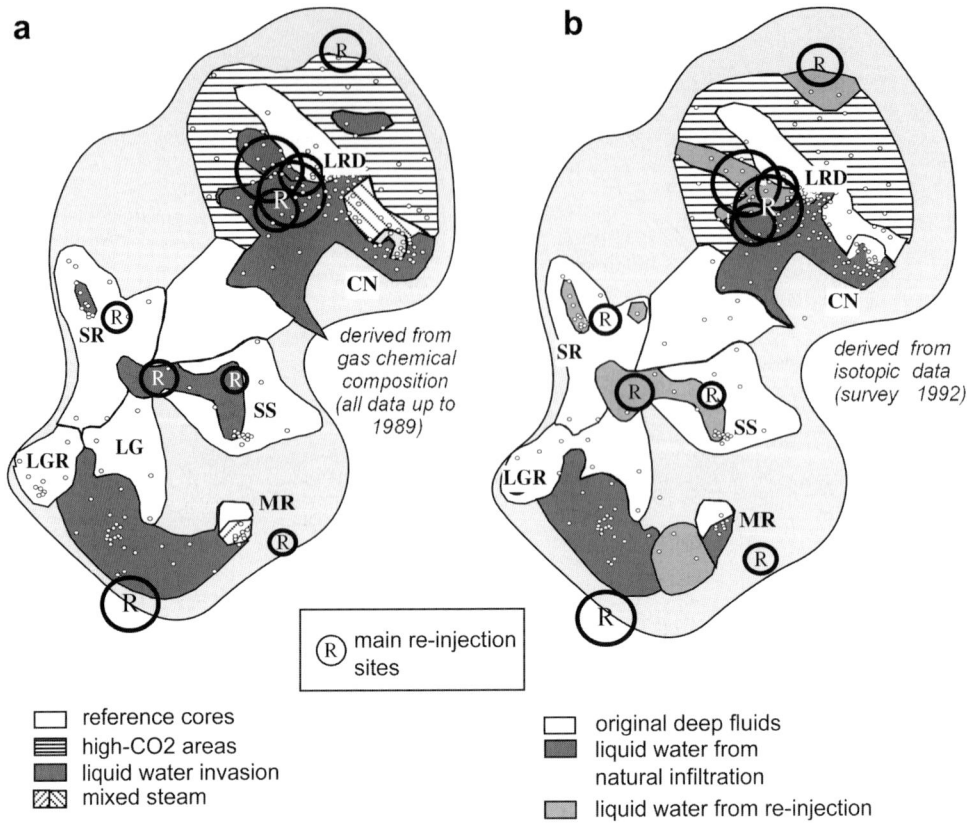

Fig. 10. Zonation of the Larderello geothermal field derived from (**a**) gas analyses, and (**b**) stable isotope values of steam produced before and after re-injection. Distribution and characterization of geothermal subunits obtained by gas analyses have been established from a data set collected before 1989, that is, 6 years after the beginning of re-injection of waste waters. White zones represent areas that produce gas mixtures with almost the same composition as that of the original gases emerging at the surface before the exploitation of the field (Scandiffio *et al.* 1995). Dashed zones produce steam affected by addition of cold water (i.e., re-injected) to the geothermal system. The zonation from the isotopes was derived from an extensive survey performed in 1992. In Fig. 10**b**, different sources of cold water are discriminated. Abbreviations: LRD = Larderello, CN = Castelnuovo, MR = Monterotondo, SS = Sasso Pisano, LGR = Lagoni Rossi, SR = Serrazzano geothermal subunits.

previously. Fractionation of isotopes of light elements between compounds is quite significant and temperature-dependent, thus allowing the use of the distribution of the stable isotopes of H, C, and O between aqueous and gaseous components as geothermometers (Richet *et al.* 1977).

Figure 11**a** illustrates the difference between the actual temperature, measured at the well head (T_{WH}), and the temperature determined from the isotopic fractionation (T_{IS}) between H_2 and H_2O. This isotope geothermometer seems to work well for this system, at least in the sense that it predicts subsurface temperatures actually encountered in deep drill holes (Panichi *et al.* 1979). Before 1979 (full symbols), the ΔT

values ($= T_{IS} - T_{WH}$) are between 10 and 30 °C; after re-injection (open symbols) the ΔT values decreased to -80°C, showing that an isotopic disequilibrium was established. The behaviour relative to the pair CO_2–CH_4 is different, as shown in Fig. 11**b**. Generally the CO_2–CH_4 isotope geothermometer indicates temperatures higher than those encountered in boreholes. This result has been taken to indicate that this geothermometer, which responds slowly, reflects temperatures at deeper levels than those penetrated by the wells (Panichi *et al.* 1977; Giggenbach 1984). In this case, ΔT between calculated and measured values was about $+40$ °C before re-injection (solid symbols), underwent a considerable decrease during the first stages of

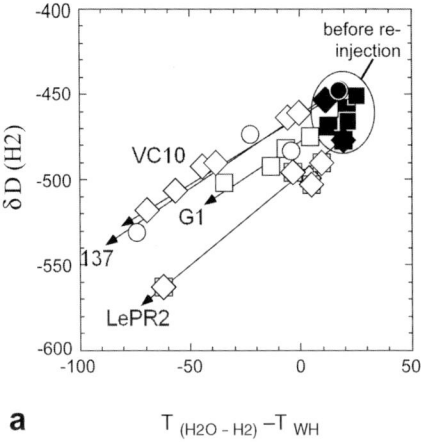

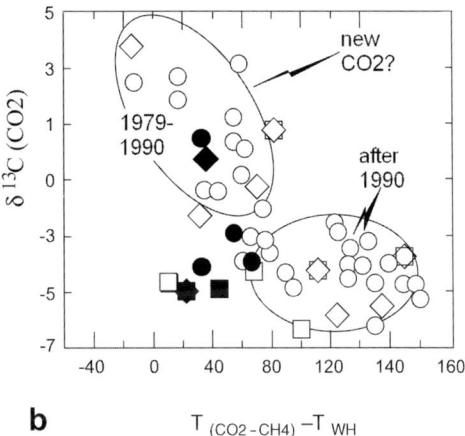

a $T_{(H2O - H2)} - T_{WH}$ **b** $T_{(CO2 - CH4)} - T_{WH}$

Fig. 11. Observed variations of isotopic temperatures before and after re-injection in selected wells (VC10, G1, 137, Le PR2). (**a**) Variations computed with the pair (H_2O-H_2) in relation to the wellhead temperatures (T_{WH}). (**b**) Isotopic temperatures are relative to the pair (CO_2-CH_4). Full symbols refer to the pre-injection data.

re-injection, and then, after 1990, increased to values of up to 150 °C. The behaviour shown in Fig. 11**b** could be related to gas samples having $\delta^{13}C_{CO_2}$ higher than 0‰ and could be explained by the addition to the system of new CO_2 formed in the presence of a boiling liquid plume in the producing reservoir.

Disequilibrium conditions at depth could also be responsible for the relationship between H_2, CO_2, and Ar shown in Fig. 12 (after Giggenbach

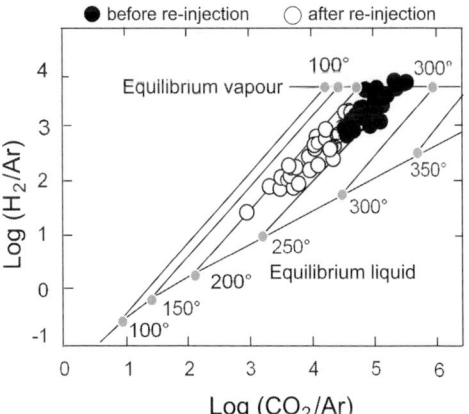

Fig. 12. Evaluation of He–Ar and CO_2–Ar equilibrium conditions. The concentrations are expressed in mmol/mol on a water-free basis. As an effect of re-injection, the data points, representative of the original fluid before re-injection, move from an almost complete equilibrium in vapour phase towards to a mixed phase, where equilibrium conditions in a pure liquid phase become more and more important. (From Giggenbach & Goguel 1989.)

& Goguel 1989). Here, empirical relationships between H_2/Ar and CO_2/Ar at different temperatures are reported for both a vapour system and a complete liquid phase. In this case, the concentration of Ar in the geothermal gas is considered almost constant and completely derived from the atmosphere; hydrogen derives from the dissociation of H_2O, and the concentrations of CO_2 are determined by a typical paragenesis of hydrothermal alteration, constituted by Ca–Al-silicate, calcite, K-feldspar and K-mica (Giggenbach 1988). The representative points of the wells selected for the already mentioned transect are compared with pre-exploitation data (Sborgi 1934) for natural manifestations and the first drillings in the Larderello field, and with unpublished data from the ENEL database for the period 1932–1950. The original composition, representing a near-vapour equilibrium condition at an average temperature of 250 °C (solid circles), shows a progressive influence of the injected water on the deep steam component. This implies an alignment along the 200 °C isotherm with compositional values that are intermediate between pure vapour and pure liquid equilibrium conditions, confirming that disequilibrium exists at depth.

Conclusions

In terms of evaluation of recovery coefficients (i.e., of the amount of re-injected water that is recovered from the production wells near the re-injection sites), the relative concentrations of Ar, He, and N_2 (or CO_2) probably represent a tool that reflects immediately the variations occurring at the re-injection sites. On the other hand, the use of the stable isotopic composition of the steam delivered at the surface may give smoothed results because the secondary vapour (from re-injected waters) consists of mixed steam from new evaporation fronts and from residual liquid plumes. Deuterium is the parameter least affected by fractionation (fractionation factor α between steam and liquid is close to 1 at boiling temperatures near 220 °C), by any kind of water–rock interaction, or by areal fluid displacement. As a consequence, δD values make possible the evaluation of the relative contributions of the original vapour and injection-derived steam. Changes in $\delta^{18}O$ may reflect recharge from different sources (from meteoric water inflow and from re-injection) into the producing aquifer. Gaseous components may be used to study the response of geothermal reservoirs to re-injection (and exploitation). CO_2, H_2, N_2, and Ar, as well as various isotopic gas-equilibria, are considered to be of interest in this respect.

Fluid extraction from the reservoir occurs simultaneously with re-injection of cold liquid and air. Atmospheric nitrogen and Ar propagate very fast through the fracture network in the reservoir rocks and arrive at producing regions with negligible thermal interference. This may be used as a simple and cheap method to estimate the three-dimensional permeability tensor.

The drawback of tracer methodologies that use stable isotope compositions of water is that some isotopic exchanges may modify the original isotopic ratios in the re-injected water. If the water remains for any length of time in the reservoir, water–rock interaction will eventually modify the oxygen isotopic composition. Moreover, isotopic re-equilibration may take place between the water and the gas species when the injected fluid comes into contact with large quantities of gas. Finally, the isotopic composition of the steam produced from injected water may differ considerably from the original composition because of the isotopic fractionation caused by boiling.

The development of a residual liquid plume in the geothermal reservoir at Larderello requires further studies because, for example, the data shown in Fig. 12 may not result from the attainment of full equilibrium by produced gases, but rather by the addition of meteoric Ar to the geothermal system. Problems may arise from possible buildup of liquid plumes inside the reservoir, which could reduce secondary heat recovery. Should this happen, the re-injection pattern will be changed: the injection in the well responsible for the plume buildup must be

reduced or halted, and new wells must be activated to look for other superheated zones of the reservoir. To monitor re-injection and particularly to locate the presence of liquid plumes, some geophysical surveys, such as microgravimetric and geoelectric measurements, should be performed periodically together with the monitoring of geochemical parameters.

In conclusion, the re-injection experiences gained at Larderello have proven that 'secondary heat recovery' is feasible in those areas characterized by a high permeability and a high degree of superheating of steam inside the reservoir. A large part of the re-injected water has been recovered as superheated steam, with a significant increase in steam flow rate and reservoir pressure. No liquid water breakthrough was observed between the re-injection and production wells, and the wellhead temperatures of most production wells remained constant, demonstrating water penetration at depth with efficient fluid–rock heat exchange. As the secondary steam contains no gas, the average gas content in the total produced fluid has decreased, with an improvement in power plant-specific consumption due to the energy saving for gas extraction from condensers (Cappetti et al. 1995).

This paper benefits largely from the collaboration with the Italian Electric National Company (ENEL) during the first experiment with the re-injection process in Larderello. I would like to thank H. Armannsson, S. Arnórsson, and B. Fritz for reviewing this manuscript and in particular, R. Gieré for his review and support. I am also grateful to technicians of the International Institute for Geothermal Researches in Pisa that supported me in sampling surveys and performing isotopic analyses.

References

ADAMS, M. C. 1995. Vapour, liquid, and two phase tracers for geothermal systems. *Proceedings of the World Geothermal Congress*, Florence, Italy, 1875–1880.

ARNÓRSSON, S. 2004. Environmental impact of geothermal energy utilization. *In*: GIERÉ, R. & STILLE, P. (eds) *Energy, Waste and the Environment: a Geochemical Perspective*, The Geological Society, London, 297–336.

BALDEYROU-BAILLY, A., SURMA, F. & FRITZ, B. 2004. Geophysical and mineralogical impacts of fluid injection in a geothermal system: the Hot Fractured Rock site at Soultz-sous-Forêts, France. *In*: GIERÉ, R. & STILLE, P. (eds) *Energy, Waste and the Environment: a Geochemical Perspective*, The Geological Society, London, 355–367.

BARELLI, A., CAPPETTI, G. & STEFANI, G. 1995. Optimum exploitation strategy at Larderello–

Valle Secolo. *Proceedings of the World Geothermal Congress*, Florence, Italy, 1779–1793.

BARELLI, A., BERTINI, G., BUONASORTE, G., CAPPETTI, G. & FIORDELISI, A. 2000. Recent deep exploration results at the margins of the Larderllo–Trvale geothermal system. *Proceedings of the World Geothermal Congress*, Japan, 965–970.

BEALL, J. J. 1993. The history of injection recovery in the units 13 and 16 area of the Geysers Steamfield. *Geothermal Resources Council Transactions*, **17**, 211–214.

BEALL, J. J., ENEDY, S. L. & BOX, W. T. JR. 1989. Recovery of injected condensate as steam in the south Geysers field. *Geothermal Resources Council Transactions*, **13**, 351–358.

BENOIT, D. 1992. A case history of injection through 1991 at Dixie Valley, Nevada. *Geothermal Resources Council Transactions*, **16**, 611–620.

BERTRAMI, R., CALORE, C., CAPPETTI, G., CELATI, R. & D'AMORE, F. 1985. A three-year recharge test by re-injection in the central area of Larderello field: analysis of production rate. *Geothermal Resources Council Transactions*, **9**, 293–298.

BOLOGNESI, L., NOTO, P. & PANICHI, C. 1990. More mileage for isotope gas geothermometers: the case of the Larderello geothermal field. *Geothermal Resources Council Transactions*, **14**, 857–861.

CALORE, C., GIANELLI, G. & PRUESS, K. 1990. Water–CO_2 version of MULKOM CODE: a tool studying the origin and the trends of CO_2 in geothermal reservoirs. *In*: PRUESS, K. (ed) *Proceedings of the 15th Workshop on Geothermal Energy*. Lawrence Berkeley Laboratory, LBL-29710, Berkeley, California, 53–60.

CAPPETTI, G., GIOVANNONI, A., RUFFILLI, C., CALORE, C. & CELATI, R. 1982. Reinjection in the Larderello geothermal field. *In: Proceedings International Conference on Geothermal Energy*. BHRA Fluid Engineering, 395–408.

CAPPETTI, G., PARISI, L., RIDOLFI, A. & STEFANI, G. 1995. Fifteen years of reinjection in the Larderello–Valle Secolo area: analysis of the production data. *Proceedings of the World Geothermal Congress*, Firenze, Italy, 18–31 May 1995, 1997–2000.

CELATI, R., NOTO, P., PANICHI, C., SQUARCI, P. & TAFFI, L. 1973. Interactions between the steam reservoir and surrounding aquifers in the Larderello geothermal field. *Geothermics*, **2**, 174–185.

D'AMORE, F., FANCELLI, R. & PANICHI, C. 1987. Stable isotope study of re-injection processes in the Larderello geothermal field. *Geochimica et Cosmochimica Acta*, **51**, 857–867.

EINARSSON, S. S., VIDES, A. & CUELLAR, G. 1975. Disposal of geothermal waste water by re-injection. *Proceedings of the 2nd U.N. Symposium on Geothermal Energy*, San Francisco, **2**, 1349–1363.

ENEDY, S., ENEDY, K. & MANEY, J. 1991. Reservoir response to injection in the southeast Geysers. *Proceedings of the 16th Workshop on Geothermal Reservoir Engineering*, Stanford University, 75–82.

GAMBILL, D. T. 1990. The recovery of injected water as steam at the Geysers. *Geothermal Resources Council Transactions*, **14**, 1655–1660.

GIGGENBACH, W. R. 1984. Mass transfer in hydrothermal alteration systems. A conceptual approach. *Geochimica et Cosmochimica Acta*, **48**, 2693–2711.

GIGGENBACH, W. R. 1988. Geothermal solute equilibria. Derivation of Na–K–Mg–Ca geoindicators. *Geochimica et Cosmochimica Acta*, **52**, 2749–2765.

GIGGENBACH, W. R. & GOGUEL, R. L. 1989. *Collection and Analysis of Geothermal and Volcanic Water and Gas Discharges*. Report N CD24014 of the Chemistry Division. Department of Scientific and Industrial Research, Petone, New Zealand, pp. 81.

GIOVANNONI, A., ALLEGRINI, G., CAPPETTI, G. & CELATI, R. 1981. First results of a re-injection experiment at Larderello. *Proceedings of the 7th Stanford Geothermal Reservoir Engineering Workshop*, 77–83.

GOYAL, K. P. & BOX, W. T. JR. 1992. Injection recovery based on production data in unit 13 and unit 16 areas of the Geysers field. *Proceedings of the 17th Stanford Geothermal Reservoir Engineering Workshop*, 103–109.

GUDMUNDSSON, J. S. 1983. Injection testing 1982 at the Svartsengi high-temperature field in Island. *Geothermal Resources Council Transactions*, **7**, 423–428.

GULATI, M. S., LIPMAN, S. C. & STROBEL, C. J. 1978. Tritium tracer survey at The Geysers. *Geothermal Resources Council Transactions*, **2**, 237–239.

HARPER, R. T. & JORDAN, O. T. 1985. Geochemical changes in response to production and re-injection for Palimpinon-I geothermal field, Negros Oriental, Philippines. *Proceedings of the New Zealand Geothermal Workshop*, University of Auckland, Auckland, 39–44.

HENLEY, R. W, TRUESDELL, A. H. & BARTON, P. B. 1984. Fluid–mineral equilibria in hydrothermal systems. *Reviews in Economic Geology*, **1**, 1–267.

MAZOR, E. 1978. Noble gases in a section across the vapour dominated geothermal field of Larderello, Italy. *Pageoph*, **117**, 262–275.

NUTI, S., CALORE, C. & NOTO, P. 1981. Use of environmental isotopes as natural tracers in a re-injection experiment at Larderello. *Proceedings of the 7th Stanford Geothermal Reservoir Engineering Workshop*, 85–89.

PANICHI, C., CELATI, R., NOTO, P., SQUARCI, P., TAFFI, L. & TONGIORGI, E. 1974. Oxygen and hydrogen isotope studies of the Larderello, Italy, geothermal system. *Proceedings of the Symposium on Isotope Techniques in Groundwater Hydrology*, **2**, 3–28. IAEA, Vienna.

PANICHI, C. & GONFIANTINI, R. 1978. Environmental isotopes in geothermal studies. *Geothermics*, **6**, 143–161.

PANICHI, C., FERRARA, G. C. & GONFIANTINI, R. 1977. Isotope geothermometry in the Larderello geothermal field. *Geothermics*, **5**, 81–88.

PANICHI, C., NUTI, S. & NOTO, P. 1979. Use of isotopic geothermometers in the Larderello geothermal field. *Isotope Hydrology*, 1978, IAEA, Vienna, **2**, 613–630.

PANICHI, C., SCANDIFFIO, G. & BACCARIN, F. 1995. Variation of geochemical parameters induced by reinjection in the Larderello area. *Proceedings of the World Geothermal Congress*, Florence, Italy, **3**, 1845–1949.

REED, M. J. & RENNER, J. L. 1995. Environmental compatibility of geothermal energy. *In*: STERRETT, F. S. (ed) *Alternative Fuels and the Environment*. Lewis Publisher, CRC Press, Boca Raton, Florida, 23–27.

RICHET, P., BOTTINGA, Y. & JAVOY, M. 1977. A review of hydrogen, carbon, nitrogen, oxygen, sulfur and chlorine stable isotope fractionation among gaseous molecules. *Annual Review of Earth and Planetary Science*, **5**, 65–110.

SANYAL, S. K., GRANADOS, E. E & MENZIES, A. J. 1995. Injection-related problems encountrered in geothermal projects and their mitigation: the United State experience. *Proceedings of the World Geothermal Congress*, Florence, Italy, **3**, 2019–2022.

SBORGI, U. 1934. *Studi e ricerche sui Gas dei Soffioni boraciferi con panicolare riguardo al loro contenuto in Elio ed altri Gas nobili*. Memorie Scienze Fisiche, Matematiche Naturali, Reale Accademia d'Italia (ed) Roma, **5**, 668–713.

SCANDIFFIO, G., PANICHI, C. & VALENTI, M. 1995. Geochemical evolution of fluids in the Larderello Geothermal field. *Proceedings of the World Geothermal Congress*, Florence, Italy, **3**, 1839–1943.

STEFANSSON, V. 1997. Geothermal re-injection experience. *Geothermics*, **26**, 99–139.

Geophysical and mineralogical impacts of fluid injection in a geothermal system: the Hot Fractured Rock site at Soultz-sous-Forêts, France

A. BALDEYROU-BAILLY, F. SURMA & B. FRITZ

Centre de Géochimie de la Surface, UMR 7517 CNRS – Université Louis Pasteur, Strasbourg, France (e-mail: armelle@illite.u-strasbg.fr)

Abstract: This paper deals with the geothermal system at Soultz-sous-Forêts, an important research site in France of the European Hot Fractured Rock project. A pilot plant is ready for electric energy production of up to 25–30 MW. Several research projects were developed in France, Switzerland, and Germany in order to study various scientific aspects of the geothermal system in parallel with the industrial development. Of primary scientific interest are the geophysical, geological, and geochemical conditions in the geothermal reservoir as well as the predictable consequences of future exploitation on the quality of the reservoir and on induced environmental problems, including: induced seismicity; rock alteration and possible formation of secondary minerals along the water pathways; and evolution of porosity and permeability in the deep aquifers. This paper focuses on two aspects: the geochemical evolution of minerals in the reservoir due to thermal gradients and fluid flow; and the evolution of physical properties of rocks during fluid injection in the wells. The petrophysical study shows that the fluid–rock interactions can be extensive in the reservoir matrix. The experimental study suggests that the fluid–mineral interactions are dynamic and that clay minerals can precipitate.

Geothermal energy has been produced for nearly a century and at present is exploited in 46 countries worldwide (Stefansson & Fridleifsson 1998). In 1998, the total energy generated in these countries was about 44 TWh/y and the direct use amounted to about 37 TWh/y (Fridleifsson 1998). Conventional electric power production is limited to systems producing fluids at near-surface temperatures above 150 °C, however, lower temperatures can be used if binary fluids are involved. The largest geothermal district heating facility in the world, the Reykjavik District Heating (serving about 152 000 people), obtains 75% of its heat from low-temperature fields (85–130 °C) and 25% from a high-temperature fields (300 °C). Geothermal energy could become one of the renewable energy types able to account for a significant contribution to relieving supply shortages.

The European Hot Fractured Rock (HFR) project at Soultz-sous-Forêts, France (Fig. 1) has been developed over the past few years and includes a considerable component of fundamental research. The concept of the HFR project is to produce a hot fluid within a hot rock. The site comprises one injection well (GPK3) and two production wells (GPK2 and GPK4), drilled to a depth of 5000 m in the granite (GPK4 is scheduled to be completed during 2004). In addition to these wells, there are several other boreholes, which have been drilled to depths ranging from 1400 (sediment/granite boundary) to 2200 m depth. The deepest one (EPS1) is the only one that has been cored.

In the HFR project it has been attempted to provide a continuous flow of water between two drillings: cold fluids (60 °C) are injected towards a deep level of the granite (5000 m, 200 °C) through the GPK3 well; these fluids then flow through the fracture network and finally are extracted as hot fluids (near 180 °C) through GPK2 and GPK4 at the surface for a future economic purpose. To produce energy, the fluid will be transformed into vapour. However, the fluid located in the wells is salt-saturated and would cause precipitation of salt during the liquid-to-vapour change. It is therefore necessary to use a heat exchanger, where the heat of the saline geothermal fluid is exchanged with a diluted water. After heating, this hot liquid water will be transformed into vapour to produce energy using turbines (Fig. 2).

To understand processes that take place in the rock at 5000 m depth, the research project

From: GIERÉ, R. & STILLE, P. (eds) 2004. *Energy, Waste, and the Environment: a Geochemical Perspective*. Geological Society, London, Special Publications, **236**, 355–367.
0305-8719/04/$15

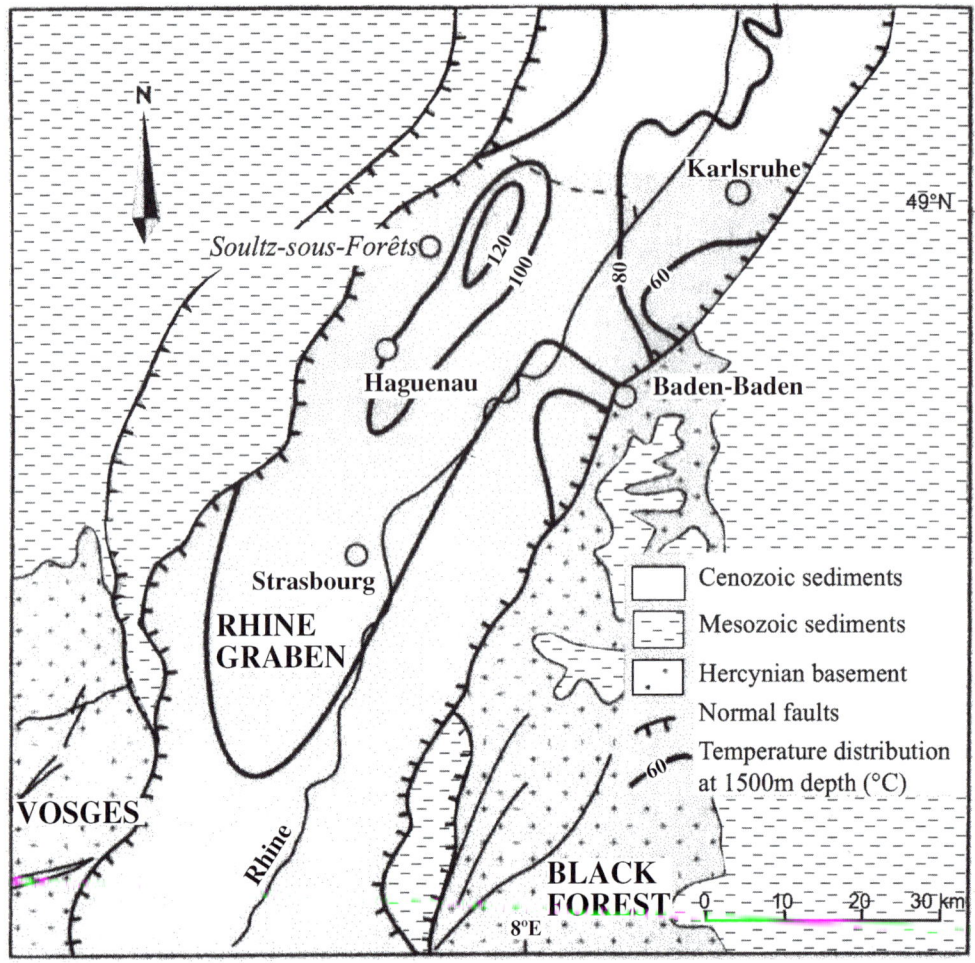

Fig. 1. Simplified geological map of the Rhine graben (Dezayes *et al.* 2000) and location of the geothermal site of the European Hot Fractured Rock project in Soultz-sous-Forêts.

combines geological, geophysical, and geochemical studies. The aim is to (1) better understand the past behaviour of the geothermal system, and (2) to predict the behaviour of the same system under heat exploitation during the planned injection and pumping phases. The results may have implications for the feasibility of the industrial project, but they also help in assessing the environmental impact of geothermal energy exploitation at Soultz-sous-Forêts.

This paper presents two aspects of the research project, namely the geophysical and the mineralogical evolution of the geothermal system as a direct consequence of fluid injection. It is a geological, mineralogical, petrophysical, and geochemical study of water–rock interactions at temperatures of up to, and higher than, 200 °C

and under pressures similar to those expected at 5000 m depth in the wells. Analytical, experimental, and modelling approaches are combined.

Geological setting

The geothermal site at Soultz-sous-Forêts is the subject of a European Hot Fractured Rock (HFR) research project. Located in the northern part of Alsace, France (Fig. 1), the geothermal system is part of the tectonic Rhine graben, which developed as a result of east–west oriented extension during the Cenozoïc. This prominent graben structure extends from Basel (Switzerland) in the South to Frankfurt am Main (Germany) in the North. It is over 300 km long and, near Soultz, 30–40 km wide. The tectonic activity led to

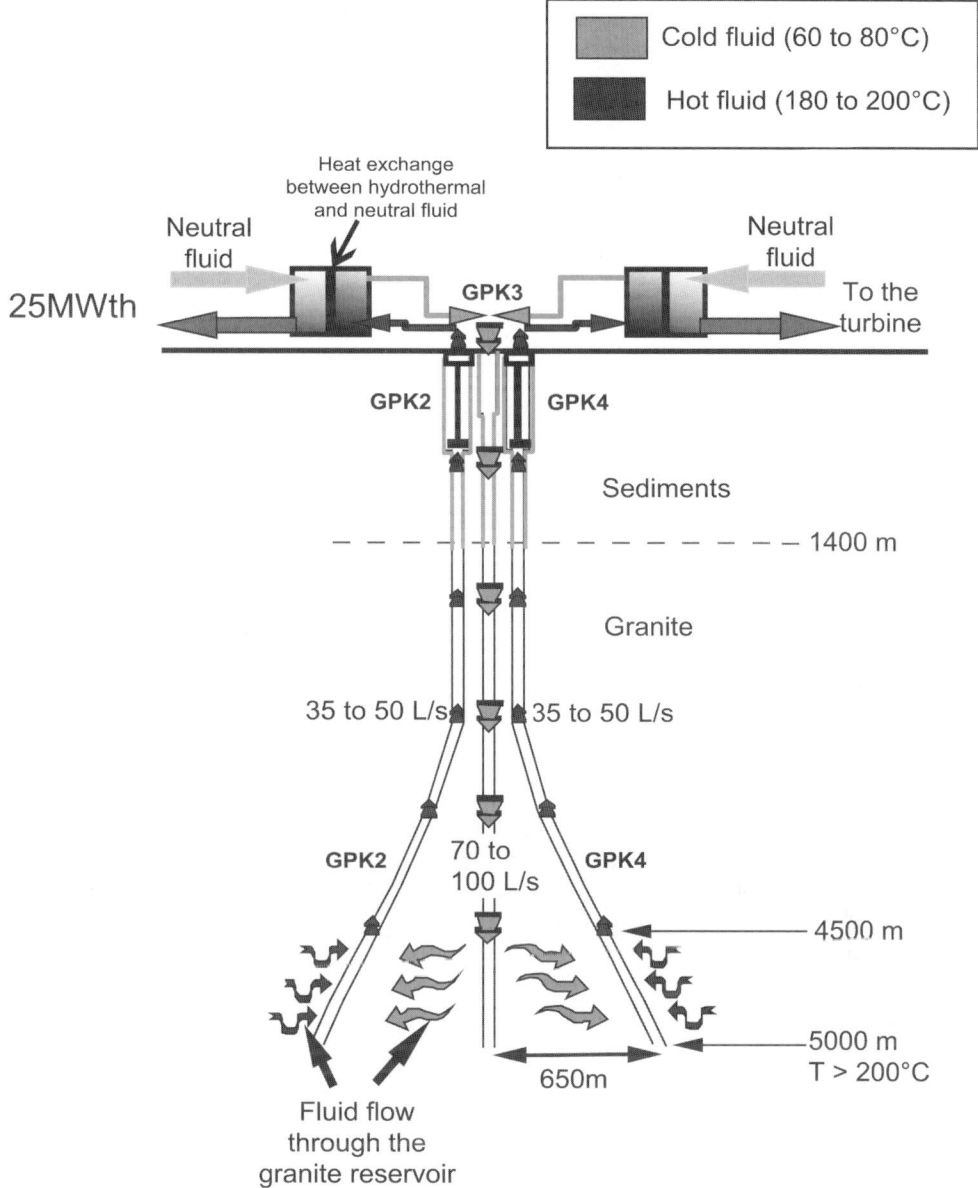

Fig. 2. Schematic view of the Soultz-sous-Forêts scientific pilot plant (from Document GEIE "EMC"/AG).

extensive fracturing of the upper continental crust (Brun *et al.* 1991). The granitic basement has a sedimentary cover, which is of Triassic to Tertiary age and, at Soultz-sous-Forêts, has a thickness of about 1400 m (Fig. 2). The sedimentary cover is well studied as it hosts an oil field that has been exploited in the past (Pechelbronn oil field). The extensional processes caused crustal thinning and induced a high geothermal gradient (Fig. 3). This tectonic setting is favourable for the development of a geothermal system, because the fractures can be used as pathways for the geothermal fluid, and because the pronounced geothermal anomaly, at least in the sedimentary cover, implies elevated temperatures at relatively shallow depths.

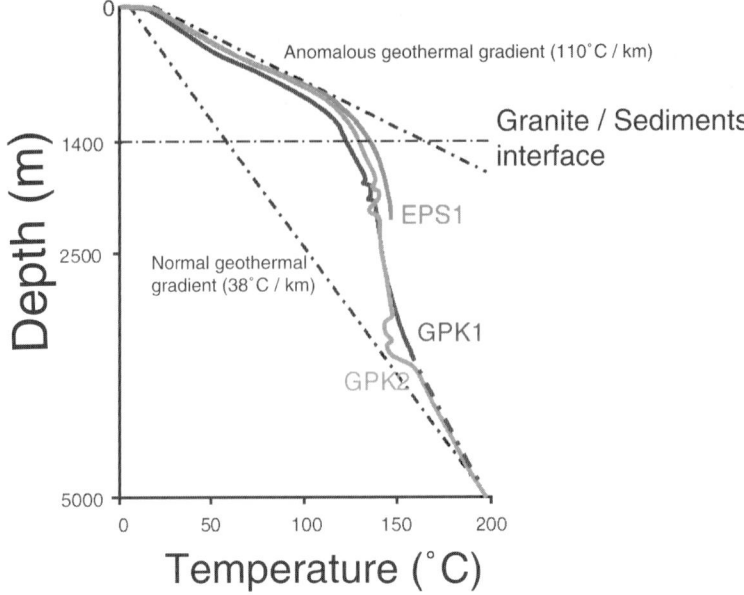

Fig. 3. Thermal gradient measured in the Soultz-sous-Forêts wells. (Modified from Pribnow *et al.* 1999).

Towards the industrial phase

Since 1987, Soultz-sous-Forêts is used as a pilot site to evaluate the potential for high-enthalpy geothermal energy production. Two major bore-holes already exist: GPK2 and GPK3 (Fig. 2). In 1997, a large-scale circulation test was performed between these two wells. The aim was to test the establishment of a fluid circulation of at least 20 L/s in the deep granitic reservoir. The intra-granitic fluid was extracted in the production well at a depth of 3876 m and with the expected flow at a temperature of 140 °C. This fluid was then rapidly cooled to 60 °C in order to simulate the effect of a heat exchanger of a surface power plant. After cooling, the fluid was re-injected in the crystalline basement through the injection well at 3590 m depth and at a horizontal distance of 450 m from the production well. This separation between injection and production wells was the largest for HFR geothermal systems tested in the world. Nevertheless the fluid flow was balanced between injection and extraction, and the circulation test was successful. A total fluid volume of 244 000 m³ circulated between the two wells during a four-month test (Jacquot 2000). The hydraulic conductivity varied between 22 and 25 L/s, and it was possible to reach a thermal output of about 10 MW, that is, twice the output of other hot dry rock sites (Los Alamos, USA; Rosenmores, UK; Hijiori, Japan).

After this first reservoir test, the project was aimed at reaching a higher temperature and a higher fluid flow. Therefore, one of the two wells was extended down to 5070 m depth. At the bottom of this well, the temperature was close to 200 °C. In the year 2001, a preliminary scientific pilot plant has been built, thus opening the pre-industrial stage of geothermal energy production. The aim of this pilot project is to inject the fluid at 5000 m depth into a central well and to pump it up via two lateral wells (Fig. 2). The last 5000 m deep well (GPK4) is currently being drilled. At the bottom, the wells are separated by a distance of about 650 m. The expected temperature and flow rate of the extracted fluid were 180 °C (at the well head) and 50 L/s, respectively, and a thermal power of 50 MW was anticipated for several years. The industrial prototype, which could emerge from this pilot plant in 2006, should reach a power of 25 MW. It would require the drilling of several wells in order to provide nearly 400 kg/s of a 200 °C hot fluid.

Monitoring electric potential during hydraulic stimulation

The measurement of streaming potential is a widely recognized method for identifying flow paths through rock (Wurmstich & Morgan 1994). This method is based on the existence of an electric double layer at the liquid–matrix interface: many minerals have negatively charged surfaces (e.g., clay minerals), and therefore there exists a diffuse layer of positive ions in the local pore solution, which are weakly bonded to the surface of the minerals. Under a local pressure gradient, these ions can be carried away from the mineral surface, yielding a negative anomaly. Thus, there is a strong relationship between the streaming potentials and the pressure gradient under which the ions move. If there is a fluid flow, there will be a variation of the streaming potential. Variations around the Soultz-sous-Forêts site have been monitored during a stimulation experiment, during which 23 000 m^3 of water were injected at 5 km depth (Marquis *et al.* 2002). These authors explain that the electrokinetic coupling parameter (relationship between the streaming potential and the pressure gradient) required to fit the surface streaming-potential data during the injection ($\sim$200 mV/bar) is in good agreement with laboratory values for a granite–water system. These measurements confirm the numerical modelling predicting the fluid flow even at 5 km depth. The modelling also shows that the chemistry of the injected fluid is controlling the magnitude of the streaming potential anomalies during the stimulation phase. The monitoring of surface streaming potential adds significantly to the local information obtained by down-hole pressure and acoustic probes and could be used in the future as a tool to monitor the dynamics of the whole geothermal reservoir. This approach

is still being developed in connection with the industrial project.

Microseismicity induced by fluid injection

In order to characterize the reservoir and to control the seismic risks and events due to fluid injections in deep wells (Zoback & Harjes 1997; Shapiro *et al.* 2003), a new seismological network has been installed around Soultz-sous-Forêts. This network of permanent stations allows the localization of a seismic event within the site and the determination of its magnitude. A seismic activity with about 7200 events has been produced during 2000 massive hydraulic stimulation experiments, including fluid injections. During the injection phase, the induced microseismicity has very low intensity, with events of <1.9 magnitude (Gérard *et al.* 1998). Several geophysical studies discuss the information about the direct seismic impact of fluid injections (Baria *et al.* 1999; Marquis *et al.* 2002; Shapiro *et al.* 2003). The spatial and temporal distributions of seismic events can be related to the fracture network (Dezayes *et al.* 1995; Genter & Traineau 1996, 1992) and to the effective permeability of the rock (Shapiro *et al.* 2003). Therefore, one can use the data for Soultz-sous-Forêts and deduce a hydraulic conductivity; the calculations yield a value of 0.05 m^2/s for the system (Fig. 4).

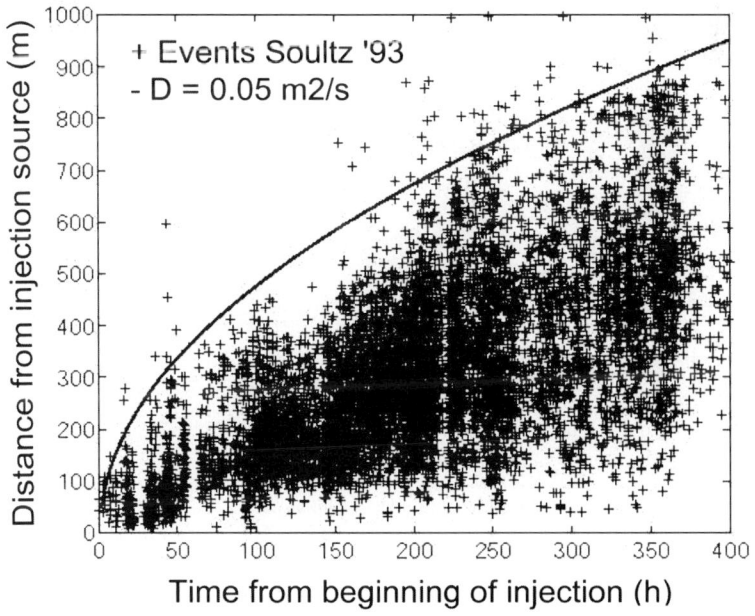

Fig. 4. Seismic events during injection in relation to time and distance from the injection source (from Shapiro *et al.* 2003).

Ancient fluid flows: chemical mobility and alteration

Mineralogical and chemical studies of the granite and of secondary mineral phases have been performed in order to identify past fluid pathways and reconstruct the alteration history of the granite (Genter 1990; Jacquemont 2002; Surma 2003). Such knowledge is important because hydrothermal alteration, induced by exploitation of the geothermal field in the near future, will have a similar and possibly important impact on fluid pathways and will significantly alter the physical properties of the reservoir. The rock system can be subdivided into three different petrographic facies (Fig. 5): (1) fresh porphyry granite; (2) hydrothermally altered granite; and (3) cataclastic granite comprising protolith, damage zone, and fault core, as has been described for various other rock types (Evans & Chester 1995; Caine *et al.* 1996; Billi *et al.* 2003). The mineralogical studies allowed

identification of different secondary minerals in different fractures, and point to the existence of several hydrothermal fluid circulation systems in space and time (Ledésert *et al.* 1996, 1999; Sausse 1999, 2001). The tectonic fracturing and the various ancient hydrothermal fluid systems transformed the granite into a excellent candidate for an induced geothermal system by creating not only a connected network of open fractures but also a connected network of pores in extended permeable zones associated with the different fractures.

The different granite facies have been studied with respect to permeability and other physical properties (Surma 2003; Surma & Géraud 2003). They are characterized by permeabilities varying between $1.1 \times 10^{-16} \, \text{m}^2$ and $10^{-18} \, \text{m}^2$ (Table 1). These values are high compared to those of other granitic rocks (usually 10^{-18} to $10^{-19} \, \text{m}^2$; Guéguen & Palciauskas 1992). The fresh porphyry granite is less porous (<1%) than the rocks of the other two facies, but there

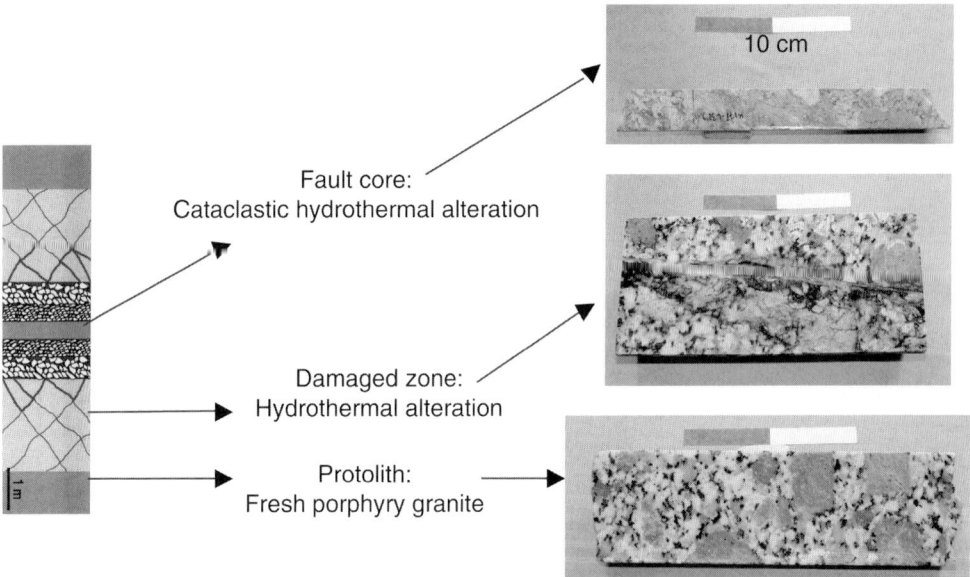

Fig. 5. Schematic diagram showing the geometry of a fracture and photographs of the associated granitic facies EPS1 cores.

Table 1. *Permeability, porosity, pore diameter and clay contents of fresh and altered granite*

Sample	Facies	Threshold (%)	Pore diam. (μm)	Permeability (m²)	Clays
K136	Fresh	0.4	0.001	1.1×10^{-17}	Chlorite
K108-2758	Hydrothermal	1.0	0.010	1.2×10^{-16}	Illite
K109-2795	Hydrothermal	5.1	0.030	1.0×10^{-18}	Illite

is a large variation of porosity in the hydrothermally altered samples due to precipitation and dissolution of secondary minerals (Table 1). These porosity variations are associated with permeability variations due to clay minerals that precipitate in the pore space and reduce connectivity and, therefore, reduce the matrix permeability. In addition, the chemical composition of fluids from one of the boreholes (Table 2) documents that the fluids are mainly brines, which might cause precipitation of additional clays and other secondary minerals in the pores and along fractures. According to Surma (2003) the main permeability is due to the presence of microfractures, but matrix permeability of the rock is not negligible because of the possible variation of the chemical composition of the fluid and the possible flow of the fluid in the matrix. Fluid pathways between fractures (main transport zone) and damage zones of the faults (secondary transport zone) were found using X-ray computer tomography (Surma 2003).

Fluid flow modelling

Our fluid flow model, which is based on seismic, petrophysical, and mineralogical data, suggests that exchange between fractures and matrix only occurs when the fractures are nearly sealed and fluid flows are low (Surma 2003). Similarly, other research predicts a possible sealing of fractures during long-term fluid injection (Jacquot et al. 1999; Jacquot 2000). Therefore, the damage zone adjacent to a fracture will play the double role of transfer and reaction zone. Fluid pathways and fluid flow are strongly related to the precipitation of secondary minerals. Slow fluid flow in the matrix induces more complete reactions in favour of long-term sealing and higher fluid flow causes less complete reactions (kinetic constraints) and, therefore, allows better retention of long-term permeability. In order to study these chemical reactions, numerical models and experiments were performed.

Mineralogical evolution during injection experiments

Thermodynamic model calculations, which take into account the current temperature gradient in the Rhine graben and the composition of the altered granite, have shown that the hydrothermal fluids present in the granite cause a new stage of alteration (Komninou & Yardley 1997). Model calculations have been performed in order to predict the formation of new

Table 2. *Chemical analyses of fluids from the production well GPK 1 at Soultz-sous-Forêts*

Sample	pH lab 25 °C	Alc. (meq/L)	Na (g/L)	K (g/L)	Ca (g/L)	Mg (mg/L)	SiO$_2$ (mg/L)	Al (mg/L)	Ba (mg/L)	Fe (mg/L)	Cl (g/L)	SO$_4$ (mg/L)	F (mg/L)
Drill head[*]	4.75	10.6	28.2	3.32	6.73	150	97	n.d.	12.3	232	58.5	215	3.9
In the borehole[*] (−1810 m)	4.75	4.29	27.2	3.21	6.76	145	97	n.d.	12.0	53	57.0	205	n.d.[‡]
In the borehole[*] (−1845 m)	4.75	3.09	28	3.28	6.96	152	94	n.d.	12.5	7.5	58.1	220	n.d.
In the borehole[*] (−1930 m)	4.75	4.18	27.9	3.40	6.93	152	93	n.d.	12.5	30	58.5	225	n.d.
In the borehole[†] (−3500 m)	4.75	7.5	27.9	2.81	7.3	112	210	n.d.	12.0	n.d.	61.0	198	n.d.

[*]Data from Pauwels et al. (1993).
[†]Data from Aquilina et al. (1997).
[‡]n.d., not determined.

hydrothermal minerals in the fractured rock system (Jacquot *et al.* 1999; Jacquot 2000). These calculations indicate that the geothermal reservoir porosity is essentially controlled by carbonate precipitation and dissolution, although smectites can also precipitate in the fracture network at low temperature (from cooled fluid). Carbonates are only minor phases in contact with the hydrothermal fluid, but dissolution and precipitation kinetics of carbonates are very fast compared to those of silicate minerals and, therefore, are of importance for the short-term evolution of the deep reservoir porosity (Durst 2002).

In addition to these thermodynamic calculations, experiments were performed to identify potential mineral precipitations along the fluid pathway (Baldeyrou *et al.* 2003). These experiments were carried out in cold-sealed vessels at 600 bar and under a defined thermal gradient of 300 to 200 °C (Fig. 6). Tube-in-tube experiments have already been carried out at higher temperatures and pressures (Robert & Goffé 1993; Vidal 1997; Poinssot *et al.* 1998). A perforated gold capsule was filled with the starting material and placed into a 15 cm long gold tube that contained only pure water. The tube was then placed in a horizontal cold-sealed vessel (Tuttle 1949). The hot and cold extremities of the tube were heated to 300 °C and 200 °C, respectively (Fig. 7). The starting material consisted of mineral mixtures (albite, K-feldspar, quartz) or crushed rock samples, which represent several granite compositions (Tables 3 and 4). At the end of the run, after 40 days, the tube was quenched from 300 °C to room temperature within five minutes of using compressed air. The gold tube was then cut longitudinally, and its internal walls were studied with a scanning electron microscope (SEM; JEOL JSM 840 with energy-dispersive analytical system [EDS]) and a scanning transmission electron microscope (STEM; Philips CM12 with EDS) in order to identify and analyse the minerals, that have crystallized along the thermal profile. If enough material was available, it was collected and studied by X-ray diffraction (XRD). To test reproducibility, two runs have been performed: MEL 1 with a starting mixture of natural albite, K-feldspar and quartz and GRAN 4 with a starting mixture of a crushed granite (Tables 3 and 4).

After the experiments, significant quantities of newly formed minerals were observed at the cold extremity of the tube, pointing to a fast material transport by diffusion from the hot to the cold end of the tube. The following spatial distribution of newly formed phases, reflecting the temperature profile, was observed in both runs (Fig. 8): quartz + K-feldspar ± plagioclase ± Mg-rich saponites (hot extremity); quartz + K-feldspar + plagioclase (middle of the tube); and alkaline or Ca-rich clays + quartz ± plagioclase (cold extremity). The cation composition of the phyllosilicates was similar in both experiments. Some newly formed quartz crystals

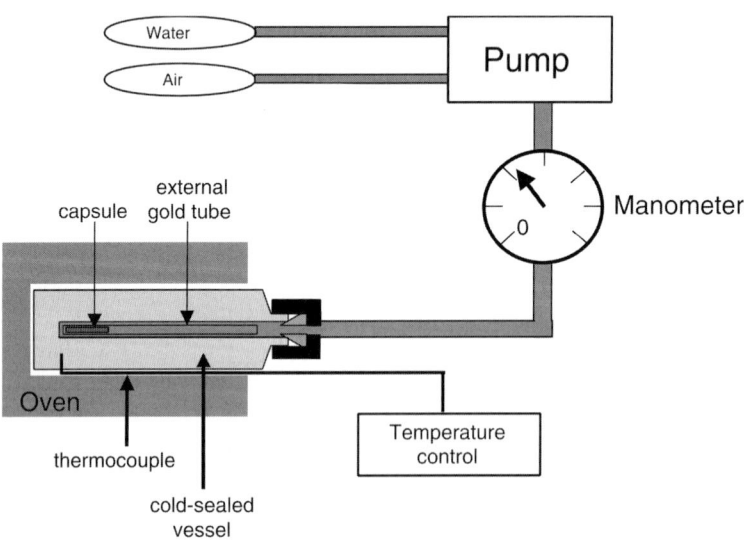

Fig. 6. Outline of a cold-sealed vessel experiment.

Temperature along the tube (°C)

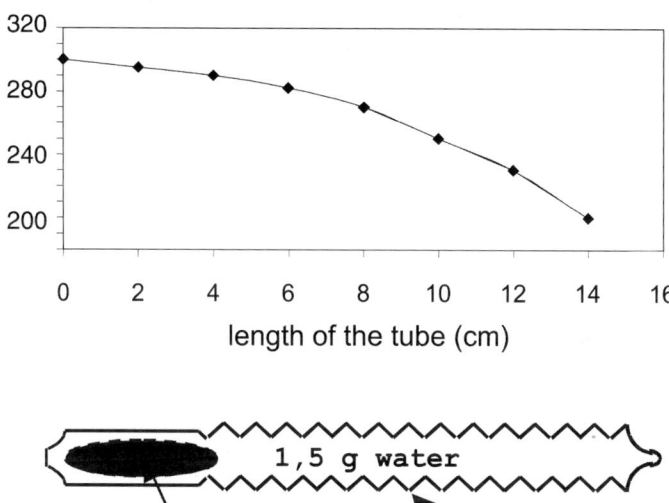

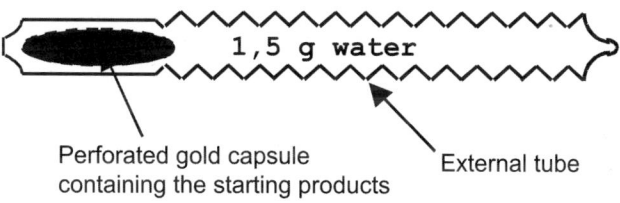

Perforated gold capsule
containing the starting products

External tube

Fig. 7. Schematic diagram showing the tube-in-tube design for the experiment and the temperature profile along the tube. The tube represents the reaction area where crystallization can occur. The capsule placed at the hot extremity of the tube contains the starting materials, which act as the source of elements precipitated in the tube (from Baldeyrou *et al.* 2003).

Table 3. *Experimental conditions and composition of the starting products*

Experimental run	*P, T* conditions	Starting mixtures (mol%)	Run duration
MEL 1	*P*: 600 bars *T* (at hot point): 300 °C	Quartz, K-Feldspar, Albite (62.5: 12.7: 24.8)	40 days
GRAN 4	*P*: 600 bars *T* (at hot point): 300 °C	Granite crush, see Table 4	40 days

Table 4. *Chemical composition of the crushed granite used as starting material for experimental run GRAN 4*

Oxides	wt%
SiO_2	69.24
TiO_2	0.48
Al_2O_3	15.29
Fe_2O_3t	3.36
MnO	0.04
MgO	0.40
CaO	2.41
Na_2O	4.59
K_2O	3.89
P_2O_5	0.30

exhibit dissolution rims, and some feldspars are overgrown by clays. These observations suggest that a thermodynamic equilibrium has not been achieved. Each part of the tube may represent a local transient equilibrium, which evolves with time (Baldeyrou-Bailly 2003).

In the run MEL 1, the proportion of quartz relative to other secondary minerals increases from the hot to the cold extremity, whereas the proportion of K-feldspar decreases. Towards the cold end of the tube, K-feldspar is gradually overgrown by Na- and K-rich clays. Albite is totally overgrown by clays at the cold extremity. According to XRD patterns (Fig. 9) and chemical compositions (Fig. 10), the clay minerals are

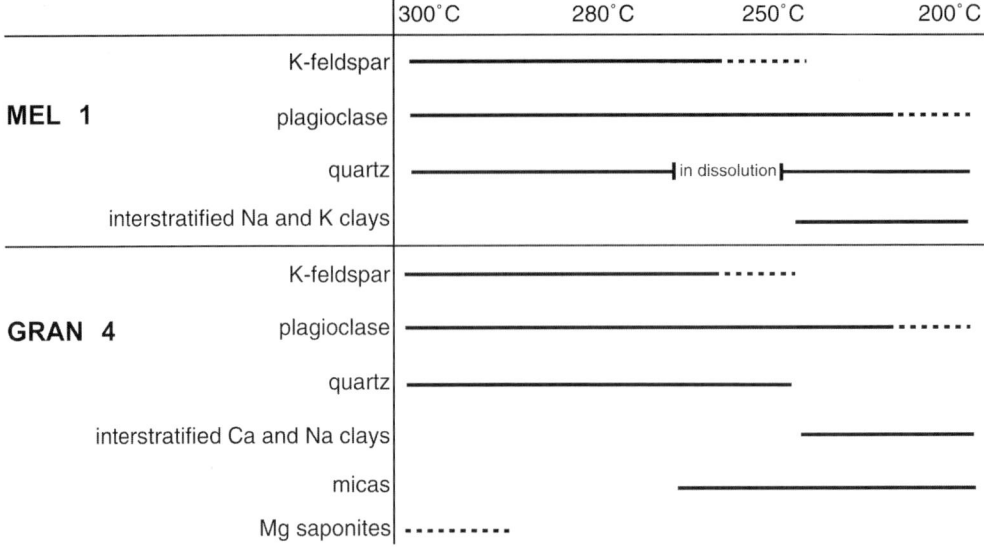

Fig. 8. Scheme showing the spatial distribution of the newly formed minerals along the temperature gradient in experimental runs MEL 1 and GRAN 4 (see also Fig. 7).

probably interstratified (20.7 Å and 14.5 Å), composed of mica and smectite layers with about 80% smectite. Figure 9 shows that the clays formed in the run GRAN 4 are interstratified clay minerals with a structure close to the one found in experiment MEL 1, and some micas at the cold extremity (10 Å peak).

The observed paragenesis contains a significant amount of feldspars. Although feldspars are not considered to crystallize in the numerical simulations (Jacquot 2000; Durst & Vuataz 2001), our experimental results suggest that formation of clay minerals is controlled by previously crystallized feldspars. Therefore, the presence of feldspars is of importance for the crystallization of secondary minerals and the evolution of porosity in the stimulated reservoir of the HFR site.

The tube-in-tube experiment is a very powerful method to determine the sequence of precipitation of secondary minerals as function of temperature for a chosen chemical system. Chemical reactions occur quickly (within 40 days) and the transport by diffusion of chemical elements is efficient. Similar crystallization sequences are observed in both experiments, suggesting that the transitions between the different mineral phases are not only controlled by the composition of the solution but also by temperature. The experimental design does not strictly correspond to the geometry encountered at the Soultz-sous-Forêts site and therefore needs to

be improved to be a real analogous experiment. Nevertheless, these experiments allowed for the first time to deduce empirically the mineralogical evolution in the Soultz-sous-Forêts granite under geothermal conditions.

Conclusions

The HFR geothermal site of Soultz-sous-Forêts is ready very soon to become a scientific pilot plant. It shows a good potential for extraction of very hot water, whose heat will be transformed into electrical energy. The expected final production level is around 25 MWe, which corresponds roughly to the demand of 25 000 people. The results of scientific studies associated with the project show that fluid–rock interactions will obviously occur in the water pathways and are controlled primarily by carbonates (result from model calculations), but also by feldspars and phyllosilicates (result from experimental study). It was further found that extensive fluid–rock interaction will occur in the rock matrix, more so than in the major fractures. The mineralogical evolution of this rock matrix during fluid flow will reduce the porosity and then the pathway of the fluid. This evolution should now be studied in greater detail in order to evaluate how long the geothermal site could be productive.

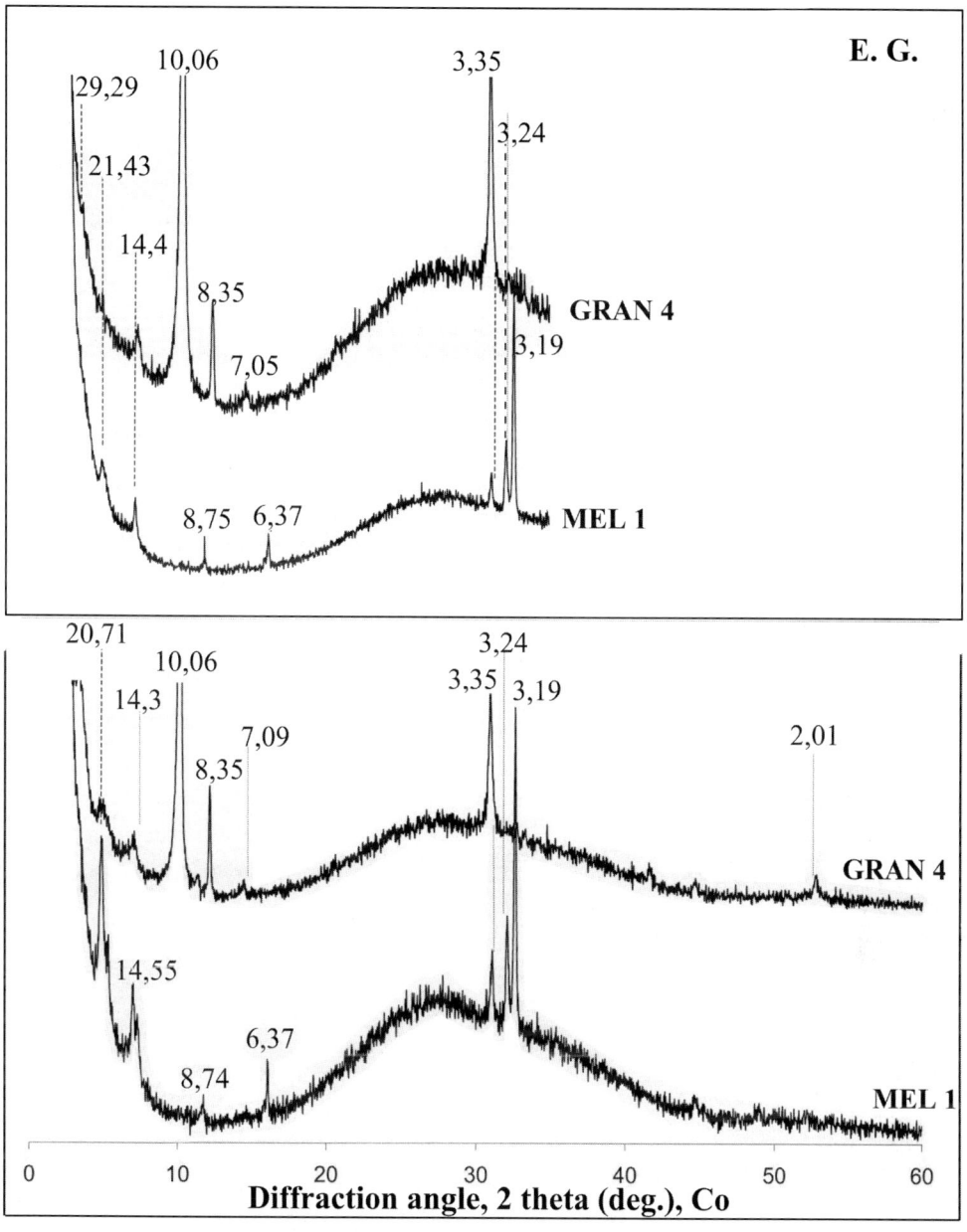

Fig. 9. X-ray diffraction patterns of newly formed minerals at the cold extremity of GRAN 4 and MEL 1 before (lower) and after (upper) treatment with ethylene glycol (E.G.). Peaks are labeled with values of d-spacings (in Å).

This project is of interest for several reasons, but primarily because the deep fluid is used as a closed loop, avoiding surface storage of brines, and because the seismicity induced by stimulation of the reservoir is continually watched in order to limit the risks. The Soultz-sous-Forêts geothermal site could be a pilot site for the future 'clean energy' projects in the field of geothermal high-energy enthalpy mining.

The authors would like to thank for financial support and a PhD grant (for F. Surma), which allowed their research: the CNRS-ADEME 'Ecodev' Program, the EC Project,

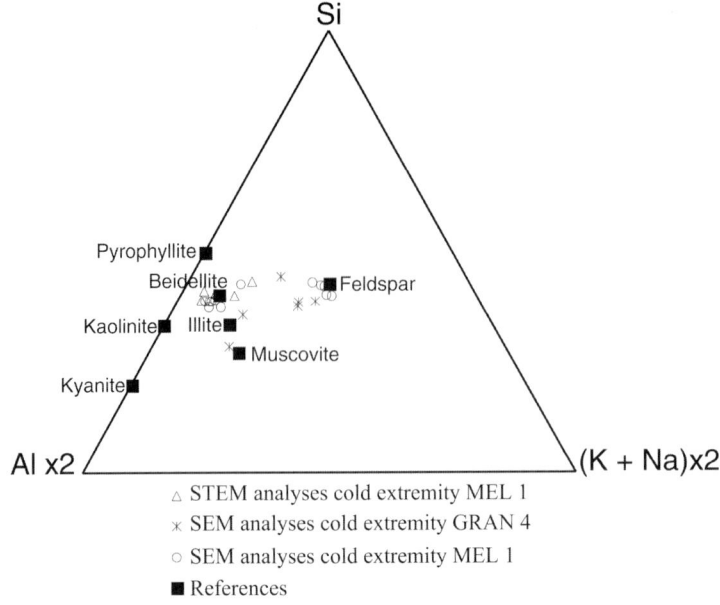

Fig. 10. Chemical compositions of the newly formed minerals as determined by EDS (SEM and STEM data). The minerals crystallized at the cold extremity of the tubes during experimental runs MEL1 and GRAN 4. Black squares are index minerals: Pyrophyllite $Si_4Al_2O_{10}(OH)_2$; Kaolinite $Si_2Al_2O_5(OH)_4$; Muscovite $KAl_3Si_3O_{10}(OH)_2$; Kyanite Al_2SiO_5; Feldspar $KAlSi_3O_8$; Beidellite $Si_{3.67}Al_{0.33}Al_2Na_{0.33}$; Illite $K_{0.7}Al_{2.7}Si_{3.3}O_{10}(OH)_2$. (Modified from Baldeyrou *et al.* 2003).

ENK562000-00301, and the Region Alsace. They would also like to thank the GEIE 'Exploitation Minière de la Chaleur' and particularly André Gérard for practical support and access to drilling and exploitation data.

References

AQUILINA, L., PAUWELLS, H., GENTER, A. & FOUILLAC, C. 1997. Water–rock interaction processes in the Triassic sandstone and the granitic basement of the Rhine Graben: geochemical investigation of a geothermal reservoir. *Geochimica et Cosmochimica Acta*, **61**, 4281–4295.

BALDEYROU-BAILLY, A. 2003. *Etude Expérimentale et Modélisation de la Stabilité des Phyllosilicates Soumis à un Fort Gradient Thermique. Test dans le Contexte du Site Géothermique de Soultz-sous-Forêts*. Thèse de l'Université Louis Pasteur, Strasbourg, 296 p.

BALDEYROU, A., VIDAL, O. & FRITZ, B. 2003. Etude expérimentale des transformations de phases dans un gradient thermique: application au granite de Soultz-sous-Forêts, France. *Comptes rendus de l'Académie des Sciences Série II Fascicule A, Paris*, **335**, 371–380.

BARIA, R., BAUMGÄRTNER, J., GÉRARD, A., JUNG, R. & GARNISH, J. 1999. European HDR Research Program at Soultz-sous-Forêts (France); 1987–1996. *Geothermics* (Special issue on HDR Academic Review), **28**, 655–669.

BILLI, A., SALVINI, F. & STORTI, F. 2003. The damage zone–fault core transition in carbonates rocks: implications for fault growth, structure and permeability. *Journal of Structural Geology*, **25**, 1779–1794.

BRUN, J. P., WENZEL, F. & ECORS TEAM 1991. Crustal-scale structure of the southern Rhine graben from ECORS-DEKORP seismic reflection data. *Geology*, **19**, 758–762.

CAINE, J. S., EVANS, J. P. & FORSTER, J. P. 1996. Fault zone architecture and permeability structure. *Geology*, **24**, 1025–1028.

DEZAYES, C., VILLEMIN, T. & PÊCHER, A. 2000. Microfracture pattern compared to core-scale fractures in the borehole of the Soultz-sous-Forêts granite, Rhine graben, France. *Journal of Structural Geology*, **22**, 723–733.

DEZAYES, C., VILLEMIN, T., TRAINEAU, H., GENTER, A. & ANGELIER, J. 1995. Analysis of fractures in boreholes of the Hot Dry Rock project at Soultz-sous-Forêts (Bas-Rhin, France). *Scientific Drilling*, **5**, 31–41.

DURST, P. 2002. *Geochemical Modelling of the Soultz-sous-Forêts Hot Dry Rock Test Site: Coupling Fluid–Rock Interactions to Heat and Fluid Transport*. PhD thesis, University of Neuchâtel, 127 p.

DURST, P. & VUATAZ, F. D. 2001. Geochemical modeling of the Soultz-sous-Forêts Hot Dry Rock system, brine rock interactions in a deep hot fractured granite reservoir. *26th Workshop on Geothermal Reservoir Engineering*, Stanford University, Stanford, California, SGP-TR-168.

EVANS, J. P. & CHESTER, F. M. 1995. Fluid rock interaction in faults of the San Andreas system:

Inferences from San Gabriel fault rock geochemistry and microstructure. *Journal of Geophysical Research*, **100**, 13007–13020.

FRIDLEIFSSON, I. B. 1998. Direct use of geothermal energy around the world. *Geo-heated Centre Quarterly Bulletin, Oregon Institute of Technology*, **19**, 4–9.

GENTER, A. 1990. Géothermie roches chaudes sèches: le granite de Soultz-sous-Forêts (Bas-Rhin, France). *Document BRGM*. **185**, 201 p.

GENTER, A. & TRAINEAU, H. 1992. Borehole EPS1, Alsace, France. Preliminary geological results from granite core analyses for hot dry rock research. *Scientific Drilling*, **3**, 205–214.

GENTER, A. & TRAINEAU, H. 1996. Analysis of macroscopic fractures in granite in the HDR geothermal well EPS-1, Soultz-sous-Forêts, France. *Journal of Volcanology and Geothermal Research*, **72**, 121–141.

Gérard, A., JUNG, R., BAUMGARTNER, J., BARIA, R., GANDY, T. & TRAN-VIET 1998. *Essais de Circulation Moyenne Durée Conduits à Soultz-sous-Forêts en 1997*. Rapport d'activité de SOCOMINE, 20 p.

GUÉGUEN, Y. & PALCIAUSKAS, V. 1992. *Introduction à la Physique des Roches*. Hermann, Paris, 299 p.

JACQUEMONT, B. 2002. *Etude des Interactions Eaux–Roches dans le Granite de Soultz-sous-Forêts. Quantifications et Modélisation des Transferts de Matière pour les Fluides*. Thèse de l'Université Louis Pasteur, Strasbourg, 181 p.

JACQUOT, E. 2000. *Modélisation Thermodynamique et Cinétique des Réactions Géochimiques entre Fluides de Bassin et Socle Cristallin: Application au Site Expérimental du Programme Européen de Recherche en Géothermie Profonde (Soultz-sous-Forêts, Bas-Rhin, France)*. Thèse de l'Université Louis Pasteur, Strasbourg, 202 p.

JACQUOT, E., FRITZ, B. & LEROY, J. 1999. Geochemical modelling of interactions between crystalline basement and basin fluid applied to the geothermal HDR plant of Soultz-sous-Forêts (Alsace, France). *Terra Abstracts*, **1**, 586. EUG X Strasbourg, France. March 28–April 1, 1999.

KOMNINOU, A. & YARDLEY, B. W. 1997. Fluid–rock interaction in the Rhine Graben: a thermodynamic model of the hydrothermal alteration observed in deep drilling. *Geochimica et Cosmochimica Acta*, **61**, 515–531.

LEDÉSERT, B., JOFFRE, J., AMBLÈS, A., SARDINI, P., GENTER, A. & MEUNIER, A. 1996. Organic matter in the Soultz HDR granitic thermal exchanger (France): natural tracer of fluid circulations between the basement and its sedimentary cover. *Journal of Volcanology and Geothermal Research*, **70**, 235–253.

LEDÉSERT, B., BERGER, G., MEUNIER, A., GENTER, A. & BOUCHET, A. 1999. Diagenetic-type reactions related to hydrothermal alteration in the Soultz-sous-Forêts granite, France. *European Journal of Mineralogy*, **11**, 731–741.

MARQUIS, G., DARNET, M., SAILHAC, P., SINGH, A. K. & GÉRARD, A. 2002. Surface electric variations induced by deep hydraulic stimulation: an example from the Soultz HDR site. *Geophysical Research Letters*, **29**, 7.1–7.4.

PAUWELS, H., FOUILLAC, C. & FOUILLAC, A. M. 1993. Chemistry and isotopes of deep geothermal saline fluids in the upper Rhine Graben: origin of compounds and water–rock interactions. *Geochimica et Cosmochimica Acta*, **57**, 2737–2749.

POINSSOT, CH., TOULHOAT, P. & GOFFÉ, B. 1998. Chemical interaction between a simulated nuclear waste glass and different backfill materials under a thermal gradient. *Applied Geochemistry*, **13**, 715–734.

PRIBNOW, D., FESCHE, W. & HÄGEDORN, F. 1999. *Heat Production and Temperature to 5 km Depth at the HDR Site in Soultz-sous-Forêts*. Geowissenschaftliche Gemeinschaftsaufgaben (G.G.A.) report, Hannover, 17 p.

ROBERT, C. & GOFFÉ, B. 1993. Zeolitisation of basalts in subaqueous freshwater settings: Field observations and experimental study. *Geochimica et Cosmochimica Acta*, **57**, 3597–3612.

SAUSSE, J. 1999. *Caractérisation et Modélisation des Écoulements Fluides en Milieu Fissure. Relation avec les Altérations Hydrothermales et Quantification des Paléocontraintes*. Thèse de l'Université Henri Poincaré, Nancy, 336 p.

SAUSSE, J. 2001. Hydromechanical properties and alteration of natural fracture surfaces in the Soultz granite (Bas-Rhin, France). *Tectonophysics*, **348**, 169–185.

SHAPIRO, S., PATZIG, R., ROTHERT, E. & RINDSCHWENTNER, J. 2003. Triggering of seismicity by pore pressure perturbations: permeability-related signatures of phenomenon. *Pure and Applied Geophysics*, **160**, 1051–1066.

STEFANSSON, V. & FRIDLEIFSSON, I. B. 1998. *Geothermal Energy – European and Worldwide Perspective. Expert hearing on Assessment and Prospects for Geothermal Energy in Europe*. Sub-committee on Technology and Energy of the Parliamentary Assembly of the Council of Europe, Strasbourg, France, p. 10.

SURMA, F. 2003. *Etude de la Porosité des Zones Endommagées autour des Failles et Rôle de l'État du Matériau sur les Propriétés d'Échange Fluide–roche: Minéralogie, Structures de Porosité, Caractéristiques Mécaniques*. Thèse de l'Université Louis Pasteur, Strasbourg, 272 p.

SURMA, F. & GÉRAUD, Y. 2003. Porosity and thermal conductivity of the Soultz-sous-Forêts granite. *Pure and Applied Geophysics*, **160**, 1125–1136.

TUTTLE, O. F. 1949. Two pressure vessels for silicate–water studies. *Geological Society of American Bulletin*, **60**, 1727–1729.

VIDAL, O. 1997. Experimental study of the thermal stability of pyrophyllite, paragonite and clays in a thermal gradient. *European Journal of Mineralogy*, **9**, 123–140.

WURMSTICH, B. & MORGAN, F. D. 1994. Modelling of streaming potential responses caused by oil well pumping. *Geophysics*, **59**, 46–56.

ZOBACK, M. D. & HARJES, J. H. 1997. In situ stress measurements to 3,5 km in the Cajon Pass scientific research borehole: implications for mechanics of crustal faulting. *Journal of Geophysical Research*, **97**, 5039–5057.

Waste heat problems and solutions in geothermal energy

LADISLAUS RYBACH[1,2] & THOMAS KOHL[1,2]

[1]*GEOWATT AG, Zürich, Switzerland*

[2]*Institute of Geophysics, ETH Hönggerberg, Zürich, Switzerland*

Abstract: All heat–power conversion systems produce waste heat, which can attain significant portions. This applies to geothermal power generation too; the waste heat fraction depends on the conversion technology. The more of the waste heat that can be utilized for some useful purpose (and consequently requiring less heat to be rejected), the better economy can be achieved, besides benign environmental effects. The best solution to avoid discharge to the atmosphere or to the hydrosphere is cascaded use. This consist of a chain of applications with stepwise decreasing temperatures, for example, from industrial uses through balneology down to fish farming. Constraints given by environmental legislation can lead to beneficiary solutions like in the case of warm tunnel waters in the Swiss Alps: these would need cooling ponds/towers before being permissible for discharge into local rivers. It is described by several specific examples how the tunnel waters can be used instead.

Geothermal energy utilization aims at taking advantage of the Earth's internal heat. The heat content of the Earth's interior is immense, and significant parts are constantly generated through the decay of naturally radioactive elements, such as uranium and thorium. The loss of heat through the Earth's surface amounts to an impressive total of 40 million MW (Chapman & Rybach 1985).

The main form of utilization is the generation of electricity from geothermal resources and the direct use of the geothermal heat in various applications like bathing, space heating/cooling, agricultural and industrial uses (see Arnórsson, this volume). In the following, the waste heat of geothermal power generation will be addressed first. Various power plant types lead to waste heat of different extents. Besides the description of generation and avoidance, environmental constraints as well as examples of waste heat utilization will be presented. In particular, geothermal use of warm tunnel waters in Switzerland will be discussed.

Power plant types and their waste heat

Fundamental to all heat–power conversion systems is that a significant portion of the heat supplied to the system must be rejected. Depending on their heat–power conversion efficiency, fossil-fueled plants waste 40–60%, nuclear-fueled plants 60–70% of the heat input; geothermal power plants make no exception here.

The heat rejected by the power cycle Q_{rej} depends on the thermal efficiency η, and is given by

$$Q_{rej} = Q_{in} - Q_w = Q_{in}(1 - \eta) \qquad (1)$$

where Q_{in} is the heat input into the cycle and Q_w is the energy converted by the cycle ($Q_w = \eta \times Q_{in}$). Figure 1 shows the ratio between Q_w for a power cycle with an efficiency of $\eta = 20\%$ set to one, and at various other thermal efficiencies. It can be clearly recognized that the heat rejected increases for less efficient power plants. Small efficiency improvements can have strong impacts on the heat rejection. Geothermal power plants differ in design and efficiency according to the pressure and temperature of the geothermal fluid extractable from the geothermal reservoir.

The simplest of the geothermal cycles is the *direct-intake non-condensing* type. Steam from the geothermal well is directly passed through a turbine and exhausted to the atmosphere without condensation. Such cycles require 15–20 kg steam per generated kWh. *Condensing plants* (with condensers at the turbine outlet) need only 6–10 kg steam per kWh, depending on the turbine inlet pressure: 6 kg for 1.5–2.0 MPa, 7–9 kg with 0.5–1.5 MPa. Turbine types for these vapour-dominated applications are typically in the capacity range 20–120 MWe, and which can rely on the relatively high enthalpy of steam

From: GIERÉ, R. & STILLE, P. (eds) 2004. *Energy, Waste, and the Environment: a Geochemical Perspective*. Geological Society, London, Special Publications, **236**, 369–380.
0305-8719/04/$15 © The Geological Society of London 2004.

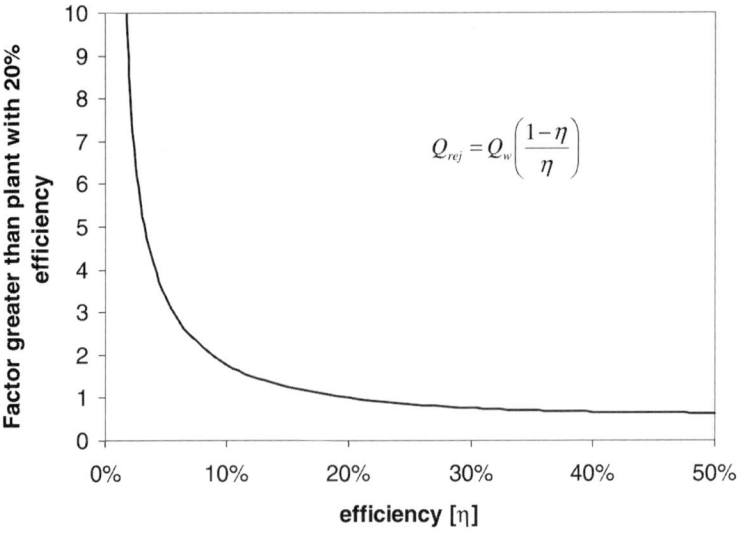

Fig. 1. Decreasing heat rejection with higher efficiencies: factor $Q_{w_{\eta=20\%}}/Q_w$ vs. thermal efficiency. (Modified from Robertson 1980). Example: at $\eta \sim 9\%$ the rejected energy is twice as much as for a plant with $\eta = 20\%$.

(around 2.8 MJ/kg) at fluid temperatures around 220 °C.

Resources with such favourable characteristics are rare and most of them already discovered. Water-dominated fields with lower temperatures and enthalpies are much more abundant. For these, *flash-steam plants* at one or more pressure stages are used to flash the geothermal water to steam, commonly in the capacity range 10–50 MWe.

The power-generating potential of a water-dominated resource depends on the geothermal fluid temperature and production flow rate (Fig. 2). The figure gives the net power output, which accounts for parasitic loads caused by the condenser and feed pump power requirements. The output power from two-phase water–steam or steam alone is much greater than the curves shown for liquid in Fig. 2.

The flash-steam systems can be used at lower geothermal fluid temperatures (<175 °C). When the fluid is flashed to steam, precipitation of solids might pose problems. Also the non-condensable gases (mainly H_2S) need to be treated ('condensing cycle', see Fig. 3). In such a design, the steam from the turbine is discharged to a condensing chamber, which is maintained at low pressure (around 10 kPa). Because of the large pressure drop across the condensing turbine more power can be generated for typical inlet conditions than with simpler design (conventional atmospheric exhaust steam turbine; for details, see Hudson 1998). Typical power plant size is 20 MWe.

If the geothermal well produces relatively low-temperature fluid (<175 °C), electricity can also be generated by means of *binary cycle plants*, albeit with low efficiency (see below). The binary system utilizes a secondary working

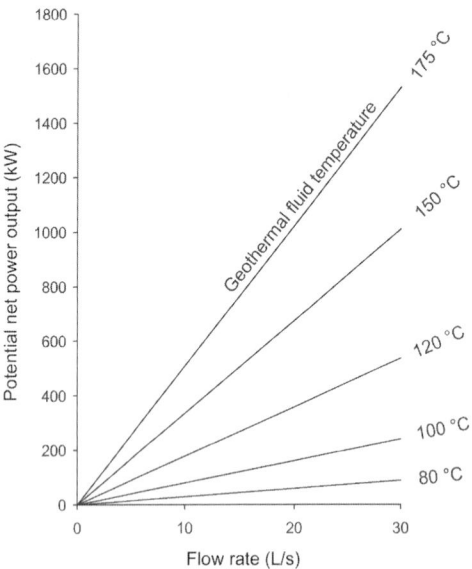

Fig. 2. Geothermal power generation capacity of a geothermal resource. Feed pump and condenser parasitic power have been accounted for. (Modified from Nichols 1986).

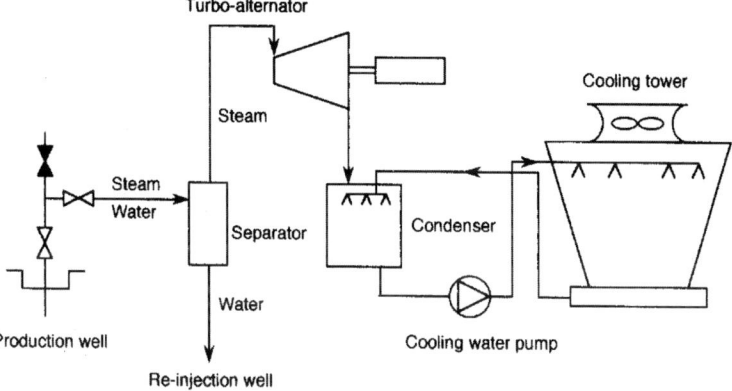

Fig. 3. Simplified schematics of *condensing cycle*-type power plant. (Modified from Hudson 1998).

fluid (e.g., isobutane, C_4H_{10}, called R600a in the refrigeration industry), which has a low boiling point and high vapour pressure at low temperature relative to steam. This secondary fluid is operated through a conventional Rankine cycle; heat is transferred from the geothermal fluid to the binary cycle via heat exchangers, where the working fluid is heated and vaporized before being expanded in the turbine (Fig. 4). By selecting the appropriate working fluid, binary systems can operate with fluid inlet temperatures as low as 85 °C. Binary plants are traditionally small, modular units varying in size from a few hundred kW to several MW (Hudson 1998).

The efficiency η of conversion of a binary cycle is estimated by a function of the inlet temperature T_f (in °C) (Hudson 1998):

$$\eta \ (\%) = 0.36(0.18 \, T_f - 10) \qquad (2)$$

that is, a geothermal fluid inlet temperature of 140 °C yields a net conversion efficiency of 5.5%, whereas at 100 °C an efficiency of 2.9% only can be achieved. In any case, significant amounts of heat will be rejected by any of the systems. Table 1 shows the heat rejected as a function of the thermal efficiency of the power cycle. Cooling towers (wet mechanical draft or air-cooled coil types) are often needed for waste heat disposal.

In comparison with other power generation technologies the geothermal waste heat generation per MW electric power is substantial (Fig. 5). The higher portion of waste heat for geothermal plants arises from lower efficiencies due to the lower cycle inlet temperatures. In the geothermal case, however, the waste heat occurs only at the production site, whereas for other energy sources (coal, oil, gas, uranium),

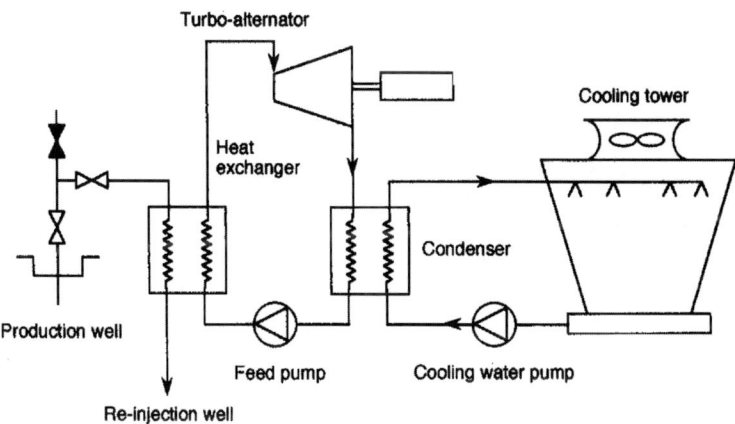

Fig. 4. Simplified schematics of *binary cycle*-type power plant. (Modified from Hudson 1998).

Table 1. *Heat rejected as a function of the thermal efficiency of the power cycle**

Thermal efficiency (%)	Heat (MWt) rejected per MWe of station capacity
5	19.00
10	9.00
15	5.07
20	4.00

*From Robertson (1980).

mining activities and transport also generate heat, which is not included in Fig. 5.

Waste heat avoidance strategies

Unless otherwise used, the waste heat due to power generation ultimately ends up in the atmosphere or hydrosphere. The latter appear to be unlimited heat sinks; however, local conditions like the microclimate can be significantly influenced. Therefore, the best solution to avoid atmospheric and hydrospheric discharge is the underground disposal of the waste fluids or use of the waste heat.

Superheating

As displayed in Fig. 1, the waste heat decreases drastically with higher efficiencies. Because the efficiency of a cycle depends mostly on the temperature range, the waste heat is reduced for higher temperatures. Since the source temperature for geothermal applications is often fixed, Kohl *et al.* (2002) developed a hybridization concept for power generation. On the basis of existing data from the experimental geothermal project at Soultz-sous-Forêts (France), these authors could demonstrate that rather high efficiency factors (~15%) are already achievable for geothermal systems of 150 °C reservoir temperatures when superheating pure water vapour. In view of the multiple economic and operational advantages such innovative concepts could develop further in the near future.

Re-injection

Re-injection into the subsurface is the most often used technique to dispose of geothermal waste heat. The cooled-down fluids, after having fulfilled their tasks in the power generation cycle (or in the direct-use application), can be re-injected into the same geothermal reservoir from which the hot fluid has been produced. The fluid re-injection can help to sustain reservoir pressure, which otherwise would decrease during production. For the re-injection, however, additional wells must be drilled and – if required by the rock permeability at depth – the fluid must

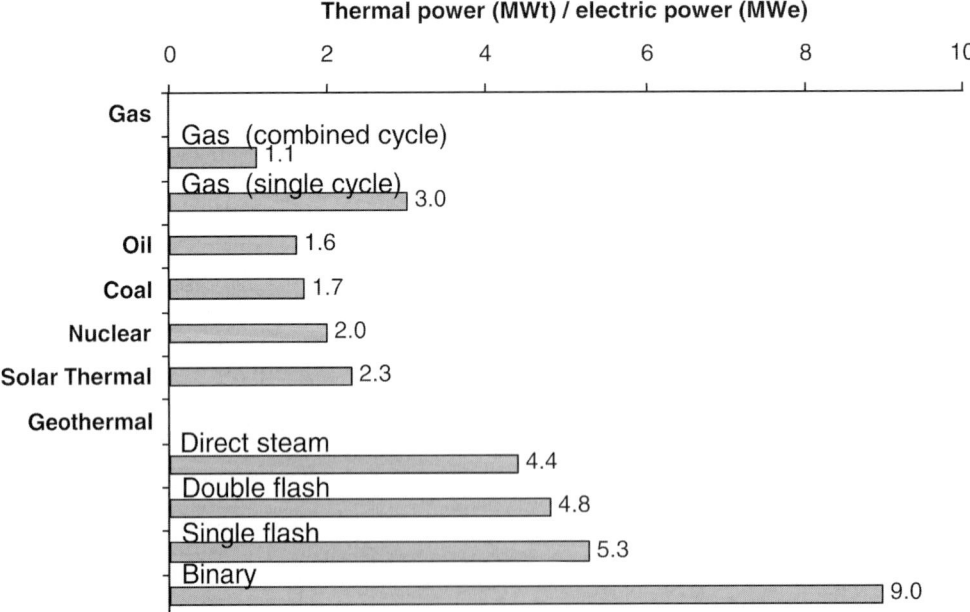

Fig. 5. Waste heat per unit electric capacity (MWt/MWe) of geothermal power plants, in comparison with other types of power generation. (Modified from DiPippo 1991).

be pressed back by pumping into the reservoir, which has economic consequences. In addition, the pumping rate must be carefully designed in order to avoid unwanted side effects like induced seismicity or thermal break-through, that is, the arrival of cooled fluid at production wells.

Co-generation

The economics of power generation can be highly improved if the heat, which otherwise would be wasted, can be sold. Therefore the combined heat–power systems ('co-generation') provide a solution with maximum economic benefits. The best way of waste heat utilization is the use of heat in a district heating system, which can take up the substantial heat load.

Unfortunately, geothermal power plants must be sited primarily by geologic criteria. The plant sites often rely on resources far remote from places where the heat could be used for space heating. Piping and pumping costs prohibit the heat transfer over greater distances. The now emerging hot-dry-rock (HDR) technology (also called enhanced geothermal systems, EGS) could site the power generation at places where heat consumers exist. Such projects are now under development in Basle and Geneva, Switzerland (Haering & Hopkirk 2001). The HDR/EGS technology operates in a closed system: water is circulated via injection and production wells through an artificially created heat

exchanger at several kilometres depth. Hybrid plants, such as those at Puna (Hawaii) and Cove Fort (Utah) use a flash steam unit followed by a binary unit to increase the resource utilization.

Recent examples of co-generation are given by the installation of small binary plants at Altheim and Bad Blumenau in Austria, which use the waste water for district heating and thermal bathing, respectively. References can be found in Pernecker & Uhlig (2002) for Altheim, and Legmann (2003) for Bad Blumenau.

Cascaded use

An important concept for making use of waste heat is the so-called cascaded use. Geothermal *direct-heat applications* can be attached to geothermal power generation systems efficiently. Applications needing temperatures not higher than 65 °C might be attached (cascaded) in series to the power plant fluid outlet line (Lund & Boyd 1999). There is a whole spectrum of possible direct-use applications suitable for stepwise reduced temperature levels, thus ensuring the minimum achievable effluent throw-away energy. The direct-use applications depend on the required fluid temperature. With decreasing temperature, the following applications can be considered for cascading: industrial use; space heating; diverse therapeutic and wellness uses; greenhouse and soil heating; de-icing of parking lots and pavements; production of domestic

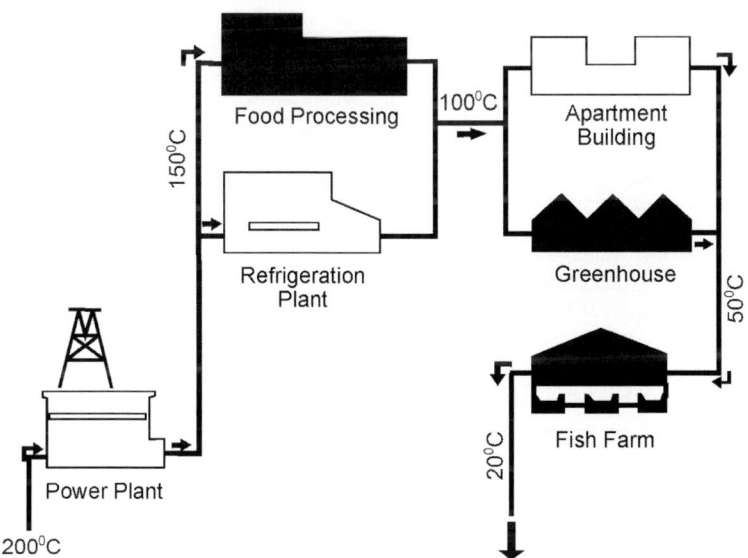

Fig. 6. Schematic flow diagram of a cascading sequence.

hot water; air-conditioning; heat pump applications; assorted low-temperature washing; low-temperature drying; and fish farming (see also Arnórsson, 2004; Banks *et al.*, 2004). Figure 6 shows a schematic cascading sequence, with progressively lower temperatures.

Examples of cascading use have existed for many years in Italy and Japan (see Minohara & Sekioka 1980; Lund 1987). Further cascading examples can be found in Dickson & Fanelli (1990, 1995).

Environmental considerations

Although the ultimate heat sink for geothermal waste heat is the Earth's atmosphere, there is no regulation to limit atmospheric discharges. Legal limitations exist, however, for discharges into surface waters. These restrictions aim at limiting the temperature rise in rivers and lakes, which could have detrimental effects on living organisms.

One such example is the maximum permissible temperature increase of $\Delta T_{reg} = +1.5\ °C$ in river waters, issued for Switzerland by the Swiss Agency for Environment, Forests and Landscape (SAEFL). This numerical value limits, at a given geothermal effluent temperature T_d, the maximum discharge flow rate into a nearby river. The calculation of the maximum permissible discharge rate R_{dmax} (L/s) must take into account the minimum flow rate R_{rmin} of the river and its minimum temperature T_{rmin} (which are usually observed during the winter season):

$$R_{dmax} = \Delta T_{reg} \times R_{rmin}/(T_d - T_{rmin} - \Delta T_{reg})$$

$$(3)$$

An example for such a situation is given by the new deep Alpine tunnels, which are being constructed in Switzerland. The AlpTransit (NEAT) project comprises two long deep (overburden up to 2.5 km) tunnels, the Gotthard and Lötschberg Base Tunnels, which drain substantial quantities of warm water from the surrounding rock. Under the influence of gravity, this warm water flows to the tunnel portals. To decrease the temperature of the tunnel waters (unless otherwise used; see below), construction of cooling ponds and/or towers would be needed at the portals because disposal into nearby rivers is limited by the authorities as described above. For example, the maximum permissible tunnel water discharges to the rivers near the portals of the Gotthard Base Tunnel are given in Table 2. It can be seen from Table 2 that much of the tunnel water outflow will not be disposable at the outflow temperature into the nearby rivers. Therefore utilization possibilities for the tunnel water heat must be sought in the vicinity of the portals. Such possibilities exist already elsewhere in Switzerland, as shown below.

Geothermal use of tunnel waters: a beneficial way to avoid disposal

Tunnels are subsurface constructions to overcome topographic barriers to transportation. They can serve various purposes (highway, railway, pipelines). Depending on local geologic conditions the tunnels can serve, especially in mountainous terrain, as heat/fluid sources: railway and road tunnels as well as major galleries drain water from their surrounding rock masses.

The piezometric surface inside mountain ranges exhibits significant relief between ridges and valleys (where the tunnel portals are located). This leads to the development of pronounced, deep-reaching groundwater circulation systems. As a result of the high piezometric level under ridges, the deep tunnels usually drain large amounts of warm water, especially in permeable fracture zones (Rybach 1995). The temperature of this inflow water depends on the overburden; in some cases it can reach more than 50 °C. The collected water flows under gravity to the portals; provided a sufficient flow rate, various types of installations at or near the tunnel portals can thus be supplied with heat.

Table 2. *Maximum disposable tunnel water flow rates R_{dmax}, for the Gotthard Base Tunnel (Switzerland), calculated by equation (3) with $\Delta T_{reg} = +1.5\ °C$*[*]

Portal	Nearby river	R_{rmin} (L/s)	T_{rmin} (°C)	T_d (°C)	Estimated outflow[†] (L/s)	R_{dmax} (L/s)
Erstfeld (N)	Schattdorf	3000	3.5	31.5	60–555	170
Bodio (S)	Pollegio	610	3.9	29.0	80–460	39

[*]From BFE (2002).
[†]From ATH (1993).

Table 3. *Geothermal potential of 15 selected tunnels in Switzerland*[*]

Tunnel	Tunnel type	Outflow (L/min)	Water temperature (°C)	Thermal power (kWt)[†]
Ascona	Highway	360	12	150
Furka[‡]	Railway	5400	16	3758
Frutigen (at Lötschberg)	Exploratory adit	800	17	612
Gotthard[‡]	Highway	7200	15	4510
Grenchenberg (south portal)	Railway	18000	10	11693
Hauenstein[‡]	Railway	2500	19	2262
Isla Bella	Highway	800	15	501
Lötschberg	Railway	731	12	305
Mappo-Morettina[‡]	Highway	983	16	684
Mauvoisin	Exploratory adit	600	20	584
Polmengo	Exploratory adit	600	20	584
Rawyl	Exploratory adit	1200	24	1503
Ricken[‡]	Railway	1200	12	501
Simplon (north portal)	Railway	1380	13	672
Vereina	Railway	2100	17	1608
Total				**29927**

[*]Data from SGA (2002).
[†]Calculated at portal, by cooling down to 6 °C.
[‡]Geothermal installation in operation.

Subsurface rock temperatures in mountain ranges depend on a large number of factors, including: cover thickness; three-dimensional topography; lithologic structure (with anisotropic thermal conductivity); hydrogeology (permeability distribution and water circulation); and transient effects like uplift/erosion and past climatic changes. Usually, the inflowing waters are in thermal equilibrium with the surrounding rock masses; that is, the water temperature equals that of the neighboring rock.

The inflow rate can change considerably with time: originally strong flows (tunnel construction phase) decrease significantly within the first few years to reach a lower, stable level, corresponding to a new hydraulic equilibrium in the circulation system (Rybach 1995).

The waters are collected and flow under gravity, from the point of tunnel culmination, towards the portals in more or less isolated conduits. Their flow rate and temperature depend on the natural inflow conditions, and on technical measures (injections to reduce inflow, insulation conditions during outflow). According to water protection regulations, waters of only limited amount and temperature can be disposed by discharge to rivers (see above). In extreme cases, the installation of cooling towers would be necessary. Therefore, the possible geothermal use of the outflowing tunnel waters is an attractive option; depending on flow rate and temperature, various types of installations at or near the tunnel portals can be supplied with heat.

Usually the outflow temperature shows no seasonal variations. Thus such waters represent an interesting potential for direct use, provided that potential heat consumers are situated at or near the portals. Contrary to the use of mine waters for space heating, as for example in Canada (Sanner 1993) or Germany (Bussmann 1994), where electrical energy is needed to lift the warm waters from the drowned mines to the surface, there is no need for pumps in the case of tunnel waters.

In the following, the tunnel water situation in Switzerland will be outlined, utilization examples will be described, and the potential of deep and long tunnels in construction will be discussed. In the concluding remarks the prerequisites of a geothermal use of tunnel waters will be addressed.

Tunnels in Switzerland

Switzerland accommodates a substantial part of the Alpine chain, and there are over 700 road or railway tunnels in the country. The cover thickness is often considerable, which can lead to rock/water inflow temperatures of 40–50 °C. An examination of the geothermal potential of Swiss tunnels screened 150 tunnels, from which 15 objects were selected for more detailed studies (Arbeitsgemeinschaft ZEWI 1995, 1996, 1998; Fig. 7). Their heat potential is estimated to be greater than 30 MWt (Table 3). The flow rates range from 360 L/min (Ascona) to

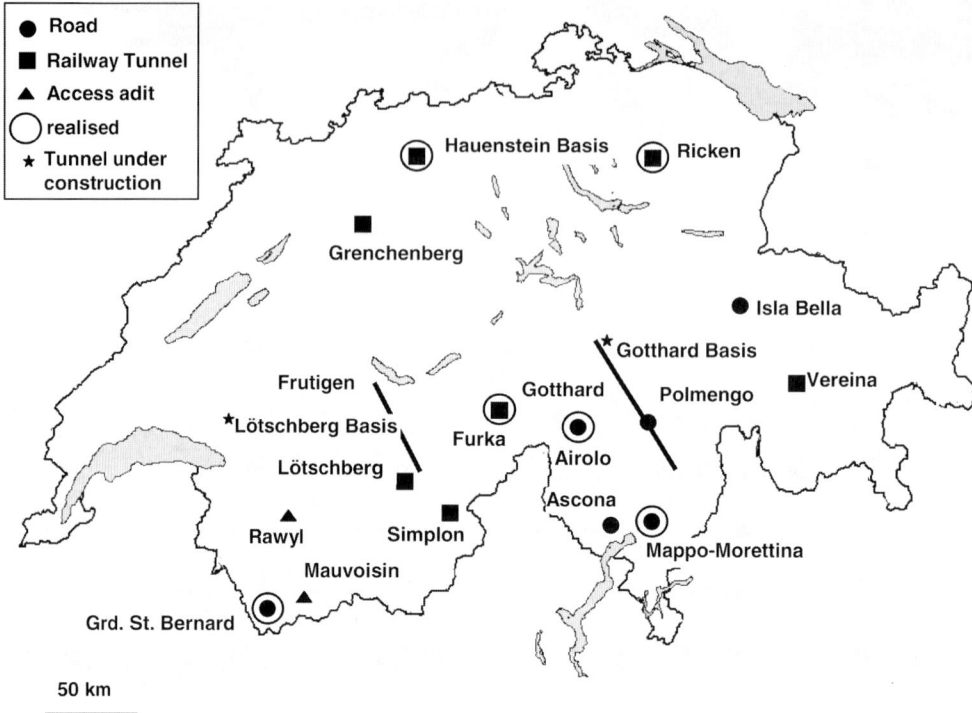

Fig. 7. Location of investigated Swiss tunnels and of geothermal direct use installations in operation.

24 000 L/min (Grenchenberg), and the outflow temperatures from 11.9 °C (Ricken) to 24.3 °C (Rawyl). Six heating systems are already in operation near the tunnel portals of St Gotthard (highway tunnel), Furka, Mappo-Morettina, Hauenstein, Ricken, and Grand St Bernard (Wilhelm & Rybach 2003).

Operating examples

The use of energy from warm tunnel waters is most advantageous in the immediate vicinity of the tunnel portals, because long transport/distribution piping is a major cost factor. Therefore, the installed heating systems are all situated at the tunnel portals. In addition to five systems where warm tunnel waters are exploited, there exists also one installation (Grand St Bernhard), where the heat content of warm tunnel air has been utilized since 1999 (for locations, see Fig. 7).

St Gotthard highway tunnel. The outflow at the south portal of the St Gotthard highway tunnel amounts to 7200 L/min with a temperature of

15 °C. The water is cooled down by 2.3 °C, and a heat pump supplies 1.9 MWt heating power to buildings of the highway service centre in Airolo (canton Ticino). This is the oldest installation, in operation since 1979. By technical upgrade (and especially with a larger cooling ΔT), the system could be improved to provide 4.0 MWt.

Furka railway tunnel. The outflow at the west portal amounts to 5400 L/min at 16 °C. The water is filtered and then piped to the nearby village Oberwald (canton Valais). There, an innovative system has been in use since 1991 (Rybach & Wilhelm 1995): the tunnel water is piped to individual heat pumps, tailored to supply various types of buildings ('cold district heating'). A total of 177 apartments and a sports centre are heated, with an installed capacity of 960 kWt (some of the buildings seen on Fig. 8).

Ricken railway tunnel. The outflow at the southern portal amounts to 1200 L/min at 12 °C. The first stage of heating installation has

Fig. 8. Apartment buildings in Oberwald (canton Valais, Switzerland) connected to the 'cold district heating' system of the Furka railway tunnel (for details see text). Photograph by J. Wilhelm.

been operational since 1998: with a heat pump the tunnel water is used in the village of Kaltbrunn (canton St Gallen) to heat a multipurpose building, a gymnastic hall, a civil shelter centre and a kindergarten. To date the installed capacity is 156 kWt.

Mappo-Morettina highway tunnel. The outflow at the northern portal amounts to 983 L/min at 16 °C. The sports and recreational centre of the community Minusio (canton Ticino) is located right at the portal, and has been heated since 1999 with the tunnel waters. During the heating season 2000/2001, the seasonal performance coefficient (SPC) of the system, including the heat pump, amounted to 4.0. This coefficient is defined as the ratio of energy produced (heat)/ energy consumed (electricity) over the entire heating season.

Hauenstein railway tunnel. The outflow at the south portal amounts to 2500 L/min at 19 °C. In the first stage, three housing complexes with a total of 150 apartments were heated and supplied with domestic hot water in the village Trimbach (canton Solothurn) from 1999. The

bivalent (dual) system consists of a heat pump for the tunnel water plus two oil burners (total installed capacity 1.0 MWt). The tunnel system including the heat pump has an SPC of 4.0. Further development stages are in planning.

The operating experience with all these installations is satisfactory to good; only some minor technical improvements, for example, exchange of sand filters, became necessary. The flow rates and temperatures of the outflowing tunnel waters show little to no seasonal variation. Therefore, this 'zero-cost' heat source can be considered as totally reliable and sustainable.

Tunnels under construction

Switzerland presently follows the traffic policy of shifting transport by heavy trucks on roads to railway transport. Towards this goal, the capacity of the main traffic axis (N–S) is now being greatly expanded by the AlpTransit project. The key elements of this project are the two railway base tunnels: the Lötschberg tunnel in the west and the Gotthard tunnel in the east (see Fig. 7). The rock overburden can reach 2.5 km, and the lengths are considerable: 57 km for the Gotthard and 35 km for the Lötschberg. Both tunnels are under construction. As a result of the thick overburden and the great lengths, the expected geothermal potential is significant.

The geothermal potential largely depends on the inflow rates and temperatures to be expected in the different tunnel sections. Whereas the rock/fluid temperatures can be predicted with relatively high accuracy (Rybach & Busslinger 1999; Busslinger & Rybach 1999; Keller 2002), the forecasts of the water inflows have a considerably range of uncertainty (ATH 1993). Therefore the estimated potentials at the portals can vary over a correspondingly wide range: Gotthard north portal: 4–23 MWt, Gotthard south portal: 4–19 MWt; Lötschberg north portal: 4–12 MWt; Lötschberg south portal: 4–7.5 MWt (BFE 2002).

The local rivers near the portals have only limited potential for tunnel water uptake. At most places cooling ponds or towers would be necessary to decrease water temperatures prior to disposal. Therefore the construction authorities show significant interest for a geothermal use of the tunnel waters. This interest is shared by the local communities, like Frutigen at the Lötschberg north portal or Biasca at the Gotthard south portal. Besides space heating, innovative uses like heating of a tropical greenhouse in Frutigen (capacity demand: 2.7 MWt) are now being implemented.

Further investigations are needed to better specify the geothermal potential of the deep Alp-Transit tunnels. There are several factors that can influence the outflow rate and temperature of tunnel waters, and these need to be addressed and quantified: (1) possible construction measures such as cement injections to seal fissures and cavities in order to prevent or reduce the water inflow at critical points along the tunnel; (2) the development of a transient, nonlinear temperature profile behind the tunnel face due to ventilation; this will lead to lower rock, and thus water, inflow temperatures with time; and (3) heat loss from conduits in which the tunnel waters are transported to the portals.

Besides the technical prerequisites for a geothermal use of tunnel waters, there are also other tasks to be completed for a successful utilization. For example, the economic viability needs to be demonstrated for attracting and motivating potential consumers. Moreover, several administrative steps have to be followed to obtain all necessary permits. Nevertheless, the various operating examples in Switzerland demonstrate that all these tasks can be done, even with a relatively low speed of development. At a given location, the addition of a capacity of 0.5–1 MWt needs about one year to implement.

Furthermore, there is still ample space for innovative solutions, especially to utilize the warm tunnel waters in summer time. The planned tropical greenhouse in Frutigen is such an example. Further options are: spa-type establishments for wellbeing, fun and pleasure, aquacultures and the like. All of these applications are already being considered.

Outlook

Significant amounts of waste heat arise from geothermal energy utilization; the amounts vary with the utilization type. The possibilities of waste heat disposal are often limited by technical or legislative barriers. For technical reasons, re-injection of used geothermal fluids into the subsurface often remains incomplete. Regulations, on the other hand, can restrict discharges of waste fluid and heat to (or dissipation) in the environment. A beneficial way of waste heat treatment is the use of the heat for purposes that can even result in economic profits. Several such options have been described above but many more possibilities are technically feasible and economically viable.

Disposal of geothermal effluent along with its heat content is not the only problem in geothermal energy utilization. The fluids can contain unpleasant constituents like non-condensable gases, among them H_2S, which requires special measures. H_2S abatement is nowadays a standard practice (see e.g., Di Pippo 1991), but it has economic consequences. Further details about chemical constituents of geothermal waste waters can be found elsewhere in this book (Panichi, 2004; Arnórsson, 2004).

There are also, however, beneficial constituents in geothermal fluids, for example, metals or even precious metals like gold (Brown & Simmons 2003). An example of metal recovery from geothermal brine is the $200 million zinc project of CalEnergy Inc. in the Imperial Valley of California, brought online in 2002. Here, the waste water from eight geothermal power plants containing 600 ppm Zn will be utilized, and 30 000 t of Zn will be extracted annually (Clutter 2000).

Certainly not all problems of geothermal waste heat have already been solved. Novel solutions and concepts will undoubtedly emerge; their success will depend, besides having positive environmental effects, on the additional economic benefits.

The helpful and constructive comments of two reviewers, Professor R. DiPippo (University of Massachusetts, Dartmouth/MS, USA) and Professor J. Lund (GeoHeatCenter, Oregon Institute of Technology, Klamath Falls/OR, USA) are gratefully acknowledged.

References

ARBEITSGEMEINSCHAFT ZEWI 1995. *Gewinnung Geothermischer Energie aus Tunnels. Bericht Phase I*. Report to the Swiss Federal Office of Energy, Berne.

ARBEITSGEMEINSCHAFT ZEWI 1996. *Gewinnung Geothermischer Energie aus Tunnels. Bericht Phase II*. Report to the Swiss Federal Office of Energy, Berne.

ARBEITSGEMEINSCHAFT ZEWI 1998. *Gewinnung Geothermischer Energie aus Tunnels. Bericht Phase III. Machbarkeitsstudien für Vereina, Grenchenberg und Sondierstollen Frutigen*. Report to the Swiss Federal Office of Energy, Berne.

ARNÓRSSON, S. 2004. Environmental impact of geothermal energy utilization. *In*: GIERÉ, R. & STILLE, P. (eds) *Energy, Waste and the Environment: a Geochemical Perspective*. The Geological Society, London, 297–336.

ATH 1993. *Bergwasserzuflüsse und Beeinflussung des Bergwasserspiegels*. Arbeitsteam Hydrogeologie, Bericht Nr. 425 bh, AlpTransit Gotthard-Basistunnel.

BANKS, D., SKARPHAGEN, H., WILTSHIRE, R. & JESSOP, C. 2004. Heat pumps as a tool for energy recovery from mining wastes. *In*: GIERÉ, R. & STILLE, P. (eds) *Energy, Waste and the Environment: a Geochemical Perspective*. The Geological Society, London, 499–513.

BFE 2002. *Wärmenutzung Tunnelwasser Basistunnel Lötschberg und Gotthard, Grundlagen Wärme-Angebot*. Report GRUNEKO to the Swiss Federal Office of Energy, Berne.

BROWN, K. & SIMMONS, S. 2003. Precious metals in high temperature geothermal systems. *Geothermics*, **32**, 619–625.

BUSSLINGER, A. & RYBACH, L. 1999. Felstemperaturprognose für tiefliegende Tunnel. *TUNNEL*, **1/99**, 24–32.

BUSSMANN, W. 1994. Ehrenfriedersdorf: Grubenwärmenutzung in Betrieb genommen. Mittgbl. *Geothermische Vereinigung e.V.*, **3**, 4–6.

CHAPMAN, D. & RYBACH, L. 1985. Heat flow anomalies and their interpretation. *Journal of Geodynamics*, **4**, 3–37.

CLUTTER, T. 2000. Mining economic benefits from geothermal brine. *Geo-Heat Center Quarterly Bulletin*, **21**, 1–3.

DIPIPPO, R. 1991. Geothermal energy – Electricity generation and environmental impact. *Energy Policy*, **October 1991**, 798–807.

DICKSON, M. H. & FANELLI, M. (eds) 1990. *Small Geothermal Resources – A Guide to Development and Utilization*. UNITAR/UNDP Centre on Small Energy Resources, Rome/Italy, 274 p.

DICKSON, M. H. & FANELLI, M. (eds) 1995. *Geothermal Energy*. John Wiley & Sons, Chichester/UK, 214 p.

HAERING, M. & HOPKIRK, R. 2001. The Swiss deep heat mining project. *Geothermal Resources Council Bulletin*, **23**, 31–33.

HUDSON, R. B. 1998. Electricity production from geothermal energy. *In*: FORJAZ, V. H., BICUDO PONTE, C. A. & POPOVSKI, K. (eds) *Proceedings of the International Summer School 'Electricity Production from Geothermal Energy'*, Ponta Delgada/PT, 4.3–4.31.

KELLER, F. 2002. Gotthardbasistunnel: Geologie zwischen Prognose und Befund. sia tec21 14–15, 7–14.

KOHL, T., SPECK, R. & STEINFELD, A. 2002. Using geothermal hybrid plants for electricity production from enhanced geothermal systems. *Geothermal Resources Council Transactions*, **26**, 315–318.

LEGMANN, H. 2003. The Bad Blumenau geothermal project: a low temperature, sustainable and environmentally benign power plant. *Geothermics*, **32**, 497–503.

LUND, J. W. 1987. Cascading of geothermal energy in Italy. *GeoHeatCenter Quarterly Bulletin*, **Summer 1987**, 13–16.

LUND, J. W. & BOYD, T. 1999. Small geothermal power project examples. *GeoHeatCenter Quarterly Bulletin*, **20**, 9–26.

MINOHARA, Y. & SEKIOKA, M. 1980. Non-electric utilization of geothermal water and stam from production wells of geothermal power stations in Japan. *GeoHeatCenter Quarterly Bulletin*, **Summer 1980**, 10–14.

NICHOLS, K. E. 1986. Wellhead power plants and operating experience at Wendel Hot Springs. *Geothermal Resources Council Transactions*, **10**, 341–346.

PANICHI, C. 2004. Geochemical impact of re-injecting geothermal waste waters: Example Larderello, Italy. *In*: GIERÉ, R. & STILLE, P. (eds) *Energy, Waste and the Environment: a Geochemical Perspective*. The Geological Society, London, 337–354.

PERNECKER, G. & UHLIG, S. 2002. Altheim Geothermal Project: District heating system in Upper Austria pays for injection well with power generated by a low-enthalpy ORC turbogenerator. *Geothermal Resources Council Bulletin*, **31**, 33–36.

ROBERTSON, R. C. 1980. Waste heat rejection from geothermal power plants. *In*: KESTIN, J. (ed) *Sourcebook on the Production of Electricity from Geothermal Energy*. US Department of Energy, Washington DC, 997 p.

RYBACH, L. 1995. Thermal waters in deep Alpine tunnels. *Geothermics*, **24**, 631–637.

RYBACH, L. & BUSSLINGER, A. 1999. Prognose der Felstemperatur Gotthard Basistunnel. *In*: LÖW, S. & WYSS, R. (eds) *Vorerkundung und Prognose der Basistunnels am Gotthard und am Lötschberg*. A.A. Balkema, Rotterdam, 257–269.

RYBACH, L. & WILHELM, J. 1995. Potential and use of warm waters from deep Alpine tunnels. Proceedings of the World Geothermal Congress, Florence, **3**, 2199–2201.

SANNER, B. 1993. Verwendung aufgelassener Kohlegruben zur Nutzung geothermischer Energie in Springhill, Kanada. *Mitteilungsblatt Geothermische Vereinigung e.V.*, **2**, 3–5.

SGA 2002. *Tunnelgeothermie*. Swiss Geothermal Association, Technische Notiz. Biel/Switzerland. 2 p.

WILHELM, J. & RYBACH, L. 2003. The geothermal potential of Swiss alpine tunnels. *Geothermics*, **32**, 557–568.

Corrosion and stability of vitrified material derived from municipal solid waste

D. PERRET[1], P. STILLE[2], J.-L. CROVISIER[2], G. SHIELDS[2,3], U. MÄDER[4],
K. SCHENK[5] & M. CHARDONNENS[5]

[1]*Ecole Polytechnique Fédérale de Lausanne, Laboratoire de Pédologie,
ENAC-IST-LPE, Lausanne, Switzerland (e-mail: didier.perret@dplanet.ch)*
[2]*ULP Ecole et Observatoire des Sciences de la Terre CNRS, Centre de Géochimie
de la Surface UMR 7517, Strasbourg, France*
[3]*James Cook University, School of Earth Sciences, Townsville;
Queensland, Australia*
[4]*Universität Bern, Mineralogisch-Petrographisches Institut; Bern, Switzerland*
[5]*Office Fédéral de l'Environnement, des Forêts et du Paysage, Division Déchets,
Bern, Switzerland*

Abstract: Advanced high-temperature technologies for the treatment of municipal solid wastes or residues from conventional incineration can produce vitrified materials with superior physical and chemical characteristics compared to conventional residues (bottom ash, fly ash, filter cake). These materials, which may contain significant metal concentrations, exhibit favourable thermodynamic characteristics and very low contaminant release when subjected to corrosion. Owing to low leachability, these materials could possibly be used as secondary raw materials for civil engineering applications, although the energy demand for their production is high. This chapter presents the physical and chemical characteristics of a range of vitrified materials originating from various high-temperature technologies; their behaviour under highly aggressive conditions of corrosion and their thermodynamic stability are also presented. It is concluded that such high-temperature materials are indeed less likely to be damaging to the environment than conventional residues.

Vitrification: an alternative route to secondary raw materials

During the last two decades, environmentally sound waste management and treatment policies have been implemented in many countries. The aim of these policies has been the reduction of waste volumes, the separation of recyclable materials and the production of energy from the incineration of the remaining municipal solid wastes (MSW). As an example, Switzerland totally changed its waste-landfilling practices of the 1970s and 1980s to global incineration. As of the year 2000, all non-recyclable MSW must be incinerated (BUWAL 1996). Today, ca. 3×10^6 t MSW/y are incinerated in Switzerland and converted into so-called bottom ash (BA: solid residues recovered in the oven after incineration; ca. 750×10^3 t/y), fly ash (FA: ashes recovered in the electrostatic filters mounted in the exhaust chimney of the incineration plant; ca. 46.5×10^3 t/y), and filter cake (FC: solid material recovered after wet washing of incineration smokes; ca. 11.6×10^3 t/y).

Unfortunately, the secondary residues (BA, FA, FC) of conventional MSW incinerators usually contain high metal concentrations; as a consequence, they cannot be re-used as secondary raw materials. Air pollution control (APC) residues, which represent a mixture of FA and FC, must be treated prior to landfilling in order to reduce the leachability of heavy metals (ABB-EAWAG-EMPA-KEZO 1990; EKESA 1992; BUWAL 1998). Ideally, heavy metal concentrations could be made sufficiently low, so that MSW residues could be used without problems. In many European countries, leachability is the criterion by which a material might be selected for re-use.

A promising alternative for reducing heavy metal concentrations and leachability are the several emerging advanced thermal treatment technologies. Operated at high combustion temperatures (i.e., between 900 and 1400 °C), these

From: GIERÉ, R. & STILLE, P. (eds) 2004. *Energy, Waste, and the Environment: a Geochemical Perspective*. Geological Society, London, Special Publications, **236**, 381–410.
0305-8719/04/$15   © The Geological Society of London 2004.

technologies are claimed by their manufacturers to produce inert high-temperature materials (HT materials) with environmentally favourable characteristics (Barin 1991, 1992; Künstler *et al.* 1994; Schumacher & Gugat 1994; Thomé-Kozmiensky 1994; Gutmann 1996; Kanczarek & Grosse-Holz 1996; Patze 1996; Stahlberg 1996). It is also claimed that metal concentrations are reduced and that the leachability of the glassy products is greatly reduced with respect to conventional products. Thus, it may be relevant from the point of view of the path 'energy → waste → environment' to consider HT technologies as plausible options for better management of our resources, as depicted on the schematic entropy diagram of Fig. 1.

The additional energy input needed for waste vitrification technologies, compared to conventional incinerators that even recover residual energy, may be considered detrimental to the net energy balance of the complete path 'resource → waste'. However, this additional energy requirement can also be considered as the additional price to pay for the efficient exploitation of natural resources. To be sustainable, HT materials should fulfil the following requirements:

(1)   The ratio [additional energy needed to produce HT materials] : [long-term environmental profit] should be as low as possible.
(2)   The leachability and long-term durability of HT materials should be such that the potential hazard to the environment is significantly reduced in comparison to conventional MSWI products.
(3)   The mechanical characteristics of HT materials should be comparable to those

of the equivalent primary raw materials extracted from natural resources.

The present chapter deals exclusively with the scientific issues of point (2). It is based on the results of an extensive study under the authority of the Swiss Agency for Environment, Forests and Landscape (SAEFL/BUWAL; Perret *et al.* 2000, 2002). The in-depth survey was performed on 23 HT materials originating from 16 different advanced thermal treatment facilities developed by 10 different companies in Switzerland, Germany, France, and Italy, and operated under realistic conditions (e.g., from wastes representative of average MSW). On the basis of the results, two options are tested to assess the long-term behaviour of HT materials:

(1)   *Landfill disposal of HT materials.* This scenario supposes that HT materials are stable enough to be considered as inert materials that can be disposed into monofills, hereafter referred to as 'glassfills', requiring no special attention (e.g., technical barriers or precautionary monitoring of leachates).
(2)   *Re-use of HT materials for civil engineering applications.* This scenario, although less conservative than the previous one, complies with the principles of a sustainable management of natural resources (Finet 1994; O'Connor *et al.* 1994; Zevenbergen *et al.* 1994a, 1994b; Yan & Neretnieks 1995; Hnat & Bartone 1996; Resce *et al.* 1996; Bottero *et al.* 1997, Cases & Thomas 1997; Kraus & Meunier 1997; Depmeier *et al.* 1997; Guyonnet *et al.* 1998; Méhu 1998). In this scenario,

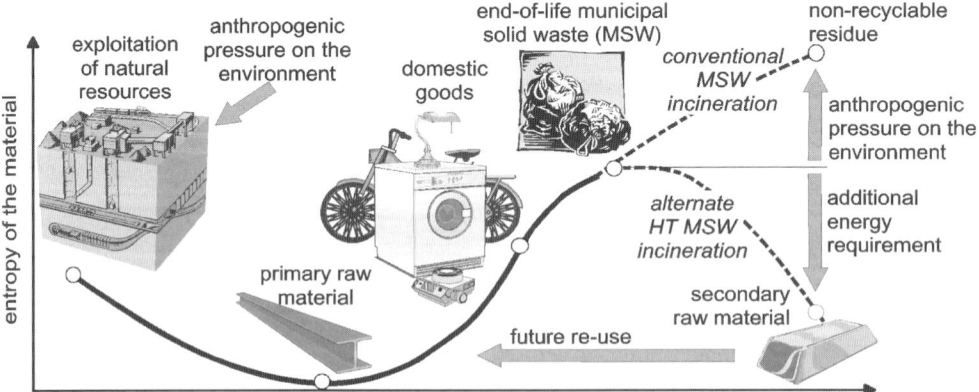

life path of domestic goods, from the natural resources to the wastes and residues

Fig. 1. Schematic representation of the evolution of the entropy of materials, from the natural resources to the production of wastes and residues, via the fabrication of domestic goods.

HT materials are re-used as foundation layers for road construction, hereafter referred to as 'glassroad'.

The estimation of the release fluxes of metals from a hypothetical 'glassfill' and a hypothetical 'glassroad' will be based on experimental results and theoretical but realistic considerations.

Characterization of HT materials

This section is primarily intended to show that the relevant combination of the intrinsic physico-chemical and microscopic characteristics of HT materials, their behaviour under aggressive conditions of corrosion, and their modelled thermodynamic stability, yields a sound composite picture of these materials. On the basis of the knowledge acquired over decades on high-level nuclear waste glasses (Vernaz & Dussosoy 1992; Bates et al. 1994; Thomassin 1995, 1996; Ewing 1996), these are the key parameters that may drastically influence the long-term durability of HT materials.

For an accurate description of the experimental set-up, the analytical procedure, the HT treatment facilities, and the detailed characteristics of each of the individual 23 HT materials studied, the reader is referred to Perret et al. (2000, 2002, 2003).

The 23 HT materials studied were produced in Swiss, German, French, and Italian HT treatment facilities, which were fed by different input materials under realistic conditions of operation (i.e., wastes and residues representative of average MSW, BA, FA, and FC). These technologies can be roughly grouped into two categories (see Fig. 2), in-line processes and post-processes:

(1) In-line processes are technologies that directly inertize municipal solid wastes. Five such HT materials were studied.
(2) Post-processes, which inertize the residues of conventional MSW incinerators, can be further grouped into two subcategories:
   - Post-processes for bottom ash, which are fed by BA (80–100 wt%, either in rough or fine fractions) and up to 20 wt% FA. Six such HT materials were studied.
   - Post-processes for fly ash, which are fed by FA (50–100 wt%) and up to 50 wt% of other waste material (filter cake, sewage sludge, cement, recycled glass, or even car shred). Twelve such HT materials were studied.

The factor governing the final size and morphology (from millimetre-sized to large irregular blocks) of HT materials is the melt quenching technique used in the different HT treatment technologies.

One non-radioactive surrogate (code: SON68) of the French high-level nuclear waste (HLW) borosilicate glass R7T7 was added to the set of samples as an analytical standard; this HT material is one of the most studied HLW glasses (Nogues 1984; Fillet 1987; Advocat 1991; Cheron et al. 1995). Two other well-defined French HT materials (codes: R2bis*, R3) produced from FA were also used for comparative purposes, as their composition and behaviour have been thoroughly studied (Colombel 1996; Colombel et al. 1997).

Physico-chemical characteristics of HT materials

Intuitively, the characteristics and behaviour of HT materials would be expected to strongly depend on the operating characteristics of the thermal treatment technologies (e.g., type of combustion chamber, melting temperature), on the feeding material (e.g., MSW, BA, FA, with other additives), and on the melt quenching rate. In some cases rough trends, such as the proportion of more volatile to less volatile elements in the HT material, were indeed observed. However, the vast majority of results did not exhibit clear differences with respect to the types of HT processes or input wastes and residues, even for samples with highly contrasting macroscopic and microscopic features (see Fig. 3).

The majority of samples (15 out of 23) were shown to contain no or less than ca. 2 vol% of crystalline inclusions and have a low specific surface area ($S_{spec} = 300–600$ cm^2/g for material ground to 100–125 µm); these samples are considered to be vitreous. The other samples contain higher amounts of crystalline components, either homogeneously or heterogeneously dispersed in a vitreous matrix; they are considered to be vitrocrystalline (labelled with an asterisk) and are expected to exhibit a lower resistance to corrosion because of their higher specific surface area. The latter typically varies between 400 and 1000 cm^2/g, but may be as high as 8200 cm^2/g. None of the HT materials studied was entirely crystalline.

X-ray diffractograms (Fig. 4) and petrographic analyses (scanning electron microscopy; also used for estimating the surface roughness and micro-heterogeneity of samples) indicate the presence of diverse silicates and oxides on the surface and in the bulk of the vitrocrystalline samples. In addition to the nearly ubiquitous quartz, other minerals were found in several samples: gehlenite, albite, diopside, portlandite, pyroxenes

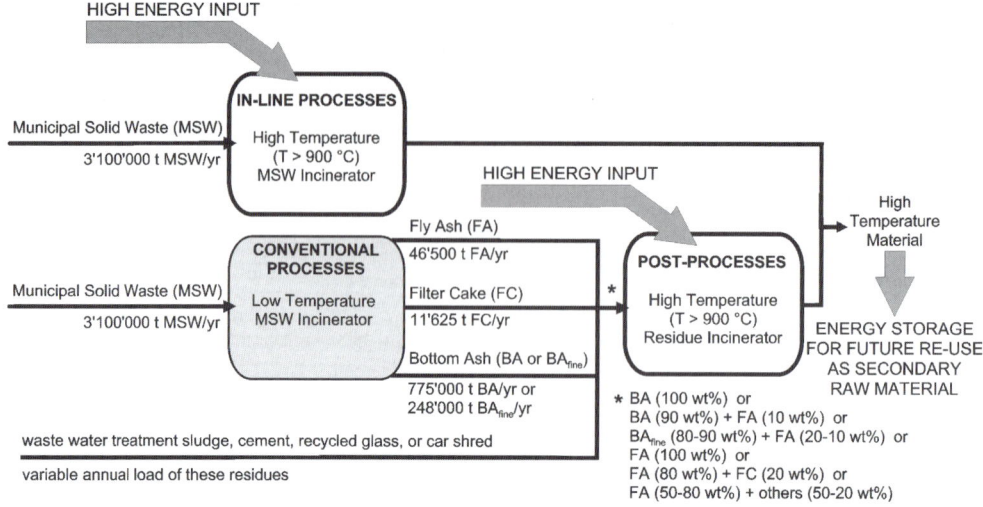

Fig. 2. Classification of HT samples on the basis of the input material from which they are produced. In-line processes directly transform MSW into HT materials. Post-processes transform residues of MSW incineration (BA, FA, FC, or mixtures of them with possible additives) into HT materials. Values are given for Switzerland (year 2002).

and spinels, but also alloys, and metallic inclusions. The presence of crystalline phases in the vitreous matrix of HT materials is explained both in terms of quenching rate and embedding of relics inherited from the input material. Many crystalline phases identified were enriched in trace metals, in particular chromium, copper, and zinc dominating most HT samples.

Although the role of crystalline phases in the leachability of HT materials is unclear and must be examined from case to case, the identified silicates and oxides are overall more resistant to corrosion than silicate glass and residues of incineration (Scholze 1991). Thus, a clear assessment of the durability of HT materials as a function of crystalline components must take into account the combined effects of their enrichment or depletion in trace metals, their individual leachability, the increase (but sometimes decrease) in overall reactivity due to local heterogeneities and increased S_{spec} (Jacquet-Francillon *et al.* 1982; Bickford & Jantzen 1984; Jantzen & Plodinec 1984; Scholze 1991; Adams 1992; Sproull *et al.* 1994; Sterpenich 1998).

Probably the most important factor affecting the long-term stability of waste glasses is their chemical composition (Scholze 1991; Gutmann 1996). Glass is an amorphous and quasi-homogeneous solid, which can accept a wide range of waste compositions and concentrations in its matrix (Zachariasen 1932; O'Keefe 1984; Dietzel 1988). Owing to its nature, glass is not in equilibrium with its environment and tends to transform into more stable phases. Glass is a

three-dimensional poorly ordered arrangement of SiO_4^{4-} tetrahedra linked together by siloxane oxygen atoms (Si–O–Si) into a network of polymeric chains (Zachariasen 1932; Stanworth 1950; Vogel 1992). Terminal silanol groups (Si–OH) host protons, alkali elements (Na$^+$, K$^+$, Li$^+$) and alkaline earths (Ca^{2+}, Mg^{2+}), while elements with higher charge (Al^{3+}, Fe^{3+}, B^{3+}, Zr^{4+}, Mn^{4+}) replace Si^{4+} in the glass matrix, acting as network-forming or network-modifying elements (Darab *et al.* 1996). As a rough guide, terminal elements decrease the stability of the glass, as opposed to matrix elements, which tend to increase its stability. The amount of the network-forming element Si will directly influence the homogeneity of the glassy matrix of HT materials, and thus their durability. On the other hand, a high concentration of the network-modifying element Ca will impair the stability of the matrix. Alternatively, elements such as Al may act either as network-forming elements (in tetrahedral sites) or as network-modifying elements (in octahedral sites). When expressed as ternary SiO_2–Al_2O_3–CaO materials, HT samples belong to the domain of compositions that form silicate structures (Perret *et al.* 2000, 2002). The vitrocrystalline samples with a high S_{spec}, a rough surface, or many mineral inclusions, are globally poorer in SiO_2 than vitreous samples.

In general (see Fig. 5), HT materials contain high amounts of Si ($[SiO_2] = 41$ wt%, ranging from an average 36.8 wt% for FA-derived HT materials to an average 48.2 wt% for BA-derived

Fig. 3. Examples of typical HT materials in their original state. Left column: vitreous samples. Right column: vitrocrystalline samples. Top rows: samples on a 10 cm^2 scaling grid. Bottom row: scanning electron micrographs (SEM in secondary electrons mode) of samples after grinding to 100–125 μm.

HT materials) and Al ($[Al_2O_3] = 16.5$ wt%, ranging from 14.5 wt% for MSW-derived samples to 18.1 wt% for FA-derived samples), whereas their Ca content ($[CaO] = 23.1$ wt%) ranges from 17.5 wt% for BA-derived samples to 27.5 wt% for FA-derived samples. It must be emphasized that HT materials prepared from fly ash (and eventual additives) are on average

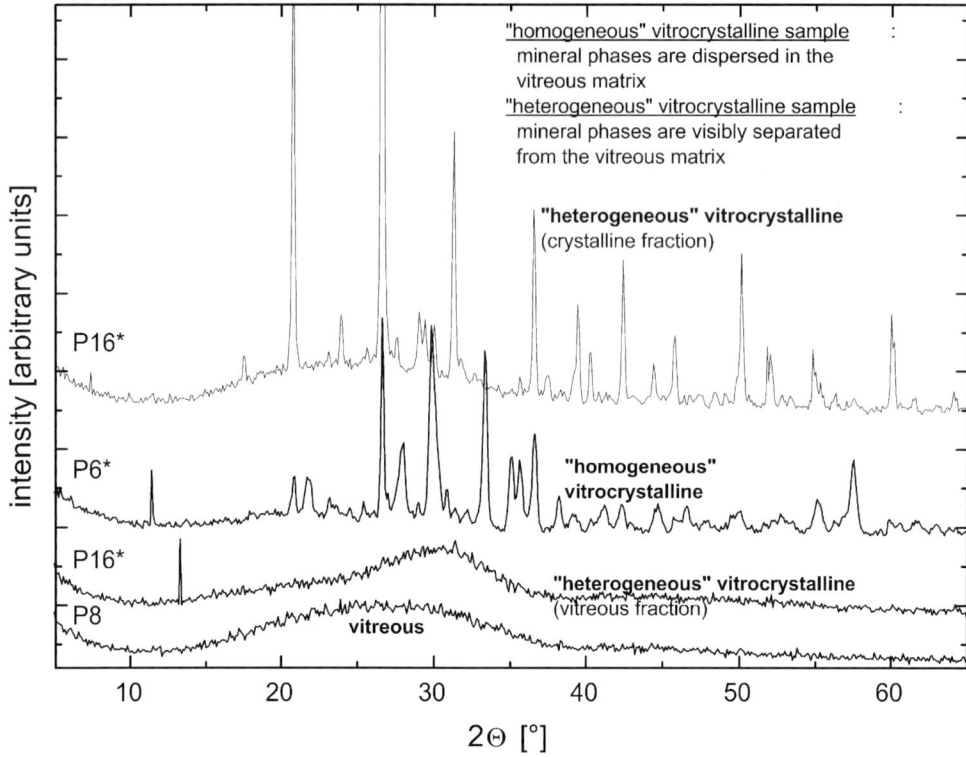

Fig. 4. Typical X-ray diffractograms of 100–125 μm ground HT materials, showing purely vitreous character (P8) or vitrocrystalline character (P6*, P16*). For most vitrocrystalline samples, mineral phases are distributed throughout the glass matrix (e.g., P6*, 'homogeneous' vitrocrystalline), but some samples (e.g., P16*, 'heterogeneous' vitrocrystalline) exhibit patches of concentrated crystalline phases visibly separated from the bulk vitreous matrix.

poorest in Si and richest in Ca (Si and Ca being inversely correlated); they would be expected to exhibit a lower durability upon corrosion. Nevertheless, the amount of Si and Al in all the samples is high ($[SiO_2 + Al_2O_3] = 45–75$ wt%), suggesting that HT materials should behave favourably when subjected to leaching, as opposed to the conventional residues (BA, FA, FC) of low-temperature incineration. In addition, the ratio $[SiO_2]:[CaO]$, which gives a rough indication of the expected stability of HT materials, is systematically greater than 1 on a weight basis, although lower for FA-derived materials (in-line processes: 2.94; post-processes for BA: 2.92; post-processes for FA: 1.4).

The minor constituents Mg, Na, K, and Fe are present in HT materials in highly variable proportions ($[MgO + Na_2O + K_2O + Fe_2O_3] = 6–25$ wt%). Among these constituents, the alkali elements are known to favour phase-separated glasses impairing resistance to leaching (Stanworth 1950; Scholze 1991), but their low proportion in the studied samples ($[Na_2O + K_2O] < 7$ wt%)

should not drastically influence the long-term stability of HT materials.

When dealing with potential secondary raw materials, one has to keep in mind that their trace metal contents should be as close as possible to that of primary raw materials, although a high metal concentration does not necessarily translate into important releases during leaching. In our case, HT treatment technologies are shown to efficiently scavenge and embed high concentrations of metals in the matrix of their corresponding HT materials (see Fig. 6). As expected, FA-derived HT materials exhibit higher concentrations of the most volatile elements Cd, Pb, Sb than MSW- or BA-derived HT materials, in agreement with the known accumulation of these metals in fly ash and depletion from bottom ash during conventional incineration (Thomé-Kozmiensky 1994).

When comparing HT materials to the Swiss Waste Management Ordinance (TVA, Technische Verordnung über Abfälle; BUWAL 1996), it appears that only 8 samples out of 23 could

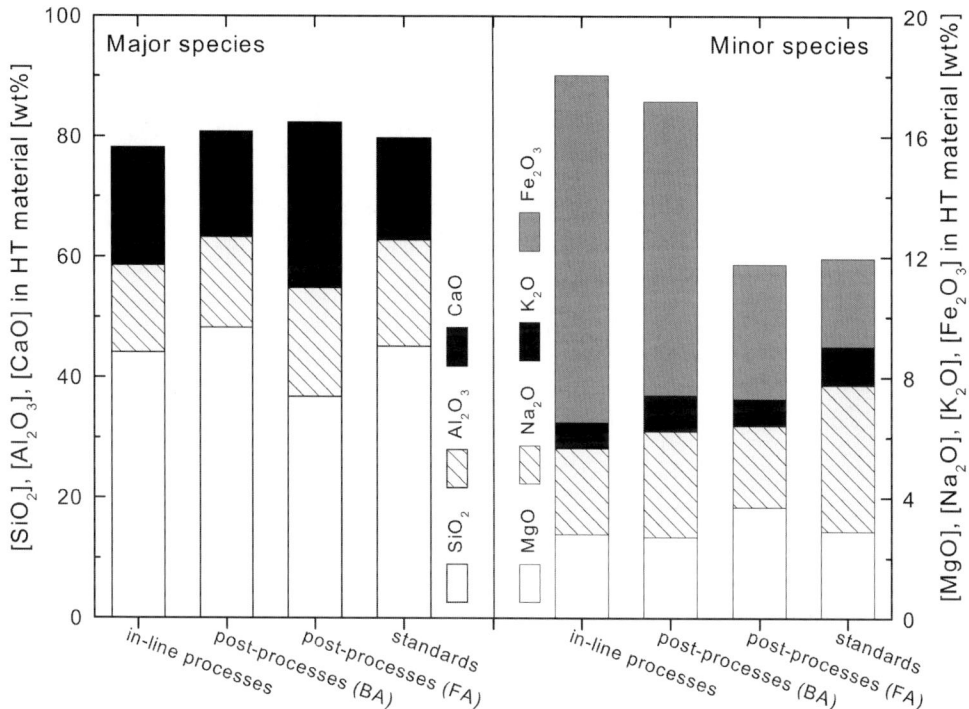

Fig. 5. Average concentrations of major and minor constituents in the different types of HT materials studied. Left: SiO$_2$ (network-forming), Al$_2$O$_3$ (network-forming or modifying), and CaO (network-modifying). Right: MgO, Na$_2$O, K$_2$O, and Fe$_2$O$_3$.

be disposed into landfills for inert materials, on the exclusive basis of their trace metal contents ([Cd]: max. 10 mg/kg; [Cu], [Ni], [Pb]: max. 500 mg/kg; [Zn]: max. 1000 mg/kg; the other metals are not regulated with respect to admissibility in landfills). The other HT materials contain between one and five metals that may exceed the levels of the Swiss limits by up to 27 times, and would thus require special post-stabilization prior to landfilling. However, Swiss legislation, which is based on concentrations and leachability of heavy metals, was designed for discriminating the conventional residues (BA, FA, FC) of low-temperature incinerators, which are known to have a poor durability with respect to leaching (ABB-EAWAG-EMPA-KEZO 1990; EKESA 1992; BUWAL 1998). Thus, it would be false to assess the critical sustainability of secondary raw materials on the exclusive basis of their metal content, without determining their behaviour upon corrosion.

Taking into account all observations on the physico-chemical characteristics of HT materials (macroscopic, microscopic and physical features, chemical composition, metal content), only three HT samples, all of them being vitrocrystalline,

would be considered as outside the range of favourable HT materials (i.e., vitreous and homogeneous, with low metal content, only few inclusions, small S_{spec}, and smooth surface) in terms of secondary raw materials. Out of the 20 other samples, 15 exhibit either two or three favourable characteristics, without obvious relationship between the physico-chemical characteristics of HT materials and their parent process or feeding material. These characteristics must be weighted together with the behaviour of HT material during corrosion, as discussed below, in order to gain a broader and more relevant picture on their long-term durability and potentials as secondary raw materials.

Dynamic behaviour of HT materials during corrosion

Glass undergoes a series of transformations over time, collectively called corrosion or alteration. Over the past decades, a consensus has emerged on the steps describing glass corrosion (Scholze et al. 1982; Adams 1984; Nogues 1984; Nogues et al. 1985; Grauer 1985; Lutze & Ewing 1988;

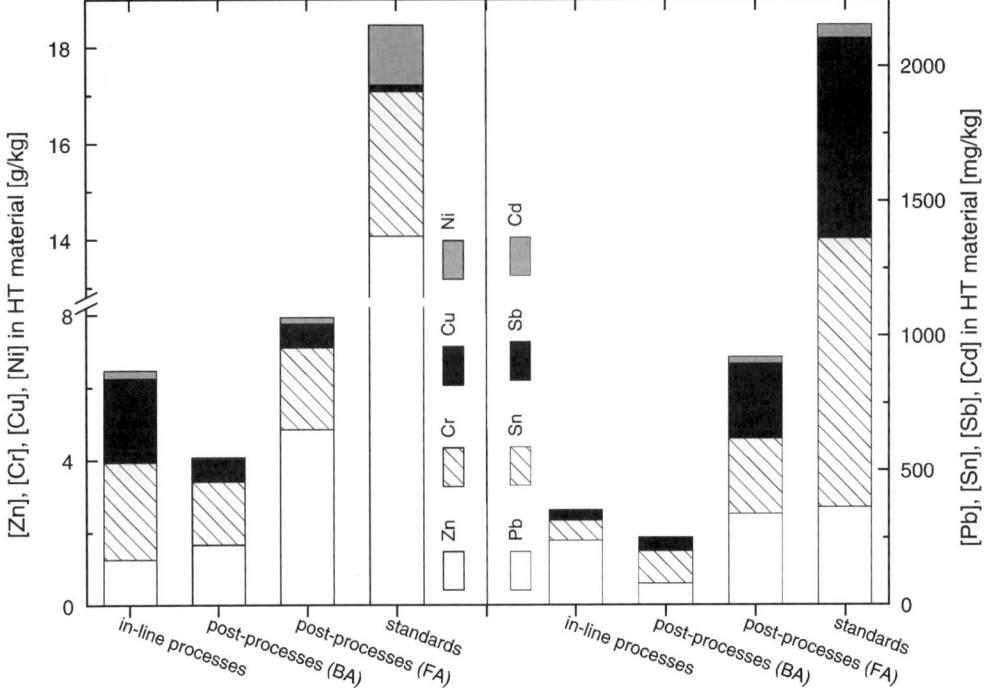

Fig. 6. Average concentrations of several potentially toxic and environmentally relevant metals in the different types of HT materials studied.

Atassi 1989; Dran *et al.* 1989; Petit *et al.* 1990; Advocat 1991; Grambow 1991; Clark & Zoitos 1992; Cunnane *et al.* 1993; Bates *et al.* 1994, 1996; Clark *et al.* 1994; Cunnane & Allison 1994; Ebert & Mazer 1994; Feng *et al.* 1994; Ewing 1996). On the basis of extensive work on the behaviour of HLW glasses, glass corrosion exhibits a sequence of the following three reactions (see Fig. 7):

(1) *Initial ion-exchange.* Unaltered glass undergoes diffusion of water molecules from its surface to its matrix. Water reacts with terminal alkali (Alk) elements in a process of ion exchange (Charles 1958; Garland & White 1985):

$$\equiv Si-O-Alk + H_2O$$
$$\rightarrow\ \equiv Si-O-H + OH^- + Alk^+$$

This attack consumes H_3O^+ from the surrounding environment and releases alkali elements from the glass (dealkalinization). The reaction is kinetically limited by diffusion of water molecules into the bulk of the glass through the gel layer (see below). The thickness of the diffusion (or reaction) layer

is 1–5 μm. This step is then followed by the following reaction (2):

(2) *Matrix hydrolysis.* Once glass starts to corrode, water molecules react with internal and external siloxane groups in a process of hydrolysis and dissolution (Zellmer & White 1985; Pederson *et al.* 1986; Bates *et al.* 1991; Scholze 1991; Clark & Zoitos 1992; Carroll *et al.* 1994):

$$\equiv Si-O-Si \equiv\ +\ OH^-$$
$$\rightarrow\ \equiv Si-O^- + HO-Si \equiv$$
(internal glass hydrolysis)
$$\equiv Si-O-Si(OH)_3 + OH^-$$
$$\rightarrow\ \equiv Si-O^- + H_4SiO_4$$
(external glass hydrolysis and dissolution)

This nucleophilic/basic attack consumes OH^- (acidification) and releases silicic acid and network-forming and network-modifying elements. During this alteration step, $\equiv Si-O^-$ groups that are formed are eventually reprotonated by surrounding water molecules. The gel layer, which thickens during hydrolysis and extensive

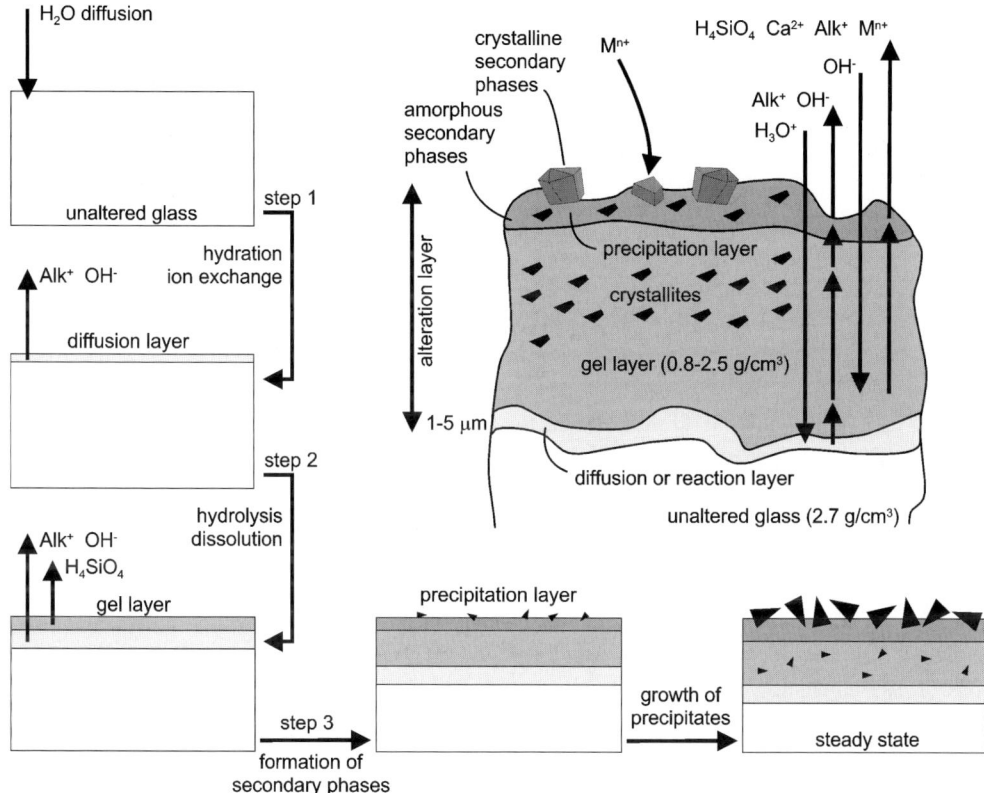

Fig. 7. Schematic of the mechanisms of corrosion taking place when a glass is exposed to water.

dissolution, is amorphous in nature (Luo *et al.* 1997).

Ion exchange (production of OH⁻ in the reaction layer) and matrix hydrolysis (consumption of OH⁻ from the reaction layer and the surrounding environment) are intimately coupled, devitrification being autocatalysed by ion exchange and both processes being enhanced by water diffusion (Grambow 1992). This step is then followed by reaction (3).

(3) *Surface precipitation of secondary phases.* Over time, ion exchange and matrix hydrolysis drive to slow supersaturation of the released species. Supersaturation occurs in the thick gel layer and in the bulk solution (Altenhein *et al.* 1981; Malow 1982; Petit *et al.* 1990). In addition, ionic species initially present in the leachant may be complexed and adsorbed at the solution–gel interface and eventually increase supersaturation (Grambow 1985; Strachan *et al.* 1985; Whitehead *et al.* 1993).

When supersaturation is reached, secondary phases start to precipitate, mostly at the surface of the altered glass, but also in solution and, to a lesser extent, in the gel layer (Grambow 1982; Flintoff & Harker 1985; Haaker *et al.* 1985; Lee & Clark 1985; Jercinovic *et al.* 1989; Petit *et al.* 1990; Bates *et al.* 1991; Shuttleworth & Monteith 1997). Indeed, the phases with the highest precipitation rate form first. Secondary phases are thermodynamically more stable than their individual constituents originally embedded in the glass matrix (Feng *et al.* 1994). These precipitates are amorphous or poorly crystallized (Murakami *et al.* 1989; Bates *et al.* 1991; Gong *et al.* 1996), although some crystalline secondary phases are frequently detected (smectites of various composition, zeolite, chrysotile, quartz, kaolinite, illite, spinel, hydrotalcite) at the surface of corroded glasses (Mottl & Holland 1978; Crovisier 1989; Evans 1989; Jercinovic *et al.* 1989; Murakami *et al.* 1989; Abrajano *et al.* 1990; Gislason *et al.* 1993; Buck *et al.* 1994; Abdelouas *et al.* 1994*a*, *b*, 1997; Gong *et al.* 1996). The diversity of precipitates (which mostly but not exclusively are oxides, oxyhydroxides, phosphates, carbonates, and silicates) reflects both the composition of the glass and the solution chemistry of the leachant (Hench *et al.* 1982; Yanagisawa & Sakai 1988; Petit *et al.* 1990; Whitehead *et al.* 1993; Feng *et al.* 1994). This ultimate step in glass corrosion is actually very difficult to model.

The time scales of these processes are in the order of days to weeks (step 1), weeks to months (step 2), and months to years (step 3). Although the two first steps are well described by refined kinetic and thermodynamic models, the long-term fate of glass constituents cannot be predicted with accuracy, mostly because of the discrepancies between models and experiments on the nature of the secondary phases formed.

A convenient way to estimate the long-term stability of glasses is to bring them under conditions of accelerated corrosion. Many glass corrosion tests have been designed to yield the detailed mechanisms of matrix corrosion and kinetics of glass dissolution. However, these tests are usually performed under conditions that are far from realistic corrosion scenarios (e.g., high temperature, high concentration of leachants, high reactive surfaces), forcing the solid–solution interface away from equilibrium and thus inducing additional uncertainties on the long-term behaviour of glasses. Of course, real-time corrosion experiments (e.g., burial of glasses in soils, landfill disposal, re-use in civil engineering applications) are of major interest, but they yield valuable results after decades and they cannot thus be considered appropriate in our case.

For our HT materials, and taking into account their morphological diversity (small grains to large blocks), a new corrosion test was designed, hereafter referred to as the Strasbourg test. The test is performed under static conditions (i.e., no flow of leachant) at high temperature (leachant = 50 mL H_2O; HT material = 50 mg, ground to 100–125 μm; no stirring; $T = 90$ °C; duration = 1, 3, 10 days), with pH evolving freely during corrosion in a closed vessel. At the end of the experiment, the leachate was analysed for pH and major, minor, and trace elements, and the HT material was recovered for microscopic examination. Under these drastic conditions of leaching ($S_{HT\ material}/V_{leachant} = 40$–$820$ m^{-1}), one-day corrosion is expected to

yield initial rates of corrosion, while steady-state conditions should eventually be approached after 10 days of corrosion.

Figure 8 illustrates the transformations induced on the surface structure of HT materials after 10 days of corrosion. The amorphous gel layer that appears during corrosion is observed on most HT samples and reveals pits, holes, and corrosion paths from case to case. For many samples, crystalline secondary phases in the 1–10 μm range cover the altered surface, and spallation/exfoliation of the gel layer is observed or at least suggested in several instances, indicating that corrosion is a discontinuous process, even under static conditions.

From a theoretical point of view, the gel layer is a barrier that reduces further hydrolysis of the silicate network, and is supposed to be more stable than the glass matrix, thus reducing the overall rate of corrosion. However, gel exfoliation may momentarily re-activate corrosion, at least locally. No clear trend was observed for the presence of the crystalline secondary phases identified at the surface of the corroded HT samples. The most abundant minerals are aluminosilicates, calcium phosphates, Fe- and Mg-rich minerals, and zeolites; their role in the scavenging or release of metals remains ambiguous, although many mineral phases identified bear traces of metals.

After one day, a direct correlation between leachant pH and release of Si in the solution can be seen (Fig. 9**a**). Later on (3- and 10-day corrosion), the increase in pH becomes much slower, suggesting that the leaching of alkaline elements is not a congruent process, and pH stabilizes rapidly during matrix dissolution. In fact, the proportion of CaO in the matrix of HT materials (11–38 wt%) directly controls the amount of Ca that will be readily leached out (see Fig. 9**b**). It can thus be expected that Ca-poor/Si-rich HT materials possess a higher durability than Ca-rich/Si-poor ones. The relationship between CaO in the matrix of HT materials, Ca^{2+} extracted into solution by the leachant, and pH is given by:

$$[CaO]_{HT\ material} + H_2O \longrightarrow Ca^{2+} + 2OH^-$$

$$[CaSiO_3]_{HT\ material} + 2H^+ + H_2O$$
$$\longrightarrow Ca^{2+} + H_4SiO_4$$

Irrespective of the physico-chemical characteristics of HT materials, the concentrations of most major, minor, and trace elements in the leachant increase, but appear to stabilize somewhat with time (Fig. 10). In addition for any given element, the release curves of different samples

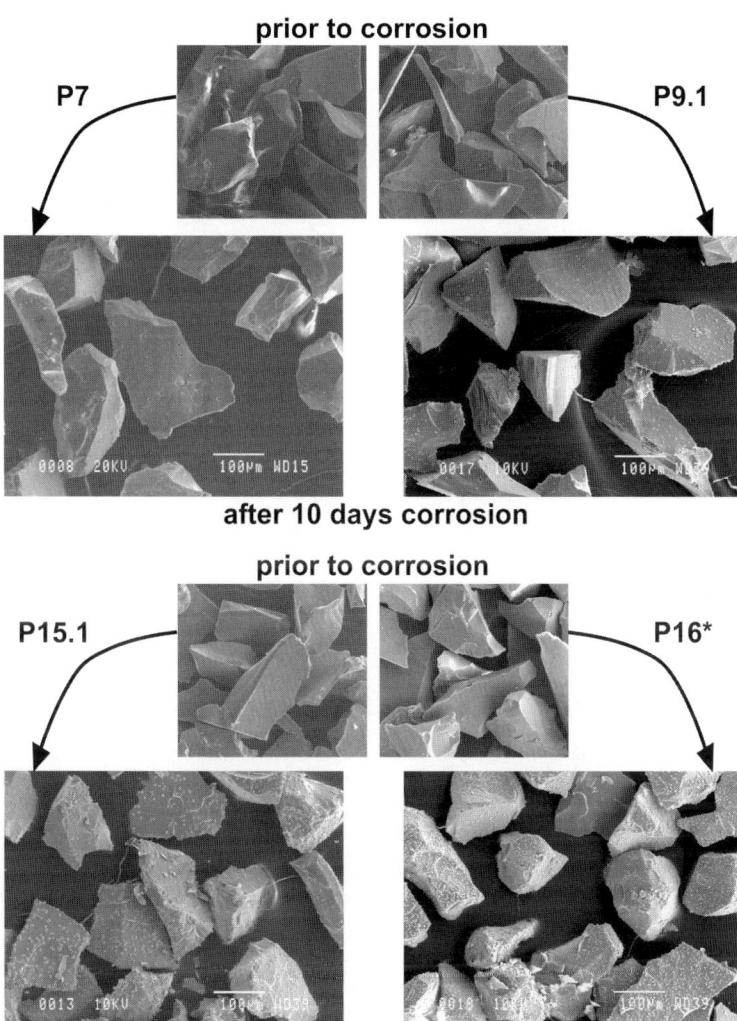

Fig. 8. Scanning electron micrographs (SEM in secondary electrons mode) of the surface features of HT materials ground to 100–125 μm prior to (small micrographs) and after 10-day corrosion (large micrographs). Corroded sample P7 exhibits pits but almost no secondary mineral phases; corroded sample P16* shows a dense cover of secondary minerals.

are not superimposed, indicating again that matrix dissolution is a complex non-congruent process, which cannot be explained in terms of simple physico-chemical characteristics of the samples. It must however be pointed out that the true release of elements, that is, their extraction from the glass matrix, cannot be determined, because many released elements will be precipitated, or trapped into the gel layer, or scavenged by the secondary minerals (Malow 1982; Crovisier *et al.* 1987, 1992; Petit *et al.* 1990;

Grambow 1994). Thus, the concentrations measured in the leachant are merely apparent releases.

Trace elements, as exemplified by Pb (Fig. 10), frequently exhibit an increase in concentration between 1 and 3 days of corrosion, followed by a decrease between 3 and 10 days, but absolute values vary strongly from sample to sample and from element to element. This observation supports the hypothesis that crystalline secondary phases forming at the surface of corroded HT materials act as trace metal scavengers.

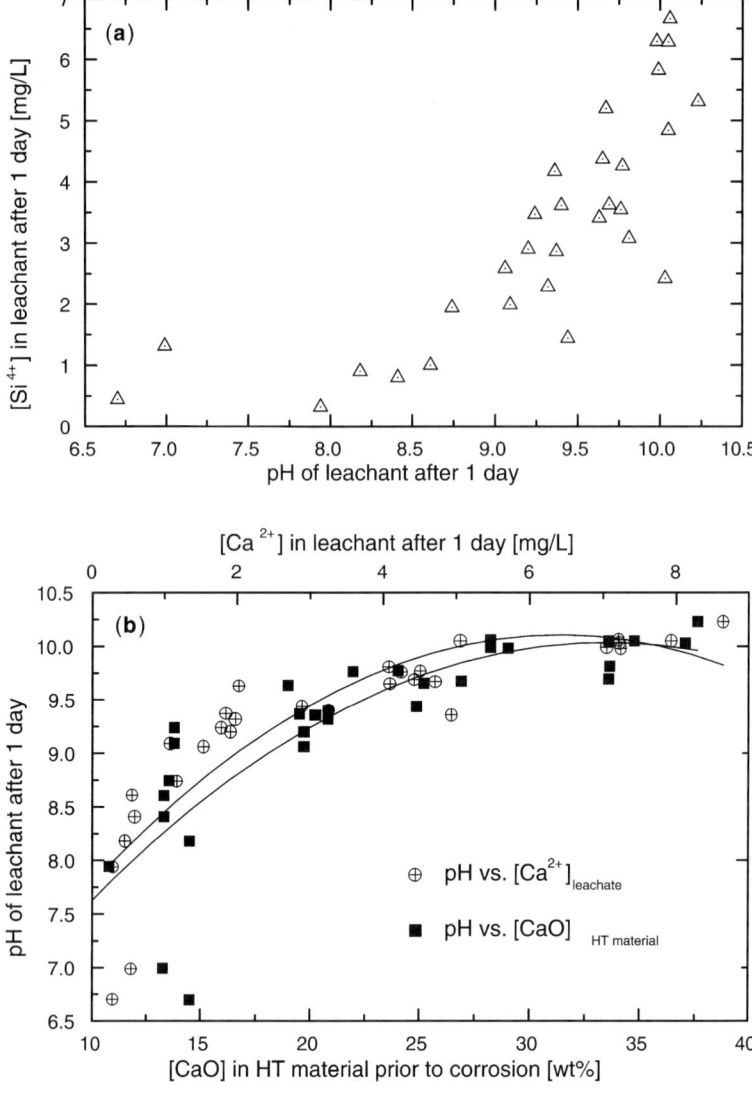

Fig. 9. (a) Concentration of Si in the leachates of the Strasbourg test after one day of corrosion at 90 °C, as a function of the pH of the leachate. (b) Relationships between the measured pH of the leachate and the concentration of CaO in HT materials, respectively the released concentration of Ca^{2+} in the leachates, after one day of corrosion.

The Strasbourg test is performed under much more drastic conditions of corrosion than the corrosion test described by the Swiss Waste Management Ordinance for disposal of residues in landfills for inert materials (coarsely ground samples reacted with CO_2-saturated water during 24 h at room temperature; BUWAL 1990). Nevertheless, we observed that the released metals never reached the maximum permissible concentrations set by Swiss legislation for inert materials (Table 1). By comparison,

BA and FA of conventional low-temperature incineration usually leads to much higher metal concentrations, commonly exceeding permissible values, when leached under the conditions of the Swiss leaching test (ABB-EAWAG-EMPA-KEZO 1990; EKESA 1992).

In order to estimate the long-term durability of HT materials, leaching experiments need to be expressed in terms of apparent normalized release rates of elements (Perret *et al.* 2003). For comparison purposes (from element to

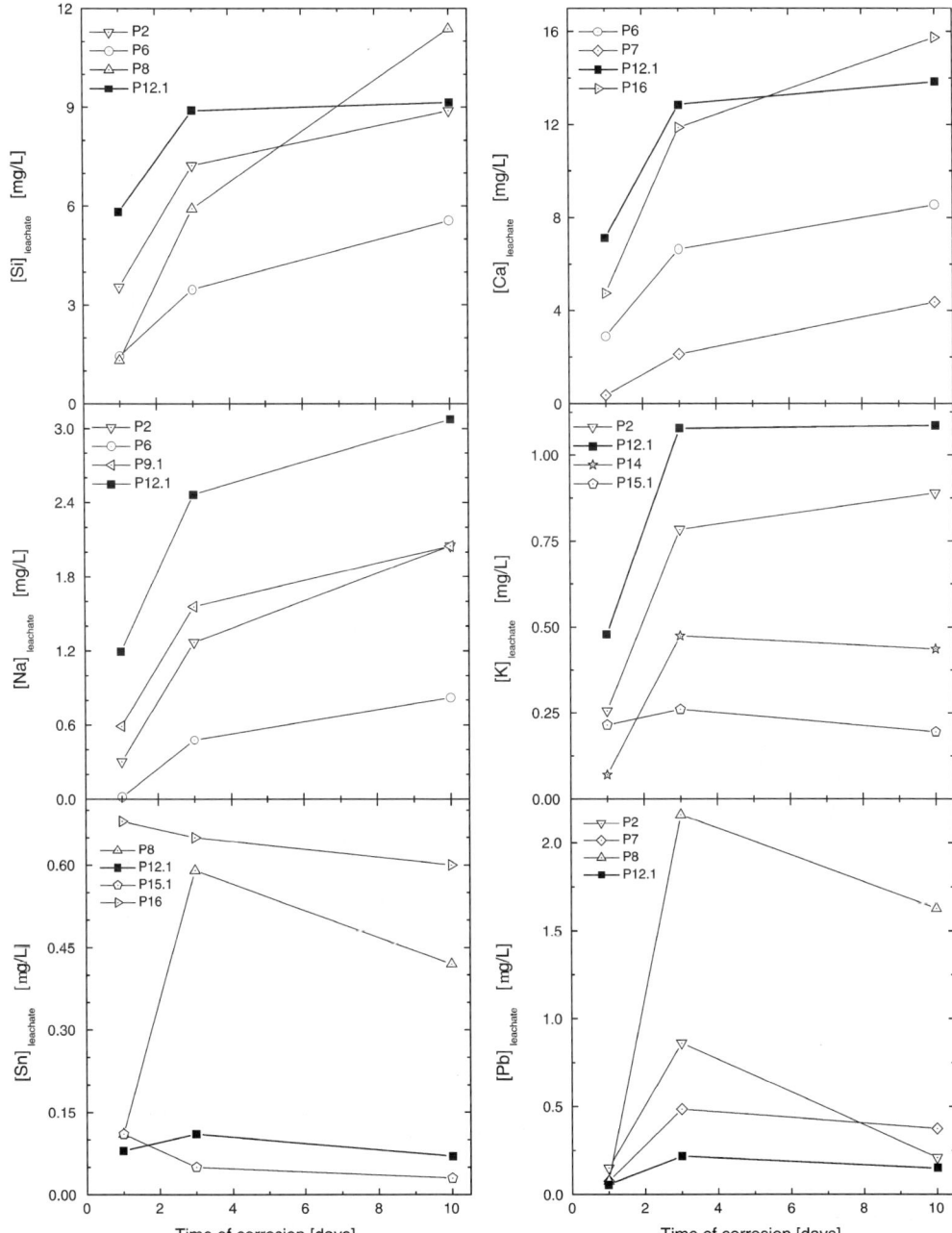

Fig. 10. Temporal evolution of the concentration of the major (Si, Ca), minor (Na, K), and trace species (Sn, Pb) in the leachates of several HT materials corroded for 1 day, 3 days, and 10 days. The temporal fingerprint is different from sample to sample and from element to element.

element, and from sample to sample), normalized rates take into account the characteristics of the HT materials (mass of sample, surface exposed to leachant), the elements of interest (concentration in the sample, apparent concentration released in the leachant), and the experimental conditions (volume of leachant, duration of corrosion). The apparent normalized

Table 1. *Measured concentrations of metals released from HT materials during the Strasbourg test* and maximum admissible concentrations of metals in the leachate of the Swiss TVA test [†]*

	Cr	Co	Ni	Cu	Zn	Cd	Sn	Pb
Strasbourg test, measured (µg/L)	≤40	≤0.4	≤22	≤19	≤113	≤2.7	≤0.7	≤2.5
TVA test, max. admissible (µg/L)	50	50	200	200	1000	10	200	100

*Maximum value after 1 day, 3 days, or 10 days.
[†]Technische Verordnung über Abfälle; BUWAL (1996).

release rate $r(i)_{norm}$ [g/m^2 × day] of element (i) is expressed as:

$$r(i)_{norm} = \frac{d\left[\dfrac{[C_i]_{leachate} \times V_{leachant}}{[C_i]_{HT\ material} \times S_{HT\ material}}\right]}{dt}$$

where $[C_i]_{leachate}$ = concentration of element (i) in the leachate [g/L], $[C_i]_{HT\ material}$ = concentration of element (i) in the HT material [g/g], $V_{leachant}$ = volume of leachant [L], $S_{HT\ material}$ = surface of HT material exposed to leachant [m^2], and dt = duration of corrosion experiment [day].

Figure 11 shows representative examples of apparent release rates of several elements for selected samples. Here again, the release mechanisms are not congruent, either from element to element in the same HT material (e.g., sample P12.1), or from sample to sample for the same element. This is indeed in agreement with observations from others, showing the complex solution and secondary mineral chemistries (Grambow 1985; Advocat 1991). For one-day corrosion, alkali elements are typically (with exceptions, e.g., P12.1) released at higher rates than network-forming and network-modifying elements, indicating that the latter will be predominantly precipitated as secondary minerals. Apparently, the normalized release rates of major elements seem to be partly influenced by the original type of residues used to

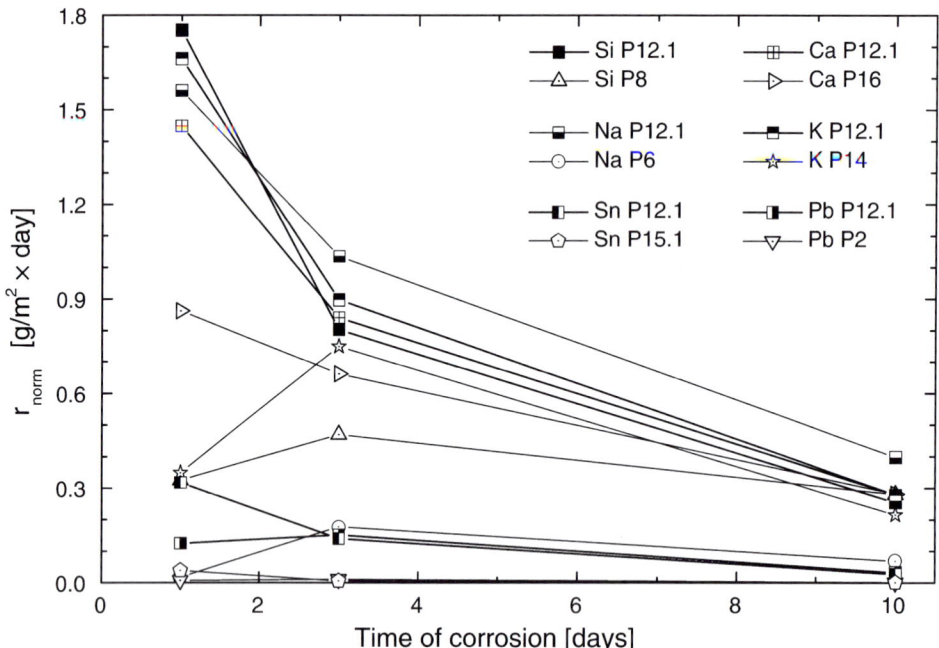

Fig. 11. Evolution of the apparent normalized release rates of various species as a function of the duration of corrosion. Although the initial rates (i.e., after one day) are different from sample to sample and from element to element, the rates measured after 10 days tend to a limiting value independent of the HT material. Indeed, steady state would be reached after a longer period of corrosion.

produce the HT materials (r\{in-line processes\} > r\{post-processes for FA\} > r\{post-processes for BA\}). On the other hand, trace elements exhibit a broad range of apparent release rates, but their calculation is partly biased by the low concentrations of metals in both HT materials and leachants.

The main conclusion from the leaching experiments performed under the drastic conditions of the Strasbourg corrosion test is that short-term releases (1 day corrosion) are distinct for each sample and can be discriminated into fast-reacting and slow-reacting materials. The chemistry of the leachates at the beginning of corrosion is partly governed by the composition of HT materials, whereby corrosion is primarily controlled by pH, which is in turn mostly related to the proportion of CaO in HT materials. However, the releases rapidly decrease and converge toward similar values, around $r(i)_{norm} = 0.2\text{--}0.25 \text{ g/m}^2 \times \text{day}$ after 10 days for major and minor elements, whatever the nature of the HT materials. For longer terms of corrosion, the rates would still decrease and reach much lower limiting values, characteristic of a steady state.

In addition, one may notice that, although most HT materials could not be disposed in landfills for inert materials on the basis of their content in trace metals, they all release very small concentrations of these metals during accelerated corrosion experiments, much below the permissible levels set by Swiss legislation. This demonstrates that HT materials pre-concentrate fairly large quantities of trace metals in their matrix, and they embed them over the long term when subjected to corrosion, whatever the technology developed to produce these materials; on this basis, HT materials comply to the conditions required for safe secondary raw materials.

Finally, the apparent normalized release rate of Si ($r(\text{Si})_{norm}$) in the HT samples was compared to the one obtained for the standard HLW glass SON68 under the same conditions of corrosion. Of the 23 HT samples tested, 6 samples exhibit $r(\text{Si})_{norm} \leq$ SON68, thus presenting a highly favourable dynamic picture with respect to their long-term durability.

While comparing the physico-chemical characteristics of HT materials and their behaviour during corrosion, samples can be roughly discriminated, but they all exhibit globally favourable features. Two samples out of the three best HT materials are vitrocrystalline. On the other hand, three HT materials containing elevated amounts of metals would not be allowed into landfills for inert materials, although their reactivity is lower than the one of the standard HLW glass SON68.

Thermodynamic characteristics of HT materials

A relevant and quantitative way of combining the physico-chemical characteristics of HT materials and their dynamic behaviour as a consequence of corrosion has been developed (Paul 1977, 1982; Newton & Paul 1980; Newton 1985) and refined later (Jantzen 1984, 1988; Jantzen & Plodinec 1984; Plodinec 1984; Plodinec & Wicks 1994). The proposed model makes use of the thermodynamic assessment of glass hydration to estimate the relative stability of glasses, and has been used successfully to predict the long-term stability of nuclear HLW glasses (Bates et al. 1994; Sproull et al. 1994; Ewing 1996). The assessment requires the determination of the overall free energy of hydration ΔG_{hydr} of the glass, which is estimated as the molar-weighted sum of the free energies of hydration $\Delta G_{hydr}(i)$ of its individual constituents (e.g., $CaSiO_3 + 2H^+ + H_2O \rightarrow Ca^{2+} + H_4SiO_4$; $SiO_2 + 2H_2O H_4SiO_4$):

$$\Delta G_{hydr} = \sum (X_i \times \Delta G_{hydr}(i))$$

where X_i = molar fraction of the constituent (i) in the glass (X_i are obtained from the chemical analysis of the glass), and $\Delta G_{hydr}(i)$ = free energy of hydration of the individual constituents (i) in the glass ($\Delta G_{hydr}(i)$ values are available in the literature for most silicates and oxides, or calculated as $\Delta G_{hydr}(i) = -RT \times \ln(K_{hydr}(i))$).

The model assumes that glass is a homogeneous three-dimensional network of tetrahedra MO_4^{4-} (M = network-forming elements Si, Al, B), with network-modifying elements (Ca, Mg, Na, K, Fe) and trace elements being dissolved into the network. According to the model, the assembly will hydrate spontaneously upon corrosion in a congruent manner, assuming that no secondary minerals form at the surface of the corroded glass. Negative values of ΔG_{hydr} correspond to materials that hydrate more spontaneously, that is, that dissolve easily in water.

It has also been observed that a fairly tight relationship exists between the calculated ΔG_{hydr} of a glass, and its apparent normalized release rate, as determined during accelerated corrosion tests (Plodinec & Wicks 1994); this relationship, although questionable, has been verified for a large palette of glasses of different origins (HLW, natural, ancient, and commercial glasses). The global thermodynamic approach thus consists of estimating the durability of a glass on the basis of its thermodynamic propensity to corrode during the initial stage of dissolution, and to relate this durability to the experimental release rate of the glass. In fact, it must be kept in mind that long-term rates of corrosion cannot be derived directly from the model of Paul, which gives only information on the initial rate of corrosion.

Although the thermodynamic model should in principle be used exclusively for glasses, we applied the concept to our set of vitreous and vitrocrystalline HT materials, assuming that the minute amounts of crystalline components present in the vitrocrystalline samples would not drastically impair the determination of their overall free energy of hydration. In addition, trace elements, for which hydration data are either unknown or inaccurate, were not included in the calculation of ΔG_{hydr}, except for Ba, Zn, Zr (known thermodynamics; non-negligible concentrations in some samples); the error in ΔG_{hydr} caused by omitting other trace elements should be negligible, except for Cr and Cu, which may be present in very high concentrations in some

samples. The model used for the computation of ΔG_{hydr} and the specific processing pertaining to our HT materials are detailed and discussed elsewhere (Advocat 1991; Linard 2000; Perret *et al.* 2000, 2003).

Figure 12 is a composite diagram showing the evolution of the calculated free energy of hydration of our HT materials over a wide range of pH. The curves obtained for all samples originating from the same family of HT treatment facilities (in-line processes for MSW wastes; post-processes for BA; post-processes for FA) have been grouped, for simplicity, into individual domains of stabilities (shaded areas).

For clarity, the schematic evolution of ΔG_{hydr} of two hypothetical glasses subjected to corrosion, one SiO_2-rich/CaO-poor (upper curve) and one SiO_2-poor/CaO-rich (lower curve), is superimposed on Fig. 12. The Si-rich glass exhibits a higher stability than the Ca-rich glass. For the former, the curve is steeper at higher pH, to

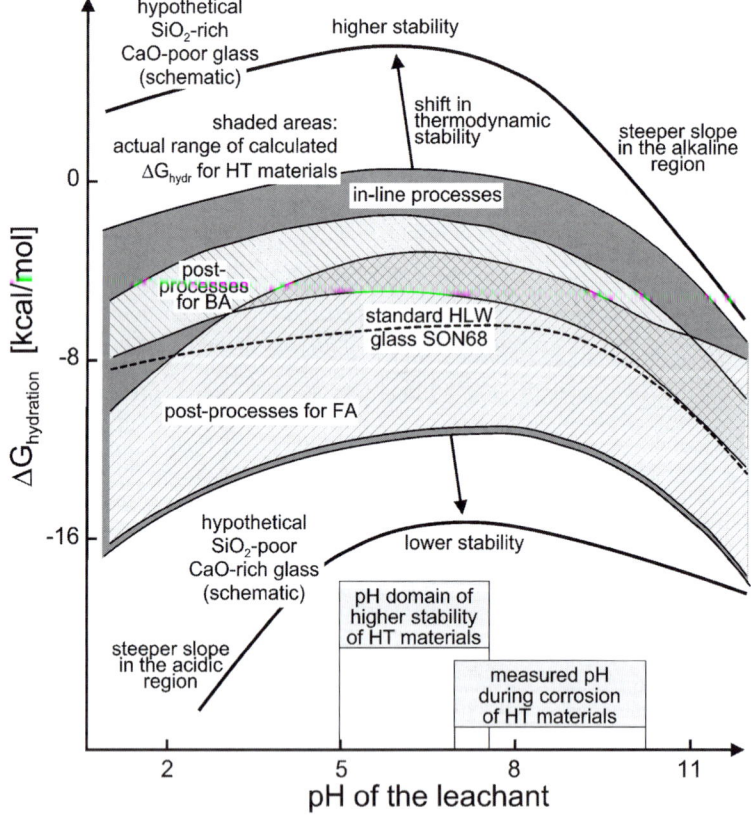

Fig. 12. Schematic curves (top, bottom) of the calculated free energy of hydration ΔG_{hydr} of two hypothetical HT materials (top: Si-rich/Ca-poor; bottom: Si-poor/Ca-rich) as a function of the pH. The shaded areas are the actual ranges of computed ΔG_{hydr} of HT materials, categorized as a function of the HT process from which they originate. For comparison, the computed curve for the standard HLW glass SON68 is also shown (dotted curve).

indicate that SiO_2-rich glasses hydrate more spontaneously in alkaline solutions; in contrast, CaO-rich glasses (steeper slope at lower pH) dissolve more easily under acidic conditions.

While the pH measured after 1–10 days of corrosion under the conditions of the Strasbourg test ranged from 7 to 10.3, the individual curves $\Delta G_{hydr} = f(pH)$ of HT materials indicate that their maximum thermodynamic stability would be reached around pH = 5–7.5. Taken collectively, HT materials originating from in-line processes are spread over a wide range of ΔG_{hydr} values ($\Delta G_{hydr} = -4.7$ kcal/mol $\pm$ 3.1); this is explained by the fact that the amount of major and minor species in these samples has a large variability. On the other hand, HT materials produced by post-processes for BA present a narrower band of stabilities ($\Delta G_{hydr} = -5.0$ kcal/mol $\pm$ 1.5), in agreement with their rather limited inter-sample variation in composition. Finally, post-processes for FA yield HT materials that are collectively less stable ($\Delta G_{hydr} = -9.6$ kcal/mol $\pm$ 2.0) than the others,

as a consequence of the enrichment in fly ashes of CaO with respect to SiO_2 (see Fig. 5).

Altogether, all HT materials exhibit stabilities similar to that of the standard HLW glass SON68, designed to be highly durable; this is indeed a highly favourable feature of the predicted long-term stability of HT materials. The overall free energy of hydration of HT materials is primarily governed by those components present in large amounts and exhibiting large $\Delta G_{hydr}(i)$ values. As expected, samples containing high amounts of network-forming components (SiO_2, Al_2O_3, Fe_2O_3, and in particular MnO_2) and low amounts of network-modifying components ($CaSiO_3$ and to a lesser extent Na_2SiO_3 and $MgSiO_3$) are characterized by the highest thermodynamic stabilities.

An extended survey of the relationship between calculated ΔG_{hydr} and experimental apparent normalized release rates $r(glass)_{norm}$ is presented in Fig. 13 (Plodinec & Wicks 1994) for a series of 115 glasses of different origins. The empirical relationship was obtained by

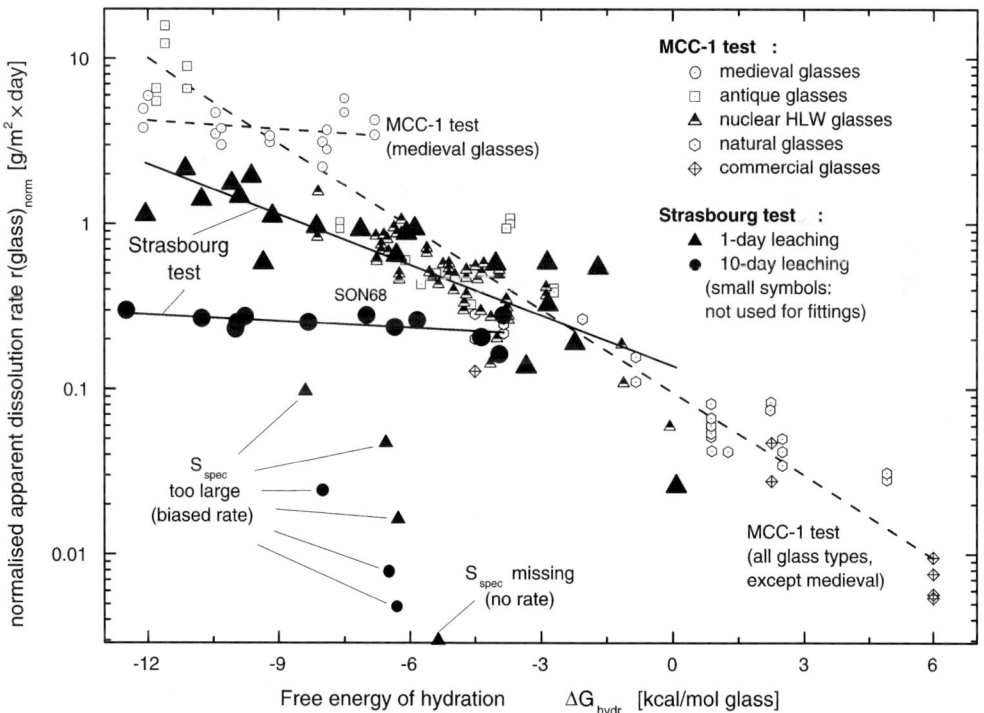

Fig. 13. Relationship between the normalized apparent dissolution rate of HT materials, $r(glass)_{norm}$ and their free energy of hydration, ΔG_{hydr}, calculated for the pH values measured after one day and 10 days of corrosion. For comparison, the literature-extracted results (Plodinec and Wicks 1994) obtained for 115 glasses of different origins corroded under the conditions of the MCC-1 test are reproduced. For simplicity, data obtained after three days of corrosion are not shown; their linear fit lies between the ones of the one-day and the 10-day corrosion experiments.

means of the MCC-1 test (Bates *et al.* 1994; Ebert & Mazer 1994), a test that is an accelerated static corrosion test performed on glass cubes in water, at 90 °C during 28 days. It was designed to assess the reactivity of nuclear HLW glasses.

Because of technical constraints (several HT samples cannot be cut into 4 cm^2 monoliths, due to their granular nature), the MCC-1 test could not be performed with our set of samples. Nevertheless, the normalized apparent dissolution rates of HT materials ($r(\text{glass})_{\text{norm}}$), approximated as the release rates of silicon ($r(\text{Si})_{\text{norm}}$) obtained by the Strasbourg test (100–125 μm powdered samples), are also shown on Fig. 13 for comparative purposes. Although not identical, the MCC-1 test and the Strasbourg test can at least be related in terms of their matching efficiencies ($S_{\text{HTmaterial}}/V_{\text{leachant}}) \times t_{\text{corrosion}} = 2.8 \text{ day}/$ cm for the MCC-1 test; 0.2–2 day/cm for the Strasbourg test). Formally, $r(\text{Si})_{\text{norm}}$ should not be taken equal to $r(\text{glass})_{\text{norm}}$, as Si, like other elements, has been shown to precipitate at the surface of our HT materials during corrosion, and is thus not the relevant tracer for describing matrix transformation; however, this approximation is acceptable in our case, because $r(\text{Si})_{\text{norm}}$ will represent the net, apparent dissolution of HT materials.

Figure 13 shows that the $r(\text{glass})_{\text{norm}}$ vs. ΔG_{hydr} relationship for HT materials corroded during one day is very close to the one obtained on various glasses by the MCC-1 test, which is performed under conditions of initial rate of corrosion. In addition, the standard HLW glass SON68 (Strasbourg test; three-day corrosion, not shown) yields similar results to the ones obtained for the other nuclear HLW glasses (MCC-1 test), confirming that both tests can be compared, at least for the initial stage of corrosion of the Strasbourg test. Although based on simplifying assumptions, the thermodynamic approach of the free energy of hydration can clearly be applied to discriminate HT materials during the initial stage of matrix hydrolysis. For the MCC-1 test and the Strasbourg test after one day, difference of 10 kcal/mol between two HT samples yields approximately one order of magnitude difference between their dissolution rates.

As already mentioned, normalized apparent release rates of major and minor elements present in HT materials converge toward similar values on longer terms of corrosion (ca. 0.2–0.25 g/m² × day; see Fig. 11). This is confirmed on Fig. 13, where the influence of the differences in ΔG_{hydr} on the release rates vanishes for 10-day leaching (quasi-flat slope). This result indicates that the corrosion of HT materials is very different during the initial stage of dissolution, but

rapidly approaches steady state as a consequence of the protective effect of secondary minerals formed at the surface of corroded HT samples, and of the increase in Si concentration in the leachate (Grambow 1985). In addition, medieval glasses (MCC-1 test) and our HT materials (Strasbourg test; 10-day corrosion) exhibit similar slopes, which may be explained by the formation of protective Mg-rich phases in both cases (Crovisier *et al.* 1987).

Medieval and antique glasses have an observed durability greater than 10^3 years (Gillies & Cox 1988; Macquet & Thomassin 1992; Sterpenich 1998), whereas the extrapolated durability of nuclear HLW glasses extends above 10^4 years (Jollivet *et al.* 1998), as determined from corrosion experiments performed under conditions of long-term rate and steady state. On this basis, and by comparison of our results (one day corrosion) and the ones obtained by the MCC-1 test, the durability of our HT materials is estimated to range at least between 10^3 years and 10^4 years.

'Glassfill' versus 'glassroad'

The description of HT materials by means of their physico-chemical and microscopic characteristics, their behaviour under aggressive conditions of accelerated corrosion, and their estimated thermodynamic stability, helps us to derive a composite, yet semi-quantitative picture of the individual qualities of HT materials. This is the aim of Fig. 14, where each HT material has been assigned a relative rank with respect to its static features (crystallinity, S_{spec}, surface roughness, density of mineral inclusions, amount of metals), dynamic features (Si loss in the leachate, normalized apparent release rate), and thermodynamic features (free energy of hydration). Details on the computing of the individual ranks are given in Perret *et al.* (2000) and briefly summarized in Fig. 14. In this figure, a distinction is made between fully characterized and partly characterized HT materials.

The grey zone in Fig. 14 encompasses the average (± standard deviation) rank for samples that were fully characterized. The standard HLW glass SON68 and the two standard HT materials R2bis and R3 are shown for comparative purposes. Several HT samples that were not fully characterized can be described as generally suitable materials (i.e., they lie within the grey zone), whereas some fully characterized HT samples lie below the grey zone. No conclusion can, however, be drawn for the incompletely characterized HT samples that lay below the grey

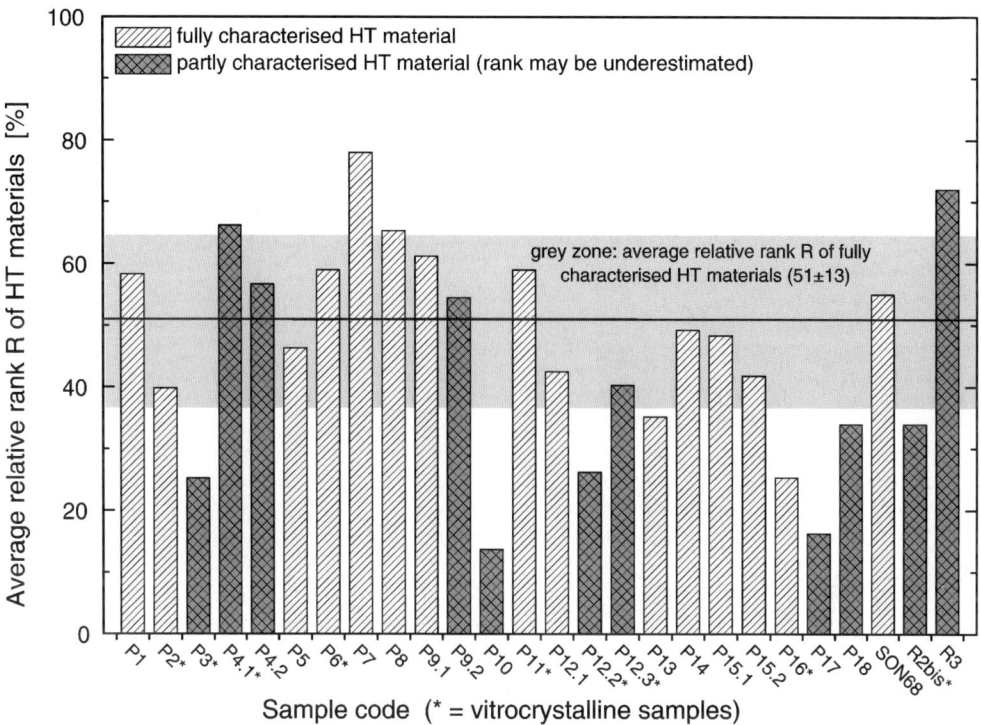

Fig. 14. Tentative semi-quantitative ranking of HT materials. The average relative rank R of each sample is the mean of their static, dynamic, and thermodynamic ranks (R = (Stat + Dyn + Thermo)/3). For the global static rank Stat each physico-chemical characteristic (vitreous/vitrocrystalline; S_{spec}; roughness; density of mineral inclusions; admissibility in inert landfills with respect to metal content) of a sample is assigned the value 100% when favourable (e.g., vitreous; $S_{spec} < S_{spec}$(SON68); etc.) or 0% when not favourable (e.g., rougher than SON68; not admissible in inert landfill; etc). The global dynamic rank Dyn is the mean of the values obtained during corrosion for $[Si]_{leachate}$, respectively $r(Si)_{norm}$ (between 0% for the worst sample and 100% for the best sample). The thermodynamic rank Thermo is the value obtained for ΔG_{hydr} (0–100% from the worst to the best sample). For partly characterized samples (i.e., samples for which several parameters are missing), the average relative rank R may be underestimated

zone, because their final rank is probably underestimated.

Taken collectively, HT materials exhibit generally favourable qualities, and can be considered as potential candidates for either disposal into specific landfills for inert-type HT materials requiring no special attention ('glassfill' scenario), or re-use for civil engineering applications ('glassroad' scenario). The 'glassfill' vs. 'glassroad' approach is based on the following considerations, assumptions, and boundary conditions (for details, see Perret *et al.* 2000):

(1) The 28 existing Swiss MSW incineration plants would be either replaced by in-line processes for MSW, or updated with post-processes for BA or FA. These plants would treat 3×10^6 t/y MSW, or 7.75×10^5 t/y BA, or 4.65×10^4 t/y FA, or combinations of them (see Fig. 2).

(2) The annual production of HT materials depends on the type of HT technologies. Taking into account the highest input → output mass transfer coefficient of each process, the following masses of HT materials would theoretically be produced in Switzerland:
   In-line processes: $400–700 \times 10^3$ t/y.
   Post-processes for BA: $200–700 \times 10^3$ t/y.
   Post-processes for FA: $30–50 \times 10^3$ t/y.
For our 'glassfill' vs. 'glassroad' scenario, it is assumed that the annual production of HT materials in Switzerland would be $50–700 \times 10^3$ t/y.

(3) The composition of HT materials can be approximated as the average composition of all HT materials tested in this study, taking into account the lower and upper concentrations of major, minor, and trace elements determined for each of the three

families of HT processes. The composition of such a hypothetical HT material is given in Table 2.

(4) The HT materials produced by these HT technologies would be either disposed into thick 'glassfills', or re-used as thin foundation layers for 'glassroad' construction. The characteristics of such hypothetical 'glassfills' and 'glassroads,' which depend on the annual production of HT materials, are given in Table 3.

(5) The resulting 'glassfills' and 'glassroads' would be exposed to realistic conditions of corrosion. Taking into account the results of the present study and long-term (90–200 days) corrosion rates given in the literature, ranging between 10^{-4} and 10^{-2} g/m^2 × day (Advocat 1991; Colombel 1996; Colombel *et al.* 1997; Jollivet *et al.* 1998; Sterpenich 1998; static and dynamic conditions of corrosion), the most realistic dissolution rate pertaining to the 'glassfill' and the 'glassroad' scenarios is $r_{\text{corrosion}} = 10^{-4}$–$10^{-3}$ g/m^2 × day.

(6) The metals released from the 'glassfills' and the 'glassroads' during corrosion would be dispersed in the environment without further control by the authorities. In this way, released concentrations and loads of metals can be directly compared with fluxes resulting from natural mechanisms and anthropogenic activities.

For the case of the 'glassfill', the annual production of HT material would result in the opening of a landfill with the dimensions 50 × 50 m^2 to 190 × 190 m^2 (typical depth of 10 m). For the case of the 'glassroad', an annual production of 50–700 × 10^3 t of HT material would be equivalent to the construction of ca. 5–70 km/y of a road with highway dimensions (ca. 10 m width; typical thickness of 50 cm). Because of their specific characteristics (different dimensions and permeabilities), the 'glassfill' would drain annually 2.6–37 × 10^3 m^3 water, while the 'glassroad' would drain 3.8–53 × 10^3 m^3 water. Thus, the scenario shall consider that an overall volume of water ranging between 3 × 10^3 m^3 and 50 × 10^3 m^3 would percolate through the pile of HT material. This is an important issue on the plausible destination of HT materials, as the investment in energy which is required in their production would generate a much more cost-effective return in the case of the 'glassroad' scenario (higher value) than in the case of the 'glassfill'.

Estimated impact of 'glassfill' and 'glassroad' on the environment

The model designed to account for all considerations discussed above expresses metal release in terms of concentrations and in terms of annual loads (i.e., masses of metals released annually). The annual loads of metals $m(M_{\text{released}})$ [kg/yr] are computed according to the following expression:

$$m(M_{\text{released}}) = [M]_{\text{HT material}} \times r_{\text{corrosion}}$$
$$\times ((m_{\text{HT material}}/\rho_{\text{HT material}})/V_{\text{grain}})$$
$$\times S_{\text{grain}} \times \Delta t_{\text{corrosion}}$$

Table 2. *Composition of a hypothetical HT material taking into account (i) the average composition of the 23 HT samples studied, and (ii) the lower and upper average concentrations of species in each of the three families of HT processes (in-line processes; post-processes for BA; post-processes for FA)*

Component	Average concentration	Minimum concentration	Maximum concentration
SiO$_2$ (wt%)	41.4	36.8	48.2
Al$_2$O$_3$ (wt%)	16.6	14.5	18.1
CaO (wt%)	23.1	17.5	27.5
MgO (wt%)	3.2	2.7	3.7
Na$_2$O (wt%)	2.9	2.7	3.5
K$_2$O (wt%)	1.0	0.9	1.2
Fe$_2$O$_3$ (wt%)	7.5	4.5	11.5
Cr (mg/kg)	2212	1746	2674
Ni (mg/kg)	157	52	222
Cu (mg/kg)	997	603	2310
Zn (mg/kg)	3237	1262	4844
Cd (mg/kg)	14	0.4	26
Pb (mg/kg)	250	81	339

Table 3. *Physical characteristics of the hypothetical 'glassfille' and 'glassroads'*

	Disposal into 'glassfill'		Re-use for 'glassroad' construction	
	Production 50 × 10³ t/y HT	Production 700 × 10³ t/y HT	Production 50 × 10³ t/y HT	Production 700 × 10³ t/y HT
Volume of granules (density: 2.6 g/cm³)	19 × 10³ m³/y	269 × 10³ m³/y	19 × 10³ m³/y	269 × 10³ m³/y
Surface of granules (diameter: 0.1–5 cm)	1.2 × 10⁹–4.6 × 10⁵ m²/y	1.6 × 10¹⁰–6.5 × 10⁶ m²/y	1.2 × 10⁹–4.6 × 10⁵ m²/y	1.6 × 10¹⁰–6.5 × 10⁶ m²/y
Apparent volume of pile of HT material (apparent density of HT pile: 2 g/cm³)	25 × 10³ m³/y	350 × 10³ m³/y	25 × 10³ m³/y	350 × 10³ m³/y
Upper surface of HT pile	2.5 × 10³ m²/y (depth: 10 m)	35 × 10³ m²/y (depth: 10 m)	50 × 10³ m²/y (depth: 0.5 m)	700 × 10³ m²/y (depth: 0.5 m)
Volume of rain through the HT pile (average rainfall: 1500 mm/y)	2.6 × 10³ m³/y (permeability: 70%)	3.7 × 10⁴ m³/y (permeability: 70%)	3.8 × 10³ m³/y (permeability: 5%)	5.3 × 10⁴ m³/y (permeability: 5%)
Flow rate through the pile of HT material	120 mL/h × m²	120 mL/h × m²	10 mL/h × m²	10 mL/h × m²

where $[M]_{HT\ material}$ = concentration of metal M in the HT material [kg/kg], $r_{corrosion}$ = apparent normalized release rate of HT material = 10^{-4}–10^{-3} [g/m² × day] = 3.65 × 10⁻⁵–3.65 × 10⁻⁴ [kg/m² × y] for realistic conditions of corrosion, $m_{HT\ material}$ = mass of HT material produced yearly and subjected to corrosion = 50–700 × 10⁶ [kg/y] depending on the type of HT process, $\rho_{HT\ material}$ = true density of HT material assumed to be glass = 2.6 [g/cm³] = 2.6 × 10³ [kg/m³], V_{grain} = volume of grains of HT material subjected to corrosion [m³], S_{grain} = surface of grains exposed to corrosion [m²], and $\Delta t_{corrosion}$ = duration of the leaching of the 'glass-fill' or 'glassroad' [y].

Figure 15 shows the relevant results obtained with the corrosion model.

It appears that the concentrations and annual masses of metals that would be released in the environment are linearly related to the proportions of these metals in the HT material and to the corrosion rate. However, the released concentrations of metals are independent of the mass of HT material produced annually, whereas the released annual masses are indeed linked to the annual production. On the other hand, the size of HT granules has a strong influence on the concentrations and masses of metals that would be released, as $m_{released} = f(1/size)$; in other words, a 1 mm granule releases 50× more metal than a 5 cm block.

Taking into consideration the upper limits of the scenarios ($[metal]_{HT\ material}$ = 5000 mg/kg; granule diameter = 1 mm; $r_{corrosion}$ = 10⁻³ g/m² × day; production = 700 kt/y), and making the simplified assumption that metals released during corrosion would disperse homogeneously in the percolating water (i.e., without advection, dispersion, concentration gradient, sorption by solid phases or biota), the maximum concentration of metal released into the environment would be on the order of 80 mg/L, for a maximum load of ca. 3 t/y. Using the lower limits of the scenarios, the minimum concentration of metal released would be on the order of 2 ng/L, for a minimum load of ca. 10 mg/y. Table 4 summarizes the concentrations and loads that would be expected within the boundaries of the scenarios, for grain sizes of 1 cm, which is a realistic size according to the actual morphologies of our set of HT samples, and taking into account the average concentrations of metals in our hypothetical HT material.

The proposed corrosion model allows one to estimate the maximum lifetime of our hypothetical average HT material subjected to realistic conditions of corrosion. The maximum lifetime represents the time necessary to corrode the

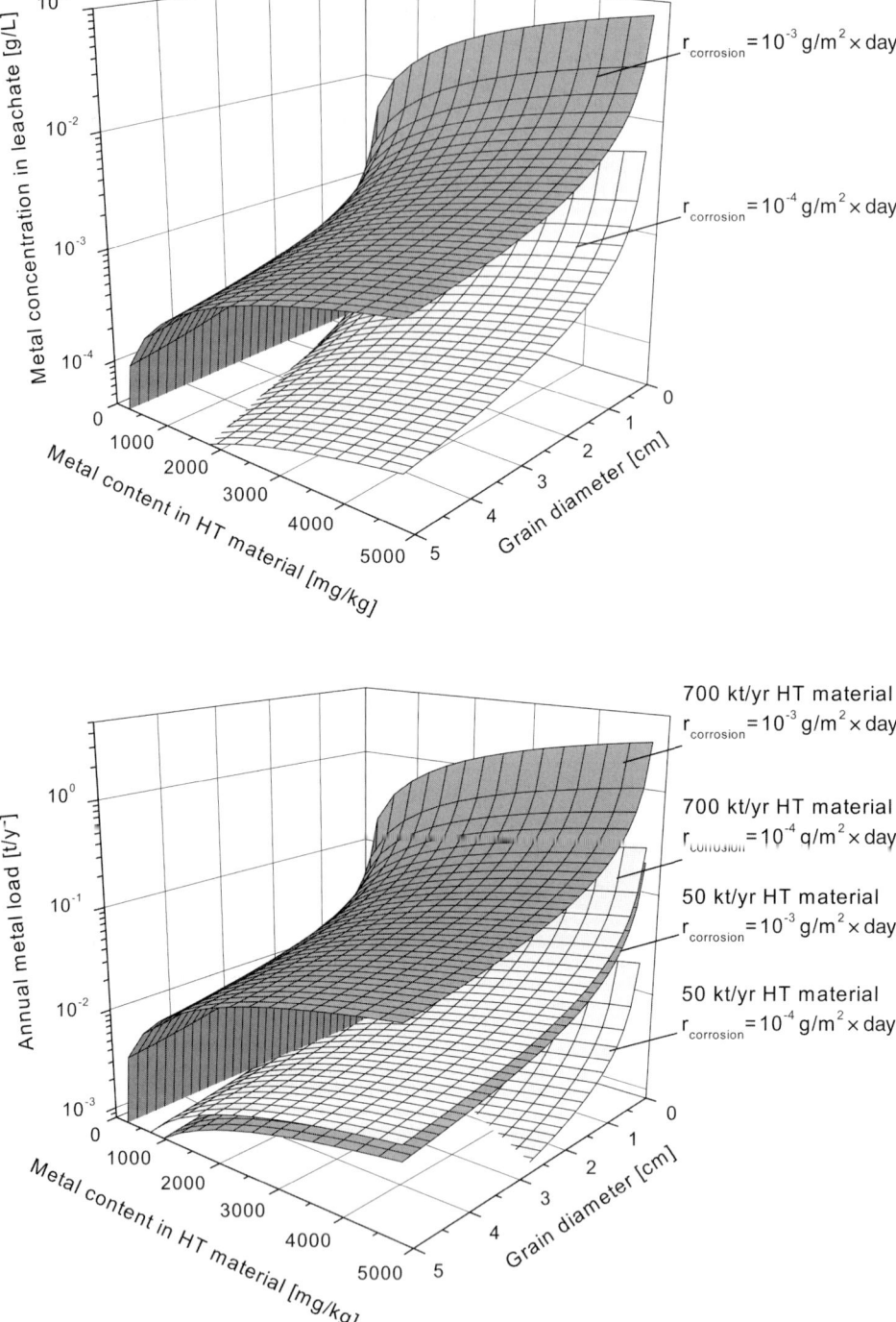

Fig. 15. Modelled concentrations (top) and absolute releases (bottom) of metals in the leachate of the corrosion experiments, as a function of the metal content in the HT materials and the size of the granules subjected to corrosion. Two realistic rates of corrosion are used for computation.

Table 4. *Estimated average impacts of the 'glassfills' and 'glassroads' on the environment, under the conditions of the corrosion scenarios discussed in the text**

	Cr	Ni	Cu	Zn	Cd	Pb
Releases for a production of 5×10^4 t/y, $r_{corrosion} = 10^{-4}$ g/m^2 × day, granule size = 1 cm (lower boundary of scenario)						
[M^{n+}] (mg/L)	0.25 (0.2–0.3)	0.02 (0.006–0.03)	0.1 (0.07–0.3)	0.36 (0.1–0.5)	0.002 (<0.003)	0.03 (0.009–0.04)
Load (kg/y)	0.9 (0.7–1.1)	0.07 (0.02–0.09)	0.4 (0.3–1)	1.4 (0.5–2)	0.006 (0.0002–0.01)	0.1 (0.03–0.2)
Releases for a production of 7×10^5 t/y, $r_{corrosion} = 10^{-3}$ g/m^2 × day, granule size = 1 cm (upper boundary of scenario)						
[M^{n+}] (mg/L)	3.6 (2.8–4.3)	0.25 (0.08–0.4)	1.6 (1–3.7)	5.2 (2–8)	0.02 (<0.04)	0.4 (0.1–0.5)
Load (kg/y)	130 (103–158)	9 (3–13)	59 (36–136)	191 (74–286)	0.8 (0.02–1.5)	15 (5–20)

*For each scenario, the minimum and maximum impacts (concentrations, annual loads; in parentheses) are calculated according to the lower and upper concentrations of metals in the hypothetical HT material, as given in Table 1.

total amount of HT material, provided that all elements are released at the same rate (i.e., congruent mechanism of dissolution) and that the rate is constant throughout the life of the HT material. Table 5 summarizes the plausible lifetimes calculated from the different scenarios.

Within realistic conditions of operation, the shortest lifetime (1200 y) represents the worst case, which is certainly difficult to envision; for this situation, HT granules would have the smallest possible size (1 mm) and would dissolve at the fastest rate (10^{-3} g/m^2 × day). Likewise, the longest lifetime (6×10^5 y) is the best case scenario, which might never happen: HT granules would have the greatest size and dissolve at the lowest corrosion rate. These two extreme lifetimes are thus not considered to be plausible. For the most realistic estimates, one would expect the hypothetical HT material to last for ca. 10^4–10^5 years, which is remarkably favourable.

Other natural and anthropogenic fluxes of metals

The impact of our hypothetical average HT material on the environment can be compared to other natural and anthropogenic sources of metals. Table 6 compares the average concentrations of metals in several materials of natural and anthropogenic origin. As expected from the operating conditions of the various HT treatment technologies, the concentration of Cr in HT materials (i.e., in our hypothetical average HT material) is much higher than in other materials, although the continental crust may contain up to 3 g/kg of this element (Table 6). Ni in HT materials is much below the permissible value for disposal of residues in landfills for inert materials, and even below the Swiss T value (tolerable concentration in excavation material used for civil engineering purposes). Cu, Zn, Cd, and Pb are less concentrated in HT materials than in the conventional residues of incineration (BA, FA, FC); except for Pb, these metals exceed the permissible value of the Swiss Waste Management Ordinance for inert landfills.

Table 7 lists the annual masses of metals immobilized in several anthropogenic materials. Data in Table 7 take into account the concentrations of metals in these materials (Table 6) and the claimed annual production of these materials in Switzerland. For the case of our hypothetical HT material, two scenarios are taken into account. It is assumed here that the annual production of HT materials would be either 7×10^5 t/y (replacement of the 28 Swiss

Table 5. *Estimated lifetimes of the hypothetical pile of HT material subjected to realistic conditions of corrosion**

	Granule size: 0.1 cm	Granule size: 1 cm	Granule size: 5 cm
Lifetime for $r_{corrosion} = 10^{-3}$ g/m^2 × day (upper boundary of scenario)	1200 y (not realistic)	**12 000 y**	**60 000 y**
Lifetime for $r_{corrosion} = 10^{-4}$ g/m^2 × day (lower boundary of scenario)	**12 000 y**	**120 000 y**	600 000 y (not realistic)

*Bold values result from the most plausible conditions of operation.

incineration plants by in-line processes for MSW to treat 3×10^6 t/y of MSW, or update with alternative post-processes for BA), or 5×10^4 t/y (update of the 28 plants with alternative post-processes for FA).

Depending on the type of HT process chosen, the annual mass of metals immobilized in the hypothetical average HT material would either be in the medium to high range, or in the low range, of masses of metals in materials produced by human activities. Cr is an exception, as a consequence of the abundant release of this element from the refractory materials used in most HT technologies. For the other metals, HT materials immobilize lower annual masses of metals than MSW and BA produced by conventional incineration; this reflects the fact that HT technologies are effectively able to separate a pool of recyclable metals, which will not end up in the HT material. The quantities of metals that would be immobilized in HT materials annually are of the same order of magnitude as those immobilized in the Swiss Portland cement.

Finally, Table 8 compares the extent of metal loss into the environment as a result of (1) the corrosion of our hypothetical HT material, (2) the anthropogenic dispersion of metal-containing materials used as fertilizers on Swiss soils, and

(3) the masses of metals transported annually by the two major Swiss rivers. Only sewage sludge falling within the Swiss limits for use as fertilizers is considered in Table 8. In the case of HT materials, two boundary conditions are presented. The most conservative case considers an annual production of 7×10^5 t HT materials being corroded at the highest plausible rate ($r_{corrosion} = 10^{-3}$ g/m^2 × day), while the second case is computed with an annual production of 5×10^4 t HT materials being corroded at the lowest plausible rate ($r_{corrosion} = 10^{-4}$ g/m^2 × day). Data are also presented as surface-weighted metal loss, that is, amount of metals released per square kilometre of soil (use of fertilizer) or 'glassfill'/'glass-road'; this distinction is necessary because, for a given mass of HT material, the upper surface of the 'glassfill' exposed to rain is smaller than its equivalent 'glassroad', as a result of their different thickness (see Table 3).

Table 8 shows that the total amount of metals that would be released by our hypothetical average HT material is distinctly lower ($\sum$(metals) = 2–400 kt/y, depending on the scenario chosen) than the total amount deposited every year in Switzerland as a result of intensive use of fertilizers ($\sum$(metals) = 10^6 kg/y for all types of fertilizers). Likewise, the total loadings

Table 6. *Concentrations of metals (expressed in mg/kg) in several natural and anthropogenic materials*

	Cr	Ni	Cu	Zn	Cd	Pb
Average HT material (see Table 1)	2212	157	997	3237	14	250
Average municipal solid wastes (MSW)	60	40	2000	1400	11	700
Average bottom ash (BA)	500	149	7880	4088	4.1	1876
Average fly ash (FA)	400	133	2000	24266	623	14933
Average filter cake (FC)	200	213	2667	3733	88	1867
Maximum concentration allowed for disposal in inert landfill (Swiss TVA limits)	—	500	500	1000	10	500
Average sludge of wastewater treatment plant	84	40	388	1110	2.4	133
Residues of automobile shredding	1350	900	16000	16000	87	7000
Swiss Portland cement (maximum concentration)	125	?	?	680	5.5	255
Swiss clinker (maximum)	320	?	136	530	1.5	112
Maximum concentration allowed in Swiss clinker	150	100	100	500	1.5	100
Upper continental crust (maximum)	2980	2000	87	130	0.3	20
Mid ocean ridge basalt (MORB; maximum)	300	140	80	70	0.1	1
Granites (maximum)	22	15	30	60	0.13	48
Unpolluted excavation material (Swiss U value)	50	50	40	150	1	50
Tolerable excavation material (Swiss T value)	250	250	250	500	5	250

Table 7. *Annual masses of metals (expressed in t/y) immobilized in anthropogenic materials produced in Switzerland*

	Cr	Ni	Cu	Zn	Cd	Pb
Average HT material; production = 700 000 t/y	1548	110	698	2266	10	175
Average HT material; production = 50 000 t/y	111	8	50	162	1	13
Average municipal solid wastes (MSW)	186	124	6200	4340	34.1	2200
Average bottom ash (BA)	386	115	6100	3200	3.4	1500
Average fly ash (FA)	19	6.2	93	1100	29	694
Average filter cake (FC)	2.3	2.5	31	43.4	1	21.7
Average sludge of wastewater treatment plant	18	8.5	82	234	0.5	28
Residues of automobile shredding	52	34	611	611	3.3	267
Swiss Portland cement (maximum concentration)	431	?	?	2344	19	879

resulting from corrosion of the HT material is much smaller than the total quantity of metals circulating via the Rhine and the Rhone ($\sum$(metals) = 10^6 kg/y).

Surface-weighted data, which are more representative of short-range contaminations, indicate that the 'glassfill' scenario is much more detrimental to the environment than every other case. The hypothetical 'glassroad', on the other hand, would have less of an impact on soils and natural waters than the use of fertilizers on soils.

In terms of loadings, it is instructive to note that the Rhine and the Rhone transport annually ca. 2.5×10^3 to 10^5 times more metal than that

which would be released during the corrosion of our hypothetical average HT material. Likewise, it appears that the Swiss soils collect annually huge amounts of metals as a consequence of agricultural activities and use of different types of fertilizers.

To conclude, it must be pointed out that the releases of metals in the 'glassfill'/'glassroad' scenarios, although realistic, do not take into account the complex mechanisms of immobilization of metals by precipitation of stable secondary solid phases, either at the surface of the corroded HT material or in solution. These mechanisms of immobilization are difficult to

Table 8. *Annual dispersions and losses of metals in the environment*[*]

	Cr	Ni	Cu	Zn	Cd	Pb
Total loadings (t/y)						
Average HT material; 7×10^5 t/y, $r = 10^{-3}$ g/m^2 × day	0.13	<0.01	<0.06	0.2	<0.001	<0.02
Average HT material; 5×10^4 t/y, $r = 10^{-4}$ g/m^2 × day	<0.001	<0.001	<0.001	<0.002	<0.001	<0.001
Average sludge of wastewater treatment plant	8.7	3.8	41	132	0.22	15.9
Average mineral fertilizers	130	4.7	7	45	1	4.3
Average compost	2	1.3	3.5	12	0.03	3.8
Average animal manures	—	—	80	420	0.4	8.3
Other average fertilizers made of wastes	3.7	4.1	6.5	37	0.3	6
Rhine River flux to the outlet of Switzerland	—	—	103	257	—	60
Rhone River flux to the outlet of Switzerland	42	93	46	264	0.9	60
Surface-weighted loadings (kg/y × km^2)						
Average HT material; 'glassfill', $r = 10^{-3}$ g/m^2 × day	3726	265	1680	5453	24	421
Average HT material; 'glassroad', $r = 10^{-4}$ g/m^2 × day	19	1.3	8.4	27	0.12	2.1
Average sludge of wastewater treatment plant	11	5	52	165	0.28	20
Average mineral fertilizers	33	1.2	1.8	11	0.25	1.1
Average compost	2	1.3	3.5	12	0.03	3.8
Average animal manures	—	—	10	53	0.05	1

[*]Values are expressed in t/y (total loadings) and in kg/y × km^2 (surface-weighted loadings). Surface-weighted loadings take into account the maximum surfaces of soils that are concerned by the use of fertilizers in Switzerland, respectively the surface of 'glassfills' or 'glassroads' necessary to account for the annual production of HT material.

model, but they would decrease the metal concentrations and loadings resulting from the corrosion of the HT materials.

Concluding remarks

The determination of the overall picture of HT materials, by a combination of physico-chemical and microscopic analyses, accelerated corrosion experiments performed under aggressive conditions, and computed spontaneity to hydrate, is a fairly simple comparative tool that allows us to estimate the relative stability of HT materials. The key results of this study show that the high-temperature technologies used to produce HT materials, or the original wastes (MSW, BA, FA, FC) used to produce HT materials, have only little influence on the final characteristics and behaviour of these materials. Taken altogether, HT materials have fairly favourable characteristics when subjected to corrosion. HT materials may contain 10–200 times more metals than some anthropogenic and natural materials (cement, clinker, continental crust, granites), but they are much cleaner overall than other residues resulting from human activities (BA, FA, FC, car shred, sewage sludge). According to the Swiss Waste Management Ordinance, most HT materials cannot be considered as inert materials and could not be disposed in landfills without requiring pre-stabilization; however, HT materials release only negligible concentrations of toxic metals under drastic conditions of corrosion, much below the Swiss permissible values used to assess the chemical stability of wastes and residues. When compared to other materials (glasses of different origins), the durability of HT materials, estimated to range between 10^4 and 10^5 years under realistic conditions of corrosion, is exceptionally high.

As a consequence, the production of HT materials and their potential re-use as foundation layers for road construction, or their immobilization into specific landfills requiring minimal care, would appear to represent an activity with low environmental impact in comparison with other natural and anthropogenic systems.

The model for the calculation of the free energy of hydration was provided by Dr T. Advocat, CEA Marcoule (France). The financial support to D. Perret, P. Stille, G. Shields and U. Mäder was provided by the Swiss Office for the Environment, Forests and Landscape. This is EOST/CGS contribution n° 2004.604-UMR7517. Valuable suggestions were provided by A. Johnson, E. Curti, D. Strachan, and R. Gieré to improve the manuscript.

References

ABB-EAWAG-EMPA-KEZO 1990. *Durchführung von Pilotversuchen zur Aufbereitung und Entsorgung von Filterstaub aus Kehrichtverbrennungsanlagen.* Schlussbericht. Baden/Dübendorf, Switzerland.

ABDELOUAS, A., CROVISIER, J.-L., LUTZE, W., MÜLLER, R. & BERNOTAT, W. 1994a. Hydrotalcite formed by alteration of R7T7 nuclear waste glass and basaltic glass in salt brine at 190°C. *In:* BARKATT, A. & VAN KONYNENBOURG, R. A. (eds) *Scientific Basis for Nuclear Waste Management XVII. Materials Research Society Symposia Proceedings*, **333**, 513–518.

ABDELOUAS, A., CROVISIER, J.-L., LUTZE, W., FRITZ, B., MOSSER, A. & MÜLLER, R. 1994b. Formation of hydrotalcite-like compounds during R7T7 nuclear waste glass and basaltic alteration. *Clays and Clay Minerals*, **42**, 526–533.

ABDELOUAS, A., CROVISIER, J.-L., LUTZE, W., GRAMBOW, B., DRAN, J.-C. & MÜLLER, R. 1997. Surface layers on a borosilicate nuclear waste glass corroded in MgCl₂ solution. *Journal of Nuclear Materials*, **240**, 100–111.

ABRAJANO, T. A., BATES, J. K., WOODLAND, A. B., BRADLEY, J. P. & BOURCIER, W. L. 1990. Secondary phase formation during nuclear waste-glass dissolution. *Clays and Clay Minerals*, **38**, 537–548.

ADAMS, P. B. 1984. Glass corrosion: A record of the past? A predictor of the future? *Journal of Non-Crystalline Solids*, **67**, 193–205.

ADAMS, P. B. 1992. Predicting corrosion. *In:* CLARK, D. E. & ZOITOS, B. K. (eds) *Corrosion of Glass, Ceramics and Ceramic Superconductors; Principles, Testing, Characterization and Applications.* Noyes, Park Ridge, USA, 29–50.

ADVOCAT, T. 1991. *Les Mécanismes de Corrosion en Phase Aqueuse du Verre Nucléaire R7T7; Approche Expérimentale, Essai de Modélisation Thermodynamique et Cinétique.* PhD thesis, University Louis Pasteur, Strasbourg, France.

ALTENHEIN, F. K., LUTZE, W. & MALOW, G. 1981. The mechanisms for hydrothermal leaching of glass and glass-ceramic nuclear waste forms. *In:* MOORE, J. G. (ed) *Scientific Basis for Nuclear Waste Management III.* Plenum Press, New York, USA, 363–370.

ATASSI, H. 1989. *Evaluation de la Résistance à la Corrosion en Solution Aqueuse de Quelques Verres Silicatés.* PhD thesis, University Louis Pasteur, Strasbourg, France.

BARIN, O. 1991. Behandlungsverfahren zur Inertisierung der Flug- und Filterstäube aus der Müllverbrennung. I. Verfestigungs-, physikalisch-chemische, Keramisierungs- Zementherstellungs-, Feuerungs- und Brennkammerverfahren. *Wissenschaft und Umwelt*, **3–4**, 159–167.

BARIN, O. 1992. Behandlungsverfahren zur Inertisierung der Flug- und Filterstäube aus der Müllverbrennung. II. Elektrothermische Verfahren und Diskussion. *Wissenschaft und Umwelt*, **1**, 93–98.

BATES, J. K., EBERT, W. L., MAZER, J. J., BRADLEY, J. P., BRADLEY, C. R. & DIETZ, N. L. 1991. The role of surface layers in glass leaching performance. *In*: ABRAJANO, T. A. & JOHNSON, L. H. (eds) *Scientific Basis for Nuclear Waste Management XIV. Materials Research Society Symposia Proceedings*, 212, 77–87.

BATES, J. K., BRADLEY, C. R. *et al.* 1994. *High-Level Waste Borosilicate Glass: A Compendium of Corrosion Characteristics.* Technical Report DOE-EM-0177, US-DOE, Washington, USA.

BATES, J. K., ELLISON, A. J. G., EMERY, J. W. & HOH, J. C. 1996. Glass as a waste form for the immobilization of plutonium. *In*: MURPHY, W. M. & KNECHT, D. A. (eds) *Scientific Basis for Nuclear Waste Management XIX. Materials Research Society Symposia Proceedings*, 412, 57–64.

BICKFORD, D. F. & JANZTEN, C. M. 1984. Devitrification behaviour of SRL defence waste glass. *In*: MCVAY, G. L. (ed) *Scientific Basis for Nuclear Waste Management VII. Materials Research Society Symposia Proceedings*, 26, 557–566.

BOTTERO, J.-Y., CHATELET, L. *et al.* 1997. Les déchets ultimes inertes peuvent-ils devenir des matériaux utiles? *In*: CASES, J.-M. & THOMAS, F. (eds) *Proceedings Procédés de Solidification et de Stabilisation des Déchets*, Nancy, France, 436–443.

BUCK, E. C., FORTNER, J. A., BATES, J. K., FENG, X., DIETZ, N. L., BRADLEY, C. R. & TANI, B. S. 1994. Analytical electron microscopy examination of solid reaction products in long-term tests of SRL 200 waste glasses. *In*: BARKATT, A. & VAN KONYNENBOURG, R. A. (eds) *Scientific Basis for Nuclear Waste Management XVII. Materials Research Society Symposia Proceedings*, 333, 585–593.

BUWAL 1990. *Eluat Test nach TVA.* BUWAL, Bern, Switzerland.

BUWAL 1996. *TVA/OTD: Technische Verordnung über Abfälle/Ordonnance sur le Traitement des Déchets, #814.015.* BUWAL, Bern, Switzerland.

BUWAL 1998. *Die Rückstände der Verbrennung; Flugasche und Filterkuchen. Umwelt-Materialien #100.* BUWAL, Bern, Switzerland.

CARROLL, S. A., BOURCIER, W. L. & PHILLIPS, B. L. 1994. Surface chemistry and durability of borosilicate glass. *In*: BARKATT, A. & VAN KONYNENBOURG, R. A. (eds) *Scientific Basis for Nuclear Waste Management XVII. Materials Research Society Symposia Proceedings*, 333, 533–540.

CASES, J.-M. & THOMAS, F. (eds) 1997. *Proceedings Procédés de Solidification et de Stabilisation des Déchets.* Nancy, France. Société Alpine de Publication, Grenoble, France.

CHARLES, R. J. 1958. Static fatigue of glass. *Journal of Applied Physics*, 29, 1549–1560.

CHERON, P., CHEVALIER, P. *et al.* 1995. Examination and testing of an active glass sample produced by COGEMA. *In*: MURAKAMI, T. & EWING, R. C. (eds) *Scientific Basis for Nuclear Waste Management XVIII. Materials Research Society Symposia Proceedings*, 353, 55–62.

CLARK, D. E., SCHULZ, R. L., WICKS, G. G. & LODDING, A. R. 1994. Waste glass alteration processes, surface layer evolution and rate limiting steps. *In*: MURAKAMI, T. & EWING, R. C. (eds)

Scientific Basis for Nuclear Waste Management XVIII. Materials Research Society Symposia Proceedings, 333, 107–122.

CLARK, D. E. & ZOITOS, B. K. 1992. *Corrosion of Glass, Ceramics and Ceramic Superconductors; Principles, Testing, Characterization and Applications.* Noyes Publications, Park Ridge.

COLOMBEL, P. 1996. *Etude du Comportement à Long Terme de Vitrifiats de REFIOM.* PhD thesis, University of Poitiers, France.

COLOMBEL, P., GODON, N., VERNAZ, E. & THOMASSIN, J.-H. 1997. Mécanismes d'altération des vitrifiats de REFIOM et analogie avec d'autres verres industriels ou naturels. *In*: CASES, J.-M. & THOMAS, F. (eds) *Proceedings Procédés de Solidification et de Stabilisation des Déchets*, Nancy, France, 450–454.

CROVISIER, J.-L. 1989. Verres basaltiques dans l'eau de mer et dans l'eau douce; essai de modélisation. PhD thesis. University of Strasbourg, Strasbourg, France.

CROVISIER, J.-L., HONNOREZ, J. & EBERHART, J.-P. 1987. Dissolution of basaltic glass in seawater: Mechanism and Rate. *Geochimica et Cosmochimica Acta*, 51, 2977–2990.

CROVISIER, J.-L., VERNAZ, E., DUSSOSSOY, J.-L. & CAUREL, J. 1992. Early phyllosilicates formed by alteration of R7T7 glass in water at 250 °C. *Applied Clay Science*, 7, 47–57.

CUNNANE, J. C. & ALLISON, J. M. 1994. High-level waste glass compendium; what it tells us concerning the durability of borosilicate waste glass. *In*: MURAKAMI, T. & EWING, R. C. (eds) *Scientific Basis for Nuclear Waste Management XVIII. Materials Research Society Symposia Proceedings*, 333, 3–14.

CUNNANE, J. C., BATES, J. K. *et al.* 1993. High-level nuclear-waste borosilicate glass. *In*: INTERRANTE, C. G. & PABALAN, R. T. (eds) *Scientific Basis for Nuclear Waste Management XVI. Materials Research Society Symposia Proceedings*, 294, 225–232.

DARAB, J. G., FENG, X., LINEHAN, J. C., SMITH, P. A. & ROTH, I. 1996. Composition–structure relationships in model Hanford low-level waste glasses. *Ceramics Transactions*, 72, 103–110.

DEPMEIER, L., TOMSCHI, U. & VETTER, G. 1997. Elutionsverhalten von Reststoffen aus der thermischen Abfallbehandlung. *Müll und Abfall*, 9, 528–533.

DIETZEL, A. H. 1988. The history of the structure of glass from the early days to present thinking. *In*: BICKFORD, D. F., BOULOS, E. N. *et al.* (eds) *Advances in the Fusion of Glass; Proceedings of the 1st International Conference on the Advances in the Fusion of Glass*, 1.1–1.7.

DRAN, J.-C., PETIT, J.-C., TROTIGNON, L., PACCAGNELLA, A. & DELLA MEA, G. 1989. Hydration mechanisms of silicate glasses: Discussion of the respective role of ion exchange and water permeation. *In*: LUTZE, W. & EWING, R. C. (eds) *Scientific Basis for Nuclear Waste Management XII. Materials Research Society Symposia Proceedings*, 127, 25–32.

EBERT, W. L. & MAZER, J. J. 1994. Laboratory testing of waste glass aqueous corrosion: effects of experimental parameters. *In*: BARKATT, A. & VAN KONYNENBOURG, R. A. (eds) *Scientific Basis for Nuclear Waste Management XVII. Materials Research Society Symposia Proceedings*, **333**, 27–40.

EKESA 1992. *Emissionsabschätzung für Kehrichtschlacke; Projekt EKESA; Schlussbericht. Auftraggebergemeinschaft für das Projekt EKESA.* Switzerland.

EVANS, R. 1989. Characterization of BNFL glasses containing simulated high-level waste. *Proceedings Testing of High-Level Waste Forms under Repository Conditions.* Cadarache, France, 106–114.

EWING, R. C. 1996. *Glass as a Waste Form and Vitrification Technology: Summary of an International Workshop.* National Research Council, National Academy Press, Washington.

FENG, X., CUNNANE, J. C. & BATES, J. K. 1994. A literature review of surface alteration layer effects on waste glass behavior. *Ceramics Transactions*, **39**, 341–352.

FILLET, S. 1987. *Mécanismes de Corrosion et Comportement des Actinides dans le Verre Nucléaire R7T7.* PhD thesis, University Montpellier II, Montpellier, France.

FINET, C. 1994. La vitrification: un procédé de banalisation des résidus de l'épuration des fumées d'incinération des ordures ménagères. *TSM*, **4**, 196–198.

FLINTOFF, J. F. & HARKER, A. B. 1985. Detailed processes of surface layer formation in borosilicate waste glass dissolution. *In*: JANTZEN, C. M., STONE, J. A. & EWING, R. C. (eds) *Scientific Basis for Nuclear Waste Management VIII. Materials Research Society Symposia Proceedings*, **44**, 147–154.

GARLAND, J. A. & WHITE, W. B. 1985. Determination of early stages of glass dissolution by pH titration. *In*: JANTZEN, C. M., STONE, J. A. & EWING, R. C. (eds) *Scientific Basis for Nuclear Waste Management VIII. Materials Research Society Symposia Proceedings*, **44**, 81–88.

GILLIES, K. J. S. & COX, G. A. 1988. Decay of medieval stained glass at York, Canterbury and Carlisle; II. Relationship between the composition of the glass, its durability and the weathering products. *Glasstechnische Berichte*, **61**, 101–107.

GÍSLASON, S. R., VEBLEN, D. R. & LIVI, K. J. T. 1993. Experimental meteoric water–basalt interactions: Characterization and interpretation of alteration products. *Geochimica et Cosmochimica Acta*, **57**, 1459–1471.

GONG, W. L., EWING, R. C., WANG, L. M., VERNAZ, E., BATES, J. K. & EBERT, W. L. 1996. Secondary phase formation and the microstructural evolution of surface layers during vapor phase alteration of the French SON68 nuclear waste glass at 200°C. *In*: MURPHY, W. M. & KNECHT, D. A. (eds) *Scientific Basis for Nuclear Waste Management XIX. Materials Research Society Symposia Proceedings*, **412**, 197–204.

GRAMBOW, B. 1982. The role of metal ion solubility in leaching of nuclear waste glasses. *In*: LUTZE, W. (ed) *Scientific Basis for Nuclear Waste Management V. Materials Research Society Symposia Proceedings*, **11**, 93–102.

GRAMBOW, B. 1985. A general rate equation for nuclear waste glass corrosion. *In*: JANTZEN, C. M., STONE, J. A. & EWING, R. C. (eds) *Scientific Basis for Nuclear Waste Management VIII. Materials Research Society Symposia Proceedings*, **44**, 16–27.

GRAMBOW, B. 1991. What do we know about nuclear waste glass performance in the repository near field? *In*: SELLIN, P., APTED, M. & GAGO, J. (eds) *Proceedings Technical Workshop on Near-Field Performance Assessment for High-Level Wastes, Madrid, Spain.* SKB Technical Report #91-59. Stockholm, Sweden, 25–49.

GRAMBOW, B. 1992. Geochemical approach to glass dissolution. *In*: CLARK, D. E. & ZOITOS, B. K. (eds) *Corrosion of Glass, Ceramics and Ceramic Superconductors; Principles, Testing, Characterization and Applications.* Noyes, Park Ridge, USA, 124–152.

GRAMBOW, B. 1994. Remaining uncertainties in predicting long-term performance of nuclear waste glass from experiments. *In*: BARKATT, A. & VAN KONYNENBOURG, R. A. (eds) *Scientific Basis for Nuclear Waste Management XVII. Materials Research Society Symposia Proceedings*, **333**, 167–180.

GRAUER, R. 1985. *Synthesis of Recent Investigations on Corrosion Behaviour of Radioactive Waste Glasses.* Technical Report 85-27. Nagra, Würenlingen, Switzerland.

GUTMANN, R. 1996. Thermal technologies to convert solid waste residuals into technical glass products. *Glastechnische Berichte*, **69**, 285–299.

GUYONNET, D., CÔME, B., OUVRY, J.-F. & BARRÈS, M. 1998. Concepts de stockage de déchets; un essai de définition dans une logique d'impact. *Déchets Sciences et Techniques*, **9**, 39–44.

HAAKER, R., MALOW, G. & OFFERMANN, P. 1985. The effect of phase formation on glass leaching. *In*: JANTZEN, C. M., STONE, J. A. & EWING, R. C. (eds) *Scientific Basis for Nuclear Waste Management VIII. Materials Research Society Symposia Proceedings*, **44**, 121–128.

HENCH, L. L., WERME, L. & LODDING, A. 1982. Burial effects on nuclear waste glass. *In*: LUTZE, W. (ed) *Scientific Basis for Nuclear Waste Management V. Materials Research Society Symposia Proceedings*, **11**, 153–162.

HNAT, J. G. & BARTONE, L. M. 1996. Recycling of boiler and incinerator ash into value added glass products. *In*: *Proceedings 17th Biennal Waste Processing Conference*, ASME, 129–142.

JACQUET-FRANCILLON, N., PACAUD, F. & QUEILLE, P. 1982. An attempt to assess the long-term crystallization rate of nuclear waste glasses. *In*: LUTZE, W. (ed.) *Scientific Basis for Nuclear Waste Management V. Materials Research Society Symposia Proceedings*, **11**, 249–259.

JANTZEN, C. M. 1984. Effects of Eh (oxidation potential) on borosilicate waste glass durability. *Advances in Ceramics*, **8**, 385–393.

JANTZEN, C. M. 1988. Prediction of glass durability as a function of glass composition and test conditions: thermodynamics and kinetics. *In*: BICKFORD, D. F., BOULOS, E. N. *et al.* (eds) *Advances in the Fusion of Glass; Proceedings of the 1st International Conference on the Advances in the Fusion of Glass*, 24.1–24.17.

JANTZEN, C. M. & PLODINEC, M. J. 1984. Thermodynamic model of natural, medieval and nuclear waste glass durability. *Journal of Non-Crystalline Solids*, **67**, 207–223.

JERCINOVIC, M. J., KASER, S. & EWING, R. C. 1989. Observations of surface layers formed on basaltic and borosilicate glass: 6 month and 1 year MIIT experiments. *In*: *Proceedings of the Workshop on Testing of High Level Waste Forms under Repository Conditions*. EUR 12017 EN, the Commission of the European Communities, 183–191.

JOLLIVET, P., NICOLAS, M. & VERNAZ, E. 1998. Estimating the alteration kinetics of the French vitrified high-level waste package in a geologic repository. *Nuclear Technology*, **123**, 67–81.

KANCZAREK, A. & GROSSE-HOLZ, G. 1996. Das Siemens-KWU-Schwel-Brenn-Verfahren. *Schriftenreihe WAR*, **88**, 77–111.

KRAUS, F. & MEUNIER, R. 1997. Propriétés d'un vitrifiat de REFIOM produit par un procédé à arc électrique. *Verre*, **3**, 22–26.

KÜNSTLER, H., KLUKOWSKI, C. & GROTEFELD, V. 1994. Der VS-Kombi-Reaktor der Firma Küpat. *Müll und Abfall, Beiheft*, **31**, 67–72.

LEE, C. T. & CLARK, D. E. 1985. Electrokinetics, adsorption and colloid study of simulated nuclear waste glasses leached in aqueous solutions. *In*: JANTZEN, C. M., STONE, J. A. & EWING, R. C. (eds) *Scientific Basis for Nuclear Waste Management VIII. Materials Research Society Symposia Proceedings*, **44**, 221–228.

LINARD, Y. 2000. *Détermination des Enthalpies Libres de Formation des Verres Borosilicatés; Application à l'Étude de l'Altération des Verres de Confinement de Déchets Radioactifs*. PhD thesis, University Paris VII, Paris, France.

LUO, J. S., EBERT, W. L., MAZER, J. J. & BATES, J. K. 1997. Simulation of natural corrosion by vapor hydration test: Seven-year results. *In*: GRAY, W. J. & TRIAY, I. R. (eds) *Scientific Basis for Nuclear Waste Management XX. Materials Research Society Symposia Proceedings*, **465**, 157–163.

LUTZE, W. & EWING, R. C. 1988. Summary and evaluation of nuclear waste forms. *In*: LUTZE, W. & EWING, R. C. (eds) *Radioactive Waste Forms for the Future*. North-Holland, Amsterdam, The Netherlands, 699–740.

MACQUET, C. & THOMASSIN, J. H. 1992. Archaeological glasses as modelling of the behaviour of buried nuclear waste glasses. *Applied Clay Science*, **7**, 17–31.

MALOW, G. 1982. The mechanisms for hydrothermal leaching of nuclear waste glasses: properties and evaluation of surface layers. *In*: LUTZE, W. (ed)

Scientific Basis for Nuclear Waste Management V. Materials Research Society Symposia Proceedings, **11**, 25–36.

MÉHU, J. 1988. Comportement à long terme et éco-compatibilité des déchets ultimes stabilisés; vers une stratégie environnementale unifiée. *Déchets Sciences et Techniques*, **9**, 54–55.

MOTTL, M. J. & HOLLAND, H. D. 1978. Chemical exchange during hydrothermal alteration of basalt by seawater. I. Experimental results for major and minor components of seawater. *Geochimica et Cosmochimica Acta*, **42**, 1103–1115.

MURAKAMI, T., BANBA, T., JERCINOVIC, M. J. & EWING, R. C. 1989. Formation and evolution of alteration layers on borosilicate and basalt glasses: Initial stage. *In*: LUTZE, W. & EWING, R. C. (eds) *Scientific Basis for Nuclear Waste Management XII. Materials Research Society Symposia Proceedings*, **127**, 65–72.

NEWTON, R. G. 1985. The durability of glass. *Glass Technology*, **26**, 21–38.

NEWTON, R. G. & PAUL, A. 1980. A new approach to predicting the durability of glasses from their chemical compositions. *Glass Technology*, **21**, 307–309.

NOGUES, J.-L. 1984. *Les Mécanismes de Corrosion des Verres de Confinement des Produits de Fission*. PhD thesis, University Montpellier II, Montpellier, France.

NOGUES, J.-L., VERNAZ, E. Y. & JACQUET-FRANCILLON, N. 1985. Nuclear glass corrosion mechanisms applied to the French LWR reference glass. *In*: JANTZEN, C. M., STONE, J. A. & EWING, R. C. (eds) *Scientific Basis for Nuclear Waste Management VIII. Materials Research Society Symposia Proceedings*, **44**, 89–98.

O'CONNOR, W. K., ODEN, L. L. & TURNER, P. C. 1994. Vitrification of municipal waste combustor residues: Physical and chemical properties of electric arc furnace feed and products. *Process Mineralogy*, **XII**, 17–37.

O'KEEFE, J. A. 1984. Natural glass. *Journal of Non-Crystalline Solids*, **67**, 1–17.

PATZE, H. U. 1996. Das Noell-Konversionsverfahren zur Umwelt-Freundlichen thermischen Verwertung von Rest- und Abfallstoffen. *Schriftenreihe WAR*, **88**, 113–129.

PAUL, A. 1977. Chemical durability of glasses. *Journal of Materials Science*, **12**, 2246–2268.

PAUL, A. 1982. *Chemistry of Glasses*. Chapman and Hall, London.

PEDERSON, L. R., BAER, D. R., MCVAY, G. L. & ENGELHARD, M. H. 1986. Reaction of soda-lime silicate glass in isotopically labelled water. *Journal of Non-Crystalline Solids*, **86**, 369–380.

PERRET, D., STILLE, P., SHIELDS, G., CROVISIER, J.-L. & MÄDER, U. 2000. *Long-Term Stability of HT Materials; Report 4. SAEFL, Section Wastes, Switzerland*. The CD-ROM 'Long-Term Stability of HT Materials. A Compendium of the Static, Dynamic, Thermodynamic Pictures of Products from the High-Temperature Treatment of Municipal Solid Wastes and Associated Residues' contains the technical report and the complete set of results; it is available upon request to the Swiss Agency for Environment, Forests and Landscape; Section Wastes; CH-3003 Berne, Switzerland.

PERRET, D., SCHENK, K., CHARDONNENS, M., STILLE, P. & MÄDER, U. (2002). Characteristics, behavior and durability of high-temperature materials. *In*: LUDWIG, C., HELLWEG, S. & STUCKI, S. (eds) *Municipal Solid Waste Management: Strategies and Technologies for Sustainable Solutions*. Springer Verlag, Berlin, Germany, 257–293.

PERRET, D., CROVISIER, J.-L. *et al.* 2003. Thermodynamic stability of waste glasses compared to leaching behaviour. *Applied Geochemistry*, **18**, 1165–1184.

PETIT, J.-C., DELLA MEA, G., DRAN, J.-C., MAGONTHIER, M.-C., MANDO, P. A. & PACCAGNELLA, A. 1990. Hydrated layer formation during dissolution of complex silicate glasses and minerals. *Geochimica et Cosmochimica Acta*, **54**, 1941–1955.

PLODINEC, M. J. 1984. Stability of radioactive waste glasses assessed from hydration thermodynamics. *In*: MCVAY, G. L. (ed) *Scientific Basis for Nuclear Waste Management VII. Materials Research Society Symposia Proceedings*, **26**, 755–762.

PLODINEC, M. J. & WICKS, G. G. 1994. Application of hydration thermodynamics to in-situ test results. *In*: BARKATT, A. & VAN KONYNENBOURG, R. A. (eds) *Scientific Basis for Nuclear Waste Management XVII. Materials Research Society Symposia Proceedings*, **333**, 145–157.

RESCE, J. L., RAGSDALE, R. G. & OVERCAMP, T. J. 1996. Vitrification of an incinerator blowdown waste containing both chloride salts and carbon. *Journal of Environmental Science and Health*, **A31**, 2381–2393.

SCHOLZE, H. 1991. *Glass, Nature, Structure, and Properties*. Springer-Verlag, New York.

SCHOLZE, H., CONRADT, R., ENGELKE, H. & ROGGENDORF, H. 1982. Determination of the corrosion mechanisms of high-level waste containing glass. *In*: LUTZE, W. (ed) *Scientific Basis for Nuclear Waste Management V. Materials Research Society Symposia Proceedings*, **11**, 173–180.

SCHUMACHER, W. & GUGAT, J.-A. 1994. EloMelt- und FosMelt-Verfahren thermische Behandlungskonzepte für Reststoffe aus der Müllverbrennung. *Müll und Abfall, Beiheft*, **31**, 152–156.

SHUTTLEWORTH, S. & MONTEITH, J. E. 1997. The use of laser microprobe-ICP-MS in the examination of nuclear waste glasses. *In*: GRAY, W. J. & TRIAY, I. R. (eds) *Scientific Basis for Nuclear Waste Management XX. Materials Research Society Symposia Proceedings*, **465**, 123–130.

SPROULL, J. F., MARRA, S. L. & JANTZEN, C. M. 1994. High-level radioactive waste glass production and product description. *In*: BARKATT, A. & VAN KONYNENBOURG, R. A. (eds) *Scientific Basis for Nuclear Waste Management XVII. Materials Research Society Symposia Proceedings*, **333**, 15–25.

STAHLBERG, R. 1996. Das Thermoselect-Verfahren. *Schriftenreihe WAR*, **88**, 131–142.

STANWORTH, J. E. 1950. *Physical Properties of Glass*. Clarendon Press, Oxford, England.

STERPENICH, J. 1998. *Altération des Vitraux Médiévaux, Contribution à l'Étude du Comportement à Long Terme des Verres de Confinement*. PhD thesis, Université Henri Poincaré, Nancy.

STRACHAN, D. M., PEDERSON, L. R. & LOKKEN, R. O. 1985. Results from the long-term interaction and modeling of SRL-131 glass with aqueous solutions. *In*: WERME, L. O. (ed) *Scientific Basis for Nuclear Waste Management IX. Materials Research Society Symposia Proceedings*, **50**, 195–202.

THOMASSIN, J.-H. 1995. *Apport des Analogues Naturels, Industriels ou Archéologiques à la Connaissance du Comportement à Long Terme des Vitrifiats de Déchets Toxiques*. Rapport final. University of Poitiers, France.

THOMASSIN, J.-H. 1996. Rapport CRP OESIP 96-1. University of Poitiers, Poitiers, France.

THOMÉ-KOZMIENSKY, K. J. 1994. *Thermische Abfallbehandlung*. EF-Verlag für Energie und Umwelttechnik, Berlin, Germany.

VERNAZ, E. & DUSSOSOY, J.-L. 1992. Current state knowledge of nuclear waste glass corrosion mechanisms: The case of R7T7 glass. *Applied Geochemistry* (Supplement Issue), **1**, 13–22.

VOGEL 1992. *Glasschemie*. Springer Verlag, Berlin, Germany.

WHITEHEAD, N. E., SEWARD, D. & VESELSKY, J. 1993. Mobility of trace elements and leaching rates of rhyolitic glass shards from some New Zealand tephra deposits. *Applied Geochemistry*, **8**, 235–244.

YAN, J. & NERETNIEKS, I. 1995. Is the glass phase dissolution rate always a limiting factor in the leaching processes of combustion residues? *The Science of the Total Environment*, **172**, 95–118.

YANAGISAWA, F. & SAKAI, H. 1988. Leaching behaviour of a simulated nuclear waste glass in groundwater of 50–240 °C. *Applied Geochemistry*, **3**, 153–163.

ZACHARIASEN, W. H. 1932. The atomic arrangement in glass. *Journal of the American Chemical Society*, **54**, 3841–3851.

ZELLMER, L. A. & WHITE, W. B. 1985. Characterization of hydrated surface layers on nuclear waste glasses by infrared reflectance spectroscopy. *In*: JANTZEN, C. M., STONE, J. A. & EWING, R. C. (eds) *Scientific Basis for Nuclear Waste Management VIII. Materials Research Society Symposia Proceedings*, **44**, 73–80.

ZEVENBERGEN, C., BRADLEY, J. P., VANDER WOOD, T., BROWN, R. S., VAN REEUWIJK, L. P. & SCHUILING, R. D. 1994a. Microanalytical investigation of mechanisms of municipal solid waste bottom ash weathering. *Microbeam Analysis*, **3**, 125–135.

ZEVENBERGEN, C., VANDER WOOD, T., BRADLEY, J. P., VAN DER BROECK, P. F. C. W., ORBONS, A. J. & VAN REEUWIJK, L. P. 1994b. Morphological and chemical properties of MSWI bottom ash with respect to the glassy constituents. *Hazardous Waste and Hazardous Materials*, **11**, 371–383.

Chemistry and mineralogy of municipal solid waste incinerator bottom ash

URS EGGENBERGER[1], KAARINA SCHENK[2] & URS MÄDER[1]

[1]Institut for Geological Sciences, University of Bern, Bern, Switzerland
(e-mail: eggenberger@geo.unibe.ch)

[2]Swiss Agency for Environment, Forests and Landscape, Division Waste,
Bern, Switzerland

Abstract: A petrographic study was conducted on a suite of bottom ash samples from 28 municipal solid waste incinerators in Switzerland. Chemical and mineralogical analyses of bottom ash from waste combustors with comparable technology show similar major oxide composition. A significant decrease of SiO_2 was observed in comparison to the chemical composition of bottom ash from 10 years ago. In contrast to major oxide contents, heavy metal concentrations vary significantly in the bottom ash samples, but without any correlation to the type of input waste or plant operating conditions. Similar types and contents of crystalline phases were observed in all samples. The content of newly formed melilite increases with decreasing bulk Si/Ca ratio, indicating that the type of Ca–Mg–Al-silicates crystallizing during incineration and cooling follow petrogenetic rules. A considerable recycling potential for ferrous and non-ferrous metals was identified in the bottom ash. Optimized mechanical metal separation technologies could reduce the waste volume, heavy metal content, H_2 production, and exothermic reactions of the bottom ash in landfills and might be economically viable.

The objectives of municipal solid waste (MSW) incineration are the reduction of waste volume (~90%) and mass (~70%), the destruction of organic compounds, the separation of volatile metals into the filter residues, the inertization of metals in a glassy and/or crystalline matrix and the use of its energy (~12 MJ/kg). Bottom ash (BA) is, with about 80%, the most abundant waste material of all MSWI residues (fly ash, air pollution control residues, mud from waste water purification). In the European Union (EU) approximately 11 Mt and in the USA 8 Mt of MSWI bottom ash are produced per year (Wiles 1996), in Switzerland approximately 100 kg of BA per year and citizen. Strategies for further volume reduction of MSWI residues and/or re-use of inorganic components are either increasing the recycling rate of waste materials through separate collection prior to incineration, or optimizing the residue quality by pyrometallurgical or mechanical separation.

The recycling rate has already reached a high level in many countries (e.g., ~45% Switzerland) and can hardly be increased. In the 1990s many investigations were therefore performed in the field of new high-temperature processes like Ash-Arc, Siemens KWU, Thermoselect,

HSR and others (Perret *et al.* 2003; Zeltner & Lichtensteiger 2001). These high-temperature processes usually operate above the liquidus temperature of a silicate melt under reducing conditions to enable pyrometallurgic separation and to produce a more homogeneous glass ceramic or glassy slag that is less soluble than BA from common MSWI systems. The high-temperature treatment processes cause a significant decrease in Pb, Cd, S, and inorganic carbon contents compared to conventional MSW grate-firing residues, but some of the processes show only minor depletion in Zn and Ni (Traber *et al.* 2002). To date none of these processes has however been able to produce solid residues with trace metal contents equal to the Earth's crust composition. Whether these high-temperature materials can be used as secondary raw materials depends on the specific procedure of thermal treatment (Zeltner & Lichtensteiger 2001; Kruspan 2000) and on the legislation of the specific countries. Technical, economic and acceptance problems are additional reasons why only a few high-temperature (HT) plants have been built worldwide. High-temperature incineration processes were developed (HSR; see Mayer & Rey 1996) or evaluated in Switzerland

From: GIERÉ, R. & STILLE, P. (eds) 2004. *Energy, Waste, and the Environment: a Geochemical Perspective*. Geological Society, London, Special Publications, **236**, 411–422.
0305-8719/04/$15 © The Geological Society of London 2004.

(Siemens KWU, Thermoselect) but no HT plant has been realized until now.

From the lifetime cycle of a MSWI plant of about 30–40 years it is obvious that at least in the next two decades conventional MSWI BA will remain the dominant residue. In some European countries (e.g., The Netherlands, Germany, or France) legislation allows MSW BA to be used as secondary raw material, mainly for road construction. Some investigations proposed the use of MSW BA in concrete or other cementitious products (e.g., Targan *et al.* 2002) or as sintering promoters in porcelainized stoneware (Barbieri *et al.* 2002).

In Switzerland, where direct landfill of MSW is prohibited, all MSW has to be burned in incinerators and the bottom ash has to be stored in special controlled landfill sites (so called 'reactor deposits'). Re-use of today's BA is no longer practised because such BA is a heterogeneous 'random product' (Lichtensteiger 1996) and elevated heavy metal concentrations in future leachates cannot be excluded (e.g., Stäubli 1992).

The Swiss Agency of Environment and Landscape decided to re-evaluate mechanical metal separation techniques to reduce the landfill volume and to improve the BA quality for deposition by exploiting metal resources. A sampling campaign in all 28 incineration plants was initiated to verify BA quality and to establish a solid data base. A common sampling procedure, sample treatment, and analytical method were prescribed in order to obtain consistent information of chemical and structural composition and the leaching behaviour of current BA. This paper is focused on the chemical and mineralogical results from the study.

Materials and methods

A sampling procedure was defined that had to be applied by all incineration plants and that had to fulfill the basic requirement of being representative. All samples had to be taken during the same season. In the most important plant, waste and firing parameters were also collected.

Six samples were taken daily during three non-consecutive days within one week to achieve an average weekly composition. The sample size of every subsample of BA had to be defined based on considerations of a particulate pollutant (e.g., heavy metals) concentration rather than that of a diffusive pollutant distribution. A sample size reduction scheme (Fig. 1) was defined according to the study of Bunge & Bunge (1999) depending on grain size and concentration of particulate pollutants. From the 18 collected subsamples,

each about 20 kg, metal pieces were separated by the plant operators and weighed separately. The mixed sample was then divided four times using cone debris sample reduction and sent to a brick factory for air drying at 105 °C to constant weight in an industrial drying oven. The grain size fractions of so-called slag (>2 mm) and ash (<2 mm) were determined according to the study of Lichtensteiger (1996), and again metal pieces were separated and weighed. The sample was then crushed twice, to 1 cm and 0.2 cm, and the sample size was again reduced according to Fig. 1, using a sample splitter. The remaining sample was ground in a tungsten carbide mill and split again. About 300 g were finely ground (<5 μm) in a tungsten carbide mill for further analyses.

Quantitative bulk-chemical analyses were carried out by X-ray fluorescence analysis (XRF), identification and quantification of crystalline phases by X-ray diffraction (XRD, using LiF as internal standard), Cu, Cd and Pb contents determined by acid digestion and atomic absorption spectroscopy (AAS), and C and S were measured as CO_2 and SO_2, respectively, using infrared spectroscopy. Polished thin sections of resin-embedded slag samples were used for examination by optical microscopy in transmitted and reflected light. Carbon-coated thin sections were used for investigations with a CamScan scanning electron microscope (SEM) equipped with an energy-dispersive spectrometer (EDS).

Results

Plant-specific parameters and general characterization of BA

Although the incinerator lines of the 28 plants differ greatly in their age (2–40 years) and capacity (9000–124 000 t/y), all incinerators show similar burning technology (reciprocating or travelling grate) and similar four-stage air pollution control systems. The plants in Switzerland together produce about 4800 TJ of electrical energy, which corresponds to the energy consumption of about 250 000 households and an additional 9600 TJ of thermal energy.

The average waste composition comprises 53% waste from households, 19% from manufactories, 18% from industrial plants, and 10% from the building material sector. In total, 3 Mt MSW are incinerated and about 650 000 tons of BA are produced per year. The relative mass of the BA are 23% and this proportion is very constant in all incinerator plants, as shown in Fig. 2.

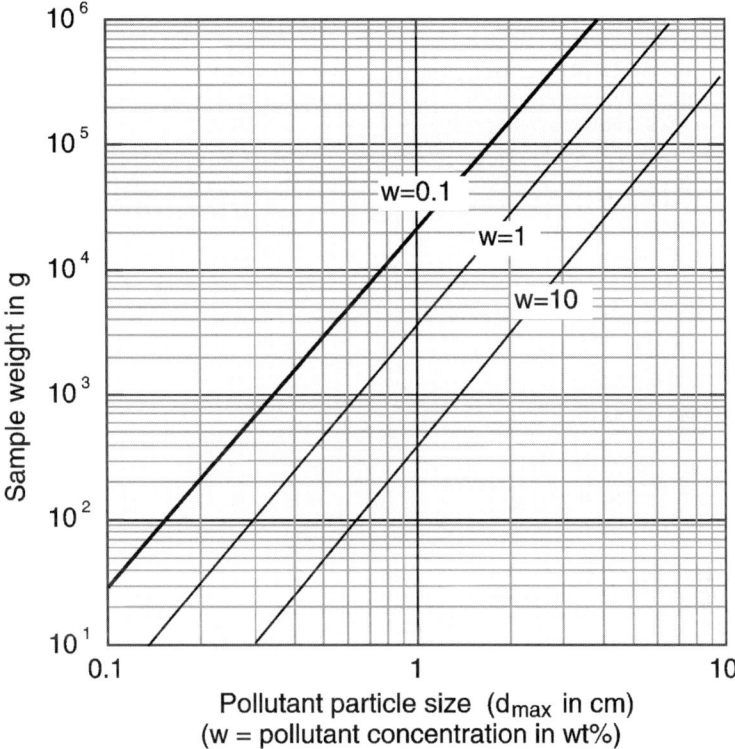

Fig. 1. Sample size reduction scheme, considering particle size of particulate pollutants (e.g., Pb as metal droplet) and pollutant concentration of the bulk sample.

Large differences exist from a macroscopic point of view regarding BA quality. Bottom ash from newer plants shows generally a smaller grain size distribution, lighter colour, and fewer unburnt organic components (Fig. 3, top). Microscopic investigations revealed a more homogeneous fabric of finer glassy and partially recrystallized particles from BA of newer plants in comparison to BA from older plants, where large differences in the mm to cm scale of regions with glassy and only partially molten components and particles were observed (Fig. 3). This is most likely due to the better homogenization of waste on the grate in newer plants.

The average macroscopic composition of all BA samples, in comparison to older studies from some of the same plants, revealed similar results to those shown in Table 1 (Lichtensteiger & Zeltner 1993). Well-established selective collection programmes for glass (80% recycling rate) lead to a lower glass fraction (2%) and also to a macroscopically visible decrease in metal content (3–4%). Because the amount of each individual component, such as glass, ceramics, stones, metals, and organic material, is rather small (<15%), its

fraction in all BA samples does not differ significantly. However, the relative amounts of slag (>2 mm) and ash (<2 mm) vary strongly throughout all samples (Fig. 4). Therefore, it is not surprising that the slag/ash ratios are also different in BA from other studies of specific incinerator plants (e.g., Chimenos et al. 1999).

It was not intended in this study to describe the BA in microscopic detail. Because only two to three intact particles bigger than 10 mm from every BA were examined, the observations may not be representative for the entire BA sample. Classification schemes were used in previous studies (e.g., Pfrang-Stotz & Schneider 1995; Chandler et al. 1997) where BA from a number of European combustors were classified into (1) glassy slags, (2) porous slags, and (3) microcrystalline slags. In some of our samples all three classes were observed in the same BA, others showed only type (1) and (3). It is not yet clear if this is a result of a sample mixture of slightly different combustion conditions during the three days of sampling or if local differences of input material and/or other conditions on the grate may be the reason. Eusden et al. (1999)

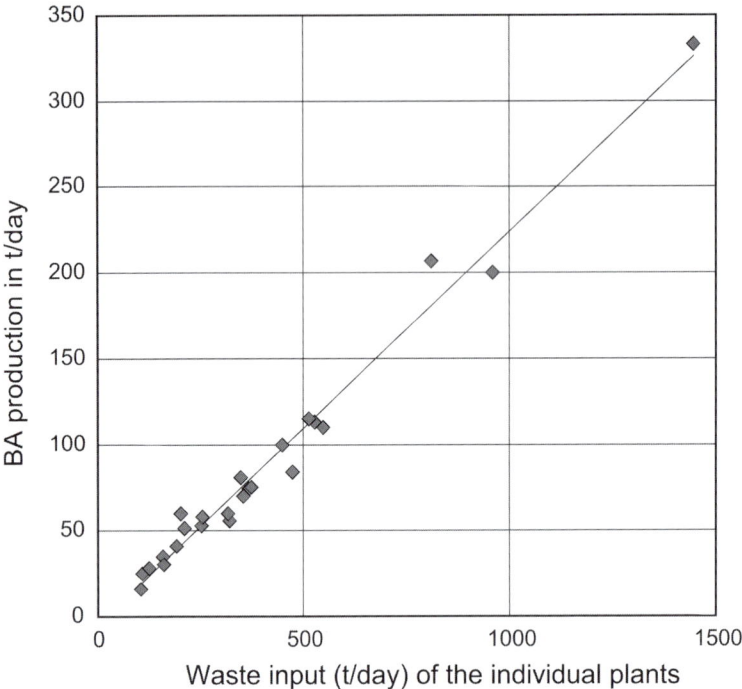

Fig. 2. Production of BA in comparison to the amount of waste input is, at 23%, very constant for all plants.

conducted a detailed petrographic study of BA from three different MSWI plants in the USA. They divided the BA constituents into two major groups, namely refractory (rock fragments, metals, etc.) and melt products (see also Lichtensteiger 1996). For the melt products Eusden *et al.* (1999) found two main types: (1) isotropic glass with complex silicate minerals produced at very high temperature (1500–1600 °C), and (2) opaque glass with metal oxides and crystallized silicate minerals from lower temperature melts (1150–1400 °C). In our samples it was not always possible to distinguish clearly between these two types of glass, but it was evident that the glassy constituents were similar in morphology and major element chemistry to some volcanic ashes, as concluded already by Zevenbergen *et al.* (1994). It is also not to be expected that the burning technology of our types of incinerators produce temperatures above 1100–1200 °C on the grate, because measured gas atmosphere temperatures are between 800 and 950 °C. The same temperature range was estimated by Pfrang-Stotz & Schneider (1995) for three German combustors.

Separated metal content

Ferrous and non-ferrous metals in BA are not only sources for potentially released heavy metals,

such as Cu, Zn, Pb, Cr, or Mn, but contribute also at different stages to the reactivity of the BA in landfills. The oxidation of ferrous metals contributes significantly to heat production, where Fe-oxides and hydroxides are formed. Oxidation of Al-metals may produce H_2 gases (Förstner 1996) and Al-hydroxides and sulphates are formed that influence pore water composition and reactivity. These reactions take place during many years as shown, for example, by Johnson *et al.* (1995). Metal fractions were separated by hand and magnetic separators at different stages of sampling and sample treatment. Silicate slag material is attached to finer grained metal pieces, and on some samples it was shown that the relative amount of metals may be overestimated by about 12–20 wt% with an increasing portion toward finer metal fractions.

Presently an average of 6 wt% is already separated by plant operators in Switzerland. However, there exist large differences between the amount of metals separately collected by the individual plant operators, covering a range of 0–15%. A further 10% of metals were separated during sample crushing and reduction. Assuming that 15% slag material is attached to the separated metals, today about 8–9% of ferrous and non-ferrous metals are still landfilled.

Further mechanical metal separation requires crushing of the BA, whereby grain size is

Fig. 3. Images of BA from two MSWI plants: on the left side BA from a recently built plant and on the right side from a very old plant. Bottom: thin section images of the corresponding BA (transmitted light, size 4 × 1.5 cm).

reduced, specific surface area enlarged and a more reactive BA is produced. Therefore it has to be demonstrated that the decrease in reactivity through metal separation balances the increase in leachability of heavy metals due to the resultant finer grain size distribution of the BA in a land-fill. It should also be tested if the increased binding capacity from improved hydraulic and pozzolanic properties allows the produc-tion of self-stabilizing blocks for deposition. Experiments where the effect of finely ground BA (<150 μm) and portlandite cement mix-tures were investigated showed that the BA

Table 1. *Macroscopic composition of BA constituents in comparison to a study from ten years ago (Lichtensteiger & Zeltner 1993)*

Source (vol%)	This study	Lichtensteiger & Zeltner, 1993
Ash	41	45
Melt	47	40
Metal	3–4	5
Ceramic	2	1–3
Glass	2	5
Stones	1	1
Organics	1	1–3

contributes only to a relatively minor chemical reactivity (Filipponi *et al.* 2003) and exhibits only a weak pozzolanic activity (Giampaolo & Lo Mastro 2002).

Chemical composition

All consecutive chemical and mineralogical data refer to BA without the amount of separated metals as described above. The chemical compo-sition of major elements shows, in terms of igneous rock classification, an ultrabasic (<45% SiO_2) rather than a basic (45–52% SiO_2) compo-sition (see also Kruspan 2000), but which is miner-alogically more akin to basalts. In the last ten years the SiO_2 content has decreased significantly, from 51.4% in 1986 (Belevi *et al.* 1992) to 42.6%, due to separate glass collection and new treatment technologies for contaminated soil material. The amount of CaO and P_2O_5 is about twice as high in BA, but SiO_2 shows lower and MgO signifi-cantly lower amounts in comparison to naturally occurring basic rocks (e.g., basalts). Such a com-parison, however, has to be carried out with caution, because naturally occurring basic melts would produce a completely different mineral assemblage. Normative calculations, such as the CIPW norm (Kelsey 1965) show that all of

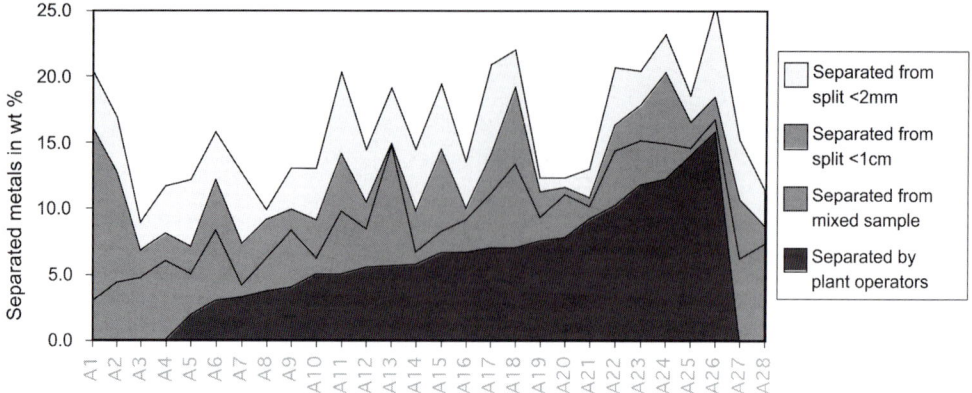

Fig. 4. Separated metal fractions from different grain size classes during sample reduction (by operators: metal >5 cm; from mixed sample: metal >1 cm; from split <1 cm: metal from 1 cm to 2 mm; from BA split <2 mm: metal <2 mm).

our BA are quartz-undersaturated. This indicates that quartz would not crystallize from natural magmas with a BA melt composition. Because BA always contains refractory components (quartz grains, ceramics, etc.), which show SiO_2 contents conspicuously higher than the whole BA composition, the partially molten or recrystallized areas must again show a more basic composition.

In general, comparison of chemical BA bulk composition with properties of naturally occurring igneous rocks may lead to some misleading interpretations; for example, the calculated viscosities of the BA according to Shaw (1972) shows at 1100 °C a range of 1.4–407 poise. However, BA from different plants do not show such big differences in the flow behaviour on the grate, because local melting processes with their adjoining solid particles are endothermal reactions and the partial melts show temperatures only a little above the solidus and remain therefore highly viscous.

In terms of a petrologic description, BA represents more closely a type of rock of detritic origin composed of phases that had undergone very different conditions of formation, and that subsequently was subject to a short low-pressure/high-temperature metamorphism where only partial melts occurred.

The use of the BA as a potential secondary raw material such as cement additive has often been discussed (Pera *et al.* 1997; Pecqueur *et al.* 2001; Targan *et al.* 2002; Filipponi *et al.* 2003). The composition of the major oxides Al + Fe, Si and Ca of the BA is in the range of substitutes already in use, such as sewage sludge or coal fly ash (Fig. 5). However, as will be shown later, elevated heavy metal contents are, from our

point of view, the limiting factor for their use as substitute material.

The variation of the chemical composition of major elements of the 28 BA samples is relatively small, as shown in Fig. 6. The box represents the interquartile distance (IQD) from the 25th to the 75th percentile and the vertical lines represent the range of composition that are not defined as outliers (bigger than $Q75 + 1.5 * IQD$, or less than $LQ-1.5 * IQD$). This indicates that the input composition of major oxides of a weekly averaged sample does not differ significantly among various incinerator plants located in very different regions, from urban to more rural regions.

The content of organic carbon has a significant influence on chemical and biological processes in landfills, also in a long-term context. Microbial degradation may favour carbonation and a decrease in pH and, therefore, accelerate the mobilization of heavy metals. Organic carbon may also influence the complexation and mobilization of metals with organic ligands (van der Sloot 1996; Meima & Comans 1999). Detailed studies of the speciation of the organic carbon in MSWI BA showed that variable amounts of carbon species are found, mainly depending on the variable degree of thermal decomposition of the organic waste material on the grate. Non-extractable organic carbon was found as an indicator for an incomplete thermal decomposition, not having reached temperatures above 600 °C (Ferrari *et al.* 2002; Rubli 2000). In this study only organic and inorganic carbon was measured. According to Swiss legislation only BA with an organic carbon content of <3 wt% is allowed to be landfilled. Today all ashes fulfill this requirement as shown in Fig. 7. However, loss on ignition (LOI) determined at

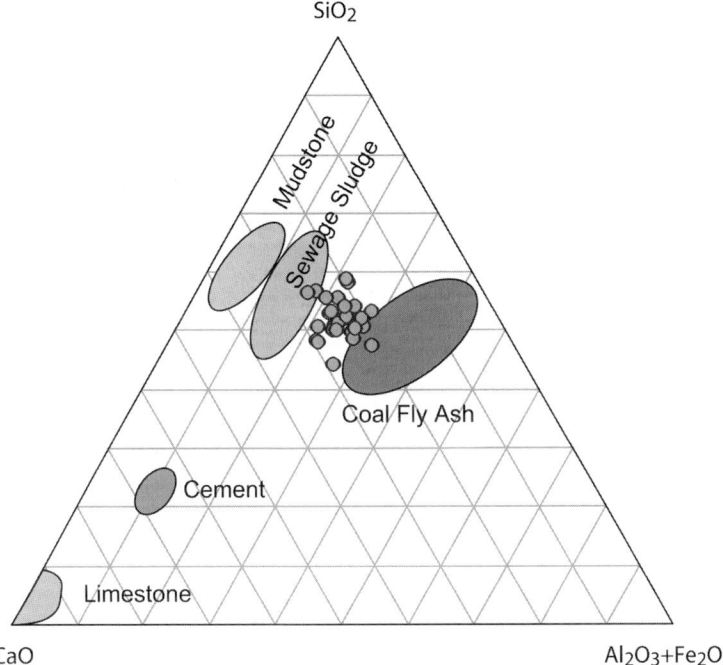

Fig. 5. $SiO_2–CaO–Al_2O_3 + Fe_2O_3$ plot of the BA in comparison to substitute or additive waste materials already used in the building material sector.

1050 °C shows a large spread representing the sum of organically and mineralogically bound CO_2 and H_2O, plus the gain (negative loss) caused by oxidation of metals. Therefore, LOI measurements, most notably above 550 °C, are not suitable as a quality standard for BA.

Trace metals show generally a much larger concentration range, particularly Cu, Zn, Pb,

and Hg (Fig. 8), in contrast to major element concentration. Although metals such as Pb, Cd, Zn, and Hg show a strong tendency to fractionate into the fly ash (Belevi, 1998), elevated median concentrations for Pb (1900 ppm) and Zn (5000 ppm) were still observed in BA. However, it has also been shown elsewhere that at temperatures up to 1400 °C a significant

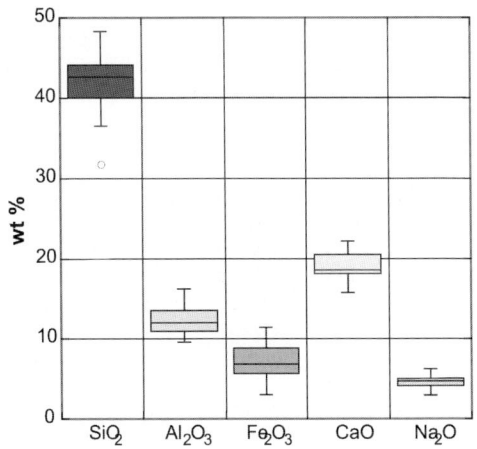

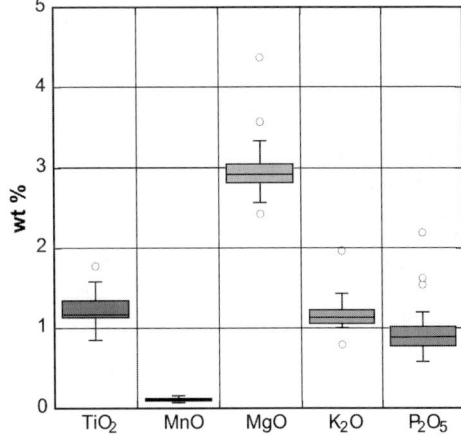

Fig. 6. Major oxide composition of the BA.

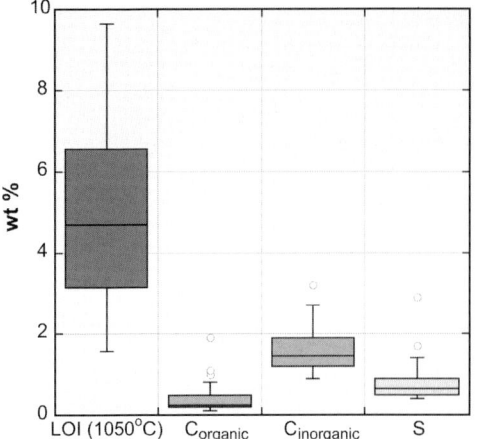

Fig. 7. Loss on ignition (LOI), carbon content (inorganic and organic) and sulphur content of the BA.

decrease in Pb and Cd content was observed compared to conventional MSWI residues, but no significant depletion in Zn. In these thermal products Zn is able to replace Ca in the newly formed melilites, which show Zn concentrations 1.5 times that of the total analysis (Traber *et al.* 2002), and also enters the glass structure comfortably.

Copper, which remains preferentially in the BA, shows a median of almost 6000 ppm and in one sample even a concentration of 3.8 wt%.

It is obvious that published Cu concentrations in BA vary strongly. Not only data from different countries, but also data from the same plant differ significantly. In Table 2 Cu, Zn, and Pb concentrations of BA from different sampling campaigns are shown. Whereas Pb and Zn concentration

ranges are comparable, the median Cu value of this study shows three times the concentration of a previous study in BA coming from the same incinerators (Ganguin 1998). It is not clear, however, if this is a consequence of the varying Cu input, a problem of sample preparation (see Zeltner 1998), or a result of different analytical techniques. We found that with certain grinding techniques, such as tungsten milling, metallic copper or Cu-bearing alloys remain as coarser grained particles (~200 μm) in the mineral powder (<5 μm) and may affect the analysis.

Comparing the BA heavy metal concentration according to the Swiss legislation for waste materials (TVA 1990) and the guidelines for non-contaminated and tolerable disposable materials (BUWAL 1999) it is evident that today's BA do not fullfil either threshold values for so called inert landfill type nor the limits for disposable materials allowed for re-use (Table 3). In comparison to naturally occurring rocks, the measured concentrations in, for example, Cu, Zn, and Pb are even orders of magnitude higher.

Comparing metal concentrations in the BA with the amount of the different types of waste (household, industrial, etc.) only one poor correlation ($R = 0.6$) was observed: a higher amount of waste coming from the construction sector caused higher CaO contents in the BA. It is likely that in this case Ca-rich CSH phases (calcium–silicate–hydrate) and gypsum account for a higher CaO content in the BA.

For none of the trace metals was a correlation found, neither for the type of waste input, nor for the firing conditions like gas temperature or oxidative conditions in the burning chamber, as theoretically described by Belevi (1998).

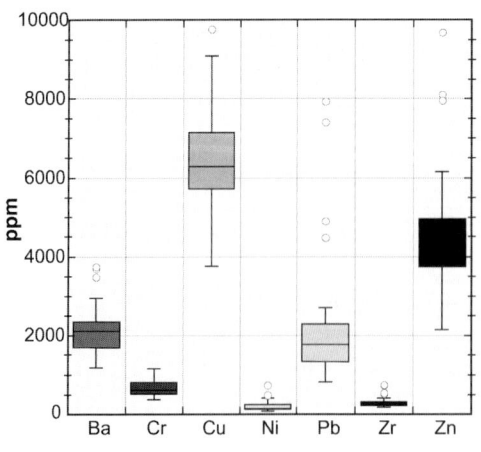

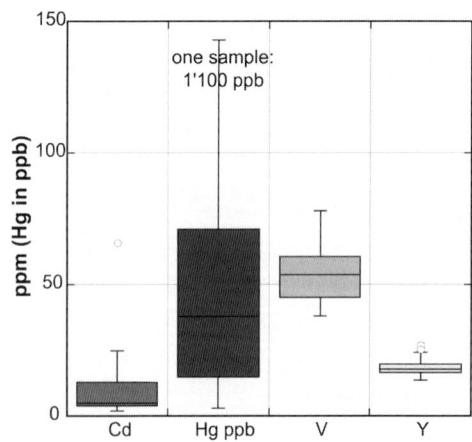

Fig. 8. Heavy metal content of the BA samples.

Table 2. *Concentration of Cu, Zn, and Pb of BA in comparison to other studies*

Source country	This study CH median	SAEFL 2002 CH mean	Ganguin 1998 CH median	Zeltner, 1998 CH mean	Knorr *et al.* 1999 D Fine fraction mean	Knorr *et al.* 1999 D Coarse fraction mean	Basalt standard BR CRPG	Granite standard GR CRPG
Cu ppm	6300	4170	2110	8300	3910	3100	72	345
Zn ppm	4800	4290	4810	4000	3920	1750	150	60
Pb ppm	1800	1670	1980	1700	1980	920	8	32

Mineralogical composition

Because BA is a product of only partially molten material, the phase assemblages do not in any way represent equlibrium conditions (Lichtensteiger 1996). At temperatures of about 1000 °C and relatively short residence times on the grate of about 30–60 minutes it is obvious that there exist large differences in the temperature that the individual particles have reached. Refractory waste products such as natural stones, building materials (tiles, bricks, fine ceramic, etc.) are not altered at all and represent the 'primary mineralogy' of these fragments, mainly quartz, feldspars, gehlenite, mullite, wollastonite, and diopside from ceramics and additionally any other naturally or industrially occurring mineral phase remaining stable under these conditions. It is assumed that finer grained carbonates and gypsum will be transformed to CaO and anhydrite respectively, whereas coarser grained carbonates will not be completely decarbonated.

In our study only small amounts of carbonates and no portlandite was found in the BA. The samples were air dried at 105 °C after sampling within 24 hours and, therefore, newly formed cabonates or portlandite formation was not expected. However, hydration and carbonation processes of wetted and disposed BA is well known as an important weathering process that also influences trace metal mobility (Johnson *et al.* 1999; Meima *et al.* 2002). Phase identification in XRD patterns from multicomponent samples like BA is relatively complex. Therefore, only major mineral phases found by microscopic or SEM investigations were quantified by XRD. The total amount of mineral phases was determined using an internal standard and the amount of glass was estimated using the broad 'glass peak' between 15 and 40° 2θ using the estimation method of Rizzoli (2001) developed for heavy clay ceramic. Other more detailed mineralogical studies (Pfrang-Stotz & Schneider 1995) showed similar amounts and types of mineral phases in the BA.

Quartz has always been found in BA and is clearly a refractory constituent. Also the higher temperature SiO_2 modification tridymite has been described in previous studies. Microscopic investigations indicate that quartz may also act as a source of Si when in contact with partial melts. There, newly formed gehlenite in the glass matrix close to quartz grains was observed. Melilites (endmembers: gehlenite–åkermanite) arc found as large phenocrysts and also as fine needles in the glass matrix. Toward smaller Si/ Ca ratios in the BA the formation of Si-poorer silicates (melilites) seems to be favoured (Fig. 9) whereas the total amount of Ca–Mg– Al-silicates remains relatively constant.

Table 3. *Legislative limiting values in Switzerland in comparison to BA content and some naturally occurring magmatic rocks*

Parameter (ppm)	TVA 1990 Inert material	Guideline disposable material Tolerable	Guideline disposable material Non-contaminated	Measured BA concentration Median	Measured BA concentration Mean	Measured BA concentration Min	Measured BA concentration Max
Pb	500	250	50	1791	2312	821	7931
Cd	10	5	1	5	10	2	66
Cr total	n.d.	250	50	609	650	376	1157
Cu	500	250	40	6290	8280	3756	38356
Ni	500	250	50	150	211	87	726
Hg	2	1	0.5	0.038	0.085	0.003	1.100
Zn	1000	500	150	4790	5115	2153	10898

n.d., not determined.

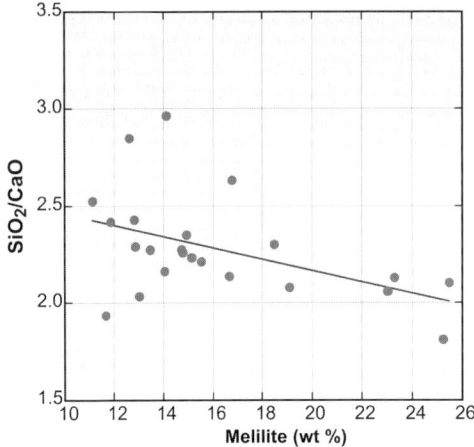

Fig. 9. Si/Ca ratio of the chemical bulk composition of BA in comparison to the amount of melilite measured by XRD.

Feldspars may be refractory as well or crystallized from partial melts. Whereas potassium feldspars are found to be mostly refractory, anorthite rich plagioclase may be newly formed. Pyroxene and spinel were identified in varying amounts but no olivine could be detected. Metals and iron oxide minerals (magnetite–hematite) are always present in BA, but have not been quantified yet. In Table 4 the amount and the ranges of measured mineral contents are summarized.

Conclusions

The formation of BA is a highly complex thermal process. Varying input materials are subject to high temperature for a short time only, whereby equilibrium conditions may occur only locally in the BA. Chemical and mineralogical analyses of BA coming from 28 MSWI plants in Switzerland with comparable technology show similar major oxide compositions and similar types and contents of mineral phases. Newly formed melilite contents increase with decreasing bulk Si/Ca ratio in the BA. This indicates that at least the type of Ca–Mg–Al-silicates crystallizing during incineration and cooling follow petrogenetic rules. Bottom ash contains many refractory phases and, therefore, the comparison with naturally occurring igneous rock classifications is only of limited applicability.

The BA is relatively well crystallized, which tends to lower hydrolysis when considering a long-term perspective. Dissolution rates are generally lowered if the crystalline fraction is higher and the growth of less resistant phases like melilites can be reduced. This is mainly the case for bulk compositions with a higher Si/Ca ratio.

For a better understanding of the reactions taking place during incineration and cooling, mineral–glass and mineral–mineral reactions have to be studied locally in detail. Currently a study is in progress using the same BA to arrive at a better understandig of these reactions and the associated trace metal distribution between the primary, secondary, and melt phases.

In contrast to major oxide contents, heavy metal concentrations vary significantly in the BA without a detectable relation to the type of input waste or plant operating conditions. The high heavy metal content of the BA exclude their use as secondary raw materials according to Swiss legislation. However, a considerable recycling potential of ferrous and non-ferrous metal content was identified in the BA. This recycling potential encouraged the Swiss Agency of Environment and Landscape to evaluate the applicability of mechanical metal separation technologies. Additionally, a reduction of heavy metal concentration, H_2 production, and exothermic reactions of BA in landfills is expected.

Table 4. *Mineralogical composition of the BA measured by XRD using internal standardization*

Mineral (wt%)	Median	Mean	Minimum	Maximum
Quartz	18	19	12	29
K-feldspar	2	2	1	4
Plagioclase	7	9	4	29
Pyroxene	7	8	4	15
Melilite	15	16	11	25
Calcite	2	2	1	3
Dolomite	1	1	0	2
Spinel	3	3	2	4
Glass	43	41	26	48

The Swiss Agency for Environment, Forests and Landscape, Division Waste, is gratefully acknowledged for financial and scientific support of this study. The authors are grateful to the companies and plant operators for their collaboration, the sampling of the BA, and the supply of the requested information.

References

BARBIERI, L., CORRADI, A., LANCELLOTTI, I. & MANFREDINI, T. 2002. Use of municipal incinerator bottom ash as sintering promoter in industrial ceramics. *Waste Management*, **22**, 859–863.

BELEVI, H. 1998. *Environmental Engineering of Municipal Solid Waste Incineration*. Verlag vdf Hochschulverlag, Habilitation, Zürich, ISBN 3 7281 2659 4.

BELEVI, H., STÄMPFLI, D. M. & BACCINI, P. 1992. Chemical behaviour of municipal solid waste incinerator bottom ash in monofills. *Waste Management & Research*, **10**, 153–167.

BUNGE, R. & BUNGE, K. 1999. Probenahme auf Altlasten: Minimal notwendige Probenmasse. *Altlasten Spektrum*, **3/99**, 174–179.

BUWAL/SAEFL 1999. *Richtline für die Verwertung, Behandlung und Ablagerung von Aushub-, Abraum- und Ausbruchmaterial (Aushubrichtlinie)*. Vollzug Umwelt Abfall und Altlasten, BUWAL, Dokumentation, CH-3003 Bern, Switzerland.

CHANDLER, A. J., EIGHMY, T. T. *et al.* 1997. *Municipal Solid Waste Combustor Residues: The International Ash Working Group, Studies in Environmental Science*. Elsevier Science, Amsterdam, 67.

CHIMENOS, J. M., SEGARRA, M., FERNANDEZ, M. A. & ESPIELL, F. 1999. Characterization of the bottom ash in municipal solid waste incinerator. *Journal of Hazardous Materials*, **64**, 211–222.

EUSDEN, J. D., EIGHMY, T. T., HOCKERT, K., HOLLAND, E. & MARSELLA, K. 1999. Petrogenesis of municipal solid waste combustion bottom ash. *Applied Geochemistry*, **14**, 1073–1091.

FERRARI, S., BELEVI, H. & BACCINI, P. 2002. Chemical speciation of carbon in municipal solid waste incinerator residues. *Waste Management*, **22**, 303–314.

FILIPPONI, P., POLETTINI, A., POMI, R. & SIRINI, P. 2003. Physical and mechanical properties of cement-based products containing incineration bottom ash. *Waste Management*, **23**, 145–156.

FÖRSTNER, U. 1996. Langzeitprognosen und naturnahe Dauerlösungen. *Geowissenschaften*, **14**, 169–172.

GANGUIN, J. 1998. Kehrichtschlacke weist Reststoffqualität auf – Untersuchungen der chemischen Zusammensetzung und des Auslaugverhaltens von Kehrichtschlacke. *Informationsbulletin des Amtes für Gewässerschutz und Abfallwirtschaft des Kantons Bern*, **2/98**, 40–47.

GIAMPAOLO, C. & LO MASTRO, S. 2002. Acid neutralisation capacity and hydration behaviour of incineration bottom ash–Portland cement mixtures. *Cement and Concrete Research*, **32**, 769–775.

JOHNSON, C. A., BRANDENBERGER, S. & BACCINI, P. 1995. Acid neutralizing capacity of municipal waste incinerator bottom ash. *Environmental Science Technology*, **28**, 142–147.

JOHNSON, C. A., KAEPPELI, M., BRANDENBERGER, S., ULRICH, A. & BAUMANN, W. 1999. Hydrological and geochemical factors affecting leachate composition in municipal solid waste incinerator bottom ash. Part II. The geochemistry of leachate from Landfill Lostorf, Switzerland. *Journal of Contaminant Hydrology*, **40**, 239–259.

KELSEY, C. H. 1965. Calculation of the C.I.P.W. norm. *Mineralogical Magazine*, **34**, 276–282.

KNORR, W., HENTSCHEL, B., MARB, C., SCHÄDEL, S., SWEREV, M., VIERLE, O. & LAY, J. P. 1999. *Rückstände aus der Müllverbrennung – Chancen für eine stoffliche Verwertung von Aschen und Schlacken. Initiativen zum Umweltschutz*, 13 ed. Deutsche Bundesstiftung Umwelt, Erich Schmidt, Berlin.

KRUSPAN, P. 2000. *Natürliche und Technische Petrogenese von Puzzolanen – ein Beitrag zur Herstellung Mineralischer Sekundärrohstoffe*. PhD thesis, ETH 13904, ETH Zürich, Switzerland.

LICHTENSTEIGER, TH. 1996. Müllschlacken aus petrologischer Sicht. *Geowissenschaften*, **14**, 173–179.

LICHTENSTEIGER, TH. & ZELTNER, CH. 1993. Wie lassen sich Feststoffqualitaeten beurteilen. *In*: BACCINI, P. & GAMPER, B. (eds) *Deponierung fester Rückstände aus der Abfallwirtschaft – Endlagerqualität am Beispiel Müllschlacke*. Verlag vdf Hochschulverlag, Zürich, 11–33.

MAYER, A. & REY, T. 1996. Innovative HSR process to transform waste incinerator slag into useful mineral additives. *Waste Management*, **16**, 27–33.

MEIMA, J. A. & COMANS, R. N. J. 1999. The leaching of trace elements from municipal solid waste incinerator bottom ash at different stages of weathering. *Applied Geochemistry*, **14**, 159–171.

MEIMA, J. A., VAN DER WEIJDEN, R. D., EIGHMY, T. & COMANS, R. N. J. 2002. Carbonation processes in municipal solid waste incinerator bottom ash and their effect on the leaching of copper and molybdenum. *Applied Geochemistry*, **17**, 1503–1513.

PECQUEUR, G., CRIGNON, C. & QUÉNÉE, B. 2001. Behaviour of cement-treated MSWI bottom ash. *Waste Management*, **21**, 229–233.

PERA, J., COUTAZ, L., AMBROISE, J. & CHABABBET, M. 1997. Use of incinerator bottom ash in concrete. *Cement and Concrete Research*, **27**, 1–5.

PERRET, D., CROVISIER, J.-L. *et al.* 2003. Thermodynamic stability of waste glasses compared to leaching behaviour. *Applied Geochemistry*, **18**, 1165–1184.

PFRANG-STOTZ, G. & SCHNEIDER, J. 1995. Comparative studies of waste incineration bottom ashes from various grate and firing systems, conducted with respect to mineralogical and geochemical methods of examination. *Waste Management & Research*, **13**, 273–292.

RIZZOLI, G. 2001. *Suitability of Carbonate-Rich Roofing Tile Raw Materials for a More Economical and Ecological Firing Technique*. PhD thesis, Institute for Geological Sciencies, University of Bern, Switzerland.

RUBLI, ST. 2000. *Thermische Abfallbehandlung – Organischer und Elementarer Kohlenstoff als Indikatoren in der Prozessoptimierung.* PhD thesis, ETH 13782, ETH Zürich, Switzerland.

SAEFL 2002. Die KVA als Instrument der Erfolgskontrolle in der Abfallwirtschaft der Schweiz. Report, Swiss Agency for Environment, Forests and Landscape, Division Waste, CH-3003 Bern, Switzerland.

SHAW, H. R. 1972. Viscosities of magmatic silicate liquids: an empirical method of prediction. *American Journal of Science*, **272**, 870–893.

STÄUBLI, B. 1992. *Emissionsabschätzung für Kehrichtschlacke. Bericht der Auftraggebergemeinschaft für das Projekt EKESA.* AGW Kanton ZH und Abt. Umweltschutz Kanton AG, 156 pp.

TARGAN, S., OLGUN, A., ERDOGAN, Y. & SEVINC, V. 2002. Effects of supplementary cementing materials on the properties of cement and concrete. *Cement and Concrete Research*, **32**, 1551–1558.

TRABER, D., MÄDER, U. & EGGENBERGER, U. 2002. Petrology and geochemistry of a municipal solid waste incinerator residue treated at high temperature. *Schweizerische Mineralogische und Petrographische Mitteilungen*, **82**, 1–14.

TVA 1990. *Technische Verordnung über Abfälle (SR 814.600).* Schweizerischer Bundesrat, Eidg. Druck- und Materialzentrale, Bern, 32 pp. World Wide Web Address: http://www.admin.ch/ch/d/sr/c814_600.html.

VAN DER SLOOT, H. A. 1996. Developments in evaluating environmental impact from utilization of bulk inert wastes using laboratory leaching tests and field verification. *Waste Management*, **16**, 65–81.

WILES, C. 1996. Municipal solid waste combustion ash: State-of-the-knowledge. *Journal of Hazardous Materials*, **47**, 325–344.

ZELTNER, C. 1998. *Petrographische Evaluation Thermischer Behandlung von Siedlungsabfällen über Schmelzprozesse.* PhD thesis, ETH 12688, ETH Zürich, Switzerland.

ZELTNER, C. & LICHTENSTEIGER, T. 2001. *Thermal Waste Treatment and Resource Management – A Petrologic Approach to Control the Genesis of Materials in Smelting Processes. Environmental Engineering and Policy.* Springer-Verlag (Electronic Journal). World Wide Web Adress: http://www.link.springer.de.

ZEVENBERGEN, C., VAN DER WOOD, T., BRADLEY, J. P., VAN DER BROEK, P. F., ORBONS, W. A. & VAN REEUWIJK, L. P. 1994. Morphological and chemical properties of MSWI bottom ash with respect to the glassy constituents. *Hazardous Waste Materials*, **11**, 371–383.

Incinerator waste as secondary raw material: examples of applications in glasses, glass–ceramics and ceramics

LUISA BARBIERI & ISABELLA LANCELLOTTI

Università di Modena e Reggio Emilia,

Dipartimento di Ingegneria dei Materiali e dell'Ambiente, Modena, Italy

(e-mail: barbieri.luisa@unimore.it)

Abstract: The recovery and beneficiation of solid waste residues (bottom and fly ash) produced during incineration of municipal wastes are very important goals from the environmental and economic point of view. Some technologies allow not only to reduce the volume of incinerator waste but also to generate products of economic value by vitrification, devitrification, and ceramic processes. Bottom ash is preferable compared to fly ash as a secondary raw material. Mixed with glass cullet, bottom ash can be easily transformed into homogeneous and inert glasses, which can be transformed into glass fibres or sintered glass–ceramics after controlled thermal treatments. Alternatively, bottom ash can be used to produce tiles, in particular glazed tiles containing vitrified bottom ash in the ceramic body, and possibly bricks. The results presented in this review are promising and in accordance with the waste minimization policy as well as resources conservation.

The increasing flux of waste and its immense impact on the environment urges the research community and the legislators to develop techniques that reduce municipal waste generation. Regarding the situation in the European Union (EU), almost 200 Mt of municipal solid waste (MSW) is generated every year (Eurostat 2000). Waste often represents an inefficient use of its material and energy resource and can lead to serious environmental pollution if managed inappropriately. There are various systems for waste management, and they vary from country to country depending on a large number of factors, including topography, population density, transportation infrastructure, socioeconomics, and environmental regulations. Most OECD (Organisation for Economic Co-operation and Development) governments, especially European governments, are shifting waste strategies from simple collection and disposal to a 'waste hierarchy' approach with an emphasis on preventing waste generation, followed by recovery systems and finally environmentally sound disposal of non-recoverable waste only. In Europe, approximately 64% of municipal waste is currently destined for landfill, 18% for incineration, and 18% for recycling, including composting. Further implementation of waste management policies, however, is expected to help to reduce landfilling and increase recycling in the future: by 2020, about 50% of municipal waste is expected to be landfilled, 17% incinerated, and 33% recycled (Zacarias-Farah & Geyer-Allèly 2003). From these data it appears evident that, among the various MSW management options, particular relevance is attributed to recycling and incineration, notwithstanding the latter produces solid residues to be disposed. In the past, recovery efforts were impeded not only by low prices for primary raw materials and sufficient landfill capacities and hence low prices for MSW disposal, but also by the chemical heterogeneity of the waste, its elevated pollutant concentrations, and reactivity of the solid residuals, as well as by a negative image of products made from waste materials. In the future, however, the development of recycling technologies will become economically sustainable. This is, among other factors, a direct consequence of the fact that refuse disposal is becoming very rare and expensive, and legal restrictions for treatment of residues are becoming more severe. Therefore, the amount of products made through recycling of components of municipal solid waste per se (e.g., glass, paper, plastic), and of incinerated MSW (e.g., solid residues), will strongly increase.

MSW incineration and its waste products

Among the methodologies used to treat MSW, incineration is a mature technology for which

From: GIERÉ, R. & STILLE, P. (eds) 2004. *Energy, Waste, and the Environment: a Geochemical Perspective*. Geological Society, London, Special Publications, **236**, 423–433.
0305-8719/04/$15   © The Geological Society of London 2004.

the environmental and safety aspects are taken into consideration at all phases of the process. The main objectives of MSW incineration are to sterilize the waste and to reduce its volume by more than 90% and its mass by about 70% before final disposal. The majority of new MSW incinerators (MSWI) are also designed for energy recovery, either in the form of electricity or steam production for industry or district heating. The concerns over air emissions from these facilities have resulted in most countries adopting very stringent air emission control regulations, which have increased the cost of constructing and operating incinerators. Although MSW incineration is capable of reducing the volume of waste by 90%, 20–30% of the original weight of the waste accumulates as ash, which requires further treatment. There are two generic ash streams from incinerators: bottom ash and fly ash, together with sludge resulting from washing steps. In most countries, the two ash streams are classified and managed differently because significant differences exist in their physical, chemical and leaching characteristics. In the EU, most countries have deemed bottom ash suitable for disposal in landfills or monofills, but some (Germany, The Netherlands, and Denmark) have already allowed the use of processed incinerator waste, in particular bottom ash, in different applications.

With respect to the management of fly ash, in addition to disposal as a hazardous material, there are four generic treatment technologies that have been used or are under development. These include solidification, chemical stabilization, ash melting or vitrification, and extraction/recovery processes. Comparing the different technologies requires not only an assessment of the costs involved, but also the characterization of the products generated from each process. Consequently, the high costs of vitrification or extraction processes may be considered acceptable in view of the potential use and added value of the generated material. Conversely, when considering the lower costs of solidification or stabilization processes, the additional weight of the solidification reagents and the potential long-term instability due to the high salt contents of the treated material must be taken into account.

Despite the emphasis on waste minimization and recycling, society will continue to generate waste requiring incineration or landfill disposal in the foreseeable future. Owing to the formation of bottom and fly ash through MSW incineration, it is important to find criteria for their utilization in an environmentally sound manner. This is in line with the policies adopted by several EU countries, such as Italy, where the utilization of the incinerator residues is not regulated. On the other hand, implementation of re-use and recovery of the wastes as secondary raw material is being promoted by means of financial support for performing feasibility studies and building pilot plants.

Characterization of MSWI residues

Bottom ash is generally defined as the material collected off the incineration grates. It is an aggregate material and constitutes 80 wt% of generated residues (Wiles 1996). Fly ash is a collective term for the finer material captured downstream of the furnace, that is, in the heat recovery and air pollution control (APC) system. In particular, APC residues consist of entrained particulates that are trapped and particles generated by acid gas scrubbers and are subsequently removed from the flue gas by fabric filters and/or electrostatic precipitators (ESP). Fly ash has a fine grain size, including the breathable fraction, and is characterized by a high concentration of heavy metal compounds. For these reasons, fly ash is considered a hazardous waste material and, at the moment, in Italy it is discarded in suitable landfills at the considerable cost of 100–150 euros/t (Falcone et al. 2003). While for bottom ash it is possible to establish an average composition, the fly ash composition differs depending on flue gas abatement systems. In the case of Italy, 41 incineration plants are operating: 28 in the North, 10 in the Middle, and 3 in the South. They incinerate about 2.6 Mt of urban wastes per year. In Table 1 the chemical compositions of ash from representative MSWI in the North are reported. Fly ash is particularly problematic because it is enriched in thermally mobile species and organochlorine compounds. Volatile metals, such as Pb and Cd, form chlorides, which are volatilized in the combustion chamber. They condensate while the flue gases are cooled down to around 200 °C and hence are transferred into the fly ash. Mercury and its compounds, however, stay mainly in the gas phase even at these temperatures and must be taken care of in the APC system to prevent their emission. Because many volatile heavy metals are rather toxic and often form water-soluble compounds, the metal volatilization is a welcome effect that removes these constituents from the bottom ash. The mineralogical compositions reported in Table 2 underline the significant differences between bottom and fly ash. Moreover, the data emphasize the higher percentages of sulphates and chlorides derived from the acid gas abatement systems in fly ash compared with bottom ash.

Table 1. *Chemical composition of MSWI bottom and fly ash from Italian incinerator plants equipped with both fabric filter and electrostatic precipitator (1), and fabric filter only (2)*

Component	Unit	Bottom ash: Range of (1) and (2) plants	Fly ash from fabric filter (1)	Fly ash from electrostatic precipitator (1)	Fly ash from fabric filter (2)
SiO_2	wt%	45–50	1.71	35.5	9.52
Al_2O_3	wt%	8–10	0.32	11.7	5.26
Na_2O	wt%	4–6	6.83	6.83	29.9
K_2O	wt%	1–1.5	46.5	9.67	4.33
CaO	wt%	16–19	1.21	18.8	16.26
MgO	wt%	2–3	0.12	2.37	2.95
BaO	wt%	2000	—	—	1126
P_2O_5	wt%	1	—	—	—
PbO	$\mu g/g$	0–2000	8359	1013	3434
TiO_2	wt%	0.7–0.9	—	—	0.85
Fe_2O_3	wt%	3–9	0.14	3.9	0.46
MnO	$\mu g/g$	0–800	—	—	426
ZnO	$\mu g/g$	1000–3000	traces	—	9800
CdO	$\mu g/g$	—	193	40	194
Cr_2O_3	$\mu g/g$	400	111	15	920
CuO	$\mu g/g$	0–1500	1275	273	800
Cl^-	$\mu g/g$	0.04	2587	1976	188100
SO_4^{2-}	wt%	1–1.5	5–5.5	7–7.5	11.59
LOI (Loss on ignition at 950 °C)	wt%	4–8	molten	6	0.001

The carbon content of MSW cannot be converted into CO_2 entirely, and due to incomplete combustion, minor amounts of CO and soot particles are found in the flue gases. The particulate carbon is known to be involved in the formation of volatile and toxic compounds especially polychlorodibenzo-dioxins and -furanes. Tests in the fully working incinerator plants revealed the presence of particulate carbon, chlorides, and Cu compounds as catalysts in the fly ash (see also Table 3).

Incinerator waste as secondary raw material

Waste management is a difficult task because of the nature of the refuse, which is so chemically heterogeneous. It is necessary to distinguish between fly and bottom ashes, which, coming from different plants, may have quite different chemical compositions. Fly ash is particularly problematic because it contains significant concentrations of heavy metals (e.g., As, Pb, Sb, Sn;) as well as trace amounts of organic pollutants (e.g., polychlordibenzo-dioxin and -furanes). In the re-utilization of this powder the environmental impact has to be taken into account. In fact, toxicity does not necessarily disappear with fly ash re-utilization and one must be careful not to create new sources of pollution. With regard to the MSWI bottom ash, the proportions of different constituents are fairly constant, (Chimenos *et al.* 1999; Ferraris *et al.* 2001) independent of the waste origin and the

Table 2. *Phase content of MSWI bottom and fly ash from an Italian incinerator plant with a double filtering system*

Waste	Crystalline phases
Bottom ash	Quartz (SiO_2), calcite ($CaCO_3$), kyanite (Al_2SiO_5), anhydrite ($CaSO_4$), enstatite ($MgSiO_3$), gehlenite ($Ca_2Al_2SiO_7$)
Fly ash from electrostatic precipitator	Halite (NaCl), sylvite (KCl), gehlenite ($Ca_2Al_2SiO_7$), anhydrite $CaSO_4$
Fly ash from fabric filter	Halite (NaCl)

Table 3. *Elementary analysis (wt%) of MSWI bottom and fly ash from Italian incinerator plants*

Waste	N	C	H	S
Bottom ash	0.03	1.70	0.15	0.65
Fly ash from fabric filter	0.04	6.5	0.05	1.85
Fly ash from electrostatic precipitator	0.06	0.60	0.06	2.75

particular incineration process. Consequently, bottom ash appears to be a suitable secondary raw material for re-use, for example, in construction. Other important aspects for bottom ash re-use are the suitability for processing and the technical performance of the final product. The suitability for processing depends on physico-chemical characteristics of the residues, such as particle size and chemical properties that may constitute a limitation for a specific process. In some cases, however, these characteristics can be adjusted so as to comply with processing requirements. Moreover, after processing of the residues the final product is acceptable only if it exhibits good technical properties. Thanks to its chemical composition, bottom ash handling is easier and it requires less treatment before being subjected to a recovery technology. Because bottom ash contains typical glass constituents, such as Si, Al, Ca, and Na, it is suitable for vitrification. Moreover, the bottom ash composition is also suitable for production of ceramic bodies. Thus, vitrification, devitrification, and the production of ceramic tiles or bricks are new promising re-use technologies. The most significant advantages of vitrification include: volume reduction of the waste; good chemical durability of the final product; and the possibility to use wastes of different origin to obtain more homogeneous batches.

Ceramics

The ceramic industry includes the manufacture of pottery and porcelain, as well as the fabrication of ceramics (e.g., bricks, tiles, stoneware). Ceramics are prepared from malleable, earthy materials, such as clay, that become rigid at high temperatures. Besides clays, which confer cohesion and plasticity, and feldspar as melting agents, ceramic pastes also include inert materials that provide structural support, necessary to retain shape during drying and firing. Quartz (SiO_2) is the most commonly used inert material and is usually supplied either as sand or schist. It is also used in the glazing of ceramic bodies in combination with stiffeners and melting agents. When a glaze layer is

applied to the ceramic body and then fired, the glaze ingredients melt and become glass-like. During thermal treatment (sintering at a maximum temperature around 1200 °C for less than 1 hour) the aforementioned raw materials are transformed into crystalline silicate phases and glassy matrices capable of fixing the wastes. The ceramic industry represents an important reference point for the recycling or re-use of several waste types; for example, urban solid waste-water sludge and sawdust (Cusidò *et al.* 1996), integrated-gasification gas-combined cycles slag (Acosta *et al.* 2002), and windshield glasses (Mortel & Fuchs 1997) for brick manufacture; MSWI fly ash and granite sawing residues for porcelainized stoneware (Hernàndez-Crespo & Rincòn 2001); wastes from the stone industry for ceramic tile manufacturing (Raigon-Pichardo *et al.* 1996; Vieira *et al.* 1999; Das *et al.* 2000); bauxite waste for ceramic glaze preparation (Yalcin & Sevinc 2000); ash from the incineration of sewage sludge for the brick and tile industry (Wiebusch & Seyfried 1997); and used catalysts from the fluidized-catalyst cracking units involved in petroleum refining for the preparation of ceramic frits (Escardino *et al.* 1995). In Italy, manufacturing of wall and floor ceramic tiles is an important part of the economy, primarily in an area of 300 km^2 in the north of the country, that is, in the provinces of Modena and Reggio Emilia, where 80% of the tiles manufactured nationally are produced. In 1999, the ceramic industries produced about 606 million m^2 of tiles, consuming of about 9 Mt of raw materials, of which 56 wt% were imported from other countries (Ukraine, Turkey, Germany, France; Assopiastrelle Indagine Statistica Nazionale 2000). Table 4 underlines the economic impact of the raw materials on the different typologies of ceramic tiles produced in Italy. It is evident that the proportion of raw material costs on the whole production cost is much higher for red and white single-firing and porcelainized stoneware tiles (11–26%) than for the double-firing ceramics (4%). These raw material costs are important and, therefore, the complementary use of alternative raw materials is of significant interest in the ceramic sector.

Table 4. *Raw material costs, expressed as percentage of the total cost of the ceramic production*

Tile typology	1998	1999	2000
Double-firing	4.08	4.61	4.31
Red single-firing	11.96	11.87	11.30
White single-firing	16.48	16.14	15.74
Unglazed porcelainized stoneware	26.18	25.20	24.91
Glazed porcelainized stoneware	26.53	24.29	23.02

Tiles. In the ceramic tile industry the production volume of especially porcelainized stoneware is becoming important. The term 'porcelainized stoneware' recalls the origin and the characteristics of the product; in fact, according to ceramic terminology, the term 'stoneware' indicates an extremely compact material, composed of different crystal phases embedded in a vitreous matrix. On the other hand, the adjective 'porcelainized' is etymologically linked to the word porcelain, the noblest ceramic material, well known and appreciated for centuries. Porcelainized stoneware is a high-sintered material obtained by fast-firing of kaolinitic ceramic bodies. It has very good technical characteristics, such as high bending strength (>27 N/mm^2) and moderate water absorption ($<0.1\%$), indicating a low open porosity (Palmonari 1989; Sezzi 2000). It is mainly because of this latter property that this material is used for interior and exterior applications and in applications where frost resistance is required. The recycling potential of MSWI solid residues in the porcelainized stoneware seems to be limited to bottom ash (Andreola *et al.* 2001). The high Na and K chloride and organic matter content of the electrostatic precipitator and fabric filter fly ashes, as detected by X-ray powder diffraction (XRD) and differential thermal analysis (DTA), respectively, causes problems for the industrial plant due to corrosive phenomena. During the ceramic process, the high amounts of soluble chloride compounds cause an important increase of the viscosity, rendering the wet-grinding step difficult. The final product will exhibit bubbles, pores, and black core, and therefore fly ash is incompatible with this kind of ceramic body. Consequently, there is no sustainable way of re-using fly ash per se without drastic modification of its quality.

Porcelainized stoneware containing up to 20 wt% of bottom ash and fired using a laboratory thermal cycle simulating industrial conditions shows a good mineralogical, thermal, and rheological compatibility with the standard ceramic body. The chemical composition and mineralogical constituents of the bottom ash (Tables 1, 2) confirm that bottom ash is suitable for mixing with the porcelainized ceramic body. The crystalline phases present in the porcelain stoneware body after firing are quartz and mullite ($Al_6Si_2O_{13}$). Introduction of bottom ash into the ceramic body causes the development of anorthite ($CaAl_2Si_2O_6$) as a new phase. This observation is in agreement with other studies on traditional ceramic bodies with added waste materials containing Ca compounds (Andreola *et al.* 1992) and with the phase diagram of the ternary $CaO-Al_2O_3-SiO_2$ system (Roth 1995). The formation of anorthite decreases the amount of mullite, since the reaction between silica, alumina, and calcium compounds to obtain anorthite,

$$2\,SiO_2 + Al_2O_3 + CaO \longrightarrow CaAl_2Si_2O_8$$

occurs at lower temperatures ($1000-1100$ °C) with respect to the characteristic temperature for mullite formation ($1100-1200$ °C). It is especially important to evaluate the rheological effect on the wet-grinding process of adding solid waste to porcelain stoneware suspensions. All suspensions containing bottom ash exhibit non-Newtonian behaviours; that is, the viscosity is not constant at a given temperature and pressure, but varies with the shear rate and shear stress applied. Therefore only 'apparent viscosity' values may be determined. This apparent viscosity increases as a function of the bottom ash amount added to the ceramic body. Rheological experiments have shown that addition of up to 20 wt% bottom ash is compatible with the porcelain stoneware wet-grinding process because under these conditions, the apparent viscosity does not exceed the limit of the industrial workability (200 mPa.s). Detailed studies have been carried out to investigate the sintering, mechanical, and aesthetic (spot resistance and colour) properties of tiles containing MSWI bottom ash compared to those of porcelainized stoneware (Andreola *et al.* 2002). It was found that introduction of this bottom ash into porcelainized stoneware does not significantly change the properties of the final products and the conditions of the firing cycle. The most evident effect is an increase in porosity, which is proportional to the waste content; this porosity increase causes a decrease in density, shrinkage, and spot resistance. Furthermore, addition of up to 10 wt% of bottom ash does not influence negatively the aesthetic characteristic of the fired tiles (Andreola *et al.* 2002).

These results suggest that introduction of bottom ash into a ceramic body might be a suitable re-use option for this waste material. Barbieri *et al.* (2001*a*) studied the possibility of including MSWI bottom ash in a ceramic body used for the manufacture of wall tiles. From the experimental results it can be concluded that the optimum firing temperature is 20 °C lower (1080 °C) than the standard-body processing temperature (1100 °C). At temperatures higher than the optimum firing temperature for traditional ceramic products, the linear shrinkage is too high, although the planarity (property related to the fact that tile should be as flat as possible) is maintained and the water absorption decreases remarkably. The water absorption test is normally conducted on ceramic tiles (EN ISO 10545-3; European norm) in order to assess the sintering process. Water absorption is related to the open porosity of the material and is minimal for compact, but high for porous materials. Introduction of the waste into this ceramic body does not change the mineral content of the fired samples. Among the phases present (quartz, anorthite, pyroxenes [e.g., diopside]), the pyroxenes become the main phases at the expense of quartz.

Because of the economic importance of finding new ceramic raw materials that favour the sintering process, the use of frits has been evaluated. Frits were obtained from glass cullet and incinerator bottom ash and were introduced as sintering promoters at a level of 3–15 wt% for the production of the porcelainized stoneware body (Barbieri *et al.* 2002). The results of the laboratory-scale tests are satisfying and allow us to conclude that, with the addition of waste-based frits, bending strength does not change significantly with respect to the standard body. The measured bending strength of >45 N/mm^2 for the waste-containing ceramic products is satisfying and conforms with the European norm (EN ISO 10545-4), which requires a strength of ≥27 N/mm^2. Spot resistance tests have been performed in order to evaluate and classify the material. Ceramic materials containing larger quantities of waste (e.g., 50 wt% bottom ash and 50% glass cullet) show even better spot resistance. The experiments further show that addition of glassy frits improves the characteristics of water absorption as a result of the absence of surface porosity, which is due to the glassy phase. Under the firing conditions used, the planarity of the tiles becomes worse and there is a modification of the colour, which becomes darker with respect to the base body. To improve the planarity of the tiles, the firing temperature has to be reduced or the porcelai-

nized stoneware body could be prepared using waste frits as melting agent together with clays and quartz. With regard to the colour change, the waste-based frits could be used to produce glazed porcelainized stoneware, because for this kind of ceramic product, while good physical and mechanical properties are required, the aesthetic characteristics are less important with respect to the unglazed one.

Bricks. Brick manufacture has also been considered as an option for re-use of MSWI waste. Because bricks are produced with a particularly simple technology, which exhibits tolerance with respect to technical parameters, and because the fired product is able to absorb heterogeneous materials, they represent a good possibility for re-using byproducts of different nature, even in large quantities. Moreover, because the brick industry is distributed across the entire Italian territory, the eventual application of this recovery technique results in shorter transport distances between incinerators and plants. The MSWI bottom ash has been added to commercial bricks in amounts ranging from 2.5 to 20 wt%. In the case of the fired bricks, there are four very important parameters to consider: loss in weight after firing (weight difference between green and fired product); apparent density; water absorption; and surface defects. It has been observed that the ash-containing samples show lower loss in weight than the standard brick. This result is probably due to the decrease in the body of the amount of clay, which is substituted by materials with lower reactivity. The apparent density of the fired bricks does not change significantly with the addition of up to 20 wt% of waste. Water absorption, strictly related to the morphology and the porosity, increases with the introduction of the incinerator ash into the body. For example, addition of 20 wt% of waste increases the water absorption to 25 wt% (compared to 15 wt% of the standard). A significant problem is related to the aesthetic aspect because of the formation of white spots as a result of salt migration towards the surface (efflorescence). Further investigations are under way in order to consider other brick types.

Glass and glass–ceramics

Vitrification is one of the most promising technological options for the transformation of MSWI residues into inert materials. The objective of this method is to generate a product that can be re-used or, at least, disposed in standard landfills with minimized risk. The inert vitreous product can, in principle, be utilized in various ways: as road-base material; in embankments; as blasting grit; to partially replace sand in concrete and in the production of construction and decorative materials, such as water-permeable blocks, tiles, pavement bricks and decorative stones for gardens (Abe *et al.* 1996; Boccaccini *et al.* 1999; Nishigaki 2000). The use of solid MSWI residues as glass-making raw materials requires temperatures up to 1000 °C for the first melt-producing reaction. However, the pyro-reactions that follow the formation of the first melt have to occur at higher melting temperatures in order to reach the technologically required formation of a homogeneous glass melt (Volf 1984). For this kind of residue, the temperature has to be higher than 1300 °C (Barbieri *et al.* 2000a, b). The high temperatures involved in the process lead to the complete destruction of organic pollutants (e.g., dioxins). Moreover, heavy metals can be incorporated in the glassy network. Because of these advantages, several companies build and install vitrification facilities around the world (especially in Japan). The disadvantage of vitrification is that it is an energy-intensive process involving relatively high costs. Moreover, the quality of the final product must be assessed with respect to long-term stability. Therefore, its use can only be fully justified if a high-quality product with optimized properties can be fabricated, which can compete with other current materials used, for example, for construction, architectural, or insulation applications. The most effective way to improve the mechanical and electrical properties in particular, and to increase the market value of the vitrified materials, is the transformation into glass–ceramics. These are fine-grained polycrystalline materials (50 vol% to 98 vol% crystalline) formed when glasses of suitable composition are heat-treated and hence undergo controlled crystallization, a phenomenon called devitrification. The three key variables in the design of a glass–ceramic are: the glass composition, which controls glass viscosity (which in turn is related to the effectiveness of nucleation and speed of crystallization); the phase assemblage (i.e., the types of crystals and the proportion of crystals to glass), which is responsible for many of the physical and chemical properties; and the microstructure (i.e., crystal size and morphology, textural relationship among crystals and glass), which is key to many mechanical and optical properties.

The procedure to obtain glass–ceramics does not differ very much from the vitrification process, as it typically only requires an additional carefully controlled thermal treatment, which induces devitrification. In particular, a partly crystalline material can be obtained by one of the following techniques:

(1) *Glass–ceramic processing.* An original glass is produced via melting and rapid quenching. This glass is then heat-treated in one or, more frequently, two stages during which nucleation takes place and crystal growth starts.

(2) *Powder sintering.* This technique is especially used where unusual product shapes are required, which cannot be obtained by the normal means of glass shaping. The advantage of this method is that glass workability characteristics do not have to be taken into account. Furthermore, because sintering is the result of atomic motion stimulated by high temperature and because diffusive processes are generally dominant and favoured by the presence of a glassy liquid phase, this method can be used for materials developing a glassy phase, that is, glass–ceramics rather than ceramics. Liquid-phase sintering is an attractive option for many of the high-performance materials because of the rapid processing cycle and excellent final properties (e.g., high density, low porosity).

Many investigations have been performed in the field of glass–ceramics and different kinds of waste materials have been used (Boccaccini *et al.* 1996; Karamanov *et al.* 1999; Romero *et al.* 1999; Barbieri *et al.* 2000a, b; Boccaccini *et al.* 2000; Francis *et al.* 2002).

In our experience, different glasses have been prepared by mixing MSWI bottom ash with glass cullet in different ratios, ranging from 10 to 100 wt% ash (Barbieri *et al.* 2001b). In DTA experiments, an exothermic peak, related to the crystallization phenomenon in the glass, develops in the 890–940 °C range and becomes sharper by increasing the bottom ash content (Fig. 1). This increase is due to the increasing amounts of Fe and alkaline-earth oxides introduced by the waste, which favour the crystallization. Glass transition and softening temperatures generally increase with increasing residue input; that is, the CaO, Al_2O_3, and Fe_2O_3 contents with respect to the alkali oxides determine the increasing refractory character of the glasses. The thermal expansion coefficient decreases by adding ash due to the decrease in the alkali content, which weakens the glassy network. The glasses exhibit good resistance to chemical attacks as assessed by standard leaching tests, which are designed to compare the durability of various glasses in solutions (Shelby *et al.*

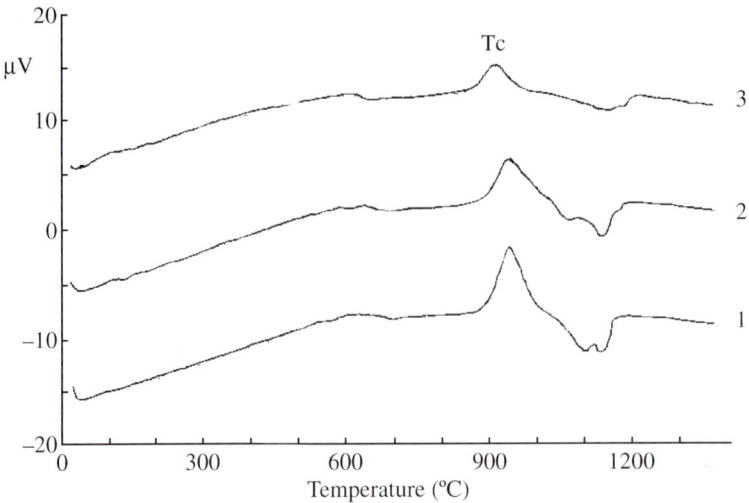

Fig. 1. DTA curves of glasses containing various amounts of MSWI bottom ash: (1) 100 wt%, (2) 90 wt%, and (3) 50 wt%. Tc = crystallization temperature.

1991). The result of the leaching experiment with distilled water during 16 days indicates that both the glass cullet (S0) and the glass based on vitrified bottom ash (S10; 100 wt% of bottom ash) release metals in quantities that are below the Italian regulation limits (Italian Official Gazette 14/04/98) (Table 5). The XRD patterns of the glass–ceramic point to the presence of wollastonite ($CaSiO_3$) as the only crystalline phase after addition of 50 wt% of bottom ash. If more than 50 wt% ash is added, crystals of pyroxenes (diopside and/or augite, general formula: $(Ca(Mg,Fe,Al)(Si,Al)_2O_6)$, and plagioclase $((Ca,Na)(Si,Al)_4O_8)$ also form. Scanning electron microscopy (SEM) shows that crystallization starts from the surface, with the formation of acicular and dendritic crystals of wollastonite and pyroxenes, respectively, dispersed in the glass matrix (Fig. 2).

These glassy materials obtained from bottom ash are also devitrified by the powder sintering technique (Barbieri et al. 2000b), showing a reduction of the time and temperature of the thermal treatment and developing the same crystalline phases typical of the conventional glass–ceramic products (wollastonite, diopside/augite) by a surface nucleation mechanism. To study sintering, density, linear shrinkage, and water absorption are monitored because these parameters allow the characterization of the sintering process. The XRD analysis as well as the experimental results shown in Fig. 3 point to 850 °C as optimal processing temperature, corresponding to high shrinkage, low water absorption, and significant degree of crystallization.

Glass fibres. Other materials with a high market value are glass fibres. Glass fibres can be divided into two broad groups depending on their geometry: (1) fibres for yarn, fabric and reinforcement, which are typically continuous, and (2) chopped fibres and discontinuous fibres used as blankets or boards for insulation and filtration. Typical fibre diameters range from 3 to 20 μm. Glass filaments are highly abrasive to each other. To minimize an abrasion-related degradation of filament strength, 'size' coatings are applied before the strand is gathered (Aubourg et al. 1991).

Table 5. *Trace element concentrations in distilled water after a 16 day leach test compared with the limit values of the Italian regulation*

Sample	Cu (mg/L)	Pb (mg/L)	Cr_{tot} (mg/L)	Zn (mg/L)
S0	0.025	0.015	–	0.02
S10	0.008	0.010	0.04	0.04
Italian limits	0.05	0.03	0.05	3

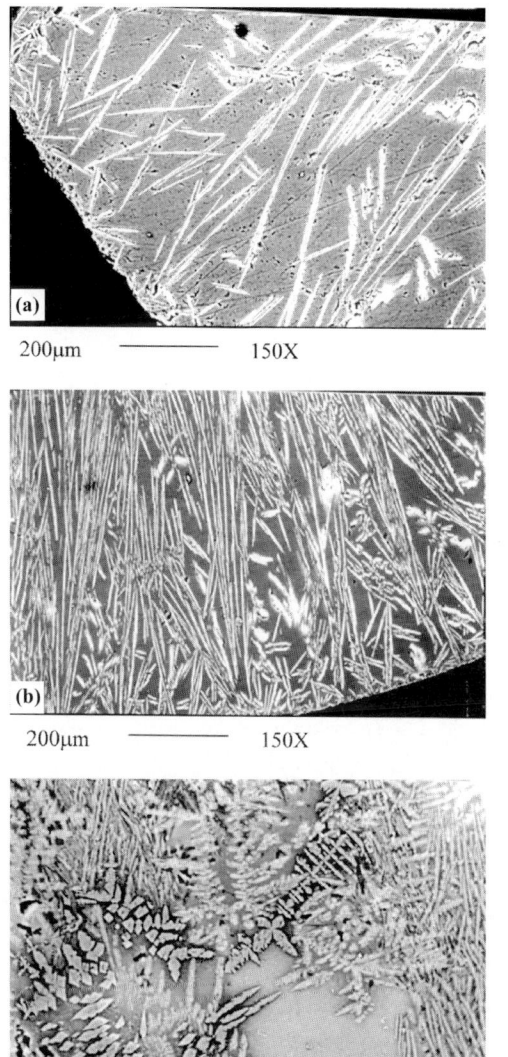

(a)

200μm ——— 150X

(b)

200μm ——— 150X

(c)

100μm ——— 300X

Fig. 2. Back scattered electron images of glass–ceramics containing various amounts of MSWI bottom ash: (**a**) 10 wt%, (**b**) 50 wt%, and (**c**) 100 wt%. Acicular are the wollastonite and dendritic (rhombic) the pyroxenes.

The MSWI bottom ash was successfully used, either vitrified alone (S10; 100 wt% of bottom ash) or with addition of 90 wt% (S1) and 50 wt% (S5) of glass cullet, and drawn into fibres using a Pt/Rh monofilament drawing apparatus (Scarinci *et al.* 2000). The choice of these glass compositions was based on their

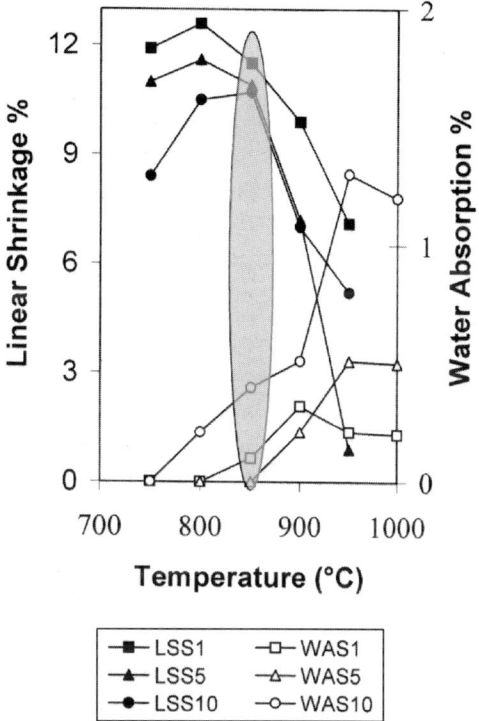

■ LSS1	□ WAS1
▲ LSS5	△ WAS5
● LSS10	○ WAS10

Fig. 3. Linear shrinkage (LS) and water absorption (WA) as a function of temperature for sintered samples containing 10 wt% (S1), 50 wt% (S5), and 100 wt% (S10) of MSWI bottom ash.

good chemical durability as determined by leaching tests. The values of tensile strength and elastic modulus are similar to values of commercial fibres (Table 6). In particular, the good effect of sizing is evident, as a strength improvement of about 40% is achieved for S5 fibres sized with γ-aminopropyl-trymethoxysilane. This behaviour results from the minor surface damage caused by handling and the lower interaction with humidity.

Conclusion

In Europe, about 64% of MSW are still land-filled. The increasing waste production necessitates a change in MSW management strategies, which depends in various countries on a large number of factors, including topography, population density, transportation infrastructure, socioeconomics, and environmental regulations. A useful option to reduce both the volume and the hazard of MSW is incineration. The MSWI generates a smaller volume of solid residues, which, however, still require further management. The re-use of incinerator waste as

Table 6. *Average tensile strength (σ) and elastic modulus (E) for glass fibres produced with different amounts of MSWI bottom ash and drawn at 1275 °C*

Sample	σ (GPa)	E (GPa)
S1	0.81 ± 0.32	58.0 ± 10.0
S5	0.82 ± 0.25	60.3 ± 8.6
S10	0.60 ± 0.26	67.1 ± 13.7
S5 with sizing	1.21 ± 0.31	71.0 ± 13.9
Commercial fibres	0.69–5.5	60–90

secondary raw material is becoming an important part of MSW management strategies. By considering physical, chemical, and mineralogical characteristics of the main solid residues, that is, bottom and fly ashes, bottom ash in particular can be used successfully for the production of glasses, glass–ceramics, and ceramic materials.

The methods and results discussed in this review underline the environmental benefits of vitrification and thus its suitability as an important process for MSWI bottom ash recovery. The possibility of transforming bottom ash into glass–ceramics is a further important development in the field of the vitrification technique. When modifying the composition of the starting batch by means of glass cullet, it is possible to obtain glasses with good chemical resistance. Moreover, this material can be drawn into fibres with qualities that are comparable with those of standard commercial fibres. Of importance is also the fact that it is possible to devitrify the amorphous material by sintering; this technique has the advantage that it reduces the time and temperature of the thermal treatment. The possibility of introducing bottom ash, especially if vitrified, into ceramic bodies without significant change of the properties of the final products and the conditions of the firing cycle is another advantage. Some technical problems could be minimized by producing glazed tiles in which the glaze can cover possible defects present on the surface of a tile.

The successful application of the presented techniques will have great advantages in waste minimization as well as resource conservation, and thus should convince both the research community and the legislators that the methods presented here are good alternatives to landfilling.

The authors thank the CNR of Italy for financial support in the project 'Innovative glass and glass–ceramic matrix composite materials (PF-MSTA II)'. They are also grateful to University of Modena and Reggio Emilia of Italy, in particular Dipartimento di Ingegneria dei Materiali e dell'Ambiente, for technical and professional support.

References

ABE, S., KAMBAYASHI, F. & OKADA, M. 1996. Ash melting treatment by rotating type surface melting furnace. *Waste Management*, **16**, 431–443.

ACOSTA, A., IGLESIAS, I., AINETO, M., ROMERO, M. & RINCÒN, J. MA. 2002. Utilisation of IGCC slag and clay steriles in soft mud bricks (by pressing) for use in building bricks manufacturing. *Waste Management*, **22**, 887–891.

ANDREOLA, F., BARBIERI, L., CORRADI, A., LANCELLOTTI, I. & MANFREDINI, T. 2001. The possibility to recycle solid residues of the municipal waste incineration into a ceramic tile body. *Journal of Material Science*, **36**, 4869–4873.

ANDREOLA, F., BARBIERI, L., CORRADI, A., LANCELLOTTI, I. & MANFREDINI, T. 2002. Utilization of municipal incinerator grate slag for porcelanized stoneware tiles manufacturing. *Journal of the European Ceramic Society*, **22**, 1457–1462.

ANDREOLA, F., BONFATTI, L. & MANFREDINI, T. 1992. Addition of exhausted lime in ceramic bodies: possibilities for an environmentally compatible tile production. Part. II Thermal behaviour of bodies. *Tile & Brick International*, **8**, 341–346.

ASSOPIASTRELLE 2000. *Indagine Statistica Nazionale, Assopiastrelle.* Modena.

AUBOURG, P. F., CRALL, C., HADLEY, J., KAVERMAN, R. D. & MILLER, D. M. 1991. Glass fibers. *In*: ASM INTERNATIONAL HANDBOOK COMMITTEE (eds) *Engineered Materials Handbook, Ceramics and Glasses.* ASM International, USA, **4**, 1027–1031.

BARBIERI, L., CORRADI BONAMARTINI, A. & LANCELLOTTI, I. 2000*a*. Alkaline and alkaline-earth silicate glasses and glass–ceramics from municipal and industrial wastes. *Journal of the European Ceramic Society*, **20**, 2477–2483.

BARBIERI, L., CORRADI, A. & LANCELLOTTI, I. 2000*b*. Bulk and sintered glass–ceramics by recycling municipal incinerator bottom ash. *Journal of European Ceramic Society*, **20**, 1637–1643.

BARBIERI, L., CORRADI, A., LANCELLOTTI, I. & PELLACANI, G. C. 2001*a*. How to transform a waste into a new marketable product with added value. *In*: AIM (eds) *Proceedings of 7th European Conference on Advanced Materials and Processes (EUROMAT 2001)*, Milan, 239–240.

BARBIERI, L., CORRADI, A. & LANCELLOTTI, I. 2001*b*. Wastes-based glasses and glass–ceramics. *Materiales de Construccion*, **51**, 197–208.

BARBIERI, L., CORRADI, A., LANCELLOTTI, I. & MANFREDINI, T. 2002. Use of municipal incinerator bottom ash as sintering promoter in industrial ceramics. *Waste Management*, **22**, 859–863.

BOCCACCINI, A. R., BÜCKER, M. & BOSSERT, J. 1996. Glass and glass–ceramics from coal fly ash and waste glass. *Tile and Brick International*, **12**, 515–518.

BOCCACCINI, A. R., KERN, H., RINCÓN, J. MA., ROMERO, M., PETITMERMET, M. & WINTERMANTEL, E. 1999. Transforming municipal incinerator fly ash in glass and glass–ceramics. *In*:

BARRAGE, A. & EDELMANN, X. (eds) *Proceedings of the R'99 World Congress on R'99 Recovery recycling re-integration*, Vol II. EMPA, Geneva, 152–157.

BOCCACCINI, A. R., SCHAWOHL, J., KERN, H., SCHUNK, B., RINCÒN, J. MA. & ROMERO, M. 2000. Sintered glass–ceramics from municipal incinerator fly ash. *Journal of Glass Technology*, **41**, 99–105.

CHIMENOS, J. M., SEGARRA, M., FERNANDEZ, M. A. & ESPIELL, F. 1999. Characterization of the bottom ash in municipal solid waste incinerator. *Journal of Hazardous Material*, **64**, 211–222.

CUSIDÒ, J. A., DEVANT, M., CELEBROVSKY, M., RIBA, J. & ARTEAGA, F. 1996. Ecobrick®: a new ceramic material for solar buildings. *Renewable Energy*, **8**, 327–330.

DAS, S. K., KUMAR, S. & RAMACHANDRARAO, P. 2000. Exploitation of iron ore tailing for the development of ceramic tiles. *Waste Management*, **20**, 725–729.

ESCARDINO, A., AMOROS, J. L., MORENO, A. & SANCHEZ, E. 1995. Utilizing the used catalyst from refinery FCC units as a substitute for kaolin in formulating ceramic frits. *Waste Management & Research*, **13**, 569–577.

EUROSTAT, 2000. *Environment Statistics: pocketbook 2000*. OPOCE, Luxembourg.

FALCONE, R., HREGLICH, S., PROFILO, B. & VALLOTTO, M. 2003. Inertization by vitrification of mixtures of municipal waste incineration slag and ash and valorisation of the obtained glass products. *Rivista della Stazione Sperimentale del Vetro*, **4**, 25–29.

FERRARIS, M., SALVO, M., SMEACETTO, F., AUGIER, L., BARBIERI, L., CORRADI, A. & LANCELLOTTI, I. 2001. Glass matrix composites from solid waste materials. *Journal of the European Ceramic Society*, **21**, 453–460.

FRANCIS, A. A., RAWLINGS, R. D., SWEENEY, R. & BOCCACCINI, A. R. 2002. Processing of coal ash into glass–ceramic products by powder technology and sintering. *Journal of Glass Technology*, **43**, 58–62.

HERNÀNDEZ-CRESPO, M. S. & RINCÒN, J. MA. 2001. New porcelainized stoneware materials obtained by recycling of MSW incinerator fly ashes and granite sawing residues. *Ceramic International*, **27**, 713–720.

ITALIAN OFFICIAL GAZETTE 14/04/98.

KARAMANOV, A., TAGLIERI, G. & PELINO, M. 1999. Iron-rich sintered glass–ceramics from industrial wastes. *Journal of the American Ceramic Society*, **82**, 3012–3016.

MORTEL, H. & FUCHS, F. 1997. *Recycling of Windshield Glasses in Fired Brick Industry*. Key Engineering Materials Trans Tech Publications, Switzerland, 132–136, 2268–2271.

NISHIGAKI, M. 2000. Producing permeable blocks and pavement bricks from molten slag. *Waste Management*, **20**, 185–192.

PALMONARI, C. 1989. *In*: RUGGERI (ed) *Il Grès Porcellanato*. Castellarano Fiandre Ceramiche S.p.A., Modena, 35–80.

RAIGON-PICHARDO, M., GARCIA-RAMOS, G. & SANCHES-SOTO, P. J. 1996. Characterisation of a waste washing solid product of mining granitic tin-bearing sands and its application as ceramic raw material. *Resources, Conservation and Recycling*, **17**, 109–124.

ROMERO, M., RAWLINGS, R. D. & RINCÒN, J. MA. 1999. Development of a new glass ceramic by means of controlled vitrification and crystallisation of inorganic wastes from urban incineration. *Journal of the European Ceramic Society*, **19**, 2049–2058.

ROTH, R. S. 1995. *Phase Diagrams for Ceramists*, The American Ceramic Society, Westerville, 293.

SCARINCI, G., BRUSATIN, G. *et al.* 2000. Vitrification of industrial and natural wastes with production of glass fibers. *Journal of the European Ceramic Society*, **20**, 2485–2490.

SEZZI, G. 2000. Grès porcellanato, riferimento obbligato. *Ceramic World Review*, **37**, 84–88.

SHELBY, J. E., LACOURSE, W. C. & CLARE, A. G. 1991. Engineering properties of oxide glasses and other inorganic glasses. *In*: ASM INTERNATIONAL HANDBOOK COMMITTEE (eds) *Engineered Materials Handbook, Ceramics and Glasses*. ASM International, USA, **4**, 845–857.

VIEIRA, M. T., CATARINO, L. *et al.* 1999. Optimisation of the sintering process of raw materials wastes. *Journal of Materials Processing Technology*, **92–93**, 97–101.

VOLF, M. B. 1984. Introduction to the chemistry of glass. *In*: *Chemical Approach to Glass – Glass Science and Technology* 7. Elsevier, The Netherlands, 17–138.

WIEBUSCH, B. & SEYFRIED, C. F. 1997. Utilization of sewage sludge ashes in the brick and tile industry. *Water Science and Technology*, **36**, 251–258.

WILES, C. C. 1996. Municipal solid waste combustion ash: State-of-the-Knowledge. *Journal of Hazardous Material*, **47**, 325–344.

YALCIN, N. & SEVINC, V. 2000. Utilization of bauxite waste in ceramic glazes. *Ceramic International*, **26**, 485–493.

ZACARIAS FARAH, A. & GEYER ALLÈLY, E. 2003. Household consumption patterns in OECD countries: trends and figures. *Journal of Cleaner Production*, **11**, 819–827.

Phosphate stabilization of municipal solid waste combustion residues: geochemical principles

T. TAYLOR EIGHMY[1] & J. DYKSTRA EUSDEN JR.[2]

[1]*Environmental Research Group, Civil Engineering Department, University of New Hampshire, Durham, NH, USA (e-mail: taylor.eighmy@unh.edu)*

[2]*Geology Department, Carnegie Science Center, Bates College, Lewiston, ME, USA (e-mail: deusden@abacus.bates.edu)*

Abstract: The use of orthophosphate (PO_4^{3-}) as a chemical stabilization agent for municipal solid waste (MSW) combustion residues is widespread in Japan and North America. The application of this technology to MSW ashes generally parallels its use with other metal contaminated wastes (e.g., soils, sediments, smelter dusts, slags, wire chopping wastes, mine tailings), especially Pb-contaminated soils. The technology relies on the fact that PO_4^{3-} forms very insoluble and stable minerals for a variety of divalent metal cations (e.g., Pb, Cd, Cu, and Zn). Extensive data from phosphate-treated contaminated soil systems suggest that stabilization involves surface immobilization reactions involving sorption, heterogeneous nucleation and surface precipitation, and/or solution phase precipitation involving homogeneous nucleation and precipitation. A geochemical basis for use of PO_4^{3-} in ash systems is presented with a focus on the wide theoretical pH distribution, pH–pE predominance and redox stability of Cd, Cu, Pb, and Zn phosphates within complex bottom ash pore water systems. Stabilization mechanisms in bottom ashes, scrubber residues, and vitrification dusts are similar to those observed in soil systems. Some longer term leaching behaviour of phosphate-stabilized ashes are presented. The roles of Ostwald ripening, solid solutions (e.g., (Pb,Ca)$_5$(PO4)$_3$OH), and kinetically controlled reaction pathways probably are more important than what is presently envisioned in phosphate-stabilized ash systems.

The stabilization of municipal solid waste (MSW) combustion or incineration residues with orthophosphate (PO_4^{3-}) is just one example of the use of phosphates to chemically stabilize wastes. Phosphates are increasingly used as chemical stabilization agents for contaminated soils (Christensen & Nielsen 1987; Rabinowitz 1993; Ruby et al. 1994; Wright et al. 1995; Cotter-Howells & Caporn 1996; Laperche et al. 1996; Berti & Cunningham 1997; Hettiarachchi et al. 2000; Hodson et al. 2000; Raicevic 2001; Basta et al. 2001; Grcman et al. 2001; Hodson et al. 2001; Matheson et al. 2001; McGowen et al. 2001; Ryan et al. 2001; Wang et al. 2001; Yang et al. 2001, 2002; Stanforth & Qiu 2001; Cao et al. 2002; Hamon et al. 2002; Theodoratos et al. 2002), contaminated sediments (Arey et al. 1999; Crannell et al. 2003a), electric arc furnace smelter dusts (Crannell et al. 1999), various smelter slags (Tickanen & Turpin 1997; Ioannidis & Zouboulis 2003), lead ore mine tailings (Xenidis et al. 1999; Eusden et al. 2002), heavy metal plating sludges (Tikanen & Turpin 1997), and metal-contaminated industrial wastewaters (Admassu

& Breese 1999; Liu & Peng 2001). They are also used in less permeable reactive barriers for heavy metal contaminated sediment capping (Crannell et al. 2003a, b) and in more permeable reactive barriers for U-contaminated groundwater contaminant plume treatment (Wright et al. 1995; Matheson et al. 2001). Phosphates are also used in the microencapsulation of low-level mixed wastes using chemically bonded phosphate ceramics (Singh et al. 1997; Wagh et al. 1999; Rao et al. 2000). Presently, there are commercial processes in use for MSW ashes, wire chopping wastes, contaminated soils, dredged materials, and contaminated groundwater. A cursory search using the online database SciFinder Scholar 2001 (from the American Chemical Society) on published patent documents pertaining to phosphate stabilization or treatment of wastes containing heavy metals shows that approximately 39 patents were issued in Japan and the USA from 1994. Many of these relate to MSW ash treatment.

The residues from MSW combustion are typically referred to as bottom or grate ashes, boiler or economizer ashes, and ashes from air pollution

From: GIERÉ, R. & STILLE, P. (eds) 2004. *Energy, Waste, and the Environment: a Geochemical Perspective*. Geological Society, London, Special Publications, **236**, 435–473.
0305-8719/04/$15 © The Geological Society of London 2004.

control operations (e.g., electrostatic precipitator (ESP) ash, dry scrubber residues, and wet scrubber solids). Combined ashes are usually mixtures of bottom ashes or grate ashes with air pollution control ashes. Dusts collected in air pollution control systems from the melting or vitrification of MSW incinerator ashes are also considered here.

During MSW combustion in a modern mass burn combustor with reciprocating grates, 1 t of MSW is converted to heat; gases such as CO_2, NO_X, SO_X, and H_2O; and about 350 kg of ash residuals partitioned into the various ash streams. For every tonne of MSW combusted, about 5 kg of grate siftings, 295 kg of grate ash, 5 kg of boiler/economizer ash, 20 kg of ESP ash, and 12 kg of dry scrubber solids are produced (Chandler et al. 1997).

Grate siftings are frequently low boiling point pure metals (Al, Pb, Zn, and Cu) that melt and drip through the reciprocating grates. Grate ash is the high-temperature lithophilic Al–Si–Fe residue that is left on the grates during combustion. When combined with grate siftings, these materials are referred to as bottom ash. Grate ash is a larger particle size (mm to cm) glassy slag similar to alkaline volcanic rock in its petrogenesis (Eusden et al. 1999). Meanwhile, ESP ashes are finer particle size (μm to mm) fly ashes similar to wollastonite- and spinel-dominated coal fly ash in their formation and mineral composition, usually with condensed volatile elements (as metal salts) on the outer surface of the particles (Eighmy et al. 1995). They are removed from hot flue gases in an ESP. Dry scrubber residues are both fly ashes and acid gas neutralization reaction products that are captured in baghouses. Lime $(Ca(OH)_2)$ or hydrated lime $(CaOH_2)$ is added to the flue gas stream to form a filter cake on the filter bag. Within the filter cake, acid gases are neutralized, forming gypsum $(CaSO_4 \cdot 2H_2O)$, calcium chloride $(CaCl_2)$, and calcium chloride hydroxide (CaClOH). Fly ash particles are also removed in the cake (Eighmy et al. 1997). Wet scrubber solids come from the use of wet scrubber towers that contain dissolved sodium carbonate (Na_2CO_3) or hydrated lime to neutralize acid gases. A filter cake is produced as the wet scrubber solution is treated to promote solids precipitation, coagulation, and separation. These residues can also contain high concentrations of metals. Vitrification dusts are collected in air pollution bag houses or ESPs downstream of high-temperature melting furnaces of MSW ashes. These residues can contain very high concentrations of more volatile metals (e.g., Pb, Zn, and Cd) (Eighmy et al. 1999).

The focus of this review is on the stabilization of the more leachable and problematic elements in MSW ashes, especially Pb, Zn, Cu, and Cd. However, given the dominance of Ca in ash systems and the role that Ca plays in phosphate mineral chemistry, this element is also included. Much of the work on phosphate stabilization has been in fundamental systems or other wastes, so the literature from these fields has great relevance to MSW ash stabilization. This review starts with phosphate stabilization principles, mineralogy and geochemistry, stabilization in natural systems, surface reactions involved with immobilization, solution phase stabilization, and use of phosphate for metal stabilization in soils. It then turns to the theory and use of phosphate in MSW incineration residues, confirmation of the stabilization mechanisms in ashes, and evaluation of longer term stability of stabilized ashes. More detailed background information of phosphate geochemistry is found in Nriagu & Moore (1984), Tardy & Veillard (1977), Lindsay (1979), and Lindsay & Vlek (1977).

Phosphate stabilization principles

To place MSW residues into a geochemical context, the types of matrix forming and leachable elements and their concentrations found in various MSW ashes are given in Table 1. There is appreciable P in ash residues, but it may not be readily available for stabilization reactions in untreated ashes. The concentrations of these elements in pore waters or low liquid-to-solid (LS) ratio leachates are given in Table 2. These concentrations are presumed to be in pseudo-equilibrium (or local equilibrium) with the ash solid phases that they contact. Such assumptions are generally sound based on geochemical modeling of solid phase control of leachates with thermodynamic equilibrium models and confirmed by spectroscopic characterization of the residues (e.g., Eighmy et al. 1995; Meima & Comans 1997, 1999). The molar concentrations in the pore waters or leachates vary widely, depending on the speciation and relative solubility of the element in the ash. Such equilibrium concentrations are useful when thermodynamically modeling the predominance and stability of metal phosphate minerals in such systems. Typically, the system is dominated by Ca, Cl, S (principally SO_4^{2-}), Na, K, and sometimes C (principally HCO_3^- and CO_3^{2-}). The systems are almost always alkaline (pH values 8.0–11.0), have high ionic strengths (0.01–1.0 molal though extremes outside this range can be observed), and can sometimes contain

Table 1. *Ranges in composition of MSW ashes[†] for some important matrix and leachable elements*[*]

Element	Bottom ash	Fly ash (ESP ash)	Dry scrubber residue (with fly ash)	Wet scrubber residue (without fly ash)
Al	0.8–2.7	1.8–3.3	0.4–3.0	0.8–1.4
Ca	0.01–3.0	1.8–3.2	2.7–9.0	2.1–5.0
Cd	0.000003–0.0001	0.0004–0.004	0.001–0.003	0.001–0.01
Cl	0.02–0.12	0.8–6.0	1.7–10.8	0.5–1.6
Cu	0.05–0.20	0.01–0.05	0.0003–0.02	0.007–0.04
F	0.01–0.05	—	—	—
Fe	0.9–2.6	0.5–1.0	0.05–1.2	0.35
K	0.02–0.4	0.5–1.5	0.1–1.0	0.02–0.2
Mg	0.02–1.0	0.4–0.8	0.2–0.6	0.08–2.9
Na	0.1–1.8	0.6–2.5	0.3–1.3	0.03–0.14
P	0.04–0.2	0.15–0.31	0.05–0.14	—
Pb	0.0005–0.03	0.02–0.12	0.01–0.04	0.0001–0.0015
S	0.03–0.15	0.3–1.4	0.04–0.8	0.8–0.2
Si	3.2–11.0	3.3–7.5	1.2–4.3	0.7–3.2
Zn	0.01–0.1	0.1–1.2	0.1–0.3	0.1–0.8

[*]From Chandler *et al.* 1997.
[†]Moles/kg dry weight.

appreciable dissolved organic carbon if the ash is not burned out or was produced under poor combustion conditions.

Chemical stabilization of waste materials offers the potential to reduce the leachability of contaminants present in the waste. As shown in Fig. 1, it involves the conversion of the available fraction (or soluble species) of an element in a waste to an unavailable fraction (insoluble species) that is so stable geochemically with respect to pH, solution phase complexation, and pE that it is considered to be stabilized and unavailable for leaching.

The stabilization reaction for essentially all ash systems likely involves solubilization of the available fraction, contact between the orthophosphate and the element to be stabilized via either (1) solution phase precipitation by

Table 2. *Matrix and leachable element composition of MSW ash leachates*[*] *in pseudoequilibrium with the ashes*

Element	Fresh bottom ash lysimeter extracted pore water and leachates (Chandler *et al.* 1997; Crannell *et al.* 2000)	Aged bottom ash extracted pore waters (Mcima 1997)	Dry and semi-dry scrubber residues low LS (0.0–0.2) column leachates (Chandler *et al.* 1997)	Wet scrubber residue low LS (0.0–0.2) column leachates (Chandler *et al.* 1997)
Al	—	4.4×10^{-6}–2.1×10^{-4}	—	—
C (CO_3^{2-})	0.001–0.01	1.0×10^{-4}–1.0×10^{-3}	1.0×10^{-3}–1.0×10^{-1}	—
Ca	—	0.01–0.03	2.75–4.0	0.15
Cd	9.0×10^{-9}	9.0×10^{-8}–9.0×10^{-6}	8.9×10^{-7}–1.7×10^{-5}	5.3×10^{-8}
Cl	0.001 – 0.13	0.003–0.3	5.1–8.3	1.85
Cu	4.7×10^{-6}	1.5×10^{-6}–2.0×10^{-4}	2.7×10^{-6}–0.001	1.4×10^{-8}
Fe	—	7.1×10^{-8}–6.0×10^{-7}	—	—
K	—	0.02–0.4	0.19–0.95	0.6
Mg	—	4.1×10^{-6}–4.6×10^{-5}	—	—
Na	—	0.04–0.5	0.4–1.6	1.0
P (PO_4^{3-})	—	0.008–0.11	—	—
Pb	9.6×10^{-8}	5.0×10^{-8}–5.0×10^{-6}	1.0×10^{-5}–0.05	9.1×10^{-9}
S (SO_4^{2-})	0.001 – 0.023	0.07–0.15	0.002–0.007	0.03
Si	—	8.9×10^{-5}–0.002	—	—
Zn	2.3×10^{-6}	1.5×10^{-6}–1×10^{-4}	3.0×10^{-7}–0.01	3.0×10^{-7}
pH	9.3	8.0–11.0	9.8–10.2	9.0
Typical I	0.01–0.2	0.01–0.5	0.1–10.0	0.1–2.0

[*]In moles/L.

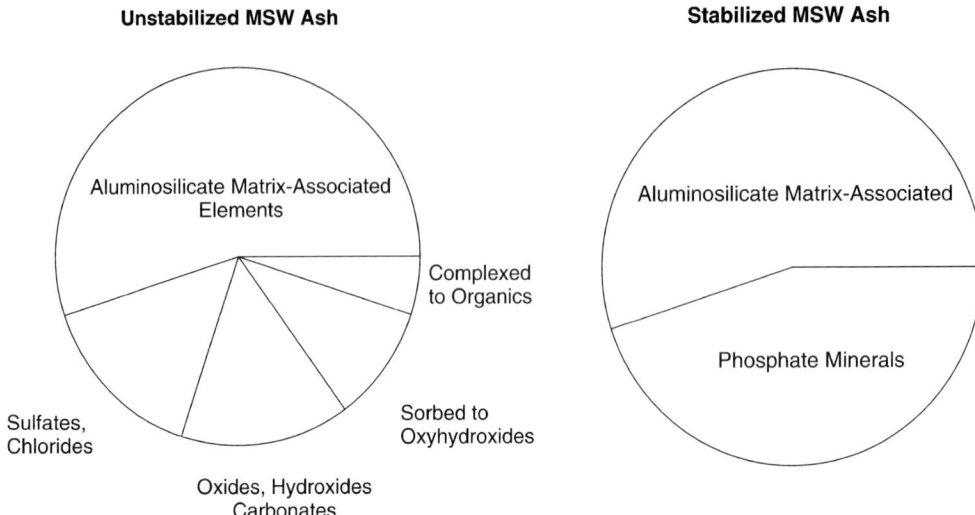

Fig. 1. Schematic of chemical stabilization principle. Soluble phases are converted to insoluble phosphate minerals.

homogeneous nucleation, crystallization, and crystal ripening or (2) surface immobilization via sorption and heterogeneous surface precipitation (and possible re-precipitation) on the surface of the solid. An example of the former mechanism is precipitation of the pure mineral $Cd_5H_2(PO_4)_4 \cdot 4H_2O$ from solutions containing 0.1 M Cd^{2+} and 0.1 M $NH_4H_2PO_4$; the crystal morphology changes from prismatic crystals to twins, aggregates and dendrites as pH of the solution changes while crystals ripen and increase in size, become more structured and ordered, and decrease in solubility (Ayati & Lundager Madsen 2000, 2001). An example of the latter mechanism is described by reactions between Pb^{2+} and hydroxyapatite $(Ca_5(PO_4)_3OH)$. As Pb sorbs to the hexagonal crystal surface, hydroxypyromorphite $(Pb_5(PO_4)_3OH)$ needles grow epitaxially off the surface as a surface precipitate, forming a different mineral surface than the host mineral (Lower *et al.* 1998a).

These stabilization reactions can occur in a variety of contexts: in a highly engineered mixing apparatus that blends the ash and the treatment agent; on a landfill site during crude mixing of the ash and a treatment agent; within a landfill, but after closure and during permanent disposal, when diffusive processes begin to control mass transport of reactants; inside a laboratory batch or column leaching vessel; or even in an extraction vessel during a regulatory leaching test (e.g., Scheckel *et al.* 2003).

The degree of success of the stabilization reaction depends not only on the system LS, pH, I, temperature, time, concentration of aqueous phase ligands, and presence of other dissolved ions that can participate in the precipitation reaction, but also on the geochemical reaction types and pathways. Geochemical modeling efforts in this review elaborate on some of these phenomena. During treatment, some phosphates are already soluble and available for contact, others will dissolve and make soluble phosphate available, and others will act as sorbent surface or crystallite seeds where surface adsorption or nucleation of the metal phosphate will occur. Some reactions are kinetically constrained, even though they may be thermodynamically preferred with respect to precipitation/dissolution equilibria.

Phosphate mineralogy and geochemistry

As shown in Fig. 2, phosphate combines with at least 38 elements to form over 300 naturally occurring phosphate minerals. The concept of the figure comes from Lindsay & Vlek (1977) with data from Nriagu (1984) through 1984. Undoubtedly, additional minerals have been identified since then. Phosphate minerals can form under a wide range of geochemical conditions: from high-temperature silicate melts, to low-temperature soil weathering systems, to diagenic processes in ocean floors (Lindsay & Vlek 1977). Metal phosphates are ubiquitous secondary minerals in the oxidized zones of lead ore deposits and as assemblages around ore bodies (Nriagu 1984). They also occur in soils, sediments, and phosphatic or phosphorite beds (Nriagu 1984). As such, they are considered

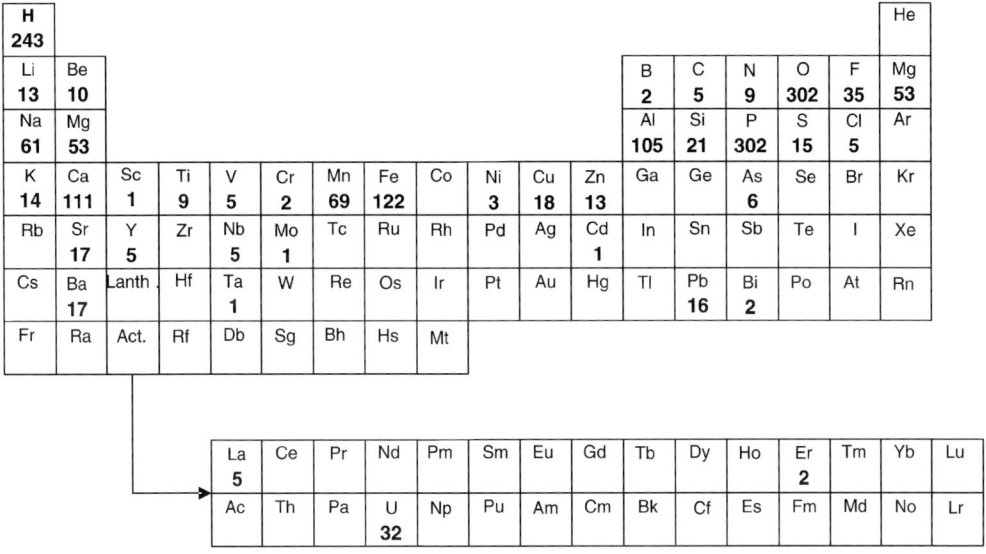

Fig. 2. Elements found in naturally occurring phosphate minerals (data derived from Nriagu, 1984). The number shown for an element indicates the number of naturally occurring phosphate minerals that contain the element. For instance, there are 16 naturally occurring phosphate minerals that contain Pb.

very stable geochemically with respect to pH, pE, and mineral authigenesis (Nriagu 1974, 1984).

The affinity between the tetrahedrally arranged orthophosphate oxyanion, PO_4^{3-}, and hexavalently coordinated metal cations lends itself to a classification of phosphate minerals in a scheme similar to silicates (SiO_4^{4-}): framework, insular, chain, and layer phosphates. Examples of this scheme, advanced by Povarennykh (1972) and further elaborated by Lindsay & Vlek (1977), include berlinite ($AlPO_4$ framework); hydroxyapatite (insular); monetite ($CaHPO_4$, chain); and vivianite ($Fe_4(PO_4)_2 \cdot 2H_2O$, layer).

The apatite group minerals are the most abundant phosphorus-bearing minerals on Earth, typically as accessory minerals in basic to acidic igneous rocks, pegmatites, hydrothermal veins and cavities, carbonates, contact and regionally metamorphosed rocks, and sedimentary rocks (Deer et al. 1996). The principal members of the apatite group include fluoroapatite ($Ca_5(PO_4)_3F$), chloroapatite ($Ca_5(PO_4)_3Cl$), hydroxyapatite, and carbonate apatite ($Ca_5(PO_4, CO_3)_3(F,OH)$) (Deer et al. 1996).

Table 3 provides a list of Pb, Cd, Cu, Zn and Ca phosphate minerals likely to be important in soils and sediments where anthropogenic contamination is likely (Nriagu 1984). These data are from the NIST 46 (version 4) database (NIST 2001), Nriagu (1976), Viellard & Tardy (1984), Nriagu (which include estimates) (1984), Rickard & Nriagu (1978), and Wagman et al. (1982). The table includes solubility product (K_{sp}) values for the indicated dissolution reaction and Gibbs free energy of formation (ΔG_f^0, kJ/mole) for the solid phase. Many of the metal phosphate minerals that form when divalent metal cations come into contact with orthophosphate are both extremely stable and insoluble (Nriagu 1974, 1984). The fluoroapatites are the most insoluble form of apatite. Generally, metal phosphates, particularly the apatite family of minerals, exhibit wide distribution in pC–pH diagrams and wide predominance in pE–pH (Pourbaix) diagrams.

The apatite family of minerals is a common feature to many of the minerals shown in Table 3. In nature, the apatite mineral structure conforms to the 6/m class of minerals with hexagonal crystal structure and the generic formula $Me_5(XO_4)_3Z$ where Me is Ca, Sr, Ba, Cd, and Pb (typically), X = P, As, V, Mn, and Cr; and Z = OH, F, Cl, and Br. In addition to carbonate apatite, chloroapatite, chloropyromorphite, fluoroaptite, fluoropyromorphite, hydroxyapatite, and hydroxypyromorphite, the family includes abukumalite (($Ca,Th,Ce)_5(PO_4,SiO_4)_3(OH,F)$), britholite (($Ca,Th,Ce)_5(PO_4,SiO_4)_3(OH,F)$, dahllite ($Ca_5(PO_4,CO_3)_3(OH,F)$, ellestadite ($Ca_5\{(Si,S,P,C)O_4\}_3(Cl,F,OH)$, fermorite ($(Ca,Sr)_5\{(As,P)O_4\}_3(F,OH)$, francolite ($Ca_5(PO_4,CO_3)_3(OH,F)$,

Table 3. *Some phosphate mineral solubility product and Gibbs free energy of formation data*

Mineral class/name	Dissolution reaction	$-pK_{sp}$	ΔG_f^0 (kJ/mole)	Ref.
Apatites				
Hydroxylapatite	$Ca_5(PO_4)_3OH + H^+ \Leftrightarrow 5Ca^{2+} + 3PO_4^{3-} + H_2O$	44.33	−6,413	G
Chloroapatite	$Ca_5(PO_4)_3Cl \Leftrightarrow 5Ca^{2+} + 3PO_4^{3-} + Cl^-$	46.89	−6,223	B
Hydroxypyromorphite	$Pb_5(PO_4)_3OH + H^+ \Leftrightarrow 5Pb^{2+} + 3PO_4^{3-} + H_2O$	62.80	−3,774	B
Chloropyromorphite	$Pb_5(PO_4)_3Cl \Leftrightarrow 5Pb^{2+} + 3PO_4^{3-} + Cl^-$	84.43	−3,791	B
$Cd_5(PO_4)_3OH$	$Cd_5(PO_4)_3OH + H^+ \Leftrightarrow 5Cd^{2+} + 3PO_4^{3-} + H_2O$	56.49	−4,004	G
$Cd_5(PO_4)_3Cl$	$Cd_5(PO_4)_3Cl \Leftrightarrow 5Cd^{2+} + 3PO_4^{3-} + Cl^-$	49.66	−3,859	B
$Zn_5(PO_4)_3OH$	$Zn_5(PO_4)_3OH + H^+ \Leftrightarrow 5Zn^{2+} + 3PO_4^{3-} + H_2O$	49.10	−4,309	C
$Zn_5(PO_4)_3Cl$	$Zn_5(PO_4)_3Cl \Leftrightarrow 5Zn^{2+} + 3PO_4^{3-} + Cl^-$	37.53	−4,137	A
$Cu_5(PO_4)_3OH$	$Cu_5(PO_4)_3OH + H^+ \Leftrightarrow 5Cu^{2+} + 3PO_4^{3-} + H_2O$	51.62	−6,279	C
$Cu_5(PO_4)_3Cl$	$Cu_5(PO_4)_3Cl \Leftrightarrow 5Cu^{2+} + 3PO_4^{3-} + Cl^-$	53.96	−3,168	A
Tertiary metal phosphates				
Low Whitlockite	$\beta\text{-}Ca_3(PO_4)_2 \Leftrightarrow 3Ca^{2+} + 2PO_4^{3-}$	28.92	−3,863	G
$Pb_3(PO_4)_2$	$Pb_3(PO_4)_2 \Leftrightarrow 3Pb^{2+} + 2PO_4^{3-}$	43.53	−2,359	G
$Zn_3(PO_4)_2$	$Zn_3(PO_4)_2 \Leftrightarrow 3Zn^{2+} + 2PO_4^{3-}$	27.11	−2,633	B
$Cu_3(PO_4)_2$	$Cu_3(PO_4)_2 \Leftrightarrow 3Cu^{2+} + 2PO_4^{3-}$	36.85	−2,051	B
$Cd_3(PO_4)_2$	$Cd_3(PO_4)_2 \Leftrightarrow 3Cd^{2+} + 2PO_4^{3-}$	32.60	−2,456	B
$Mg_3(PO_4)_2$	$Mg_3(PO_4)_2 \Leftrightarrow 3Mg^{2+} + 2PO_4^{3-}$	24.38	−3,538	B
Tetra metal phosphates				
Hilgenstockite	$Ca_4O(PO_4)_2 + 2H^+ \Leftrightarrow 4Ca^{2+} + 2PO_4^{3-} + H_2O$	17.36	−4,588	B
$Pb_4O(PO_4)_2$	$Pb_4O(PO_4)_2 + 2H^+ \Leftrightarrow 4Pb^{2+} + 2PO_4^{3-} + H_2O$	36.86	−2,582	B
Other phosphate minerals				
$AlPO_4$	$AlPO_4 \Leftrightarrow Al^{3+} + PO_4^{3-}$	17.00	−1,601	B
Monetite	$CaHPO_4 \Leftrightarrow Ca^{2+} + PO_4^{3-} + H^+$	19.09	−1,681	E

Mineral	Reaction			Ref.
Brushite	$CaHPO_4 \cdot 2H_2O \Leftrightarrow Ca^{2+} + PO_4^{3-} + 2H_2O + H^+$	18.93	-2.154	B
Cornetite	$Cu_3PO_4(OH)_3 + 3H^+ \Leftrightarrow 3Cu^{2+} + PO_4^{3-} + 3H_2O$	5.94	-1.567	C
Libethenite	$Cu_2PO_4OH + H^+ \Leftrightarrow 2Cu^{2+} + PO_4^{3-} + H_2O$	14.00	-1.204	C
Pseudomalachite	$Cu_5(PO_4)_2(OH)_4 + 4H^+ \Leftrightarrow 5Cu^{2+} + 2PO_4^{3-} + 4H_2O$	19.83	-2.771	C
Corkite	$PbFe_3(PO_4)(OH)_6SO_4 + 6H^+ \Leftrightarrow Pb^{2+} + 3Fe^{3+} + PO_4^{3-} + SO_4^{2-} + 6H_2O$	28.66	-3.388	C
Spencerite	$Zn_4(PO_4)_2(OH)_2AlPO_4 \cdot 3H_2O + 2H^+ \Leftrightarrow 4Zn^{2+}Al^{3+} + 3PO_4^{3-} + 5H_2O$	24.77	-3.953	C
Zn Rockbridgite	$ZnFe_4(PO_4)_3(OH)_5 + 5H^+ \Leftrightarrow Zn^{2+} + 4Fe^{3+} + 3PO_4^{3-} + 5H_2O$	68.55	-4.799	C
Scholzite	$CaZn_2(PO_4)_2 \cdot 2H_2O \Leftrightarrow Ca^{2+} + 2Zn^{2+} + 2PO_4^{3-} + 2H_2O$	34.10	-3.553	C
Tarbuttite	$Zn_2(PO_4)OH + H^+ \Leftrightarrow 2Zn^{2+} + PO_4^{3-} + H_2O$	12.55	-1.621	C
Faustite	$ZnAl_6(PO_4)_4(OH)_8 \cdot 4H_2O + 8H^+ \Leftrightarrow Zn^{2+} + 6Al^{3+} + 4PO_4^{3-} + 12H_2O$	65.70	-10.355	C
Plumbogummite	$PbAl_3(PO_4)_2(OH)_5 \cdot H2O + 5H^+ \Leftrightarrow Pb^{2+} + 3Al^{3+} + 2PO_4^{3-} + 6H_2O$	29.36	-5.108	C
Hinsdalite	$PbAl_3(PO_4)(SO_4)(OH)_6 + 6H^+ \Leftrightarrow Pb^{2+} + 3Al^{3+} + PO_4^{3-} + SO_4^{2-} + 6H_2O$	15.10	-4.753	D
Tsumebite	$CuPb_2(PO_4)(OH)_3 \cdot 3H_2O + 3H^+ \Leftrightarrow Cu^{2+} + 3Pb^{2+} + PO_4^{3-} + 6H_2O$	9.36	-2.478	D
Cadmium phosphate	$Cd_5H_2(PO_4)_4 \cdot 4H_2O + 2H^+ \Leftrightarrow 5Cd^{2+} + 4PO_4^{3-} + 4H_2O$	30.9	—	F

A, Nriagu 1976; B, Vielard & Tardy 1984; C, Nriagu 1984; D, Rickard & Nriagu 1978; E, Wagman et al. 1982; F, Ayati & Lundager Madsen 2001; G, NIST Standard Reference Database 46, version 4.0.

mimetite $(Pb_5(AsO_4)_3Cl)$, svabite $(Ca_5(AsO_4)_3$ (F,Cl,OH), vanadinite $(Pb_5(VO_4)_3Cl)$, and wilkeite $(Ca_5(SiO_4, PO_4,SO_4)_3(O,OH,F))$ (Nriagu & Moore 1984). It is possible to make synthetic hydroxyapatite, fluoroapatite and chloroapatite minerals that conform to the same mineral structure and formula as natural apatites, although some can have distorted habits (McConnell 1973).

In hydroxyapatites, the calcium can typically be substituted with divalent, hexavalently coordinated cations with ionic radii between 0.69 and 1.35 Å. This includes the elements Ba (1.34 Å), Cd (0.97 Å), Co (0.72 Å), Cu (0.72 Å), Mg (0.66 Å), Mn (0.80 Å), Ni (0.69 Å), Pb (1.20 Å), and Sr (1.12 Å). The orthophosphate can typically be substituted with oxyanion-forming elements with ionic radii between 0.29 and 0.60 Å. This includes As (0.46 Å) and V (0.59 Å). In chloroapatites, the calcium can typically be substituted with divalent, hexavalently coordinated cations with ionic radii between 0.80 and 1.35 Å. This includes the elements Ba (1.34 Å), Cd (0.97 Å), Mn (0.80 Å), Pb (1.20 Å), and Sr (1.12 Å). The orthophosphate can typically be substituted with oxyanion-forming elements with ionic radii between 0.29 and 0.60 Å. This includes As (0.46 Å) and V (0.59 Å). These isomorphic substitutions can be complete for elements like Pb, Sr, and Ba as evidenced by minerals like chloropyromorphite (Nriagu 1984) or partial for elements like Zn, Cu, Ni, and Ca, forming solid solutions.

Solid solutions within the apatite family are readily synthesized in the laboratory. Some examples include $(Ca,Zn,Pb)_5(PO_4)_3OH$ (Panda et al. 1991), $(Ca,Cd,Pb)_5(PO_4)_3OH$ (Mahapatra et al. 1995), $(Ca,Sr,Cu)_5(PO_4)_3OH$ (Pujari & Patel 1989), or other apatite solid solutions containing various quantities of Cd, Mg, Zn, Cd or Y (Ergun et al. 2001). They are also found naturally (Botto et al. 1997). Unlike well-ordered naturally occurring minerals, these solid solutions may actually be the more typical form of the mineral in stabilized ash systems given the system complexity, rapid precipitation kinetics, and wide prevalence of available divalent cations.

Evidence of metal phosphates in natural and anthropogenically contaminated systems

It is useful to examine metal phosphate geochemical principles and behaviour in perturbed natural systems to elucidate the role of phosphate in sequestering metals. Early research efforts predicted that phosphate minerals are likely controlling solids for metals in soil systems; depending upon ion activity products (IAPs), soil type, pH, CO_2 (g) partial pressures, etc. (Nriagu 1972, 1973a, b, 1974, 1984; Lindsay & Moreno 1960; Jurniak & Inouye 1962; Santillan-Medrano & Jurniak 1975; Street et al. 1977; Ruby et al. 1994; Xu & Schwartz 1994; Xu et al. 1994). More recently, spectroscopic techniques have been employed to identify important phosphate minerals in various contaminated or altered natural systems. These methods have allowed for visual examination of fine precipitates or sorption surfaces, determination of the chemical state of the metal in the phosphate mineral structure, and on the phosphate mineral surface and the actual coordination mechanism and local bonding environment on the phosphate mineral surface.

In native soil systems, Morin et al. (2001) used electron microprobe analysis (EMPA), scanning electron microscopy-energy dispersive spectroscopy (SEM-EDS), synchrotron-based X-ray microfluorescence (μ-SXRF), and extended X-ray absorption fine structure (EXAFS) to characterize soils in Pb-mineralized sandstones from Largentière, France. At this site, natural soil development occurred upon a Pb–Zn strata bound deposit. Plumbogummite was the dominant Pb phase in the soil profile, though other species (Pb^{2+}–Mn–(hydro)oxide surface complexes, Pb^{2+}-surface organic complexes) were observed.

In anthropogenically impacted systems, Cotter-Howells & Thornton (1991) used SEM-EDS to characterize soil grains from a Derbyshire, UK, mining village where Pb mining historically occurred. Many of the soil grains were chloropyromorphite. EXAFS was used to identify pyromorphite in mine-waste-contaminated soils in follow-on studies (Cotter-Howells et al. 1994).

Davis et al. (1993) used EMPA to look at soil grains in Butte, Montana, impacted by Pb ore mining. Their modelled paragenetic sequences showed galena (PbS) and Pb-oxide particles weathering to anglesite ($PbSO_4$) and chloropyromorphite solid solutions (($Pb_{0.2}Ca_{0.8})_5(PO_4)_3Cl$). The most dominant phosphate minerals were drugmanite ($Pb_2Fe(PO_4)_2(OH) \cdot H_2O$), plumbogummite, hinsdalite, and corkite ($PbFe_3(PO_4)$ ($SO_4)(OH)_6$).

Davis et al. (1999) used EMPA, pore water analyses, and geochemical modeling to show that overbank deposits of mine tailings from Butte, Montana, that were treated with lime or revegetated had significant relative levels of plumbogummite that controlled Pb in pore waters.

Ostergren et al. (1999) used EXAFS to study Pb speciation in galena-dominated lead ore mine tailings in Leadville, Colorado. Fine-grained solids (<100 μm and <10 μm) from the Hamm's tailings pile showed dominant Pb speciation as adsorbed Pb, pyromorphites, and hydrocerrusites ($Pb_3(CO_3)_2(OH)_2$).

Overview of immobilization mechanisms

The general depiction of chemical reactivity versus time offered by McBride (1994) is a useful framework for discussion of chemical stabilization through metal phosphate precipitation. As shown in Fig. 3, possible stabilization mechanisms can involve surface sorption processes to existing particulate surfaces that result, through heterogeneous nucleation, in formation of surface precipitates. It can also involve the formation of crystallites and clusters that result, through homogeneous nucleation, in discrete precipitates. Both the surface precipitates and discrete precipitates can further ripen and become well ordered. Some of these reactions happen in the order of minutes (sorption), others in the order of millennia (recrystallization of surface precipitates or nucleated crystals into more ordered crystals). Spectroscopic techniques exist to distinguish between sorption and precipitation (Corey 1981; Farley et al. 1985; Sposito 1986).

The types of metal immobilization mechanisms that can exist when orthophosphate (e.g., as dilute H_3PO_4) is added to an MSW ash are likely to be varied and dependent upon time, concentration of a sorbate or precipitant, concentration of a sorbent or co-precipitant, pH, temperature, reaction kinetics, mixing, effective diffusivity, and so on. The reaction system is initially defined by soluble solid phases (e.g., $PbCl_2$, $ZnSO_4$, and $CdCO3$), sorbed or surface-complexed metal phases, and relatively unavailable phases within the aluminosilicate framework. The addition of a source of phosphate such as rock apatite (or perhaps phosphoric acid to help dissolve the more available fraction) will initiate a sequence of events culminating in metal phosphate precipitation. Using the mechanisms outlined in Fig. 3, it is likely that sorption on particle surfaces (e.g., apatite), solid solution formation, and subsequent heterogeneous nucleation and precipitation is one dominant removal mechanism. Homogeneous nucleation and precipitation of new, discrete phases is the other likely immobilization reaction.

Surface immobilization: sorption mechanisms

Apatites have been demonstrated to be highly effective sorbent surfaces. They can have specific

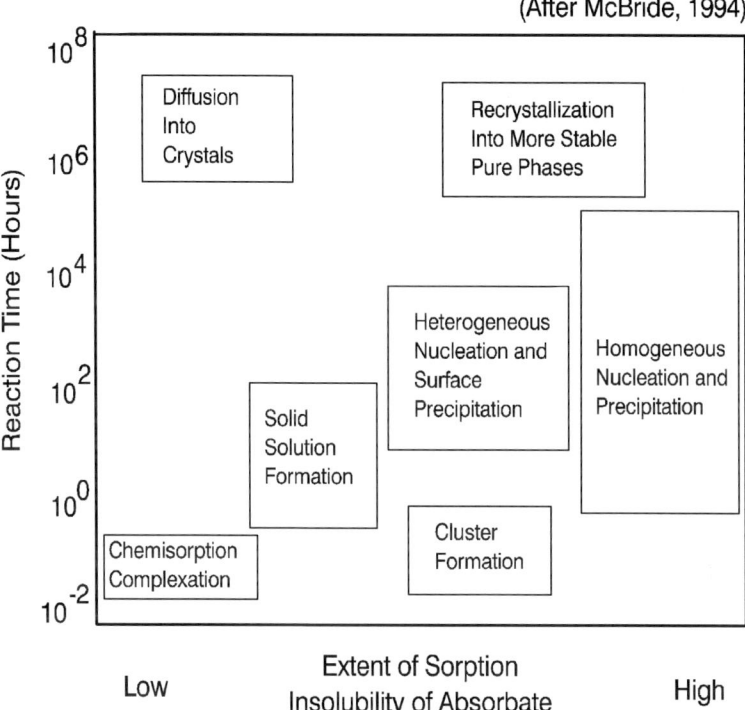

Fig. 3. Mechanisms for removal of metals from solid-phase-dominated aqueous systems (after McBride 1994).

surface areas (SSA) that vary depending on the form of the mineral (recently precipitated to well crystallized), purity, and so on. (Nicol & Clark 1969). Reported values range from 9.1 to 87.2 m^2/g (Ma 1996) with most values between 25 and 55 m^2/g.

Apatite surface properties have been characterized with respect to their role as sorptive surfaces (Wu et al. 1991; Somasundaran & Wang 1984; Chander & Fuerstenau 1984; Leyva et al. 2001). The point of zero charge (PZC), as measured by titration, electrophoresis, or streaming potential varies from pH values of 7 to 10 for hydroxyapatite and from 4 to 12 for fluoroapatite, and is a function of (1) the presence of CO_2, (2) ionic strength, and (3) time/aging of the mineral.

There appear to be two surface functional groups on fluoroapatite (with conditional constants corrected for coulombic energy of the charged surface) as determined by potentiometric titration under nitrogen (Wu et al. 1991):

$$\equiv PO^- + H^+ \ = \ \equiv POH$$
$$\log \beta^s_{101}(\text{int.}) = 6.6 \pm 0.1 \qquad (1)$$

$$\equiv CaOH_2^+ - H^+ \ = \ \equiv CaOH$$
$$\log \beta^s_{-110}(\text{int.}) = -9.7 \pm 0.1 \qquad (2)$$

A number of researchers have attempted to identify the precise sorption or ion exchange mechanism on apatites (Mandjiny et al. 1995, 1998; Fedoroff et al. 1999; Vega et al. 1999; Monteil-Rivera et al. 1999, 2000; Levya et al. 2001; Perrone et al. 2001; Cheung et al. 2002; Fuller et al. 2002; Duc et al. 2003). Some of these studies are highlighted here.

Leyva et al. (2001) studied Sb(III) adsorption to hydroxyapatite. They conducted adsorption isotherms in closed vessels at Sb concentrations of 0.05–50 mg/L, constant I (0.01 M), constant solid phase concentrations of 10 g/dm³ at pH values between 5 and 8. The hydroxyapatite was characterized by X-ray diffraction (XRD), SEM-EDS, X-ray photoelectron spectroscopy (XPS), and infrared (IR) spectroscopy. Langmuir adsorptions models revealed Γ_{max} of 6.7×10^{-8} mol/m² and $K_{ads} = 1.5 \times 10^3$ dm³/mol. As Sb adsorption occurred, the isoelectric point (pH_{iep}) of the hydroxyapatite changed from 4.0 to 12.0. The decline in the pH_{iep} during sorption as well as the absence of precipitation of any new Sb(III) phases, led the authors to conclude that either the following ion exchange or hydrolysable cation chemisorption conceptual processes were occurring:

$$\equiv Ca^{2+} + SbO^+ \iff \equiv SbO^+ + Ca^{2+} \quad (3)$$

or

$$\equiv OH + SbO^+ \iff \equiv O-SbO + H^+ \quad (4)$$

$$\equiv O_3P-OH^+ + SbO^+ \iff \equiv O_3P-O-SbO^+ + H^+ \quad (5)$$

$$\equiv O-SbO + H_2O \iff \equiv O-SbO(OH)^- + H^+ \quad (6)$$

$$\equiv O_3P-O-SbO^+ + H_2O \iff \equiv O_3P-O-Sb(OH) + OH^- \quad (7)$$

Perrone et al. (2001) modelled Ni(II) adsorption to synthetic carbonate fluoroapatite (Ca_{10} ((PO_4)$_5$(CO_3))(OH,F). The solid phase had a pH_{IEP} of 6.3 and a ZPC of 6.4 with an SSA of 8.8 m^2/g, an estimated sorption site density of 3.1 sites/nm^2. They conducted 8-day isotherms in closed vessels at Ni concentrations of 5×10^{-10} to 1×10^{-8} M, constant I (0.05, 0.1 or 0.5 M), constant solid phase concentrations of 10 g/dm^3 at pH values of 4 to 12. As Ni sorption occurred, no significant release of Ca was seen. Sorption was reversible. Rather than precisely characterize surface functional groups, they elected to describe their sorbent surfaces using acid–base reactions for the average behaviour of all sites involved in protonation and deprotonation. Potentiometric titration data were used to estimate the constants with the FITEQL computer code:

$$\equiv OH + H^+ \iff SOH_2^+ \quad \log K^+ = 5.98 \quad (8)$$

$$\equiv OH \iff SO^- + H^+ \quad \log K^- = -7.22 \quad (9)$$

They used the constant capacitance model (surface capacitance of 18 F/m^2) to fit the following three sorption reactions to observed absorption edge data:

$$\equiv OH + Ni^{2+} \iff \equiv OHNi^{2+} \quad \log K_1 = 3.5 \quad (10)$$

$$\equiv OH + Ni^{2+} \iff \equiv ONi^+ + H^+$$
$$\log K_2 = -3.2 \quad (11)$$

$$\equiv OH + Ni^{2+} + H_2O \iff \equiv ONi(OH) + 2H^+$$
$$\log K_3 = -11.5 \quad (12)$$

Fuller et al. (2002) used EXAFS to study the sorption/surface precipitation of U(VI) to hydroxyapatites used in permeable reactive barriers. At low sorbed concentrations (<4700 mg/kg U), spectral fitting and interpretation suggested that bidentate edge sharing and/or monodentate corner sharing between U(VI) O_x (x = 5 to 6) polyhedra and CaO_y polyhedra (where y = 6 O, 8 O/F atoms) at the mineral surface. Spectral fitting also indicated bidentate linkages between the U(VI) equatorial shell and phosphate tetrahedra. Solid solution formation at the surface was ruled out.

Earlier sorption studies involved more traditional isotherm and kinetic studies that allowed estimation of maximum coverage (Γ_{max}), sorption constants (K_{ads}), and the observation of desorption. Occasionally, studies showed the effects of sorption on solution Ca^{2+}, H^+, and PO_4^{3-} activities over time that suggested initial exchange mechanisms resulting in apatite constituent release and then precipitation mechanisms resulting in removal or burial of the metal cations as metal phosphate minerals. These early works include the efforts of Brudervold et al. (1963), Ingram et al. (1992), Christoffersen et al. (1988), Dalas & Koutsoukos (1989), Lazic & Vukovic (1991), Misra et al. (1975), Pak & Bartter (1967), Kukura et al. (1973), Suzuki & colleagues (Suzuki et al. 1981, 1982, 1984), Misra et al. (1975), Misra & Bowen (1984), Middelburg & Comans (1991), and Xu et al. (1994). However, there was divergence in the literature about the relative role of ion exchange, matrix diffusion, and surface precipitation. More recently, the application of advanced surface spectroscopies has helped to clarify the roles of these mechanisms.

Surface immobilization: spectroscopic evidence of surface precipitation

Application of spectroscopic and surface analysis techniques has allowed the observation of heterogeneous nucleation of more insoluble metal phosphate precipitates on the surface of the apatite crystal or the homogeneous nucleation of precipitates from solution (Jeanjean et al. 1996; Chen et al. 1997a, b; Da Rocha et al. 2002; Lusvardi et al. 2002; Mavropoulos et al. 2002). These more recent works clearly suggest that a range of phenomena occur: some type of sorption/ion exchange reaction at low sorbate concentrations that happens quickly and some dissolution/surface precipitation or re-precipitation reaction that occurs at higher metal concentrations and may be slow because of the rate limiting step of apatite dissolution and or coverage of available surface sites by surface precipitates that blind the original surface.

The presence of the metal cation and an appropriate anion (OH^-, Cl^-, F^-, etc.) in contact with the surface of the apatite is sufficient to cause partial/complete dissolution of the apatite and precipitation of the more stable pyromorphite. A typical set of stoichiometric dissolution and precipitation reactions would be (Lower et al. 1998b):

$$Ca_5(PO_4)_3OH_{(s)} + 7H^+ \longleftrightarrow$$
$$5Ca^{2+} + 3H_2PO_4^- + H_2O \quad (13)$$
$$5Pb^{2+} + 3H_2PO_4^- + H_2O \longleftrightarrow$$
$$Pb_5(PO_4)_3OH_{(s)} + 7H^+ \quad (14)$$

At some distance close to the mineral surface, the saturation index (SI, IAP/K_{sp}) of pyromorphite must be greater than zero to allow local super saturation and precipitation. The question is whether this is a heterogeneous nucleation/surface precipitation reaction or a dissolution homogeneous nucleation reaction.

Lower et al. (1998a) conducted studies with atomic force microscopy (AFM), optical microscopy, EDS and XRD to study the reactions between $Pb(NO_3)_2$ solutions (100 mg/L Pb), buffered to pH 6, and hydroxyapatite in the AFM fluid cell. This allowed examination of the transformation of hydroxyapatite to hydroxypyromorphite. They used a synthetic, poorly crystalline hydroxyapatite with hexagonal morphology. A pulsed flow system was used to watch transformations over 3 to 12 h periods. Transport-limited dissolution of hydroxyapatite and of homogeneous nucleation of hydroxypyromorphite was seen. Hydroxypyromorphite needles were found in close contact with the hydroxyapatite surface, suggesting PO_4^{3-} diffusion from the hydroxyapatite to the hydroxypyromorphite might be involved in the overall rate limitation.

Lower et al. (1998b) can shed more light on this phenomenon. They again used AFM, SEM, TEM, SEM-EDS, electron diffraction, and XRD to study the reactions between 0.5 and 500 mg/L of Pb with hydroxyapatite at pH 6 and a reaction temperature of 22 °C. A commercial hydroxyapatite was used at sorbent concentrations of 0.5 g/L. Reactions were observed over a 2 h period. At high initial Pb concentrations, Pb solution concentrations dropped from 500 mg/L to <100 mg/L. At concentrations of 0.5–100 mg Pb/L, after reaction, Pb levels dropped to less than 15 µg/L. In both cases, hydroxyapatite dissolved and hydroxypyromorphite formed. The authors applied some nucleation and crystal growth theory developed

by Berner (1981) to hypothesize the saturation state of hydroxypyromorphite, Ω_{HYP}:

$$\Omega_{HYP} = [Pb^{2+}]^5[PO_4^{3-}]^3[OH]^-/K_{sp,HYP} \quad (15)$$

At the higher suspected Ω_{HYP} ($[Pb_{aq}] \sim 10–100$ mg/L), small nuclei or aggregates of poorly crystalline hydroxypyromorphite formed homogeneously in solution. At suspected intermediate Ω_{HYP} ($[Pb_{aq}] > 100$ mg/L), large, euhedral needles of hydroxypyromorphite precipitated homogeneously from solution. At suspected low levels of Ω_{HYP} ($[Pb_{aq}] < 10$ mg/L), a needle-like Pb-containing phase grew heterogeneously on the surface of the hydroxyapatite. The needles sometimes exhibited preferential orientation and several intergrowths could be seen with AFM. The phases may have been hydroxypyromorphite, but at concentrations below detection by XRD. Sorbed Pb, amorphously precipitated Pb, or Pb substituted into the hydroxyapatite may also have been present.

Valsami-Jones et al. (1998) conducted similar studies with a synthetic, microcrystalline hydroxyapatite in either a Pb or Cd buffered solution. Using AFM, hydroxypyromorphite was observed to grow epitaxially on the hydroxyapatite surface in a clear example of heterogeneous nucleation. Cd removal differed, with the likely formation of a Ca–Cd phosphate solid solution.

Maneki et al. (2000a) used synthetic hydroxyapatite and natural fluoroapatite and chloroapatite to study rate limitations to heterogeneous nucleation. The former had an SSA of 59.1 m^2/g while the latter had 3.0 and 4.0 m^2/g, respectively. Dissolution and sorption experiments were conducted. Sorption reactions were run for 14 days at pH 4.2, sorbent loading of 1 g/L, and Pb and Cl concentrations 38.7 and 13.1 mg/L, respectively. Dissolution was linear (zeroth order). After correction for SSA, dissolution rates were greatest in the chloroapatite, followed by the fluoroapatite, and then the hydroxyapatite. In the presence of Cl and Pb, all three apatites formed chloropyromorphite. The rate of uptake (not corrected for SSA) was greatest in the hydroxyapatite, followed by the fluoroapatite, and then the chloroapatite. Rate control was governed by apatite dissolution. Heterogeneous nucleation did not prevent further apatite dissolution.

Manecki et al. (2000b) used the same three apatites to study both homogeneous and epitaxial heterogeneous nucleation. For this, AFM, SEM, optical microscopy, EDS, Fourier-transformed infrared (FTIR) spectroscopy, and XRD were used to characterize the apatites and the reaction products. Heterogeneous nucleation on the fluoroapatite and chloroapatite showed strong epitaxy. Epitaxial crystals of pyromorphite on chloroapatite and hydroxyapatite grew away from the apatite surface, the only source of phosphate. The morphologies of the epitaxial crystals differed from the homogeneously nucleated crystals; a spiral growth mechanism was observed.

Laperche & Traina (1998) studied Pb uptake on hydroxyapatite at low initial solution concentration of Pb (103 mg/L). For this, EXAFS was used to characterize the local coordination environment of Pb on the apatite. The baseline corrected, Fourier-transformed EXAFS spectra revealed k-values at >3 Å, suggesting that Pb was not randomly sorbed. Radial structure functions (RSF) showed three intense peaks, characteristic of pyromorphite.

One useful way of describing the ability of apatites to take up and immobilize contaminants is the linear distribution coefficient, K_d, the mass of contaminant sorbed to the mass of apatite ($M_{i(ads)}$) relative to the mass of contaminant in solution at equilibrium ($M_{i(soln)}$):

$$M_{i(ads)} = K_d M_{i(soln)} \qquad (16)$$

Table 4 summarizes some experimentally determined K_d values for a variety of apatites and contaminants. Some values are very large ($>10,000$), indicating the propensity for apatites to sorb and/or induce surface precipitation reactions. These values are useful in terms of understanding the broad affinity that apatites have for various solutes. However, they reflect the operationally defined conditions particular to each partition study (pH, I, solid-to-liquid ratio, and apatite type).

Solution phase immobilization: dissolution, precipitation, solid solution formation

In systems where all components are initially soluble, it is simple to form pyromorphites (Nriagu 1974, 1984; Ma 1996) or ternary metal apatites where Pb^{2+}, Cd^{2+}, Cu^{2+}, and Zn^{2+} isomorphically substitute for Ca^{2+} and form solid

Table 4. *Reported K_d values for various apatites and various contaminants*

Element	Range of K_d (mL/g) and apatite type*	Reference
Ba(II)	222 (North Carolina)	Knox *et al.* 2003
Cd(II)	30–200,000 (HAP, Florida, North Carolina, Apatite II	Middleburg & Comans 1991; Wright *et al.* 1995; This work
Co(II)	470 (North Carolina)	Knox *et al.* 2003
Cu(II)	1,100–16,400	This work
Pb(II)	3,900–533,000 (Florida, North Carolina, HAP, Bone Char, Apatite II)	Kaplan *et al.* 2002; Knox *et al.* 2003; Wright *et al.* 1995; This work
Ni(II)	10,000 (Carbonate FAP)	Perrone *et al.* 2001
Zn(II)	550–175,000 (North Carolina, Apatite II)	Wright *et al.* 1995; This work
As(III)	95–110 (North Carolina)	This work
As(V)	2–100 (HAP, Florida, Bone Char)	Thomson *et al.* 2003
Cr(III & VI)	5–18,000 (North Carolina, Apatite II)	Kaplan *et al.* 2002
Cr(VI)	5.3–11.4 (Florida, North Carolina)	This work
Se(IV)	3–6 (FAP) 9-900 (HAP)	Duc *et al.* 2003
Se(VI)	1.1–650 (Florida, HAP, Bone Char)	Thomson *et al.* 2003
Am(III)	270–46,500 (Florida, HAP, Bone Char)	Thomson *et al.* 2003
Eu(III)	313,200 (North Carolina)	Knox *et al.* 2003
Pu(III)	270–10,000,000 (North Carolina, HAP, Bone Char, Apatite II)	Knox *et al.* 2003; Thomson *et al.* 2003; Wright (unpublished)
Tc(VII)	0.11–36.88 (Florida, HAP, Bone Char)	Thomson *et al.* 2003
U(VI)	177–268,000 (North Carolina, Florida, Natural, HAP, Bone Char)	Knox *et al.* 2003; Thomson *et al.* 2003

*HAP ($Ca_5(PO_4)_3OH$), FAP ($Ca_5(PO_4)_3F$), Carbonate FAP ($Ca_{10}((PO_4)_5CO_3)OH,F$). Florida is a mined natural apatite mixture of FAP and HAP (IMC Agrico, Mulberry, FL). North Carolina is mined natural Carbonate FAP (Aurora Phosphate Mine, Lee, NC). Bone char is apatite made from bone char. Apatite II is an FAP and HAP mixture made from fish bones.

solutions. These precipitation reactions can be very fast (seconds to minutes) depending on stoichiometric availabilities, pH ($11-12$ is optimum), and mixing energy. As the nanoprecipitates nucleate and crystallize, they become more ordered and ripen according to Ostwald's process with concomitant reduction in solubility approaching that of large, well-ordered crystals (Mahapatra *et al.* 1995).

Ayati and Lundager Madsen (2000, 2001) studied the formation of Cd, Cu, and Pb phosphates from solutions individually and in the presence of Ca during carefully controlled titrations of 0.1 M metal nitrate solutions with 0.1 M $NH_4H_2PO_4$ at constant temperatures (5, 37 °C) for reactions of up to $6-10$ days. The precipitates were examined with light microscopy, XRD, and IR spectroscopy. For Cd titrations, Cd_5H_2 $(PO_4)_4 \cdot 4H_2O$ formed (pK_{sp} of 30.9). As pH changed, crystal morphology changed from prismatic crystals to twins, aggregates and dendrites. For Pb titrations, $PbHPO_4$ (lower pH) and hydroxypyromorphite formed. The crystal structure of the former was elongated, tabular and irregular. The crystal structure of the latter was formed of very fine crystals. For Cu titrations, $Cu_3PO_4(OH)_3$ and $CuNH_4PO_4$ formed at 5 °C and Cu_2PO_4OH formed at 37 °C. The presence of Ca at different concentrations influenced all three titrations. For Cd, solubility measurements decreased and mixed crystals were observed. For Pb, the presence of Ca did not affect solubility or IR spectra. For Cu, the presence of Ca increased crystal size markedly.

In systems where a solid phase such as phosphate rock is added as the source of phosphate, the dissolution of the solid phase is required before homogeneous nucleation can occur (Ma *et al.* 1993, 1994*a, b*, 1995; Wright *et al.* 1995; Ma 1996; Laperche *et al.* 1996, 1997). The dissolution rate plays a role in the degree of immobilization. More kinetically constrained sources (e.g., phosphate rock) take longer to dissolve and thus can limit the degree of immobilization (Ma 1996, Sugiyama *et al.* 2002). In simple Pb solutions, hydroxyapatite can dissolve relatively quickly (on the order of minutes) and precipitate hydroxypyromorphite. However, in more complex solutions containing soil particles, the reaction is slower, on the order of hours (Ma 1996).

In the presence of high concentrations ($12\times$ Pb concentration) of SO_4^{2-}, NO_3^-, Cl^-, or F^-, hydroxyapatite will still completely dissolve and form hydroxypyromorphite, chloropyromorphite or fluoropyromorphite. In the presence of high concentrations ($7\times$ Pb concentration) of Al, Cd, Cu, Fe(II), Ni, or Zn, hydroxyapatite

will still readily dissolve and form hydroxypyromorphite, although some inhibition occurs (Ma 1996), likely through competition during solid solution formation. Carbonate (CO_3^{2-}) plays an indirect role in the dissolution/precipitation process. At the high pH values where carbonate is present, the rate of dissolution of hydroxyapatite is lower (Ma 1996).

The influence of Na^+, K^+, calcite, and EDTA on complete hydroxyapatite dissolution and hydroxypyromorphite formation was also examined (Ma 1996). Only high EDTA concentrations (EDTA/Pb molar ratios up to 5) significantly affected the ability of dissolving hydroxyapatite from sequestering hydroxypyromorphite; the strong bond between the EDTA ligand and the Pb^{2+} was fairly competitive with the tendency to precipitate out as hydroxypyromorphite.

Zhang, Ryan and co-workers (Zhang *et al.* 1997, 1998; Zhang & Ryan 1998, 1999; Scheckel & Ryan 2002) have studied the dissolution of Pb minerals (galena, cerrusite, and anglesite) or Pb sorbed to goethite (α-FeOOH) in the presence of synthetic hydroxyapatites and the subsequent formation of Pb phosphates. Both fixed and dynamic systems were used to look at transformations, with pH and P/Pb molar ratios as variables. In all cases, the initial Pb mineral of phase was transformed from a more bioavaliable form to the less bioavailable, geochemically stable pyromorphite. Generally, the rate of formation of pyromorphites was limited by the rate of hydroxyapatite dissolution. Transformation rates were on the order of days.

Use of phosphate in contaminated soil stabilization

The use of phosphate has been widely evaluated and subjected to field trials for Pb-contaminated soils. Most treatment systems involve excavation, pug milling of the soil with the stabilization agent, and either replacement or landfill disposal. Occasionally, for larger sites and deeper contamination, *in situ* mixing with large augers is used.

A variety of phosphate sources have been used: soluble P (e.g., H_3PO_4 or solutions made from the dissolution of soluble phosphates such as $(NH_4)_2HPO_4$, Na_2HPO_4, NaH_2PO_4, K_2HPO_4, KH_2PO_4,), mineral P (e.g., $Ca(H_2PO_4)_2 \cdot H_2O$, Na_2HPO_4, KH_2PO_4), hydroxyapatite, bone meal, rock phosphate or rock apatite (typically $Ca_5(PO_4)_3(F,OH)$ or $Ca_5(PO_4)_3(F,CO_3)$), and Apatite II, a more amorphous apatite made from fish bone (e.g., Wang *et al.* 2001; Sugiyama *et al.* 2002). Three of these studies are illustrated further.

Laperche *et al.* (1996) studied the conversion in stirred reactors of soil Pb phases such as litharge (PbO), and cerrusite (PbCO$_3$) in soils contaminated with paint spills (Pb = 37,026 mg/kg) to hydroxypyromorphite after the addition of hydroxyapatite. These soil Pb fractions (100 mg) were added with hydroxyapatite (20–40 mg) in 100 mL of H$_2$O at pH 5 and 7.7. Subsequently XRD and SEM identified the formation of hydroxypyromorphite (and possible solid solutions) after 9 to 14 days.

Ryan *et al.* (2001) studied the effects of hydroxyapatite addition to soils impacted by Pb from smelter operations. Dialysis experiment were conducted where the soil and hydroxyapatite solids were placed in separate dialysis bags suspended in 0.01 M NaNO$_3$. Chloropyromorphite formed on the dialysis membrane containing the soil. The dissolution of solid-phase soil Pb was the rate-limiting step for pyromorphite formation. EXAFS showed that after the 240 day incubation the hydroxyapatite treatment caused a change in the average, local molecular bonding environment of soil Pb.

Cao *et al.* (2002) used three treatments (H$_3$PO$_4$, H$_3$PO$_4$/Ca(H$_2$PO$_4$)$_2$, H$_3$PO$_4$/phosphate rock) at a P/Pb molar ratio of 4 to look at transformation of soil Pb in test plots in soils at a battery-recycling yard. A selective extraction method, based on Tessier's Sequential Chemical Extraction Procedure, (Tessier *et al.* 1979), was used to evaluate re-speciation of the soil Pb. After 220 to 320 days of incubation at typical soil moisture levels, 10–50% of the available soil Pb (exchangeable, carbonate, Fe/Mn oxyhydroxide, and organic fractions) converted to the residual fraction. Chloropyromorphite was identified in the soils samples using Visual MINTEQ-modelled activity-ratio diagrams. H$_3$PO$_4$/phosphate rock was the best treatment.

Phosphates have also been used as either impermeable or permeable reactive barriers for containment or remediation of contaminated groundwater. Fuller *et al.* (2002) looked at the mechanisms of U(VI) immobilization by apatites with EXAFS. Matheson *et al.* (2001) examined mechanisms of U(VI) removal in apatite-based permeable reactive barriers. Gauglitz *et al.* (1992) conducted laboratory studies on the efficacy of hydroxyapatite to remove U and Th from brine solutions. The minerals saleeite (Mg(UO$_2$)$_2$(PO$_4$)$_2$·9H$_2$O) and metaatunite (Ca(UO$_2$)$_2$(PO$_4$)$_2$·6H$_2$O) were observed. The concept of using apatite as a backfill material around repositories was explored. Apatite barriers made of Apatite II have been successfully

demonstrated at a number of contaminated groundwater sites, reducing levels of U, Cd, and Zn to very low levels (Bostick *et al.* 2000; US EPA 2000; Conca *et al.* 2002). Apatites are also being evaluated for use as less permeable reactive barriers in confined aquatic disposal cells for contaminated sediments (Crannell *et al.* 2003*a, b*).

Use of phosphate in ceramic-based stabilization systems

Low-temperature treatment of low-level mixed wastes has also been accomplished by solidification/stabilization with chemically bonded phosphate ceramics (CBPC). These are made by hydrothermal chemical reaction rather than by sintering. Chemical bonding develops when acid phosphates react with oxides to form crystalline orthophosphate (Singh *et al.* 1997). The ceramic matrix stabilizes the wastes by microencapsulation. The low temperature of the reaction allows volatile radionuclides to be treated (Singh *et al.* 1997).

Wagh *et al.* (1999) studied the use of CBPC for stabilization of combustion residue of high transuranic Pu wastes. A CBPC matrix formed by the reaction of MgO and KH$_2$PO$_4$ was used to microencapsulate Pu at a 5 wt% loading. The ceramic waste forms had compressive strengths twice that of conventional cement grout controls. The connected porosity was 50% less than cement grout. Waste forms displayed high leaching resistance for both hazardous metals and Pu. PuO$_2$ was the dominant species in the waste form.

Rao *et al.* (2000) solidified soil and ash containing spiked levels of Cd, Cr, Pb, Ni, and Hg with CBPC binder. The volume of ceramic matrix was reduced by uniaxial or harmonic compression, while CBPC binding and compaction achieved 39–47% volume reductions at waste loadings of 80 wt%. Compressive strengths of final waste forms ranged from 1,500 to 2,000 psi. Regulatory leaching (using the Toxicity Characteristics Leaching Procedure) showed that all heavy metals except Hg passed TCLP limits using the CBPC.

Use of phosphate stabilization for MSW incineration residues

Phosphate-based chemical stabilization systems are presently used in the USA, Canada, and Japan (Lyons 2000; Ecke *et al.* 2000). In the USA, the technology was first used in 1987 (Lyons 2000). It is based on the use of agricultural grade H_3PO_4, which is added to pug mills treating the scrubber residue before it is added with the bottom ash to create a treated or amended combined ash waste stream. It has been used at over 30 facilities in the USA and Canada. As of 2000, over 9 million tonnes of ash were treated (mostly combined ash and scrubber residue). In most cases, the use of the technology allows the ash to pass the regulatory leaching tests and observed levels of Pb and Cd in landfill leachates are low. In studies conducted by the US EPA (Kosson et al. 1993), the USA-based phosphate stabilization process was evaluated along with other ash stabilization and solidification technologies. Generally, the phosphate-based system did comparatively well in stabilizing Pb (and other metals) in the residues. In the Dutch Total Availability Leaching Test, the phosphate stabilization process reduced lead leaching by a factor of 10 (500 mg/kg to 46 mg/kg).

In Japan in 2000, the technology was used at over 120 units (Ecke et al. 2000). H_3PO_4 was used at a dose of about 10% agricultural grade phosphate to the wet weight of the ash stream (Ecke et al. 2002). It was applied to scrubber residues that were managed separately from bottom ash. Treatment costs roughly range from 7000 to 14 000 Y/tonne (ca. $70–140/ tonne). Some modifications to the process included the addition of high specific surface area lime (Uchida et al. 1996). Typically, the technology allows the ash waste stream to pass the Japanese 13 regulatory leaching test and observed levels of metals in landfill leachates are low (Shimaoka et al. 1998). The technology is also viewed as a long term insurance about the chemical stabilization of metals within the disposed ash.

The process is being considered for use in Europe (Sabbas et al. 2003). In Sweden, some preliminary tests have been conducted on its ability to treat scrubber residue (Kullberg 1995). In Denmark, pilot-scale plants have been built and tested for treating ESP ashes and scrubber residues from two large incineration facilities near Copenhagen (Hjelmar et al. 2000a, b). The process involves the use of both H_3PO_4 and $CO_2(g)$ to promote carbonation. Both phosphate and phosphate/carbonate are able to reduce regulatory leaching of metals.

In France, a number of researchers have explored the washing, phosphate stabilization, and sintering or calcination of residues (Derie 1996; Iretskya et al. 1999; Nzihou & Sharrock 2002; Piantone et al. 2003). The formation of more crystalline hydroxyapatite reaction products helps reduce leachability of metals incorporated into apatite minerals. Calcination is done upwards of 900 °C. Detailed studies have shown that Pb and Cd, originally associated with silicate and carbonate phases, are respeciated into phosphate minerals with much greater resistance to leaching.

Calcium phosphate minerals in thermodynamically modelled MSW bottom ash leachate systems

For waste materials such as MSW ashes that contain appreciable concentrations of Ca^{2+} (e.g., 0.01–9.0 moles/kg dry weight, 0.01–4.0 M local equilibrium solutions), it is useful to review Ca^{2+} and PO_4^{3-} crystallization and precipitation chemistry developed from medical and marine science literature (Nancollas 1984; van Capellen & Berner 1989). Ca^{2+} can represent a significant PO_4^{3-} sink. It can also form one of the more stable calcium phosphate minerals, hydroxyapatite, which in turn can act as a sorbent surface for other elements. When Ca^{2+} and PO_4^{3-} are titrated in solution at generally neutral pH values, a variety of phases form, with the kinetically favoured minerals forming early and then transforming into successively more thermodynamically stable forms. In a simple system, the reaction sequence generally involves the elusive $Ca_9(PO_4)_6$ (non-stoichiometric, amorphous calcium phosphate), then $CaHPO_4 \cdot 2H_2O$ (brushite); $CaHPO_4$ (monetite); $Ca_8H_2(PO_4)_6 \cdot 5H_2O$ (octacalcium phosphate), β-$Ca_3(PO_4)_2$ (there are two forms of this mineral, the β form (low whitlockite, ΔG_f^0 of -3884 kJ/mole) and the α form (high whitlockite, ΔG_f^0 of -3875 kJ/mole)); and ultimately hydroxyapatite (Morse & Casey 1988). The sequence is influenced by IAPs, pH, ionic strength, reaction kinetics, the presence of precursor substrates or 'seed', and the presence of inhibitors like Mg^{2+} (Koutsoukos et al. 1980; Nancollas 1984; Christoffersen et al. 1988, 1989; van Capellen and Berner 1989).

The fairly wide distribution of calcium phosphates in complex solutions containing typical dissolved constituents in equilibrium bottom ash leachates can be explored using thermodynamic modeling. Such calculations are useful to discern the role of solution phase complexation of the major cations, typically via OH^-, Cl^-, SO_4^{2-}, HCO_3^-, and CO_3^{2-}.

The stability calculation software STABCAL (Dr. Hsin-Hsiung Huang, Metallurgical Engineering Department, Montana Technological University, http://www.mtech.edu/hhuang/stabcal.

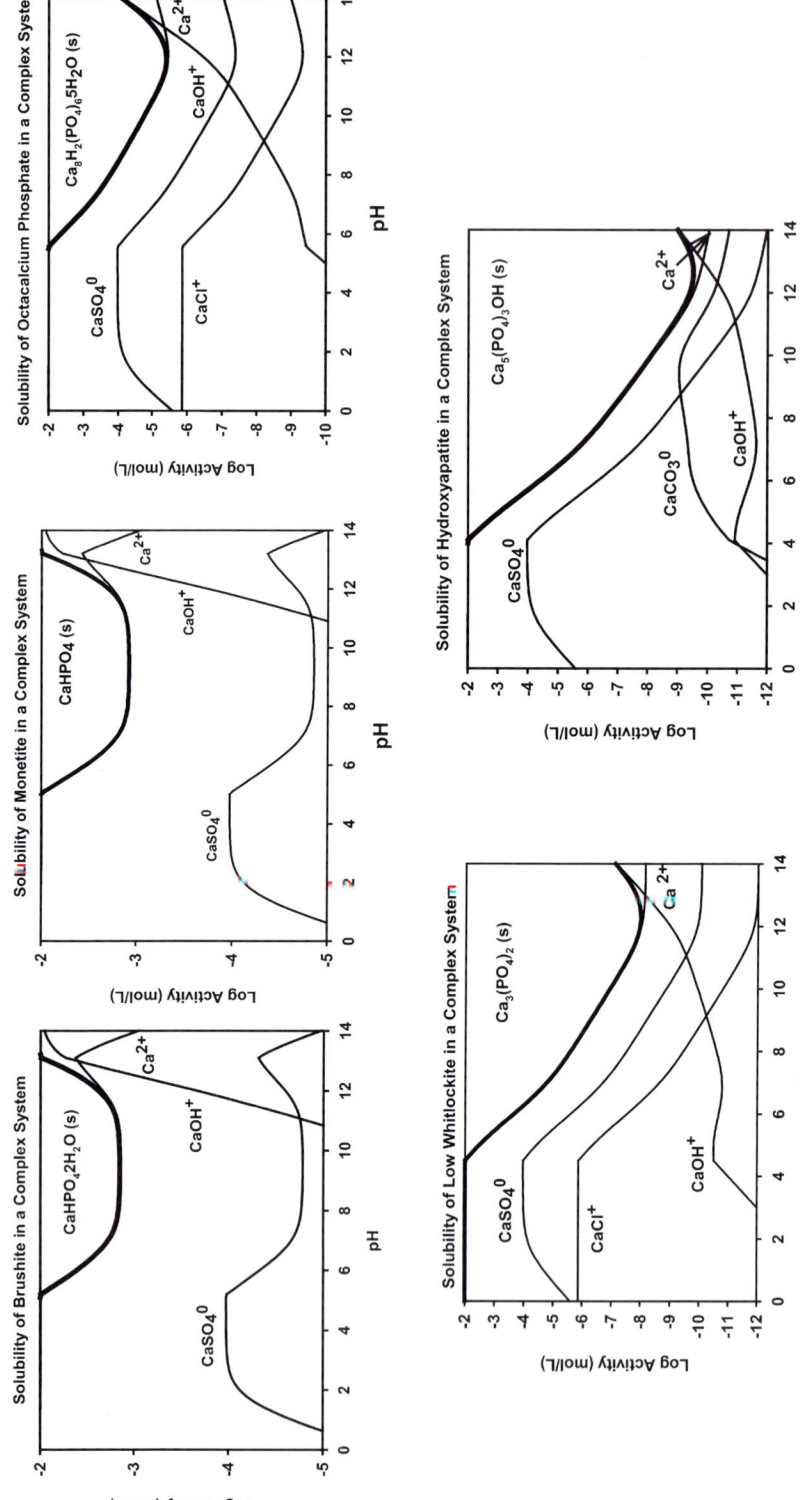

Fig. 4. Calcium phosphate mineral pC–pH distribution diagrams. In all cases, an ionic strength of 0.1 molal was assumed and activity coefficients were calculated using the extended Debye–Hückel equation. Temperatures were set at 25 °C. Note the differences in scale of the *y*-axes of the graphs. In all cases, the systems were modelled as follows (unless otherwise noted): $C_{T,P} = 1 \times 10^{-2}$ M, $C_{T,Cl} = 1 \times 10^{-3}$ M, $C_{T,S} = 1 \times 10^{-3}$ M. For brushite, the system included the components $CaHPO_4 \cdot 2H_2O(s)$, Ca^{2+}, $CaOH^+$, $CaSO_4^0$, and $CaCl^+$. For monetite, the system included the components: $CaHPO_4(s)$, Ca^{2+}, $CaOH^+$, $CaSO_4^0$, and $CaCl^+$. For octacalcium phosphate, the system included the components: $Ca_8H_2(PO_4)_6 \cdot 5H_2O(s)$, Ca^{2+}, $CaOH^+$, $CaSO_4^0$, $CaCl^+$ and $CaCl_2^0$. For hydroxyapatite, the system included the components: $Ca_5(PO_4)_3OH(s)$, $CaCO_3(s)$, $CaSO_4 \cdot 2H_2O(s)$, Ca^{2+}, $CaOH^+$, $CaSO_4^0$, $CaCl^+$, $CaCl_2^0$ and $CaCO_3^0$.

htm) was used to construct pC–pH distribution diagrams (Fig. 4) for brushite, monetite, octacalcium phosphate, low whitlockite, and hydroxyapatite. The MINTEQA2 thermodynamic database (Allison et al. 1990) was used as a source of ΔG_f^0 values. This database of free energy of formation values is in general agreement with the values listed in Table 3. The modelling exercise used typical equilibrium concentrations for MSW bottom ash leachates as shown in Table 2. Fluoride was not included in the modelling exercise as levels are very low in ash (Chandler et al. 1997) and little information is available about levels in leachates. The inclusion of $C_{T,C}$ was done only for the hydroxyapatite example as hydroxyapatite is the most likely calcium phosphate mineral in mineralization sequence.

Brushite has a very wide distribution field, from about pH 5 to pH 13, with minimum Ca^{2+} solubilities at around 1×10^{-3} M between pH 7 and 11. At the solute concentrations modelled, only the $CaOH^+$ complex influences the solubility of the solid phase. Monetite also has a very wide distribution field, from about pH 5 to pH 13, with minimum Ca^{2+} solubilities at around 1×10^{-3} M between pH 7 and 11. At the solute concentrations modelled, only the $CaOH^+$ complex influences the solubility of the solid phase. Octacalcium phosphate has a very wide distribution field, from about pH 5 to pH 14, with minimum Ca^{2+} solubilities at around 1×10^{-5} M around pH 12. At the solute concentrations modelled, only the $CaOH^+$ complex influences the solubility of the solid phase. Low whitlockite has a very wide distribution field, from about pH 4 to pH 14, with minimum Ca^{2+} solubilities at around 1×10^{-8} M around pH 12. At the solute concentrations modelled, only the $CaOH^+$ complex influences the solubility of the solid phase. Hydroxyapatite has a very wide distribution field, from about pH 4 to pH 14, with minimum solubilities at around 5×10^{-9} M around pH 12. At the solute concentrations modelled, only the $CaOH^+$ and $CaCO_3^0$ ligands influence the solubility of the solid phase.

The role of ligand concentrations ($C_{T,S}$, $C_{T,C}$, and $C_{T,Cl}$) on the general distribution of hydroxyapatite was explored (data not shown). The general distribution field for hydroxyapatite does not significantly change relative to 100-fold increases or decreases in these ligand concentrations. Further, the distribution of hydroxyapatite relative to the distribution of calcite and gypsum requires high concentrations of $C_{T,S}$ and $C_{T,C}$ ($>100\times$) before these solids exhibit any appreciable distribution (data not shown).

STABCAL was also used to construct pE–pH stability fields for a Ca–P–S–Cl system where hydroxyapatite, gypsum and calcite ($CaCO_3$) were used as components (Fig. 5). The NBS thermodynamic database (Wagman et al. 1982) was used as a source of thermodynamic data. The modeling exercise used typical equilibrium concentrations for MSW bottom ash leachates as shown in Table 2. As shown in Fig. 5, hydroxyapatite has a wide stability field. It always has predominance over oldhamite (CaS) at all three $C_{T,P}$ levels. It also has wide predominance over gypsum. Only at the lowest $C_{T,P}$ level (1×10^{-9} M) does calcite appear in the predominance diagram. These data suggest that apatite is extremely stable at high alkalinities where carbonates might prevail or at low Eh values where sulphides traditional are controlling solids.

Divalent metal phosphates in thermodynamically modelled MSW bottom ash leachate systems

The fairly wide distribution of metal phosphates in complex solutions containing typical dissolved constituents in equilibrium ash leachates was also explored using STABCAL. pC–pH distribution diagrams (see Fig. 6) were prepared for chloropyromorphite, hinsdalite ($PbAl_3(PO_4)_2$ $(OH)_5$), plumbogummite, tricadmium diphosphate ($Cd_3(PO_4)_2$), tricopper diphosphate ($Cu_3(PO_4)_2$), and hopeite ($Zn_3(PO_4)_2 \cdot 5H_2O$). These are some of the more commonly encountered metal phosphate minerals in phosphate-stabilized waste systems (Jurniak & Inouye 1962; Santillan-Medrano & Jurniak 1975; Street et al. 1977; Eighmy et al. 1997, 1999f; Ayati & Lundager Madsen 2000; Crannell et al. 2000; Basta et al. 2001; McGowen et al. 2001; Eusden et al. 2002). The MINTEQA2 thermodynamic database (Allison et al. 1990) was used as a source of ΔG_f^0 values. This database is in general agreement with the values listed in Table 3. There are slight differences for plumbogummite ($-5,216$ vs. $-5,108$ kJ/mole) and hinsdalite ($-4,679$ vs. $-4,753$ kJ/mole). The modelling exercise used typical equilibrium concentrations for MSW bottom ash leachates as shown in Table 2. $C_{T,P}$ values were selected to produce a reasonable predominance for each of the indicated solids. $C_{T,C}$ was used for chloropyromorphite as a means of illustrating its relative influence on distribution. It should be noted that labile organic matter is capable of complexing divalent metals in ash systems (e.g., Meima et al. 1999). It can impact the general distribution of Pb in modelled phosphate systems (Sauve et al. 1998).

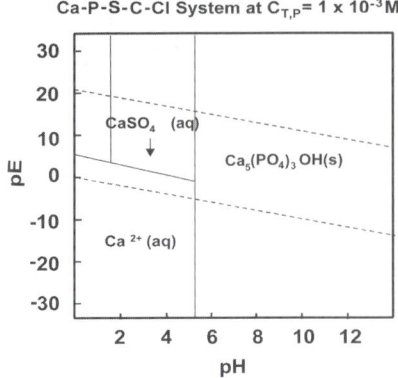

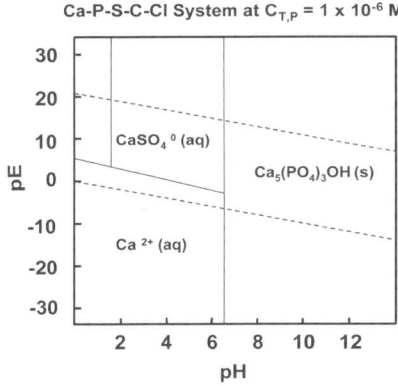

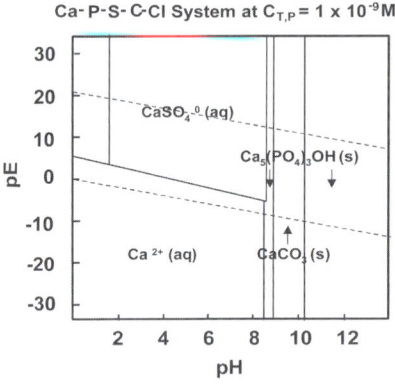

Fig. 5. Calcium phosphate mineral pE–pH predominance diagrams. In all cases, an ionic strength of 0.1 molal was assumed and activity coefficients were calculated using the extended Debye–Hückel equation. Temperatures were set at 25 °C. The following system was modeled: $C_{T,Ca} = 1 \times 10^{-3}$ M, $C_{T,Cl} = 1 \times 10^{-3}$ M, $C_{T,S} = 0.1$ M, $C_{T,C} = 1 \times 10^{-3}$ M; and included the components: $Ca_5(PO_4)_3OH(s)$, $CaCO_3(s)$, $CaSO_4 \cdot 2H_2O(s)$, $CaS(s)$, $S(s)$, Ca^{2+}, $CaOH^+$, $CaSO_4^0$, $CaCl^+$, $CaCl_2^0$, $CaCO_3^0$, PO_4^{3-}, HPO_4^{2-}, $H_2PO_4^-$, $H_3PO_4^0$, S^{2-}, HS^-, H_2S^0, SO_4^{2-}, HSO_4^-, Cl^-, Cl_2^0, CO_3^{2-}, HCO_3^-, and $H_2CO_3^0$. The $C_{T,P}$ varied from 1×10^{-3} M to 1×10^{-9} M.

Chloropyromorphite has a very wide distribution field, from about pH 4 to pH 11, with minimum solubilities at around 3×10^{-10} M around pH 7. At the solute concentrations modelled, the $Pb(OH)_3^-$, $Pb(OH)_4^{2-}$, $PbCO_3^0$, and $Pb(CO_3)_2^{2-}$ complexes influence the solubility of the solid phase. The role of carbonates on the predominance is significant from pH 7.5 to 9. Hinsdalite has a narrow distribution field, from about pH 5 to pH 7, with minimum solubilities at around 1×10^{-8} M around pH 6. At the solute concentrations modelled, the $Pb(OH)^+$ and $PbSO_4^0$ complexes influence the solubility of the solid phase. Plumbogummite has a broad distribution field, from about pH 4 to pH 9, with minimum solubilities at around 1×10^{-20} M around pH 6. At the solute concentrations modelled, the $Pb(OH)^+$, $Pb(OH)_2^0$, and $Pb(OH)_3^-$ complexes influence the solubility of the solid phase. Tricadmium diphosphate has a very wide distribution field, from about pH 6 to pH 12, with minimum solubilities at around 1×10^{-6} M around pH 10. At the solute concentrations modelled, the $Cd(OH)_2^0$, $Cd(OH)_3^-$, $Cd(OH)_4^{2-}$, $CdSO_4^0$, and $CdCl^+$ complexes influence the solubility of the solid phase. Tricopper diphosphate has a very wide distribution field, from about pH 4 to pH 10, with minimum solubilities at around 1×10^{-6} M around pH 6.5. At the solute concentrations modelled, the $Cd(OH)_2^0$ complex influences the solubility of the solid phase. Hopeite has a wide distribution field, from about pH 6 to pH 10, with minimum solubilities at around 1×10^{-5} M around pH 8. At the solute concentrations modelled, the $ZnOH^+$, $Zn(OH)_2^0$, $Zn(OH)_3^-$, $Zn(SO_4)^{2-}$, and $ZnCl^+$ complexes influence the solubility of the solid phase.

STABCAL was also used to construct pE–pH stability fields for chloropyromorphite, hinsdalite, plumbogummite, tricadmium diphosphate, tricopper diphosphate, and hopeite (Fig. 7). These diagrams allow for estimation of stability with respect to pH and to the presence of insoluble sulphides. The NBS thermodynamic database (Wagman *et al.* 1982) was used as a source of thermodynamic data. The total concentrations chosen for each metal were selected to produce a stability region for the metal phosphate solid. In some cases, this was a very low total concentration (e.g., $C_{T,Pb} = 1 \times 10^{-10}$ M for Pb). In other cases, the total metals concentration was high (e.g., $C_{T,Cd} = 1 \times 10^{-3}$ M for Cd). The modelling exercise used typical equilibrium concentrations for MSW bottom ash leachates as shown in Table 2.

Chloropyromorphite has a large predominance field at a very low $C_{T,Pb}$ (1×10^{-10} M). It also has a fairly large predominance field over

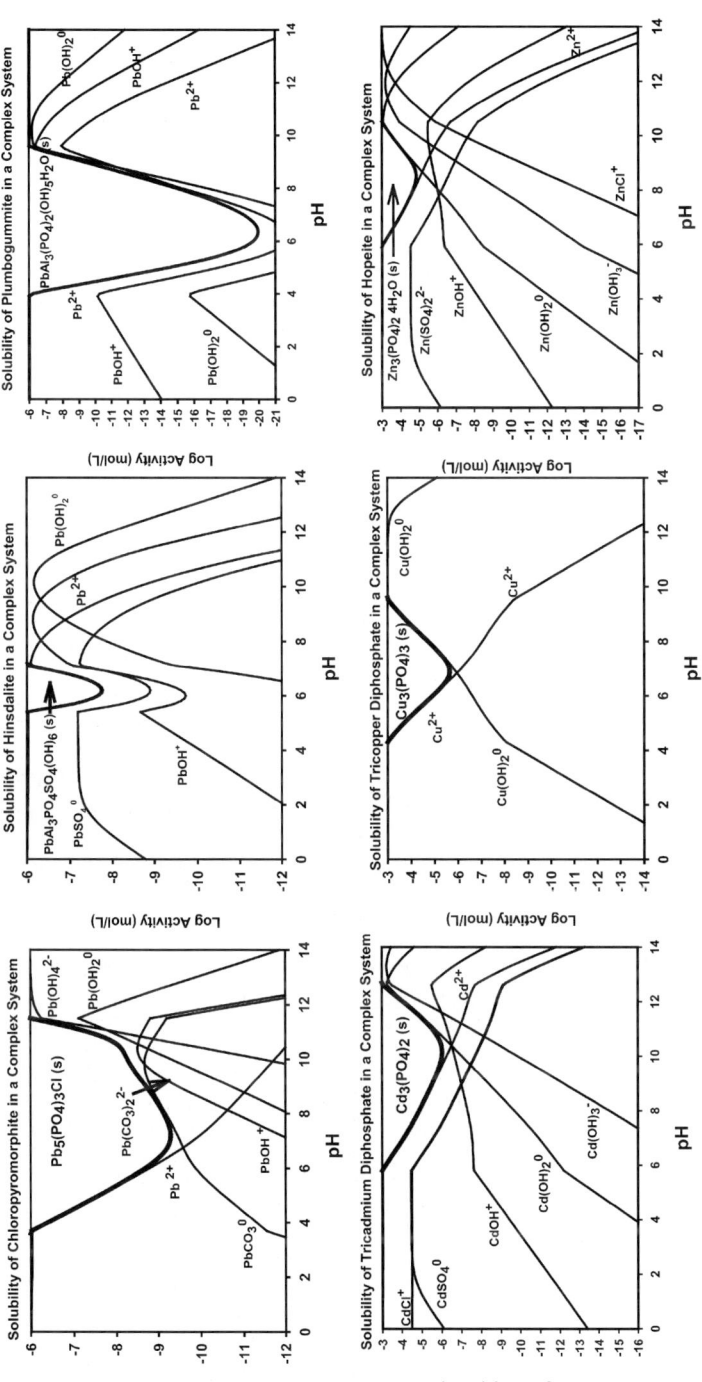

Fig. 6. pH–pC distribution diagrams for various metal phosphate minerals. In all cases, an ionic strength of 0.1 molal was assumed and activity coefficients were calculated using the extended Debye–Hückel equation. Temperatures were set at 25 °C. Note the differences in scale of the y-axes of the graphs. For chloropyromorphite, the following system was modelled: $C_{T,P} = 1 \times 10^{-4}$ M, $C_{T,Cl} = 1 \times 10^{-3}$ M, $C_{T,S} = 1 \times 10^{-3}$ M; and included the components: $Pb_5(PO_4)_3Cl(s)$, Pb^{2+}, $PbOH^+$, $Pb(OH)_2^0$, $Pb(OH)_3^-$, $Pb(OH)_4^{2-}$, $PbCl^+$, $PbCl_2^0$, $PbSO_4^0$, $Pb(SO_4)_2^{2-}$, $Pb(CO_3)_2^{2-}$, and $PbCO_3^0$. For hinsdalite, the following system was modelled: $C_{T,P} = 5 \times 10^{-6}$ M and $C_{T,Al} = 1 \times 10^{-3}$ M; and included the components: $PbAl_3(PO_4)_2(OH)_5(s)$, Pb^{2+}, $PbOH^+$, $Pb(OH)_2^0$, $Pb(OH)_3^-$, $Pb(OH)_4^{2-}$. For plumbogummite, the following system was modelled: $C_{T,P} = 1 \times 10^{-5}$ M, $C_{T,Al} = 1 \times 10^{-3}$ M; and included the components: $PbAl_3(PO_4)_2(OH)_5 \cdot H_2O(s)$, Pb^{2+}, $PbOH^+$, $Pb(OH)_2^0$, $Pb(OH)_3^-$, Pb_2OH^{3+}, $Pb_3(OH)_4^{2+}$, and $Pb(OH)_4^{2-}$. For tricadmium diphosphate, the following system was modelled: $C_{T,P} = 1 \times 10^{-3}$ M, $C_{T,Cl} = 1 \times 10^{-3}$ M, and $C_{T,S} = 1 \times 10^{-3}$ M; and included the components: $Cd_3(PO_4)_2(s)$, Cd^{2+}, $CdOH^+$, $Cd(OH)_2^0$, $Cd(OH)_3^-$, Cd_2OH^{3+}, $CdCl^+$, $CdCl_2^0$, $CdOHCl^0$, $CdSO_4^0$, and $Cd(SO_4)_2^{2-}$. For tricopper diphosphate, the following system was modelled: $C_{T,P} = 1 \times 10^{-3}$ M, $C_{T,Cl} = 1 \times 10^{-3}$ M, and $C_{T,S} = 1 \times 10^{-3}$ M; and included the components: $Cu_3(PO_4)_2(s)$, Cu^{2+}, $CuOH^+$, $Cu(OH)_2^0$, $Cu(OH)_3^-$, $Cu(OH)_4^{2-}$, and $Cu_2(OH)_2^{2+}$. For hopeite, the following system was modelled: $C_{T,P} = 1 \times 10^{-3}$ M, and $C_{T,Cl} = 1 \times 10^{-3}$ M; and included the components: $Zn_3(PO_4)_2 \cdot 5H_2O(s)$, Zn^{2+}, $ZnOH^+$, $Zn(OH)_2^0$, $Zn(OH)_3^-$, $Zn(OH)_4^{2-}$, $ZnCl^+$, $ZnCl_2^0$, $ZnCl_3^-$, $ZnOHCl^0$, $ZnSO_4^0$, and $Zn(SO_4)_2^{2-}$.

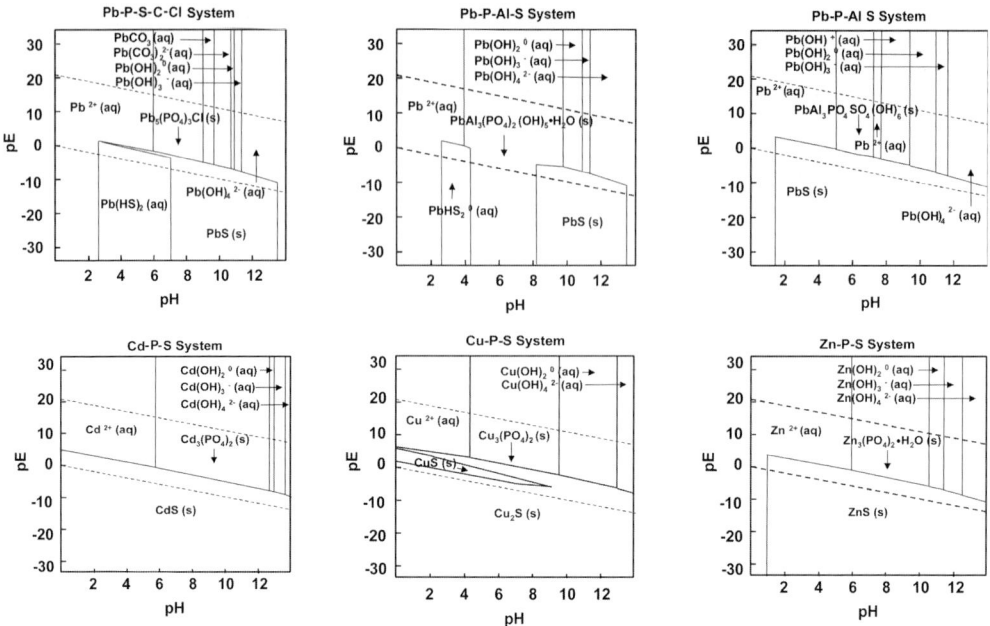

Fig. 7. pE–pH predominance diagrams for various metal phosphate minerals. In all cases, an ionic strength of 0.1 molal was assumed and activity coefficients were calculated using the extended Debye–Hückel equation. Temperatures were set at 25 °C. For chloropyromorphite, the following system was modelled: $C_{T,Pb} = 1 \times 10^{-10}$ M, $C_{T,P} = 1 \times 10^{-2}$ M, $C_{T,Cl} = 1 \times 10^{-3}$ M, $C_{T,S} = 1 \times 10^{-3}$ M, and $C_{T,C} = 1 \times 10^{-3}$ M; and included the components: $Pb_5(PO_4)_3Cl(s)$, $PbCO_3(s)$, $PbSO_4(s)$, $PbS(s)$, $S(s)$, Pb^{2+}, $PbOH^+$, $Pb(OH)_2^0$, $Pb(OH)_3^-$, $Pb(OH)_4^{2-}$, Pb_2OH^{3+}, $Pb_3(OH)_4^{2+}$, $PbCl^+$, $PbCl_2^0$, $PbCl_3^-$, $PbCl_4^{2-}$, $Pb(HS)_2^0$, $Pb(HS)_3^-$, $PbSO_4^0$, $Pb(SO_4)_2^{2-}$, $Pb(CO_3)_2^{2-}$, $PbCO_3^0$, $PbHCO_3^+$, PO_4^{3-}, HPO_4^{2-}, $H_2PO_4^-$, $H_3PO_4^0$, S^{2-}, S_2^{2-}, HS^-, H_2S^0, SO_4^{2-}, HSO_4^-, Cl^-, Cl_2^0, CO_3^{2-}, HCO_3^-, and $H_2CO_3^0$. For hinsdalite, the following system was modelled: $C_{T,Pb} = 1 \times 10^{-8}$ M, $C_{T,P} = 1 \times 10^{-3}$ M, $C_{T,S} = 1 \times 10^{-3}$ M, and $C_{T,Al} = 1 \times 10^{-3}$ M; and included the components: $PbAl_3PO_4SO_4(OH)_6(s)$, $PbSO_4(s)$, $PbS(s)$, Pb^{2+}, $PbOH^+$, $Pb(OH)_2^0$, $Pb(OH)_3^-$, $Pb(OH)_4^{2-}$, $PbSO_4^0$, $Pb(SO_4)_2^{2-}$, Al^{3+}, $AlOH^{2+}$, $Al(OH)_2^+$, $Al(OH)_3^0$, $AlSO_4^+$, $Al(SO_4)_2^-$, PO_4^{3-}, HPO_4^{2-}, $H_2PO_4^-$, $H_3PO_4^0$, S^{2-}, HS^-, H_2S^0, SO_4^{2-}, and HSO_4^-. For plumbogummite, the following system was modelled: $C_{T,Pb} = 1 \times 10^{-8}$ M, $C_{T,P} = 1 \times 10^{-3}$ M, $C_{T,S} = 1 \times 10^{-3}$ M, and $C_{T,Al} = 1 \times 10^{-3}$ M; and included the components: $PbAl_3PO_4SO_4(OH)_6 \cdot H_2O(s)$, $PbSO_4(s)$, $PbS(s)$, Pb^{2+}, $PbOH^+$, $Pb(OH)_2^0$, $Pb(OH)_3^-$, $Pb(OH)_4^{2-}$, $PbSO_4^0$, $Pb(SO_4)_2^{2-}$, $Pb(HS)_2^0$, $Pb(HS)_3^-$, Al^{3+}, $AlOH^{2+}$, $Al(OH)_2^+$, $Al(OH)_3^0$, $Al(OH)_4^-$, $AlSO_4^+$, $Al(SO_4)_2^-$, PO_4^{3-}, HPO_4^{2-}, $H_2PO_4^-$, $H_3PO_4^0$, S^{2-}, HS^-, H_2S^0, SO_4^{2-}, and HSO_4^-. For tricadmium diphosphate, the following system was modelled: $C_{T,Cd} = 1 \times 10^{-3}$ M, $C_{T,P} = 1 \times 10^{-3}$ M, and $C_{T,S} = 1 \times 10^{-3}$ M; and included the components: $Cd_3(PO_4)_2(s)$, $CdSO_4(s)$, $CdS(s)$, Cd^{2+}, $CdOH^+$, $Cd(OH)_2^0$, $Cd(OH)_3^-$, $Cd(OH)_4^{2-}$, $CdSO_4^0$, $Cd(SO_4)_2^{2-}$, PO_4^{3-}, HPO_4^{2-}, $H_2PO_4^-$, $H_3PO_4^0$, S^{2-}, HS^-, H_2S^0, SO_4^{2-}, and HSO_4^-. For tricopper diphosphate, the following system was modelled: $C_{T,Cu} = 1 \times 10^{-3}$ M, $C_{T,P} = 1 \times 10^{-3}$ M, and $C_{T,S} = 1 \times 10^{-3}$ M; and included the components: $Cu_3(PO_4)_2(s)$, $CuSO_4(s)$, $CuSO_4 \cdot 5H_2O(s)$, $CuS(s)$, $Cu_2S(s)$, $S(s)$, Cu^+, Cu^{2+}, $CuOH^+$, $Cu(OH)_2^0$, $Cu(OH)_3^-$, $Cu(OH)_4^{2-}$, $Cu_2(OH)_2^{2+}$, $CuSO_4^0$, $Cu(SO_4)_2^{2-}$, PO_4^{3-}, HPO_4^{2-}, $H_2PO_4^-$, $H_3PO_4^0$, S^{2-}, HS^-, H_2S^0, SO_4^{2-}, and HSO_4^-. For hopeite, the following system was modelled: $C_{T,Zn} = 1 \times 10^{-3}$ M, $C_{T,P} = 1 \times 10^{-3}$ M, and $C_{T,S} = 1 \times 10^{-3}$ M; and included the components: $Zn_3(PO_4)_2 \cdot 4H_2O(s)$, $ZnSO_4(s)$, $ZnSO_4 \cdot H_2O(s)$, $ZnSO_4 \cdot 6H_2O(s)$, $ZnS(s)$, Zn^{2+}, $ZnOH^+$, $Zn(OH)_2^0$, $Zn(OH)_3^-$, $Zn(OH)_4^{2-}$, $ZnSO_4^0$, $Zn(SO_4)_2^{2-}$, PO_4^{3-}, HPO_4^{2-}, $H_2PO_4^-$, $H_3PO_4^0$, S^{2-}, HS^-, H_2S^0, SO_4^{2-}, and HSO_4^-.

galena, and completely dominates cerrusite. Carbonate, hydroxide and sulphide complexes with Pb are also present in the stability field. These data suggest that chloropyromorphite is very stable at moderate pH values (6–9) and at high $C_{T,C}$ values where carbonates might prevail. It is also stable at fairly low pE values. Hinsdalite has a fairly large predominance field over galena at $C_{T,Pb} = 1 \times 10^{-8}$ M. Hydroxide and sulphide complexes with Pb are also

present in the stability field. These data suggest that hinsdalite is very stable over a large range in pH values (4–10). It is also stable at fairly low pE values. Plumbogummite has a modest predominance field between pH 5 and 7.5 at $C_{T,Pb} = 1 \times 10^{-8}$ M. It has predominance over galena. Hydroxide complexes with Pb are also present in the stability field. Tricadmium diphosphate has a wide predominance field between pH 6 and 12 at $C_{T,Cd} = 1 \times 10^{-3}$ M. It has predominance over greenockite (CdS). Hydroxide complexes with Cd are also present in the stability field. Tricopper diphosphate has a wide predominance field between pH 4 and 9.5 at $C_{T,Cu} = 1 \times 10^{-3}$ M. It has some predominance over chalcocite (Cu_2S) and wide predominance over covellite (CuS). Hydroxide complexes with Cu are also present in the stability field. Hopeite has a wide predominance field between pH 4 and 9.5 at $C_{T,Zn} = 1 \times 10^{-3}$ M. It has some predominance over sphalerite (CuS). Hydroxide complexes with Zn are also present in the stability field.

Confirmation of phosphate stabilization mechanisms in phosphate-stabilized MSW ashes

Three detailed studies are reviewed here that examine the precise nature of the chemical stabilization of metals in MSW incineration bottom ash (Crannell et al. 2000), scrubber residue (Eighmy et al. 1997), and vitrification dust (Eighmy et al. 1999).

The general experimental approach adopted for each residue was similar and is briefly outlined here. The bottom ash and scrubber residue came from a 1500 t/day modern, massburn, reciprocating grate system equipped with $Ca(OH)_2$ venturi scrubbers with fabric filters. The vitrification dust came from a bag house system of a surface melting facility in Japan that treats MSW ashes.

The bottom ash was treated with 0.4 moles of PO_4^{3-} (from industrial-grade phosphoric acid) per kg dry weight of ash. The scrubber residue was treated with 1.2 moles of PO_4^{3-} per kg of scrubber residue. The vitrification dust was treated with 0.4 moles of PO_4^{3-} per kg of residue. Process mixing in all cases used an LS of 0.4. Mixing was done for 10 minutes in a Hobart mixer using a tined paddle (108 rpm) with a planetary orbit (48 rpm).

All ashes were subjected to a variety of analyses in either their untreated state or after treatment. Further, the ashes were then subjected to rigorous leaching in the Dutch Total Availability leaching procedure, producing four samples for characterization: (1) untreated and unleached, (2) untreated and leached, (3) treated and unleached, and (4) treated and leached. Presumably, solid-phase reaction products would remain in the treated and leached residue after leaching. Some of the spectroscopic analyses included XRD, magic-angle spinning nuclear magnetic resonance (MAS-NMR), and XPS. The reduction in leachability was evaluated

with both the Dutch Total Availability leaching procedure and pH-dependent leaching in pH stat systems.

XRD was used to identify crystalline mineral phases in the samples. A Rigaku–Geigerflex goniometer was used (copper X-ray source, 45 kV, 35 mA, 1575 W). Samples were run in triplicate and a fourth run was conducted after tungsten was added to one of the replicates as an external standard. The data were evaluated for possible crystalline phases using the PC-based search and match program MICRO-ID (Materials Data Corp., Livermore, CA).

MAS NMR was used to monitor the ^{31}P isotropic chemical shifts and chemical shift anisotropy tensors of the component species in the scrubber residue and vitrification dust samples. The Fe content in the bottom ash made spectra acquisition difficult. The spectra were taken on a 400 MHz Chemagnetics Infinity NMR spectrometer with a 9.4 Tesla magnet corresponding to a ^{31}P NMR frequency of 161.9 MHz. Chemical shifts were referenced to an external standard sample of $NH_4H_2PO_4$; (δ (^{31}P) = 0 ppm with respect to 85% H_3PO_4). The analysis of ^{31}P NMR spectra of inorganic phosphates uses isotropic chemical shift and individual chemical shift tensor elements (principal axis system) obtained from Simplex and gradient search implementation of the Herzfeld–Berger method. Numerous primary standards were run to generate reference spectra.

In order to identify and quantify possible chemical phases as well as to quantify elements in the samples, XPS was used. A Perkin-Elmer Physical Electronics Division 5100 hybrid XPS was used. Broadband Mg kα X-rays at 10 eV and 300 W were used for excitation. For energy referencing, the entire system was calibrated to the gold 84.0 $4f_{7/2}$ binding energy (BE). Correction for peak shift due to static charge buildup on the sample system was achieved through the adventitious carbon reference method using a C 1s BE of 284.8 eV as a conducting reference. The NIST database was used for peak BE identification. Numerous metal phosphate reference spectra were obtained (Eighmy et al. 1998, 1999a–e) to assist in Ca $2p_{3/2}$ and Pb $4f_{7/2}$ peak identification.

The Dutch Total Availability Leaching test (NEN 7341) was used to operationally quantify the elemental mass fraction available for leaching in the samples. The procedure involves two sequential extractions; the first were conducted at a pH of 7.0 and LS of 100. For examination of mass fractions available for leaching, 8 g of sample was added to 800 mL of distilled, deionized water and stirred in a capped Teflon vessel.

Concentrated HNO_3 was added to bring the solutions to pH 7.0 and then 3 N HNO_3 was used to maintain pH throughout the leaching period. After three hours, the leachates were filtered through a 0.22 μm polycarbonate filter. The second extraction entailed leaching the samples and filters from the first extraction at pH 4.0 for four hours at an LS of 100. Concentrated HNO_3 was added to bring the solution to pH 4.0 then 3 N HNO_3 was used to control pH. The leachates were filtered and analysed using graphite furnace atomic absorption spectrometry, ion chromatography and low-level colorimetric assays.

The pH-dependent leaching procedure is a means of determining the equilibrium leaching behaviour in the samples over a range of pH values relevant to regulatory leaching and landfill disposal. Each extraction was done at an LS of 10 so as to ensure solid-phase control. The leaching tests were conducted in a pH-stat for 24 hours, a duration typically used for achieving pseudo-equilibrium. Typically, 80 g of sample were placed into the Teflon vessel to which 800 mL of distilled, deionized water was added. Then 3 N HNO_3 was used to control the pH at various set points (e.g., 4, 6, and 8). The leachates were filtered and analysed using graphite furnace atomic absorption spectrometry, ion chromatography and low-level colorimetric assays.

The geochemical equilibrium model MINTEQA2 (Allison et al. 1990) was used to determine which solid phase controls leachate composition as a function of pH. MINTEQA2 was modified to include all the phosphate mineral phases shown in Table 3. The likelihood of solid solution formation required further modification to MINTEQA2 to use idealized solid solutions as possible controlling solids. A simplistic zero heat of mixing and ideal site substitution model was assumed (Davis et al. 1996; Greenwood 1977). Standard free energies of formation for the solid solutions ($\Delta G_{f,ij}^0$) between end members i and j were used to determine theoretical K_{sp} values (see Table 5) for the solid solutions using:

$$\Delta G_{f,ij}^0 = x\Delta G_{f,i}^0 + (1-x)\Delta G_{f,j}^0$$
$$+ nRT[x \ln x + (1-x)\ln(1-x)]$$

$$(19)$$

where x is the mole fraction of end member i, $\Delta G_{f,i}^0$ and $\Delta G_{f,j}^0$ are the free energies of formation of end members i and j, respectively, n is the number of sites in the mineral undergoing substitution (e.g., 3, 4, or 5 depending on the mineral solid solution series being calculated; e.g., 5 for

apatites), R is the universal gas constant, and T is degrees Kelvin. No attempts were made to evaluate the likelihood of these solid solutions with respect to theoretical (e.g., $K_{sp,i}/K_{sp,j}$) or experimental distribution coefficients. These K_{sp} values were entered into the MINTEQA2 database.

A typical suite of X-ray diffractograms is shown in Fig. 8 for bottom ash samples. Diffraction peaks differ between sample treatments. With bottom ash, a large amorphous background signal is present. Thirty to 40 peaks are selected for analysis in the search match software. As shown in Tables 6 to 8, a number of metal phosphates were found in the treated samples and the treated and leached samples for the bottom ash, scrubber residue, and vitrification dust samples. Apatite family and tertiary metal phosphates are common to both the treated and unleached samples and the treated and leached samples for all three ashes.

Typical ^{31}P MAS-NMR spectra for the treated and unleached vitrification dust and treated and leached vitrification dust samples are shown in Fig. 9. In the treated and unleached sample, a major component (62.4%) resonance is present at -7.5 ppm with sidebands spaced at ± 4 and ± 8 kHz. A minor component (37.6%) resonance is present at -34.6 ppm with sidebands spaced at ± 4, ± 8, and ± 12 kHz. The major component is likely either hopeite or $CaZn(PO_4)_2$. The minor component is likely a pyrophosphate. As shown in the plot, the residuals trace suggests that other components may be present. In the treated and leached sample, a major component (49.4%) resonance is present at -11.5 ppm with sidebands spaced at ± 4, ± 8, and ± 12 kHz. A second major component (50.7%) resonance is present at -34.6 ppm with sidebands spaced at ± 4, ± 8, and ± 12 kHz. The first component is likely either hopeite or $CaZn(PO_4)_2$. The second major component is a simple orthophosphate with a more negative counter ion than Ca or Zn. As shown in the plot, the residuals trace suggests that other components are present. As shown in Tables 7 and 8, similar Ca phosphate phases were seen in both dry scrubber residues and vitrification dusts; the phases were present after aggressive leaching, and there is generally agreement between the XRD and MAS-NMR data.

Typical Pb $4f_{7/2}$ XPS spectra for the four fractions of the vitrification dust sample are shown in Fig. 10. Component curves are shown under the actual curves that were fit to the spectra. In some cases, two species of Pb were fit. In others, three species were identified. As shown in Tables 6 to 8, similar Pb phosphate phases

Table 5. *Solubility products and ΔG_f^0 for various ideal solid solutions*

Mineral	Dissolution reaction	$-pK_{sp}$	ΔG_f^0 (kJ/mole*)
Hydroxyapatites			
$Ca_4Pb(PO_4)_3OH$	$Ca_4Pb(PO_4)_3OH + H^+ \Leftrightarrow 4Ca^{2+} + Pb^{2+} + 3PO_4^{3-} + H_2O$	49.10	−5,812
$Ca_3Pb_2(PO_4)_3OH$	$Ca_3Pb_2(PO_4)_3OH + H^+ \Leftrightarrow 3Ca^{2+} + 2Pb^{2+} + 3PO_4^{3-} + H_2O$	53.17	−5,306
$Ca_2Pb_3(PO_4)_3OH$	$Ca_2Pb_3(PO_4)_3OH + H^+ \Leftrightarrow 2Ca^{2+} + 3Pb^{2+} + 3PO_4^{3-} + H_2O$	56.86	−4,798
$CaPb_4(PO_4)_3OH$	$CaPb_4(PO_4)_3OH + H^+ \Leftrightarrow Ca^{2+} + 4Pb^{2+} + 3PO_4^{3-} + H_2O$	60.18	−4,288
$Ca_4Cd(PO_4)_3OH$	$Ca_4Cd(PO_4)_3OH + H^+ \Leftrightarrow 4Ca^{2+} + Cd^{2+} + 3PO_4^{3-} + H_2O$	47.85	−5,858
$Ca_3Cd_2(PO_4)_3OH$	$Ca_3Cd_2(PO_4)_3OH + H^+ \Leftrightarrow 3Ca^{2+} + 2Cd^{2+} + 3PO_4^{3-} + H_2O$	50.66	−5,399
$Ca_2Cd_3(PO_4)_3OH$	$Ca_2Cd_3(PO_4)_3OH + H^+ \Leftrightarrow 2Ca^{2+} + 3Cd^{2+} + 3PO_4^{3-} + H_2O$	53.09	−4,936
$CaCd_4(PO_4)_3OH$	$CaCd_4(PO_4)_3OH - H^+ \Leftrightarrow Ca^{2+} + 4Cd^{2+} + 3PO_4^{3-} + H_2O$	55.14	−4,472
$Ca_4Cu(PO_4)_3OH$	$Ca_4Cu(PO_4)_3OH + H^+ \Leftrightarrow 4Ca^{2+} + Cu^{2+} + 3PO_4^{3-} + H_2O$	46.91	−5,710
$Ca_3Cu_2(PO_4)_3OH$	$Ca_3Cu_2(PO_4)_3OH + H^+ \Leftrightarrow 3Ca^{2+} + 2Cu^{2+} + 3PO_4^{3-} + H_2O$	48.78	−5,101
$Ca_2Cu_3(PO_4)_3OH$	$Ca_2Cu_3(PO_4)_3OH + H^+ \Leftrightarrow 2Ca^{2+} + 3Cu^{2+} + 3PO_4^{3-} + H_2O$	50.26	−4,490
$CaCu_4(PO_4)_3OH$	$CaCu_4(PO_4)_3OH + H^+ \Leftrightarrow Ca^{2+} + 4Cu^{2+} + 3PO_4^{3-} + H_2O$	51.41	−3,877
$Ca_4Zn(PO_4)_3OH$	$Ca_4Zn(PO_4)_3OH + H^+ \Leftrightarrow 4Ca^{2+} + Zn^{2+} + 3PO_4^{3-} + H_2O$	46.33	−5,919
$Ca_3Zn_2(PO_4)_3OH$	$Ca_3Zn_2(PO_4)_3OH - H^+ \Leftrightarrow 3Ca^{2+} + 2Zn^{2+} + 3PO_4^{3-} + H_2O$	47.62	−5,521
$Ca_2Zn_3(PO_4)_3OH$	$Ca_2Zn_3(PO_4)_3OH + H^+ \Leftrightarrow 2Ca^{2+} + 3Zn^{2+} + 3PO_4^{3-} + H_2O$	48.53	−5,119
$CaZn_4(PO_4)_3OH$	$CaZn_4(PO_4)_3OH + H^+ \Leftrightarrow Ca^{2+} + 4Zn^{2+} + 3PO_4^{3-} + H_2O$	49.07	−4,716
Chloroapatites			
$Ca_4Pb(PO_4)_3Cl$	$Ca_4Pb(PO_4)_3Cl \Leftrightarrow 4Ca^{2+} + Pb^{2+} + 3PO_4^{3-} + Cl^-$	55.48	−5,743
$Ca_3Pb_2(PO_4)_3Cl$	$Ca_3Pb_2(PO_4)_3Cl \Leftrightarrow 3Ca^{2+} + 2Pb^{2+} + 3PO_4^{3-} + Cl^-$	63.36	−5,259
$Ca_2Pb_3(PO_4)_3Cl$	$Ca_2Pb_3(PO_4)_3Cl \Leftrightarrow 2Ca^{2+} + 3Pb^{2+} + 3PO_4^{3-} + Cl^-$	70.87	−4,772
$CaPb_4(PO_4)_3Cl$	$CaPb_4(PO_4)_3Cl \Leftrightarrow Ca^{2+} + 4Pb^{2+} + 3PO_4^{3-} + Cl^-$	78.00	−4,284
$Ca_4Cd(PO_4)_3Cl$	$Ca_4Cd(PO_4)_3Cl \Leftrightarrow 4Ca^{2+} + Cd^{2+} + 3PO_4^{3-} + Cl^-$	48.52	−5,756
$Ca_3Cd_2(PO_4)_3Cl$	$Ca_3Cd_2(PO_4)_3Cl \Leftrightarrow 3Ca^{2+} + 2Cd^{2+} + 3PO_4^{3-} + Cl^-$	49.45	−5,286
$Ca_2Cd_3(PO_4)_3Cl$	$Ca_2Cd_3(PO_4)_3Cl \Leftrightarrow 2Ca^{2+} + 3Cd^{2+} + 3PO_4^{3-} + Cl^-$	50.01	−4,813

(continued)

Table 5. *Continued*

Mineral	Dissolution reaction	$-pK_{sp}$	ΔG_f^0 (kJ/mole*)
	$CaCd_4(PO_4)_3Cl \Leftrightarrow Ca^{2+} + 4Cd^{2+} + 3PO_4^{3-} + Cl^-$	50.18	−4,338
	$Ca_4Cu(PO_4)_3Cl \Leftrightarrow 4Ca^{2+} + Cu^{2+} + 3PO_4^{3-} + Cl^-$	49.43	−5,618
	$Ca_3Cu_2(PO_4)_3Cl \Leftrightarrow 3Ca^{2+} + 2Cu^{2+} + 3PO_4^{3-} + Cl^-$	51.26	−5,009
	$Ca_2Cu_3(PO_4)_3Cl \Leftrightarrow 2Ca^{2+} + 3Cu^{2+} + 3PO_4^{3-} + Cl^-$	52.71	−4,398
	$CaCu_4(PO_4)_3Cl \Leftrightarrow Ca^{2+} + 4Cu^{2+} + 3PO_4^{3-} + Cl^-$	53.80	−3,785
	$Ca_4Zn(PO_4)_3Cl \Leftrightarrow 4Ca^{2+} + Zn^{2+} + 3PO_4^{3-} + Cl^-$	46.06	−5,812
	$Ca_3Zn_2(PO_4)_3Cl \Leftrightarrow 3Ca^{2+} + 2Zn^{2+} + 3PO_4^{3-} + Cl^-$	44.52	−5,397
	$Ca_2Zn_3(PO_4)_3Cl \Leftrightarrow 2Ca^{2+} + 3Zn^{2+} + 3PO_4^{3-} + Cl^-$	42.61	−4,980
	$CaZn_4(PO_4)_3Cl \Leftrightarrow Ca^{2+} + 4Zn^{2+} + 3PO_4^{3-} + Cl^-$	40.32	−4,560
Tertiary Metal Phosphates			
	$Ca_3Pb(PO_4)_2 \Leftrightarrow 2Ca^{2+} + Pb^{2+} + 2PO_4^{3-}$	34.62	−3,367
	$CaPb_2(PO_4)_2 \Leftrightarrow Ca^{2+} + 2Pb^{2+} + 2PO_4^{3-}$	39.49	−2,865
	$Ca_2Cd(PO_4)_2 \Leftrightarrow 2Ca^{2+} + Cd^{2+} + 2PO_4^{3-}$	30.97	−3,399
	$CaCd_2(PO_4)_2 \Leftrightarrow Ca^{2+} + 2Cd^{2+} + 2PO_4^{3-}$	32.20	−2,930
	$Ca_2Cu(PO_4)_2 \Leftrightarrow 2Ca^{2+} + Cu^{2+} + 2PO_4^{3-}$	32.39	−3,264
	$CaCu_2(PO_4)_2 \Leftrightarrow Ca^{2+} + 2Cu^{2+} + 2PO_4^{3-}$	35.04	−2,660
	$Ca_2Zn(PO_4)_2 \Leftrightarrow 2Ca^{2+} + Zn^{2+} + 2PO_4^{3-}$	29.11	−3,458
	$CaZn_2(PO_4)_2 \Leftrightarrow Ca^{2+} + 2Zn^{2+} + 2PO_4^{3-}$	28.46	−3,048
Other Apatites			
	$Ca_3PbO(PO_4)_2 + 2H^+ \Leftrightarrow 3Ca^{2+} + Pb^{2+} + 2PO_4^{3-} + H_2O$	23.21	−4,092
	$Ca_2Pb_2O(PO_4)_2 + 2H^+ \Leftrightarrow 2Ca^{2+} + 2Pb^{2+} + 2PO_4^{3-} + H_2O$	28.32	−3,592
	$CaPb_3O(PO_4)_2 + 2H^+ \Leftrightarrow Ca^{2+} + 3Pb^{2+} + 2PO_4^{3-} + H_2O$	32.97	−3,090

*A simplistic zero heat of mixing and ideal site substitution model was assumed (Davis *et al.* 1996). Standard free energies of formation for the solid solutions ($\Delta G_{f,ij}^0$) between end members i and j were used to determine theoretical K_{sp} values.

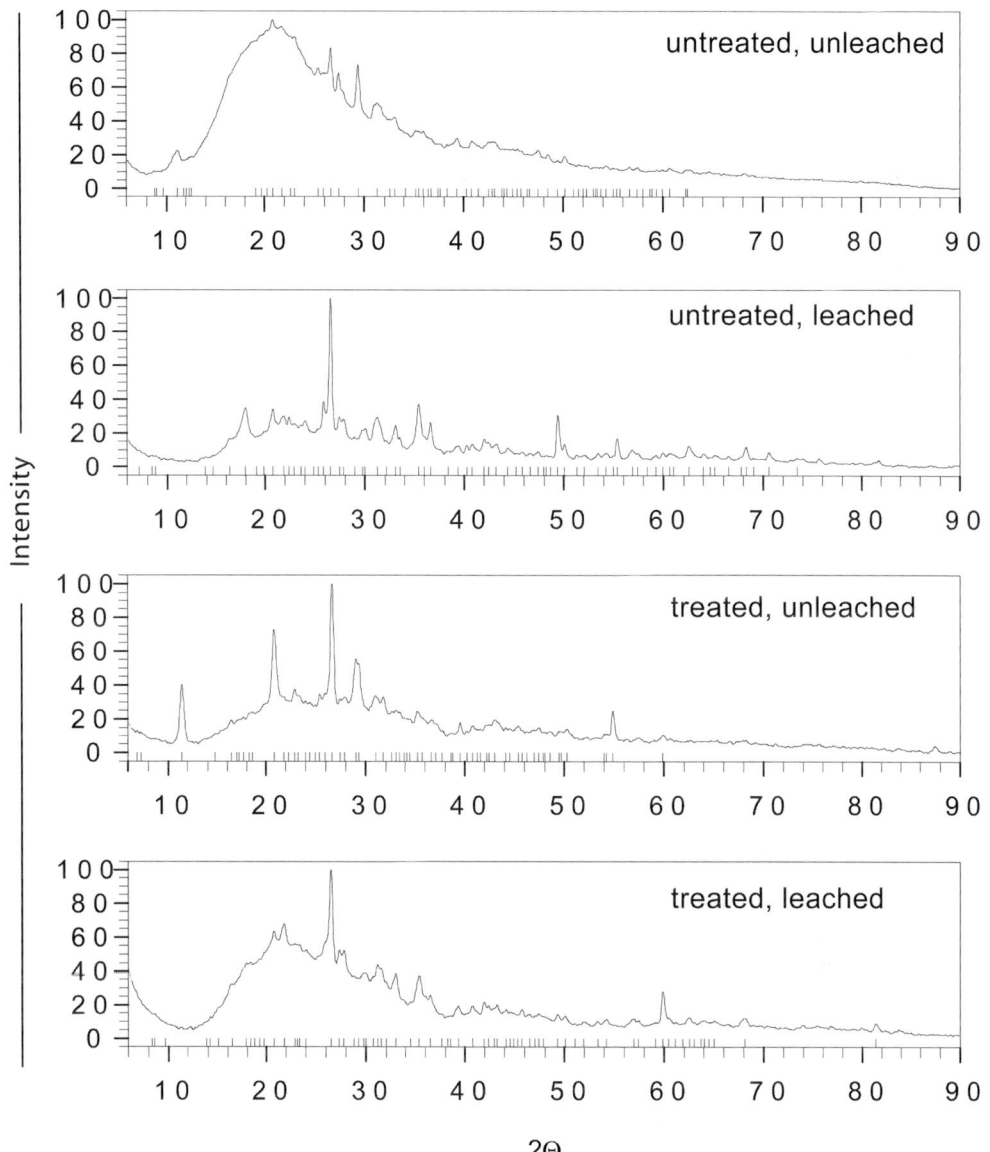

Fig. 8. X-ray diffraction spectra for MSW bottom ash treated 0.4 moles of PO_4^{3-}/kg of ash. The four fractions were produced using the presence or absence of treatment and the use of the Dutch Total Availability leaching protocol (from Crannell *et al.* 2000).

were seen in bottom ash, scrubber residues, and vitrification dusts; the phases were present after aggressive leaching, and there is generally agreement between the XRD and XPS data.

A typical pH-dependent leaching plot for Pb is shown in Fig. 11 for vitrification dusts. The plots show the measured concentration of Pb in the pH-dependent leaching test leachate as a function of pH for both the untreated sample and

the treated sample. A one- to two-order reduction in leachability is seen. The modelling of solid-phase control suggests that chloropyromorphite is the best candidate, although the ideal solid solution phase $(Ca_4,Pb)(PO_4)_3Cl$ is also a likely candidate. As shown in Tables 6 to 8, similar metal phosphate candidate controlling solids were seen in all three ash types; the phases were present after aggressive leaching,

Table 6. *Identified phosphate mineral reaction products in phosphate-stabilized MSW bottom ash**

	Spectroscopic method				pH-dependent controlling solids
	XRPD		XPS		
	Unleached	Leached	Unleached	Leached	
Ca	$Ca_5(P,Si,S)_3O_{12}(Cl,OH,F)$ $CaZn_2(PO_4)_2$ $Ca_3(PO_4)_2$ $Ca_5(PO_4)_3OH$	$Ca_8Pb_2(PO_4)_6(OH)_2$ $CaZn_2(PO_4)_2$ $Ca_3(PO_4)_2 \cdot xH_2O$ $Ca_5(PO_4)_3OH$	$Ca_5(PO_4)_3OH$ $Ca_5(PO_4)_3Cl$ $Ca_3(PO_4)_2$ $Ca_2P_2O_7$ $CaHPO_4$ $CaHPO_4 \cdot 2H_2O$	$Ca_5(PO_4)_3OH$ $Ca_5(PO_4)_3Cl$ $(Ca,Pb)_3(PO_4)_2$ $Ca_2P_2O_7$ $CaHPO_4$	$CaHPO_4$ (pH 4,6,8) $CaHPO_4 \cdot 2H_2O$ (pH 4,6,8) $Ca_3(PO_4)_2$ (pH 4) $Ca_4O(PO_4)_2$ (pH 6)
Cd	$CdTi_4(PO_4)_6$	$Cd_5(PO_4)_3Cl$	—	—	—
Cu	$Cu_4O(PO_4)_2$ $Cu_5(PO_4)_2(OH)_4$	$Cu_2(PO_4)OH$ $Cu_3(PO_4)(OH)_3$	—	—	$(Cu_2,Ca)(PO_4)_2$ (pH 4,6,8) $(Cu,Ca_4)(PO_4)_3Cl$ (pH 6,8) $(Cu,Ca_2)(PO_4)_2$ (pH 4)
Pb	$Pb_4O(PO_4)_2$ $Pb_3(PO_4)_2$ $Pb_5(CrO_4)_2(PO_4)_2 \cdot H_2O$	$Pb_5(PO_4)_3Cl$ $Pb_5(PO_4)_3OH$ $Pb_4O(PO_4)_2$ $Pb_2(Fe,Al)(PO_4)_2(OH) \cdot H_2O$ $PbFe_3(PO_4)(SO_4)(OH)_6$	—	$Pb_5(PO_4)_3Cl$	$(Pb_2,Ca)(PO_4)_2$ (pH 6,8) $PbHPO_4$ (pH 6) $(Pb_2,Ca_3)(PO_4)_3OH$ (pH 6) $Pb_5(PO_4)_3Cl$ (pH 4)
Zn	$Zn_3(PO_4)_2 \cdot 4H_2O$	$Zn_3(PO_4)_2$ $ZnHPO_4 \cdot H_2O$	—	—	$CaZn_2(PO_4)_2 \cdot 2H_2O$ (pH 6) $(Zn,Ca_2)(PO_4)_2$ (pH 6) $(Zn_3,Ca_2)(PO_4)_3OH$ (pH 6)

*From Crannell *et al.* 2000.

Table 7. *Identified phosphate mineral reaction products in phosphate-stabilized MSW dry scrubber residues**

| | Spectroscopic method | | | | | | pH-dependent controlling solids |
| | XRPD | | MAS-NMR | | XPS | | |
	Unleached	Leached	Unleached	Leached	Unleached	Leached	
Ca	$Ca_5(P,Si,S)_3O_{12}(Cl,OH,F)$ $\alpha\text{-}CaZn_2(PO_4)_2$ $Ca_{10}(PO_4)_3(CO_3)_3(OH)_2$	$Ca_3Mg_2(PO_4)_2$ $Ca_3(PO_4)_2 \cdot xH_2O$ $\alpha\text{-}CaZn_2(PO_4)_2$ $Ca_5(P,Si,S)_3O_{12}(Cl,OH,F)$	$CaHPO_4 \cdot 2H_2O$ $CaHPO_4$ $Ca_5(PO_4)_3OH$ $\alpha\text{-}CaZn_2(PO_4)_2$ $\alpha\text{-}Ca_2P_2O_7$	$CaHPO_4 \cdot 2H_2O$ $\alpha\text{-}Ca_2P_2O_7$	$Ca_5(PO_4)_3OH$ $Ca_5(PO_4)_3Cl$ $\alpha\text{-}Ca_2P_2O_7$ $Ca_8H_2(PO_4)_6 \cdot 5H_2O$ $CaHPO_4$	$Ca_5(PO_4)_3OH$ $Ca_5(PO_4)_3Cl$ $\alpha\text{-}Ca_2P_2O_7$ $\beta\text{-}Ca_3(PO_4)_2$	$CaHPO_4$ (pH 4,6,8) $CaHPO_4 \cdot 2H_2O$ (pH 4,6,8) $Ca_5(PO_4)_3Cl$ (pH 4,6,8) $Ca_5(PO_4)_3OH$ (pH 4,6,8)
Cd	$Cd_3(PO_4)_2$	$Cd_5(PO_4)_3Cl$ $Cd_3(PO_4)_2$	—	—	—	—	$(Cd,Ca_2)(PO_4)_2$ (pH 6) $(Cd,Ca_4)(PO_4)_3Cl$ (pH 6)
Cu	—	$CuFe_2(PO_4)_2(OH)_2$ $Cu_3(PO_4)_2(OH)_3$	—	—	—	—	$(Cu,Ca_4)(PO_4)_3OH$ (pH 6)
Pb	$Pb_4O(PO_4)_2$ $PbAl_3PO_4(OH)_6SO_4$ $Pb_5(PO_4)_3OH$ $Pb_5(PO_4)_3Cl$ $PbAl_3(PO_4)_2(OH)_5 \cdot H_2O$ $\beta\text{-}Pb_9(PO_4)_6$ $PbFePO_4SO_4(OH)_6$	$Pb_4O(PO_4)_2$ $Pb_5(PO_4)_3OH$ $Pb_5(PO_4)_3Cl$ $\beta\text{-}Pb_9(PO_4)_6$	—	—		$Pb_5(PO_4)_3Cl$	$Pb_5(PO_4)_3Cl$ (pH 4,6,8) $(Pb,Ca_4)(PO_4)_3OH$ (pH 6) $(Pb_2,Ca)(PO_4)_2$ (pH 8) $(Pb_3,Ca_2)(PO_4)_3OH$ (pH 8)
Zn	$Zn_3(PO_4)_2 \cdot 2H_2O$ $ZnHPO_4 \cdot H_2O$	$Zn_3(PO_4)_2$	$\alpha\text{-}CaZn_2(PO_4)_2$	—	—	—	$Zn_5(PO_4)_3OH$ (pH 4,6,8) $CaZn_2(PO_4)_2 \cdot 2H_2O$ (pH 4,6,8) $(Zn,Ca_2)(PO_4)_2$ (pH 6) $(Zn_2,Ca_3)(PO_4)_3OH$ (pH 6)

*From Eighmy et al. 1997.

Table 8. *Identified phosphate mineral reaction products in phosphate-stabilized MSW ash vitrification dust*[*]

	Spectroscopic Method						pH-dependent controlling solids
	XRPD		MAS-NMR		XPS		
	Unleached	Leached	Unleached	Leached	Unleached	Leached	
Ca	$CaHPO_4 \cdot 2H_2O$	$\alpha\text{-}CaZn_2(PO_4)_2$	$\alpha\text{-}CaZn_2(PO_4)_2$	$\alpha\text{-}CaZn_2(PO_4)_2$	—	—	$Ca_5(PO_4)_3Cl$ (pH 5,7,9) $(Pb,Ca_4)(PO_4)_3Cl$ (pH 5,7,9) $\beta\text{-}Ca_3(PO_4)_2$ (pH 5,7,9) $(Pb,Ca_2)(PO_4)_2$ (pH 5,7,9) $(Cu,Ca_2)(PO_4)_2$ (pH 5,7,9)
Cd	—	—	—	—	—	—	$(Cd,Ca_2)(PO_4)_2$ (pH 7,9) $(Cd,Ca_4)(PO_4)_3Cl$ (pH 7,9) $(Cu,Ca_2)(PO_4)_2$ (pH 5,7)
Cu	$Cu_2(PO_4)OH$ $Cu_3Al_4(PO_4)_3(OH)_9 \cdot 4H_2O$	$Cu_6(NO_3)_2(PO_4)(OH)_7$ $(Cu,Zn)_6(PO_4)_2(OH)_6 \cdot H_2O$	—	—	—	—	
Pb	$Pb_5(PO_4)_3Br$ $KNaPb_8(PO_4)_6$	$\beta\text{-}Pb_9(PO_4)_6$ $Pb_5(PO_4)_3OH$ $KPb_4(PO_4)_3$ $PbHPO_4 \cdot H_2O$	$Pb_5(PO_4)_3Cl$ $Pb_3(PO_4)_2$	—	—	$Pb_5(PO_4)_3Cl$	$(Pb,Ca_4)(PO_4)_3Cl$ (pH 5,7) $Pb_5(PO_4)_3Cl$ (pH 9) $(Pb_2,Ca)(PO_4)_2$ (pH 9)
Zn	$KZn_2H(PO_4)_2 \cdot 2H_2O$	$Zn_3(PO_4)_2 \cdot 4H_2O$ $(Cu,Zn)_6(PO_4)_2(OH)_6 \cdot H_2O$ $Zn_3(PO_4)_2$	$\alpha\text{-}CaZn_2(PO_4)_2$ $Zn_3(PO_4)_2$	$\alpha\text{-}CaZn_2(PO_4)_2$ $Zn_3(PO_4)_2$	—	—	$CaZn_2(PO_4)_2 \cdot 2H_2O$ (pH 5,7,9) $(Zn_2,Ca_3)(PO_4)_3OH$ (pH 7,9) $(Zn,Ca_2)(PO_4)_2$ (pH 7,9) $(Zn_3,Ca_2)(PO_4)_3OH$ (pH 7,9)

[*] From Eighmy *et al.* 1999.

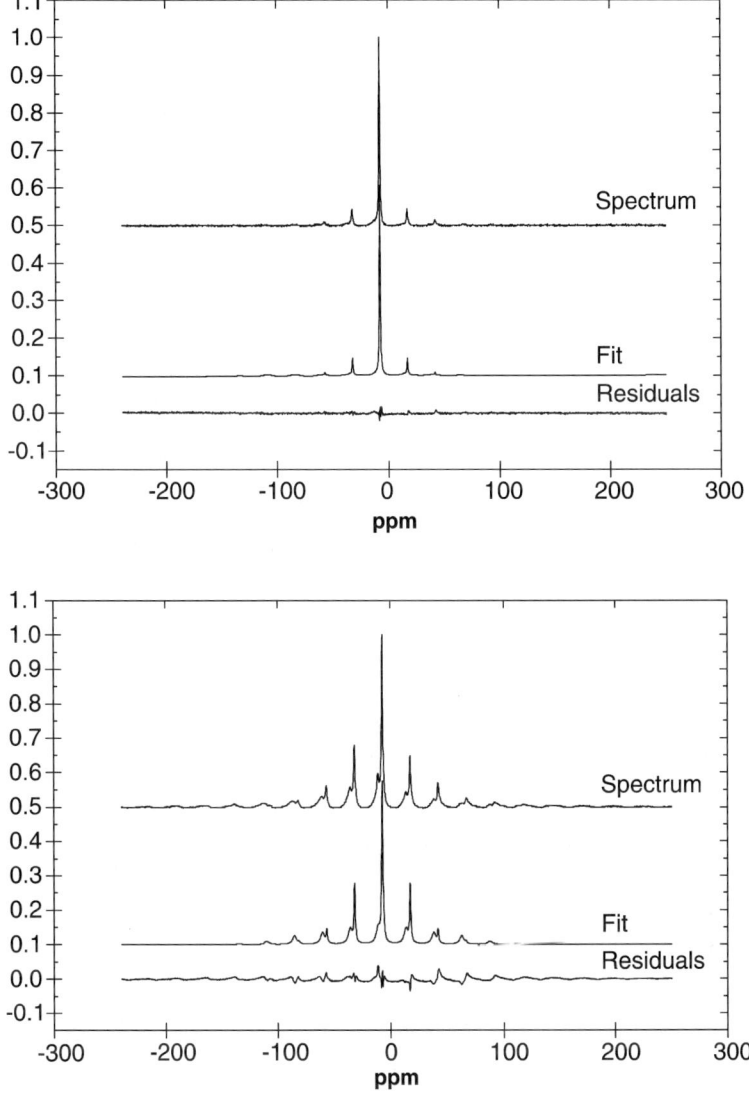

Fig. 9. ^{31}P MAS-NMR spectra for treated and unleached (upper plot) and treated and leached (lower plot) MSW ash vitrification dust treated with 0.4 moles PO_4^{3-}/kg of ash (from Eighmy *et al.* 1998).

and there is generally agreement between the spectroscopic and geochemical modelling data.

The sum of these observations suggests that metal phosphate crystalline mineral phases, some of which are solid solutions, are present in the samples. These phases may be discrete or surface precipitates. Some work on scrubber residues using depth profiling techniques with secondary ion mass spectrometry (SIMS) suggest that surface precipitates are present for these more soluble and fine-grained residues (Eighmy *et al.* 1997).

The observed metal phosphate phases agree with thermodynamic models of the ash system described here. These phases control leaching in pH-stat systems and are present after aggressive leaching designed to remove available or leachable fractions. These phases are also similar to ones observed in soil, sediment, smelter dust, industrial wastewater, and slag systems.

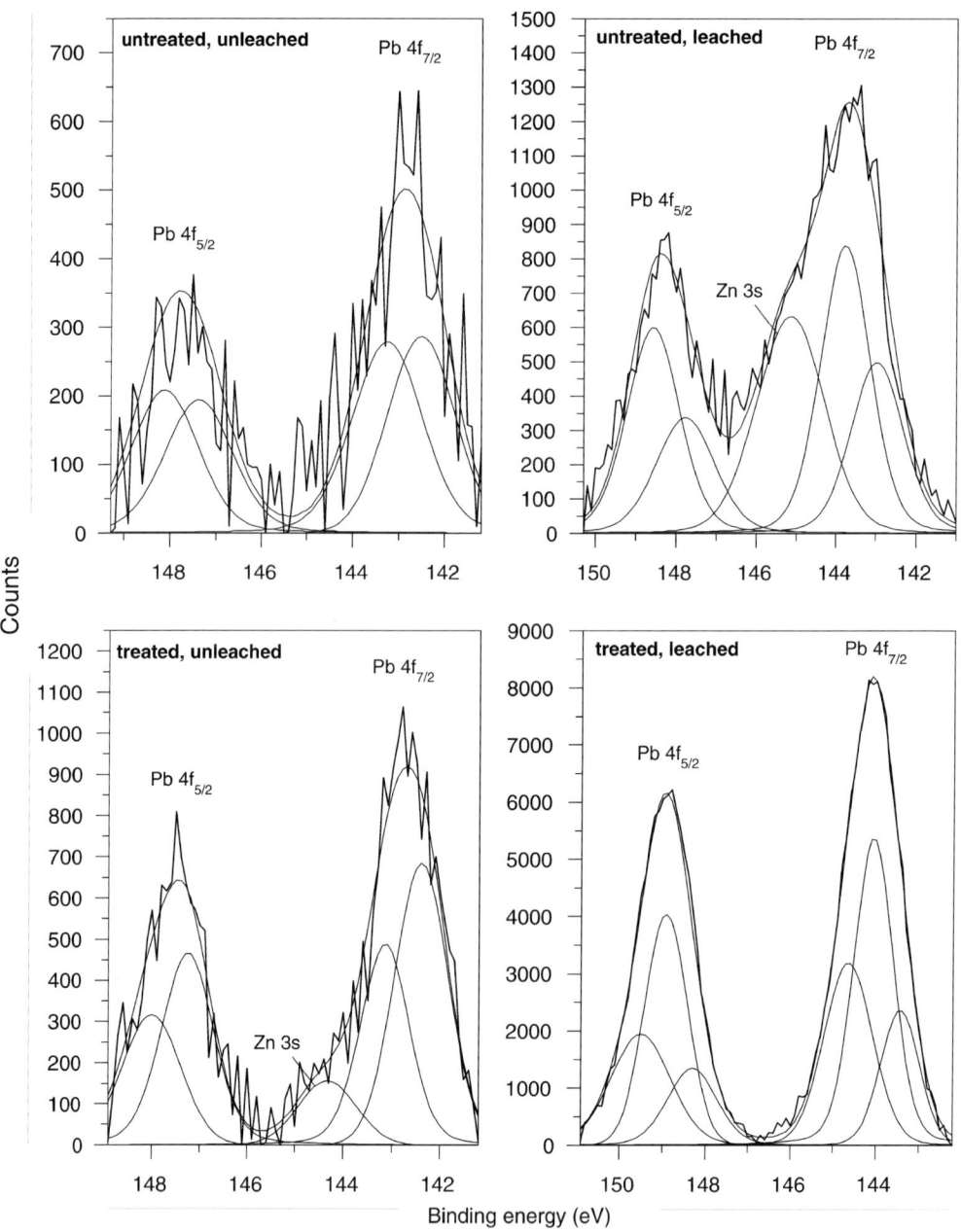

Fig. 10. Pb $4f_{7/2}$ high-resolution XPS spectra for the four fractions of MSW ash vitrification dust treated with 0.4 moles PO_4^{3-}/kg of ash (from Eighmy *et al.* 1998).

Evaluation of longer term performance of phosphate-stabilized ashes

The longer term efficacy of the phosphate stabilization of combined bottom ash and scrubber residues has been evaluated since 1988 through the use of mini-landfills. Residues amended with 0.6 moles PO_4^{3-}/kg of dry combined ash have been maintained in large outdoor lysimeters (ca. 2 m × 2 m × 3 m) designed with leachate collection systems. Approximately 6.2 t of wet treated combined ash (4.03 t dry) and 8.05 t of untreated

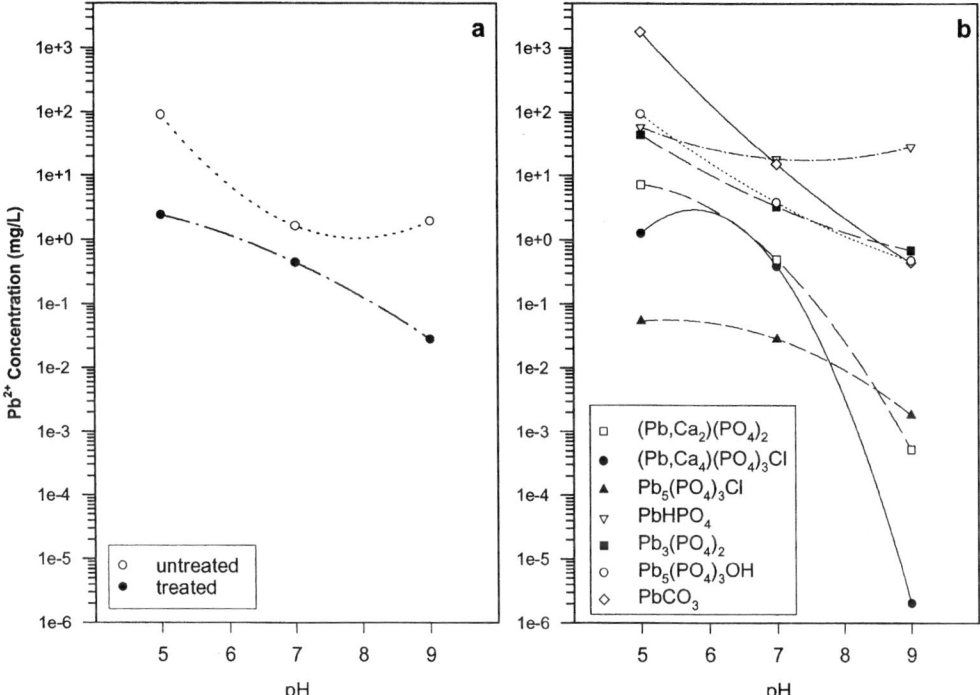

Fig. 11. pH-dependent leaching plots and geochemical modelling plots for MSW ash vitrification dust treated with 0.4 moles PO_4^{3-}/kg of ash (from Eighmy *et al.* 1998). The left-hand plot is the observed leaching behaviour of the untreated and treated ashes. The right-hand plot is the results of geochemical modelling with MINTEQA2 to identify candidate controlling solids for the treated sample leaching behaviour.

combined ash (5.47 t dry) were placed in their respective outdoor lysimeters located at the University of New Hampshire. The lysimeters have been subjected to seasonal (spring through fall/autumn) natural precipitation and infiltration for over 14 years.

Cumulative leaching plots (cumulative mass of element leached per mass dry weight of ash vs. cumulative mass of leachate collected vs. mass dry weight of ash) for Pb, Cu, and Zn are shown in Fig. 12. The plots are constructed using measured element leachate concentrations and volumes of leachate collected. In many cases, the detection limit of the element was used as the concentration, so that the traces are conservative. Over time, the pH of leachates has dropped from about 12 to about 9. Data for Cd are not shown because of the high number of detection limit values that were observed in both the unamended and amended lysimeters.

Generally, the phosphate stabilization process reduced the leachability of the elements from the treated ash. Percentage reductions in leaching were 90% for Pb, 44% for Cu, and 33% for Zn.

However, as pH drops, the ability of natural attenuation mechanisms in the ash (e.g., carbonation, sorption to hydrous ferric oxides) becomes more dominant and leaching is reduced.

Conclusions

Phosphate is widely used as a chemical stabilization agent for MSW combustion residues in Japan and North America and is under consideration for use in parts of Europe. The application of this technology to MSW ashes generally parallels its application to contaminated soils. Metal phosphates (notably Cd, Cu, Pb and Zn) frequently have wide pH distribution, pH–pE predominance, and redox stability within complex ash pore water systems. Stabilization mechanisms identified in other contaminated systems (e.g., soils), involving a combination of sorption, heterogeneous nucleation, and surface precipitation, or solution-phase precipitation are generally observed in ash systems.

There are opportunities to further explore the geochemistry of phosphate stabilization of

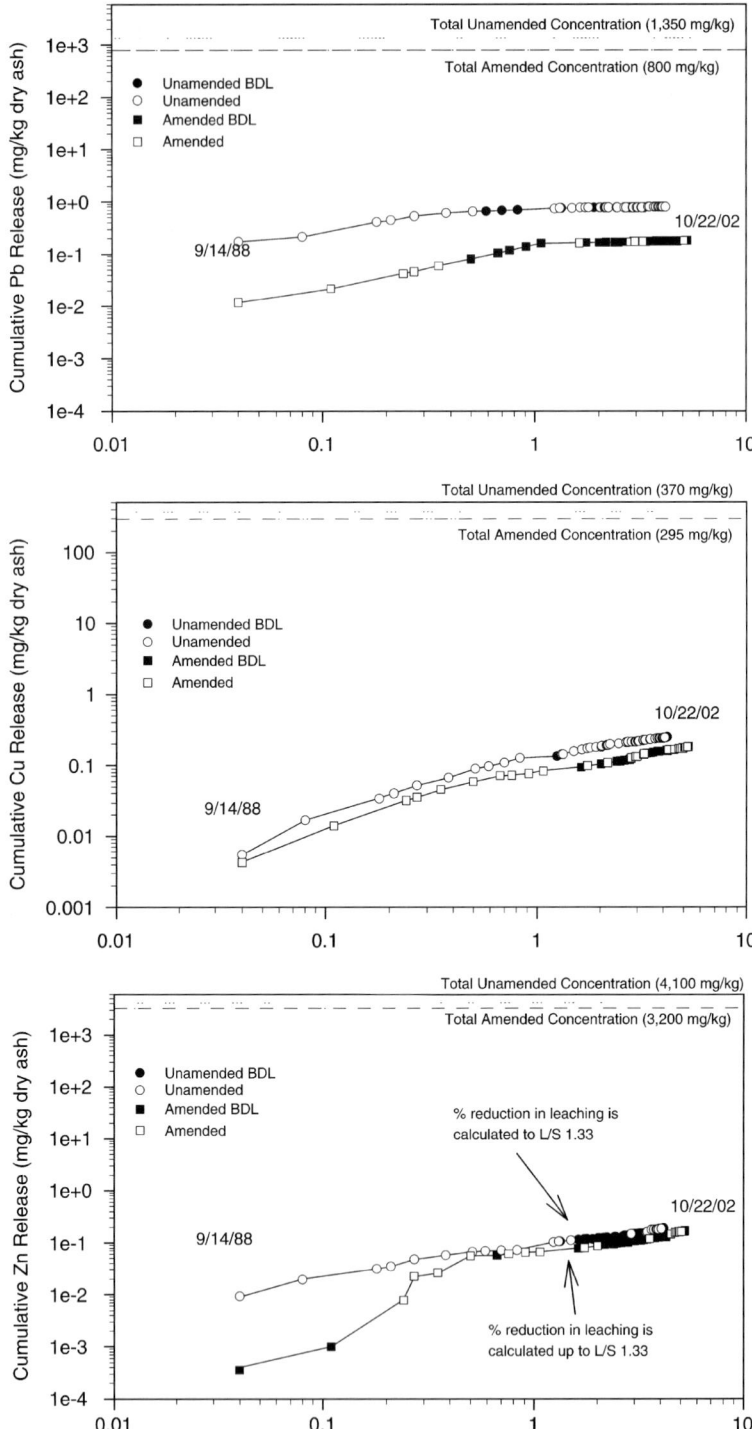

Fig. 12. Cumulative release plots for the long-term leaching behaviour of Pb, Cu, and Zn leaching from combined bottom ash and scrubber residue amended with 0.6 moles PO_4^{3-}/kg of ash (unpublished data). When below detection limit (BDL) values were obtained for analytes, the detection limit value was used in the calculation of cumulative leaching.

MSW ashes. The role of solid solutions (e.g., $(Pb,Ca)_5(PO_4)_3OH$ probably plays a much more important role than presently envisioned and the spectroscopic identification of such phases as well as more precise estimations of their stabilities, aqueous solubilities, stoichiometric saturation states, ideal or non-ideal behaviour, and thermodynamic properties (e.g., excess free energy) will help to further understand the stabilization process. The reaction pathways that occur when phosphate is added to a highly solubilized matrix with a cornucopia of elements would benefit from reaction path and kinetic modelling and from careful laboratory observations. The role of Ostwald ripening, crystal aging, and changes in solubility of discrete phases as crystallites nucleate, grow, and mature is also an important area of investigation, particularly as it relates to changes in solubility and bioavailability after stabilization is initiated, but before final disposition of the ash. This is particularly true for systems where mixing is initially incomplete (e.g., mixing in landfills) or in systems where ashes are treated and then immediately subjected to regulatory leaching tests. The long-term stability of metal phosphates in strongly reducing environments where sulphides are present deserves further theoretical, laboratory, and field study. Finally, the role of strong complexants like humic, fulvic, and low molecular weight carboxylic acids in influencing the stability of metals phosphates (e.g., Lang & Kaupenjohann 2003) deserves further exploration. All of these issues are also germane to the use of phosphate stabilization for many other waste systems.

We thank the series editors (R. Gieré, P. Stille) for the opportunity to prepare a review chapter for this book. We thank the three reviewers (R. Comans, S. Traina, A. Wagh) for their constructive comments. We thank our colleagues L. Butler, F. Cartledge, B. Crannell, C. Francis, L. Gallagher, J. Krzanowski, D. Oblas, and L. Shaw for their excellent collaborations. Some of the work presented herein has been supported by Wheelabrator Environmental Systems, Kurita Water Industries Ltd, the National Oceanic and Atmospheric Administration and the Federal Highway Administration.

References

ADMASSU, W. & BREESE, T. 1999. Feasibility of using natural fishbone apatite as a substitute for hydroxyapatite in remediating aqueous heavy metals. *Journal of Hazardous Materials*, **69**, 187–196.

ALLISON, J. D., BROWN, D. S. & NOVO-GRADAC, K. K. 1990. *MINTEQA2/PRODEFA2, A Geochemical Assessment Model for Environmental Systems: Version 3.0 User's Manual.* Environmental Research Laboratory, US EPA, Athens, Georgia.

AREY, J. S., SEAMAN, J. C. & BERTSCH, P. M. 1999. Immobilization of uranium in contaminated sediments by hydroxyapatite addition. *Environmental Science & Technology*, **33**, 337–342.

AYATI, M. & LUNDAGER MADSEN, H. E. 2000. Crystallization of some heavy-metal phosphates alone and in the presence of calcium ion. *Journal of Crystal Growth*, **208**, 579–591.

AYATI, M. & LUNDAGER MADSEN, H. E. 2001. Solubility product of the cadmium phosphate $Cd_5H_2(PO_4)_2 \cdot 4H_2O$ at 37 °C. *Journal of Chemical Engineering Data*, **46**, 113–116.

BASTA, N. T., GRADWOHL, R., SNETHEN, K. L. & SCHRODER, J. L. 2001. Chemical immobilization of lead, zinc and cadmium in smelter-contaminated soils using biosolids and rock phosphate. *Journal of Environmental Quality*, **30**, 1222–1230.

BERNER, R. A. 1981. Kinetics of weathering and diagenesis. *In*: LASAGA, A. C. & KIRKPATRICK, R. J. (eds) *Kinetics of Geochemical Processes*. Mineralogical Society of America, Washington, DC, 111–134.

BERTI, W. R. & CUNNINGHAM, S. D. 1997. In-place inactivation of Pb in Pb-contaminated soils. *Environmental Science & Technology*, **31**, 1359–1364.

BOSTICK, W. D., STEVENSON, R. J., JARABEK, R. J. & CONCA, J. L. 2000. Use of apatite and bone char for the removal of soluble radionuclides in authentic and simulated DOE groundwater. *Advances in Environmental Research*, **3**, 488–498.

BOTTO, I. L., BARONE, V. L., CATIGLIONI, J. L. & SCHALAMUK, I. B. 1997. Characterization of natural substituted pyromorphite. *Journal of Materials Science*, **32**, 6549–6553.

BRUDERVOLD, F., STEADMAN, L. T., SPINELLI, M. A., AMDUR, B. H. & GRON, P. 1963. A study of zinc in human teeth. *Archives in Oral Biology*, **8**, 135–144.

CAO, X., MA, L. Q., CHEN, M., SINGH, S. P. & HARRIS, W. G. 2002. Impacts of phosphate amendments on lead biogeochemistry at a contaminated site. *Environmental Science & Technology*, **36**, 5296–5304.

CHANDER, S. & FUERSTENAU, D. W. 1984. Solubility and interfacial properties of hydroxyapatite: a review. *In*: MISRA, D. N. (ed) *Adsorption and Surface Chemistry of Hydroxyapatite*. Plenum Press, New York, 28–49.

CHANDLER, A. G., EIGHMY, T. T. *et al.* 1997. *Municipal Solid Waste Incinerator Residues*. Elsevier Science BV, Amsterdam.

CHEN, X., WRIGHT, J. V., CONCA, J. L. & PEURRUNG, L. M. 1997a. Evaluation of heavy metal remediation using mineral apatite. *Water, Air & Soil Pollution*, **98**, 57–78.

CHEN, X.-B., WRIGHT, J. V., CONCA, J. L. & PEURRUNG, L. M. 1997b. Effects of pH on heavy metal sorption on mineral apatite. *Environmental Science & Technology*, **31**, 624–631.

CHEUNG, C. W., PORTER, J. F. & McKAY, G. 2002. Removal of Cu(II) and Zn(II) ions by sorption onto bone char using batch agitation. *Langmuir*, **18**, 650–656.

CHRISTENSEN, T. H. & NIELSEN, B. G. 1987. Retardation of lead in soils. *In*: *Heavy Metals in The Environment*. CEP Consultants, Edinburgh, UK, 319–321.

CHRISTOFFERSEN, J., CHRISTOFFERSEN, M. R., LARSEN, R., ROSTRUP, E., TINGSGAARD, P., ANDERSEN, O. & GRANDJEAN, P. 1988. Interaction of cadmium ions with calcium hydroxyapatite crystals: a possible mechanism contributing to the pathogenesis of cadmium-induced bone diseases. *Calcified Tissue International*, **42**, 331–339.

CHRISTOFFERSEN, J., CHRISTOFFERSEN, M. R., KIBALCZYC, W. & ANDERSEN, F. A. 1989. A contribution to the understanding of the formation of calcium phosphates. *Journal of Crystal Growth*, **94**, 767–777.

CONCA, J., STRIETELMEIER, LU, N., WARE, S. D., TAYLOR, T. P., KASZUBA, J. & WRIGHT, J. 2002. Treatability study of reactive materials to remediate groundwater contaminated with radionuclides, metals, and nitrates in a four-component permeable reactive barrier. *In*: *Handbook of Groundwater Remediation Using Permeable Reactive Barriers*. Elsevier Science, New York, 221–252.

COREY, R. B. 1981. Adsorption vs. precipitation. *In*: ANDERSON, M. A. & RUBIN, A. J. (eds) *Adsorption of Inorganics at Solid–Liquid Interfaces*. Ann Arbor Science, Ann Arbor, Michigan, 161–182.

COTTER-HOWELLS, J. & THORNTON, I. 1991. Sources and pathways of environmental lead to children in a Derbyshire mining village. *Environmental Geochemical Health*, **13**, 127–135.

COTTER-HOWELLS, J. D., CHAMPNESS, P. E., CHARNOCK, J. M. & PATTRICK, R. A. D. 1994. Identification of pyromorphite in mine-waste contaminated soils by ATEM and EXAFS. *European Journal of Soil Science*, **45**, 393–402.

COTTER-HOWELLS, J. & CAPORN, S. 1996. Remediation of contaminated land by formation of heavy metal phosphates. *Applied Geochemistry*, **11**, 335–342.

CRANNELL, B. S., EIGHMY, T. T., KRZANOWSKI, J. R. & EUSDEN, J. D. JR. 1999. Phosphate stabilization mechanisms for heavy metals in electric arc furnace smelter dusts. *In*: NIKOLAIDIS, N., ERKEY, C. & SMETS, B. (eds) *Hazardous Wastes and Hazardous Materials*. Technomic Publishing, Lancaster, Pennsylvania, 561–570.

CRANNELL, B. S., EIGHMY, T. T., KRZANOWSKI, J. E., EUSDEN, J. D. JR., SHAW, E. L. & FRANCIS, C. A. 2000. Heavy metal stabilization in municipal solid waste combustion bottom ash using soluble phosphate. *Waste Management*, **20**, 135–148.

CRANNELL, B. S., EIGHMY, T. T., BUTLER, L., EMERY, E. & CARTLEDGE, F. 2003*a*. Use of phosphate to stabilize heavy metals in contaminated sediments. *In*: PEDERSON, J. & ADAMS, E. (eds) *Dredged Material Management: Options and Environmental Concerns*. MIT Sea Grant Publications, Cambridge, Massachusetts, 175–178.

CRANNELL, B. S., EIGHMY, T. T. *et al.* 2003*b*. Reactive barriers for containment of metals-contaminated dredged materials: diffusion studies. *In*: EIGHMY, T. T. (ed) *Beneficial Use of Recycled Materials in Transportation Applications*. Air & Waste Management Assoc., Sewickley, Pennsylvania, 377–388.

DALAS, E. & KOUTSOUKOS, P. G. 1989. Crystallization of hydroxyapatite from aqueous solutions in the presence of cadmium. *Journal of the Chemical Society, Faraday Transactions 1: Physical Chemistry in Condensed Phases*, **85**, 3159–3164.

DA ROCHA, N. C. C., DE CAMPOS, R. C., ROSSI, A. M., MOREIRA, E. L., BARBOSA, A. D. F. & MOURE, G. T. 2002. Cadmium uptake by hydroxyapatite synthesized in different conditions and submitted to thermal treatment. *Environmental Science & Technology*, **36**, 1630–1635.

DAVIS, A., DREXLER, J. W., RUBY, M. V. & NICHOLSON, A. 1993. Micromineralogy of mine wastes in relation to lead bioavailability, Butte, Montana. *Environmental Science & Technology*, **27**, 1415–1425.

DAVIS, A., RUBY, M. V., BLOOM, M., SCHOOF, R., FREEMAN, G. & BERSTROM, P. D. 1996. Mineralogic constraints on the bioavailability of arsenic in smelter impacted soils. *Environmental Science & Technology*, **30**, 392–399.

DAVIS, A., EARY, L. E. & HELGEN, S. 1999. Assessing the efficacy of lime amendment to geochemically stabilize mine tailings. *Environmental Science & Technology*, **33**, 2626–2632.

DEER, W. A., HOWIE, R. A. & ZUSSMAN, J. 1996. *An Introduction to Rock-Forming Minerals*. Addison-Wesley, Boston.

DERIE, R. 1996. A new way to stabilize fly ash from municipal incinerators. *Waste Management*, **16**, 711–716.

DUC, M., LEFEVRE, G. *et al.* 2003. Sorption of selenium anionic species on apatites and iron oxides from aqueous solutions. *Journal of Environmental Radioactivity*, **70**, 61–72.

ECKE, H., SAKANAKURA, H., MATSUTO, T., TANAKA, N. & LAGERKVIST, A. 2000. State-of-the-art treatment processes for municipal solid waste incineration residues in Japan. *Waste Management & Research*, **18**, 41–51.

ECKE, H., MENAD, N. & LAGERKVIST, A. 2002. Treatment-oriented characterization of dry scrubber residue from municipal solid waste incineration. *Journal of Material Cycles & Waste Management*, **4**, 117–126.

EIGHMY, T. T., EUSDEN, J. D. JR., KRZANOWSKI, J. E., DOMINGO, D., STÄMPFLI, D., MARTIN, J. R. & ERICKSON, P. M. 1995. Comprehensive approach towards understanding element speciation and leaching behavior in municipal solid waste electrostatic precipitator ash. *Environmental Science & Technology*, **29**, 629–646.

EIGHMY, T. T., CRANNELL, B. S. *et al.* 1997. Heavy metal stabilization in municipal solid waste combustion dry scrubber residue using soluble phosphate. *Environmental Science & Technology*, **31**, 3330–3338.

EIGHMY, T. T., SHAW, E. L., EUSDEN, J. D. JR. & FRANCIS, C. A. 1998. Environmentally important phosphate minerals (I.) Chloropyromorphite (Pb$_5$(PO$_4$)$_3$Cl) by XPS. *Surface Science Spectra*, **5**, 122–129.

EIGHMY, T. T., KINNER, A. E., SHAW, E. L., EUSDEN, J. D. JR. & FRANCIS, C. A. 1999a. Hinsdalite (PbAl$_3$PO$_4$SO$_4$(OH)$_6$) characterized by XPS: An environmentally important secondary mineral. *Surface Science Spectra*, **6**, 184–192.

EIGHMY, T. T., KINNER, A. E., SHAW, E. L., EUSDEN, J. D. JR. & FRANCIS, C. A. 1999b. Hydroxlapatite (Ca$_5$(PO$_4$)$_3$OH) characterized by XPS: An environmentally important secondary mineral. *Surface Science Spectra*, **6**, 193–201.

EIGHMY, T. T., KINNER, A. E., SHAW, E. L., EUSDEN, J. D. JR. & FRANCIS, C. A. 1999c. Plumbogummite (PbAl$_3$(PO$_4$)$_2$(OH)$_5 \cdot$H$_2$O) characterized by XPS: An environmentally important secondary mineral. *Surface Science Spectra*, **6**, 202–209.

EIGHMY, T. T., KINNER, A. E., SHAW, E. L., EUSDEN, J. D. JR. & FRANCIS, C. A. 1999d. Chloroapatite (Ca$_3$(PO$_4$)$_3$Cl) characterized by XPS: An environmentally important secondary mineral. *Surface Science Spectra*, **6**, 210–218.

EIGHMY, T. T., KINNER, A. E., SHAW, E. L., EUSDEN, J. D. JR. & FRANCIS, C. A. 1999e. Whitlockite (β-Ca$_3$(PO$_4$)$_2$) characterized by XPS: An environmentally important secondary mineral. *Surface Science Spectra*, **6**, 219–227.

EIGHMY, T. T., CRANNELL, B. S. et al. 1999f. Characterization and phosphate stabilization of dusts from the vitrification of MSW combustion residues. *Waste Management*, **18**, 513–524.

ERGUN, C., WEBSTER, T. J., BIZIOS, R. & DOREMUS, R. H. 2001. Hydroxylapatite with substituted magnesium, zinc, cadmium, and yttrium. I. Structure and microstructure. *Journal of Biomedical Materials Research*, **59**, 305–311.

EUSDEN, J. D. JR., EIGHMY, T. T., HOCKERT, K., HOLLAND, E. & MARSELLA, K. 1999. Petrogenesis of municipal solid waste combustion bottom ash. *Applied Geochemistry*, **14**, 25–43.

EUSDEN, J. D., GALLAGHER, L. et al. 2002. Petrographic and spectroscopic characterization of phosphate-stabilized mine tailings from Leadville, Colorado. *Waste Management*, **22**, 117–135.

FARLEY, K. J., DZOMBAK, D. A. & MOREL, F. M. M. 1985. A surface precipitation model for the sorption of cations on metal oxides. *Journal of Colloid and Interface Science*, **106**, 226–242.

FEDOROFF, M., JEANJEAN, J., ROUCHAUD, J. C., MAZEROLLES, L., TROCELLIER, P., MAIRELES-TORRES, P. & JONES, D. J. 1999. Sorption kinetics and diffusion of cadmium in calcium hydroxyapatites. *Solid State Sciences*, **1**, 71–84.

FULLER, C. C., BARGAR, J. R., DAVIS, J. A. & PIANA, M. J. 2002. Mechanisms of uranium interactions with hydroxyapatite: implications for groundwater remediation. *Environmental Science & Technology*, **36**, 158–165.

GAUGLITZ, R., HOLTERDORF, M., FRANKE, W. & MARX, G. 1992. Immobilization of heavy metals by hydroxylapatite. *Radiochimica Acta*, **58–59**, 253–257.

GRCMAN, H., PERSOLJA, J., LOBNIK, F. & LESTAN, D. 2001. Modifying lead, zinc and cadmium bioavailability in soil by apatite and EDTA addition. *Fresenius Environmental Bulletin*, **10**, 727–730.

GREENWOOD, H. J. 1977. Entropy, randomness, and activity. In: GREENWOOD, H. J. (ed) *Short Course in Application of Thermodynamics to Petrology and Ore Deposits*. Mineralogical Association of Canada, Toronto, 38–46.

HAMON, R. E., McLAUGHLIN, M. J. & COZENS, G. 2002. Mechanisms of attenuation of metal availability in in situ remediation treatments. *Environmental Science & Technology*, **36**, 2991–2996.

HETTIARACHCHI, G. M., PIERZYNSKI, G. M. & RANSOM, M. D. 2000. In situ stabilization of soil lead using phosphorus and manganese oxide. *Environmental Science & Technology*, **34**, 4614–4619.

HJELMAR, O., BIRCH, H. & HANSEN, J. B. 2000a. Treatment of APC residues from MSW incinerators. In: DHIR, R. K., DYER, T. D. & PAINE, K. A. (eds) *Sustainable Construction: Use of Incinerator Ash*. Thomas Telford Publishing, London, 185–194.

HJELMAR, O., BIRCH, H. & HANSEN, J. B. 2000b. Further development of a process for treatment of APC residues from MSW incinerators. In: WOOLLEY, G. R., GOUMONS, J. J. J. M. & WAINWRIGHT, P. J. (eds) *Waste Materials in Construction*. Pergamon Press, Amsterdam, 872–883.

HODSON, M. E., VALSAMI-JONES, E. & COTTER-HOWELLS, J. D. 2000. Bonemeal additions as a remediation treatment for metal contaminated soil. *Environmental Science & Technology*, **34**, 3501–3507.

HODSON, M. E., VALSAMI-JONES, E., COTTER-HOWELLS, J. D., DUBBIN, W. E., KEMP, A. J., THORNTON, I. & WARREN, A. 2001. Effect of bone meal (calcium phosphate) amendments on metal release from contaminated soils – a leaching column study. *Environmental Pollution*, **112**, 233–243.

INGRAM, G. S., HORAY, C. P. & STEAD, W. J. 1992. Interaction of zinc with dental mineral. *Caries Research*, **26**, 248–253.

IOANNIDIS, T. A. & ZOUBOULIS, A. I. 2003. Detoxification of a highly toxic lead-loaded industrial solid waste by stabilization using apatites. *Journal of Hazardous Materials*, **97**, 173–191.

IRETSKAYA, S., NZIHOU, A., ZAHRAOUI, C. & SHARROCK, P. 1999. Metal leaching from MSW fly ash before and after chemical and thermal treatments. *Environmental Progress*, **18**, 144–148.

JEANJEAN, J., McGRELLIS, S., ROUCHARD, J. C., FEDOROFF, M., RONDEAU, A., PEROCHEAU, S. & DUBIS, A. 1996. A crystallographic study of the sorption of cadmium on calcium hydroxyapatites: incidence of cationic vacancies. *Journal of Solid State Chemistry*, **126**, 195–201.

JURNIAK, J. J. & INOUYE, T. S. 1962. Some aspects of zinc and copper phosphate formation in aqueous systems. *Soil Science Society of America Proceedings*, **26**, 144–147.

KAPLAN, D., KNOX, A. & COFFEY, C. 2002. *Reduction of Contaminant Mobility at the TNX Outfall Delta Through the Use of Apatite and Zero-Valent Iron as Soil Amendments*. Westinghouse Savannah River Company, Savannah River, Georgia, WSRC-TR-2002–00370.

KNOX, A. S., KAPLAN, D. I., ADRIANO, D. C., HINTON, T. G. & WILSON, M. D. 2003. Apatite and phillipsite as sequestering agents for metals and radionuclides. *Journal of Environmental Quality*, **32**, 515–525.

KOSSON, D. S., KOSSON, T. T. & VAN DER SLOOT, H. 1993. *Evaluation of Solidification/Stabilization Treatment Processes for Municipal Waste Combustion Residues*. US Environmental Protection Agency, EPA/600/SR-93–167, Washington, DC.

KOUTSOUKOS, P., AMJAD, Z., TOMSON, M. B. & NANCOLLAS, G. H. 1980. Crystallization of calcium phosphates. A constant composition study. *Journal of the American Chemical Society*, **102**, 1553–1557.

KUKURA, M, BELL, L. C., POSNER, A. M. & QUIRK, J. P. 1973. Kinetics of isotope exchange on hydroxyapatite. *Soil Science Society of America Journal*, **37**, 364–366.

KULLBERG, S. 1995. *Fosfatstabilisering av Rökgasaska fron Avfallsförbränning*. Stiftelsen REFORSK FoU 124, Stockholm, Sweden.

LANG, F. & KAUPENJOHANN, M. 2003. Effect of dissolved organic matter on the precipitation and mobility of the lead compound chloropyromorphite in solution. *European Journal of Soil Science*, **54**, 139–147.

LAPERCHE, V., TRAINA, S. J., GADDAM, P. & LOGAN, T. J. 1996. Chemical and mineralogical characterizations of Pb in a contaminated soil: reactions with synthetic apatite. *Environmental Science & Technology*, **30**, 3321–3326.

LAPERCHE, V., LOGAN, T. J., GADDAM, P. & TRAINA, S. J. 1997. Effect of apatite amendments on plant uptake of lead from contaminated soil. *Environmental Science & Technology*, **31**, 2745–2753.

LAPERCHE, V. & TRAINA, S. J. 1998. Immobilization of Pb by hydroxylapatite. *In*: JENNE, E. (ed) *Adsorption of Metals by Geomedia*. Academic Press, New York, 255–276.

LAZIC, S. & VUKOVIC, Z. 1991. Ion exchange of strontium on synthetic hydroxyapatite. *Journal of Radioanalytical and Nuclear Chemistry*, **149**, 161–168.

LEYVA, A. G., MARRERO, J., SMICHOWSKI, P. & CICERONE, D. 2001. Sorption of antimony onto hydroxyapatite. *Environmental Science & Technology*, **35**, 3669–3675.

LINDSAY, W. L. 1979. *Chemical Equilibria in Soils*. John Wiley & Sons, New York.

LINDSAY, W. L. & MORENO, E. C. 1960. Phosphate phase equilibria in soils. *Soil Science Society of America Proceedings*, **24**, 177–182.

LINDSAY, W. L. & VLEK, P. L. 1977. Phosphate minerals. *In*: DIXON, J. B., WEED, S. B., KITTRICK, J. A., MILFORD, M. H. & WHITE, J. L.

(eds) *Minerals in Soil Environments*. Soil Science Society of America, Madison, Wisconsin, 639–672.

LIU, Y. & PENG, M.-S. 2001. Applications of mineral apatites in the treatment of wastewater. *Anquan Yu Huanjing Xuebao*, **1**, 9–12.

LOWER, S. K., MAURICE, P. A. & TRAINA, S. J. 1998a. Simultaneous dissolution of hydroxylapatite and precipitation of hydroxypyromorphite: direct evidence of homogeneous nucleation. *Geochimica Cosmochimica Acta*, **62**, 1773–1780.

LOWER, S. K., MAURICE, P. A., TRAINA, S. J. & CARLSON, E. H. 1998b. Aqueous Pb sorption by hydroxylapatite: applications of atomic force microscopy to dissolution, nucleation, and growth studies. *American Mineralogist*, **83**, 147–158.

LUSVARDI, G., MALAVASI, G., MENABUE, L. & SALADINI, M. 2002. Removal of cadmium ion by means of synthetic hydroxyapatite. *Waste Management*, **22**, 853–857.

LYONS, M. R. 2000. The use of soluble phosphates to stabilize heavy metals in fly ash. *In*: DHIR, R. K., DYER, T. D. & PAINE, K. A. (eds) *Sustainable Construction: Use of Incinerator Ash*. Thomas Telford Publishing, London, 87–96.

MA, L. Q. 1996. Factors influencing the effectiveness and stability of aqueous lead immobilization by hydroxyapatite. *Journal of Environmental Quality*, **25**, 1420–1429.

MA, Q. Y., TRAINA, S. J., LOGAN, T. J. & RYAN, J. A. 1993. *In situ* lead immobilization by apatite. *Environmental Science & Technology*, **27**, 1803–1810.

MA, Q. Y., LOGAN, T. J., TRAINA, S. J. & RYAN, J. A. 1994a. Effects of NO_3^-, Cl^-, F^-, SO_4^{2-}, and CO_3^{2-} on Pb^{2+} immobilization by hydroxyapatite. *Environmental Science & Technology*, **28**, 408–418.

MA, Q. Y., TRAINA, S. J., LOGAN, T. J. & RYAN, J. A. 1994b. Effects of aqueous Al, Cd, Cu, Fe(II), Ni, and Zn on Pb immobilization by hydroxyapatite. *Environmental Science & Technology*, **28**, 1219–1228.

MA, Q. Y., LOGAN, T. J. & TRAINA, S. J. 1995. Lead immobilization from aqueous solutions and contaminated soils using phosphate rocks. *Environmental Science & Technology*, **29**, 1118–1126.

MAHRAPATRA, P. P., SARANGI, D. S. & MISHRA, B. 1995. Kinetics of nucleation of lead hydroxylapatite and preparation of solid solutions of calcium–cadmium–lead hydroxyapatite: an x-ray and IR study. *Journal of Solid State Chemistry*, **116**, 8–14.

MANECKI, M., MAURICE, P. A. & TRAINA, S. J. 2000a. Kinetics of aqueous Pb reaction with apatites. *Soil Science*, **165**, 920–933.

MANECKI, M., MAURICE, P. A. & TRAINA, S. J. 2000b. Uptake of aqueous Pb by Cl-, F-, and OH-apatites: mineralogic evidence for nucleation mechanisms. *American Mineralogist*, **85**, 932–942.

MANDJINY, S., ZOUBOULIS, A. I. & MATIS, K. A. 1995. Removal of cadmium from dilute solutions by hydroxyapatite. I. sorption studies. *Separation Science and Technology*, **30**, 2963–2978.

MANDJINY, S., MATIS, K. A. *et al.* 1998. Calcium hydroxyapatites: evaluation of sorption properties for cadmium ions in aqueous solution. *Journal of Materials Science*, **33**, 5433–5439.

MATHESON, L. J., GOLDBERG, W. C. & BOSTICK, W. D. 2001. Laboratory batch and column studies to evaluate Apatite II removal of soluble uranium from contaminated groundwater. *American Chemical Society National Meeting, American Chemical Society, Division of Environmental Chemistry*, **41**, 109–113.

MAVROPOULOS, E., ROSSI, A. M., COSTA, A. M., PEREZ, C. A. C., MOREIRA, J. C. & SALDANHA, M. 2002. Studies on the mechanisms of lead immobilization by hydroxyapatite. *Environmental Science & Technology*, **36**, 1625–1629.

MCBRIDE, M. B. 1994. *Environmental Chemistry of Soils*. Oxford University Press, New York.

MCCONNELL, D. 1973. Apatite: Its crystal chemistry, mineralogy, utilization, and geologic and biologic occurrences. *In*: HERAUSGEGBEN, V., FRÉCHETTE, V. D., KIRSCH, H., SAND, L. B. & TROJER, F. (eds) *Applied Mineralogy*, Springer-Verlag, New York, 22–32.

MCGOWEN, S. L., BASTA, N. T. & BROWN, G. O. 2001. Use of diammonium phosphate to reduce heavy metal solubility and transport in smelter-contaminated soil. *Journal of Environmental Quality*, **30**, 493–500.

MEIMA, J. A. 1997. *Geochemical Modelling and Identification of Leaching Processes in MSWI Bottom Ash*. PhD dissertation, Utrecht University, Utrecht, The Netherlands.

MEIMA, J. A. & COMANS, R. N. J. 1997. Geochemical modeling of weathering reactions in municipal solid waste incinerator bottom ash. *Environmental Science and Technology*, **31**, 1269–1276.

MEIMA, J. A. & COMANS, R. N. J. 1999. The leaching of trace elements from municipal solid waste incinerator bottom ash at different stages of weathering. *Applied Geochemistry*, **14**, 159–171.

MEIMA, J. A., VAN ZOMMEREN, A. & COMANS, R. N. J. 1999. Complexation of Cu with dissolved organic carbon in municipal solid waste incinerator bottom ash leachates. *Environmental Science & Technology*, **33**, 1424–1429.

MIDDLEBURG, J. J. & COMANS, R. N. J. 1991. Sorption of cadmium on hydroxyapatite. *Chemical Geology*, **90**, 45–53.

MISRA, D. N., BOWEN, R. L. & WALLACE, B. M. 1975. Adhesive bonding of various materials to hard tooth tissues. VII nickel and copper ions on hydroxyapatite; role of ion exchange and surface nucleation. *Journal of Colloid and Interface Science*, **51**, 36–43.

MISRA, D. N. & BOWEN, R. L. 1984. Interactions of zinc ions with hydroxylapatite. *In*: MISRA, D. N. (ed) *Adsorption and Surface Chemistry of Hydroxyapatite*. Plenum Press, New York, 179–185.

MONTEIL-RIVERA, F., MASSET, S., DUMONCEAU, J., FEREROFF, M. & JEANJEAN, J. 1999. Sorption of selenite ions on hydroxyapatite. *Journal of Materials Science Letters*, **18**, 1143–1145.

MONTEIL-RIVERA, F., FEDOROFF, M., JEANJEAN, J., MINEL, L., BARTHES, M.-G. & DUMONCEAU, J. 2000. Sorption of selenite (SeO_3^{2-}) on hydroxyapatite: an exchange process. *Journal of Colloid and Interface Science*, **221**, 291–300.

MORIN, G., JUILLOT, F., ILDEFONSE, P., CALAS, G., SAMAMA, J.-C., CHEVALLIER, P. & BROWN, G. E. JR. 2001. Mineralogy of lead in a soil developed on a Pb-mineralized sandstone (Largentiere, France). *American Mineralogist*, **86**, 92–104.

MORSE, J. W. & CASEY, W. N. 1988. Ostwald processes and mineral paragenesis in sediments. *American Journal of Science*, **288**, 537–560.

NANCOLLAS, G. H. 1984. The nucleation and growth of phosphate minerals. *In*: NRIAGU, J. O. & MOORE, P. B. (eds) *Phosphate Minerals*. Springer-Verlag, Berlin, 137–154.

NICOL, S. K. & CLARKE, A. J. 1969. Some electrochemical properties of the calcium hydrozyapatite/solution interface. *In*: FEARNHEAD, R. W. & STACK, M. V. (eds) *Tooth Enamel II: Its Composition, Properties, and Fundamental Structure*. John Wright & Sons, Bristol, UK, 179–186.

NIST 2001. NIST *Standard Reference Database 46, Version 4.0*, NIST, Gaithersburg, Maryland.

NRIAGU, J. O. 1972. Lead orthophosphates. I: solubility and hydrolysis of secondary lead orthophosphate. *Inorganic Chemistry*, **11**, 2499–2503.

NRIAGU, J. O. 1973a. Lead orthophosphates. II: stability of chloropyromorphite at 25 °C. *Geochimica Cosmochimica Acta*, **37**, 367–377.

NRIAGU, J. O. 1973b. Lead orthophosphates. III: stabilities of fluoropyromorphite and bromopyromorphite at 25 °C. *Geochimica Cosmochimica Acta*, **37**, 1735–1743.

NRIAGU, J. O. 1974. Lead orthophosphates. IV: formation and stability in the environment. *Geochimica Cosmochimica Acta*, **38**, 887–898.

NRIAGU, J. O. 1976. Phosphate–clay mineral relations in soils and sediments. *Canadian Journal of Earth Sciences*, **13**, 717–736.

NRIAGU, J. O. 1984. Formation and stability of base metal phosphates in soils and sediments. *In*: NRIAGU, J. O. & MOORE, P. B. (eds) *Phosphate Minerals*. Springer-Verlag, Berlin, 318–329.

NRIAGU, J. O. and MOORE, P. B. 1984. *Phosphate Minerals*. Springer-Verlag, Berlin, 442 p.

NZIHOU, A. & SHARROCK, P. 2002. Calcium phosphate stabilization of fly ash with chloride extraction. *Waste Management*, **22**, 235–239.

OSTERGREN, J. D., BROWN, G. E. JR., PARKS, G. A. & TINGLE, T. N. 1999. Quantitative speciation of lead in selected mine tailings from Leadville, CO. *Environmental Science & Technology*, **33**, 1627–1636.

PAK, C. Y. C. & BARTTER, F. C. 1967. Ionic interaction with bone material, I. Evidence for isoionic calcium exchange with hydroxyapatite. *Biochimica Biophysica Acta*, **141**, 401–409.

PANDA, A., SAHU, B., PATEL, P. N. & MISHRA, B. 1991. Hydroxylapatite solid solutions: preparation, infrared and lattice constant measurements. *Transition Metal Chemistry*, **16**, 476–477.

PERRONE, J., FOUREST, B. & GIFFAUT, E. 2001. Sorption of nickel on carbonate fluoroapatites. *Journal of Colloid and Interface Science*, **239**, 303–313.

PIANTONE, P., BODÉNAN, F., DERIE, R. & DEPELSENAIRE, G. 2003. Monitoring the stabilization of municipal solid waste incineration fly ash by phosphatation: mineralogical and balance approach. *Waste Management*, **23**, 225–243.

POVARENNYKH, A. S. 1972. *Crystal Chemical Classification of Minerals*. Plenum Press, New York.

PUJARI, M. & PATEL, P. N. 1989. Strontium–copper–calcium hydroxyapatite solid solutions: preparation, infrared, and lattice constant measurements. *Journal of Solid State Chemistry*, **83**, 100–104.

RABINOWITZ, M. B. 1993. Modifying soil lead bioavailability by phosphate addition. *Bulletin of Environmental Contamination and Toxicology*, **51**, 438–444.

RAICEVIC, S. 2001. Remediation of uranium contaminated water and soil using phosphate-induced metal stabilization (PIMS). *Hemijska Industrija*, **55**, 277–280.

RAO, A. J., PAGILLA, K. R. & WAGH, A. S. 2000. Stabilization and solidification of metal-laden wastes by compaction and magnesium phosphate-based binder. *Journal of the Air & Waste Management Association*, **50**, 1623–1631.

RICKARD, D. T. & NRIAGU, J. O. 1978. Aqueous chemistry of lead. *In*: NRIAGU, J. O. (ed) *The Biochemistry of Lead in the Environment*. Elsevier/North-Holland Biomedical Press, Amsterdam, 219–284.

RUBY, M. V., DAVIS, A. & NICHOLSON, A. 1994. *In situ* formation of lead phosphates in soils as a method to immobilize lead. *Environmental Science & Technology*, **28**, 646–654.

RYAN, J. A., ZHANG, P., HESTERBERG, D., CHOU, J. & SAYERS, D. E. 2001. Formation of chloropyromorphite in a lead-contaminated soil amended with hydroxyapatite. *Environmental Science & Technology*, **35**, 3798–3803.

SABBAS, T., POLETTINE, A. *et al.* 2003. Management of municipal solid waste incineration residues. *Waste Management*, **23**, 61–88.

SANTILLAN-MEDRANO, J. & JURINAK, J. J. 1975. The chemistry of lead and cadmium in soil: solid phase formation. *Soil Science Society of America Proceedings*, **39**, 851–856.

SAUVE, S., MCBRIDE, M. & HENDERSHOT, W. 1998. Lead phosphate solubility in water and soil suspensions. *Environmental Science & Technology*, **32**, 388–393.

SCHECKEL, K. G. & RYAN, J. A. 2002. Effects of aging and pH on dissolution kinetics and stability of chloropyromorphite. *Environmental Science & Technology*, **36**, 2198–2204.

SCHECKEL, K. G., IMPELLITTTERI, C. A., RYAN, J. A. & MCEVOY, T. 2003. Assessment of a sequential extraction procedure for perturbed lead-contaminated samples with and without phosphorus amendments. *Environmental Science & Technology*, **37**, 1892–1898.

SHIMAOKA, T., OKU, K. *et al.* 1998. Stability of hazardous heavy metals in chemically immobilized fly ash in landfill. *Haikibutsu Gakkai Ronbunshi*, **9**, 264–273.

SINGH, D., WAGH, A. S., CUNNANE, J. C. & MAYBERRY, J. L. 1997. Chemically bonded phosphate ceramics for low-level mixed-waste stabilization. *Journal of Environmental Science and Health, Part A: Environmental Science and Engineering & Toxic and Hazardous Substance Control*, **A32**, 527–541.

SOMASUNDARAN, P. & WANG, Y. H. C. 1984. Surface chemical characteristics and adsorption properties of apatite. *In*: MISRA, D. N. (ed) *Adsorption and Surface Chemistry of Hydroxyapatite*. Plenum Press, New York, 129–149.

SPOSITO, G. 1986. Distinguishing adsorption from surface precipitation. *In*: DAVIS, J. A. & HAYES, K. F. (eds) *Geochemical Processes at Mineral Surfaces*. American Chemical Society, Washington, DC, 217–228.

STANFORTH, R. & QIU, J. 2001. Effect of phosphate treatment on the solubility of lead in contaminated soil. *Environmental Geology*, **41**, 1–10.

STREET, J. J., LINDSAY, W. L. & SABEY, B. R. 1977. Solubility and plant uptake of cadmium is soils amended with cadmium and sewage sludge. *Journal of Environmental Quality*, **6**, 72–77.

SUGIYAMA, S., ICHII, T., HAYASHI, H. & TOMIDA, T. 2002. Lead immobilization by non-apatite-type calcium phosphates in aqueous solutions. *Inorganic Chemistry Communications*, **5**, 156–158.

SUZUKI, T., HATSUSHIKA, T. & HAYAKAWA, Y. 1981. Synthetic hydroxyapatites as inorganic cation-exchangers. *Journal of the Chemical Society Faraday Transactions*, **77**, 1059–1062.

SUZUKI, T., HATSUSHIKA, T. & MIYAKE, M. 1982. Synthetic hydroxyapatite as inorganic cation exchangers, Part 2. *Journal of the Chemical Society Faraday Transactions*, **78**, 3605–3611.

SUZUKI, T., ISHIGAKI, K. & MIYAKE, M. 1984. Synthetic hydroxyapatites as inorganic cation exchangers. Part 3 – Exchanger characteristics of lead ions (Pb^{2+}). *Journal of the Chemical Society Faraday Transactions*, **80**, 3157–3165.

TARDY, Y. & VEILLARD, P. 1977. Relationships among Gibbs Free Energies and enthalpies of formation of phosphates, oxides and aqueous ions. *Contributions to Mineralogy and Petrology*, **63**, 75–88.

TESSIER, A., CAMPBELL, P. G. C. & BISSON, M. 1979 Sequential extraction procedure for the speciation of particulate trace metals. *Analytical Chemistry*, **51**, 844–51.

THEODORATOS, P., PAPASSIOPI, N. & XENIDIS, A. 2002. Evaluation of monobasic calcium phosphate for the immobilization of heavy metals in contaminated soils from Lavrion. *Journal of Hazardous Materials*, **94**, 135–146.

THOMSON, B. M., SMITH, C. L., BUSCH, R. D., SIEGEL, M. D. & BALDWIN, C. 2003. Removal of metals and radionuclides using apatite and other natural sorbents. *Journal of Environmental Engineering*, **129**, 492–499.

TICKANEN, L. D. & TURPIN, P. D. 1997. Treatment of heavy metal-bearing wastes using a buffered phosphate stabilization system. *In*: WUKASCH, R. F. & DALTON, C. S. (eds) *Proceedings of the 51st Industrial Waste Conference*. CRC Press, Boca Raton, Florida, 627–635.

UCHIDA, T., ITOH, I. & HARADA, K. 1996. Immobilization of heavy metals contained in incinerator fly ash by application of soluble phosphate – treatment and disposal cost reduction by combined use of 'high specific surface area lime'. *Waste Management*, **16**, 475–481.

US EPA 2000. *Field Demonstration of Permeable Reactive Barriers to Remove Dissolved Uranium From Groundwater, Fry Canyon, Utah*. EPA 402-C-00–001, US EPA, Washington, DC.

VALSAMI-JONES, E., RAGNARSDOTTIR, K. V., PUTNIS, A., BOSBACH, D., KEMP, A. J. & CRESSEY, G. 1998. The dissolution of apatite in the presence of aqueous metal cations at pH 2–7. *Chemical Geology*, **151**, 215–233.

VAN CAPPELLEN, P. & BERNER, R. A. 1989. Marine apatite precipitation. *In*: MILES, D. (ed) *Water–Rock Interaction*. Balkema, Rotterdam, 707–710.

VEGA, E. D., PEDREGOSA, J. C. & NARDA, G. E. 1999. Interaction of oxyvanadium (IV) with crystalline calcium hydroxyapatite: surface mechanism with no structural modification. *Journal of Physics and Chemistry of Solids*, **60**, 759–766.

VIELLARD, P. & TARDY, Y. 1984. Thermochemical properties of phosphates. *In*: NRIAGU, J. O. & MOORE, P. B. (eds) *Phosphate Minerals*. Springer-Verlag, Berlin, 171–198.

WAGH, A. S., STRAIN, R., JEONG, S. Y., REED, D., KRAUSE, T. & SINGH, D. 1999. Stabilization of Rocky Flats Pu-contaminated ash within chemically bonded phosphate ceramics. *Journal of Nuclear Materials*, **265**, 295–307.

WAGMAN, D. D., EVANS, W. H. *et al.* 1982. The NBS tables of chemical thermodynamic properties: selected values for inorganic and C_1 and C_2 organic substances in SI units. *Journal of Physical Chemical Reference Data*, **11**, 2-1–2-392.

WANG, Y. M., CHEN, T. C., YEH, K. J. & SHUE, M. F. 2001. Stabilization of an elevated heavy metal contaminated site. *Journal of Hazardous Materials*, **88**, 63–74.

WRIGHT, J. V., PEURRUNG, L. M., MOODY, T. E., CONCA, J. L., CHEN, X., DIDZEREKIS, P. P. & WYSE, E. 1995. *In situ Immobilization of Heavy Metals in Apatite Mineral Formulations*. Technical Report to the Strategic Environmental Research and Development Program, Department of Defense, Pacific Northwest Laboratory, Richland, Washington.

WU, L., FORSLING, W. & SCHINDLER, P. 1991. Surface complexation of calcium minerals in aqueous solution 1. Surface protonation at fluoroapatite–water interfaces. *Journal of Colloid and Interface Analysis*, **147**, 178–185.

XENIDIS, A., STOURAITI, C. & PASPALIARIS, I. 1999. Stabilization of oxidic tailings and contaminated soils by monocalcium phosphate monohydrate addition: the case of Montevecchio (Sardinia, Italy). *Journal of Soil Contamination*, **8**, 681–697.

XU, T. & SCHWARTZ, F. W. 1994. Lead immobilization by hydroxyapatite in aqueous solutions. *Journal of Contaminant Hydrology*, **15**, 187–206.

XU, T., SCHWARTZ, F. W. & TRAINA, S. J. 1994. Sorption of Zn^{2+} and Cd^{2+} on hydroxyapatite surfaces. *Environmental Science & Technology*, **28**, 1472–1480.

YANG, J., MOSBY, D. E., CASTEEL, S. W. & BLANCHAR, R. W. 2001. Lead immobilization using phosphoric acid in a smelter-contaminated urban soil. *Environmental Science & Technology*, **35**, 3553–3559.

YANG, J., MOSBY, D. E., CASTEEL, S. W. & BLANCHAR, R. W. 2002. In vitro lead bioaccessibility and phosphate leaching as affected by surface application of phosphoric acid in lead-contaminated soil. *Archives of Environmental Contamination and Toxicology*, **43**, 399–405.

ZHANG, P. & RYAN, J. A. 1998. Formation of pyromorphite in anglesite–hydroxyapatite suspensions under varying pH conditions. *Environmental Science & Technology*, **32**, 3318–3324.

ZHANG, P. & RYAN, J. A. 1999. Transformation of Pb(II) from cerrusite to chloropyromorphite in the presence of hydroxyapatite under varying conditions of pH. *Environmental Science & Technology*, **33**, 625–630.

ZHANG, P., RYAN, J. A. & BRYNDZIA, L. T. 1997. Pyromorphite formation from goethite adsorbed lead. *Environmental Science & Technology*, **31**, 2673–2678.

ZHANG, P., RYAN, J. R. & YANG, J. 1998. In vitro soil Pb solubility in the presence of hydroxyapatite. *Environmental Science & Technology*, **32**, 2763–2768.

Environmental impact of energy recovery from waste tyres

RETO GIERÉ[1,2], SARA T. LAFREE[2], LORAN E. CARLETON[2] & JODY K. TISHMACK[3]

[1]*Institut für Mineralogie, Petrologie und Geochemie, Universität Freiburg, Freiburg, Germany*
(e-mail: giere@uni-freiburg.de)

[2]*Department of Earth and Atmospheric Sciences, Purdue University, West Lafayette, IN, USA*
[3]*Building Services & Grounds, Purdue University, West Lafayette, IN, USA*

Abstract: Accumulation of millions of worn automotive tyres poses a considerable environmental problem. As an important part of the solid waste stream in today's society, worn tyres have traditionally been discarded in landfills or stored in stockpiles. Over the past several decades, however, innovative alternatives to disposal have been developed, partly as a result of high tipping fees charged by landfill operators. Because of their high heat content and their low levels of moisture and nitrogen compared to coal, tyres are ideally suited for energy recovery through combustion. Utilization of waste tyres as supplemental or alternate fuel in various industrial combustion facilities, thus, has become one of the most important alternatives to disposal. Combustion processes, however, generate gaseous pollutants and solid waste materials, which must be disposed of or re-used as secondary raw materials. It is therefore important to characterize these combustion products in order to assess the environmental impacts of energy recovery from scrap tyres. Studies have shown that substantial reductions of some environmental pollutants can be achieved by partially replacing conventional fuels with waste tyres. On the other hand, using tyres as fuel may lead to considerable increases in the levels of other pollutants. Most notable among the effects of tyre combustion are, relative to conventional fuels, a pronounced decrease in the emission of nitrogen oxides into the atmosphere, and a generally significant increase in atmospheric zinc emissions as well as in the zinc contents of the solid combustion products. The geochemical effects on solid and gaseous combustion products are more or less pronounced depending on fuel composition, conditions of combustion, type of facility, and effectiveness of air pollution control devices. Thus, the use of tyre fuel has environmental impacts that must be weighed against the benefits of reducing the large volume of waste tyres in the global waste stream.

Every year, millions of worn tyres are discarded worldwide. In 2001 alone, Americans disposed of 281 million car, truck, bus, and aeroplane tyres, equivalent to approximately one discarded tyre per person (RMA 2002). The large amounts of tyres removed from vehicles each year make proper waste management vital. Traditionally, discarded tyres have been treated as waste and placed in landfills or stockpiles; however, in many cases they are also illegally dumped. In 1990, approximately 65.5% of the 278 million tyres discarded in the USA were disposed of in landfills or added to existing stockpiles; only 34.5% were reused, recycled or recovered for various commercial and industrial applications (Blumenthal 1993). Since then, it has been recognized that old tyres represent a resource rather than a waste material, and thus, various solutions for re-using or recycling tyres have been, and are currently being, developed (Fig. 1). As a result, the number of tyres being disposed of in landfills has decreased dramatically over the past decade, reaching 10% in 2001 (Table 1).

One of the main reasons for this decrease is that scrap tyres are an important source of energy. In the USA, for example, energy recovery from tyres increased from 11% of the total number of scrap tyres in 1990 to approximately 40% in 1994, and still remains at approximately this level now (Fig. 2). Today, energy recovery through combustion is the single most important market for discarded tyres in the USA (Table 1, Fig. 3). This procedure is also extensively applied in many other countries, including Japan and Korea (Jang *et al.* 1998; JATMA 2002), Canada (Polasek & Jervis 1994), and several European nations (Giugliano *et al.* 1999; BUWAL 2001; Mukherjee *et al.* 2003; NARRA 2003). It is therefore important to assess the environmental impacts of energy recovery from scrap tyres.

This chapter will present a short overview of worn tyre management practices, with data for the USA, followed by a discussion of the effects of energy recovery through tyre combustion on both solid waste products and atmospheric emissions. In addition to reviewing

From: GIERÉ, R. & STILLE, P. (eds) 2004. *Energy, Waste, and the Environment: a Geochemical Perspective*. Geological Society, London, Special Publications, **236**, 475–498.
0305-8719/04/$15   © The Geological Society of London 2004.

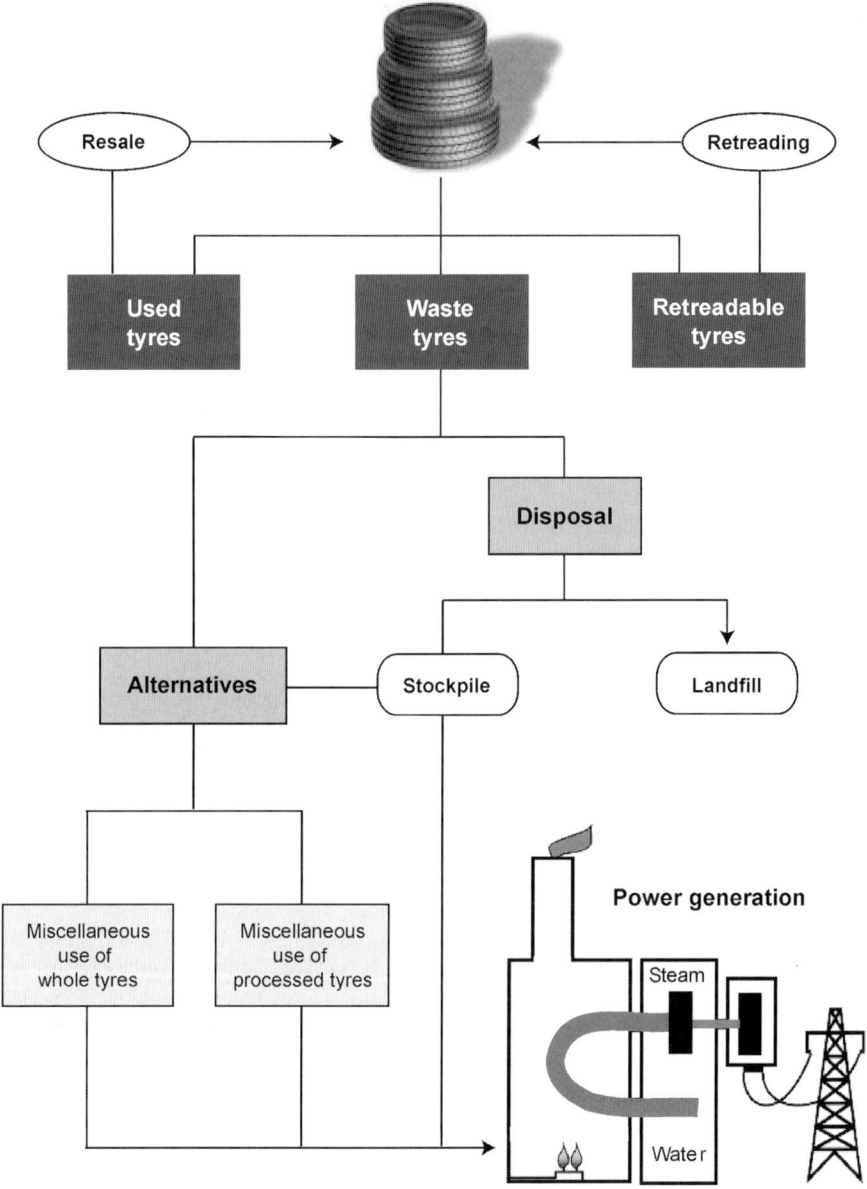

Fig. 1. Schematic diagram showing the life-cycle of tyres including current tyre waste management strategies.

literature data, we also discuss a case study of tyre co-combustion at the Purdue University power plant, where both solid waste products and atmospheric emissions have been chemically characterized.

Used and retreadable tyres

Substantial numbers of tyres are still usable on vehicles after they have been removed from initial service. In 2001, approximately 30 million used tyres were resold either in the USA or in other countries (Table 1), whereby it is estimated that 15 million were exported, representing about 7% of the market for discarded tyres (Fig. 3b).

Another important pathway for worn tyres is recycling by retreading (Fig. 1). The estimated number of retreaded tyres in the USA ranges

Table 1. *Discarded tyre disposition in 2001 for the USA**

General application	Millions of tyres	Specific application
Total worn tyres accumulated	281	
Total discarded tyres used[†]	218	
Tyre fuel	115	Total tyre fuel
	53	Cement kilns
	19	Pulp and paper mills
	18	Electric utilities
	11	Industrial boilers
	14	Tyre-to-energy facilities
Rubber products	41	Total rubber products
	33	Ground rubber
	8	Cut, punched, or stamped
Civil engineering applications	40	
Miscellaneous applications	7	
Resale	30	Total resale
	15	Export for resale
	15	Domestic resale
Retreading	16	
Disposal in landfills	28	Landfill or monofill

*Data from RMA (2002).
[†]Excludes domestic resale and retreated tyres; for details, see RMA (2002).

from 16.4 million (Table 1) to 24.2 million (TRIB 2003) for the year 2001. The consumer price for retreaded tyres may be considerably less than that of new tyres, because retreading a worn tyre requires up to 70% less oil when compared to manufacturing a new tyre (EPA 2003). Retreading relatively expensive tyres, such as truck tyres, which amount to ~15% of all discarded tyres in the USA, is an economic and widely practised option (Jang *et al.* 1998; Blumenthal 1993). Moreover, 80% of the tyres used by the commercial aviation industry are retreaded (TRIB 2003). Retreading offers the best strategy for value recovery from tyres,

requiring the least new material and energy (Amari *et al.* 1999).

Disposal of waste tyres

Landfills

In the past, landfills have been the preferred method of waste tyre disposal because burial eliminates the fire hazard associated with above-ground storage, as well as the unsightliness of tyre piles. However, problems with this method have become increasingly apparent. Because of their composition, tyres are neither decomposable, nor can they easily be compacted (Blumenthal 1993). Therefore, tyres occupy large volumes of landfills, contributing to the already prevalent problem of landfill crowding.

The large amount of empty space (75% of the volume is void; Jang *et al.* 1998), that results from the shape of tyres is often filled by air or other gases, causing the tyre to be buoyant relative to its surroundings when buried. Therefore, tyres disposed of in landfills tend to rise to the surface, disrupting the protective final cover of closed landfills (Blumenthal 1993). This disruption is potentially hazardous because precipitation can then easily enter the landfill, increasing the likelihood for contaminant leaching and subsequent transport into surrounding groundwaters or soils. Additionally, a breach in the landfill cover allows insects, birds, and rodents to interfere with landfill processes, and further

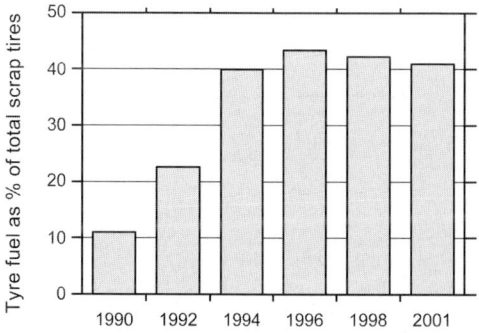

Fig. 2. Evolution of tyre fuel consumption since 1990, shown as percentage of the total amount of tyres discarded in the USA (data from RMA 2002).

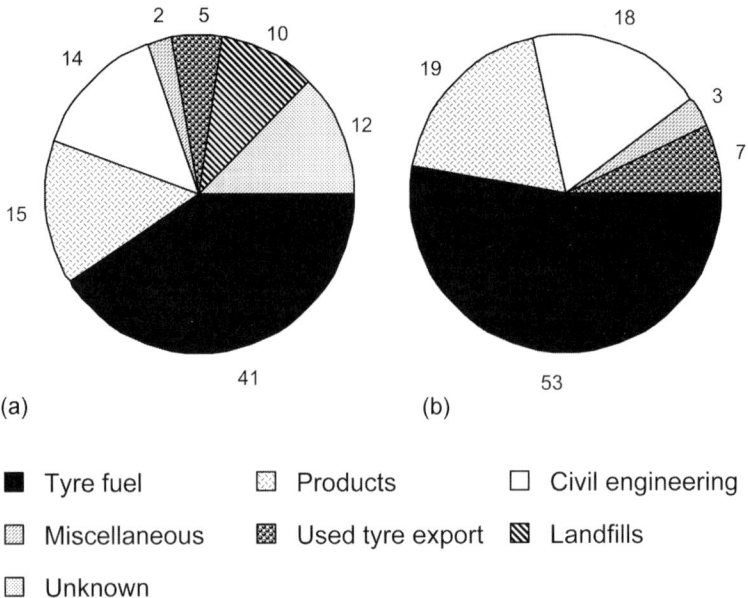

Fig. 3. Tyre data from 2001 for the USA: (**a**) Disposition of tyres, shown as percentage of the total number of discarded tyres (281 million tyres); (**b**) Markets for tyres, shown as percentage of the total number of discarded tyres used (218 million tyres) (data from Table 1).

allows for the escape of landfill gases (e.g., methane), which could otherwise be used as a source of energy. Owing to these complications, many landfill operators refuse tyre disposal or charge high tipping fees (Ohio Air Quality Development Authority 1991; EPA 1999). The regulatory trend is to ban whole tyres from landfills (Blumenthal 1993; EPA 1999), but the placement of shredded tyres in monofills has become a prominent alternative for long-term storage of scrap tyres in the recent past (RMA 2002). Today in the USA, about 10% of all worn tyres are still disposed of in landfills (Fig. 3a). Moreover, tyre shreds are now widely used in landfill engineering as well.

Stockpiles

The increasing cost or unavailability of landfill disposal as well as the existence of sham recycling operations can contribute to the growth of tyre stockpiles. The hazards associated with stockpiled tyres are equal to, if not greater than, those associated with landfill disposal. The most significant problem of tyre stockpiles is that they are prone to fires. Once tyre fires are ignited, they are extremely difficult to extinguish because of the almost continuous source of oxygen contained within the void space of the tyres. Moreover, quenching the tyre fires with water increases the production of pyrolytic oil, which can be transported by the water into the environment, where it aggravates soil and groundwater contamination (Jang *et al.* 1998). Uncontrolled tyre burning also releases significant amounts of hazardous gases and unburned hydrocarbons, which create thick black smoke plumes (Blumenthal 1993; EPA 1997). Moreover, extinguishing fires with water creates steam clouds, which serve to entrain even more particulate matter (Lemieux, personal communication 2003).

Outdoor tyre stockpiles also represent a breeding ground for insects and rodents. Rainwater collects in the open space of the impermeable tyres, and, together with trapped windblown pollen and dust, creates an ideal environment for the development of mosquito larvae. The warm, dark recesses of the tyres also provide a supportive environment for rodent habitation. Both rodents and mosquitoes pose a potential threat to human health due to their ability to transmit various diseases (Ohio Air Quality Development Authority 1991).

As a result of stockpile abatement programmes, the number of stockpiled tyres in the USA has decreased from 700 or 800 million in 1994 to approximately 300 million at the end of 2001 (RMA 2002). These abatement programmes take advantage of various alternative methods to tyre disposal.

Alternatives to waste tyre disposal

Several approaches have been made to reduce the number of landfilled, stockpiled, or illegally dumped tyres. Markets now exist for about 78% of all waste tyres (Fig. 3a), up from ~34.5% in 1990 (Blumenthal 1993). These markets continue to grow due to innovative approaches, which include improvements in design and material to increase the lifetime of a tyre, as well as various alternatives to tyre disposal (Table 2).

Utilization of whole waste tyres

Whole waste tyres can be utilized for many practical purposes including: crash barriers for both automobiles and boats; breakwaters to protect shorelines and harbours; playground equipment; erosion control in mountainous areas (e.g., construction of retaining walls); and artificial reefs to encourage growth of marine populations and natural reef systems (CIWMB 1992; Jang et al. 1998).

Utilization of processed waste tyres

Processing of discarded tyres by shredding, cutting, stamping, or grinding is becoming an increasingly important part of the waste tyre cycle. Shredding can reduce the tyre volume by up to 75%, thus significantly reducing transportation costs (Jang et al. 1998). Shredded tyres are used extensively in many civil engineering applications, which in the USA consumed 40 million, or 14% of all old tyres accumulated in 2001 (Table 1, Fig. 3a). In some other countries, civil engineering applications consume most

Table 2. *Alternatives to waste tyre disposal*[*]

	Advantages	Disadvantages
Reuse, retreading	Extended tyre life	Limited market
	Conserves resources	Eventual disposal required
Whole tyre utilization		
Bumpers, crash barriers, and erosion control	Effective Low cost	Eventual disposal required
Artificial reefs	Increase fish habitation	Expensive Potential instability
Processed tyre utilization		
Road base fill	Good filling agent	Potential leachate
Landfill operations and construction	Wide variety of applications Cost-efficient	Cutting or shredding required
Highway construction	Reduced road noise Increased durability	Current implementation limited and costly
Reclaimed rubber products	Unlimited possible products	Eventual disposal required
Pyrolysis/gasification	Energy-efficient Low air emissions Potentially saleable products	Potential for fugitive emissions Economically marginal
Tyres as energy source		
Cement kilns	Conserve natural resources Reduce NO_x emissions Complete tyre combustion Large market Ash incorporated into product	Increase in particulate emissions Increase in Zn emissions
Pulp and paper mills	Conserve natural resources	Large transportation costs for tyres Limited use Necessary to implement better emission controls Increased maintenance costs Tyres must be de-wired
Utility plants	Decrease SO_2 and NO_x emissions Conserve fossil fuels	Increase in particulates released into atmosphere Increase Zn concentration in ash
Dedicated tyre-to-energy facilities	Use whole tyres Conserve fossil fuels	Start-up costs very high Continuous tyre supply needed

[*]Sources: CIWMB 1992; Amari et al. 1999; RMA 2002; and other references cited in text.

of the discarded tyres (e.g., Finland: 70%; Mukherjee *et al.* 2003). Shredded tyres are used for landfill construction and operation (e.g., gas venting and leachate collection systems, daily covers, cap closures); road bases; rubber-modified asphalt concrete; highway embankments; drainage systems; erosion control; and as a substitute for wood chips and gravel (Blumenthal 1993).

Another promising utilization of shredded tyres is pyrolysis, that is, the thermal degradation in the absence of oxygen. This process converts waste tyres into secondary products of significant value, for example, various chemicals, oils, and residual char, which can be used as smokeless fuel, carbon black, or activated charcoal (Williams & Taylor 1993; Mastral *et al.* 1999*a*). However, pyrolysis is an expensive method of tyre utilization, and the resulting oils may contain high concentrations (up to 10 wt%; Williams & Taylor 1993) of polycyclic aromatic hydrocarbons (PAH), some of which are carcinogenic and/or mutagenic (Lee *et al.* 1981). These high PAH concentrations limit the use of the tyre-derived oil as a substitute for petroleum-derived fuels, because the oil would represent a health hazard when combusted.

A large amount of scrap tyres is also processed into ground rubber, or crumb rubber, either by mechanical grinding with an abrasive at ambient temperature or by cryogenic fracturing after cooling with liquid nitrogen (EPA 1994). The most important use of ground rubber is as an asphalt modifier in highway construction. Ground rubber is mixed with asphalt to produce reinforced road paving materials. These materials, known as rubberized asphalt or asphalt rubber, are characterized by reduced road noise, shorter breaking distances, and, due to increased durability of the roadway, lower maintenance costs (CIWMB 1992; Blumenthal 1993; RMA 2002). Ground rubber can also be used to fabricate athletic field surfaces, carpet underlayment, and railroad crossing beds. Discarded tyres are further processed to reclaim rubber, yielding a raw material to make various commercial products, including new tyres. In the USA, the grinding method accounted for 15% of all discarded tyre usage in 2001, whereas products made through cutting, punching, or stamping only accounted for about 4% (Table 1).

Tyres as energy source

A promising and relatively new way to utilize both whole and processed old tyres (Fig. 1) is to recover the energy contained in this waste material. As an energy source, tyres represent an attractive alternative to other fuels due to their high heating value. On average, the energy content of tyres ranges between 27 and 39 MJ/kg; the value varies because of the varying compositions of tyres combusted. This energy content is relatively high when compared to the average energy contained in bituminous coal, and is more than twice the heating value of municipal solid waste or chipped wood waste (Table 3). On the other hand, less than 40% of the energy contained in tyres is recoverable through combustion, as demonstrated by life-cycle energy budget calculations that consider both the energy consumed during production of the tyre and the energy recovered from their use as fuel (Amari *et al.* 1999). Another important advantage of tyre fuel is that it typically contains less than 2 wt% moisture, which is considerably less than coal and negligible compared to wood waste (Table 3). Owing to these advantages, the market for tyres as fuel has grown in the past decade (Fig. 2), and is the single most important end market for discarded tyres in the USA (Fig. 3**b**) and other countries, for example, Switzerland (50%; BUWAL 2001) and Japan (61% by weight; JATMA 2002). Waste tyres are a typical example of an opportunity fuel (Johnson *et al.* 1997), that is, a byproduct of other industrial processes with a significant potential in reducing the costs of generating electricity.

Although tyres can be used as an *alternate* fuel, they are most often utilized as a *supplemental* fuel. Tyre fuel exists either in shredded form (known as TDF, or tyre-derived fuel) or as whole tyres. Tyre-derived fuel consists of tyre chips, usually no larger than 5 cm on a side (Blumenthal 1993). The size reduction procedure is itself an energy-intensive process, and costs increase as the particle size decreases (Atal & Levendis 1995; Amari *et al.* 1999). The cost of cryogenic grinding of tyres can be as much as five times higher than that of pulverizing coal (Atal & Levendis 1995). Whole tyres or TDF

Table 3. *Comparison of typical moisture contents and heating values of select fuels**

Fuel	Moisture (wt%)	Heating value (MJ/kg)
Waste tyres	<2	27–39
Bituminous coal	3–10	26–30
Petroleum coke	0.5–1	29–35
No. 6 fuel oil	—	42–43
Municipal solid waste	15–40	8.1–13
Chipped wood waste	10–60	9.3–13

*Adapted from CIWMB (1992), with additional data from Table 4.

are combusted together with traditional fuels, predominantly coal, in cement kilns, pulp and paper mills, and electric utility or industrial boilers (Table 1). In addition, whole tyres are used as the sole fuel source in dedicated tyre-to-energy facilities, which, however, only combusted ~12% of all tyre fuel used in the USA in 2001 (Fig. 4). Tyres may also be combusted in waste-to-energy facilities designed for the incineration of municipal solid waste (Amari *et al.* 1999).

Tyres as supplemental fuel

Cement kilns. Cement kilns use whole tyres or TDF to supplement energy generated through the combustion of coal. These facilities represent by far the most important tyre fuel market (Table 1, Fig. 4). The recovered energy is used to produce the clinker by heating a mixture of finely ground calcareous, argillaceous, and siliceous materials to 1500–1600 °C. The high flame temperature (up to 2000 °C) and long gas residence times (several seconds at >1000 °C) allow for near complete combustion of the tyres (Ohio Air Quality Development Authority 1991; Giugliano *et al.* 1999; Karell & Blumenthal 2001). Moreover, unlike in other facilities, the reinforcing steel wires do not need to be removed from the tyre prior to combustion, because they serve as a source of supplemental Fe for the cement, thus reducing raw material needs (Barlaz *et al.* 1993; Giugliano *et al.* 1999). Similarly, the ash resulting from tyre combustion is beneficial to the cement-making process, as it is incorporated into the product material rather than discarded (Blumenthal 1993; Giugliano *et al.* 1999). Despite these facts, scrap tyres cannot completely replace other fuels because of their high contents of Zn (see below), which would adversely affect clinker quality and setting times (Karell & Blumenthal 2001).

Pulp and paper mills. The second largest consumers of tyre fuel are pulp and paper mills (Table 1, Fig. 4). Because the production of pulp and paper is an energy-intensive process, these facilities typically have their own boilers and turbines to meet electrical needs. Pulp and paper mills use TDF instead of whole tyres to supplement wood waste as fuel. The wood waste, also referred to as hog fuel, consists of chipped bark and other unusable tree parts (Barlaz *et al.* 1993). The use of TDF helps maintain constant combustion conditions in the stoker grate boiler system, which are not easily achieved using wood waste alone (Ohio Air Quality Development Authority 1991). The fuel-feeding process of these types of boilers, however, requires that the TDF be almost entirely free of wires (Jones *et al.* 1990). This requirement increases the costs of fuel significantly.

Electric utility and industrial boilers. The use of tyres in electric utility and industrial boilers comprises ~26% of the market for tyre fuel (Fig. 4). Although whole tyres can be used in some industrial boilers, most facilities combust TDF mixed with coal in a variety of concentrations ranging from as little as 2 to as much as 40% TDF by weight (e.g., Tesla 1994; EPA 1997; Ohio Air Quality Development Authority 1991). There are many different types of boilers that can utilize TDF combustion to augment energy generation (see Amari *et al.* 1999), the most popular of which is the cyclone-fired boiler. This type of boiler typically requires TDF pieces to be smaller than 2.5 cm × 2.5 cm, and the tyre chips must be de-wired prior to combustion (Ohio Air Quality Development Authority

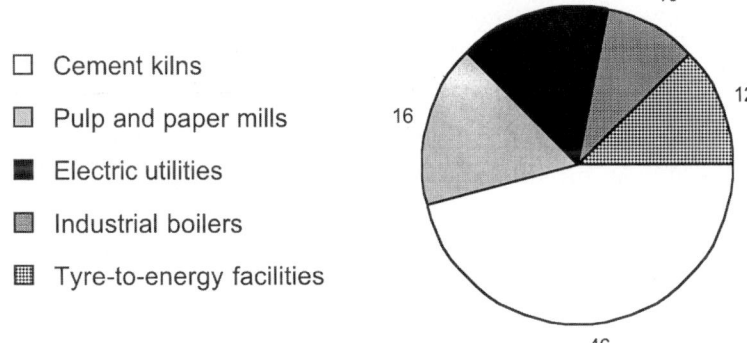

Fig. 4. Markets for tyre fuel consumed in the USA during 2001 (data from Table 1).

1991). De-wiring increases the costs of energy recovery by 25–50% (Amari *et al.* 1999). Some power companies have also conducted successful test burns with crumb-rubber TDF in their pulverized-coal boilers (Jones 1998).

Tyres as alternate fuel

The most direct process to use tyres for energy recovery is to combust whole tyres. Facilities that combust discarded tyres as the sole energy source are termed dedicated tyre-to-energy facilities. This type of facility is relatively new and appears to offer a promising and environmentally more suitable method for recovering energy from waste tyres. Only two such facilities are currently in operation in the USA, due to the complexity and cost associated with the construction and operation of these plants (RMA 2002). Nevertheless, combustion in tyre-to-energy facilities accounts for 12% of the end market of discarded tyres. At full capacity, one of the US facilities (in Stirling, Connecticut) combusts about 30 000 tyres daily, which amounts to more than 10 million tyres per year (Makansi 1992).

Composition of tyres

The production of modern car tyres uses more than 100 raw materials, most of which are based on petroleum products. Tyres consist of natural and synthetic rubber, typically styrene-butadiene (SBR); reinforcing fillers (e.g., carbon black, silica, clay, calcium carbonate); reinforcing fibres (e.g., nylon, polyester, rayon, steel wires); extenders or softeners (e.g., petroleum process oils); vulcanizing agents and accelerators (e.g., zinc oxide, stearic acid, organic sulphur compounds); and various antioxidants (JATMA 2002). The amounts of these materials vary widely among manufacturers and types of tyres, but a typical passenger car tyre contains, in addition to natural and synthetic rubber, ~3–5 wt% textile fibres, 10–17 wt% steel wire, 1 wt% ZnO, and 21–31 wt% carbon black (Amari *et al.* 1999; BUWAL 2001; JATMA 2002). As a result of these rather large ranges, the chemical composition of tyres is highly variable, thus complicating the quantitative assessment of the environmental impact of tyre combustion.

A typical automobile tyre consists of several parts (Fig. 5). The tread and sidewall make up the outside of the tyre. The tread is the portion of the tyre that comes into contact with the road, whereas the sidewall, or wall, is the side of the tyre that connects the tread and the bead. The tread is reinforced from underneath by the belt, which comprises a set of fabrics and/or wires. The plies, found in association with the belt, are layers of rubber-coated cords, which are typically made of fabric, polymer, fibreglass, or steel. The set of belts and plies is often collectively referred to as the cord, or cord body, of the tyre. The cords of the tyre wrap around the bead, the part of the tyre that rests on the tyre rim. The bead consists of both rubber bead filler and a series of steel bead wires. The last major part of the tyre, the innerliner, is a low-permeability

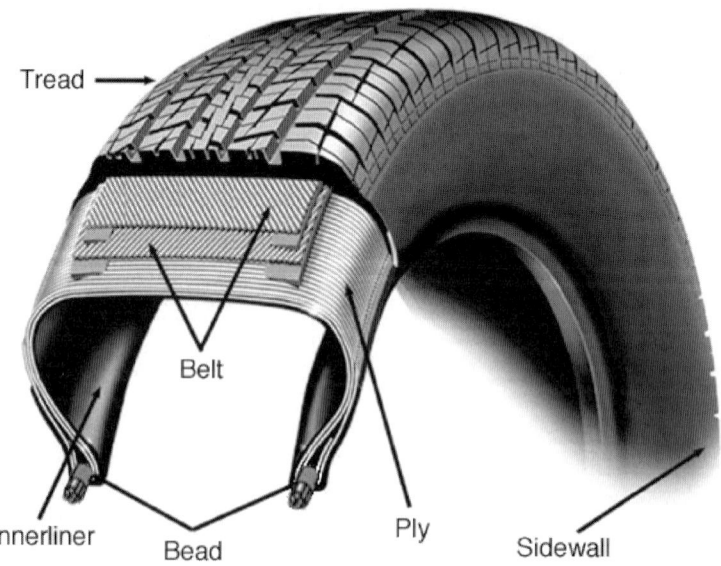

Fig. 5. Schematic diagram showing the most important parts of a car tyre. Modified from original diagram posted on the World Wide Web (http://tiresafety.com).

rubber. This rubber is laminated to the inside of tubeless tyres in order to improve the tyres' air retention capabilities.

The ash content of tyres and TDF varies within a large range, depending on composition. Steel-belted tyres exhibit considerably higher ash contents than textile-based tyres (Table 4). The available data show that, compared to typical coal, tyres and TDF contain significantly less N and, in general, less moisture (see also Table 3). On the other hand, tyres are distinctly richer in volatiles than coal, and their S content is similar to that of medium-S coal. Only a few studies report Cl concentrations, but these indicate that waste tyres are generally richer in Cl than typical coal (Tables 4, 5). Most investigations on waste tyres, unfortunately, do not provide analytical data for elements other than those shown in Table 4. Table 5 lists available chemical compositions of scrap tyres. The data reveal that tyres may contain considerable quantities of trace elements, in addition to the high amounts of Zn and, for steel-belted tyres, Fe (14–16 wt%; Makansi 1992; BUWAL 2001). Of particular interest is the relatively volatile element Zn, whose concentration ranges mostly between 1 and 2 wt% (Polasek & Jervis 1994), but may be as high as 4.3 wt% in certain truck tyres (Smolders & Degryse 2003). Because of the high Zn content of tyres, there is concern about diffuse Zn contamination of the environment from tyre wear. Several studies, for example, have documented steep Zn concentration gradients in roadside soils (see discussion in Smolders & Degryse 2003).

Table 5 further documents that major differences in elemental concentrations do not only exist between different types of tyres, but also between the tread and the wall of a tyre. Most of the elements are preferentially contained in the tyre walls relative to the tread. The wall is particularly enriched in Al and Ti, as well as in the volatile elements Cl, Br (in truck tyres), and Sb (Table 5). Tyre fuel consists most likely predominantly of wall material, and therefore, it is expected that tyre fuel will exhibit concentrations at the upper end of the range listed in Table 5. Of further note are the environmentally critical and volatile elements Hg and Pb, which range between 8 ppb and 0.4 ppm and 5 ppb and 65 ppm, respectively.

Solid waste products from tyre combustion

A large portion of combustion products from industrial and utility operations is regarded as waste, which is directly disposed of in surface impoundments and landfills. The interest in the chemical composition of combustion products is largely due to concerns about the potential release of toxic elements through leaching processes in ash disposal sites (see chapters by Tishmack & Burns, Spears & Lee, and Donahoe in this volume). It is therefore necessary to know the characteristics of the discarded combustion products in order to assess possible environmental impacts. The following section concentrates on the bulk chemical composition of the solid combustion products generated by burning of pure tyres or blends of tyres and coal.

Solid waste from pure tyre combustion

The ash generated through combustion of pure tyres, that is, without an additional fuel, varies in composition depending on the kind and manufacturer of the tyre, and on the combustion conditions. Unfortunately, there are only very few chemical analyses available for pure tyre ash. To our knowledge, the only investigations reporting TDF or tyre ash compositions are those of Granger & Clark (1991) and Polasek & Jervis (1994). These authors have shown that one of the principal TDF or tyre ash components is ZnO (Table 6), which is added during the vulcanization process. Polasek and Jervis (1994) stated that ZnO may account for as much as 75 wt% of the ash, but did not publish this specific analysis.

Other major constituents of TDF and tyre ash include Al_2O_3, SiO_2, CaO, TiO_2, and Fe_2O_3 (Table 6). Polasek & Jervis (1994) observed that the concentrations of Al_2O_3 and TiO_2 may be as high as 23 wt% and 33 wt%, respectively. Granger & Clark (1991) reported SO_3 contents of 15 wt% in ash produced through combustion of de-wired TDF (Table 6).

Solid waste from tyre co-combustion

As more industrial facilities are beginning to use tyres or TDF as a supplemental fuel, studies of the chemical composition of the solid waste resulting from co-combustion of tyres or TDF with other fuels (e.g., coal, natural gas, wood waste, or municipal solid waste) are becoming increasingly important. Owing to the large compositional variability observed for both tyres (Tables 4 and 5) and other fuels, it is difficult to compare data sets collected at different combustors. Direct comparison of various data sets is further complicated by differences in combustion conditions and ash collection procedures at different industrial and utility facilities. These difficulties are already a problem when comparing the chemical composition of pure coal combustion products generated at different facilities.

Table 4. *Composition of various tyres, TDF, and coal*

Fuel	Specifics	Source	Energy content (MJ/kg)	Moisture (wt%)	Ash (wt%)	Volatiles (wt%)	H (wt%)	C (wt%)	Fixed C (wt%)	N (wt%)	O (wt%)	S (wt%)	Cl (ppm)
Tyre	Fibreglass	Pope 1991	32.5	—	11.7	—	6.6	75.8	—	0.2	4.4	1.29	—
Tyre	Steel-belted	Pope 1991	26.7	—	25.2	—	5.0	64.2	—	0.1	4.4	0.91	—
Tyre	Nylon	Pope 1991	34.6	—	7.2	—	7.0	78.9	—	<0.1	5.4	1.51	—
Tyre	Polyester	Pope 1991	34.3	—	6.5	—	7.1	83.5	—	<0.1	1.7	1.20	—
Tyre	Kevlar-belted	Pope 1991	39.2	—	2.5	—	7.4	86.5	—	<0.1	2.1	1.49	—
Tyre	De-wired	Mastral *et al.* 2000	38.6	0.9	3.8	67.3	8.3	88.6	31.1	0.40	—	1.40	—
Tyre	Car	Karell & Blumenthal 2001	36.9	—	3.9	—	7.6	89.5	—	0.27	<0.01	1.88	700
Tyre	Truck	Karell & Blumenthal 2001	34.8	—	5.5	—	7.5	89.6	—	0.25	<0.01	2.09	600
Tyre	—	Burger 1991	33.7	0.8	10	65	—	78	—	0.2	—	1.4	1500
TDF	—	Karell & Blumenthal 2001	36.5	—	4.2	—	7.6	89.5	—	0.27	<0.01	1.92	700
TDF	De-wired	Karell & Blumenthal 2001	36.0	0.6	4.8	66.6	7.1	83.9	28.0	0.24	2.2	1.23	1490
TDF	Pulverized	Biagini *et al.* 2002	36.9	0.9	5.1	63.7	7.0	78.9	30.2	0.6	5.6	1.85	—
TDF	1.25 cm, rubber 'fuzz'	Granger & Clark 1991	32.1	2.3	16.5	—	6.3	69.7	—	0.45	3.4	1.30	700
TDF	5 cm, de-wired	Granger & Clark 1991	32.6	1.0	8.7	—	6.7	72.2	—	0.36	9.7	1.23	900
TDF	5 cm, with metal	Granger & Clark 1991	31.1	0.8	23.2	—	5.8	67.0	—	0.25	1.6	1.33	300
TDF	With metal	Gaglia *et al.* 1991	37.8	0.6	4.8	—	7.1	83.9	—	0.24	2.2	1.23	—
TDF	Crumb rubber	Hutchinson *et al.* 1991	37.9	0.2	4.2	64.8	—	—	28.9	—	—	1.90	—
TDF	2.5 cm, with metal	Hutchinson *et al.* 1991	34.6	2.4	13.2	60.7	—	—	22.4	—	—	1.30	—
TDF	2.5 cm, de-wired	Hutchinson *et al.* 1991	36.5	1.3	6.8	—	—	—	—	—	—	1.40	—
TDF	1.25 cm, de-wired	Hutchinson *et al.* 1991	36.2	0.8	9.5	60.9	—	—	27.3	—	—	1.50	—
TDF	3.75 cm, with metal	Howe 1991	29.2	8.6	14.8	54.3	6.0	62.0	22.4	0.25	7.3	1.19	—
TDF	1.88 cm, with metal	Phalen *et al.* 1991	36.7	0.4	8.3	62.7	6.8	77.4	28.6	0.37	5.1	1.62	—
TDF	—	Jones *et al.* 1990	36.0	0.6	4.8	—	7.1	83.9	—	0.24	2.2	1.23	—

TDF	Crumbs, <0.64 cm, de-wired	Lemieux 1994; Miller et al. 1998	37.3	0.8	7.2	65.5	7.2	76.0	26.4	0.34	7.1	1.75	3100*
TDF	Ground	Levendis et al. 1998a	29.0	—	26.0	52.3	5.3	60.9	21.7	0.28	7.1	2.46	—
TDF	2.5 cm, not de-wired	Unpublished data†	—	—	—	—	—	83.3	—	—	—	1.69	2938
TDF	—	University of Iowa‡	33.6	6.6	4.7	60.4	6.6	77.1	28.4	0.50	3.2	1.39	—
Coal	Sub-bituminous	Unpublished data§	25.9	14.6	7.2	33.6	—	75.4	43.9	—	—	1.47	215
Coal	Bituminous	Levendis et al. 1998a	29.2	—	13.7	34.4	4.7	71.9	51.9	1.36	7.0	1.36	—
Coal	Bituminous	Karell & Blumenthal 2001	31.5	—	7.8	—	5.1	75.8	—	1.50	8.2	1.60	—
Coal	Bituminous	Hower et al. 2001	28.4	2.7	14.0	38.4	4.8	64.3	—	1.43	—	4.37	—
Coal	—	University of Iowa‡	28.1	8.5	9.4	32.4	4.6	67.7	49.7	1.37	7.3	1.14	—

*Represents total halogens, calculated as Cl.
†TDF used in test burn conducted by the Purdue University power plant (this study).
‡Data compiled from original University of Iowa data set provided by Schwarzhoff & Milster (personal communication 2003).
§Illinois Basin coal from Southern Indiana, used in Purdue University power plant test burn (this study).

However, test burns that were conducted at a single facility to study the effects of adding TDF to a base fuel reveal chemical trends that may be similar to those observed at other facilities, and thus could be useful for assessing ash disposal and groundwater protection strategies.

In experiments to study the potential for emissions of hazardous air pollutants from combustion of TDF, Lemieux (1994) co-fired natural gas and wire-free crumb rubber at $\sim900-1000\ °C$ in a rotary kiln incinerator simulator (composition of crumb rubber given in Tables 4 and 5). The resulting fly ash is characterized by high contents of Al_2O_3, SiO_2, CaO, and ZnO (Table 6). The solid combustion product is thus similar to the ash produced through combustion of pure tyres or TDF (cf. data of Granger & Clark 1991 and Polasek & Jervis 1994 in Table 6), documenting that co-firing TDF with natural gas does not affect the ash composition to any significant extent. Using X-ray diffractometry, Lemieux (1994) identified cristobalite/quartz (both SiO_2), mullite ($Al_6Si_2O_{13}$), and willemite ($ZnSiO_4$) as major phases in this fly ash.

Table 7 lists the results of test burns at two coal-combusting power plants: a western Kentucky power plant, which burned pure Illinois Basin coal (Table 4) and two blends of this coal + TDF (99 wt% coal + 1 wt% TDF; 97 wt% coal + 3 wt% TDF) in a cyclone boiler (Hower et al. 2001); and the Purdue University power plant, which used a stoker boiler to combust a comparable Illinois Basin coal as well as a blend containing 95 wt% of this coal and 5 wt% TDF (Table 4; this study).

The Kentucky data (Hower et al. 2001) reveal considerable changes in the concentration of several components in the bulk fly ash with the addition of TDF. The most drastic changes are observed for Zn and Pb in the fly ash resulting from combustion of the 3 wt% TDF blend: relative to pure coal, the Zn content of the fly ash more than doubled, whereas that of Pb increased from undetectable levels to more than 300 ppm (Table 7). Notable increases in concentration are also observed for Na_2O, Al_2O_3, SiO_2, and TiO_2, whereas most other components of the bulk fly ash, including Cr, Ni, and Cu, show reduced concentrations (Fig. 6). For the bulk bottom ash samples, pronounced increases in concentration were only observed for CaO and Fe_2O_3. When compared to the pure coal and the blend with 3 wt% TDF, there are some inconsistencies in the chemical trends observed for the blend containing 1 wt% TDF (Table 7); these may result from heterogeneities in the fuel.

The Purdue University data (this study) revealed that the bulk fly ash from the coal +

Table 5. Concentrations of various elements (in ppm) in select tyres or parts thereof, and in TDF

Element	Typical car tyre tread Mean	Std. dev.	Typical car tyre wall Mean	Std. dev.	Typical truck tyre tread Mean	Std. dev.	Typical truck tyre wall Mean	Std. dev.	Bulk car tyre	TDF	TDF (de-wired)	TDF (de-wired)	Car tyre	TDF (not de-wired) Mean	Std. dev.
Be	—	—	—	—	—	—	—	—	—	0.4	—	—	—	<0.5	—
Na	610	160	330	85	510	140	530	130	—	—	—	—	—	<70	35
Mg	32	—	230	160	44	43	688	670	—	—	—	—	—	500	35
Al	420	120	5800	1500	340	90	33000	9000	—	318	3100	1490	—	2520	61
Cl	690	220	1300	420	410	140	1900	690	—	—	—	—	—	2938	0
K	180	70	520	130	1700	360	—	—	—	—	—	—	—	332	0
Ca	330	240	650	440	160	100	2000	1000	—	—	—	3780	—	1550	82
Sc	0.011	0.008	0.6	0.1	—	—	2.2	0.4	—	—	—	—	—	<10	—
Ti	195	84	3500	1200	42	23	64000	18000	—	0.4	—	—	—	1480	35
V	0.8	0.4	4.8	1.6	0.9	0.3	12	6	—	—	—	—	—	<5	—
Cr	4	2	8	4	3.8	1.7	22	13	90	1	<5	97	0.025	<68	—
Mn	1.9	—	7.4	2.3	2.9	1.1	5.5	4.2	—	11	—	—	—	<80	—
Fe	210	70	790	210	—	—	1300	300	160000	1605	295	3210	—	4400	176
Co	2.9	0.8	2.4	0.7	2.6	0.4	1.9	0.6	—	182	—	—	—	63	9
Ni	—	—	—	—	—	—	—	—	80	2	—	—	—	22	1
Cu	—	—	—	—	—	—	—	—	200	70	—	—	0.013	38	2
Zn	17000	3000	20000	4000	12000	1400	19000	4000	8030	12137	21400	15200	15300	10900	849
As	—	—	0.6	0.3	—	—	—	—	—	0.8	—	—	0.65	0.9	0.0
Se	—	—	—	—	—	—	—	—	—	3.6	—	—	—	<0.1	—
Br	1.8	1	3	1.6	2	1.5	1400	560	—	—	—	—	—	—	—
Mo	—	—	—	—	—	—	—	—	—	1.5	—	—	—	<2	—
Cd	—	—	—	—	—	—	—	—	10	0.4	<2	6	<2	0.1	0.0
In	—	—	—	—	—	—	—	—	—	10.8	—	—	—	—	—
Sb	0.06	0.06	46	15	0.18	0.9	11	3.3	—	3.4	—	—	—	3.6	0.3
Ba	17.2	—	17.3	—	—	—	—	—	—	0.4	—	—	—	11.6	0.8
Sm	0.1	0.03	0.4	0.1	0.07	0.03	1.2	0.3	—	—	—	—	—	0.2	0.0
Hg	—	—	—	—	—	—	—	—	—	0.4	—	—	<0.1	0.008	0.001
Pb	—	—	—	—	—	—	—	—	50	38	51	65	0.005	6	2
Source	Polasek & Jervis 1994		Polasek & Jervis 1994		Polasek & Jervis 1994		Polasek & Jervis 1994		BUWAL 2001	Hutchinson et al. 1991	Lemieux 1994; Miller et al. 1998	Karell & Blumenthal 2001	Mukherjee et al. 2003	Unpublished data*	

*Average of three analyses of TDF, used in test burn conducted by the Purdue University power plant (this study).

Table 6. *Composition of TDF and tyre combustion ash*

	Ash from TDF rubber 'fuzz' (1.25 cm)*	Ash from TDF (de-wired, 5 cm)*	Ash from TDF (with metal, 5 cm)*	Ash from typical car tyre[†]		Ash from wire-free crumb rubber (<0.64 cm)[‡]	
				Mean	Std. dev.	Sample TB3	Sample TB6
Major element oxides (wt%)							
Na_2O	1.07	1.10	0.13	—	—	0.92	1.6
MgO	0.73	1.35	0.10	0.5	0.2	1.4	1.6
Al_2O_3	6.99	9.09	1.93	18	3	5.3	3.6
SiO_2	18.21	22.00	5.16	—	—	68	68
P_2O_5	0.56	1.03	0.10	—	—	0.002	0.0002
SO_3	8.35	15.38	0.99	—	—	0.001	0.001
K_2O	0.55	0.92	0.14	0.5	0.2	0.70	0.70
CaO	5.99	10.64	0.56	0.4	0.2	4.3	4.0
TiO_2	6.01	2.57	0.14	28	5	0.25	0.14
Fe_2O_3	30.93	1.45	0.35	1.0	0.1	1.2	1.2
ZnO	20.60	34.50	5.14	22	2	3.0	6.5
Metal[§]	—	—	85.28	—	—	—	—
Trace elements (ppm)							
Cl	—	—	—	86	43	—	—
Sc	—	—	—	7.2	0.9	—	—
V	—	—	—	46	9.9	—	—
Cr	—	—	—	107	26	20	100
Mn	—	—	—	160	18	—	—
Co	—	—	—	19	3	50	100
Ni	—	—	—	—	—	70	30
Cu	—	—	—	—	—	20	1
Br	—	—	—	280	77	—	—
Sr	—	—	—	—	—	20	100
Zr	—	—	—	—	—	100	100
Sb	—	—	—	7	1	—	—
La	—	—	—	33	5	—	—
Sm	—	—	—	3.8	0.6	—	—
Pb	—	—	—	—	—	10	9

*Data from Granger & Clark (1991). Major element oxides normalized to 100 wt%. Ash represents residue from combustion of pure TDF, whose fuel characteristics are listed in Table 4.
[†]Data from Polasek & Jervis (1994), for pure tyre ash. Major element oxides add up to only 71 wt%; no further details available.
[‡]Data from Lemieux (1994), collected after co-firing natural gas and TDF crumbs. Major element oxides add up to 85 and 87 wt%, respectively. Ash represents fly ash from combustion of TDF crumbs, whose fuel characteristics are listed in Tables 4 and 5. Lemieux (1994) suspected that some elements (Al, Si, Zr) may have originated from the rotary kiln insulation.
[§]Metal (reported as Fe_2O_3) was removed from sample before the analysis was performed.

TDF mixture was distinctly richer in SO_3, Zn, Ge, and As than the fly ash from pure coal. Most pronounced are the changes in the contents of SO_3 and Zn, which increased by factors of more than 2 and 16, respectively (Fig. 6). This result can be explained by the average concentrations of these volatile components in the two fuels used in the test burn: the coal + TDF blend contains 2.1 wt% S and 183 ppm Zn, whereas the pure coal only contains 1.5 wt% S (Table 4) and 36 ppm Zn (Gieré, unpublished data). The levels of most other analysed trace metals in the bulk fly ash decreased with the addition of TDF. For the bottom ash, many of the patterns observed for the fly ash are reversed (Table 7).

Overall, the Kentucky and Purdue studies show similar trends. The most important change resulting from the addition of TDF to coal is the increase in the Zn content of the fly ash. In contrast, most other trace elements have lower concentrations in the fly ash resulting from the combustion of the blend. Despite many similarities, two notable differences between the two test burns are observed. SO_3 and Pb showed opposite trends in the Purdue fly ash compared to the Kentucky fly ash. In both studies, enrichments of many trace elements

Table 7. *Comparison of bulk solid waste compositions from two coal-combusting facilities*

	Hower et al. 2001 100% Coal Cyclone 1649 °C — BA (n=1)	FA (n=1)	Hower et al. 2001 99% Coal+1% TDF Cyclone 1649 °C — BA (n=1)	FA (n=1)	Hower et al. 2001 97% Coal+3% TDF Cyclone 1649 °C — BA (n=1)	FA (n=1)	Purdue University 100% Coal Stoker 1500 °C — BA (n=1)	FA† (n=2)	Purdue University 95% Coal+5% TDF Stoker 1500 °C — BA (n=3)	Std. dev.	FA† (n=2)
C (wt%)	0.63	1.40	1.79	21.36	3.17	0.81	2.0	15.2	2.1	0.6	16.2
Major element oxides (wt%), normalized to 100 wt% on a C-free basis											
Na_2O	0.12	1.04	0.38	6.94	0.42	2.10	0.29	0.86	0.26	0.02	0.79
MgO	0.67	0.46	0.64	0.32	0.60	0.19	0.56	0.64	0.51	0.02	0.69
Al_2O_3	20.90	17.16	20.02	21.72	18.95	20.61	29.3	23.2	27.3	0.09	22.4
SiO_2	48.07	29.26	44.38	30.26	44.36	35.14	47.9	42.8	46.2	1.2	41.3
P_2O_5	0.27	0.17	0.24	0.02	0.22	0.05	0.13	1.48	0.16	0.03	1.47
SO_3	0.41	1.04	0.57	0.37	0.79	0.79	0.32	7.65	0.8	0.1	16.4
K_2O	2.71	3.11	2.75	2.82	2.80	3.18	1.74	2.67	1.52	0.06	2.58
CaO	4.85	8.87	5.25	2.19	5.93	3.48	0.54	1.03	0.55	0.03	1.30
TiO_2	1.08	1.12	1.13	1.40	1.10	1.45	1.22	1.68	1.25	0.02	1.72
Fe_2O_3	20.92	37.76	24.63	33.96	24.84	33.00	18.0	18.0	21.4	1.6	11.2
Trace elements (ppm)‡											
V		702	174	233	252	410	281	614	307	6	548
Cr		283	125	156	80	212	220	487	262	70	382
Mn		179	199	99	262	135	100	200	152	50	150
Ni		397	67	244	62	233	358	1765	291	22	1185
Cu		298	39	166	52	230	118	945	118	9	680
Zn		2420	90	1840	275	5750	68	3475	270	28	56400
Ga		107	<0.6	46	<0.6	98	20	587	19	1	411
Ge		278	13	149	<0.9	265	19	1220	23.5	0.3	1925
As		468	<1	627	9	302	2	650	4	5	700
Rb		<0.7	70	<0.7	160	<0.7	91	124	83	7	107
Sr		275	259	183	146	238	600	574	671	89	476
Y		<1.2	<1.2	<1.2	<1.2	<1.2	92	111	80	2	85
Zr		340	192	126	211	230	242	222	234	8	195
Mo		<1.3	<1.3	<1.3	<1.3	<1.3	8	112	8	1	81
Pb		<1.8	20	<1.8	<1.8	327	22	2045	31	10	1605

*BA and FA represent bottom ash and fly ash, respectively.
†Purdue University FA collected from an electrostatic precipitator hopper.
‡Detection limits for the Hower *et al.* (2001) data set extracted from Wong & Robertson (1993).

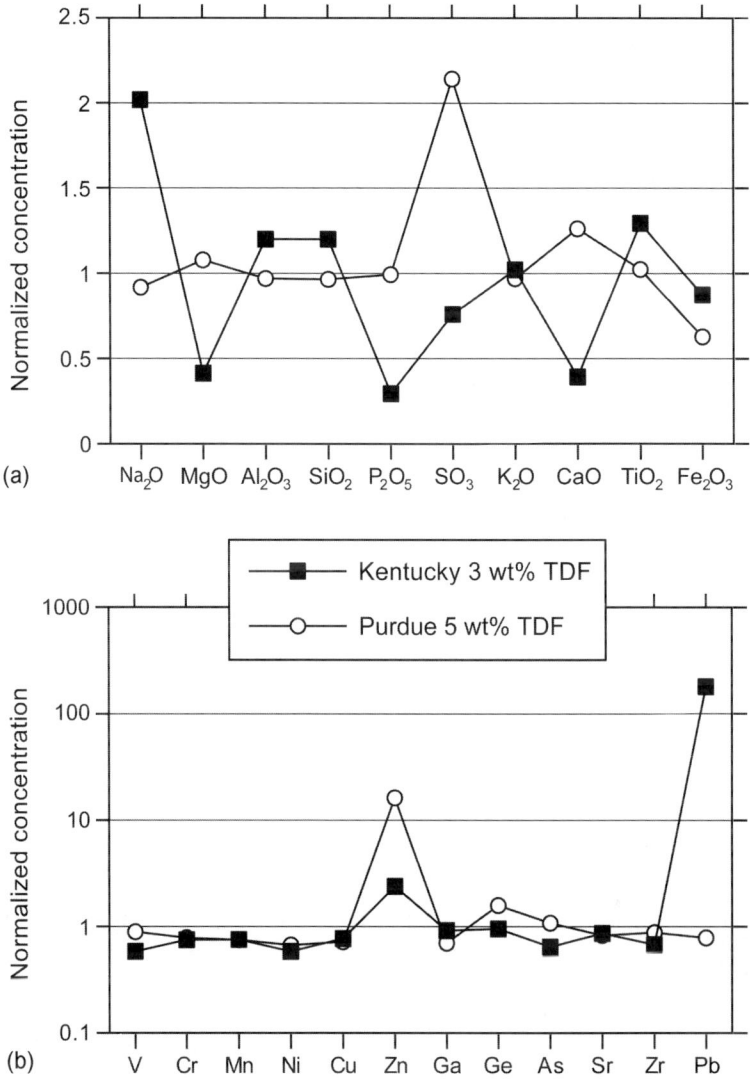

Fig. 6. Comparison of normalized concentrations of chemical components of fly ash from two boilers. Values represent concentrations in fly ash from combustion of coal/tyre blends normalized to the respective concentrations in fly ash from combustion of pure coal (data from Table 7).

are observed in the fly ash compared to the bottom ash (Table 7). This result is not surprising, however, because the bottom ash accumulates directly from the boiler, which is much hotter than the collection points for the fly ash, and therefore, the relatively volatile elements (e.g., S, Zn, As, and Pb) are partitioned into the fly ash due to temperature. Moreover, the smaller particles of the fly ash are more effective at capturing volatile elements than the larger ones

in the bottom ash, because they exhibit relatively large surface areas (e.g., Hower *et al.* 2001).

Atmospheric emissions from tyre combustion

Emissions from pure tyre combustion

As discussed above, the chemical composition of tyres differs greatly among various manufacturers, tyre types, and tyre parts. Therefore, results of emission tests from pure tyre combustion may vary between different studies. However, several general trends in emissions have been reported from laboratory-scale studies at carefully controlled conditions: compared to coal combustion, burning of pure tyres under the same conditions typically produces much smaller amounts of NO_x $(NO + NO_2)$ emitted into the atmosphere, but higher emissions of CO, PAHs, and submicrometre particulate matter (Levendis *et al.* 1996, 1998*a, b*; Courtemanche & Levendis 1998; Mastral *et al.* 1999*b*, 2000). The quantitative effects, however, vary greatly and depend on various factors, including fuel composition and particle size, boiler type, and conditions of combustion.

The influence of fuel composition on atmospheric emissions is best documented for CO_2, SO_2, and NO_x. In experiments using an electrically heated drop-tube furnace, Levendis *et al.* (1998*a, b*) observed that the CO_2 yields from tyre combustion were considerably lower than those from coal combustion, reflecting the difference in C contents of the specific fuels used (60.9 wt% vs. 71.9 wt% C, respectively; see Table 4). Moreover, the particular batch of tyres used by these authors had an S content of 2.46 wt%, which is unusually high for tyres (see Table 4), and generated higher SO_2 emissions than the bituminous coal reference material (1.36 wt% S). Finally, the NO_x emissions generated during combustion of tyres are much lower than those resulting from coal combustion because tyres typically contain considerably less N than coal (Table 4). Despite this fact, a larger fraction of the fuel-bound N is converted to NO_x when tyres are combusted rather than coal (Levendis *et al.* 1996). This conversion, however, does not compensate for the lower N content of tyres and thus, the NO_x emissions from tyre combustion remain low relative to those of coal. The influence of fuel particle size has been studied by Levendis *et al.* (1996), who showed that tyre particles combust considerably faster than coal particles of similar size.

Another key factor determining the emission levels is the bulk equivalence ratio (Φ) in the furnace, a parameter that describes the relative amounts of fuel and air. Levendis and co-workers (1996, 1998*b*) have shown that for both coal and tyre (and for various blends thereof) the specific emissions, that is, the mass of a certain pollutant relative to the mass of fuel burned, vary systematically when the combustion environment changes from fuel-lean (low Φ) to fuel-rich (high Φ) conditions. Their results document that, at a given temperature, the specific emissions of NO_x, SO_2, and CO_2 decrease moderately to strongly with increasing Φ, whereas those of CO and PAH increase. The observed changes are drastic for CO and PAH, whose specific emissions were insignificant at fuel-lean conditions, but increased exponentially with increasing Φ in fuel-rich conditions.

Emission levels are also strongly influenced by the gas temperature in the furnace. Between 1000 and 1300 °C, for example, an increase in temperature leads to a decrease in emissions of PAH, but to an increase in CO and NO_x emissions (Levendis *et al.* 1998*a*; Courtemanche & Levendis 1998). The observed increase in CO has been attributed to partial oxidation of hydrocarbons. Temperature-dependence of PAH emissions has also been reported below 1000 °C: in laboratory experiments using an atmospheric fluidized-bed reactor, Mastral *et al.* (1999*b*, 2000) have shown that within the studied range (650–950 °C), total PAH emissions were highest at 750 °C and 950 °C for coal and tyres, respectively, whereby the temperature effect was much more pronounced for the tyre fuel. In both cases, the main contribution to the total PAHs was from the most volatile compounds, that is, those consisting of three (coal) or three and four (tyre) rings. During tyre combustion, most of the emitted PAHs are associated with particulate matter, especially at higher temperatures, whereas coal combustion more frequently produces PAH emissions in the gas phase (Mastral *et al.* 1999*b*).

The PAH emissions from combustion processes are a particular concern because they are organic compounds, which can be more dangerous than inorganic compounds, even though they are emitted in smaller quantities. During the combustion process, PAHs are formed by two mechanisms. Pyrolysis first leads to partial fragmentation of organic compounds, producing small compounds, mainly free radicals, which are highly reactive and thus only exist for a short period of time. Subsequent pyrosynthesis recombines these intermediary compounds into the relatively stable PAHs, which consist of two or more aromatic rings (Mastral *et al.* 1999*b*). One cause for the high PAH emissions from tyre combustion, relative to coal combustion, may be the large amount of carbon black in tyres, which, when only partially burned, appears to provide particle surfaces favourable for PAH deposition (Mastral *et al.* 1999*b*).

A larger-scale tyre combustion experiment was performed by Lemieux & Ryan (1993) in order to collect emission data from a simulated open waste tyre fire. In addition to identifying a large number of organic compounds, including

PAHs, in the emitted gases, these authors also studied particulate matter, in which they observed elevated levels of Zn and Pb.

Uncontrolled open fires in tyre dumps may burn for several months (e.g., Rhinehard tyre fire in Winchester, Virginia), generating many hazardous products of incomplete combustion, which are released directly into the atmosphere (EPA 1997). Owing to safety concerns and other factors (e.g., meteorological conditions, fire-fighting activities), it is difficult to collect air samples from the smoke plumes of these fires. The available field data revealed potentially hazardous levels of several PAHs and CO, as well as relatively high concentrations of Pb and Zn in such plumes (Ohio Air Quality Development Authority 1991; EPA 1997).

Emissions from tyre co-combustion

Besides examining the emissions resulting from pure coal and pure tyre combustion, Levendis *et al.* (1998b) also studied two different coal/ TDF blends (75 wt% coal + 25 wt% TDF; 50 wt% coal + 50 wt% TDF). The emissions of NO_x, SO_2, CO, CO_2, and particulate matter resulting from the combustion of the two blends were in between those of the parent materials, confirming the general trends observed for pure coal and pure tyre (see above). Surprisingly, the PAH emissions from the two fuel blends were only slightly greater than those from pure coal, that is, much lower than expected based on the compositions of the blends. This result possibly suggests that the combustion of the coal in the fuel blends supports the oxidative destruction of PAHs produced during the burning of the tyre particles (Levendis *et al.* 1998b).

In another experiment, de-wired TDF was combusted together with natural gas as the primary fuel in a rotary kiln incinerator simulator (Lemieux 1994). This study examined organic gases, along with emissions of particulate matter and metal aerosols. It revealed that adding TDF to natural gas increased emissions of PAH, particulate matter, and metals, most notably those of Zn, As, Sb, and Pb. Another small-scale study with natural gas and TDF has demonstrated that TDF can be a fuel that is technically viable for reburning, that is, a special method of significantly reducing NO_x emissions (Miller *et al.* 1998).

Besides these laboratory experiments there are several emission studies on controlled tyre co-combustion in large-scale industrial facilities, including cement and lime kilns, utility boilers, and pulp and paper mills (Ohio Air Quality Development Authority 1991; EPRI 1991; EPA

1997; Giugliano *et al.* 1999). Comparison between data sets is often difficult because the combustion conditions were either very different between the various experiments or not reported in detail. Inter-facility comparisons are further complicated by the fact that different units are given for the emission levels. Moreover, most of these studies did not report the composition of the tested fuels. Despite these complications, it is observed that emissions of NO_x generally decreased with increasing amounts of TDF in the fuel blend (e.g., Horvath 1991), whereas emissions of some metals, particularly Zn, increased. In most coal-burning facilities, emissions of particulate matter also augmented with the addition of TDF. Co-firing of TDF with coal resulted in either enhanced or reduced SO_2 emissions, depending on the S content of the coal and TDF used. Overall, the results indicate that properly designed industrial facilities combusting solid fuels can supplement their normal fuels with 10–20 wt% TDF and still comply with environmental regulations for atmospheric emissions (EPA 1997). Dedicated tyre-to-energy facilities, however, are designed to burn exclusively tyres and, due to the high boiler temperatures (about 1400 °C) and sophisticated air pollution control systems, achieve emission rates that are much lower for most compounds than those from other solid fuel combustors (Ohio Air Quality Development Authority 1991; EPA 1997).

Below, we discuss test burns in industrial facilities that combusted both pure coal and coal/TDF blends. These fuels were burned at identical conditions in each combustor: a stoker boiler at the University of Iowa (EPA 1997; Schwarzhoff & Milster, personal communication 2003), a stoker boiler at Purdue University (this study), and two different cement kilns (Carrasco *et al.* 1998; Mukherjee *et al.* 2003).

The University of Iowa conducted test burns with three different fuels (pure coal, and coal/ TDF blends of 96/4 wt% and 92/8 wt%), whereas the Purdue University study compared pure coal and a coal/TDF blend of 95/5 wt% (Table 8). In the Iowa test burn results, it is noted that the 96/4 wt% blend did not produce data that are entirely consistent with the trend exhibited by the blend containing 8 wt% TDF. This inconsistency is probably due to heterogeneity within each of the two fuel components or due to insufficient mixing of the two fuels. The Purdue data correspond both qualitatively and quantitatively more closely with the 8 wt% TDF Iowa data than they do with the 4 wt% TDF data (Figs 7 and 8; Table 8). Both the University of Iowa and the Purdue University

Table 8. *Comparison of average emissions resulting from combustion of various fuel blends in stoker boilers*

| | University of Iowa* | | University of Iowa* | | University of Iowa* | | Purdue University | | Purdue University | |
| | 100% Coal | | 96% Coal + 4% TDF | | 92% Coal + 8% TDF | | 100% Coal | | 95% Coal + 5% TDF | |
Fuel blend (wt%)	n = 3	Std. dev.	n = 3	Std. dev.	n = 3	Std. dev.	n = 3	Std. dev.	n = 3	Std. dev.
Gas temperature (°C)	202	1	199	5	199	0.6	183.4	0.7	195	3
Total particulate matter (kg/h)	14	2	10	2	13	1	13.6	0.9	13	2
Metals (g/h)										
Be	0.04	0.01	0.03	0.01	0.2	0.1	0.20	0.03	0.7	0.2
Mg	9	5	8	1	20	13	—	—	—	—
Al	—	—	—	—	—	—	60	12	700	169
Ca	—	—	—	—	—	—	11	4	40	10
V	—	—	—	—	—	—	0.9	0.4	6	1
Cr	0.9	0.2	0.8	0.1	2	2	0.91	0.00	3.9	0.9
Mn	—	—	—	—	—	—	48.1	—	40	12
Co	—	—	—	—	—	—	1.1	0.3	4	1
Ni	1.4	0.5	1.8	0.3	4	2	4.5	0.4	14	3
Cu	2.1	0.6	2.5	0.2	9	4	3.3	0.7	14	2
Zn	40	15	164	9	1600	360	15	2	2,400	741
As	3	2	2.1	0.7	12	2	6	2	26	4
Se	—	—	—	—	—	—	7	2	11	4
Cd	<0.19	—	<0.188	—	0.5	0.1	0.04	0.00	0.24	0.07
Sb	—	—	—	—	—	—	0.45	—	1.5	0.3
Te	—	—	—	—	—	—	<0.9	—	1.8	0.4
Ba	1.2	0.6	0.9	0.2	3	2	0.2	0.1	0.7	0.3
Hg	1.42	0.05	1.2	0.2	1.2	0.4	0.45	0.00	0.45	0.00
Tl	—	—	—	—	—	—	0.45	0.00	1.7	0.5
Pb	2.0	0.8	2.3	0.2	10	4	8	2	20	7
Gases (kg/h) [†]										
CO	5.0	0.4	8	5	6.0	0.2	5	2	3.0	0.5
NO$_x$	82	9	78	3	70	22	63.2	0.5	67	3
SO$_2$	320	16	291	2	290	27	390	22	360	12
HCl	5	1	6.3	0.9	7	3	0.69	0.06	1.1	0.3
Fluoride[‡]	0.00080	0.00003	0.00064	0.00007	0.00077	0.00000	0.20	0.01	0.22	0.03
Organics										
THC (mg/s)	265	—	265	—	189	—	120	83	20	23
Dioxins/Furans (ng/s)[§]	18	4	10	1	6	3	<19	—	<17	—

* Average data for the University of Iowa published by EPA (1997). Data presented here compiled from original University of Iowa data set provided by Schwarzhoff & Milster (personal communication 2003).
[†] At 7% O$_2$ by volume.; [‡] Purdue University dioxins/furans data not reported at 7% O$_2$ by volume.; [§] Purdue University dioxins/furans data reported at 7% O$_2$ by volume.
[‡] University of Iowa tested for fluoride as HF. University of Iowa fluoride data not reported at 7% O$_2$ by volume.

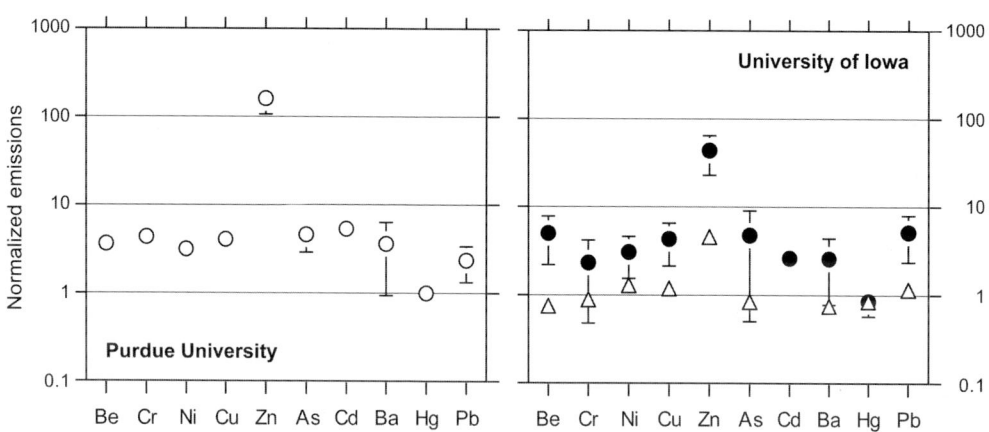

Fig. 7. Comparison of normalized metal emissions from two stoker boilers. Values represent emissions from combustion of coal/tyre blends normalized to the respective emissions from combustion of pure coal. Purdue University data set is for 5 wt% TDF. Symbols for University of Iowa data set: △, 4 wt% TDF; ●, 8 wt% TDF. Data without error bars have standard deviations that are smaller than the symbols. Data from Table 8.

studies document that combustion of coal/TDF blends with ≥5 wt% TDF rather than pure coal leads to significantly enhanced average emissions of Be, Ni, Cu, Zn, As, Cd, and Pb (Fig. 7). Most dramatic are the increases observed for Zn, whose average emission rate was multiplied by a factor of more than 43 when ≥5 wt% TDF were added to coal (Fig. 7). In the Purdue test burn, Zn emissions averaged 2.4 kg/h for the 95/5 wt% blend, compared to 15 g/h for the combustion of pure coal (Table 8). The Purdue study further reveals significantly increased emissions in most other analysed metals, including Al, Cr,

and Sb (Table 8). For example, combusting the 95/5 wt% blend led to Al emissions that were 12 times larger than those from combustion of pure coal. On the other hand, emissions of Hg and total particulate matter remained constant in both studies (Table 8; Fig. 7). Total hydrocarbon (THC) emissions decreased in both stoker boiler experiments when the coal + TDF mixture was used as fuel (Fig. 8). The high Cl content (nearly 3000 ppm) of the TDF used in the Purdue test, the only comparative study for which Cl concentrations in the fuel are available, is responsible for the pronounced increase in HCl

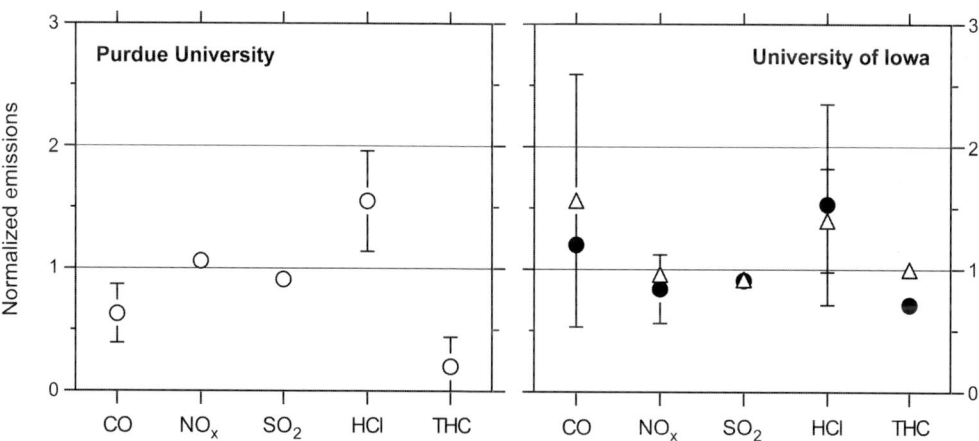

Fig. 8. Comparison of normalized gas emissions from two stoker boilers. Values represent emissions from combustion of coal/tyre blends normalized to the respective emissions from combustion of pure coal. Purdue University data set is for 5 wt% TDF. Symbols for University of Iowa data set: △, 4 wt% TDF; ●, 8 wt% TDF. Data without error bars have standard deviations that are smaller than the symbols. Data from Table 8.

emissions (Fig. 8) from the blend relative to those from pure coal, which contains only 215 ppm Cl (Table 4). Additionally, the Iowa study showed a notable reduction in dioxin/furan emissions with increasing TDF percentage.

The results reported by Carrasco *et al.* (1998) revealed that nearly all studied metal emissions, measured at the exit of a cement kiln stack, were significantly higher when a blend of 80 wt% coal + 20 wt% TDF was combusted instead of pure coal. Especially notable are increased emissions in Cr, Mn, Cu, Zn, and Pb (Table 9). The exception to this trend is Hg, which exhibited a 30% reduction in its emission rate when the coal + TDF mixture was burned. The data further document reductions in NO_x and organic compound emissions, including PAHs, where the most drastic decrease was observed for dioxins and furans. On the other hand, emissions of CO, SO_2, and HCl increased considerably with the addition of TDF (Table 9). The total particulate emissions from combustion of the blend were only slightly greater than those from pure coal. Carrasco *et al.* (1998) used their data to model atmospheric dispersion of the emitted contaminants in the vicinity of the cement kiln, and found that the environmental impact was most pronounced within an area with a 500 m radius around the pollution source. Their calculations revealed that the ground-level maximum 24 h concentrations of particulate matter, CO, SO_2, Fe, Zn, Pb, and HCl increased significantly when TDF was added to the fuel, whereas the concentrations of NO_x and several organic compounds decreased. These findings are in agreement with the values measured at the kiln stack exit, but the relative changes are different, because they are influenced also by the direction of the dominant winds and by the presence of obstacles in the flow path (e.g., buildings). It should be noted that Giugliano *et al.* (1999) found in their study of TDF co-firing in a cement kiln that the metal emissions, including those of volatile elements, such as Zn and Pb, were not affected by the addition of TDF to petroleum coke. This somewhat surprising result was attributed to the high particle-removal efficiency of the air pollution control system. On the other hand, Giugliano *et al.* (1999) observed, like Carrasco *et al.* (1998), that addition of TDF led to a considerable reduction in NO_x emissions.

Table 9. *Emissions from combustion of fuel blends in a cement kiln**

	Emissions 100% Coal	Emissions 80% Coal + 20% TDF[†]	Normalized emissions[‡]
	$n = 1$	$n = 1$	
Gas temperature (°C)	292	273	—
Total particulate matter (mg/m^3)	99.2	106.5	1.07
Metals (μg/m^3)			
Al	745.2	1284.0	1.72
Cr	48.0	315.5	6.57
Mn	27.5	87.3	3.17
Fe	873.0	1862.4	2.13
Cu	3.93	11.2	2.85
Zn	185.1	1700.1	9.18
Hg	55.1	37.7	0.68
Pb	120.8	334.8	2.77
Gases (mg/m^3)			
CO	155.4	228.0	1.47
NO_x	1754.4	1483.9	0.84
SO_2	696.8	905.2	1.30
HCl	9.68	15.80	1.63
Organics (μg/m^3)			
Chlorobenzene	1.72	0.88	0.51
PAH	83.5	68.4	0.82
Naphthalene	76.3	68.3	0.90
Dioxins/Furans	0.0010	0.0004	0.40

*Data collected at the kiln stack exit (from Carrasco *et al.* 1998).
[†]Percentage is probably by weight, but was not specified in the original study.
[‡]Normalized emissions = Emissions (80% Coal + 20% TDF)/Emissions (100% Coal).

Mukherjee *et al.* (2003) investigated atmospheric emissions during a test burn in a cement kiln in Finland. They compared the heavy metal emissions resulting from combustion of coal/petcoke with those arising from burning a blend consisting of 10 wt% shredded car tyres and 90 wt% coal/petcoke. The diagrams presented by Mukherjee *et al.* (1993) show that adding car tyres led to a significant increase in the total emissions of V, Cr, Mn, Fe, Ni, Cu, Zn, and Pb. These trends are similar to those noted by Carrasco *et al.* (1998). On the other hand, the emissions of Hg and Tl decreased considerably as a result of adding tyres to the fuel. Mukherjee *et al.* (2003) further observed that this decrease was accompanied by a shift of Hg and Tl from the gaseous phase to particulates and argued that this effect is environmentally beneficial because particulate matter can be controlled more easily. The study of Mukherjee *et al.* (2003), unfortunately, presented mainly diagrams and reported actual emission values for Fe and Hg only; it will therefore not be discussed further (normalized emissions: Fe = 3.12; Hg = 0.11; cf. Carrasco *et al.* 1998 data in Table 9).

Comparing the cement kiln results of Carasco *et al.* (1998) to the Iowa and Purdue stoker boiler data is difficult because of the differences in types of facilities, compounds analysed, composition of TDF and coal, and percentage of TDF used in the fuel blends. Moreover, emission units of mass per time (Table 8) are less suitable for inter-facility comparison than mass per heat input (e.g., kg/MJ; EPA 1997), which were not available. Nevertheless, there are similar emission trends for a number of compounds analysed in all three test burns: the average metal emissions, when normalized to the emissions from pure coal combustion, are on approximately the same level (Fig. 7, Table 9), despite the fact that the cement kiln used a much higher percentage of TDF in the fuel blend. The exceptions to this trend are Zn and Hg. Zinc appears to be emitted at significantly higher levels when coal + TDF mixtures were burned in the two stoker boilers. This result is difficult to interpret because the Zn contents of the coal and TDF used in the actual experiment are only known for the Purdue test burn (coal: 36 ppm; coal + TDF: 183 ppm; TDF: 10 900 ppm; unpublished data). In the two stoker boiler experiments, the Hg emissions are practically unaffected by the addition of TDF to coal (Fig. 7), whereas a considerable reduction is observed in the cement kiln. Similarities between the three test burns are also observed for the gaseous emissions (Fig. 8, Table 9), but the effect of adding TDF to coal is less pronounced for the gases than for most metals. The reduction in CO emissions displayed

for the Purdue experiment appears to be in contrast to the results from the other studies. On the other hand, the Carrasco *et al.* (1998) data show a marked increase in SO₂ emissions, in contrast to both stoker boiler experiments. This discrepancy probably reflects differences in the S content of the fuel. Reduced SO₂ emissions have also been observed by Giugliano *et al.* (1999) for TDF co-combustion in a cement kiln.

Discussion

The relatively large heating value of tyres makes the use of TDF or whole tyres for combustion in utilities, industry, and manufacturing the most promising method for waste tyre management.

The most obvious advantage to using tyres as a fuel source is the partial alleviation of many of the environmental problems associated with traditional disposal of scrap tyres in stockpiles and landfills. The eradication of tyre piles and landfill waste will reduce many of the risks to human health linked to the buildup of tyre waste. The removal of outdoor tyre stockpiles provides great aesthetic benefits as well.

Tyre combustion not only helps eliminate a waste problem, but also conserves the limited amount of fossil fuel supplies. Because the energy extracted from tyres per mass unit is greater than that of most other fuels, tyre combustion can prove to be a financially beneficial alternative to fossil fuel combustion (e.g., McGowin 1991; Lamarre 1995). In some cases, the use of tyre fuel can decrease the cost of energy production by up to 40% (Ohio Air Quality Development Authority 1991). However, several factors may reduce the economic benefits of using this fuel. The proximity of the tyre source to the facility weighs heavily on the economic benefits garnered by the operating party. As the distance increases, the cost effectiveness of combusting tyres for energy decreases (Jones 1998). It is for this reason that the large tyre-to-energy facility in Modesto, California, is located adjacent to the world's largest tyre dump (Makansi 1992). This measure guarantees a continuous supply of scrap tyres, which is an important requirement for tyre-to-energy facilities. Many newer power plants, however, lie at a reasonably large distance from urban centres, and therefore, at a fairly large distance from major waste tyre sources, thus increasing transportation costs.

In addition to collection and transportation costs, there are major expenses associated with the preparation of the tyres for combustion or with modifications to the fuel feeding systems of power plants (e.g., Goddard 1992; Lamarre

1995). The costs associated with shredding tyres into usable chips increases as the size of the chip decreases. Energy production costs further increase when de-wired tyres are used due to the added expense of wire removal, which is commonly achieved via magnetic or gravity separation (Ohio Air Quality Development Authority 1991; EPA 1994). This financial burden is especially prevalent for facilities using cyclone-fired boilers and for paper and pulp mills, as they typically require small, de-wired tyre chips. Therefore, the most cost-effective process for utilizing tyres as fuel is one that uses whole, unaltered tyres. However, dedicated tyre-to-energy facilities combusting whole tyres are costly to build, even when modifying a pre-existing combustion facility. Other cost drawbacks to tyre combustion are those associated with the installation and maintenance of additional air pollution control devices.

Several studies investigating the environmental effects of controlled tyre combustion have been conducted. It is evident that atmospheric emissions can be greatly reduced if proper air pollution control systems are installed. Laboratory and field data provide evidence indicating that concentrations of some environmental pollutants, especially NO_x, may decrease due to tyre combustion, whereas others increase compared to pure coal combustion. Zinc is an example of an element that increases in both solid combustion products and atmospheric emissions. The geochemical impact of higher Zn contents in fly and bottom ash on leaching processes in disposal sites remains to be tested.

Combustion of tyres as alternate fuel, for example in tyre-to-energy facilities, produces ash with particularly high Zn contents. Such ash might represent a valuable resource for Zn recovery. Makansi (1992) reported that the Stirling tyre-to-energy facility in Connecticut had plans to sell the fly ash to metal refineries for extraction of the Zn. It should be noted that the highest amounts of Zn found in pure tyre ash are 60 wt% (Polasek & Jervis 1994). This concentration is similar to the Zn content of pure sphalerite (ZnS), the most important Zn ore. However, even if the Zn contents of tyre ash are lower (Table 6), recovery of Zn from TDF or tyre ash may be more economical than extraction from sphalerite ore because it does not involve mining for sphalerite and because the bulk ash is a fairly homogeneous and fine-grained material. Moreover, the use of tyre ash as a source of Zn would be an environmentally beneficial process for three reasons: (1) it would not create sulphide waste with its associated environmental problems (e.g., Sidenko

et al. 2001); (2) it would help conserve natural resources; and (3) it would take advantage of a waste material rather than discard it. From an environmental standpoint, specially designed tyre-to-waste facilities therefore represent a promising option for dealing with one of the greatest disposal challenges for solid waste and at the same time contribute to a sustainable energy cycle.

The authors would like to express their gratitude to Dr J. Hower (University of Kentucky) and Dr P. Lemieux (US EPA) for their thorough reviews and very helpful suggestions. J. Schwarzhoff and F. Milster are thanked for kindly providing the complete set of original data from the TDF test burns carried out at the University of Iowa power plant. We thank Purdue University for giving us permission to collect samples during the test burn and for financial support. Additional funding was provided by a grant from the Indiana Academy of Science.

References

AMARI, T., THEMELIS, N. J. & WERNICK, I. K. 1999. Resource recovery from used rubber tires. Resources Policy, 25, 179–188.

ATAL, A. & LEVENDIS, Y. A. 1995. Comparison of the combustion behaviour of pulverized waste tyres and coal. Fuel, 74, 1570–1581

BARLAZ, M. A., ELEAZER, W. E. & WHITTLE, D. J. 1993. Potential to use waste tires as supplemental fuel in pulp and paper mill boilers, cement kilns and in road pavement. Waste Management and Research, 11, 463–480.

BIAGINI, E., TOGNOTTI, L., MALLOGGI, S. & PASINI, S. 2002. Co-combustion of coal and tire residue in a pilot plant: a simplified modelling approach for scale-up predictions of char oxidation. Combustion Science and Technology, 174, 129–150.

BLUMENTHAL, M. H. 1993. Tires. In: LUND, H. F. (ed) The McGraw-Hill Recycling Handbook. McGraw-Hill, New York, 18.1–18.64.

BURGER, C. 1991. Illinois scrap tire management study summary. Proceedings: 1991 Conference on Waste Tires as a Utility Fuel. Electric Power Research Institute, Palo Alto, CA. EPRI GS-7538, 3/1–3/10.

BUWAL 2001. Information zur Entsorgung von Altreifen. Swiss Agency for the Environment, Forests and Landscape. World Wide Web Address: http://www.buwal.ch/abfall/docu/pdf/altpneu_d.pdf. (in German)

CARRASCO, F., BREDIN, N., GNINGUE, Y. & HEITZ, M. 1998. Environmental impact of the energy recovery of scrap tires in a cement kiln. Environmental Technology, 19, 461–474.

CIWMB 1992. Tires as a Fuel Supplement: Feasibility Study. Report to the Legislature, State of California, California Integrated Waste Management Board.

COURTEMANCHE, B. & LEVENDIS, Y. A. 1998. A laboratory study on the NO, NO_2, SO_2, CO and CO_2

emissions from the combustion of pulverized coal, municipal waste plastics and tires. *Fuel*, **77**, 183–196.

DONAHOE, R. J. 2004. Secondary mineral formation in coal combustion byproduct disposal facilities: implications for trace element sequestration. *In*: GIERÉ, R. & STILLE, P. (eds) *Energy, Waste, and the Environment: a Geochemical Perspective.* Geological Society, London, Special Publications, **236**, 641–658.

EPA 1994. *Handbook, Recycling and Reuse of Material Found on Superfund Sites.* US Environmental Protection Agency, EPA-625/R-94-004.

EPA 1997. *Air Emissions from Scrap Tire Combustion.* US Environmental Protection Agency, EPA-600/R-97-115.

EPA 1999. *State Scrap Tire Programs. A Quick Reference Guide: 1999 Update.* US Environmental Protection Agency, EPA-530/B-99-002.

EPA 2003. *Municipal Solid Waste: Tires.* US Environmental Protection Agency. World Wide Web Address: www.epa.gov/epaoswer/non-hw/muncpl/tires.htm.

EPRI 1991. *Proceedings: 1991 Conference on Waste Tires as a Utility Fuel.* Electric Power Research Institute, Palo Alto, CA, EPRI GS-7538.

GAGLIA, N., LUNDQVIST, R., BENFIELD, R. & FAIR, J. 1991. Design of a 470,000 lb/hr coal/tire-fired circulating fluidized bed boiler for united development group. *Proceedings: 1991 Conference on Waste Tires as a Utility Fuel.* Electric Power Research Institute, Palo Alto, CA, EPRI GS-7538, 13/1–13/20.

GIUGLIANO, M., CERNUSCHI, S., GHEZZI, U. & GROSSO, M. 1999. Experimental evaluation of waste tires utilization in cement kilns. *Journal of the Air and Waste Management Association*, **49**, 1405–1414.

GODDARD, H. C. 1992. Incentives for solving the scrap tire problem through existing markets. *Journal of Hazardous Materials*, **29**, 165–177.

GRANGER, J. E. & CLARK, G. A. 1991. Fuel characterization of coal/shredded tire blends. *Proceedings: 1991 Conference on Waste Tires as a Utility Fuel.* Electric Power Research Institute, Palo Alto, CA, EPRI GS-7538, 5/1–5/14.

HORVATH, M. 1991. Results of the Ohio Edison whole-tire burn test. *Proceedings: 1991 Conference on Waste Tires as a Utility Fuel.* Electric Power Research Institute, Palo Alto, CA, EPRI GS-7538, 6/1–6/19.

HOWE, W. C. 1991. Fluidized bed combustion experience with waste tires and other alternate fuels. *Proceedings: 1991 Conference on Waste Tires as a Utility Fuel.* Electric Power Research Institute, Palo Alto, CA, EPRI GS-7538, 10/1–10/20.

HOWER, J. C., ROBERTSON, J. D. & ROBERTS, J. M. 2001. Petrology and minor element chemistry of combustion by-products from the co-combustion of coal, tire-derived fuel, and petroleum coke at a western Kentucky cyclone-fired unit. *Fuel Processing Technology*, **74**, 125–142.

HUTCHINSON, W., EIRSCHELE, G. & NEWELL, R. 1991. Experience with tire-derived fuel in a cyclone-fired

utility boiler. *Proceedings: 1991 Conference on Waste Tires as a Utility Fuel.* Electric Power Research Institute, Palo Alto, CA, EPRI GS-7538, 8/1–8/8.

JANG, J.-W., YOO, T.-S., OH, J.-H. & IWASAKI, I. 1998. Discarded tire practices in the United States, Japan and Korea. *Resources, Conservation and Recycling*, **22**, 1–14.

JATMA 2002. *Tire Industry of Japan 2002.* The Japan Tyre Manufacturers Association Inc. World Wide Web Address: www.jatma.or.jp.

JOHNSON, S. A., GIERMAK, E. A. & KHANA, R. D. 1997. Advantages of repowering with solid-fuel combustion technologies. *American Power Conference*, **50**, 1192–1199.

JONES, C. 1998. Opportunity fuels: can they go the distance? *Power*, **142**, 25–26.

JONES, R. M., KENNEDY, J. M. & HEBERER, N. L. 1990. Supplementary firing of tire-derived fuel (TDF) in a combination fuel boiler. *Tappi Journal*, **73**, 107–113.

KARELL, M. A. & BLUMENTHAL, M. H. 2001. Air regulatory impacts of the use of tire-derived fuel. *Environmental Progress*, **20**, 80–86.

LAMARRE, L. 1995. Tapping the tire pile. *Electric Power Research Institute Journal*, **20/5**, 28–34.

LEE, M. L., NOVOTNY, M. & BARTLE, K. D. 1981. *Analytical Chemistry of Polycyclic Aromatic Compounds.* Academic Press, New York.

LEMIEUX, P. M. 1994. *Pilot-Scale Evaluation of the Potential for Emissions of Hazardous Air Pollutants from Combustion of Tire-Derived Fuel.* US Environmental Protection Agency, EPA-600/R-94-070.

LEMIEUX, P. M. & RYAN, J. V. 1993. Characterization of air pollutants emitted from a simulated scrap tire fire. *Journal of the Air and Waste Management Association*, **43**, 1106–1115.

LEVENDIS, Y. A., ATAL, A., CARLSON, J. B., DUNAYEVSKIY, Y. & VOUROS, P. 1996. Comparative study on the combustion and emissions of waste tire crumb and pulverized coal. *Environmental Science and Technology*, **30**, 2742–2754.

LEVENDIS, Y. A., ATAL, A. & CARLSON, J. B. 1998a. On the correlation of CO and PAH emissions from the combustion of pulverized coal and waste tires. *Environmental Science and Technology*, **32**, 3767–3777.

LEVENDIS, Y. A., ATAL, A., COURTEMANCHE, B. & CARLSON, J. B. 1998b. Burning characteristics and gaseous/solid emissions of blends of pulverized coal with waste tire-derived fuel. *Combustion Science and Technology*, **131**, 147–185.

MAKANSI, J. 1992. Tires-to-energy plant takes highroad in managing discharges. *Power*, **April 1992**, 152r.

MASTRAL, A. M., ÁLVAREZ, R., CALLÉN, M. S., CLEMENTE, C. & MURILLO, R. 1999a. Characterization of chars from coal-tire copyrolysis. *Industrial and Engineering Chemistry Research*, **38**, 2856–2860.

MASTRAL, A. M., CALLÉN, M. S., MURILLO, R. & GARCÍA, T. 1999b. Combustion of high calorific value waste material: organic atmospheric pollution. *Environmental Science and Technology*, **33**, 4155–4158.

MASTRAL, A. M., CALLÉN, M. S. & GARCÍA, T. 2000. Fluidized bed combustion (FBC) of fossil and non-fossil fuels. A comparative study. *Energy and Fuels*, **14**, 275–281.

MCGOWIN, C. R. 1991. Alternate fuel cofiring with coal in utility boilers. *Proceedings: 1991 Conference on Waste Tires as a Utility Fuel*. Electric Power Research Institute, Palo Alto, CA, EPRI GS-7538, 1/1–1/9.

MILLER, C. A., LEMIEUX, P. M. & TOUATI, A. 1998. Evaluation of tire-derived fuel for use in nitrogen oxide reduction by reburning. *Journal of the Air and Waste Management Association*, **48**, 729–735.

MUKHERJEE, A. B., KÄÄNTEE, U. & ZEVENHOVEN, R. 2003. The effects of switching from coal to alternative fuels on heavy metals emissions from cement manufacturing. *In*: SAJWAN, K. S., ALVA, A. K. & KEEFER, R. F. (eds) *Chemistry of Trace Elements in Fly Ash*. Kluwer Academic/Plenum Publishers, New York, 45–61.

NARRA 2003. *Tire Recycling in Europe*. North American Recycled Rubber Association, Whidby, Ontario. World Wide Web Address: www.recycle.net/recycle/assn/narra/europe.html

OHIO AIR QUALITY DEVELOPMENT AUTHORITY 1991. *Air Emissions Associated with the Combustion of Scrap Tires for Energy Recovery*. Malcolm Pirnie Inc., Columbus, OH.

PHALEN, J., LIBAL, A. S. & TAYLOR, T. E. 1991. Manitowoc coal/tire chip-cofired circulating fluidized bed combustion project. *Proceedings: 1991 Conference on Waste Tires as a Utility Fuel*. Electric Power Research Institute, Palo Alto, CA, EPRI GS-7538, 12/1–12/14.

POLASEK, M. & JERVIS, R. E. 1994. Elements in car and truck tires and their volatilization upon incineration. *Journal of Radioanalytical and Nuclear Chemistry, Articles*, **179**, 205–209.

POPE, K. M. 1991. Tires to energy in a fluidized bed combustion system. *Proceedings: 1991 Conference on Waste Tires as a Utility Fuel*. Electric Power Research Institute, Palo Alto, CA, EPRI GS-7538, 11/1–11/9.

RMA 2002. *U.S. Scrap Tire Markets 2001*. Rubber Manufacturers Association, Washington, DC. World Wide Web Address: www.rma.org.

SIDENKO, N. V., GIERÉ, R., BORTNIKOVA, S. B., PAL'CHIK, N. A. & COTTARD, F. 2001. Mobility of heavy metals in self-burning waste heaps of the zinc smelting plant in Belovo (Kemerovo Region, Russia). *Journal of Geochemical Exploration*, **74**, 109–125.

SMOLDERS, E. & DEGRYSE, F. 2003. Fate and effect of zinc from tire debris in soil. *Environmental Science and Technology*, **36**, 3706–3710.

SPEARS, D. A. & LEE, S. 2004. Geochemistry of leachates from coal ash. *In*: GIERÉ, R. & STILLE, P. (eds) *Energy, Waste, and the Environment: a Geochemical Perspective*. Geological Society, London, Special Publications, **236**, 619–639.

TESLA, M. R. 1994. Scrap tire process turns waste into fuel. *Power Engineering*, **98**, 43–44.

TISHMACK, J. K. & BURNS, P. E. 2004. The chemistry and mineralogy of coal and coal combustion products. *In*: GIERÉ, R. & STILLE, P. (eds) *Energy, Waste, and the Environment: a Geochemical Perspective*. Geological Society, London, Special Publications, **236**, 223–246.

TRIB 2003. Tire Retread Information Bureau, Pacific Grove, CA. World Wide Web Address: www.retread.org.

WILLIAMS, P. T. & TAYLOR, D. T. 1993. Aromatization of tyre pyrolysis oil to yield polycyclic aromatic hydrocarbons. *Fuel*, **72**, 1469–1474.

WONG, A. S. & ROBERTSON, J. D. 1993. Multi-elemental analysis of coal and its by-products by simultaneous proton-induced gamma-ray/X-ray emission analysis. *Journal of Coal Quality*, **12**, 146–150.

Heat pumps as a tool for energy recovery from mining wastes

DAVID BANKS[1], H. SKARPHAGEN[2], R. WILTSHIRE[3], & C. JESSOP[3]

[1]*Holymoor Consultancy, Chesterfield, Derbyshire, UK, (e-mail: david@holymoor.co.uk)*
[2]*Båsum Boring AS, Nydalen, Oslo, Norway*
[3]*Building Research Establishment Ltd (BRE), Garston, Watford, UK*

Abstract: Heat pumps extract heat energy from a low-temperature source and transfer it to a higher temperature sink, usually via a closed loop of volatile 'refrigerant' fluid in a compression/expansion cycle. They can be efficiently used for space heating (and cooling), extracting heat from seawater, rivers, lakes, groundwater, rocks, sewage, or mine water. Electrical energy powers the heat pump's compressor. The ratio of total heat output to electrical energy input, called the coefficient of performance, typically ranges from 3.0 to 6.0. The use of mine water for space heating or cooling purposes has been demonstrated to be feasible and economic in applications in Scotland, Canada, Norway, and the USA. Mine water is an attractive energy resource due to: (1) the high water storage and water flux in mine workings, representing a huge renewable enthalpy reservoir; (2) the possibility of re-branding a potentially polluting environmental liability as a 'green' energy resource; and (3) the development of many mine sites as commercial/industrial parks with large space heating/cooling requirements. The exothermic nature of the pyrite oxidation reaction (>1000 kJ/mol) implies added benefits if closed-loop systems can harness the chemical energy released in mine-waste tips. An appreciation of geochemistry also assists in identifying and solving possible problems with precipitation reactions occurring in heat pump systems.

International conventions on emission of greenhouse gases have attempted to commit individual nations to decreased use of fossil fuels or, at the very least, decreased net emissions of carbon dioxide. It is recognized that no single new technology is likely to provide a complete solution to such issues and that a combination of policies is required, including: the development of renewable or non-fossil energy resources (biomass, wind, wave, solar, nuclear); increased efficiency of energy consumption; and alternative disposal routes for CO_2, other than mere atmospheric emission.

However, such courses of action are often significantly more expensive than continued use of fossil fuel. They have their own adverse environmental or aesthetic impacts (e.g., waste from nuclear power, visual impact of wind turbines). This paper discusses the use of heat pumps as a technology for efficiently transferring environmental heat (from the air, from water or from the ground) to the interior of buildings for purposes of space- or water-heating. In total, around 100 million heat pumps are installed world wide, with an annual capacity of 1300 TWh energy. They are estimated to result in a global saving in annual CO_2 emissions of 0.13 Gt, in the context of a global total CO_2 emission of 22 Gt (Bouma 2002). Specifically, this paper discusses the exploitation of *ground source heat* via heat pumps, and is largely based on a paper presented at a conference in Newcastle by Banks *et al.* (2003). This established and demonstrated technology has altered the pattern of energy usage in certain nations, such as Sweden, who have dared to invest in it. Ground source heat pumps essentially use electrical energy very efficiently to transfer renewable energy stored in rocks, water, or geogenic wastes to the user. This transferred energy may ultimately be:

- geothermal (derived from the Earth's natural geothermal gradient);
- solar (derived by solar heat warming up near-surface rocks and groundwater);
- chemical (derived from, for example, sulphide oxidation).

A recent Canadian assessment (Caneta Research 1999) concluded that, 'There is unlikely to be a potentially larger mitigating effect on greenhouse gas emissions and the resulting global warming impact of buildings from any other current, market-available single technology, than from ground source heat pumps.'

From: GIERÉ, R. & STILLE, P. (eds) 2004. *Energy, Waste, and the Environment: a Geochemical Perspective*. Geological Society, London, Special Publications, **236**, 499–513.
0305-8719/04/$15   © The Geological Society of London 2004.

Ground source heat

Groundwater has a huge capacity to store heat (4181 J/kg/K). Rocks and minerals have a lesser (around 800 J/kg/K; Mellon 2001), but still significant, heat capacity. They also have a certain thermal conductivity and these properties allow them to act as enormous subsurface heat storage and exchange reservoirs. Table 1 provides some examples of specific heat capacities and thermal conductivities. This heat, stored in the geological environment, can be extracted, manipulated and utilized.

At shallow depths, rocks and groundwater have an approximately constant temperature that roughly corresponds to the annual average air temperature. In Scandinavia, this may be as little as 4–7 °C, while in the southern UK temperatures of 10–12 °C are more typical. In the immediate subsurface (down to a few metres depth), some seasonal fluctuation will occur, with heat from the sun tending to warm the subsoil in the summer. The large heat capacity

of the geological environment means that any seasonal variations are highly damped and that the subsoil is generally warmer than the air in winter. Conversely, it is generally cooler than the air in summer. At shallow depths, the stored heat energy is thus largely solar in origin. The Earth's surface acts, in effect, as an enormous solar collector.

At greater depth, stored heat energy in rocks and groundwater begins to have a larger geothermal component. The geothermal gradient varies from location to location at the Earth's surface and also varies with rock type (and thermal conductivity). Higher gradients are observed in areas of volcanic activity and crustal thinning. In general, however, gradients of 1–3 °C/100 m are common, reflecting a typical European geothermal heat flux of around 0.04–0.1 W/m². Thus, at the base of a typical British 100 m deep borehole, temperatures of around 13–14 °C may be expected.

Heat pumps may be used to extract some of this stored heat. The use of this 'ground source heat' was first proposed in 1912 in Switzerland, while

Table 1. *The thermal conductivity and specific volumetric heat capacity of selected rocks and minerals**

	Thermal conductivity (W/m/K)	Specific heat capacity (MJ/m³/K)
Rocks		
Coal	*0.3*	*1.8*
Limestone	1.5–3.0 (*2.8, massive limestone*)	*2.3*
Shale	1.5–3.5 (*2.1*)	*2.3*
Wet clay	*1.6*	*2.4*
Basalt	*1.7*	*2.4*
Diorite	1.7–3.0 (*2.6*)	*2.9*
Sandstone	2.0–6.5 (*2.3*)	*2.0*
Gneiss	2.5–4.5 (*2.9*)	*2.1*
Arkose	*2.9*	*2.0*
Granite	3.0–4.0 (*3.4*)	*2.4*
Quartzite	5.5–7.5 (*6.0*)	*2.1*
Minerals		
Plagioclase	1.5–2.3	—
Mica	2.0–2.3	—
K-feldspar	2.5	—
Olivine	3.1–5.1	—
Quartz	7.7	—
Other		
Air	0.024	1.29×10^{-3} at 1 atm
Glass	0.8	—
Concrete	0.8 (*1.6*)	*1.8*
Ice	1.7 (*2.2*)	*1.9*
Water	*0.6*	4.18
Copper	390	3.5

*Data from inter alia Sundberg (1991), Banks & Robins (2002), Halliday & Resnick (1978). Italics show recommended values cited by Eskilson *et al.* (2000). Note that thermal conductivity increases with quartz content, and that most rock materials have a specific heat capacity of around 800 J/kg/K, or slightly over 2 MJ/m³/K (compare with copper at only 386 J/kg/K).

the first applied research on heat pumps was carried out in the 1940s in the UK and USA (Rawlings & Sykulski 1999). This led, in 1948, to the installation in the UK of 12 prototype heat pumps with ground collectors, 9 kW output and a coefficient of performance (see below) as high as 3. It was not before the oil crisis of the 1970s, however, that the commercial use of ground source heat pumps (GSHPs) became widespread. Sweden, already with around 1000 GSHPs by the end of the 1970s, paved the way for much of the recent development of GSHPs, along with the USA, Austria, Switzerland, and Germany (partly due to government subsidies to assist in introducing and demonstrating the technology). In Sweden, for example, some 50 000 ground source heat plants were installed between 1980 and 1986, yielding about 500 MW of heat (Albu et al. 1997). Currently (Bouma 2002), over 90% of new Swedish houses are built with installed heat pumps, and 75% of such residential heat pumps are ground-sourced. Swedish heat pumps generate an estimated 7 TWh/y to district heating and a further 6 TWh/y to residential heating. In the USA, some 28 000 new GSHP units were installed annually in 1994, but by 1999 this total had risen to 50 000. In Canada, around 30 000 GSHPs were being installed annually throughout the 1990s, with 20% of this going to the commercial and institutional (e.g., school) sector (Bouma 2002). In the UK, although air-sourced (i.e., withdrawing heat from ambient air temperature) heat pumps are widely used (some 600 000 installed units), GSHP technology is almost unknown. Rawlings & Sykulski (1999) could find a mere ten examples in the UK, although the concept is rapidly gaining momentum, to the point where 150 new GSHP units were commissioned in 2001 (Table 2). Currently,

around 500 000 GSHPs are believed to be installed world-wide, providing an estimated 6.7 GW energy, of which 4.8 GW are installed in the USA (Bouma 2002). These figures compare favourably with other 'alternative' energy sources. For example, the total-world installed wind energy capacity was 7.2 GW in 2002, whereas solar energy was <1 GW (BP 2003). However, these values are still of minor significance compared to the 'main players': consumption of coal accounted for 3190 GW, natural gas for 3040 GW, and oil for 4690 GW of power equivalent on average in 2002, with hydroelectric power weighing in at 790 GW, and nuclear power at 813 GW (BP 2003).

Heat pumps

A heat pump is an electromechanical device that takes heat from a low-temperature medium (e.g., groundwater at, say, 7 °C in Norway) and transfers it to a high-temperature space-heating medium via a compression–expansion cycle. The space-heating medium may be air at, say, 25 °C (common in the USA) or the hot water of a household central heating system at 50–60 °C (common in Europe). The electricity consumed by the heat pump is used to 'push' the heat 'up' the temperature gradient (by powering the compressor). Many people find the concept of extracting heat from a cool medium intuitively problematic, yet the same people are comfortable enough with the concept of a domestic refrigerator. The refrigerator is, in fact, a heat pump: heat is extracted from a cool medium (the cool-box interior) and transferred to a warmer one (kitchen air). There are a number of different heat pump concepts (Fig. 1):

Open-loop systems. Groundwater is pumped from a borehole (or mine shaft) and circulated directly through the heat pump, which extracts heat directly from the water. This method is obviously appropriate where a significant water yield, with a suitable quality, can be achieved and maintained.

Closed-loop systems. In closed-loop systems, groundwater abstraction is not required. Such systems can be used in poorly permeable formations, such as hard rocks or clays. Closed-loop systems may be *direct circulation* where the heat pump refrigerant is pumped through a closed loop installed in the ground. More frequently they utilize *indirect* circulation. Here, water, usually with added anti-freeze (e.g., ethylene glycol or ethanol), is circulated through a closed polyethene hose system (the *collector*

Table 2. *New ground source heat pumps sold in 2001, and annual growth rate in market, for selected European countries**

Country	Annual ground source heat pump sales (units/y)	Market growth rate for GSHPs (%)
UK	150	>100
Czech Republic	350	25
Poland	500	5
Norway	650	10
Switzerland	2800	6
Sweden	27000	6
Europe (total)	41000	

*After Bouma 2002.

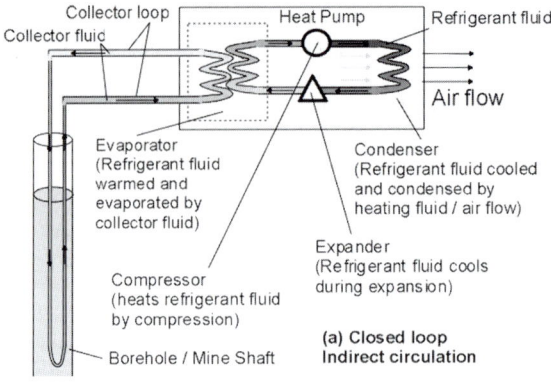

Collector loop
Collector fluid
Heat Pump
Refrigerant fluid
Air flow

Evaporator
(Refrigerant fluid
warmed and
evaporated by
collector fluid)

Condenser
(Refrigerant fluid cooled
and condensed by
heating fluid / air flow)

Expander
(Refrigerant fluid cools
during expansion)

Compressor
(heats refrigerant fluid
by compression)

Borehole / Mine Shaft

**(a) Closed loop
Indirect circulation**

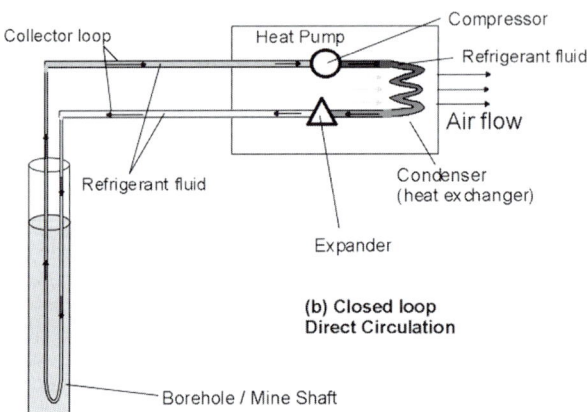

Collector loop
Compressor
Heat Pump
Refrigerant fluid
Air flow

Refrigerant fluid

Condenser
(heat exchanger)

Expander

**(b) Closed loop
Direct Circulation**

Borehole / Mine Shaft

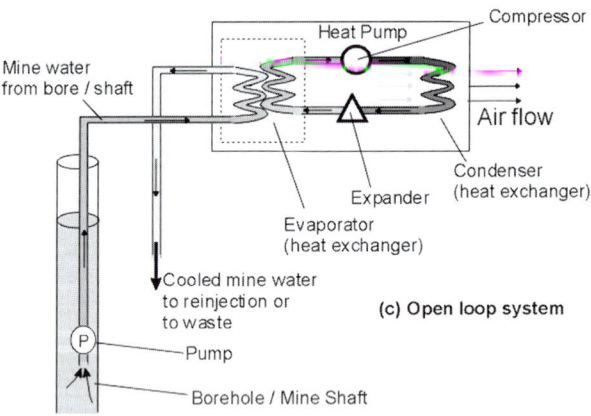

Compressor
Heat Pump
Mine water
from bore / shaft
Air flow

Condenser
(heat exchanger)

Expander

Evaporator
(heat exchanger)

Cooled mine water
to reinjection or
to waste

(c) Open loop system

P

Pump

Borehole / Mine Shaft

Fig. 1. Schematic diagrams of (**a**) a vertically installed, closed-loop, indirect circulation water-to-air heat pump. The collector fluid may be a solution based on glycol or ethanol mixture; the refrigerant may be hydrofluorocarbon (although use of many compounds may be limited by environmental legislation), hydrocarbon or ammonia-based (modified after Banks & Robins 2002). This type of system is well-suited to small-scale installations, installations in poor aquifers or aquitards (e.g., crystalline rocks, clays), or where water is of incompatible quality with a heat exchanger; (**b**) a closed-loop, direct circulation water-to-air heat pump. Here the refrigerant is circulated directly downhole; (**c**) an open-loop heat pump system, where the groundwater or mine water is passed directly through the heat pump. This is ideal for large-scale applications based on water from a good aquifer or mine system, where water quality is favourable. The water, once used, is either pumped to a suitable waste recipient, or re-injected to the aquifer or mine system. It may alternatively be used for 'grey water' applications, such as toilet flushing, if the water quality is suitable.

loop) in the ground. This may be installed *vertically* (down a borehole or shaft) or *horizontally* (e.g., below a grassed lawn, yard, or in a mine gallery). The fluid is warmed to the subsurface temperature and, on its return to the surface, may be sent through a heat pump.

The ratio of the total heat output of the heat pump to the quantity of electric energy input to the heat pump is called the *heat effect* or *co-efficient of performance* (COP). This value depends on the temperature gradient, but is typically around 3–4 for an output temperature of 30 °C (and can be higher). Canada specifies a minimum COP of 3.0 for all ground source heat pumps rated below 35 kW (Bouma 2002).

As GSHPs are essentially transferring solar or geothermal energy, they can be regarded as environmentally sustainable, provided that the flux of energy extracted does not exceed the replenishment of energy in the ground from solar input or geothermal gradient. The only environmental drawback to heat pumps is that they require a certain electrical energy input to extract the ground source heat. However, the net energy benefit is huge, and savings in energy consumption in conventional heating outweigh the electricity consumed by heat pumps. Use of GSHPs powered by electricity generated by wind turbines can, assuming typical COP values, deliver heat to three to four times as many houses than if the heat power was generated by wind turbines alone.

Mines and mine wastes as sources for ground heat

It is possible to base heat pump systems on a variety of environmental media: seawater, unfrozen rivers, lakes, groundwater, rocks, or sewage. The one proviso for the efficient use of heat pumps is that the point of use must be relatively close to the source.

In ore- and coalfields, abandoned mines may be a particularly attractive energy resource, for several reasons. First, flooded, abandoned mines contain huge volumes of water at a relatively constant temperature. In other words, they are major reservoirs of stored enthalpy, which can be seasonally manipulated as sources or sinks for heat energy. Secondly, a mine system contains extensive areas (tunnels, working faces, mine waste, goaf) where flowing groundwater is in contact with rocks. The mine is thus a major heat-exchanger between a storage medium (rocks and minerals) and a transport medium (flowing groundwater). Thirdly, where flooded mines overflow at the surface, gravity drainage from mines may yield considerable quantities of

water at a constant temperature that can be used in heat pumps with minimal consumption of energy for pumping. Fourthly, even where gravity drainage does not exist, boreholes can be drilled into old mine workings. Open- or closed-loop systems can then be installed in these boreholes or existing shafts. The large quantities of water travelling through disused workings ensure that the available energy is renewed by convection as well as by thermal conduction. The size and complexity of some mine workings mean that different parts of a mine system may have differing thermal properties and can be used for extracting heat and for cooling purposes, respectively. Fifthly, in deep mines, there may be a substantial geothermal component to the temperature, as well as the solar component, adding to the efficiency of the system. Finally, and crucially, many mining areas in Western Europe and North America are gradually being abandoned, and are often the recipients of grants to stimulate new 'brownfield' development and industry. New commercial and industrial parks, with demands for environmentally friendly and economically efficient space heating and cooling are thus currently being planned or constructed in the vicinity of flooded mines in these areas. The developments may even occur at the mine head sites themselves, where mine-waste tips may be located or where existing shafts may facilitate access for heat pump systems to mine water reservoirs.

Groundwater or mine water may not only be used for heating, but also to cool offices, communal spaces, or areas with a high density of heat-generating equipment (e.g., computers, telephone exchanges). This is often most efficiently done by direct 'free cooling'. Here, heat migrates from a building interior to circulating groundwater via a system of heat exchangers with high surface area. Use of chilled beams or panels can result in the establishment of convection cells in a room. The 'direction' of heat pumps can also be reversed, leading to 'active chilling' of a space, such that heat is transferred from a building interior to an environmental recipient (air, water, rock). 'Waste' heat from space-cooling applications may be re-used to heat other parts of a building complex. Alternatively, it may be re-injected to the groundwater body (Morgan 1997), possibly to be re-extracted and be used later (e.g., in winter).

Geochemistry and mine water heat pumps

Geochemistry is relevant to the concept of using heat pumps in mine systems from three perspectives: (1) the generation of contaminant fluxes by sulphide oxidation in mine systems, which may require treatment – recovery of energy can assist in offsetting the costs of treatment; (2) exothermic sulphide oxidation reactions can boost the efficiency of heat extraction systems; and (3) mine water may have a tendency to precipitate oxyhydroxide (or other chemical) deposits, which will require special consideration to avoid fouling of circulation pipes and exchangers.

Mine water: hydrochemical liability or environmental asset?

Many metalliferous ore deposits and most coal-fields are characterized by the presence of sulphide minerals, such as pyrite (FeS_2), galena (PbS), or sphalerite (ZnS). When exposed to water and oxygen, these sulphides have a tendency to oxidize, releasing dissolved metals, sulphate and, in the case of pyrite, acid (equations 1 and 2).

$$2\,FeS_2 + 7\,O_2 + 2\,H_2O$$
$$= 2\,Fe^{2+} + 4\,SO_4^{2-} + 4\,H^+ \quad (1)$$
$$ZnS + 2\,O_2 = Zn^{2+} + SO_4^{2-} \quad (2)$$

Table 3 illustrates the composition of four mine waters. It will be noted that sulphide ore mines tend to generate more aggressive mine water than coal mines, and that recently flooded mines tend to generate poorer water qualities than long-

abandoned mines (Banks *et al.* 1997*b*). Of the mines detailed in Table 3, the Dunston and Morlais mine waters are being or planned to be treated by passive settlement/wetland systems (Banks & Banks 2001; Younger, 2004). Kongens mine, Røros, is being assessed for possible treatment using electrolysis, active treatment and/or anaerobic cells (Banks *et al.* 1997*b*; Iversen & Knudsen 1997; Ettner 2002). The San José mine (Banks *et al.* 2002) area is the subject of intense planning for remediation of mine water and mine wastes: a favoured option for mine water treatment involves evaporation.

Further oxidation of ferrous iron (Fe^{2+}) to ferric (Fe^{3+}) may take place, followed by hydrolysis and precipitation of Fe^{3+}-oxyhydroxides:

$$2\,H^+ + 2\,Fe^{2+} + \frac{1}{2}O_2 = 2\,Fe^{3+} + H_2O \quad (3)$$

$$2\,Fe^{3+} + 6\,H_2O = 2\,Fe(OH)_3 + 6\,H^+ \quad (4)$$

Because of the elevated concentrations of potentially ecotoxic metals and acid in mine water, and because of the potential for precipitation of benthos-smothering Fe, Al, Mn-oxyhydroxides (and other phases) in downstream recipient water-courses, water draining from mines is often regarded as a major environmental problem. Considerable effort and expense are incurred by treating

Table 3. *Composition of four selected mine waters to illustrate possible range in composition*

	Mine			
	Dunston (Fender)	Morlais	Kongens Gruve	San José
Location	Chesterfield	nr. Llangennech	Røros	Oruro
Region	Derbyshire	South Wales		Altiplano
Country	UK	UK	Norway	Bolivia
Type of mine	Long-abandoned coal mine; overflowing shaft	Coal mine, flooded by 1986; overflowing shaft	Cu–Zn sulphide mine; overflowing via adit	Pumped Ag–Sn mine
Source	Banks *et al.* (1997*a*) Banks & Banks (2001)	Unpublished Coal Authority data	Iversen & Knudsen (1997)	Banks *et al.* (2002)
Flow rate (L/s)	*c.* 20	*c.* 100–200	5.8 (average)	8
Temperature (°C)	9.4	14.2	n/a	20.8
pH	6.3	6.9	2.7	1.47
Alkalinity (meq/L)	3.74	6.07	0	0
Cl^- (mg/L)	26	25	n/a	32670
SO_4^{2-} (mg/L)	210	455	901	8477
Ca^{2+} (mg/L)	64.5	91.8	47.8	1780
Na^+ (mg/L)	51.4	155	n/a	17256
Fe (mg/L)	10.6	26.6	134	2460
Al (mg/L)	<0.045	<0.01	33.1	559
Mn (mg/L)	1.26	0.93	n/a	27.4
Zn (mg/L)	<0.007	<0.002	36.3	79.4

such waters, either by active or passive methods (see Younger *et al.* 2002; Younger, 2004).

However, some of these mine waters have significant potential for energy recovery via use of heat pumps. For example, the Morlais mine water of South Wales has an estimated discharge of at least 100 L/s. The specific heat capacity of water is around 4181 J/L/°C or 1.16 kWh/m^3/°C. If the Morlais mine water's temperature could be lowered by using a heat pump by 5 °C (from 14 to 9 °C), a heat flux (power) of:

$$100 \, \text{L/s} \times 4.2 \, \text{kJ/L/}^{\circ}\text{C} \times 5 \, ^{\circ}\text{C} = 2100 \, \text{kW}$$

could be removed. Assuming a COP of 3.5, the energy input to drive an (admittedly rather large) heat pump array would be around 840 kW, yielding a total effect of 2.94 MW for heating purposes. A similar calculation for the 20 L/s Dunston mine water would yield a total heat effect of 470 kW (Banks *et al.* 2003). Clearly, these are theoretical calculations: one would need a heat pump array of adequate capacity and efficiency to harness this energy and, not least, one would need nearby consumers to consume the heat energy produced. Abandoned mine sites are, however, extremely attractive areas for new commercial and industrial developments. The effective harnessing of energy from mine water, either from naturally overflowing mines or pumped shafts/boreholes, can turn an environmental liability (mine water) into an environmental asset (a clean energy source). Provision of energy can also help in recouping the costs of mine water treatment.

Harnessing exothermic reactions

Sulphide oxidation is typically a highly exothermic reaction, and is typically microbiologically catalysed by bacteria such as *Thiobacillus ferrooxidans*. Norwegian miners were familiar with the phenomenon of the *kisbrann* or spontaneous combustion of sulphide minerals. The continued exothermic oxidation of pyrite minerals is evidenced by the excessive down-mine temperatures in, for example, the Norwegian Killingdal Cu–Zn–Fe–S mine (Banks 1994). At the Richmond Mine, Iron Mountain, California (Banfield & Gihring 2003), in-mine temperatures of 35–50 °C are reached. Within mine-waste tips and tailings deposits of sulphide mines, interior temperatures in excess of several tens of degrees have been recorded (e.g., over 50 °C at San José mine in Oruro; Banks *et al.* 2002), with values of up to 70 °C in extreme cases (Wels *et al.* 2003). Even at Killingdal,

where the annual average air temperature is around 0 °C, exothermic reactions maintain the interior temperature of mine-waste tips at around 10 °C or above (Iversen 1998). Spontaneous combustion (often initiated by pyrite oxidation (Stevens 1869), and maintained by combustion of coal) is also a recognized phenomenon in coal mines, coal stores, and coal spoil tips (Younger, 2004). Sulphide oxidation is also known to initiate the combustion of oil shales on the UK's Dorset Coast (West 2001).

An indication of the degree of exothermicity of sulphide oxidation reactions can be gained by comparing the enthalpy of formation (ΔH_f°), that is, a measure of the energy locked up in each chemical species, relative to native elements. The difference in enthalpies of formation of all reactants and all products defines the enthalpy (heat released or absorbed) of the reaction. Thermodynamic data on sulphide minerals, such as pyrite, are notoriously varied and disputed, and the values in Table 4 must be treated with caution. Nevertheless, depending on whether one defines the reaction as ending in an aqueous solution (equation 5), an intermediate secondary sulphate (e.g., melanterite – equation 6) or in complete oxidation to an oxyhydroxide (equation 7), the calculated reaction enthalpy (ΔH_r°) released is of the order of at least 1000 kJ/mol.

$$\text{FeS}_2 + \frac{7}{2}\text{O}_2 + \text{H}_2\text{O} = \text{Fe}^{2+} + 2\,\text{SO}_4^{2-} + 2\,\text{H}^+ \quad (5)$$

$$\Delta H_{f(\text{Reactants})}^{\circ} = -178 - 286 = -464 \, \text{kJ/mol}$$
$$\Delta H_{f(\text{Products})}^{\circ} = -90 - 1818 = -1908 \, \text{kJ/mol}$$
$$\Delta H_r^{\circ} - -1444 \, \text{kJ/mol}$$

$$\text{FeS}_2 + \frac{7}{2}\text{O}_2 + 8\text{H}_2\text{O} = \text{FeSO}_4 \cdot 7\,\text{H}_2\text{O}$$
$$+ \text{SO}_4^{2-} + 2\,\text{H}^+ \quad (6)$$

$$\Delta H_{f(\text{Reactants})}^{\circ} = -178 - 2288 = -2466 \, \text{kJ/mol}$$
$$\Delta H_{f(\text{Products})}^{\circ} = -3013 - 909 = -3922 \, \text{kJ/mol}$$
$$\Delta H_r^{\circ} = -1456 \, \text{kJ/mol}$$

$$\text{FeS}_2 + \frac{15}{4}\text{O}_2 + \frac{7}{2}\text{H}_2\text{O} = \text{Fe(OH)}_3$$
$$+ 2\,\text{SO}_4^{2-} + 4\,\text{H}^+ \quad (7)$$

$$\Delta H_{f(\text{Reactants})}^{\circ} = -178 - 1001 = -1179 \, \text{kJ/mol}$$
$$\Delta H_{f(\text{Products})}^{\circ} = -823 - 1818 = -2641 \, \text{kJ/mol}$$
$$\Delta H_r^{\circ} = -1462 \, \text{kJ/mol}$$

Table 4. *Enthalpies of formation of selected species involved in sulphide oxidation reactions*

Reactants		Products	
Species	kJ/mol	Species	kJ/mol
$O_2(g)$	0^d	$Pb^{2+}(aq)$	-1.7^b
O_2 (aq)	-12^b	$Fe^{3+}(aq)$	-49^a
H_2O (l)	-286^f	$Fe^{2+}(aq)$	-90^a
H_2S (g)	-21^f	$Zn^{2+}(aq)$	-154^b
Covellite, CuS	-53^b	Anglesite, $PbSO_4$	-920^b
Chalcocite, Cu_2S	-80^b	$FeSO_4$	-932^a
Galena, PbS	-100^b	Rozenite, $FeSO_4.4H_2O$	-2129^a
Pyrrhotite, FeS	-100^b	Bianchite, $ZnSO_4.6H_2O$	-2778^a
Troilite, FeS	-102^e	Melanterite, $FeSO_4.7H_2O$	-3013^a
Marcasite, FeS_2	-172^e	Coquimbite, $Fe_2(SO_4)_3.9H_2O$	-5288^a
Pyrite, FeS_2	-167^e	Römerite, $Fe^{II}Fe_2^{III}(SO_4)_4.14H_2O$	-7730^a
Pyrite, FeS_2	-178^b	Halotrichite, $FeAl_2(SO_4)_4.22H_2O$	-11041^a
Sphalerite, ZnS	-206^b		
		$Fe(OH)_3$ crystalline	-823^b
		$Al(OH)_3$	-1276^b
		$H^+(aq)$	0^d
		SO_4^{2-} (aq)	-909^b

*Enthalpies are cited in kJ/mol at 25 °C and 1 bar. Note: pyrrhotite is cited as an idealized FeS phase. In reality, pyrrhotite has a formula $Fe_{1-x}S$, where x ranges from 0 to 0.17. The pure FeS mineral, troilite, does not exist in nature on the Earth's surface and is found in meteorites.
[a]Hemingway *et al.* (2002); [b]USNBS (1982); [c]Cox *et al.* (1989); [d]base state; [e]Chase (1998); [f]Moore (1972).

Wels *et al.* (2003) cite an exothermic energy release upon pyrite oxidation of 1409 kJ/mol pyrite, equating to 11.7 kJ/g pyrite. Of course, the rate of release of heat in kW from a waste tip will depend on physicochemical factors such as grain size (exposed surface area), permeability of the mass to O_2, and also on the activity of the bacteria that catalyse the redox reactions involving Fe and S.

For other minerals, such as pyrrhotite (and, in Table 4, an ideal composition of FeS is assumed, rather than the more realistic $Fe_{1-x}S$), galena or sphalerite, the following equations may apply:

$$FeS + O_2 = Fe^{2+} + 2SO_4^= \quad (8)$$

$$\Delta H^\circ_{f(Reactants)} = -100 - 0 = -100\,kJ/mol$$
$$\Delta H^\circ_{f(Products)} = -90 - 1818 = -1908\,kJ/mol$$
$$\Delta H^\circ_r = -1808\,kJ/mol$$

$$ZnS + O_2 = Zn^{2+} + 2SO_4^= \quad (9)$$

$$\Delta H^\circ_{f(Reactants)} = -206 - 0 = -206\,kJ/mol$$
$$\Delta H^\circ_{f(Products)} = -154 - 1818 = -1972\,kJ/mol$$
$$\Delta H^\circ_r = -1766\,kJ/mol$$

In the case of coals, the energy released upon thermal oxidation will depend on the sulphide content, the calorific content of the coal (typically ranging from 14 MJ/kg (lignite) to > 33 MJ/kg (anthracite); Wood *et al.* 1983), and the extent of oxidation (whether incomplete to CO, or complete to CO_2).

The energy released by exothermic reactions (such as equations 5 to 9) could be efficiently recovered from mine-waste tips using heat pumps or exchangers. Closed-loop coils of suitable materials could be buried within the wastes and water circulated as a collector fluid. The collector fluid could be passed via a heat pump to boost temperature or, in extreme cases where temperature was already sufficiently high, used directly for space heating via heat exchangers. In Oruro, Bolivia, for example, 'hot' mine-waste deposits (sometimes >50 °C) occur in the immediate proximity of residential dwellings and public buildings in the high, seasonally cold Altiplano city (with average monthly temperatures ranging from 5 to 13 °C), with its significant energy demands and a limited energy supply.

Fouling of heat pump/exchanger elements

In the context of use of mine water for space heating and cooling purposes, concern is often expressed as to the possibility of fouling of pipework and exchanger elements by precipitation of secondary minerals such as Fe-oxyhydroxides (equation 4) or other hydroxide or, conceivably, carbonate minerals. This concern is valid, but several methods are available for dealing with the problem:

(1) Use closed-loop systems, which might be installed 'down-mine', or within settlement basins at a minewater treatment plant. Here the collector fluid circulates in a closed-loop and never comes into contact with the mine water. Likewise, mine water never enters the heat pump or heat exchanger elements. The problem is thus wholly averted, although there are limitations in capacity to closed-loop systems. For larger capacity operations, open systems will be preferred. Furthermore, the exterior of such closed-loop systems may become heavily fouled, necessitating periodic removal for cleaning.

(2) Where open-loop systems are unavoidable, mine water should be circulated through heat pumps and exchangers in such a way as to minimize contact with atmospheric O_2, which promotes oxidation of Fe^{2+} to poorly soluble Fe^{3+}. Also consideration should be given to pressurized systems, in which down-mine pressures are maintained as far as possible, to hinder degassing of, for example, CO_2. Such degassing may elevate pH and promote precipitation of Fe-oxyhydroxides or carbonate scales.

(3) A range of mechanical techniques are available to de-scale pipes and exchanger surfaces. High pressure jetting of the interior of a pipe circuit may be sufficient to remove loose chemical precipitates. The use of acids in jetting might also be considered to redissolve hydroxide precipitates, provided that their compatibility with pipe circuit and exchanger components has been considered. Pigging involves the insertion of a cylindrical tool (a 'pig') into a pipe. Different 'pigs' may be used for inspection, scouring or abrasion of the interior surface of the pipe. The addition to minewater pipelines of phosphate-based dispersing agents to break up precipitates, or reducing agents (such as sodium dithionate) to scavenge oxygen, are also being trialled (Dudeney *et al.* 2002).

Sustainability and kinetics

The question naturally arises as to the sustainability of mine-based heat pump systems: are

we removing more heat from the ground than is naturally replenished by solar energy, the geothermal gradient or exothermic reactions? In other words, can we overabstract a heat reservoir?

Underground mines

The physics of thermal conduction and storage are, in fact, directly analogous to those of groundwater flow. Thermal conductivity (k_T) and hydraulic conductivity (k) are analogous, as are heat capacity and storage coefficient; and temperature (T) and hydraulic head (h). Indeed, heat flow (H) is estimated by an analogous equation to Darcy's Law:

$$H = -k_T \times A \times dT/dx \qquad (10)$$

$$Q = -k \times A \times dh/dx \quad \text{(Darcy's Law)}, \qquad (11)$$

where Q is groundwater flow, A is cross-sectional area, and x is a distance coordinate).

The sustainable yield of groundwater from an aquifer depends ultimately on recharge from rainfall and other sources. It also depends on how efficiently the aquifer and the borehole can capture that recharge (i.e., develop a 'well catchment area'), and thus on aquifer properties such as hydraulic conductivity and storage. Similarly, the sustainable heat yield of a shallow borehole-based heat pump system will depend on input of solar energy to the geological environment (and also geothermal gradient and exothermic reaction rates, typically in that order of importance), thermal conductivity and heat capacity. However, many of these parameters exhibit relatively low variability (Table 1), thus permitting the application of standard dimensioning 'rules of thumb' for small heat pump systems in hard, crystalline rock aquifers.

The physics of subsurface heat flow differs from that of groundwater flow in one important aspect, however. In aquifer strata, heat is not transported just by thermal conduction in proportion to the temperature gradient and thermal conductivity. It can also be transported by convection; that is, heat stored in groundwater is transported by groundwater flow. Thus, a heat pump system installed in a flowing system (an aquifer or a large mine) will often be able to extract greater sustainable heat yields (and develop a greater thermal catchment) than those installed in low-permeability crystalline rock, due to subsurface heat flow by convection as well as conduction.

In most flooded mine systems (including all the case studies mentioned below), the ultimate

source of heat will be either solar energy input to
the geological environment from the surface or,
in deeper mines, the geothermal gradient. The
energy input from exothermic oxidation reactions will typically be relatively low: the bacterially catalysed sulphide oxidation reaction
requires a large flux of oxidizing species (in practical terms, the presence of gas-phase O_2) to progress at a significant rate. Sulphide oxidation will
thus only typically be significant in terms of heat
flux above, or within the zone of fluctuation of,
the water table. Below the water table in
flooded mines, the rate of sulphide oxidation is
severely limited by the solubility of O_2. Indeed,
flooding of mines, or underwater disposal of
mine wastes, is a recognized method of suppressing acid rock drainage generation (Banks *et al.*
1997*b*). Thus, the design of heat pump systems
for underground flooded mine systems requires,
first and foremost, a good understanding of the
hydraulics and groundwater throughput of the
mine system. This will typically be mine-
specific, and a considerable volume of literature
is available concerning such hydraulic characterization (see, for example, Younger & Robins
2002; Younger *et al.* 2002).

Mine-waste tips

The use of heat pumps to harness the heat generated by exothermic reactions in mine-waste tips,
however, remains largely undemonstrated. Here,
clearly, a thorough understanding of geochemical processes and, in particular, kinetics, will
be important to be able to assess sustainable
rates of heat extraction. However, as the sulphide
oxidation reaction generates heat in proportion to
generation of contaminants (metals, acid, sulphate; e.g., equations 1 and 2), it should be possible to apply the results of the large volume of
research on the kinetics of generation of acid
rock drainage (e.g., Strömberg & Banwart
1994, 1999; Banwart *et al.* 2002) to sulphide-
dominated mine-waste tips. The primary controls
on rate of the pyrite oxidation reaction are
(Banwart *et al.* 2002; Banks *in press*):

- mass of pyrite present in waste rock;
- specific surface area of pyrite (i.e., grain
  size);
- concentrations of reactants (O_2, Fe^{3+}) – the
  abiotic rate of reaction increases as concentrations increase;
- bacterial activity (sulphide oxidation reactions are bacterially catalysed), which may,
  in turn, be influenced by nutrient concentrations and temperature; while abiotic
  pyrite oxidation typically has a rate of 10^{-9}

to 10^{-10} mol/m^2/s at pH 7 (Younger *et al.*
2002), bacterially mediated pyrite oxidation
may be 25–34 times faster than abiotic oxidation. Bacterially mediated Fe^{2+} oxidation
(equation 3) may be 10^6 times faster than
abiotic oxidation;
- temperature (abiotic rate decreases with
  temperature);
- pH (abiotic rate decreases as pH decreases).
  However, *Thiobacillus ferrooxidans* thrives
  at a pH range of 1.5–3.0 and, at low pH
  values, bacterially mediated oxidation will
  overwhelmingly dominate abiotic oxidation.

Thus, the *a priori* estimation of sulphide oxidation rates (and thus of heat generation) in
many coarse metals mine-waste tips will be
almost impossible, requiring a quantification
not only of three-phase (gas, water, solid) geochemical processes but also of microbiological
activity, and the flux of O_2 through the wastes
(convection as well as diffusion). In such cases,
the most practical approach may simply be
reverse modelling: inferring rates of reaction
from fluxes of reaction products (e.g., $SO_4^=$)
from the waste tip. This should, however, be tem-
pered by an understanding that extraction of heat
will change the temperature within a waste tip,
which will in turn affect abiotic reaction rate,
microbiological activity, and the convection of
air through coarse-grained wastes.
    In the case of mine waste comprising finer
grained tailings, a theoretical approach to calculating exothermic reaction rates can be tractable,
as the diffusion of O_2 into the wastes may be a
rate-limiting step, and a process that can be simulated by relatively straightforward models.

Longevity

For heat pumps based on flooded mine systems,
where the heat abstracted is replenished seasonally by inputs from solar radiation and geothermal gradient, the longevity of the reservoir is,
in principle, limitless. The sustainability of
the operation will only be limited by the durability of the infrastructure (e.g., biofouling,
clogging or physical instability of boreholes,
longevity of collector coils).
    For heat pumps based on the exothermic
oxidation reactions taking place in mine-waste
tips, the longevity of the operation will be
limited by the mass of reactant (sulphide or
organic carbon/coal) within the waste. However,
if the principles above are applied and the reaction rates within the waste deduced, the longevity
of the exothermic reactants can be estimated
(see, e.g., Strömberg & Banwart 1994, 1999).
In practice, however, many mine-waste tips are

estimated to have a sulphide oxidation longevity of decades or centuries (Younger *et al.* 2002).

Case studies

Four case studies will be presented to demonstrate that mine-water-based heat pumps are not 'pie-in-the-sky', but technology that has been demonstrated to be functional, economically advantageous, and environmentally friendly. Further details may be found in the paper by Banks *et al.* (2003).

Metals mines: 1, Park Hills, Missouri, USA

At this site (Heat Pump News 1996; CADDET 2000), water is extracted from workings between 11 and 133 m below ground level in abandoned and flooded Pb mines and has been used to heat and cool the Municipal Building (area 753 m^2) since 1995. The minewater, at a temperature of 13.9 °C, is abstracted at 4.7 L/s via a 122 m deep well and is passed through a plate heat exchanger, which transfers heat to a closed-loop water system. This, in turn, feeds nine water-to-air heat pumps for space heating. The mine water is re-injected via a second well. In the summer, the building can be cooled by reversing the whole system. The system cost US$132 400 (GBP£72 800 at current rates) to install, some $22 200 more than a conventional system. The annual savings, compared with a conventional system, are estimated as US$4800, resulting in a payback time of 4.6 years.

Metals mines: 2, Folldal and Kongsberg, Norway

Folldal is located in Central Norway (Fig. 2) and has operated a mine since 1748. Mining for Cu, Zn, and S continued for almost 200 years, and a community became established around the mine. After 1941, however, the main production shifted to other mines, including Hjerkinn, some distance away in the Dovre Mountain massif.

At the original Folldal mine, the underground Wormshall cavern is still used for concerts and banquets. Since October 1998, it has been heated by a heat pump system based on mine water. A 600 m long, 50 mm diameter closed-loop collector pipe is installed in a 600 m deep flooded shaft. This feeds a heat pump running off 4.6 kW electrical energy and delivering warm air at 22 °C, with a space-heating effect of 18 kW (Åge Kristoffersen, Folldal mine, personal communication). The

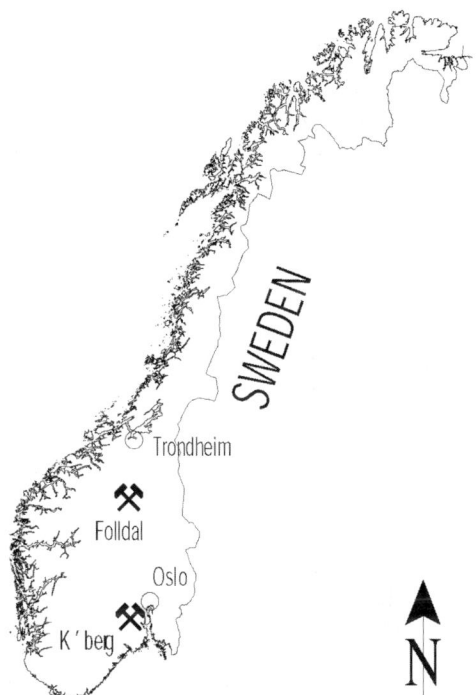

Fig. 2. Overview map of Norway, showing locations of Folldal mine and Kongsberg mine, together with the two major cities of Trondheim and Oslo.

total installation cost is reported to be some 150 000 NoK (GBP£12 000).

A similar system is being planned by researchers at Buskerud College (Hostvedt *et al.* 2002) at the Kongsberg Silver Mine of southern Norway (Fig. 2), which has a working history stretching back several centuries. Since 1957, however, it has only been active as a tourist site, comprising a mining museum and an underground banqueting and concert hall (the 'Festsalen', located some 342 m below ground level). Most of the various mine complexes of Kongsberg are linked at the Christian VII adit level at 200 m above sea level. This also represents the lowest mine drainage level and the workings below it are flooded (Fig. 3). Even in summer, the air temperature around the Festsalen does not exceed 6 °C, despite a rock temperature of c. 9 °C, as density-driven air currents enter the mine via an upland entrance. The Festsalen thus requires heating in order to be a comfortable venue.

Fortuitously, the Festsalen is located near the flooded main mine shaft, at the base of which temperatures are believed to be as high as 16.4 °C. A proposal for an open-loop heat

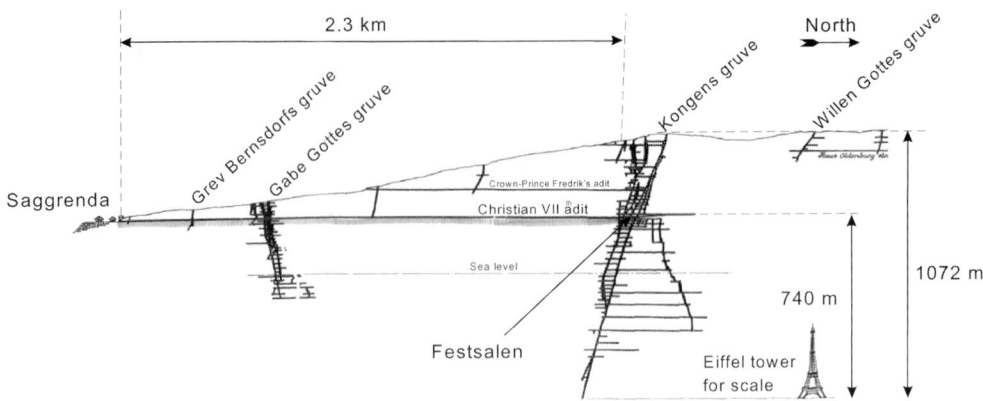

Fig. 3. Cross-section through the Kongsberg mine complex, showing the location of the Festsalen in Kongens Gruve (not to be confused with the mine of the same name at Røros, Table 3), approximately at the coincidence of the main lift shaft and the Christian VII adit (after Hostvedt *et al.* 2002).

pump scheme was rejected due to concerns over compatibility of water quality with heat exchangers. Planning has continued on the basis of an indirect closed loop system (Fig. 4). It is proposed that an anti-freeze fluid would circulate through a closed polyethene collector loop with a pipe length of 130–250 m installed in the main shaft. This system is dimensioned to be able to collect some 12 kW (of the mine's peak demand of 15 kW) heat energy, raising the temperature of the collector fluid to some 10 °C. This temperature would be elevated by passage through a water–water heat pump, to 35 °C. Of this output, 20% would serve a water–air

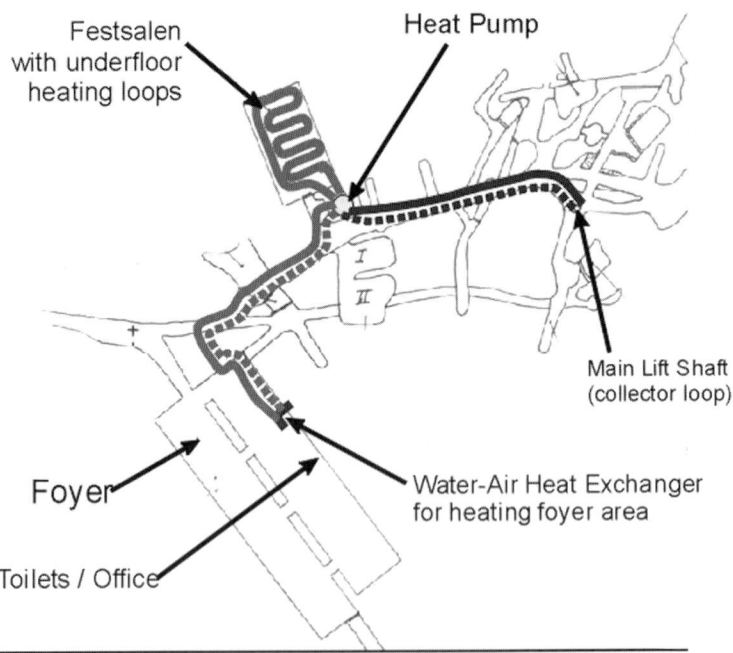

Fig. 4. Schematic plan of the closed-loop scheme to heat the Kongsberg Festsalen. The collector loop is installed in the main lift shaft, and the heat pump outside the Festsalen. The heat pump feeds an underfloor heating loop in the Festsalen and a water–air exchanger in the peripheral areas (after Hostvedt *et al.* 2002).

exchanger, heating the Festsalen's access areas and vestibule, and 80% would heat the Festsalen itself via underfloor loops. The scheme is estimated to cost NoK 81 950 (or c. GBP£7000, of which the heat pump accounts for NoK 42 800, or GBP£3500), and to result in energy savings of NoK 39 000 (or GBP£3100) per year. The payback time for the total cost of the scheme is thus around 2.3 years.

There are also tentative proposals to extend a collector hose from another part of the mine to satisfy the heating requirements of the Mining Museum's topside complex and the State Forestry School at Saggrenda (Fig. 3).

Coal mines: 1, Springhill, Nova Scotia, Canada

At this site (CADDETT 1992, 1997), an industrial plastic packaging plant with an effective area of 13 500 m^2, was retrofitted with 11 heat pumps. These used coal-mine water as a source for all the plant's cooling and heating requirements. Mine water, with a temperature of 18 °C, is pumped from 140 m depth at a rate of 4 L/s. The GSHPs extract heat and lower the water temperature to 13 °C. The mine water is returned to a shallower level (30 m depth) in the mine system. In the summer, the heat pumps are reversed and heat is extracted from the warm air of the building and transferred to the mine water, which is returned to the mine system at 23 °C. The cost of the system was reported to be some Can$110 000 (GBP£47 000 at current rates). A conventional propane-based heating system was estimated to cost Can$70 000 (GBP£30 000). The heat-pump based system resulted, however, in a reported annual energy saving of Can$45 000 (GBP£19 000). These figures imply a payback time of the capital difference of around only 1 year.

Coal mines: 2, Shettleston and Lumphinnans, Scotland

The firm John Gilbert Architects (1999, 2001a, b) has been involved in the design of two mine water-based heat pump systems in Scotland. The first was completed in 1999 in Shettleston, Glasgow, serving 16 newly built dwellings. The heating system uses water from 100 m depth in a flooded coal mine underlying the housing site. The 12 °C mine water circulates through a water-to-water heat pump, and is thereafter returned to the groundwater body at 3 °C. Alternatively, this 'waste' mine water can be used as 'grey water' for flushing toilets. The

heated water output from the heat pump at 55 °C is then transferred to a thermal storage tank, with a side loop to 36 m^2 solar collector panels, which further contribute to the stored enthalpy. This storage tank supplies both central heating systems and hot water immersion heaters for individual dwellings. The total annual heating cost per home is estimated at only GBP£19–30, while the hot water cost is estimated as GBP£55–60.

Another similar scheme is operational at Ochil View, Lumphinnans, Fife, where the heat pump system was retrofitted to 18 dwellings. Water is pumped from a flooded coal mine beneath the site at 170 m depth and utilized via water-to-water heat pumps in a similar manner to Shettleston (although without the solar collectors to boost energy input). The project was completed in August 2001.

Furthermore, a major new town development around the former 1000 m deep Monktonhall Colliery in Midlothian, Scotland, is in the planning stage. Mine water is being seriously assessed as a possible source of heating and cooling requirements for the town, whose peak heat energy demand could reach 30 MW (ENDS 2002).

Conclusions

Wastes from both coal and metals mining (mine water, waste rock, and tailings) represent potential reservoirs of heat energy that can be utilized with the assistance of modern heat-pump technology. Heat pumps based on mine water and mine waste are an extremely attractive proposition because: (1) they can represent a potentially genuinely competitive energy source, especially for large buildings/industrial estates (although this obviously depends on energy policy and pricing in the country concerned); and (2) they convert an environmental liability (a mine water) to an environmental asset (a renewable energy source).

It is, however, reasonable to ask why, if mine waters and wastes represent such an attractive source for heat pumps, larger numbers of such systems have not been installed. In great part, the authors would argue that this is due to lack of awareness on the part of developers and potential users, and their risk-averse attitudes to new technology that may not be regarded as having been adequately 'proven'. Additionally, relatively low energy prices in certain countries (e.g., electricity, until recently, in Norway; mains gas in the UK) have discouraged developers from adopting poorly known alternative technologies. Other potential disadvantages that may be cited by sceptics include: (1) the fact that such systems (like other 'alternative' technologies, such as solar cells or panels) require an initially large capital investment, rendering them unattractive in the short term to small-scale users (although still competitive for large users, with payback times as short as one to two years); and (2) heat exchangers may be liable to fouling by precipitates from mine water in open-loop systems (but technological solutions are available). Closed-loop systems will be more robust in this respect, though these are limited in the energy output, and even these may require periodic removal of mine-water-related precipitates.

Moreover, mine-water-based heat pump systems have been installed in Canada, the USA, Scandinavia, and Scotland, and have been demonstrated to be a realistic alternative space-heating and -cooling solution.

Finally, it should be noted that regulations surrounding the implementation of ground source heat pumps are unclear. Any organization considering the installation of a mine-based heat pump should contact the relevant mining, environmental, and water authorities to obtain the necessary drilling, abstraction, and discharge permits. In the UK, this includes the Environment Agency and the Coal Authority.

The authors wish to extend their thanks to the three reviewers who contributed useful suggestions regarding this paper, namely: Dr C. Nuttall (TNEI Services), Dr E. Reusser (ETH Zürich) and Dr S. Fritz (Purdue University, USA). Dr R. Gieré is also thanked for his enthusiastic suggestions and careful editing.

References

ALBU, M., BANKS, D. & NASH, H. 1997. *Mineral and Thermal Groundwater Resources*. Chapman and Hall, London.

BANFIELD, J. & GIHRING, T. 2003. *Sulfide Mineral Weathering and Acid Mine Research*. Earth & Planetary Science Group, UC Berkeley. World Wide Web Address: http://www.seismo.berkeley.edu/~jill/amd/AMDresearch.html.

BANKS, D. 1994. The abandonment of the Killingdal sulphide mine, Norway: a saga of acid mine drainage and radioactive waste disposal. *Mine Water and the Environment*, **13**, 35–48.

BANKS D. 2004. Geochemical processes controlling minewater pollution. *Proc. 2nd Advanced IMAGE-TRAIN Study Course, Groundwater management in mining areas*, 23–27 June 2003, Pécs, Hungary. Umweltbundesamt GmbH, Vienna, Austria.

BANKS, S. B. & BANKS, D. 2001. Abandoned mines drainage: impact assessment and mitigation of discharges from coal mines in the UK. *Engineering Geology*, **60**, 31–37.

BANKS, D. & ROBINS, N. 2002. *An Introduction to Groundwater in Crystalline Bedrock*. Norges geologiske undersøkelse, Trondheim.

BANKS, D., BURKE, S. P. & GRAY, C. G. 1997*a*. Hydrogeochemistry of coal mine drainage and other ferruginous waters in north Derbyshire and south Yorkshire, UK. *Quarterly Journal of Engineering Geology*, **30**, 257–280.

BANKS, D., YOUNGER, P. L., ARNESEN, R.-T., IVERSEN, E. R. & BANKS, S. B. 1997*b*. Mine water chemistry: the good, the bad and the ugly. *Environmental Geology*, **32**, 157–174.

BANKS, D., HOLDEN, W. *et al.* 2002. Contaminant source characterisation of the San José mine, Oruro, Bolivia. *In*: YOUNGER, P. L. & ROBINS, N. S. (eds) *Mine Water Hydrogeology and Geochemistry*. Geological Society Special Publication, London, **198**, 215–239.

BANKS, D., SKARPHAGEN, H., WILTSHIRE, R. & JESSOP, C. 2003. Mine water as a resource: space heating and cooling via use of heat pumps. *Land Contamination and Reclamation*, **11**, 191–198.

BANWART, S. A., EVANS, K. A. & CROXFORD, S. 2002. Predicting mineral weathering rates at field scale for mine water risk assessment. *In*: YOUNGER, P. L. & ROBINS, N. S. (eds) *Mine Water Hydrogeology and Geochemistry*. Geological Society Special Publication, London, **198**, 137–157.

BOUMA, J. 2002. Heat pumps – better by nature. *IEA Heat Pump Centre Newsletter*, **20** (2), 10–27.

BP 2003. *Statistical Review of World Energy in 2003*. British Petroleum. World Wide Web Address: http://www.bp.com.

CADDET 1992. *Geothermal Mine Water as an Energy Source for Heat Pumps*. Centre for the Analysis and Dissemination of Demonstrated Energy Technologies Result 112 (CA 91.003/3D.H03), October 1992, Sittard, Netherlands.

CADDET 1997. *Geothermal Mine Water as an Energy Source for Heat Pumps*. Centre for Analysis and Dissemination of Demonstrated Energy Technologies project reference CA-1991-003. World Wide Web Address: http://www.caddet.org/infostore/details.php?id = 1659

CADDET 2000. *Geothermal System Saves Energy at a Municipal Building Using Water from Flooded Mine*. Centre for Analysis and Dissemination of Demonstrated Energy Technologies project reference US-2000-507. World Wide Web Address: http://www.caddet.org/infostore/details.php?id = 2789

CANETA RESEARCH 1999. *Global Warming Impacts of Ground Source Heat Pumps Compared to other Heating/Cooling Systems*. Report prepared for NRCan's (Ottawa) Renewable & Electrical Energy Division, April 1999, Caneta Research, Mississauga.

CHASE, M. W. 1998. *NIST-JANAF Thermochemical Tables*, 4th edn. *J. Phys. Chem. Ref. Data*, Monograph **9**, 1–1951. Cited in US National Institute of Standards and Technology (NIST) Webbook, http://webbook.nist.gov.

COX, J. D., WAGMAN, D. D. & MEDVEDEV, V. A. 1989. *CODATA Key Values for Thermodynamics*. Hemisphere, New York.

DUDENEY, B., DEMIN, O. & TARASOVA, I. 2002. Control of ochreous deposits in mine water treatment. *In*: NUTTALL, C. A. (ed) *Proceedings of the Conference on Mine Water Treatment: a Decade of Progress*, Newcastle, 11–13 November 2002, 195–202.

ENDS 2002. Study into use of mine water for district heating. *ENDS Report*, **328**, 13.

ESKILSON, P., HELLSTRÖM, G., CLAESSON, J., BLOMBERG, T. & SANNER, B. 2000. Earth energy designer v. 2 (software parameter database). Blocon Software, Sweden.

ETTNER, D. C. 2002. Våtmarksrensing av gruvevann. Erfaringer fra Titania og Kongens gruve på Røros *[Wetland treatment of mine water; experiences from Titania and King's mine at Røros - in Norwegian]*. *Vann*, **37**, 142–147.

HALLIDAY, D. & RESNICK, R. 1978. *Physics*, 3rd edn. Wiley, New York, 1131 pp.

HEAT PUMP NEWS 1996. City hall gets heat from disused mine. *IEA Heat Pump Newsletter*, **14** (3), 5.

HEMINGWAY, B. S., SEAL, R. R. & MING CHOU, I. 2002. *Thermodynamic Data for Modeling Acid Mine Drainage Problems: Compilation and Estimation of data for Selected Soluble Iron-Sulfate Minerals*. United States Geological Survey Open File Report 02–161.

HOSTVEDT, K., PAULSEN, S. T., RENSKOUG, H. & AURE, O. 2002. Varmepumpe i sølvgruvene/ geothermal energy in Kongsberg silver mines. Hovedoppgave Thesis EMR2002-16, Department of Engineering, Høgskolen i Buskerud, Norway.

IVERSEN, E. R. 1998. *Vinddrevet Forurensningskontroll av Gruveavrenning [Wind-powered monitoring of pollution from mine run-off - in Norwegian]*. Norsk Institutt for Vannforskning (NIVA) Årsberetning (Annual Report) 1998, (also at World Wide Web Address: http://www.niva.no/Aarsberetninger/98aarbok/IVE_98.htm).

IVERSEN, E. R., KNUDSEN, C. H. 1997. *Kjemisk Rensing av Gruvevann fra Kongens Gruve i Nordgruvefeltet, Røros [Chemical treatment of minewater from the King's Mine in the northern mining field, Røros - in Norwegian]*. Norsk Institutt for Vannforskning (NIVA), Report 3632–97, Project O-96099, Oslo.

JOHN GILBERT ARCHITECTS 1999. *Sustainable Housing, Glenalmond Street, Shettleston*. John Gilbert Architects, Glasgow.

JOHN GILBERT ARCHITECTS 2001*a*. *How Geothermal Heating Works*. John Gilbert Architects, Glasgow.

JOHN GILBERT ARCHITECTS 2001*b*. *A Technical Report on Ochil View, Lumphinnans*. John Gilbert Architects, Glasgow.

MELLON, M. T. 2001. Thermal inertia and rock abundance. *Proceedings of the Data User's Workshop on Exploring Mars with TES*, 13th–15 November 2001, Tempe, Arizona, USA.

MOORE, W. J. 1972. *Physical Chemistry*. Longman, London, 977 pp.

MORGAN, H. 1997. Breaking new ground in refrigeration technology. *Refrigeration and Air Conditioning*, **June 1997**, 48–50.

RAWLINGS, R. H. D. & SYKULSKI, J. R. 1999. Ground source heat pumps: a technology review. *Building Services Engineering Research and Technology*, **20**, 119–129.

STEVENS, R. W. 1869. *On the Stowage of Ships and their Cargoes: with Information Regarding Freights*. C Longman, London. Cited on World Wide Web Address: http://energyconcepts.tripod.com/energyconcepts/weathering_of_coal.htm.

STRÖMBERG, B. & BANWART, S. A. 1994. Kinetic modelling of geochemical reactions at the Aitik mining waste rock site in northern Sweden. *Applied Geochemistry*, **9**, 583–595.

STRÖMBERG, B. & BANWART, S. A. 1999. Experimental study of acidity-consuming processes in mining waste rock: some influences of mineralogy and particle size. *Applied Geochemistry*, **14**, 1–16.

SUNDBERG, J. 1991. *Termiska Egenskaper i Jord Och Berg [Thermal properties of soils and rock - in Swedish]*. Statens geotekniska institut Report, Linköping, Sweden.

USNBS 1982. United States National Bureau of Standards tables of molar thermodynamic properties. *J. Phys. Chem. Ref. Data*, **11** (Supp. 2). World Wide Web Addresses: http://www.ualberta.ca/ ~ jplambec/che/data/p00404.htm and http://www.ucdsb.on.ca/tiss/stretton/chem1/data3.htm.

WELS, C., LEFEBVRE, R. & ROBERTSON, A. 2003. An overview of prediction and control of air flow in acid-generating waste-rock dumps. *Enviromine Online Journal*, World Wide Web Address: http://www.infomine.com/technology/enviromine/publicat/Airflow_Wels_Lefebvre_Robertson.pdf.

WEST, I. M. 2001. *Burning Beach, Burning Cliffs and the Lyme Volcano: Oil Shale Fires; Geology of the Dorset Coast*. School of Ocean and Earth Sciences, Southampton University, Report Version U05.10.01, World Wide Web Address: http://www.soton.ac.uk/ ~ imw/kimfire.htm.

WOOD, G. H., KEHN, T. M., CARTER, M. D. & CULBERTSON, W. C. 1983. Coal resource classification system of the U.S. Geological Survey. *US Geological Survey Circular*, **891**, 65 pp. http://pubs.usgs.gov/circ/c891/index.htm.

YOUNGER, P. L. & ROBINS, N. S. (eds.) 2002. *Mine Water Hydrogeology and Geochemistry*. Geological Society Special Publication, London, **198**, 396 pp.

YOUNGER, P. L., BANWART, S. A. & HEDIN, R. S. 2002. *Mine Water: Hydrology, Pollution, Remediation*. Kluwer, Dordrecht, 442 pp. London

YOUNGER, P. L. 2004. Environmental impacts of coal mining and associated wastes: a geochemical perspective. *In*: GIERÉ, R. & STILLE, P. (eds) *Energy, Waste, and the Environment: a Geochemical Perspective*. Geological Society, London, Special Publications, **236**, 169–209.

Modelling near- and far-field processes in nuclear waste management

J. BRUNO, D. ARCOS, E. CERA, L. DURO & M. GRIVÉ

Enviros. Passeig de Rubí, Valldoreix, Spain (e-mail: jbruno@enviros.biz)

Abstract: In this chapter we present some of the recent advances concerning process understanding and modelling of radionuclide migration in nuclear waste disposal systems. The present geochemical modelling approaches used to quantify the processes concerning spent fuel dissolution, radionuclide interactions with the canister materials and the bentonite buffer are thoroughly discussed. Finally, some applications to natural analogue studies of spent nuclear fuel disposal, as testing ground for concepts and models developed for waste management systems, are presented.

Radioactive waste management is a quite mature field of application of basic geoscientific disciplines. As we will discuss in forthcoming sections, the long-term performance and henceforth the safety of radioactive waste disposal systems, deeply relies on the basic principles that control the release, mobility, and transport of the chemical elements in the geosphere. In the context of radioactive waste disposal, the waste matrix constitutes the innermost of the barriers that may control the release and ulterior transport of radionuclides through the groundwater systems.

Consequently, it is of paramount importance to understand the processes that control the stability of the waste matrix and the potential trace component release when exposed to the action of groundwater. Most of this process understanding is based on well-controlled experiments performed in laboratory conditions and for limited periods of time. At the same time, the performance of radioactive waste repositories must be predicted over geological time-scales. Therefore, bridging the gap between the experimental and the prediction time-scales requires the development of a proper process understanding based upon well-founded scientific principles (thermodynamic laws) and testing of the subsequent thermodynamic models in situations that mimic the expected repository materials and processes in geological systems (natural analogues). We must also consider that, regardless of the long time-scales involved, some of the key geochemical processes are kinetically controlled and consequently kinetics must be considered.

In this chapter, we will describe the state of the art regarding the current research on waste/water interactions with special emphasis on UO_2 spent fuel disposal in saturated groundwater conditions and with a specific outlook onto the findings concerning uraninite stability in natural systems.

The problem setting

Radioactive wastes arise in many different forms and from a wide range of activities. The main streams come from plants and processes associated with nuclear power production and research, and unfortunately also from widespread military applications. There are other industrial applications also producing minor radioactive waste volumes. Categorization schemes are normally based on the following attributes:

- the type of process generating the waste, e.g., reactor operation, reprocessing activities, decommissioning and medical procedures;
- the radioactivity and thermal activity levels;
- the physical and chemical forms of the waste;
- the disposal route.

The most usual route of waste classification is by radioactivity and thermal emission, mainly between high-level nuclear waste (HLNW) and low-level nuclear waste (LLNW). Depending on the countries there are other categories, such as intermediate-level nuclear waste (ILNW) and more recently another category has been introduced in order to avoid unnecessary saturation of LLNW repositories; these are the so-called very low activity nuclear wastes (VLNW).

The discussions in this chapter will be mainly centered on the disposal of HLNW, which con-

From: GIERÉ, R. & STILLE, P. (eds) 2004. *Energy, Waste, and the Environment: a Geochemical Perspective*. Geological Society, London, Special Publications, **236**, 515–528.
0305-8719/04/$15

sists of spent nuclear fuel produced at power stations or the products of immobilization of the highly active liquid wastes arising from fuel reprocessing. These wastes typically contain radioactive isotopes (radionuclides) with decay periods over 30 years and that generate significant heat of decay.

It is clear that the disposal of HLNW requires a high level of effective isolation for geological time-scales. In this context deep geological disposal has arisen as the most accepted option and there are already operational repositories of this type (waste isolation pilot plant, WIPP) in the USA, and in Finland and Sweden the plans are well advanced for the siting and construction of such facilities.

The isolation and safety functions of HLNW deep geological repositories are based upon the multibarrier concept, where a number of containment and isolation barriers are put in place. A schematic view of the multibarrier HLNW concept is given in Fig. 1. The main barriers of the system are: the waste matrix itself; a metallic container (either corrosion resistant like Cu or Ti, or based upon stainless steel); a buffer material (normally bentonite); and finally the host rock itself (essentially granite or clay, although salt domes are also being considered).

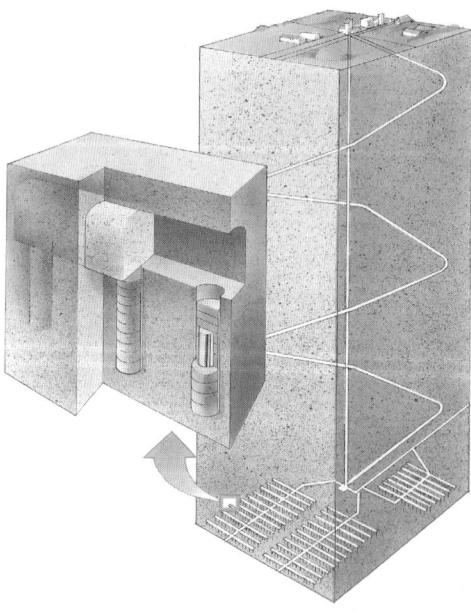

Fig. 1. Schematic multibarrier HLNW concept of deep geological repositories (with permission of SKB).

In this chapter we will pay most attention to the isolation function of the innermost of the barriers, the waste matrix, and its potential interactions with the contacting water. In addition, and because of the similarities in the processes involved, we will also discuss the key processes that control the mobility of some of the critical components of waste in groundwaters. These key processes are bentonite/groundwater interactions, which can exert a large influence on the processes controlling the master pH/pe variables, iron corrosion processes responsible for poising the redox potential of the system and the interactions between the waste matrix itself and the contacting fluids, which produce radiolysis reaction processes.

Near-field waste–water interactions

The near field of the repository includes the engineered barrier system (EBS, i.e., canister and buffer) and the waste form. Also included in the near field is the interface between the buffer and the host rock, denoted as excavation disturbed zone (EDZ). In terms of waste/water interactions, the geochemical evolution of the near field is essential as it controls the composition of the fluids that will eventually contact the waste.

Two master variables, pH and pe, can be used to define the limits of stability of the UO_2 spent fuel matrix under repository conditions. The Pourbaix diagram depicted in Fig. 2 clearly indicates the stability space of UO_2 and consequently the desired chemical conditions that ensure the correct performance of the waste matrix.

The chemical composition of the fluids potentially contacting the spent fuel matrix is mainly controlled by the water–rock interactions between groundwater and the amount (both in terms of mass and surface) of bentonite buffer material, and to a lesser extent, by further interaction with the canister material. In this context, we can propose that the bentonite/groundwater interactions are key in controlling the alkalinity of the system and consequently in buffering the pH of the system. The redox capacity of the system, although initially set by the amount of Fe(II) minerals present in the bentonite clay material, is mainly controlled by the corrosion processes in the container, particularly those involving metallic iron. We will briefly discuss the main processes arising at the bentonite/groundwater interface and the role of bentonite as a chemical buffer. The

$[U]_{tot} = 0.1 \mu M, 25°C, I=0$

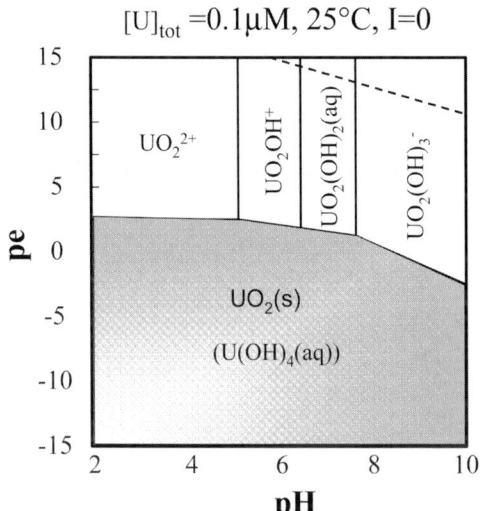

Fig. 2. Predominance diagram of the uranium system showing the influence of pe and pH in the system in the absence of carbonate. In brackets are the aqueous species in equilibrium with $UO_2(s)$. (Database taken from Grenthe *et al.* (1992) and Silva *et al.* (1995).)

mineralogical composition of some representative bentonite materials thought to be used as a buffer material in engineered barriers in repositories is given in Table 1. The main characteristic of the mineralogical composition of these bentonite rocks is the high content of clay minerals such as montmorillonite and/or smectite,

which determines the exchange and swelling capacity of the bentonite.

The alkalinity buffer capacity of bentonite is mainly provided by the following processes:

(1) Dissolution of the calcite content in bentonite, according to the following reaction:

$$CaCO_3(s) + H^+ \Leftrightarrow Ca^{2+} + HCO_3^- \quad (1)$$

(2) Ion exchange reactions between cations present in groundwater and in the interlayer space of the montmorillonite structure. These are mainly the replacement of $2Na^+$ by Ca^{2+} and Mg^{2+} present in groundwaters. This triggers the dissolution of carbonates according to equilibria like the one exemplified by equation (1). The ion-exchange reactions can be described by:

$$2\,NaX + Ca^{2+} \Leftrightarrow CaX_2 + 2\,Na^+ \quad (2)$$
$$2\,NaX + Mg^{2+} \Leftrightarrow MgX_2 + 2\,Na^+ \quad (3)$$

(3) Surface protonation/deprotonation reactions at the edge of the silanol and aluminol sites (>SOH) of montmorillonite, which can be exemplified by the following reactions:

$$>SOH + H^+ \Leftrightarrow >SOH_2^+ \quad (4)$$
$$>SOH \Leftrightarrow >SO^- + H^+ \quad (5)$$

Table 1. *Mineralogical composition of the representative bentonites FEBEX, MX80, and Kunigel V1*

Mineralogical composition	FEBEX*	MX-80[†]	Kunigel V1[‡]
Mineral (wt%)			
Montmorillonite	92	75	46—49
Kaolinite	—	<1	—
Quartz	4	15.2	38—39
Feldspar	—	5—8	2.7—5.5
K-Feldspar	Trace	—	—
Plagioclase	2	—	—
Organic matter (as CO_2)	0.35	0.4	—
Carbonate	0.6	1.4	4.1—6.4
Calcite	—	0.7	2.1—2.6
Dolomite	—	—	2.0—3.8
Siderite	—	0.7	—
Soluble sulphates (gypsum)	0.14	23.5	0.24
Low soluble sulphates (barite, celestite)	0.12	—	—
Sulphide (py)	0.02	0.3	0.5—0.7
Chloride	0.13	1.35	0.014
Iron oxides	0.105	10.6	—

*Fernández *et al.* (2001).
[†]Bradbury & Baeyens (2002).
[‡]Tachi *et al.* (2001).

(4) Finally, the montmorillonite weathering reactions also contribute in a kinetic fashion to the pH buffering capacity of the bentonite barrier according to the following reaction:

$$Na_{0.33}(Mg_{0.33}Al_{1.67})Si_4O_{10}(OH)_2 + 6\,H^+$$
$$+ 4\,H_2O$$
$$\Rightarrow 0.33\,Mg^{2+} + 0.33\,Na^+$$
$$+ 1.67\,Al^{3+} + 4\,H_4SiO_4 \qquad (6)$$

(5) The redox buffering capacity of the bentonite material is provided by the Fe(II)-containing accessory minerals, particularly Fe(II)-carbonates and pyrite. The key reactions are the following:

$$FeCO_3(s) + 2.5\,H_2O + 0.5\,O_2$$
$$\Leftrightarrow Fe(OH)_3(s) + H^+ + HCO_3^- \qquad (7)$$
$$FeS_2(s) + 3.75\,O_2(g) + 3.5\,H_2O$$
$$\Leftrightarrow Fe(OH)_3(s) + 2\,SO_4^{2-} + 4\,H^+ \qquad (8)$$

The effect of these processes in controlling the pH and pe of the bentonite pore water is indicated in Figs 3, 4 and 5, where the expected evolution of these parameters is calculated with the PHREEQC geochemical code (Parkhust & Appelo 1999) for two different bentonite groundwater systems. The results presented in these figures correspond to different reactive transport models where main geochemical processes considered in the bentonite are cation exchange and dissolution/precipitation reactions of bentonite accessory minerals. These near-field models have evolved from relatively simple mix-tank models (Bruno *et al.* 1999) to more complex and realistic models (Arcos *et al.* 2002*a, b*), where diffusion is the main transport mechanism and a thermal gradient was also implemented in order to simulate the heating effect of the spent fuel on the near field of a repository. In Fig. 3, the time-dependent diagrams for the evolution of the Äspö groundwater/MX80 bentonite interactions are depicted. Simulations are calculated at different initial pO$_2$(g) of the system. The input of Ca(II) due to the large concentrations in Äspö groundwater induces not only the exchange of 2 Na by Ca in the bentonite, but also the precipitation of calcite. The calcite buffers the alkalinity of the system and keeps the pH values above 8, ensuring the chemical integrity of the bentonite material (Fig. 3a). In terms of the redox evolution (Fig. 3b), the simulations indicate that when extremely high oxygen fugacities (log $fO_2 = -0.22$ bar, given by

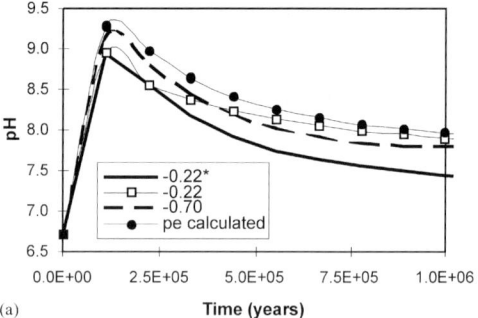

(a)

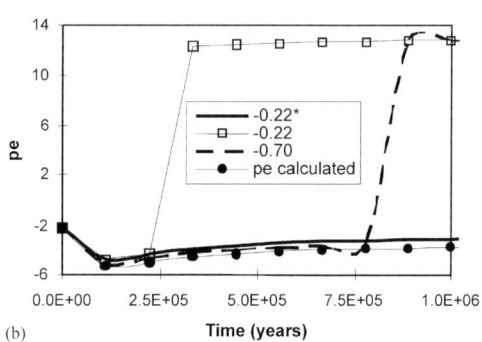

(b)

Fig. 3. Simulations calculated with the PHREEQC geochemical code (Parkhust & Appelo 1999): (a) time-dependent diagram for the pH evolution of the Äspö ground water/bentonite interaction; (b) time-dependent diagram for the pe evolution of the Äspö groundwater/bentonite interaction. Curves correspond to different initial partial oxygen pressures. Initial calcite and pyrite contents are 0.3 wt% and 0.01 wt% respectively, except for the curve of log $fO_2 = -0.22^*$ where calcite and pyrite contents are 1.4 wt% and 0.3 wt%, respectively. pe calculated stands for the cases where the oxygen fugacity is obtained from the groundwater redox potential (Bruno *et al.* 1999).

potential infiltration of oxygenated ice melting water) are assumed the average pyrite content is depleted after 300 000 years. However, if the pyrite content is assumed to be in the higher range of what has been measured (wt% = 0.3) the reducing capacity of the buffer bentonite system is maintained even in these very unlikely dramatic circumstances. Atmospheric fugacities (log $fO_2 = -0.70$ bar) would only have an impact if the post-closure period is extended for some 750 000 years, which is a rather unlikely possibility (Bruno *et al.* 1999), given that the post-closure period may extend from 10 000 up to 1 000 000 years maximum (SKI 2001).

In Fig. 4 the evolution of pH and Eh (mV) for a Spanish granitic groundwater in contact with FEBEX bentonite is shown, together with the spatial distribution of the bentonite blocks.

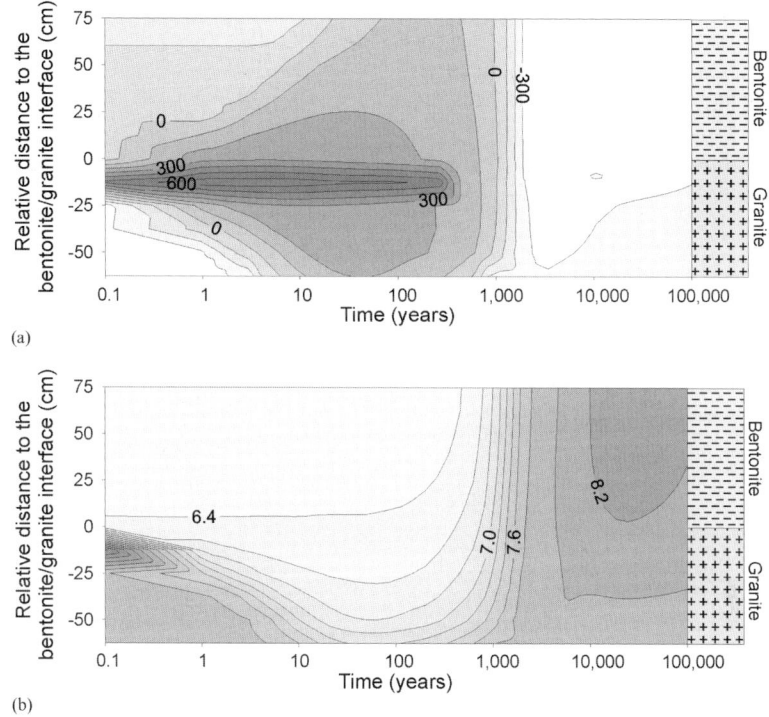

Fig. 4. Simulations calculated with the PHREEQC geochemical code (Parkhust & Appelo 1999): (a) time-dependent evolution of Eh (mV) for a Spanish granite groundwater in contact with FEBEX bentonite; (b) time-dependent evolution of pH for a Spanish granite groundwater in contact with FEBEX bentonite (Arcos *et al.* 2000*a*).

Initially, O_2 diffuses through the bentonite and granitic domains, controlling the redox state of the system. Once O_2 is exhausted, granitic groundwater controls the redox state of the system. The results of these calculations performed with the PHREEQC geochemical code (Parkhust & Appelo 1999) clearly indicate that there is a substantial variability in pH/pe space along the temporal and spatial evolution of the near field of a repository. This has clear consequences for the subsequent interactions with the Fe canister material and finally with the spent fuel matrix.

Conditioning of the near field geochemistry by the canister corrosion processes

The fluids that have evolved as a result of the bentonite–groundwater interactions will contact the canister on their travel towards the spent fuel matrix. Most of the proposed canister materials in different countries have in common the presence of Fe in the system, either as cast iron (Sweden, Finland) or as stainless steel (France, Spain). While the bentonite–groundwater processes have

a large influence on the chemical composition of the near-field fluids, Fe corrosion will be the process responsible for poising the redox potential of the system. As we have previously seen, the resulting bentonite pore water will be mainly anoxic and consequently the anaerobic corrosion of iron will be the dominant process. Under these conditions the main reactions are:

(1) Initial corrosion of Fe(s) to form metastable $Fe(OH)_2$(s):

$$Fe(s) + 2 H_2O \Leftrightarrow Fe(OH)_2(s) + H_2(g) \tag{9}$$

(2) The subsequent formation of magnetite from the metastable $Fe(OH)_2$(s), describing the Schikorr reaction:

$$3 Fe(OH)_2(s) \Leftrightarrow Fe_3O_4(s) + H_2(g) + 2 H_2O \tag{10}$$

(3) In the presence of carbonates, $Fe(OH)_2$(s) may also evolve towards the formation of

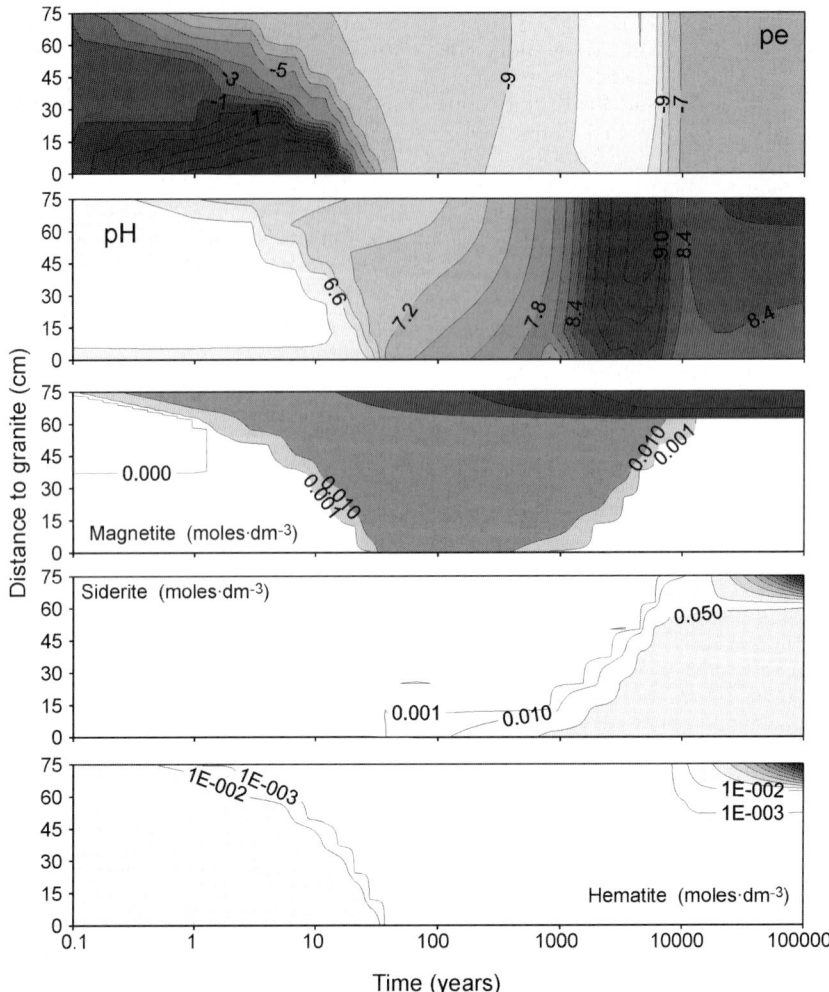

Fig. 5. Evolution of some relevant parameters of the bentonite–steel canister system with time (Arcos *et al.* 2000*b*).

siderite, $FeCO_3(s)$:

$$Fe(OH)_2(s) + HCO_3^- + H^+$$
$$\Leftrightarrow FeCO_3(s) + 2\,H_2O \quad (11)$$

(4) Another important intermediate in the corrosion of Fe(s) is the so-called 'green rust', which consists of layered Fe(II)–Fe(III) hydroxides, where the positive charge of the isomorphic substitution of Fe(II) by Fe(III) is balanced by anions like Cl^-, SO_4^{2-}, and CO_3^{2-}.

All these processes contribute to controlling the redox state and the alkalinity of the system. The typical evolution of the bentonite–steel

canister system is given in Fig. 5. We can see that there are two different stages. First is the iron corrosion stage (up to 5000 years) where the oxidation front from the EDZ (O_2 diffusion) is neutralized by the reductive front from the canister. In the final stage (from 5000 years), once the iron corrosion process ends, the processes governing the system will be the same as when iron corrosion is not considered. During the corrosion stage, magnetite and hematite precipitate in the whole bentonite domain. Once the corrosion finishes, magnetite and hematite precipitation is restricted to the bentonite–canister boundary and they are dissolved in the rest of the bentonite domain to precipitate siderite.

In summary, the setting for the stability of spent fuel is quite varying depending on the geochemical evolution of the near field, but, by and large, it remains under reducing conditions and with a moderate alkalinity from the onset of the repository. This would in principle ensure the stability of the main component of the spent fuel matrix, $UO_2(s)$. However, as we will see in the following section, UO_2 spent fuel, when contacting water, constitutes a dynamic redox system.

UO_2 spent nuclear fuel, a fascinating chemical system

The interactions between the waste matrix and the contacting fluids in the near field constitute the source term of the ulterior complex models used in the performance assessment of spent fuel repositories.

The spent fuel matrix is a ceramic material with a fascinating chemical composition and a large degree of phase heterogeneity. The physical state and chemical composition of spent fuel largely depends on the burn-up of the fuel once it is taken out of the reactor. In Fig. 6 we indicate the dependence of the chemical composition on the burn-up for a series of PWR fuels. However, the fact that remains constant is that UO_2 constitutes the major component of spent fuel, ranging within a total of 95–98% in weight (see Fig. 7).

In order to appraise the processes that control the interactions between such a complex material and the result of the groundwater interactions

with the components of the near field, there is a need for a substantial amount of data and models. The data are provided by various types of laboratory experiments with actual spent fuel, unirradiated UO_2 and also from chemical analogues to spent fuel, like SIMFUEL. This is a material that reproduces the chemical composition of spent nuclear fuel without being radioactive, with obvious simplifications for experimental design. Another important field of data comes from natural analogue studies, particularly from uranium ore deposits where some of the key constituents and processes expected in a repository have occurred over geological time-scales.

The inherent radioactive characteristics of the spent nuclear fuel condition determine many of the key processes to be studied. Owing to its energy content, spent fuel relaxes by transferring alpha, beta, and gamma radiation to water when contacting it. This originates what is known as radiolysis reactions. The key processes occurring at the spent fuel water interface are depicted in Fig. 8.

The sequence of reactions is the following:

(1)   Water radiolysis:

$$H_2O + (\alpha\beta\gamma) \text{ radiolysis} \implies O_2 + H_2$$
$$+ H_2O_2 + \cdots \quad (12)$$
$$H_2O_2 \implies 0.5\,O_2 + H_2O \quad (13)$$

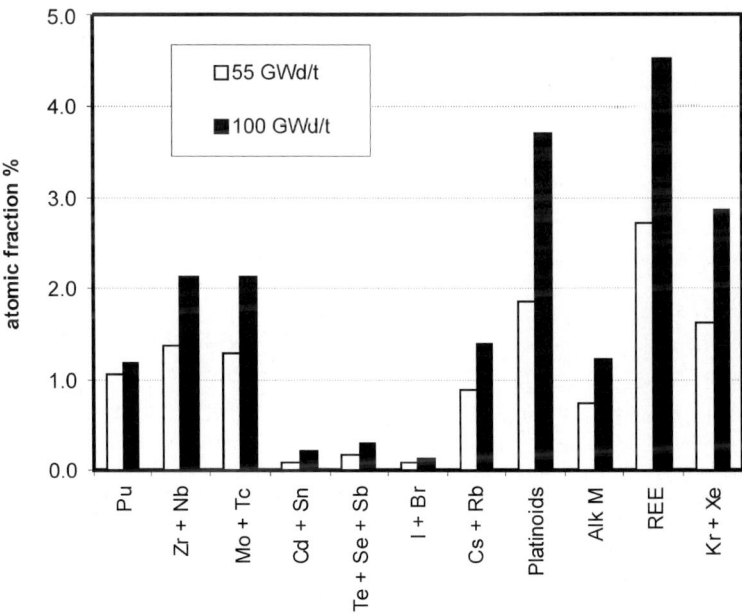

Fig. 6. Dependence of the chemical composition on the burn-up for a series of PWR fuels (Data taken from Poinssot *et al.* 2001).

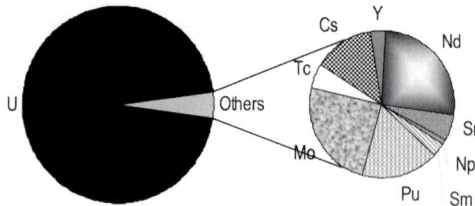

Fig. 7. Major and minor components of nuclear spent fuel in wt%.

Depending on the water composition other radical species are formed, such as carbonate and chloride radicals. This imposes net oxidizing conditions at the water–fuel interface because the generated oxidants, molecular oxygen and hydrogen peroxide, predominate under α radiation, and other radical species like $OH^{-\bullet}$ or $CO_3^{2-\bullet}$ are more active than the generated reductants, mainly molecular hydrogen. This is why we propose that the spent fuel–water interface is a dynamic redox system, independently of the conditions imposed on the near field (Merino *et al.* 2001).

(2) Oxidation of the UO_2 matrix by the oxidants being produced at the interface. This is a surface controlled process and therefore it should be described according to the following reaction:

$$> UO_2 \equiv UO_2 + 0.33\,H_2O$$
$$\Rightarrow\; > UO_2 \equiv UO_{2.33} + 0.67\,H^+ + 0.67\,e^-$$
$$(14)$$

where $> UO_2 \equiv UO_2$ stands for the unaltered surface of the UO_2 spent fuel and $> UO_2 \equiv UO_{2.33}$ exemplifies the oxidation of the surface up to the threshold for the breakdown of the cubic uraninite structure of UO_2.

(3) Reduction of the oxidants radiolitically generated in the system. The main oxidant species produced by α-radiolysis of water are O_2 and H_2O_2. The reduction reactions can be expressed as follows:

$$O_2 + 4\,H^+ + 4\,e^- \;\Leftrightarrow\; 2\,H_2O \qquad (15)$$
$$H_2O_2 + 2\,H^+ + 2\,e^- \;\Leftrightarrow\; 2\,H_2O \qquad (16)$$

(4) Dissolution of the oxidized portion of the surface of the spent fuel, that is, the release of uranium(VI) to the solution. This is described by the following expression:

$$> UO_2 \equiv UO_{2.33} + 0.67\,H^+$$
$$\Rightarrow\; > UO_2 \equiv (UO_2)_{0.67} + 0.33\,UO_2^{2+}$$
$$+ 0.33\,H_2O \qquad (17)$$

The U(VI) aqueous speciation will depend on the actual chemical composition of the contacting fluid. According to what we have discussed in the previous section, the U(VI) speciation will be mainly controlled by hydroxo or carbonato complexes. Hence the dissolution reaction will be described by the following set of reactions.

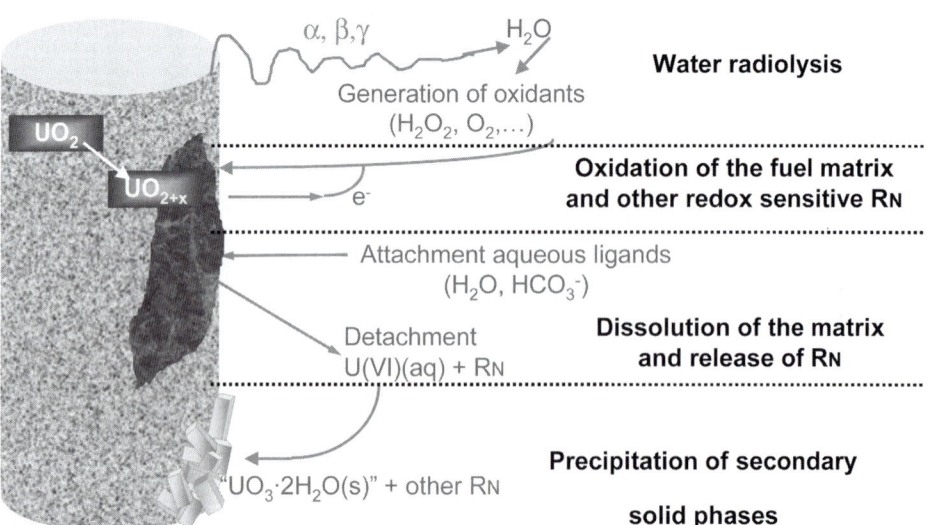

Fig. 8. Key processes occurring at the spent fuel water interface (R_N = radionuclides).

In the absence of carbonates:

$$> UO_2 \equiv UO_{2.33} + 0.33 H_2O$$
$$\Rightarrow \quad > UO_2 \equiv (UO_2)_{0.67}$$
$$+ 0.33 UO_2(OH)_2(aq) \quad (18)$$

In the presence of carbonates at typical groundwater alkalinities:

$$> UO_2 \equiv UO_{2.33} + 0.67 HCO_3^-$$
$$\Rightarrow \quad > UO_2 \equiv (UO_2)_{0.67}$$
$$+ 0.33 UO_2(CO_3)_2^{2-}$$
$$+ 0.33 H_2O \quad (19)$$

(5) The released U(VI) from the UO_2 matrix will continue to be dissolved until saturation with secondary U(VI) solid phases is reached. The observations from both laboratory and natural systems would indicate that the kinetically preferred phase is hydrated schoepite. This will be denoted as $UO_2(OH)_2(s)$ for the sake of description of the model, although the correct notation would be $UO_3 \cdot xH_2O$, with x oscillating between 0 and 2. Depending on the presence of carbonates in the contacting solution, the reactions can be described as:

$$UO_2(OH)_2(aq) \Leftrightarrow UO_2(OH)_2(s) \quad (20)$$
$$UO_2(CO_3)_2^{2-} + 2 H_2O$$
$$\Leftrightarrow UO_2(OH)_2(s) + 2 HCO_3^- \quad (21)$$

(6) Recently, there have been observations indicating that in the presence of excess hydrogen peroxide a uranium(VI) peroxide, studtite ($UO_4 \cdot 4H_2O$), is formed (McNamara et al. 2002). This has been investigated by using a combination of Atomic Force Microscopy (AFM) and X-ray diffraction analysis. The reaction appears to be surface controlled; studtite only forms when $UO_2(s)$ is present and consequently this would suggest the following reaction scheme:

$$> UO_2 \equiv UO_2 + H_2O_2$$
$$\Rightarrow \quad > UO_2 \equiv UO_4 + 2 H^+ + 2 e^-$$
$$(22)$$

for the oxidation reaction of the UO_2 surface to studtite, and

$$> UO_2 \equiv UO_4 + 2 H^+ + 2 e^- + 2 HCO_3^-$$
$$\Rightarrow \quad > UO_2 + UO_2(CO_3)_2^{2-} + 2 H_2O$$
$$(23)$$

for the reductive dissolution reaction of the surface-generated studtite in the presence of carbonates (crystal water has been omitted in the surface reaction).

When describing the evolution of the geochemistry of the near field, we have observed that the anaerobic corrosion of iron generates substantial amounts of hydrogen gas. The partial pressure of hydrogen in the near field will build up until the lithostatic pressure of the repository is reached. This is calculated to be some 10–12 MPa. This large hydrogen overpressure is bound to have some effect on the rates and extent of the radiolytic oxidation of the UO_2 surface. The preliminary experimental observations would indicate that at $p_{H_2} > 0.1$ MPa the reduction of U(VI) to U(IV) is observed (Rovira et al. 2003). The reduction process is particularly efficient in the presence of corroding Fe and magnetite. In the case of spent fuel the observations would indicate that the radiolytic dissolution of UO_2 is totally stopped in presence of high hydrogen partial pressures.

This is a process that is at present being integrated in the current models of spent fuel stability in repository conditions, but some work is still necessary in order to ascertain the reactions and mechanisms that control the reductive passivation of the UO_2 surface and the inhibition of the radiolytic production of oxidants.

Lessons from natural analogue studies

There are inherent scale limitations in the time and space dimensions covered by laboratory studies. The applicability of the near field geochemical models derived from laboratory observations have to be applied to long-term, large-scale situations like the ones involved in the safety assessment of nuclear waste repositories. Hence, there is a need to test the models developed from laboratory investigations in field situations that are related to the ones to be encountered in repository systems.

In this context, natural analogue studies have brought some key evidence in the validation of the concepts and parameters used in modelling uranium release from repository systems. As an example, we investigated the solubility behaviour of several uraninite samples from three uranium ore deposits that have been thoroughly investigated as analogues to spent fuel disposal: Jachimov (Czech Republic), Cigar Lake (Canada), and Oklo (Gabon). The uraninite samples were thoroughly characterized with a variety of analytical techniques, including X-ray diffraction (XRD) optical and electronic microscopy, scanning electron microscopy (SEM), environmental scanning electron microscopy (ESEM), and transmission electron microscopy (TEM). These uraninite samples were put in contact with groundwater solutions of varying acidity. Experimental data were rationalized in terms of an overall solubility model that integrated the two master variables, pH and pe. As can be seen in Fig. 9, the measured data for the various uraninite samples can be well reproduced by a common thermodynamic model that integrates the variability of the redox properties of the surface of the mineral phase together with the variability of the alkalinity of the contacting solutions. The solubility model includes all the U(IV) and U(VI) species calculated for different values of oxygen fugacities, as explained in Casas *et al.* (1998). This is a rewarding result as it indicates that it is possible to describe the complexity of the natural materials by the appropriate thermodynamic parameters. This is further strengthened when including the experimental data from other investigators in this common model, as shown in Fig. 10.

Once this common thermodynamic framework is established for the solubility of UO_2 under nominally reducing conditions, we have to ascertain the most probable pathway for the oxidative alteration of UO_2 spent fuel in geological repository conditions. There is a large body of evidence on the processes involved in the oxidative alteration of natural uraninites and unirradiated UO_2. Long-term unsaturated tests performed by Wronckiewicz *et al.* (1992) on groundwater from Yucca Mountain (the so-called J-13 groundwater), indicated that the formation of schoepite, as described by process (20) and (21), occurs, but is a transient event and that the alteration proceeds towards the precipitation of

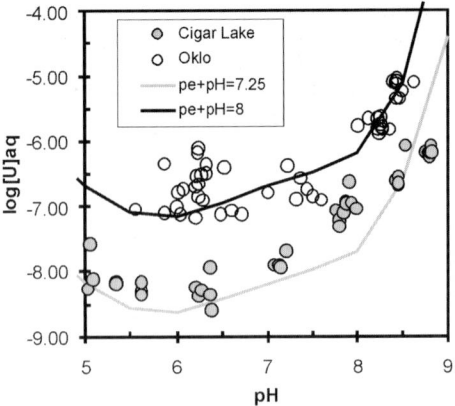

Fig. 9. Experimental solubilities as total uranium concentration in solution for experiments on dissolution of uraninite samples from Oklo and Cigar Lake. Solid lines correspond to the calculated solubilities. Calculations performed with PHREEQC geochemical code (Parkhust & Appelo 1999) and uranium database taken from Grenthe *et al.* (1992) and Bruno & Puigdomènech (1989).

becquerelite $(Ca(UO_2)_6O_4(OH)_6 \cdot 8H_2O)$, soddyite $((UO_2)_2SiO_4 \cdot 2H_2O)$, uranophane $(Ca(UO_2)_2 [SiO_3(OH)]_2 \cdot 5H_2O)$ and other Ca–U(VI)-silicates. The oxidative alteration pathway of uraninite has been well known since the seminal work by Frondel (1956), which indicates the sequence:

Uraninite $\Rightarrow$ Schoepite (K–Ca–U(VI) oxyhydroxides)
$\Rightarrow$ Ca–U(VI) silicates,

depending on the silica to phosphate ratio.

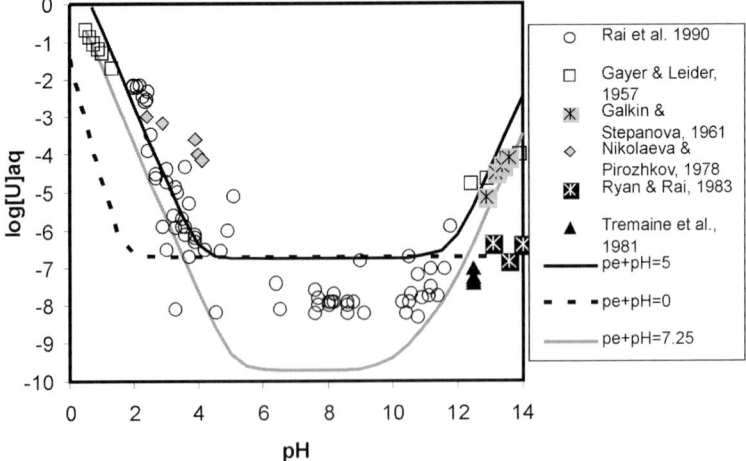

Fig. 10. Uranium dioxide solubility data obtained under nominally reducing conditions, corresponding to values extracted from the literature as given in the figure. Thermodynamic database for uranium from Grenthe *et al.* (1992) and Bruno & Puigdomènech (1989).

Hence, as a continuation of the radiolytic oxidation scheme previously presented, we propose that the alteration of UO_2 in geological time-scales proceeds as follows: full oxidation of the UO_2 surface to U(VI) oxyhydroxides that would include Ca^{2+} and/or K^+ as accompanying cations:

$$6\,UO_2 + O_2 + Ca^{2+} + 14\,H_2O$$
$$\Rightarrow Ca(UO_2)_6O_4(OH)_6 \cdot 8\,H_2O(becquerelite)$$
$$+ 6\,H^+ \qquad (24)$$

The final alteration product strongly depends on the phosphate/silica ratio in the contacting groundwaters. However, in the near field and because of the massive presence of bentonite, the supply of silica will be unlimited and, therefore, the final alteration will most probably be:

$$Ca(UO_2)_6O_4(OH)_6 \cdot 8H_2O(becquerelite)$$
$$+ 6\,SiO_2 + 2\,Ca^{2+}$$
$$\Rightarrow 3\,Ca[(UO_2)(SiO_3OH)]_2$$
$$\cdot 5H_2O(uranophane)$$
$$+ 4\,H^+ + H_2O \qquad (25)$$

Application of the collected thermodynamic data to model the oxidative alteration pathway of UO_2 under repository conditions by using the PHREEQC code (Parkhurst & Appelo 1999) is given in Fig. 11a and b. Once the thermodynamic framework is set for the geochemical evolution of the repository system, we have to take into consideration that for many of the processes involved, there will be some kinetic constraints. This is illustrated by Table 2, where a comparison of the expected lifetime for some of the phases expected in the repository system is made.

It is clear that in addition to thermodynamic models, kinetic mass transfer models can bring about some additional information that is required for a better definition of the system. In this context, natural analogues provide some of the required scale and time-frames necessary for the testing of kinetic mass transfer models and the Cigar Lake ore deposit is probably the better constrained for such an exercise.

The kinetic mass transfer model developed to take into consideration the geochemical evolution of the Cigar Lake ore deposit was mainly done by simulating the evolution of the Al–Si system in the Cigar Lake ore deposit system. To this aim the system formed by kaolinite, gibbsite and illite as main aluminosilicate solid phases was considered and kinetics for the dissolution–precipitation processes were taken from the open scientific literature (Nagy *et al.*

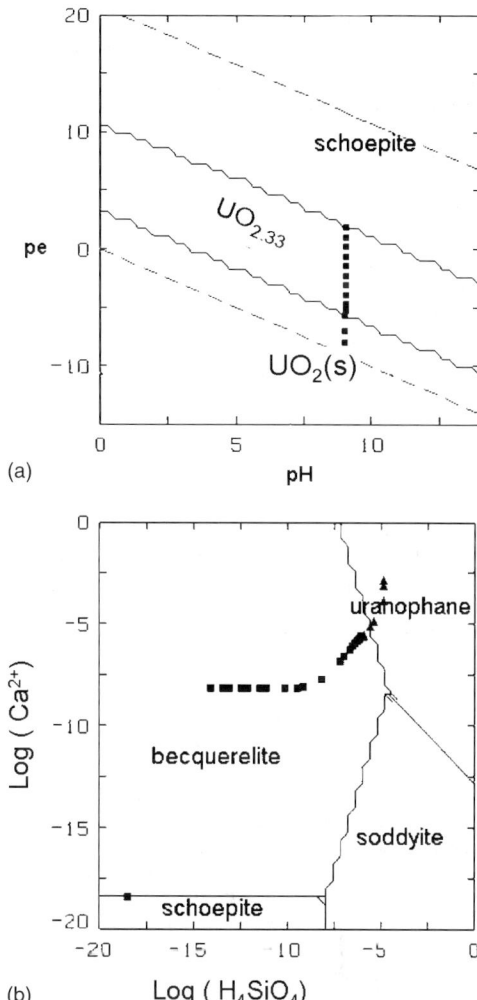

(a)

(b)

Fig. 11. (a) Thermodynamic reaction pathway for the initial oxidative alteration of the spent fuel matrix at pH 8, calculated by using the PHREEQC code (adapted from Bruno *et al.* 1995). (b) Thermodynamic reaction pathway for the alteration of schoepite in granitic/bentonite groundwater at pH 8, calculated by using the PHREEQC code (adapted from Bruno *et al.* 1995 with permission).

1991; Nagy & Lasaga 1992). Two different codes were used in the modelling, the mass-balance-based code NETPATH (Plummer *et al.* 1991) and the box-modelling code STEADYQL (Furrer *et al.* 1989). Details of the modelling can be found in Bruno *et al.* (1997). A transect of the ore deposit with the surrounding mineralogy and the relevant boreholes used in the modelling is shown in Fig. 12. The observed and simulated geochemical evolution of the system is also shown in Fig. 12, where three different

Table 2. *Rate constants and estimated lifetimes of some minerals and solid phases of importance in a HLNW repository**[*]

Mineral	log Rate (mole/m²/s)	Molar volume (cm³/mole)	Lifetime of 1 mm crystal (years)
Quartz	-13.39	22.69	34 million
Kaolinite	-13.28	99.52	6 million
Muscovite	-13.07	140.71	2.6 million
Epidote	-12.61	139.2	923 000
Albite	-12.26	100.07	575 000
Gibbsite	-11.45	31.96	276 000
Pyrite	-12.65	23.94	5.9 million
Fe(III)-oxyhydroxide	-12.56	22.36	5.1 million
Calcite	-8.33	36.93	184
Uraninite	-11.25	24.62	229 000
Schoepite	-9.2	32	1570
Uranophane	-11.95	150	188 000

[*]From Lasaga *et al.* (1994).

evolutionary pathways have been represented. The three evolutionary pathways represent quite well the geochemical and hydrogeochemical data collected in the site (Winberg & Stevenson 1994). The suggested evolutionary pathways were then used to model the evolution of the uranium system in the ore deposit and the calculated pH and pe values were compared to the measured ones. As may be seen in Fig. 13a and b, there is a quite satisfactory agreement, which independently confirms the goodness of the geochemical model proposed. Black dots in Fig. 13a and b stand for the experimental values of pH and

pe, while the solid line represents the calculated evolution of the system as O_2 is added. As the redox potential increases, the pH of the system decreases as a consequence of the hydrolysis of the uranium dissolved. The values obtained reproduce fairly well the values of pe and pH experimentally observed as water evolves from reduced to oxidized zones (Bruno *et al.* 1997).

The major findings of this exercise indicate that, in spite of the long residence times of groundwaters in the Cigar Lake ore deposit system (over 10 000 years), the Al–Si system appears to be kinetically controlled and the

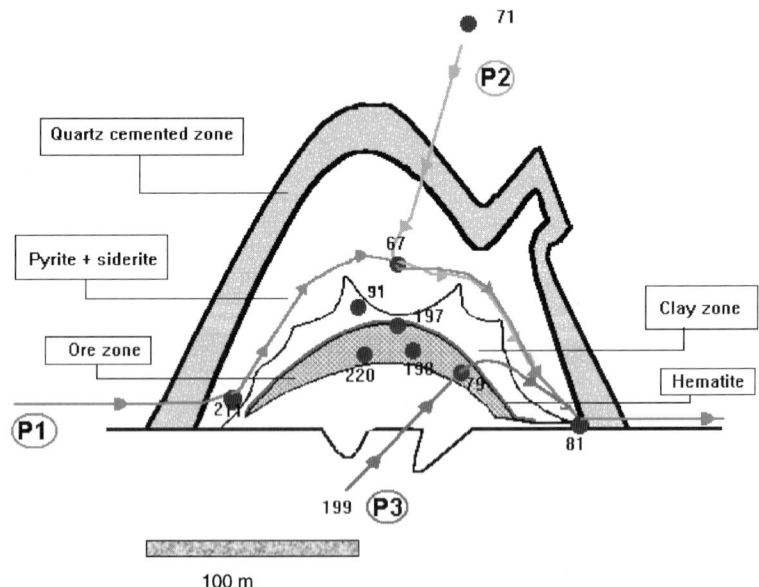

Fig. 12. Transect of the ore deposit with the surrounding mineralogy and sampling points (black circles) and flow paths tested in the calculations at the Cigar Lake uranium deposit (arrows).

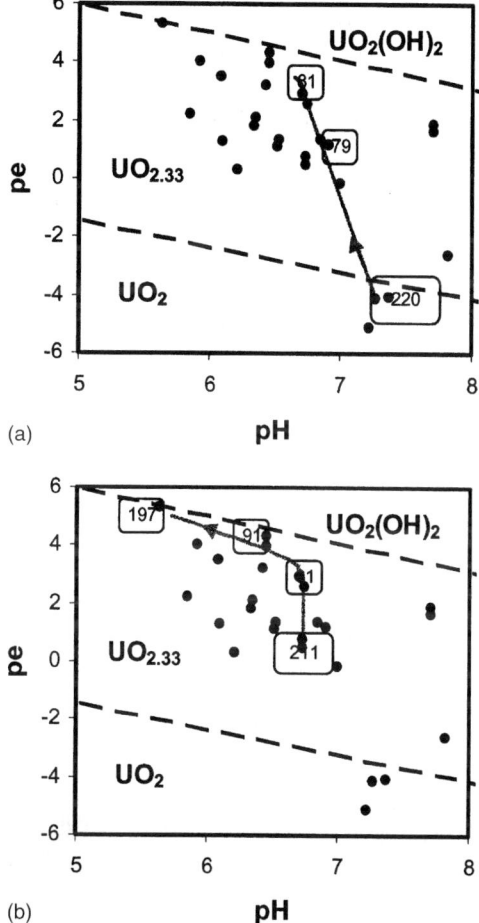

(a)

(b)

Fig. 13. Evolution of the pe and pH of waters along the flow paths suggested from the hydrogeological analysis of the Cigar Lake site. Thermodynamic database for uranium from Grenthe *et al.* (1992) and Bruno & Puigdomènech (1989). (a) Uranium modelling. Equilibrium assumption. Evolution from borehole 220. (b) Uranium modelling. Equilibrium assumption. System evolving from borehole 211.

kinetic mass transfer models really could help to unravel the key processes and parameters controlling uranium mobility in this geochemical system.

Summary

In this chapter we have attempted to give an overview of current developments concerning waste/water interactions with special emphasis on UO_2 spent fuel disposal in saturated groundwater conditions and with a specific look at the findings concerning uraninite stability in natural systems. The outcome of this should be to demonstrate that the combination of thorough laboratory studies, comprehensive geochemical modelling and its application to natural analogue studies bring some confidence about the predictions that have to be made in order to assess the future behaviour of repository systems. It would be desirable that similar strategies are pursued when developing safety and performance criteria for waste management systems arising from other energy sources.

References

ARCOS, D., BRUNO, J., DURO, M. & Grivé, M. 2000*a*. *Desarrollo de un Modelo Geoquímico de Campo Próximo.* Empresa Nacional de Residuos Radioactivos, S.A., Madrid, Spain, Enresa. Publicación técnica 04/00, 60 pp.

ARCOS, D., BRUNO, J., BENBOW, S. & TAKASE, H. 2000*b*. *Behaviour of Bentonite Accessory Minerals During the Thermal Stage.* Svensk Kärnbränslehantering, AB, Stockholm, Sweden, SKB Technical Report TR-00-06, 38 pp.

BRADBURY, M. H. & BAEYENS, B. 2002. *Porewater Chemistry in Compacted Resaturated MX-Bentonite: Physicochemical Characterisation and Geochemical Modelling.* Paul Scherrer Institut, Villigen, Switzerland, PSI Bericht Nr. 02-10, 42 pp.

BRUNO, J. & PUIGDOMÈNECH, I. 1989. Validation of the SKBU1 Uranium thermodynamic database for its use in geochemical calculations with EQ3/6. *Materials Research Society Symposium Proceedings,* **127**, 887–896.

BRUNO, J., CASAS, I., CERA, E. & DURO, L. 1997. Development and application of a model for the long-term alteration of UO_2 spent nuclear fuel. Test of equilibrium and kinetic mass transfer models in the Cigar Lake ore deposit. *Journal of Contaminant Hydrology,* **26**, 19–26.

BRUNO, J., ARCOS, D. & DURO, L. 1999. *Processes and Features Affecting the Near Field Hydrochemistry. Groundwater–Bentonite Interaction.* Svensk Kärnbränslehantering, AB, Stockholm, Sweden, SKB TR-99-29, 56 pp.

BRUNO, J., CASAS, I., CERA, E., SWING, R. C., FINCH, R. C. & WERME, L. O. 1995. The assessment of the long-term evolution of the spent nuclear fuel matrix by kinetic, thermodynamic and spectroscopic studies of uranium minerals. *Materials Research Society Symposium Proceedings,* **353**, 633–639.

CASAS, I., DE PABLO, J. *et al.* 1998. The role of pe, pH and carbonate on the solubility of UO_2 and uraninite under nominally reducing conditions. *Geochimica et Cosmochimica Acta,* **62**, 2223–2231.

FERNÁNDEZ, A. M., CUEVAS, J. & RIVAS, P. 2001. Pore water chemistry of the FEBEX bentonite. *Materials Research Society Symposium Proceedings,* **663**, 573–588.

FRONDEL, C. 1956. Mineral composition of Gummite. *American Mineralogist*, **41**, 539–568.

FURRER, G., WESTALL, J. & SOLLINS, P. 1989. The study of soil chemistry through quasi-steady state models: I. Mathematical definition of model. *Geochimica et Cosmochimica Acta*, **53**, 595–601.

GALKIN, N. P. & STEPANOV, M. A. 1961. The solubility of uranium(IV) hydroxide in sodium hydroxide. *Atomic Energy (USSR)*, **8**, 258–261.

GAYER, K. H. & LEIDER, H. 1957. The solubility of uraium(IV) hydroxide in solutions of sodium hydroxide and perchloric acid at 25 °C. *Canadian Journal of Chemistry*, **35**, 5–7.

GRENTHE, I., FUGER, J., KONINGS, R. J. M., LEMIRE, R. J., MULLER, A. B., NGUYEN-TRUNG, C. & WANNER, H. 1992. Chemical thermodynamics of uranium. *In*: WANNER, H. & FOREST I. (eds) *Chemical Thermodynamics*. NEA OECD, Elsevier, The Netherlands, **1**, 715 pp.

LASAGA, A. C., SOLER, J. M., GANOR, J., BURCH, T. E. & NAGY, K. L. 1994. Chemical weathering rate laws and global geochemical cycles. *Geochimica et Cosmochimica Acta*, **58**, 2361–2386.

MCNAMARA, B., HANSON, B. & BUCK, E. 2002. Observation of studtite and metastudtite on commercial spent nuclear fuel. *Materials Research Society Symposium*, **757**, 401–406.

MERINO, J., CERA, E., BRUNO, J., ERIKSSEN, T., QUIÑONES, J. & MARTÍNEZ-ESPARZA, A. 2001. Long term modelling of spent fuel oxidation/dissolution under repository conditions. ICEM'01. Session 23, V. *The 8th International Conference on Radioactive Waste Management and Environmental Remediation*. 30 September–4 October 2001, Bruges (Brugge), Belgium.

NAGY, K. L. & LASAGA, A. C. 1992. Dissolution and precipitation kinetics of gibbsite at 80°C and pH 3. The dependence on solution saturation state. *Geochimica et Cosmochimica Acta*, **56**, 3093–3111.

NAGY, K. L. BLUM, A. E. & LASAGA, A. C. 1991. Dissolution and precipitation kinetics of kaolinite at 80°C and pH 3. The dependence on solution saturation state. *American Journal of Science*, **291**, 649–686.

NIKOLAEVA, N. M. & PIROZHKOV, A. V. 1978. The solubility product of U(IV) hydroxide at elevated temperatures. *Izvestiya Sibirskogo Otdeleniya Akademii Nauk SSSR, Seriya Khimicheskikh Nauk*, **5**, 82–88.

PARKHURST, D. L. & APPELO, C. A. J. 1999. *User's Guide to PHREEQC (Version 2). A Computer Program for Speciation, Batch-Reaction, One-Dimensional Transport and Inverse Geochemical Calculations*, USGS, 312 pp. Water Resources Investigations Report 99–4259. U.S. Department of the Inaterior U.S. Geological Survey, Denver, Colorado.

PLUMMER, L. N. PRESTEMON, E. C. & PARKHURST, D. L. 1991. *An Interactive Code (NETPATH) for Modeling Net Geochemical Reactions Along a Flow Path*. U.S. Geological Survey, Water Resources Investigations Report 91-4078.

POINSSOT, C., TOULHOAT, P. *et al.* 2001. *Synthesis on the Long Term Behaviour of the Spent Nuclear Fuel. Volume I. Direction des Technologies de l'Information*. CEA Rapport CEA-R-5958 (E).

RAI, D., FELMY, A. R. & RYAN, J. L. 1990. Uranium (IV) hydrolysis constants and solubility product of $UO_2 \cdot xH_2O(am)$. *Inorganic Chemistry*, **29**, 260–264.

ROVIRA, M., DE PABLO, J., EL AAMRANI, S., DURO, L., GRIVÉ, M. & BRUNO, J. 2003. *Study of the Role of Magnetite in the Immobilisation of U(VI) by Reduction to U(IV) under the Presence of $H_2(g)$ in Hydrogen Carbonate Medium*. Svensk Kärnbränslehantering, AB, Stockholm, Sweden, SKB TR-03-04, 44 pp.

RYAN, J. L. & RAI, D. 1983. The solubility of U(IV) hydrous oxide in sodium hydroxide solutions under reducing conditions. *Polyhedron*, **2**, 947–952.

SILVA, R. J., BIDOGLIO, G., RAND, M. H., ROBOUCH, P. B., WANNER, H. & PUIGDOMENECH, I. 1995. Chemical thermodynamics of americium. *In*: *Chemical Thermo-dynamics 2*. NEA OECD, Nuclear Energy Agency, Organisation for Economic Cooperation Development, Ed. North Holland Elsevier Science Publishers, B.V., Amsterdam, The Netherlands.

SKI 2001. SKI Report 01 : 4–SSI Report 2001 : 03. SKI and SSI's Joint Review of SKB's Safety Assessment Report, SR 97 Review Report (May 2001) ISSN 1104-1374; ISSN 0282-4434; ISRN SKI-R–01/4–SE. Swedish Nuclear Power Inspectorate, Stockholm, Sweden.

TACHI, Y., SHIBUTANI, T., SATO, H. & YUI, M. 2001. Experimental and modeling studies on sorption and diffusion of radium in bentonite. *Journal of Contaminant Hydrology*, **47**, 171–186.

TREMAINE, P. R. CHEN, J. D., WALLACE, G. J. & BOIVIN, W. A. 1981. Solubility of uranium(IV) oxide in alkaline aqueous solutions to 300°C. *Journal of Solution Chemistry*, **3**, 221–230.

WINBERG, A. & STEVENSON, D. 1994. Hydrogeological modelling. *In*: CRAMER, J. & SMELLIE, J. (eds) *Final Report of the AECL/SKB Cigar Lake Analog Study*. Svensk Kärnbränslehantering, AB, Stockholm, Sweden, SKB TR 94-04, 104–142.

WRONKIEWICZ, D. J., BATES, J. K., GERDING, T. J., VELECKIS, E. & TANI, B. 1992. Uranium release and secondary phase formation during unsaturated testing of UO_2 at 90 °C. *Journal of Nuclear Materials*, **190**, 107–127.

Colloid influence on the radionuclide migration from a nuclear waste repository

HORST GECKEIS

Institut für Nukleare Entsorgung, Forschungszentrum Karlsruhe – in der Helmholtz Gemeinschaft, Karlsruhe, Germany (e-mail: geckeis@ine.fzk.de)

Abstract: Colloid formation is discussed as a possible pathway for the radionuclide release from a nuclear waste repository. An assessment of the colloid relevance on radionuclide migration requires insight into the possible colloid generation mechanisms, their stability and mobility under given groundwater conditions. In various experiments dedicated to the investigation of nuclear waste form behaviour in contact with groundwater, colloidal species are observed mainly for the tri- and tetravalent actinides even in rather high ionic strength solutions where destabilization of colloids is expected. Experimental evidence of laboratory and field studies suggests colloid instability in saline groundwater with time. Groundwater of low ionic strength and high pH enhances colloid stability, as demonstrated in various laboratory and field experiments. The results of an *in situ* colloid characterization study at the Äspö hard rock laboratory in Sweden are discussed as an example. The mechanism of radionuclide interaction with colloids and notably the reversibility of the radionuclide–colloid binding are other key issues of colloid studies. Kinetic stabilization of the radionuclide binding to colloids may lead to a considerable enhancement of the colloid-mediated radionuclide migration. Substantial kinetic inhibition of actinide dissociation from humic colloids has been established by studying the behaviour of natural humic colloid borne U, Th and rare earth elements. Such behaviour might be explained by the incorporation of these elements into inorganic nanoparticles stabilized by humic coating. Spectroscopic evidence for the occurrence of actinide incorporation into colloidal structures is briefly discussed and the importance of considering the kinetics involved in all kinds of colloidal processes is emphasized. The enhanced migration of tri- and tetravalent actinide ions has been observed recently in various *in situ* dipole tests at the Grimsel hard rock laboratory in Switzerland. Such experimental findings underline the necessity of further studies on the colloid influence on actinide mobility. Moreover, an improved understanding of colloid–rock interaction mechanisms is required, which is essential for the description and prediction of colloid filtration processes.

The reliable long-term safety assessment of a nuclear waste repository requires the quantification of all processes that may affect the isolation of the nuclear waste from the biosphere. The colloid-mediated radionuclide migration is discussed as a possible pathway for radionuclide release. As soon as groundwater has access to the nuclear waste, a complicated interactive network of physical and chemical reactions is initiated, and may lead to: (1) radionuclide mobilization; (2) radionuclide retardation by surface sorption and co-precipitation reactions; and (3) radionuclide immobilization by mineralization reactions, that is, the inclusion of radionuclides into thermodynamically or kinetically stabilized solid host matrices.

Ninety years ago, Paneth observed for the first time the occurrence of colloidal species for low-solubility radionuclides and coined the term *radiocolloid* (Paneth 1914). Since the 1980s, many efforts have been undertaken to assess the role of such colloids on the geochemical behaviour of radionuclides and particularly of the nuclear-waste-derived radionuclides (Kim *et al.* 1984). Knowledge of the aquatic chemistry of radionuclides with regard to their interaction with dissolved ligands and with mineral surfaces has clearly increased during recent years. Considering the colloid chemistry of nuclear-waste-derived radionuclides, the situation is much worse. Today, neither the colloid generation mechanisms in the near field of a nuclear waste repository nor the colloid migration processes in the far field are yet fully understood in a quantifiable way (Honeyman 1999). The assessment of colloid relevance for the safety of nuclear waste disposal still appears to be rather difficult. To some extent this may be

From: GIERÉ, R. & STILLE, P. (eds) 2004. *Energy, Waste, and the Environment: a Geochemical Perspective*. Geological Society, London, Special Publications, **236**, 529–543.
0305-8719/04/$15   © The Geological Society of London 2004.

the consequence of the difficulties connected with the accurate and sensitive analysis of colloids in natural and laboratory-derived samples and with a number of possible artifacts. Another major problem is the variety of possible colloid types exhibiting quite different geochemical properties. Polyvalent actinides such as Pu form nm-sized polymerized species upon hydrolysis under oversaturation conditions, a behaviour that is also observed for elements like Fe and Si. Clay minerals are known to occur dispersed in groundwater in sizes down to several tens of nanometers and the macromolecular humic/fulvic acids produced through degradation of all kinds of organic matter in the biosphere are also frequently found as constituents of various natural waters. Colloid formation of radionuclides by polymerization or by interaction with different types of groundwater colloids may result in species of quite variable geochemical behaviour and mobility.

The potentially important role of colloidal species in the geochemical behaviour of the polyvalent actinides has nevertheless been stated by various authors (e.g., Kim 1991; Kersting *et al.* 1999). The present paper discusses the role of colloids on the release of radionuclides from a nuclear waste repository with regard to the processes leading to: (1) colloid generation and stability; (2) radionuclide interaction with aquatic colloids; and (3) colloid-borne radionuclide migration.

Colloid generation processes and colloid stability

In general, generation of aquatic colloids in natural groundwater can be expected in chemically or mechanically perturbed aquifer systems (e.g., Ryan & Gschwend 1994). Therefore, it is reasonable to assume that in a nuclear waste repository system, consisting of the nuclear waste forms, container materials and backfill material artificially placed into the respective hostrock, the contact with groundwater may create hydraulic and geochemical gradients for a certain time period. In such systems, the possible colloid generation and interaction processes are manifold (Fig. 1).

In a first step, waste-form components start to dissolve when in contact with intruding groundwater and release radionuclides (Fig. 1, left). Subsequent nucleation due to oversaturation with regard to stable secondary phases or as a consequence of redox reactions may lead to colloid formation. Radionuclide sorption to already present groundwater colloids or colloids generated during container corrosion (Fig. 1, middle) and detachment of waste-form particles from the solid surface or surface alteration layers are further possible colloid-generating mechanisms. Colloid formation upon nuclear waste form contacted with groundwater has been observed in various studies (summarized in Ahn 1996; Geckeis *et al.* 1998). Colloidal Np, Pu, and U species have been found to

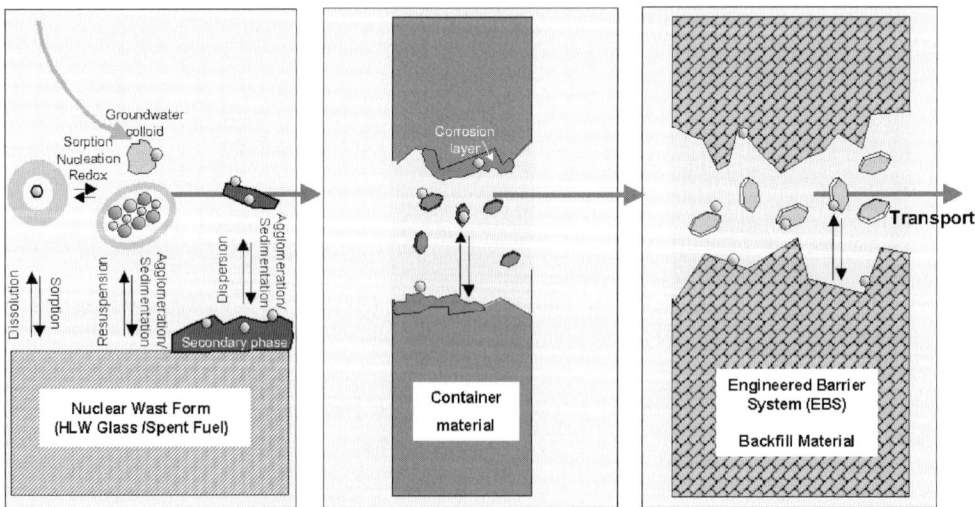

Fig. 1. Potential colloid generation processes in a nuclear waste repository near field (for explanation, see text) (HLW, high-level waste; EBS, engineered barrier system).

contribute considerably to the release from vitrified nuclear waste. They consist mainly of secondary smectite, phosphate and silicate phases containing various amounts of radionuclides (Bates *et al.* 1992; Feng *et al.* 1994; Buck & Bates 1999). The concentration of these colloids appears to correlate with the ionic strength (I) of the leaching solution. The salinity of the leachate increases in closed-system experiments (batch experiments) at extended corrosion periods due to the increasing concentration of dissolved glass components. As a consequence, the simultaneous decrease of the colloid population is stated. Colloidal phases generated from concrete and cemented waste have been identified to consist mainly of calcium silicate hydrates (CSH) and $CaCO_3$ (Wieland & Spieler 2001; Fujita *et al.* 2003; Swanton & Vines 2003). Various radionuclides are known to interact with these phases and thus may be mobilized. The relatively low colloid concentrations found by the authors in their investigations are explained by the high concentrations of Ca acting as coagulant and, thus, exerting a destabilizing effect. In spent fuel dissolution experiments, colloidal species are mainly found for the tri- and tetravalent actinides. The exact chemical composition of these colloids is presently unknown (Geckeis *et al.* 1998). No significant colloid impact is noted for Cs and Sr. Colloids exist at generally higher concentrations in low-mineralized granitic groundwater as compared to saline solutions, taken to simulate the conditions in a rock salt repository (Fig. 2).

It is interesting to note that even at the highly saline conditions of 5 mol/kg NaCl (Fig. 2) colloidal radionuclide species are still observed. In the majority of experiments carried out in simulated salt brines, the appearance of colloid-borne actinides and rare earth elements (REE) is found to be transient. A subsequent decrease in their concentration indicates that these colloids are unstable. In Fig. 3, the time-dependent behaviour of dissolved and colloidal (determined by ultra-filtration) Am(III) and Eu(III) released from a spent fuel pellet is shown to be accompanied by a pH increase and a decrease in U concentration. Both facts suggest the co-precipitation of REE and trivalent actinides with secondary U-containing phases (see also Quinones *et al.* 1996). The example demonstrates that colloids may be generated as a consequence of geochemical variations even under high ionic strength conditions. The transient colloid concentration reveals that the observation of increased colloid concentrations at a certain point in time does not necessarily allow conclusions with regard to their relevance in the long term. Thus, the kinetics of colloid processes have to be considered.

The generation of colloids from bentonite discussed as a barrier and backfill material (Fig. 1,

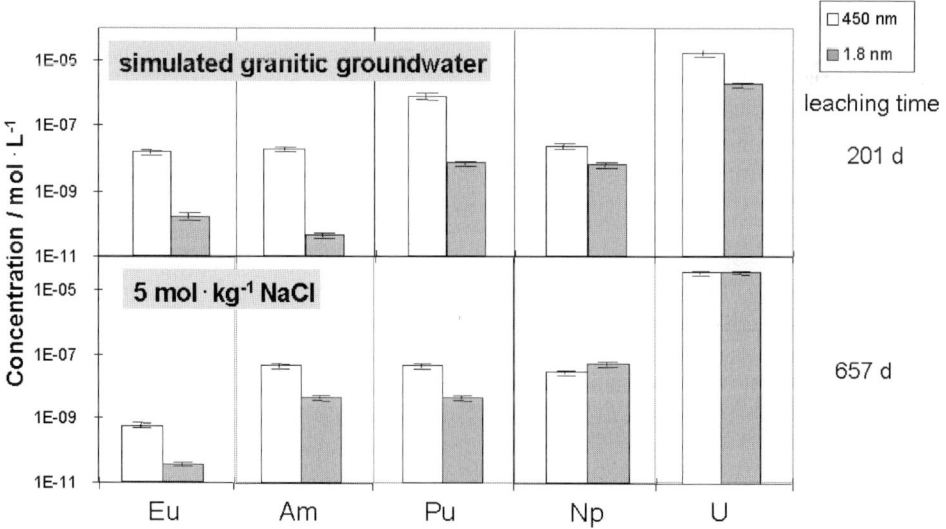

Fig. 2. Radioelement concentrations in solutions contacted with powdered spent fuel (UO_2; burn-up: 50 MW d/kg U; γ-dose rate: ~10 Mrad/h; solid surface/solution volume ratio: 1000/m at 25 °C) after sequential filtration (filter pore size: 450 nm → white bars; filter pore size: 1.8 nm → grey bars); solutions consist of concentrated brine (5 mol/kg NaCl) and simulated granitic groundwater ($I = 2.8 \times 10^{-3}$ mol/L, pH ~ 8) (Geckeis *et al.* 1998).

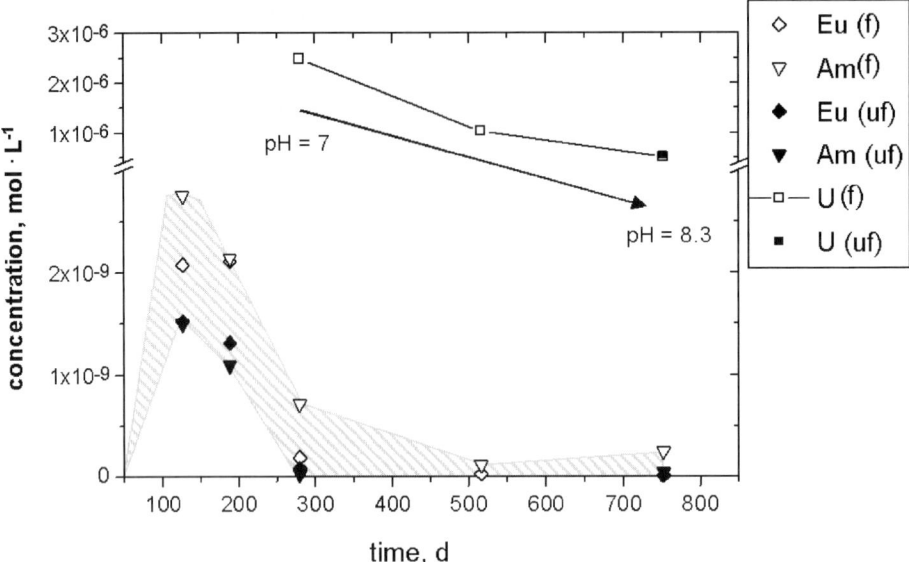

Fig. 3. Evolution of Am(III), Eu(III) and U concentrations with time in spent fuel pellet leaching experiments (leachate: 5 mol/kg NaCl solution; anaerobic conditions); radionuclides found in ultrafiltered samples (uf: filter pore size: 1.8 nm) are considered as 'truly dissolved'; radionuclide concentrations found in filtered samples (f: filter pore size: 450 nm) are attributed to 'truly dissolved + colloidal' species; the grey shaded area marks the fraction of colloidal radioelement species in solution; the black arrow indicates the pH increase in solution during the leaching experiment (Geckeis *et al.* 1998).

right part), especially in granitic repository concepts, has been the subject of recent experimental studies (Missana & Adell 2000; Schäfer *et al.* 2000; Missana *et al.* 2002). Mechanical erosion induced by water flow is assumed to play the dominant role, leading to the washout of smectite-like clay colloids with sizes ranging from 10 to 1000 nm. Such colloids may contribute to the mobilization of those radionuclides with a high sorption affinity to clay colloids. Stability of smectite clay colloids derived from bentonite barriers has been studied under various conditions. Surface charge is one parameter controlling colloid stability and quite often is assessed by analysing the ζ-potential. It is defined as the electrostatic potential measured at the plane of hydrodynamic slippage outside the colloidal surface and is usually measured by electrokinetic methods. A clearly negative ζ-potential of -30 to -50 mV has been found over wide ranges in pH and ionic strength for the smectite clay colloids. Hence, strong electrostatic repulsion would prevent close approach of colloids and thus colloid agglomeration. Results of agglomeration experiments reveal, however, a considerable colloidal stability only at an ionic strength of $\leq 10^{-3}$ mol/L and at pH > 7. This behaviour is explained by the different protonation/deprotonation characteristics of silanol and aluminol

groups. Deprotonation of the silanol groups at quite low pH (~ 2) and permanent charge induced by isomorphic substitution in octahedral and tetrahedral sheets of the clay minerals is responsible for the overall negative charges of the smectite colloid surface. The aluminol sites at the smectite edges, however, start to deprotonate at pH $\sim 8-9$, creating local patches with positive charges at lower pH. Agglomeration is thus induced at near-neutral pH due to edge to plane attraction. Interaction of groundwater with natural bentonite leads to the establishment of porewater solutions with considerably higher ionic strength ~ 0.1 mol/L (Bradbury & Baeyens 2003). Such processes should inherently delimit the colloid concentration over the long term in the near field by destabilising them. Stable bentonite colloid dispersions can, therefore, be expected to play an important role at the interface of the bentonite with groundwater of low ionic strength.

A major outcome of the abovementioned studies is the importance of pH and notably of the salinity of the groundwater controlling colloid concentrations and, consequently, the relevance of colloids for radionuclide transport. The pH-dependent colloid stability varies considerably for different colloid types. Experimental data for the relationship of the stability ratio W

with pH are plotted in Fig. 4 for smectite (Missana & Adell 2000) and ZrO_2 colloids (Bitea *et al.* 2003*a*). The latter are taken as a potential homologue to colloidal PuO_2 due to their similar isolectric point (pH_{iep} for ZrO_2: 8.2; for PuO_2: 8.5–9). The definition of W is as follows:

$$W = \frac{\left[\left(\frac{dr_h}{dt}\right)_{t\to0}\right]_{(f)}}{\left(\frac{dr_h}{dt}\right)_{t\to0}}$$

where r_h corresponds to the hydrodynamic radius of the colloid and t represents the time; W is determined from the initial agglomeration rate $(dr_h/dt)_{t\to0}$ in the fast agglomeration regime (index: f) where no electrostatic repulsion occurs divided by the rate observed under stabilizing conditions. At $pH \to pH_{iep}$ and at high ionic strength, W approximates a value of 1. While for the smectite colloids destabilization is found under acidic conditions, the opposite is true for ZrO_2 colloids showing high stability at low pH. For both colloid types, however, W drops to 1 for a wide pH range when increasing the ionic strength to 0.1 mol/L. Even under destabilizing conditions close to the pH_{iep} and at high ionic strength, the agglomeration rate strongly depends on the colloid concentration. This relationship has been validated for ZrO_2

colloid trace concentrations ranging from 3.4×10^{11} to 2.6×10^9 particle number/L by applying the laser-induced breakdown detection (LIBD) arrangement (Bitea *et al.* 2003*a*). This method has been developed as a colloid-selective analytical tool of ultimate sensitivity, ranging to the ng/L concentration level for particles as small as 5 nm (e.g., Walther *et al.* 2002). Such sensitivity is particularly required for colloid characterization in groundwater of low colloid population. While at the highest investigated colloid concentration, the agglomeration half-life is at 21.3 min, it increases to more than two days for the lowest concentration. This finding, again, demonstrates the necessity of considering the kinetics involved in all kinds of colloidal processes, and shows that colloids at trace concentrations may exist for some time even under destabilizing conditions.

Colloid analysis in natural groundwater supports the findings on colloid stability obtained in laboratory experiments. A recent experimental study on various granitic groundwaters at the Äspö hard rock laboratory (HRL) in Sweden demonstrates the necessity of artifact-free *in situ* colloid characterization (Hauser *et al.* 2003). Groundwater samples with a broad variation in salinity and concentration of natural organic matter due to their origination from different sources are studied by the application of a mobile LIBD arrangement. The Äspö groundwater is

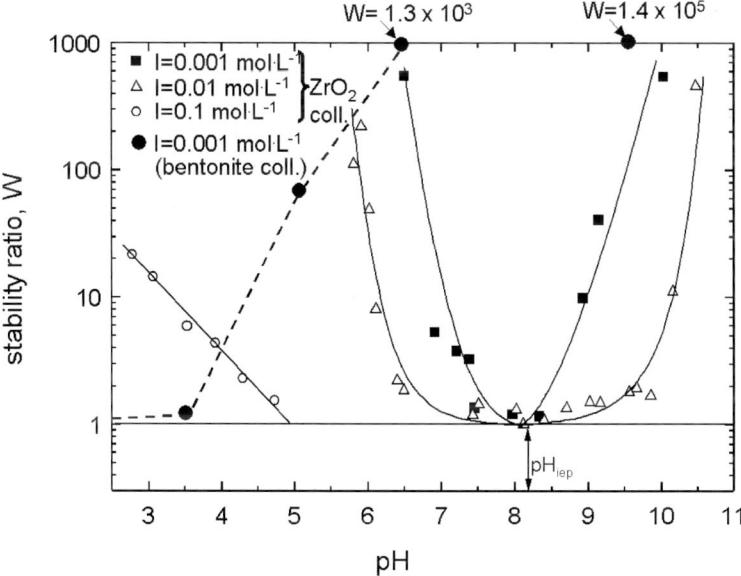

Fig. 4. Stability ratio (W) determined for ZrO_2 (Bitea *et al.* 2003*a*) and smectite colloids (Missana & Adell 2000) as a function of pH and ionic strength.

characterized by quite high concentrations of Fe(II), dissolved CO_2 and high hydraulic pressures. Off-site colloid analysis becomes almost impossible due to artificial colloid generation by degassing processes and oxygen accessing the sample. These processes result in precipitation of carbonate and Fe(III) phases and, thus, lead to erroneous results on colloid concentrations. Extensive reviews on analytical strategies for colloid characterization emphasizing the possible interferences are already available (e.g., McCarthy & Degueldre 1993). On-line analysis of the water samples in a flow-through pressure cell under *in situ* conditions allows for the minimization of those artifacts. A clear correlation of increasing colloid concentration with decreasing groundwater salinity is observed *in situ* at the Äspö HRL (Fig. 5a).

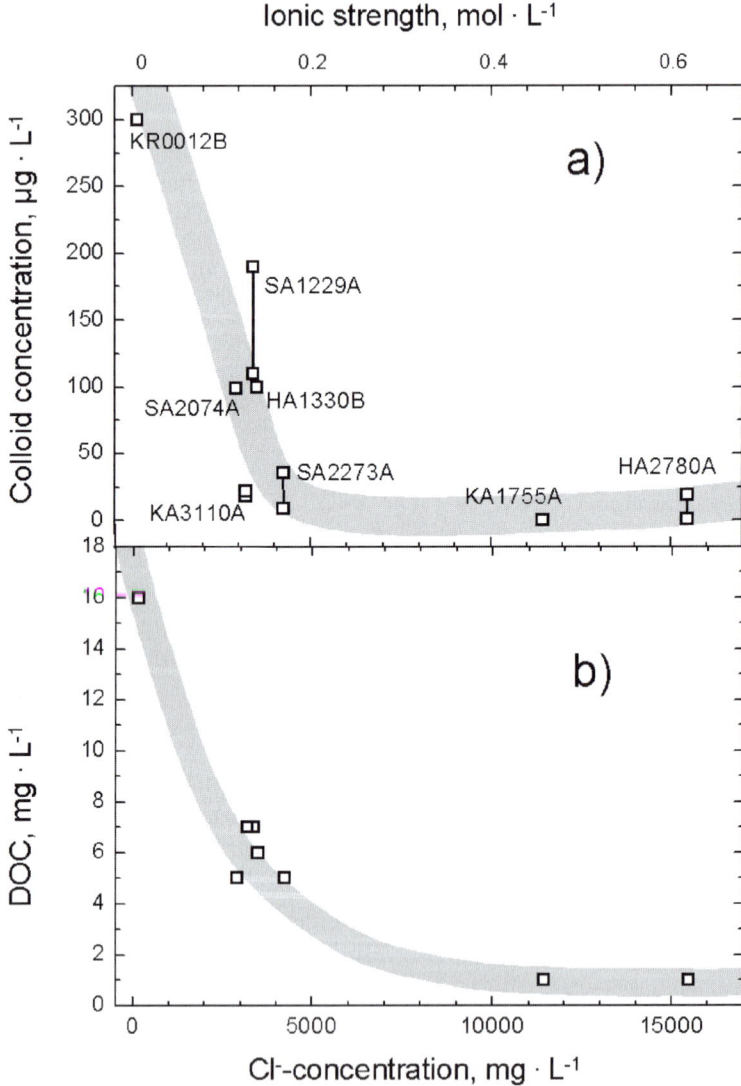

Fig. 5. Colloid concentrations determined under *in situ* conditions at the Äspö hard rock laboratory, Sweden. Analysis of colloids has been performed on line by (**a**) using a mobile laser-induced breakdown detection (LIBD) arrangement, and (**b**) DOC analysis of collected samples in the laboratory (Hauser *et al.* 2003). Bars in the upper diagram represent colloid concentration ranges detected during the campaign. Colloid concentrations and DOC are plotted against salinity expressed as the groundwater Cl^- concentration (lower *x*-axis) and ionic strength (upper *x*-axis).

Colloid concentrations drop to almost zero at $I > 0.2$ mol/L. The same trend is found for humic/fulvic acids represented in Fig. 5b as dissolved organic carbon (DOC). The studies, however, also make clear that the variation of the chemical conditions with the creation of oxidizing conditions by, for example, radiolysis may temporarily generate high colloid populations even under conditions of high ionic strength.

Correlation of colloid concentration with groundwater chemistry has also been reported by Degueldre et al. (1996a, b, 2000). Qualitatively, the influence of ionic strength on colloid abundance can be explained by the increasing tendency of colloids to surface attachment at increasing salinity. Degueldre and co-workers, however, also emphasize the influence of DOC counteracting colloid destabilization. Humic/fulvic acids have been reported to form during chemical and microbial degradation of organic ingredients in low-level nuclear waste (Caron et al. 1996). They are known to be available as inherent constituents of the backfill material or the host rock (Pirlet 2001) and as dissolved species in groundwater (Kim et al. 1984; Buckau et al. 2000). Humic/fulvic colloids are known to be potentially stable in groundwater for long time periods (Buckau et al. 2000). The polyelectrolyte-like organic colloids are able to complex polyvalent actinide ions (Kim et al. 1993) and may interact with inorganic colloid surfaces (e.g., Kretzschmar et al. 1998). The hydrophilic humic/fulvic acid coating with negatively charged deprotonated carboxylate groups exerts strong repulsive forces and thus increases the stability of such colloids even at higher ionic strength.

The behaviour of colloidal Th-oxyhydroxide species investigated in the course of solubility studies (Neck et al. 2002; Bitea et al. 2003b) appears to deviate from the general trend of colloid destabilization at increased ionic strength. Mostly very small colloidal species with sizes of 2–10 nm have been determined in those studies by the application of LIBD. They are found to be stable at quite high salinity (0.5 mol/L NaCl) for up to 489 days at pH ∼ 4. Colloidal Th concentrations in those experiments were in a range of 1.3×10^{-5} to 2.5×10^{-3} mol/L Th (i.e., 3×10^{-3} to 0.58 g/L). Based on these results, it has been argued that under certain conditions polynuclear oxide/hydroxide colloids might even be the thermodynamically stable species. Nothing, however, is known at present about their exact nature and their mobility, and further characterization studies are required. As the colloidal

Th-oxyhydroxide species readily dissolve upon dilution below the solubility limit, it is not very likely that such actinide(IV) colloids play a role away from the source in the far field of a repository. In the near field of a repository, however, they may be predominant species controlling the solubility of tetravalent actinide species such as U(IV) and Pu(IV) and thus the source term. Unusual stability at high ionic strength has been also reported for amorphous SiO_2 colloids (Iler 1979; Healy 1994) which also cannot be explained solely by electrostatic repulsion. Formation of oligomeric or polymeric silicate species at the colloid–water interface are thought to exert additional steric stabilization by preventing close approach of those particles.

Interaction of radionuclides with colloids

Mechanisms of radionuclide–colloid interaction relate to the critical questions regarding colloid relevance on radionuclide migration. Contardi et al. (2001) applied a three-phase model in order to account for the colloid influence on radionuclide migration at the Yucca mountain site. Based on various assumptions (reversible attachment of radionuclides to colloids; no colloid filtration), it is found that colloids significantly influence the retardation factor only at quite high colloid concentrations (≥ 1 mg/L). Furthermore, sorption coefficients for the radionuclide–colloid interaction must be high ($K_d > 10^4$ mL/g; K_d, ratio of sorbed to dissolved radionuclide mass: (mg/g)/(mg/mL)) as found, for example, for the tri- and tetravalent actinides. For weakly sorbing radionuclides, such as ^{90}Sr(II), transport of dissolved species will dominate their migration. Owing to its high affinity to the frayed edge sites of clay minerals (Bradbury & Baeyens 2000), Cs(I) speciation and migration may also be influenced by the presence of clay colloids. Colloid-mediated Cs migration has indeed been observed in column experiments related to the radionuclide contaminated site at Hanford, USA (Zhuang et al. 2003). Cs desorption from the mainly clay-type colloidal matter has been noted during passage through a porous medium, which points to the reversible binding of Cs to the colloids. Consequently, the assessment of the colloid relevance on radionuclide migration has up to now mostly been done by considering the radionuclide–colloid binding as being readily reversible. A similar three-phase approach has been applied by Meier et al. (2003) for the case of humic colloid mediated radionuclide transport in a porous sediment aquifer, and similar conclusions are drawn. Thus, one could come to the conclusion that colloidal species may only slightly reduce the radionuclide retardation and will not affect the long-term safety of a repository considerably unless colloid concentrations are relatively high (Ryan & Elimelech 1996; Contardi et al. 2001).

In a number of recent studies, however, it is found that the desorption of radionuclides from inorganic or organic colloids is subject to considerable kinetic hindrance. The question arises, whether radionuclide/colloid binding can be considered as reversible with respect to the relevant time-scale. Delayed desorption of Pu from hematite and goethite colloids has been observed, for example, by Lu et al. (1998). The pronounced hindrance of actinide ion desorption from humic colloids has led to the recognition that migration of colloid-borne metal ions usually cannot be described by a thermodynamic equilibrium approach (Artinger et al. 1998). Kinetic modeling appeared to be necessary to account for the experimental observations (Schuessler et al. 2000, 2001; Warwick et al. 2000). A closer inspection of the actinide desorption from humic colloids reveals that at least two kinetic modes exist. The first mode involves the complexation of polyvalent metal ions with purified humic/fulvic acids, where the subsequent dissociation rates decrease with increasing contact time (Choppin & Clark 1991; Geckeis et al. 2002). Such behaviour is explained by agglomeration or re-orientation of the humic macromolecules, thus 'wrapping' the actinide ion, or by the transfer of the metal ion to energetically more favourable binding sites. Values for the half-life of the dissociation reaction may range up to several days. A second kinetic mode is found for the desorption reaction of the naturally occurring humic colloid-borne REE, U and Th from the humic colloid matter in a natural groundwater. Those elements can only partly be desorbed from the humic matter within 200 days using a chelating ion-exchanger resin. Within that time period, added radioisotopes do not undergo any exchange with the naturally abundant humic colloid-bound metal ions (Geckeis et al. 2002). Application of flow-field flow fractionation (FFFF) allows a closer inspection of natural humic groundwater colloids from the Gorleben site in Germany. Colloids are fractionated in FFFF according to their size in a thin ribbon-like channel delimited by permeable walls (Beckett & Hart 1993). The elution profile obtained by monitoring the colloid concentration in the channel outflow is called fractogram and can be converted into a size distribution. Fractograms of a Eu(III) humate solution, prepared from purified humic acid (HA), and of natural groundwater humic colloids obtained by inductively coupled plasma mass spectrometric (ICP-MS) and spectrophotometric detection of the FFFF effluent reveal distinct differences (Fig. 6). The naturally abundant REE and actinides are situated in colloids of a size >17 nm, whereas Eu(III) bound as humate complex is located in colloids of <3 nm corresponding to the size range where the humic acid colloids are found. The natural humic colloids obviously contain entities of larger size hosting primarily the polyvalent metal ions. Such observation is explained by the presence of inorganic colloids stabilized by a humic/fulvic acid coating, which have indeed been visualized by atomic force microscopy (AFM) (Plaschke et al. 2002). The 'quasi irreversible' incorporation of actinide ions into nanocrystalline colloids may represent a possible way for the significant kinetic stabilization of colloid-borne radionuclides and, thus, to a colloid-mediated transport of a certain radionuclide fraction over long distances.

Whether or not such type of 'quasi irreversible' radionuclide binding to colloids at the long time-scales pertinent to nuclear waste disposal is relevant, is certainly difficult to assess. Some general considerations, however, can be made by examining the mechanisms of radionuclide interaction with colloids. The application of spectroscopic tools, such as X-ray absorption spectroscopy (XAS) and time-resolved laser fluorescence spectroscopy (TRLFS), allows such mechanistic insight (e.g., Charlet & Manceau 1993; Geckeis et al. 1999). The outcome of a number of TRLFS studies on the sorption of the trivalent actinide Cm(III) onto colloidal solids is schematically represented in Fig. 7. The fluorescent Cm(III) exhibits a geochemical behaviour consistent with other trivalent actinides, for example, Am(III) (Rabung et al. 2000), allowing TRLFS investigations even at trace concentrations (10^{-8} mol/L). Sorption to smectite colloids is found to proceed via outer-sphere complexation at pH < 5 and at low ionic strength. Under these conditions, the hydration sphere of the Cm^{3+} aquo ion remains unaffected upon the interaction with the smectite surface (Stumpf et al. 2001a). Such information becomes available from both the unchanged fluorescence lifetime ($\tau = 68$ μs) and peak position of the Cm(III) fluorescence spectra ($\lambda = 593.8$ nm), even though almost complete metal ion sorption is found in batch sorption experiments. The outer-sphere complexation is explained as an interaction of Cm^{3+} with permanent charges at the smectite surface. At increasing pH, a clear red shift of the Cm(III) emission spectrum combined with an increase

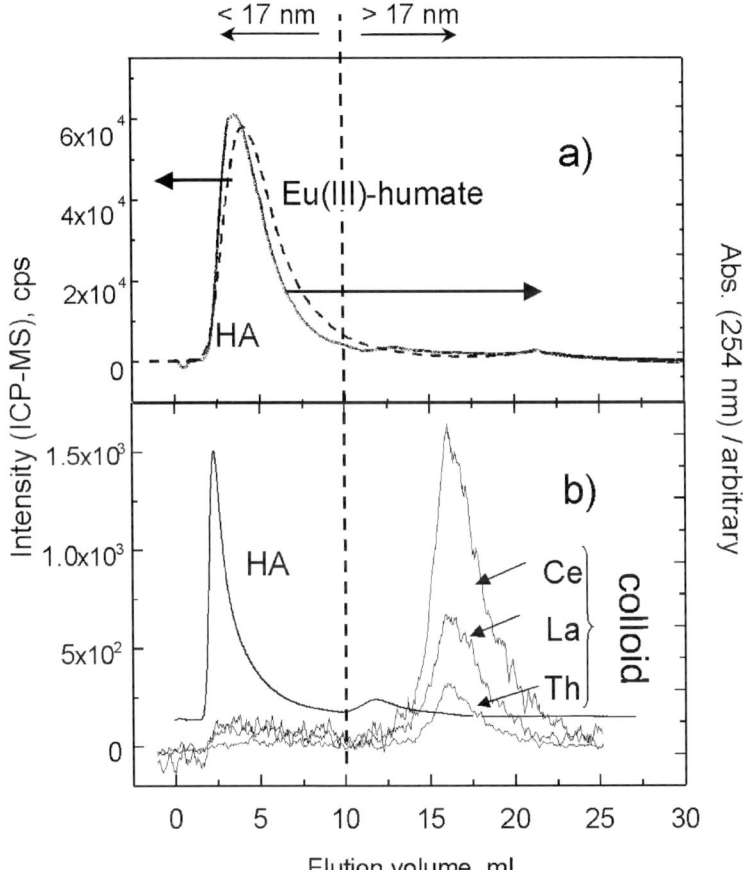

Fig. 6. Colloid size fractionation obtained by FFFF of (**a**) a Eu(III) humate solution ([HA]: 30 mg/L; Eu(III): 10^{-6} mol/L; pH 8.0) and (**b**) a humic/fulvic acid-containing groundwater ([DOC]: 81.9 mg/L; pH 7.7). Vertical axes depict the response of the ICP-MS detector (cps: counts per second; left axis) and of the photometric detector (UV/Vis; $\lambda = 254$ nm; right axis) (Geckeis *et al.* 2002).

of the fluorescence life-time ($\tau = 110$ μs) indicates the formation of inner-sphere complexation with colloid-surface hydroxyl groups. Via an empirical correlation (Kimura & Choppin 1994), the fluorescence life-time can be used to determine the number of coordinating H_2O molecules in the first coordination sphere. It decreases from eight to nine characteristic to the aquo ion down to four to five for the surface sorbed species (Fig. 7). The TRLFS spectra are quite comparable with the results obtained from studies on the Cm(III) sorption on colloidal γ-alumina (Stumpf *et al.* 2001*b*) and, thus, the surface complexation to aluminol groups at the smectite edge sites is suggested at increased pH. Inner-sphere complexation with carboxylate functional groups is also stated by spectroscopic methods for actinide interaction

with humic/fulvic colloids (Kim *et al.* 1993). Both types of actinide sorption reactions can be considered reversible within reasonably short time-scales as the surface-sorbed species are still in direct contact with the surrounding aquatic environment.

Much stronger kinetic stabilization can be expected for processes leading to the inclusion of radionuclide ions into the colloid structure (Fig. 7, lower part). Spectroscopic indications for such processes have indeed been found again by TRLFS for the Cm(III) interaction with colloidal and particulate amorphous silica, calcite and CSH phases (Chung *et al.* 1998; Stumpf & Fanghänel 2002; Tits *et al.* 2003). The incorporation of actinide ions into colloidal precursor clay phases has been recently investigated as a possible mechanism in natural

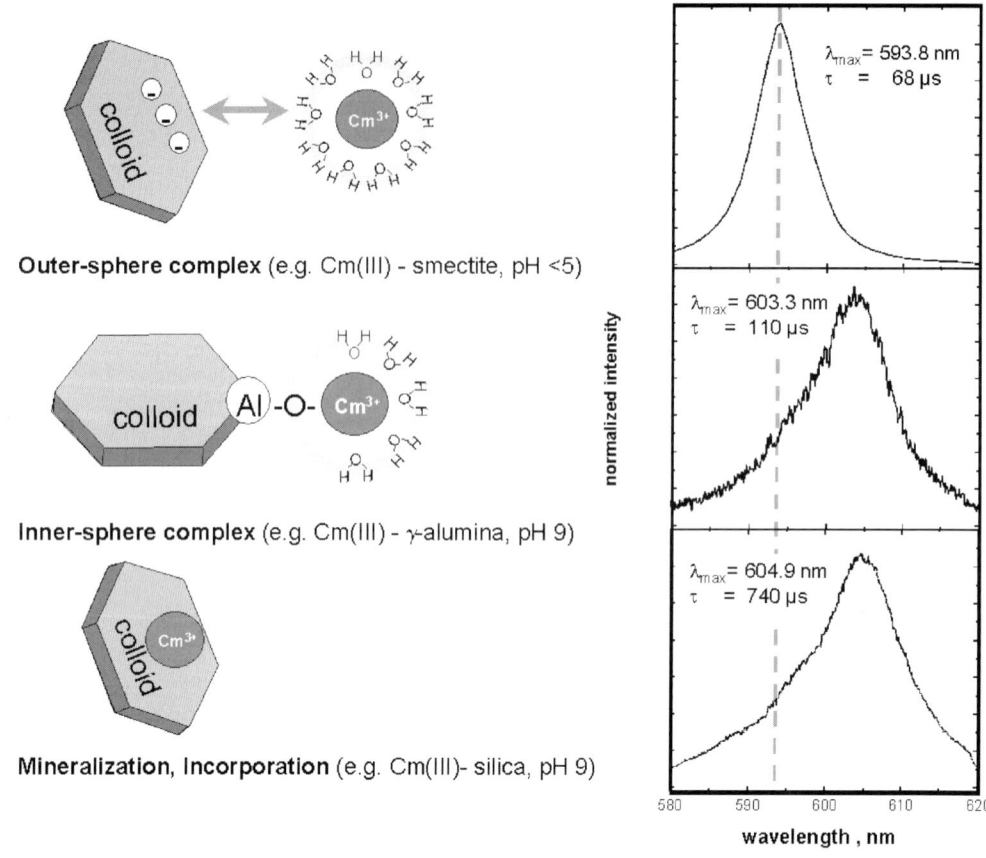

Fig. 7. Potential mechanisms of actinide (represented by Cm(III)) interaction with colloids as interpreted from laser fluorescence spectroscopy (TRLFS) experiments. Spectra are taken from Stumpf *et al.* (2001*a, b*) and Chung *et al.* (1998).

systems (Kim *et al.* 2003). Under given conditions, almost complete displacement of the H_2O molecules from the first coordination sphere is found. Incorporation of trace metals into mineral phases by solid–solution formation or non-ideal incorporation processes (e.g., occlusion) is a common observation in nature. It is, therefore, permissible to assume similar behaviour in the case of colloidal matter. For such compounds, a very strong kinetic stabilization of the actinide–colloid binding for long timescales cannot be excluded, but in most cases has not yet been examined in detail.

Colloid mobility

The mobility of humic/fulvic colloid-borne tri-, tetra-, penta- and hexavalent actinide ions has been clearly observed in sand column experiments (Kim *et al.* 1994; Artinger *et al.*

1998, 2002, 2003). In those studies, the colloid-containing groundwater had been previously equilibrated with the sandy sediment until the humic/fulvic acid concentrations of inflowing and outflowing water were identical. Under these conditions, the actinide dissociation rate from the colloids plays the dominant role in controlling the fraction of mobile colloid-borne actinides.

In a number of dynamic laboratory experiments carried out in porous and fractured media, colloid-facilitated radionuclide migration has been investigated without equilibrating the stationary phase with colloid-containing groundwater. Vilks and co-workers studied colloid migration (Vilks & Bachinski 1996) and the colloid influence on radionuclide migration in granitic fractures in block scale experiments (Vilks & Baik 2001). It was shown that Am(III) was carried to a small extent by natural colloids (19.3 mg/L) consisting of mainly organic particles and aluminosilicates. Bentonite colloids appeared to be mobile to only a small extent under given conditions (pH 8.3, $I = 1.3 \times 10^{-2}$ mol/L). The major fraction of those colloids underwent filtration processes. Experiments with functionalized latex and silica nanoparticles in such a system showed the colloid migration being different from that of conservative non-retarded tracers with respect to recovery and arrival times. Different pathways of conservative tracer and colloid migration due to size exclusion and filtration phenomena are considered responsible for that observation. A decrease of colloid mobility with decreasing flow rates is observed. Field-scale experiments indicated a rather high mobility of silica colloids in a granitic fracture over distances of 17 m (Vilks et al. 1997).

In contrast to the findings of Vilks and co-workers, a quite high mobility for bentonite colloids has been reported in the various in situ migration studies performed at the Grimsel test site in Switzerland. Under the high-pH and low-ionic-strength conditions (pH 9.6, $I = 10^{-3}$ mol/L) of the groundwater, colloid dispersions have been found to be extremely stable even over months. Such a finding is in agreement with colloid studies performed in various groundwater

samples from this area (Degueldre et al. 1996b). The colloid-borne lanthanides and actinides Tb(III), Am(III) and Th(IV), Pu(IV) are found to be transported in dipole experiments over distances up to 5 m with quite high recoveries ranging from 30% to 80% (Hauser et al. 2002; Möri et al. 2003). Dipoles have been established in a granitic shear zone filled with secondary minerals. Breakthrough concentrations for the migration of actinides (U(VI), Np(V), Pu(IV), Th(IV), Am(III)), and the fission products Cs(I) and Sr(II) in a 2.23 m long dipole are plotted over the elution time in Fig. 8 (Geckeis et al. 2003). Such breakthrough curves reflect the mobility of individual species by comparing them with the non-retarded or so-called conservative tracer I-131 injected as iodide. More than 70% of the colloids (injection concentration: 20 mg/L) and the adherent Am(III) and Pu(IV) are recovered. Owing to its comparatively strong sorption to smectite and clay in general (Bradbury & Baeyens 2000; Nakano et al. 2003), a small fraction of <1% Cs(I) is also found to be transported with bentonite colloids while the major part experiences a quite extensive retardation. As a result of size or charge exclusion from fracture walls and pores, the maximum of the colloid breakthrough curves appears slightly ahead of the I⁻ peak.

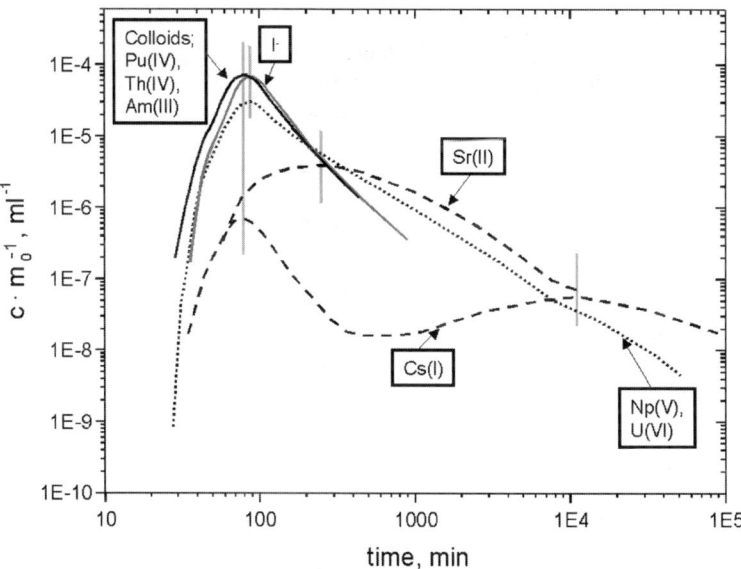

Fig. 8. Radionuclide migration studied in a granitic shear zone at the Grimsel test site, Switzerland (injection flow rate: 10 mL/min; extraction flow rate: 150 mL/min, dipole distance: 2.3 m). Am(III), Pu(IV) and Th(IV) are co-eluted with the colloids; grey vertical lines indicate maxima of breakthrough curves (Geckeis et al. 2003). In order to allow a direct comparison of breakthrough curves, the colloid and radionuclide concentrations (c in mg/mL) in the extracted water samples are normalized to the total injected mass of individual colloid or radionuclide tracers (m_0 in mg).

Under given conditions, colloids do not have an impact on the migration of Sr(II), U(VI) and Np(V), which display only weak or no interaction with the colloids. While Sr(II) migrates as a weak sorbing tracer, the oxidized actinides U(VI) and Np(V) are partly eluted together with the conservative tracer, that is, without retardation. A second part of U(VI) and Np(V) appears slightly retarded probably due to the weak reversible interaction with surfaces of the fracture infill.

The partly contradictory outcome of the migration experiments in granitic systems can be explained to a certain extent by the different groundwater chemistry. The high pH and weakly mineralized Grimsel groundwater is certainly predestined for high colloid stability and mobility, thus being responsible for the high colloid recoveries in the investigated system. The groundwater of low ionic strength ($\sim 2 \times 10^{-3}$ mol/L) and high pH (8.4) may also account for the colloid-mediated radionuclide migration in a shallow aquifer over a distance of 1.3 km observed by Kersting et al. (1999). They studied the propagation of different radionuclides, including Pu, from a nuclear test facility at the Nevada site through a subsurface aquifer. Their results are currently considered as a strong indication that radionuclides hitherto assumed to be immobile can be mobilized by colloid formation and migrate over long ranges.

The studies on colloid-facilitated radionuclide migration discussed above demonstrate that colloids may indeed act as mobile carriers for the transport of radionuclides. Notably under those groundwater conditions favouring colloid stability, the colloid mobility is for obvious reasons found to be quite high. However, there are indications that the colloid migration under low groundwater flow velocity conditions considerably favours colloid attachment and thus filtration of colloids in porous and fractured media (Vilks & Bachinski 1996). Therefore, different flow velocities and flow path geometries may also contribute to the different outcome of the experiments in granitic fractures discussed above. In order to obtain a complete picture of the colloid-mediated radionuclide migration processes, more knowledge on the interaction of the colloids with sediment and rock surfaces is required. The currently available theories on colloid filtration still appear to be inappropriate to describe the processes in heterogeneous natural systems (Ryan & Elimelech 1996). The existence of extensive chemical and physical heterogeneity of natural solids is invoked as a main reason for the failure of classical electrostatic surface interaction models as

represented by the DLVO theory (named after the scientists who first published that approach: B. Derjaguin, L. Landau, E. Vervey, and T. Overbeek), which describes the attractive and repulsive forces between surfaces. Modified modelling approaches taking such heterogeneities into account are under development (Song et al. 1994). The enlightenment of respective colloid–rock interaction processes would certainly enhance the reliability of colloid transport modelling efforts.

Conclusions

The present paper shows that it is still necessary to fill some gaps before a reasonable assessment of colloid relevance for the long-term safety of a nuclear waste repository can be given. As a matter of fact, a nuclear waste repository deep underground, with water penetrating, has to be considered as a chemically and mechanically disturbed system more or less far from geochemical equilibrium. The generation of colloidal species, therefore, cannot be excluded. A quantification of a 'colloidal source term' in a repository system of pronounced heterogeneity with respect to waste components and geochemical processes will to be rather difficult. In particular for the tri- and tetravalent actinides, the generation of colloidal species will be of high relevance due to their tendency to undergo polynucleation in solution and to interact strongly with aquatic colloids. In the presence of significant clay colloid concentrations, Cs isotopes may also be prone to colloid formation. An important criterion influencing the colloid relevance for a nuclear waste repository is the stability of colloidal dispersions, which is clearly decreased in high ionic strength groundwater as concluded from laboratory and in situ colloid studies. Hydrophilic and steric stabilization mechanisms may, however, counteract this tendency and have to be studied in future investigations.

Mechanisms that may lead to the 'quasi irreversible' binding of radionuclides to colloids belong to the key uncertainties of the assessment of the colloid problem. The kinetics of the dissociation of colloid-bound radionuclides are not yet understood. Radionuclide incorporation into stable colloids may enhance the colloid-mediated radionuclide release considerably. It is clear that only the investigation of the interaction mechanisms by spectroscopic methods is able to unravel the relevance of such processes. In order to allow the description of colloid-facilitated radionuclide migration, it is furthermore required to improve our understanding of the colloid interaction with geomatrix surfaces. The quantification of colloid migration will strongly depend on the availability of applicable models for colloid filtration.

The careful reviews of T. Missana, P. Vilks, E. Hoehn, and the Editors are gratefully acknowledged.

References

AHN, T. M. 1996. *Long-Term Kinetic Effects and Colloid Formations in Dissolution of LWR Spent Fuels*. US Nuclear Regulatory Commission, NUREG-1564.

ARTINGER, R., KIENZLER, B., SCHÜSSLER, W. & KIM, J. I. 1998. Effects of humic substances on the ^{241}Am migration in a sandy aquifer: column experiments with Gorleben groundwater/sediment systems. *Journal of Contaminant Hydrology*, 35, 261–275.

ARTINGER, R., RABUNG, TH. *et al.* 2002. Humic colloid-borne migration of uranium in sand columns. *Journal of Contaminant Hydrology*, 58, 1–12.

ARTINGER, R., BUCKAU, G., ZEH, P., GERAEDTS, K., VANCLUYSEN, J., MAES, A. & KIM, J. I. 2003. Humic colloid mediated transport of tetravalent actinides and technetium. *Radiochimica Acta*, 91, 743–750.

BATES, J. K., BRADLEY, J. P., TEETSOV, A., BRADLEY, C. R. & BUCHHOLTZ TEN BRINK, M. 1992. Colloid formation during waste form corrosion: Implications for nuclear waste disposal. *Science*, 256, 469–471.

BECKETT, R. & HART, B. T. 1993. Use of field flow fractionation techniques to characterize aquatic particles, colloids, and macromolecules. *In*: BUFFLE, F. & VAN LEEUWEN, H. P. (eds) *Environmental Particles*, vol. 2. Lewin Publishers, London, 162–205.

BITEA, C., WALTHER, C., KIM, J. I., GECKEIS, H., RABUNG, TH., SCHERBAUM, F. J. & CACUCI, D. G. 2003a. Time-resolved observation of ZrO$_2$– colloid agglomeration. *Colloids & Surfaces A*, 215, 55–66.

BITEA, C., MÜLLER, R., NECK, V., WALTHER, C. & KIM, J. I. 2003b. Study of the generation and stability of thorium(IV) colloids by LIBD combined with ultrafiltration. *Colloids & Surfaces A*, 215, 63–70.

BUCK, E. C. & BATES, J. K. 1999. Microanalysis of colloids and suspended particles from nuclear waste glass alteration. *Applied Geochemistry*, 14, 635–653.

BUCKAU, G., ARTINGER, R., FRITZ, P., GEYER, S., KIM, J. I. & WOLF, M. 2000. Origin and mobility of humic colloids in the Gorleben aquifer system. *Applied Geochemistry*, 15, 183–191.

BRADBURY, M. H. & BAEYENS, B. 2000. A generalised sorption model for the concentration dependent uptake of caesium by argillaceous rocks. *Journal of Contaminant Hydrology*, 42, 141–163.

BRADBURY, M. H. & BAEYENS, B. 2003. Porewater chemistry in compacted re-saturated MX-80 bentonite. *Journal of Contaminant Hydrology*, 61, 329–338.

CARON, F., ELCHUK, S. & WALKER, Z. H. 1996. HPLC characterization of dissolved organic matter from low-level radioactive waste leachates. *Journal of Chromatography*, 739, 281–294.

CHARLET, L. & MANCEAU, A. 1993. Structure, formation and reactivity of hydrous oxide particles: Insights from X-ray absorption spectroscopy. *In*: BUFFLE, J. & VAN LEEUWEN, H. P. (eds) *Environmental Particles*, Vol. 2. Lewis, Boca Raton, 117–164.

CHOPPIN, G. R. & CLARK, S. B. 1991. The kinetic interactions of metal ions with humic acids. *Marine Chemistry*, 36, 27–38.

CHUNG, K. H., KLENZE, R., PARK, K. K., PAVIET-HARTMANN, P. & KIM, J. I. 1998. A study of the surface sorption process of Cm(III) on silica by time-resolved laser fluorescence spectroscopy (I). *Radiochimica Acta*, 82, 215–219.

CONTARDI, J. S., TURNER, D. R. & AHN, T. M. 2001. Modeling colloid transport for performance assessment. *Journal of Contaminant Hydrology*, 47, 323–333.

DEGUELDRE, C., PFEIFFER, H. R., ALEXANDER, W., WERNLI, B. & BRUETSCH, R. 1996a. Colloid properties in granitic groundwater systems. I. Sampling and characterization. *Applied Geochemistry*, 11, 677–695.

DEGUELDRE, C., GRAUER, R., LAUBE, A., OESS, A. & SILBY, H. 1996b. Colloid properties in granitic groundwater systems. II. Stability and transport study. *Applied Geochemistry*, 11, 697–710.

DEGUELDRE, C., TRIAY, I., KIM, J. I., VILKS, P., LAAKSOHARJU, M. & MIEKELEY, N. 2000. Groundwater colloid properties: a global approach. *Applied Geochemistry*, 15, 1043–1051.

FENG, X., BUCK, E. C., MERTZ, C., BATES, J. K., CUNNANE, J. C. & CHAIKO, D. J. 1994. Characteristics of colloids generated during corrosion of nuclear waste glass in groundwater. *Radiochimica Acta*, 66/67, 197–205.

FUJITA, T., SUGIYAMA, D., SWANTON, S. W. & MYATT, B. J. 2003. Observation and characterization of colloids derived from leached cement hydrates. *Journal of Contaminant Hydrology*, 61, 3–16.

GECKEIS, H., GRAMBOW, B., LOIDA, A., LUCKSCHEITER, H., SMAILOS, E. & QUINONES, J. 1998. Formation and stability of colloids under simulated near field conditions. *Radiochimica Acta*, 82, 123–128.

GECKEIS, H., KLENZE, R. & KIM, J. I. 1999. Solid–water interface reactions of actinides and homologues: Sorption onto mineral surfaces. *Radiochimica Acta*, 87, 13–21.

GECKEIS, H., RABUNG, TH., NGO MANH, TH., KIM, J. I. & BECK, H. P. 2002. Humic colloid-borne natural polyvalent metal ions: Dissociation experiment. *Environmental Science and Technology*, 36, 2946–2952.

GECKEIS, H., GEYER, F. W. *et al.* 2003. GTS V/CRR: Tracer tests #28–#35 (January–March 2002); raw data report. *In*: MÖRI, A. (ed) *Nagra Internal Report*. Nagra, Wettingen, Switzerland, NIB 02-18.

HAUSER, W., GECKEIS, H., KIM, J. I. & FIERZ, TH. 2002. A mobile laser-induced breakdown detection system and 1st application for the *in-situ* monitoring of colloid migration. *Colloids & Surfaces*, **203**, 37–45.

HAUSER, W., GÖTZ, R., GECKEIS, H. & KIENZLER, B. 2003. *In-situ* colloid detection in granite groundwater along the Äspö hard rock laboratory access tunnel. *In*: LAAKSOHARJU, M. (ed) *Äspö Hard Rock Laboratory: Status Report of the Colloid Investigation Conducted at the Äspö HRL During the Years 2000–2003*. Swedish Nuclear Fuel and Waste Management Co. (SKB), Stockholm International Progress Report, IPR-03-38.

HEALY, T. W. 1994. Stability of aqueous silica sols. *In*: BERGNA, H. E. (ed) *The Colloid Chemistry of Silica, Advances in Chemistry Series 234*. American Chemical Society, Washington, 147–159.

HONEYMAN, B. D. 1999. Colloidal culprits in contamination. *Nature*, **397**, 23–24.

ILER, R. K. 1979. *The Chemistry of Silica*. John Wiley & Sons, New York.

KERSTING, A. B., EFURD, D. W., FINNEGAN, D. L., ROKOP, D. J., SMITH, D. K. & THOMPSON, J. L. 1999. Migration of plutonium in ground water at the Nevada Test Site. *Nature*, **397**, 56–59.

KIM, J. I., BUCKAU, G., BAUMGÄRTNER, F., MOON, H. C. & LUX, D. 1984. Colloid generation and the actinide migration in Gorleben groundwaters. *In*: MCVAY, G. L. (ed) *Scientific Basis for Nuclear Waste Management VII*. North-Holland, New York, 31–40.

KIM, J. I. 1991. Actinide colloid generation in groundwater. *Radiochimica Acta*, **52/53**, 71–81.

KIM, J. I., RHEE, D. S., WIMMER, H., BUCKAU, G. & KLENZE, R. 1993. Complexation of trivalent actinide ions (Am^{3+}, Cm^{3+}) with humic acid: a comparison of different experimental methods. *Radiochimica Acta*, **62**, 35–43.

KIM, J. I., DELAKOWITZ, B., ZEH, P., KLOTZ, D. & LAZIK, D. 1994. A column experiment for the study of colloidal radionuclide migration in Gorleben aquifer systems. *Radiochimica Acta*, **66/67**, 165–171.

KIM, M. A., PANAK, P. J., YUN, J. I., KIM, J. I., KLENZE, R. & KÖHLER, K. 2003. Interaction of actinides with aluminosilicate colloids in statu nascendi: Part I: generation and characterization of actinide(III)–pseudocolloids. *Colloids & Surfaces*, **216**, 97–108.

KIMURA, T. & CHOPPIN, G. R. 1994. Luminescence study on determination of the hydration number of Cm(III). *Journal of Alloys and Compounds*, **213/214**, 313–317.

KRETZSCHMAR, R., HOLTHOFF, H. & STICHER, H. 1998. Influence of pH and humic acid on coagulation kinetics of kaolinite: A dynamic light scattering study. *Journal of Colloid and Interface Science*, **202**, 95–103.

LU, N., COTTER, C. R., KITTEN, H. D., BENTLEY, J. & TRIAY, I. R. 1998. Reversibility of sorption of plutonium-239 onto hematite and goethite colloids. *Radiochimica Acta*, **83**, 167–173.

MCCARTHY, J. F. & DEGUELDRE, C. 1993. Sampling and characterization of colloids and particles in groundwater for studying their role in contaminant transport. *In*: BUFFLE, J. & VAN LEEUWEN, H. P. (eds) *Environmental Particles*, Vol. 2. CRC Press, Inc., Boca Raton, 247–315.

MEIER, H., ZIMMERHACKL, E. & ZEITLER, G. 2003. Modeling of colloid-associated radionuclide transport in porous groundwater aquifers at the Gorleben site, Germany. *Geochemical Journal*, **37**, 325–350.

MISSANA, T. & ADELL, A. 2000. On the applicability of DLVO theory to the prediction of clay colloids stability. *Journal of Colloid and Interface Science*, **230**, 150–156.

MISSANA, T., ALONSO, U. & TURRERO, M. J. 2002. Generation and stability of bentonite colloids at the bentonite/granite interface of a deep geological radioactive waste repository. *Journal of Contaminant Hydrology*, **61**, 17–31.

MÖRI, A., ALEXANDER, W. R. *et al.* 2003. The colloid and radionuclide retardation experiment at the grimsel test site: Influence of bentonite colloids on radionuclide migration in a fractured rock. *Colloids & Surfaces*, **217**, 33–47.

NAKANO, M., KAWAMURA, K. & ICHIKAWA, Y. 2003. Local structural information of Cs in smectite hydrates by means of an EXAFS study and molecular dynamics simulations. *Applied Clay Science*, **23**, 15–23.

NECK, V., MÜLLER, R., BOUBY, M., ALTMAIER, M., ROTHE, J., DENECKE, M. & KIM, J. I. 2002. Solubility of amorphous Th(IV) hydroxide – application of LIBD to determine the solubility product and EXAFS for aqueous speciation. *Radiochimica Acta*, **90**, 485–494.

PANETH, F. 1914. Colloidal solutions of radioactive substances. *Kolloid-Zeitschrift*, **13**, 1–4.

PIRLET, V. 2001. Overview of actinides (Np, Pu, Am) and Tc release from waste glasses: influence of solution composition. *Journal of Nuclear Materials*, **298**, 47–54.

PLASCHKE, M., RÖMER, J. & KIM, J. I. 2002. Characterization of Gorleben groundwater colloids by atomic force microscopy. *Environmental Science and Technology*, **36**, 4483–4488.

QUINONES, J., GRAMBOW, B., LOIDA, A. & GECKEIS, H. 1996. Coprecipitation phenomena during spent fuel dissolution. Part 1: Experimental procedure and initial results on trivalent ion behavior. *Journal of Nuclear Materials*, **238**, 38–43.

RABUNG, T., STUMPF, T., GECKEIS, H., KLENZE, R. & KIM, J. I. 2000. Sorption of Am(III) and Eu(III) onto γ-Alumina: Experimental results and modeling. *Radiochimica Acta*, **88**, 711–716.

RYAN, J. N. & ELIMELECH, M. 1996. Colloid mobilization and transport in groundwater. *Colloids & Surfaces*, **107**, 1–56.

RYAN, J. N. & GSCHWEND, P. M. 1994. Effects of ionic strength and flow rate on colloid release: Relating kinetics to intersurface potential energy. *Journal of Colloid and Interface Science*, **164**, 21–34.

SCHÄFER, TH., BAUER, A., BUNDSCHUH, T., RABUNG, TH., GECKEIS, H. & KIM, J. I. 2000. Colloidal stability of inorganic colloids in natural and synthetic groundwater. Applied mineralogy in research, economy, technology and culture. *Proceedings of the 6th International Congress ICAM 2000*, Göttingen, July 13–21. A.A. Balkema, Rotterdam, Vol. 2, 675–678.

SCHUESSLER, W., ARTINGER, R., KIENZLER, B. & KIM, J. I. 2000. Conceptual modeling of the humic colloid-borne americium(III) migration by a kinetic approach. *Environmental Science and Technology*, **34**, 2608–2611.

SCHUESSLER, W., ARTINGER, R., KIM, J. I., BRYAN, N. D. & GRIFFIN, D. 2001. Numerical modeling the humic colloid borne Americium (III) migration in column experiments using the transport/speciatrion code K1D and the KICAM model. *Journal of Contaminant Hydrology*, **47**, 311–322.

SONG, L., JOHNSON, P. R. & ELIMELECH, M. 1994. Kinetics of colloid deposition onto heterogeneously charged surfaces in porous media. *Environmental Science and Technology*, **28**, 1164–1171.

STUMPF, TH., BAUER, A., COPPIN, F. & KIM, J. I. 2001a. Time-resolved laser fluorescence spectroscopy study of the sorption of Cm(III) onto smectite and kaolinite. *Environmental Science and Technology*, **35**, 3691–3694.

STUMPF, TH., RABUNG, TH., KLENZE, R., GECKEIS, H. & KIM, J. I. 2001b. Spectroscopic study of Cm(III) sorption onto γ-alumina. *Journal of Colloid and Interface Science*, **238**, 219–224.

STUMPF, TH. & FANGHÄNEL, TH. 2002. A time-resolved laser fluorescence spectroscopy (TRLFS) study of the interaction of trivalent actinides (Cm(III)) with calcite. *Journal of Colloid and Interface Science*, **249**, 119–122.

SWANTON, S. W. & VINES, S. 2003. Equilibrium leach tests: Colloid generation and the association of

radionuclides with colloids under simulated repository conditions. *Colloids & Surfaces*, **217**, 71–79.

TITS, J., STUMPF, TH., RABUNG, TH., WIELAND, E. & FANGHÄNEL, TH. 2003. Uptake of trivalent actinides (Cm(III)) and lanthanides (Eu(III)) by Calcium silicate hydrates: a wet chemistry and time-resolved laser fluorescence spectroscopy (TRLFS) study. *Environmental Science and Technology*, **37**, 3568–3573.

VILKS, P. & BACHINSKI, D. B. 1996. Colloid and suspended particle migration experiment in a granite fracture. *Journal of Contaminant Hydrology*, **21**, 269–279.

VILKS, P., FROST, L. H. & BACHINSKI, D. B. 1997. Field-scale colloid migration experiments in a granitic fracture. *Journal of Contaminant Hydrology*, **26**, 203–214.

VILKS, P. & BAIK, M. H. 2001. Laboratory migration experiments with radionuclides and natural colloids in a granitic fracture. *Journal of Contaminant Hydrology*, **47**, 197–210.

WALTHER, C., BITEA, C., HAUSER, W., KIM, J. I. & SCHERBAUM, F. J. 2002. Laser induced breakdown detection for the assessment of colloid mediated radionuclide migration. *Nuclear Instruments and Methods in Physics Research Section B*, **195**, 374–388.

WARWICK, P. W., HALL, A., PASHLEY, V., BRYAN, N. D. & GRIFFIN, D. 2000. Modelling the effect of humic substances on the transport of europium through porous media: a comparison of equilibrium and equilibrium/kinetic models. *Journal of Contaminant Hydrology*, **42**, 19–34.

WIELAND, E. & SPIELER, P. 2001. Colloids in the mortar backfill of a cementitious repository for radioactive waste. *Waste Management*, **21**, 511–523.

ZHUANG, J., FLURY, M. & JIN, Y. 2003. Colloid-facilitated Cs transport through water-saturated Hanford sediment and Ottawa sand. *Environmental Science and Technology*, **37**, 4905–4911.

Mechanisms of uranyl sorption

M. DEL NERO[1], A. FROIDEVAL[1], C. GAILLARD[1], G. MIGNOT[2], R. BARILLON[1], I. MUNIER[1] & A. OZGÜMÜS[1]

[1]*Institut de Recherches Subatomiques, UMR 7500 ULP/CNRS/IN2P3, Strasbourg, France*
(e-mail: mireille.delnero@ires.in2p3.fr)
[2]*Commissariat à l'Energie Atomique, DAM/RCE, Bruyères-le-Châtel, France*

Abstract: Detailed knowledge of the reactions at the water/colloid/mineral interface is crucial to model accurately actinide behaviour in nature. In this paper, we review current knowledge of the sorption of actinides and of the mechanisms of sorption, with a particular focus on uranyl. Of major interest is the influence of the aqueous uranyl species (e.g., carbonate complexes, polynuclear species, colloids) on the uranyl sorption species. We present extended X-ray absorption fine structure (EXAFS) and X-ray photoelectron spectroscopy (XPS) studies on the coordination of uranyl onto an amorphous Al phase and onto quartz, respectively. Our XPS investigations show that two components having uranyl ions in very distinct coordination environments co-exist on quartz at high uranyl surface coverage, independently of the presence of uranyl carbonate complexes or uranyl colloids in solution. One component corresponds to polynuclear surface species and/or schoepite-like surface precipitates. In the case of similar uranyl concentrations and of high carbonate solution concentrations, polymeric uranyl species are formed on quartz, whereas no such surface species occurs on the Al phase. Uranyl is found on the Al phase as mononuclear uranyl carbonato surface complexes only. These results are of importance because they suggest that mineral surface characteristics strongly control the uranyl surface species in aquifers.

Actinides (An) originating from nuclear explosions and spent fuels have long-lived isotopes with large production amounts. Some hundreds of years after discharge from a reactor, the fuel radiotoxicity is dominated by long-lived nuclides of plutonium (^{238}Pu, ^{239}Pu, ^{240}Pu) and americium (^{241}Am, ^{243}Am). Neptunium (^{237}Np) and uranium also contribute to the long-term fuel radiotoxicity. Curium is produced in a small quantity, but its active nuclide in fuel, ^{244}Cm, is of concern for several human generations. These An are thus present and future radiological hazards for the environment. They are proposed to be deposited in geological media, at high-level waste (HLW) repositories, which are potential sources for An dissemination in case of groundwater intrusion. Actinides have already been introduced into the environment by reprocessing of spent nuclear fuel, bomb testing, and reactor accidents. Mining and reprocessing of natural U have also led to generation of large amounts of U-mill tailings over the world. Migration of U (e.g., Morrison *et al.* 1995) and of other An in groundwaters is of great concern at several contaminated areas. The importance of An as contaminants depends not only on their concentration and half-lives, but also on their bioavailability and on their mobility in natural waters, that is, on their chemical state and form. Knowledge of their chemical behaviour in the environment is thus of major interest for remediation of polluted areas and for safety assessment of repository sites.

Actinides undergo a variety of chemical reactions in the environment. An accurate prediction of their migration behaviour requires the development of geochemical codes that permits handling of numerous cooperative or competitive processes, all dependent on the An oxidation state and the chemical conditions in a given aquatic system. Possible reactions are dissolution/precipitation of An- or An-bearing compounds, redox reactions, hydrolysis, polynucleation, complexation with organic and inorganic ligands, (pseudo-)colloid generation, and sorption onto minerals. Potentially mobile species are complex ions, true colloids, and pseudo-colloids formed by An sorption onto natural colloids present in waters. Sorption onto rock-forming minerals or large soil particles may immobilize all these species. Thus, the interactions at the water/colloid/mineral interface may play a crucial role in the geo-cycling of An. Such interactions are still poorly understood due to the existence of

From: GIERÉ, R. & STILLE, P. (eds) 2004. *Energy, Waste, and the Environment: a Geochemical Perspective.* Geological Society, London, Special Publications, **236**, 545–560.
0305-8719/04/$15 © The Geological Society of London 2004.

many potentially sorbed An species in natural waters and to the variety of sorption processes. Acquiring further knowledge of the mechanisms of An sorption is thus of fundamental importance for the selection of accurate model sorption reactions and for improving predictive capabilities of geochemical codes.

This paper is devoted to the sorption of uranyl, which exhibits a complex aqueous and surface chemistry. We review briefly the sorption behaviour of An in the environment, and illustrate the variety of environmental processes using published data of uranyl sorption in the Bangombé natural reactor zone. After summarizing the general findings of the mechanisms of An sorption, we then focus particularly on the current knowledge of the mechanisms of uranyl sorption. A major area of research is the influence of the aqueous uranyl speciation on the uranyl surface species. Spectroscopic data of U(VI) sorbed onto silica and alumina minerals are examined and used to discuss the role of aqueous uranyl polynuclear species, $UO_2(OH)_2$ colloids and uranyl–carbonate complexes. The influence of the mineral surface properties on the mechanisms of sorption is also discussed.

Sorption processes of actinides in the environment

Overview

A major problem for the modelling of the sorption of An in the environment is the complexity of the aqueous chemistry of such polyvalent metal ions. The most important property regulating the chemistry of an An is its oxidation state. Uranium, Pu and Np have three possible oxidation states in nature, namely the tetravalent, pentavalent, and hexavalent states, whereas Am and Cm occur only as trivalent elements. After release into solution, the An ions tend to decrease their charge by complexation. The tendency for hydrolysis, polynucleation and generation of true colloids through successive oxygen bridging follows the effective charge of the bare An ion in the order: $An^{4+} > AnO_2^{2+} > An^{3+} > AnO_2^+$. The effective charges of AnO_2^{2+} and AnO_2^+ are equal to 3.3 and 2.3, respectively (Choppin 1988a). The tendency for surface hydrolysis also follows the effective charge of the bare An ion (Silva & Nitsche 1995) leading possibly to considerable adsorption of some An onto oxide and silicate phases in the absence of complexing anions. Tetravalent, hexavalent, and trivalent An thus have low solubilities, but a strong tendency to generate potentially mobile pseudo-colloids and true colloids (e.g., Kim 1991; Silva & Nitsche

1995). Laboratory studies have indeed assessed that colloids are initially formed from oversaturated Th(IV) or U(VI) solutions as microcrystalline or amorphous hydroxides and hydrous oxides of very small sizes (Rothe *et al.* 2002; Bitea *et al.* 2003; Froideval *et al.* 2003). Inorganic pseudo-colloids of Pu(IV), U(VI), Am, and rare earth elements (REE), as well as true Pu(IV) colloids, are recognized as An carriers in aquifers (Runde *et al.* 2002; Geckeis *et al.* 2003; Möri *et al.* 2003). The distinction between mobile and non-mobile colloids can be made for individual aquifers only, because such phases may migrate through porous spaces and rock fractures or may be filtered by such systems (Kim 1991). Many anions present in natural waters (e.g., carbonates, humates, phosphates, silicates) stabilize An as complex ions. Silicate and phosphate ions are of particular interest because they are expected to form poorly soluble minerals with trivalent and hexavalent An (Silva & Nitsche 1995). Several kinds of silicate and phosphate minerals are reported to form in nature under specific chemical conditions, such as uranyl silicates (Pérez del Villar *et al.* 1997; Kienzler *et al.* 2001; Bruno *et al.* 2002) and lanthanide phosphates (Janeczek & Ewing 1996; Jensen *et al.* 2000; Bruno *et al.* 2002). Carbonate ions and humates are of cardinal importance, due to their strong complexing properties, their omnipresence in ground and surface waters, respectively, and their strong influence on An mobility. Choppin & Allard (1985) and Choppin (1988b) have shown that the strength of the humate complexation varies with the humic acid composition and increases with the degree of ionization of the humic acid up to pH 7. The constant values of humic complex formation reported by these authors for An at pH 7 are much larger than the carbonate complexation constants (Silva & Nitsche 1995). They suggest the following trend in the strength of humate complexation for An: $Pu^{4+} > Am^{3+} > UO_2^{2+} > NpO_2^+$. Several studies have shown that in humic-rich waters the concentrations of trivalent, tetravalent, and hexavalent elements, including uranyl, increase with the dissolved organic content (Kim *et al.* 1987; Choppin 1992). The binding of metal ions to humic molecules finally leads to the formation of humic colloids, which may either remain in solution or attach onto inorganic colloids/particulates in soils (Choppin 1988b, 1992). The geochemical reactions between humic substances, An, and minerals or inorganic colloids constitute a major field of current research. In natural waters at near-neutral pH, carbonate and humate complexation are competitive among one another

(Kim 1991), particularly for hexavalent and trivalent An. The trend in carbonate complexation is in the order: $An^{4+} > AnO_2^{2+} > An^{3+} > AnO_2^+$ (Kim 1991; Silva & Nitsche 1995). Uranyl carbonate complexes account indeed for dissolved uranyl in several marine and groundwaters (Choppin 1989). Even neptunium, in its highly stable pentavalent state, is potentially mobile not only as neptunyl cation NpO_2^+, but also as neptunyl carbonate complexes in specific ecosystems. The carbonate complexation of REE, which are possible homologues to trivalent An, competes against sorption and is partly responsible for the fractionation between light REE (LREE) and heavy REE (HREE) elements in some rock profiles (Fairhurst et al. 1995; Johannesson et al. 1997; Takahashi et al. 1999). Although carbonate complexation increases An mobility in the environment, recent spectroscopic studies point to the formation in specific cases of An surface carbonate complexes, for example, for uranyl onto hematite (Swanton et al. 1998; Bargar et al. 2000). Thus, many potentially adsorbed An species co-exist in natural waters as aqueous or colloidal species. Moreover, a variety of sorption processes such as ion exchange, diffusion, surface complexation, surface polymerization, surface precipitation and co-precipitation may occur in nature and will be illustrated in the following sections for uranyl.

Uranyl sorption processes in the Bangombé natural reactor zone

Studying sorption/desorption processes of U in rocks from natural analogue sites may provide valuable insight into the variety of sorption processes and the host mineral phases responsible for the long-term immobilization of U. Because it is located at a very shallow depth (12 m below the surface), the natural reactor of Bangombé is of particular interest to elucidate the role of present-day weathering in the mobility of uranium. The unique location and the geochemical features of the Bangombé reactor zone (e.g., Gauthier-Lafaye et al. 1996, 2004; Salah 2000; Stille et al. 2003; see also Chapter 7 of this volume) make it a valuable site to improve understanding of the U retention processes under various geochemical conditions. The uraninite reactor zone is partially preserved due to the redox buffering capacity of the organic matter associated with Fe(II)/Fe(III) minerals (Madé et al. 2000). Above the reactor, the sequence is composed of black shales containing ferrous minerals like pyrite or chlorite,

whose alteration by oxidizing meteoric waters led to the formation of weathered pelites (uppermost oxidizing zone). Fe- and Mn-crusts occur between the reactor zone and the black shales and correspond to local oxidizing conditions. The boundaries of the oxidizing zones are sharp (Gauthier-Lafaye et al. 1996; Salah et al. 1999; Salah 2000). Nevertheless, gains of U occur in the weathered rocks surrounding the reactor, particularly in the Fe-, P-, and Mn-rich zones, pointing to efficient retention of U under oxidizing conditions (Salah et al. 1999; Salah 2000).

Precipitation of uranyl silicates is a long-term trapping mechanism at Bangombé (Salah et al. 1999; Jensen et al. 2000; Salah 2000). Apart from precipitation, several sorption processes contribute significantly to the long-term uranyl retention at Bangombé. Figure 1 reports experimental results for uranyl adsorption/desorption processes in rock samples collected in the weathered zones around the natural nuclear reactor of Bangombé (Del Nero et al. 1999a). The method used was the U isotope exchange technique of Payne & Waite (1991). In such experiments, the isotope ^{233}U is added to ^{238}U-containing subsamples brought in contact with aqueous solutions at different pH. A linear relation between the amounts of natural ^{238}U desorbed and the percentages of ^{233}U adsorbed gives the fraction of total uranium in the sample governed by sorption/desorption equilibrium (Fig. 1). Samples 6.2 and 6.8 in Fig. 1 were collected in the weathered pelites (at 6.2 and 6.8 m depth). They contain illite, kaolinite and Fe-oxyhydroxides as accessory minerals. Sample 10.3 was collected in the P-rich Fe-crust (at 10.3 m depth). The exchange experiments show that adsorption is a major uranyl retention process in the uppermost clayey layers containing Fe-oxyhydroxides as accessory minerals, whereas the fraction of total uranium adsorbed at mineral surfaces in the Fe-crust is small (Fig. 1, Table 1). Thus, uranyl is potentially accessible to weathering solutions in the clayey layers only. Experiments using the sequential extraction procedure of Yanase et al. (1991) revealed that the extraction of Fe-oxyhydroxides and of minor P-rich phases by a citrate–dithionate–bicarbonate solution (CDB) could remove a large part of U in the Fe-rich crust (Table 1). A possible long-term trapping mechanism of U(VI) at Bangombé is thus U incorporation into the structure of Fe-oxyhydroxides. Several studies of minerals or rocks have highlighted the potential role of U incorporation into Fe-oxyhroxides for the long-term retention of uranyl in the environment. Payne et al. (1994) have shown that U, initially bound

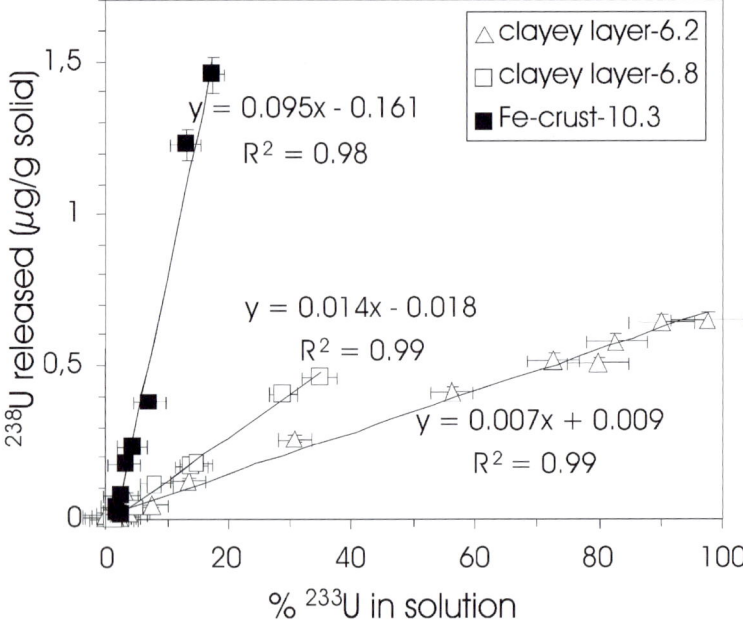

Fig. 1. Amounts of ^{238}U released from solid samples vs. the percentages of ^{233}U remaining in solution in U isotope exchange experiments performed by Del Nero *et al.* (1999*a*) using weathered pelites (6.2 and 6.8) and an Fe-crust (10.3) from Bangombé.

at the ferrihydrite surface, is incorporated into the solid during the transformation (aging) of ferrihydrite to crystalline minerals. Bruno *et al.* (1995) have pointed out that U uptake during formation of Fe-oxyhydroxides under well-defined experimental conditions may result from $Fe(OH)_3-UO_2(OH)_2$ co-precipitation processes. Bruno *et al.* (1998) successfully described the behaviour of U(VI) at the El Berrocal site by including such a co-precipitation process in their modelling. Another possible long-term trapping mechanism at Bangombé suggested by the extraction experiments is the association of uranyl with ferric phosphates, identified by Salah (2000) as surface precipitates on the Fe-

oxyhydroxides. The precise role of phosphate ligands on the mechanisms of U sorption remains so far poorly characterized. Payne *et al.* (1998) have shown that P added to a solution substantially increases the uptake of uranyl by ferrihydrite at low pH values. They suggested that ternary surface complexes involving both U and P formed at the ferrihydrite surface. They further pointed out the possibility of surface precipitation of ferric phosphates having a stronger affinity for U(VI) than the original ferrihydrite. Murakami *et al.* (1997) studied samples of the Koongarra deposit, Australia, and proposed a surface precipitation process for the formation of uranyl phosphate micro-crystals

Table 1. *Results of the U isotope exchange experiments (Fig. 1) and of the sequential extraction experiments of Del Nero et al. (1999a), using Bangombé samples collected in the weathered pelites (6.2 and 6.8) and in the Fe-crust (10.3)*

Sample	Total U content (ppm)	% of total U as sorbed U	% of total U extracted/in residue			
			Extractant solutions			Residue
			Morgan	Tamm	CDB	
6.2	4.1 ± 0.2	22 ± 4	8 ± 4	29 ± 4	—	75 ± 9
6.8	6.2 ± 0.3	23 ± 3	5 ± 3	18 ± 10	18 ± 3	57 ± 10
10.3	309 ± 16	3 ± 1	2 ± 1	3 ± 2	40 ± 5	33 ± 4

within veins of Fe^{3+}-minerals. The sorptive behaviour of uranyl at Bangombé also reveals the crucial role of phosphate ligands on the long-term retention of U under oxidizing conditions. Thus, studies at natural analogue sites have revealed that several processes may be relevant to the long-term immobilization of uranyl in the environment. Nevertheless, the mechanisms of many sorption processes, such as surface polymerization/precipitation, remain so far poorly understood.

Uranyl sorption mechanisms

General findings on actinide sorption mechanisms

Adsorption is the general term for the attachment of species at mineral–solution interfaces due to either chemical binding (inner-sphere complexation) or electrostatic attraction of charged aqueous species (outer-sphere complexation). In the presence of water, the surfaces of oxides and of silicates are covered by different types of surface hydroxyl groups, coordinated to one metal atom (terminal hydroxyls) or to two or three metal atoms (bridging hydroxyls) in the crystal lattice. Outer-sphere surface complexation results from the attachment of charged aqueous species to surface hydroxyl groups of opposite charge. In such a case, the hydration sphere of the metal ion remains unaffected by adsorption. Inner-sphere complex formation involves terminal hydroxyl groups, which behave as Lewis bases (i.e., they share the lone pair of electrons on the hydroxyl oxygen) and interact with Lewis acids, such as metal ions. The direct coordination of the metal ion to the surface hydroxyl groups results in strong chemical binding and in short distances between metal ion and hydroxyl oxygen. Surface spectroscopic studies using extended X-ray absorption fine structure (EXAFS) have provided clear evidence for the sorption of An via inner-sphere surface complex formation, at least in a first step. Combes et al. (1992) reported neptunyl mononuclear surface complexes onto goethite. Uranyl may form mononuclear, inner-sphere bidentate surface complexes onto silica, alumina and ferrihydrite (Reich et al. 1998; Sylwester et al. 2000; Bargar et al. 2000) and mononuclear, inner-sphere monodendate surface complexes onto hematite (Swanton et al. 1998). Studies using time-resolved laser-induced fluorescence spectroscopy (TRLFIS) have also shown that inner-sphere surface complexation of Cm(III) takes place at the surface of alumina (Stumpf et al. 2001).

Chemical reactions at hydroxyl sites – such as protonation/deprotonation, ligand exchange, formation of surface complexes – generate a surface charge and a surface potential (Stern layer). The accumulation of electrolyte solution counter-ions in a diffuse ion swarm (Gouy layer) at a certain distance from the surface (double-layer theory) compensates this surface charge. Several surface complexation models (SCM) are available, such as the constant capacitance model (CCM), the diffuse layer model (DLM), and the triple layer model (TLM) (Davis et al. 1978; Bowden et al. 1980; Stumm et al. 1980; van Riemsdijk et al. 1986; Schindler & Stumm 1987; Hayes et al. 1991). They differ from one to another in the microscopic description of the double-layer interface and in the relation used between surface charge and surface potential (e.g., Davis & Kent 1990; Goldberg 1992). All SCM include reactions for inner-sphere surface complexation and ionization of hydroxyl groups; only the TLM considers reactions for outer-sphere surface complexation. In the DLM, the equations related to the ionization of the surface hydroxyls and to the creation of a surface charge and potential are the following:

$$S-OH + H^+ \xrightarrow{K^{int}_+} S-OH_2^+ \tag{1}$$

$$S-OH \xrightarrow{K^{int}_-} S-O^- + H^+ \tag{2}$$

$$K^{int}_+ = K^c_+ \exp(+F\psi_0/RT)$$
$$= \frac{[S-OH_2^+]}{[S-OH]\{H^+\}} \exp(+F\psi_0/RT) \tag{3}$$

$$K^{int}_- = K^c_- \exp(-F\psi_0/RT)$$
$$= \frac{[S-O^-]\{H^+\}}{[S-OH]} \exp(-F\psi_0/RT) \tag{4}$$

$$-\sigma_0 = \sigma_d = -0.1174\sqrt{I}\sinh(zF\psi_0/2RT) \tag{5}$$

$$\psi_0 = \psi_d \tag{6}$$

A general formulation of the equilibrium, involving a monodendate, inner-sphere surface complex, and of the associated constants, is as follows:

$$S-OH + xM^{m+} + yH_2O$$

$$\xrightarrow{K^{int}_{species}} (S-OM_xOH_y)^{(xm-y-1)^+} + (y+1)H^+ \tag{7}$$

$$K_{species}^{int} = K_{species}^c \exp\left((xm - y - 1)F\psi_0/RT\right) \tag{8}$$

$$K_{species} = \frac{[(S-OM_xOH_y)^{(xm-y-1)+}]\{H^+\}^{(y+1)}}{[S-OH]\{M^{m+}\}^x} \tag{9}$$

In the equations above, brackets and square brackets stand respectively for aqueous ion activities and for ion surface concentrations, S−OH is a surface hydroxyl group, M^{m+} is a metal ion, $\sigma_0(\sigma_d)$ is the surface charge at the surface 0-plane (d-plane or diffuse layer), $\psi_0(\psi_d)$ is the potential at the 0-plane (d-plane), and K^{int} and K^c are intrinsic and conditional constants, respectively. The surface complexation models are pseudo-thermodynamic models; the constants associated to surface complexation reactions are model-dependent ('intrinsic constants'). Nevertheless, SCM are valuable predictive tools because they allow the description of metal ion adsorption by taking into account (1) the mineral surface properties and the aqueous phase characteristics, (2) the competitive adsorption of the solute ions, and (3) the competition between surface and aqueous ligands for ion coordination. For example, SCM models have been widely used to interpret the pH-dependent macroscopic sorption of An in terms of inner-sphere complex formation (e.g., Hsi & Langmuir 1985; Laflamme & Murray 1987; Girvin *et al.* 1991; Payne & Waite 1991; Zachara & McKinley 1993; Waite *et al.* 1994; McKinley *et al.* 1995; Turner *et al.* 1996; Labonne-Wall *et al.* 1997; Del Nero *et al.* 1998, 1999*a*; Kohler *et al.* 1999; Gabriel *et al.* 2001).

Typically, the pH-dependent macroscopic sorption of An onto (hydr-)oxides displays a sharp increase in a narrow pH range (sorption edge) in the absence of aqueous ligands (Fig. 2). For a given mineral, the location of the sorption edge of an An at trace level concentrations is highly dependent on the total concentration of hydroxyl groups (i.e., the mineral/solution ratio or the mineral specific surface area) and on the An oxidation state. The ionic strength has generally little or no influence, suggesting inner-sphere surface complexation. The tendency of metals to form inner-sphere surface complexes follows their tendency to form aqueous hydroxide complexes (Silva & Nitsche 1995). There is indeed an analogy between the surface complexation reaction (7) and the following reaction

describing the formation of aqueous hydrolysis species:

$$xM^{m+} + yH_2O \leftrightarrow (M_x(OH)_y)^{(xm-y)} + yH^+ \tag{10}$$

Assuming a correlation between surface complexation and aqueous hydrolysis exists, the trend in strengths of surfaces complexes for An in different oxidation states onto a given mineral would be in the order: $An^{4+} > AnO_2^{2+} > An^{3+} > AnO_2^+$. Several authors have provided evidence for linear relations between the first hydrolysis constant of metals and the intrinsic constant associated to the formation of surface species of metals as $S-OM^{(m-1)+}$; for example: $\gamma-Al_2O_3$ (Hachiya *et al.* 1984), amorphous silica (Schindler & Stumm 1987), hydrous ferric oxides (Dzombak & Morel 1990), aluminum (hydr-)oxides and kaolinite (Del Nero *et al.* 1997, 1999*a*).

Experimental and modelling studies reporting self-consistent data sets for uranyl (McKinley *et al.* 1995; Turner & Sassman 1996; Turner *et al.* 1996) and for americium (Degueldre *et al.* 1994) support a stronger affinity of An for aluminol and ferrinol surface sites than for silanol surface sites. This finding is consistent with the Lewis base properties of terminal surface hydroxyls for a series of metal oxides, expressed qualitatively by the valence coordination number ratio (VNCR) (Hayes & Katz 1996; Geckeis & Rabung 2002). The VCNR value is the ratio of the valence charge of the lattice metal atom to the number of oxygen atoms coordinated by the lattice metal. For example, the four oxygen atoms coordinated to Si in quartz share equally a charge of 4; that is, each oxygen experiences an effective charge of +1 (VNCR value: 1). In Al (ferric) oxides, the six oxygen atoms coordinated to Al (Fe) share a charge of 3, giving a VNCR value of 0.5. In other words, the attraction exerted by the lattice metal ion on the hydroxyl oxygen electrons is greater for silanol surface groups than for aluminol (ferrinol) surface groups. A silanol surface group has thus a lower tendency to donate its oxygen electrons to form inner-sphere surface complexes than a surface aluminol (ferrinol) group. By contrast, the ability of a surface hydroxyl to give up a proton roughly correlates with higher VNCR values. The greater the attraction of the metal lattice on surface oxygen electrons, the weaker is the attraction of the surface oxygen on the proton. This explains the difference in the surface-protonation behaviour of silica and alumina minerals, which exhibit over a wide

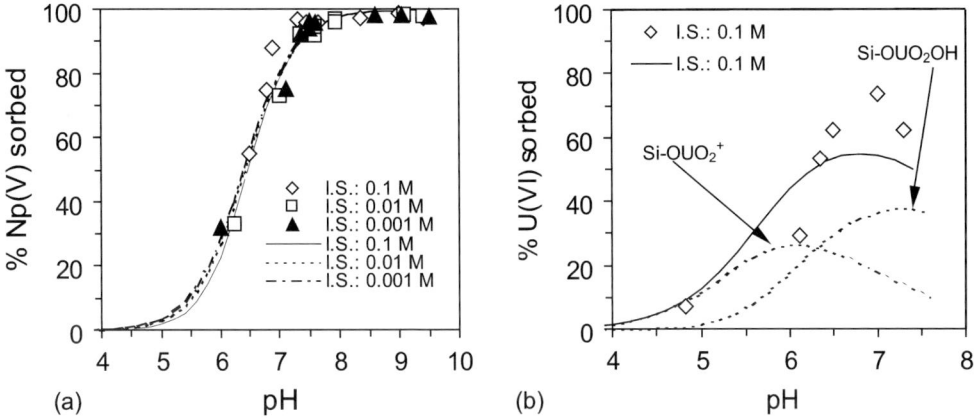

Fig. 2. Sorption edges and DLM fitting curves of (**a**) neptunyl on alumina gel, and (**b**) uranyl on quartz at different ionic strengths (I.S.) using experimental conditions and model parameters listed in Table 2. Uranyl data are from Mignot (2001).

pH range overall negative surface charges and overall positive surface charges, respectively (Figs 3 and 4). Thus, each oxide or silicate has its own surface protonation characteristics, depending on the bonding in the bulk crystal structure connected to the surface. The point of zero charge (pHpzc) of a mineral refers to the pH at which the negative surface charge equals the positive surface charge at the mineral surface. Values of pHpzc are available in the literature for a variety of (hydr-)oxides and clays (e.g., Goldberg 1992). In the absence of potentially-

determining ions other than H^+ and OH^-, Al and Fe (hydr-)oxides display a high pHpzc value, whereas Mn-oxides, silica minerals and clays, such as kaolinite or montmorillonite, have a low pHpzc value. The point of zero charge shifts to lower (higher) pH values with increasing inner-sphere complexation of anions (cations). Modelling An sorption edges requires knowledge of the acid/base properties of minerals (equations (1) to (9) for the DLM). For example, an accurate description of the pH-dependent neptunyl (uranyl) sorption on alumina gel (quartz) is achieved by taking into account in DLM both inner-sphere neptunyl

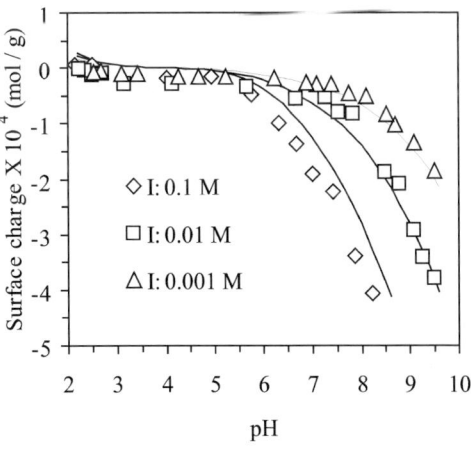

Fig. 3. Experimental surface charges on silica gel of Del Nero *et al.* (1999*b*), as measured by adsorption of counter-ions from $NaClO_4$ electrolytes at different ionic strength (I) and DLM fitting curves from parameters given in Table 2.

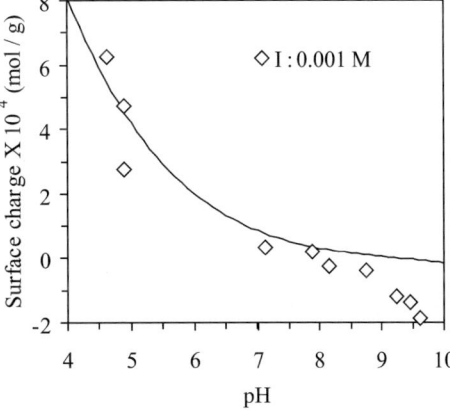

Fig. 4. Experimental surface charges on alumina gel, as measured by adsorption of counter-ions from 0.01 M NaCl electrolyte, and DLM simulated curves using parameters listed in Table 2.

(uranyl) surface complexation and ionization reaction of surface aluminol (silanol) groups (Figs 2 to 4, Table 2). Sverjensky & Sahai (1996) developed a method to estimate the intrinsic protonation constants for use in SCM from theoretical considerations and known properties of solids. The acid/base properties of mineral surfaces are, however, generally determined by fitting simulated surface charges, using SCM, to experimentally determined surface charges (Fig. 3). The experiments classically performed are potentiometric titrations of mineral suspensions. Alternatively, experiments are conducted as described by Del Nero *et al.* (1999*b*) to measure the retention by the mineral surface of the electrolyte anions and cations, as counter-ions to the positive and negative surface charges, respectively (Figs 3 and 4). The interpretation of surface charge experiments is complex, because reactions other than surface protonation also involve protons during the titrations. In addition, electrolyte ions adsorb as outer-sphere surface complexes and contribute significantly to the surface charge at high ionic strength. Moreover, for clays, the surface charge results from reactions occurring on a variety of distinct surface sites and there is so far no generally accepted way to model acidities of the surface components from bulk measurements (Wieland & Stumm 1992; Brady *et al.* 1996; Huertas *et al.* 1998). The capability of SCM to describe the pH- and ionic strength-dependence of the experimental mineral surface charges varies from one model to another (Hayes *et al.* 1991). A drawback of the fitting procedure may also be the high number of fitting parameters, particularly for the TLM, which considers additional binding of electrolyte ions.

Actinide sorption may deviate from linearity when the An concentration increases, leading to changes in the shape and location of sorption edges. Possible explanations for non-linear sorption are the progressive saturation of high surface affinity sites on minerals and/or the occurrence of reactions such as surface polymerization, surface precipitation, or homogenous precipitation. The latter may occur even when not anticipated in experiments, because poorly crystalline, small solids may control dissolved An concentrations (e.g., Diaz Arocas & Grambow 1998). The presence of aqueous ligands also strongly affects the shape of the sorption curves and often causes a decrease of An sorption at basic pH. A reliable modelling of the effect of An concentration and carbonate ligands on sorption edges requires knowledge of the structure of the An surface species.

Investigations of uranyl carbonate complexes on an amorphous Al phase using EXAFS

Several modelling and experimental studies suggested that the decrease of the An sorption with increasing aqueous carbonate concentration is due to competition between surface hydroxyl groups and aqueous carbonate anions for the coordination of An ions and/or between aqueous An species and carbonate species for binding at surface sites (Hsi & Langmuir 1985; Sanchez *et al.* 1985; Laflamme & Murray 1987; Waite *et al.* 1994; Del Nero *et al.* 1998, 1999*a*; Kohler *et al.* 1999). Modelling suggested also that carbonate ligands participate in the formation of ternary surface–An–carbonate complexes (Hsi & Langmuir 1985; Waite *et al.* 1994; Turner & Sassman 1996; Kohler *et al.*

Table 2. *DLM adsorption constants optimized from experimental sorption data shown in Fig. 2 and experimental and model parameters*

Experiments	Modelling	
	Surface complex	K^{DLM}
1 g/L alumina gel in 0.1, 0.01 and 0.001 M NaCl	$AlOH_2^+$	7.5*
solutions containing 0.5 μmol/L Np(V)	AlO^-	−10.2*
Specific surface area of mineral: 90 m²/g	$AlONpO_2^0$	−2.9 ± 0.1
Surface site concentration: 345 μmol/L[‡]		
25 g/L quartz in 0.1 M NaNO₃ solutions containing	$SiOH_2^+$	1.0 ± 0.5[†]
0.1 μmol/L U(VI)	SiO^-	−7.2 ± 0.2[†]
Specific surface area of mineral: 0.065 m²/g	$SiOUO_2^+$	−0.25 ± 0.04
Surface site concentration: 6.2 μmol/L[‡]	$SiOUO_2OH^0$	−6.24 ± 0.18

[*]Values of Sverjensky & Sahai (1996) for gibbsite used to describe surface charge experiments on alumina gel (Fig. 4).
[†]Values optimized from surface charge experiments on silica gel (Fig. 3).
[‡]Calculated using a site density value of 2.31 sites/nm recommended by Davis & Kent (1990).

1999; Del Nero *et al.* 1999*a*; Gabriel *et al.* 2001). Spectroscopic studies using EXAFS have confirmed such surface complexes for uranyl, but, to our knowledge, the studies have so far been limited to hematite (Swanton *et al.* 1998; Bargar *et al.* 2000). Spectroscopic studies are thus greatly needed to characterize directly the structure of the possible U(VI)–carbonato surface complexes onto other mineral surfaces. Particularly, probing the formation of such surface species on Al phases is of primary importance for predicting uranyl mobility in the environment, due to the ubiquitous occurrence of carbonate ions under aquifer conditions and to the strong scavenging property of such phases for uranium. The aim of the present EXAFS study was to gain insights into the coordination of U(VI) on an amorphous alumina phase ($Al_2O_3 \cdot nH_2O$) under conditions relevant to groundwaters, that is, at near-neutral pH values and in the presence of carbonate ligands.

Procedure. The solid used is a Merck 'alumina gel.' The product is predominantly amorphous but its X-ray diffraction pattern exhibits small peaks characteristics of gibbsite. Its surface charge determined by counter-ions adsorption is reported in Fig. 4. Batch experiments were performed using 5 g/L solid contacted with 10^{-4} M uranyl carbonated solutions at pH $\sim$ 7

(total carbonate concentration, expressed as $C = 10^{-2}$ M) or with CO_2-free uranyl solutions in an N_2 atmosphere. The uranyl-bearing solid samples were conditioned as wet pastes (leading possibly to contaminations by atmospheric CO_2). Uranium L_{III}-edge EXAFS spectra were recorded in fluorescence mode at the Rossendorf beamline of the European Synchrotron Radiation Facility (ESRF), using a Si(111) double-crystal monochromator. Energy calibration was obtained by measuring the K-edge absorption of yttrium metal. The EXAFS spectra were analysed according to the standard procedure using the program FEFFIT and theorical scattering phases and amplitudes were calculated with the code FEFF (Rehr & Albers 2000).

Results and discussion. Fourier transforms (FT), not corrected for phase shift, of the k^3-weight XAFS spectra are presented in Fig. 5. They exhibit two distinct major peaks in the shorter distance region. The first one, at about 1.5 Å, corresponds to the U–O_{ax} single scattering interactions, the second one, at about 2 Å, corresponds to the U–O_{eq} interactions. All samples also display a weak contribution at 2.7 Å. Table 3 reports the best-fit structural parameters. No splitting of the equatorial oxygen shell was observed. Attributing the small contribution at 2.7 Å (on the FT) to Al neighbours resulted in bad fitting parameters and no evidence for

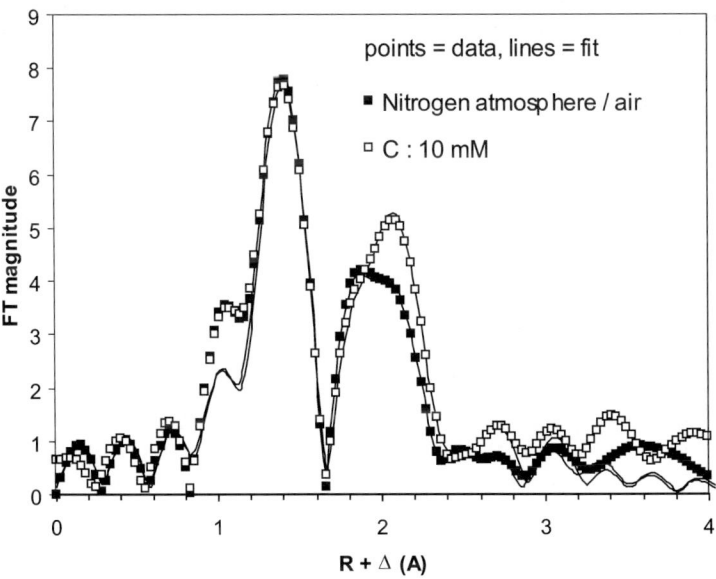

Fig. 5. Fourier transforms, not corrected for phase shift, of the k^3-weight XAFS spectra for uranyl ions sorbed on alumina gel using the experimental conditions given in Table 3.

Table 3. *Best fit structural parameters of the uranium L_{III}-edge EXAFS spectra shown in Fig. 5 and experimental and spectroscopic sample conditions*

Experiments: 5 g/L alumina gel in 10^{-4} M uranyl solutions at pH ~ 7		Coordination shells: coordination number (N), distance (R in Å), Debye–Waller factor (σ^2 in Å^2)								
Carbonate concentration (expressed C)	U loading (μmol U/m^2)	U–O$_{ax}$			U–O$_{eq}$			U–C		
		N	R	σ^2	N	R	σ^2	N	R	σ^2
N$_2$ atm./air*	$\sim$0.2	2.0†	1.80	0.002	5.2	2.44	0.007	1.7	2.94	0.002
10^{-2} M	$\sim$0.15	2.0†	1.81	0.002	5.3	2.46	0.005	2.3	2.92	0.001

*Sample under nitrogen atmosphere during experiments and re-equilibrated with air during conditioning.
†Value held constant in the fits.

U–Al contribution was found, even at longer distances. The contribution at 2.7 Å actually corresponds to the U–C single scattering contribution. A surprising feature is that the presence of carbon is observed for samples maintained under an N$_2$ atmosphere during the batch experiments. This indicates either carbonate impurities on the alumina gel, which could not be removed by our mineral pretreatments, and/or carbonate contamination of the samples during the sorption experiments or the conditioning for EXAFS analysis. Uranium is found surrounded by an average of two carbon atoms at a distance of 2.92–2.94 Å (Table 3). From such a distance between U and C, we suggest formation of a bidentate surface complex. A bidentate coordination results indeed in distances close to 2.9 Å, whereas monodentate coordination would result in larger distances ($\sim$3.2 Å). The spectral features lead us to conclude that an outer-sphere complexation mechanism is possibly responsible for the formation of surface–U(VI)–carbonato species at near-neutral pH values. Studies using EXAFS of U(VI) sorbed onto hematite in the presence of carbonates reported U–Fe interactions and inner-sphere surface complexation (Bargar *et al.* 2000). By contrast, EXAFS data on uranyl sorbed onto hematite at pH 7 and in equilibrium with 2% CO$_2$ atmosphere point to formation of outer-sphere surface complexes (Swanton *et al.* 1998). Further spectroscopic studies are currently in progress to confirm the surface complexation mode and to understand the influence of chemical parameters on the mechanisms of uranyl surface complexation onto alumina minerals.

XPS studies in uranyl–quartz systems. Using X-ray photoelectron spectroscopy (XPS), Froideval *et al.* (2003) found two kinds of pH-dependent uranyl species at high surface coverage onto quartz (initial aqueous uranyl concentration: [U]$_{aq,i}$ = 10^{-4} M). A uranyl surface component exhibiting the U 4f XPS lines characteristics of a reference meta-schoepite was interpreted as polynuclear surface oligomers and/or as amorphous schoepite-like (surface) precipitates. This result confirmed the possible formation at mineral surfaces of polynuclear uranyl species, which has long been disputed in the literature. The position of the U 4f XPS lines of the second component indicated uranyl ions in coordination environments very distinct to that of uranyl ions in U(VI) oxide hydrate-like compounds, probably as mononuclear uranyl surface species. Mononuclear surface species were *a priori* unexpected under conditions where polynuclear species and UO$_2$(OH)$_2$ colloids predominate in solution. The findings of Froideval *et al.* (2003) supported the conclusion of Chisholm-Brause *et al.* (2001) on the predominance of mineral surface characteristics (i.e., of the nature and abundance of hydroxyl surface sites) over aqueous uranyl speciation on the uranyl sorption species. Nevertheless, confirming such a conclusion requires spectroscopic investigations under a wide range of experimental conditions.

Although EXAFS is a powerful technique to determine the chemical environment of an atom, a limitation is that it observes the average of all structures present for an adsorbing ion. In XPS measurements, the energy of the emitted photoelectron for a given atom is related to its coordination environment. The U 4f XPS binding energy value is thus dependent on the U(VI) chemical environment (i.e., on the number of equatorial oxygen atoms bound to a central U atom) and/or on the U(VI) structural environment (i.e., on the length of the U–O$_{eq}$ bonds). The deconvolution of XPS spectra may thus provide valuable insights into co-existing uranyl surface species having distinct coordination environments. We present here the deconvolution of U 4f XPS spectra taken from Mignot (2001) and recorded for uranyl/quartz systems containing carbonate ligands.

Procedure. A deconvolution procedure was applied to the U 4f spectra of Mignot (2001) and to a U 4f spectrum of Froideval *et al.* (2003). The batch experiments performed by the authors to obtain the samples suitable for spectroscopy are as follows. The partitioning of uranyl between quartz, colloidal and aqueous phases is determined using a centrifugation/ ultracentrifugation procedure. Quartz (25 g/L) was brought in contact with CO_2-free 5×10^{-5} M uranyl solution or with 10^{-4} M uranyl solutions at total carbonate concentrations of 10^{-3} M or 10^{-2} M (expressed as C). The samples were adjusted to desired final pH values (expressed as pH_F) and shaken for three days at 298 K. The U(VI)-containing quartz fractions to be analysed by XPS were collected after centrifugation of the samples. Blank experiments were also conducted for the possibility of formation of uranyl compounds from aged uranyl solutions containing no quartz. The XPS measurements were performed using a VSW-HA 150 spectrometer with a monochromatized Al K$_\alpha$ X-ray source (hν = 1486.6 eV). The procedures are described in detail by Mignot (2001) and Froideval *et al.* (2003).

Results and discussion. Uranyl colloids are formed in the blank and sorption samples at

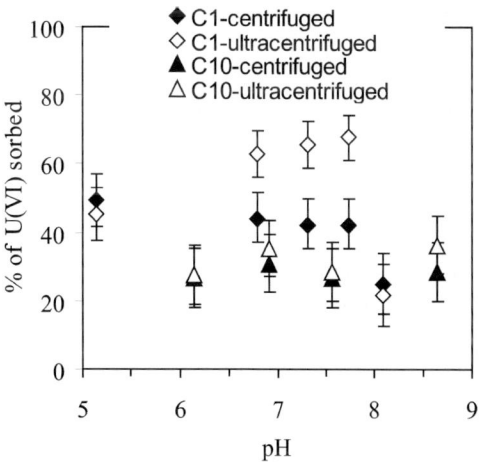

Fig. 6. Percentage of uranyl sorbed on quartz at a total carbonate concentration of 10^{-3} M (series C1) or 10^{-2} M (series C10) obtained after sample centrifugation and ultracentrifugation (after Mignot 2001).

near-neutral pH except at $C = 10^{-2}$ M (Fig. 6). The uranyl colloids are possibly uranyl solubility-controlling phases similar to the amorphous shoepite of Torrero *et al.* (1994). Nevertheless, quartz is responsible for the removal of uranyl from solution in the low pH (pH $\sim$ 5) and/or highly carbonated samples. It increases significantly the percentage of U(VI) collected by centrifugation in the other samples. Such features indicate that sorption processes on quartz intervene in the retention of U(VI) in the samples analysed by XPS. Table 4 reports the spectroscopic sample conditions. The procedure of deconvolution of the U 4f peaks is similar to that described by Froideval *et al.* (2003). All the spectra are successfully deconvoluted by (1) taking parameters determined for the U 4f peaks of a 'metaschoepite' reference sample for a high binding energy component, and by (2) adding only one component having U 4f lines at lower binding energies (Fig. 7). The component at low binding energy displays XPS lines characteristics that are similar to those for uranyl ions sorbed onto quartz at very low pH (Froideval *et al.* 2003). Comparing the position of the U 4f XPS lines with EXAFS structural parameters published for uranyl ions onto minerals provide insight into the uranyl coordination environments. Studies using EXAFS reported similar U–O equatorial parameters for U atoms in metaschoepite and in polynuclear surface oligomers/surface precipitates onto silica and onto hydrobiotite (Hudson *et al.* 1999; Sylwester *et al.* 2000). As suggested by Froideval *et al.* (2003), we attribute the high binding energy component having an U(VI) oxide hydrate character in our XPS spectra to surface polymers and/or to shoepite-like surface precipitates. The formation of polynuclear uranyl surface complexes has long been controversially discussed. On the one hand, EXAFS analyses indicated the formation of polynuclear uranyl surface species on Al-oxides, silica and silicates brought in contact with near-neutral solutions having aqueous uranyl concentrations between $\sim$3 $\times$ 10^{-6} M and $\sim$4 $\times$ 10^{-5} M (Hudson *et al.* 1999; Sylwester *et al.* 2000). On the other hand, EXAFS analyses provided evidence for the existence of mononuclear bidentate surface complexes not only for uranyl at acidic pH on various minerals (Waite *et al.* 1994; Reich *et al.* 1998; Sylwester *et al.* 2000) but also at near-neutral pH on ferric oxyhydroxides where aqueous polynuclear species might predominate (Reich *et al.* 1998; Bargar *et al.* 2000). The spectroscopic investigation of Chisholm-Brause *et al.* (2001) of uranyl sorbed onto montmorillonite shed new light on such discrepancies by suggesting

Table 4. *Experimental and spectroscopic sample conditions of the U(VI)-containing quartz analysed by XPS*

Sample	Initial aqueous U concentration (μM)	Carbonate concentration (mM)	pH	U loading (μmol U/g)	U loading (μmol U/m^2)
C0-5*	50	N$_2$-atmosphere	5.5	~0.2	~3.1
C1-7[†]	100	1	6.8	~0.4	~6.2
C1-8[†]	100	1	8.1	~0.25	~3.8
C10-6[†]	100	10	6.2	~0.3	~4.6

*Froideval *et al.* (2003); [†]Mignot (2001).

additional polymeric species formed only at high uranyl surface coverage, independently of the aqueous uranyl speciation. A major finding of our work is that polymeric uranyl species may form at a high uranyl surface coverage onto silica minerals, and in addition to other uranyl surface components. In our EXAFS study on the uranyl species sorbed onto an amorphous Al phase, no evidence for polymeric uranyl surface species at high carbonate solution

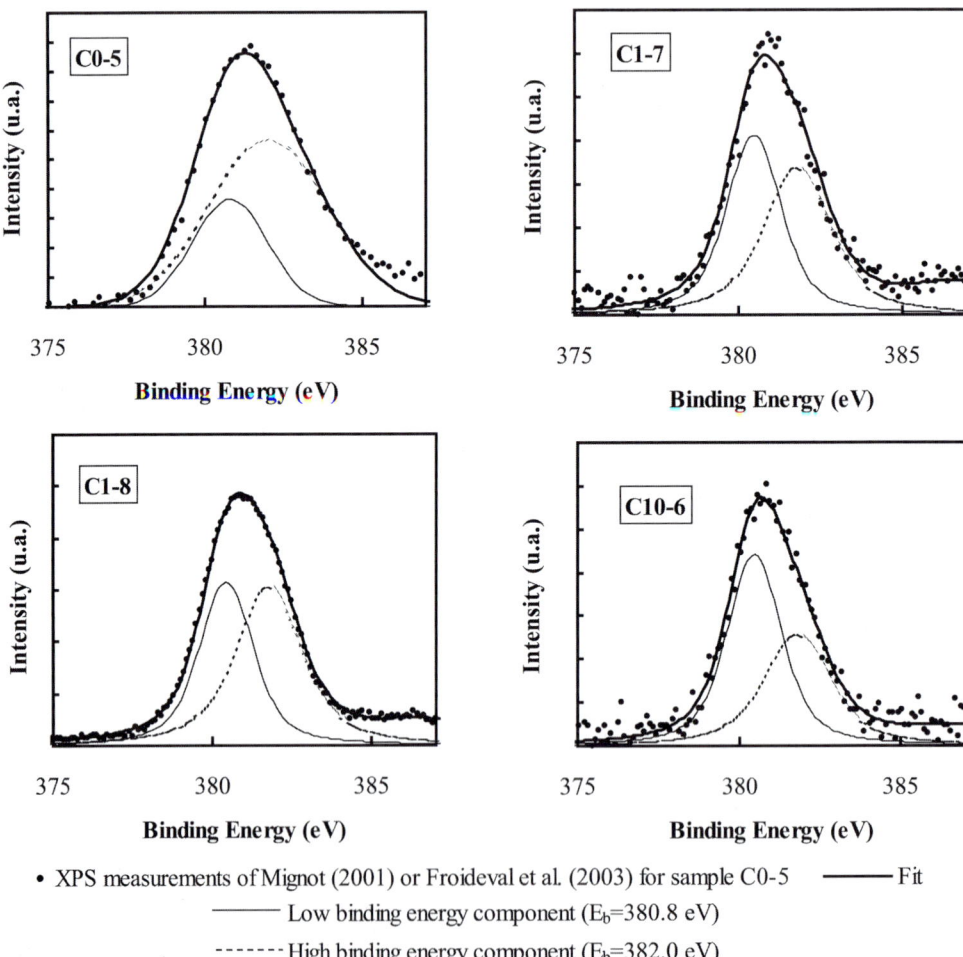

• XPS measurements of Mignot (2001) or Froideval et al. (2003) for sample C0-5   ——— Fit
——— Low binding energy component (E$_b$=380.8 eV)
------ High binding energy component (E$_b$=382.0 eV)

Fig. 7. Fitting curves showing the contribution of two uranyl components with distinct binding energies compared with the U 4F XPS spectra of U(VI)-containing quartz samples described in Table 4. u.a., arbitary units.

concentration ($C = 10^{-2}$ M) was found. This supports the predominance of surface coverage (i.e., of the surface mineral characteristics such as the abundance of surface hydroxyls) over aqueous uranyl speciation on the uranyl sorption species. The component having U 4f lines at low binding energy values occurs possibly as mononuclear uranyl surface complexes. Indeed, either the $U–O_{eq1}$ bond distance (Waite et al. 1994; Bargar et al. 2000) or the $U–O_{eq2}$ bond distance (Reich et al. 1998) is higher for U atoms in mononuclear surface complexes than for U atoms in polynuclear surface species and/or (surface-) schoepite-like precipitates (Hudson et al. 1999; Sylwester et al. 2000). This implies lower U 4f XPS binding energies for uranyl ions in mononuclear surface species than in polynuclear species/schoepite-like precipitates. We can, however, draw no conclusion on the presence of uranyl carbonate species on quartz, due to the scarcity of EXAFS data for ternary uranyl/mineral/carbonate solution systems. Further spectroscopic studies are currently in progress to investigate uranyl on quartz at lower surface coverage and to determine the nature of the uranyl species in the low binding energy component.

Conclusions

Studies of natural analogue sites provide information on the variety of sorption processes relevant to the retention of uranyl in the environment. The sorption/desorption behaviour of uranyl from weathered rock horizons at Bangombé point to several processes of long-term uranyl immobilization, that is, surface complexation onto clays, co-precipitation with ferric oxyhydroxides and/or incorporation into ferric phosphates occurring as coatings on ferric oxyhydroxides. Modelling uranyl sorption, however, requires detailed knowledge of the mechanisms of sorption and of the affinity of many potentially adsorbed uranyl species for mineral surfaces. In particular, probing the formation of uranyl carbonate complexes onto minerals is of cardinal importance due to the omnipresence of carbonate ions in aquifers. The EXAFS study presented here provides new data on the coordination of U(VI) on an alumina mineral in the presence of carbonate ligands. It confirms that uranyl carbonato surface complexes may occur on alumina minerals. Such a finding is of importance for predicting uranyl mobility in groundwaters due to the widespread occurrence of Al (hydr-)oxides in aquifer systems. A current research field concerns the formation of polynuclear uranyl surface species onto minerals,

which has long been a subject of debate in literature. The XPS work presented here shows that two components having uranyl ions in very distinct coordination environments co-exist at high surface coverage on quartz, independently of the presence of uranyl colloids and/or uranyl carbonate complexes in solutions. One component corresponds to polynuclear surface species and/or schoepite-like surface precipitates. At similar carbonate solution concentrations ($C = 10^{-2}$ M), uranyl polymeric surface compounds occur on quartz whereas no evidence for such species could be found in our EXAFS study of uranyl on Al gel. These spectroscopic results support the predominance of the uranyl surface coverage (i.e., of the mineral surface characteristics such as the nature and abundance of surface hydroxyls) over the aqueous uranyl speciation on the sorption mechanisms. They further suggest that mineral surface characteristics are important factors controlling the uranyl surface species in aquifers. Spectroscopic work is currently in progress to confirm such a conclusion. Our paper illustrates that, whereas natural analogues provide general information on processes and host phases relevant to long-term uranyl retardation in the environment, surface spectroscopic studies are irreplaceable to obtain detailed knowledge of the mechanisms of sorption.

The experimental studies of the neptunyl/alumina gel system were performed in the framework of the contract 'GLASTAB' (N°FIKW-CT-2000-00007) of the 5th Euratom Programme. We thank the reviewers and P. Stille for improvement of the manuscript.

References

BARGAR, J. R., REITMEYER, R., LENHART, J. J. & DAVIS, J. A. 2000. Characterization of U(VI)–carbonato ternary complexes on hematite: EXAFS and electrophoretic mobility measurements. *Geochimica et Cosmochimica Acta*, **64**, 2737–2749.

BITEA, C., MULLER, R., NECK, V., WALTHER, C. & KIM, J. I. 2003. Study of the generation and stability of thorium(IV) colloids by LIBD combined with ultrafiltration. *Colloids and Surfaces A: Physicochemical and Engineering Aspects*, **217**, 63–70.

BOWDEN, J. W., NAGARAJAH, S., BARROW, N. J., POSNER, A. M. & QUIRK, J. P. 1980. Describing the adsorption of phosphate, citrate and selenite on a variable charge mineral surface. *Australian Journal of Soil Research*, **18**, 49–60.

BRADY, P. V., CYGAN, R. T. & NAGY, K. L. 1996. Molecular controls on kaolinite surface charge. *Journal of Colloid Interface Science*, **183**, 356–364.

BRUNO, J., de PABLO, J., DURO, L. & FIGUEROLA, E. 1995. Experimental study and modeling in the $U(VI)-Fe(OH)_3$ surface precipitation/coprecipitation equilibria. *Geochimica et Cosmochimica Acta*, **59**, 4113–4123.

BRUNO, J., DURO, L. *et al.* 1998. Estimation of the concentrations of trace metals in natural systems. The application of codissolution and coprecipitation approaches to El Berrocal (Spain) and Poços de Caldas (Brazil). *Chemical Geology*, **151**, 277–291.

BRUNO, J., DURO, L. & GRIVÉ, M. 2002. The applicability of thermodynamic geochemical models to simulate trace element behaviour in natural waters. Lessons learned from natural analogue studies. *Chemical Geology*, **190**, 371–393.

CHOPPIN, G. R. 1988a. Chemistry of actinides in the environment. *Radiochimica Acta*, **43**, 82–83.

CHOPPIN, G. R. 1988b. Humics and radionuclide migration. *Radiochimica Acta*, **44/45**, 23–28.

CHOPPIN, G. R. 1989. Soluble rare earth and actinide species in seawater. *Marine Chemistry*, **28**, 19–26.

CHOPPIN, G. R. 1992. The role of natural organics in radionuclide migration in natural aquifer systems. *Radiochimica Acta*, **58/59**, 113–120.

CHOPPIN, G. R. & ALLARD, B. 1985. *In:* FREEMAN, A. G. & KELLER, C. (eds) *Handbook on the Physics and Chemistry of the Actinides.* Elsevier, 407–429.

CHISHOLM-BRAUSE, C. J., BERG, J. M., MATZNER, R. A. & MORRIS, D. E. 2001. Uranium(VI) sorption on montmorillonite as a function of solution chemistry. *Journal of Colloid and Interface Science*, **233**, 38–49.

COMBES, J. M., CHISHOLM-BRAUSE, C. J. *et al.* 1992. EXAFS spectroscopic study of neptunium(V) sorption at the α-FeOOH/water interface. *Environmental Science and Technology*, **26**, 376–382.

DAVIS, J. A., JAMES, R. O. & LECKIE, J. O. 1978. Surface ionization and complexation at the oxide/water interface. I. Computation of electrical double layer properties in simple electrolytes. *Journal of Colloid and Interface Science*, **63**, 480–499.

DAVIS, J. A. & KENT, D. B. 1990. Surface complexation modeling in aqueous geochemistry. *Reviews in Mineralogy*, **23**, 177–260.

DEGUELDRE, C., ULRICH, H. G. & SILBY, H. 1994. Sorption of ^{243}Am onto montmorillonite, illite and hematite colloids. *Radiochimica Acta*, **65**, 173–179.

DEL NERO, M., MADÉ, B., BONTEMS, G. & CLÉMENT, A. 1997. Adsorption of neptunium(V) on hydrargilite. *Radiochimica Acta*, **76**, 219–228.

DEL NERO, M., BEN SAÏD, K., MADÉ, B., CLÉMENT, A. & BONTEMS, G. 1998. Effect of pH and carbonate concentration in solution on the sorption of neptunium(V) by hydrargilite: Application of the non-electrostatic model. *Radiochimica Acta*, **81**, 133–141.

DEL NERO, M., SALAH, S., MIURA, T., CLÉMENT, A. & GAUTHIER–LAFAYE, F. 1999a. Sorption/desorption processes of uranium in clayey samples of the Bangombé natural reactor zone, Gabon. *Radiochimica Acta*, **87**, 135–149.

DEL NERO, M., ADVOCAT, T., JOLLIVET, P. & BONTEMS, G. 1999b. Sorption of neptunium (V) on an alteration gel of alumino-borosilicate glasses and on synthetic silicate gels. *Proceedings 2nd International Symposium on Nuclear Fuel Cycle Safety Engineering Research Facility NUCEF'98*, Hitachinaka, Ibaraki, Japan, 584–595.

DIAZ AROCAS, P. & GRAMBOW, B. 1998. Solid–liquid phase equilibria of U(VI) in NaCl solutions. *Geochimica et Cosmochimica Acta*, **62**, 245–263.

DZOMBAK, D. A. & MOREL, F. M. M. 1990. *Surface Complexation Modeling. Hydrous Ferric Oxide.* John Wiley & Sons, New York.

FAIRHURST, A., WARWICK, P. & RICHARDSON, S. 1995. The influence of humic acid on the adsorption of europium onto inorganic colloids as a function of pH. *Colloids Surfaces A: Physicochemical and Engineering Aspects*, **99**, 187–199.

FROIDEVAL, A., DEL NERO, M., BARILLON, R., HOMMET, J. & MIGNOT, G. 2003. pH-dependence of uranyl retention in a quartz/solution system: an XPS study. *Journal of Colloid and Interface Science*, **266**, 221–235.

GABRIEL, U., CHARLET, L., SCHLÄPFER, C. W., VIAL, J. C., BRACHMANN, A. & GEIPEL, G. 2001. Uranyl surface speciation on silica particles studied by time-resolved laser-induced fluorescence spectroscopy. *Journal of Colloid and Interface Science*, **239**, 358–368.

GAUTHIER-LAFAYE, F., HOLLIGER, P. & BLANC, P.-L. 1996. Natural fission reactors in the Franceville basin, Gabon: a review of the conditions and results of a 'critical event' in a geologic system. *Geochimica et Cosmochimica Acta*, **60**, 4831–4852.

GAUTHIER-LAFAYE, F., STILLE, P. & BROS, R. 2004. Special cases of natural analogues: The Gabon and Cigar Lake U ore deposits. *In:* GIERÉ, R. & STILLE, P. (eds) *Energy, Waste, and the Environment: a Geochemical Perspective.* Geological Society, London, Special Publications, **236**, 123–134.

GECKEIS, H. & RABUNG, T. 2002. Solid-water interface reactions of polyvalent metal ions at iron oxide–hydroxide surfaces. *In:* HUBBARD, A. (ed) *Encyclopedia of Surface and Colloid Science.* Dekker Inc., 4737–4748.

GECKEIS, H., NGO MANH, T. H., BOUBY, M. & KIM, J. I. 2003. Aquatic colloids relevant to radionuclide formation: characterization by size fractionation and ICP-mass spectrometric detection. *Colloids and Surfaces A: Physicochemical and Engineering Aspects*, **217**, 101–108.

GIRVIN, D. C., AMES, L. L., SCHWAB, A. P. & MCGARRAH, J. E. 1991. Neptunium adsorption on synthetic amorphous iron oxyhydroxide. *Journal of Colloid and Interface Science*, **141**, 67–78.

GOLDBERG, S. 1992. Use of surface complexation models in soil chemical systems. *In:* SPARKS, D. L. (ed) *Advances in Agronomy, 47.* Academic Press, Inc., San Diego, 233–329.

HACHIYA, K., SASAKI, M., SARUTA, M., MIKAMI, N. & YASUNAGA, T. 1984. Static and kinetic studies of adsorption–desorption of metal ions on a γ-Al$_2$O$_3$ surface. 1. Static study of adsorption–desorption. *Journal of Physical Chemistry*, **88**, 23–27.

HAYES, K. F., REDDEN, G., ELA, W. & LECKIE, J. O. 1991. Surface complexation models: an evaluation of model parameter estimation using FITEQL and oxide mineral titration data. *Journal of Colloid and Interface Science*, **142**, 448–469.

HAYES, K. F. & KATZ, L. E. 1996. Application of X-ray absorption spectroscopy for surface complexation modelling of metal ion sorption. *In*: BRADY, P. V. (ed) *Physics and Chemistry of Mineral Surfaces*. CRC Press, Boca Raton, 147–223.

HSI, C.-K. D. & LANGMUIR, D. 1985. Adsorption of uranyl onto ferric oxyhydroxides: application of the surface complexation site-binding model. *Geochimica et Cosmochimica Acta*, **49**, 1931–1941.

HUDSON, E. A., TERMINELLO, L. J. *et al.* 1999. The structure of U^{6+} sorption complexes on vermiculite and hydrobiotite. *Clays and Clay Minerals*, **47**, 439–457.

HUERTAS, F. J., CHOU, L. & WOLLAST, R. 1998. Mechanism of kaolinite dissolution at room temperature and pressure: Part I. Surface speciation. *Geochimica et Cosmochimica Acta*, **62**, 417–731.

JANECZEK, J. & EWING, R. 1996. Florencite-(La) with fissiogenic REE from a natural fission reactor at Bangombé, Gabon. *American Mineralogist*, **81**, 1263–1269

JENSEN, K. A., JANECZEK, J., EWING, R. C., STILLE, P., GAUTHIER-LAFAYE, F. & SALAH, S. 2000. Crandallites and coffinite: retardation of nuclear fission products at the Bangombé nuclear fission reactor. *Material Research Society Symposium Proceedings*, **608**, 525–532.

JOHANNESSON, K. H., STETZENBACH, K. J. & HODGE, V. F. 1997. Rare earth elements as geochemical tracers of regional groundwater mixing. *Geochimica et Cosmochimica Acta*, **64**, 3605–3618.

KIENZLER, B., LUCKSCHEITER, B. & WILHELM, S. 2001. Waste form corrosion modeling: comparison with experimental results. *Waste Management*, **21**, 741–752.

KIM, J. I., BUCKAU, G. & KLENZE, R. 1987. *In*: COME, B. & CHAPMAN, N. (eds) *Natural Analogues in Radioactive Waste Disposal*. Graham and Trotman, London, 289.

KIM, J. I. 1991. Chemical behaviour of transuranic elements in the natural environment. *Proceedings 3rd International Symposium on Advanced Nuclear Energy Research – Global Environment and Nuclear Energy*, March 1991, Mito, Japan, 1–13.

KOHLER, M., HONEYMAN, B. D. & LECKIE, J. O. 1999. Neptunim(V) sorption on hematite (alpha-Fe$_2$O$_3$) in aqueous suspension: the effect of CO$_2$. *Radiochimica Acta*, **85**, 33–48.

LABONNE-WALL, N., MOULIN, V. & VILAREM, J.-P. 1997. Retention properties of humic substances onto amorphous silica: consequences for the sorption of cations. *Radiochimica Acta*, **79**, 37–49.

LAFLAMME, B. D. & MURRAY, J. W. 1987. Solid/solution interaction: the effect of carbonate alkalinity on adsorbed thorium. *Geochimica et Cosmochimica Acta*, **51**, 243–250.

MADÉ, B., LEDOUX, E., SALIGNAC, L., LE BOURSICAUD, B. & GURBAN, I. 2000. Modélisation du transport réactif de l'uranium autour du réacteur nucléaire naturel de Bangombé (Oklo, Gabon). *Comptes Rendus de L'Académie des Sciences*, **331**, 587–594.

MCKINLEY, J. P., ZACHARA, J. M., SMITH, S. C. & TURNER, G. 1995. The influence of uranyl hydrolysis and multiple site-binding reactions on adsorption of U(VI) to montmorillonite. *Clays and Clay Minerals*, **43**, 586–598.

MIGNOT, G. 2001. *Rétention de l'Uranium(VI) sur le Quartz et la Kaolinite. Expériences et Modélisations*. PhD thesis, Paris XI, France, 218.

MÖRI, A., ALEXANDER, W. R. *et al.* 2003. The colloid and radionuclide retardation experiments at the Grimsel Test Site: influence of bentonite colloids on radionuclide migration in a fractured rock. *Colloids and Surfaces A: Physicochemical and Engineering Aspects*, **217**, 1–3, 33–47.

MORRISON, S. J., TRIPATHI, V. S. & SPANGLER, R. R. 1995. Coupled reaction/transport modeling of a chemical barrier for controlling uranium(VI) contamination in groundwater. *Journal of Contaminant Hydrology*, **17**, 347–363.

MURAKAMI, T., OHNUKI, T., Isobe, H & SATO, T. 1997. Mobility of uranium during weathering. *American Mineralogist*, **82**, 888–899.

PAYNE, T. E. & WAITE, T. D. 1991. Surface complexation modeling of uranium sorption data obtained by isotope exchange techniques. *Radiochimica Acta*, **52/53**, 487–493.

PAYNE, T. E., DAVIS, J. A. & WAITE, T. D. 1994. Uranium retention by weathered schists. The role of iron minerals. *Radiochimica Acta*, **66/67**, 297–303.

PAYNE, T. E., LUMPKIN, G. R. & WAITE, T. D. 1998. Uranium(VI) adsorption on model minerals: controlling factors and surface complexation modeling. *In*: JENNE, E. (ed) *Adsorption of Metals by Geomedia*. Academic Press, San Diego, 75–97.

PÉREZ DEL VILLAR, L., PELAYO, M. *et al.* 1997. Mineralogical and geochemical evidence of the migration/retention processes of U and Th in fracture fillings from the El Berrocal granitic site (Spain). *Journal of Contaminant Hydrology*, **26**, 45–60.

REHR, J. J. & ALBERS, R. C. 2000. Theoretical approaches to X-ray absorption fine structure. *Review in Modern Physics*, **72**, 621–654.

REICH, T., MOLL, H. *et al.* 1998. An EXAFS study of uranium(VI) sorption onto silica gel and ferrihydrite. *Journal of Electron Spectroscopy and Related Phenomena*, **96**, 237–243.

ROTHE, J., DENECKE, M. A., NECK, V., MULLER, R. & KIM, J. I. 2002. XAFS investigation of the structure of aqueous thorium(IV) species, colloids, and solid thorium(IV) oxide/hydroxide. *Inorganic Chemistry*, **41**, 249–258.

RUNDE, W., CONRADSON, S. D., EFURD, W. D., LU, N., VANPELT, C. E. & TAIT, C. D. 2002. Solubility and sorption of redox-sensitive radionuclides (Np, Pu) in J-13 water from the Yucca Mountain Site:

comparison between experiment and theory. *Applied Geochemistry*, **17**, 837–853.

SALAH, S. 2000. *Weathering Processes at the Natural Reactor of Bangombé (Gabon). Identification and Geochemical Modeling of the Retention and Migration Mechanisms of Uranium and Rare Earth Elements.* PhD thesis, Strasbourg, France.

SALAH, S., GAUTHIER-LAFAYE, F. & DEL NERO, M. 1999. Behaviour of REE and U in the weathering sequence of Bangombé. *In*: LOUVAT, D., MICHAUD, I. & VON MARAVIC H. (eds) *EUR Report 19116.* Commission of the European Communities, Luxembourg, 255–268.

SANCHEZ, A. L., MURRAY, J. W. & SIBLEY, T. H. 1985. The adsorption of plutonium(IV) and (V) on goethite. *Geochimica et Cosmochimica Acta*, **49**, 2297–2307.

SCHINDLER, P. W. & STUMM, W. 1987. The surface chemistry of oxides, hydroxides, and oxide minerals. *In*: STUMM, W. (ed) *Aquatic Surface Chemistry.* John Wiley & Sons, New York, 83–107.

SILVA, R. J. & NITSCHE, H. 1995. Actinide environmental chemistry. *Radiochimica Acta*, **70/71**, 377–396.

STILLE, P., GAUTHIER-LAFAYE, F. *et al.* 2003. REE mobility in groundwater proximate to the natural fission reactor at Bangombé (Gabon). *Chemical Geology*, **198**, 289–304.

STUMM, W., KUMMERT, R. & SIGG, L. 1980. A ligand exchange model for the adsorption of inorganic and organic ligands at hydrous oxide interfaces. *Croatica Chemica Acta*, **53**, 291–312.

STUMPF, TH., RABUNG, TH., KLENZE, R., GECKEIS, H. & KIM, J. I. 2001. Spectroscopic study of Cm(III) sorption onto γ-alumina. *Journal of Colloid and Interface Science*, **238**, 219–224.

SVERJENSKY, D. A. & SAHAI, N. 1996. Theoretical prediction of single-site surface-protonation equilibrium constants for oxides and silicates in water. *Geochimica et Cosmochimica Acta*, **60**, 3773–3797.

SWANTON, S. W., BASTON, G., COWPER, M. M. & CHARNOCK, J. M. 1998. EXAFS study of uranium(VI) sorbed to hematite. *Proceedings Workshop on Speciation, Techniques and Facilities for Radioactive Materials at Synchroton Light Sources*, Grenoble, France, Nuclear Energy Agency.

SYLWESTER, E. R., HUDSON, E. A. & ALLEN, P. G. 2000. The structure of uranium(VI) sorption complexes on silica, alumina and montmorillonite. *Geochimica et Cosmochimica Acta*, **64**, 2431–2438.

TAKAHASHI, Y., MINAI, Y, AMBE, S., MAKIDE, Y. & AMBE, F. 1999. Comparison of adsorption behaviour of multiple inorganic ions on kaolinite and silica in the presence of humic acids using the multiple sorption tracer technique. *Geochimica et Cosmochimica Acta*, **63**, 815–836.

TORRERO, M. E., CASAS, I., DE PABLO, J., SANDINO, M. C. A. & GRAMBOW, B. 1994. A comparison between unirradiated $UO_2(s)$ and schoepite solubilities in 1 M NaCl medium. *Radiochimica Acta*, **66/67**, 29–35.

TURNER, D. R. & SASSMAN, S. A. 1996. Approaches to sorption modeling for high-level waste performance assessment. *Journal of Contaminant Hydrology*, **21**, 311–332.

TURNER, G. D., ZACHARA, J. M., MCKINLEY, J. P. & SMITH, S. C. 1996. Surface-charge properties and UO_2^{2+} adsorption of a subsurface smectite. *Geochimica et Cosmochimica Acta*, **60**, 3399–3414.

VAN RIEMSDIJK, W. H., BOLT, G. H., KOOPAL, L. K. & BLAAKMEER, J. 1986. Electrolyte adsorption on heterogeneous surfaces: Adsorption models. *Journal of Colloid and Interface Science*, **109**, 219–228.

WAITE, T. D., DAVIS, J. A., PAYNE, T. E., WAYCHUNAS, G. A. & XU, N. 1994. Uranium(VI) adsorption to ferrihydrite. *Geochimica et Cosmochimica Acta*, **58**, 5465–5478.

WIELAND, E. & STUMM, W. 1992. Dissolution kinetics of kaolinite in acidic aqueous solutions at 25 °C. *Geochimica et Cosmochimica Acta*, **56**, 3339–3355.

YANASE, N., NIGHTINGALE, T., PAYNE, T. & DUERDEN, P. 1991. Uranium distribution in mineral phases of rock by sequential extraction procedure. *Radiochimica Acta*, **52/53**, 387–393.

ZACHARA, J. M. & MCKINLEY, J. P. 1993. Influence of hydrolysis on the sorption of metal cations by smectites: importance of edge coordination reactions. *Aquatic Sciences*, **55**, 250–261.

Development and application of the Nagra/PSI Chemical Thermodynamic Data Base 01/01

TRES THOENEN[1], URS BERNER[2], ENZO CURTI[2], WOLFGANG HUMMEL[2] & F. J. PEARSON[3]

[1]*Paul Scherrer Institut, Waste Management Laboratory, 5232 Villigen PSI, Switzerland*
(e-mail: tres.thoenen@psi.ch)
[2]*Paul Scherrer Institut, Waste Management Laboratory, 5232 Villigen PSI, Switzerland*
[3]*Ground-Water Geochemistry, 411 East Front St. New Bern, NC 28560-4916, USA*

Abstract: The Nagra/PSI Chemical Thermodynamic Data Base is a critical selection of thermodynamic data used to support performance assessments for planned radioactive waste repositories in Switzerland. For this purpose, the data base is focused on actinides and fission products, but also includes additional data for major elements occurring in common groundwaters. Recently, a peer-reviewed update was carried out (from data base version 05/92 to 01/01) involving major revisions for most actinides and fission products. Altogether, more than 70% of the data base contents have been revised. Most of the thermodynamic data for U, Np, Pu, Am, and Tc were adopted from the reviews by the NEA TDB project. In contrast, data for Th, Sn, Eu, Pd, Al, as well as the solubility and metal complexation of sulphides and silicates, were extensively reviewed. Less effort was put into the review of data for Zr, Ni, and Se as these elements are currently being reviewed in Phase II of the NEA TDB project. The data for metal complexation with organic ligands from data base version 05/92 were not included in this update. They will be reconsidered in a future update, after completion of the NEA TDB Phase II review of organic ligands. The chemical consistency of the selected data was checked with empirical rules related to the periodic properties of the elements and with correlations based on charge/size relations. As a result of the update, major gaps in the data base could be identified, especially with respect to missing carbonate complexes. In some systems, for example, Th(IV)–H$_2$O and U(IV)–H$_2$O, conflicting experimental data cannot be described by a unique set of thermodynamic constants and a pragmatic approach closely reproducing measured solubility data was chosen for application to performance assessment. The electronic version of the Nagra/PSI Chemical Thermodynamic Data Base 01/01 and information concerning its full documentation is available on the PSI web site.

Equilibrium thermodynamics is one of the pillars supporting the safety analyses of radioactive waste repositories. Thermodynamic constants are used for modelling reference porewaters, calculating radionuclide solubility limits, deriving case-specific sorption coefficients, and analysing experimental results. It is essential to use the same data base in all instances of the modelling chain in order to ensure internally consistent results.

The original Nagra Thermochemical Data Base (version 05/92; Pearson & Berner 1991; Pearson *et al.* 1992) was based on data reviews and experimental studies published through 1990. It was used to support safety studies for a planned Swiss repository for low- and intermediate-level radioactive waste (Nagra 1993, 1994*a*) and for a planned repository of high-level waste in the crystalline basement (Nagra 1994*b*). Since then, the data base has been updated, resulting in the Nagra/PSI Chemical Thermodynamic Data Base 01/01 (Hummel *et al.* 2002), henceforth referred to as Nagra/PSI TDB 01/01. This data base has been used in a recent performance assessment for a proposed Swiss repository in the Opalinus Clay formation (Nagra 2002*a*), which is planned to host spent fuel (SF), vitrified high-level waste (HLW) deriving from the reprocessing of spent fuel, and long-lived intermediate-level waste (ILW).

Update procedure

The update project was guided by the following criteria, which have to be met when using thermodynamic data in performance assessment:

(1) *Accuracy.* Quantitative estimates of radionuclide solubility and speciation need reliable thermodynamic data, that is,

From: GIERÉ, R. & STILLE, P. (eds) 2004. *Energy, Waste, and the Environment: a Geochemical Perspective*. Geological Society, London, Special Publications, **236**, 561–577.
0305-8719/04/$15 © The Geological Society of London 2004.

equilibrium constants. These data cannot yet be calculated ab initio, so they must be derived from experimental studies of appropriate chemical systems. The documentation of the statistical precision of thermodynamic data gained in such studies is usually good. However, the data might be affected by systematic errors and therefore, the accuracy of the data is often lower than the estimated precision. Differences in the chemical speciation models used to evaluate the experimental data in various investigations can be the source of significant systematic errors in the resulting equilibrium constants. It is very difficult to give a reliable estimate of such errors. In our review project we followed the practice of the Thermodynamic Data Base Project of OECD's Nuclear Energy Agency (NEA TDB; see Grenthe et al. 1992; Silva et al. 1995; Rard et al. 1999; Lemire et al. 2001) in trying to estimate the accuracy of the recommended thermodynamic quantities. Also following the practice of the NEA TDB project, we have thoroughly documented our data selection procedures for each recommended quantity in Hummel et al. (2002).

(2) *Completeness.* There are two philosophies concerning the purpose and contents of thermodynamic data bases. (a) An ideal 'puristic' data base would include only 'generally accepted' and 'well-established' thermodynamic quantities that are based on carefully evaluated and sound experimental data. However, due to numerous gaps in chemical knowledge, a puristic data base will always remain incomplete. The effect of this incompleteness must be evaluated if such a data base is used for predictions in complex chemical environments. (b) An ideal 'ready-for-use' data base would contain all the information necessary to describe the geochemical system under study in sufficient detail. However, the completeness of a 'ready-for-use' data base strongly depends on the properties of the investigated system, since a certain amount of site-specific chemical information is always required to identify gaps in the description of safety-relevant processes, and expert judgement is needed to fill these gaps, for example, by applying the methods discussed by Grenthe et al. (1997a). Failure to identify such gaps and assess their consequences may afflict predictive models with serious errors.

In the course of the update process of the Nagra/PSI TDB 01/01 we decided to go beyond the puristic concept and identified several important cases of gaps in chemical knowledge that needed to be filled. Two types of gaps were distinguished. (a) Insufficient process understanding: a growing number of experimental data for specific chemical systems cannot be interpreted by a unique and self-consistent set of thermodynamic constants (see the discussion of the ThO_2-H_2O and UO_2-H_2O systems below). In these cases we chose a pragmatic approach that is well suited for the performance assessment of radioactive waste repositories by including parameters in the data base that are not thermodynamic parameters sensu stricto, but make it possible to reproduce the relevant experimental observations. (b) Insufficient experimental data: several important equilibrium constants are either uncertain or simply unknown. We decided to include uncertain data and to provide estimates for missing data whenever their omission in modelling calculations might lead to more erroneous results than their inclusion (see discussion below).

(3) *Applicability.* The distinction between equilibrium and non-equilibrium systems is highly case-specific, especially in the case of redox reactions. For most potential redox reactions involving nitrogen, sulphur, or radionuclides in different oxidation states the question as to whether equilibrium thermodynamics is applicable calls for system-specific information beyond the realm of thermodynamics. Qualitative kinetic information is needed to decide whether a thermodynamically feasible reaction actually takes place in a specific environment within a given time frame. Although we have included several redox equilibrium constants in the data base, we cannot guarantee that any of these apply to a specific geochemical situation. For the purpose of performance assessment modelling, it has to be decided for every redox sensitive radionuclide whether a particular thermodynamic equilibrium applies or whether a fixed oxidation state has to be assumed.

Data base version 05/92 distinguished between two types of data, 'core data' and 'supplemental data'. Core data are for elements commonly found as major solutes in natural waters. These data are well established and have not been changed to any significant degree in the update. Supplemental data comprise actinides, fission products, Mn, Fe, Si, and Al (see Fig. 1). The update from data base version 05/92 to 01/01 involved major revisions for most of the supplemental data.

The revision of thermodynamic data for Tc and for the actinides U, Np, Pu, and Am was based on the reviews of the NEA TDB project (Grenthe *et al.* 1992; Silva *et al.* 1995; Rard *et al.* 1999; Lemire *et al.* 2001) and we adopted most of the recommended values. Reasons for rejecting some of the NEA recommendations are documented in detail (Hummel *et al.* 2002). Thermodynamic data for Th, Sn, Eu, Pd, Al, and for the solubility and metal complexation of sulphides and silicates were extensively reviewed, while data for Zr, Ni, and Se were examined less rigorously, since these elements are currently under review in Phase II of the NEA TDB project. Data base version 05/92 contained equilibrium constants for protonation and metal complexation of oxalic (ox), citric (cit), nitrilotriacetic (nta), and ethylenediaminetetraacetic (edta) acid. For the Nagra/PSI TDB 01/01 we decided to remove all data referring to organic ligands and to wait for the completion of the NEA TDB Phase II review of ox, cit, edta, and isosaccharinic (isa) acid. The complexation of metals with these ligands will then be considered in a future update.

In total, more than 70% of the data base contents have been updated. It is remarkable that, unlike most other data base updates, the Nagra/PSI TDB 01/01 includes fewer compounds and complexes than the previous version. This is the result of our efforts to remove data for hypothetical, questionable or irrelevant compounds and complexes. In addition, data of dubious quality were also removed.

The Nagra/PSI TDB 01/01 and its entire documentation (Hummel *et al.* 2002) has undergone a peer review by an independent expert, according to Nagra's QA procedures. The complete peer review comment records may be obtained on request from Nagra.

Structure, contents, and availability of the data base

Since the Nagra/PSI TDB 01/01 was primarily developed for use with computer codes that apply the law of mass action (LMA) algorithm, for example, PHREEQC (Parkhurst & Appelo 1999) and MINEQL (Westall *et al.* 1976), the main effort in the data base update was put into the critical evaluation of equilibrium constants. The selected constants refer to standard state conditions, that is, infinite dilution (zero ionic strength) at 25 °C and 1 bar. Equilibrium constants studied in the laboratory are usually determined in ionic background media, the most popular being $NaClO_4$ and KNO_3 at high concentrations (both are of no relevance in environmental modelling). Hence, all experimental data had to be extrapolated to zero ionic strength as a critical part of the data review procedure. For this purpose, activity coefficients were calculated with the Specific Ion Interaction Theory (SIT) (Grenthe *et al.* 1997b), following the practice of the NEA TDB project. Ideally, the same method should be used for the extrapolation of standard state data contained in the data base to the environmental conditions under investigation, but this is usually not possible, since the SIT equation is presently not implemented in geochemical codes used in performance assessment. Furthermore, SIT parameters are missing for most of the species present in natural waters. As a pragmatic remedy to this unfortunate situation, the Nagra/PSI TDB 01/01 also contains parameters for the calculation of

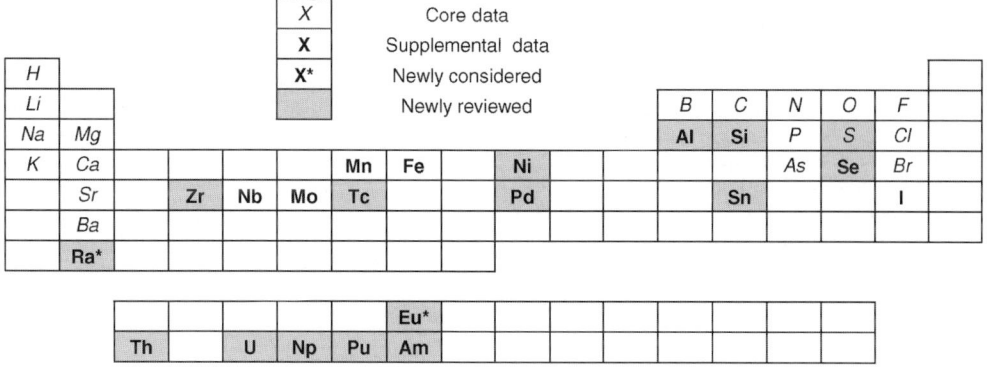

Fig. 1. Periodic table indicating the elements considered in the Nagra/PSI TDB 01/01.

activity coefficients with the extended Debye–Hückel and the WATEQ Debye–Hückel equation (see below), which are both implemented in PHREEQC.

The essential thermodynamic data in the Nagra/PSI TDB 01/01 are equilibrium constants ($\log_{10} K^{\circ}$) for the formation reactions of so-called *product species*, which comprise aqueous product species, solids, and gases. Each formation reaction involves a single product species, which is related to at least one of the aqueous master species. Two types of master species are distinguished. The *primary master species* consist of the minimal set of aqueous species required to formulate all product species. They are thus the basic building blocks for setting up formation reactions. For reasons of convenience in formulating a number of reactions, there are also several *secondary master species*, the formation reactions of which are written entirely in terms of *primary master species*.

Besides equilibrium constants, additional thermodynamic data were included, if available, although little emphasis was put on their completeness. The data for *primary master species* comprise the standard molar thermodynamic properties of formation from the elements ($\Delta_f G_m^{\circ}$: standard molar Gibbs energy of formation; $\Delta_f H_m^{\circ}$: standard molar enthalpy of formation; $\Delta_f S_m^{\circ}$: standard molar entropy of formation), the standard molar entropy (S_m°), the standard molar isobaric heat capacity ($C_{P,m}^{\circ}$), the coefficients $\Delta_f a$, $\Delta_f b$, and $\Delta_f c$ for the temperature-dependent molar isobaric heat capacity equation

$$\Delta_f C_{P,m}(T) = \Delta_f a + \Delta_f bT + \Delta_f c/T^2$$

and finally the *a* and *b* parameters for calculation of the activity coefficient γ according to the WATEQ Debye–Hückel equation (Truesdell & Jones 1974, see also Parkhurst & Appelo 1999)

$$\log_{10} \gamma = \frac{-A z^2 \sqrt{I}}{1 + Ba\sqrt{I}} + bI$$

where A and B are constants for a given temperature, z is the charge, and I is the ionic strength. Note that for $b = 0$, this equation is identical with the extended Debye–Hückel equation.

Data for *secondary master species* and *product species* include the stoichiometry and $\log_{10} K^{\circ}$ of the formation reactions, the standard Gibbs energy of reaction ($\Delta_r G_m^{\circ}$), the standard enthalpy of reaction ($\Delta_r H_m^{\circ}$), the standard entropy of reaction ($\Delta_r S_m^{\circ}$), the standard isobaric heat capacity

of reaction ($\Delta_r C_{P,m}^{\circ}$), the coefficients $\Delta_r a$, $\Delta_r b$, and $\Delta_r c$ for the temperature-dependent equation of the standard isobaric heat capacity of reaction

$$\Delta_r C_{P,m}(T) = \Delta_r a + \Delta_r b T + \Delta_r c/T^2$$

and the coefficients A, B, C, D, E for the temperature-dependent equation of the equilibrium constant

$$\log_{10} K(T) = A + BT + C/T + D\log_{10} T$$
$$+ E/T^2.$$

In addition, data for the thermodynamic properties of the species themselves are given as mentioned above for the *primary master species* (of course, no aqueous activity coefficients are required for solids and gases).

The Nagra/PSI TDB 01/01 was set up and maintained with the data base management program PMATCHC (Pearson *et al.* 2001). It allows both interactive and batch operation for data entry, manipulation, and output in user-specified formats. Performing calculations that maintain internal consistency among data entered by the user in various forms, it provides messages to the user whenever data inconsistencies occur that the program cannot resolve internally. PMATCHC is available for download (together with an electronic version of the Nagra/PSI TDB 01/01 for use with PMATCHC) at *http://les.web.psi.ch/Software/PMATCHC* (versions for Windows and Linux). PMATCHC was also used for the preparation of a PHREEQC version of the Nagra/PSI TDB 01/01, the data file is available for download at *http://les.web.psi.ch/TDBook*.

An additional version (Nagra/PSI TDB 01/01 GEMS) was prepared by Thoenen & Kulik (2003) for use with the geochemical code GEM-Selektor (Karpov *et al.* 1997), which is based on a Gibbs energy minimization (GEM) algorithm. For this purpose, PMATCHC was used for deriving a consistent set of values for $\Delta_f G_m^{\circ}$ from the values of $\log_{10} K^{\circ}$ given by the Nagra/PSI TDB 01/01. GEM-Selektor including Nagra/PSI TDB 01/01 GEMS is available for download at *http://les.web.psi.ch/Software/GEMS-PSI* (versions for Mac OS X, Windows and Linux).

Chemical consistency of selected data

The values of equilibrium constants of aqueous complexes and solids are expected to reveal systematic patterns when different groups of metals and ligands are compared. Some general patterns can be deduced from the position of the element in the periodic table (i.e., principal element groups, Irving–Williams series of transition metals, lanthanide systematics). Many empirical and semi-empirical correlations have been proposed relating aqueous complex and solid formation constants to various parameters, such as ionization potentials, electronegativity, charge/radius ratio, and protonation constants (Grenthe *et al.* 1997*a*). Such correlations are important in assessing the quality of new data and may be used to estimate unknown or experimentally inaccessible constants (Brown & Wanner 1987; Neck & Kim 2000).

We did not use such correlations to estimate formation constants for the Nagra/PSI TDB 01/01 in any systematic way. Our selection of recommended values is almost exclusively based on experimental findings. Therefore, chemical patterns and correlations revealed in the selected data may serve, at least to some extent, as an independent measure of the chemical consistency of the data base.

The obvious first step in checking for the consistency of the data is comparing equilibrium constants of cations with the same charge and similar ionic radius. Comparison of the formation constants of complexes and solids of tetravalent actinides (Th^{4+}, U^{4+}, Np^{4+}, Pu^{4+}), Zr^{4+}, and Sn^{4+} reveals that the selected data are very similar, which is to be expected from a chemical point of view, and none of the formation constants appears to be improbable (for details see Hummel *et al.* 2002). Similar pictures of chemical consistency emerge from the trivalent Np^{3+}, Pu^{3+}, Am^{3+}, and Eu^{3+} complexes and solids, and from the hexavalent UO_2^{2+}, NpO_2^{2+}, and PuO_2^{2+} complexes and solids (Hummel *et al.* 2002).

These data comparisons were used by Hummel & Berner (2002) to estimate values for the missing formation constants of the mixed carbonate hydroxide complexes of U^{4+}, Np^{4+} and Pu^{4+}, and for various missing complexation constants of Np^{3+} and Pu^{3+}. The estimated constants were used as limiting values in performance assessment but are not included in the Nagra/PSI TDB 01/01.

Chemical patterns based on periodic properties of elements

The internal geochemical consistency of the selected thermodynamic constants can be further tested by means of well-established empirical rules related to periodic properties of the elements. One of the most popular and useful schemes for the semi-quantitative prediction of stability constants is that of classifying cations and ligands into 'hard' and 'soft' (or 'A' and 'B') groups (Schwarzenbach 1961; Pearson 1963; Grenthe *et al.* 1997*a*, p. 86, fig. III.7). The stability of metal complexes and solid compounds with ionic bonding character ('hard' or 'A' metals) increases with decreasing cation radius and increasing charge. Moreover, the well-established series of acid dissociation constants serves as an additional guide to predict the sequence of increasing complex stability for a given metal (Schwarzenbach 1961). The pK_a° values of these acids may be regarded as formation constants of the proton (a hard cation) with the anions (ligands) of these acids. By analogy, the sequence of stability constants for complexes of other hard cations can be predicted from the sequence of increasing pK_a°, that is,

$$Cl^- < NO_3^- < SO_4^{2-} < H_2PO_4^- < F^-$$
$$< HCO_3^- < HPO_4^{2-} < SiO(OH)_3^- < CO_3^{2-}$$
$$< PO_4^{3-} < OH^-.$$

The histogram in Fig. 2 gives a panoramic view of the $\log_{10} K^\circ$ values of the formation constants for mononuclear single ligand complexes included in the Nagra/PSI TDB 01/01. On the *x*-axis, the cations (with hard or borderline character) are listed in order of decreasing metal 'hardness' (decreasing ionic charge and increasing ionic radius). On the *y*-axis, the ligands are listed in order of increasing hardness. From the empirical rules mentioned above, one expects that, as a general trend, the $\log_{10} K^\circ$ values decrease in the *x*-direction and increase in the *y*-direction. The overall picture confirms this prediction, but there are exceptions.

In general, the formation constants of metals with a specific ligand follow the predicted trend and increase in step with increasing metal hardness. Mg, Al, and Mn complexes, however, do not follow this trend. The stability constants of Mg and Al are well established and thus, their anomalous behaviour may indicate the limitation of this simple correlation when comparing 'A' metals (Mg, Al) with d-transition elements (Fe). The data for Mn need to be thoroughly reviewed before any conclusions can be drawn.

It is also worth mentioning that the constants for the metal silicate complexes fit very well into the relative order of ligand stability, even though they have been classified as uncertain data (see discussion below) because of

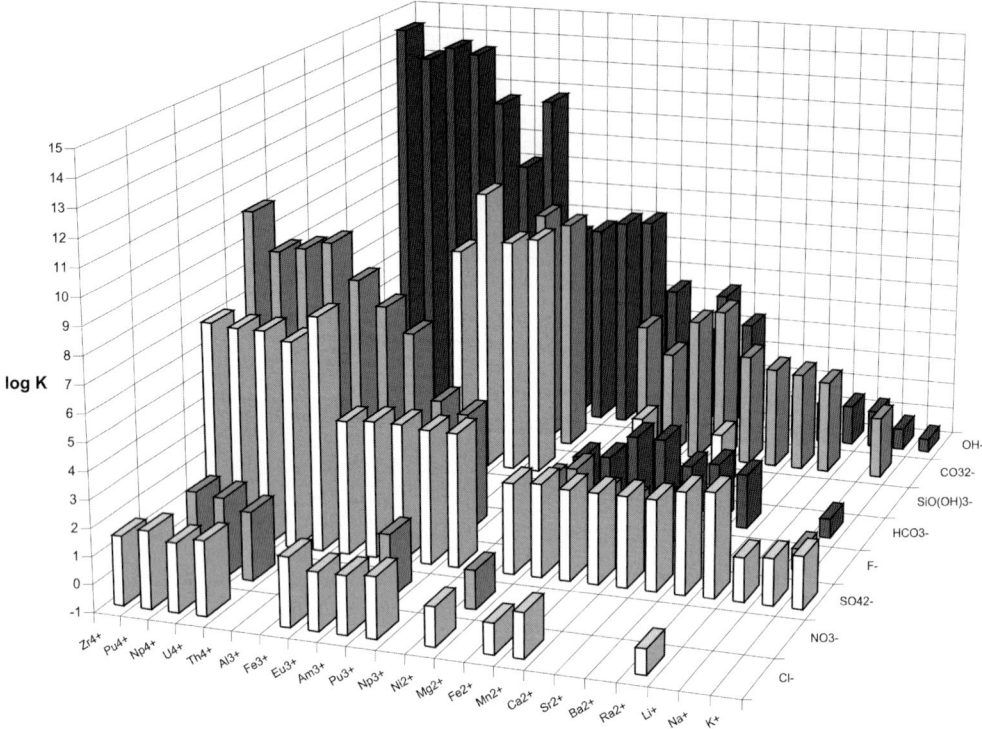

Fig. 2. Histogram of the formation constants of selected 1 : 1 complexes. Formation constants generally increase with increasing metal hardness (increasing charge, decreasing ionic radius) and with increasing ligand hardness. Metal hardness increases from K^+ to Zr^{4+} and ligand hardness from Cl^- to OH^-. The most stable complexes are formed between hard metals and hard ligands, the weakest complexes are formed betweeen soft metals and soft ligands.

unresolved ambiguities concerning the stoichiometry of the complexes.

More systematic anomalies appear when comparing complexation constants of a given cation with different ligands (compare with the sequence of acid dissociation constants given above). (1) Monocarbonato complexes are generally more stable than monohydroxo complexes (up to 2 log units). This may be due to a chelate effect, leading to the preferred formation of bidentate carbonate complexes. (2) Monosulphato complexes of alkalis, alkaline earths, and transition metals are more stable than the corresponding monofluorido complexes (by about 1 to 2 log units). The order reverts to normal for the tetravalent actinides. The reason for this behaviour is unclear.

Empirical correlations based on charge/size relations

A convenient way for comparing the complexation of different hard cations with a specific ligand is plotting their complexation constants against their ionic indices (z^2/r, square of charge divided by ionic radius, after Nieboer & Richardson 1980), which permits consideration of changes in complex stability due to both ionic radius and ionic charge. The plot of the formation constants of monofluorido complexes versus the ionic index reveals an excellent linear correlation (Fig. 3b), indicating an overall consistency of the selected constants. Figure 3a shows a reasonably well-defined linear trend for the monohydroxo complexes. Only the constants for $AlOH^{2+}$ and $Fe(III)OH^{2+}$ are found to deviate considerably. Since they are well determined, it is not possible to ascribe these anomalies to poor data quality. Rather, they reflect a particular chemical behaviour that cannot be rationalized in terms of simple empirical rules. On the other hand, the constants for trivalent and tetravalent actinides fit reasonably well into the linear trend defined by all metals.

The monosulphato complexes display a different behaviour (Fig. 3c). Although the constants generally increase with decreasing ionic radius,

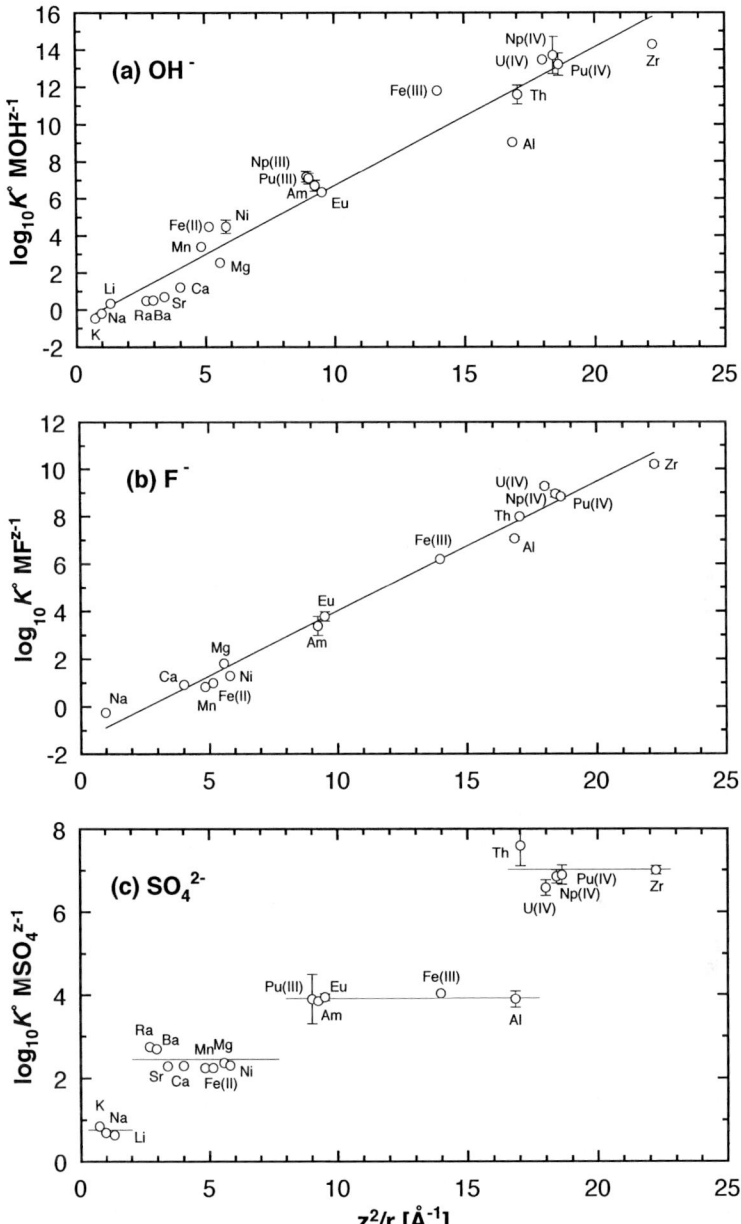

Fig. 3. Correlations between ionic index (z^2/r) and complexation constants $\log_{10} \beta_1^\circ$ for (**a**) the monohydroxo, (**b**) the monofluorido, and (**c**) the monosulphato complexes of hard cations and the borderline cation Ni(II).

they are nearly independent from ionic radius for metals of a given charge. This is particularly evident in the case of trivalent cations. It is not yet clear whether this is an artefact due to experimental problems or whether this represents the genuine chemical behaviour of sulphate. It is noteworthy, however, that the charge distribution in the sulphate ion is not point-like (in contrast to the hydroxide and fluoride ion, where this approximation can be made) and varies as a function of the coordination mode of the ligand.

In spite of their empirical nature, the correlations shown in Fig. 3 indicate a good overall chemical consistency of the three types of complexes considered.

Conflicting data

Parameters describing a particular thermodynamic equilibrium system are derived from experimental quantities obtained by a variety of methods, for example, calorimetry, potentiometry, and solubility studies. In the ideal case, critical examination of well-studied systems reveals high-quality experimental data that lead to a unique set of thermodynamic constants, which are internally consistent, not only formally, but also from a chemical point of view. In the course of our reviews, however, we encountered several cases of conflicting experimental data that resisted any attempt to cast them into a unique set of thermodynamic parameters. The following summarizes the conflicting data and our pragmatic solutions.

In the system Th(IV)–H_2O three sets of thermodynamic quantities can be derived from experimental data: (1) the hydrolysis constants $\log_{10} \beta_1^\circ$ and $\log_{10} \beta_4^\circ$ of $ThOH^{3+}$ and $Th(OH)_4(aq)$, respectively, have been determined potentiometrically by several authors over a wide range of ionic strength (for references see Hummel et al. 2002); (2) the thermodynamic properties of $ThO_2(cr)$ have been determined by calorimetry, and thus a solubility product $\log_{10} K_{s,0}^\circ$ (cr) for

$$ThO_2(cr) + 4\,H^+ \iff Th^{4+} + 2\,H_2O$$

can be derived from the available thermochemical data (Rai et al. 2000); (3) several solubility studies of $ThO_2(s)$ have been reported; however, they reveal a very peculiar behaviour, such that these data cannot be brought into consistency with the other two data sets.

A synopsis of several solubility studies (Baes et al. 1965; Moon 1989; Felmy et al. 1991; Bundschuh et al. 2000; Rai et al. 2000) is shown in Fig. 4. At pH < 5 the solubility measured at a given pH for crystalline $ThO_2(cr)$ is lower than the solubility of amorphous $ThO_2(am)$ by several orders of magnitude. At pH > 6 both sets of experimental data converge and the solubility of $ThO_2(s)$ becomes independent of pH.

At pH < 3 solubility data for $ThO_2(cr)$ agree fairly well with the solubility predicted by calorimetric data. On the other hand, solubility data of amorphous $ThO_2(am)$ indicate a solubility product more than 10 orders of magnitude

higher than that of $ThO_2(cr)$ (see Fig. 4). In both cases, combining the solubility products $\log_{10} K_{s,0}^\circ$ (cr) and $\log_{10} K_{s,0}^\circ$ (am), respectively, with the independently determined hydrolysis constant $\log_{10} \beta_4^\circ$ of $Th(OH)_4(aq)$ results in predicted Th concentrations far from any measured values in neutral and alkaline solutions. In the case of $ThO_2(cr)$ the concentration of dissolved Th(IV) is predicted to fall below any detection limit to [Th] < 10^{-16} M, whereas for equilibrium with $ThO_2(am)$ concentrations of [Th] > 10^{-7} M are predicted (see question marks in Fig. 4).

All measured solubility data for $ThO_2(s)$ at pH > 6 have been found in the range 10^{-7} M > [Th] > 10^{-9} M. A mean value of $10^{-8.5}$ M is represented by $\log_{10} K_{s,4}^\circ$ (s) = −8.5 for

$$ThO_2(s) + 2\,H_2O \iff Th(OH)_4(aq).$$

If this constant is combined with the hydrolysis constant $\log_{10} \beta_4^\circ = -18.4$ for $Th(OH)_4(aq)$, a solubility product $\log_{10} K_{s,0}^\circ(s) = 9.9$ for $ThO_2(s)$ is calculated that lies between the values for $ThO_2(am)$ and $ThO_2(cr)$ (see Fig. 4). This set of equilibrium constants now describes the measured solubilities at pH > 6, but fails to account for the solubility variation of more than 10 orders of magnitude at lower pH.

Recently, solubility measurements have been combined by Rothe et al. (2002) with XAFS investigations on crystalline, anhydrous $ThO_2(cr)$, microcrystalline $ThO_2 \cdot xH_2O(s)$, amorphous $ThO_n(OH)_{4-2n} \cdot xH_2O(am)$, aqueous Th(IV) solutions and colloidal suspensions, all at pH <3.7. The EXAFS spectra of microcrystalline $ThO_2 \cdot xH_2O(s)$ are clearly different from those of amorphous $ThO_n(OH)_{4-2n} \cdot xH_2O(am)$ and of bulk, anhydrous $ThO_2(cr)$. Therefore, the structure of these microcrystalline particles must be different as well.

The system U(IV)–H_2O (Fig. 5) shows a behaviour similar to Th(IV)–H_2O. At pH < 3 the solubility measured for freshly precipitated $UO_2(am)$ and the solubility of $UO_2(cr)$ derived from calorimetric data differ by nine orders of magnitude. All measured solubility data for $UO_2(s)$ > 5 have been found in the range 10^{-7} M > [U(IV)] > 10^{-10} M. A mean value of 10^{-9} M is represented by $\log_{10} K_{s,4}^\circ$ (s) = −9 for

$$UO_2(s) + 2\,H_2O \iff U(OH)_4(aq).$$

This behaviour is not restricted to Th(IV) and U(IV) as similar patterns have been found for all tetravalent actinides, An(IV). A more detailed discussion and comparison of An(IV) solubility and hydrolysis is given by Neck & Kim (2001). These authors conclude from solubility data

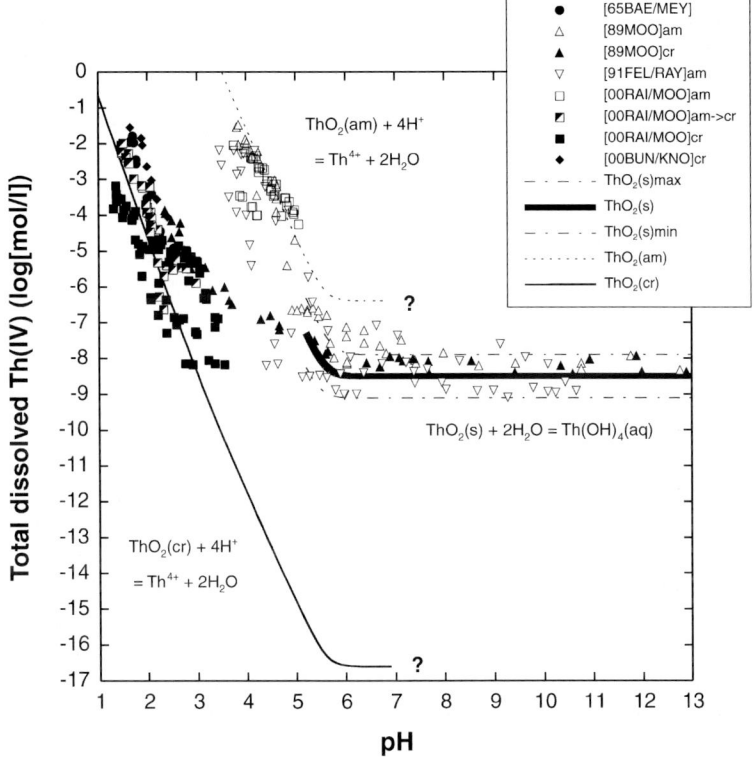

Fig. 4. Solubility data for the system Th(IV)–H₂O. The thick solid line was calculated from thermodynamic data selected for the Nagra/PSI TDB 01/01. Dot-dashed lines represent the estimated uncertainty. The dotted line was calculated from data for ThO₂(am), and the thin solid line from data for ThO₂(cr). Solubility data from: [65BAE/MEY]: Baes *et al.* (1965); [89MOO]: Moon (1989); [91FEL/RAY]: Felmy *et al.* (1991); [00RAI/MOO]: Rai *et al.* (2000); [00BUN/KNO]: Bundschuh *et al.* (2000).

measured for ThO₂(cr) and UO₂(cr) that the crystalline dioxide is the solubility limiting solid only at very low pH, where the non-hydrolysed An⁴⁺ is the predominant aqueous species. They postulate that the bulk crystalline dioxide must be covered with an amorphous surface layer as soon as the An⁴⁺ ion undergoes hydrolysis reactions, and the dissolution of AnO₂(cr) seems to become quasi-irreversible. According to Neck & Kim (2001) further investigations are needed to verify this hypothesis and to ascertain the chemical form of the solubility limiting solid in natural systems. Despite this promising qualitative model, we have to conclude that at present the systems ThO₂–H₂O and UO₂–H₂O are not understood in terms of quantitative equilibrium thermodynamics. The experimental data sets cannot be described by a consistent quantitative model without ad hoc assumptions.

As a pragmatic solution to this dilemma, we decided to rely on measured solubilities of

Th(IV) and U(IV) in neutral and alkaline solutions (the pH region of interest in performance assessment). Hence, the thermodynamic constants selected for the data base update (Hummel *et al.* 2002) do not refer to well-defined thorianite, ThO₂(cr), and uraninite, UO₂(cr), used in calorimetric measurements, but to the still poorly defined solids ThO₂(s) and UO₂(s) encountered in solubility studies at pH > 6 and pH > 5, respectively. Therefore, the thermodynamic constants selected in the data base update cannot be used to represent the widely varying solubilities of ThO₂ and UO₂ at pH < 5.

Uncertain data

All thermodynamic constants in the Nagra/PSI TDB 01/01 are uncertain to varying degrees. We estimated these uncertainties wherever possible and, as a consequence, most of the updated

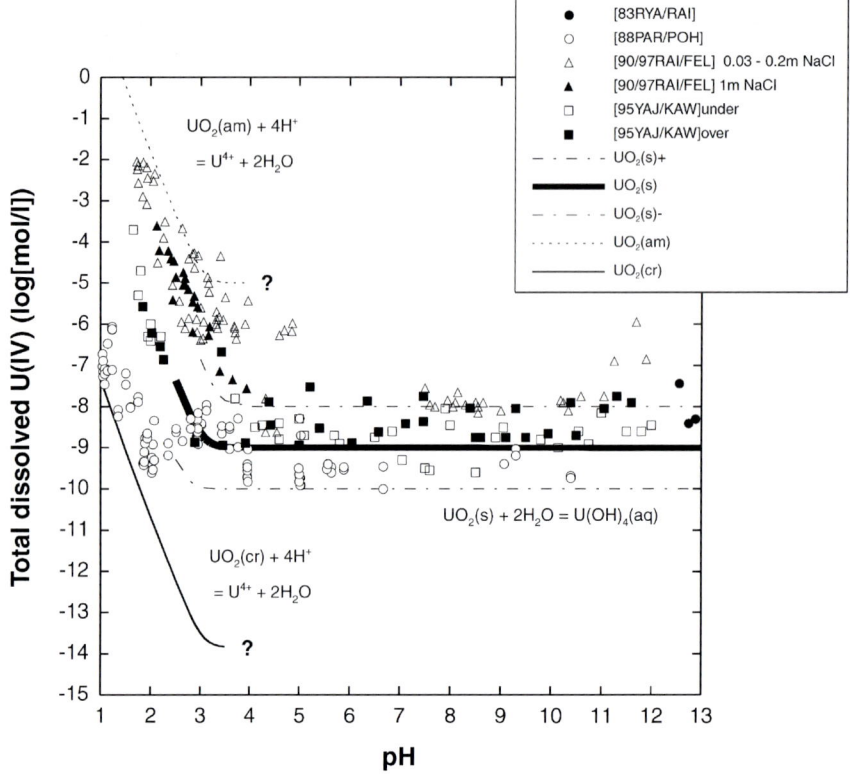

Fig. 5. Solubility data for the system U(IV)–H_2O. The thick solid line was calculated from thermodynamic data selected for the Nagra/PSI TDB 01/01. Dot-dashed lines represent the estimated uncertainty. The dotted line was calculated from data for UO_2(am), and the thin solid line from data for UO_2(cr). Solubility data from: [83RYA/RAI]: Ryan & Rai (1983); [88PAR/POH]: Parks & Pohl (1988); [90/97RAI/FEL]: Rai *et al.* (1990, 1997); [95YAJ/KAW]: Yajima *et al.* (1995).

constants are associated with an uncertainty range. However, simply comparing and ranking the selected values according to the stated uncertainties does not tell the complete story. At least two additional classes of 'especially uncertain' data can be discerned: 'limiting values' and 'placeholders'.

(1) Some of the selected stability constants are given as limiting values only ($<$ or $\leq$), or as approximate values ($\approx$) without uncertainty estimates. This concerns the aqueous species $Ni(CO_3)_2^{2-}$, $NiHCO_3^+$, $PdCl_2(OH)_2^{2-}$, TcO^{2+}, $NpO_2(OH)_3^-$, $NpO_2(OH)_4^{2-}$, and PuO_2OH(aq). Because such approximate data cannot be handled by conventional speciation codes, we were forced to treat these uncertain equilibrium constants as exact for the electronic versions of the data base and the $\approx$, $<$, and $\leq$ signs had to be dropped. Should any of these species turn out to be of importance

in the result of a speciation calculation, the approximate nature of the result should be kept in mind.

(2) Some stability constants have been included in the data base as 'placeholders': The redox equilibrium between Se(0) and Se(-II)

$$Se(cr) + 2H^+ + 2e^- \iff H_2Se(aq)$$

is not well constrained by experimental data and the equilibrium constant included in the data base is only a rough estimate. Since this equilibrium is of considerable importance in the field of radioactive waste disposal, its omission in modelling would probably cause more erroneous results than the inclusion of a rough estimate for the equilibrium constant.

The stability constant for $ThHPO_4^{2+}$ included in the data base is not of sufficient quality to be considered a recommended value. It should rather be thought of as a placeholder for missing data on Th phosphate complex formation. It is intended to serve as a flag in speciation calculations, warning modellers about the possible importance of phosphate complex formation.

To some extent, the aqueous metal silicate complexes constitute an entire class of placeholders. There is ample evidence that silica forms strong complexes with several metals, but the stoichiometry of such complexes has not yet been established. Different speciation schemes have been proposed to interpret experimental data, for example, 1 : 2 complexes, chelates, or mixed hydroxide silicate complexes. Despite these ambiguities we decided to include several metal silicate complexes in the data base as guidelines for modellers. Should such complexes turn out to be of crucial importance in particular systems, additional experimental studies would be called for.

Missing data

In an ideal world, a data base should be complete in terms of all relevant compounds and complexes. In practice this goal can only be approximated. There is a delicate balance between including high-quality data only and filling gaps with estimated values as placeholders as discussed above. A guideline to keep this balance is provided by the question: 'Are the missing data of any importance to the envisioned application of the data base?'

In the case of Th(IV) and U(IV) hydrolysis we selected values for $ThOH^{3+}$, $Th(OH)_4(aq)$, and UOH^{3+} derived from potentiometric studies, and in addition an estimated value for $U(OH)_4(aq)$. This estimate is of crucial importance for modelling the system $UO_2(s)-H_2O-CO_2$ above pH 4. However, we decided not to include, just for the sake of completeness, estimated values for the complexes $Th(OH)_2^{2+}$, $Th(OH)_3^+$, $U(OH)_2^{2+}$, and $U(OH)_3^+$. These complexes would change the Th(IV) and U(IV) solubility curves only marginally and at a pH low enough to lie outside the range of applicability of our selected parameter set (see Figs 4 and 5, and the detailed discussion in Neck & Kim 2001).

However, a few cases of serious data gaps have been identified:

(1) In the case of tin, no meaningful value for the Sn(IV)/Sn(II) redox equilibrium could be derived from experimental data and no estimate was possible. As a consequence,

the tin system is redox decoupled in the Nagra/PSI TDB 01/01 and two *primary master species* were selected for tin, one for Sn(II) and one for Sn(IV). It is therefore not possible to model the speciation of tin as a function of the oxidation potential.

(2) Solubility studies indicate the formation of strong Zr carbonate complexes (Pouchon *et al.* 2001). This is not a surprise when considering the behaviour of other tetravalent metals. However, the available experimental data are not sufficient to derive the stoichiometry of the limiting carbonate complex, that is, to discern between a tetra- and a pentacarbonato complex. This prevents the derivation of any meaningful stability constant and no value can be recommended at present.

(3) In the case of ferric iron, solubility studies also indicate the formation of strong Fe(III) carbonate complexes (Bruno *et al.* 1992). A detailed review of the experimental data revealed an unresolved ambiguity concerning the nature of the solubility limiting phase (Hummel 2000; Bruno & Duro 2000). This ambiguity results in uncertainties of several orders of magnitude for Fe(III) carbonate equilibrium constants and no equilibrium constants can be recommended, although the experimental data suggest that Fe(III) carbonate complexes predominate in some carbonate-rich groundwaters. However, for exploring the potential effects of Fe(III) carbonate complexation in modelling exercises, estimates for an equilibrium constant are provided (Hummel *et al.* 2002), although not formally included in the Nagra/PSI TDB 01/01.

Future data needs

The most obvious future data needs concern the missing, uncertain, and conflicting data identified above. Additional experimental investigations are needed in the case of Fe(III) and Zr(IV) carbonate complexation, and in the case of the Sn(IV)/Sn(II) and the Se(0)/Se(-II) redox couples. The molecular structure of metal silicate complexes needs clarification in order to remove ambiguities in the speciation scheme of these complexes. A rather challenging topic concerns the supposed transformation of crystalline tetravalent actinide oxides, $AnO_2(cr)$, to solids with an amorphous surface layer as soon as the An^{4+} ion hydrolyses. The consequences of such a reaction for thermodynamic equilibria, solubility, and sorption of actinides are largely unexplored.

Ternary complexes must be considered in models of environmental systems. However, there is no chance to explore experimentally the huge number of potentially forming ternary complexes. Modellers must first estimate which ternary complexes might be important for a specific system, and only then should experimental investigations be started. Hummel & Berner (2002) have shown, for example, that mixed hydroxide carbonate complexes of U(IV), Np(IV), and Pu(IV) are of particular interest in carbonate-rich groundwaters.

Review work for future updates of our data base should focus on iron compounds and complexes. The iron system is thought to be of crucial importance for characterizing the redox behaviour of radioactive waste repositories. Preliminary applications have indicated that the lack of data for the iron system is a source of major uncertainties associated with the definition of an oxidation potential. Hence, there is little use in developing sophisticated redox models for radionuclides as long as the dominant redox processes in a repository are poorly known.

Another field with a large potential for improvements concerns aluminosilicate minerals, which are of great importance in determining the chemistry of water in many types of rock. In backfill clays, aluminosilicates are responsible for the retention (sorption, incorporation) of trace elements and may affect both oxidation potential (incorporation of $Fe(II)/Fe(III)$) and pH (hydrolysis of silicate and/or exchange of H^+). Related classes of compounds (i.e., calcium silicates and calcium aluminates) form the chemical backbone of cementitious materials. The thermodynamic properties of these substances are still largely unexplored.

Application of the data base in performance assessment

The proposed Swiss repository for SF, HLW, and ILW is situated in the Opalinus Clay of the Zürcher Weinland in northern Switzerland, where an exploratory borehole was drilled near the village of Benken (Nagra 2002*a*). The Opalinus Clay formation consists of a well-consolidated clay shale, which is suitable for the construction of small, unlined tunnels and larger, lined tunnels at depths of several hundred metres. The engineered barrier system includes the waste containers and the backfill of construction, operation, and emplacement tunnels. The SF and HLW steel canisters will be placed on highly compacted bentonite blocks in SF/HLW emplacement tunnels. The resulting voids will be backfilled with granular bentonite. The ILW drums (containing wastes in cementitious or bituminous matrices) will be incorporated into concrete emplacement containers with cementitious mortar used to fill the void space around the drums. After emplacement of the ILW containers in their emplacement tunnels the void regions will be backfilled with a cementitious mortar. Construction and operation tunnels will be backfilled with a bentonite–sand mixture.

The porewater chemistry in the Opalinus Clay hostrock, in the bentonite backfill, and in the cementitious backfill will strongly affect the mobility of radionuclides via sorption and solubility equilibria. Because water samples could not be directly collected from the borehole at Benken it was necessary to develop the reference water chemistry for Opalinus Clay (Benken reference water) by geochemical modelling (see Nagra 2002*b* and references therein). The modelling approach assumed that, with the exception of some free or non-reacting constituents (chloride, bromide, and sulphate, whose concentrations must be specified, and pCO_2, which must be fixed), the porewater composition and speciation can be modelled adequately by saturation with minerals occurring in the Opalinus Clay (calcite, dolomite, siderite, rhodochrosite, fluorite, kaolinite, quartz, and UO_2) and by cation exchange with the clay minerals.

This Benken reference water was used as a starting point for deriving reference porewaters for the bentonite and the cementitious backfill. The composition of the reference bentonite porewater was calculated by Curti & Wersin (2002) from the reaction between Benken reference water and MX-80 Bentonite (75 wt% montmorillonite). The chemical processes considered include gas exchange (CO_2), complexation reactions in solution, saturation equilibria for accessory minerals in the bentonite (calcite, gypsum, quartz, and kaolinite), and cation exchange and surface complexation reactions with montmorillonite. Cement porewater compositions were calculated from the equilibrium reaction between Benken reference water and the cement model phases portlandite, brucite, calcite, and fluorite (see Nagra 2002*b* and references therein). All these reference porewater compositions were based on thermodynamic data from the Nagra/PSI TDB 01/01, with the exception of ion-exchange and surface complexation parameters, which had to be adopted from other sources.

Radionuclide concentration limits (solubility limits) for bentonite porewater were determined by Berner (2002a) using the following modelling procedure. Increasing amounts of the element in question are added (in ionic form, with Na^+ or Cl^- as counter-ions) to a given mass of reference bentonite porewater. The maximum concentration is obtained at incipient precipitation of the most stable solid. Berner (2002b) determined concentration limits for cement porewaters in a similar way. The modelling in both studies was based on the Nagra/PSI TDB 01/01.

Bradbury & Baeyens (2002a, b) provided sorption data bases for Opalinus Clay and MX-80 bentonite, respectively, based on numerous experiments with varying aqueous solution compositions. Since sorption can be influenced by the complexation of radionuclides with various ligands, corrections had to be made to account for the differences in speciation between the experimental solutions and the Opalinus Clay and bentonite porewaters. For this purpose, the Nagra/PSI TDB 01/01 was used to model the speciation of all safety-relevant radionuclides in the experimental solutions and in the Opalinus Clay and bentonite porewaters.

The modelling chain described in this section, linking the reference porewaters, radionuclide concentration limits, and sorption data bases, is summarized in Fig. 6.

Outlook and conclusions

Presently, thermodynamic data bases for environmental chemistry are far from being complete. We believe that many built-in data bases of geochemical codes that include an impressive number of data for aqueous species and solid phases for most elements may easily produce incorrect results if used without criticism. Indeed, one of the main lessons learnt during our update exercise is that completeness and reliability of the data are mutually exclusive. On the other hand, reducing the data base to a small number of 'best' thermodynamic data severely limits its field of applicability. Thus, in order to model specific systems of fundamental relevance for radioactive waste disposal we were forced to make compromises and had to include estimated constants.

The Nagra/PSI data base represents a work-in-progress, which requires further updates. In addition to satisfying the data needs discussed above, further development of the data base is planned to proceed along the following lines:

Temperature extrapolations

For about 60% of the formation reactions in the Nagra/PSI TDB 01/01 only equilibrium constants for use at a standard temperature of 25 °C are given, while additional data ($\Delta_r H_m^\circ$, $\Delta_r S_m^\circ$, $\Delta_r C_{P,m}^\circ$, etc.) required for extrapolation to higher temperatures are missing. The temperatures relevant to radioactive waste disposal may extend to more than 100 °C, which is definitely outside the range of applicability of these constants. Since it cannot be expected that all the missing experimental data for temperature extrapolation will be available in the near future, a pragmatic solution to this problem is required. A promising approach is the one-term extrapolation for isocoulombic reactions. In such reactions, an equal number of like-charged species appears on both sides of the reaction. The fundamental assumption of the isocoulombic principle is that the thermodynamic properties of like-charged species have similar responses to changes in ionic strength and temperature (see Gu et al. 1994 and references therein). From this follows that $\Delta_r G_m^\circ$ of isocoulombic reactions is nearly independent from temperature, and $\log_{10} K^\circ$ (T) can be calculated directly from $\log_{10} K^\circ$ (25 °C), without additional thermodynamic data. With the proper choice of the isocoulombic formulation, the one-term extrapolation for a reaction is reliable up to 200–300 °C (Kulik 2002). The first steps in extending the Nagra/PSI TDB 01/01 with the one-term extrapolation of isocoulombic reactions have been taken by Thoenen et al. (2002).

Solid solutions

An important subject of ongoing experimental and modelling studies is the behaviour of solid solutions in aqueous environments. Although it is not yet a standard procedure in studies related to radioactive waste disposal, the quantitative description of solid solution systems has become a practicable method in the last decade. Every geochemist is aware of the fact that almost all minerals formed in complex natural environments are impure phases and, therefore, solid solution formation is the rule rather than the exception. Solid solutions only slightly affect the solubility of the major components (e.g., Ca^{2+} in calcite) but have a large influence on the equilibrium concentration of trace components, like most of the safety-relevant radionuclides (e.g., Ni, Sr, Ra in calcite). Thus, modelling equilibrium concentrations with the usual assumption of saturation equilibrium with any single-component solid included in the data base will frequently lead to overestimated radionuclide concentrations. This situation is generally accepted in safety

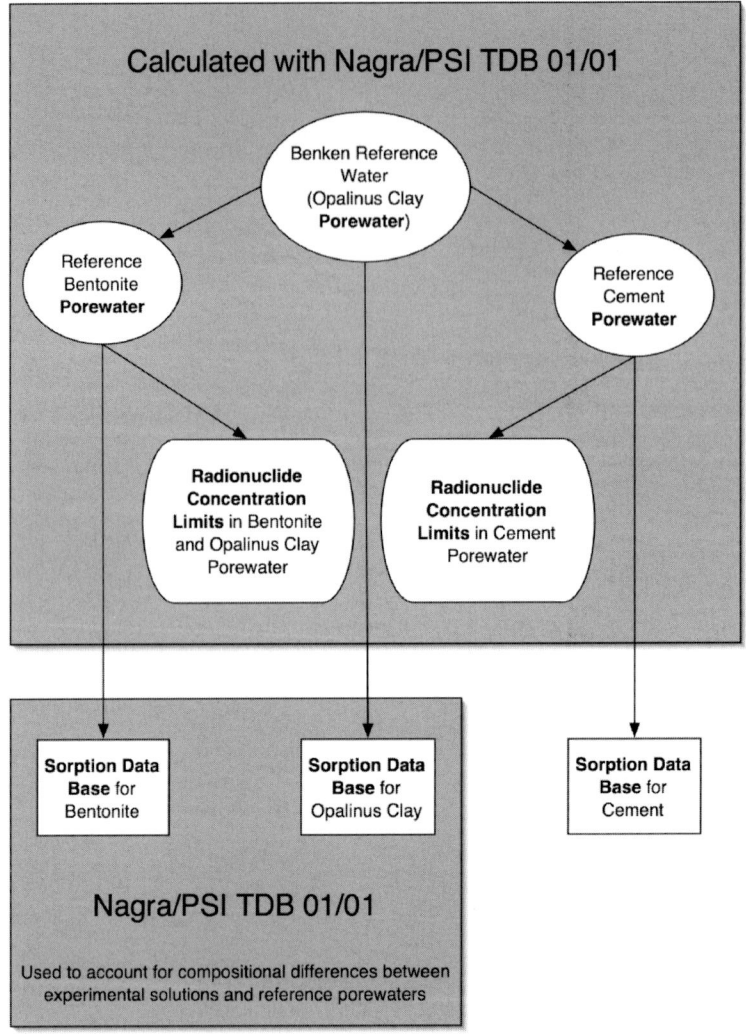

Fig. 6. The Nagra/PSI TDB 01/01 was used in this modelling chain for the performance assessment of a proposed Swiss repository for spent fuel, vitrified high-level waste, and long-lived intermediate-level waste.

assessments, since the errors made are on the safe side; that is, the results are said to be conservative. However, the results may be far from realistic. Hence, considering solid solutions may greatly reduce existing uncertainties and overconservatisms in the understanding of trace element behaviour. A future extension of the Nagra/PSI data base should include the model parameters necessary for a consistent description of important solid solution systems. The feasibility of such an extension has been demonstrated by Berner & Curti (2002) in a study on aqueous radium solubility controlled by sulphate solid solutions.

Impact of data uncertainty on modelling results

At present, none of the conventional geochemical codes can actually cope with uncertainties of thermodynamic parameters and the uncertainties specified in the data bases are treated as mere decorations. However, disregarding uncertainties in geochemical modelling may lead to serious misinterpretations of modelling results or to completely wrong conclusions. This calls for a rigorous analysis of the propagation of uncertainties in geochemical modelling. To this end, work is in progress to include the treatment of thermodynamic parameter uncertainty in the Gibbs energy minimization algorithm (Chudnenko *et al.* 2004).

To conclude, we see the recent update of the Nagra/PSI data base as a small, but important, step towards completeness and reliability of the large body of thermodynamic data needed to calculate chemical equilibrium in the complex geochemical systems occurring within or in the vicinity of radioactive waste disposal sites. The most important achievement in this exercise was probably the elimination of a conspicuous number of thermodynamic data not supported by experimental evidence or of dubious origin. This 'sieving' procedure resulted in a reduced, but at least transparent and self-consistent data base. Future extensions can now be built on this well-documented basis.

Partial financial support by the National Cooperative for the Disposal of Radioactive Waste (Nagra) and the comments of two anonymous reviewers are gratefully acknowledged.

References

BAES, C. F., JR., MEYER, N. J. & ROBERTS, C. E. 1965. The hydrolysis of thorium(IV) at 0 and 95 °C. *Inorganic Chemistry*, **4**, 518–527.

BERNER, U. 2002*a*. *Project Opalinus Clay: Radionuclide Concentration Limits in the Near-Field of a Repository for Spent Fuel and Vitrified High-Level Waste*. Nagra Technical Report NTB 02-10, Nagra, Wettingen, Switzerland. Also issued as PSI Bericht Nr. 02-22, Paul Scherrer Institut, Villigen, Switzerland.

BERNER, U. 2002*b*. *Radionuclide Concentration Limits in the Cementitious Near-Field of an ILW Repository*. Nagra Technical Report NTB 02-22, Nagra, Wettingen, Switzerland. Also issued as PSI Bericht Nr. 02-26, Paul Scherrer Institut, Villigen, Switzerland.

BERNER, U. & CURTI, E. 2002. *Radium Solubilities from SF/HLW Wastes Using Solid Solution and Co-Precipitation Models*. PSI Technical Report TM-44-02-04, Paul Scherrer Institut, Villigen, Switzerland.

BRADBURY, M. H. & BAEYENS, B. 2002*a*. *Near-Field Sorption Data Bases for Compacted MX-80 Bentonite for Performance Assessment of a High Level Radioactive Waste Repository in Opalinus Clay Host Rock*. Nagra Technical Report NTB 02-18, Nagra, Wettingen, Switzerland. Also issued as PSI Bericht Nr. 03-07, Paul Scherrer Institut, Villigen, Switzerland.

BRADBURY, M. H. & BAEYENS, B. 2002*b*. *Far-Field Sorption Data Bases for Performance Assessment of a HLW Repository in an Undisturbed Opalinus Clay Host Rock*. Nagra Technical Report NTB 02-19, Nagra, Wettingen, Switzerland. Also issued

as PSI Bericht Nr. 03-08, Paul Scherrer Institut, Villigen, Switzerland.

BROWN, P. L. & WANNER, H. 1987. *Predicted Formation Constants Using the Unified Theory of Metal Ion Complexation*. OECD Nuclear Energy Agency, Paris.

BRUNO, J., STUMM, W., WERSIN, P. & BRANDBERG, F. 1992. On the influence of carbonate in mineral dissolution: 1. The thermodynamics and kinetics of hematite dissolution in bicarbonate solutions at T = 25 °C. *Geochimica et Cosmochimica Acta*, **56**, 1139–1147.

BRUNO, J. & DURO, L. 2000. Reply to W. Hummel's comment on and correction to 'On the influence of carbonate in mineral dissolution: 1. The thermodynamics and kinetics of hematite dissolution in bicarbonate solutions at T = 25 °C' by J. BRUNO, W. STUMM, P. WERSIN, and F. BRANDBERG. *Geochimica et Cosmochimica Acta*, **64**, 2173–2176.

BUNDSCHUH, T., KNOPP, R., MÜLLER, R., KIM, J. I., NECK, V. & FANGHÄNEL, T. 2000. Application of LIBD to the determination of the solubility product of thorium(IV)-colloids. *Radiochimica Acta*, **88**, 625–629.

CHUDNENKO, K. V., KARPOV, I. K., KULIK, D. A., BERNER, U. R., HUMMEL, W. & ARTIMENKO, M. V. 2004. *GEM Uncertainty Space Approach for Sensitivity Analysis of Solid-Aqueous Chemical Equilibrium Models: A Pilot Study*. PSI Technical Report TM-44-04-01, Paul Scherrer Institut, Villigen, Switzerland.

CURTI, E. & WERSIN, P. 2002. *Assessment of Porewater Chemistry in the Bentonite Backfill for the Swiss SF/HLW Repository*. Nagra Technical Report NTB 02-09, Nagra, Wettingen, Switzerland.

FELMY, A. R., RAI, D. & MASON, M. J. 1991. The solubility of hydrous thorium(IV) oxide in chloride media: development of an aqueous ion-interaction model. *Radiochimica Acta*, **55**, 177–185.

GRENTHE, I., FUGER, J., KONINGS, R. J. M., LEMIRE, R. J., MULLER, A. B., NGUYEN-TRUNG, C. & WANNER, H. 1992. *Chemical Thermodynamics of Uranium*. Elsevier, Amsterdam.

GRENTHE, I., HUMMEL, W. & PUIGDOMÈNECH, I. 1997*a*. Chemical background for the modelling of reactions in aqueous systems. *In*: GRENTHE, I. & PUIGDOMÈNECH, I. (eds) *Modelling in Aquatic Chemistry*. OECD Nuclear Energy Agency, Paris, 69–129.

GRENTHE, I., PLYASUNOV, A. V. & SPAHIU, K. 1997*b*. Estimation of medium effects on thermodynamic data. *In*: GRENTHE, I. & PUIGDOMÈNECH, I. (eds) *Modelling in Aquatic Chemistry*. OECD Nuclear Energy Agency, Paris, 325–426.

GU, Y., GAMMONS, C. H. & BLOOM, M. S. 1994. A one-term extrapolation method for estimating equilibrium constants of aqueous reactions at elevated temperatures. *Geochimica et Cosmochimica Acta*, **58**, 3545–3560.

HUMMEL, W. 2000. Comment on 'On the influence of carbonate in mineral dissolution: 1. The thermodynamics and kinetics of hematite dissolution in bicarbonate solutions at T = 25 °C' by J. BRUNO, W. STUMM, P. WERSIN, and F. BRANDBERG. *Geochimica et Cosmochimica Acta*, **64**, 2167–2171.

HUMMEL, W., BERNER, U., CURTI, E., PEARSON, F. J. & THOENEN, T. 2002. *Nagra/PSI Chemical Thermodynamic Data Base 01/01*. Nagra Technical Report NTB 02-16, Nagra, Wettingen, Switzerland. Also published by Universal Publishers/uPublish.com Parkland, Florida, USA. World Wide Web Address: http://www.upublish.com.

HUMMEL, W. & BERNER, U. 2002. Application of the Nagra/PSI TDB 01/01: Solubility of Th, U, Np and U. Nagra Technical Report NTB 02-12, Nagra, Wettingen, Switzerland.

KARPOV, I. K., CHUDNENKO, K. V. & KULIK, D. A. 1997. Modeling chemical mass transfer in geochemical processes: Thermodynamic relations, conditions of equilibria, and numerical algorithms. *American Journal of Science*, **297**, 767–806.

KULIK, D. A. 2002. Minimising uncertainty induced by temperature extrapolations of thermodynamic data: A pragmatic view on the integration of thermodynamic databases into geochemical computer codes. *Proceedings of the Workshop on 'The Use of Thermodynamic Databases in Performance Assessment'*, 29–30 May 2001, Barcelona, Spain. Organisation for Economic Cooperation and Development OECD, Paris, France, 125–137.

LEMIRE, R. J., FUGER, J. et al. 2001. *Chemical Thermodynamics of Neptunium & Plutonium*. Elsevier, Amsterdam.

MOON, H. C. 1989. Equilibrium ultrafiltration of hydrolyzed thorium(IV) solutions. *Bulletin of the Korean Chemical Society*, **10**, 270–272.

NAGRA 1993. *Beurteilung der Langzeitsicherheit des Endlagers SMA am Standort Wellenberg (Gemeinde Wolfenschiessen, NW)*. Nagra Technical Report NTB 93-26, Nagra, Wettingen, Switzerland.

NAGRA 1994a. *Bericht zur Langzeitsicherheit des Endlagers SMA am Standort Wellenberg*. Nagra Technical Report NTB 94-06, Nagra, Wettingen, Switzerland.

NAGRA 1994b. *Kristallin-I Safety Assessment Report*. Nagra Technical Report NTB 93-22, Nagra, Wettingen, Switzerland.

NAGRA 2002a. *Project Opalinus Clay: Safety Report. Demonstration of Disposal Feasibility for Spent Fuel, Vitrified High-Level Waste and Long-Lived Intermediate-Level Waste (Entsorgungsnachweis)*. Nagra Technical Report NTB 02-05, Nagra, Wettingen, Switzerland.

NAGRA 2002b. *Projekt Opalinuston – Synthese der Geowissenschaftlichen Untersuchungsergebnisse. Entsorgungsnachweis für Abgebrannte Brennelemente, Verglaste Hochaktive Sowie Langlebige Mittelaktive Abfälle*. Nagra Technical Report NTB 02-03, Nagra, Wettingen, Switzerland.

NECK, V. & KIM, J. I. 2000. An electrostatic approach for the prediction of actinide complexation constants with inorganic ligands – application to carbonate complexes. *Radiochimica Acta*, **88**, 815–822.

NECK, V. & KIM, J. I. 2001. Solubility and hydrolysis of tetravalent actinides. *Radiochimica Acta*, **89**, 1–16.

NIEBOER, E. & RICHARDSON, D. H. S. 1980. The replacement of the nondescript term 'heavy metal' by a biologically and chemically significant classification of metal ions. *Environmental Pollution (Ser. B)*, **1**, 3–26.

PARKHURST, D. L. & APPELO, C. A. J. 1999. *User's Guide to PHREEQC (Version 2) – A Computer Program for Speciation, Batch-Reaction, One-Dimensional Transport, and Inverse Geochemical Calculations*. Water-Resources Investigations Report 99-4259, US Geological Survey, Denver, Colorado, USA.

PARKS, G. A. & POHL, D. C. 1988. Hydrothermal solubility of uraninite. *Geochimica et Cosmochimica Acta*, **52**, 863–875.

PEARSON, F. J., JR. & BERNER, U. 1991. *Nagra Thermochemical Data Base I. Core Data*. Nagra Technical Report NTB 91-17, Nagra, Wettingen, Switzerland.

PEARSON, F. J., JR., BERNER, U. & HUMMEL, W. 1992. *Nagra Thermochemical Data Base II. Supplemental Data 05/92*. Nagra Technical Report NTB 91-18, Nagra, Wettingen, Switzerland.

PEARSON, F. J., THOENEN, T., DMYTRIYEVA, S., KULIK, D. A. & HUMMEL, W. 2001. *PMATCHC: A Program to MAnage ThermoCHemical data, written in C++*. PSI Technical Report TM-44-01-07, Paul Scherrer Institut, Villigen, Switzerland.

PEARSON, R. G. 1963. Hard and soft acids and bases. *Journal of the American Chemical Society*, **85**, 3533–3539.

POUCHON, M. A., CURTI, E., DEGUELDRE, C. & TOBLER, L. 2001. The influence of carbonate complexes on the solubility of zirconia: new experimental data. *Progress in Nuclear Energy*, **38**, 443–446.

RAI, D., FELMY, A. R. & RYAN, J. L. 1990. Uranium(IV) hydrolysis constants and solubility product of $UO_2 \cdot xH_2O$(am). *Inorganic Chemistry*, **29**, 7852–7865.

RAI, D., FELMY, A. R., STERNER, S. M., MOORE, D. A., MASON, M. J. & NOVAK, C. F. 1997. The solubility of Th(IV) and U(IV) hydrous oxides in concentrated NaCl and $MgCl_2$ solutions. *Radiochimica Acta*, **79**, 239–247.

RAI, D., MOORE, D. A., OAKES, C. S. & YUI, M. 2000. Thermodynamic model for the solubility of thorium dioxide in the $Na^+ - Cl^- - OH^- - H_2O$ system at 23 °C and 90 °C. *Radiochimica Acta*, **88**, 297–306.

RARD, J. A., RAND, M. H., ANDEREGG, G. & WANNER, H. 1999. *Chemical Thermodynamics of Technetium*. Elsevier, Amsterdam.

ROTHE, J., DENECKE, M. A., NECK, V., MÜLLER, R. & KIM J. I. 2002. XAFS investigation of the structure of aqueous thorium(IV) species, colloids, and solid thorium(IV) oxide/hydroxide. *Inorganic Chemistry*, **41**, 249–258.

RYAN, J. L. & RAI, D. 1983. The solubility of uranium (IV) hydrous oxide in sodium hydroxide solutions under reducing conditions. *Polyhedron*, **2**, 947–952.

SCHWARZENBACH, G. 1961. The general, selective, and specific formation of complexes by metallic cations. *In*: EMELÉUS, H. J. & SHARPE, A. G. (eds) *Advances in Inorganic Chemistry and Radiochemistry*, **3**. Academic Press, New York, 257–285.

SILVA, R. J., BIDOGLIO, G., RAND, M. H., ROBOUCH, P. B., WANNER, H. & PUIGDOMÈNECH, I. 1995. *Chemical Thermodynamics of Americium.* Elsevier, Amsterdam.

THOENEN, T., BERNER, U., HUMMEL, W. & KULIK, D. 2002. *Equilibrium Constants at 50 °C for Solids and Aqueous Species Determining the Solubility of Am, Pu, Np, U, Th, and Tc in the Reference Bentonite Porewater.* PSI Technical Report TM-44-02-05, Paul Scherrer Institut, Villigen, Switzerland.

THOENEN, T. & KULIK, D. 2003. *Nagra/PSI Chemical Thermodynamic Data Base 01/01 for the GEM-Selektor (V.2-PSI) Geochemical Modeling Code: Release 28-02-03*. PSI Technical Report TM-44-03-04, Paul Scherrer Institut, Villigen, Switzerland.

TRUESDELL, A. H. & JONES, B. F. 1974. WATEQ, A computer program for calculating chemical equilibria of natural waters. *Journal of Research, U.S. Geological Survey*, **2**, 233–274.

WESTALL, J., ZACHARY, J. L. & MOREL, F. M. M. 1976. *MINEQL, a Computer Program for the Calculation of Chemical Equilibrium Composition of Aqueous Systems*. Technical Note 18, Department of Civil and Environmental Engineering, Massachusetts Institute of Technology, Cambridge, MA, USA.

YAJIMA, T., KAWAMURA, Y. & UETA, S. 1995. Uranium(IV) solubility and hydrolysis constants under reduced conditions. *Materials Research Society Proceedings*, **353**, 1137–1142.

Towards a consistent rate law: glass corrosion kinetics near saturation

JONATHAN P. ICENHOWER[1], S. SAMSON[1], A. LÜTTGE[2] & B. P. McGRAIL[1]

[1]*Pacific Northwest National Laboratory, Applied Geology and Geochemistry Group, Richland, WA, USA (e-mail: jonathan.icenhower@pnl.gov)*

[2]*Rice University, Department of Geology and Geophysics, Houston, TX, USA*

Abstract: Although glass corrosion resistance has been tested with laboratory methods for decades, investigators are now just beginning to understand the reaction phenomena at or close to saturation with respect to the rate-limiting phase(s). Near saturation, the phenomena that govern element release rates include alkali–hydrogen (species) exchange, differential reactivity of phase-separated glass, and accelerated corrosion rates due to precipitation of key secondary phases. These phenomena were not anticipated by early models of glass dissolution and are incompletely quantified in current rate representations. This review discusses the two over-arching models for glass reactivity, diffusion and surface reaction control, and demonstrates the importance of glass reactivity in terms of glass composition and micro-heterogeneity of the glass. Our conclusion is that surface reaction control best describes the release of elements to solution, but that models based on current interpretations of transition state theory (TST) must be modified to account for reported anomalies in behaviour near saturation.

Glass has been studied for millennia, and our forbearers used the existing technology and science to create vitreous objects with desirable colour, durability, patterning, and quality. At the dawn of recorded human history the 'technology' of the times typically included religious rituals so that the glass objects could be fashioned with the approval of the gods. For example, we have this glass-making 'recipe' from a clay tablet found in the Assyrian Temple of Nabu from the seventh century BCE:

'When thou settest out the ground-plan of a furnace for 'minerals' thou shalt seek out a favorable day in a fortunate month, and thou shalt set out the ground-plan of the furnace … The wood which thou shalt burn underneath the furnace shall be styrax, thick, decorticated billets which have not lain exposed in bundles but have been kept in leather coverings, cut in the month of Ab … Thou shalt mix them [components of the frit] together and put them down in the furnace … then thou shalt keep a good smokeless fire burning until it liquefies: then thou shalt pour it on burnt brick.' (Horton 1929)

Modern techniques to fashion glass do not call upon fortune or the favour of gods, but upon the insights of glass structure garnered from modern science. Production of homogeneous unflawed glass resulted in the invention of the microscope and telescope, which presaged revolutionary discoveries on vastly different scales, from the immense (astronomy) to the minute (biology).

In more modern times, use of manufactured glass has expanded to a variety of demanding environments and the chemical durability of glass has become another highly prized quality. Commonplace glass objects, such as windows, laboratory equipment, vessels for preserving materials are subject to corrosion, reaction, or decay, optical and electrical components, and insulation may be subject to chemically harsh settings. Therefore, the corrosion resistance of a glass item is an important consideration in its manufacture. More relevant to this review, recognition of glass as a candidate waste form for disposal of hazardous and radioactive waste has sparked a vigorous research effort to understand the chemical durability of glass, yet there are still many divisions among investigators over the interpretation of experiments aimed at quantifying corrosion resistance. These controversies have fostered a large number of reviews of glass dissolution kinetics, and, almost by ritual, every two to three years a new review is offered. Among the many excellent reviews are those of Bourcier (1991, 1994), Casey & Bunker (1990), Vernaz & Dussossoy (1992), Barkatt *et al.* (1986), Hench *et al.* (1986), Bunker *et al.* (1988), Werme *et al.* (1990), and Vernaz *et al.*

From: GIERÉ, R. & STILLE, P. (eds) 2004. *Energy, Waste, and the Environment: a Geochemical Perspective*. Geological Society, London, Special Publications, **236**, 579–594.
0305-8719/04/$15 © The Geological Society of London 2004.

(2001). This burgeoning set of reviews serves as a starting point for our discussion of glass corrosion resistance, although our review differs from those noted above by emphasizing dissolution behaviour near saturation with respect to potential rate-limiting phases.

The reasons for this shift are multifarious and include the observation that rates acquired from field studies (typically on natural vitreous materials, such as obsidian) (Yokoyama & Banfield 2002) do not accord with those obtained from laboratory experiments. In general this observation parallels results on tests of periodic solids (i.e., minerals); rates derived from the field on common rock forming minerals are up to a factor of $1000\times$ slower compared to laboratory rates (Velbel 1993). There has been much discussion of this topic in the literature (e.g., Sverdrup & Warfvinge 1995) and the difference is ascribed to differences in surface area, temperature, saturation state, and manner of aqueous solution contact (partial versus full) and the presence of corrosion resistant coatings of other minerals (such as aluminum or iron oxyhydroxides) on the solid of interest (Hochella & Banfield 1995; White 1995). Although no single factor can be presented as the reason for such a large discrepancy, the influence of solution saturation state is a clear target for consideration. The second reason for studying reaction rates near saturation is that it goes to the heart of the debate surrounding reaction mechanisms. For example, it can easily be shown that dissolution kinetics of glass and minerals near saturation do not conform to expectations founded upon current interpretations of transition state theory (TST), as elucidated more fully below. The third reason for studying reaction rates near solution saturation is that the relative importance of reaction mechanisms changes near saturation. Because the pore fluids in many of the waste disposal settings under consideration are at a near-saturated state (Murphy & Palaban 1994; Bacon *et al.* 2000), understanding the corrosion durability of glass under these conditions takes on a high importance for predicting long-term release of radionuclides or other elements to the environment.

In this review we will outline the major themes in glass dissolution kinetics and point out the papers that best describe these models. In pursuit of these themes, the reader will note that we accept or reject certain fundamental ideas embedded in the two principal models presented below. This should hardly come as a surprise, given the wide division in the two chief models and the disparate outcomes predicted from each. However, in pointing out our

underlying bias, it is our aim to educate the reader and allow him or her to arrive at his/her own conclusions based on the facts presented. If this admitted bias disturbs some, it should be pointed out that this quote from Jacques Barzun is most appropriate: 'It does not follow that bias cannot be guarded against, that all biases distort equally, or that controlled bias remains as bad as propaganda' (Barzun 2000). Barzun concludes that the essence of objectivity lies not in the eradication of opinion, but in the open display of conflicting ideas: 'One has then the duty to report the informed judgment of others' (Barzun 2000). It is this body of 'informed judgement' that is presented below.

Specifically, the major topics covered by this review are: (1) a brief discussion of models of element release rates based on diffusion and on TST, and (2) glass–water reactions that dominate near equilibrium. We will discuss these themes with the assumption of some general understanding of chemical kinetics, but the concepts should be comprehensible to readers from outside this field as well.

The two over-arching models for glass/water reactivity

Diffusion through a reaction layer

Early glass/water experiments demonstrated the relatively quick release of alkali ions to solution at the beginning of the test. Another observation noted in these early days was the presence of a macroscopically visible alteration layer forming on the surface of the reacting glass. More modern instruments, such as Rutherford backscattering spectroscopy (RBS) (Bunker *et al.* 1983; Baer *et al.* 1984; Pederson *et al.* 1986), nuclear reaction analysis (NRA) (Lanford *et al.* 1979; Dran *et al.* 1988; Pederson *et al.* 1990), X-ray photoelectron spectroscopy (XPS) (McGrail *et al.* 2001*a*), and nuclear magnetic resonance (NMR) (Bunker *et al.* 1988; Tsomaia *et al.* 2003), by themselves or in combination, have been used to characterize and quantify the chemical composition of such reaction layers. However, these sophisticated techniques, instead of clarifying the role of reaction layers, have revealed a complex chemistry, both temporally and spatially (laterally and in terms of depth). Because of these complications, attempts to quantify the role of reaction layers in the reactivity of glass are hampered.

Models for the development of reaction layers have centered on ion exchange between hydrogen species (OH^-, H_3O^+, H^+) or water (H_2O)

and cations in the glass (e.g., Li^+, Na^+, K^+). Doremus (1977a, b, 1981, 1983, 2000) pioneered the model that diffusion of H_2O into glass is the *primary* rate-limiting mechanism, at least during the initial stages of dissolution. More complex models were later developed in which diffusion-control through a 'transport' or 'passivating' layer was identified as the chief mechanism for release of dissolved species to solution (Leturcq *et al.* 1999; Jégou *et al.* 2000; Gin & Mestre 2001; Advocat *et al.* 2001; Gin *et al.* 2001a, b; Linard *et al.* 2001; Vernaz *et al.* 2001). This emphasis gained considerable currency in some circles, especially when the observation that TST-based models (see below) may not be adequate for solutions near saturation with respect to the rate-limiting solid.

Note, however, that even proponents of this idea disagree on the long-term effects of diffusion control on element release rates through a passivating surface layer. While a more traditional model holds that element release decreases as the square root of time, new long-term experiments (up to five years) have exhibited evidence for a constant rate of release (Gin & Frugier 2003). All of this underscores the dilemma faced by stewards of waste repositories; one diffusion model indicates a low, but constant release of elements with time, the other a low and decreasing rate. Comparing the two models reveals that the gap between predicted element releases becomes significantly large over time (Gin & Frugier 2003).

Another important consequence of the constant rate of release diffusion model is that it mimics many of the features that have commonly been attributed to surface reaction (matrix dissolution) control. If one were to account for changes in surface area over time, the predicted long-term dissolution rate due to surface reaction control would also yield constant element release. In surface reaction controlled models, the invariant release rate with respect to time is considered to be the natural consequence of the system achieving steady-state conditions. Other features of experiments commonly cited as evidence for surface reaction control, such as relatively high experimental activation energies (60–70 kJ/mol), could be explained as easily by the diffusion-control model. These findings show how similar the observations are between proponents of the two models: it is only the *interpretation* of the mechanism that differs.

Transition state theory

Eyring (1935a, b) developed 'Transition State Theory', or 'Absolute Rate Theory', in the 1930s.

He was motivated by the theoretical constructs of collision theory (CT), in which 'correction factors' of up to 10^{-8} had to be applied to force theory to comply with experiment (Eyring 1935b). The reason why CT failed to accurately predict reaction rates is that one would have to estimate the number of collisions between reacting molecules based on observed momentum transfers. The estimated cross-section area of the reacting molecules is likely to be incorrect since momentum transfer can occur without products forming due to a lack of proper orientation or relative velocities of the reactant molecules. However, TST-based models have been able to avoid this problem by considering only the momentum of the so-called 'activated complex' on the potential energy surface in the region near the transition to products. By activated complex, we mean an aggregate of atoms at a higher potential energy than reactants or products that have a statistically defined probability of decaying back to reactants or proceeding to products. Figure 1 illustrates the relationship between reactants, products, and the activated complex along the reaction coordinate. Therefore, statistical mechanical constructs can be devised to calculate the activities of activated complexes and their velocity over the activation energy barrier.

Investigations in succeeding years demonstrated that TST was able to correctly square theory with experiment for gas phase reactions by allowing investigators to map out the potential energy surface of simple molecular reactions (Golden 1979; Karplus *et al.* 1965). In contrast, application to solid/aqueous solution systems has not been universally accepted. Aagaard & Helgeson (1982) developed the most comprehensive and detailed treatment of TST for such systems. The current application of TST to reactions describing the dissolution of a solid into aqueous solution can be written:

$$r_+ = \vec{k}_o a_{H^+}^{-\eta_{H^+}} \exp\left(\frac{-E_a}{RT}\right) \times \left(1 - \frac{Q}{K}\right)^\sigma \prod_{i=j}^j a_j$$

(1)

where r_+ is the dissolution rate, $\vec{k}_o$ is the reaction constant, E_a is the activation energy, RT is the product of the gas constant and absolute temperature, Q is the ion activity product, K is the equilibrium constant of the rate-determining solid, and σ is the ratio of the order of the elementary reaction to that of the overall reaction and is generally considered to be unity for theoretical reasons. The last term, $\prod_{i=j}^i a_j$, is the

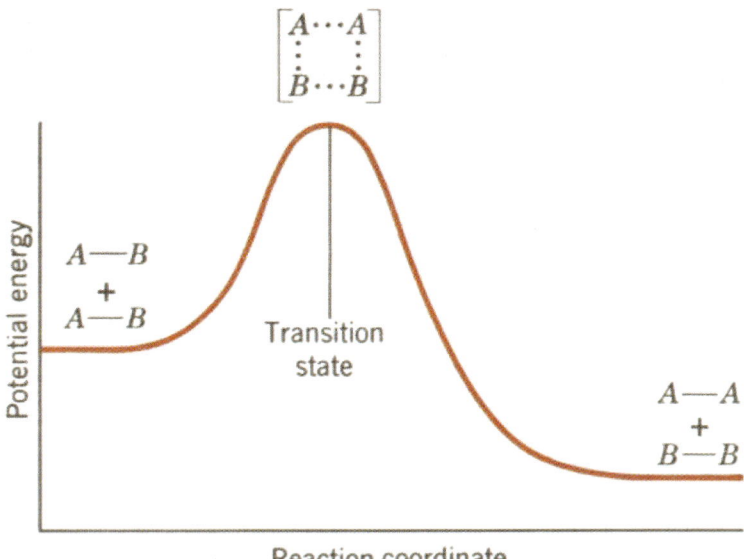

Fig. 1. Illustration of the relationship between reactants (designated as 2·A–B), products (A–A and B–B), and the activated complex. According to transition state theory, reaction kinetics is limited by the irreversible decay of the activated complex minus the rate at which the activated complex reversibly breaks down to reactants.

activity product of all rate catalyzing or inhibiting species, aside from H^+.

For the purposes of this review, the key concept embedded in equation (1) is the chemical affinity term, which is expressed as $1 - Q/K$. The chemical affinity of a system is related to the free energy of the reaction and is a measure of the degree of departure from equilibrium (i.e., $f(\Delta G) = 1 - Q/K$). The form of the chemical affinity term indicates that as the concentrations of dissolved elements build up in solution, the system approaches saturation in a rate-limiting solid and, the overall dissolution reaction slows down, and, at equilibrium, the rate would be zero. In the case of glass dissolution, there are many circumstances, which are reviewed below, where the rate behaviour does not comply with these expectations.

Some of these difficulties can be traced to unproven, and perhaps incorrect, assumptions implicit within the Aagaard & Helgeson (1982) model. For example, reaction rates at the glass/water interface are modelled as sequential. As we discuss below, there is good reason to consider reactions as taking place concurrently. In a system that is close to equilibrium, the identity of the dominant dissolution mechanism among a set of concurrent elementary reactions may change as conditions shift. Another potential pitfall is the assumption that the 'principle of

detailed balancing' applies to glass/water reactions. In the principle of detailed balancing, the equilibrium constant, K, is equal to the ratio of the forward to reverse rate (i.e., k_+/k_-). While this may be true for systems that come to equilibrium, it is clearly untrue for glass/water reactions because equilibrium between glass and water is never realized (e.g., Bourcier 1994). Further, it is also untrue for a large set of minerals that formed under high temperature and pressure conditions: these phases cannot be in equilibrium with solution at near surface conditions. In summary, these assumptions, although unverified at the time, represented a good starting point to evaluate dissolution kinetics, but may require amendment based on the data presented in the next section.

Reactions near equilibrium

There are a number of mechanisms that pose potential problems to predicting dissolution rate kinetics as the system approaches saturation. Part of this conundrum originates from current models of glass corrosion kinetics that cannot yet incorporate these unanticipated phenomena into a mathematical equation that is consistent with the constraints of thermodynamics or kinetics. These phenomena include: (1) alkali–hydrogen exchange; (2) dissimilar reactivity of phase-separated glass; and (3) acceleration or deceleration of rates as secondary corrosion products precipitate.

Alkali–hydrogen exchange kinetics

Models of glass corrosion that include alkali–hydrogen exchange are some of the earliest and most long-lived in the literature, beginning with papers by Rana & Douglas (1961a, b) and Douglas & El-Shamy (1967). These papers noted that at the start of the experiment, release of alkalis to solution occurred to a greater extent than could be explained by stoichiometric dissolution alone. Further measurements demonstrated that, for experiments in which excess alkalis are released to solution, the contacting solution pH increased, resulting in a more rapid dissolution of glass. The explanation given for these observations is ion exchange (IEX) between Na^+ in glass and hydrogen species in solution. Accordingly, for every mole of Na^+ released to solution by IEX reactions, one mole of H^+ or H_3O^+ must be incorporated into glass. Therefore, IEX reactions are driven by chemical potential differences between glass and solution:

$$\mu^i_{glass} > \mu^i_{solution} \qquad (2)$$

where μ^i_{glass} and $\mu^i_{solution}$ are the chemical potentials of an element, i, in glass and aqueous solution, respectively. One expectation from this chemical potential inequality is that as the concentration of alkali elements increases in solution, the chemical potential difference will diminish, resulting in slower alkali release rates.

A number of experiments, however, demonstrated that alkali release rates do not decrease, even in solutions with high ionic strength. For example, Pederson et al. (1993) reported that IEX rates in experiments with aluminosilicate glasses were not dependent upon the concentration of Na up to 6.11 molal. Even simple alkali silicate glasses demonstrated no decrease in IEX rates, provided that the solution was kept at the isoelectric point of the glass (pH 2.3). Apparently, the chemical potential difference of Na^+ in glass and solution is insensitive to ionic strength, even in the case of highly concentrated brines. However, this was not the only surprising observation that was revealed in more recent sets of experiments.

Sodium-rich glass compositions tested in single-pass flow-through (SPFT) apparatus at Pacific Northwest National Laboratory support the low-activity waste (LAW) glass disposal programme. The LAW glass compositions are non-radioactive analogues of waste that is currently being held in large underground single- and double-shell tanks. The glass chemistries are characterized by a high molar ratio of Na to the sum of trivalent cations (such as Al, B, and Fe), but release of Na and B are at the same rate in dilute solutions. However, as the concentration of dissolved silica in solution increases, either by addition of Si to the input solution or by slowing the flow rate, release of Na^+ is anomalously fast with respect to B (McGrail et al. 2001a, b). Figure 2 illustrates this behaviour; congruent dissolution is indicated in dilute solutions, but as the activity of silicic acid $[H_4SiO_4(aq)]$ increases, a difference in rate between B and Na release begins to surface. Both the B and Na rates decrease, but the decrease in Na release rate is not strong and becomes constant towards amorphous silica saturation. Boron release rates are more strongly affected, with a sharp decrease in rate appearing even as small quantities of silicic acid are added to solution. Surprisingly, as in the case of Na, boron release rates also become independent of silicic acid activity. At conditions of amorphous silica saturation, Na release rates are up to a factor of $50\times$ faster than that of B (McGrail et al. 2001b), depending on glass composition, temperature, and pH conditions.

Clearly, the rate of release of Na^+ is not due solely to stoichiometric dissolution of the glass; rather, Na^+ release is a consequence of two reactions occurring in parallel: (1) matrix dissolution (i.e., the breakup of the polymerized glass structure, as evidenced by B release), and (2) alkali–hydrogen exchange reactions. In dilute solutions matrix dissolution is the dominant reaction such that the slower IEX reaction cannot be detected. However, as matrix dissolution slows as a consequence of higher concentrations of dissolved silica, the IEX reaction becomes dominant and the apparent rate of Na^+ release increases over that of B. The rate of the IEX reaction can be computed by subtracting out the matrix rate from the apparent Na rate (e.g., McGrail et al. 2001b, 2002). These results underscore the importance of conducting tests over a range of solution saturation states.

Although the operation of parallel reactions explains why B and Na rates are different, there is no immediate explanation for why B rates become independent of solution saturation state as Si builds up in solution. Models of glass dissolution fashioned from TST arguments do not anticipate these results. According to TST-based models, the glass dissolution rate should depend solely on the solution saturation state

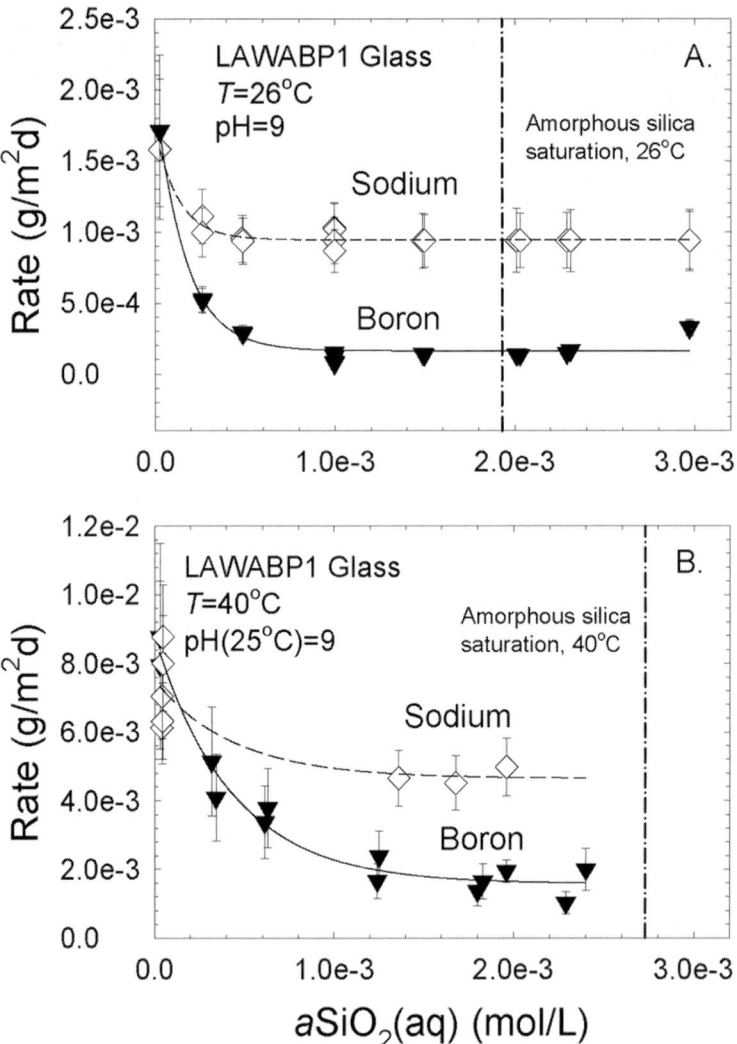

Fig. 2. Plot of normalized rate vs. the activity of silicic acid for the LAWABP1 (see Table 1) glass composition at two temperatures (26 and 40 °C). Rates are all computed at steady-state conditions. Boron and Na release rates are identical at low silica activities, then decrease, and become constant at or near saturation with respect to amorphous silica (vertical dot-dashed line). Note that the B rate decreases more than the Na rate. This behaviour can be rationalized as competition between two concurrent reactions: alkali–hydrogen exchange and matrix dissolution (see text). Error bars represent 2-σ experimental uncertainties.

(Aagaard & Helgeson 1982), provided that solution pH and temperature remain constant. However, the solution saturation state is difficult to define because glass cannot be in equilibrium with aqueous solution. Accordingly, the identity of the rate-limiting solid for glass is controversial, but for the moment, we will consider the simplest of the potential candidates.

Grambow, who perhaps wrote the most commonly referenced paper in the glass corrosion literature (Grambow 1985), proposed that a state of 'micro-equilibrium' exists (between Si–O–Si polymers and solution) and can be written as follows:

$$2\,H_2O + SiO_2(am) \Leftrightarrow H_4SiO_4(aq) \qquad (3)$$

Grambow (1985) based this proposal on the tenet of TST that says that with any set of reactions releasing element *i* to solution, the slowest reaction rate will be rate limiting. He noted that the Si–O bond is the least likely to break out of all the bonds present in glass, and that a layer relatively enriched in Si forms on the surface of the glass. The redistribution of Si on the surface, including condensation (re-polymerization) reactions, is an expression of the importance of Si–O bonding during glass reactions and has been the subject of numerous papers (e.g., Baer *et al.* 1984; Pederson *et al.* 1986, 1990). Direct measurement of silica-like surface species was carried out on sodium borosilicate glass by Bunker *et al.* (1988).

The constant rate of reaction at high activities of silicic acid can be explained by IEX reactions. Exchange reactions between a monovalent alkali cation M^+ (e.g., Na^+) and H^+ species can be written in the following manner:

$$\equiv Si-O-M + H^+ \rightarrow \equiv Si-OH + M^+ \quad (4)$$

or

$$\equiv Si-OM + H_3O^+$$
$$\rightarrow \equiv Si-OH + M^+ + H_2O \quad (5)$$

where $\equiv Si$ symbolizes a Si atom bound to the glass structure by three bonds. Note that either equation (4) or (5) produces a silanol group and increases the *local* pH, thus catalysing the irreversible hydrolysis reaction:

$$\begin{array}{c} OH \\ | \\ -Si-OH + OH^- \rightarrow H_4SiO_4(aq) \\ | \\ OH \end{array} \quad (6)$$

The reaction in equation (6) is irreversible since glass cannot precipitate from solution. Equations (4) or (5) affect the overall rate at which equation (6) proceeds because they impact the rate at which Si–OH groups are produced. However, three silanol groups must be formed before the last anchoring bond is broken, releasing a silicic acid (H_4SiO_4) molecule into solution.

The above reactions occur only when Na^+ is vulnerable to release via IEX reactions. Therefore, the *structure* of the glass (molecular arrangement of glass species) clearly impacts the overall hydrolysis rate. In order to visualize the impact of glass structure on IEX rates, consider Fig. 3. This figure is a plot of the normalized Na IEX rate, in which the contribution of Na from matrix dissolution, Na_{matrix}, is subtracted out (i.e., $Na_{matrix} - Na_{IEX}$), versus the amount of 'excess'

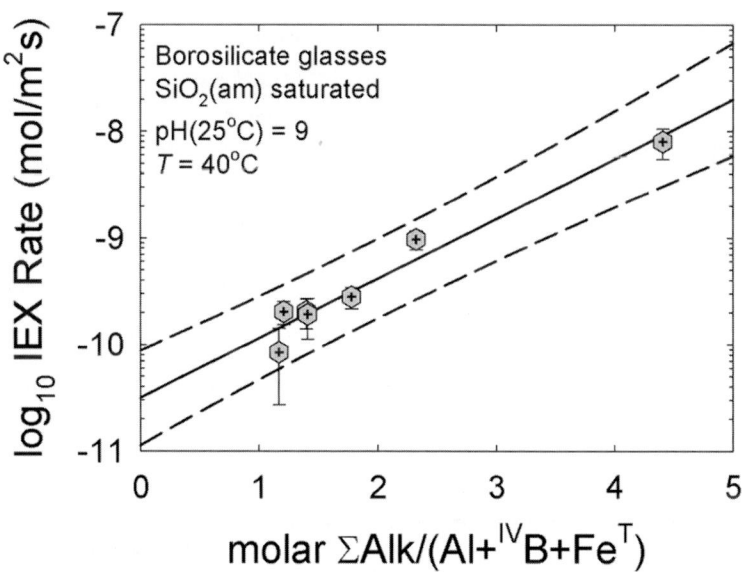

Fig. 3. Plot of log_{10} normalized ion-exchange rate at amorphous silica saturation vs. the amount of 'excess' alkalis (Na, K), denoted by the molar ratio $\Sigma Alk/(Al + {}^{IV}B + Fe^T)$. All boron is treated as four-fold coordinated (${}^{IV}B$) and total iron (Fe^T) is regarded as ferric. The ion-exchange rate subtracts out the contribution of alkalis to solution from matrix dissolution. As the amount of 'excess' alkali increases, the ion-exchange rate increases. This increase in rate reflects the increasing amount of alkalis in non-bridging oxygen (NBO) configurations. Error bars represent 2-σ experimental uncertainties and the dashed lines signify the prediction interval.

Na in glass, as indexed by the molar ratio $Na^+/(Al + B + Fe)$. This ratio assumes that: (1) all B is in four-fold coordination, and (2) that all Fe is trivalent. The first assumption is undoubtedly false, yet the correlation between ratio and the IEX rate in Fig. 3 is very good. What the molar ratio really expresses is that Na ions, in excess of what can be charge-compensated for by the sum of trivalent Al, B, and Fe, must be relegated to non-bridging oxygen (NBO) positions (sometimes referred to as 'sites'). In other words, not all Na^+ is incorporated into the matrix structure, but is in a different bonding environment: one that is energetically more susceptible to IEX reactions (Pederson *et al.* 1986, 1990; McGrail *et al.* 2001a). This observation is especially important for Na-rich glass compositions, such as those contemplated for disposal of Hanford, USA, tank waste (Table 1), and high-level waste from Japan. The highly sodic glass compositions typical of the bulk of US wastes may be one reason why US and European scientists differ in their interpretations of glass corrosion mechanisms.

Finally, our observations regarding the long-term impact of alkali ion exchange on glass dissolution now provide a mechanistic basis for the empirical 'residual' rate of reaction appended to the TST rate law articulated by Grambow (1985). The 'residual' rate was appended to prevent calculated glass dissolution rates from dropping to zero under silica-saturated conditions, which is not in accord with experimental observations.

As we have demonstrated, IEX reactions occur independent of the saturation state of the aqueous solution, raise the local solution pH, and so cause additional glass dissolution via reaction (6). The 'residual' rate for alkali-rich glass compositions, therefore, is simply the net rate of glass dissolution controlled by the rate at which the ion-exchange reaction proceeds.

Dissolution of phase-separated glass

By virtue of its ability to act as a fluxing agent, even in small concentrations, boron is included in most waste glass compositions worldwide. However, this positive property is somewhat lessened by the propensity of B to exist in two different forms in melt and glass. Four-fold coordinated B (^{IV}B), manifested as BO_4 units, is the desired moiety since these molecules co-polymerize with the Si–O–Al framework. Investigations using different techniques indicate that three-fold coordinated B (BO_3 or ^{III}B) can also exist in borosilicate glasses (Bray 1978; Mozzi & Warren 1970; Konijnendijk & Stevels 1975, 1976). The BO_3 units can readily be identified by spectroscopic techniques as predominantly three-membered boroxyl rings (B_3O_6), and it is easy to understand how such a configuration would have difficulty fitting into the relatively rigid SiO_4^{4-}- or AlO_4^{5-}-based framework. As a result of this maladaptation, phase separation, resulting in BO_3- and BO_4-rich domains, occurs in many borate and

Table 1. *Target chemical compositions of typical glasses tested at Pacific Northwest National Laboratory*

	LAWA33 (wt%)	LAWABP1 (wt%)	LAWA44 (wt%)	LD6-5412 (wt%)	HLP-9 (wt%)	MAGNOX (wt%)	HLP-31 (wt%)	Na–B–Si (wt%)
Al_2O_3	11.97	10.00	6.20	12.00	6.84	7.72	4.00	n.a.[†]
B_2O_3	8.85	9.25	8.90	5.00	12.00	27.87	12.00	20.23
CaO	n.a.	n.a.	1.99	4.00	0.01	0.01	n.a.	n.a.
Cl	0.58	0.58	0.65	n.a.	0.27	n.a.	0.32	n.a.
Cr_2O_3	0.02	0.02	0.02	n.a.	0.07	0.92	0.09	n.a.
F	0.04	0.04	0.01	n.a.	0.01	n.a.	0.01	n.a.
Fe_2O_3	5.77	2.50	6.98	n.a.	5.38	3.36	3.36	n.a.
K_2O	3.10	2.20	0.50	1.46	0.40	0.02	0.47	n.a.
La_2O_3	n.a.	2.00	n.a.	n.a.	n.a.	0.67	n.a.	n.a.
MgO	1.99	1.00	1.99	n.a.	1.47	2.99	0.92	n.a.
Na_2O	20.00	20.00	20.00	20.00	19.56	7.36	23.00	14.21
P_2O_5	0.08	0.08	0.03	0.19	0.05	0.36	0.06	n.a.
SO_3	0.10	0.10	0.10	n.a.	0.07	0.12	0.08	n.a.
SiO_2	38.25	41.89	44.55	55.91	47.98	35.95	52.00	65.56
TiO_2	2.49	2.49	1.99	n.a.	2.93	0.01	1.83	n.a.
ZnO	4.27	2.60	2.96	n.a.	1.47	0.65	0.92	n.a.
ZrO_2	2.49	5.25	2.99	n.a.	1.47	0.68	0.92	n.a.
Others*	n.a.	n.a.	n.a.	n.a.	n.a.	11.31	n.a.	n.a.
Total	100.00	100.00	99.86	98.56	99.98	100.00	99.98	100.00

*'Others' refers to BaO, CeO_2, Cs_2O, HfO_2, Li_2O, MoO_3, Nd_2O_3, NiO, RbO_2, RuO_2, Pr_2O_3, Sm_2O_3, SrO, TeO_2, and Y_2O_3.
[†]n.a. = not analysed.

borosilicate glass compositions (Charles & Wagstaff 1968; Haller *et al.* 1970). In the case of borosilicate glass compositions, there is a tendency for glass to separate into two chemically distinct regions: one rich in alkali-metaborate – the other in silicon (Bray 1978). Accordingly, the thermodynamic properties of glasses prone to phase separation display large values of configurational entropy as the various cations (Si, Al, B) compete for oxygen atoms and their moieties jostle each other for space (Hervig & Navrotsky 1985).

What is particularly problematic about this phase separation process from a chemical corrosion point of view is that the B-rich phase (which can be modeled as a sodium-metaborate component; $Na_2O \cdot 2B_2O_3$ or $Na_2B_4O_7$) is highly water-soluble. As dissolution of sodium-metaborate is independent of dissolved silica activity:

$$Na_2B_4O_7 + 7\,H_2O = 4\,H_3BO_3(aq)$$

$$+ 2\,Na^+ + 2\,OH^- \quad (7)$$

we should expect to find unequal dissolution behaviour in phase-separated glass with only the Si-rich regions (modelled as a reedmergnerite ($NaBSi_3O_8$) component) responding to differences in chemical affinity of the system based on dissolved silica activity. Areas of fast dissolution corresponding to alkali-metaborate regions should form etch pits as the chemical affinity of the system approaches saturation with respect to amorphous silica.

Such 'pits' have recently been observed in a reacted phase-separated glass using vertical scanning interferometry (VSI) techniques, and an image of the pitted surface is shown in Fig. 4. The glass composition is the simple Na–B–Si glass included in Table 1; note that the glass is relatively B-rich (molar B > Na). The glass specimen was a 10 mm × 10 mm × 2 mm coupon that was run in an SPFT apparatus for 24 h at 90 °C at pH 9. The coupon was examined *ex situ* by VSI techniques and height differences were recorded with a resolution down to 1 nm. For a further discussion of the VSI techniques, the reader is referred to Lüttge *et al.* (1999).

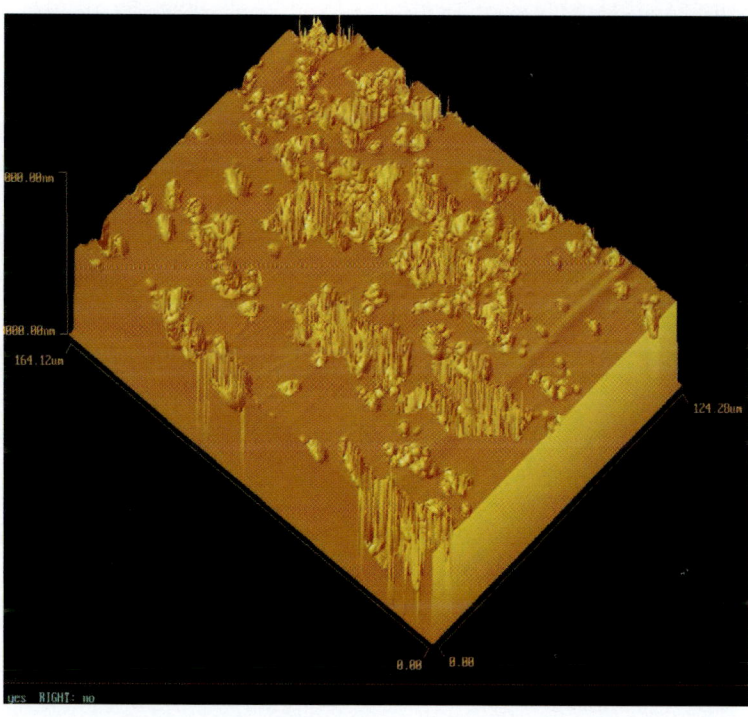

Fig. 4. An image (124 × 164 μm) illustrating height differences on a $Na_2O–B_2O_3–SiO_2$ glass that is phase-separated (composition 'Na–B–Si' in Table 1). Height differences were measured using the vertical scanning interferometer (VSI) technique. The 'etch pits' correspond to domains rich in water-soluble sodium-metaborate. Thus, the figure indicates differential rates of dissolution of phase-separated domains consisting of sodium-metaborate and Si-richer (and, therefore, more corrosion resistant) matrix.

Figure 5 illustrates the effect that glass–glass phase separation can have on dissolution rates of alkali borosilicate glass compositions. On this diagram, we have plotted the normalized $\log_{10}$ rates as a function of silicic acid activity for both homogeneous and phase-separated glass. One notable feature of this diagram is that the phase-separated glass dissolution rates (plots of MAGNOX and HLP-31 data) are a factor of $\sim 10 \times$ faster than those of the homogeneous glasses. Note also that the release rates of Na and B from the phase-separated glasses are, unlike in the case of the homogeneous glasses (LAWA33 and HLP-9 data), independent of the activity of silicic acid activity. The reason that the phase-separated glass compositions do not display evidence for silica-saturation control can be seen in equation (7). If the sodium-metaborate region dissolves rapidly with respect to the Si-richer matrix, then the release of Na and B to solution should show no dependence upon silicic acid activity. As pre-dicted, the release rates of Na and B are faster than for their homogeneous counterparts. In addi-tion, the phase-separated glasses display an equal *mass* release rate of Na and B, which is predicted through the stoichiometry of the glass (McGrail *et al.* 2002). This behaviour is in keeping with a chemical affinity control over glass dissolu-tion. Another relevant point is that the rate of release of Na and B from phase-separated glass depends on the relative proportions of reedmerg-nerite- and sodium-metaborate- rich regions that make up the phase-separated glass. A phase-separated glass made up predominately of a reed-mergnerite-like phase will dissolve relatively slowly. In contrast, a phase-separated glass con-taining a high proportion of sodium-metaborate glass, especially if such domains are intercon-nected, will dissolve more rapidly. Again, the theme of parallel reactions, one of which will be dominant, is reiterated.

The tendency of B to promote phase sepa-ration in glass specimens predicts that a wide

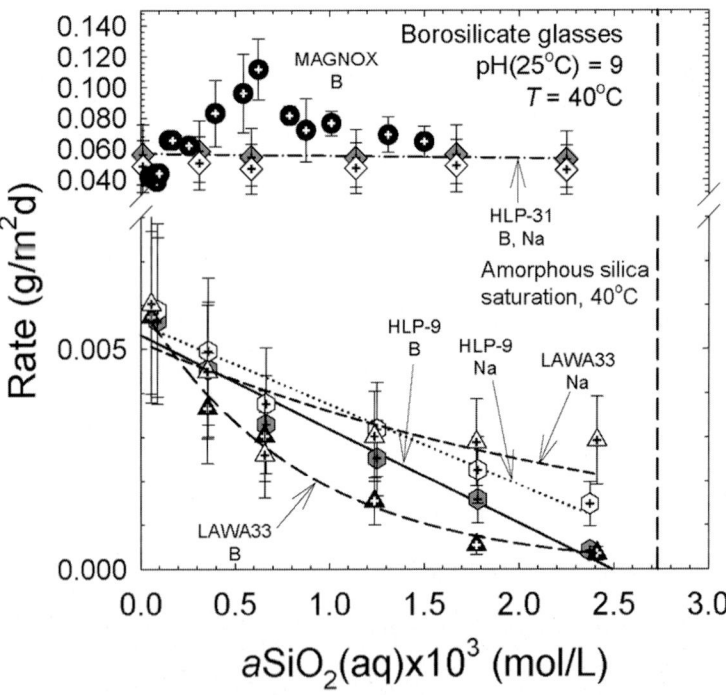

Fig. 5. Plot of $\log_{10}$ B and Na normalized release rates vs. the activity of silicic acid for both phase-separated and physically homogeneous glass specimens. All rates were plotted at steady-state conditions. MAGNOX and HLP-31 represent phase-separated whereas LAWA33 and HLP-9 represent homogeneous glass specimens. The behaviour of homogeneous glass includes an inverse relationship between rates and silicic acid activity and a difference between B and Na rates as silicic acid activity increase. The magnitude of the difference between B and Na rates is related the amount of 'excess' Na (see Fig. 3). Relatively faster element release rates ($\sim 10 \times$), identical release rates of Na and B, and independence from activity of silicic acid appears to characterize phase-separated glass.

variety of borosilicate glasses should be phase-separated. We have tested numerous glass compositions that show consistently high dissolution rates and do not respond to increasing concentrations of silica activity (McGrail *et al.* 2001*b*). These findings are in accord with the expectation that many borosilicate glass compositions, especially those that have low Al_2O_3 concentrations, are phase-separated. However, both compositionally simple and complex glass specimens can show evidence for phase separation. For example, ample evidence exists that both the >15 oxide component MAGNOX and the simple three-component $(Na_2O–B_2O_3–SiO_2)$ glass are phase-separated (McGrail *et al.* 2002). This evidence includes NMR data, secondary electron microscope observations of acid-reacted glass, and direct documentation of etch pit formation caused by dissolution of the soluble sodium-metaborate glass (see Fig. 4 and Icenhower *et al.* 2003).

However, this begs the question of why other investigators have not detected phase separation over a wide range of borosilicate compositions. The answer seems to be that the very fine scale of phase separation typically complicates detection efforts. Typical micro-heterogeneity of the glass is on the order of 10 to 500 Å (Konijnendijk & Stevels 1976; Porai-Koshits *et al.* 1982), which means that many modern techniques used to detect phase separation (>100 Å) are not necessarily sensitive enough.

Influence of secondary mineral precipitation

Over the last 20 years investigators have reported that precipitation of secondary phases can accelerate the corrosion rate of glass. This is because precipitation can cause a sudden drop in the activity of a key aqueous species. In other words, we hypothesize that the rate is affected through the chemical affinity of the system.

Bates & Steindler (1983) first reported the linkage between dissolution rates and secondary mineral precipitation, and a number of more recent papers (Van Iseghem & Grambow 1988; Strachan & Croak 2000) have reported similar findings. In these studies, precipitation of the zeolite phase, analcime, $Na(AlSi_2O_6)\cdot H_2O$, consumed silicic acid, causing the corrosion of the glass to accelerate back up to the maximum or forward rate of reaction. Again, these findings argue against control of element release via a thickening reaction layer. The mineral herschelite, $(Na, K)AlSi_2O_6\cdot 3H_2O$ (Fig. 6), covers the surface of many of the glass specimens that we have worked with (along with analcime and kaolinite), and, like analcime, controls the activities of silicic acid and Al in solution. Because zeolite minerals can completely coat the glass surface,

Fig. 6. Scanning electron micrograph picture of borosilicate glass (LAWA33, see Table 1) reacted with solution at an elevated temperature. The hexagonal phase is herschelite, $(Na, K)AlSi_2O_6\cdot 3H_2O$.

why should the rates of reaction increase? Why should zeolite minerals not also behave as a diffusion barrier? The simplest way to interpret these results is to invoke changes in the chemical affinity of the system triggered by the precipitation of silica-consuming phases.

If secondary minerals affect the dissolution rate of glass, then the composition of glass itself becomes important because the chemistry of the glass ultimately determines which phase or phases will precipitate. Strachan & Croak (2000) made a set of empirical plots of the effects that glass composition exerts on the corrosion resistance of glass. These plots were based on their original research and that of Van Iseghem & Grambow (1988), which showed that a critical concentration of Na and, especially, Al, is needed to stabilize analcime. Therefore, Al – and not just Si – were shown to adversely affect dissolution. Furthermore, their investigation found that a critical volume of analcime must form before the rates can accelerate to the forward rate. This last finding explains why precipitation of analcime by itself is not sufficient for controlling rates; a threshold volume of analcime must form in order to control the dissolved Si-activity of contacting water.

Although these investigations are immensely helpful for deciphering dissolution behaviour of borosilicate glass, they are, in the end, only empirical relationships. One of the more difficult aspects of investigating glass corrosion behaviour is that phases that should form, using thermodynamic criteria, often do not (at least on the time-scale of laboratory experiments). This has to do with the inherent precipitation kinetics of secondary phases; typically, thermodynamically metastable phases form before their fully crystalline, and thermodynamically stable, counterparts. Note that this is especially true in experiments performed at high temperatures; typically, more crystalline phases are favoured by higher temperatures and can obfuscate interpretations of glass corrosion performance under (lower) repository temperatures where the metastable phases may persist over the entire time of interest with respect to repository performance.

Another difficulty presented by precipitation of secondary phases is that other aqueous species, such as those of Al, can be affected, as stated above. Under certain conditions the activity of $Al(OH)_4^-$ in mildly alkaline solutions can drop to near zero due to precipitation of zeolite phases. The question is how to represent this change in aluminate activity using a rate equation based on chemical affinity concepts. For example, in equation (1), the activity of the

aluminate ion can be accommodated a number of different ways, but they all ultimately cause problems for long-term prediction of glass corrosion rates.

If the aluminate ion is a component in the chemical affinity term, then the activity of dissolved silica, by itself, cannot describe the change in rate with changes in chemistry of the contacting fluid. A number of attempts have been made to explicitly include Al activity in the chemical affinity term. Gin (1996) suggested that glass dissolution could be modelled using a 'mixed' Si/Al term for the ion activity product (Q):

$$f(\Delta G) = \left(1 - \frac{Q}{K}\right)$$

$$= \left(1 - \left[\frac{a_{Al(OH)_4^-}^x \times a_{H_4SiO_4}^y}{K_g}\right]\right) \quad (8)$$

where the terms $a_{Al(OH)_4^-}^x$ and $a_{H_4SiO_4}^y$ are the activities of the aluminate ion and silicic acid, respectively. The exponents of the activities (x and y) represent the stoichiometry of the rate-limiting reaction. The term in the denominator, K_g, represents the fictive glass equilibrium constant. As noted above in the 'alkali–hydrogen kinetics' discussion, concepts such as K_g as applied to glass are not valid, but aid in developing the concepts of solution composition control on glass dissolution. Bourcier *et al.* (1994) proposed that the solution is in equilibrium with an amorphous gel and the saturation state could be defined using the components Si, Fe, Al, Ca, and Mg. This view is similar to that of Daux *et al.* (1997), in which the aqueous activities of Si, Fe, and Al should be used to define Q for basaltic glass. Oelkers advocated a rate law for basalt and aluminosilicate materials in general that is based upon the activity of Al^{3+} (Oelkers 2001; Oelkers & Gislason 2001). This model is echoed in models for aluminosilicate glass compositions containing variable amounts of Al (Hamilton *et al.* 2001). Advocat *et al.* (1998) went further than the others, suggesting that the affinity term include *all* glass components.

Although all of these ideas have merit, they suffer from the same problem: saturation with respect to certain solid phases causes one or more activities to decrease to near zero, causing the entire $1 - Q/K$ term to go to unity. In other words, the model implies that rates should accelerate to the maximum or forward rate, despite how much silicic acid is present in solution, contrary to experimental results. As long as the glass is physically homogeneous (i.e., no phase separation), some decrease in the rate must occur, even with relatively sodic glass compositions (see above section on 'ion exchange kinetics'). Therefore, models based upon multicomponent ion activity products do not allow for an accurate picture of the rate dependence.

Equally problematic is representing Al as a rate inhibitor – thereby removing the activity of Al from the ion activity quotient term to the $\prod_i a_i^{n_i}$ term (see equation 1). Yet, precipitation of an Al-bearing phase, such as zeolite, can dramatically lower the activity of Al in solution. Because inhibitor species will have a negative exponent, the term gets very large at low activities of Al, which again causes the rate to increase, despite high activity of silicic acid in solution. McGrail *et al.* (2001*b*) reported a number of cases of this phenomenon. These findings make plain that a more mathematically stable form of the rate equation must be fashioned.

Conclusions

We are only beginning to understand the dissolution kinetics of alkali borosilicate glass under the context of conditions most likely to persist in subsurface environments. Because of typically slow pore water recharge rates, especially in relatively dry repository settings (e.g., Hanford, Washington; Yucca Mountain, Nevada; and Los Alamos, New Mexico, USA), aqueous solutions in contact with glass will be extremely concentrated in dissolved glass components and subject to the chemical affinity of the system. However, this chapter vividly illustrates that the conventional means of linking element release rates to solution saturation state fail for a number of circumstances. Failure is likely related to assumptions of consecutive, rather than concurrent, reactions and the idea that chemical affinity solely explains glass dissolution behaviour. The data presented in this review indicate that element release rates are governed by competing reactions that: (1) limit release through the dominant reaction, and (2) the identity of the dominant reaction changes as the system approaches saturation. The dominance of 'secondary' reactions, unanticipated by earlier investigators, govern rates in concentrated solution and include: (1) the magnitude of excess alkalis in glass, (2) dissolution of micro-heterogeneous borosilicate glass, and (3) the identity and volume of precipitating phases on the surface of glass. The fact that failure to conform to conventional rate laws will occur for common, as opposed to exotic, conditions germane to waste glass disposal is reason for further serious contemplation. Fortunately, there is an expanding list of detailed dissolution kinetics studies that have proven their value by withstanding scrutiny from an informed and perceptive group of peers. These works provide a sound basis for evaluating a spectrum of glass corrosion behaviour.

The failure of models based on application of TST rate laws to glass/water systems does not mean, however, that diffusion through a 'leach layer' is by default the answer to this dilemma. Clearly, the set of recently reported data on glass corrosion resistance shows that it is not an 'either-or' situation between affinity- and diffusion-based rate laws. Finding a mathematically stable form of the rate equation appears to be more worthy of pursuit.

We thank L. Royack and D. Wellman for their aid in constructing diagrams. We thank E. Rodriguez and J. Steele for their unflagging dedication to the laboratory work. Conversations, formal and informal, on many of the key concepts in this paper with D. Strachan, J. Vienna, E. Pierce, D. Wellman, and D. Rimstidt are noted with appreciation. Last, we gratefully acknowledge the incisive and detailed reviews of K. Lemmens, P. Frugier, and D. Rimstidt, whose comments greatly improved the manuscript.

References

AAGAARD, P. & HELGESON, H. C. 1982. Thermodynamic and kinetic constraints on reaction rates among minerals and aqueous solutions. I. Theoretical considerations. *American Journal of Science*, **282**, 237–285.

ADVOCAT, T., CHOUCHAN, J. L., CROVISIER, J. L., GUY, C., DAUX, V., JÉGOU, S. & VERNAZ, E. 1998. Borosilicate nuclear waste glass alteration kinetics: Chemical inhibition and affinity control. *Materials Research Society Symposium Proceedings*, **506**, 63–70.

ADVOCAT, T., JOLLIVET, P., CROVISIER, J. L. & DEL NERO, M. 2001. Long-term alteration mechanisms in water for SON68 radioactive borosilicate glass. *Journal of Nuclear Materials*, **298**, 55–62.

BACON, D. H., WHITE, M. D. & MCGRAIL, B. P. 2000. *Subsurface Transport Over Reactive Multiphases (STORM): A General, Coupled Nonisothermal Multiphase Flow, Reactive Transport, and Porous Medium Alteration Simulator, Version 2, User's Guide*. Pacific Northwest National Laboratory, PNL-13108, Richland, WA.

BAER, D. R., PEDERSON, L. R. & MCVAY, G. L. 1984. Glass reactivity in aqueous solutions. *Journal of Vacuum Science Technology*, **A 2 (2)**, 738–743.

BARKATT, A., GIBSON, B. C. *et al.* 1986. Mechanisms of defense waste glass dissolution. *Nuclear Technology*, **73**, 140–164.

BARZUN, J. 2000. *From Dawn to Decadence: 1500 to the Present. 500 Years of Western Cultural Life*. Harper Collins Publishers, Inc., New York, 800 pp.

BATES, J. K. & STEINDLER, M. J. 1983. Alteration of nuclear waste glass by hydration. *In*: BROOKINS, D. G. (ed) *Scientific Basis for Nuclear Waste Management VI*, **15**. Elsevier, New York, 83–90.

BOURCIER, W. L. 1991. Overview of chemical modeling of nuclear waste glass dissolution. *Materials Research Society Symposium Proceedings*, **212**, 3–18.

BOURCIER, W. L. 1994. Waste glass corrosion modeling: Comparison with experimental results. *Materials Research Society Symposium Proceedings*, **333**, 69–82.

BOURCIER, W. L., CARROLL, S. A. & PHILLIPS, B. L. 1994. Constraints on the affinity term for modeling long-term glass dissolution rates. *Materials Research Society Symposium Proceedings*, **333**, 507–512.

BRAY, P. J. 1978. NMR studies of borates. *In*: PYE, L. D., FRECHETTE, V. D. & KREIDL, N. F. (eds) *Borate Glasses: Structure, Properties, Applications*. Plenum, New York, 321–352.

BUNKER, B. C., ARNOLD, G. W., BEAUCHAMP, E. K. & DAY, D. E. 1983. Mechanisms for alkali leaching in mixed-Na–K silicate glasses. *Journal of Non-Crystalline Solids*, **58**, 295–322.

BUNKER, B. C., TALLANT, D. R., HEADLEY, T. J., TURNER, G. L. & KIRKPATRICK, R. J. 1988. The structure of leached sodium borosilicate glass. *Physics and Chemistry of Glasses*, **29**, 106–120.

CASEY, W. H. & BUNKER, B. 1990. Leaching of mineral and glass surfaces during dissolution. *In*: HOCHELLA, M. F. & WHITE, A. F. (eds) *Mineral–water interface geochemistry*, **23**. Mineralogical Society of America, Washington, 397–426.

CHARLES, R. J. & WAGSTAFF, F. E. 1968. Metastable immiscibility in the B_2O_3–SiO_2 system. *Journal of the American Ceramics Society*, **51**, 16–20.

DAUX, V. C. G., ADVOCAT, T., CROVISIER, J.-L. & STILLE, P. 1997. Kinetic aspects of basaltic glass dissolution at 90 °C: Role of aqueous silicon and aluminum. *Chemical Geology*, **142**, 109–126.

DOREMUS, R. H. 1977a. Diffusion in glasses and melts. *Journal of Non-Crystalline Solids*, **25**, 261–292.

DOREMUS, R. H. 1977b. Interdiffusion of hydrogen and alkali ions in a glass surface. *Journal of Non-Crystalline Solids*, **19**, 137–144.

DOREMUS, R. H. 1981. Time dependence of the reaction of water with glass. *Nuclear and Chemical Waste Management*, **2**, 119–123.

DOREMUS, R. H. 1983. Diffusion-controlled reaction of water with glass. *Journal of Non-Crystalline Solids*, **55**, 143–147.

DOREMUS, R. H. 2000. Diffusion of water in rhyolite glass: Diffusion–reaction model. *Journal of Non-Crystalline Solids*, **261**, 101–107.

DOUGLAS, R. W. & El-SHAMY, T. M. M. 1967. Reactions of glasses with aqueous solutions. *Journal of the American Ceramic Society*, **50**, 1–8.

DRAN, J.-C., DELLA MEA, G., PACCAGNELLA, A., PETIT, J.-C. & TROTIGNON, L. 1988. The aqueous dissolution of alkali silicate glasses: Reappraisal of mechanisms by H and Na depth profiling with high energy ion beams. *Physics and Chemistry of Glasses*, **29**, 249–255.

EYRING, H. 1935a. The activated complex and the absolute rate of chemical reactions. *Chemical Reviews*, **17**, 65–77.

EYRING, H. 1935b. The activated complex in chemical reactions. *Journal of Chemical Physics*, **3**, 107–115.

GIN, S. 1996. Control of R7T7 nuclear glass alteration kinetics under saturation conditions. *Materials Research Society Symposium Proceedings*, **412**, 189–196.

GIN, S. & MESTRE, J. P. 2001. SON 68 nuclear glass alteration kinetics between pH 7 and pH 11.5. *Journal of Nuclear Materials*, **295**, 83–96.

GIN, S. & FRUGIER, P. 2003. SON68 glass dissolution kinetics at high reaction progress: Experimental evidence of the residual rate. *Materials Research Society Symposium Proceedings*, **757**, 175–182.

GIN, S., RIBET, I. & COULLIARD, M. 2001a. Role and properties of the gel formed during nuclear waste glass alteration: Importance of gel formation kinetics. *Journal of Nuclear Materials*, **298**, 1–10.

GIN, S., JOLLIVET, P., MESTRE, J. P., JULLIEN, M. & POZO, C. 2001b. French SON 68 nuclear glass alteration mechanisms on contact with clay media. *Applied Geochemistry*, **16**, 861–881.

GOLDEN, D. M. 1979. Experimental and theoretical examples of the value and limitations of Transition State Theory. *The Journal of Physical Chemistry*, **83**, 108–113.

GRAMBOW, B. 1985. A general rate equation for nuclear waste glass corrosion. *Materials Research Society Symposium Proceedings*, **44**, 15–27.

HALLER, W., BLACKBURN, D. H., WAGSTAFF, F. E. & CHARLES, R. J. 1970. Metastable immiscibility surface in the system Na_2O–B_2O_3–SiO_2. *Journal of the American Ceramic Society*, **53**, 34–39.

HAMILTON, J. P., BRANTLEY, S. L., PANTANO, C. G., CRISCENTI, L. J. & KUBICKI, J. D. 2001. Dissolution of nepheline, jadeite and albite glasses: Toward better models for aluminosilicate dissolution. *Geochemica et Cosmochimica Acta*, **65**, 3683–3702.

HENCH, L. L., CLARK, D. E. & HARKER, A. B. 1986. Nuclear waste solids. *Journal of Materials Science*, **21**, 1457–1478.

HERVIG, R. L. & NAVROTSKY, A. 1985. Thermochemistry of glasses in the system Na_2O–B_2O_3–SiO_2. *Journal of the American Ceramic Society*, **11**, 284–298.

HOCHELLA, M. F. & BANFIELD, J. F. 1985. Chemical weathering of silicates in nature: A microscopic perspective with theoretical considerations. *In*: WHITE, A. F. & BRANTELY, S. L. (eds) *Chemical Weathering Rates of Silicate Minerals*. **31**, Mineralogical Society of America, Washington, DC, 353–406.

HORTON, H. H. 1929. *Outline of Science. Part II. Man's Material Achievements*. Funk and Wagnalls Co., New York, NY, pp 207.

ICENHOWER, J. P., LÜTTGE, A. *et al.* 2003. Results of vertical scanning interferometry (VSI) of dissolved borosilicate glass: Evidence for variable surface features and global surface retreat. *Materials*

Research Society Symposium Proceedings, **757**, 119–126.

JÉGOU, C., GIN, S. & LARCHÉ, F. 2000. Alteration kinetics of a simplified nuclear glass in an aqueous medium: Effects of solution chemistry and of protective gel properties on diminishing the alteration rate. *Journal of Nuclear Materials*, **280**, 216–229.

KARPLUS, M., PORTER, R. N. & SHARMA, R. D. 1965. Exchange reactions with activation energy. I. Simple barrier potential for (H, H_2). *The Journal of Chemical Physics*, **43**, 3259–3287.

KONIJNENDIJK, W. H. & STEVELS, J. M. 1975. The structure of borate glasses studied by Raman scattering. *Journal of Non-Crystalline Solids*, **18**, 307–331.

KONIJNENDIJK, W. H. & STEVELS, J. M. 1976. The structure of borosilicate glasses studied by Raman scattering. *Journal of Non-Crystalline Solids*, **20**, 193–224.

LANFORD, W. A., DAVIS, K., LAMARCHE, P., LAURSEN, T., GROLEAU, R. & DOREMUS, R. H. 1979. Hydration of soda-lime glass. *Journal of Non-Crystalline Solids*, **33**, 249–266.

LETURQ, G., BERGER, G., ADVOCAT, T. & VERNAZ, E. 1999. Initial and long-term dissolution rates of aluminosilicate glasses enriched with Ti, Zr and Nd. *Chemical Geology*, **160**, 39–62.

LINARD, Y., ADVOCAT, T., JÉGOU, C. & RICHET, P. 2001. Thermochemistry of nuclear waste glasses: Application to weathering studies. *Journal of Non-Crystalline Solids*, **289**, 135–143.

LÜTTGE, A., BOLTON, E. W. & LASAGA, A. C. 1999. An interferometric study of the dissolution kinetics of anorthite: The role of reactive surface area. *American Journal of Science*, **299**, 652–678.

McGRAIL, B. P., ICENHOWER, J. P., *et al.* 2001a. The structure of $Na_2O-Al_2O_3-SiO_2$ glass: Impact of sodium ion exchange in H_2O and D_2O. *Journal of Non-Crystalline Solids*, **296**, 10–26.

McGRAIL, B. P., ICENHOWER, J. P., MARTIN, P. F., SCHAEF, H. T., O'HARA, M. J., RODRIGUEZ, E. A. & STEELE, J. L. 2001b. *Waste Form Release Data Package for the 2001 Immobilized Low-Activity Waste Performance Assessment*. Pacific Northwest National Laboratory, PNNL-13043, Rev. 2, Richland, WA.

McGRAIL, B. P., ICENHOWER, J. P. & RODRIGUEZ, E. A. 2002. Origins of discrepancies between kinetic rate law theory and experiments in the $Na_2O-B_2O_3-SiO_2$ system. *Materials Research Society Symposium Proceedings*, **713**, 537–546.

MOZZI, R. L. & WARREN, B. E. 1970. The structure of vitreous boron oxide. *Journal of Applied Crystallography*, **3**, 251–257.

MURPHY, W. M. & PALABAN, R. T. 1994. *Geochemical Investigations Related to the Yucca Mountain Environment and Potential Nuclear Waste Repository*. Southwest Research Institute, NUREG/CR-6288, San Antonio, TX.

OELKERS, E. H. 2001. General kinetic description of multioxide silicate mineral and glass dissolution. *Geochimica et Cosmochimica Acta*, **65**, 3703–3719.

OELKERS, E. H. & GISLASON, S. R. 2001. The mechanism, rates and consequences of basaltic glass dissolution: I. An experimental study of the dissolution rates of basaltic glass as a function of aqueous Al, Si and oxalic acid concentration at 25 °C and pH = 3 and 11. *Geochimica et Cosmochimica Acta*, **65**, 3671–3681.

PEDERSON, L. R., BAER, D. R., McVAY, G. L. & ENGELHARD, M. H. 1986. Reaction of soda lime silicate glass in isotopically labeled water. *Journal of Non-Crystalline Solids*, **86**, 369–380.

PEDERSON, L. R., BAER, D. R., McVAY, G. L., FERRIS, K. F. & ENGELHARD, M. H. 1990. Reaction of silicate glasses in water labeled with D and ^{18}O. *Physics and Chemistry of Glasses*, **31**, 177–182.

PEDERSON, L. R., McGRAIL, B. P., McVAY, G. L., PETERSEN-VILLALOBOS, D. A. & SETTLES, N. S. 1993. Kinetics of alkali silicate and aluminosilicate glass reactions in alkali chloride solutions: Influence of surface charge. *Physics and Chemistry of Glasses*, **34**, 140–148.

PORAI-KOSHITS, I. A., GOLUBKOV, V. V., TITOV, A. P. & VASILEVSKAYA, T. N. 1982. The microstructure of some glasses and melts. *Journal of Non-Crystalline Solids*, **49**, 143–156.

RANA, M. A. & DOUGLAS, R. W. 1961a. The reaction between glass and water. Part 1. Experimental methods and observations. *Physics and Chemistry of Glasses*, **2**, 179–195.

RANA, M. A. & DOUGLAS, R. W. 1961b. The reaction between glass and water. Part 2. Discussion of the results. *Physics and Chemistry of Glasses*, **2**, 196–204.

STRACHAN, D. M. & CROAK, T. L. 2000. Compositional effects on long-term dissolution of borosilicate glass. *Journal of Non-Crystalline Solids*, **272**, 22–33.

SVERDRUP, H. & WARFVINGE, P. 1995. Estimating field weathering rates using laboratory kinetics. *In*: WHITE, A. F. & BRANTLEY, S. F. (eds) *Chemical Weathering Rates of Silicate Minerals*, **31**. Mineralogical Society of America, Washington, 485–541.

TSOMAIA, N., BRANTLEY, S. L., HAMILTON, J. P., PATANO, C. G. & MUELLER, K. T. 2003. NMR evidence for formation of octahedral and tetrahedral Al and repolymerization of the Si network during dissolution of aluminosilicate glass and crystal. *American Mineralogist*, **88**, 54–67.

VAN ISEGHEM, P. & GRAMBOW, B. 1988. The long-term corrosion and modeling of two simulated Belgian reference high-level waste glasses. *Materials Research Society Symposium Proceedings*, **112**, 631–639.

VELBEL, M. A. 1993. Constancy of silicate-mineral weathering-rate ratios between natural and experimental weathering: Implications for hydrologic control of differences in absolute rates. *Chemical Geology*, **105**, 89–99.

VERNAZ, E. & DUSSOSSOY, J. L. 1992. Current state of knowledge of nuclear waste glass corrosion mechanisms: The case of R7T7 glass. *Applied Geochemistry* (Supplementary Issue), **1**, 13–22.

VERNAZ, E., GIN, S., JÉGOU, C. & RIBET, I. 2001. Present understanding of R7T7 glass alteration kinetics and their impact on long-term behavior modeling. *Journal of Nuclear Materials*, **298**, 27–36.

WERME, L. B., BJÖRNER, I. K., *et al.* 1990. Chemical corrosion of highly radioactive borosilicate nuclear waste glass under simulated repository conditions. *Journal of Materials Research*, **5**, 1130–1146.

WHITE, A. F. 1995. Chemical weathering rates of silicate minerals in soils. *In*: WHITE, A. F. & BRANTLEY, S. F. (eds) *Chemical Weathering Rates of Silicate Minerals, Mineralogical Society of America*, **31**. Washington, 407–461.

YOKOYAMA, T. & BANFIELD, J. F. 2002. Direct determinations of the rates of rhyolite dissolution and clay formation over 52,000 years and comparison with laboratory measurements. *Geochimica et Cosmochimica Acta*, **66**, 2665–2681.

Cement stabilization of heavy-metal-containing wastes

C. ANNETTE JOHNSON

Swiss Federal Institute of Environmental Science and Technology (EAWAG), Dübendorf, Switzerland (e-mail: johnson@eawag.ch)

Abstract: Cement has found wide usage in the stabilization of heavy-metal-containing wastes as cement minerals can substantially reduce heavy metal solubility as a result of precipitation, adsorption to the surfaces and incorporation. The solubility of some heavy metal cations is limited by the precipitation of hydroxides, while that of some oxyanions is limited by the formation of Ca salts. Only ions that are sufficiently soluble in basic media will be incorporated in or sorbed to hydrated cement minerals to a significant degree. Heavy metal cations may sorb quite strongly to calcium silicate hydrate (C–S–H). The cations diffuse into the C–S–H particles where they are probably sorbed to the silicate chains. Pure phases of oxyanion-substituted ettringite ($3CaO \cdot Al_2O_3 \cdot 3(CaSO_4) \cdot 32H_2O$, an AFt phase) and monosulphate ($3CaO \cdot Al_2O_3 \cdot CaSO_4 \cdot 12H_2O$, an AFm phase) phases and solid solutions with SO_4^{2-} and OH^- have been synthesized. Some thermodynamic data are available for the pure phases. For most elements an approximate range of solubility has recently become known. However, it is not possible to predict solubility from the available data.

Wastes are being increasingly used in the production of cements. They may be added as raw meal substitutes, fuels or as mineral additives. Despite the benefits derived from the reduction of fuel usage or the consumption of raw materials, the impact of increases in heavy metal concentrations to the environment due to the use of wastes is not well understood. Although metals may not be leached to a significant degree from primary structures, they may be leached after demolition if concrete is recycled, used as unbound aggregate or landfilled (van der Sloot 2001). Such a broad view is taken in 'cradle to grave' life-cycle assessments that are being developed in Europe for a wide range of construction products (e.g., Eyerer & Reinhardt 2000). Cement is also used worldwide to 'stabilize' hazardous and nuclear wastes in order to reduce the mobility of contaminants, namely heavy metals metalloids and radionuclides. Here the long-term assessment of the effectiveness of such a treatment is essential. The contaminant leachability is reduced by a lowering of the permeability of the wastes and by the binding of ions in the cementitious matrix. The geochemical properties of the metal/metalloid/radionuclide ion and the binding mechanism control the porewater concentrations. Recognition of the importance of understanding binding in the cementitious matrix (Spence 1993) has led to an increase in research activity over the last decade. The investigations have fallen broadly into two categories.

Leaching experiments have been carried out on ground hydrated cement pastes doped with metal ions, on cement-stabilized wastes and on cements and mortars containing elevated metal concentrations derived from wastes. Such experiments have been collated in the MONOLITH database (Stegemann *et al.* 2001). The results of such experiments scatter widely and are generally very difficult to compare. This may be a consequence of experimental procedure including the type of cement, the water : cement ratio for the initial hydration, hydration time, the form and the time in which the metal ion is added, the water : solid ratio of the leaching test, the time for the leaching process and so on. There are so many variables that it is difficult to compare leaching experiments. Nevertheless the following observations have been made. Some ions appear to be solubility-controlled, that is, discrete solid phases are precipitated. Other, more soluble ions exhibit a direct correlation between solid and aqueous phase concentrations. Last, but not least, the dissolved concentration is pH-dependent (e.g., Trussel & Batchelor 1996; van der Sloot *et al.* 2001; Sanchez *et al.* 2002). The underlying mechanisms have been discussed in Spence (1993) and more recently by Glasser (1997); others will be discussed here in depth later in the chapter.

While progress has been made in this complex system, the investigation of heavy-metal-binding to single cement minerals has brought great

From: GIERÉ, R. & STILLE, P. (eds) 2004. *Energy, Waste, and the Environment: a Geochemical Perspective*. Geological Society, London, Special Publications, **236**, 595–606.
0305-8719/04/$15 © The Geological Society of London 2004.

advances in recent years. This chapter focuses on these model investigations.

Minerals available for binding in the cement matrix

The hydrated Portland cement matrix contains calcium silicate hydrate (C–S–H, 50 wt%), port-landite ($Ca(OH)_2$, 20 wt%) and Ca aluminates. The most important Ca aluminates are ettringite ($3CaO·Al_2O_3·3CaSO_4·32H_2O$, 4 wt%), mono-sulphate ($3CaO·Al_2O_3 · CaSO_4·12H_2O$, 7 wt%) and Ca carboaluminate ($3CaO·Al_2O_3·CaCO_3·11H_2O$, 7 wt%) (Taylor 1997). Together they used to make up almost 90 wt% of the mineral phases in the hydrated OPC cement paste. Now, supplementary materials, such as fly ash, slag or limestone are added in significant quan-tities (specifications allow 10%). The additions alter bulk compositions and so the secondary phases (Glasser 1993). Other crystalline phases, such as hibschite ($Ca_3Al_2SiO_4(OH)_8$), hydro-garnet ($Ca_3Al_2(OH)_{12}$), strätlingite ($Ca_2Al_2SiO_2$ $(OH)_{10}$) or layered double hydroxides ($Mg_2Al(OH)_7$), while still minor, are of increased importance. Portlandite has not been found to play an important role in the binding process. To date C–S–H, ettringite and mono-sulphate are considered to have the greatest potential for heavy metal binding, due partially to their abundance but also because of their structure.

Calcium silicate hydrate

The structure of C–S–H has been discussed in detail in various excellent publications (Glasser 1993; Richardson & Groves 1993; Taylor 1997) and will only be discussed here in terms of its potential for binding metal and metalloid ions.

The most commonly accepted model of the C–S–H is that suggested by Taylor (1997). He suggests that C–S–H is likely to have two kinds of local structure present in different layers, one related to 14 Å tobermorite ($Ca_5Si_{5.5}O_{17}H_2·8H_2O$) and the other to jennite ($Ca_9(Si_6O_{18}H_2)$ $(OH)_8·6H_2O$). The basic structure of both crystalline minerals is one of alternating CaO layers and SiO_2 chains. Figure 1 shows the

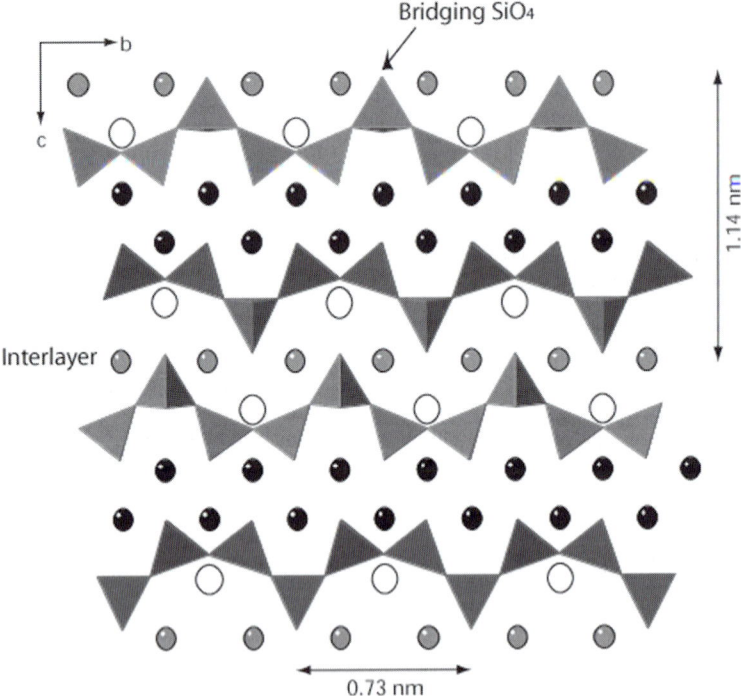

Fig. 1. Schematic representation showing two layers of 11 Å tobermorite adapted from Ziegler (2000) after Hamid (1981). The tetrahedra represent SiO_4 units, the grey and black spheres represent Ca atoms in the interlayers and main layers respectively and the open circles represent H_2O molecules.

structure of a single layer of 11Å tobermorite (those of the aforementioned minerals have not been exactly determined), which is thought to be most likely to represent that of C–S–H in a mature cement paste. The figure shows a layer structure in which the central layer has the empirical composition CaO_2 and is flanked on both sides by parallel rows of infinite SiO_4 chains that are polymerized to form sheets by bridging SiO_4 groups. In immature cement pastes these bridging groups are largely missing and dimers predominate. They still make up about 40% of the chains after 20–30 years. As the paste matures, the bridging silicates form pentamers or higher polymers. It is currently thought that these bridging positions may be occupied by AlO_4 groups and that the difference in charge is balanced in the interlayers by an alkali cation or Ca^{2+}. It is also thought that anions such as OH^-, SO_4^{2-}, or CO_3^{2-} may be incorporated into the interlayers to introduce the layered or brucite or hydrotalcite-like structures. The suggested contribution of these additional structures is thought to be dependent on the system and the composition of the grain from which the C–S–H was derived. Glasser (1993) has suggested that the nanostructure of the C–S–H gel is a stack of irregular C–S–H crystalline domains of more or less two-dimensional order of a few nanometres in size with a large volume of micropores with unsatisfied surface charges that may be available for binding.

From the above description, it becomes apparent that with such disorder, the sorption of both heavy metal cations and oxyanionic species is favoured. Owing to the lack of knowledge of the structure of the C–S–H gel itself, however, it is unlikely that an exact binding mechanism could be discerned at present.

Ettringite

The crystal structure of ettringite is shown in Fig. 2. It is often refered to in terms of cement nomenclature as an AFt phase (A = Al_2O_3, perhaps substituted by F (Fe_2O_3); t = trisulphate). It is made of columns composed of $Al(OH)_6$ octahedra alternating with triangular groups of edge-sharing CaO_8 polyhedra (4Os coming from H_2O) in hexagonal array. In the channels SO_4^{2-} ions balance the charge. Water is also present in the interchannels.

A short review by Taylor (1997) illustrates that bivalent cations may be exchanged for Ca^{2+}, trivalent cations and $Si(OH)_4$ for Al^{3+} and a number of anions have been exchanged for SO_4^{2-} ions. To date most investigations of ettringite as a heavy metal or metalloid binding agent have considered the exchange of anions for SO_4^{2-}. This will be discussed in detail later in the text.

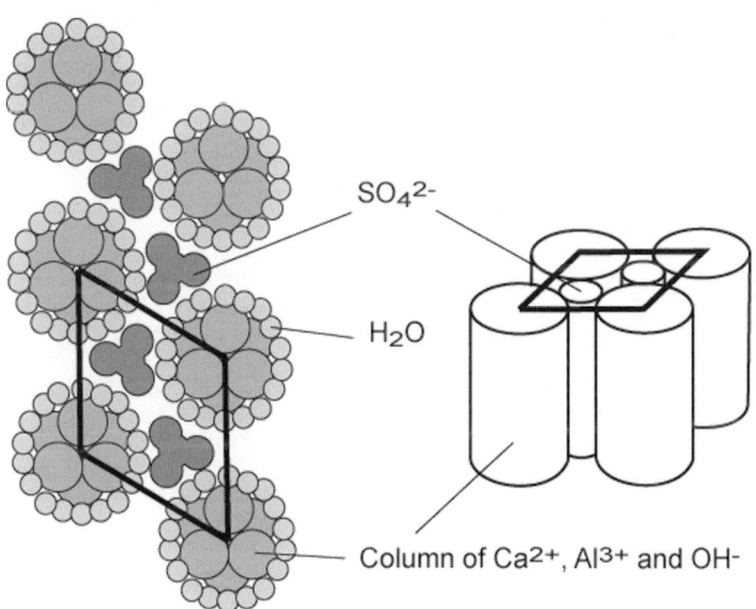

Fig. 2. Schematic representation of ettringite adapted from Myneni *et al.* (1997) after Moore & Taylor (1970).

Monosulphate and similar minerals

This class of minerals forms hexagonal platelets (Fig. 3) and are referred to as AFm (m = monosulphate) phases in cement nomenclature. The structure is derived from $Ca(OH)_2$ but one in three of the Ca^{2+} cations is replaced by an Al^{3+} or a Fe^{3+} cation. The charge is balanced by anions in the interlayer. Monosulphate ($3CaO \cdot Al_2O_3 \cdot CaSO_4 \cdot 12H_2O$) and $3CaO \cdot Al_2O_3 \cdot CaCO_3 \cdot 12H_2O$ are common phases in cement pastes. Strätlingite (or gehlenite hydrate) is also an AFm mineral with aluminosilicate groups in the interlayer. Again, as for ettringite, SO_4^{2-} or CO_3^{2-} groups may be exchanged for other anions.

Binding mechanisms in the cement matrix

In general terms, the binding of heavy metals and metalloid ions in the cement matrix is quite well understood. There appear to be three basic types of binding mechanism (e.g., Cocke & Mollah 1993; Cougar et al. 1996; Glasser 1997). A metal ion may be:

(1) Precipitated in the alkaline cement matrix as an oxide, mixed oxide or as other solid phases;
(2) Sorbed or precipitated onto the surfaces of cement minerals; or
(3) Incorporated into hydrated cement minerals.

The solubility of heavy-metal-containing solid phases (1) does seem to be a limiting factor with regard to the second and third mechanisms (Glasser 1994), so only ions that do not precipitate as oxides will be incorporated in or sorbed to hydrated cement minerals to a significant degree.

Figure 4 illustrates the effect of different binding mechanisms on the porewater concentrations of metals and metalloid species in relation to the solid-phase concentrations. First, if a certain component were very soluble and not bound to the solid in any way, the pore water concentration at equilibrium would equal the total concentration in the system. However,

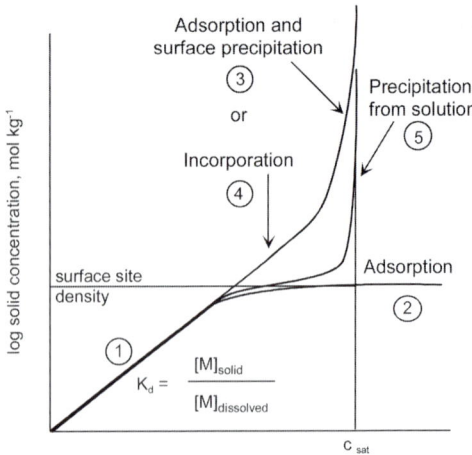

Fig. 4. Schematic adsorption isotherms adapted from Stumm (1992). The dissolved concentration at saturation is represented by c_{sat}.

as soon as sorption takes place, by mechanism (2) or (3), a relationship between the aqueous and the solid phase is established. This relationship may be described by the distribution ratio (①). The distribution ratio is dependent on time, pH, porewater and cement composition.

If a species sorbs to a crystalline mineral surface, in an ideal case, no more sorption occurs when all the available surface sites are occupied and concentrations in the solid phase remain constant (②). In reality surface precipitation may occur, resulting in a sorption isotherm schematically shown in Fig. 4 (③).

Incorporation in cement minerals will lead to a similar relationship, which may be described by a distribution ratio with the exception that uptake may be much greater than that of surface sorption (④). Such mechanisms may apply to C–S–H, AFt, and AFm phases. The mechanism of incorporation may be by isomorphic substitution of a particular species within a crystal lattice. A good example here is the exchange of SO_4^{2-} in ettringite for another anion. Another possibility is the adsorption to sites within a crystal structure, as may occur at silicate sites within C–S–H.

Finally, if a solid phase is precipitated, the concentration of the component will remain constant in the aqueous phase irrespective of its concentration in the solid phase as long as the porewater composition remains constant (⑤). This limiting case is easy to interpret and model. From Fig. 4 it can be seen that should precipitation occur at low dissolved concentrations,

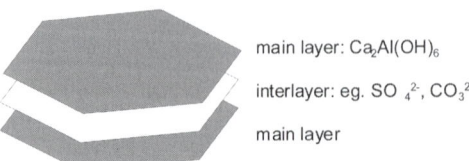

main layer: $Ca_2Al(OH)_6$

interlayer: eg. SO_4^{2-}, CO_3^{2-}

main layer

Fig. 3. Schematic representation of AFm (after Taylor 1997).

and the vertical line shifts to the left, other mechanisms of binding will be less significant with respect to controlling the porewater concentration of a given ion. It is therefore very helpful to examine the solubility of the components under consideration.

The solubility of heavy metal and metalloid components

Heavy metal cations precipitate readily as hydroxides or carbonates in alkaline media. Dissolved carbonate content will be limited by calcite precipitation or by conversion of hydroxyl AFm to carbonate AFm. Hydroxide ions, on the other hand, are abundant. Here only the solids that may be present under oxic conditions will be discussed. Figure 5a shows the total dissolved heavy metal cation concentrations that would prevail if hydroxide precipitation were to be the dominant solubility-controlling process. Figure 5b shows the solubility of Ca metallate species, as these are likely to act as solubility-controlling phases for oxyanionic species.

From these figures it is easy to see that Co, Ni, Cd, AsO_4, MoO_4, and WO_4 are more likely to be solubility controlled than Pb, Zn, Cr(III), SeO_3, SeO_4, CrO_4, and AsO_3. It should be pointed out, however, that there is still much to be learned about the formation and thermodynamic stability of mixed hydroxide phases. One commonly observed phase, Ca zincate ($CaZn_2(OH)_6$) (e.g., Cocke & Mollah 1993), is thermodynamically more stable than $Zn(OH)_2$

above pH 12.5 (Ziegler & Johnson 2001). Another group of minerals is that of the hydrotalcite-like minerals, the layered double hydroxides (LDH, $M_2^{2+}M^{3+}1/yX^{y-}(OH)_6$ where X is an anion): cobalt, Ni and Zn can form such minerals (Johnson & Glasser 2003). To date there has been one report of a Ni/Al LDH (where X is probably OH) occurring in a cementitious matrix (Scheidegger et al. 2000).

Binding mechanisms of ions to specific cement minerals

Because of the complexity of hydrated cement pastes and the variety of possibilities for binding, a study of the binding of metal and metalloid ions to specific cement minerals is advantageous. The binding to single components of the hydrated cement paste can be compared to the leachability of an ion bound in a hydrated cement paste and deductions made as to the dominant mechanism that limits the solubility in the porewater.

Only a few studies of the binding of metal and metalloid ions to cement minerals have been carried out to date. The amorphous nature of C–S–H makes the investigation of binding mechanisms difficult and only Pb(II) and Zn(II) have been investigated in detail. Set against this, there is more information regarding the binding of oxyanions to ettringite and monosulphate, partially because these minerals are crystalline and can thus be characterized by standard methods. An outline of the studies is given below.

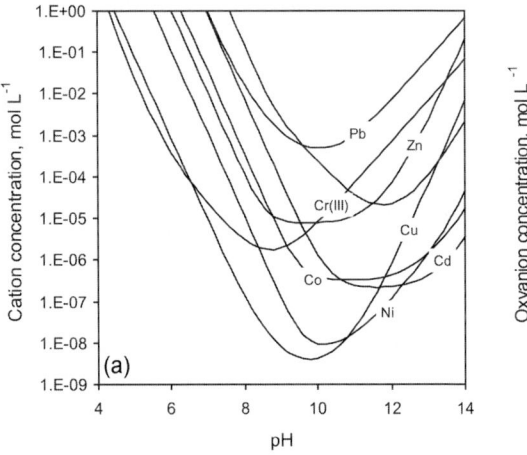
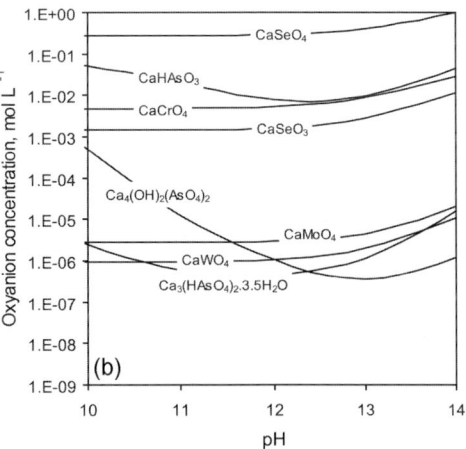

Fig. 5. Solubility of heavy metal cations (**a**) and anions (**b**) in equilibrium with their corresponding hydroxides and Ca metallates respectively as a function of pH. Conditions assumed were $I = 0.15$ M and $[Ca] = 0.02$ M. The dashed line indicates the solubility curve for $CaZn_2(OH)_6$.

The binding of ions to CSH

Studies of Zn(II) binding to C–S–H to date are in agreement that it appears to bind to the silicate layers (Moulin *et al.* 1999; Ziegler *et al.* 2001*b*; Tommaseo 2002). Each of these three research groups carried out extended X-ray absorption fine structure (XAFS) measurements and determined that Zn was tetrahedrally coordinated to O in the first atomic shell and that Si and not Ca was to be found in the second shell. The authors agree that Zn is bound to the silicate chains, although opinions diverge as to whether the binding is at the bridging position or at the end of the silica chains. In addition, Moulin *et al.* (1999) found a change in spectra on addition of Zn(II) and Pb(II) when using ^{29}Si nuclear magnetic resonance. They attributed this to the formation of Zn–O–Si and Pb–O–Si bonds. It should also be noted that there does not appear to be an exchange of Ca for Zn cations. Ziegler *et al.* (2001*a*) found, using electron probe micro analyses (EPMA), that the Ca : Si ratio in the solid phase did not change significantly even when Zn was incorporated into C–S–H particles at more than 3 wt%.

Sorption isotherms obtained by Ziegler *et al.* (2001*a*) are shown in Fig. 6. A number of features can be discerned. First, there is a direct relationship between solid and dissolved concentrations at low Zn(II) concentrations. The extent of sorption is pH-dependent. Ziegler *et al.* (2001*a*) also reported that sorption is time-dependent, as Zn(II) diffuses into the C–S–H particles from solution. At high concentrations, Zn(II) precipitates as β_2-Zn(OH)$_2$ or CaZn$_2$(OH)$_6$. There is a good agreement, for a given pH, between the K_d values (Fig. 4) determined from experiments in which Zn(II) is sorbed to C–S–H and those in which C–S–H containing Zn(II) is precipitated directly from solution (Johnson & Kersten 1999; Ziegler 2000). This would indicate that Zn(II) occupies the same kind of sorption site, independent of the method of synthesis. Ziegler *et al.* (2001*a*) report a maximum Zn(II) sorption at a Si : Zn ratio of approximately 6 : 1.

The sorption of Pb(II) to C–S–H has been investigated by Moulin (1999) and Pointeau (2000). In fact XAFS studies indicate that, as for Zn(II), Pb(II) is adsorbed to the silicate chains within the C–S–H structure (Rose *et al.* 2000). Figure 7 shows sorption isotherms for C–S–H of different Ca : Si ratios. It appears that sorption is greater at a lower Ca : Si ratio.

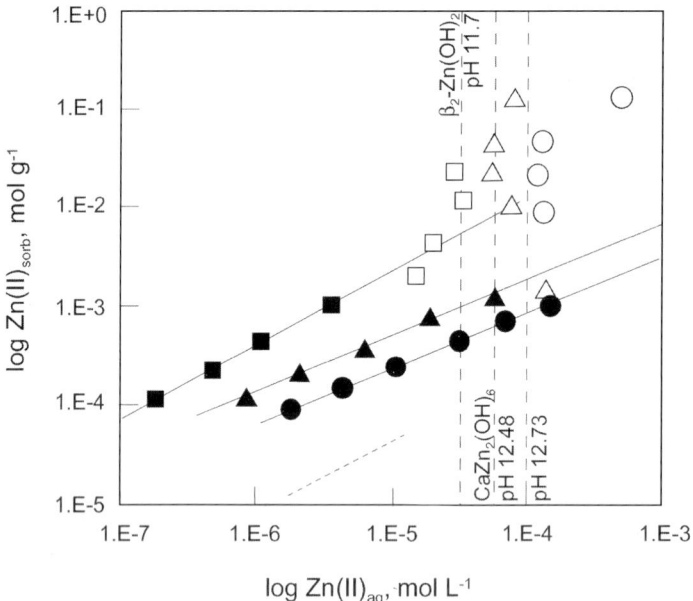

Fig. 6. Sorption isotherms of Zn(II) binding to C–S–H (Ca : Si = 1 : 1) at pH 11.7 (■), 12.48 (▲), and 12.78 (●) after 87 days equilibration. Open symbols represent experiments in which Zn solid phases were detected by X-ray diffraction (XRD). The figure is adapted from Ziegler (2000). The dashed line represents data from Moulin *et al.* (1999) with Ca : Si = 1.7 : 1, pH 12 and an equilibration time of 72 h.

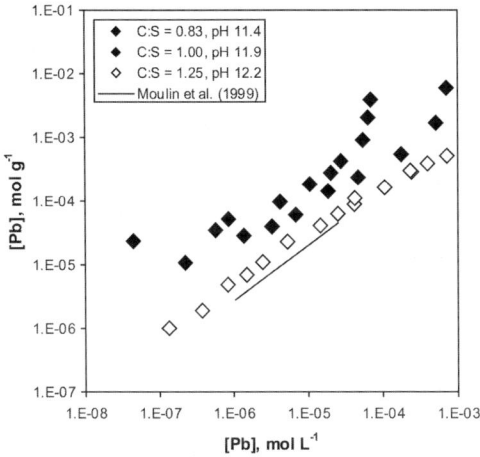

Fig. 7. Sorption isotherms of Pb(II) as a function of Ca : Si ratio, adapted from Pointeau (2000). The equilibration time was 24 h. The dotted line represents data from Moulin *et al.* (1999) for C–S–H with a Ca : Si ratio of 1.7 : 1, pH 12 and an equilibration time of 72 h.

However, the pH varies as well and might be the cause of the difference in sorption isotherms. Further investigations would be necessary. The sorption isotherms determined by Moulin *et al.* (1999) appear to be slightly lower for both Zn(II) (Fig. 6) and Pb(II) (Fig. 7), but the agreement is quite good.

The recent detailed investigations on Zn(II) and Pb(II) have advanced our understanding of heavy metal cation binding to C–S–H. However there are still a number of important questions that require attention. We need to understand the relationship, if any between Ca : Si ratio and metal uptake. We also need to understand how to interpret the sorption process, whether as a solid solution (Tommaseo & Kersten 2002) or as precipitation within the C–S–H particles or as a sorption process suggested by Glasser (1993) to the 'surfaces' of the crystalline domains.

Further XAFS studies by Bonhoure *et al.* (2003) suggests that Sn(IV) sorbs to silicate groups within C–S–H. Preliminary data (Rigo *et al.* 2000) suggest that Cu(II) is taken up by C–S–H, but it is not yet clear whether such a process influences the porewater concentrations of the relatively insoluble cations such as Cd(II). Moulin (1999) finds that Pb(II), Zn(II), Cu(II) and Cr(III) all have a strong affinity for the hydration products of alite ($3CaO \cdot SiO_2$). If, however, the four cations are added separately to a suspension containing unhydrated alite, then Cu precipitates as CuO.

The role of C–S–H in the binding of anions remains to be studied. While Baur & Johnson (2003*a*) found that uptake of SeO_4^{2-} was insignificant, the sorption of SeO_3^{2-} was stronger, in keeping with its greater propensity for surface adsorption. Under extreme conditions CrO_4^{2-} can bind to C–S–H within a short space of time, although how this sorption process develops with time is unclear (Omotoso *et al.* 1998*a* 1998*b*; Moulin 1999).

The incorporation of species in AFt and AFm phases

The AFt phases containing SO_4^{2-}, OH^-, SO_3^{2-}, NO_3^- and CO_3^{2-}, $B(OH)_4^-$ but also SeO_4^{2-} and CrO_4^{2-} have been synthesized (Pöllmann *et al.* 1989; Hassett *et al.* 1990; Motzet & Pöllman 1999; Perkins & Palmer 2000; Baur & Johnson 2003*b*). Some AFt phases (SO_3^{2-}, NO_3^- SeO_4^{2-}, CO_3^{2-}) could only be synthesized using the 'saccharate' method in which Ca is solubilized as a saccharate complex, while others were readily formed (SO_4^{2-}, CrO_4^{2-}) in suspensions of CaO, Na aluminate and a Na salt of the appropriate anion (known as the 'paste reaction'; Atkins *et al.* 1991). Hydroxide-AFt was found to be sensitive to carbonation and Cl^--AFt could not be formed at all. Borate forms mixed phases with OH^- (Csetenyi & Glasser 1993). Available solubility products are listed in Table 1. The lowest value, that is, the most stable phase, is obtained for SO_4^{2-}-AFt, followed by CrO_4^{2-}-AFt and SeO_4^{2-}-AFt.

Solid solutions in AFt phases of OH^- and $B(OH)_4^-$, of SO_4^{2-} and $B(OH)_4^-$ and of SO_4^{2-} and CrO_4^{2-} have been synthesized (Csetenyi & Glasser 1993; Pöllmann *et al.* 1993). Evidence of $Si(OH)_4$ substitution was determined by Atkins *et al.* (1993) in SiO_2-modified OPC, but the crystal size was much finer in comparison to Aft phases in unmodified OPC and it was believed that CO_3^{2-} substitution had also taken place. Halides also substitute for SO_4^{2-} to form $3CaO\cdot(Al_2O_3\cdot3CaX_2\cdot30H_2O$ (where X represents Cl^- or F^-) according to Bensted (1977), although the AFm phases such as Friedel's salt ($3CaO\cdot Al_2O_3\cdot CaCl_2\cdot10H_2O$) appear to be favoured in OPC (Suryavanshi *et al.* 1995).

Kumarathasan *et al.* (1990) found that substitution for SO_4^{2-} in AFt is usually partial. The maximum amount of AsO_4^{3-}, for example, was only 10% and approximately one-sixth of SO_4^{2-} could be exchanged for VO_4^{3-}. They found that MoO_4^{2-}-AFt could not be synthesized and powellite ($CaMoO_4$) precipitated instead. This could be related to the relative insolubility

Table 1. *Thermodynamic solubility data for AFt and AFm phases*

$3CaO \cdot Al_2O_3 \cdot 3CaX \cdot yH_2O$			$3CaO \cdot Al_2O_3 \cdot CaX \cdot yH_2O$		
X	y	K_{so}	X	y	K_{so}
SO_4^{2-}	32	57.45[a]	SO_4^{2-}	12	72.57[a]
CrO_4^{2-}	29	60.54[b]	CrO_4^{2-}	15	71.62[c]
SeO_4^{2-}	37.5	61.29[d]	SeO_4^{2-}	13–14	73.40[d]
$B(OH)_4^-$ ($\times 4$), OH^- ($\times 2$)	30	90.02[e]	$B(OH)_4^-$, OH^-	11.5	88.69[e]
$B(OH)_4^-$ ($\times 2$), OH^- ($\times 4$)	30	102.39[e]			
			MoO_4^{2-}	10–14	69.48[f]
			OH^- ($\times 2$)	12	104.37[g]

[a]Damidot & Glasser (1993).
[b]Perkins & Palmer (2000).
[c]Perkins & Palmer (2001).
[d]Baur & Johnson (2003b).
[e]Csetenyi & Glasser (1993).
[f]Kindness et al. (1994) (one data point, to be confirmed).
[g]Zhang (2000).

of the Ca metallates of these oxyanions. Borate, SeO_4^{2-} and CrO_4^{2-} substituted more extensively, but it appeared that Na had to be incorporated for charge balance because of an Al^{3+} deficiency, particularly for the CrO_4^{2-}-AFt phase. The occupancy of the channels appeared to be variable. The thermodynamic stability of ettringite is significantly greater than that of CrO_4^{2-}-AFt and SeO_4^{2-}-AFt, indicating that the SO_4^{2-} ion has a size, electronegativity and possibly a geometry that is particularly suited to the stabilization of the AFt structure and making an exchange for another anion less favourable (Zhang 2000).

Monosulphate and other AFm phases have received relatively little attention until recently. In a PhD thesis, Zhang (2000) describes the equilibration of the anions $B(OH)_4^-$, CrO_4^{2-}, MoO_4^{2-}, SeO_4^{2-} and SO_4^{2-} with hydrocalumite $(Ca_4Al_2(OH)_{12}(OH)_2 \cdot 6H_2O)$ in mixtures between OH^- and the appropriate oxyanion from 0 to 100%. The samples were equilibrated for 260 days. The author noted that with increasing amounts of oxyanion, the formation of hydrogarnet $(Ca_3Al_2(OH)_{12} \cdot 6H_2O)$ decreased. (Hydrogarnet is observed with time, as hydrocalumite is thermodynamically unstable with respect to hydrogarnet). Borate/OH^--AFm was the dominant species in the $B(OH)_4^-$ series, with $B(OH)_4^-$-AFt precipitating at high $B(OH)_4^-$ concentrations. Similar observations were made for SeO_4^{2-}. Chromate/OH^--AFm was the dominant phase of the chromate series and a pure CrO_4^{2-}-AFm was obtained as an end-member of the solid solution series. The thermodynamic stability of CrO_4^{2-}-AFm is reflected in the solubility products (Table 1). Molybdate did form a mixed AFm phase with OH^-, but its range of stability was limited. Powellite $(CaMoO_4)$ precipitation

was observed at high Mo concentrations. Kindness et al. (1994) observed a mixture of powellite and MoO_4^{2-}-AFm in OPC. Hydrocalumite and monosulphate formed two immiscible phases with only very limited solid solution formation at either end of the series. Glasser (1999) obtain similar results for OH^- and SO_4^{2-} and they suggest that extensive solid solution is most likely to occur between anions of the same charge and shape, while a difference in shape limits solid solution formation. These authors also point out that the anionic content of AFm is very sensitive to the composition of the porewater.

In conclusion it may be said that work is still in progress with regard to understanding anion substitution in AFm and AFt phases.

If the solubilites of the AFm and AFt phases are compared (Fig. 8), it can be seen that under the modelling conditions, AFm phases result in lower dissolved oxyanion concentrations. Selenate and chromate appear to form AFm phases readily (Baur 2002; Zhang 2000). From a perspective of oxyanion immobilization, it would therefore be preferable that the conditions within a cement matrix should favour AFm formation. However, wastes containing high concentrations of SO_4^{2-} form ettringite in favour to monosulphate (e.g., Baur et al. 2001) when mixed with cement. This may unwittingly lead to a higher solubility of oxyanions. It should be pointed out, however, that our understanding is incomplete and we are not yet in a position to tailor the formation of AFm and AFt phases in cement-waste mixes. It is fortunate that AFm phases are predominant in hydrated cement pastes. However, one important question in this respect is why monosulphate is persistent in hydrated cement pastes, even though it is

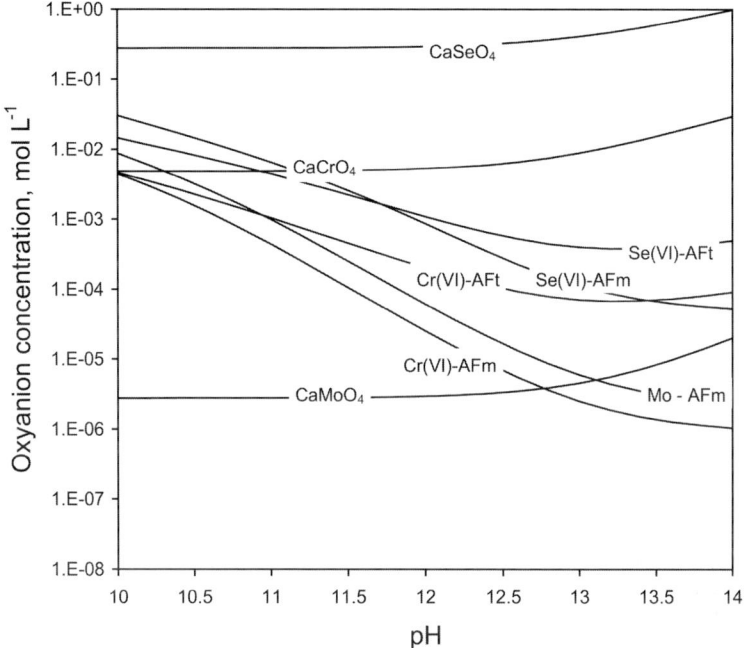

Fig. 8. Solubility of Ca metallates, AFm and AFt phases of MoO_4^{2-}, SeO_4^{2-}, and CrO_4^{2-}.

thermodynamically metastable with respect to ettringite below 40 °C (Turriziani 1964; Lea 1970; Atkins *et al.* 1991; Damidot & Glasser 1993). Thus we are not yet in the position to predict the solubility of oxyanions that are bound as solid solution in AFm and AFt phases or the long-term mobility of oxyanions. Ochs *et al.* (2002), for example, observed that Se(VI) uptake in fresh cement pastes was high, followed by a decrease within a year. The cause of this observation most probably has to do with a change in mineralogy, particularly with respect to AFt and AFm phases of SO_4^{2-} and SeO_4^{2-}.

Finally it should be noted that in real systems, solid solutions between SO_4^{2-} or OH^- and a given contaminant oxyanion is likely to be the rule, whether AFt or AFm phases dominate. Little is known about the thermodynamics of such mixed solids and this certainly deserves further attention.

Surface sorption

Adsorption to the surfaces of ettringite and monosulphate is an alternative immobilization mechanism that has been examined for SeO_3^{2-}, SeO_4^{2-} and of AsO_4^{3-} (Myneni *et al.* 1997; Baur 2002; Baur & Johnson 2002a). The studies estimated a maximum surface site concentration of approximately 0.03–0.1 mol/kg

for synthetic ettringite from the surface area and the crystal unit structure. Baur & Johnson (2002a) estimate a surface Ca and Al concentration of 0.1 mol/kg and a surface SO_4^{2-} concentration of 0.015 mol/kg for monosulphate.

Myneni *et al.* (1997) report that the sorption of AsO_4^{3-} to ettringite was strong. Below surface saturation, over 90% was sorbed to the surface, but when higher concentrations of As(V) were added, precipitation of an As(V) phase occurred. Baur & Johnson (2002a) also observed sorption to surface saturation for SeO_3^{2-} to ettringite, and they found that the SeO_4^{2-} sorbed much less strongly. Selenite sorbed similarly strongly to monosulphate, but SeO_4^{2-} sorbed more strongly and X-ray diffraction evidence could be used to show that SeO_4^{2-} was incorporated into the monosulphate structure. For SeO_3^{2-} they reported K_d values of 0.18 and 0.38 m³/kg for ettringite and monosulphate respectively, whilst for SeO_4^{2-} the K_d values were 0.03 and 2.06 m³/kg, respectively. The reported K_d values for SeO_3^{2-} are in agreement with values determined with cement pastes at pH 12.4 (Johnson *et al.* 2000). These investigations show that even if a given oxyanion does not incorporate within the structure of an AFm or an AFt phase, it may still sorb to the surface. Distinguishing between adsorption or

incorporation can only be done using spectroscopic and microscopic techniques.

Outlook

Enormous progress has been made in the last 10
years and for most elements an approximate
range of solubility is known. However, it is not
possible, particularly in the case of the soluble
species, to predict solubility from the available
data. Prediction is important for the formulation
of waste products stabilized with cement, for
example, or for setting limits on solid-phase concentrations. In order to predict ionic concentrations, binding mechanisms must be better
quantified. It is important to be able to quantify
the sorption of cations to C–S–H. For oxyanions,
AFt and AFm phases are important for controlling
solubility, but it is not yet clearly understood what
the propensity of each species is, to form either
pure AFt and AFm phases as opposed to the formation of solid solutions with SO_4^{2-} in ettringite
and monosulphate. The relative stability of
the AFt or AFm phases in general, and ettringite
and monosulphate in particular, needs to be
addressed. Lastly, it would be useful to know to
what extent heavy metal cations substitute for
Ca^{2+} and Al^{3+} in the Ca sulphoaluminate
hydrates and whether this process is important
relative to binding in the C–S–H matrix.

Cemsuisse are acknowledged for their financial support.
F. Glasser and U. Mäder are thanked for their thorough
reviews.

References

ATKINS, M., MACPHEE, D., KINDNESS, A & GLASSER,
    F. P. 1991. Solubility properties of ternary and quater-
    nary compounds in the $CaO–Al_2O_3–SO_3–H_2O$
    system. Cement and Concrete Research, 21, 991–998.
BAUR, I. 2002. The Immobilisation of Heavy Metals
    and Metalloids in Cement-Stabilized Wastes:
    A Study Focussing on the Selenium Oxyanions
    SeO_3^{2-} and SeO_4^{2-}. PhD thesis, Swiss Federal Insti-
    tute of Science and Technology (ETHZ), Zürich,
    Switzerland.
BAUR, I., LUDWIG, C. & JOHNSON, C. A. 2001. Leach-
    ing behaviour of cement-stabilized incinerator
    ashes: A comparison of field and laboratory
    measurements. Environmental Science and Tech-
    nology, 35, 2817–2822.
BAUR, I. & JOHNSON, C. A. 2003a. Sorption of selenite
    and selenate to cement minerals. Environmental
    Science and Technology, 37, 3442–3447.
BAUR, I. & JOHNSON, C. A. 2003b. The solubility of
    selenate ettringite and selenate monosulfate.
    Cement and Concrete Research, 33, 1741–1748.

BENSTED, J. 1977. Chloroaluminates and the role of
    calcium chloride in accelerated hardening of Portland
    cement. World Cement Technology, 8, 171–175.
BONHOWE, I., WIELAND, E., SHEIDEGGER, A. M.,
    OCHS, M. & KUNZ, D. 2003. EXAFS study of
    Sn(IV) immobilization by hardened cement paste
    and calcium silicate hydrates. Environmental
    Science and Technology, 37, 2184–2191.
COCKE, D. L. & MOLLA, M. Y. A. 1993. The chemistry
    and leaching mechanisms of hazardous substances
    in cementitious solidification/stabilization systems.
    In: SPENCE, R. D. (ed) Chemistry and Microstruc-
    ture of Solidified Waste Forms. Lewis, Boca Raton,
    187–242.
COUGAR, M. D. L., SCHEETZ, B. E. & ROY, D. M.
    1996. Ettringite and C–S–H Portland cement
    phases for waste ion immobilization: A review.
    Waste Management, 16, 295–303.
CSETENYI, L. J. & GLASSER, F. P. 1993. Borate substi-
    tuted ettringites. Materials Research Society Sym-
    posium Proceedings, 294, 273–278.
DAMIDOT, D. & GLASSER, F. P. 1993. Thermodynamic
    investigation of the $CaO–Al_2O_3–CaSO_4–H_2O$
    system at 25 °C and the influence of Na_2O.
    Cement and Concrete Research, 23, 221–228.
EYERER, P. & REINHARDT, H. W. 2000. Ökologische
    Bilanzierung von Baustoffen und Gebäuden.
    Wege zur ganzheitlichen Bilanzierung. BauPraxis,
    Verlag Birkhölzer.
GLASSER, F. P. 1993. Chemistry of cement-solidified
    waste forms. In: SPENCE, R. D. (ed) Chemistry
    and Microstructure of Solidified Waste Forms.
    Lewis, Boca Raton, 1–39.
GLASSER, F. P. 1994. Immobilisation potential of
    cementitious materials. In: GOUMANS, J. J. J. M.,
    VAN DER SLOOT, H. A & AALBERS, TH. G. (eds)
    Environmental Aspects of Construction with
    Waste Materials. Elsevier, Amsterdam, 77–86.
GLASSER, F. P. 1997. Fundamental aspects of cement
    solidification and stabilisation. Journal of Hazar-
    dous Materials, 52, 151–170.
GLASSER, F. P. 1999. Scientific basis for nuclear waste
    management XXII. In: WRONKIEWIECZ, D. J. &
    LEE. J. H. (eds) The Solubility Limited Source
    Term for Cement Conditioned Wastes. Materials
    Research Society Proceedings, 556, 1225–1236.
HAMID, S. A. 1981. The crystal structure of 11-Å
    natural tobermorite $Ca_{2.25}[Si_3O_{7.5}(OH_{1.5})]\cdot H_2O$.
    Zeitschrift für Kristallographie, 154, 189–198.
HASSETT, D. J., McCARTHY, G. J., KUMARATHASAN,
    P. & PFLUGHOEFT-HASSETT, D. 1990. Synthesis
    and characterization of selenate and sulphate-
    selenate ettringite structure phases. Materials
    Research Bulletin, 25, 1347–1354.
JOHNSON, C. A. & KERSTEN, M. 1999. A possible
    mechanism of Zn binding to Ca silicate hydrates.
    Environmental Science and Technology, 33,
    2296–2298.
JOHNSON, C. A. & GLASSER, F. P. 2002. Synthesis,
    characterisation and solubilities of hydrotalcites
    $(M_2Al(OH)_6(CO_3)_{0.5}$ where M = Mg, Zn, Co,
    Ni). Clays and Clay Minerals, 51, 1–8.
JOHNSON, E. A., RUDIN, M. J., STEINBERG, S. M. &
    JOHNSON, W. H. 2000. The sorption of selenite

on various cement formulations. *Waste Management*, **20**, 509–516.

KINDNESS, A., LACHOWSKI, E. E., MINOCHA, A. K. & GLASSER, F. P. 1994. Immobilisation and fixation of molybdenum(VI) by portland cement. *Waste Management*, **14**, 97–102.

KUMARATHASAN, P, MCCARTHY, G. J., HASSETT, D. J. & PFLUGHOEFT-HASSETT, D. F. 1990. Oxyanion substituted ettringites: Synthesis and charactersiation and their potential role in immobilization of As, B, Cr, Se and V. *Materials Research Society Symposium Proceedings*, **178**, 83–104.

LEA, F. M. 1970. *The Chemistry of Cement and Concrete* (3rd ed). Arnold, London, UK.

MOORE, A. W. & TAYLOR, H. F. W. 1970. Crystal structure of ettringite. *Acta Crystallographica Section B: Structural Science*, **26**, 386–393.

MOTZET, H. & PÖLLMANN, H. 1999. Synthesis and characterisation of sulfite-containing AFm phases in the system $CaO-Al_2O_3-SO_2-H_2O$. *Cement and Concrete Research*, **29**, 1005–1011.

MOULIN, I. 1999. *Lead, Copper, Zinc, Chromium(III) and (VI) Speciation in Cement Hydrated Phases*. PhD thesis, Université Aix-Marseille III, France.

MOULIN, I., STONE, W. E. E., SANZ, J., BOTTERO, J.-Y., MOSNIER, F. & HAEHNEL, C. 1999. Lead and zinc retention during hydration of tri-calcium silicate, a study by sorption isotherms and ^{29}Si NMR spectroscopy. *Langmuir*, **15**, 2829–2835.

MYNENI, S. C. B., TRAINA, S. J., LOGAN, T. J. & WAYCHUNAS, G. A. 1997. Oxyanion behaviour in alkaline environments: sorption and desorption of arsenate in ettringite. *Environmental Science and Technology*, **31**, 1761–1768.

OCHS, M., LOTHENBACH, B. & GIFFAUT, E. 2002. Uptake of oxo-anions by cements through solid–solution formation: experimental evidence and modelling. *Radiochimica Acta*, **90**, 639–646.

OMOTOSO, O. E., IVEY, D. G. & MIKULA, R. 1998a. Hexavalent chromium in tricalcium silicate Part 1: Quantitative X-ray diffraction analysis of crystalline hydration products. *Journal of Materials Science*, **33**, 507–513.

OMOTOSO, O. E., IVEY, D. G. & MIKULA, R. 1998b. Hexavalent chromium in tricalcium silicate Part II: Effects of CrVI on the hydration of tricalcium silicate. *Journal of Materials Science*, **33**, 515–522.

PERKINS, R. B. & PALMER, C. D. 2000. Solubility of $Ca_6[Al(OH)_6]_2(CrO_4)_3 \cdot 26H_2O$, the chromate analog of ettringite; 5–75 °C. *Applied Geochemistry*, **15**, 1203–1218.

PERKINS, R. B. & PALMER, C. D. 2001. Solubility of chromate hydrocalumite ($3CaO\ Al_2O_3\ CrO_4 \cdot nH_2O$; 5–75 °C. *Cement and Concrete Research*, **31**, 983–992.

POINTEAU, I. 2000. *Etude Mechanistique et Modelisation de la Retention de Radionucleides par les Silicates de Calcium Hydrates (CSH) des Ciments*. PhD thesis. Universite de Reims Champagne-Ardenne, France.

PÖLLMANN, H., KUZEL, H.-J. & WENDA, R. 1989. Compounds with ettringite structure. *Neues Jahrbuch für Mineralogie*, **160**, 133–158.

PÖLLMANN, H., KUZEL, H.-J. & WENDA, R. 1993. Solid solution of ettringites. Part II: Incorporation of $B(OH)_4^-$ and CrO_4^{2-} in $3CaO \cdot Al_2O_3 \cdot 3CaSO_4 \cdot 32H_2O$. *Cement and Concrete Research*, **23**, 422–430.

RICHARDSON, I. G. & GROVES, G. W. 1993. The incorporation of minor and trace elements into calcium silicate hydrate (C–S–H) gel in hardened cement pastes. *Cement and Concrete Research*, **23**, 131–138.

RIGO, E., GIES, H. *et al.* 2000. Beitrag zur Immobilisierung von Blei-, Zink- und Kupferionen in schlecht geordneten C–S–H-Phasen. *Zement-Kalk-Gips International*, **53**, 414–423.

ROSE, J., MOULIN, I. *et al.* 2000. X-ray absorption spectroscopy study of immobilization processes for heavy metals in calcium silicate hydrates: 1. Case of lead. *Langmuir*, **16**, 9900–9906.

SANCHEZ, F., GERVAIS, C., GARRABRANTS, A. C., BARNA, R. & KOSSON, D. S. 2002. Leaching of inorganic contaminants from cement-based waste materials as a result of carbonation during intermittent wetting. *Waste Management*, **22**, 249–260.

SCHEIDEGGER, A. M., WIELAND, E., SCHEINOST, A. C., DÄHN, R. & SPIELER, P. 2000. Spectroscopic evidence for the formation of layered Ni–Al double hydroxides in cement. *Environmental Science and Techology*, **34**, 4545–4548.

SPENCE, R. D. (ed) 1993. *Chemistry and Microstructure of Solidified Waste Forms*. Lewis Publishers, Boca Raton.

STEGEMANN, J. A., BUTCHER, E. J. *et al.* 2001. MONOLITH – *A Database and Interface for Cement-Based Products, Developed Under Contract No. BRPR-CT97-0570*. Commission of the European Community, Brussels, Belgium.

STUMM, W. 1992. *Chemistry of the Solid–Water Interface*. John Wiley & Sons, New York.

SURYVANSHI, A. K., SCANTLEBURY, J. D. & LYON, S. B. 1995. The binding of chloride ions by sulphate resistant portland cement. *Cement and Concrete Research*, **25**, 581–592.

TAYLOR, H. F. W. 1997. *Cement Chemistry*, 2nd ed. Thomas Telford, London, p. 459.

TOMMASEO, C. E. 2002. *EXAFS – Untersuchungen zur Rolle von Silizium bei der Sorption von Umweltrelevanten Schwermetallen (Zn, As, Pb) in Speichermineralien (FeOOH, CSH)*. PhD thesis, Johannes Gutenberg-University, Mainz, Germany.

TOMMASEO, C. E. & KERSTEN, M. 2002. Aqueous solubility diagrams for cementitious waste stabilization systems. 4. Mechanism of zinc immobilization by calcium silicate hydrate. *Environmental Science and Technology*, **36**, 2919–2925.

TRUSSEL, S. & BATCHELOR, B. 1996. Chemical characterization of pore water of a solidified hazardous waste. *In*: GILLIAM, T. M. & WILES, C. C. (eds) *Stabilization/Solidification of Hazardous, Radioactive and Mixed Wastes*. ASTM,

West Conshohocken, Pennsylvania, Vol. 3, ASTM STP 1240.

TURRIZIANI, R. 1964. The calcium aluminate hydrates and related compounds. *In*: TAYLOR, H. F. W. (ed) *The Chemistry of Cements*. Academic Press, London, Vol. 1, 233–286.

VAN DER SLOOT, H., HOEDE, D. *et al.* 2001. *Environmental Criteria for Cement Based Products ECRICEM Phase I: Ordinary Portland Cements*. Petten, The Netherlands, ECN-C–01-069.

ZHANG, M. 2000. *Incorporation of Oxyanionic B, Cr, Mo, and Se into Hydrocalumite and Ettringite: Application to Cementitious Systems*. PhD thesis, University of Waterloo, Waterloo, Canada.

ZIEGLER, F. 2000. *Heavy Metal Binding in Cement-Based Waste Materials: An Investigation of the Mechanism of Zn Sorption to Calcium Silicate Hydrate*. Diss. ETH No. 13 569, ETH Zürich, Switzerland.

ZIEGLER, F. & JOHNSON, C. A. 2001. The solubility of calcium zincate ($CaZn_2(OH)_6 \cdot 2H_2O$). *Cement and Concrete Research*, **31**, 1327–1332.

ZIEGLER, F., GIERÉ, R. & JOHNSON, C. A. 2001a. The sorption mechanisms of zinc to calcium silicate hydrate: Sorption studies and microscopic investigations. *Environmental Science and Technology*, **35**, 4556–4561.

ZIEGLER, F., SCHEIDEGGER, A. M., JOHNSON, C. A., DÄHN, R. & WIELAND, E. 2001b. The sorption mechanisms of zinc to calcium silicate hydrate: An X-ray absorption fine structure (XAFS) investigation. *Environmental Science and Technology*, **35**, 1550–1555.

Hydrological and geochemical factors controlling leachate composition in incinerator ash landfills

C. ANNETTE JOHNSON & KARIM C. ABBASPOUR

Swiss Federal Institute of Environmental Science and Technology (EAWAG), Dübendorf, Switzerland

(e-mail: johnson@eawag.ch)

Abstract: The hydrological and geochemical processes that control leachate composition are discussed for two examples, the first being a landfill containing municipal solid waste incinerator (MSWI) bottom ash, Landfill Lostorf, and the second a lysimeter containing blocks of cement-stabilized MSWI air pollution control (APC) residues. In both examples the leachate consists of a component that takes months to years to pass through the landfill and rainwater that passes through the landfilled material via preferential flow paths. The composition of the leachate is relatively constant and is only diluted with rainwater during rain events. The concentration of major leachate components (Na^+, K^+, Cl^- and OH^-) is controlled by diffusion processes and by the precipitation of solubility-controlling phases (Ca(II), Al(III), SO_4^{2-} and Si(IV)). The observed concentrations of some relatively insoluble heavy metal and metalloid species may be explained by the precipitation of secondary phases. More soluble species may be adsorbed on surfaces of or incorporated in minerals, but the mechanisms remain to be elucidated.

One of the main environmental concerns with regard to landfilling is the leaching and transport of contaminants from the landfill into groundwater. Thus leachate containment and treatment is the central feature of the modern landfill. An understanding of the leaching process is essential for the long-term assessment of landfill safety as well as for the design of new and the treatment of old landfills. The leaching process is dependent upon two factors: first the landfill hydrology, and secondly the geochemistry and reactivity of the landfill material. Landfill hydrology depends on the permeability, the homogeneity, and the physical characteristics of the landfilled material. The flow paths and interaction times determine the extent of the leaching process. The composition of the leachate depends on the composition of the waste, on diffusion rates and water residence times and water mixing processes within the landfill.

The aim of this chapter is to present a conceptual model of the leaching process in landfills based on two exemplary field studies at a municipal solid waste incinerator (MSWI) bottom ash landfill, Landfill Lostorf, and a lysimeter investigation, Landfill Teuftal, containing cement-stabilized MSWI air pollution control (APC) residues.

The field sites

Landfill Lostorf

Landfill Lostorf, a MSWI bottom ash monofill near Buchs AG, Switzerland, is situated in a disused gravel pit (Johnson *et al.* 1998). A liner consisting of 0.8 m opalinous clay supporting a gravel drainage layer (0.2 m) serves to collect the leachate via high-density polyethylene (HDPE) tubing into a shaft. Two geotextiles separate the MSWI bottom ash from the drainage layer and the drainage layer from the clay liner. The landfill has a depth of 6 m and has been successively filled to this depth from east to west in discrete stages (at 6 to 9 month intervals). The landfill has three compartments with separate leachate drainage. This is achieved by the topography of the liner that is inclined at a gradient of around 4% towards the compartment boundary. The experiments were carried out on the oldest compartment (5850 m^2) of the landfill with MSWI bottom ash produced in 1991. The surface of the landfill was not covered or cultivated until 2000, so there was direct contact between the atmosphere and the disposed ash throughout the duration of the experiments. Drilling in the landfill has shown that although

From: GIERÉ, R. & STILLE, P. (eds) 2004. *Energy, Waste, and the Environment: a Geochemical Perspective*. Geological Society, London, Special Publications, **236**, 607–617.
0305-8719/04/$15   © The Geological Society of London 2004.

the ash is in general unsaturated, ponding does exist. The extent of these formations is unknown.

Teuftal lysimeter

The lysimeter was located next to the landfill Teuftal in Mühleberg (Canton Berne). It was constructed in the winter of 1990/1991 (AGW 1991; Ochs *et al.* 1999) and dismantled in December 1997. The lysimeter had two compartments, one containing cement-stabilized APC residues in the form of cubic blocks with an edge length of 0.5 m. This compartment was approximately 1.5 m deep with an area of 15–16 m². The space around the edge and between the cement blocks (around 1 cm) was filled with sand. The compartment was covered with geotextile, gravel (0.8 m), and humus (0.3 m) layers. The cement-stabilized APC residue blocks were composed of APC residues (41%), Portland cement (22%), water (32%), and NaOH (3%).

The hydrology of incinerator ash deposits

A landfill can be seen as a spatially heterogeneous porous and, by design, unsaturated system. Flow paths and physical mechanisms of runoff generation play a crucial role for the hydrology and ultimately leaching reactions within a landfill. Figure 1 illustrates the flow

paths through a landfill. In simple terms water entering the landfill appears to flow through the landfilled material in two ways. The first is through the pore channels, commonly referred to as matrix flow. This flow depends on water content and the connectivity of water within these channels; the greater the connectivity, the greater the flow. This flow may be described by the Richards equation. It is described as a function of water content of the material (θ, $L^3 L^{-3}$), pressure head (h, L), and hydraulic conductivity (K, LT^{-1}). Assuming a vertical positive downward coordinate (z), the water content as a function of time (t, T), may be described as follows:

$$\frac{\partial \theta}{\partial t} = \frac{\partial}{\partial z}\left(K(h)\frac{\partial h}{\partial z} - K(h) \right) \qquad (1)$$

This approach has been used in the well-known model HELP (Hydrological Evaluation of Landfill Performance, Schroeder *et al.* 1994) and a number of complementary models (Nixon *et al.* 1997). These models mostly assume the landfilled material to be idealized layers with homogenous properties. One such model, HYDRUS, has been used to model flow through Landfill Lostorf, but it was found that it could not fully catch the dynamics of flow, particularly after rain events (Johnson *et al.* 2001). Water passes through the

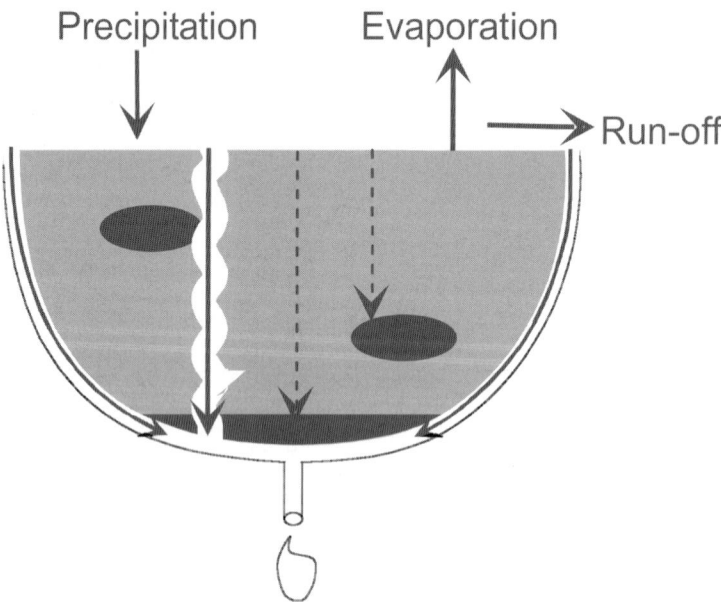

Fig. 1. Simplified representation of flow through a landfill. Light and dark grey areas represent unsaturated and saturated parts of the landfill, respectively. Solid and dashed lines represent preferential and matrix flow, respectively.

landfill in this way quite slowly and may have an average residence time of years. Salts present within the landfilled material may diffuse out and become concentrated in this water. The important physical factors that control the extent of the leaching process are diffusivity, the diffusion path length and the residence time of the water. This will be discussed later.

The second type of flow is that through preferential paths within and at the edge of the landfilled material. The amount of water that passes through preferential flow paths is dependent on the net input of precipitation to the landfill and on the water content of the landfilled material. There is evidence to suggest that landfills take up water until a water content is reached at which preferential flow is triggered, and that preferential flow is the cause of discharge during the initial period after landfilling. In landfill lysimeter tests, Maloszewski et al. (1995) have employed ^{18}O and ^{2}H isotope tracers to determine the extent of preferential flow. They found that the residence or transit time of water through their municipal solid waste and sewage sludge mix lysimeters (1.5–4 m high, 5 m in diameter) was 3–6 years. In addition they found that individual rain events required 3–7 weeks to be expressed in the leachate of the lysimeter and the proportion of preferential flow varied between 1 and 40% of the total flow.

Tracers for the determination of mixing processes

The mixing equation often used to distinguish waters of different origin (Pinder & Jones 1969) can be expressed as

$$Q_1/Q_{tot} = (c_{dis} - c_{rain})/(c_{t=0} - c_{rain}) \quad (2)$$

where Q_1 and Q_{tot} are the matrix flow discharge and the total or leachate discharge (L/min), respectively, and c the tracer concentration of the discharge during a rain event (c_{dis}), the rainwater (c_{rain}) and the discharge prior to the rain event ($c_{t=0}$), respectively. There are important general requirements for the application of the mixing equation for a particular rain event. The tracer concentrations of the rainwater and the leachate have to differ significantly and the rainwater has to have either a constant tracer content throughout the event or the temporal variations have to be negligible in comparison with those in total runoff. Only discrete rain events can be analysed. Lastly, the pre-event concentration in the outflow has to be representative of the leachate during the rain event.

Ash deposits are generally high in readily leachable salts. Leachate composition is generally quite saline and so on-line measurements of electrical conductivity can be used to distinguish between the components of leachate discharge, that is, matrix flow and preferential. Another tracer that has been used in the Lostorf study is oxygen isotopic composition. The values in the leachate drainage remain fairly constant as they are a result of the mixing of rainwaters over several years, whereas values in individual rain events can differ quite significantly (Johnson et al. 1998). In fact oxygen isotopic composition is by far the more reliable tracer, but electrical conductivity is very easily measured. In summer months it was found that both tracer measurements agreed quite well, whereas in winter months, when the landfill was partially saturated, the electrical conductivity measurements were generally less reliable.

Landfill Lostorf

The hydrological characteristics of this MSWI bottom ash landfill are clearly seen in Fig. 2. An increase in total or leachate discharge in response to rainfall is observed within hours. Within 30 to 100 hours approximately 50% of water from a rain event has been discharged. In winter months over 90% of rainfall was expressed in the landfill discharge, whereas between 9% and 40% was observed in summer months, depending on rainfall intensity. This is because of increased evaporation and a reduction of water content in the landfill. Over a hydrological year approximately 50% of incident rainfall was observed in the discharge, evaporation (and surface runoff) accounting for the remaining 50%. The discharge was less than 4 L/min for approximately 50% of the year in 1996. The average residence time of the water is estimated to be roughly three years.

The electrical conductivity is a mirror image of the total discharge. This illustrates the dilution of the matrix drainage caused by the preferential flow just after rain events. It is interesting to note that the matrix drainage increases as a result of a rain event. This increase may be regarded as a flushing process or possibly piston-flow as a result of an increase in pressure head. The electrical conductivity recovers after each rain event but there is a slight downward trend, typical for the wet winter months. After prolonged dry periods the electrical conductivity may be as high as 15 mS/cm. Looking more closely at the different discharge components for the May 1995 rain event (Fig. 3), it can be seen that the maximum of the matrix discharge is slightly before that of

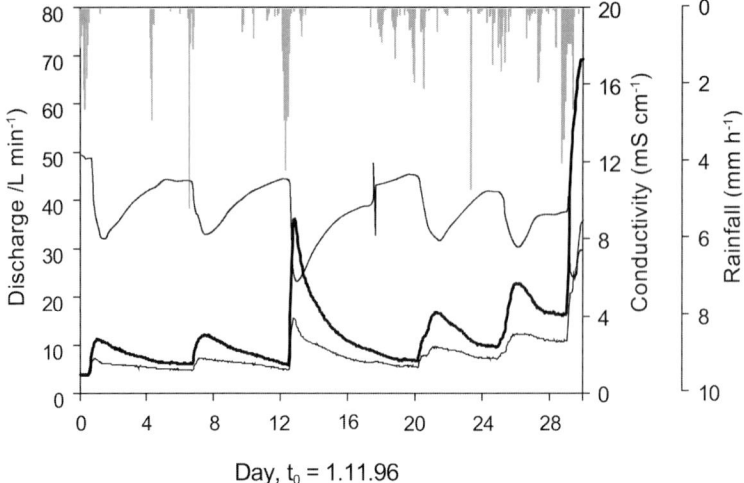

Fig. 2. The response of leachate drainage (——), electrical conductivity (– – –) and matrix discharge (——) to rainfall at Landfill Lostorf in November 1996.

the discharge through preferential paths. This indicates that a piston-flow type mechanism may be in operation. It should be noted that for this rain event, agreement between the two tracers, electrical conductivity and oxygen isotope ratio measurements was very good.

Teuftal lysimeter

Discharges from this lysimeter only occurred after heavy rainfall and lasted for short periods.

Only four events were monitored at this site in 1996. Discharge rates were between 0.01 and 2 L/h. Approximately 20% of incident rainfall passed through the landfilled material (Ochs *et al.* 1999; Ludwig *et al.* 2000), much being lost by evapotranspiration and runoff. An average residence time of water within the sand between the blocks and assuming that the water content of the cement blocks remains constant has been roughly estimated to be 6 months. A typical example of discharge resulting from a

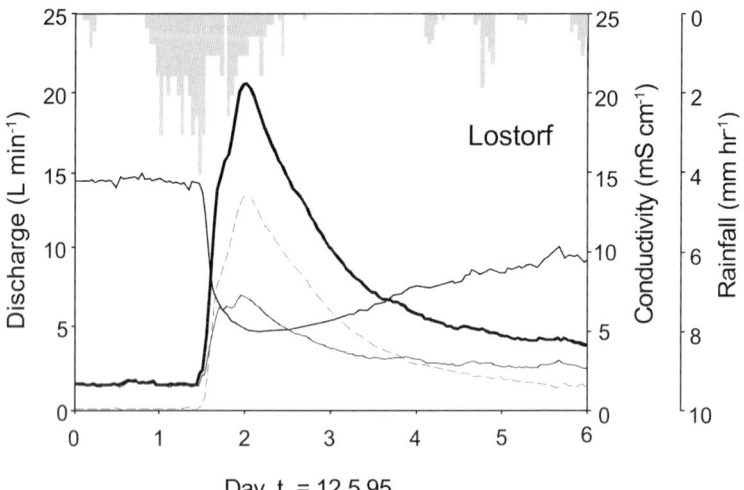

Fig. 3. The response of leachate drainage (——), electrical conductivity (– – –), matrix discharge (——) and preferential flow (——) to rainfall at Landfill Lostorf during a rain event in May 1995.

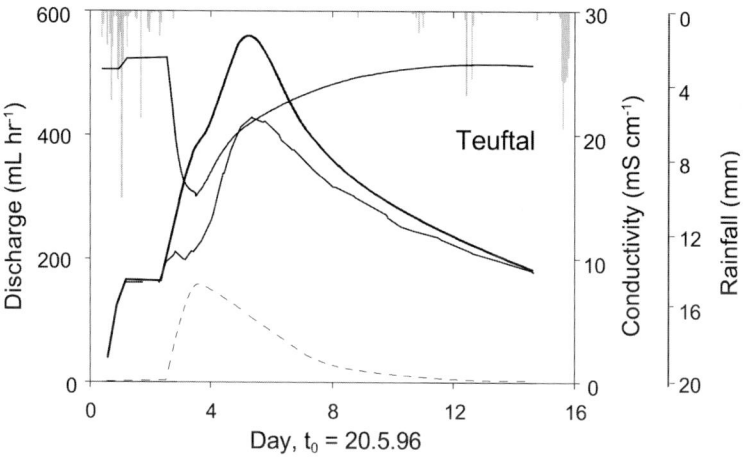

Fig. 4. The response of leachate drainage (——), electrical conductivity (– – –), matrix discharge (——) and preferential flow (— —) to rainfall at Lysimeter Teuftal during a rain event in May 1996.

rain event is shown in Fig. 4. A peak in discharge is observed four days after the rain event. The rainwater, or preferential flow, reaches the drainage outlet approximately two days earlier than the matrix drainage. The characteristic decrease in electrical conductivity is clearly seen, as is the recovery of electrical conductivity after the rain event.

Modelling flow in Landfill Lostorf

In a recent paper two empirical (neural networks and linear storage models) and two numerical methods were used to model flow through Landfill Lostorf (HYDRUS, with no preferential flow and MACRO with preferential flow; Johnson *et al.* 2001). The models were calibrated using an eight-month data set from 1996 and validated on a three-month data set from winter 1994/1995. The data sets comprised hourly values of rainfall, evaporation (estimated from the Penman–Monteith relationship), drainage discharge and electrical conductivity. Predicted and measured discharges were compared. It was found that all model approaches would have benefited from a more exact knowledge of initial water content. Nevertheless, the empirical models calibrated on the 1996 data set were able to model the response of the discharge to rainfall very well. However, the linear storage model in particular could not cope with the non-linearity of the system that was caused by seasonal changes in water content of the MSWI bottom ash. The most useful modelling approach was the numerical one that included preferential flow (MACRO). It was necessary to

fit 11 parameters. Fitting such a large number of parameters results in covariances and it is unlikely that they have meaning individually. However a parameter set generated with the Lostorf data set has been successfully used to model flow in another landfill, Landfill Seckenberg, Switzerland (Abbaspour *et al.* 2004). It should be noted that none of the models could model preferential flow. Electrical conductivity has been successfully used by Abbaspour *et al.* (2004) to distinguish preferential flow modeled by MACRO, but the next challenge will be to better model fast flow through landfills during rain events.

Geochemical factors controlling leachate composition

Leachate composition is dominated by geochemical processes within the pores of the waste matrix, diffusion to mobile water within larger pores and ultimately mixing with preferential flow (Fig. 5). Geochemical factors, such as precipitation of new phases and sorption to surfaces play an important role in controlling the dissolved concentrations of potential contaminants, while hydrological processes cause fluctuations in concentration. It is important to be able to interpret leachate compositions (for our examples see Table 1) in order to assess the impact on the environment, not forgetting that leachate composition will change with time as a result of leaching and weathering processes within the landfill. In this chapter it is not possible to discuss all processes in detail. Instead the processes are shortly

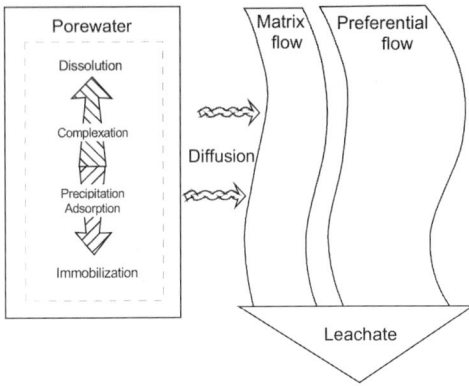

Fig. 5. Conceptual model of hydrological and geochemical processes controlling leachate composition.

introduced and illustrated with examples from Landfill Lostorf and Lysimeter Teuftal.

Influence of hydrological processes on leachate composition

Leachate appears to be of more or less constant composition under conditions not affected by rain events. In winter, when the landfills have a higher water content overall, concentrations may be lower due to dilution, but the effect is not dominant (Johnson *et al.* 1999). This is illustrated by the example shown in Fig. 6. The concentrations of several leachate components are shown for Landfill Lostorf as a function of time during a rain event. It should be noted that

these are not logarithmic plots. The concentrations of the components Mo and Cd follow the same dilution pattern as the major components that contribute to the electrical conductivity of the leachate discharge. (Electrical conductivity is a measure of the sum of the dissolved ions, mainly Na^+, K^+, Cl^-, Ca^{2+}, SO_4^{2-} and OH^-.) Other components that appear to follow this trend are V, (redox state unknown), Mn(II) and Zn(II). Concentrations of Cu(II), WO_4^{2-} and $Sb(OH)_6^-$ increase as electrical conductivity decreases, as do Cr(VI) concentrations (not shown). The most probable reason for such increases are probably geochemical in nature and related to changes in leachate composition.

In the case of Lysimeter Teuftal, most dissolved heavy metal and metalloid concentrations are directly correlated to electrical conductivity. The plot in Fig. 7 shows the relative dilution of different components during a rain event in comparison to that of electrical conductivity. The trend is independent of the speciation of these components. It should be noted that there are three notable exceptions, namely Ni(II), Co(II), and Cu(II). The common property of these cationic species is that they are extremely insoluble under the highly basic conditions encountered in the leachates of cement-stabilized residues (pH 12.8). It is possible that precipitation is the cause of such behaviour.

Diffusion processes

Species within the pores of the waste matrix dissolve in pore water and diffuse to mobile water

Table 1. *Average parameter values and concentrations (mol/L) of all samples from Landfill Lostorf (13—194 samples) and Lysimeter Teuftal (65 samples) during the experimentation period from 28.11.94 to 15.11.96 and 1.5.96 to 5.8.96 respectively*

	MSWI bottom ash Lostorf	Cement-stabilized APC residues Teuftal		MSWI bottom ash Lostorf	Cement-stabilized APC residues Teuftal
I	0.10	0.18			
pH	10.28	12.8			
DOC	8.61×10^{-4}	—			
Alk	1.39×10^{-3}	—			
Ca	8.23×10^{-3}	8.49×10^{-5}	Cd	1.17×10^{-8}	1.22×10^{-8}
Na	4.45×10^{-2}	1.19×10^{-1}	Mn	8.53×10^{-8}	2.45×10^{-8}
K	1.18×10^{-2}	2.73×10^{-2}	Mo	5.44×10^{-6}	1.67×10^{-5}
Mg	6.26×10^{-4}	—	Pb	1.30×10^{-8}	1.85×10^{-7}
Al	2.87×10^{-5}	1.74×10^{-3}	Sb	2.65×10^{-7}	—
B	2.19×10^{-4}	—	W	6.13×10^{-7}	4.93×10^{-6}
Si	1.35×10^{-4}	2.33×10^{-3}	Cu	1.59×10^{-6}	2.34×10^{-7}
Cl	4.71×10^{-2}	1.36×10^{-2}	Zn	8.68×10^{-8}	6.41×10^{-6}
SO4	1.24×10^{-2}	2.42×10^{-2}	Cr	2.09×10^{-7}	3.71×10^{-6}
CO3	7.32×10^{-4}	4.58×10^{-3}	V	4.27×10^{-7}	5.41×10^{-6}

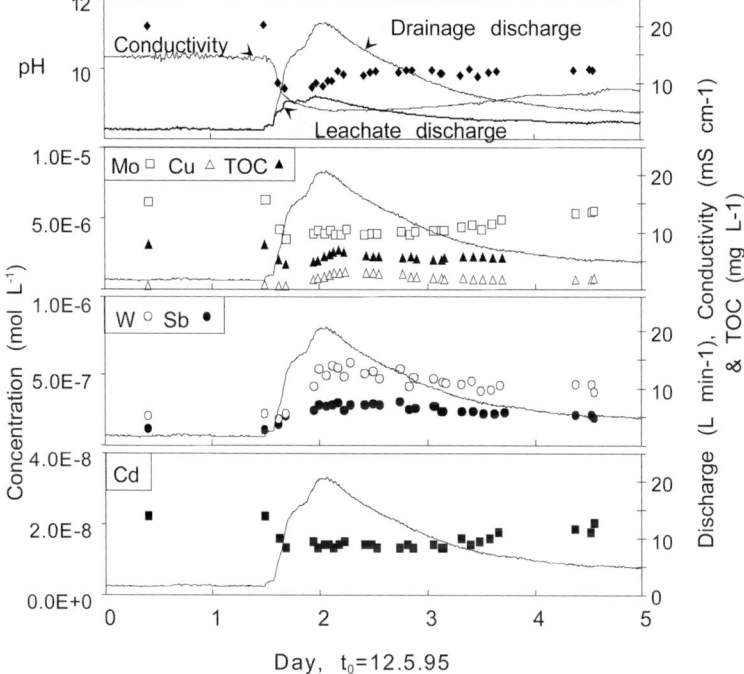

Fig. 6. Concentrations of selected leachate components during a rain event in May 1995 at Landfill Lostorf.

within within the waste matrix. The cumulative fraction (CFR) of the amount of a particular species that has dissolved from the waste matrix at time t (M_t) depends on the surface area (S) relative to volume (V), on the effective diffusion coefficient (D_e) of the ionic species and the total mass of that species (M) in the solid phase (equation 3; Crank 1990).

$$CFR = \frac{M_t}{M} = \frac{2S}{\sqrt{\pi}V}\sqrt{D_e \times t} \qquad (3)$$

It is not always a trivial matter to assess whether the dissolved leachate concentration of a particular species is diffusion controlled or controlled by geochemical processes. The interest in discerning between these processes is twofold. On the one hand it is important to eliminate artifacts in the experimental determination of D_e, which may be lowered by geochemical processes that remove a particular species from solution. On the other hand, it is important to know how a change in hydrology may affect leachate composition since the residence time of water greatly influences leachate concentrations of species that are diffusion controlled while there may be less influence on those species whose concentrations are controlled by geochemical mechanisms. Figure 8 illustrates how the dissolved concentrations in leachate from Lysimeter Teuftal of some species may be diffusion controlled, while others are most probably solubility controlled.

Fig. 7. The correlation of changes in dissolved heavy metal concentrations to changes in electrical conductivity for the May 1996 rain event.

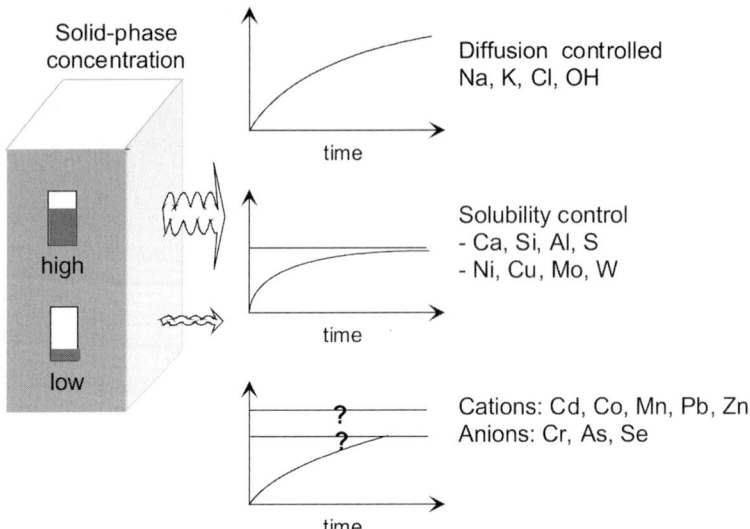

Fig. 8. Illustration of possible mechanisms controlling the dissolved concentrations of leachate components in Lysimeter Teuftal (adapted from Baur, personal communication).

For the majority of heavy metal and metalloid species it is unclear which mechanism prevails.

Geochemical factors

Major components appear to be either diffusion or solubility controlled. Ions such as Na^+, K^+, Cl^- and OH^- are diffusion controlled. Their concentration depends wholly on diffusion rates, diffusion path lengths, and the residence time of the leachate. On the other hand, major components, such as $Ca(II)$, $Al(III)$, SO_4^{2-} and most probably $Si(IV)$, are solubility controlled in both Landfill Lostorf and Lysimeter Teuftal leachates. In the former case (see Fig. 9), gypsum ($CaSO_4 \cdot 2H_2O$), ettringite ($3CaO \cdot Al_2O_3 \cdot 3CaSO_4 \cdot 32H_2O$) and gibbsite ($Al(OH)_3$) appear to be solubility-limiting phases, while the solubility of Si does not appear to be limited by crystalline or amorphous SiO_2, Ca silicate hydrate ($1.7CaO \cdot SiO_2 \cdot 1.4H_2O$ or C–S–H) or imogolite ($Al_2SiO_3(OH)_4$). It is possible that Si forms clay precursors of unknown composition at the surface of glassy materials in bottom ash, as observed by Zevenberg *et al.* (1996). In the cement-stabilized APC residues of Lysimeter Teuftal porewater composition is dominated by the cement phases and the pH value of the leachate is approximately 13. Solubility-controlling phases for $Ca(II)$, $Al(III)$, SO_4^{2-} and $Si(IV)$ are C–S–H, portlandite ($Ca(OH)_2$) and ettringite (Baur *et al.* 2001).

There are possibly a greater number of factors controlling heavy metal and metalloid species concentration (Fig. 10). While ion exchange can be eliminated in systems with leachates containing high salt concentrations, experience shows that each element may be controlled by a unique mechanism or combination of mechanisms that depends on the geochemical properties of the species under consideration within the particular matrix of the waste material.

Precipitation of (hydr)oxides and carbonates appears to control the concentrations of some cationic species. In Lysimeter Teuftal $Ni(II)$ and $Cu(II)$ appear to be precipitated as hydroxides, while in leachates of Landfill Lostorf $Cd(II)$ concentrations appear to be controlled by the precipitation of $CdCO_3$. In both systems MoO_4^{2-} and WO_4^{2-} appear to precipitate as Ca metallates (e.g., Fig. 11).

There are probably several mineral phases, particularly for the highly alkaline systems, that remain to be discovered. Mixed hydroxides may control solubility. Calcium zincate ($CaZn_2(OH)_6$), for example, is thermodynamically more stable than $Zn(OH)_2$ above pH 11.5 and may be important in cementitious systems. Another group of minerals is that of the hydrotalcite-like minerals, the layered double hydroxides (LDH, $M^{2+}_2M^{3+}1/yX^{y-}(OH)_6$ where X is an anion). Cobalt, Ni and Zn can form such minerals (Johnson & Glasser 2003) under neutral to alkaline conditions. For the majority of species, however, solubility-limiting phases do not appear to control dissolved concentrations.

Complexation with strong complexing agents may enhance the solubility of a metal ion.

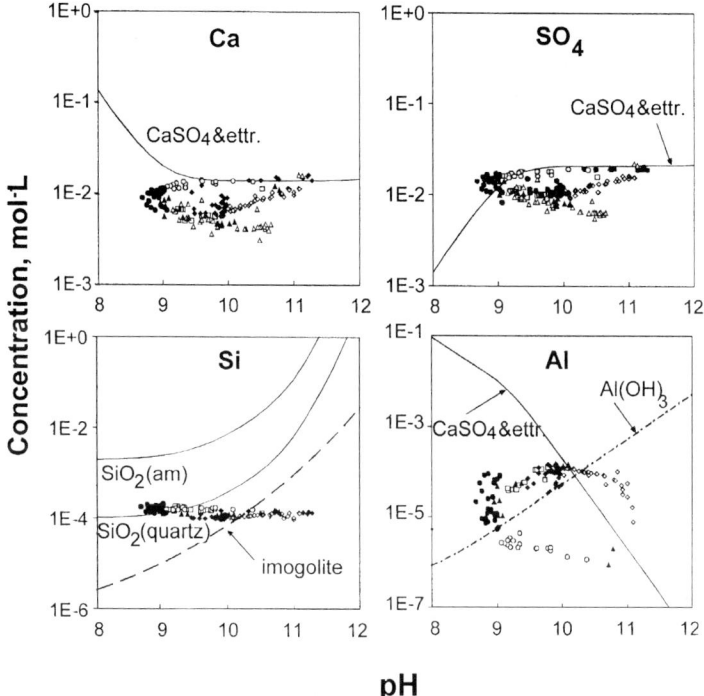

Fig. 9. Possible solubility controlling phases for dissolved concentrations of Ca(II), Al(III), SO_4^{2-} and Si(IV) in Landfill Lostorf leachate. The symbols represent measured concentrations for different rain events and the curves represent model calculations for the conditions given in Table 1.

The most striking example is Cu(II) mobilization by organic ligands. In Landfill Lostorf leachate dissolved Cu(II) concentrations are to a large extent correlated to total organic carbon (TOC,

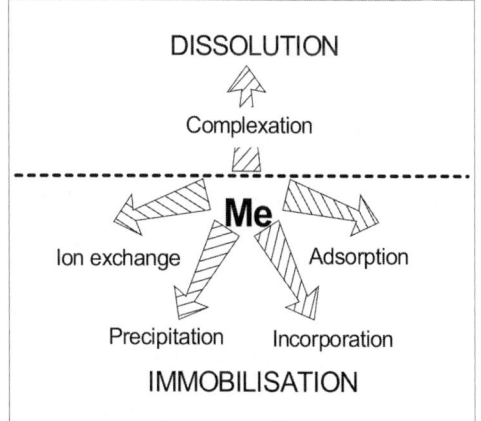

Fig. 10. Illustration of possible mechanisms controlling the dissolved concentrations of heavy metals and metalloid species.

Fig. 6). Meima *et al.* (1999) determined the binding constants for Cu(II) with dissolved organic carbon from MSWI bottom ash. These authors found that 90–100% of Cu(II) was organically bound. Using these constants, the dissolved concentrations of Cu(II) can be successfully modelled for the Landfill Lostorf data (Fig. 11). Freyssinet *et al.* (2002) also elevated Cu(II) concentrations in leachate of their lysimeter study, but ascribed these elevated concentrations to complexation with Cl^-.

Adsorption to mineral surfaces such as Fe and Al (hydr)oxides has long been known to be an important process that limits the mobility of heavy metals and metalloid species in aqueous systems (e.g., Stumm 1992). The sorption of ionic species in MSWI bottom ash has been recently studied in detail by Meima & Comans (1998, 1999). These authors used a sequence of 'selective' chemical extractions to determine sorbent concentration, namely Fe and Al (hydr)oxides. Their model calculations suggested that Zn(II) and MoO_4^{2-} sorbed to Fe (hydr)oxides, while Pb(II) and Cu(II) appeared to have a greater affinity for Fe (hydr)oxides. The sorption of Cd(II) was found to be very weak. The interpretation of

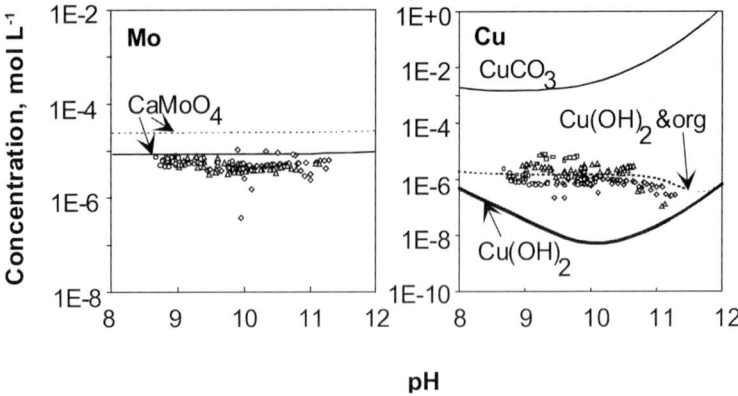

Fig. 11. Possible solubility controlling phases for dissolved concentrations of Ca(II), Al(III), SO_4^{2-} and Si(IV) in Landfill Lostorf leachate. The symbols represent measured concentrations for different rain events and the curves represent model calculations for the conditions given in Table 1.

adsorption processes is difficult mainly because the determination of concentration of surface sites available for adsorption is very difficult in such complex solid mixtures.

Incorporation in existing mineral phases may also limit the solubility of dissolved species. Isomorphic substitution of ions in a crystal lattice for other ions, known as solid solution, is thought to be an important process for some species. For example, Fruchter *et al.* (1990) suggest that CrO_4^{2-} substitutes for SO_4^{2-} in $BaSO_4$, thus lowering the solubility of the former ion. Chromate may also substitute for SO_4^{2-} in ettringite (Kumarathasan *et al.* 1990). There are many possibilities for the formation of solid solutions for species that are soluble in the alkaline pH range.

Studies on binding mechanisms sometimes appear to give quite contradictory results. The most simple mechanisms, such as precipitation of a solid phase, are more open to determination than those that are dependent on an understanding of the solid phase, for example with regard to sorption capacity. Certainly it is important to further our understanding in this field.

Conclusions

Geochemical processes appear to control the concentration ranges of a number of leachate parameters, while hydrological factors cause fluctuations in leachate composition due to mixing with rainwater that passes through the landfill as preferential flow. The modelling of these mixing processes still requires improvement, although great advances have been made. One of the most difficult remaining challenges is to

understand the geochemical processes that control the dissolved concentrations of soluble heavy metal and metalloid cations and oxyanions in these complex alkaline systems. Only with a better understanding of these processes will we be able to improve long-term predictions of contaminant mobility.

Kanton Aargau is gratefully acknowledged for financial support. C. Ludwig and I. Baur are acknowledged for their work on Lysimeter Teuftal.

References

ABBASPOUR, K. C., JOHNSON, C. A. & VAN GENUCH-TEN, M. TH. 2004. Estimation of uncertain flow and transport parameters by a sequential uncertainty fitting procedure: SUFI-2. *Vadose Zone Journal*, 2004, *to appear*.

AGW 1991. *Immobilisierung von Rauchgasreinigungsrückständen aus Kehrichtverbrennungsanlagen, Schlussbericht zum Projekt IMRA April 1991.* Projektleitung AGW (Amt für Gewässerschutz und Wasserbau), Zürich, Switzerland.

BAUR, I., LUDWIG, C. & JOHNSON, C. A. 2001. Leaching behaviour of cement-stabilised incinerator ashes: A comparison of field and laboratory measurements. *Environmental Science and Technology*, **35**, 2817–2822.

CRANK, J. 1990. *The Mathematics of Diffusion.* Clarendon Press, Oxford, UK.

FREYSSINET, PH., PIANTONE, P., AZAROUAL, M., ITARD, Y., CLOZEL-LELOUP, B., GUYONNET, D. & BAUBRON, J. C. 2002. Chemical changes and leachate mass balance of municipal solid waste bottom ash submitted to weathering. *Waste Management*, **22**, 159–172.

FRUCHTER, J. S., RAI, D. & ZACHARA, J. M. 1990. Identification of solubility-controlling solid phases in a large fly ash field lysimeter. *Environmental Science and Technology*, **24**, 1173–1179.

JOHNSON, C. A., RICHNER, G. A., VITVAR, T., SCHITTLI, N. & EBERHARD, M. 1998. Hydrological and geochemical factors affecting leachate composition in municipal solid waste incinerator bottom ash. Part I: The hydrology of Landfill Lostorf, Switzerland. *Journal of Contaminant Hydrology*, **33**, 361–376.

JOHNSON, C. A., KAEPPELI, M., BRANDENBERGER, S., ULRICH, A. & BAUMANN, W. 1999. Hydrological and geochemical factors affecting leachate composition in municipal solid waste incinerator bottom ash. Part II: The geochemistry of leachate from Landfill Lostorf, Switzerland. *Journal of Contaminant Hydrology*, **40**, 239–259.

JOHNSON, C. A., SCHAAP, M. K. & ABBASPOUR, K. 2001. Modelling of flow through a municipal solid waste incinerator ash landfill. *Journal of Hydrology*, **243**, 55–72.

JOHNSON, C. A. & GLASSER F. P. 2003. Synthesis, characterisation and solubilities of hydrotalcites $(M_2Al(OH)_6(CO_3)_{0.5}$ where $M = Mg$, Zn, Co, Ni). *Clays and Clay Minerals*, **51**, 1–8.

LUDWIG, C., JOHNSON, C. A., KAEPPELI, M., ULRICH, A. & RIEDIKER, S. 2000. Hydrological and geochemical factors controlling the weathering of cemented MSWI filter ashes: A lysimeter field study. *Journal of Contaminant Hydrology*, **42**, 253–272.

KUMARATHASAN, P., MCCARTHY, G. J., HASSETT, D. J. & PFLUGHOEFT-HASSETT, D. F. 1990. Oxyanion substituted ettringites: Synthesis and characterisation and their potential role in immobilization of As, B, Cr, Se and V. *Materials Research Society Symposium Proceedings*, **178**, 83–104.

MALOSZEWSKI, P., MOSER, H., STICHLER, W. & TRIMBORN, P. 1995. Isotope hydrology investigations in large refuse lysimeters. *Journal of Hydrology*, **167**, 149–166.

MEIMA, J. A. 1997. *Geochemical Modelling and Identification of Leaching Processes in MSWI Bottom Ash*. PhD thesis, University of Utrecht, The Netherlands, pp. 217.

MEIMA, J. A., & COMANS, R. N. J. 1998. Application of surface complexation/precipitation modeling to contaminant leaching from weathered municipal solid waste incinerator bottom ash. *Environmental Science and Technology*, **32**, 688–693.

MEIMA, J. A., & COMANS, R. N. J. 1999. The leaching of trace elements from MSWI bottom ash at different stages of weathering. *Applied Geochemistry*, **14**, 159–171.

MEIMA, J. A., VAN ZOMEREN, A. & COMANS, R. N. J. 1999. Complexation of Cu with dissolved organic carbon in municipal solid waste incinerator bottom ash. *Environmental Science and Technology*, **33**, 1424–1429.

NIXON, W. B., MURPHY, R. J. & STESSEL, R. I. 1997. An emprical approach to the performance assessment of solid waste landfills. *Waste Management and Research*, **15**, 607–626.

OCHS, M., STÄUBLI, B. & WANNER, H. 1999. Eine Versuchsdeponie für verfestigte Rückstände aus der Kehrichtverbrennungsanlage. *Müll und Abfall*, **1999**(4), 194–205.

PINDER, G. F. & JONES, J. F. 1969. Determination of the groundwater component of peak discharge chemistry of total runoff. *Water Resouces Research*, **5**, 438–445.

SCHROEDER, P. R., DOZIER, T. S., ZAPPI, P. A., MCENROE, B. M., SJOSTROM, B. M. & PEYTON, R. L. 1994. *The Hydrological Evaluation of Landfill Performance (HELP) Model: Engineering Documentation for Version 3*, EPA/600/R-94/168b. US EPA, Cinncinnati, Ohio.

STUMM, W. 1992. *Chemistry of the Solid–Water Interface*. John Wiley & Sons, New York.

ZEVENBERG, C., VAN REEUWIK, P., BRADLEY, J. P., BLOEMEN, P. & COMANS, R. N. J. 1996. Mechanism and conditions of clay formation during natural weathering of MSWI bottom ash. *Clays and Clay Minerals*, **44**, 546–552.

Geochemistry of leachates from coal ash

D. A. SPEARS[1] & S. LEE[2]

[1]Centre for Analytical Sciences, Department of Chemistry, University of Sheffield, Sheffield, UK
(e-mail: d.a.spears@sheffield.ac.uk)

[2]Department of Environmental Engineering, The Catholic University of Korea, Wonmi-Gu,
Buchon-City, Gyonggi-Do, Korea

Abstract: The combustion of coal around the world for power generation produces huge volumes of fly ash. In Europe alone this amounted to about 40 Mt in 2000 of which less than 50% was utilized. The waste ends in lagoons, ash mounds, and landfill sites. Coal ashes have high concentrations of many trace elements, some of which are of environmental concern. Although the origin of elements in coals is not considered in this chapter, other aspects of the geochemistry are, and in particular the location of elements within the coal as this influences the behaviour of elements during combustion. During combustion many elements are volatile and are concentrated on the surfaces of the ash particles. Analyses of input coal and combustion residues from Eggborough power station (UK) demonstrate retention of the majority of elements in the solid combustion products, and analyses of size-fractionated fly ash have enabled the percentage surface association to be calculated, which for elements such as As and Mo is considerable. A consideration of the general leaching behaviour leads to the conclusion that it is the surface-associated elements that are most susceptible to leaching in the aqueous environment. The pH is an important control on trace element mobility in water, and in leachates from fly ash ranges from 3.3 to 12.3. High-sulphur coals generate acidic leachates, but not exclusively as the laboratory and field data demonstrate in case studies on UK coals. Batch and column leaching tests on fly ashes are reviewed and data presented for fresh fly ash and weathered fly ashes from two UK mounds dating back 17 and 40 years. The weathered ashes do have lower leachate concentrations than the fresh ash, but in spite of their ages they would not be considered to be inert. The batch leaching tests are of value in simulating high liquid-to-solid ratios encountered in ash lagoons, whereas the column leaching tests relate more closely to ash mounds. Finally, the results of field studies are reviewed and data presented for samples from boreholes in the two ash mounds. Analyses of the ash and extracted porewaters demonstrate depth-related changes due to reaction of the ash with the infiltrating water and whether or not equilibrium was established. Calculations demonstrate that the porewaters achieve saturation with respect to gypsum at depth in the boreholes. Infiltration over the years has led to detectable changes in the concentrations of some of the major elements in the bulk ash, such as Ca, and this enables realistic infiltration rates to be calculated. However, there are not comparable changes in the concentrations of the trace elements in the ash because the rate at which elements are being removed in solution is not sufficiently great.

When coal is combusted a number of ash products are produced. In a conventional coal-fired power station the ash that enters the flue gas stream is referred to as the fly ash or pulverized fuel ash. This is volumetrically the most important fraction and although considerable progress has been made in utilizing this material, nevertheless there is an excess production, much of which ends in lagoons or, on a longer term basis, in landfill sites. The potential impacts of fly ash on surface water and groundwater therefore have to be considered both in the short and long term. The annual European production of fly ash in 2000 was 38.96×10^6 t, of which

46% was directly utilized (ECOBA 2003). The remainder was disposed of in landfill sites and to reclaim and restore land in old industrial areas. The fly ash consists of glass and high-temperature phases and in this respect is analogous to a volcanic glass. However, the leachates from volcanic glass are not of concern whereas those from fly ash are. This is because volatile elements during combustion have condensed on the surfaces of the fly ash particles (Clarke & Sloss 1992) and are potentially mobile in the aqueous environment. The volatile elements include many trace elements as well as major elements. The major elements contribute to the

From: GIERÉ, R. & STILLE, P. (eds) 2004. *Energy, Waste, and the Environment: a Geochemical Perspective*. Geological Society, London, Special Publications, **236**, 619–639.
0305-8719/04/$15 © The Geological Society of London 2004.

bulk composition of the porewaters, including pH, but the trace elements are more of a potential problem because concentrations in leachates may be unusually high compared with most aqueous environments and hence the emphasis in this chapter.

Bottom ash from power stations is less of a problem compared with fly ash for the contamination of natural waters firstly because the proportions of fly ash to bottom ash are approximately three to one and a greater proportion of the bottom ash is used (ECOBA 2003). Secondly, the volatile elements are depleted compared with fly ash (Clarke & Sloss 1992). Other combustion residues include fluidized-bed boiler ashes and the products from flue gas desulphurization (FGD). The non-regenerable FGD systems commonly use limestone, slaked lime, or a mixture of slaked lime and alkaline fly ash that are sprayed into the flue gases to remove SO_2 (Clarke & Sloss 1992). Although 90 wt% of the product is used to replace natural gypsum in plasters and wallboards, there is currently a small excess production in Europe of that is disposed of in landfill and equivalent sites (ECOBA 2003). Because the FGD plant treats the cooled flue gases volatile elements are concentrated and there will be similarities with fly ash.

The geochemistry of leachates from fly ashes is a function of the composition of the fly ash, which depends on the composition of the input coal and the combustion behaviour. The latter aspect is dealt with earlier in this book and hence the impact of combustion on the element distributions in the fly ashes is only briefly covered. However, the inorganic geochemistry of coals is not covered elsewhere and is therefore discussed more fully here. In this paper we will also refer to work done in our laboratories on the characterization of coal and combustion residues from Eggborough power station (National Grid Ref. SE 575245) and fly ash from Drax power station (National Grid Ref. SE 655277). The power stations are adjacent and are approximately 50 km northeast of Sheffield. At the time of sampling Upper Carboniferous bituminous coals from local mines were in use. Fly ash from Drax was subjected to various leaching tests in the laboratory and an ash mound at Drax was sampled to study the effects of natural leaching over a period of 17 years. Additional samples were taken from an ash mound at the now decommissioned Meaford power station (National Grid Ref. SJ 896373) to extend the period of weathering to 40 years. The latter site is some 75 km to the southwest of Sheffield. Samples from the ash mounds are referred to in the text as weathered.

The geochemistry of coals and coal ashes

In Goldschmidt's work on the laws governing the distribution of the chemical elements in nature, especially in rocks and minerals (reviewed in Goldschmidt 1954), the analysis of coal ashes figured prominently. One of the first samples he analysed, with the newly developed optical spectrograph in the 1920s, was a coal ash. High concentrations of Ge were found that were greatly in excess of crustal averages and this created an interest that led to many other investigations over the years.

Goldschmidt (1954) listed some 20 trace elements that could be concentrated in coal ashes at many times the crustal average with concentration factors as high as 20 000. It is this possibility of high trace element concentrations that has to be considered when leachates originate from fly ashes. This approach of comparing with crustal averages is still used, but comparisons made using average coal compositions based on worldwide compilations are more revealing, as for example in the work of Shao et al. (2003). The ash in bituminous coals and anthracites is mainly derived from the minerals, both detrital and diagenetic, with the organic matter and the connate fluids also contributing. The detrital minerals mainly consist of quartz and clay minerals, whereas the diagenetic minerals are primarily the sulphides (most commonly pyrite), clay minerals and carbonates. In anthracites and bituminous coals the concentrations of major elements expressed on an ash basis are broadly similar to non-calcareous mudrocks, unless ash contents are low (≤ 5 wt%), in which case the connate fluids (Na and Cl) and the diagenetic minerals (Fe in pyrite) make a proportionally greater contribution. Typically in an ash the concentrations of SiO_2, Al_2O_3 and Fe_2O_3 exceed 70 wt% and the resulting fly ash is classed as Type F in American Society of Testing Materials (ASTM) standard C618 (for further details, see Mantz 1998). In an ash with less than 70 wt% of these oxides, but greater than 50 wt%, the other oxide that has increased is usually CaO. This is a Class-C fly ash and is usually referred to as a high-lime or high-Ca ash. Carbonates and organic matter are the sources for the Ca and their concentrations are generally greatest in lignites and sub-bituminous coals. In the USA, high-lime fly ashes characterize the products of many power plants in the west and mid-western states burning low-sulphur coals (Adriano et al. 1980).

In the last 20 years or so, many workers have established that most of the trace elements are dominantly associated with the mineral matter. For example, statistical analysis of the trace element data for two physically fractionated UK coals showed that Cr, Ga, Rb, Sr, Y, Zr, Nb, Ba, La, Ce, Sm, Th, and U are dominantly associated with the silicate fraction, Mo, Se, Ni, As, Pb, Sb, Cu, and Zn are concentrated in pyrite, and Ge, Br, and V are present in the organic matter (Spears 2001). A multinational collaborative research project, supported by the International Energy Association, reached broad conclusions on the likely locations for a number of environmentally important trace elements in four World traded coals (Davidson 2000). However, most elements are not exclusively present in only one fraction. This means that as the proportions of the fractions vary so the bulk composition changes. The variation could be temporal, that is, through a seam or a succession, or lateral. A major fraction of the coal may have relatively low trace element contents but still makes the greatest contribution to the overall coal composition simply because of its abundance. Although in combustion the emphasis is on the bulk composition, rather than the composition of the fractions, the location of the elements within fractions does have an influence on the combustion behaviour of the elements. Knowledge of element locations in the coal also helps in understanding why coals differ in their trace element compositions.

Trace element analyses of coals from around the world have been compiled by a number of authors, notably by Swaine (1990). Such an undertaking is not without problems however. Not only have analytical methods evolved but also the number of elements of interest has increased. The COALQUAL database maintained by the US Geological Survey (USGS) is the largest database containing quality information on US coals (USGS 2003a). Representative coals from all the UK coalfields are held in a coal bank and their analyses are in the literature (Spears & Zheng 1999), but the samples were collected in the 1980s and since then there has been a 70% reduction in UK coal output and many of the mines have now closed. The samples may be a unique archive but they are not representative of the current coal production and it is that which is of immediate environmental concern. There is a need for reliable analytical data on current coal production from around the World. To meet this requirement the USGS has embarked on an ambitious programme to collect and comprehensively analyse coals from all coal-producing countries. The World Coal Quality Inventory (WoCQI) will satisfy the need for a comprehensive database. This will be accessible in a number of formats including the World Wide Web, as for example with the data acquired for Chinese coals (USGS 2003b).

Trace elements in combustion

The liberation of elements from the combustion residues into the aqueous environment, and thus the composition of the leachates will be related to the element distribution in the ashes, hence the importance of the combustion behaviour of the trace elements. In addition to the earlier Chapter in this book, a very good summary of the combustion behaviour of trace elements is provided by Clarke & Sloss (1992). Most of the trace elements are associated with the particulate matter. Some elements show no fractionation between coarse and fine residues, others are depleted in the coarse residues (furnace bottom ash) and are concentrated in the fly ash, particularly the fine-grained particles, whereas the most volatile elements are depleted in all the solid phases. Many authors have observed an increase in the concentrations of a number of trace elements with decreasing grain size of the fly ash particles (reviewed in Clarke & Sloss 1992). This is attributed to the condensation of the volatile elements on the surfaces of particles as the temperatures fall in the waste gas stream. Surface-associated elements are accessible for leaching in the groundwater and hence the interest in this mode of occurrence. It is noteworthy that the combustion behaviour of some elements may vary significantly depending on combustion conditions and coal composition. Thus, As may be retained as involatile calcium arsenate if Ca is present as more than a minor element in the coal (≥ 0.1 wt%), and Hg as condensed $HgCl_2$ on the surfaces of ash particles if HCl is present in the gas stream (Quick & Irons 2002).

The emphasis in the present work is on leachates derived from fly ash disposal, but it must not be forgotten that fine ash particles may escape from power plants. The transport and dispersion of particulate emissions from tall stacks has been actively researched over the last 20–30 years. A review by Carras (1995) records that plumes have been observed to remain as coherent units up to distances of at least 1800 km and thus deposit pollution far from source. However, maximum particulate surface loadings are likely to be found adjacent to old power plants with limited fine-particle fly ash entrapment. Evans et al. (1980) estimated that in a period of 23 years the cumulative stack ash load at a distance of 1.7 km from an 83 MW plant was $2.75 \, \text{kg/m}^2$. This is a significant amount and clearly there will be important environmental impacts on the soil system and possibly on the groundwater. The submicron-sized fly ash particles have the highest trace element concentrations and it is these particles that are most likely to escape into the atmosphere even with improved retention systems. Fly ash particles have been identified in sediments at some distance from power plants. Based on lake sediments the spatial and temporal distributions of carbonaceous fly ash particles have been studied (Rose *et al.* 1999) and shown to agree with that of the S emissions.

Case study

In Eggborough power plant it was found that Zn, Pb, Mo, Cu and As are concentrated in the fly ash compared with the bottom ash, whereas for the other trace elements studied (Ba, Cr, Nb, Rb, Sr, V, Y, Zr) there was very little fractionation (Martinez-Tarazona & Spears 1996, tables 3 and 4). Mass balance calculations show that for the elements studied only S and As are depleted in the combustion ashes and all the other elements analysed appear to have been retained. Not included within the analyses were Hg, Cl, and F, and loss of these elements would be anticipated based on their volatility (Sloss & Davidson 2001).

The analyses of the Eggborough fly ash included size-fractionated material (Martinez-Tarazona & Spears 1996, table 5). Enrichment factors were calculated as a ratio of the concentration of an element in a specific fraction to the concentration in the whole fly ash. Concentrations were also normalized against the Al concentration on the basis that this element is involatile. In the finest grain sizes there was significant enrichment for As, Mo, Pb, Cu, and Zn with enrichment factors close to two, and minor enrichment for Ba, Ni, Cr, V, and P (Martinez-Tarazona & Spears 1996, fig. 3). Shown in Table 1 (column A) is the ratio of element concentration in the fine fraction to that in the coarse fraction calculated from the 1996 data. Absolute values depend on the grain size of the fraction but the ratio does provide a sensitive measure of the element volatility. The first nine elements were identified in the 1996 work, but possibly Cr is more volatile than was originally thought. The ash size fractions were obtained by air elutriation, and a Coulter Counter was used to determine the grain size distribution for each ash fraction to check the size fractionation. Subsequently, in as yet unpublished work, we used the grain size data to calculate the average

Table 1. *Size-fractionated Eggborough power station fly ash*[*,†]

A ($<$2.5 µm/$>$65 µm size)		B (surface%)	
As	13.79	Mo	90
Mo	12.32	As	81
Zn	6.56	Cu	61
Cu	6.51	Zn	48
Pb	6.30	Pb	48
Cr	2.94	V	28
Ni	2.61	Cr	25
V	2.14	Ni	22
Ba	1.49	Ba	13
Sr	1.46	Sr	9
Nb	1.45	Nb	7
Na	1.26	Rb	6
Y	1.17	Y	4
Mn	1.09	Mn	0
Ti	1.07	Zr	0
Fe	1.05		
Zr	1.01		
Rb	1.00		
K	1.00		
Ca	1.00		
Al	1.00		
Mg	0.94		
Si	0.78		

[*]Column A: The ratio of element concentration in the $<$2.5 µm size fraction over the concentration in the $>$65 µm size fraction. Column B: Percentage of element surface associated.
[†]Analyses from Martinez-Tarazona & Spears 1996, table 5.

surface area of each fraction and have found that, for many of the elements, there is a positive linear relationship between the surface area and the concentration of an element in that fraction. From these relationships we have calculated the actual contribution that the surface-associated elements make to the bulk composition of the fly ash. These values are shown for the bulk fly ash in Table 1, column B, expressed as a percentage of the whole ash composition. The calculations are based on analyses of eight fractions, whereas the values in column A are based on two fractions. Although the order of the elements in the two columns in Table 1 is similar, more credence should be given to the percent surface-associated elements (column B) because of the greater number of data points. Nevertheless there are errors, but Ba and Sr probably do show a minor surface association, whereas lower elements in the order probably do not.

Discussion

The elements not present on the grain surfaces are presumably contained within the particles. In the fly ash from Eggborough, glass dominates with minor amounts of quartz, mullite, hematite and magnetite (Martinez-Tarazona & Spears 1996). Substitutions in quartz and mullite are limited, although Fe^{3+} and Ti^{4+} may replace Al^{3+} in the mullite structure to an appreciable extent (Deer et al. 1966). It is possible that in fly ash much of the mullite may be derived directly from the breakdown of kaolinite, which contains few cations other than Si and Al. Hower et al. (1999) separated a size-fractionated fly ash into a magnetic and a non-magnetic fraction. Chromium was found to be significantly higher in the magnetic (magnetite-rich) fraction, with Ni less so. In magnetite ($Fe^{2+}Fe_2^{3+}O_4$) a range of substitutions is possible, with Cr and V replacing the Fe^{3+}, and Ni, Co, and Zn replacing the Fe^{2+}. The other major Fe mineral is hematite and limited substitution of Fe^{2+}, Mn^{2+}, and Ti^{4+} may occur.

Direct analysis of individual ash particles is difficult because of the fine grain size. Gieré et al. (2003) used a combination of scanning and transmission electron microscopy (SEM and TEM) and electron probe microanalysis (EPMA) to show that most individual particles are fairly homogeneous, but a pronounced compositional variation exists among particles with similar physical and structural attributes. Several trace elements, including U, are partioned into Fe-rich particles, and small Ca-rich phases present on the grain surfaces may be important hosts for elements such as V and Zn. In US fly ashes Zielinski & Finkelman (1997) had also noted that U was contained within particles. The direct analysis of fly ash particles has also been attempted by Spears (2004) using laser-ablation inductively coupled plasma mass spectrometry (LA-ICP-MS). Conclusions reached were that there is a major association with the ash particle surfaces of As, U, Pb, Tl, Mo, and Se, the glass is an important location for V, Cr, Cu, and Zn and although magnetite may contain higher concentrations of Cr and V the glass is the major source. For three of the surface-associated elements we believe from the size fraction analyses that the particle surfaces are the main location for the elements in the fly ash. The LA-ICP-MS results suggest that this also applies to U, Tl, and Se.

Unburnt carbon is another minor component of fly ash. Ideally its content should be less than 1 wt%, but concentrations of 3–4 wt% or higher may be recorded if the combustion efficiency falls. Any trace elements contained within the unburnt carbon are likely to have condensed from the gas stream rather than inherited from the original organic matter. Loss of Hg from the gas stream due to unburnt carbon has been recorded (Sakulpitakphon et al. 2000).

In conclusion, although some trace elements are concentrated in minerals such as magnetite, it is the association with the surfaces of the ash particles that dominates.

General leaching behaviour of fly ash

As discussed in the previous section, trace elements are essentially retained in the solid combustion products and, because many are present on the surfaces of the particles, they are potentially leachable. Our data show the elements Mo, As, Cu, Zn, Pb, U, Tl, and Se will be readily accessible for leaching. A significant fraction of the V, Cr, and Ni, and a minor proportion of the Ba and Sr will also be potentially leachable because of the surface association, but most of these elements appear to be located in particles and will be released more slowly as the dissolution of the glass and other phases takes place. Rubidium, Y, Zr, Mn, and Nb are contained almost entirely within the particles and dissolution is potentially slower. The extent to which elements are leached also depends on their speciation and solubility in the porewaters, and the pH exerts a major control. In oxidizing solutions, elements such as, Cd, Cu, Mn, Ni, Pb, and Zn form hydrated cations that adsorb onto mineral surfaces at higher pH values and desorb at lower pH values. In contrast, the elements As, U, Mo, Se, and V, under similar Eh conditions, form oxyanions that adsorb onto mineral surfaces at low pH values and desorb at higher values (Jones 1995).

The major primary minerals present in fly ash are quartz, mullite, magnetite, and hematite. All of these phases are relatively unreactive in the weathering environment and release of elements into solution will be extremely slow. Most of the trace elements are located elsewhere, as discussed above, and may therefore be associated with more reactive phases. The glass in the fly ash is more reactive than the primary minerals by at least an order of magnitude. The glass essentially forms from sedimentary minerals in the coal, and in typical UK coals with ash contents of about 17 wt% most of the minerals are detrital. The ash therefore has a composition approximating to igneous materials. Volcanic glass is unstable in the weathering environment, but the rate of alteration is relatively slow on a non-geological timescale. In New Zealand, less than 5% clay has formed in volcanic ashes in 3000 years and only 5–10% clay in ashes that range in age from 3000–10 000 years (Lowe 1986). Zielinski & Finkelman (1997) have argued that the rate of alteration of the fly ash particles is slow in the natural environment, which is consistent with the low concentrations of U in the pore waters in ash disposal sites. However, diffusion-controlled hydration of the glass and alkali ion exchange generates high-pH conditions in a relatively closed system (Fisher & Schminke 1984). Hydration reactions involving the univalent and divalent cations also occur in the minerals and, in a closed system, alkaline conditions will again result. Mineral reactions dominate in soils lacking organic matter and porewater values are typically alkaline. In other soils organic growth and decomposition leads to porewater values that are typically acidic, not only from the presence of important complexing carboxylic and phenolic acids, but also from dissolved CO_2.

Based on published data it would appear that the majority of fly ashes (70%) produce a neutral to alkaline leachate (Mattigod et al. 1990). In the eastern USA, the high-S coals are generally associated with low pH values in fly ash leachates, whereas the western coals are those that generate more alkaline conditions (Adriano et al. 1980). The oxide analyses of the fly ash, including the S content, have been used to predict the pH of fly ash leachate (van der Sloot et al. 1982). The acidic leachate has generally been attributed to either adsorption of SO_2 or condensation of sulphuric acid from the gas stream onto the surfaces of particles. A different mechanism has been proposed by Fishman et al. (1999), namely reaction of the sulphuric acid in the gas stream with the fly ash to produce a S-bearing phase visible under the SEM and tentatively identified as an Al–K-sulphate. Hydrolysis of the Al produces low pH values (observed and calculated reductions in pH agree within 0.2 pH units). Their work was based on paired samples taken from different positions within the fly ash collection system over a period of time. Fly ash exposed to exhaust gases for a longer period has a preferential concentration of the sulphate and this material produces an acid leachate, whereas the other produces an alkaline leachate. The lesson to be learnt from this is that it is not only the composition of the input coal controlling the pH of the leachate but also the combustion regime. The pH of the leachate from fly ash exerts an important control on mobility of trace elements as noted above, but if the pH is alkaline, which it is for the majority of fly ashes, then opportunities arise to counter the effects of acid weathering mine discard.

Thermodynamic modelling of the behaviour of elements during cooling in the gas stream (Thompson & Argent 2002) predicts the formation of an alkali sulphate-based melt, with the majority of the other elements also forming sulphates, implying a complex melt and solid solution(s) based on sulphates. At the present, information on the activity of cations in mixed sulphate melts is lacking and identification of actual phases is difficult because of the problem of concentrating the sulphate fraction. It is therefore difficult to predict with certainty exactly how a fly ash sample will behave in water and hence the need for standardized leaching tests.

Carbon compounds are extremely important in the weathering environment as an energy source for microbial action. The destruction of the organic matter creates acidic conditions in porewaters, as we have noted for soils, but in addition reactions lead to lower Eh values. The early diagenetic zones in modern sediments are well documented and with restricted water movement, as in clay-rich sediments, sulphate reduction occurs (Potter 2003). The permeability in ash deposits is higher and porewaters are unlikely to become anaerobic, and Eh values will remain high. Although unburnt carbon is present in fly ashes this is not a good substrate for microbial growth because the more reactive organic molecules were eliminated during coalification (Bustin & Wust 2003) and combustion. However, organic matter added to facilitate plant growth and soil formation could influence porewater compositions. It is possible then, that conditions could become anaerobic if porewater flow was restricted, not unlike the situation that exists in muds during diagenesis and, in a comparable manner, the mobility of elements could change with changes in the elements' oxidation states. For example, the presence of reduced S species in solution would lead to a number of elements precipitating as sulphides. We have already seen that this is an efficient process in the earlier references to the trace element concentrations in pyrite in the coal. A more comprehensive discussion of the oxidations states, speciation and mobilities of the key trace elements derived from fly ash in a greater range of conditions is to be found in Jones (1995).

The leaching behaviour of fly ash is of considerable interest in horticulture, and the literature is extensive. In addition to the reclamation of fly ash disposal sites, fly ash is used extensively as a soil conditioner. Acidic agricultural soils benefit from the addition of alkaline fly ash, and essential plant nutrients present in the ash, such as B, Ca, Mg, K, S, Mn, and Zn are also added improving both chemical and physical properties (Clark et al. 1995; Sikka & Kansal 1995). In coal mining areas, acid mine drainage and acid weathering of mine discard are major problems. Remediation using alkaline weathering fly ash is a cost-effective solution that has been widely adopted (see Daniels *et al.* 2002). In some unpublished work we examined the reactions between weathered fly ash and the acid leachate from municipal landfill sites. We found that the concentrations of a number of the trace elements were greatly reduced in the leachates, but this was attributable to an increase in pH and consequent precipitation rather than reaction with any products of the fly ash weathering. The use of fly ash as a soil conditioner, resulting in both increased pH and element concentrations, may thus create problems and could include plant nutrient imbalances, possible phytotoxic levels of B and elevated levels of As, Mo, and Se in plant tissue (Adriano *et al.* 2002). Although B phytotoxicity is a concern, this element is mobile in the soil environment and is rapidly lost from the system. Continued application of high loads of fly ash to the soil could lead to bioaccumulation of Mo and Se in particular, and as they have a very narrow range of safe threshold concentrations in animal nutrition (Tolle *et al.* 1983) the plants grown should not be part of the food chain.

Laboratory-based leaching studies

Laboratory-based leaching tests to determine the possible impact of fly ash on the aqueous environment have been widely used. The extractability of various elements in short-term extraction studies is influenced by several factors such as the nature of ash, the solution to solid ratio, the duration of the experiment, and the intensity and method of agitation during extraction (Mattigod *et al.* 1990). Therefore, the results of leaching experiments do not accurately reflect the natural weathering condition in the field (see also Donahoe, this volume). Elemental leaching from fly ash columns in the laboratory, which is closer to field conditions compared with batch leaching methods, may in some cases provide qualitative information for assessing the effects of kinetic factors on mobilization of the elements from weathering of fly ashes (Mattigod *et al.* 1990). Notwithstanding such restrictions and the difficulty in simulating the natural environment, the results from laboratory-based studies are extremely useful in providing some measure of the potential release of elements from fly ash into solution. If a major disposal site is planned then it might be possible to initiate field trials if sufficient time is available, but usually this is not possible and laboratory tests are essential.

This section mainly outlines the laboratory-based leaching of two UK coal fly ashes from Drax and Meaford power stations that we have also studied in the field (Lee & Spears 1995, 1997, 1998) Previous experimental studies on the leaching of fly ash are also reviewed. Samples were investigated from ash mounds at the two power stations to provide information on the long-term leaching of fly ash under natural conditions. The samples from the Drax mound have been subjected to weathering for 17 years and the material from Meaford for about 40 years. In addition the Meaford ash was first pumped into lagoons, dewatered and then transferred to the mound, whereas at Drax the fly ash was transferred directly to the mound without a prior leaching stage.

The data produced from our laboratory work was also processed using the geochemical modelling program WATEQ4F (Ball *et al.* 1987). Such an approach is useful in identifying potential solubility-controlling solid phases that might limit the aqueous concentrations in the fly ash leachate (Mattigod 1983; Roy & Griffin 1984; Ainsworth & Rai 1987; Rai *et al.* 1988; Lee & Spears 1997, 1998; Mudd & Kodikara 2000). Calculation of the chemical speciation of the leachate also provides important information on the mobility and potential toxicity of the elements liberated from the fly ash. It is well known that precipitation/dissolution, complex formation, adsorption/desorption, and redox reactions control the mobilization of various elements from fly ash (see also Donahoe, this volume). Geochemical equilibrium approaches have been conducted toward evaluating the chemical composition of the fly ash leachate as reflection of precipitation/dissolution reactions during the water–waste interaction (Talbot *et al.* 1978; Mattigod 1983; Roy & Griffin 1984; Ainsworth & Rai 1987; Rai *et al.* 1988; Lee & Spears 1995, 1997, 1998; Mudd & Kodikara 2000).

Previous studies on leaching tests

Major and trace elements in fly ash are released to varying extent to the surrounding environment on contact with water. Many studies have focused on the extractability of the elements in fly ash using batch leaching tests with various reactants (Brown *et al.* 1976; Theis & Wirth 1977; Talbot *et al.* 1978; Theis *et al.* 1978; Elseewi *et al.* 1980; Kopsik & Angino 1981; Roy *et al.* 1981; James *et al.* 1982; van der Sloot *et al.* 1982; Mattigod 1983; Roy & Griffin 1984; Wadge *et al.* 1986; Ainsworth & Rai 1987; Rai *et al.* 1988; Grisafe *et al.* 1988; Hjelmar 1990; Fishman *et al.* 1999; Benito *et al.* 2001; Choi *et al.* 2002; Praharaj *et al.* 2002). The agents used in the studies include distilled and demineralized water, hot water, dilute or strong acids, alkalies and other solvents. The major factors that affect the extractability of the elements are: the solution to solid ratio, the type and concentrations of the reactants, and the duration and temperature of the extraction (Mattigod *et al.* 1990). The results of various water and acid extractions reported in the literature are summarized in Table 2.

The results of the *batch leaching* experiments show large variations in the water-soluble concentrations of all major and trace elements, depending on the controlling factors mentioned above. In general, stronger acid extractions mobilize a greater fraction of the elements in the fly ash, sometimes extracting nearly all the available Al, Ca, Fe, Mg, K, and Na. Major fractions of the Na and S are extracted by distilled water, whereas less than 10 wt% of the total Al, Fe, Mg, and K are extracted. Calcium, Na, S are the major elements in the extracts, ranging from tens to thousands of mg/L, whereas the concentrations of K and Mg are up to several hundred mg/L in some extracts. The pH of the water extract from the fly ash is extremely variable and ranges from 3.3 to 12.3. The trace elements that are known to be enriched on the surface of the fly ash particles, such as As, B, Cd, Cr, Cu, Mo, Zn, show higher concentrations in extracts (Theis & Wirth, 1977; Hansen & Fisher 1980; Ainsworth & Rai 1987). The concentrations of trace elements are also higher in acid extracts than in water extracts. Acid extracts may have value in determining the location of elements in the fly ash, but the extraction conditions are unlikely to be encountered in the field.

Some studies used the *column leaching* method (Jones & Lewis 1960; Brown *et al.* 1976; Dudas 1981; van der Sloot *et al.* 1982; Warren & Dudas 1984; Ainsworth & Rai 1987; Fruchter *et al.* 1988, 1990; Hjelmar 1990) or *field lysimeters* (Brown *et al.* 1976; Dudas 1981; Ainsworth & Rai 1987; Hjelmar 1990; Mudd & Kodikara 2000). Mattigod *et al.* (1990) summarized the leaching trend from continuous column leaching tests into two categories. Calcium, Na, K, and S are in the first group with the elements showing higher concentrations in the initial leachate, and thereafter the concentrations decline rapidly before reaching a steady state. In contrast, the elements in the second group, such as Al and K, are very low in concentration in the early leaching stages, but increase with the progress of leaching. The fly ash columns have been leached with water (Dudas 1981; Lee 1994) or dilute acid (Warren & Dudas 1984). Some column experiments have been conducted over several years (Brown *et al.* 1976; Dudas 1981; Fruchter *et al.* 1990; Hjelmar 1990). Aluminium and Si fall into the second group, and Mg, K, and Fe fall into the first group according to the study of van der Sloot *et al.* (1982), but Mg, K, and Na are in the second group in the data of Dudas (1981) and Warren & Dudas (1984).

The continuous column leaching test and the field lysimeter are better methods for understanding the longer-term leaching behaviour than the batch leaching test. On the other hand, the latter provides information on a system with a high solution to solid ratio appropriate for lagoons and also helps characterize the ash in terms of element distributions. A combination of laboratory experiments and field-based studies probably represents the ideal approach to understanding the leaching behaviour of a specific fly ash.

Table 2. *Concentrations of elements in the extract from batch leaching tests (extracted by water or acids) for pH 3.3–12.34*

Element	Concentration (mg/L)
Al	0.12–62
Ca	67–1468
Mg	<0.05–118
Na	1.87–2008
K	0.72–191
Si	0.05–46
S	33–3583
Fe	<0.005–3
As	<0.08–14
Ba	0.05–2.0
B	0.1–109
Cd	<0.01–1.8
Cr	0.02–44
Pb	<0.05–38
Mn	<0.01–290
Zn	<0.01–121
Mo	0.01–6.8
Ni	<0.01–8.5
Se	<0.05–0.4
V	0.003–1.1

Sources: Dreesen *et al.* (1977); Adriano *et al.* (1980); Hansen & Fisher (1980); Roy *et al.* (1981); Ainsworth & Rai (1987); Choi *et al.* (2002).

Case study

Batch leaching tests. The results of batch leaching with deionized water using the UK fly ashes from Drax and Meaford power stations are shown in Table 3. Three different fly ash samples were used in the study; Drax fresh,

Table 3. *Analytical result of batch leaching test with deionized water**

Treatment	Drax fresh ash				Drax weathered ash				Meaford weathered ash			
	dfd1	dfd2	dfd3	dfd4	dwd1	dwd2	dwd3	dwd4	mwd1	mwd2	mwd3	mwd4
pH	9.3	10.0	7.1	8.8	7.36	7.6	7.7	8.1	7.1	7.4	7.3	7.7
EC (mS/s)	1130	150.8	242	70	215	102.4	61	102.9	193	111.5	60.5	65.3
Ca	173.1	19.3	35.3	10	26.6	10.5	9.4	8.6	21.1	10.3	9.4	8.8
Na	42.1	4.9	2.9	1.5	4.3	2.6	1.8	1.1	3.6	4.2	2.1	3.1
K	21.3	2.8	1.9	0.4	2.4	0.5	0.3	0.1	3.6	0.4	0.7	0.2
Mg	0.99	1.50	2.12	1.40	1.98	0.78	0.57	0.47	1.27	0.71	0.62	0.55
Fe	0.25	0.04	0.04	0.1	0.02	0.02	0.01	0.01	0.14	0.03	0.08	0.06
Al	3.67	1.37	0.07	1.72	0.60	0.79	0.80	1.07	0.37	0.28	0.75	0.74
B	2.77	0.93	0.52	0.24	0.18	0.08	0.06	0.03	0.18	0.15	0.08	0.08
Ba	0.3	0.29	0.89	0.25	0.54	0.4	0.24	0.17	0.28	0.72	0.23	0.45
Cd	<0.01	<0.01	<0.01	<0.01	0.05	<0.01	<0.01	<0.01	0.16	0.01	0.01	<0.01
Co	0.05	0.02	0.02	0.02	<0.02	<0.02	<0.02	<0.02	<0.02	<0.02	<0.02	<0.02
Cr	0.30	0.06	0.04	0.03	0.01	<0.01	<0.01	<0.01	0.06	<0.01	<0.01	0.35
Cu	0.10	0.02	0.01	0.01	<0.01	<0.01	<0.01	<0.01	0.05	<0.01	<0.01	<0.01
Li	0.74	0.10	0.06	0.02	0.03	<0.01	0.01	<0.01	0.02	<0.01	<0.01	<0.01
Pb	0.15	0.10	0.05	0.05	<0.01	<0.01	<0.01	<0.01	<0.01	<0.01	<0.01	1.41
Si	1.55	1.86	3.38	1.56	1.25	1.08	1.16	1.15	1.55	1.78	1.69	1.26
Sr	0.48	0.1	0.17	0.05	0.14	0.06	0.04	0.04	0.17	0.09	0.07	0.07
V	0.36	0.26	0.21	0.24	0.11	0.12	0.11	0.09	0.07	0.09	0.09	0.07
Zn	0.18	0.02	0.15	0.02	0.06	0.06	0.03	0.01	0.12	0.05	0.04	0.02
Ni	0.18	0.02	<0.01	0.02	<0.01	<0.01	<0.01	<0.01	0.03	<0.01	<0.01	<0.01
As	0.14	0.12	0.49	0.34	0.31	0.29	0.30	0.22	0.18	0.17	0.18	0.18
Hg	0.22	0.03	0.03	0.14	0.01	<0.01	0.01	<0.01	0.01	0.03	<0.01	<0.01
Se	0.15	0.15	0.11	0.14	0.05	<0.01	<0.01	0.01	0.07	0.06	0.08	0.04
Mo	1.12	0.23	0.09	0.04	<0.01	<0.01	<0.01	<0.01	<0.01	<0.01	<0.01	0.03
Cl$^-$	6.10	2.30	0.46	0.34	7.31	5.21	1.65	0.54	3.20	3.61	1.50	2.90
NO$_3^-$	0.75	0.73	7.23	2.59	24.29	1.22	1.57	0.57	2.37	0.52	0.59	1.30
SO$_4^{2-}$	492.43	34.64	22.10	19.7	17.52	3.99	1.50	1.26	21.08	4.90	2.30	1.41

*Concentrations in mg/L unless indicated; df, Drax fresh ash; dw, Drax weathered ash; mw, Meaford weathered ash; 1, 2, 3, and 4 denote successive treatment with deionized water; EC, electric conductivity.

Drax weathered, and Meaford weathered ash. The method used in the batch leaching tests was based on the German standard (DIN 38414, Part 4) for the examination of water, wastewater and sludge. The ash samples were mixed with deionized water with a ratio of solution to solid of 500 mL to 50 g in 1 L bottles and then shaken for 24 h in an end-over-end shaker. With this water to solid proportion the results are more representative of wet ash disposal. At the end of leaching the solutions were decanted and filtered through a 0.45 μm membrane filter. This process was repeated four times. In the case of Drax fresh ash, most of the total water-soluble Ca, Na, K, and SO_4^{2-} is liberated in the first treatment. Other elements exhibiting this pattern are B, Cr, Cu, Li, Ni, Hg, Mo, and Cl^-. This trend is consistent with a surface association. However, there are other elements that differ markedly in behaviour in that comparable amounts are dissolved gradually with each leaching. Falling into this group are Mg, Al, Ba, Si, V, As and Se. One interpretation would be that these elements are present in less readily soluble components of the ash, such as the glass. However our data on the combustion behaviour of As in particular (Table 1) is not entirely consistent with this possibility. Elements intermediate in leaching behaviour are Fe, Co, Pb, Sr, Zn, and Zr.

Most of the readily soluble surface-associated elements may have been depleted in the Drax and Meaford weathered ashes, compared with the fresh ash from Drax, because of their prolonged contact time with water. The Drax ash samples have been deposited 17 years and the Meaford samples some 40 years. Loss of the readily soluble elements from the weathered fly ashes would therefore be anticipated. The results, shown in Table 3, clearly demonstrate that this is the case. The concentrations of Ca, Na, K, SO_4^{2-}, B, Cr, Cu, Li, Ni, Hg, and Mo are all much lower, with some of the trace elements below the detection limits. The only exception is NO_3^-, and the field studies, discussed later, show that fertilizers are probably responsible.

Column leaching tests. The design and running time of column leaching tests are closer to conditions found in dry ash disposal where ash is emplaced in mounds or landfills and is subjected to infiltration. The tests are generally closer to natural conditions compared with batch leaching tests, and therefore, the data generated from the column leaching test with deionized water should provide more practical data by being closer to natural weathering reactions. In the column leaching tests, the liquid to solid

proportions differ from the batch leaching procedure, with 1000 mL of solution flowing through around 900–1000 g of fly ash sample. The column leaching tests using Drax and Meaford weathered ash also extend the degree of weathering investigated in the field study, providing an extended database for the long-term behaviour of the ash.

Three fly ash samples, Drax fresh, Drax weathered and Meaford weathered, in their natural moisture condition were lightly packed into acrylic columns (5 cm diameter, 40 cm long) to achieve a uniform bulk density, comparable to values in the literature (dry bulk density: 1.01–1.43 mg/m^3, Mattigod *et al.* 1990). Then, 1000 mL of deionized water were passed through each column at a flow rate of 25 mm/day, and 50 mL of solution samples were collected every two days.

Concentration changes with solutions passing through the columns are shown in Fig. 1. The pH values are higher for the fresh ash, above 9, whereas weathered ashes are lower, around 8. For the Drax fresh fly ash, Na and K show a rapid decline after the initial leaching stage (Fig. 1a). However, Ca and SO_4^{2-}, Mg and Al concentrations in the leaching of fresh ash do not decrease rapidly with increasing volume, but remain relatively uniform throughout the leaching procedure. The leaching behaviour of Ca is somewhat different from the pattern of Ca reviewed by Mattigod *et al.* (1990). This may indicate the presence of solubility-controlling solid phase involving Ca and S, and gypsum was indeed identified as the control by equilibrium calculations using the field data (Lee & Spears 1997). Aluminium concentrations are low for most solution samples and no initial rapid decline was observed. Most trace elements including B, Li, Mo, and Cr also decrease rapidly in the initial leachates and continue to decline throughout the leaching period (Fig. 1a). Arsenic, Mn, Ni, Zn, Hg, and Se concentrations are low, recording ranges of a few to several tens of μg/L, and the Hg, Se, and Zn concentrations fell below detection levels during the experiment. Arsenic and Se are known to be volatile and therefore surface-enriched. The As and Se concentrations are higher for the first couple of leachate samples and then nearly constant for the remainder.

Leaching of Drax and Meaford weathered ashes produced much lower initial concentrations for the readily soluble elements, in particular for Na and K (Fig. 1b). Boron, Cr, Ni, and Mo are also present at much lower concentrations in the weathered ash leachates. The higher concentrations for the Ca and S in the initial leachate

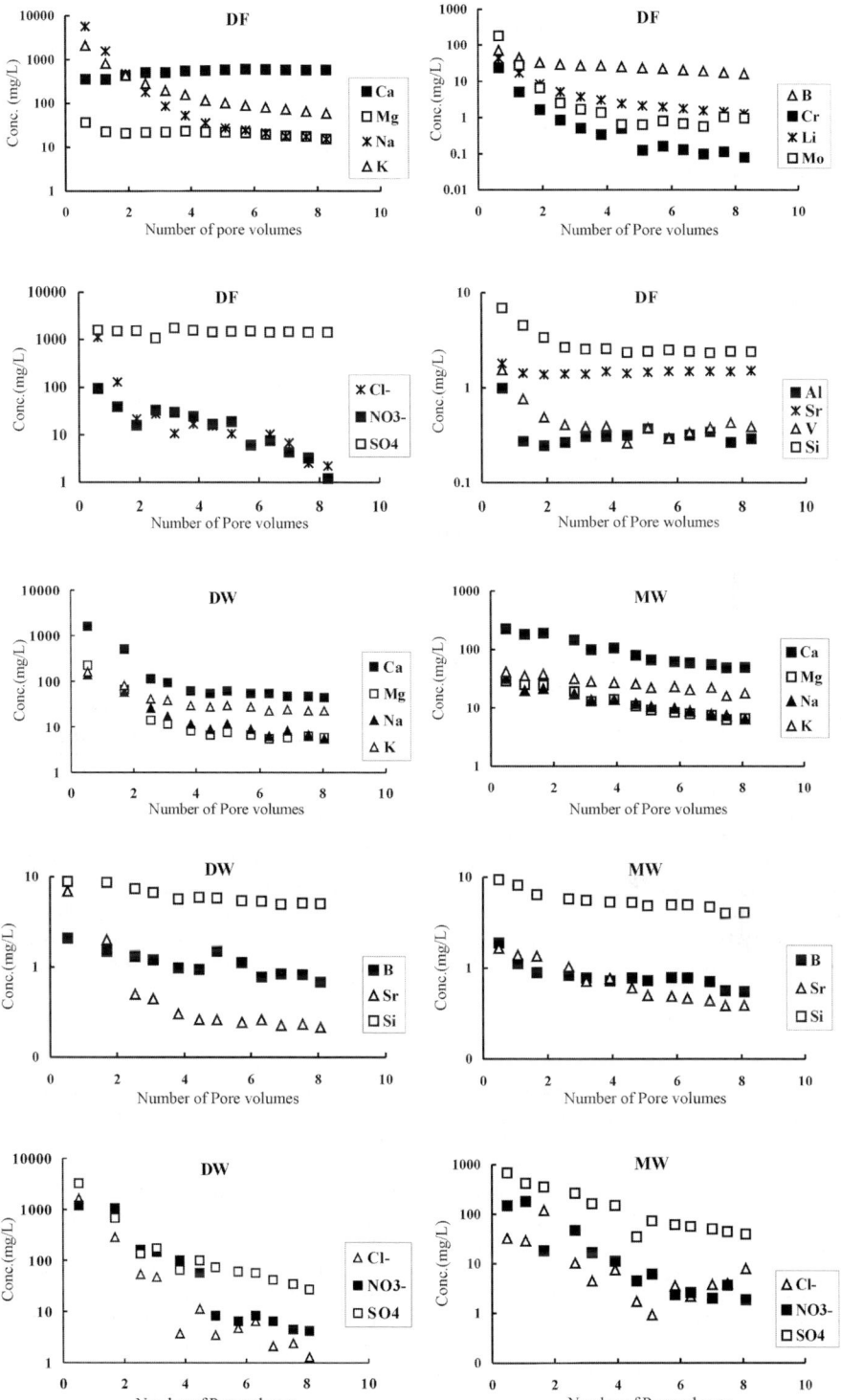

Fig. 1. Variations in elemental concentrations in fresh Drax, weathered Drax and weathered Meaford fly ash eluates treated with deionized water in column leaching tests (DF: Drax fresh, DW: Drax weathered, MW: Meaford weathered).

seem to be from the dissolution of soluble secondary precipitates, such as gypsum, that may have been formed during the dissolution of fresh fly ash. A range of trace elements, including Cd, Co, Cu, and Hg, is detectable in all the leachate samples, although the concentrations for some elements are close to detection limits. Silicon, mainly associated with the glass fraction of the ash particle, is continuously released and the constant liberation of the trace elements from the weathered ash indicates the trace elements are also incorporated within the aluminosilicate matrices, as well as associated with the particle surface. In this case, it is not only the disposal of fresh ashes in landfill sites that is of environmental concern, but also the prolonged interaction of weathered ash. The fly ashes that we have referred to as weathered have been exposed for up to 40 years and yet some fraction of the mobile elements still remains. This suggests that either weathering of the surface-associated elements is less effective than we would have predicted or some of the elements are also incorporated into less soluble components of the ash. In either case, the findings from the leaching tests using Drax and Meaford weathered fly ashes demonstrates that the weathered fly ashes are not inert, and that further release of elements into the environment is possible.

Comparison of leaching behaviour between batch and column leaching tests. Results from the batch and column leaching tests broadly agree, and the same conclusions can be drawn for the column leaching tests as for the batch leaching tests. Lithium, Mo and B were readily soluble in both tests, and the other trace elements also revealed comparable leaching patterns to one another. The experiments are inter-related, in term of the specific factors controlling the solubility of elements. The column and batch leaching experiments are complementary to each other, and also to field observations.

Batch leaching tests are generally taken as an initial step to evaluate toxicity of the ashes because the batch tests represent the best leaching conditions and generate the highest quantity of leachates (Benito *et al.* 2001). Serial batch leaching tests also often successfully simulate ash pond conditions (Praharaj *et al.* 2002), whereas the column leaching tests simulate dry ash disposal (Benito *et al.* 2001).

Geochemical modelling

The relationship between the solution concentrations in the fly ash leachates and specific chemical reactions in waste/water system has been examined from a fundamental thermo-chemical perspective by some authors, including Ainsworth & Rai (1987), Rai *et al.* (1988), Fruchter *et al.* (1988, 1990), Lee (1994), Lee & Spears (1997, 1998), and Mudd & Kodikara (2000). This approach involves the concept of waste as a mixture of discrete solid phases that control the concentrations of elements in the solution through the processes of dissolution/precipitation and adsorption/desorption (Rai *et al.* 1988). Once an element has achieved an equilibrium concentration, its concentration in aqueous solution is determined by specific solubility-controlling solid phases, regardless of the location, depth and age of the sample collected. This approach is an appropriate way of dealing with the longer-term weathering reactions of coal ash, in which equilibrium reactions are a primary control. The analytical data from the leaching tests were processed using WATEQ4F (Ball *et al.* 1987).

Several solubility-controlling phases were predicted from the analytical data from the column and batch leaching tests using water. Gypsum was identified as a concentration-limiting phase for Ca^{2+} and SO_4^{2-} in the eluates from the laboratory leaching tests and also in the field studies (Lee & Spears 1997). Other solubility-controlling solid phases identified in the leaching tests include quartz and wairakite $(CaAl_2(SiO_3)_4 \cdot 2H_2O)$ for SiO_2, $Al(OH)_3$ for Al, $Fe(OH)_2(am)$ for Fe, tenorite (CuO) or malachite $(Cu(OH)_2CO_3)$ for Cu, and $PbCO_3$ for Pb. These findings agree well with the previous studies (e.g., Rai *et al.* 1988; Fruchter *et al.* 1990). However, it should be noted that the presence of the predicted solubility-controlling solid phases has not been confirmed by either X-ray diffraction (XRD) or SEM. One reason for this is that the concentrations of these phases may be very low and consequently difficult to detect (see also Donahoe, this volume).

If phases that control solubility are known to be present then the calculations can be used to predict the chemical composition of leachates with some confidence. The geochemical modelling undertaken in our study suggests that both in the fresh and weathered ash, reactions of many elements are generally controlled by a solubility-controlling phase involving secondary reaction products. It is, however, necessary to identify these minerals to make the geochemical modelling a practical and feasible tool in understanding the reaction of fly ash quantitatively and in modelling the natural weathering of fly ash.

Field-based leaching studies

The elevated concentrations of soluble salts, such as those of Ca, Mg, Na, K, and SO_4 and trace elements in fly ash have led to concerns regarding disposal, but few field studies have been made due to the difficulties and expense of full-scale monitoring of ash mounds (Carlson & Adriano 1993). More information on the potential environmental influence of fly ash disposal on groundwater is available from laboratory studies, as discussed above, but these may not be fully representative of natural conditions. One method of fly ash disposal is to pump it into lagoons and most field-based studies have been conducted on such disposal sites (Dreesen et al. 1977; Theis et al. 1978; Theis & Richter 1979). With such a wet ash disposal method the ratio of water to solids is high and the emphasis of the published work has been on the early reactions. The work by Theis & Richter (1979) detailed the movement of a seepage plume in the groundwater from an ash disposal pond, and $Cr(OH)_3$, $Cu_2(OH)_2CO_3$ and $Pb(OH)_2$ were identified as possible solubility-controlling solids for Cr, Cu, and Pb. A study by Hardy (1981) showed elevated concentrations of Ca, K, Fe, and S (as SO_4^{2-}) and As, B, Mn, Mo, Ni, Sr, and Zn in groundwater samples collected from an area bordering a fly ash settling pond compared with uncontaminated groundwater. In monitoring a fly ash mound for three years, Simsiman et al. (1987) observed large B, Na, and SO_4^{2-} plumes in the adjacent groundwater system. Sakata (1987) reported Ca and S (as SO_4^{2-}) as major ions in the leachate of weathered fly ash from mounds and showed that infiltration had also taken place into the underlying soil layer. In their study of a soil amended with fly ash, focussing mainly on the plants, Adriano et al. (2002) noted that soluble salts were leached downwards from the soil within a period of four years, although groundwater samples did not show detectable changes in elemental concentrations with most of the trace elements at or below detection limits. A germane observation (Adriano et al. 2002) on amending soils with fly ash was that unweathered fly ash should be allowed to weather to enable leaching of harmful constituents. The results of groundwater studies do demonstrate that elements added to the groundwater will be dispersed in a similar manner to elements added to surface waters. Groundwater flow is also considered in a hypothetical case study (Hansen et al. 2002) that combines information on leachate generation from fly ash with the integration of speciation and fluid flow models.

Fly ash is extensively used in civil engineering and is considered relatively impermeable with typical permeability values in the range 2×10^{-5} to 3×10^{-7} m/s, but even at these rates appreciable infiltration can take place through the ash and into an underlying aquifer over a period of several years. In the technical specifications issued by industry it is claimed that the low permeability and pozzolanic (self-cementing) properties make fly ash a suitable material for containing contaminants on brown field sites (old industrial sites). Fly ash may well be extremely valuable in this role, but possibly more because of reaction potential than low permeability. Furthermore, the indigenous coals in the UK produce Type-F (not Type-C) fly ashes, and it is the latter with their high Ca contents that are pozzolanic. In the case study described below there was a need to demonstrate that water ingress did take place into the ash mounds and that all the precipitation was not lost as runoff.

Case study

In our field-based work (Lee & Spears 1995, 1997, 1998) samples were investigated from the ash mounds at Drax and Meaford power stations. At the well-managed and carefully engineered Drax mound, the water content of the ash when tipped is less than 5 wt%. The fly ash is transported directly to the mound and tipped laterally onto a permeable layer of coarse furnace bottom ash containing field drains that overlie impermeable boulder clay. The discharge from the field drains has a chemical signature of the ash demonstrating infiltration through the mound. There is a water-filled ditch around the mound (supporting fish) that is fed by the discharge from the drains and the surface runoff from the mound, which then flows into natural watercourses. The mound is progressively covered with topsoil, typically 0.3 m, and returned to agricultural use. Because of the lateral construction of the mound it is possible to sample fly ash of a known age, and the ash that is described here was 17 years old. At Meaford the fly ash was pumped into a series of lagoons, eventually being transferred to an off-site mound covered with topsoil and returned to agricultural use. The fly ash sampled at Meaford is believed to be about 40 years old.

A hand-auger drill was used to obtain samples at 30 cm depth intervals down to 5.0 m. At these depths the fly ash is unsaturated in both sites. Porewater samples were extracted from the fly ash samples by centrifugation, following the method of Edmunds & Bath (1976) and analysed by inductively coupled plasma spectrometry (ICP), atomic absorption spectrometry (AAS), and ion chromatography. Because the porewaters were obtained by centrifuge extraction, the

HCO_3^- was not determined. Ion balance calculations show, however, that this is a minor component in the system. The chemical composition of the fly ash was determined by X-ray fluorescence (XRF). The purpose in studying a sequence of samples from a borehole is to observe changes with depth reflecting continuing reaction between fly ash and infiltrating porewaters. The porewaters are sensitive to the weathering reactions of the fly ash, whereas changes in the bulk composition will only be detected if the weathering changes are major and there have been significant losses from the system. At both sites the water content remains relatively constant with depth in the boreholes. At Drax, for example, the mean water content in the upper part of the borehole is 12.3 ± 2.1 wt% ($n = 10$) and in the lower part is 11.9 ± 1.1 wt% ($n = 10$). If the water content did vary, possibly through evaporation, depth-related trends might have an explanation other than reaction between infiltrating porewater and fly ash.

At both the Drax and Meaford sites, whole rock and clay fractions were analysed by XRD. Clay minerals were found in the fly ash at Drax, but the suite is the same as in the topsoil, and there is little doubt that the clay minerals are present in the fly ash because of translocation from the overlying soil resulting from rainwater infiltration However, the possibility cannot be excluded that some of the kaolinite might be neoformed from the fly ash and this would be consistent with porewater compositions falling in the stability field of kaolinite. With this possible exception, secondary weathering products were not identified in the fly ash at either site. We have been able to demonstrate that there was movement of the clay minerals through the fly ash because the suite of clay minerals could only have originated in the topsoil, but this does raise the possibility that there has also been movement of similar sized submicron fly ash particles through the fly ash that could contribute towards depth-related concentration changes. Although this size of fly ash particle will have high trace element concentrations, as discussed earlier, it is a relatively insignificant fraction of the whole and therefore any movement should not lead to depth-related concentration changes. We have not detected such changes in the trace element concentrations in the bulk ash even due to reaction with infiltrating porewaters, as we discuss later.

The composition of porewaters in the fly ash mounds provides a sensitive record of reactions between infiltrating porewaters and fly ash. The depth trend of element concentrations is of particular interest. Increasing concentrations with depth are indicative of continued reaction between ash and the infiltrating porewater as a function of contact time. As pore fluids move downwards, the older porewater is located at a greater depth compared with younger porewaters, and the greater contact time results in higher solution concentrations. If concentrations reach constant values with depth this suggests equilibrium has been achieved, but confirmation is required from equilibrium calculations. Elemental concentrations decreasing with depth could be due to an element originating from an external source, with concentrations decreasing either by reaction or by dilution.

The major cations present in the porewaters are Ca, Mg, K, and Na, and anions SO_4^{2-}, Cl^-, and NO_3^-. These are the major ions that would be expected in the leaching of fly ashes based on the batch and column leaching described earlier, with the exception of Cl^- and NO_3^-. In Fig. 2 are plotted the depth variations of major elements in the porewaters from the Drax and Meaford ash disposal sites. Concentrations show systematic changes with depth, and there are similarities between the two sites, but even where trends are similar absolute values differ. Sulphate concentrations increase progressively with depth in both sites. This is a consequence of the accumulating effect of reaction of the fly ash with porewaters as they infiltrate. The initial values are higher for the Drax mound, which is consistent with the younger age of this material and presumably less weathered condition. If sulphate shows one trend, Cl^- and NO_3^- concentrations in the Drax site show a very different trend, with values high near surface and decreasing with depth, although there is a minor spike between 2.5 and 3.0 m. There is no question that the sulphate is derived from the fly ash, which is known to be deficient in nitrate, and the different trend for NO_3^- and Cl^- is a further indication that they originate elsewhere. The near-surface concentrations are very high, particularly for NO_3^-, and fertilizer is the most likely source. Possibly the minor spike noted in the NO_3^- and Cl^- depth trends could be a record of an earlier application of fertilizer. The overall decrease with depth is attributed to mixing and dispersion. The Cl^- and NO_3^- concentrations in the Meaford porewater also generally decrease with depth, but concentrations are lower and this is probably due to less fertilizer being used at this site.

Sodium and K in the Drax porewaters decrease with depth initially and then increase. The near-surface values are thought to be due to fertilizers, and below this zone the fly ash contribution

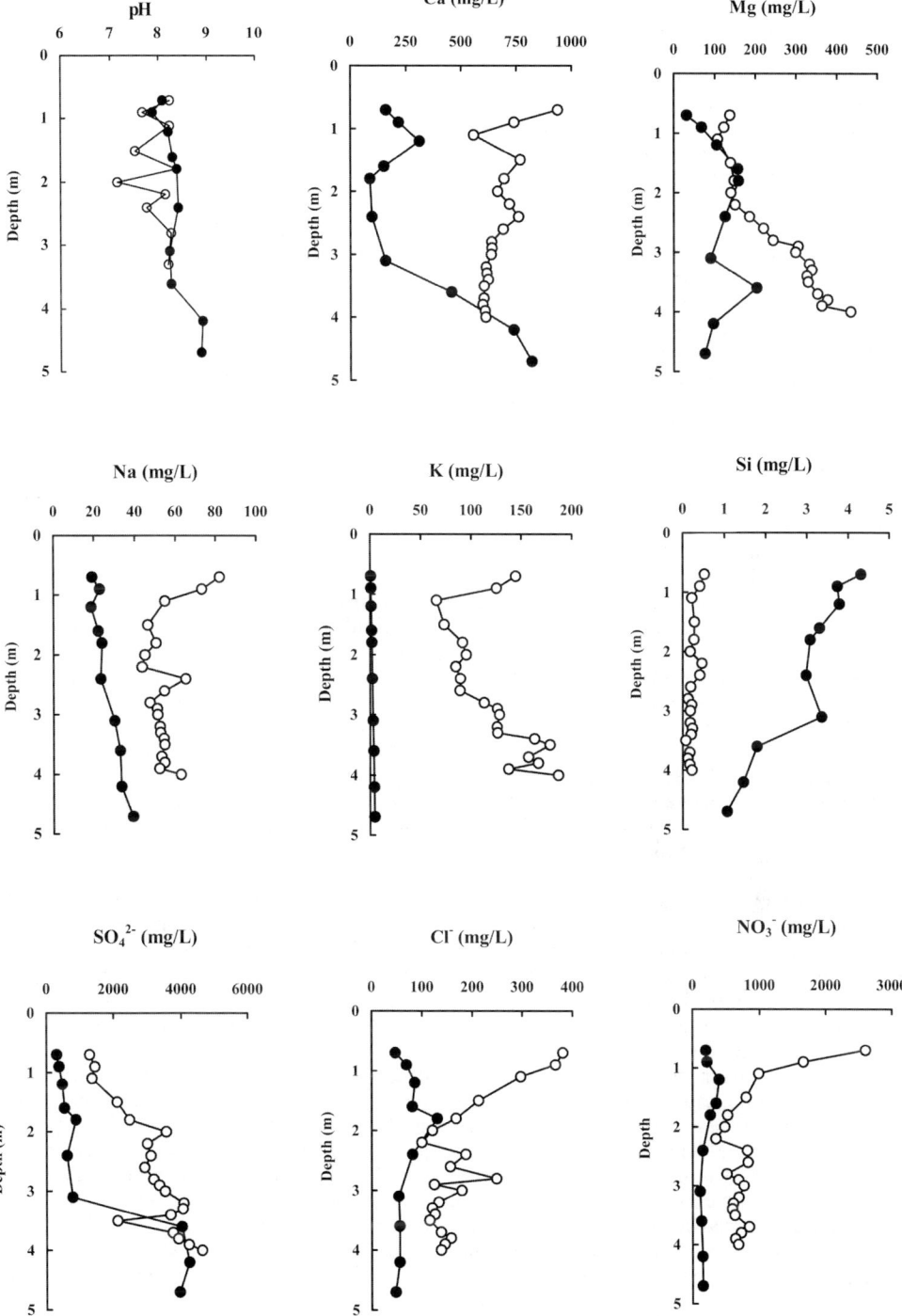

Fig. 2. Variation of major elements in porewaters as a function of depth in the Barlow (open circles) and Meaford (solid circles) ash disposal mounds.

becomes more apparent. The depth trend is thus a combination of surface and internally derived material. In the Meaford porewaters, Na and K increase with depth but the lower values agree with a more weathered material compared with Drax. Magnesium concentrations increase with depth in the Drax porewaters, whereas the lower values in the Meaford do not show a consistent depth trend. The Ca concentrations are high and relatively constant in the Drax porewaters and equilibrium may have been achieved. Calculations should confirm whether or not this is the case and indicate a possible controlling solid phase. In the Meaford porewaters, the Ca concentrations are lower, as with other elements, but increase with depth and approach values comparable to those recorded at Drax. Silicon is one of the few elements higher in the Meaford porewaters and concentrations decrease with depth. In the Meaford site the pH values are marginally higher (8.38 ± 0.33, $n = 10$) than at Drax (7.96 ± 0.40, $n = 10$), which is a possible explanation for the difference in the silica concentrations in the two sites, but there is not a decrease in pH values with depth at Meaford (Fig. 2) to account for the decrease in silica concentrations with depth.

Nearly all trace elements show depth-related trends, with minor fluctuations (Fig. 3). The elements that increase with depth in porewaters from both sites are B, Cr, Li, Mo, Pb, and Ni. Concentrations are mostly higher in the Drax porewaters in a comparable manner to the major ions. Arsenic and Se are similar to each other in the porewaters from the two sites. Concentrations are lower than the other elements considered here and both elements show a general increase with depth in the Drax porewaters whereas at Meaford there is not a clear depth trend. The Ba concentrations decrease with depth in the Drax samples in a comparable manner to the NO_3^- and Cl^- depth trends and there is also a similar reversal of the depth trend between 2 and 3 m. The fact that the trends appear similar suggests that at least some of the Ba may be derived from the surface. In the Meaford samples Ba does not decrease with depth, but neither do NO_3^- and Cl^-, and therefore there is further support for a possible surface-derived component in the Drax porewaters. The Ba concentrations at Meaford fluctuate and do not vary consistently with depth making interpretation difficult. The Sr porewater concentrations in the Meaford samples do show a more consistent depth relationship suggesting dissolution from the ash, whereas at Drax, below the surface zone, values are approximately constant, which could

an equilibrium value. A final point is that average concentrations of some of the trace elements in the porewaters are greater than World Health Organization (WHO) guidelines on drinking water quality, with B, Pb, Mo, As, and Ba present in the porewaters at concentrations greater by a factor of ten.

There are changes in the concentrations of some of the mobile major elements in the fly ashes as a result of infiltration. The result of such leaching should be a reduction in bulk concentrations in the fly ashes. A comparison of the fresh ash composition with an average for the weathered fly ash from Drax shows decreases in SO_3 from 0.35 to 0.14 wt%, Na_2O from 1.56 to 1.02 wt%, CaO from 1.90 to 1.39 wt%, and MgO from 1.73 to 1.48 wt%. These values are consistent with the major ion composition of the porewaters. Using the S data it is possible to calculate approximate infiltration rates through the fly ash. The total amount of S (μg) removed from the fly ash is known from the bulk analyses, as is the S concentrations in the porewaters ($\mu g/mL$), and from which can be obtained the total volume passing through unit area in the total time and thus, the infiltration rate per year. This calculation gives a rate of 42 cm/y, which is a reasonable value given that it is approximately half of the potential infiltration based on the rainfall averages for the period. A similar calculation for the Meaford site based on S and Ca gave 25 and 29 cm/y (Lee & Spears 1997). Although the error margins are high, the calculations do demonstrate that changes in bulk ash compositions and elemental porewater concentration are consistent with realistic infiltration rates.

The Fe^{2+}/Fe^{3+} ratio in the fly ash is a sensitive indicator of the extent of weathering. In the fresh fly ash samples from Drax the ratio of wt% FeO to total Fe (expressed as wt% Fe_2O_3) is 0.342 and in the weathered ash from the Drax mound the ratio is 0.326. This is a small difference that suggests that although the oldest fly ash has been subjected to weathering for 17 years there has been little oxidation of iron. It is also instructive to compare the trace element concentrations in fresh ash with weathered ash. Trace elements are passing into solution in the porewaters and, if the rate at which this is happening and the time period are large enough, there will be detectable changes in the bulk composition. The analyses in Lee & Spears (1998, Table 3) show little difference with the fresh ash composition that can be attributed to leaching, including elements such as Pb, Zn, and Cu that have a significant surface association in the fly ash. A similar conclusion is reached when

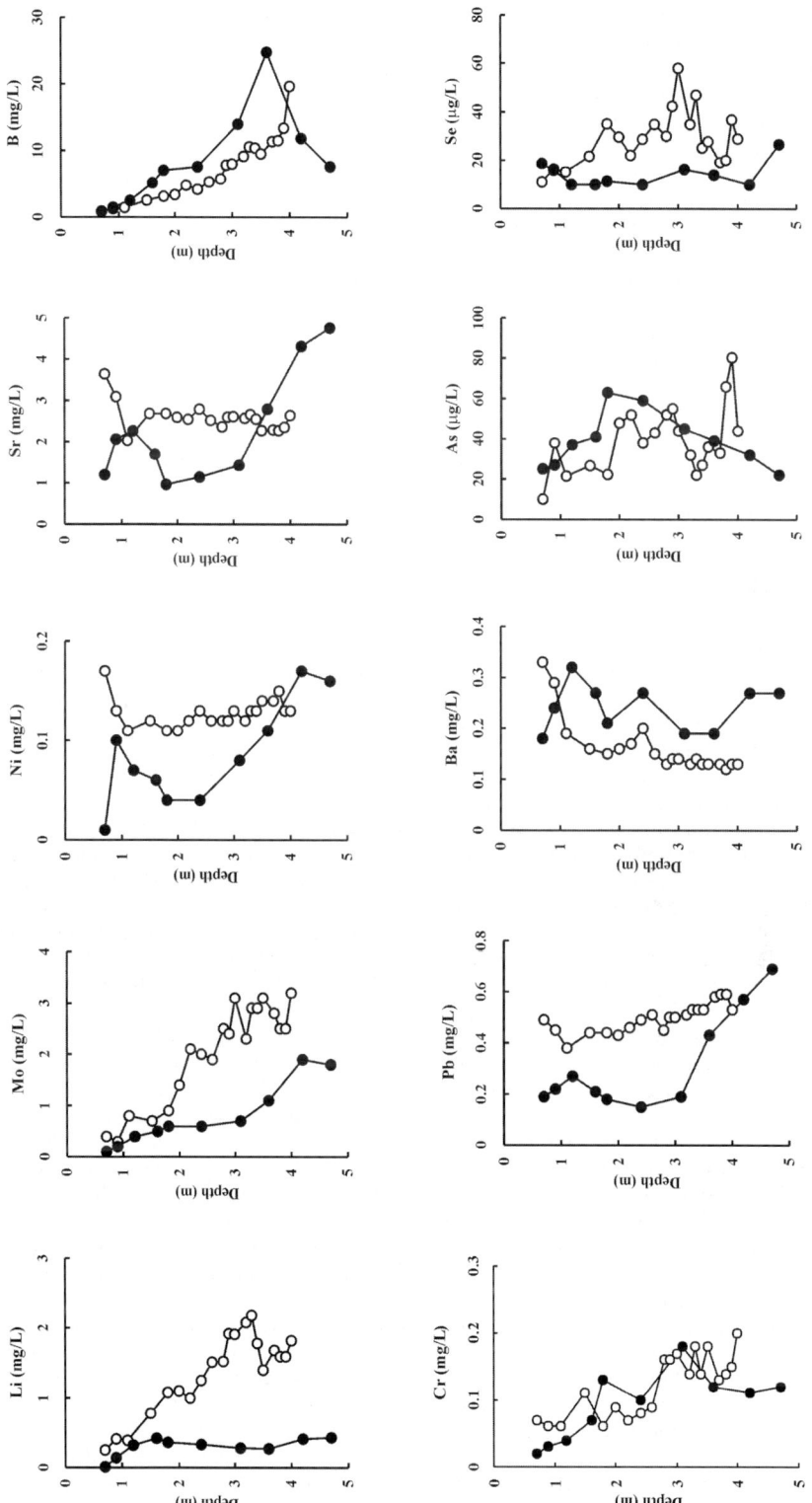

Fig. 3. Variation of trace elements in porewaters as a function of depth in the Drax (open circles) and Meaford (solid circles) ash disposal mounds.

samples from the boreholes are compared on a depth basis. The trace element data are inherently more variable, but this only partly explains why differences cannot be proved. The prime reason is that a relatively small volume of water has passed through the ash since emplacement. Using the calculated infiltration values and the observed Pb concentrations in the porewaters, the Pb concentration in the fly ash would be lowered in total by an average of less than 10 ppm, which could well be difficult to demonstrate. Essentially the mass in the solid is very much greater than the mass in solution. What this means is that porewaters emanating from fly ash mounds pose less of a threat than ash ponds or other situations, where the proportion of water to ash is much greater. Although some of the trace elements are present at concentration levels in excess of WHO drinking water guidelines in the ash mounds, the impact on an underling aquifer could be relatively minor because of the infiltration rate. At the Drax ash mound impermeable boulder clay protects the underlying aquifer and the infiltration through the mound escapes into surface drainage ditches that, as we noted earlier, support an aquatic fauna. If the infiltration did pass into the aquifer its impact on the water quality might well have been no greater.

Several elements, including Ca, Ba, and Sr have depth profiles in the porewaters; that is, the concentrations do achieve more constant values with depth, suggesting attainment of equilibrium concentrations. Porewater analytical data were processed using a computer modelling code WATEQ4F (Ball et al. 1987) to detect any potential solid phases controlling the solubility of specific elements. Calculated activities are compared with the activities in equilibrium with respect to a specific solid phase, or a single ion activity product is compared with the equilibrium constant of a specific solid phase to determine the saturation state of the porewater. Both Drax and Meaford porewaters are undersaturated with respect to gypsum near surface, but are saturated at depth. At Drax this takes place at about 1 m, whereas at Meaford the depth is about 4 m (Lee & Spears 1998, fig. 3), which is consistent with the more weathered nature of the ash at Meaford. Calculations suggest that Ba and Sr activities are linked to the presence of solubility-controlling solid phases. Felmy et al. (1993) suggested co-precipitation of Ba and Sr in a mixed sulphate as the possible solid phase controlling the Ba and Sr activities. Other predicted solubility-controlling solid phases are $Fe(OH)_3$ for Fe, $Al(OH)_3$ for Al, and CuO for Cu (Lee & Spears 1997). The inferred solubility-controlling

solid phases have not been identified directly by either XRD or SEM, but this is probably because of the minor amounts involved.

Conclusions

Many trace elements are present in coal ashes at concentration levels far in excess of average crustal values. Although new uses are being found for fly ashes, there remains an excess production that ends in lagoons, ash mounds and landfill sites. In the ashes produced in coal-fired power plants trace elements are present in minerals such as quartz, mullite and magnetite, but their overall contribution to the bulk is minimal. The trace elements present in the glass are quantitatively more important, but for the more volatile elements it is surface association that is particularly important, because this is the major location in the ash for some elements and they are potentially leachable.

Based on analyses of size-fractionated fly ash, which confirmed the high trace element concentrations in the finest fractions, a relationship between element concentration and surface area has been observed and used to calculate the percentage surface association, which is greater than approximately 50% for Mo, As, Cu, Zn, and Pb. Results from a LA-ICP-MS study suggest that U, Tl, and Se may also be in this group, although U has been shown to be present within the glass. In leachates from fly ashes Ca and SO_4^{2-} are usually the dominant ions, with lesser concentrations of Na, Mg, K, and Cl^-. The pH of most leachates is alkaline, which would be predicted from the reaction of glass and minerals with porewaters, but a significant number are acidic, and this is associated with the use of high-S coals and the combustion conditions. Higher concentrations of elements such as Cd, Cu, Mn, Ni, Pb, and Zn are present in solution under acid conditions, whereas under alkaline conditions concentrations of As, U, Mo, Se, and V are higher.

Batch and column leaching tests on fresh and weathered UK fly ashes conform to the general pattern of leachate generation. The batch leaching tests, with the high solution to solid ratios, are more akin to conditions in lagoons, and the columns to conditions in ash mounds. Although the weathered ashes come from ash mounds dating back 17 and 40 years, a proportion of the mobile trace elements still remains. Borehole samples from two UK ash mounds show depth-related concentration changes in the porewaters that can be attributed to reaction with the fly ash. Constant values with depth are an indication of equilibrium, and calculations confirm that solutions achieve saturation with respect to gypsum at depth. Infiltration rates can be calculated from changes in the Ca and S content of the ash, but in general there are few detectable changes in element concentrations due to leaching and this is due to the relatively low infiltration rate through the ash. The potential impact on an underlying aquifer would appear to be relatively minor.

DAS would like to acknowledge the recent support of BCURA and the earlier support from the National Power/PowerGen Joint Environmental Programme. The views expressed in this Chapter, however, are those of the authors and not necessarily those of the power generators. SL would like to acknowledge the financial support from Korea Science Foundation (Fund No. Ro1-2000-000-00057-0). The authors very much appreciate the thoughtful reviews by Drs A. Page, O.M. Saether and K.S. Reddy and also the painstaking editorial work of R. Gieré.

References

ADRIANO, D. C., PAGE, A. L., ELSEEWI, A. A. & CHANG, A. C. 1980. Utilization and disposal of fly ash and other coal residues in terrestrial ecosystem: A review. *Journal of Environmental Quality*, **9**, 333–344.

ADRIANO, D. C., WEBER, J., BOLAN, N. S., PARAMASIVAM, S., KOO, B.-J. & SAJWAN, K. S. 2002. Effects of high rates of coal fly ash on soil, turf grass and groundwater quality. *Water, Air, and Soil Pollution*, **139**, 365–385.

AINSWORTH, C. C. & RAI, D. 1987. *Chemical Characterization of Fossil Fuel Wastes*. EPRI Rep. EA-5321, Electric Power Research Institute, Palo Alto, CA.

BALL, J. W., NORDSTROM, D. K. & ZACHMAN, D. K. 1987. *WATEQ4F – A Personal Computer FORTRAN Translation of the Geochemical Model WATEQ2 With Revised Database*. US Geological Survey, Open-file report 87–50. US Geological Survey, National Center, Reston, VA.

BENITO, Y., RUIZ, M., COSMEN, P. & MERINO, J. L. 2001. Study of leaches obtained from the disposal of fly ash from PFBC and AFBC processes. *Chemical Engineering Journal*, **84**, 167–171.

BROWN, J., RAY, N. J. & BALL, M. 1976. The disposal of pulverised fuel ash in water supply catchment areas. *Water Resources*, **10**, 1115–1121.

BUSTIN, R. M. & WUST, R. A. J. 2003. Maturation, organic. *In*: MIDDLETON, G. V. (ed) *Encyclopedia of Sediments and Sedimentary Rocks*. Kluwer Academic Publishers, Dordrecht, 424–429.

CARLSON, C. L. & ADRIANO, D. C. 1993. Environmental impacts of coal combustion residues. *Journal of Environmental Quality*, **22**, 227–247.

CARRAS, J. N. 1995. The transport and dispersion of plumes from tall stacks. *In*: SWAINE, D. J. & GOODARZI, F. (eds) *Environmental Aspects of Trace Elements in Coal*. Kluwer Academic Publishers, Dordrecht, 146–177.

CHOI, S.-K., LEE, S., SONG, Y.-K. & MOON, H.-S. 2002. Leaching characteristics of selected Korean fly ashes and its implications for the groundwater composition near the ash disposal mound. *Fuel*, **81**, 1083–1090.

CLARK, R. B., ZETO, S. K., RICTCHEY, K. D., WEDDELL, R. R. & BALIGAR, V. C. 1995. *Coal combustion By-Product Use in Acid Soil: Effects on Maize Growth and Soil pH and Electrical Conductivity*. ASA Special Publications, Madison, WI, vol. 58.

CLARKE, L. B. & SLOSS, L. L. 1992. *Trace Elements – Emissions from Coal Combustion and Gasification*. IEA Coal Research, London, 111 pp.

DANIELS, W. L., STEWART, B., HAERING, K. & ZIPPER, C. 2002. *The Potential for Beneficial Reuse of Coal Fly Ash in Southwest Virginia Mining Environments*. Virginia Cooperative Extension Virginia Tech, Publication number 460-134.

DAVIDSON, R. M. 2000. *Modes of Occurrence of Trace Elements in Coal*. IEA Coal Research, London, pp. 36.

DEER, W. A., HOWIE, R. A. & ZUSSMAN, J. 1966. *An Introduction to the Rock Forming Minerals*. Longmans, London, pp. 528.

DIN 38414 1984. *German Standard Methods for the Examination of Water, Waste Water and Sludge: Sludge and Sediments (Groups)*. Fachgruppe Wasserchemie in der GDCH, Normausschuss Wasserwessen im DIN (eds), VCH, Weinheim, Part 4. 641–658.

DONAHOE, R. J. 2004. Secondary mineral formation in coal combustion byproduct disposal facilities: implications for trace element sequestration. *In*: GIERÉ, R. & STILLE, P. (eds) *Energy, Waste and the Environment: a Geochemical Perspective*. Geological Society, London, Special Publications, **236**, 641–658.

DREESEN, D. R., GLADNEY, E. S., OWENS, J. W., PERKINS, B. L., WIENKE, C. L. & WANGEN, L. E. 1977. Comparison of levels of trace elements extracted from fly ash and levels found in effluent waters from a coal fired power plant. *Environmental Science and Technology*, **10**, 1017–1019.

DUDAS, M. J. 1981. Long-term leachability of selected extracts from fly ash. *Environmental Science and Technology*, **15**, 840–843.

ECOBA 2003. World Wide Web Address: http://www.ecoba.com/docs/ccps/product.

EDMUNDS, W. M. & BATH, A. H. 1976. Centrifuge extraction and chemical analysis of interstitial waters. *Environmental Science and Technology*, **10**, 467–472.

ELSEEWI, A. A., PAGE, A. L. & GRIMM, S. 1980. Chemical characterization of fly ash aqueous systems. *Journal of Environmental Quality*, **9**, 424–428.

EVANS, D. W., WIENER, J. G. & HORTON, J. H. 1980. Trace element input from a coal-burning power plant to adjacent terrestrial environments. *Journal of Air Pollution Control Association*, **30**, 567–573.

FELMY, A. R., RAI, D. & MOORE, D. A. 1993. The solubility of (Ba, Sr)SO$_4$ precipitates: Thermodynamic equilibrium and reaction path analysis. *Geochimica et Cosmochimica Acta*, **57**, 4345–4363.

FISHER, R. V. & SCHMINCKE, H. U. 1984. *Pyroclastic Rocks*. Springer-Verlag, Berlin.

FISHMAN, N. S., RICE, C. A., BREIT, G. N. & JOHNSON, R. D. 1999. Sulfur-bearing coatings on fly ash from a coal-fired power plant: composition, origin, and influence on ash composition. *Fuel*, **78**, 187–196.

FRUCHTER, J. S., RAI, D., ZACHARA, J. M. & SCHMIDT, R. L. 1988. *Leachate Chemistry at the Montour Fly Ash Test Cell*, EPRI-EA-5922. Electric Power Research Institute, Palo Alto, CA.

FRUCHTER, J. S., RAI, D. & ZACHARA, J. M. 1990. Identification of solubility-controlling solid phases in a large fly ash field lysimeter. *Environmental Science and Technology*, **24**, 1173–1179.

GOLDSCHMIDT, V. M. 1954. *Geochemistry*. Oxford University Press, Oxford, 454 pp.

GIERÉ, R., CARLETON, L. E. & LUMPKIN, G. R. 2003. Micro- and nanochemistry of fly ash from a coal-fired power plant. *American Mineralogist*, **88**, 1853–1865.

GRISAFE, D. A., ANGINO, E. E. & SMITH, S. M. 1988. Leaching characteristics of a high-calcium fly ash as a function of pH: a potential source of selenium toxicity. *Applied Geochemistry*, **3**, 601–608.

HANSEN, L. D. & FISHER, G. L. 1980. Elemental distribution in coal fly ash. *Environmental Science and Technology*, **14**, 1111–1117.

HANSEN, Y., NOTTEN, P. J. & PETRIE, J. G. 2002. The environmental impact of ash management in coal-based power generation. *Applied Geochemistry*, **17**, 1131–1141.

HARDY, M. A. 1981. *Effects of Coal-Fly Ash Disposal on Water Quality in and around the Indiana Dunes National Lake-Shore, Indiana*. Water-Resources Investigations, 81–16. USGS Indianapolis.

HJELMAR, O. 1990. Leachate from land disposal of coal fly ash. *Water Management and Research*, **8**, 429–449.

HOWER, J. C., RATHBONE, R. F., ROBERTSON, J. D., PETERSON, G. & TRIMBLE, A. S. 1999. Petrology, mineralogy, and chemistry of magnetically-separated sized fly ash. *Fuel*, **78**, 197–203.

JAMES, W. D., GRAHAM, C. C., GLASCOCK, M. D. & HANNA, A.-S. G. 1982. Water-leachable boron from coal ashes. *Environmental Science and Technology*, **16**, 195–197.

JONES, D. R. 1995. The leaching of major and trace elements from coal ash. *In*: SWAINE, D. J. & GOODARZI, F. (eds) *Environmental Aspects of Trace Elements in Coal*. Kluwer Academic Publishers, Dordrecht, 221–262.

JONES, L. H. & LEWIS, A. V. 1960. Weathering of fly ash. *Nature*, **185**, 404–405.

KOPSICK, D. A. & ANGINO, E. E. 1981. Effect of leachate solutions from fly and bottom ash on groundwater quality. *Journal of Hydrology*, **54**, 341–356.

LEE, S. 1994. *The Long-Term Weathering of PFA and its Implications for Groundwater Pollution*. PhD thesis, University of Sheffield.

LEE, S. & SPEARS, D. A. 1995. The long-term weathering of PFA and implications for groundwater composition. *Quarterly Journal of Engineering Geology*, **28**, S1–S16.

LEE, S. & SPEARS, D. A. 1997. Natural weathering of pulverised fuel ash and porewater evolution. *Applied Geochemistry*, **12**, 367–376.

LEE, S. & SPEARS, D. A. 1998. Potential contamination of groundwater by pulverised fuel ash. *In*: MATHER, J., BANKS, D., DUMPLETON, S. & FERMOR, M. (eds) *Groundwater contaminants and their Migration*. Geological Society of London, Special Publications, **128**, 51–61.

LOWE, D. J. 1986. Controls on the rate of weathering and clay mineral genesis in airfall tephras: a review and New Zealand case study. *In*: COLEMAN, S. M. & DETHIER, D. P. (eds) *Rates of Chemical Weathering of Rocks and Minerals*. Academic Press, New York, 265–330.

MANTZ, O. E. 1998. Coal fly ash: A retrospective and future look. *Energeia*, **9/2**, 1–3 (CAER, University of Kentucky).

MARTINEZ-TARAZONA, M. R. & SPEARS, D. A. 1996. The fate of trace elements and bulk minerals in pulverised coal combustion in a power station. *Fuel Processing Technology*, **47**, 79–92.

MATTIGOD, S. V. 1983. Chemical composition of aqueous extracts of fly ash: Ionic speciation as a controlling factor. *Environmental Technology Letters*, **4**, 485–490.

MATTIGOD, S. V., RAI, D., EARY, L. E. & AINSWORTH, C. C. 1990. Geochemical factors controlling the mobilization of inorganic constituents from the fossil fuel combustion residue: I Review of the major elements. *Journal of Environmental Quality*, **19**, 188–201.

MUDD, G. & KODIKARA, J. 2000. Field studies of the leachability of aged brown coal ash. *Journal of Hazardous Materials*, **76**, 159–192.

POTTER, P. E. 2003. Mudrocks. *In*: MIDDLETON, G. V. (ed) *Encyclopedia of Sediments and Sedimentary Rocks*. Kluwer Academic Publishers, Dordrecht, 451–459.

PRAHARAJ, T., POWELL, M. A., HART, B. R. & TRIPATHY, S. 2002. Leachability of elements from sub-bituminous coal fly ash from India. *Environment International*, **27**, 609–615.

QUICK, W. J. & IRONS, R. M. A. 2002. Trace element partitioning during the firing of washed and untreated power station coal. *Fuel*, **81**, 665–672.

RAI, D., MATTIGOD, S. V., EARY, L. E. & AINSWORTH, C. C. 1988. Fundamental approach for prediction pore-water composition in fossil fuel combustion wastes. *In*: McCARTHY, G. J., GLASSER, F. P., ROY, D. M. & HEMMINGS, R. T. (eds) *Fly Ash and Coal Conversion Byproducts: Characterization, Utilization and Disposal* (Symposium Proceedings V113). Materials Research Society, Pittsburg, PA, 317–324.

ROSE, N. L., HARLOCK, S. & APPLEBEY, G. G. 1999. The spatial and temporal distribution of spheroidal carbonaceous fly ash particles (SCP) in the sediment records of European mountain lakes. *Water, Air and Soil Pollution*, **113**, 1–32.

ROY, W. R., THIERY, R. G., SCHULLER, R. M. & SULOWAY, J. J. 1981. *Coal Fly Ash: A Review of the Literature and Proposed Classification System with Emphasis on Environmental Impacts*, Environmental Geology Note No.96. Illinois State Geological Survey, Champaign. IL.

ROY, W. R. & GRIFFIN, R. A. 1984. Illinois basin coal fly ashes. 2. Equilibria relationships and qualitative modeling of ash–water reactions. *Environmental Science and Technology*, **18**, 739–742.

SAKATA, M. 1987. Movement and neutralization of alkaline leachate at coal ash disposal. *Environmental Science and Technology*, **21**, 771–777.

SAKULPITAKPHON, T., HOWER, J. C., TRIMBLE, A. S., SCHRAM, W. H. & THOMAS, G. A. 2000. Mercury capture by fly ash: Study of the combustion of a high-mercury coal at a utility boiler. *Energy and Fuels*, **14**, 727–733.

SHAO, L., JONES, T., GAYER, R., DAI, S., LI, S. & JIANG, Y. 2003. Petrology and geochemistry of the high-sulphur coals from the Upper Permian carbonate coal measures in the Hesham Coalfield, southern China. *International Journal of Coal Geology*, **55**, 1–26.

SIKKA, R. & KANSAL, B. D. 1995. Effect of fly ash application on yield and nutrient composition of rice, wheat and soil pH and available nutrient status of soils. *Bioresource Technology*, **51**, 199–203.

SIMSIMAN, G. V., CHESTERS, G. & ANDERSON, A. W. 1987. Effect of ash disposal ponds in groundwater at coal-fired power plant. *Water Resources*, **21**, 417–426.

SLOSS, L. L. & DAVIDSON, R. M. 2001. Partioning of potentially hazardous trace elements in coal combustion. *Proceedings of 18th Annual International Pittsburgh Coal Conference*. CD Rom.

SPEARS, D. A. & ZHENG, Y. 1999. Geochemistry and origin of elements in some UK coals. *International Journal of Coal Geology*, **38**, 161–179.

SPEARS, D. A. 2001. Trace elements in some UK coals: quantitative distribution within the coal. *Proceedings of 18th Annual International Pittsburgh Coal Conference*. CD Rom.

SPEARS, D. A. 2004. The use of laser ablation inductively coupled mass spectrometry (LA-ICP-MS) for the analysis of fly ash. *Fuel*, **83**, 1765–1770.

SWAINE, D. J. 1990. *Trace Elements in Coal*. Butterworths, London, 294 pp.

TALBOT, R. W., ANDERSON, M. A. & ANDERS, W. A. 1978. Qualitative model of heterogeneous equilibria in a fly ash pond. *Environmental Science and Technology*, **12**, 1056–1062.

THEIS, T. L. & RICHTER, R. O. 1979. Chemical speciation of heavy metals in power plant ash pond leachate. *Environmental Science and Technology*, **13**, 219–224.

THEIS, T. L. & WIRTH, J. L. 1977. Sorptive behaviour of trace metals on fly ash in aqueous systems. *Environmental Science and Technology*, **11**, 1096–1100.

THEIS, T. L., WESTRICK, J. D., HSU, C. L. & MARLEY, J. J. 1978. Field investigation of trace metals in groundwater from fly ash disposal. *Journal of Water Pollution Control Federation*, **50**, 2457–2469.

THOMPSON, D. & ARGENT, B. B. 2002. Thermodynamic equilibrium study of trace element mobilisation under pulverised fuel combustion conditions. *Fuel*, **81/3**, 345–361.

TOLLE, D. A., ARTHUR, M. F. & VAN VORIS, P. 1983. Microcosm/field comparison of trace element uptake in crops grown in fly ash-amended soil. Ecological effects of soil amended with waste products. *Science of the Total Environment*, **31**, 243–261.

USGS 2003a. World Wide Web Address: http://www.energy.er.usgs.gov/products/databases/coalqual/intro.htm.

USGS 2003b. World Wide Web Address: http://www.pubs.usgs.gov/openfile/of00-047/.

VAN DER SLOOT, H. A., WIJKSTRA, J., VAN DAL, A., DAS, H. A., SLANNA, J., DEKKERS, J. J. & WALS, J. D. 1982. *Leaching of Trace Elements from Coal Solid Waste*. ECN-120. Netherlands Energy Research Foundation, Petten.

WADGE, A., HUTTON, M. & PETERSON, P. J. 1986. The concentration and particle size relationships of selected trace elements in fly ashes from U.K coal-fired power plants and a refuse incinerator. *Science of the Total Environment*, **54**, 13–27.

WARREN, C. J. & DUDAS, M. J. 1984. Weathering processes in relation to leachate properties of alkaline fly ash. *Journal of Environmental Quality*, **13**, 530–538.

ZIELINSKI, R. A. & FINKELMAN, R. B. 1997. *Radioactive Elements in Coal and Fly Ash: Abundance, Forms and Environmental Significance*. USGS, Central Region Energy Resources Team, Fact Sheet FS-163-97. U.S. Geological Survey National Center, Reston, VA.

Secondary mineral formation in coal combustion byproduct disposal facilities: implications for trace element sequestration

RONA J. DONAHOE

Department of Geological Sciences, The University of Alabama, Tuscaloosa, AL, USA
(e-mail: rdonahoe@wgs.geo.ua.edu)

Abstract: Coal combustion byproducts (CCBs) are high-volume wastes produced by the electrical power industry and typically disposed of in landfills and lagoon impoundments. In the disposal environment, meteoric water or groundwater may percolate through and interact with the ash materials, producing leachate that contains elevated levels of many trace elements. Various geochemical reactions control the release of solutes and the formation of secondary minerals in CCB disposal facilities during weathering. Concern about the potential release of trace elements into the environment has motivated a large number of studies aimed at predicting the maximum concentrations of elements in leachate solutions. Secondary minerals formed during weathering of CCBs have the potential to limit the mobility of trace elements in an ash disposal facility. Geochemical modelling has been used by many investigators to predict the equilibrium concentrations of solutes in CCB leachate solutions and the stable secondary minerals that will form in weathered ash. Unfortunately, basic kinetic, thermodynamic and adsorption data are lacking for many solid phases, particularly those that may contain trace elements. In addition, secondary solid phases are very difficult to identify by direct analytical methods due to their low abundances and/or amorphous character, so it is often not possible to directly determine the identity and compositions of secondary phases in ash disposal environments. Despite these difficulties, numerous secondary solid phases have been directly observed and/or predicted to form via weathering reactions in CCBs. A tabulation of all secondary phases that have formed or have the potential to form in the CCB disposal environment is given, along with the relevant references. The potential for secondary phases to sequester trace elements via precipitation, adsorption and co-precipitation is discussed. Secondary phases with the greatest potential to limit the mobility of trace elements are amorphous Fe-oxyhydroxide and amorphous aluminosilicate phases, which are metastable precursors to Fe-oxide and clay minerals, respectively.

Coal combustion byproducts (CCBs) predominantly consist of solid phases formed at high temperatures. These high-temperature phases are typically metastable under the low-temperature weathering conditions characteristic of disposal environments, and will convert to thermodynamically stable phases over time. The new minerals formed can strongly influence the physical and chemical properties of altered ash materials. Mineral transformation reactions that are deleterious to the physical properties of CCBs used in engineering and construction applications may be favourable for decreasing the leachability of some hazardous elements. Diagenetic alteration and secondary mineral formation therefore have important implications for both the re-use of ash materials and the long-term mobility of chemical elements in the disposal environment.

Primary minerals in CCBs consist of residual phases that survive the combustion process, and high-temperature phases that are formed during coal combustion. The mineralogy and composition of CCBs is determined by the composition and mineralogy of the feed coal (McCarthy *et al.* 1999; Kolker *et al.* 2000) and by the details of the combustion process itself, including any additives for emissions control (Wu & Chen 1987; Hower *et al.* 1999a; Steenari *et al.* 1999). Many studies have shown that trace elements are unevenly distributed in CCB materials (e.g., Natusch *et al.* 1975; see summary in Smith 1980). Although volatile trace elements (As, Cd, Cu, Pb, Hg, Se, V, Zn) are assumed to be concentrated on small particles and particle surfaces by many investigators, Hulett *et al.* (1980) reported significant fractions of As, Ba, Hg, Pb, Se, and Sr to be incorporated into the glassy fly ash matrix. The distribution and speciation of trace elements in CCB materials will influence their leachability and thus their possible

From: GIERÉ, R. & STILLE, P. (eds) 2004. *Energy, Waste, and the Environment: a Geochemical Perspective*. Geological Society, London, Special Publications, **236**, 641–658.
0305-8719/04/$15 © The Geological Society of London 2004.

incorporation by secondary minerals. For detailed information about the chemistry and mineralogy of CCB wastes, the reader is referred to Tishmack & Burns (this volume).

Secondary minerals are defined here as phases that form as the result of chemical weathering reactions between the CCB wastes and water-based solutions that come into contact with the ash material. The degree of contact between ash and water is, in part, a function of the disposal method used. For wet ash disposal, ash is slurried with effluent water and sent by pipe to an ash lagoon. Large volumes of water containing solutes dissolved from the ash are produced by the high liquid-to-solid (L/S) ratios (10:1 to 20:1) employed (Hansen *et al.* 2002). In contrast, dry ash disposal impounds dry ash in above-grade containment areas (Hansen *et al.* 2002), reducing the contact between ash and the weathering environment. These systems have low L/S ratios and, therefore, limited contact between the ash and weathering solutions occurs. It has been observed that dry deposits do not generate secondary phases, even over periods of several years (Zevenbergen *et al.* 1994; McCarthy *et al.* 1997). This observation shows the importance of the fluid phase in accomplishing conversion of metastable primary minerals into stable secondary phases.

Although a number of secondary minerals have been predicted to form in weathered CCB materials, few have been positively identified by physical characterization methods. Secondary phases in CCB materials may be difficult or impossible to characterize due to their low abundance and small particle size. Conventional mineral identification methods such as X-ray diffraction (XRD) analysis fail to identify secondary phases that are less than 1–5% by weight of the CCB or are X-ray amorphous. Scanning electron microscopy (SEM) and transmission electron microscopy (TEM), coupled with energy dispersive spectroscopy (EDS), can often identify phases not seen by XRD. Additional analytical methods used to characterize trace secondary phases include infrared (IR) spectroscopy, electron microprobe (EMP) analysis, differential thermal analysis (DTA), and various synchrotron radiation techniques (e.g., micro-XRD, X-ray absorption near-edge spectroscopy [XANES], X-ray absorption fine-structure [XAFS]).

This chapter will review the formation of secondary minerals in CCB materials and discuss the ability of secondary phases to sequester trace metals in the ash pond environment.

Secondary mineral formation

Geochemical controls

High-temperature phases produced by coal combustion alter in contact with water and air to produce new, secondary minerals that are more stable in the low-temperature environment typical of a CCB disposal facility. Initial dissolution of glass and other soluble primary coal ash components often present on the surfaces of glassy ash particles (e.g., Dudas 1981; Fishman *et al.* 1999) provide high levels of solutes to the weathering solution. If formed quickly enough, secondary solid phases in weathered CCB wastes will affect the long-term leaching rates of trace elements that have the potential to contaminate surface and groundwaters.

The secondary phases formed in a disposal facility will be determined by the primary mineralogy, grain size distribution and permeability of the CCB materials, and by the composition, redox state, and pH of the weathering solution. Because the L/S ratio influences the rate and extent of weathering reactions, the chemical characteristics of the weathering solution and therefore the composition of the secondary phases may also be determined by the hydrology of the ash pond environment. The hydrology of the disposal environment is dependent upon the disposal method used (wet vs. dry) as well as the physical properties (grain size, porosity and permeability) of the ash.

Secondary mineral formation in CCB disposal facilities takes place via a variety of geochemical processes. Important alteration reactions include hydration, hydrolysis, carbonation, oxidation, adsorption/desorption and a variety of mineral dissolution/precipitation reactions that collectively convert metastable primary phases into stable secondary phases. Early diagenesis of CCB materials takes place when ash is slurried with power plant effluent water and piped to the ash impoundment. These initial reactions involve simple hydration and rapid solution of soluble elements. However, more complex mineral transformation reactions typically take place over periods of months to years when infiltrating meteoric water or groundwater contacts the CCBs impounded in ash lagoons or landfills. The principal geochemical variables and processes responsible for the formation of secondary minerals in coal ash waste are summarized in Fig. 1. Table 1 gives examples of important geochemical reactions resulting in secondary mineral formation in CCB wastes.

Solution parameters determining the progression and outcome of coal ash alteration by

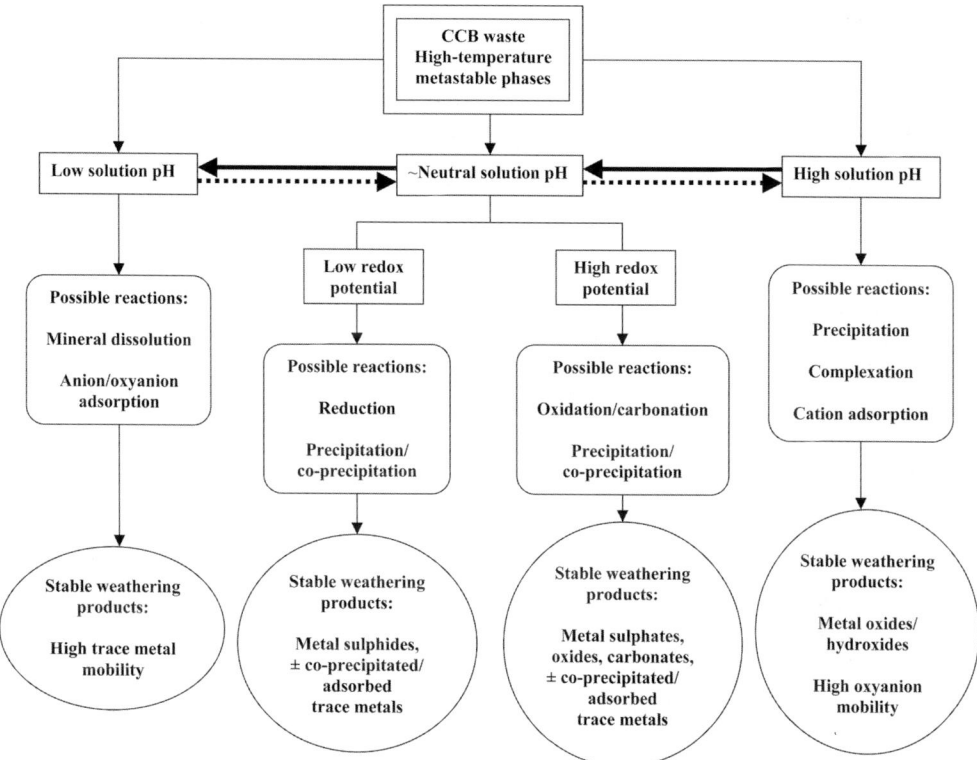

Fig. 1. Schematic diagram showing major pathways, reactions and stable weathering products for CCB alteration. Heavy solid arrow indicates changes caused by progressive natural weathering of alkaline CCBs at high L/S (liquid/ solid) ratios. Heavy dashed arrow indicates changes caused by progressive alteration of CCBs by acidic leachate solutions at low L/S ratios.

Table 1. *Important geochemical reactions resulting in the formation of secondary minerals in CCB wustes*

Reaction	Examples	
Hydration	Anhydrite to gypsum:	$CaSO_4 + 2H_2O \rightarrow CaSO_4 \cdot 2H_2O$
	Lime to portlandite:	$2CaO + H_2O \rightarrow 2Ca(OH)_2$
Solution	Dissolution of gypsum:	$CaSO_4 \rightarrow Ca^{2+} + SO_4^{2-}$
Hydrolysis	Dissolution of glass: (schematic reaction)	Me-Al-Si-O glass $+ H_2O + H^+ \rightarrow Me^+ +$ $Al(OH)_3(am) + SiO_2(am)$
Carbonation	Portlandite to calcite:	$Ca(OH)_2 + CO_2 \rightarrow CaCO_3 + H_2O$
Oxidation	Magnetite to hematite or maghemite	$2Fe_3O_4 + 0.5O_2 \rightarrow 3Fe_2O_3$
Precipitation	HFO:	$Fe^{3+} + 3OH^- \rightarrow Fe(OH)_3$
	Ettringite:	$3CaSO_4 + Ca_3Al_2O_6 + 32H_2O \rightarrow Ca_6Al_2(SO_4)_3(OH)_{12} \cdot 26H_2O$
Adsorption	Heavy metals on oxides:	$\equiv$Fe-OH $+ Me^{2+} \rightarrow \equiv$Fe-OMe$^+ + H^+$
Co-precipitation	Ideal solid solution between $Fe(OH)_3$ and $Me(OH)_2$	$Fe(OH)_3(s) + Me^{2+} \rightarrow Me(OH)_2(s) + Fe^{3+} + OH^-$

Me, metal cation; HFO, hydrous ferric oxide or ferric hydroxide; am, amorphous.

chemical weathering include pH and redox potential. The solubility and dissolution rate of primary minerals in CCB wastes may be dependent on both parameters. In addition the leachability/mobility of trace elements derived from CCBs is determined by the aqueous concentration and speciation of the metal, which are also functions of pH and redox potential. The pH values of ash leachates reported in the literature show a wide range (4.5–12) and change over time due to mineral buffering reactions (Zevenbergen et al. 1999b; Hansen et al. 2002). It should be noted, however, that the lower pH values in the reported range are produced by acid laboratory extraction solutions and are therefore not typical of weathering CCB in a disposal facility. Fossil fuel combustion byproduct leachates can also show a range of redox potentials from oxic to anoxic. Redox potential can be determined by the characteristics of the waste material itself and/or by those of the disposal environment (van der Sloot et al. 1994). Many trace elements exhibit highly different mobility under oxidizing versus reducing conditions (van der Sloot et al. 1994, Stumm & Morgan 1996).

Solutes released via dissolution of primary solid phases and surface precipitates have several possible fates. If aqueous ligands necessary to form solid precipitates with a solute are absent or in low concentration, the pore water solutions may remain undersaturated with respect to secondary solid phases and the solute may simply remain in solution. Aqueous complexation and various chemical parameters can increase the solubility of secondary solid phases, thus increasing the mobility and total concentration of solutes. Solute species may also be adsorbed onto charged solid surfaces as inner-sphere and outer-sphere complexes (Sposito 1984). Solid surfaces can become charged due to pH-dependent reactions taking place at the outer surfaces of oxide/hydroxide minerals (pH-dependent surface charge) or by isomorphous substitution of ions having different valence in the structure of a mineral (permanent structural charge). Adsorption will cause a decrease in element mobility compared to the aqueous phase. In general, species held as inner-sphere complexes are less mobile than those forming outer-sphere complexes on sorbent phases. The adsorption capacities of different sorbent phases show dramatic variation, as does the reversibility of solute adsorption (Stumm & Morgan 1996). Finally, solute species can be incorporated into the structure of mineral phases via precipitation or co-precipitation (i.e., the solute exists in solid solution

with a primary precipitate) reactions. The formation of secondary solid phases limits the mobility of constituent species to the solubility of the solid phase. However, solutions that are supersaturated with respect to one or more secondary solid phases may not precipitate the phases for long periods due to kinetic constraints on the nucleation and growth of the mineral.

Solution pH is a primary factor determining the solubility and mobility of trace elements in weathering systems. Many hydrous metal oxides are amphoteric (i.e., show both acidic and basic behaviour) due to the progressive hydrolysis of the metal cation with increasing solution pH. Metal mobility is controlled by the solubility of the hydrous oxides, and shows minimum values at pH ~ 7–10 (Stumm & Morgan 1996). In the presence of complexing ligands (e.g., sulphate, chloride, bicarbonate), aqueous speciation of trace elements is a function of ligand availability and solution pH. Complexation strongly affects the solubility of phases containing the trace element. As described above, surface charges of sorbent phases are, at least in part, pH-dependent. Metal cations and hydroxycations can be adsorbed onto negatively charged surfaces, while metal oxyanions and hydroxyanions can be adsorbed onto positively charged surfaces (Sposito 1984). Changes in solution pH can therefore be expected to change the adsorption behaviour and solubility of trace elements, and consequently their mobility.

Solution pH is also an important control on the identity and composition of secondary phases that form in weathered CCB wastes. Coal ash pore-water solutions often have extremely high initial pH values (>10), largely due to dissolution of highly reactive, small (<1 μm) particles of lime (CaO) and other alkaline earth oxides that are common as coatings on larger glass particles (Dudas 1981; Fishman et al. 1999). Glass undergoes rapid dissolution at high pH values (Hay 1966). After complete dissolution of portlandite (Dudas 1981), solution pH may drop by 3–4 units due to carbonation, hydrolysis and mineral precipitation reactions buffering the alkalinity (Zevenbergen et al. 1999b). The higher average pH values of weathering CCB wastes lead to precipitation of aluminosilicates with lower Si/Al ratios than those formed in natural settings such as weathering of siliceous volcanic ashes in open hydrologic systems (Hay & Sheppard 1981) and weathering of volcanic ash in saline, alkaline lakes (Mariner & Surdam 1970) where solution pH is generally <10. For example, Zevenbergen et al. (1996, 1998) described allophane (amorphous hydrous aluminum silicate of variable composition) and a well-ordered primitive clay that formed, respectively, in 4- and 12-year-old weathered MSWI bottom ash. They found that solution pH controlled the Si/Al ratio of the precipitating hydrous aluminosilicate phase. The dependence of the Si/Al ratio of precipitated aluminosilicate phases on solution pH was previously noted by Mariner & Surdam (1970) and Donahoe & Liou (1985).

Although Eh (redox potential) can affect both the adsorption/desorption reactions that can occur as well as the secondary phases that can form during CCB weathering, Theis & Richter (1979) stated that their studies of ash disposal ponds showed that oxidizing conditions rather consistently prevail in the active leaching zone (i.e., the unsaturated zone above the water table).

Warren & Dudas (1984) inferred the presence of two glass compositions in unreacted fly ash particles, using the results of column leaching tests as well as differences in the levels of acid ammonium oxalate-extractable Al and Si between weathered and fresh ash. Hulett & Weinberger (1980) indicated that elements such as Fe, Ca, and Mg, which are excluded from isomorphous substitution in mullite, migrate toward the surfaces of molten ash particles through distances of $1-2$ μm. This agrees with the ~ 100 nm depth of metal-enriched surface layers estimated by Natusch et al. (1975) and the suggestion of Elseewi et al. (1980) that soluble B migrates from the inner portion to the outer surface of fly ash particles. Warren & Dudas (1984) used these observations to support the presence of reactive surface glass rich in Ca, Mg, Fe and oxalate-extractable Al, Si, and Fe. Increases in concentrations of Na and K during progressive leaching were attributed to the selective leaching of the alkali elements from the less soluble silica-rich interior glass of the ash particles (Warren & Dudas 1984). Other studies have identified several different glass compositions within a single coal combustion residue (e.g., Mukhopadhyay et al. 1996; Gieré et al. 2003), which can be expected to exhibit solubility differences.

Janssen-Jurkovicová et al. (1994) presented a conceptual model for CCB weathering, which breaks the process into four phases. Figure 2 illustrates their model for weathering of alkaline fly ash (i.e., fly ash producing alkaline water extracts; defined as fly ash having Ca/S > 2.5 by Ainsworth & Rai 1987), modified using observations of Warren & Dudas (1984). In phase 1, oxides and soluble salts present on ash particle surfaces dissolve in contact with water. Hydrolysis of metal oxides (especially CaO) results in a rapid rise of pH to values of approximately 11. During phase 2, the more soluble glass at the outer surfaces of ash particles dissolves in response to the high solution pH, releasing cations. Amorphous aluminosilicate and Fe-hydroxide crusts precipitate on the surfaces of ash particles due to high solution supersaturation states. Solid phase precipitation and carbonation (CO_2 provided by air in the leaching zone and organic decay) reactions eventually buffer the pH to lower values (~ 8). In phase 3, the formation of amorphous crusts continues and their initial transformation into more crystalline phases begins. In addition, the less reactive interior glass of ash particles dissolves, releasing cations. Diffusion of glass components through amorphous crusts may limit the rate of glass dissolution. Phase 4 would involve continued aging of amorphous phases into thermodynamically stable phases such as Fe-oxides, clay minerals, and zeolites. Although CCB pore-waters and leachate solutions are typically supersaturated with respect to a variety of clay and zeolite minerals, their formation in weathered CCB materials has been documented only rarely (Warren & Dudas 1985; Fruchter et al. 1990; Janssen-Jurkovicová et al. 1994; Butler & Pflughoeft-Hassett 1995; Vassilev & Vassileva 1996; McCarthy et al. 1997; Zevenbergen et al. 1999a, b; Anthony et al. 2002). A number of important secondary mineral formation reactions are kinetically controlled, and therefore may not reach thermodynamic equilibrium for long time periods (Janssen-Jurkovicová et al. 1994).

Secondary phases

Two different approaches have been taken by researchers to determine the secondary mineralogy of CCBs: (1) direct observation, which is accomplished via analysis of weathered ash materials, and (2) prediction, based on chemical equilibrium solubility calculations for ash pore-waters and/or experimental ash leachate or extractant solutions. Because the secondary phases are typically present in very low abundance, their characterization by direct analysis is difficult. On the other hand, predictions based on chemical equilibrium modelling or laboratory leaching experiments may not be reliable indicators of element leachability or accurately indicate the secondary phases that will form under field conditions (Eighmy et al. 1994; Janssen-Jurkovicová et al. 1994).

Eary et al. (1990) reviewed previous literature on the formation of secondary solid phases in fossil fuel wastes. At the time of their review, few primary phases containing minor elements had been directly observed by previous studies

Phase 1: Fresh fly ash. Dissolution of soluble salts and oxides on particle surfaces.

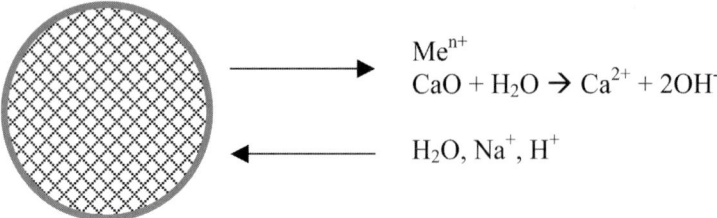

$$Me^{n+}$$
$$CaO + H_2O \rightarrow Ca^{2+} + 2OH^-$$

$$H_2O, Na^+, H^+$$

Phase 2: Dissolution of outer glass from ash particles. Formation of amorphous crusts.

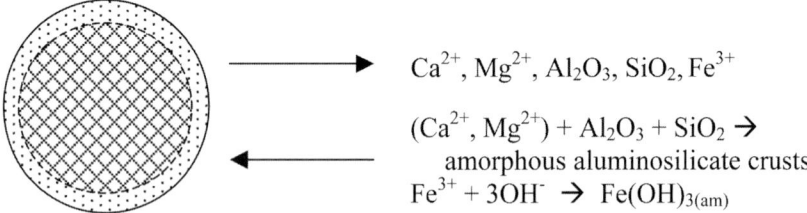

$$Ca^{2+}, Mg^{2+}, Al_2O_3, SiO_2, Fe^{3+}$$

$$(Ca^{2+}, Mg^{2+}) + Al_2O_3 + SiO_2 \rightarrow$$
$$\text{amorphous aluminosilicate crusts}$$
$$Fe^{3+} + 3OH^- \rightarrow Fe(OH)_{3(am)}$$

Phase 3: Continuing formation of amorphous crusts, aging of crusts. Diffusion-limited dissolution of inner glass from ash particles.

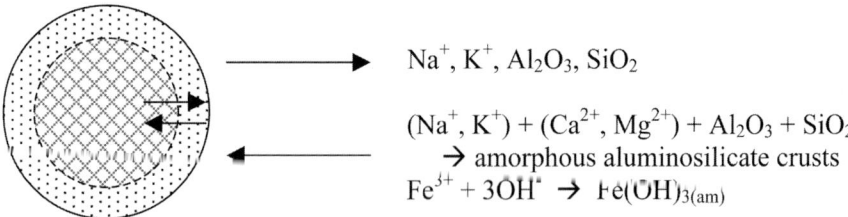

$$Na^+, K^+, Al_2O_3, SiO_2$$

$$(Na^+, K^+) + (Ca^{2+}, Mg^{2+}) + Al_2O_3 + SiO_2$$
$$\rightarrow \text{amorphous aluminosilicate crusts}$$
$$Fe^{3+} + 3OH^- \rightarrow Fe(OH)_{3(am)}$$

Phase 4: Transformation of amorphous crusts.

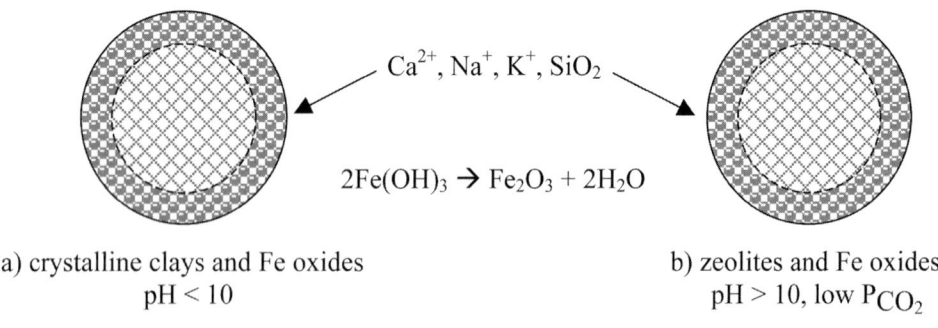

$$Ca^{2+}, Na^+, K^+, SiO_2$$

$$2Fe(OH)_3 \rightarrow Fe_2O_3 + 2H_2O$$

a) crystalline clays and Fe oxides
pH < 10

b) zeolites and Fe oxides
pH > 10, low P_{CO_2}

Fig. 2. Model for alkaline fly ash weathering (am, amorphous) (modified from Janssen-Jurkovicová *et al.* 1994).

due to their low abundances. The exceptions were As_2O_3, $BaSO_4$, and $SrCO_3$, which were identified in fly ash (Turner 1981; Bauer & Natusch 1981; Mattigod & Ervin 1983). In addition, Cr, Mn, Ni, Se, V, and Zn were found in primary spinels by Hulett *et al.* (1980) and later studies (e.g., Ainsworth *et al.* 1993; Hower *et al.* 1999*b*). Eary *et al.* (1990) listed no secondary phases in fossil fuel wastes that had been previously identified by direct analytical methods. Even now, most secondary phases containing trace elements have been inferred (not directly observed) to form, based on predictive models. Table 2 summarizes secondary minerals (including the idealized chemical formulae) that have been observed or predicted to form in the CCB disposal environment as a result of weathering. A discussion of studies that have characterized secondary minerals in weathered CCB wastes by direct analysis is presented first, followed by a discussion of papers that have used theoretical modelling to predict secondary mineral formation in disposal facilities.

Observed secondary minerals. Mukhopadhyay *et al.* (1996) determined that the fly ash produced by combustion of eastern Canadian coal had a higher percentage of non-crystalline Fe-oxides than the bottom ash. The very fine (< 1 μm) particles of CCB materials are quite important in determining the overall chemical and physical properties of the ash due to their large surface areas (Nathan *et al.* 1999). Although the majority of Fe phases in CCBs are crystalline, even small quantities of non-crystalline Fe-oxyhydroxides can have a profound effect on ash properties due to their much higher solubilities, surface areas and adsorption capacities. In order for amorphous Fe-oxyhydroxides to form by chemical weathering, a more soluble source of Fe than the primary crystalline Fe-oxides must dissolve. Most of the glass composing CCB materials contains Fe (Warren & Dudas 1984; Mukhopadhyay *et al.* 1996; Nathan *et al.* 1999). Warren & Dudas (1984) found that soluble surface glass contained more Ca, Mg, and Fe than the underlying, less soluble, alkali-rich glass. Two of the three types of glasses identified by Mukhopadhyay *et al.* (1996) in eastern Canadian feed coal combustion residues contained Fe as a major element. Both glass and mullite in South African and Columbian coal fly ash were reported to contain Fe by Nathan *et al.* (1999). Dissolution of soluble Fe-containing phases could provide sufficient Fe to supersaturate the solution with respect to various Fe-oxyhydroxide phases. Amorphous Fe-oxyhydroxides have been observed to form in artificially weathered CCB

material (Warren & Dudas 1985) and in naturally weathered CCBs (Vassilev & Vassileva 1996; Zevenbergen *et al.* 1999*a*, *b*). The formation of small amounts of secondary amorphous Fe-oxyhydroxides in weathered ash could dramatically increase the adsorption of trace elements mobilized through glass dissolution.

Vassilev & Vassileva (1996) studied the mineralogical composition of solid waste products from coal combustion, including lagooned ashes, from 11 Bulgarian thermoelectric power stations. They classified the minerals into three categories: (1) 'primary' (minerals and phases present in the original coal and undergoing little to no phase transition during combustion), (2) 'secondary' (mineral phases formed during combustion; equivalent to primary minerals, as defined in this chapter), and (3) 'tertiary' (minerals and phases formed during the transport and storage of fly and bottom ashes. Primary minerals, as defined by Vassilev & Vassileva (1996) are equivalent to entrained or residual minerals of other authors. Tertiary minerals (equivalent to secondary minerals, as defined in this chapter and generally used in geochemistry) identified by Vassilev & Vassileva (1996) in Bulgarian CCB materials included talc, maghemite, hematite, limonite, portlandite, brucite, Al-hydroxides, sulphates (gypsum, anhydrite, Fe-sulphates, Mg-sulphates, Na/K-sulphates, barite), and carbonates (calcite, dolomite, Mn-rich calcite, cerussite, witherite). Changes identified due to weathering of fly and bottom ash in the depositories were typically solution and redistribution of some sulphates, carbonates and chlorides, an increase in hematite and maghemite contents due to magnetite oxidation, hematite limonitization, hydration of anhydrite and Ca–Mg silicates, and encrustation of ash particles by fine weathered material (for all formulae, see Table 2).

Clay mineral formation has been observed in weathered coal ash materials in several studies (Warren & Dudas 1985; Zevenbergen *et al.* 1999*a*, *b*). Zevenbergen *et al.* (1999*a*) compared the geochemical properties of fresh fly ash and those of fly ash that had been weathered for eight years; both samples were derived from the same site in India. The weathering was found to have a marked effect on the chemical properties of the ash, increasing the specific surface area, the cation exchange capacity (CEC), and oxalate-extractable Al, Fe, and Si, and decreasing the pH. They attributed these changes in the ash properties to the observed rapid and widespread formation of amorphous clay by alteration of aluminosilicate glass in a high pH environment. Zevenbergen *et al.* (1999*b*) showed the widespread formation of

Table 2. *Secondary minerals observed and/or predicted to form in CCB materials exposed to chemical weathering*

Phase	Composition	Observed	Predicted
Böhmite	$AlO(OH)$		30
Gibbsite	$Al(OH)_3$		1,2,4,15,28,30,33
Al-hydroxide (am)*	$Al(OH)_3$	7	1,3,12,15, 28,30,33
Hydrous Al-sulphate	$Al(OH)SO_4$		15,24,28
Imogolite/allophane	Hydrous aluminum silicate	5,8	
Halloysite	$Al_2Si_2O_5(OH)_4$		30
Illite	$K_{0.6}Mg_{0.25}Al_{2.3}Si_{3.5}O_{10}(OH)_2$		30
Smectite	$(Ca,Na)_{0.3}Al_2Si_4O_{10}(OH)_2 \cdot nH_2O$		20,30
Arsenolite	As_2O_3	31	
Barite	$BaSO_4$	7	11,12,15,31,35
Witherite	$BaCO_3$	7	12
Ba-arsenate	$Ba_3(AsO_4)_2$		14,31
Ba-chromate/sulphate	$Ba(Cr,S)O_4$		1,19
Ba/Sr-sulphate	$(Ba,Sr)SO_4$		1,12,15
Gypsum	$CaSO_4 \cdot 2H_2O$	3,4,5,6,7,10,22,23, 27,29,30,37,38	1,2,15,33
Calcite (aragonite)	$CaCO_3$	3,5,7,8,9,10,22,23,30,37	12,25
Vaterite	$CaCO_3$	3	
Dolomite	$CaMg(CO_3)_2$	7	25,33
Ca-aluminate	$Ca_3Al_2O_6$	24,34	
Ca-silicate	$CaSiO_3$	24	28
Portlandite	$Ca(OH)_2$	4,7,9,22,24,30,34	32
Ettringite	$Ca_6Al_2(SO_4)_3(OH)_{12} \cdot 26H_2O$	3,4,9,10,21,22,27,29,30,36	26
Thaumasite	$Ca_6Si_2(SO_4)_2(CO_3)_2(OH)_{12} \cdot 24H_2O$	4,10,21,36	
Anhydrite	$CaSO_4$	7	16,30,35
Hannebachite	$CaSO_3 \cdot 0.5H_2O$	4,10,23,27,37	15
Tobermorite	$Ca_9Si_{12}O_{30}(OH)_6 \cdot 4H_2O$	4,10	
Whitlockite	$Ca_9(Mg,Fe)H(PO_4)_7$		30
Apatite	$Ca_5(PO_4)_3OH$		30
Powellite	$CaMoO_4$		15,35
Otavite	$CdCO_3$		16
Cr-hydroxide	$Cr(OH)_3$		17,18,33
Tenorite	CuO		1,12,33
Malachite	$Cu_2(OH)_2CO_3$		5,17
Fe-(oxy)hydroxide	$Fe(OH)_3$, $FeOOH$	5,7,8	1,16,30,33
Maghemite	Fe_2O_3	7	
Hematite	Fe_2O_3	7	30
Fe-sulphate	$FeSO_4$	7	
Ferric arsenate	$FeAsO_4$		13
Fe/Cr-hydroxide	$(Fe,Cr)(OH)_3$		12,15,18,28
Fe-vanadate	$FeVO_4$		13
Pb-hydroxide	$Pb(OH)_2$		17
Anglesite	$PbSO_4$		12
Hydrocerussite	$Pb_3(CO_3)_2(OH)_2$		12
Magnesio-ferrite	$MgFe_2O_4$		30
Magnesite	$MgCO_3$		30
Talc	$Mg_3Si_4O_{10}(OH)_2$	7	
Brucite	$Mg(OH)_2$	7,32	30,33
Mg-sulphates	$MgSO_4$, etc.	7	
Rhodochrosite	$MnCO_3$		33
Mn-rich calcite	$(Mn,Ca)CO_3$	7	
Pyrolusite	MnO_2		13,31
Zeolite Na–P1	$Na_6Al_6Si_{10}O_{32} \cdot 12H_2O$	4,10	
Phillipsite	$NaKAl_2Si_6O_{16} \cdot 6H_2O$		30
Laumontite	$CaAl_2Si_4O_{12} \cdot 4H_2O$		26,30
Wairakite	$CaAl_2Si_4O_{12} \cdot 2H_2O$		1,30,33
Nosean-haüyne	$(Na,Ca)_8Al_6Si_6O_{24}(SO_4)_{1-2} \cdot H_2O$	4,10	
Thenardite	Na_2SO_4	4	
Na/K,Ca-sulphates	e.g., $Na_2Ca_5(SO_4)_6 \cdot 3H_2O$	4,7,10	

(Continued)

Table 2. *Continued*

Phase	Composition	Observed	Predicted
Nickel carbonate hexihydrate	NiCO$_3$·6H$_2$O		33
Cerussite	PbCO$_3$	5,7	
Pb-hydroxide	Pb(OH)$_2$		5,17,33
Amorphous silica	SiO$_2$		30,35
Quartz	SiO$_2$		30,33
Al-Si phase (am)	Undetermined	30	
Mg–Ca–Na–K–Al–Si phase (am)	Undetermined	30	
Strontianite	SrCO$_3$		12,31,33
Celestite	SrSO$_4$		5,12,33,35
Willemite	Zn$_2$SiO$_4$		30,33

*am, amorphous.

1. Fruchter *et al.* (1990); 2. Jones (1995); 3. Anthony *et al.* (2002); 4. McCarthy *et al.* (1997); 5. Warren & Dudas (1985); 6. Humenick *et al.* (1983); 7. Vassilev & Vassileva (1996); 8. Zevenbergen *et al.* (1999*a, b*); 9. Steenari *et al.* (1999); 10. Butler & Pflughoeft-Hassett (1995); 11. Mattigod & Ervin (1983); 12. Fruchter *et al.* (1988); 13. Theis & Wirth (1977); 14. Turner (1981); 15. Ainsworth & Rai (1987); 16. Roy & Griffin (1984); 17. Theis & Richter (1979); 18. Rai & Szelmeczka (1990); 19. Amonette & Rai (1987); 20. Hollman *et al.* (1992); 21. Sahu *et al.* (2002); 22. Vempati *et al.* (1995); 23. Liem *et al.* (1982); 24. Malek & Roy (1985); 25. Talbot *et al.* (1978); 26. Mattigod (1983); 27. McCarthy *et al.* (1983); 28. Rai *et al.* (1988); 29. Simons & Jeffery (1960); 30. Janssen-Jurkovicová (1994); 31. Eary *et al.* (1990); 32. Warren & Dudas (1984); 33. Garavaglia & Caramuscio (1994); 34. Jones & Lewis (1960); 35. Rai *et al.* (1987*a*); 36. McCarthy (1997); 37. Rai *et al.* (1989); 38. Natusch *et al.* (1975).

non-crystalline hydrous aluminosilicates in two different fly ash deposits weathered for 8 and 14 years, respectively. The non-crystalline clay formed in the weathered fly ash was found to have constant major element composition, regardless of on what surface it was deposited. This observation was interpreted to suggest that the clay formed through precipitation, rather than by *in situ* transformation of aluminosilicate glass.

Clay mineral formation has also been observed in weathered municipal solid waste incinerator (MSWI) materials (Zevenbergen *et al.* 1993, 1994, 1996). Although the trace element distributions of CCB and MSWI ash can be quite different, it is likely that the general process for MSWI ash weathering is similar to CCB weathering because both waste types have glass-dominated compositions produced by high-temperature combustion and because geochemical modelling predicts very similar secondary phases to form from MSWI and CCB leachates (Comans & Meima 1994; Eighmy *et al.* 1994). Electron microscopy was utilized by Zevenbergen *et al.* (1993, 1994) to characterize the formation of allophane and well-ordered clay (illite) formed in MSWI bottom ash weathered for 12 years. Zevenbergen *et al.* (1996) suggested that alternate wetting/drying of bottom ash at the disposal site may greatly accelerate the clay formation process. As for their later coal ash studies, the authors found the CEC and bulk trace element (Cu, Ni, Pb, Zn) concentrations of the weathered MSWI ash increased steadily with increasing depth. This was attributed to the neo-formation of clays via weathering of glass.

Because the composition of the clay was independent of glass substrate composition, Zevenbergen *et al.* (1994) proposed that local hydrochemical conditions (rather than ash composition) was the primary control on secondary mineral formation. This conclusion suggests that at least some weathering behaviours of CCB and MSWI wastes are analogous.

In their classic study of CCB diagenesis, Warren & Dudas (1985) proposed that weathering of CCB materials is analogous to that of volcanic ash because both materials contain aluminosilicate glass as a dominant constituent. After 90 days of sequential acidic leaching of five columns packed with alkaline Modic (Roy & Griffin 1982) fly ash, the weathered residues were characterized by a variety of analytical techniques (XRD, DTA, IR spectroscopy, SEM, TEM). The major weathering reactions identified in the non-acidified (4th and 5th) coal ash columns were replacement of gypsum by calcite (possibly aragonite), formation of amorphous Fe coatings, and precipitation of an amorphous aluminosilicate that was described as proto-imogolite. Zevenbergen *et al.* (1999*b*) found that after only 10 years, the non-crystalline hydrous aluminosilicate content of weathered fly ash was higher than that of 250-year-old volcanic ash, due to the dissolution of CaO and resulting higher initial pH of the weathering solution.

McCarthy *et al.* (1999) studied CCB diagenesis in core material recovered from four disposal landfills or civil engineering works. In three of the sites, where sufficient moisture infiltration occurred, ash materials experienced significant diagenesis. Two of the sites showed indications of long-term alteration after burial. One of these sites contained primarily Class-F fly ash (produced by burning anthracite or bituminous coal) and other conventional boiler byproducts (bottom ash and slag) emplaced for up to 19 years. Primary constituents of the unaltered ash were quartz, mullite, magnetite, hematite, and low-Ca aluminosilicate glass. Interestingly, ettringite was present, although it is more typically found with calcite and gypsum as alteration products of high-Ca CCBs (Butler & Pflughoeft-Hassett 1995; Solem-Tishmack et al. 1995; McCarthy et al. 1997; Steenari et al. 1999; Anthony et al. 2002). The alteration products of Class-C fly ash (produced by lignite or sub-bituminous coal combustion) materials were minor amounts of ettringite and other hydration phases, for example, monosulphoaluminate (nominally $Ca_4Al_2SO_4(OH)_{12} \cdot 6H_2O$) and strätlingite ($Ca_2Al_2SiO_7 \cdot 8H_2O$).

Secondary mineral formation has also been studied in CCBs produced by other combustion methods and those using flue gas desulphurization (FGD) processes. Although circulating fluidized-bed combustion (CFBC) ashes are produced at significantly lower temperatures than pulverized fuel ashes, mineral transformations taking place within working CFBC ash–water systems show similarities to those taking place in disposal lagoons. For example, conversion of anhydrite to gypsum, ettringite formation and, ultimately, carbonation of all non-sulphate CaO was observed in wet, high-Ca CFBC ashes exposed to the air (Anthony et al. 2002). Calcite, gypsum and hannebachite have been identified in weathering FGD sludges (Liem et al. 1982; McCarthy et al. 1983; Rai et al. 1989). Ettringite and thaumasite were observed by Sahu et al. (2002) to form as secondary minerals in FGD ash reacting with water. In addition to these minerals, an undetermined Ca–Al–Si–S phase was also found.

McCarthy et al. (1997) used XRD to characterize CCBs at four landfill sites. Ettringite was found at all sites; thaumasite at three of the sites. Ettringite, gypsum, and portlandite formed quickly (within 2–5 years) via hydration reactions, while thaumasite precipitation was the result of carbonation of lime, dissolution of the resulting calcite, and gypsum dissolution. A high-Na ash produced two unique diagenetic assemblages that included tobermorite, zeolite Na–P1 and nosean–haüyne or thenardite and anhydrite after 9–12 years. Two sites in which the ash remained largely dry after initial hydration by conditioning water showed no significant diagenesis after 2 and 5 years, respectively.

Predicted secondary minerals. As discussed above, direct evidence for the formation of secondary trace element phases in weathered ash is rare. Most evidence suggesting precipitation of trace element phases by weathering processes is therefore indirect. For example, geochemical modelling of pore fluids in CCB disposal facilities and ash leachate solutions indicates that aqueous trace element concentrations are controlled by a variety of secondary phases. Table 2 lists numerous secondary trace element phases predicted, but not yet observed, to form in weathering CCB systems.

The use of chemical modelling to predict the formation of secondary phases and the mobility of trace elements in the CCB disposal environment requires detailed knowledge of the primary and secondary phases present in CCBs, thermodynamic and kinetic data for these phases, and the incorporation of possible adsorption/desorption reactions into the model. As noted above, secondary minerals are typically difficult to identify due to their low abundance in weathered CCB materials. In many cases, appropriate thermochemical, adsorption/desorption and kinetic data are lacking to quantitatively describe the processes that potentially affect the leaching behaviour of CCBs. This is particularly true for the trace elements. Laboratory leaching studies vary in the experimental conditions used (e.g., the type and concentration of the extractant solution, the L/S ratio, and other parameters such as temperature and duration/intensity of agitation), and therefore may not adequately simulate the weathering environment (Rai et al. 1988; Eary et al. 1990; Spears & Lee, 2004).

Secondary phases predicted by thermochemical models may not form in weathered ash materials due to kinetic constraints or non-equilibrium conditions. It is therefore incorrect to assume that equilibrium concentrations of elements predicted by geochemical models always represent *maximum* leachate concentrations that will be generated from the wastes, as stated by Rai et al. (1987a, b; 1988) and often repeated by other authors. In weathering systems, kinetic constraints commonly prevent the precipitation of the most stable solid phase for many elements, leading to increasing concentrations of these elements in natural solutions and precipitation of metastable amorphous phases. Over time, the metastable phases convert to thermodynamically stable phases by a process explained by the Guy-Lussac-Ostwald (GLO) step rule, also known as Ostwald ripening (Steefel & Van Cappellen 1990). The importance of time (i.e., kinetics) is often overlooked due to a lack of kinetic data for mineral dissolution/

precipitation reactions pertinent to weathering systems. In addition, kinetic data determined in the laboratory are not applicable in most cases to natural weathering systems. Janssen-Jurkovicová et al. (1994) showed that a number of major weathering reactions of importance to CCB diagenesis are kinetically controlled.

Numerous laboratory leachability tests have been devised to estimate the maximum amount of soluble elements available from granular materials. It is beyond the scope of this chapter to review the different leachability tests in detail. In general, leachability can be assessed through batch experiments or column experiments. It is generally accepted that column experiments better simulate the leaching environment of a typical CCB disposal facility, which is characterized by infiltrating meteoric water reacting with ash in the unsaturated zone. Column tests can be categorized by the type of leachate solution utilized: acidic leachates (dilute solutions of acetic acid, HCl, HNO_3 and H_2SO_4) and water leachates (DDI = distilled, deionized water, synthetic precipitation and synthetic groundwater solutions). Some studies of leachates/effluents from fly ash ponds qualitatively confirm column leaching results, in that Ca and sulphate are the major ions and other element concentrations are much lower (Theis et al. 1978; Dodd et al. 1981; Groenewold et al. 1985). However, in other cases column tests may show little correlation to leaching observed in a natural environment (Janssen-Jurkovicová et al. 1994).

The experimental approach to understanding the long-term mobility of trace elements in the CCB disposal environment has several inherent problems: (1) the leachate/extractant solution chemistry is specific to the particular waste sample and experimental parameters used; (2) the chemical reactions that occur and the secondary phases formed in these short-term experiments do not necessarily represent those typical of the CCB disposal environment during natural weathering; and (3) in addition, solubility, kinetic and adsorption data are lacking for many phases of potential interest (e.g., glass, mullite, ettringite). Despite these problems, the experimental approach has been used to set approximate boundaries on the concentrations of solute species and the formation of secondary minerals in weathering CCB materials.

A number of studies have compared the compositions of extractant and/or leachate solutions with equilibrium solubilities of potential secondary phases (Talbot et al. 1978; Henry & Knapp 1980; Hulett et al. 1980; Mattigod 1983; Roy & Griffin 1984; Ainsworth & Rai 1987; Rai et al. 1987a; Eary et al. 1990; Fruchter et al. 1990; Mattigod et al. 1990; Garavaglia & Caramuscio 1994; Meima & Comans 1998; Zevenbergen et al. 1999a). Zevenbergen et al. 1999a found that leachates derived from both weathered and fresh ash at the same site were approximately saturated with respect to gibbsite and amorphous aluminosilicate. The authors proposed that amorphous aluminosilicate phases may transform into more stable minerals (zeolites, smectite, or halloysite) over the longer term, if pH and leaching rates are favourable. Although Elseewi et al. (1980) indicated that aqueous systems of fly ash do not represent true equilibrium, they used only 24 h of equilibration in their batch extraction experiments. In contrast, Fruchter et al. (1990) showed that equilibrium was quickly achieved (7 days or less) for Al, Ba, Ca, Cr, Cu, Fe, S, Si, and Sr concentrations in leachate collected from a large fly ash field lysimeter, and determined likely solubility-controlling solid phases for these elements using chemical equilibrium modeling. Because many secondary solids (e.g., hydroxides, carbonates, sulphates and sulphides) have rapid precipitation kinetics, chemical equilibrium generally can be assumed (Eary et al. 1990).

Many authors have suggested that secondary solid phases control the concentrations of certain elements leached from fly ash and can be predicted using thermochemical modeling (Talbot 1978; Mattigod 1983; Rai et al. 1987a, b, 1988; Fruchter et al. 1990; Mattigod et al. 1990; Garavaglia & Caramuscio 1994). Mattigod et al. (1990) used chemical equilibrium calculations to model the concentrations of elements in fly ash leachate solutions generated through column experiments. Rai et al. (1987a, b, 1988) have shown that thermodynamic calculations can be used to predict equilibrium element concentrations in ash pore-waters and leachates. If the dissolution/precipitation reactions are rapid, element concentrations will be determined by equilibrium thermodynamics, and adsorption/desorption reactions can be ignored for many elements. However, when no secondary solids can be identified as controlling the aqueous concentrations of some trace elements (e.g. As, B, Cd, Mo, Sb, Se, V, and U; Fruchter et al. 1990; Garavaglia & Caramuscio 1994), many authors propose that adsorption/desorption reactions control their mobility. Because thermodynamic data for trace element solids and solid solutions typically are lacking, this assumption may not be appropriate (i.e., the geochemical model is only as good as the data base). For example, B and Mo are not included in the MINTEQA2 (Allison et al.

1991) database; the MINTEQ database used by PHREEQC (Parkhurst 1995) does not contain Co or Mo.

Trace element sequestration

Studies of fresh ash produced by coal combustion have shown that many trace elements (As, B, Bi, Cd, Cr, Cu, Ge, Hg, Mo, Pb, Ni, Se, Sr, Tl, V, W, Zn) are enriched in the fly ash compared to the bottom ash (Hansen & Fisher 1980; Eary et al. 1990; Mukhopadhyay et al. 1996; Karayigit et al. 2001). For example, Mukhopadhyay et al. (1996) reported 10–20 times enrichment of most trace elements in the fly ash compared to the feed coal and association of As with crystalline Fe–O and Fe–S phases in the bottom ash from a power plant in Nova Scotia fed by eastern Canadian coal. Elements enriched in fly ash are typically those more easily volatilized. Because fly ash particles also have smaller sizes and therefore greater reactivity than bottom ash, the probability of metal leaching is correspondingly greater. Ainsworth & Rai (1987) and Rai et al. (1988) found that most of the Cu, Mo, Se, Sr, and V in fly ash was readily soluble.

The speciation of trace elements in solid phases determines their mobility and toxicity. Spectroscopic techniques such as XANES and XAFS, can be used to determine directly the oxidation and structural state of elements in coal combustion byproducts. For example, Huggins et al. (2000) used these synchrotron techniques to determine that Cr and As occur predominately in the less toxic oxidation states Cr(III) and As(V) in CCBs. In addition, they found As, Cd, Cr, Ni, and Zn were present primarily as oxidized species (i.e., as oxides, sulphates, arsenates, etc.) in unweathered CCBs.

Considerable literature has examined the occurrence of trace elements in unweathered CCB wastes. For example, Lauf et al. (1982) and Mattigod (1982) suggested that trace metals Cr, Cu, Mn, Ni, V, and Zn are incorporated into Fe-oxides during combustion. Kolker et al. (2000) inferred that Ni, Cr, and Co occur in Fe-oxide phases due to their enrichment in the magnetic fraction of coal fly ash, relative to the bulk ash. Hulet et al. (1980) made a similar observation for Cr, Cu, Mn, Ni, V, and Zn, but proposed that these metals exist as isomorphous substitutions in ferrospinel ($Fe_{3-x}Me_xO_4$). Arsenic is distributed more uniformly at low levels throughout unweathered fly ash, either in solid solution in the glass phase (McGee et al. 1995; Mukhopadhyay et al. 1996), or possibly

as a $Ca_3(AsO_4)_2$ condensate on the ash particle surfaces (EPRI 1998). Van der Hoek et al. (1994) successfully described the leaching of As and Se from acidic fly ash using a simplified surface complexation model with hematite. Leaching of fly ash by HF showed Fe, Co, Ni, and Cr were primarily present in silicate phases and in Fe-oxides exposed by glass dissolution (Kolker et al. 2000).

Leaching of trace metals from CCB materials is of potential concern for contamination of natural water supplies. For example, Groenewold et al. (1985) found elevated levels of As, Cr, Mo, Pb, and Se in groundwater below a lignite fly ash landfill in North Dakota, and Shende et al. (1994) found stockpiled coal bottom ash contributed leachable metals to adjoining rivers. However, it also has been observed that weathered CCBs have the ability to retain many metals (Janssen-Jurkovicová et al. 1994; Steenari et al. 1999).

The most comprehensive review of the geochemical controls on element mobilization from fossil fuel combustion residues was published by Eary et al. (1990). Although combustion conditions affect the leachability of elements from fresh coal ash by determining the chemical form of the elements and the primary phases present in the ash produced (Wu & Chen 1987), the principal factors influencing the mobility of elements in weathering CCB disposal facilities are geochemical. The leachability of CCB trace elements is affected by the aqueous chemistry of weathering solutions, the distributions of these elements in the solid wastes, and their incorporation into, or adsorption onto, secondary phases. The incorporation of trace elements into secondary solids formed by weathering is likely the major long-term control on trace element mobility in CCB wastes (Eary et al. 1990).

Steenari et al. (1999) reported that certain minor and trace metal species present in coal ash were retained in the weathered ash materials. They attributed this to (1) the alkaline pH of the pore solution, which would create negatively charged surfaces on oxide and aluminosilicate mineral surfaces, making the ash materials effective sorbents for metal cations, and (2) the formation of calcite during curing of wetted ash, which lowered Mn and Zn levels through their incorporation in the calcite structure as solid solutions and as precipitates on calcite surfaces. Many metal carbonates and some metal hydroxides precipitate with increasing pH, which could also be a mechanism for immobilization of trace elements. Hollis et al. (1988) suggested possible inclusion of B in secondary $CoCO_3$ in alkaline soils, which led Eary et al.

(1990) to speculate that a similar sequestration mechanism might exist in CCB wastes.

Janssen-Jurkovicová *et al.* (1994) found significant immobilization of Ba, Mo, V, As, and Se in alkaline fly ash subjected to natural weathering for eight years, compared to fresh fly ash leached in laboratory column tests. They suggested that the positive correlations between Ba, Mo, V, and the major elements Al and Si indicate that these trace elements are associated with secondary clay-like materials. Positive correlations between As, Se, and the major elements Ca, Mg, and Fe were suggested to indicate possible co-precipitation or adsorption of As and Se on Fe-oxyhydroxides and/or carbonates. Important parameters specific to fly ash aging, such as CEC, surface area and permeability, were observed to hardly change during a three-week column test, while large changes were seen for fly ash after four years of weathering in a field lysimeter. Janssen-Jurkovicová *et al.* (1994) concluded that standard column leachability tests designed to assess the long-term mobility and availability of trace elements in CCBs are too short in duration to identify certain immobilization processes that are important under natural weathering conditions.

Taylor (1990) described rather extensive substitution of ions in the ettringite structure; Stokely & Newland (1992) suggested structural substitution as one mechanism for the retention of trace elements by wetted and cured ash materials. Hassett & Thompson (1994) stated that arsenate, borate, chromate, selenate, vanadate, and molybdate may substitute for sulphate in the ettringite structure. Vempati *et al.* (1995) presented XRD evidence suggesting that arsenate can be incorporated into ettringite through replacement of sulphate or hydroxyl groups. Table 2 shows the ideal, or stoichiometric end-member composition of ettringite. It is also possible for structural Ca to be replaced by other divalent metal cations such as, Zn, Cd, Pb, and Ni, and for structural Al to be replaced by a variety of metal cations, some of which have different charge. In addition, isomorphous substitution of metal cations of differing charge will result in the development of permanent structural charge, which would increase the ion-exchange capacity of ettringite (Vempati *et al.* 1995). Thus, it is possible for ettringite to sequester metal cations and oxyanions via several mechanisms.

Hassett & Thompson (1994) stated that ettringite formation requires the presence of constituent cations and anions and pH above 11.5. These conditions are met upon contact of high-Ca CCBs with moderate amounts of water, generating high aqueous concentrations of Ca, sulphate, and Al and solution pH values typically in excess of 12. However, ettringite is not thermodynamically stable in contact with aqueous solutions at $pH < 10.7$ (Zhang & Reardon 2003). In most cases, studies in which substantial quantities of ettringite formed and persisted were for high-Ca CCBs where the wastes were tested under saturated, closed system conditions (cured pastes and/or batch experiments). These conditions are not applicable to most ash pond disposal sites, which are hydrologically open and subject to long-term leaching by meteoric waters. For example, at four of five CCB disposal sites studied by Howell *et al.* (2003), pore-water pH values ranged from 5.75 to 8.26, averaging 6.93, 7.01, 7.18, and 7.94, respectively, for ash pond sites closed for 34, 37, 49, and 33 years. At only one studied site did pore-water pH values exceed 10, and averaged 10.04. It should be noted that the more alkaline site has been closed for only 23 years and contains CCB wastes from sub-bituminous coal. Ettringite was not found in weathered ash materials at any of these sites. While ettringite and other cementitious secondary minerals may be effective in sequestering significant amounts of certain trace elements in high-Ca CCBs over periods as long as several years, other secondary minerals are more likely to control trace element leaching over much longer time periods.

Theis & Richter (1979) showed that adsorption onto hydrous Fe-oxide was the major solubility control for Cd, Ni, and Zn in soils surrounding two ash disposal ponds, and stated that certain fly ashes are capable of causing the deposition of secondary Fe-oxides. The ash pond leachate conditions favoured the precipitation of Pb and Cr hydroxides, Pb carbonate, and precipitation of Cu as malachite. The leaching of As (as arsenate) and Se (as selenite) was found to be controlled by adsorption onto amorphous Fe-oxide by van der Hoek & Comans (1994).

Clay mineral formation in CCB materials is likely to be very important to long-term metals sequestration by the weathered ash. In addition to metal adsorption, clay minerals provide a physical barrier to heavy metal leaching through encapsulation of metal-rich particles, and they act as a chemical barrier by incorporating metals into their structure (Zevenbergen *et al.* 1999*b*). For example, Zevenbergen *et al.* (1994) found direct evidence for co-precipitation of Ca molybdate with, and incorporation of Cu into, neo-formed clays in weathered MSWI bottom ash. Zevenbergen *et al.* (1999*b*) used TEM/EDS to analyse the trace element contents of secondary clay and ferrihydrite in weathered coal fly ash. Although the data showed that detectable heavy metals (Cr, Ni, Ti, Zn) were preferentially partitioned into Fe-oxyhydroxide compared to the non-crystalline clay, the much greater abundance of clay led them to conclude that non-crystalline aluminosilicates are likely to be the major mechanism for heavy metal retention by weathered CCB materials.

Conclusions

The most common secondary minerals (gypsum, portlandite, ettringite) found in weathered coal combustion byproducts are those produced by cementitious reactions that rapidly take place when the CCB material is brought into contact with moderate amounts of water. Although ettringite has been shown to accommodate via substitution rather extensive amounts of trace element oxyanions in its structure, it is doubtful that this is a significant mechanism for trace element sequestration in CCB disposal facilities that are open to water infiltration, due to its instability at pH < 10.7. Solem-Tishmack et al. (1995) reported that ettringite formed in high-Ca CCB pastes survived leaching by rainwater and groundwater for up to five years in instrumented test cells, but questioned its long-term stability under more neutral aqueous solution conditions.

Curing of wetted coal ash results in carbonation to form calcite and possibly other metal carbonates. Calcite surfaces have been shown by Zachara et al, (1991) to preferentially sorb certain metals through adsorption followed by recrystallization (Zn, Co, Ni) or surface precipitation (Cd, Mn), but the total adsorption capacity of calcite for these metals is minor compared to their adsorption by other phases.

Other secondary phases such as amorphous clays and Fe-oxyhydroxides form in weathered CCB materials as a result of glass dissolution and subsequent precipitation reactions that take place over longer time periods (months to years). These phases have adsorption capacities greatly exceeding those of other secondary phases shown to form in weathered CCBs. Because glass is the most abundant phase in CCB materials, clays are likely to be the more abundant alteration products, and therefore exert a major control on trace element mobility, compared to Fe-oxyhydroxide phases. In addition, the rate of clay mineral formation from hydrous, amorphous aluminosilicate may be greatly accelerated (by as much as a factor of 25) if the weathering ash is subjected to repeated wetting/drying cycles (Zevenbergen et al. 1999b).

Secondary mineral formation in CCB materials occurs naturally as the result of chemical weathering reactions between the ash and percolating weathering solutions. As shown above, many of these secondary phases have high surface areas and substantially increase the CEC of the weathered ash compared to fresh ash. For this reason, Zevenbergen et al. (1999b) suggest that isolation of CCB materials from contact with weathering solutions may be counterproductive.

In summary, many studies have indicated the widespread formation of a variety of secondary phases in weathered CCB waste materials. Natural weathering processes therefore appear to play an important role in the sequestration of trace elements in ash disposal environments. Additional study is needed to identify and determine the chemical compositions of these secondary phases and to obtain pertinent thermodynamic, kinetic and adsorption data that can be used to model the mobility of trace elements in these complex weathering systems.

The author is indebted to J. Schexnayder and S. Bhattacharyya for their help with the literature search for this chapter. Thanks are also due to J. R. Howell and J. Redwine of Southern Company Services, Inc., who first interested the author in CCB diagenesis and metals sequestration. The author is grateful to J. Hower, E. J. Anthony, and R. Gieré for their helpful comments on the manuscript.

References

AINSWORTH, C. C. & RAI, D. 1987 Chemical Characterization of Fossil Fuel Wastes, EPRI EA-5321. Electric Power Research Institute, Palo Alto, CA.

AINSWORTH, C. C., MATTIGOD, S. V., RAI, D. & AMONETTE, J. E. 1993. Detailed Physical, Chemical and Mineralogical Analyses of Selected Coal and Oil Combustion Ashes. Electric Power Research Institute, Palo Alto, CA.

ALLISON, J. D., BROWN, D. S. & NOVO-GRADAC, K. J. 1991. MINTEQA2/PRODEFA2, A Geochemical Assessment Model for Environmental Systems: Version 3.0 Users Manual, EPA/600/3-917021. US Environmental Protection Agency, Athens, GA.

AMONETTE, J. & RAI, D. 1987. Ba(S,Cr)O₄ Solid Solution as a Possible Phase Controlling Cr(VI) Levels in Soils. Agronomy Abstracts, ASA, CSSA and SSSA, Madison, WI, 165.

ANTHONY, E. J., BULEWICZ, E. M., DUDEK, K. & KOZAK, A. 2002. The long term behaviour of CFBC ash–water systems. Waste Management, 22, 99–111.

BAUER, C. F. & NATUSCH, D. F. S. 1981. Identification and quantification of carbonate compounds in coal fly ash. Environmental Science and Technology, 15, 783–788.

BUTLER, R. D. & PFLUGHOEFT-HASSETT, D. F. 1995. Diagenesis and Leaching Characteristics of Aged

Coal Conversion Solid Residues from Mine Disposal Environments. US Bureau of Mines Contract Number CO388026 Research Report, Washington, DC.

COMANS, R. N. J. & MEIMA, J. A. 1994. Modelling Ca-solubility in MSWI bottom ash leachates. *In*: GOUMANS, J. J. J. M., VAN DER SLOOT, H. A. & AALBERS, TH. G. (eds) *Environmental Aspects of Construction with Waste Materials*. Elsevier Science BV, Amsterdam, 103–110.

DODD, D. J. R., GOLOMB, A., CHAN, H. T. & CHARTIER, D. 1981. A comparative field and laboratory study of fly ash leaching characteristics. *In*: CONWAY, R. A. & MALLOY, B. C. (eds) *Hazardous Solid Waste Testing: First Conference*, ASTM Special Technical Publication 760. ASTM, Philadelphia, PA, 164–185.

DONAHOE, R. J. & LIOU, J. G. 1985. An experimental study on the process of zeolite formation. *Geochimica et Cosmochimica Acta*, **49**, 2349–2360.

DUDAS, M. J. 1981. Long-term leachability of selected elements from fly ash. *Environmental Science and Technology*, **15**, 840–843.

EARY, L. E., RAI, D., MATTIGOD, S. V. & AINSWORTH, C. C. 1990. Geochemical factors controlling the mobilization of inorganic constituents from fossil fuel combustion residues: II. Review of the minor elements. *Journal of Environmental Quality*, **19**, 202–214.

EIGHMY, T. T., EUSDEN, J. D., JR., MARSELLA, K., HOGAN, J., DOMINGO, D., KRZANOWSKI, J. E. & STAMPFLI, D. 1994. Particle petrogenesis and speciation of elements in MSW incineration bottom ashes. In: GOUMANS, J. J. J. M., VAN DER SLOOT, H. A. & AALBERS, TH. G. (eds) *Environmental Aspects of Construction with Waste Materials*. Elsevier Science BV, Amsterdam, 111–135.

ELSEEWI, A. A., PAGE, A. L. & GRIMM, S. R. 1980. Chemical characterization of fly ash aqueous systems. *Journal of Environmental Quality*, **9**, 424–428.

EPRI (ELECTRIC POWER RESEARCH INSTITUTE) 1998. *Identification of Arsenic Species in Coal Ash Particles*. EPRI, Palo Alto, CA, TR-109002.

FISHMAN, N. S., RICE, C. A., BREIT, G. N. & JOHNSON, R. D. 1999. Sulfur-bearing coatings on fly ash from a coal-fired power plant: composition, origin and influence on ash alteration. *Fuel*, **78**, 187–196.

FRUCHTER, J. S., RAI, D., ZACHARA, J. M. & SCHMIDT, R. L. 1988. *Leachate Chemistry at the Montour Fly Ash Test Cell*. EPRI EA-5922. Electric Power Research Institute, Palo Alto, CA.

FRUCHTER, J. S., RAI, D. & ZACHARA, J. M. 1990. Identification of solubility-controlling solid phases in a large fly ash field lysimeter. *Environmental Science and Technology*, **24**, 1173–1179.

GARAVAGLIA, R. & CARAMUSCIO, P. 1994. Coal fly-ash leaching behaviour and solubility controlling solids. *In*: GOUMANS, J. J. J. M., VAN DER SLOOT, H. A. & AALBERS, TH. G. (eds) *Environmental Aspects of Construction with Waste Materials*. Elsevier Science BV, Amsterdam, 87–102.

GIERÉ, R., CARLETON, L. E. & LUMPKIN, G. R. 2003. Micro- and nanochemistry of fly ash from a coal-fired power plant. *American Mineralogist*, **88**, 1853–1865.

GROENEWOLD, G. J., HASSETT, D. J., KOOB, R. D. & MANZ, O. E. 1985. Disposal of western fly ash in the northern Great Plains. *Materials Research Society Symposium Proceedings*, **86**, 213–226.

HANSEN, L. D. & FISHER, G. L. 1980. Elemental distribution in coal fly ash particles. *Environmental Science and Technology*, **14**, 1111–1117.

HANSEN, Y., NOTTEN, P. J. & PETRIE, J. G. 2002. The environmental impact of ash management in coal-based power generation. *Applied Geochemistry*, **17**, 1131–1141.

HASSETT, D. J. & THOMPSON, J. S. 1994. Formation of ettringite in disposed low-rank coal fly ash: Implications for impact on groundwater chemistry. *Proceedings of the North Dakota Water Quality Symposium*, 30–31 March, 1994, North Dakota State University Extension Service, 122.

HAY, R. L. 1966. Zeolites and zeolitic reactions in sedimentary rocks. *Geological Society of America Special Paper*, **85**, 53–64.

HAY, R. L. & SHEPPARD, R. A. 1981. Zeolites in open hydrologic systems. *In*: MUMPTON, F. A. (ed) *Mineralogy and Geology of Natural Zeolites. Mineralogical Society of America, Reviews in Mineralogy*, Washington, DC, **4**, 93–102.

HENRY, W. M. & KNAPP, K. T. 1980. Compound forms of fossil fuel fly ash emissions. *Environmental Science and Technology*, **14**, 450–456.

HOLLIS, J. F., KEREN, R. & FAL, M. 1988. Boron release and sorption by fly ash as affected by pH and particle size. *Journal of Environmental Quality*, **17**, 181–184.

HOLLMAN, G. G., NOORDERWIER, M. A. & JANSSEN-JURKOVICOVÁ, M. 1992. Voortgangsrapportage van het Project Veraarding van Vliegas voor het onderzoeksjaar 1992. *Proceedings of the Project 'Soil formation in fly ash deposits.'* MvD-1992-3 Rijks Universiteit Utrecht.

HOWELL, J., REDWINE, J. & DONAHOE, R. 2003. *Secondary Mineral Formation in Weathered Ash: Implications for Metals Sequestration*. EPRI 1005261. Electric Power Research Institute, Palo Alto, CA.

HOWER, J. C., ROBL, T. L. & THOMAS, G. A. 1999*a*. Changes in the quality of coal combustion by-products produced by Kentucky power plants, 1978 to 1997: consequences of Clean Air Act directives. *Fuel*, **78**, 701–712.

HOWER, J. C., RATHBONE, R. F., ROBERTSON, J. D., PETERSON, G. & TRINBLE, A. S. 1999*b*. Petrology, mineralogy and chemistry of magnetically-separated sized fly ash. *Fuel*, **78**, 197–203.

HUGGINS, F. E., SHAH, N., HUFFMAN, G. P. & ROBERTSON, J. D. 2000. XAFS spectroscopic characterization of elements in combustion ash and fine particulate matter. *Fuel Processing Technology*, **65–66**, 203–218.

HULETT, L. D. & WEINBERGER, A. J. 1980. Some etching studies of the microstructure and composition of large aluminosilicate particles in fly ash from coal-burning power plants. *Environmental Science and Technology*, **14**, 965–970.

HULETT, L. D., WEINBERGER, A. J., NORTHCUTT, K. J. & FERGUSON, M. 1980. Chemical species in fly ash from coal-burning power plants. *Science*, **210**, 1356–1358.

HUMENICK, J. M., LANG, M. & JACKSON, K. F. 1983. Leaching characteristics of lignite ash. *Journal of the Water Pollution Control Federation*, **55**, 310–316.

JANSSEN-JURKOVICOVÁ, M., HOLLMAN, G. G., NASS, M. M. & SCHUILING, R. D. 1994. Quality assessment of granular combustion residues by a standard column test: Prediction versus reality. In: GOUMANS, J. J. J. M., VAN DER SLOOT, H. A. & AALBERS, TH. G. (eds) *Environmental Aspects of Construction with Waste Materials*. Elsevier Science BV, Amsterdam, 161–178.

JONES, D. R. 1995. The leaching of major and minor trace elements from coal ash. *In*: SWAINE, D. J. & GOODARZI, F. (eds) *Environmental Aspects of Trace Elements in Coal*. Kluwer Academic Publishers, Netherlands, 221–262.

JONES, L. H. & LEWIS, A. V. 1960. Weathering of fly ash. *Nature*, **185**, 404–405.

KARAYIGIT, A. I., ONOCAK, T., GAYER, R. A. & GOLDSMITH, S. 2001. Mineralogy and geochemistry of feed coals and their combustion residues from the Cayirhan power plant, Ankara, Turkey. *Applied Geochemistry*, **16**, 911–919.

KOLKER, A., FINKELMAN, R. B., AFFOLTER, R. H. & BROWNFIELD, M. E. 2000. The composition of coal combustion by-products: examples from a Kentucky power plant. *In*: VORIES-KIMERY, C. & THROGMORTON, D. (eds) *The Use and Disposal of Coal Combustion By-Products at Coal Mines: A Technical Interactive Forum*. US Department of Interior, Office of Surface Mining. Alton, IL, United States and Southern Illinois University, Coal Research Center, Carbondale, IL, 15–24.

LAUF, R. J., HARRIS, L. A. & RAWISTON, S. S. 1982. Pyrite framboids as the source of magnetite spheres in fly ash. *Environmental Science and Technology*, **16**, 218–220.

LIEM, H., SANDSTROEM, M., WALLIN, T., CARNE, A., RYDEVIK, U., THURENIUS, B. & MOBERG, P. O. 1982. Studies on the leaching and weathering processes of coal ashes. *In: International Conference on Coal Fired Power Plants and the Aquatic Environment*, CONF-8208123, Water Quality Institute, Hoersholm, Denmark, 338–366.

MALEK, R. A. I. & ROY, D. M. 1985. Electrokinetic phenomena and surface characteristics of fly ash particles. *Materials Research Society Symposium Proceedings*, **86**, 41–50.

MARINER, R. H. & SURDAM, R. C. 1970. Alkalinity and formation of zeolites in saline alkaline lakes. *Science*, **170**, 977–980.

MATTIGOD, S. V. 1982. Characterization of fly ash particles. *In: Scanning Electron Microscopy/1982/2*. SEM Inc., Chicago, 611–617.

MATTIGOD, S. V. 1983. Chemical composition of aqueous extracts of fly ash: Ionic speciation as a controlling factor. *Environmental Technology Letters*, **4**, 485–490.

MATTIGOD, S. V. & ERVIN, J. O. 1983. Scheme for density separation and identification of compound forms in size-fractionated fly ash. *Fuel*, **62**, 927–931.

MATTIGOD, S. V., RAI, D., EARY, L. E. & AINSWORTH, C. C. 1990. Geochemical factors controlling the mobilization of inorganic constituents from fossil fuel combustion residues: I. Review of the major elements. *Journal of Environmental Quality*, **19**, 188–201.

MCCARTHY, G. J. 1997. Residues from coal conversion and utilization: Advanced mineralogical characterization and disposed byproduct diagenesis. Research Report DE-F622-96PC96207, US Department of Energy, Morgantown, VA.

MCCARTHY, G. J., SWANSON, K. D., SCHIELDS, P. J. & GROENEWOLD, G. H. 1983. *Mineralogical Controls on Toxic Element Contamination of Groundwater from Buried Electrical Utility Solid Wastes. 1. Solid Waste Mineralogy. 2. Literature Review of Fly Ash Mineralogy*, PB-83-265116. North Dakota State University, Fargo, ND.

MCCARTHY, G. J., BUTLER, R. D., GRIER, D. G., ADAMEK, S. D., PARKS, J. A. & FOSTER, H. J. 1997. Long-term stability of landfilled coal combustion by-products. *Fuel*, **76**, 697–703.

MCCARTHY, G. J., GRIER, D. G. *et al.* 1999. Coal combustion by-product diagenesis II. *Proceedings, 1999 International Ash Utilization Symposium*, Lexington, Kentucky, Center for Applied Energy Research.

MCGEE, J. J., FINKELMAN, R. B. & PONTOLILLO, J. 1995. Microanalysis of hazardous air pollutants in fly ash and bottom ash from a coal-burning power plant. *Geological Society of America, Abstracts with Programs*, **27/6**, 139.

MEIMA, J. A. & COMANS, R. N. J. 1998. Application of surface complexation/precipitation modeling to contaminant leaching from weathered municipal solid waste incinerator bottom ash. *Environmental Science and Technology*, **32**, 688–693.

MUKHOPADHYAY, P. K., LAJEUNESSE, G. & CRANDLEMIRE, A. L. 1996. Mineralogical speciation of elements in feed coal and their combustion residues from an eastern Canadian coalfield and power plant. *International Journal of Coal Geology*, **32**, 279–312.

NATHAN, Y., DVORACHEK, M., PELLY, I. & MIMRAN, U. 1999. Characterization of coal fly ash from Israel. *Fuel*, **78**, 205–213.

NATUSCH, D. F. S., BAUER, C. F. *et al.* 1975. Characterization of trace elements in fly ash. *In: Proceedings of the International Conference on Heavy Metals in the Environment*. Vol. II, Part 2, Toronto, Ontario, Canada, 553–575.

PARKHURST, D. L. 1995. *Users Guide to PHREEQC – A Computer Program for Speciation, Reaction-Path, Advective-Transport, and Inverse Geochemical Calculations*. US Geological Survey Water Resources Investigations Report 95-4227, Reston, VA.

RAI, D. & SZELMECZKA, R. W. 1990. Aqueous behavior of chromium in coal fly ash. *Journal of Environmental Quality*, **19**, 378–382.

RAI, D., AINSWORTH, C. C., EARY, L. E., MATTIGOD, S. V. & JACKSON, D. R. 1987*a*. *Inorganic and Organic Constituents in Fossil Fuel Combustion Residues*, Vol. 1, A critical review, EPRI EA-5176. Electric Power Research Institute, Palo Alto, CA.

RAI, D., EARY, L. E., MATTIGOD, S. B., AINSWORTH, C. C. & ZACHARA, J. M. 1987*b*. Leaching behavior of fossil fuel wastes: Mineralogy and geochemistry of calcium. *Materials Research Society Symposium Proceedings*, **86**, 3–15.

RAI, D., MATTIGOD, S. V., EARY, L. E. & AINSWORTH, C. C. 1988. Fundamental approach for predicting pore-water composition in fossil fuel combustion wastes. *Materials Research Society Symposium Proceedings*, **113**, 317–324.

RAI, D., ZACHARA, J. M., MOORE, C. A., MCFADDEN, K. M. & RESCH, C. T. 1989. *Field Investigation of a Flue Gas Desulfurization (FGD) Sludge Disposal Pit*, EPRI EA-5923. Electric Power Research Institute, Palo Alto, CA.

ROY, W. R. & GRIFFIN, R. A. 1982. A proposed classification system for coal fly ash in multidisciplinary research. *Journal of Environmental Quality*, **11**, 563–568.

ROY, W. R. & GRIFFIN, R. A. 1984. Illinois basin coal fly ashes. 2. Equilibrium relationships and qualitative modeling of ash–water reactions. *Environmental Science and Technology*, **18**, 739–742.

SAHU, S., BROWN, S. A. & LEE, R. J. 2002. Thaumasite formation in stabilized coal combustion by-products. *Cement and Concrete Composites*, **24**, 385–391.

SHENDE, A., JUWARKAR, A. S. & DURA, S. S. 1994. Use of fly ash in reducing heavy metal toxicity to plants. *Resource Conservation and Recycling*, **12**, 221–228.

SIMONS, H. S. & JEFFERY, J. W. 1960. An X-ray study of pulverized fuel ash. *Journal of Applied Chemistry*, **10**, 328–336.

SMITH, R. D. 1980. The trace element chemistry of coal during combustion and the emissions from coal-fired plants. *Progressive Energy Combustion Science*, **6**, 53–119.

SOLEM-TISCHMACK, J. K., MCCARTHY, G. J., DOCKTOR, B., EYLANDS, K. E., THOMPSON, J. S. & HASSETT, D. J. 1995. High-calcium coal combustion by-products: engineering properties, ettringite formation and potential application in solidification and stabilization of selenium and boron. *Cement and Concrete Research*, **25**, 658–670.

SPEARS, D. A. & LEE, S. 2004. Geochemistry of Leachates from coal ash. In: Gieré, R. & Stille, P. (eds) *Energy, Waste and the Environment: a Geochemical Perspective*. Geological Society, London, Special Publications, **236**, 619–639.

SPOSITO, G. 1984. *The Surface Chemistry of Soils*. Oxford University Press, New York.

STEEFEL, C. I. & VAN CAPPELLEN, P. V. 1990. A new kinetic approach to modeling water–rock interaction: The role of nucleation, precursors, and

Ostwald ripening. *Geochimica et Cosmochimica Acta*, **54**, 2657–2677.

STEENARI, B. M., SCHELANDER, S. & LINDQVIST, O. 1999. Chemical and leaching characteristics of ash from combustion of coal, peat and wood in a 12 MW CFB – a comparative study. *Fuel*, **78**, 249–258.

STOKELY, M. E. & NEWLAND, L. 1992. Pozzolana-induced reductions in the environmental availability of cadmium from fly and bottom ash. *Chemosphere*, **24**, 1591–1596.

STUMM, W. & MORGAN, J. J. 1996. *Aquatic Chemistry. Chemical Equilibria and Rates in Natural Waters*, 3rd edn. Wiley Interscience, New York.

TALBOT, R. W., ANDERSON, M. A. & ANDREN, A. W. 1978. Qualitative model of heterogeneous equilibria in a fly ash pond. *Environmental Science and Technology*, **12**, 1056–1062.

TAYLOR, H. F. W. 1990. *Cement Chemistry*. Academic Press, New York, NY.

THEIS, T. L., WESTRICK, J. D., HSU, C. L. & MARLEY, J. J. 1978. Field investigation of trace metals in groundwater from fly ash disposal. *Journal of the Water Pollution Control Federation*, **50**, 2457–2469.

THEIS, T. L. & RICHTER, R. O. 1979. Chemical speciation of heavy metals in power plant ash pond leachate. *Environmental Science and Technology*, **13**, 219–224.

THEIS, T. L. & WIRTH, J. L. 1977. Sorptive behavior of trace metals on fly ash in aqueous systems. *Environmental Science and Technology*, **11**, 1096–1100.

TISHMACK, J. K. & BURNS, P. E. 2004. The chemistry and mineralogy of coal and coal combustion products. In: Gieré, R. & Stille, P. (eds) *Energy, Waste and the Environment: a Geochemical Perspective*. Geological Society, London, Special Publications, **236**, 223–246.

TURNER, R. R. 1981. Oxidation state of arsenic in coal ash leachate. *Environmental Science and Technology*, **15**, 1062–1066.

VAN DER HOEK, E. E. & COMANS, R. N. J. 1994. Speciation of As and Se during leaching of fly ash. *In*: GOUMANS, J. J. J. M., VAN DER SLOOT, H. A. & AALBERS, TH. G. (eds) *Environmental Aspects of Construction with Waste Materials*. Elsevier Science BV, Amsterdam, 467–475.

VAN DER HOEK, E. E., BONOUVRIE, P. A. & COMANS, R. N. J. 1994. Sorption of As and Se on mineral components of fly ash: relevance for leaching processes. *Applied Geochemistry*, **9**, 403–412.

VAN DER SLOOT, H. A., HOEDE, D. & COMANS, R. N. J. 1994. The influence of reducing properties on leaching of elements from waste materials and construction materials. *In*: GOUMANS, J. J. J. M., VAN DER SLOOT, H. A. & AALBERS, TH. G. (eds) *Environmental Aspects of Construction with Waste Materials*. Elsevier Science BV, Amsterdam, 483–489.

VASSILEV, S. V. & VASSILEVA, C. G. 1996. Mineralogy of combustion wastes from coal-fired power stations. *Fuel Processing Technology*, **47**, 261–280.

VEMPATI, R. K., MOLLAH, Y. A., CHINTHALA, A. K. & COCKE, D. L. 1995. Solidification/stabilization of toxic metal wastes using coke and coal combustion by-products. *Waste Management*, **15**, 433–440.

WARREN, C. J. & DUDAS, M. J. 1984. Weathering processes in relation to leachate properties of alkaline fly ash. *Journal of Environmental Quality*, **13**, 530–538.

WARREN, C. J. & DUDAS, M. J. 1985. Formation of secondary minerals in artificially weathered fly ash. *Journal of Environmental Quality*, **14**, 405–410.

WU, E. J. & CHEN, K. Y. 1987. *Chemical Form and Leachability of Inorganic Trace Elements in Coal Ash*, EPRI EA-5115. Electic Power Research Institute, Palo Alto, CA, 200 pp.

ZACHARA, J. M., COWAN, C. E. & RESCH, C. T. 1991. Sorption of metals on calcite. *Geochimica et Cosmochimica Acta*, **55**, 1549–1562.

ZEVENBERGEN, C., BRADLEY, J. P., VANDER WOOD, T., BROWN, R. S., VAN REEUWIJK, L. P. & SCHUILING, R. D. 1993. Weathering as a process to control the release of toxic constituents from MSW bottom ash. *In*: ARNOULD, M. & COME, M. B. (eds) *Geology and Confinement of Toxic Wastes: Geoconfine, Proceedings of the International Symposium*, Balkema, Rotterdam, 591–595.

ZEVENBERGEN, C., BRADLEY, J. P., VANDER WOOD, T., BROWN, R. S., VAN REEUWIJK, L. P. & SCHUILING, R. D. 1994. Microanalytical investigation of mechanisms of municipal solid waste bottom ash weathering. *Microbeam Analysis*, **3**, 125–135.

ZEVENBERGEN, C., VAN REEUWIJK, L. P., BRADLEY, J. P., BLOEMEN, P. & COMANS, R. N. J. 1996. Mechanism and conditions of clay formation during natural weathering of MSWI bottom ash. *Clays and Clay Minerals*, **44**, 546–552.

ZEVENBERGEN, C., VAN REEUWIJK, L. P., BRADLEY, J. P., COMANS, R. N. J. & SCHUILING, R. D. 1998. Weathering of MSWI bottom ash with emphasis on the glassy constituents. *Journal of Geochemical Exploration*, **62**, 293–298.

ZEVENBERGEN, C., BRADLEY, J. P., VAN REEUWIJK, L. P. & SHYAM, A. K. 1999a. Clay formation during weathering of alkaline coal fly ash. *Proceedings, 1999 International Ash Utilization Symposium*, Lexington, Kentucky, Center for Applied Energy Research.

ZEVENBERGEN, C., BRADLEY, J. P., VAN REEUWIJK, L. P., SHYAM, A. K., HJELMAR, O. & COMANS, R. N. 1999b. Clay formation and metal fixation during weathering of coal fly ash. *Environmental Science and Technology*, **33**, 3405–3409.

ZHANG, M. & REARDON, E. J. 2003. Removal of B, Cr, Mo, and Se from wastewater by incorporation in hydrocalumite and ettringite. *Environmental Science and Technology*, **37**, 2947–2952.

Index